Mitochondrial Biology and Experimental Therapeutics

Paulo J. Oliveira
Editor

Mitochondrial Biology and Experimental Therapeutics

Springer

Editor
Paulo J. Oliveira
CNC - Center for Neuroscience and Cell Biology
University of Coimbra
Coimbra, Portugal

The chapter on comics is copyrighted and Springer does not have rights to the Portuguese translation.

ISBN 978-3-030-10358-3 ISBN 978-3-319-73344-9 (eBook)
https://doi.org/10.1007/978-3-319-73344-9

Softcover re-print of the Hardcover 1st edition 2018

Printed on acid-free paper

This Springer imprint is published by the registered company Springer International Publishing AG part of Springer Nature.
The registered company address is: Gewerbestrasse 11, 6330 Cham, Switzerland

Foreword

I am honored to have been asked by the Editor to write a brief introduction to this new book on *Mitochondrial Biology and Experimental Therapeutics*. What makes this book stand out among the many other writings on this topic is its focus on the bioenergetics and molecular features of the mitochondrion that can be exploited to pharmacologically target assorted mitochondrial disease states. The chapters are categorized and sequenced to first provide an overview of the structure and function of the mitochondrion in order to set the stage for describing how mitochondrial bioenergetics is a dynamic process that influences mitochondrial replication and capacity, which then regulates cell homeostasis by way of triggering one or more cell death pathways. The book continues then to describe distinct forms of mitochondrionopathies associated with various organ disease states, including metabolic, cardiovascular, and neurodegenerative disorders, some of which may be drug-induced. The thrust of the book follows wherein the individual authors describe strategies for identifying and designing specific small molecule drugs, bioenergetic interventions, or gene-directed targeting to alter mitochondrial competence, thereby restoring the bioenergetic capacity of diseased or aging tissue.

There are numerous published treatises describing the rapid growth of interest in mitochondrial-mediated disease and toxicities. This book, however, takes the conversation to the next level by compiling a long list of internationally recognized experts to lend their individual insights into developing strategies for a new era of mitochondrial therapeutics. The Editor and authors should be complimented for their contributions to organizing this introduction into the next chapter of mitochondrial medicine.

University of Minnesota
Duluth, MN, USA

Kendall B. Wallace

Contents

Part I
Mitochondrial Biology

Introduction: Mitochondria, the Cell Furnaces

Paulo J. Oliveira

Abstract In the motion picture "The Phantom Menace", the Star Wars episode that caused so much controversy for fans of the series, Qui-Gon Jinn describes the source of the Jedi Force to Anakin Skywalker. "Without the midi-chlorians, life could not exist, and we would have no knowledge of the Force. They continually speak to us, telling us the will of the Force. When you learn to quiet your mind, you'll hear them speaking to you…". The Star Wars lore suggests that midi-chlorians are intelligent microscopic life forms that live symbiotically inside the cells of all living things, with higher numbers present in individuals with the ability to feel the Force. Does this sound familiar? Yes, it is a crude description of mitochondria, the former symbiotic bacteria that are now the cell powerhouses, among other critical roles in cell metabolism, calcium and redox signaling, and regulation of cell death processes. With multiple shapes and forms, mitochondria are an amazingly complex organelle, which never ceases to surprise the researchers with new and important functions in the context of cell metabolism. Moreover, the idea that mitochondria are an attractive drug target has been gaining traction in the last years, with two molecules claimed to target mitochondria already in the market.

Keywords Mitochondria · Furnace · Disease · Therapy · Metabolism

In the motion picture "The Phantom Menace", the Star Wars episode that raised so much controversy for fans of the series, Qui-Gon Jinn describes the source of the Jedi force to Anakin Skywalker. "Without the midi-chlorians, life could not exist, and we would have no knowledge of the Force. They continually speak to us, telling us the will of the Force. When you learn to quiet your mind, you'll hear them speaking to you…"

P. J. Oliveira (✉)
CNC – Center for Neuroscience and Cell Biology, UC-Biotech, University of Coimbra, Cantanhede 3060-197, Portugal
e-mail: pauloliv@cnc.uc.pt

P. J. Oliveira (ed.), *Mitochondrial Biology and Experimental Therapeutics*,
https://doi.org/10.1007/978-3-319-73344-9_1

The Star Wars lore (as referenced in http://starwars.wikia.com/wiki/Midi-chlorian/Legends) suggests that midi-chlorians are intelligent microscopic life forms that live symbiotically inside the cells of all living things, with higher numbers present in individuals with the ability to feel the Force. Although fans of the Star Wars series felt betrayed because George Lucas detailed something that should have remained mysterious, the truth is that the whole story is reminiscent of what occurs with mitochondria, the intracellular organelles that are described in a simplistic form as the powerhouses of the cells (which is far from the truth). The word mitochondrion (singular) is derived from the Greek μίτος, mitos, meaning "thread", and χονδρίον, chondrion, meaning "granule" and its origins are traced to a prokaryotic organism (Gray et al. 2001; Degli Esposti 2014).

Decades of mitochondrial research have shed some light on the truly complete and yet complex nature of mitochondria. It is no surprise that a simple Pubmed search using the word "mitochondria" yielded at the time of writing (July, 2017) 2,477 papers in 1980, which increased to 8,455 in 2016. One can then do the same exercise with other intracellular organelles and the trend will not look similar. What makes mitochondrial research so interesting that the number of publications increases year after year?

Instead of being static structures, mitochondria present different forms and shapes, and can indeed move long distances in longer cells, such as neurons, attached to the cell cytoskeleton by specific proteins, including Miro and Milton (Schwarz 2013). Moreover, responding to environmental factors (e.g., starvation) and intrinsic factors (e.g., oxidative stress), mitochondria undergo fusion and fission cycles, which are intrinsically related to energy production and mechanisms of cell quality control, including mitophagy, which regulates unwanted or damaged mitochondrial bodies (Galluzzi et al. 2017; McWilliams and Muqit 2017).

From a major site of energy production, stored as adenosine triphosphate (ATP), through oxidative phosphorylation, to control of cytosolic calcium, regulation of cell death, intermediate metabolism, and production of reactive oxygen species (ROS), to name just a few of the cellular roles of mitochondria, it is clear why these organelles are thus considered the "powerhouses" of cells. More than being simply "powerhouses", mitochondria act as factories and regulators of cellular dynamics. In fact, even the regulation of gene expression in the nucleus is regulated by mitochondria. Epigenetics at DNA and histone level depends on different metabolites that are produced by mitochondria, including acetyl-coA, acetyl-carnitine and ATP, which closely links the dynamics of mitochondrial operation with cycles of gene transcription (Schell and Rutter 2017; Matilainen et al. 2017; Wallace and Fan 2010).

Although they represent a perfect nightmare for many Biochemistry students, mitochondria are the site of several metabolic pathways that go beyond energy production. Fatty acid beta-oxidation and the Krebs cycle are two examples of perfectly orchestrated metabolic pathways, finely tuned to cell metabolism and cell requirements, which ultimately lead to the production of ATP. But more than that, several key steps in the Krebs cycle can be used for biosynthetic pathways leading to the production of different cell components. Examples are the production of sterols and fatty acids from citrate, purines and some amino acids from alpha-ketoglutarate, or amino acids, purines and pyrimidines from oxaloacetate. Some intermediates regulate cell epigenetics, as described above (Sajnani et al. 2017; de Castro Fonseca et al. 2016; Salminen et al. 2014).

Reminiscent of their bacterial origin, mitochondria have multiple copies of their own genome, mtDNA, which code for 13 subunits of the respiratory chain, plus tRNA and rRNAs. Although much is known already about the organization, sequence and dynamics of the mitochondrial genome, some surprises are still in store. Open-reading frames were identified in the mitochondrial genome, which code for previously unknown peptides. Examples include humanin, a 24 amino acid peptide coded from the 16S ribosomal RNA gene, MT-RNR2 (Voigt and Jelinek 2016; Paharkova et al. 2015), and MOTS-c, a 16 amino acid peptide derived from an open-reading frame of the 12S rRNA-c (Lee et al. 2016; Lee et al. 2015). Interestingly, both were described to modulate cell metabolism (Kim et al. 2017), demonstrating that mitochondrial signaling is far from being understood.

With so many "moving parts", it is only natural that the mitochondrial furnace may be affected by multiple internal and external factors. Although production of ROS is part of the normal mitochondrial activity, resulting in multiple downstream physiological effects, including regulation of gene expression, in some cases that same production can be increased without being counteracted by the intrinsic network of mitochondrial antioxidant defenses (Scialo et al. 2017; Kuksal et al. 2017). This can result in progressive damage to mitochondrial structures, from ion channels and respiratory chain complexes to membrane lipids and mtDNA. Progressive damage to the mitochondrial structure, if not counteracted by selective elimination of defective organelles (mitophagy) and consequent mitochondrial biogenesis, can lead to metabolic collapse in the cell. This is actually observed in multiple pathologies in which primary or secondary mitochondrial dysfunction plays a critical role. Diseases include multiple cancer types (Barbosa et al. 2012) and neurodegenerative (Wilkins et al. 2017; Grimm and Eckert 2017), cardiovascular (Silva et al. 2016b), renal (Duann and Lin 2017), and hepatic (Grattagliano et al. 2011) diseases. Diabetes and obesity-related complications also show a very important mitochondrial component, and it is not surprising that several of the described pathologies are associated with the aging process. Multiple publications have demonstrated progressive mitochondrial loss of fitness, mediated by or resulting in increased oxidative stress (Devarshi et al. 2017; Rovira-Llopis et al. 2017).

The field of mitochondrial toxicology has also been very active in recent decades. After an initial burst of activity in the 80s, with many papers focusing on the role of mitochondrial disruption in environmental pollutant-mediated toxicity (e.g., pesticides (Peixoto et al. 2003)), pharmaceutical companies soon appreciated how the prediction of mitochondrial liabilities could prevent delays and unnecessary costs in the process of drug development. In fact, several clinical-used drugs show mitochondrial toxicity, which can mediate off-target effects and lead to withdrawal from the market if the problem is not detected in early pre-clinical development (Atienzar et al. 2016; Pereira et al. 2012; Nadanaciva and Will 2011a, b). Drugs such as nefazodone, an anti-depressant withdrawn from the market (Silva et al. 2016a), or doxorubicin, an anti-cancer agent still in widespread use (Carvalho et al. 2014; Carvalho et al. 2009), showed mitochondrial toxicity in humans and animal models. Moreover, other toxicants, such as dioxins, parabens, cigarette-containing particles, ethanol or even heavy metals, interact with the mitochondrial respiratory chain, causing a decrease in ATP-generating capacity and an increase in oxidative stress (Pereira et al. 2012). The problems go beyond "simple" inhibition of mitochondrial ATP generation. As described above, mitochondrial metabolites (including ATP and acetyl-coA) regulate cell epigenetics, so it is hardly surprising that mitochondrial interactions with xenobiotics during pregnancy, or even pre-conception, can cause a lasting effect on the progeny (Janssen et al. 2015; Barua and Junaid 2015; Jimenez-Chillaron et al. 2015). Moreover, there is still a lack of knowledge on regulation of mtDNA dynamics by methylation of the genome. How and where mtDNA can be methylated, and the consequences in terms of replication and transcription are now only beginning to be understood. How this still poorly understood phenomenon is affected in different pathologies, aging, and xenobiotic-induced toxicity still remains an open question, despite some pioneer publications (Stoccoro et al. 2017; Novielli et al. 2017; Saini et al. 2017; Pollack et al. 1984).

Mutations in mtDNA, or in nuclear genes that directly or indirectly affect mitochondrial function (e.g., in subunits of the respiratory chain, ion transporters, proteins involved in mtDNA replication, or mitochondrial antioxidant enzymes), can also result in a disease phenotype, which can be clinically observed after birth, often leading to early death or a progressive loss of tissue function during adulthood (Craven et al. 2017; Bacalhau et al. 2017; Tavares et al. 2013).

The previous paragraphs demonstrate well how the mitochondrial furnace must be coupled to the metabolic needs of the cell, and how different protective strategies must be developed by the cell to protect its energy sources. Even with so much knowledge available on mitochondrial function, signaling, and cross-talk with other organelles, there are still several surprises. A pre-print, non-peer reviewed observation from Chretien et al., using a temperature-sensitive fluorescent probe targeted to mitochondria, indicated that active organelles operate at temperatures close to 50 °C, which contradicts many dogmas in Biochemistry books (Chretien et al. 2017). Although this work has not been yet published in a peer-reviewed publication at the time of writing*, this clearly shows that our favorite organelle (or furnace) still has some surprises enclosed in its double membrane system.

Acknowledgements We are extremely thankfully to Alexandra Holy, MBA/MPP, Mills College, Oakland, CA, USA for proofreading English language in this text.Work in the author's laboratory is funded by FEDER funds through the Operational Programme Competitiveness Factors—COMPETE and national funds by FCT—Foundation for Science and Technology under research grants PTDC/DTP-FTO/2433/2014, POCI-01-0145-FEDER-016659, and POCI-01-0145-FEDER-007440.

---

*Note added in proof: This publication has just been peer-reviewed and published: Chrétien et al., PLoS Biol 16(1):e2003992.

References

Atienzar FA, Blomme EA, Chen M, Hewitt P, Kenna JG, Labbe G, Moulin F, Pognan F, Roth AB, Suter-Dick L, Ukairo O, Weaver RJ, Will Y, Dambach DM (2016) Key challenges and opportunities associated with the use of in vitro models to detect human DILI: integrated risk assessment and mitigation plans. Biomed Res Int 2016:9737920. https://doi.org/10.1155/2016/9737920

Bacalhau M, Pratas J, Simoes M, Mendes C, Ribeiro C, Santos MJ, Diogo L, Macario MC, Grazina M (2017) In silico analysis for predicting pathogenicity of five unclassified mitochondrial DNA mutations associated with mitochondrial cytopathies' phenotypes. Eur J Med Genet 60(3):172–177. https://doi.org/10.1016/j.ejmg.2016.12.009

Barbosa IA, Machado NG, Skildum AJ, Scott PM, Oliveira PJ (2012) Mitochondrial remodeling in cancer metabolism and survival: potential for new therapies. Biochim Biophys Acta 1826(1):238–254. https://doi.org/10.1016/j.bbcan.2012.04.005

Barua S, Junaid MA (2015) Lifestyle, pregnancy and epigenetic effects. Epigenomics 7(1):85–102. https://doi.org/10.2217/epi.14.71

Carvalho C, Santos RX, Cardoso S, Correia S, Oliveira PJ, Santos MS, Moreira PI (2009) Doxorubicin: the good, the bad and the ugly effect. Curr Med Chem 16(25):3267–3285

Carvalho FS, Burgeiro A, Garcia R, Moreno AJ, Carvalho RA, Oliveira PJ (2014) Doxorubicin-induced cardiotoxicity: from bioenergetic failure and cell death to cardiomyopathy. Med Res Rev 34(1):106–135. https://doi.org/10.1002/med.21280

de Castro Fonseca M, Aguiar CJ, da Rocha Franco JA, Gingold RN, Leite MF (2016) GPR91: expanding the frontiers of Krebs cycle intermediates. Cell Commun Signal 14:3. https://doi.org/10.1186/s12964-016-0126-1

Chretien D, Benit P, Ha H-H, Keipert S, El-Khoury R, Chang Y-T, Jastroch M, Jacobs H, Rustin P, Malgorzata R (2017) Mitochondria are physiologically maintained at close to 50 C. bioRxiv. https://doi.org/10.1101/133223

Craven L, Alston CL, Taylor RW, Turnbull DM (2017) Recent advances in mitochondrial disease. Annu Rev Genomics Hum Genet 18:257–275. https://doi.org/10.1146/annurev-genom-091416-035426

Degli Esposti M (2014) Bioenergetic evolution in proteobacteria and mitochondria. Genome Biol Evol 6(12):3238–3251. https://doi.org/10.1093/gbe/evu257

Devarshi PP, McNabney SM, Henagan TM (2017) Skeletal muscle nucleo-mitochondrial crosstalk in obesity and Type 2 diabetes. Int J Mol Sci 18(4):E831. https://doi.org/10.3390/ijms18040831

Duann P, Lin PH (2017) Mitochondria damage and kidney disease. Adv Exp Med Biol 982:529–551. https://doi.org/10.1007/978-3-319-55330-6_27

Galluzzi L, Baehrecke EH, Ballabio A, Boya P, Bravo-San Pedro JM, Cecconi F, Choi AM, Chu CT, Codogno P, Colombo MI, Cuervo AM, Debnath J, Deretic V, Dikic I, Eskelinen EL, Fimia GM, Fulda S, Gewirtz DA, Green DR, Hansen M, Harper JW, Jaattela M, Johansen T, Juhasz G, Kimmelman AC, Kraft C, Ktistakis NT, Kumar S, Levine B, Lopez-Otin C, Madeo F, Martens S, Martinez J, Melendez A, Mizushima N, Munz C, Murphy LO, Penninger JM, Piacentini M, Reggiori F, Rubinsztein DC, Ryan KM, Santambrogio L, Scorrano L, Simon AK, Simon HU, Simonsen A, Tavernarakis N, Tooze SA, Yoshimori T, Yuan J, Yue Z, Zhong Q, Kroemer G (2017) Molecular definitions of autophagy and related processes. EMBO J 36(13):1811–1836. https://doi.org/10.15252/embj.201796697

Grattagliano I, Russmann S, Diogo C, Bonfrate L, Oliveira PJ, Wang DQ, Portincasa P (2011) Mitochondria in chronic liver disease. Curr Drug Targets 12(6):879–893

Gray MW, Burger G, Lang BF (2001) The origin and early evolution of mitochondria. Genome Biol 2(6):REVIEWS1018

Grimm A, Eckert A (2017) Brain aging and neurodegeneration: from a mitochondrial point of view. J Neurochem 143(4):418–431. https://doi.org/10.1111/jnc.14037

Janssen BG, Byun HM, Gyselaers W, Lefebvre W, Baccarelli AA, Nawrot TS (2015) Placental mitochondrial methylation and exposure to airborne particulate matter in the early life environment: an ENVIRONAGE birth cohort study. Epigenetics 10(6):536–544. https://doi.org/10.1080/15592294.2015.1048412

Jimenez-Chillaron JC, Nijland MJ, Ascensao AA, Sardao VA, Magalhaes J, Hitchler MJ, Domann FE, Oliveira PJ (2015) Back to the future: transgenerational transmission of xenobiotic-induced epigenetic remodeling. Epigenetics 10(4):259–273. https://doi.org/10.1080/15592294.2015.1020267

Kim SJ, Xiao J, Wan J, Cohen P, Yen K (2017) Mitochondrial derived peptides as novel regulators of metabolism. J Physiol 595(21):6613–6621. https://doi.org/10.1113/JP274472

Kuksal N, Chalker J, Mailloux RJ (2017) Progress in understanding the molecular oxygen paradox - function of mitochondrial reactive oxygen species in cell signaling. Biol Chem 398(11):1209–1227. https://doi.org/10.1515/hsz-2017-0160

Lee C, Zeng J, Drew BG, Sallam T, Martin-Montalvo A, Wan J, Kim SJ, Mehta H, Hevener AL, de Cabo R, Cohen P (2015) The mitochondrial-derived peptide MOTS-c promotes metabolic homeostasis and reduces obesity and insulin resistance. Cell Metab 21(3):443–454. https://doi.org/10.1016/j.cmet.2015.02.009

Lee C, Kim KH, Cohen P (2016) MOTS-c: a novel mitochondrial-derived peptide regulating muscle and fat metabolism. Free Radic Biol Med 100:182–187. https://doi.org/10.1016/j.freeradbiomed.2016.05.015

Matilainen O, Quiros PM, Auwerx J (2017) Mitochondria and epigenetics - crosstalk in homeostasis and stress. Trends Cell Biol 27(6):453–463. https://doi.org/10.1016/j.tcb.2017.02.004

McWilliams TG, Muqit MM (2017) PINK1 and Parkin: emerging themes in mitochondrial homeostasis. Curr Opin Cell Biol 45:83–91. https://doi.org/10.1016/j.ceb.2017.03.013

Nadanaciva S, Will Y (2011a) Investigating mitochondrial dysfunction to increase drug safety in the pharmaceutical industry. Curr Drug Targets 12(6):774–782

Nadanaciva S, Will Y (2011b) New insights in drug-induced mitochondrial toxicity. Curr Pharm Des 17(20):2100–2112

Novielli C, Mando C, Tabano S, Anelli GM, Fontana L, Antonazzo P, Miozzo M, Cetin I (2017) Mitochondrial DNA content and methylation in fetal cord blood of pregnancies with placental insufficiency. Placenta 55:63–70. https://doi.org/10.1016/j.placenta.2017.05.008

Paharkova V, Alvarez G, Nakamura H, Cohen P, Lee KW (2015) Rat humanin is encoded and translated in mitochondria and is localized to the mitochondrial compartment where it regulates ROS production. Mol Cell Endocrinol 413:96–100. https://doi.org/10.1016/j.mce.2015.06.015

Peixoto F, Vicente JA, Madeira VM (2003) The herbicide dicamba (2-methoxy-3,6-dichlorobenzoic acid) interacts with mitochondrial bioenergetic functions. Arch Toxicol 77(7):403–409. https://doi.org/10.1007/s00204-003-0456-9

Pereira CV, Nadanaciva S, Oliveira PJ, Will Y (2012) The contribution of oxidative stress to drug-induced organ toxicity and its detection in vitro and in vivo. Expert Opin Drug Metab Toxicol 8(2):219–237. https://doi.org/10.1517/17425255.2012.645536

Pollack Y, Kasir J, Shemer R, Metzger S, Szyf M (1984) Methylation pattern of mouse mitochondrial DNA. Nucleic Acids Res 12(12):4811–4824

Rovira-Llopis S, Banuls C, Diaz-Morales N, Hernandez-Mijares A, Rocha M, Victor VM (2017) Mitochondrial dynamics in type 2 diabetes: pathophysiological implications. Redox Biol 11:637–645. https://doi.org/10.1016/j.redox.2017.01.013

Saini SK, Mangalhara KC, Prakasam G, Bamezai RNK (2017) DNA Methyltransferase1 (DNMT1) isoform3 methylates mitochondrial genome and modulates its biology. Sci Rep 7(1):1525. https://doi.org/10.1038/s41598-017-01743-y

Sajnani K, Islam F, Smith RA, Gopalan V, Lam AK (2017) Genetic alterations in Krebs cycle and its impact on cancer pathogenesis. Biochimie 135:164–172. https://doi.org/10.1016/j.biochi.2017.02.008

Salminen A, Kauppinen A, Hiltunen M, Kaarniranta K (2014) Krebs cycle intermediates regulate DNA and histone methylation: epigenetic impact on the aging process. Ageing Res Rev 16:45–65. https://doi.org/10.1016/j.arr.2014.05.004

Schell JC, Rutter J (2017) Mitochondria link metabolism and epigenetics in haematopoiesis. Nat Cell Biol 19(6):589–591. https://doi.org/10.1038/ncb3540

Schwarz TL (2013) Mitochondrial trafficking in neurons. Cold Spring Harb Perspect Biol 5(6):a011304. https://doi.org/10.1101/cshperspect.a011304
Scialo F, Fernandez-Ayala DJ, Sanz A (2017) Role of mitochondrial reverse electron transport in ROS signaling: potential roles in health and disease. Front Physiol 8:428. https://doi.org/10.3389/fphys.2017.00428
Silva AM, Barbosa IA, Seabra C, Beltrao N, Santos R, Vega-Naredo I, Oliveira PJ, Cunha-Oliveira T (2016a) Involvement of mitochondrial dysfunction in nefazodone-induced hepatotoxicity. Food Chem Toxicol 94:148–158. https://doi.org/10.1016/j.fct.2016.06.001
Silva FS, Simoes RF, Couto R, Oliveira PJ (2016b) Targeting mitochondria in cardiovascular diseases. Curr Pharm Des 22(37):5698–5717
Stoccoro A, Siciliano G, Migliore L, Coppede F (2017) Decreased methylation of the mitochondrial D-loop region in late-onset Alzheimer's disease. J Alzheimer's Dis 59(2):559–564. https://doi.org/10.3233/JAD-170139
Tavares MV, Santos MJ, Domingues AP, Pratas J, Mendes C, Simoes M, Moura P, Diogo L, Grazina M (2013) Antenatal manifestations of mitochondrial disorders. J Inherit Metab Dis 36(5):805–811. https://doi.org/10.1007/s10545-012-9567-x
Voigt A, Jelinek HF (2016) Humanin: a mitochondrial signaling peptide as a biomarker for impaired fasting glucose-related oxidative stress. Physiol Rep 4(9):e12796. https://doi.org/10.14814/phy2.12796
Wallace DC, Fan W (2010) Energetics, epigenetics, mitochondrial genetics. Mitochondrion 10(1):12–31. https://doi.org/10.1016/j.mito.2009.09.006
Wilkins HM, Weidling IW, Ji Y, Swerdlow RH (2017) Mitochondria-derived damage-associated molecular patterns in neurodegeneration. Front Immunol 8:508. https://doi.org/10.3389/fimmu.2017.00508

Mitochondria: Where Are They Coming From?

Mauro Degli Esposti

Abstract This chapter summarizes the evidence sustaining the concept that mitochondria derive from a particular type of bacteria, the alpha proteobacteria. Once considered a hypothesis, this concept is now a scientific fact and therefore should not be viewed in dubitative terms. What remains uncertain is from which bacterial taxa the proto-mitochondria originated. However, the quest to find the living bacterial relatives of our mitochondria is narrowing to a limited set of alpha proteobacteria with ample metabolic versatility. The results of this quest are likely to throw new light on various aspects of mitochondrial physiology and biochemistry.

Keywords Proto-mitochondria · Alpha proteobacteria · Endo-symbiotic theory · Evolution · Metabolism

1 Introduction

The mitochondrial organelle defines eukaryotic cells (Lane 2006; Lane and Martin 2010) and thus it is at least as old as the first proto-eukaryotic organism, often referred to as LECA—Last Eukaryotic Common Ancestor (Burki 2014). Among all cellular compartments in eukaryotic cells, only mitochondria have retained structural membrane components, RNA elements and a circular DNA chromosome that are directly related to bacteria. These features clearly indicate that our mitochondria derive from bacteria (Lane and Martin 2010; Gray 2015; Lazcano and Peretó 2017), a concept that should now be considered to be a fact, since it is supported by clear evidence, and not in dubitative terms as often reported in the biomedical literature. Moreover, phylogenetic and biochemical data point to a specific group of bacteria that gave origin to the

M. D. Esposti (✉)
Italian Institute of Technology, Genoa, Italy

Center for Genomic Sciences, UNAM Campus of Cuernavaca, Cuernavaca, Morelos, Mexico
e-mail: mauro.degliesposti@iit.it

P. J. Oliveira (ed.), *Mitochondrial Biology and Experimental Therapeutics*,
https://doi.org/10.1007/978-3-319-73344-9_2

proto-mitochondria, and hence to our mitochondrial organelles: the alpha proteobacteria (Williams et al. 2007; Atteia et al. 2009; Müller et al. 2012; Thiergart et al. 2012; Degli Esposti 2014, 2016). Alpha proteobacteria include genera well known for their symbiotic or pathogenic relationships with plants (*Rhizobium*, *Agrobacterium*) and animals (*Brucella*, *Rickettsia*), as well as an increasing variety of metabolically versatile organisms that often dominate entire ecosystems. Recent surveys have shown that alpha proteobacteria contain basically all the metabolic pathways known to be present in bacteria (Louca et al. 2016; Degli Esposti and Martinez Romero 2017). The big question is then: from which lineage of such a metabolically versatile class of bacteria did proto-mitochondria come from? Of course, we cannot trace the unique event of symbiosis that led to the establishment of proto-mitochondria inside LECA, because such an event occurred nearly two billion years ago (Müller et al. 2012). So long ago that it is nearly impossible to consider the lucky perspective of finding missing links or close relatives among today's bacteria. Hence, it is a very difficult quest.

As most other prokaryotes, alpha proteobacteria constantly exchange chunks of DNA in the environment and thus their genome has been progressively re-shaped by the process of Lateral Gene Transfer (LGT) during the last two billions years (Müller et al. 2012; Thiergart et al. 2012). This process has not touched the progressively eroded genome of the proto-mitochondrial endosymbionts, which has been insulated from the environment, and hence other bacteria, by its intracellular location (Thiergart et al. 2012; Ku et al. 2015). In any case, the few genes retained in mitochondrial DNA code for membrane bioenergetic proteins that are rarely exchanged by LTG in the environments inhabited by bacteria, because these proteins generally form their core energy metabolism (Degli Esposti 2014). In the relatively rare cases in which LGT has been shown to act on these proteins, it actually produces duplicate copies of entire operons for energy conservation in the same genome, as for instance with complex I (Degli Esposti and Martinez Romero 2016). Given the above, the sequences of the longest mitochondrially encoded proteins, from ND5 of complex I to cytochrome *b* of complex III and COX1 of complex IV, have retained signatures that are shared by homologous proteins from bacteria, especially alpha proteobacteria, as shown in Fig. 1 (cf. Degli Esposti 2014, 2016). Consequently, such proteins contain a strong phylogenetic signal which, in principle, could be exploited to identify the likely living ancestors of proto-mitochondria (Thiergart et al. 2012; Degli Esposti 2014, 2016).

I will briefly review here the problems that render this possibility hard and controversial, as well as the evidence that point to a specific group of extant bacteria that may well be considered to include the living ancestors of our mitochondria.

2 Problems and Controversies Regarding the Precise Bacterial Origin of Mitochondria

For a long time, I have been fascinated by the quest for the living relatives of the ancestral bacteria from which mitochondria came from. The more I have learnt about this topic, the more problems I have found that render this quest so terribly

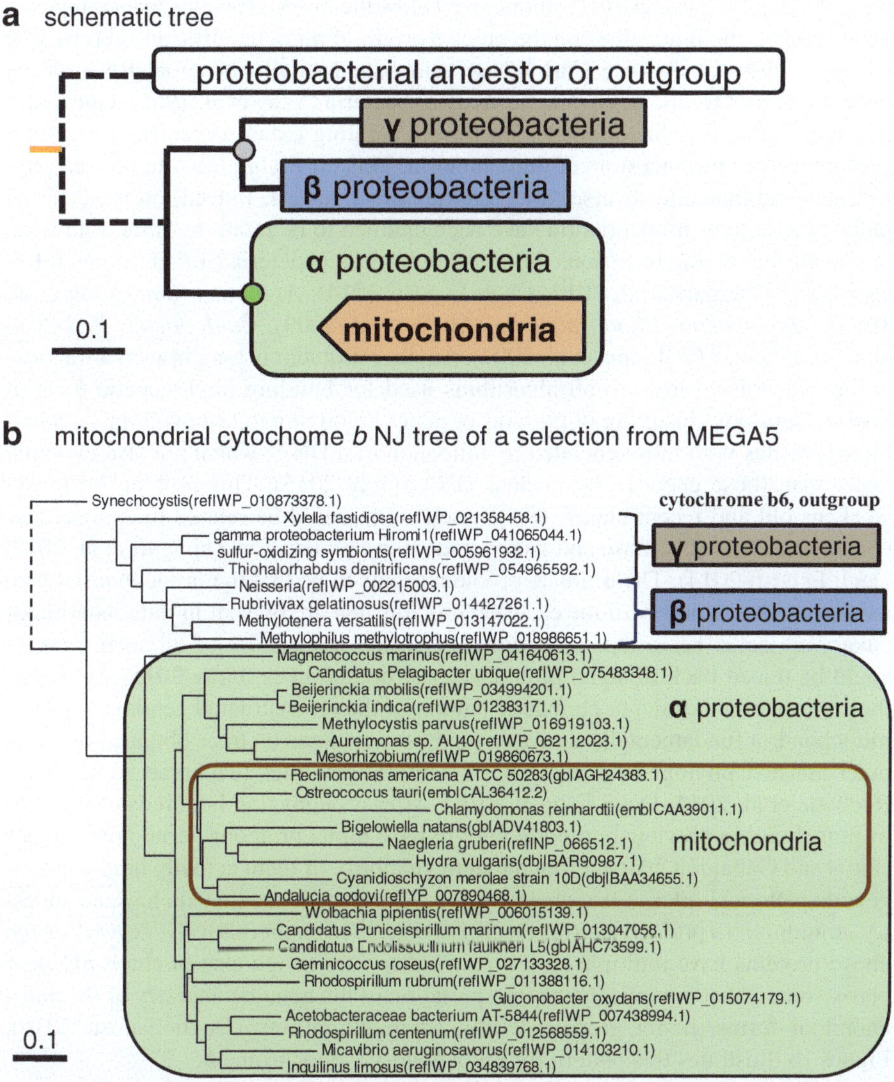

Fig. 1 Phylogenetic pattern of mitochondrial proteins. (**a**) Schematic representation of typical phylogenetic trees for mitochondrial proteins involved in energy metabolism (Degli Esposti 2016). Mitochondrial sequences are embedded within the clade of alpha proteobacterial homologues, which is in sister position of the clade containing the homologous proteins from either gamma or beta proteobacteria. (**b**) Neighbour Joining tree obtained from BLASTsearches of alpha proteobacterial cytochrome *b* proteins that were extended to a wide variety of proteobacterial and eukaryotic taxa (cf. Degli Esposti 2016). Equivalent tree topologies have been found for other subunit of complex III encoded by nuclear DNA (Degli Esposti 2016) as well as in other mitochondrially encoded proteins such as COX1 and COX3 (cf. Degli Esposti 2014)

difficult (Degli Esposti 2016). I summarize below the major problems that complicate the efforts of obtaining clues on the precise origin of mitochondria, an exercise that has been undertaken since Carl Woese established small ribosomal RNA as the benchmark tool for the phylogenetic study of bacteria (Yang et al. 1985). One major problem is the abovementioned long time separating extant organisms, and their genomes, from the ancestors of mitochondria. This inevitably leads to blurred phylogenies and hence to diverse, often contradictory results. Indeed, on the basis of phylogenetic trees mitochondria have been claimed to be relatives of the following organisms belonging to various orders of alpha proteobacteria (for additional references, see Thiergart et al. 2012; Degli Esposti 2014): *Agrobacterium* (Yang et al. 1985), *Rhodobacter*, *Rhodospirillum* (Atteia et al. 2009), *Caulobacter*, *Rickettsia* and *Pelagibacter* (Williams et al. 2007). Another problem is long branch attraction, a phenomenon intrinsic to all algorithms used for building phylogenetic trees. It leads to spurious clustering of proteins or genes belonging to fast evolving endocellular parasites with those encoded by mitochondrial DNA, which are also evolving faster than those encoded by nuclear DNA (Gray 2015). This problem seriously weakens old and recent claims that mitochondria could be related to endocellular parasites such as *Rickettsia* and its relatives (for a review, see Thiergart et al. 2012; Degli Esposti 2014). The ultimate problem seems to be that the mitochondrial proteome, i.e. the assembly of the ca. 1000 proteins that are present in mitochondria of eukaryotic cells, has a mosaic nature, since only about 20% of all such proteins could be traced back to alpha proteobacterial homologues (Gray 2015). However, the concept of a mosaic, or chimaeric nature for the reconstructed genome of proto-mitochondria fundamentally derives from heavy reliance on trees obtained by computer-assisted phylogenetic analysis of multiple genomes (Thiergart et al. 2012; Rochette et al. 2014; Degli Esposti 2016). After re-analyzing hundreds of proteins of mitochondria previously reported not to have alpha proteobacterial homologues (Pittis and Gabaldón 2016), I found that the majority of them actually have a phylogenetic pattern similar to that shown in Fig. 1, and therefore they do have an ancestry among alpha proteobacteria (Degli Esposti 2016). Alternatively, several of the above proteins have multiple isoforms in eukaryotes (often also in chloroplasts or peroxisomes) that complicate the interpretation of the genuine ancestry of the mitochondrial forms, if not properly evaluated in its entirety (Degli Esposti 2016). Figure 1b illustrates this pattern for mitochondrial cytochrome *b*.

3 Integrated Approaches to Identify Possible Relatives of Proto-mitochondria

The key approach to investigate which extant alpha proteobacterium may be really related to proto-mitochondria is to analyze a series of proteins and metabolic traits that can be definitively assigned to the ancestral lineage from which proto-mitochondria originated (Degli Esposti 2014). Such traits pivot on the adaptation to the predominantly anoxic conditions that were present in archaeal oceans at the time

Table 1 Proteins and traits that are shared by mitochondria and alpha proteobacteria

Protein and metabolic trait	Present in other proteobacteria	References
UbiL, ubiquinone biosynthesis	No	Pelosi et al. (2016)
CoQ9, ubiquinone biosynthesis	No	MDE, unpublished[a]
AOX, alternative ubiquinol oxidase, energy conservation	No	Atteia et al. (2004)
CtaBCtaE (Cox10Cox3) synteny, energy conservation	No	Degli Esposti (2014)
Rhodoquinone, energy conservation	A few beta proteobacteria	Hiraishi and Hoshino (1984), Müller et al. (2012)
CoQ4, ubiquinone biosynthesis	A few beta and gamma proteobacteria	MDE, unpublished[a]

[a]Unpublished observations by the Author

in which LECA evolved (Müller et al. 2012). Some of the ancestral traits that enabled this adaptation may be still used by contemporary bacteria with facultatively anaerobic metabolism. Certain proteins may thus represent scattered heritage of these metabolic traits, while the taxonomic distribution of proteins involved in energy conservation may reveal intriguing common characters between mitochondria and their possible ancestors (Degli Esposti 2014). Along this rationale, I present in Table 1 a preliminary list of the proteins that are shared by mitochondria and, predominantly, alpha proteobacteria. Intriguingly, several of these proteins shared by mitochondria and alpha proteobacteria are involved in the biosynthesis of the membrane quinones, ubiquinone and rhodoquinone (Aussel et al. 2014; Kawamukai 2015). Of note, rhodoquinone is present only in two related species of the Rhodospirillaceae family of alpha proteobacteria, *Rhodospirillum rubrum* and *Pararhodospirillum* (formerly *Rhodospirillum*) *photometricum* (Hiraishi and Hoshino 1984), as well as in mitochondria of invertebrates adapted to anaerobiosis (Müller et al. 2012).

4 Conclusion

This chapter provides an overview of current knowledge on the origin of mitochondria. There is no doubt, really, that mitochondrial organelles came from bacteria (Lane 2006; Gray 2015). And we also know which class of bacteria generated proto-mitochondria: the alpha proteobacteria. This is a conclusion that is very hard to dispute, or even doubt nowadays. Therefore, it can be considered a scientific fact. What remains uncertain is the kind of alpha proteobacteria that may have close relatives to the original proto-mitochondria. However, the information briefly presented and discussed here allows us to exclude all strictly aerobic and parasitic alpha proteobacteria from these potential relatives. Indeed, the quest for the living relatives of mitochondria can now be narrowed to early branching, metabolically versatile and facultatively anaerobic organisms of the Rhodospirillales and Rhizobiales orders

(Atteia et al. 2009; Thiergart et al. 2012; Degli Esposti 2014), especially their genera that maintain genes for traits of strictly anaerobic energy metabolism such as FeFe-hydrogenases, e.g. *Phaeospirillum* and *Pleomorphomonas* (Degli Esposti et al. 2016).

Even if the controversy regarding the origin of mitochondria is not resolved yet, and will probably continue for years, detailed analysis of the ever-expanding number of metagenomic organisms (e.g. Degli Esposti et al. 2016) will probably reveal the closest possible relatives to the ancestors of mitochondria. My current bet is that non-photosynthetic members of the family Rhodospirillaceae will turn out to be the closest relatives to proto-mitochondria which we can possibly envisage using a combination of approaches, not just phylogenetic trees.

Acknowledgements I thank Esperanza Martinez Romero (E.M.R.) for her support. This work was sponsored in part by CONACyT grants No. 263876 and 253116 to E.M.R.

References

Atteia A, van Lis R, van Hellemond JJ, Tielens AG, Martin W, Henze K (2004) Identification of prokaryotic homologues indicates an endosymbiotic origin for the alternative oxidases of mitochondria (AOX) and chloroplasts (PTOX). Gene 330:143–148

Atteia A, Adrait A, Brugière S, Tardif M, van Lis R, Deusch O, Dagan T, Kuhn L, Gontero B, Martin W, Garin J, Joyard J, Rolland N (2009) A proteomic survey of Chlamydomonas reinhardtii mitochondria sheds new light on the metabolic plasticity of the organelle and on the nature of the alpha-proteobacterial mitochondrial ancestor. Mol Biol Evol 26(7):1533–1548. https://doi.org/10.1093/molbev/msp068

Aussel L, Pierrel F, Loiseau L, Lombard M, Fontecave M, Barras F (2014) Biosynthesis and physiology of coenzyme Q in bacteria. Biochim Biophys Acta 1837(7):1004–1011. https://doi.org/10.1016/j.bbabio.2014.01.015

Burki F (2014) The eukaryotic tree of life from a global phylogenomic perspective. Cold Spring Harb Perspect Biol 6:a016147

Degli Esposti M (2014) Bioenergetic evolution in proteobacteria and mitochondria. Genome Biol Evol 6(12):3238–3251. https://doi.org/10.1093/gbe/evu257

Degli Esposti M (2016) Late mitochondrial acquisition, really? Genome Biol Evol 8(6):2031–2035. https://doi.org/10.1093/gbe/evw130

Degli Esposti M, Martinez Romero E (2016) A survey of the energy metabolism of nodulating symbionts reveals a new form of respiratory complex I. FEMS Microbiol Ecol 92(6):fiw084. https://doi.org/10.1093/femsec/fiw084

Degli Esposti M, Martinez Romero E (2017) The functional microbiome of arthropods. PLoS One 12(5):e0176573. https://doi.org/10.1371/journal.pone.0176573

Degli Esposti M, Cortez D, Lozano L, Rasmussen S, Nielsen HB, Martinez Romero E (2016) Alpha proteobacterial ancestry of the [Fe-Fe]-hydrogenases in anaerobic eukaryotes. Biol Direct 11:34. https://doi.org/10.1186/s13062-016-0136-3

Gray MW (2015) Mosaic nature of the mitochondrial proteome: implications for the origin and evolution of mitochondria. Proc Natl Acad Sci U S A 112:10133–10138. https://doi.org/10.1073/pnas.1421379112

Hiraishi A, Hoshino Y (1984) Distribution of rhodoquinone in Rhodospirillaceae and its taxonomic implication. J Gen Appl Microbiol 30:435–448

Kawamukai M (2015) Biosynthesis of coenzyme Q in eukaryotes. Biosci Biotechnol Biochem 80(1):23–33. https://doi.org/10.1080/09168451.2015.1065172

Ku C, Nelson-Sathi S, Roettger M, Garg S, Hazkani-Covo E, Martin WF (2015) Endosymbiotic gene transfer from prokaryotic pangenomes: inherited chimerism in eukaryotes. Proc Natl Acad Sci U S A 112(33):10139–10146. https://doi.org/10.1073/pnas.1421385112

Lane N (2006) Power, sex, suicide: mitochondria and the meaning of life. Oxford University Press, Oxford

Lane N, Martin W (2010) The energetics of genome complexity. Nature 467(7318):929–934. https://doi.org/10.1038/nature09486

Lazcano A, Peretó J (2017) On the origin of mitosing cells: a historical appraisal of Lynn Margulis endosymbiotic theory. J Theor Biol 434(17):80–87. https://doi.org/10.1016/j.jtbi.2017.06.036

Louca S, Parfrey LW, Doebeli M (2016) Decoupling function and taxonomy in the global ocean microbiome. Science 353(6305):1272–1277. https://doi.org/10.1126/science.aaf4507

Müller M, Mentel M, van Hellemond JJ, Henze K, Woehle C, Gould SB, Yu RY, van der Giezen M, Tielens AG, Martin WF (2012) Biochemistry and evolution of anaerobic energy metabolism in eukaryotes. Microbiol Mol Biol Rev 76(2):444–495. https://doi.org/10.1128/MMBR.05024-11

Pelosi L, Ducluzeau AL, Loiseau L, Barras F, Schneider D, Junier I, Pierrel F (2016) Evolution of ubiquinone biosynthesis: multiple proteobacterial enzymes with various regioselectivities to catalyze three contiguous aromatic hydroxylation reactions. mSystems 1(4). pii: e00091-16

Pittis AA, Gabaldón T (2016) Late acquisition of mitochondria by a host with chimaeric prokaryotic ancestry. Nature 531:101–104. https://doi.org/10.1038/nature16941

Rochette NC, Brochier-Armanet C, Gouy M (2014) Phylogenomic test of the hypotheses for the evolutionary origin of eukaryotes. Mol Biol Evol 31:832–845. https://doi.org/10.1093/molbev/mst272

Thiergart T, Landan G, Schenk M, Dagan T, Martin WF (2012) An evolutionary network of genes present in the eukaryote common ancestor polls genomes on eukaryotic and mitochondrial origin. Genome Biol Evol 4:466–485

Williams KP, Sobral BW, Dickerman AW (2007) A robust species tree for the alphaproteobacteria. J Bacteriol 189:4578–4586

Yang D, Oyaizu Y, Oyaizu H, Olsen GJ, Woese CR (1985) Mitochondrial origins. Proc Natl Acad Sci U S A 82(13):4443–4447

Mitochondrial Dynamics: A Journey from Mitochondrial Morphology to Mitochondrial Function and Quality

David Sebastián and Antonio Zorzano

Abstract Mitochondria were considered in the past as static and isolated organelles inside the cell. However, currently it is clear that they change in shape, and they are continuously moving along the cell forming an interconnected and highly dynamic network. The sum of all these processes are referred to as mitochondrial dynamics. In addition to the control of mitochondrial morphology and movement, a large number of evidences have turned mitochondrial dynamics into a key factor controlling mitochondrial function and quality, having an important role in respiration, oxidative metabolism, Ca^{2+} homeostasis, mitochondrial quality control, autophagy and apoptosis. The importance of mitochondrial dynamics in physiology is reflected by the increasing number of pathologies associated with its dysregulation, such as neuropathies, neurodegenerative diseases, atherosclerosis, metabolic diseases, sarcopenia and aging. Therefore, the factors regulating mitochondrial dynamics and the understanding of the molecular mechanisms by which mitochondrial dynamics regulates key aspects of mitochondrial biology is of great importance. In this chapter, we will focus on the proteins involved in mitochondrial dynamics, their regulation and their impact in the control of mitochondrial function and quality.

Keywords Mitochondria · Dynamics · Morphology · Neurodegenerative diseases · Mitochondrial quality control

---

D. Sebastián · A. Zorzano (✉)
Institute for Research in Biomedicine (IRB Barcelona), The Barcelona Institute of Science and Technology, Barcelona, Spain

Departament de Bioquímica i Biomedicina Molecular, Facultat de Biologia, Universitat de Barcelona, Barcelona, Spain

Centro de Investigación Biomédica en Red de Diabetes y Enfermedades Metabólicas Asociadas (CIBERDEM), Instituto de Salud Carlos III, Madrid, Spain
e-mail: antonio.zorzano@irbbarcelona.org

P. J. Oliveira (ed.), *Mitochondrial Biology and Experimental Therapeutics*,
https://doi.org/10.1007/978-3-319-73344-9_3

1 Mitochondrial Fusion and Fission: Shaping the Mitochondrial Network

Mitochondrial dynamics is the process by which mitochondria move along the cytoskeleton, and change their morphology, distribution and connectivity by fusion and fission events. Mitochondrial fusion leads to formation a more elongated mitochondrion (or mitochondrial network) through tethering of two adjacent mitochondria, whereas mitochondrial fission generates two daughter mitochondria (or more fragmented mitochondrial network) by the division of an existing mitochondrion (Fig. 1). Therefore, the balance between these two processes determines the mitochondrial network of the cell, and their regulation is carried out by mitochondrial fusion and fission proteins, which in turn are subjected to transcriptional, post-transcriptional and post-translational regulation (Liesa et al. 2009).

1.1 Mitochondrial Fusion

1.1.1 Mitochondrial Fusion Proteins

Mitochondrial fusion is a sequential process in which first the outer and then the inner mitochondrial membrane of two adjacent mitochondria are fused. In mammals outer mitochondrial membrane (OMM) fusion is regulated by mitofusin 1 (MFN1) and mitofusin 2 (MFN2), and inner mitochondrial membrane (IMM)

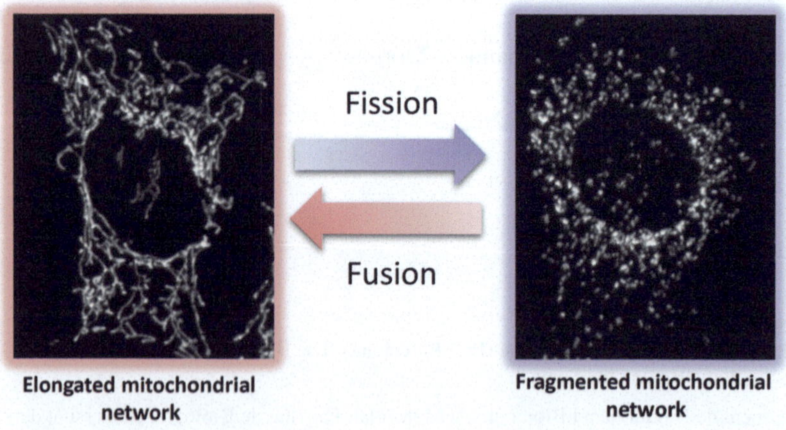

Fig. 1 Mitochondrial network in cells. Mitochondria can organize in either an elongated (left) or a more fragmented mitochondrial network (right), regulated by fusion and fission of mitochondria. In the picture, mouse embryonic fibroblasts (MEF) were incubated with Mitotracker Deep Red which stains specifically mitochondria, and then images were taken by using a confocal microscope

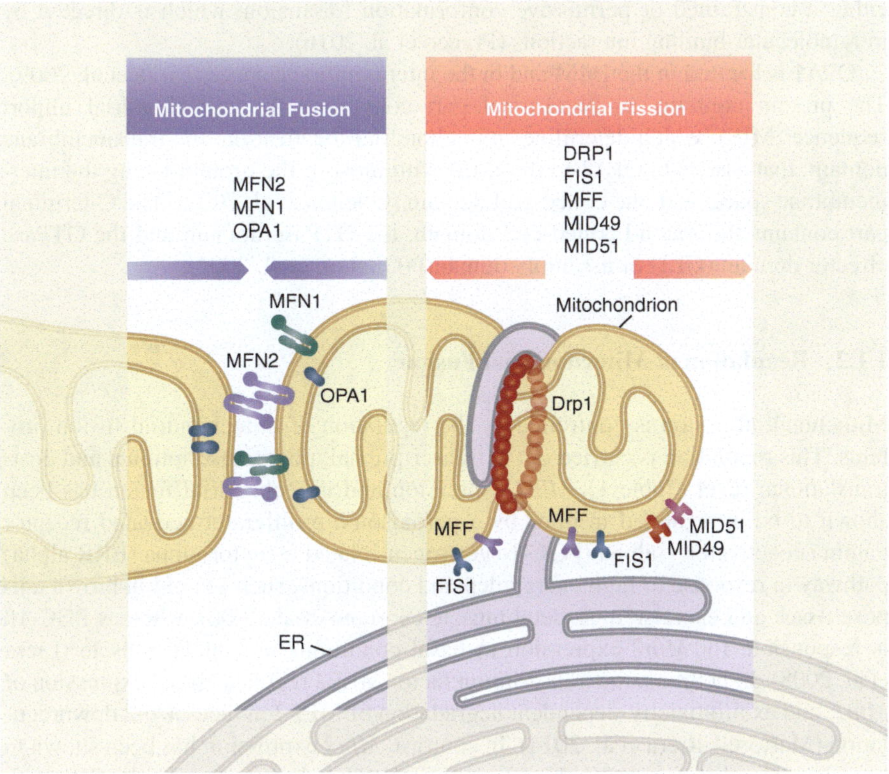

Fig. 2 Proteins involved in mitochondrial dynamics. Mitochondrial outer membrane fusion is mediated by mitofusin 2 (MFN2) and mitofusin 1 (MFN1), and inner mitochondrial membrane is fused by the action of OPA1. Mitochondrial fission is mediated by dynamin-related protein 1 (DRP1), which is recruited to the mitochondrial membrane by different receptors: fission homolog protein 1 (FIS1), mitochondrial fission factor (MFF), and mitochondrial dynamics proteins of 49 kDa and 51 kDa (MID49 and MID51). The endoplasmic reticulum (ER) interacts with mitochondria marking the constriction sites in which fission will occur

fusion is controlled by optic atrophy gene 1 (OPA1) (Liesa et al. 2009) (Fig. 2). All of them are large GTPases, and therefore, mitochondrial fusion is dependent on GTP hydrolysis (Liesa et al. 2009).

MFN1 and MFN2 are located in the OMM, being MFN2 also present in endoplasmic reticulum (ER) and enriched in the contact sites between ER and mitochondria (MAMs) (De Brito and Scorrano 2008). Both MFN1 and MFN2 are composed of a C-terminal part containing a transmembrane domain and a coiled-coil domain (also called heptad-repeat domain 2, HR2), and an N-terminal part containing the GTPase domain and another coiled-coil domain (HR1) (Santel et al. 2003; Rojo et al. 2002). OMM fusion requires the formation of homo-oligomeric and hetero-oligomeric complexes between MFN1 and MFN2, mediated by the HR2 and dependent on GTP hydrolysis (Koshiba et al. 2004). Structurally, mitofusins can adopt either a constrained or permissive conformation for fusion, which is directed by intramolecular binding interactions (Franco et al. 2016).

OPA1 is located in the IMM and in the intermembrane space (Liesa et al. 2009). The protein contains an N-terminal part containing the mitochondrial import sequence (MIS), which determines mitochondrial localization, the transmembrane domain, that associate OPA1 to the IMM with most of the protein facing the intermembrane space, and one coiled-coil domain (Olichon et al. 2002). The C-terminal part contains the second coiled-coil domain, the GTPase domain and the GTPase effector domain (GED or assembly domain) (Olichon et al. 2002).

1.1.2 Regulation of Mitochondrial Fusion

Mitochondrial fusion is controlled by the regulation of mitochondrial fusion proteins. This regulation is carried out at transcriptional, post-transcriptional and post-translational level (Table 1). *Mfn2* expression and mitochondrial fusion has been shown to be upregulated in mice by a peroxisome proliferator-activated receptor gamma coactivator 1-alpha (PGC-1α)/estrogen-related receptor-alpha (ERR alpha) pathway in response to high energy demand conditions, such as cold in brown adipose tissue and exercise in skeletal muscle (Soriano et al. 2006), whereas PGC-1β is responsible for *Mfn2* expression in basal conditions in skeletal muscle (Liesa et al. 2008). In neurons, the transcription factor MEF2 regulates basal expression of *Mfn2*, and excitotoxicity-dependent degradation of MEF2 induces *Mfn2* downregulation (Martorell-Riera et al. 2014). In contrast, *Mfn2* expression has been shown to be reduced by different miRNAs, such as miR-106b in breast cancer cells (Wu et al. 2016a), miR-214 in neuroblastoma cells (Bucha et al. 2015), and mi2R-761 in hepatoma cells (Zhou et al. 2016). In addition to this transcriptional and post-transcriptional regulation, MFN2 is also subjected to post-translational regulation. Hence, MFN2 can be ubiquitinated and marked for proteasomal degradation by different E3-ubiquitin ligases, such as MARCH5 (Nakamura et al. 2006; Yonashiro et al. 2006), PARKIN (Gegg et al. 2010), MUL1 (Peng et al. 2016), and HUWE1 (Leboucher et al. 2012), with a subsequent inhibition of mitochondrial fusion. MFN2 can be also phosphorylated by JNK and PINK1 in response to cellular stress signals and then be tagged for binding with the E3 ubiquitin ligases HUWE1 and PARKIN respectively, which trigger its degradation by the proteasome (Leboucher et al. 2012; Chen and Dorn 2013).

Mfn1 expression is also regulated by miRNAs. Hence, after oxidative or genotoxic stress, miR-140 increases and negatively regulates *Mfn1* translation in cardiomyocytes, inducing mitochondrial fragmentation (Li et al. 2014a). In addition, miR-19b has been found to repress *Mfn1* in osteosarcoma cells (Li et al. 2014b). MFN1 can be also ubiquitinated by PARKIN, thereby promoting its degradation (Gegg et al. 2010). Post-translationally, MFN1 can be phosphorylated by extracellular-signal-regulated kinase (ERK) in its HR1 domain, leading to the inhibition of its mitochondrial fusion activity (Pyakurel et al. 2015).

Table 1 Regulation of mitochondrial fusion and fission proteins

Mitochondrial dynamics protein	Type of regulation	Regulator	Physiological consequence on mitochondrial morphology
MFN2	Transcription	PGC1α/ERRα (8); PGC1β/ERRβ (9); MEF2 (10)	Increased fusion
	Translation	miR-106b (11); miR-214 (12); miR-761 (13)	Reduced fusion
	Ubiquitination	MARCH5 (14, 15); PARKIN (16); MUL1 (17); HUWE1 (18)	Reduced fusion
	Phosphorylation	JNK (18); PINK1 (19)	Reduced fusion
MFN1	Translation	miR-140 (20); miR-19b (21)	Reduced fusion
	Ubiquitination	PARKIN (16)	Reduced fusion
	Phosphorylation	ERK1/2 (22)	Reduced fusion
OPA1	Transcription	NF-kB (23)	Increased fusion
	Cleavage	OMA1 (26, 27); YME1L (26, 27)	Reduced fusion
	OPA1 processing	Prohibitin 2 (28)	Reduced fusion and aberrant cristae morphology
DRP1	Translation	miR-30 (32); miR-499 (36)	Reduced fission
	Phosphorylation	PKA (33–35); Pim-1 (37); GSK3β	Reduced fission
		CaMKIα (38); ROCK1 (39); CDK1 (40); ERK1/2 (41); PKCδ (42)	Increased fission
	Dephosphorylation	Calcineurin (35)	Increased fission
	Sumoylation	MAPL (43)	Increased fission
	Desumoylation	SENP5 (44)	Reduced fission
	Ubiquitination	PARKIN (45)	Reduced fission
		MARCH5 (46)	Increased fission
	O-linked-N-Acetylglucosamination	N-acetyl-glucosaminidase (47)	Increased fission
	S-nitrosylation	Nitric oxide (48)	Increased fission

Expression of *Opa1* has been shown to be regulated by NF-kB in response to insulin in cardiomyocytes, through a signaling pathway involving Akt and mTOR (Parra et al. 2014). In addition, OPA1 is subjected to complex post-transcriptional and post-translation regulation. The *Opa1* gene encodes for eight alternative splice mRNA variants in humans and four in mice (Delettre et al. 2001), which are also regulated by proteolytic cleavage (Ishihara et al. 2006). Thus, alternative splicing of OPA1 produces various long-forms (L-OPA1), which are cleaved to short-forms (S-OPA1) by the action of specific mitochondrial proteases in response to dissipation of mitochondrial membrane potential or induction of apoptosis, leading to mitochondrial fragmentation (Duvezin-Caubet et al. 2006). Two mitochondrial proteases have been shown to cleave OPA1: overlapping activity with the m-AAA protease (OMA1), and YME-like protein 1 (YME1L) (Anand et al. 2014). In addition, the inner membrane prohibitins also play a role in regulating OPA1 processing, since cells lacking prohibitin 2 (PHB2) show a selective loss of L-OPA1 and mitochondrial fragmentation (Merkwirth et al. 2008).

1.2 Mitochondrial Fission

1.2.1 Mitochondrial Fission Proteins

The central player in mitochondrial fission is the cytosolic soluble GTPase dynamin-related protein 1 (DRP1) (Fig. 2). DRP1 consists of a GTPase domain, a middle domain, a variable domain and a GTPase effector domain (GED). Mitochondrial fission requires the recruitment of DRP1 to the OMM, where it accumulates forming dotted structures in the future fission sites and constricts the mitochondrial tubule to mediate membrane fission (Smirnova et al. 2001). It has been suggested that the interaction of ER with these constriction sites in mitochondria is an early step to mark the sites for DRP1 recruitment and mitochondrial fission (Friedman et al. 2011). The recruitment of DRP1 to mitochondria depends on several mitochondria-bound proteins, which act as a DRP1 receptors in the OMM, including FIS1, MFF, MiD49 and MiD51 (Loson et al. 2013) (Fig. 2).

1.2.2 Regulation of Mitochondrial Fission

To date, the main protein shown to be involved in the regulation of mitochondrial fission is DRP1 (Table 1). Expression of *Drp1* has been shown to be modulated by miRNAs, such as miR-30. Hence, in response to induction of apoptosis, miR-30 levels decrease with the subsequent upregulation of p53, which in turn activates *Drp1* expression (Li et al. 2010).

DRP1 is also subjected to different post-translational modifications regulating its activity and translocation to mitochondria, including phosphorylation, SUMOylation, ubiquitination, O-GlcNAcylation and nitrosylation. Serine 637 (S637) in human DRP1 is subjected to phosphorylation and dephosphorylation by PKA and calcineurin respectively, regulating its translocation to mitochondria and its GTPase activity (Kim et al. 2011; Cribbs and Strack 2007; Cereghetti et al. 2008). Hence, phosphorylation at S637 inhibits DRP1 translocation and activity and dephosphorylation of this residue causes the opposite effect. In this regard, miR-499 is involved in the repression of calcineurin, which has been shown to dephosphorylate DRP1 and to inhibit its pro-fission activity (Cereghetti et al. 2008). Therefore, miR-499 induced downregulation of calcineurin leads to a protection of mitochondrial fragmentation in response to ischemia in cardiomyocytes (Wang et al. 2011). In addition to PKA, other kinases have been shown to phosphorylate DRP1 at S637, such as Pim-1, which decreases DRP1 protein levels and translocation to mitochondria (Din et al. 2013), and CaMKIα and ROCK1 in response to calcium overload or hyperglycemia, leading to increase in DRP1 activity and promoting mitochondrial fission (Han et al. 2008; Wang et al. 2012). Other residues in DRP1 can be also phosphorylated, such as S693 by GSK3β, leading to a decrease in DRP1 activity, and S616 by CDK1, ERK1/2 and PKCδ, which promote DRP1 activity and mitochondrial fission (Taguchi et al. 2007; Yu et al. 2011; Qi et al. 2011).

DRP1 protein stability is also regulated by SUMOylation. Thus, the mitochondria-anchored SUMO E3-ligase MAPL, SUMOylates and stabilizes DRP1, inducing mitochondrial fission, and removal of SUMO-1 from DRP1 by the SUMO protease SENP5 leads to inhibition of mitochondrial fission (Braschi et al. 2009; Zunino et al. 2007). Ubiquitination of DRP1 is carried out by PARKIN and MARCH5. PARKIN-mediated ubiquitination induces DRP1 degradation leading to a suppression of mitochondrial fragmentation in neurons, whereas MARCH5-induced ubiquitination promotes mitochondrial fission by regulating the subcellular trafficking of DRP1 (Lutz et al. 2009; Karbowski et al. 2007). DRP1 can also be O-GlcNAcylated on threonine 585 (T585) and T586 in response to high glucose treatment or N-acetyl-glucosaminidase inhibition, leading to reduced S637 phosphorylation, enhanced DRP1 activity and increased mitochondrial fragmentation (Gawlowski et al. 2012). Lastly, nitric oxide, via S-nitrosylation of DRP1, has been shown to induce mitochondrial fission and induce neuronal damage (Cho et al. 2009).

2 Mitochondrial Dynamics Regulates Mitochondrial Function and Quality

In addition to the regulation of mitochondrial morphology, studies performed during the past decade have demonstrated an important physiological role of mitochondrial dynamics in many other cellular functions. In this regard, genetic approaches oriented to repress or overexpress mitochondrial dynamics proteins have revealed that this process plays a crucial role in the control of mitochondrial function and maintenance of mitochondrial quality in a wide range of cells and tissues (Sebastian et al. 2017). Therefore, all the factors mentioned in the previous section and known to regulate mitochondrial fusion and fission proteins, could also regulate mitochondrial function and quality. These observations have led to the demonstration that alterations in mitochondrial dynamics are linked to several human diseases, such as neuropathies (Charcot-Marie-Tooth Syndrome and Autosomal Dominant Hereditary Optic Atrophy), neurodegenerative diseases, atherosclerosis, obesity, type 2 diabetes, muscle atrophy and sarcopenia (Sebastian et al. 2017).

2.1 Regulation of Mitochondrial Function by Mitochondrial Dynamics

The maintenance of a correct balance between mitochondrial fusion and fission is crucial for sustaining a correct mitochondrial function. Hence, *Mfn2* downregulation in mammalian cells, mouse liver or mouse skeletal muscle induced mitochondrial dysfunction, characterized by a decrease in oxygen consumption linked to ATP synthesis, increased proton leak, reduced mitochondrial membrane potential, increased ROS production, and reduced coenzyme Q levels, thereby leading to reduced glucose, pyruvate and fatty acid oxidation (Bach et al. 2003; Sebastian et al. 2012; Chen et al. 2005; Mourier et al. 2015). Conversely, overexpression of either full-length or a carboxyl-terminal truncated variant of MFN2 in HeLa, muscle or liver cultured cells increases glucose oxidation and mitochondrial membrane potential (Pich et al. 2005; Segales et al. 2013). However, ablation of *Mfn1* in MEF cells or mouse liver lead to an increase in ATP-linked and maximal respiration (Kulkarni et al. 2016). *Opa1* downregulation in different cultured cells leads to a reduction in basal and ATP-linked respiration, together with a decrease in mitochondrial membrane potential and an increase in proton leak (Chen et al. 2005; Buck et al. 2016; Zhang et al. 2011). In addition, OPA1 is also involved in the control of mitochondrial cristae shape, respiratory efficiency and the assembly of respiratory chain supercomplexes in mouse (Cogliati et al. 2013). Finally, *Drp1* loss-of-function leads also to an impairment in respiration and an increase in proton leak, and these effects have been shown to be either dependent or independent of mitochondrial fission (Zepeda et al. 2014; Zhang et al. 2017). Altogether, these data reflect the importance of a correct balance between mitochondrial fusion and fission in cells in order to maintain a correct mitochondrial function.

2.2 Regulation of Mitochondrial Quality by Mitochondrial Dynamics

Degradation of damaged mitochondria is a key process for maintaining a healthy and functional mitochondrial population inside the cell. In this regard, selective elimination of mitochondria by autophagy (also named mitophagy) have emerged of vital importance. In the last years, it has been demonstrated that mitochondrial dynamics has a key role in autophagy and mitophagy. Hence, MFN2 and DRP1 have been documented to be necessary for autophagy, although the mechanisms involved are still unclear (Sebastian et al. 2016; Zhao et al. 2012; Hailey et al. 2010; Lin et al. 2015). In addition to these effects on autophagy, mitochondrial dynamics is also involved in mitophagy. Mitochondrial fission is needed for the segregation of damaged mitochondria from the mitochondrial network, which is needed for mitophagy (Twig et al. 2008). In addition, PARKIN, which is recruited to mitochondria after induction of mitophagy, ubiquitinates MFN1 and MFN2 promoting their degradation and then inhibiting mitochondrial fusion (Gegg et al. 2010; Poole et al. 2010). MFN2 could also act as a PARKIN receptor in mitochondria, facilitating its phosphorylation, activation by PINK1 and promoting mitophagy in cardiomyocytes (Chen and Dorn 2013). Phosphorylation and inhibition of mitochondrial translocation of DRP1 is necessary for the preservation of an elongated mitochondrial network and protection from mitophagy after nutrient deprivation (Rambold et al. 2011; Gomes et al. 2011). Moreover, DRP1 has been shown to be necessary in mediating different forms of mitophagy, such as after hypoxic/ischemic injury in the brain, and after hypoxia in HeLa cells interacting with the mitophagy receptor FUNDC1 (Ikeda et al. 2015; Zuo et al. 2014; Wu et al. 2016b). Therefore, all these data strongly suggest that mitochondrial dynamics regulates mitochondrial quality by modulation of autophagy and mitophagy.

References

Anand R, Wai T, Baker MJ, Kladt N, Schauss AC, Rugarli E, Langer T (2014) The i-AAA protease YME1L and OMA1 cleave OPA1 to balance mitochondrial fusion and fission. J Cell Biol 204:919–929

Bach D, Pich S, Soriano FX, Vega N, Baumgartner B, Oriola J, Daugaard JR, Lloberas J, Camps M, Zierath JR, Rabasa-Lhoret R, Wallberg-Henriksson H, Laville M, Palacin M, Vidal H, Rivera F, Brand M, Zorzano A (2003) Mitofusin-2 determines mitochondrial network architecture and mitochondrial metabolism. A novel regulatory mechanism altered in obesity. J Biol Chem 278:17190–17197

Braschi E, Zunino R, McBride HM (2009) MAPL is a new mitochondrial SUMO E3 ligase that regulates mitochondrial fission. EMBO Rep 10:748–754

Bucha S, Mukhopadhyay D, Bhattacharyya NP (2015) Regulation of mitochondrial morphology and cell cycle by microRNA-214 targeting Mitofusin2. Biochem Biophys Res Commun 465:797–802

Buck MD, O'Sullivan D, Klein Geltink RI, Curtis JD, Chang CH, Sanin DE, Qiu J, Kretz O, Braas D, van der Windt GJ, Chen Q, Huang SC, O'Neill CM, Edelson BT, Pearce EJ, Sesaki H, Huber TB, Rambold AS, Pearce EL (2016) Mitochondrial dynamics controls T cell fate through metabolic programming. Cell 166:63–76

Cereghetti GM, Stangherlin A, Martins De Brito O, Chang CR, Blackstone C, Bernardi P, Scorrano L (2008) Dephosphorylation by calcineurin regulates translocation of Drp1 to mitochondria. Proc Natl Acad Sci U S A 105:15803–15808

Chen Y, Dorn GW 2nd (2013) PINK1-phosphorylated mitofusin 2 is a Parkin receptor for culling damaged mitochondria. Science 340:471–475

Chen H, Chomyn A, Chan DC (2005) Disruption of fusion results in mitochondrial heterogeneity and dysfunction. J Biol Chem 280:26185–26192

Cho DH, Nakamura T, Fang J, Cieplak P, Godzik A, Gu Z, Lipton SA (2009) S-nitrosylation of Drp1 mediates beta-amyloid-related mitochondrial fission and neuronal injury. Science 324:102–105

Cogliati S, Frezza C, Soriano ME, Varanita T, Quintana-Cabrera R, Corrado M, Cipolat S, Costa V, Casarin A, Gomes LC, Perales-Clemente E, Salviati L, Fernandez-Silva P, Enriquez JA, Scorrano L (2013) Mitochondrial cristae shape determines respiratory chain supercomplexes assembly and respiratory efficiency. Cell 155:160–171

Cribbs JT, Strack S (2007) Reversible phosphorylation of Drp1 by cyclic AMP-dependent protein kinase and calcineurin regulates mitochondrial fission and cell death. EMBO Rep 8:939–944

De Brito OM, Scorrano L (2008) Mitofusin 2 tethers endoplasmic reticulum to mitochondria. Nature 456:605–610

Delettre C, Griffoin JM, Kaplan J, Dollfus H, Lorenz B, Faivre L, Lenaers G, Belenguer P, Hamel CP (2001) Mutation spectrum and splicing variants in the OPA1 gene. Hum Genet 109:584–591

Din S, Mason M, Volkers M, Johnson B, Cottage CT, Wang Z, Joyo AY, Quijada P, Erhardt P, Magnuson NS, Konstandin MH, Sussman MA (2013) Pim-1 preserves mitochondrial morphology by inhibiting dynamin-related protein 1 translocation. Proc Natl Acad Sci U S A 110:5969–5974

Duvezin-Caubet S, Jagasia R, Wagener J, Hofmann S, Trifunovic A, Hansson A, Chomyn A, Bauer MF, Attardi G, Larsson NG, Neupert W, Reichert AS (2006) Proteolytic processing of OPA1 links mitochondrial dysfunction to alterations in mitochondrial morphology. J Biol Chem 281:37972–37979

Franco A, Kitsis RN, Fleischer JA, Gavathiotis E, Kornfeld OS, Gong G, Biris N, Benz A, Qvit N, Donnelly SK, Chen Y, Mennerick S, Hodgson L, Mochly-Rosen D, Dorn GW II (2016) Correcting mitochondrial fusion by manipulating mitofusin conformations. Nature 540:74–79

Friedman JR, Lackner LL, West M, Dibenedetto JR, Nunnari J, Voeltz GK (2011) ER tubules mark sites of mitochondrial division. Science 334:358–362

Gawlowski, T., Suarez, J., Scott, B., Torres-Gonzalez, M., Wang, H., Schwappacher, R., Han, X., Yates, J. R. 3rd, , Hoshijima, M. & Dillmann, W. 2012. Modulation of dynamin-related protein 1 (DRP1) function by increased O-linked-beta-N-acetylglucosamine modification (O-GlcNAc) in cardiac myocytes. J Biol Chem, 287, 30024-30034.

Gegg ME, Cooper JM, Chau KY, Rojo M, Schapira AH, Taanman JW (2010) Mitofusin 1 and mitofusin 2 are ubiquitinated in a PINK1/parkin-dependent manner upon induction of mitophagy. Hum Mol Genet 19:4861–4870

Gomes LC, Di Benedetto G, Scorrano L (2011) During autophagy mitochondria elongate, are spared from degradation and sustain cell viability. Nat Cell Biol 13:589–598

Hailey DW, Rambold AS, Satpute-Krishnan P, Mitra K, Sougrat R, Kim PK, Lippincott-Schwartz J (2010) Mitochondria supply membranes for autophagosome biogenesis during starvation. Cell 141:656–667

Han XJ, Lu YF, Li SA, Kaitsuka T, Sato Y, Tomizawa K, Nairn AC, Takei K, Matsui H, Matsushita M (2008) CaM kinase I alpha-induced phosphorylation of Drp1 regulates mitochondrial morphology. J Cell Biol 182:573–585

Ikeda Y, Shirakabe A, Maejima Y, Zhai P, Sciarretta S, Toli J, Nomura M, Mihara K, Egashira K, Ohishi M, Abdellatif M, Sadoshima J (2015) Endogenous Drp1 mediates mitochondrial autophagy and protects the heart against energy stress. Circ Res 116:264–278

Ishihara N, Fujita Y, Oka T, Mihara K (2006) Regulation of mitochondrial morphology through proteolytic cleavage of OPA1. EMBO J 25:2966–2977

Karbowski M, Neutzner A, Youle RJ (2007) The mitochondrial E3 ubiquitin ligase MARCH5 is required for Drp1 dependent mitochondrial division. J Cell Biol 178:71–84

Kim H, Scimia MC, Wilkinson D, Trelles RD, Wood MR, Bowtell D, Dillin A, Mercola M, Ronai ZA (2011) Fine-tuning of Drp1/Fis1 availability by AKAP121/Siah2 regulates mitochondrial adaptation to hypoxia. Mol Cell 44:532–544

Koshiba T, Detmer SA, Kaiser JT, Chen H, Mccaffery JM, Chan DC (2004) Structural basis of mitochondrial tethering by mitofusin complexes. Science 305:858–862

Kulkarni SS, Joffraud M, Boutant M, Ratajczak J, Gao AW, Maclachlan C, Hernandez-Alvarez MI, Raymond F, Metairon S, Descombes P, Houtkooper RH, Zorzano A, Canto C (2016) Mfn1 deficiency in the liver protects against diet-induced insulin resistance and enhances the hypoglycemic effect of metformin. Diabetes 65:3552–3560

Leboucher GP, Tsai YC, Yang M, Shaw KC, Zhou M, Veenstra TD, Glickman MH, Weissman AM (2012) Stress-induced phosphorylation and proteasomal degradation of mitofusin 2 facilitates mitochondrial fragmentation and apoptosis. Mol Cell 47:547–557

Li J, Donath S, Li Y, Qin D, Prabhakar BS, Li P (2010) miR-30 regulates mitochondrial fission through targeting p53 and the dynamin-related protein-1 pathway. PLoS Genet 6:e1000795

Li J, Li Y, Jiao J, Wang J, Qin D, Li P (2014a) Mitofusin 1 is negatively regulated by microRNA 140 in cardiomyocyte apoptosis. Mol Cell Biol 34:1788–1799

Li X, Wang FS, Wu ZY, Lin JL, Lan WB, Lin JH (2014b) MicroRNA-19b targets Mfn1 to inhibit Mfn1-induced apoptosis in osteosarcoma cells. Neoplasma 61:265–273

Liesa M, Borda-d'Agua B, Medina-Gomez G, Lelliott CJ, Paz JC, Rojo M, Palacin M, Vidal-Puig A, Zorzano A (2008) Mitochondrial fusion is increased by the nuclear coactivator PGC-1beta. PLoS One 3:e3613

Liesa M, Palacin M, Zorzano A (2009) Mitochondrial dynamics in mammalian health and disease. Physiol Rev 89:799–845

Lin JR, Shen WL, Yan C, Gao PJ (2015) Downregulation of dynamin-related protein 1 contributes to impaired autophagic flux and angiogenic function in senescent endothelial cells. Arterioscler Thromb Vasc Biol 35:1413–1422

Loson OC, Song Z, Chen H, Chan DC (2013) Fis1, Mff, MiD49, and MiD51 mediate Drp1 recruitment in mitochondrial fission. Mol Biol Cell 24:659–667

Lutz AK, Exner N, Fett ME, Schlehe JS, Kloos K, Lammermann K, Brunner B, Kurz-Drexler A, Vogel F, Reichert AS, Bouman L, Vogt-Weisenhorn D, Wurst W, Tatzelt J, Haass C, Winklhofer KF (2009) Loss of parkin or PINK1 function increases Drp1-dependent mitochondrial fragmentation. J Biol Chem 284:22938–22951

Martorell-Riera A, Segarra-Mondejar M, Munoz JP, Ginet V, Olloquequi J, Perez-Clausell J, Palacin M, Reina M, Puyal J, Zorzano A, Soriano FX (2014) Mfn2 downregulation in excitotoxicity causes mitochondrial dysfunction and delayed neuronal death. EMBO J 33:2388–2407

Merkwirth C, Dargazanli S, Tatsuta T, Geimer S, Lower B, Wunderlich FT, Von Kleist-Retzow JC, Waisman A, Westermann B, Langer T (2008) Prohibitins control cell proliferation and apoptosis by regulating Opa1-dependent cristae morphogenesis in mitochondria. Genes Dev 22:476–488

Mourier A, Motori E, Brandt T, Lagouge M, Atanassov I, Galinier A, Rappl G, Brodesser S, Hultenby K, Dieterich C, Larsson NG (2015) Mitofusin 2 is required to maintain mitochondrial coenzyme Q levels. J Cell Biol 208:429–442

Nakamura N, Kimura Y, Tokuda M, Honda S, Hirose S (2006) MARCH-V is a novel mitofusin 2- and Drp1-binding protein able to change mitochondrial morphology. EMBO Rep 7:1019–1022

Olichon A, Emorine LJ, Descoins E, Pelloquin L, Brichese L, Gas N, Guillou E, Delettre C, Valette A, Hamel CP, Ducommun B, Lenaers G, Belenguer P (2002) The human dynamin-related protein OPA1 is anchored to the mitochondrial inner membrane facing the inter-membrane space. FEBS Lett 523:171–176

Parra V, Verdejo HE, Iglewski M, Del Campo A, Troncoso R, Jones D, Zhu Y, Kuzmicic J, Pennanen C, Lopez-Crisosto C, Jana F, Ferreira J, Noguera E, Chiong M, Bernlohr DA, Klip A, Hill JA, Rothermel BA, Abel ED, Zorzano A, Lavandero S (2014) Insulin stimulates mitochondrial fusion and function in cardiomyocytes via the Akt-mTOR-NFkappaB-Opa-1 signaling pathway. Diabetes 63:75–88

Peng J, Ren KD, Yang J, Luo XJ (2016) Mitochondrial E3 ubiquitin ligase 1: a key enzyme in regulation of mitochondrial dynamics and functions. Mitochondrion 28:49–53

Pich S, Bach D, Briones P, Liesa M, Camps M, Testar X, Palacin M, Zorzano A (2005) The Charcot-Marie-Tooth type 2A gene product, Mfn2, up-regulates fuel oxidation through expression of OXPHOS system. Hum Mol Genet 14:1405–1415

Poole AC, Thomas RE, Yu S, Vincow ES, Pallanck L (2010) The mitochondrial fusion-promoting factor mitofusin is a substrate of the PINK1/parkin pathway. PLoS One 5:e10054

Pyakurel A, Savoia C, Hess D, Scorrano L (2015) Extracellular regulated kinase phosphorylates mitofusin 1 to control mitochondrial morphology and apoptosis. Mol Cell 58:244–254

Qi X, Disatnik MH, Shen N, Sobel RA, Mochly-Rosen D (2011) Aberrant mitochondrial fission in neurons induced by protein kinase C{delta} under oxidative stress conditions in vivo. Mol Biol Cell 22:256–265

Rambold AS, Kostelecky B, Elia N, Lippincott-Schwartz J (2011) Tubular network formation protects mitochondria from autophagosomal degradation during nutrient starvation. Proc Natl Acad Sci U S A 108:10190–10195

Rojo M, Legros F, Chateau D, Lombes A (2002) Membrane topology and mitochondrial targeting of mitofusins, ubiquitous mammalian homologs of the transmembrane GTPase Fzo. J Cell Sci 115:1663–1674
Santel A, Frank S, Gaume B, Herrler M, Youle RJ, Fuller MT (2003) Mitofusin-1 protein is a generally expressed mediator of mitochondrial fusion in mammalian cells. J Cell Sci 116:2763–2774
Sebastian D, Hernandez-Alvarez MI, Segales J, Sorianello E, Munoz JP, Sala D, Waget A, Liesa M, Paz JC, Gopalacharyulu P, Oresic M, Pich S, Burcelin R, Palacin M, Zorzano A (2012) Mitofusin 2 (Mfn2) links mitochondrial and endoplasmic reticulum function with insulin signaling and is essential for normal glucose homeostasis. Proc Natl Acad Sci U S A 109:5523–5528
Sebastian D, Sorianello E, Segales J, Irazoki A, Ruiz-Bonilla V, Sala D, Planet E, Berenguer-Llergo A, Munoz JP, Sanchez-Feutrie M, Plana N, Hernandez-Alvarez MI, Serrano AL, Palacin M, Zorzano A (2016) Mfn2 deficiency links age-related sarcopenia and impaired autophagy to activation of an adaptive mitophagy pathway. EMBO J 35:1677–1693
Sebastian D, Palacin M, Zorzano A (2017) Mitochondrial dynamics: coupling mitochondrial fitness with healthy aging. Trends Mol Med 23:201–215
Segales J, Paz JC, Hernandez-Alvarez MI, Sala D, Munoz JP, Noguera E, Pich S, Palacin M, Enriquez JA, Zorzano A (2013) A form of mitofusin 2 (Mfn2) lacking the transmembrane domains and the COOH-terminal end stimulates metabolism in muscle and liver cells. Am J Physiol Endocrinol Metab 305:E1208–E1221
Smirnova E, Griparic L, Shurland DL, Van Der Bliek AM (2001) Dynamin-related protein Drp1 is required for mitochondrial division in mammalian cells. Mol Biol Cell 12:2245–2256
Soriano FX, Liesa M, Bach D, Chan DC, Palacin M, Zorzano A (2006) Evidence for a mitochondrial regulatory pathway defined by peroxisome proliferator-activated receptor-gamma coactivator-1 alpha, estrogen-related receptor-alpha, and mitofusin 2. Diabetes 55:1783–1791
Taguchi N, Ishihara N, Jofuku A, Oka T, Mihara K (2007) Mitotic phosphorylation of dynamin-related GTPase Drp1 participates in mitochondrial fission. J Biol Chem 282:11521–11529
Twig G, Elorza A, Molina AJ, Mohamed H, Wikstrom JD, Walzer G, Stiles L, Haigh SE, Katz S, Las G, Alroy J, Wu M, Py BF, Yuan J, Deeney JT, Corkey BE, Shirihai OS (2008) Fission and selective fusion govern mitochondrial segregation and elimination by autophagy. EMBO J 27:433–446
Wang JX, Jiao JQ, Li Q, Long B, Wang K, Liu JP, Li YR, Li PF (2011) miR-499 regulates mitochondrial dynamics by targeting calcineurin and dynamin-related protein-1. Nat Med 17:71–78
Wang W, Wang Y, Long J, Wang J, Haudek SB, Overbeek P, Chang BH, Schumacker PT, Danesh FR (2012) Mitochondrial fission triggered by hyperglycemia is mediated by ROCK1 activation in podocytes and endothelial cells. Cell Metab 15:186–200
Wu H, Li Z, Wang Y, Yang P, Li H, Wu C (2016a) MiR-106b-mediated Mfn2 suppression is critical for PKM2 induced mitochondrial fusion. Am J Cancer Res 6:2221–2234
Wu W, Lin C, Wu K, Jiang L, Wang X, Li W, Zhuang H, Zhang X, Chen H, Li S, Yang Y, Lu Y, Wang J, Zhu R, Zhang L, Sui S, Tan N, Zhao B, Zhang J, Li L, Feng D (2016b) FUNDC1 regulates mitochondrial dynamics at the ER-mitochondrial contact site under hypoxic conditions. EMBO J 35:1368–1384
Yonashiro R, Ishido S, Kyo S, Fukuda T, Goto E, Matsuki Y, Ohmura-Hoshino M, Sada K, Hotta H, Yamamura H, Inatome R, Yanagi S (2006) A novel mitochondrial ubiquitin ligase plays a critical role in mitochondrial dynamics. EMBO J 25:3618–3626
Yu T, Jhun BS, Yoon Y (2011) High-glucose stimulation increases reactive oxygen species production through the calcium and mitogen-activated protein kinase-mediated activation of mitochondrial fission. Antioxid Redox Signal 14:425–437
Zepeda R, Kuzmicic J, Parra V, Troncoso R, Pennanen C, Riquelme JA, Pedrozo Z, Chiong M, Sanchez G, Lavandero S (2014) Drp1 loss-of-function reduces cardiomyocyte oxygen dependence protecting the heart from ischemia-reperfusion injury. J Cardiovasc Pharmacol 63:477–487
Zhang Z, Wakabayashi N, Wakabayashi J, Tamura Y, Song WJ, Sereda S, Clerc P, Polster BM, Aja SM, Pletnikov MV, Kensler TW, Shirihai OS, Iijima M, Hussain MA, Sesaki H (2011)

The dynamin-related GTPase Opa1 is required for glucose-stimulated ATP production in pancreatic beta cells. Mol Biol Cell 22:2235–2245

Zhang H, Wang P, Bisetto S, Yoon Y, Chen Q, Sheu SS, Wang W (2017) A novel fission-independent role of dynamin-related protein 1 in cardiac mitochondrial respiration. Cardiovasc Res 113:160–170

Zhao T, Huang X, Han L, Wang X, Cheng H, Zhao Y, Chen Q, Chen J, Xiao R, Zheng M (2012) Central role of mitofusin 2 in autophagosome-lysosome fusion in cardiomyocytes. J Biol Chem 287:23615–23625

Zhou X, Zhang L, Zheng B, Yan Y, Zhang Y, Xie H, Zhou L, Zheng S, Wang W (2016) MicroRNA-761 is upregulated in hepatocellular carcinoma and regulates tumorigenesis by targeting Mitofusin-2. Cancer Sci 107:424–432

Zunino R, Schauss A, Rippstein P, Andrade-Navarro M, McBride HM (2007) The SUMO protease SENP5 is required to maintain mitochondrial morphology and function. J Cell Sci 120:1178–1188

Zuo W, Zhang S, Xia CY, Guo XF, He WB, Chen NH (2014) Mitochondria autophagy is induced after hypoxic/ischemic stress in a Drp1 dependent manner: the role of inhibition of Drp1 in ischemic brain damage. Neuropharmacology 86:103–115

Mitochondria and Ageing

Jose Viña and Consuelo Borras

Abstract Mitochondria are the major sites of oxygen utilisation for energy production in cells. Indeed, all the reactions of the Krebs' Cycle take place in mitochondria and they produce NADH and succinate, which are then oxidised in the respiratory chain. Experiments dating back to the early part of the twentieth century seemed to indicate that at a high rate of oxygen consumption (referred to gram of body weight) was normally associated with a low maximum lifespan. Thus, it was thought that it was the rate of oxygen utilisation that was related to "the rate of living". However, more recent data pointed out that birds are unique because they combine high rates of oxygen consumption with a high maximum lifespan. It would later be pointed out that the maximal lifespan is more correlated with the rate of free radical production by mitochondria rather than the rate of oxygen utilisation. These experiments were performed under the general scheme of the free radical theory of ageing. Still, more than 300 theories have been postulated to explain ageing and this can indicate that none of them is completely satisfactory to explain a complex phenomenon such as ageing. We postulate in this chapter that the free radical theory of ageing could be revisited and that it is the age-associated derangement of the free radical signalling network that is central to understand ageing.

Keywords Free radicals · Oxidants · Longevity · Antioxidants · Frailty

1 The Free Radical Theory of Ageing. The Mitochondrial Free Radical Theory of Ageing as Proposed by Miquel

Mitochondria are the major sites of oxygen utilisation for energy production in cells. Indeed, all the reactions of the Krebs' Cycle take place in mitochondria and they produce NADH and succinate, which are then oxidised via complexes 1 and 2 of the

J. Viña (✉) · C. Borras
Department of Physiology, Faculty of Medicine and Dentistry, University of Valencia, Valencia, Spain

Center for Biomedical Network Research on Frailty and Healthy Aging (CIBERFES), CIBER-ISCIII, Valencia, Spain
e-mail: jose.vina@uv.es

P. J. Oliveira (ed.), *Mitochondrial Biology and Experimental Therapeutics*,
https://doi.org/10.1007/978-3-319-73344-9_4

respiratory chain. Experiments dating back to the early part of the twentieth century seemed to indicate that at a high rate of oxygen consumption (referred to gram of body weight) was normally associated with a low maximum lifespan. Thus, it was thought that it was the rate of oxygen utilisation that was related to "the rate of living" (Pearl 1928). However, Barja and co-workers (Barja de Quiroga 1999) pointed out that birds are unique because they combine high rates of oxygen consumption with a high maximum lifespan. It would later be pointed out that the maximal lifespan is more correlated with the rate of free radical production by mitochondria rather than the rate of oxygen utilisation. These experiments were performed under the general scheme of the free radical theory of ageing.

This theory was first postulated by Denham Harman who proposed that "ageing and the degenerative diseases associated with it could be attributed basically to the deleterious side attacks of free radicals on cell constituents and on the connective tissues" (Harman 1956). An important antecedent of this critical paper was that of Gershman and co-workers who proposed that oxygen poisoning and X irradiation have a mechanism in common (Gershman et al. 1954). They are both caused by deleterious effects of free radicals. Therefore, back in the 1950s, the essential bases for the free radical theory of ageing were proposed.

Importantly, Harman himself already mentioned that mitochondria are essential to his free radical theory of ageing (Harman 1972). However, it was Jaime Miquel in 1980 who clearly formulated the mitochondrial free radical theory of ageing (Miquel et al. 1980). What Miquel stated was that mitochondria are the origins of a substantial part of the free radical production by cells and that because the close vicinity of mitochondrial DNA and the sites of radical productions, mitochondrial DNA should be considered as an essential target of the free-radical-induced damage to cells.

It is outside the scope of this review to dwell on the number of theories that have tried to explain ageing; the reader is referred to a review on this topic by the authors of this paper in IUBMB Life (Vina et al. 2007). Suffice it to say that more than 300 theories have been postulated to explain ageing and this can indicate that none of them is completely satisfactory to explain a complex phenomenon such as ageing. We postulated that the free radical theory of ageing could be revisited and that it is the age-associated derangement of the free radical signalling network that is central to understand ageing (Vina et al. 2013). However, mitochondria are central to most of theories proposed for ageing that have high explanatory power.

2 Mitochondrial Disruption of Cell Signalling in Ageing

Mitochondria were originally thought of as organelles that produce energy and they were also considered as a source of free radicals that would literally cause damage to cells. Around the turn of the century, however, the idea that free radicals had a physiological function and were not just there to damage cells began to take shape. A critical paper by Wolf Droge published in physiological reviews in 2002 summarised the knowledge in this field at the time (Droge 2002). What Droge pointed out was that "at moderate concentrations" nitric oxide superoxide anion and other reactive oxygen species were not causative of damage, but would play an important role as regulatory mediators in signalling processes. This author went on to say that "many of the ROS mediated responses actually protect cells against oxidative stress and re-establish redox homeostasis" (Droge 2002). Therefore, the concept of oxidative stress as first pointed out by Helmut Sies was then extended to the concept of redox homeostasis. Another paper published in 2004 by Enrique Cadenas had the title "Mitochondrial function in ageing: coordination with signalling and transcriptional pathways" (Yin et al. 2004). The title itself underpinned the importance of mitochondria in generating signals that are free radicals in chemical nature, but that are essential for normal cell functioning. It is not the aim of this chapter to review the role of free radicals produced by mitochondria as signalling molecules. But what we want to point out is that in ageing, the whole network of the free radical based cell signalling pathways is deranged.

In the last 15 years, the very concept of the free radical theory of ageing as postulated by Harman has been challenged. On some occasions, ageing has been associated not with an increased oxidative damage, but with a reductive damage. In our opinion, many of the criticisms of the free radical theory of ageing are quite correct in that if one considers the strict concept of free radicals "causing damage" to cells. We have proposed the cell signalling disruption theory of ageing. The major postulate of this theory is that, as stated before, reactive oxygen species generated by mitochondria are responsible for an altered cell signalling network. It is not just molecular damage to structures like double bonds that is responsible for the altered cell function in ageing tissues, but the altered signalling causing a cascade of event leading to disrupted function in old animals and even persons (Vina et al. 2013).

3 Mitochondrial DNA Is More Susceptible to Damage than Nuclear DNA

One critical aspect of the free radical-associated damage was put forward by Britton Chance and his colleagues who proposed the idea that because radicals are so reactive they would do much of the damage at sites near their production (Boveris and Chance 1973). It was therefore important to try and show which the major sites of free radical production in ageing were.

DNA damage has been observed in a large number of cell lines from mammals exposed to oxidative stress (Halliwell and Auroma 1991). This damage includes double and single chain breaks, deletions, base changes, oxidative damage, and even chromosome aberrations. The main molecular mechanisms involved are the direct reaction of hydroxyl radicals and carbonyl compounds with DNA and the activation of nucleases (Halliwell and Auroma 1991). The superoxide anion and H_2O_2 do not react with DNA unless there are transition metal ions that allow the formation of hydroxyl radicals. Hydroxyl radicals are capable of attacking the deoxyribose, purines and pyrimidines, generating numerous products, such as 8-hydroxydesoxyguanosine (8-oxodG), thymidine glycols and 8-hydroxyadenosine (Halliwell and Auroma 1991).

Bruce Ames and co-workers calculated that reactive oxygen species modify approximately 10,000 bases of DNA per cell (Ames et al. 1993). DNA repairing enzymes are able to repair the vast majority of these lesions, but not all. Therefore, DNA lesions that go un-repaired, such as 8-oxo-dG accumulate with age.

As stated before, according to the mitochondrial theory of ageing, mitochondria are the main source of free radical production. This implies that they are also the main target of free radicals, as they are very unstable molecules (Miquel et al. 1992). Therefore, it is well known that mitochondrial DNA is much more oxidised with age than nuclear DNA (Richter et al. 1988). Our group, in 1996, showed that oxidative damage to mitochondrial DNA correlates with oxidation of mitochondrial glutathione (García de la Asunción et al. 1996).

Moreover, mitochondrial DNA (mtDNA) is especially susceptible to oxidative damage and mutations because it lacks protective histones (Johns 1995). Thus, the formation of 8-oxoDG in mitochondrial DNA increases as the rate of hydroperoxide production increases by mitochondria (Giulivi et al. 1995). Suter and Richter have reported that oxidized bases are present in moderate amounts in 16.3 kb mitochondrial DNA molecules but are found in large numbers in mitochondrial DNA fragments (Suter and Richter 1999). These results, together with the discovery of endonucleases related to mitochondrial oxidative damage, demonstrate the existence of a mitochondrial DNA repair system (Suter and Richter 1999; Shen et al. 1995; Croteau et al. 1999).

According to the mitochondrial aging theory, Barja and Herrero found that the oxidative damage associated with mtDNA is inversely related to the maximal survival of mammals, while oxidative damage to nuclear DNA is not (Barja and Herrero 2000). In addition, several studies have reported that levels of oxidative damage to mtDNA are several times higher than those produced in nuclear DNA, and that mutations in mtDNA are also more frequent than in nuclear DNA (Richter et al. 1988; Shigenaga et al. 1994; Suter and Richter 1999; Barja and Herrero 2000). However, Anson et al. pointed out that when oxidative damage to mtDNA purified from isolated mitochondria is measured, it is observed that it has been over-estimated (Anson et al. 2000). In fact, they observed similar levels of 8-oxodG in nuclear and mitochondrial DNA. Our group published a method of isolation of mtDNA that does not require previous isolation of mitochondria (Asunción et al. 1996). Using this method we obtained levels of 8-oxodG three to nine times higher in mtDNA than nuclear DNA in the eight species studied (Sastre et al. 1998). Nevertheless, these results should be confirmed using different methods of isolation of mtDNA.

Oxidative lesions to mtDNA accumulate with age in human and rodent tissues. (Halliwell and Auroma 1991; Asunción et al. 1996; Ames et al. 1993).

Two characteristics were thought to occur that supported this conclusion. The first is that it was believed that mitochondrial DNA is much less protected against free radical attack than nuclear DNA because of a lower number of histones and other DNA-associated proteins. This is still thought to be the case. On the other hand, it was also thought that mitochondrial DNA does not have repair mechanisms and therefore the damage could be less reversible than nuclear DNA. This is not thought to be the case anymore. We know now that mitochondrial DNA contains repair mechanisms to counteract oxidative damage. In any case, it still holds true that oxidative damage to mitochondrial DNA is (at least as determined by measuring 8-hydroxy-2-deoxy-guanosine) ten times higher than nuclear DNA damage.

The mtDNA repair system is unable to counteract the amount of ROS generated in mitochondria throughout life. Point mutations and deletions in mtDNA are produced in the tissues of old animals (Gadaleta et al. 1992; Lezza et al. 1994; Lee et al. 1997). In humans the deletions in mtDNA increase more than 10,000 times with age (Lezza et al. 1999). According to the Miquel's hypothesis, the highest percentage of mtDNA deletion is observed at the end of life in postmitotic tissues such as brain, heart and muscle (Lezza et al. 1999). Point mutations and aberrant forms of mtDNA from postmitotic cells are also related to degenerative diseases associated with aging.

Age-associated mtDNA deletions display mosaic distribution. This supposes a localized distribution of the deletions even in the same tissue, so that some cells possess a greater percentage of deletions than others. Thus, a difference of two or three orders of magnitude for the deletion at the 4977 bp level—the most common deletion—is observed in different regions of the brain (Corral-Debrinski et al. 1992; Cortopassi et al. 1992).

Damage to mtDNA may affect mitochondrial gene transcription (Kristal et al. 1994). In fact, it has been reported that there is a decline associated with age of mitochondrial transcript levels in some rat tissues and in Drosophila (Gadaleta et al. 1990; Calleja et al. 1993). In addition, since mtDNA lacks introns, any mutation will affect DNA coding sequences (Johns 1995). Thus, Lezza et al. observed a correlation between the decrease in oxidative phosphorylation capacity and the increase in the percentage of current deletions in mtDNA during aging (Lezza et al. 1994). Therefore, it has been suggested that mutations in mitochondrial DNA may contribute significantly to the aging process and to the development of neurodevelopmental diseases.

Furthermore, we have more recently proposed that there is also a relationship between mitochondrial and nuclear DNA damage. We found that mtDNA fragments are inserted into nuclear DNA contributing to aging and related diseases by alterations in the nucleus (Caro et al. 2010). Consequently, mitochondria can be a major trigger of aging but the final target could also be the nucleus.

4 Mitochondria Are Damaged Inside Cells

We provided the first evidence that mitochondria are damaged inside cells (see Fig. 1) (Sastre et al. 1996). The question in the early '90s was whether mitochondria were damaged when isolated from old tissues because they were more fragile and therefore they were damaged in the isolation procedures or whether they were already damaged inside cells. By using metabolic as well as flow cytometry studies we demonstrated that mitochondria were damaged inside cells and not during isolation. The metabolic approach consisted of checking that those metabolic pathways that involved mitochondria were more affected in aged tissues than those that involved only extra mitochondrial compartments. For instance, the rate of gluconeogenesis from lactate

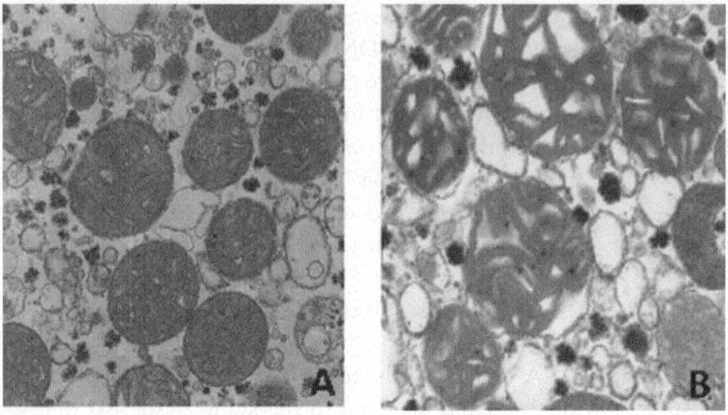

Fig. 1 Electron microscopy images of young (**a**) and old (**b**) mitochondria

and pyruvate was much more affected than that from glycerol, the first involving mitochondria and the second not (Sastre et al. 1996).

Our findings were independently confirmed almost simultaneously by the group of Bruce Ames who showed the mitochondrial decay in ageing and postulated that this was majr characteristic of the ageing process (Ames et al. 1995, Hagen et al. 1997).

Once it was clear that free radicals produced by mitochondria were responsible for much of the damage that was associated with ageing, a number of laboratories set out to test with newer techniques the old idea that the rate of respiration was responsible for "the rate of ageing". Prominent among these laboratories was that of Gustavo Barja in Madrid. Barja et al. compared the rate of oxygen consumption in tissues, for instance muscle from pigeons and rats, two species whose individuals are approximately the same body weight (Barja 1998; Barja de Quiroga 1999). There was a very significant difference between the longevity of the maximal longevity in these two species: rats live up to 3 or 4 years whereas pigeons can live more than 25. What Barja and his colleagues observed was that longevity did not correlate with the oxygen consumption by mitochondria but with the rate of free radical production by these organelles (Barja et al. 1994). These results have strong correlative evidence in favour of the free radical theory of ageing. These experiments have been confirmed in many laboratories and some of the criticisms of the free radical theory of ageing are based on assumptions that are not strictly based on the idea that the rate of production of radicals is what determines the rate of ageing.

Damage to DNA as well as to other critical components of the mitochondria like lipids, results in two important pathophysiological facts. The first is that since mitochondria are damaged, this may explain in a large part the energy collapse that occurs with ageing. The second conclusion is that oxidant production by mitochondria from old animals will be increased when compared with that of young ones. The old idea that "2% of all oxygen consumption in organisms is converted to free radicals rather than to water and energy" is not true. It may serve as an "average value", but the fact is that when electrons flow through the respiratory chain and active oxidative phosphorylation takes place, the rate of radical production in mitochondria can be as low as 10% that of damaged mitochondria when energy production is far less. This idea which is based on clear chemical considerations has far-reaching biological conclusions. For instance, the old idea that in exercised muscle, more free radicals are produced because the cells consume more oxygen is wrong. When an active respiratory chain and oxidative phosphorylation takes place, fewer radicals are produced. The inverse can be said of ageing mitochondria. When mitochondria from old animals use oxygen less efficiently, more radicals are produced. Therefore, the "inefficient work" of mitochondria from old animals explains both the energy collapse that occurs in aged cells and the increased production of oxygen radicals that not only alter the cell signalling network but also cause damage to susceptible molecules like the double bonds in unsaturated fatty acids or the deoxyguanosine residues in mitochondrial DNA.

5 Sex Differences in Free Radical Production and Its Relationship with Longevity

Differences in longevity between sexes offer interesting possibilities to understand ageing. Animals with a very similar genetic background may differ in their average life span for a much as 10%. Of paramount importance is the fact that these differences is also occur in humans. However, in order to find proof that this is not due to sociological changes or peculiarities between societies (i.e. whether women smoke more of less than men etc.) one must understand the differences in longevity in animal species.

We and others tried to explain why Wistar rat females live longer than males, the maximal life span being approximately 10% (Borras et al. 2003). In a similar fashion, using Fisher 344 rats, i.e. the same species but a different strain, in which longevity is also higher in females than in males, the group of Leeuwenburgh observed that males produce more reactive oxygen species than females (Jang et al. 2004). However, the higher longevity of females as compared with males is not a universal phenomenon. In other species of rodents such as mice, some strains, for instance the C57BL6, show a higher longevity of males when compared with females (Ali et al. 2006). Moreover, a variation of this strain which was used by Leeuwenburgh's group, i.e. C57B16J mice show no differences in longevity between sexes (Sanz et al. 2007). In contrast, the Swiss albino mouse shows an increased longevity in females when compared with males (Navarro et al. 2004). So not only do we observe different sex specific longevity in different species, but also in different strains of the same species. This offers a unique opportunity to study comparative ageing, i.e. whether there are hormonal differences, different sensitivity or reactivity to hormones or different fundamental molecular mechanisms of ageing. Since we are dealing with the same species, only different sexes, we are faced with optimal animal models to understand fundamental gerontological aspects as well as the hormonal regulation of ageing (Borras et al. 2003; Ali et al. 2006; Sanz et al. 2007)

In many species in which females live longer than males, the former produce approximately half the amount of mitochondrial peroxide than the latter (Borras et al. 2003).

In early 2000s, we measured peroxide production by mitochondria from female and from male Wistar rats and found that females produce approximately half the amount of peroxides than males. This was completely reversed when rats were ovariectomised, thus tracing the differential gender effect on radical production to ovarian hormones. We then measured the free radical production of mitochondria from ovariectomised females which had been treated with estrogens. In this case, estradiol was able to reverse the effect of ovariectomy and the rate of peroxide production was similar to that of females. The decrease in radical production resulted in a significantly lower damage to mitochondrial DNA in females than in males. In fact, the level of 8-oxo-deoxyguanosine was up to four-fold higher in males than in females (Borras et al. 2003).

However, Ali et al. showed that in those strains of mice like the C57BL6 they were using in which males live longer than females it is males that produce fewer oxidants than females (Ali et al. 2006). The authors measured not only oxidant production by determining dihydroethidium oxidation in brain but they also measured the EPR signals in brain mitochondria of their mice. Thus, in this strain of mice in which males live longer than females it is males that produce fewer radicals. Far from contradicting results fromour laboratory, these results nicely confirm our results. The claim by all of us is that the sex that lives longer produces fewer radicals independently of whether it is males or females who live longer. The critical test will be to see why estrogens promote the expression of antioxidant genes in a given species and do not promote that expression in other species or strains. A third confirmation of this hypothesis came from the laboratory of Christiaan Leeuwenburgh (Sanz et al. 2007). These researchers studied a particular strain derived from the C57B6J in which males and females live the same. Not surprisingly, these authors did not observe changes in either oxygen consumption, complex 1 and complex 3 oxidant production, protein carbonyls, oxidised DNA or other indicators of oxidative stress. Thus, in those strains in which females live the same as males, there is no difference in oxidant production or in oxidative stress associated with sex. The overall conclusion is that estrogens promote a lower rate of production of radicals in those species or strains in which females live longer than males. On the other hand, in those animals in which males live longer than females it is males that produce fewer radicals and those in which longevity is the same in both sexes, radical production is also similar.

6 Mitochondrial Diseases

It has long been recognised that mitochondria are essential, and therefore when altered, can cause diseases many of which are very severe. These are known as mitochondrial diseases. It is not within the scope of this chapter which is aimed more towards understanding the role of mitochondria in ageing, to describe in detail these mitochondrial diseases. Tissues most susceptible to mitochondrial-driven disease states are those with higher metabolic demand, i.e. brain, eye, liver, heart, and skeletal muscle. Mitochondrial disease states include:

1. The mitochondrial myopathies, a group of neuromuscular diseases that includes Kearns-Sayre syndrome, mitocondrial encephalopathy lactic acidosis and strokes, myoclonic epilepsy with ragged red fibers, and mitocondrial neuro-gastrointestinal encephalomyopathy that have genetic origins (Schapira 2006).
2. Disorders of mitochondrial electron transport chain that affect its assembly and/or stability and function. They involve both genetic factors and cofactor deficiencies (coenzyme Q10) that can lead to decreased ATP production and increased free-radical production. This free radical production leads to neurodegenerative diseases such as Alzheimer's disease, Parkinson disease, Huntington's disease, and amyotrophic lateral sclerosis (Johri and Beal 2012).

Leber hereditary optic neuropathy, which involves visual failure caused by the degeneration of retinal ganglion cells, is the most common disease with mtDNA mutations with a prevalence of approximately 12 cases per 100,000 in the population (Schapira 2006).

7 Toxicological Aspects: The Treatment of AIDS with Zidovudine Causes Mitochondrial Pathology that Explains Muscle Damage Associated with AIDS Treatment

Moreover, mitochondria are involved in not only genetic diseases but also pharmacological and toxicological diseases. For instance, patients who suffer from HIV infection, who are successfully treated with a cocktail of drugs including Zidovudine (AZT) frequently suffer from muscle myopathies. These were attributed to free radical damage associated with increased production of mitochondrial reactive oxygen species. We were interested in this pathology because first of all HIV patients can be considered as an accelerated model of ageing and secondly because mitochondria were clearly involved in this myopathy which is, it is important to underline, associated with the treatment and not with the primary disease. In any case, we observed that patients who suffer from HIV infection have a decreased activity of the cystathionase activity, an important pathway in the biosynthesis of thiols from methionine (de la Asuncion et al. 1998). Cystathionase itself is lower in patients suffering from HIV infection (Martin et al. 2001) and this renders patients more susceptible to oxidative insults. When these susceptible patients are treated with Zidovudine, mitochondrial damage clearly occurs, and this can be seen from both biochemical and molecular analysis. Histological evidence can also be observed. These alterations can, we observed, be successfully treated when patients (or experimental animals) who are treated with Zidovudine, receive high doses of vitamins C and E. We mention these toxicological aspects of the mitochondrial function because it is now becoming a trend to think that vitamin supplementation is never useful. It can be so if the patients are under specific oxidative damage, especially associated with mitochondria as is the case in Zidovudine treatment.

8 Alzheimer's Disease

On a different note, one of the most prevalent and devastating age-associated diseases is Alzheimer's. Early work from George Perry and Mark Smith showed that many of the manifestations of Alzheimer's disease can be traced to free radical damage (Perry et al. 1998). What Smith and Perry did was analysing the brains of Alzheimer's patients and seeing that in the areas more damaged in the disease, one could observe oxidative damage even by histochemical analysis.

Our understanding of the pathophysiology of Alzheimer's changed dramatically when it was appreciated that amyloid beta caused damage from inside the cells (Selkoe 1991) and not only from the plaques as was Alzheimer's original observation. In fact, some researchers believe that amyloid beta when it is in the plaques is in quite an inert form and it is the soluble Abeta what causes damage. Pioneer work by Catarina Oliveira at the University of Coimbra in Portugal showed that mitochondria were affected by amyloid beta and that in fact, cells that were artificially depleted of mitochondria were much more resistant to damage associated with amyloid beta than those containing normally functioning mitochondria (Pereira et al. 1998). The work of Oliveira prompted us to study the rate of radical production by mitochondria in the presence and absence of amyloid beta. We observed that the toxic peptide causes an increase in the rate of oxygen production by mitochondria and that this could be prevented when mitochondria were co-incubated with amyloid beta and heme (Lloret et al. 2008). The protective effect of heme was based on the observation by Atamna and co-workers that amyloid beta strongly binds to iron in heme and that this can explain the lower rate of respiration in mitochondria in the presence of Alzheimer's peptide (Atamna and Frey 2004). As stated before, a lower rate of oxygen consumption by mitochondria is usually associated with an increased production of radicals. This was the case and therefore the increased rate of oxygen production could contribute to damage to the mitochondria and to other organelles and explain the low energy production in areas of brain affected by Alzheimer's disease.

The altered free radical production in mitochondria of neurons in the presence of amyloid beta, prompted us to study the role of this increased free radical production in the cell signalling associated with Alzheimer's. For instance, we observed that RCAN 1 which is an adaptive enzyme that is upregulated in the presence of chronic oxidative stress (Davies et al. 2007) and that is an inhibitor of calcineurin could contribute to understanding the relationship between amyloid beta and Tau, originally thought of as two distinct and independent hallmarks of Alzheimer's disease (Lloret et al. 2015). Indeed, an upregulation of RCAN 1, will inhibit the dephosphorylation of phospho-Tau and lead to increase its levels of phosphor-tau (see Fig. 2). Altered radical production by mitochondria not only in normal ageing but also age-associated disease such as Alzheimer's can lead to pathophysiological hallmarks of the disease and of the ageing process itself.

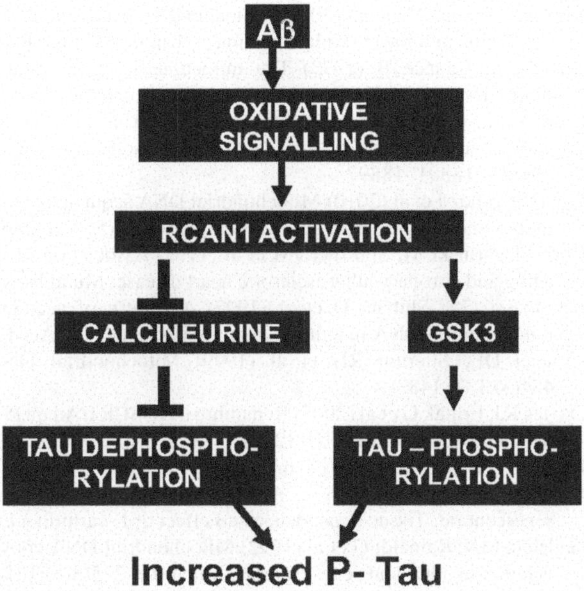

Fig. 2 Abeta and p-Tau interaction model

References

Ali SS, Xiong C et al (2006) Gender differences in free radical homeostasis during aging: shorter-lived female C57BL6 mice have increased oxidative stress. Aging Cell 5(6):565–574

Ames BN, Shigenaga MK et al (1993) Oxidants, antioxidants, and the degenerative diseases of aging. Proc Natl Acad Sci U S A 90:7915–7922

Ames BN, Shigenaga MK et al (1995) Mitochondrial decay in aging. Biochim Biophys Acta 1271(1):165–170

Anson RM, Hudson E et al (2000) Mitochondrial endogenous oxidative damage has been overestimated. FASEB J 14(2):355–360

Asunción JG, Millan A et al (1996) Mitochondrial glutathione oxidation correlates with age-associated oxidative damage to mitochondrial DNA. FASEB J 10:333

de la Asuncion JG, del Olmo ML et al (1998) AZT treatment induces molecular and ultrastructural oxidative damage to muscle mitochondria. Prevention by antioxidant vitamins. J Clin Invest 102(1):4–9

Atamna H, Frey WH 2nd (2004) A role for heme in Alzheimer's disease: heme binds amyloid beta and has altered metabolism. Proc Natl Acad Sci U S A 101(30):11153–11158

Barja G (1998) Mitochondrial free radical production and aging in mammals and birds. Ann N Y Acad Sci 854:224–238

Barja de Quiroga C (1999) Mitochondrial oxygen radical generation and leak: sites of produccion in states 4 an 3, organ specificity and relation to aging and longevity. J Bioenerg Biomembr 31(4):347–366

Barja G, Herrero A (2000) Oxidative damage to mitochondrial DNA is inversely related to maximum life span in the heart and brain of mammals. FASEB J 14(2):312–318

Barja G, Cadenas S et al (1994) Low mitochondrial free radical production per unit O2 consumption can explain the simultaneous presence of high longevity and high aerobic metabolic rate in birds. Free Radic Res 21(5):317–327

Borras C, Sastre J et al (2003) Mitochondria from females exhibit higher antioxidant gene expression and lower oxidative damage than males. Free Radic Biol Med 34(5):546–552
Boveris A, Chance B (1973) The mitochondrial generation of hydrogen peroxide. General properties and effect of hyperbaric oxygen. Biochem J 134(3):707–716
Calleja M, Pena P et al (1993) Mitochondrial DNA remains intact during Drosophila aging, but the levels of mitochondrial transcripts are significantly reduced. J Biol Chem 268(25):18891–18897
Caro P, Gomez J et al (2010) Mitochondrial DNA sequences are present inside nuclear DNA in rat tissues and increase with age. Mitochondrion 10(5):479–486
Corral-Debrinski M, Shoffner JM et al (1992) Association of mitochondrial DNA damage with aging and coronary atherosclerotic heart disease. Mutat Res 275(3-6):169–180
Cortopassi GA, Shibata D et al (1992) A pattern of accumulation of a somatic deletion of mitochondrial DNA in aging human tissues. Proc Natl Acad Sci U S A 89(16):7370–7374
Croteau DL, Stierum RH et al (1999) Mitochondrial DNA repair pathways. Mutat Res 434(3):137–148
Davies KJ, Ermak G et al (2007) Renaming the DSCR1/Adapt78 gene family as RCAN: regulators of calcineurin. FASEB J 21(12):3023–3028
Droge W (2002) Free radicals in the physiological control of cell function. Physiol Rev 82(1):47–95
Gadaleta MN, Petruzzella V et al (1990) Reduced transcription of mitochondrial DNA in the senescent rat. Tissue dependence and effect of L-carnitine. Eur J Biochem 187(3):501–506
Gadaleta MN, Rainaldi G et al (1992) Mitochondrial DNA copy number and mitochondrial DNA deletion in adult and senescent rats. Mutat Res 275(3-6):181–193
García de la Asunción J, Millan A et al (1996) Mitochondiral glutathione oxidation correlates with age-associated oxidative damage to mitochondrial DNA. FASEB J 10:333–338
Gershman R, Gilbert DL, Nye SW, Dwyer P, Fenn WO (1954) Oxygen poisoning and X irradiation: a mechanism in common. Science 119:623–626
Giulivi C, Boveris A et al (1995) Hydroxyl radical generation during mitochondrial electron transfer and the formation of 8-hydroxydesoxyguanosine in mitochondrial DNA. Arch Biochem Biophys 316(2):909–916
Hagen TM, Yowe DL, Bartholomew JC, Wehr CM, Do KL, Park JY, Ames BN (1997) Mitochondrial decay in hepatocytes from old rats: membrane potential declines, heterogeneity and oxidants increase. Proc Natl Acad Sci U S A 94(7):3064–3069
Halliwell B, Auroma OI (1991) DNA damage by oxygen derived species. Its mechanism of action and measurement in mammalian systems. FEBS Lett 281:9–19
Harman D (1956) Aging: a theory based on free radical and radiation chemistry. J Gerontol 2:298–300
Harman D (1972) The biological clock: The mitocondria. J Am Geriatr Soc 20(4):145–147
Jang YM, Kendaiah S et al (2004) Doxorubicin treatment in vivo activates caspase-12 mediated cardiac apoptosis in both male and female rats. FEBS Lett 577(3):483–490
Johns DR (1995) Seminars in medicine of the Beth Israel Hospital, Boston. Mitochondrial DNA and disease. N Engl J Med 333(10):638–644
Johri A, Beal MF (2012) Mitochondrial dysfunction in neurodegenerative diseases. J Pharmacol Exp Ther 342(3):619–630
Kristal BS, Chen J et al (1994) Sensitivity of mitochondrial transcription to different free radical species. Free Radic Biol Med 16(3):323–329
Lee CM, Weindruch R et al (1997) Age-associated alterations of the mitochondrial genome. Free Radic Biol Med 22(7):1259–1269
Lezza AM, Boffoli D et al (1994) Correlation between mitochondrial DNA 4977-bp deletion and respiratory chain enzyme activities in aging human skeletal muscles. Biochem Biophys Res Commun 205(1):772–779
Lezza AM, Mecocci P et al (1999) Mitochondrial DNA 4977 bp deletion and OH8dG levels correlate in the brain of aged subjects but not Alzheimer's disease patients. FASEB J 13(9):1083–1088
Lloret A, Badia MC et al (2008) Gender and age-dependent differences in the mitochondrial apoptogenic pathway in Alzheimer's disease. Free Radic Biol Med 44(12):2019–2025

Lloret A, Fuchsberger T et al (2015) Molecular mechanisms linking amyloid beta toxicity and Tau hyperphosphorylation in Alzheimers disease. Free Radic Biol Med 83:186–191
Martin JA, Sastre J et al (2001) Hepatic gamma-cystathionase deficiency in patients with AIDS. JAMA 285(11):1444–1445
Miquel J (1992) An update on the mitochondrial-DNA mutation hypothesis of cell aging. Mutat Res 275(3-6):209–216
Miquel J, Economos AC et al (1980) Mitochondrial role in cell aging. Exp Gerontol 15(6):575–591
Navarro A, Gomez C et al (2004) Beneficial effects of moderate exercise on mice aging: survival, behavior, oxidative stress, and mitochondrial electron transfer. Am J Physiol Regul Integr Comp Physiol 286(3):R505–R511
Pearl R (1928) The rate of living. University of London Press, London
Pereira C, Santos MS et al (1998) Mitochondrial function impairment induced by amyloid beta-peptide on PC12 cells. Neuroreport 9(8):1749–1755
Perry G, Castellani RJ et al (1998) Reactive oxygen species mediate cellular damage in Alzheimer disease. J Alzheimers Dis 1(1):45–55
Richter C, Park JW et al (1988) Normal oxidative damage to mitochondrial and nuclear DNA is extensive. Proc Natl Acad Sci U S A 85:6465–6467
Sanz A, Hiona A et al (2007) Evaluation of sex differences on mitochondrial bioenergetics and apoptosis in mice. Exp Gerontol 42(3):173–182
Sastre J, Pallardo FV et al (1996) Aging of the liver: age-associated mitochondrial damage in intact hepatocytes. Hepatology 24(5):1199–1205
Sastre J, Millan A et al (1998) A Ginkgo biloba extract (EGb 761) prevents mitochondrial aging by protecting against oxidative stress. Free Radic Biol Med 24(2):298–304
Schapira AH (2006) Mitochondrial disease. Lancet 368(9529):70–82
Selkoe DJ (1991) Amyloid protein and Alzheimer's disease. Sci Am 265(5):68–71. 74-6, 78
Shen CC, Wertelecki W et al (1995) Repair of mitochondrial DNA damage induced by bleomycin in human cells. Mutat Res 337(1):19–23
Shigenaga MK, Hagen TM et al (1994) Oxidative damage and mitochondrial decay in aging. Proc Natl Acad Sci U S A 91:10771–10778
Suter M, Richter C (1999) Fragmented mitochondrial DNA is the predominant carrier of oxidized DNA bases. Biochemistry 38(1):459–464
Vina J, Borras C et al (2007) Theories of ageing. IUBMB Life 59(4-5):249–254
Vina J, Borras C et al (2013) The free radical theory of aging revisited: the cell signaling disruption theory of aging. Antioxid Redox Signal 19(8):779–787
Yin F, Sancheti H et al (2004) Mitochondrial function in ageing: coordination with signalling and transcriptional pathways. J Physiol 594(8):2025–2042

The Mitochondrial Permeability Transition Pore

Claudia Morganti, Massimo Bonora, Luigi Sbano, Giampaolo Morciano, Giorgio Aquila, Gianluca Campo, Mariusz R. Wieckowski, Carlotta Giorgi, and Paolo Pinton

Abstract The mitochondrial permeability transition (MPT) consists of an abrupt increase in the permeability of the inner mitochondrial membrane to low molecular weight solutes, resulting in the osmotic breakout of the organelle. MPT drives cell death and provides an etiological contribution to several human disorders characterized by the acute loss of post-mitotic cells. These conditions include ischemia/reperfusion injury, cancer and neurodegenerative disorders. However, precise knowledge of the structure and regulators of the supramolecular entity that induces MPT, the so-called *permeability transition pore complex* (PTPC), is lacking and this constitutes a substantial obstacle in the development of MPT-targeting agents with clinical applications. Here we report the current evidences about molecular structure and regulatory components of PTPC. In particular we pay attention on new two proteins which recently were added to the list of PTPC components: the mitochondrial F_1F_O ATP synthase, particularly and the SPG7 paraplegin matrix AAA peptidase subunit. At least a detailed overview of MPT contribution to pathological condition is provided, focusing on the idea that to develop therapeutic drugs, it will be fundamental to understand the molecular composition of the PTPC.

Keywords Mitochondrial permeability transition · Permeability transition pore complex · F1FO ATP synthase · Mitochondrial disorders

C. Morganti · M. Bonora · L. Sbano · G. Morciano · C. Giorgi · P. Pinton (✉)
Section of General Pathology, Department of Morphology, Surgery and Experimental Medicine, University of Ferrara, Ferrara, Italy

Laboratory for Technologies of Advanced Therapies (LTTA), University of Ferrara, Ferrara, Italy
e-mail: paolo.pinton@unife.it

G. Aquila
Department of Morphology, Surgery and Experimental Medicine, University of Ferrara, Ferrara, Italy

G. Campo
Cardiovascular Institute, University of Ferrara, Ferrara, Italy

M. R. Wieckowski
Department of Biochemistry, Nencki Institute of Experimental Biology, Warsaw, Poland

P. J. Oliveira (ed.), *Mitochondrial Biology and Experimental Therapeutics*,
https://doi.org/10.1007/978-3-319-73344-9_5

1 Mitochondrial Permeability Transition

The concept of the *mitochondrial permeability transition* (MPT) refers to a sudden and irreversible increase in the permeability of the inner mitochondrial membrane (IMM) to small solutes up to 1.5 kDa, leading to the progressive dissipation of mitochondrial transmembrane potential ($\Delta\psi_m$). This unregulated passage of water into the mitochondrial matrix results in the osmotic breakdown of the organelle (Bonora et al. 2015); this implies the suspension of mitochondrial functions, including ATP production.

In the 1990s, the mechanism by which the MPT ultimately initiates a form of regulated cell death (RCD) that often (but not always) manifests with necrotic morphological features was shown (Galluzzi et al. 2015; Crompton and Costi 1990). The MPT also constitutes a central checkpoint in the apoptotic pathway; it can generate apoptotic waves. For example, when outer mitochondrial membrane (OMM) rupture occurs upon mitochondrial swelling, the release of proteins involved in the effector phase of apoptosis (such as cytochrome C, AIF, SMAC/DIABLO and EndoG) into the cytosol is inevitable. This notion is supported by various studies in which isolated mitochondria and living cells treated with MPT-inducing stimuli clearly show apoptotic-like features, and MPT targeting protects against cell death (Marchetti et al. 1996; Kroemer 1997).

A frequently discussed problem involves what determines the induction of necrosis or apoptosis upon MPT onset. The answer may reside in the ATP levels of the cell (Eguchi et al. 1997); when cellular ATP content is high, apoptosis can initiate and proceed, but when the energy level is low or insufficient, the necrotic pathway prevails. In addition, and related to the ATP level, necrosis may be caused by a prolonged and irreversible MPT event experienced by all mitochondria in the cell. Conversely, the MPT may affect only one or few mitochondria (Duchen et al. 1998) allowing the cell to either recover completely or initiate the apoptotic pathway. However, it has recently been demonstrated how transient MPT events (called tPTPs or MitoWinks) are also correlated with normal physiologic manifestations that allow the resetting of mitochondria (Lu et al. 2016) that is essential for cell survival.

For these reasons, the MPT is a significant event in different human pathologies (e.g., reperfusion injuries, neurodegeneration, and cancer). However, precise knowledge of the structure and mode of action of the supramolecular entity that induces MPT, the so-called *permeability transition pore complex* (PTPC), is lacking, and this constitutes a substantial obstacle in the development of MPT-targeting agents with clinical applications.

The best-characterized MPT is triggered by the accumulation of Ca^{2+} ions in the cytosol and then in the mitochondria (Izzo et al. 2016). Thus, besides the accumulation of mitochondrial Ca^{2+}, major MPT stimulators include reactive oxygen species (ROS), inorganic phosphate, intracellular alkalinization, long-chain fatty acids, atractyloside and carboxyatractyloside. The latter two inhibit members of the adenine nucleotide translocase (ANT) protein family by locking them into a cytoplasmic-side open conformation (Brenner and Grimm 2006).

Conversely, MPT inhibitors include ATP and ADP, NADH and NAD^+, glutamate, bongkrekic acid, which locks ANT family members into a matrix-side open conformation, 5-isothiocyanato-2-[2-(4-isothiocyanato-2-sulfophenyl) ethenyl]benzene-1-sulfonic acid (DIDS), an inhibitor of voltage-dependent anion channel (VDAC), and cyclosporine A (CsA), which targets peptidylprolyl isomerase F (PPIF, best known as cyclophilin D, CyPD) (Martel et al. 2012).

The MPT-inhibitory potential of CsA has been documented so extensively, in vitro and in vivo, that this molecule is currently considered the gold standard method for the confirmation of presumed instances of MPT (Kepp et al. 2011).

2 Mitochondrial Permeability Transition Pore Complex: Molecular Structure

Despite the intense experimental interest generated by MPT throughout the last two decades, the precise molecular composition of the PTPC remains elusive.

Accordingly, the first PTPC model proposed at the end of 1990s was a supramolecular entity assembled at the juxtaposition of the inner and outer mitochondrial membranes, composed by VDAC, ANT, and regulatory components including hexokinase 1 (HK1) and creatine kinase mitochondrial 1 (CKMT1) (Beutner et al. 1996, 1998). Moreover, CyPD was supposed to have a central role in the PTPC due to its interacting partners (Crompton et al. 1998), which include VDAC and ANT, and its pharmacological profile (Tanveer et al. 1996).

This model was substantially challenged by genetic approaches. Indeed, the genetic co-inactivation of three distinct VDAC isoforms (Vdac1, Vdac2 and Vdac3) failed to protect murine fibroblasts from MPT induction by hydrogen peroxide (an MPT inducer) and did not influence the ability of their mitochondria to undergo MPT in response to Ca^{2+} (Baines et al. 2007).

Similarly, the simultaneous knockout of the genes coding for two distinct ANT isoforms, namely, Slc25a4 (encoding Ant1) and Slc25a5 (encoding Ant2), failed to abolish the ability of murine hepatocytes to succumb to several MPT inducers, including the Ca^{2+} ionophore Br-A23187, in a CsA-inhibitable manner. Moreover, mitochondria isolated from $Slc25a4^{-/-}Slc25a5^{-/-}$ hepatocytes retained the ability to undergo MPT in vitro upon exposure to a depolarizing agent (Kokoszka et al. 2004).

The only component to survive genetic analysis was CyPD (Baines et al. 2005; De Marchi et al. 2006), confirming its role as a modulator of the PTPC. Indeed, it is unlikely that CyPD, which is mainly localized within the mitochondrial matrix, would constitute the pore-forming component of the PTPC. Therefore, CyPD is currently viewed as the major gatekeeper of the MPT, regulating the opening of the PTPC but not lining up the pore that physically allows for the entry of low-molecular-weight solutes into the mitochondrial matrix. Additionally, CyPD played a central role in the identification of (or attempts to identify) the channel-forming components of the PTPC core, primarily through the identification of CyPD-interacting proteins.

For instance, in 2008, the phosphate carrier PiC was shown to bind CyPD and ANT1, an interaction that was potentiated by MPT-inducing conditions and inhibited by CsA (Leung et al. 2008).

Inorganic phosphate has been known since 1965 as an MPT-promoting metabolite (Tedeschi et al. 1965), suggesting that the PTPC would possess a specific binding site. In physiological conditions, inorganic phosphate is transported across the inner mitochondrial membrane by members of the SLC protein family, including SLC25A3 (best known as PHC or PiC) and SLC25A24 (also known as APC1) (Palmieri 2004).

A high-throughput genetic screen showed that PiC overexpression promotes apoptotic cell death and that a small-interfering RNA-mediated depletion of PiC has cytoprotective effects (Alcala et al. 2008). Later, Baines and coworkers demonstrated that PiC is not a core component of the PTPC, although in its absence, the MPT occurred more slowly (Gutierrez-Aguilar et al. 2014; Kwong et al. 2014). Although the ability of PiC to influence mitochondrial dynamics may be involved in this process (Pauleau et al. 2008), the exact molecular mechanisms by which PiC promotes cell death under some circumstances remain to be elucidated.

Concerning APC1, it is known that it can respond to increases in cytosolic Ca^{2+} levels, favoring the mitochondrial uptake of ATP and ADP and consequently inhibit MPT (Traba et al. 2012).

Recently, thanks to monitoring MPT in living cells using fluorescence-imaging-based techniques (Bonora et al. 2016), two proteins were added to the list of PTPC components: the mitochondrial F_1F_O ATP synthase, particularly the c subunit of the F_O domain (which in humans is encoded by three genes, ATP5G1, ATP5G2 and ATP5G3), and the SPG7 paraplegin matrix AAA peptidase subunit (Giorgio et al. 2013; Bonora et al. 2013; Alavian et al. 2014; Shanmughapriya et al. 2015) (in Fig. 1 a model of PTPC is reported).

The mitochondrial F_1F_O ATP synthase is a multiprotein complex consisting of a globular domain that protrudes into the mitochondrial matrix (F_1 domain) and an inner mitochondrial membrane-embedded domain (F_O domain); the domains are interconnected by a central and a lateral stalk (Yoshida et al. 2001). Mammalian ATP synthases contain 15 different subunits: α, β, γ, δ, ε, a, b, c, d, e, f, g, A6L, F6 and O (also known as oligomycin sensitivity-conferring protein, OSCP). These subunits form a fully functional holoenzyme with a total molecular weight of ~ 600 kDa.

The interest in the mitochondrial F_1F_O ATP synthase as the possible molecular identity of the PTPC is the result of assays screening for potential CyPD binding partners. The screen identified CyPD as co-migrating with the mitochondrial F_1F_O ATP synthase in blue native gels (Giorgio et al. 2009) and the subunit OSCP (oligomycin sensitivity conferring protein) as a binding site (Giorgio et al. 2013).

In this study, Giorgio et al. proposed that the PTPC forms from dimers of the F_1F_O ATP synthase (Giorgio et al. 2013). Indeed, the mitochondrial F_1F_O ATP synthase dimers excised and extracted from blue native gels and reconstituted into lipid bilayers have been reported to provoke currents that are consistent with the known electrophysiological properties of the PTPC. However, no PTPC-like currents were observed after the addition of monomeric F_1F_O ATP synthase that was extracted from the same

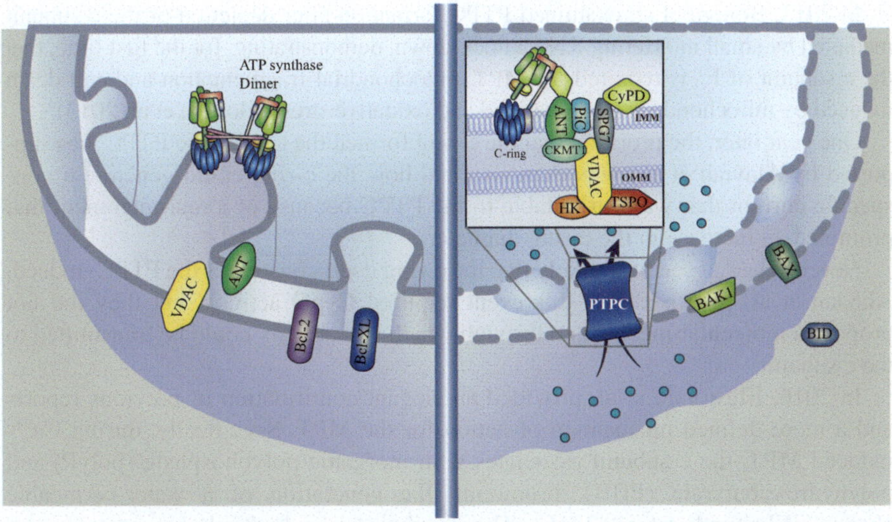

Fig. 1 Hypothetical PTPC molecular structure. MPT is mediated by the opening of a supramolecular entity, called PTPC, assembled at the juxtaposition between mitochondrial membranes. Structural and functional studies show that multiple mitochondrial and cytosolic proteins intervene in the formation or regulation of the PTPC, yet the actual pore-forming unit of the complex remains elusive. These proteins include VDAC, ANT, HK, CYPD, PiC, TSPO, CKMT1; in the text a detailed explanation is reported

blue native gel. These data have not been reproduced by independent investigators and appear to contradict several studies demonstrating, for instance, the cytoprotective effects of endogenous molecules that promote F_1F_O ATP synthase dimerization (Garcia et al. 2006). Interestingly, in 2016, Gerle highlighted that a straightforward alternative explanation could be the loss of PTPC-specific subunits during extraction from the excised gel bands or reconstitution into the black membrane. Indeed, bovine F_1F_O ATP synthase comprises 17 different subunits of which the two F_O subunits DAPIT and 6.8 kDa are easily lost during extraction of this fragile multisubunit membrane complex from the inner mitochondrial membrane (Gerle 2016). However, robust evidence excluding a key role for F_1F_O ATP synthase dimers in the MPT has not yet been provided. Only recently, it was reported that MPT induction is linked to F_1F_O ATP synthase dimers dissociation and that stabilizing F_1F_O ATP synthase dimers by genetic approaches inhibits PTPC opening (Bonora et al. 2017).

An alternative proposal for a pore-forming entity within the components that make up the mitochondrial F_1F_O ATP synthase highlights its proton-transporting c-ring.

F_1F_O ATP synthase c-rings consist of multiple copies of c subunits, varying between species (8–15), that are arranged as a circle (Pogoryelov et al. 2012). As describe above, the c subunits of F_1F_O ATP synthase are encoded by three genes, ATP5G1, ATP5G2 and ATP5G3, and it has been shown that cells depleted of ATP5G1 or ATP5G3 exhibit reduced sensitivity to MPT-driven RCD (Bonora et al. 2013; Alavian et al. 2014).

In 2013, Bonora et al. examined PTPC formation after depletion of the c-subunit mediated by small interfering RNA knock-down, demonstrating, for the first time, that the c subunit of F_O is required for MPT, mitochondrial fragmentation and cell death induced by mitochondrial Ca^{2+} overload and oxidative stress (Bonora et al. 2013).

One year later, the proposal that the c-ring formed the core of the PTPC was supported by Alavian et al., who demonstrated how the c-ring could generate a non-specific current that was attributable to the PTPC because of a rearrangement that promoted an increase in the c-ring diameter.

Nevertheless, it seems unlikely that the c-ring itself constitutes the PTPC. Indeed, Alavian et al. proposed that this event required CypD activity, but they did not propose a molecular mechanism through which this activity could be transmitted to the c subunit.

In 2016, Elustondo et al. provided an elegant confirmation of previous reports and a more defined mechanism of action for the MPT. Specifically, during Ca^{2+}-induced MPT, the c subunit associates with inorganic polyphosphate (polyP) and polyhydroxybutyrate (PHB), promoting the generation of a water-permeable channel (Elustondo et al. 2016). The c subunit is a hydrophobic protein with properties that are similar to those of lipids, and it is not expected to be able to form water-filled pores in its c-ring. Therefore, these data suggest that the c subunit is responsible for forming the calcium-dependent channel with the help of polyP possibly serving as the hydrophilic coating of the pore (Morciano et al. 2017).

Despite the increasing evidence for the pivotal roles of c subunits in the MPT (Halestrap 2014), a study in March 2017, by the Walker group showed the MPT in the absence of a c subunit (He et al. 2017). Specifically, they generated a clonal cell, HAP1-A12 (near-haploid human cell), in which ATP5G1, ATP5G2, and ATP5G3 were disrupted. They reported that the HAP1-A12 cells were incapable of producing the c subunit, but they preserved the characteristic properties of the PTPC. These data are the results of a single cell clone, so clonal adjustment cannot be excluded. Indeed, they reported that HAP1-A12 cells assembled a vestigial ATP synthase, with intact F_1-catalytic and peripheral stalk domains and supernumerary subunits e, f, and g but without membrane subunits ATP6 and ATP8. The authors did not exclude the possibility that the PTPC could be associated with the ATP synthase complex, but they speculated that the most likely components available to form the pore were the b, e, f, and g subunits (He et al. 2017). Although these data should be confirmed and strictly challenged before excluding the c subunits from the list of PTPC components, the creation of a cell clone characterized by ATP5G1, ATP5G2, and ATP5G3 deletion will improve experimental research regarding the involvement of c-subunits in the MPT.

Finally, SPG7, an integral protein of the inner mitochondrial membrane with metalloprotease activity, has recently been identified as a PTPC component (Shanmughapriya et al. 2015). A phenotypic screen based on the mitochondrial Ca^{2+} retention capacity (CRC) of digitonin-permeabilized cells after treatment with siRNAs designed to suppress translation of a set of mitochondrial proteins was used to identify regulators of the PTPC. The screen identified 13 proteins whose suppression caused desensitization of the PTPC to Ca^{2+}, among which well-known modulators that do not take part in PTPC core formation, such as CyPD, were revealed. Stable depletion of CyPD, VDAC1, or SPG7 appeared to be equally effective in protecting cultured human cells from hydrogen peroxide-dependent PTPC opening. Moreover, it was shown that SPG7 could be co-immunoprecipitated with CyPD in a complex that also included VDAC1. The interaction between CyPD and SPG7 depends on the C-terminus of SPG7 (but not on its catalytic activity) and is sensitive to CsA. Indeed, a CyPD mutant lacking the seven highly conserved residues that constitute the CsA binding site is unable to bind to SPG7. Conversely, the transmembrane domain of SPG7 is responsible for VDAC1 binding and does not depend on CyPD–SPG7 interactions. Finally, deletion of SPG7 resembles the deletion of CyPD in terms of resistance to MPT-inducing stimuli (Shanmughapriya et al. 2015). Thus, SPG7 might constitute a key regulator of MPT. However, these findings have been obtained in cultured cells only, and they have not yet been reproduced by independent investigators.

2.1 Regulatory Components

Several proteins have been shown to regulate the activity of the core PTPC, including cytosolic and mitochondrial proteins.

The translocator protein (18 kDa) (TSPO), a protein of the outer mitochondrial membrane, constitutes the benzodiazepine-binding component of the so-called peripheral benzodiazepine receptor, an oligomeric complex involving VDAC and ANT (Mcenery et al. 1992).

The physiological role of TSPO involves steroid biosynthesis regulation (Mukhin et al. 1989). Moreover, several studies have implicated TSPO in the MPT. Indeed, the ability of a series of endogenous (e.g., protoporphyrin IX) (Pastorino et al. 1994) and exogenous (e.g., PK11195, Ro5–4864, diazepam) (Hirsch et al. 1998; Chelli et al. 2001) TSPO agonists to cause an MPT has been reported. Although their roles in modulating the MPT are clear, the effects of TSPO ligands are variable, ranging from cytoprotective to cytotoxic (Kugler et al. 2008; Shargorodsky et al. 2012; Campanella et al. 2008; Decaudin et al. 2002).

Various kinases have been shown to interact with the core PTPC, such as CKMT1, HK1, HK2, glycogen synthase kinase 3β (GSK3β) and protein kinase Cε (PKCε) (Verrier et al. 2004).

Some of these kinases, including CKMT1, HK1 and HK2, do not phosphorylate protein substrates, implying that their MPT-modulatory activity originates either from their physical interaction with core PTPC components or from their ability to catalyze metabolic reactions.

CKMT1 is localized to the mitochondrial intermembrane space, and it can bind to VDAC1 and ANT1 (Beutner et al. 1996, 1998). Additionally, CKMT1 phosphorylates creatine to generate phosphocreatine, a reaction that is tightly coupled to oxidative phosphorylation and of consequence to the availability of ATP and ADP (Wallimann et al. 1998; Dolder et al. 2003). It remains to be formally demonstrated whether the MPT-modulatory activity of CKMT1 originates from its physical interaction with the PTPC components or its catalytic activity.

HKs catalyze the rate-limiting step of glycolysis, converting glucose into glucose-6-phosphate in an ATP-dependent manner (Wilson 2003). Both HK1 and HK2 interact with VDAC isoforms (Pastorino and Hoek 2008). These interactions are associated with an optimal flux through glycolysis and with major cytoprotective effects (Pastorino and Hoek 2003).

Conversely, PKCε and GSK3β exert MPT-modulatory functions by phosphorylating core PTPC components (Pastorino et al. 2005; Baines et al. 2003).

PKCε has been reported to phosphorylate VDAC1, promoting HK2 binding and consequent PTPC inhibition (Baines et al. 2003).

Activation of GSK3β has been reported to disrupt the binding of HK2 to mitochondria by phosphorylating VDAC1, resulting in an enhancement of chemotherapy-induced MPT-related cytotoxicity (Pastorino et al. 2005). Moreover, the activation of GSK3β has also been linked to the MPT-triggering phosphorylation of CyPD (Chiara et al. 2012). However, inactivation of GSK3β caused by phosphorylation on Ser9 has been shown to inhibit the PTPC by physically disrupting the ANT1/CyPD interaction (Nishihara et al. 2007). In addition, the activation of several upstream signal transducers, such as AKT1, mammalian target of rapamycin (mTOR), protein kinase A and protein kinase cGMP-dependent type I (PRKG1), has been reported to converge with the inactivation of GSK3β, mediating MPT-inhibitory effects (Juhaszova et al. 2004; Takuma et al. 2001; Pediaditakis et al. 2010).

Interestingly, multiple components of the MOMP-regulatory machinery have been shown to physically and functionally interact with core components of the PTPC, suggesting a tight relationship between the two RCD processes, including mutually regulatory crosstalk.

For instance, BCL-2 and BCL-2-like 1 (BCL-2L1, best known as BCL-XL) have been shown to inhibit MPT by regulating the open state of VDAC1 (Shimizu et al. 1999; Vander Heiden et al. 1999), but the MPT-regulating activity of anti-apoptotic BCL-2 family members remains questionable.

Instead, BAX, BAK1 and BCL-2-like 11 (BCL-2L11, a BH3-only protein best known as BID) reportedly promote MPT-driven apoptosis by interacting with ANT1 and/or VDAC1 (Marzo et al. 1998; Zamzami et al. 2000; Narita et al. 1998). Similarly, BCL-2-associated agonist of cell death (BAD, another BH3-only protein) can trigger a VDAC1-dependent, BCL-XL-responsive mechanism of the MPT. Indeed, the MPT appears to result from BAD-dependent displacement of BCL-XL from VDAC1 rather than from a physical BAD/VDAC1 interaction (Roy et al. 2009).

Furthermore, in 2012, a pool of p53 localized to the mitochondrial matrix was shown to participate in the MPT; in response to oxidative stress, p53 accumulated in the mitochondrial matrix and triggered PTPC opening and necrosis through a physical interaction with CyPD (Vaseva et al. 2012).

3 Pathological Relevance

Throughout the last two decades, several studies have implicated the MPT as a major etiological determinant in a wide variety of acute and chronic disorders characterized by an unwarranted loss of postmitotic cells. These conditions include ischemia/reperfusion injury, cancer and neurodegenerative disorders (in Fig. 2 a schematic summary of involvement of MPT in pathologies is reported).

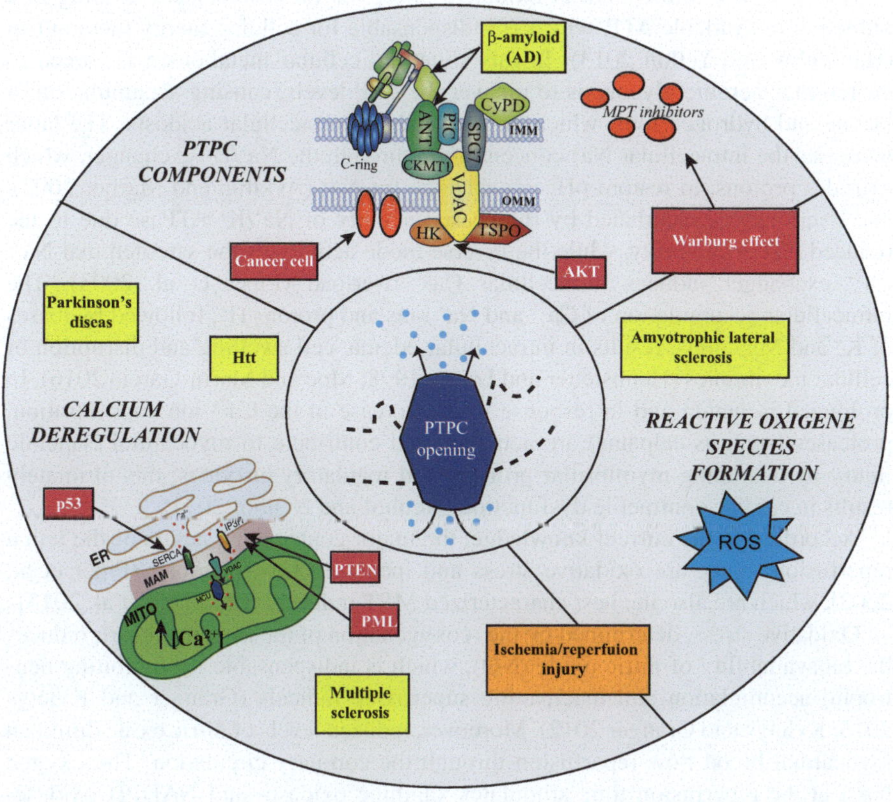

Fig. 2 Involvement of MPT in pathologies. The MPT is been identify as a major etiological determinant in a wide variety of acute and chronic pathologies. The PTPC opening could be caused by (1) changes in PTPC components and regulators, (2) calcium deregulation and/or (3) ROS formation increment. Here it is shown a summary of the most common alterations (deeply described in the text) reported in ischemia/reperfusion injury (orange box), cancer (red boxes) and neurodegenerative disorders (yellow boxes) which link MPT to disease onset

3.1 Ischemia/Reperfusion Injury

In 2012, the World Health Organization estimated 7.4 million deaths from coronary heart diseases (CHDs) worldwide, particularly in low- and middle-income countries. Accounting for nearly 13% of all deaths globally, CHDs have since been in the spotlight of cardiovascular research. Although the risk factors that promote CHDs have been identified (Akhabue et al. 2014), not all players involved in this pathology are well characterized. Generally, the pathophysiological effects of CHD are imputable to the debilitating consequences of coronary occlusions, followed by additional damage due to reperfusion, which as a whole, is known as ischemia-reperfusion injury (Kawajiri et al. 2011; Araszkiewicz et al. 2013; Frank et al. 2012).

In cardiomyocytes, the deprivation of oxygen following myocardial infarction (MI) results in a mitochondrial oxidative phosphorylation blockade, leading to a reduction in available ATP, which is indispensable for cellular energy metabolism (Hausenloy and Yellon 2013). In this situation, cellular metabolism is forced to switch to anaerobic glycolysis to recover the ATP levels, causing accumulation of lactate and hydrogen ions, which in turn leads to intracellular acidosis. The latter increases the intracellular Na^+ concentration through the Na^+/H^+-exchanger, which extrudes protons, to restore pH, in exchange for Na^+ (Avkiran and Marber 2002). Na^+ overload is exacerbated by the ceased activity of Na^+/K^+ ATPase due to the reduced ATP availability, while the reverse mode activity of the sarcolemmal Na^+/Ca^{2+} exchanger induces intracellular Ca^{2+} overload (Piper et al. 2003). The intracellular accumulation of Ca^{2+} and Na^+ ions and protons H^+, followed by losses of K^+ and Mg^{2+} ions, results in intracellular edema, cell swelling and disruption of cellular membranes (Haunstetter and Izumo 1998; Moe and Marin-Garcia 2016). In prolonged ischemia and in response to an increase in the Ca^{2+} ion concentration, proteases (such as calpains) are activated and contribute to myocardial ischemic injury by degrading myofibrillar proteins and regulatory enzymes; this ultimately results in cardiac contractile dysfunction (Neuhof and Neuhof 2014).

According to the current knowledge, the major contributing actors of the lethal reperfusion injury are oxidative stress and increased Ca^{2+} overload (Piper et al. 2003), which are also the best characterized MPT inducers (Morciano et al. 2015).

Oxidative stress, determined by the reoxygenation of the ischemic heart, reduces the bioavailability of nitric oxide (NO), which is indispensable for inhibiting neutrophil accumulation and inactivating superoxide radicals (Granger and Kvietys 2015; Kvietys and Granger 2012). Moreover, reduced levels of nitric oxide diminish myocardial blood flow reperfusion through the coronary circulation. The oxygen burst at the reperfusion time stimulates xanthine oxidase- and NADPH oxidase-dependent reactive oxygen species (ROS) formation. Intracellular Ca^{2+} overload, initiated during ischemia, is exacerbated at the time of reperfusion due to the oxidative stress-mediated impairment of the sarcoplasmic reticulum and damage to the plasma membrane (Hausenloy and Yellon 2013). In addition, the recovery of oxygen levels following the reperfusion re-energizes the mitochondria, enhancing their ability to accumulate Ca^{2+} ions (Lemasters 1999; Giorgi et al. 2012). Upon reperfusion, the rapid restoration of intracellular pH, favored by the reactivation of the Na^+-H^+ exchanger and the consequent washout of lactic acid, stimulates the opening of the PTPC. As demonstrated in 1995 by Griffiths and Halestrap (1995), despite the presence of pro-opening factors (such as Ca^{2+}, inorganic phosphate, oxidative stress and ADP), the PTPC remains closed during ischemia due to low pH levels (<7.0), since H^+ inhibits Ca^{2+} binding to the PTPC trigger site (Lemasters et al. 1996). At the reperfusion time, the opened PTPC can be permeated by protons and any molecule less than 1.5 kDa. This event favors mitochondrial membrane potential dissipation, uncoupling oxidative phosphorylation and ATP depletion, whose generation is also prevented by reversal of the ATPase (Halestrap and Richardson 2015). In this way, energy metabolism is impaired, leading to further intensification of Ca^{2+} deregulation and PTPC opening.

In the last few decades, the PTPC has emerged as a promising therapeutic target. In 2002, Hausenloy et al. confirmed PTPC opening at the onset of reperfusion and demonstrated that exclusive administration of CsA at the onset of reperfusion could limit the infarct size (Hausenloy et al. 2002). One year later, using the immunosuppressant sanglifehrin-A, it was shown that the mPTP-opening inhibition was effective in limiting the infarct size in isolated perfused rat hearts if performed in the first few minutes of reperfusion (Hausenloy et al. 2003). Long-term cardioprotective effects of PTPC inhibition were shown in a study by Gomez et al., in which the PTPC inhibitor Debio-025, a CsA analog, was administered to mice that underwent IRI. They have demonstrated that inhibition of the PTPC at the reperfusion time improves functional recovery and mortality in mice at 30 days (Gomez et al. 2007). In 2014, a clinical study, conducted in patients undergoing aortic valve surgery, showed that cyclosporine administration at the time of reperfusion protects against reperfusion injury by reducing the levels of cardiac troponin I (Chiari et al. 2014).

Moreover, ischemia induces the release of cell membrane receptor ligands, such as adenosine, generated by the breakdown of ATP (Leung et al. 2014) and bradykinin, which induces production of ROS and NO (Sharma et al. 2015). Subsequently, the activation of a set of kinases known as RISKs (reperfusion injury salvage kinases), including Akt, Erk1/2, PKG, PKC-ε and p70s6K, is triggered. By activating the Akt/eNOS pathway, ischemic preconditioning results in S-nitrosylation of multiple mitochondrial proteins; S-nitrosylation is a modification that is thought to protect sensitive sites from subsequent ROS-induced oxidation (Sun et al. 2015). These proteins include CypD, which can be nitrosylated at cysteines 103, 156 and 203 (Gutierrez-Aguilar and Baines 2015). The RIS kinases deliver an inactivating phosphorylation onto GSK3β, which is constitutively active and phosphorylates, among other substrates, CyPD, favoring its interaction with the PTPC (Rasola et al. 2010). A cardioprotective role is attributed to hydrogen sulfide (H_2S), which can function through the Akt/GSK3β axis (Andreadou et al. 2015). In this case, ischemia causes a decrease in endogenous H2S production, which can be antagonized by preconditioning protocols. Recently, mitochondrial calpain was proposed to contribute to the onset of the MPT after IRI, following its activation by a matrix Ca^{2+} concentration increase (Shintani-Ishida and Yoshida 2015).

Finally, another regulatory pathway has been shown to involve CyPD acetylation/deacetylation (Bochaton et al. 2015). Deacetylation of Lys166 of CyPD by mitochondrial deacetylase SIRT3 (Hafner et al. 2010) favors CyPD interaction with PTPC components and thus, the MPT. In addition, SIRT3 downregulation sensitizes mitochondria to PTPC opening, and SIRT3-mediated deacetylation of CyPD is reportedly enhanced downstream of postconditioning (Bochaton et al. 2015). Interestingly, the knockdown of SIRT4, another mitochondrial isoform that lacks deacetylase activity but has ADP ribosyl-transferase activity, has been reported to have the opposite effect, increasing resistance to PTPC induction (Verma et al. 2013).

3.2 Cancer

Apoptosis is recognized as a hallmark of cancer, and loss of its control is required for the development and progression of the pathology (Hanahan and Weinberg 2011). Considering the role of the MPT in cell death induction, it is logical to speculate that alterations to the PTPC exist in cancer.

Experimental data that corroborate this hypothesis have been collected and are discussed below.

As previously mentioned, the best-known MPT inducer is intra-mitochondrial calcium, which is provided to mitochondria by the endoplasmic reticulum through MAMs (Marchi et al. 2014; Patergnani et al. 2011; Giorgi et al. 2015c).

Cancer develops several mechanisms to inhibit toxic Ca^{2+} signaling. Two oncosuppressors, PML (Bernardi and Pandolfi 2014) and PTEN (Pulido et al. 2014), in cooperation with protein phosphatase 2A, sustain the transfer between the ER and mitochondria through the mitochondrial Ca^{2+} uniporter (MCU) complex by regulating the phosphorylation state of the channel responsible for Ca^{2+} release, the inositol-3-phosphate receptor (IP3R) (Bononi et al. 2013; Giorgi et al. 2010). A different mechanism, but a similar result, occurs with the loss (or mutation) of the master oncosuppressor p53. Indeed, p53 stabilization leads to stimulation of sarco/endoplasmic reticulum Ca^{2+}-ATPase (SERCA) and maintains elevated levels of Ca^{2+} in the endoplasmic reticulum ($[Ca^{2+}]$er). Loss of p53, PML and PTEN (as well as several of their oncogenic mutations) leads to a decrease in $[Ca^{2+}]$er resulting in reduced signaling to the mitochondria (Giorgi et al. 2015a, b). In contrast, oncogenes can display the opposite effect. Indeed, tumor progression can be sustained by the accumulation of a series of changes in the Ca^{2+} regulatory machinery that decrease the cytotoxic Ca^{2+} signal. An in vitro mimicking of tumor transformation induced by activation of H-RAS is accompanied by a progressive reduction in the amount of intracellular Ca^{2+} that is transferable to the mitochondria. This effect can be counteracted by a controlled increase in the extracellular Ca^{2+} level that causes an increase in the intracellular Ca^{2+} level and impedes H-RAS-induced transformation (Rimessi et al. 2015). For example, Bcl-2 reduces the Ca^{2+} content in the ER (Pinton et al. 2000), and the mitogenic kinase AKT strongly inhibits the IP3R.

Interestingly, AKT and PTEN are both members of the PI3K pathway, one the most studied survival signaling pathways, which plays a critical role in resistance to anticancer therapies (Slomovitz and Coleman 2012; Fresno Vara et al. 2004; Wolin 2013).

AKT can also phosphorylate and consequently inhibit GSK3β. This kinase has a significant stimulatory effect on the PTPC and can act as a tumor suppressor in several tumors. Indeed, as already mentioned, its inactivation by AKT promotes the association between HK2 and VDAC, causing inhibition of PTPC and an increase in cell survival (Juhaszova et al. 2004). In addition, the WNT, PKC and ERK pathways regulate GSK3β phosphorylation, which potentially connects the MPT to signals involved with growth factors, G protein-coupled receptor ligands and the extracellular matrix (Graber et al. 1999; Kang et al. 2007).

These data suggest that kinase cascades that are significantly activated in cancers can keep the PTPC closed, especially when acting on HK2.

Apart from their phosphorylation statuses, different PTPC components are differentially expressed in cancer cell lines and tumor models (Brenner and Grimm 2006; Fulda et al. 2010). VDAC isoforms are significantly higher in malignant tumor cells (Shinohara et al. 2000); ANT-2 is upregulated in renal tumors and transformed hepatocytes (Faure Vigny et al. 1996). HK2 is upregulated in multiple tumors (Shinohara et al. 1991; Rempel et al. 1996; Azoulay-Zohar et al. 2004; Gudnason et al. 1984), and there is a positive correlation between tumorigenesis and the expression level of TSPO (Beinlich et al. 2000; Maaser et al. 2001). Such alterations in gene expression would lead to the erroneous conclusion that the MPT should be favored in tumors. Nonetheless, tumor cells also contain alterations to the expression levels of the BCL-2 family members that keep the MPT probability low. Indeed, antiapoptotic members, such as BCL-XL and Mcl-1, are overexpressed in cancer (Quinn et al. 2011). BCL-XL has been shown to negatively regulate PTPC opening by directly interacting with VDAC (Arbel et al. 2012) from the cytoplasmic side, while a mitochondrial-matrix-located BCL-XL interacts with the β-subunit of F_1F_O ATP synthase inhibiting PTPC opening (Beinlich et al. 2000). Additionally, Mcl-1 plays a role in inhibiting the MPT (Thomas et al. 2013) even though the mechanism is not clearly elucidated. Furthermore, it is proposed that antiapoptotic members BCL-2, BCL-XL and Mcl-1 interfere with the proapoptotic interactions formed by BAX and BAK, which are known positive regulators of the MPT (Narita et al. 1998; Brenner et al. 2000; Karch et al. 2013).

Upregulation in HK2 expression can also be linked to the Warburg effect, a well-known metabolic hallmark of cancer. This term ("Warburg effect") usually refers to the atypical increase in glucose uptake and lactic fermentation observed in tumor masses regardless of the aerobic environment in which they are observed (Warburg et al. 1927; Boland et al. 2013). Glucose is converted to pyruvate, which is further reduced to lactate and completes the lactic fermentation process. The findings reported by Warburg (and confirmed by other groups) result in two main consequences for the MTP: (1) an increase in glucose uptake that allows for the continued synthesis of ATP, which impedes the depletion of adenine nucleotides and Pi accumulation and (2) the accumulation of lactate and lowering of pH as described for the MPT inhibitor. This suggests that the Warburg effect may result in the accumulation of PTPC inhibitors.

Another hypothesis, which is poorly investigated, is that the Warburg effect would simply make the MPT inefficient toward stressing cells sufficiently to induce cell death. Indeed, the large dependence on glycolytic metabolism would allow cells to continue surviving even with the loss of the mitochondrial functions induced by the MPT (at least to some extent). The observation that osteosarcoma cells with overt Warburg effects also display signs of the MPT supports this hypothesis.

The Warburg effect causes alterations in mitochondrial redox potential, ultimately changing ROS generation (Locasale and Cantley 2011). Indeed, ROS, the second most important MPT inducer, appears at higher levels in tumor cells, where they promote several other hallmarks of cancer, such as proliferation, invasion and metastasis (Gupta et al. 2012; Yang et al. 2013). However, ROS also reduces the PTPC threshold for Ca^{2+} and should increase tumor cell sensitivity to MPT induction, leading to an apparent paradox.

Nonetheless, several cancer cell types display higher levels of antioxidants, which could inhibit ROS toxicity. Specifically, increased levels of superoxide dismutase (SOD2) and thioredoxin reductase 2 (TRX-2) were observed in mitochondria from cancer samples (Dvorakova et al. 2002; Biaglow and Miller 2005; Pani et al. 2004). These data, together with the impaired Ca^{2+} signals already discussed, may allow the ROS level to increase without alerting the cell's regulatory mechanisms. Evading these regulatory mechanisms would then allow the tumor cell to undergo the cancer-promoting changes induced by a high ROS level.

Finally, a chaperone network could be an additional adaption mechanism for allowing cancer cells to escape MPT induction. Indeed, the PTPC relies on the activity of a well-known chaperone, CyPD, as its inhibition by CsA dramatically reduces the probability of PTPC opening.

3.3 *Neurodegenerative Diseases*

Mitochondria participate in various fundamental cellular processes, including energy production, regulation of cell death, and metabolism. There is no protagonist of life whose cellular fate is so strongly dependent on mitochondrial functions as the neuronal cell. The involvement of mitochondrial dysfunctions in neuronal damage associated with neurodegenerative diseases and brain damage has become increasingly relevant.

As already mentioned, the MPT is activated in response to pro-apoptotic stimuli, such as ROS and Ca^{2+} overload, that are common important pathological features of multiple diseases of the nervous central system (Martin et al. 2009). Despite the origin of its discovery (Crofts and Chappell 1965), the PTPC function in neurodegenerative diseases was uncovered rather recently (Du et al. 2008; Gautier et al. 2012).

Here, we describe neurodegenerative diseases that are characterized by dysfunctions in Ca^{2+} and ROS homeostasis that induce PTPC.

Alzheimer's disease (AD) is the most common form of dementia related to aging, and it caused by chronic neurodegenerative processes. AD is characterized by the accumulation and deposition of amyloid plaques formed by the β-amyloid peptide (Aβ), a cleavage product of the amyloid precursor protein (Haass and Selkoe 2007), and by phosphorylation of the tau protein in the brain (Rao et al. 2014). It has been shown that Aβ oligomers alter intracellular Ca^{2+} homeostasis, accelerating global neuropathological cascades (Demuro et al. 2010). Moreover, Aβ can be imported into the mitochondria (Hansson Petersen et al. 2008), where it promotes PTPC opening upon binding to CyPD. Consistent with this notion, neurons derived from CyPD knockout mice are protected from cell death induced by amyloid-dependent PTP opening, and the CyPD deficiency is associated with alleviation of neuronal cell death and improvements in cognitive function in AD mice (Du et al. 2008). Interestingly, Elkamhawy et al. reported that CyPD selective inhibitors, which are novel quinazoline-urea analogs, show protective effects in neuronal cells, blocking amyloid-dependent PTP opening (Elkamhawy et al. 2014). Furthermore, the selective loss of the oligomycin sensitivity conferring protein (OSCP) subunit of the F_1F_O-ATP synthase and the physical interaction of OSCP with Aβ constitute the major OSCP alterations in the brains of AD patients and the AD mouse model. Loss of OSCP leads to a reduction in ATP production, which increases oxidative stress and activation of MPT, whereas its restoration reportedly ameliorates Aβ-mediated mouse and human neuronal mitochondrial impairments, opening up new therapeutic implications for AD based on the stabilization of the OSCP protein (Beck et al. 2016). Other critical components of the PTPC appear to have a role in AD development. ANT can bind directly to Aβ, reducing the ATP/ADP exchange rate and energy metabolism in AD (Singh et al. 2009). VDAC can form complexes with Aβ and phosphorylated tau in AD mouse models, although its involvement in PTPC formation is unclear (Crompton et al. 1998; Zheng et al. 2004; Baines et al. 2007).

The second most common disease among the neurodegenerative pathologies and the most frequent movement disorder is Parkinson's disease (PD) (Bernardi et al. 2015). It is caused by the death of dopaminergic neurons in the mesencephalic region known as the "substantia nigra pars compacta", which is characterized by the presence of α-synuclein and aggregates of protein included in neuronal cells known as Lewy bodies (Rasheed et al. 2017). Dopaminergic neurons have a peculiar mechanism for controlling intracellular Ca^{2+} fluctuations and the Ca^{2+} storage capacity of their mitochondria that involves an autonomous pace-making activity that relies on voltage-dependent L-type Ca^{2+} channels that modulate the release of the neurotransmitter dopamine (Winklhofer and Haass 2010). Accordingly, these neurons are particularly sensitive to mitochondrial Ca^{2+} perturbations, and it has been proposed that the increased sensitization of PTPC opening is a major cause of neurodegeneration in PD patients and a mouse model of the pathology (Luth et al. 2014; Martin et al. 2014). Furthermore, an altered mitochondrial Ca^{2+} storage capacity, impaired respiratory complex I, and altered mitophagy, which exacerbate the sensitization of the MPT, are observed when the complex I activity is suppressed as in the case of patients with PD (Beal 2000; Greenamyre et al. 2001; Seaton et al. 1998) and when the PINK1 Ser/Thr kinase is inactivated by RNA interference-mediated downregulation or by PINK-1 knockout in mice (Rasola and Bernardi 2011; Valente et al. 2004; Kawajiri et al. 2011). Moreover, there is a reduced capacity for this mechanism to buffer intracellular ROS in dopaminergic neurons, which could induce an increase in the PTPC activity (Brundin et al. 2008). Interestingly, a reduction in the degeneration of dopaminergic neurons has been observed in patients treated with dopamine agonists rather than with L-dopa. This effect could be due to a neuroprotective effect of the direct inhibition of the MPT, as demonstrated after treatment with pramipexole, ropinirole and rasagiline in vitro (Sayeed et al. 2006; Youdim et al. 2005; Schonfeld et al. 2013).

Huntington's disease (Forte et al. 2007) is caused by autosomal dominant mutations that lead to translation of an expanded polyglutamine tract in the gene encoding the huntingtin (Htt) protein. Patients display progressive uncontrolled movements, psychiatric disturbances, and dementia; this disease is invariably lethal (Landles and Bates 2004). Choo et al. demonstrated that mitochondrial dysfunctions, caused by pathogenic Htt mutations, depend on PTPC opening after Ca^{2+} overload in isolated mitochondria from an HD mouse model (Choo et al. 2004), and these effects are inhibited by the addition of CsA and ADP (Milakovic et al. 2006). Therefore, neurons with the mutated Htt protein reportedly show an altered Ca^{2+} homeostasis, mtDNA damage and mitochondrial fragmentation that could trigger the MPT (Quintanilla et al. 2017). Interestingly, a significant association of Htt with the mitochondria in synaptic extracts (Hamilton et al. 2015) has been observed; this may indicate a possible interaction with several elements of the PTPC that affects mitochondrial function.

Amyotrophic lateral sclerosis (ALS) is a neurodegenerative disease characterized by motor neuron degeneration, especially in the brain and spinal cord, which rapidly leads to the loss of voluntary muscle control and subsequent paralysis and mortality (Lezi and Swerdlow 2012). Some forms of ALS are inherited and caused by mutations in the superoxide dismutase-1 (SOD1) gene (Martin et al. 2009). Transgenic mice expressing the mutant SOD1 were shown to display mitochondrial alterations, including swollen megamitochondria with cristae remodeling and matrix vacuolization, respiratory chain inhibition, a reduced mitochondrial Ca^{2+} buffering capacity and fragmentation, and an elevated generation of ROS (Leal and Gomes 2015; Kawamata and Manfredi 2010). Interestingly, regulatory components of the PTPC, such as ANT and CyPD, were found to be highly expressed in the same mouse model, indicating that the MPT might be the triggering cause of neuronal cell death (Martin et al. 2009).

Multiple sclerosis is the most common chronic inflammatory disease of the central nervous system, where the myelin sheath of neurons is destroyed by endogenous myelin-associated antigens, such as myelin oligodendrocyte glycoprotein, proteolipoprotein, and myelin basic protein (Stys et al. 2012). During an active inflammatory attack in MS, large quantities of glutamate, an essential excitatory neurotransmitter, are produced by activated immune cells, such as macrophages and microglia. Overstimulation of glutamate receptors leads to mitochondrial Ca^{2+} overload and a consequent bioenergetics dysfunction and MPT stimulation (Su et al. 2009). Experimental autoimmune encephalomyelitis is widely used as an animal model of multiple sclerosis. For example, decreased levels of axonal damage and a marked protection from the inflammatory response have been observed in CyPD knockout mice compared to wild-type mice (Forte et al. 2007). Moreover, to highlight the central role of the PTPC in neurodegeneration, experimental autoimmune encephalomyelitis was performed in p66Shc-null mice, where ROS-induced PTP opening was abrogated (Savino et al. 2013).

Although neurodegenerative diseases constitute a large portion of the research focus, it is also important to underline the role of mitochondria in neuronal degeneration and brain damage caused by acute and chronic abuse of alcohol, which is a serious public health problem. Recently, an in vitro study by Lamarche and co-workers (Lamarche et al. 2013) revealed a close correlation between stimulation of the PTPC and neuronal damage induced by chronic administration of ethanol. Moreover, CsA is reportedly able to attenuate ethanol withdrawal-induced cell death in the HT22 cultured hippocampal cell line (Jung et al. 2009).

4 Conclusions

Along with the recognition that the MPT has a critical role in multiple pathophysiological scenarios, there has been strong interest in the therapeutic potential of the MPT. MPT pharmacological inhibitors could be used to prevent the cell death caused by PTPC opening. Conversely, pharmacological activators of the MPT could be used to selectively kill neoplastic cells based on their intrinsically elevated levels of stress. To develop therapeutic drugs, it will be fundamental to understand the molecular composition of the PTPC. The recent key discoveries surrounding the composition of the PTPC, particularly the F_1F_O ATP synthase, have opened new perspectives into the molecular definition of its role in pathophysiology and will rapidly enhance the understanding of pore structure and function, which will bring about the design and validation of PTP-active compounds to treat cancer and cardiac and degenerative diseases.

Acknowledgements P.P. is grateful to Camilla degli Scrovegni for continuous support. P.P. is supported by the Italian Ministry of Education, University and Research (COFIN no. 20129JLHSY_002, FIRB no. RBAP11FXBC_002, and Futuro in Ricerca no. RBFR10EGVP_001), the Italian Cystic Fibrosis Research Foundation (19/2014) and Telethon (GGP15219/B). P.P. and C.G. are supported by local funds from the University of Ferrara and the Italian Association for Cancer Research (IG-18624 and MFAG-13521). M.R.W. is supported by the National Science Center, Poland (grant 2014/15/B/NZ1/00490).

References

Akhabue E, Thiboutot J, Cheng JW, Vittorio TJ, Christodoulidis G, Grady KM, Lerakis S, Kosmas CE (2014) New and emerging risk factors for coronary heart disease. Am J Med Sci 347:151–158

Alavian KN, Beutner G, Lazrove E, Sacchetti S, Park HA, Licznerski P, Li H, Nabili P, Hockensmith K, Graham M, Porter GA Jr, Jonas EA (2014) An uncoupling channel within

the c-subunit ring of the F1FO ATP synthase is the mitochondrial permeability transition pore. Proc Natl Acad Sci U S A 111:10580–10585
Alcala S, Klee M, Fernandez J, Fleischer A, Pimentel-Muinos FX (2008) A high-throughput screening for mammalian cell death effectors identifies the mitochondrial phosphate carrier as a regulator of cytochrome c release. Oncogene 27:44–54
Andreadou I, Iliodromitis EK, Rassaf T, Schulz R, Papapetropoulos A, Ferdinandy P (2015) The role of gasotransmitters NO, H2S and CO in myocardial ischaemia/reperfusion injury and cardioprotection by preconditioning, postconditioning and remote conditioning. Br J Pharmacol 172:1587–1606
Araszkiewicz A, Grygier M, Lesiak M, Grajek S (2013) The impact of ischemia-reperfusion injury on the effectiveness of primary angioplasty in ST-segment elevation myocardial infarction. Postepy Kardiol Interwencyjnej 9:275–281
Arbel N, Ben-Hail D, Shoshan-Barmatz V (2012) Mediation of the antiapoptotic activity of Bcl-xL protein upon interaction with VDAC1 protein. J Biol Chem 287:23152–23161
Avkiran M, Marber MS (2002) Na(+)/H(+) exchange inhibitors for cardioprotective therapy: progress, problems and prospects. J Am Coll Cardiol 39:747–753
Azoulay-Zohar H, Israelson A, Abu-Hamad S, Shoshan-Barmatz V (2004) In self-defence: hexokinase promotes voltage-dependent anion channel closure and prevents mitochondria-mediated apoptotic cell death. Biochem J 377:347–355
Baines CP, Song CX, Zheng YT, Wang GW, Zhang J, Wang OL, Guo Y, Bolli R, Cardwell EM, Ping P (2003) Protein kinase Cepsilon interacts with and inhibits the permeability transition pore in cardiac mitochondria. Circ Res 92:873–880
Baines CP, Kaiser RA, Purcell NH, Blair NS, Osinska H, Hambleton MA, Brunskill EW, Sayen MR, Gottlieb RA, Dorn GW, Robbins J, Molkentin JD (2005) Loss of cyclophilin D reveals a critical role for mitochondrial permeability transition in cell death. Nature 434:658–662
Baines CP, Kaiser RA, Sheiko T, Craigen WJ, Molkentin JD (2007) Voltage-dependent anion channels are dispensable for mitochondrial-dependent cell death. Nat Cell Biol 9(5):550
Beal MF (2000) Energetics in the pathogenesis of neurodegenerative diseases. Trends Neurosci 23:298–304
Beck SJ, Guo L, Phensy A, Tian J, Wang L, Tandon N, Gauba E, Lu L, Pascual JM, Kroener S, Du H (2016) Deregulation of mitochondrial F1FO-ATP synthase via OSCP in Alzheimer's disease. Nat Commun 7:11483
Beinlich A, Strohmeier R, Kaufmann M, Kuhl H (2000) Relation of cell proliferation to expression of peripheral benzodiazepine receptors in human breast cancer cell lines. Biochem Pharmacol 60:397–402
Bernardi R, Pandolfi PP (2014) A dialog on the first 20 years of PML research and the next 20 ahead. Front Oncol 4:23
Bernardi P, Rasola A, Forte M, Lippe G (2015) The mitochondrial permeability transition pore: channel formation by F-ATP synthase, integration in signal transduction, and role in pathophysiology. Physiol Rev 95:1111–1155
Beutner G, Ruck A, Riede B, Welte W, Brdiczka D (1996) Complexes between kinases, mitochondrial porin and adenylate translocator in rat brain resemble the permeability transition pore. FEBS Lett 396:189–195
Beutner G, Ruck A, Riede B, Brdiczka D (1998) Complexes between porin, hexokinase, mitochondrial creatine kinase and adenylate translocator display properties of the permeability transition pore. Implication for regulation of permeability transition by the kinases. Biochim Biophys Acta 1368:7–18
Biaglow JE, Miller RA (2005) The thioredoxin reductase/thioredoxin system: novel redox targets for cancer therapy. Cancer Biol Ther 4:6–13
Bochaton T, Crola-Da-Silva C, Pillot B, Villedieu C, Ferreras L, Alam MR, Thibault H, Strina M, Gharib A, Ovize M, Baetz D (2015) Inhibition of myocardial reperfusion injury by ischemic postconditioning requires sirtuin 3-mediated deacetylation of cyclophilin D. J Mol Cell Cardiol 84:61–69

Boland ML, Chourasia AH, Macleod KF (2013) Mitochondrial dysfunction in cancer. Front Oncol 3:292

Bononi A, Bonora M, Marchi S, Missiroli S, Poletti F, Giorgi C, Pandolfi PP, Pinton P (2013) Identification of PTEN at the ER and MAMs and its regulation of Ca(2+) signaling and apoptosis in a protein phosphatase-dependent manner. Cell Death Differ 20:1631–1643

Bonora M, Bononi A, De Marchi E, Giorgi C, Lebiedzinska M, Marchi S, Patergnani S, Rimessi A, Suski JM, Wojtala A, Wieckowski MR, Kroemer G, Galluzzi L, Pinton P (2013) Role of the c subunit of the FO ATP synthase in mitochondrial permeability transition. Cell Cycle 12:674–683

Bonora M, Wieckowski MR, Chinopoulos C, Kepp O, Kroemer G, Galluzzi L, Pinton P (2015) Molecular mechanisms of cell death: central implication of ATP synthase in mitochondrial permeability transition. Oncogene 34:1475–1486

Bonora M, Morganti C, Morciano G, Giorgi C, Wieckowski MR, Pinton P (2016) Comprehensive analysis of mitochondrial permeability transition pore activity in living cells using fluorescence-imaging-based techniques. Nat Protoc 11:1067–1080

Bonora M, Morganti C, Morciano G, Pedriali G, Lebiedzinska-Arciszewska M, Aquila G, Giorgi C, Rizzo P, Campo G, Ferrari R, Kroemer G, Wieckowski MR, Galluzzi L, Pinton P (2017) Mitochondrial permeability transition involves dissociation of F1Fo ATP synthase dimers and C-ring conformation. EMBO Rep 18(7):1077–1089

Brenner C, Grimm S (2006) The permeability transition pore complex in cancer cell death. Oncogene 25:4744–4756

Brenner C, Cadiou H, Vieira HL, Zamzami N, Marzo I, Xie Z, Leber B, Andrews D, Duclohier H, Reed JC, Kroemer G (2000) Bcl-2 and Bax regulate the channel activity of the mitochondrial adenine nucleotide translocator. Oncogene 19:329–336

Brundin P, Li JY, Holton JL, Lindvall O, Revesz T (2008) Research in motion: the enigma of Parkinson's disease pathology spread. Nat Rev Neurosci 9:741–745

Campanella M, Szabadkai G, Rizzuto R (2008) Modulation of intracellular Ca2+ signalling in HeLa cells by the apoptotic cell death enhancer PK11195. Biochem Pharmacol 76:1628–1636

Chelli B, Falleni A, Salvetti F, Gremigni V, Lucacchini A, Martini C (2001) Peripheral-type benzodiazepine receptor ligands: mitochondrial permeability transition induction in rat cardiac tissue. Biochem Pharmacol 61:695–705

Chiara F, Gambalunga A, Sciacovelli M, Nicolli A, Ronconi L, Fregona D, Bernardi P, Rasola A, Trevisan A (2012) Chemotherapeutic induction of mitochondrial oxidative stress activates GSK-3alpha/beta and Bax, leading to permeability transition pore opening and tumor cell death. Cell Death Dis 3:e444

Chiari P, Angoulvant D, Mewton N, Desebbe O, Obadia JF, Robin J, Farhat F, Jegaden O, Bastien O, Lehot JJ, Ovize M (2014) Cyclosporine protects the heart during aortic valve surgery. Anesthesiology 121:232–238

Choo YS, Johnson GV, Macdonald M, Detloff PJ, Lesort M (2004) Mutant huntingtin directly increases susceptibility of mitochondria to the calcium-induced permeability transition and cytochrome c release. Hum Mol Genet 13:1407–1420

Crofts AR, Chappell JB (1965) Calcium ion accumulation and volume changes of isolated liver mitochondria. reversal of calcium ion-induced swelling. Biochem J 95:387–392

Crompton M, Costi A (1990) A heart mitochondrial Ca2(+)-dependent pore of possible relevance to re-perfusion-induced injury. Evidence that ADP facilitates pore interconversion between the closed and open states. Biochem J 266:33–39

Crompton M, Virji S, Ward JM (1998) Cyclophilin-D binds strongly to complexes of the voltage-dependent anion channel and the adenine nucleotide translocase to form the permeability transition pore. Eur J Biochem 258:729–735

De Marchi U, Basso E, Szabo I, Zoratti M (2006) Electrophysiological characterization of the Cyclophilin D-deleted mitochondrial permeability transition pore. Mol Membr Biol 23:521–530

Decaudin D, Castedo M, Nemati F, Beurdeley-Thomas A, De Pinieux G, Caron A, Pouillart P, Wijdenes J, Rouillard D, Kroemer G, Poupon MF (2002) Peripheral benzodiazepine receptor ligands reverse apoptosis resistance of cancer cells in vitro and in vivo. Cancer Res 62:1388–1393

Demuro A, Parker I, Stutzmann GE (2010) Calcium signaling and amyloid toxicity in Alzheimer disease. J Biol Chem 285:12463–12468

Dolder M, Walzel B, Speer O, Schlattner U, Wallimann T (2003) Inhibition of the mitochondrial permeability transition by creatine kinase substrates. Requirement for microcompartmentation. J Biol Chem 278:17760–17766

Du H, Guo L, Fang F, Chen D, Sosunov AA, Mckhann GM, Yan Y, Wang C, Zhang H, Molkentin JD, Gunn-Moore FJ, Vonsattel JP, Arancio O, Chen JX, Yan SD (2008) Cyclophilin D deficiency attenuates mitochondrial and neuronal perturbation and ameliorates learning and memory in Alzheimer's disease. Nat Med 14:1097–1105

Duchen MR, Leyssens A, Crompton M (1998) Transient mitochondrial depolarizations reflect focal sarcoplasmic reticular calcium release in single rat cardiomyocytes. J Cell Biol 142:975–988

Dvorakova K, Payne CM, Tome ME, Briehl MM, Vasquez MA, Waltmire CN, Coon A, Dorr RT (2002) Molecular and cellular characterization of imexon-resistant Rpmi8226/I myeloma cells. Mol Cancer Ther 1:185–195

Eguchi Y, Shimizu S, Tsujimoto Y (1997) Intracellular ATP levels determine cell death fate by apoptosis or necrosis. Cancer Res 57:1835–1840

Elkamhawy A, Lee J, Park BG, Park I, Pae AN, Roh EJ (2014) Novel quinazoline-urea analogues as modulators for Abeta-induced mitochondrial dysfunction: design, synthesis, and molecular docking study. Eur J Med Chem 84:466–475

Elustondo PA, Nichols M, Negoda A, Thirumaran A, Zakharian E, Robertson GS, Pavlov EV (2016) Mitochondrial permeability transition pore induction is linked to formation of the complex of ATPase C-subunit, polyhydroxybutyrate and inorganic polyphosphate. Cell Death Dis 2:16070

Faure Vigny H, Heddi A, Giraud S, Chautard D, Stepien G (1996) Expression of oxidative phosphorylation genes in renal tumors and tumoral cell lines. Mol Carcinog 16:165–172

Forte M, Gold BG, Marracci G, Chaudhary P, Basso E, Johnsen D, Yu X, Fowlkes J, Rahder M, Stem K, Bernardi P, Bourdette D (2007) Cyclophilin D inactivation protects axons in experimental autoimmune encephalomyelitis, an animal model of multiple sclerosis. Proc Natl Acad Sci U S A 104:7558–7563

Frank A, Bonney M, Bonney S, Weitzel L, Koeppen M, Eckle T (2012) Myocardial ischemia reperfusion injury: from basic science to clinical bedside. Semin Cardiothorac Vasc Anesth 16:123–132

Fresno Vara JA, Casado E, De Castro J, Cejas P, Belda-Iniesta C, Gonzalez-Baron M (2004) PI3K/AKT signalling pathway and cancer. Cancer Treat Rev 30:193–204

Fulda S, Galluzzi L, Kroemer G (2010) Targeting mitochondria for cancer therapy. Nat Rev Drug Discov 9:447–464

Galluzzi L, Bravo-San Pedro JM, Vitale I, Aaronson SA, Abrams JM, Adam D, Alnemri ES, Altucci L, Andrews D, Annicchiarico-Petruzzelli M, Baehrecke EH, Bazan NG, Bertrand MJ, Bianchi K, Blagosklonny MV, Blomgren K, Borner C, Bredesen DE, Brenner C, Campanella M, Candi E, Cecconi F, Chan FK, Chandel NS, Cheng EH, Chipuk JE, Cidlowski JA, Ciechanover A, Dawson TM, Dawson VL, De Laurenzi V, De Maria R, Debatin KM, Di Daniele N, Dixit VM, Dynlacht BD, El-Deiry WS, Fimia GM, Flavell RA, Fulda S, Garrido C, Gougeon ML, Green DR, Gronemeyer H, Hajnoczky G, Hardwick JM, Hengartner MO, Ichijo H, Joseph B, Jost PJ, Kaufmann T, Kepp O, Klionsky DJ, Knight RA, Kumar S, Lemasters JJ, Levine B, Linkermann A, Lipton SA, Lockshin RA, Lopez-Otin C, Lugli E, Madeo F, Malorni W, Marine JC, Martin SJ, Martinou JC, Medema JP, Meier P, Melino S, Mizushima N, Moll U, Munoz-Pinedo C, Nunez G, Oberst A, Panaretakis T, Penninger JM, Peter ME, Piacentini M, Pinton P, Prehn JH, Puthalakath H, Rabinovich GA, Ravichandran KS, Rizzuto R, Rodrigues CM, Rubinsztein DC, Rudel T, Shi Y, Simon HU, Stockwell BR, Szabadkai G, Tait SW, Tang HL, Tavernarakis N, Tsujimoto Y, Vanden Berghe T, Vandenabeele P, Villunger A, Wagner EF et al (2015) Essential versus accessory aspects of cell death: recommendations of the NCCD 2015. Cell Death Differ 22:58–73

Garcia JJ, Morales-Rios E, Cortes-Hernandez P, Rodriguez-Zavala JS (2006) The inhibitor protein (IF1) promotes dimerization of the mitochondrial F1F0-ATP synthase. Biochemistry 45:12695–12703

Gautier CA, Giaime E, Caballero E, Nunez L, Song Z, Chan D, Villalobos C, Shen J (2012) Regulation of mitochondrial permeability transition pore by PINK1. Mol Neurodegener 7:22

Gerle C (2016) On the structural possibility of pore-forming mitochondrial FoF1 ATP synthase. Biochim Biophys Acta 1857:1191–1196

Giorgi C, Ito K, Lin HK, Santangelo C, Wieckowski MR, Lebiedzinska M, Bononi A, Bonora M, Duszynski J, Bernardi R, Rizzuto R, Tacchetti C, Pinton P, Pandolfi PP (2010) PML regulates apoptosis at endoplasmic reticulum by modulating calcium release. Science 330:1247–1251

Giorgi C, Baldassari F, Bononi A, Bonora M, De Marchi E, Marchi S, Missiroli S, Patergnani S, Rimessi A, Suski JM, Wieckowski MR, Pinton P (2012) Mitochondrial Ca(2+) and apoptosis. Cell Calcium 52:36–43

Giorgi C, Bonora M, Missiroli S, Poletti F, Ramirez FG, Morciano G, Morganti C, Pandolfi PP, Mammano F, Pinton P (2015a) Intravital imaging reveals p53-dependent cancer cell death induced by phototherapy via calcium signaling. Oncotarget 6:1435–1445

Giorgi C, Bonora M, Sorrentino G, Missiroli S, Poletti F, Suski JM, Galindo Ramirez F, Rizzuto R, Di Virgilio F, Zito E, Pandolfi PP, Wieckowski MR, Mammano F, Del Sal G, Pinton P (2015b) p53 at the endoplasmic reticulum regulates apoptosis in a Ca2+-dependent manner. Proc Natl Acad Sci U S A 112:1779–1784

Giorgi C, Missiroli S, Patergnani S, Duszynski J, Wieckowski MR, Pinton P (2015c) Mitochondria-associated membranes: composition, molecular mechanisms, and physiopathological implications. Antioxid Redox Signal 22:995–1019

Giorgio V, Bisetto E, Soriano ME, Dabbeni-Sala F, Basso E, Petronilli V, Forte MA, Bernardi P, Lippe G (2009) Cyclophilin D modulates mitochondrial F0F1-ATP synthase by interacting with the lateral stalk of the complex. J Biol Chem 284:33982–33988

Giorgio V, Von Stockum S, Antoniel M, Fabbro A, Fogolari F, Forte M, Glick GD, Petronilli V, Zoratti M, Szabo I, Lippe G, Bernardi P (2013) Dimers of mitochondrial ATP synthase form the permeability transition pore. Proc Natl Acad Sci U S A 110:5887–5892

Gomez L, Thibault H, Gharib A, Dumont JM, Vuagniaux G, Scalfaro P, Derumeaux G, Ovize M (2007) Inhibition of mitochondrial permeability transition improves functional recovery and reduces mortality following acute myocardial infarction in mice. Am J Physiol Heart Circ Physiol 293:H1654–H1661

Graber HU, Friess H, Zimmermann A, Korc M, Adler G, Schmid R, Buchler MW (1999) Bak expression and cell death occur in peritumorous tissue but not in pancreatic cancer cells. J Gastrointest Surg 3:74–80. discussion 81

Granger DN, Kvietys PR (2015) Reperfusion injury and reactive oxygen species: the evolution of a concept. Redox Biol 6:524–551

Greenamyre JT, Sherer TB, Betarbet R, Panov AV (2001) Complex I and Parkinson's disease. IUBMB Life 52:135–141

Griffiths EJ, Halestrap AP (1995) Mitochondrial non-specific pores remain closed during cardiac ischaemia, but open upon reperfusion. Biochem J 307(Pt 1):93–98

Gudnason V, Ingvarsson S, Jonasdottir A, Andresdottir V, Egilsson V (1984) Isoenzyme pattern and subcellular localization of hexokinases in human breast cancer and nonpathological breast tissue. Int J Cancer 34:63–66

Gupta SC, Hevia D, Patchva S, Park B, Koh W, Aggarwal BB (2012) Upsides and downsides of reactive oxygen species for cancer: the roles of reactive oxygen species in tumorigenesis, prevention, and therapy. Antioxid Redox Signal 16:1295–1322

Gutierrez-Aguilar M, Baines CP (2015) Structural mechanisms of cyclophilin D-dependent control of the mitochondrial permeability transition pore. Biochim Biophys Acta 1850:2041–2047

Gutierrez-Aguilar M, Douglas DL, Gibson AK, Domeier TL, Molkentin JD, Baines CP (2014) Genetic manipulation of the cardiac mitochondrial phosphate carrier does not affect permeability transition. J Mol Cell Cardiol 72:316–325

Haass C, Selkoe DJ (2007) Soluble protein oligomers in neurodegeneration: lessons from the Alzheimer's amyloid beta-peptide. Nat Rev Mol Cell Biol 8:101–112

Hafner AV, Dai J, Gomes AP, Xiao CY, Palmeira CM, Rosenzweig A, Sinclair DA (2010) Regulation of the mPTP by SIRT3-mediated deacetylation of CypD at lysine 166 suppresses age-related cardiac hypertrophy. Aging (Albany NY) 2:914–923

Halestrap AP (2014) The C ring of the F1Fo ATP synthase forms the mitochondrial permeability transition pore: a critical appraisal. Front Oncol 4:234

Halestrap AP, Richardson AP (2015) The mitochondrial permeability transition: a current perspective on its identity and role in ischaemia/reperfusion injury. J Mol Cell Cardiol 78:129–141

Hamilton J, Pellman JJ, Brustovetsky T, Harris RA, Brustovetsky N (2015) Oxidative metabolism in YAC128 mouse model of Huntington's disease. Hum Mol Genet 24:4862–4878

Hanahan D, Weinberg RA (2011) Hallmarks of cancer: the next generation. Cell 144:646–674

Hansson Petersen CA, Alikhani N, Behbahani H, Wiehager B, Pavlov PF, Alafuzoff I, Leinonen V, Ito A, Winblad B, Glaser E, Ankarcrona M (2008) The amyloid beta-peptide is imported into mitochondria via the TOM import machinery and localized to mitochondrial cristae. Proc Natl Acad Sci U S A 105:13145–13150

Haunstetter A, Izumo S (1998) Apoptosis: basic mechanisms and implications for cardiovascular disease. Circ Res 82:1111–1129

Hausenloy DJ, Yellon DM (2013) Myocardial ischemia-reperfusion injury: a neglected therapeutic target. J Clin Invest 123:92–100

Hausenloy DJ, Maddock HL, Baxter GF, Yellon DM (2002) Inhibiting mitochondrial permeability transition pore opening: a new paradigm for myocardial preconditioning? Cardiovasc Res 55:534–543

Hausenloy DJ, Duchen MR, Yellon DM (2003) Inhibiting mitochondrial permeability transition pore opening at reperfusion protects against ischaemia-reperfusion injury. Cardiovasc Res 60:617–625

He J, Ford HC, Carroll J, Ding S, Fearnley IM, Walker JE (2017) Persistence of the mitochondrial permeability transition in the absence of subunit c of human ATP synthase. Proc Natl Acad Sci U S A 114:3409–3414

Hirsch T, Decaudin D, Susin SA, Marchetti P, Larochette N, Resche-Rigon M, Kroemer G (1998) PK11195, a ligand of the mitochondrial benzodiazepine receptor, facilitates the induction of apoptosis and reverses Bcl-2-mediated cytoprotection. Exp Cell Res 241:426–434

Izzo V, Bravo-San Pedro JM, Sica V, Kroemer G, Galluzzi L (2016) Mitochondrial permeability transition: new findings and persisting uncertainties. Trends Cell Biol 26:655–667

Juhaszova M, Zorov DB, Kim SH, Pepe S, Fu Q, Fishbein KW, Ziman BD, Wang S, Ytrehus K, Antos CL, Olson EN, Sollott SJ (2004) Glycogen synthase kinase-3beta mediates convergence of protection signaling to inhibit the mitochondrial permeability transition pore. J Clin Invest 113:1535–1549

Jung ME, Wilson AM, Ju X, Wen Y, Metzger DB, Simpkins JW (2009) Ethanol withdrawal provokes opening of the mitochondrial membrane permeability transition pore in an estrogen-preventable manner. J Pharmacol Exp Ther 328:692–698

Kang BH, Plescia J, Dohi T, Rosa J, Doxsey SJ, Altieri DC (2007) Regulation of tumor cell mitochondrial homeostasis by an organelle-specific Hsp90 chaperone network. Cell 131:257–270

Karch J, Kwong JQ, Burr AR, Sargent MA, Elrod JW, Peixoto PM, Martinez-Caballero S, Osinska H, Cheng EH, Robbins J, Kinnally KW, Molkentin JD (2013) Bax and Bak function as the outer membrane component of the mitochondrial permeability pore in regulating necrotic cell death in mice. Elife 2:e00772

Kawajiri S, Saiki S, Sato S, Hattori N (2011) Genetic mutations and functions of PINK1. Trends Pharmacol Sci 32:573–580

Kawamata H, Manfredi G (2010) Mitochondrial dysfunction and intracellular calcium dysregulation in ALS. Mech Ageing Dev 131:517–526

Kepp O, Galluzzi L, Lipinski M, Yuan J, Kroemer G (2011) Cell death assays for drug discovery. Nat Rev Drug Discov 10:221–237

Kokoszka JE, Waymire KG, Levy SE, Sligh JE, Cai J, Jones DP, Macgregor GR, Wallace DC (2004) The ADP/ATP translocator is not essential for the mitochondrial permeability transition pore. Nature 427:461–465

Kroemer G (1997) The proto-oncogene Bcl-2 and its role in regulating apoptosis. Nat Med 3:614–620

Kugler W, Veenman L, Shandalov Y, Leschiner S, Spanier I, Lakomek M, Gavish M (2008) Ligands of the mitochondrial 18 kDa translocator protein attenuate apoptosis of human glioblastoma cells exposed to erucylphosphohomocholine. Cell Oncol 30:435–450

Kvietys PR, Granger DN (2012) Role of reactive oxygen and nitrogen species in the vascular responses to inflammation. Free Radic Biol Med 52:556–592

Kwong JQ, Davis J, Baines CP, Sargent MA, Karch J, Wang X, Huang T, Molkentin JD (2014) Genetic deletion of the mitochondrial phosphate carrier desensitizes the mitochondrial permeability transition pore and causes cardiomyopathy. Cell Death Differ 21:1209–1217

Lamarche F, Carcenac C, Gonthier B, Cottet-Rousselle C, Chauvin C, Barret L, Leverve X, Savasta M, Fontaine E (2013) Mitochondrial permeability transition pore inhibitors prevent ethanol-induced neuronal death in mice. Chem Res Toxicol 26:78–88

Landles C, Bates GP (2004) Huntingtin and the molecular pathogenesis of Huntington's disease. Fourth in molecular medicine review series. EMBO Rep 5:958–963

Leal SS, Gomes CM (2015) Calcium dysregulation links ALS defective proteins and motor neuron selective vulnerability. Front Cell Neurosci 9:225

Lemasters JJ (1999) The mitochondrial permeability transition and the calcium, oxygen and pH paradoxes: one paradox after another. Cardiovasc Res 44(3):470

Lemasters JJ, Bond JM, Chacon E, Harper IS, Kaplan SH, Ohata H, Trollinger DR, Herman B, Cascio WE (1996) The pH paradox in ischemia-reperfusion injury to cardiac myocytes. EXS 76:99–114

Leung AW, Varanyuwatana P, Halestrap AP (2008) The mitochondrial phosphate carrier interacts with cyclophilin D and may play a key role in the permeability transition. J Biol Chem 283:26312–26323

Leung CH, Wang L, Nielsen JM, Tropak MB, Fu YY, Kato H, Callahan J, Redington AN, Caldarone CA (2014) Remote cardioprotection by transfer of coronary effluent from ischemic preconditioned rabbit heart preserves mitochondrial integrity and function via adenosine receptor activation. Cardiovasc Drugs Ther 28:7–17

Lezi E, Swerdlow RH (2012) Mitochondria in neurodegeneration. Adv Exp Med Biol 942:269–286

Locasale JW, Cantley LC (2011) Metabolic flux and the regulation of mammalian cell growth. Cell Metab 14:443–451

Lu X, Kwong JQ, Molkentin JD, Bers DM (2016) Individual cardiac mitochondria undergo rare transient permeability transition pore openings. Circ Res 118:834–841

Luth ES, Stavrovskaya IG, Bartels T, Kristal BS, Selkoe DJ (2014) Soluble, prefibrillar alpha-synuclein oligomers promote complex I-dependent, Ca2+-induced mitochondrial dysfunction. J Biol Chem 289:21490–21507

Maaser K, Hopfner M, Jansen A, Weisinger G, Gavish M, Kozikowski AP, Weizman A, Carayon P, Riecken EO, Zeitz M, Scherubl H (2001) Specific ligands of the peripheral benzodiazepine receptor induce apoptosis and cell cycle arrest in human colorectal cancer cells. Br J Cancer 85:1771–1780

Marchetti P, Castedo M, Susin SA, Zamzami N, Hirsch T, Macho A, Haeffner A, Hirsch F, Geuskens M, Kroemer G (1996) Mitochondrial permeability transition is a central coordinating event of apoptosis. J Exp Med 184:1155–1160

Marchi S, Patergnani S, Pinton P (2014) The endoplasmic reticulum-mitochondria connection: one touch, multiple functions. Biochim Biophys Acta 1837:461–469

Martel C, Huynh le H, Garnier A, Ventura-Clapier R, Brenner C (2012) Inhibition of the mitochondrial permeability transition for cytoprotection: direct versus indirect mechanisms. Biochem Res Int 2012:213403

Martin LJ, Gertz B, Pan Y, Price AC, Molkentin JD, Chang Q (2009) The mitochondrial permeability transition pore in motor neurons: involvement in the pathobiology of ALS mice. Exp Neurol 218:333–346

Martin LJ, Semenkow S, Hanaford A, Wong M (2014) Mitochondrial permeability transition pore regulates Parkinson's disease development in mutant alpha-synuclein transgenic mice. Neurobiol Aging 35:1132–1152

Marzo I, BRENNER C, Zamzami N, Jurgensmeier JM, Susin SA, Vieira HL, Prevost MC, Xie Z, Matsuyama S, Reed JC, Kroemer G (1998) Bax and adenine nucleotide translocator cooperate in the mitochondrial control of apoptosis. Science 281:2027–2031

Mcenery MW, Snowman AM, Trifiletti RR, Snyder SH (1992) Isolation of the mitochondrial benzodiazepine receptor: association with the voltage-dependent anion channel and the adenine nucleotide carrier. Proc Natl Acad Sci U S A 89:3170–3174

Milakovic T, Quintanilla RA, Johnson GV (2006) Mutant huntingtin expression induces mitochondrial calcium handling defects in clonal striatal cells: functional consequences. J Biol Chem 281:34785–34795

Moe GW, Marin-Garcia J (2016) Role of cell death in the progression of heart failure. Heart Fail Rev 21:157–167

Morciano G, Giorgi C, Bonora M, Punzetti S, Pavasini R, Wieckowski MR, Campo G, Pinton P (2015) Molecular identity of the mitochondrial permeability transition pore and its role in ischemia-reperfusion injury. J Mol Cell Cardiol 78:142–153

Morciano G, Bonora M, Giorgi C, Pinton P (2017) Other bricks for the correct construction of the mitochondrial permeability transition pore complex. Cell Death Dis 8:e2698

Mukhin AG, Papadopoulos V, Costa E, Krueger KE (1989) Mitochondrial benzodiazepine receptors regulate steroid biosynthesis. Proc Natl Acad Sci U S A 86:9813–9816

Narita M, Shimizu S, Ito T, Chittenden T, Lutz RJ, Matsuda H, Tsujimoto Y (1998) Bax interacts with the permeability transition pore to induce permeability transition and cytochrome c release in isolated mitochondria. Proc Natl Acad Sci U S A 95:14681–14686

Neuhof C, Neuhof H (2014) Calpain system and its involvement in myocardial ischemia and reperfusion injury. World J Cardiol 6:638–652

Nishihara M, Miura T, Miki T, Tanno M, Yano T, Naitoh K, Ohori K, Hotta H, Terashima Y, Shimamoto K (2007) Modulation of the mitochondrial permeability transition pore complex in GSK-3beta-mediated myocardial protection. J Mol Cell Cardiol 43:564–570

Palmieri F (2004) The mitochondrial transporter family (SLC25): physiological and pathological implications. Pflugers Arch 447:689–709

Pani G, Colavitti R, Bedogni B, Fusco S, Ferraro D, Borrello S, Galeotti T (2004) Mitochondrial superoxide dismutase: a promising target for new anticancer therapies. Curr Med Chem 11:1299–1308

Pastorino JG, Hoek JB (2003) Hexokinase II: the integration of energy metabolism and control of apoptosis. Curr Med Chem 10:1535–1551

Pastorino JG, Hoek JB (2008) Regulation of hexokinase binding to VDAC. J Bioenerg Biomembr 40:171–182

Pastorino JG, Simbula G, Gilfor E, Hoek JB, Farber JL (1994) Protoporphyrin IX, an endogenous ligand of the peripheral benzodiazepine receptor, potentiates induction of the mitochondrial permeability transition and the killing of cultured hepatocytes by rotenone. J Biol Chem 269:31041–31046

Pastorino JG, Hoek JB, Shulga N (2005) Activation of glycogen synthase kinase 3beta disrupts the binding of hexokinase II to mitochondria by phosphorylating voltage-dependent anion channel and potentiates chemotherapy-induced cytotoxicity. Cancer Res 65:10545–10554

Patergnani S, Suski JM, Agnoletto C, Bononi A, Bonora M, De Marchi E, Giorgi C, Marchi S, Missiroli S, Poletti F, Rimessi A, Duszynski J, Wieckowski MR, Pinton P (2011) Calcium signaling around Mitochondria Associated Membranes (MAMs). Cell Commun Signal 9:19

Pauleau AL, Galluzzi L, Scholz SR, Larochette N, Kepp O, Kroemer G (2008) Unexpected role of the phosphate carrier in mitochondrial fragmentation. Cell Death Differ 15:616–618

Pediaditakis P, Kim JS, He L, Zhang X, Graves LM, Lemasters JJ (2010) Inhibition of the mitochondrial permeability transition by protein kinase A in rat liver mitochondria and hepatocytes. Biochem J 431:411–421

Pinton P, Ferrari D, Magalhaes P, Schulze-Osthoff K, Di Virgilio F, Pozzan T, Rizzuto R (2000) Reduced loading of intracellular Ca(2+) stores and downregulation of capacitative Ca(2+) influx in Bcl-2-overexpressing cells. J Cell Biol 148:857–862

Piper HM, Meuter K, Schafer C (2003) Cellular mechanisms of ischemia-reperfusion injury. Ann Thorac Surg 75:S644–S648

Pogoryelov D, Klyszejko AL, Krasnoselska GO, Heller EM, Leone V, Langer JD, Vonck J, Muller DJ, Faraldo-Gomez JD, Meier T (2012) Engineering rotor ring stoichiometries in the ATP synthase. Proc Natl Acad Sci U S A 109:E1599–E1608

Pulido R, Baker SJ, Barata JT, Carracedo A, Cid VJ, Chin-Sang ID, Dave V, Den Hertog J, Devreotes P, Eickholt BJ, Eng C, Furnari FB, Georgescu MM, Gericke A, Hopkins B, Jiang X, Lee SR, Losche M, Malaney P, Matias-Guiu X, Molina M, Pandolfi PP, Parsons R, Pinton P, Rivas C, Rocha RM, Rodriguez MS, Ross AH, Serrano M, Stambolic V, Stiles B, Suzuki A, Tan SS, Tonks NK, Trotman LC, Wolff N, Woscholski R, Wu H, Leslie NR (2014) A unified nomenclature and amino acid numbering for human PTEN. Sci Signal 7:pe15

Quinn BA, Dash R, Azab B, Sarkar S, Das SK, Kumar S, Oyesanya RA, Dasgupta S, Dent P, Grant S, Rahmani M, Curiel DT, Dmitriev I, Hedvat M, Wei J, Wu B, Stebbins JL, Reed JC, Pellecchia M, Sarkar D, Fisher PB (2011) Targeting Mcl-1 for the therapy of cancer. Expert Opin Investig Drugs 20:1397–1411

Quintanilla RA, Tapia C, Perez MJ (2017) Possible role of mitochondrial permeability transition pore in the pathogenesis of Huntington disease. Biochem Biophys Res Commun 483:1078–1083

Rao VK, Carlson EA, Yan SS (2014) Mitochondrial permeability transition pore is a potential drug target for neurodegeneration. Biochim Biophys Acta 1842:1267–1272

Rasheed MZ, Tabassum H, Parvez S (2017) Mitochondrial permeability transition pore: a promising target for the treatment of Parkinson's disease. Protoplasma 254:33–42

Rasola A, Bernardi P (2011) Mitochondrial permeability transition in Ca(2+)-dependent apoptosis and necrosis. Cell Calcium 50:222–233

Rasola A, Sciacovelli M, Pantic B, Bernardi P (2010) Signal transduction to the permeability transition pore. FEBS Lett 584:1989–1996

Rempel A, Mathupala SP, Griffin CA, Hawkins AL, Pedersen PL (1996) Glucose catabolism in cancer cells: amplification of the gene encoding type II hexokinase. Cancer Res 56:2468–2471

Rimessi A, Patergnani S, Bonora M, Wieckowski MR, Pinton P (2015) Mitochondrial Ca(2+) remodeling is a prime factor in oncogenic behavior. Front Oncol 5:143

Roy SS, Madesh M, Davies E, Antonsson B, Danial N, Hajnoczky G (2009) Bad targets the permeability transition pore independent of Bax or Bak to switch between Ca2+-dependent cell survival and death. Mol Cell 33:377–388

Savino C, Pelicci P, Giorgio M (2013) The P66Shc/mitochondrial permeability transition pore pathway determines neurodegeneration. Oxid Med Cell Longev 2013:719407

Sayeed I, Parvez S, Winkler-Stuck K, Seitz G, Trieu I, Wallesch CW, Schonfeld P, Siemen D (2006) Patch clamp reveals powerful blockade of the mitochondrial permeability transition pore by the D2-receptor agonist pramipexole. FASEB J 20:556–558

Schonfeld P, Siemen D, Kreutzmann P, Franz C, Wojtczak L (2013) Interaction of the antibiotic minocycline with liver mitochondria—role of membrane permeabilization in the impairment of respiration. FEBS J 280:6589–6599

Seaton TA, Cooper JM, Schapira AH (1998) Cyclosporin inhibition of apoptosis induced by mitochondrial complex I toxins. Brain Res 809:12–17

Shanmughapriya S, Rajan S, Hoffman NE, Higgins AM, Tomar D, Nemani N, Hines KJ, Smith DJ, Eguchi A, Vallem S, Shaikh F, Cheung M, Leonard NJ, Stolakis RS, Wolfers MP, Ibetti J, Chuprun JK, Jog NR, Houser SR, Koch WJ, Elrod JW, Madesh M (2015) SPG7 is an essential and conserved component of the mitochondrial permeability transition pore. Mol Cell 60:47–62

Shargorodsky L, Veenman L, Caballero B, Pe'er Y, Leschiner S, Bode J, Gavish M (2012) The nitric oxide donor sodium nitroprusside requires the 18 kDa Translocator Protein to induce cell death. Apoptosis 17:647–665

Sharma R, Randhawa PK, Singh N, Jaggi AS (2015) Bradykinin in ischemic conditioning-induced tissue protection: evidences and possible mechanisms. Eur J Pharmacol 768:58–70

Shimizu S, Narita M, Tsujimoto Y (1999) Bcl-2 family proteins regulate the release of apoptogenic cytochrome c by the mitochondrial channel VDAC. Nature 399:483–487
Shinohara Y, Shima A, Kamida M, Terada H (1991) Uncoupling protein is expressed in liver mitochondria of cold-exposed and newborn rats. FEBS Lett 293:173–174
Shinohara Y, Ishida T, Hino M, Yamazaki N, Baba Y, Terada H (2000) Characterization of porin isoforms expressed in tumor cells. Eur J Biochem 267:6067–6073
Shintani-Ishida K, Yoshida K (2015) Mitochondrial m-calpain opens the mitochondrial permeability transition pore in ischemia-reperfusion. Int J Cardiol 197:26–32
Singh P, Suman S, Chandna S, Das TK (2009) Possible role of amyloid-beta, adenine nucleotide translocase and cyclophilin-D interaction in mitochondrial dysfunction of Alzheimer's disease. Bioinformation 3:440–445
Slomovitz BM, Coleman RL (2012) The PI3K/AKT/mTOR pathway as a therapeutic target in endometrial cancer. Clin Cancer Res 18:5856–5864
Stys PK, Zamponi GW, Van Minnen J, Geurts JJ (2012) Will the real multiple sclerosis please stand up? Nat Rev Neurosci 13:507–514
Su KG, Banker G, Bourdette D, Forte M (2009) Axonal degeneration in multiple sclerosis: the mitochondrial hypothesis. Curr Neurol Neurosci Rep 9:411–417
Sun J, Nguyen T, Aponte AM, Menazza S, Kohr MJ, Roth DM, Patel HH, Murphy E, Steenbergen C (2015) Ischaemic preconditioning preferentially increases protein S-nitrosylation in subsarcolemmal mitochondria. Cardiovasc Res 106:227–236
Takuma K, Phuagphong P, Lee E, Mori K, Baba A, Matsuda T (2001) Anti-apoptotic effect of cGMP in cultured astrocytes: inhibition by cGMP-dependent protein kinase of mitochondrial permeable transition pore. J Biol Chem 276:48093–48099
Tanveer A, Virji S, Andreeva L, Totty NF, Hsuan JJ, Ward JM, Crompton M (1996) Involvement of cyclophilin D in the activation of a mitochondrial pore by Ca2+ and oxidant stress. Eur J Biochem 238:166–172
Tedeschi H, Hegarty HJ, James JM (1965) Osmotic reversal of phosphate-induced mitochondrial swelling. Biochim Biophys Acta 104:612–615
Thomas RL, Roberts DJ, Kubli DA, Lee Y, Quinsay MN, Owens JB, Fischer KM, Sussman MA, Miyamoto S, Gustafsson AB (2013) Loss of MCL-1 leads to impaired autophagy and rapid development of heart failure. Genes Dev 27:1365–1377
Traba J, Del Arco A, Duchen MR, Szabadkai G, Satrustegui J (2012) SCaMC-1 promotes cancer cell survival by desensitizing mitochondrial permeability transition via ATP/ADP-mediated matrix Ca(2+) buffering. Cell Death Differ 19:650–660
Valente EM, Abou-Sleiman PM, Caputo V, Muqit MM, Harvey K, Gispert S, Ali Z, Del Turco D, Bentivoglio AR, Healy DG, Albanese A, Nussbaum R, Gonzalez-Maldonado R, Deller T, Salvi S, Cortelli P, Gilks WP, Latchman DS, Harvey RJ, Dallapiccola B, Auburger G, Wood NW (2004) Hereditary early-onset Parkinson's disease caused by mutations in PINK1. Science 304:1158–1160
Vander Heiden MG, Chandel NS, Schumacker PT, Thompson CB (1999) Bcl-xL prevents cell death following growth factor withdrawal by facilitating mitochondrial ATP/ADP exchange. Mol Cell 3:159–167
Vaseva AV, Marchenko ND, Ji K, Tsirka SE, Holzmann S, Moll UM (2012) p53 opens the mitochondrial permeability transition pore to trigger necrosis. Cell 149:1536–1548
Verma M, Shulga N, Pastorino JG (2013) Sirtuin-4 modulates sensitivity to induction of the mitochondrial permeability transition pore. Biochim Biophys Acta 1827:38–49
Verrier F, Deniaud A, Lebras M, Metivier D, Kroemer G, Mignotte B, Jan G, Brenner C (2004) Dynamic evolution of the adenine nucleotide translocase interactome during chemotherapy-induced apoptosis. Oncogene 23:8049–8064
Wallimann T, Dolder M, Schlattner U, Eder M, Hornemann T, O'Gorman E, Ruck A, Brdiczka D (1998) Some new aspects of creatine kinase (CK): compartmentation, structure, function and regulation for cellular and mitochondrial bioenergetics and physiology. Biofactors 8:229–234
Warburg O, Wind F, Negelein E (1927) The metabolism of tumors in the body. J Gen Physiol 8:519–530

Wilson JE (2003) Isozymes of mammalian hexokinase: structure, subcellular localization and metabolic function. J Exp Biol 206:2049–2057
Winklhofer KF, Haass C (2010) Mitochondrial dysfunction in Parkinson's disease. Biochim Biophys Acta 1802:29–44
Wolin EM (2013) PI3K/Akt/mTOR pathway inhibitors in the therapy of pancreatic neuroendocrine tumors. Cancer Lett 335:1–8
Yang Y, Karakhanova S, Werner J, Bazhin AV (2013) Reactive oxygen species in cancer biology and anticancer therapy. Curr Med Chem 20:3677–3692
Yoshida M, Muneyuki E, Hisabori T (2001) ATP synthase—a marvellous rotary engine of the cell. Nat Rev Mol Cell Biol 2:669–677
Youdim MB, Bar Am O, Yogev-Falach M, Weinreb O, Maruyama W, Naoi M, Amit T (2005) Rasagiline: neurodegeneration, neuroprotection, and mitochondrial permeability transition. J Neurosci Res 79:172–179
Zamzami N, El Hamel C, Maisse C, Brenner C, Munoz-Pinedo C, Belzacq AS, Costantini P, Vieira H, Loeffler M, Molle G, Kroemer G (2000) Bid acts on the permeability transition pore complex to induce apoptosis. Oncogene 19:6342–6350
Zheng Y, Shi Y, Tian C, Jiang C, Jin H, Chen J, Almasan A, Tang H, Chen Q (2004) Essential role of the voltage-dependent anion channel (VDAC) in mitochondrial permeability transition pore opening and cytochrome c release induced by arsenic trioxide. Oncogene 23:1239–1247

Mitochondrial Regulation of Cell Death

Dawei Liu, Jean-Luc Perfettini, and Catherine Brenner

Abstract Mitochondria are multifaceted organelles exerting vital as well as lethal functions within eukaryotic cells. When fueled with substrates and oxygen, mitochondria govern metabolic pathways, regulate calcium fluxes and are deeply involved in redox homeostasis. In stress conditions, notably when calcium and redox balances are altered, mitochondria sense cellular damages and ultimately, can orchestrate some phylogenetically-conserved forms of cell death such as intrinsic apoptosis, parthanatos as well as mitochondrial permeability transition-mediated necrosis. In contrast, they do not influence other cell death modalities such as necroptosis and ferroptosis. The execution of these mitochondria-dependent lethal processes involves the expression of mitochondria or nucleus-encoded proteins such as BCL-2 family members, VDAC, ANT, cytochrome *c*, Smac/Diablo, as well as Omi/HtrA2. In addition, mitochondria can also influence the cell fate through fusion/fission of the mitochondrial network and mitophagy to eliminate damaged mitochondria. Here, we will review and discuss basic knowledge on the role of mitochondria in the complex regulation of cell death.

Keywords ROS · Ant · BCL-2 · Calcium · Energetic metabolism · Permeability Transition · VDAC

D. Liu · C. Brenner (✉)
INSERM UMR-S 1180-LabEx LERMIT, Université Paris-Sud, Université Paris-Saclay, Châtenay Malabry, France

J.-L. Perfettini
Cell Death and Aging Team, Gustave Roussy, Villejuif, France

Laboratory of Molecular Radiotherapy, INSERM U1030, Gustave Roussy, Villejuif, France
Université Paris Sud - Paris Saclay, Villejuif, France

P. J. Oliveira (ed.), *Mitochondrial Biology and Experimental Therapeutics*,
https://doi.org/10.1007/978-3-319-73344-9_6

1 Introduction

Mitochondria are intracellular organelles, which are structured into four compartments and two membranes allowing a variety of functions beneficial for their maintenance as well as for the cell. This symbiosis implies the exchange of metabolites, ions such as calcium (Ca^{2+}), cofactors and transport of proteins between the inside and the outside of mitochondria, e.g. the cytosol, the nucleus or the endoplasmic reticulum. Thus, mitochondria rely on the import of cytosolic substrates to produce aerobically ATP by the respiratory chain in the inner membrane (IM) and to export ATP to energy demanding cellular sites. These metabolic reactions imply a pool of mitochondrial enzymes and cofactors such as NAD^+/NADH, FAD^+/FADH, $NADP^+$/NADPH and glutathione (GSH), distinct from a (similar) cytoplasmic pool to maintain the redox homeostasis (Lewis et al. 2014).

The vital functions of mitochondria require a strict compartmentation and the regulation of the permeability of mitochondrial membranes, the outer membrane (OM) being semi-permeable and the IM being almost impermeable. Accordingly, breaking the permeability barrier of mitochondria leads to cell death (Kroemer et al. 2007). In the 1990s, mitochondrial membrane permeabilisation (MMP) has been demonstrated to sense cellular damages and to govern cell death execution in a variety of cellular models, leading to the proposal that mitochondria control cell death in certain diseases such as cancer, ischemia/reperfusion and infection (Brenner and Kroemer 2000; Reed 2000; Martinou and Green 2001; Kroemer et al. 2007). The use of eukaryotic unicellular models (e.g. *S. cerevisiae*, *C. elegans*), cell-free systems and isolated mitochondria-based techniques, contributed largely to unravel the mechanisms of MMP (Zamzami et al. 1995; Petit et al. 1998). Briefly, under stimulation by pro-apoptotic proteins of the BCL-2 family such as BAX, BID or other stress signals (e.g. calcium (Ca^{2+}), reactive oxygen species (ROS)), it was demonstrated that mitochondria lost their transmembrane potential ($\Delta\Psi m$) and release lethal factors such as cytochrome *c* (Cyt *c*), Omi/HtrA2, Smac/DIABLO and in some particular cell death models, apoptosis inducing factor (AIF), from the intermembrane space (IMS) into the cytosol (Susin et al. 1996; Du et al. 2000; Van Loo et al. 2002; Garrido et al. 2006). In general, these proteins acquire novel function into the cytosol and interact with new partners to activate specific lethal signaling cascades of the so-called intrinsic apoptosis pathway. These events can be amplified by the concomitant initiation of other death pathways such as the extrinsic apoptosis pathway mediated by the activation of cell surface death receptors (e.g. TNF, Fas) and *vice versa*. Another consequence of MMP is that mitochondria become dysfunctional and, once a megachannel called the permeability transition pore (PTP) opens in the IM, mitochondria release molecules with a molecular weight inferior to 1.5 kDa such as calcium (Ca^{2+}) and produce more reactive oxygen species (ROS) in a process called ROS-induced ROS release (Ichas et al. 1997; Marzo et al. 1998). Moreover, in parallel to the loss of $\Delta\Psi m$, mitochondria consume ATP due to the reverse functioning of the F1Fo-ATPase (Di Lisa et al. 2007). The use of fluorescent probes such as TMRM, DiOC(6)3, calcein/cobalt, JC-1 combined with siRNA and/

or pharmacological inhibitors led to the demonstration that MMP is irreversible at this stage (for review: (Kroemer et al. 2007)). P53 and protein kinases such as GSK3β can also be translocated to mitochondria to regulate PTP opening and cell death (Juhaszova et al. 2004; Vaseva et al. 2012; Martel et al. 2012).

Likewise, cell death has been recently reclassified in several modalities including extrinsic and intrinsic apoptosis, mitochondrial permeability transition pore (MPT)-mediated necrosis, necroptosis, parthanatos, ferroptosis, and autosis to form the large group of regulated cell deaths (RCD) (Galluzzi et al. 2015). More recently, another cell death modality, name entosis emerged and has been described (Martins et al. 2017). Entosis is characterized by cell-in-cell structures corresponding to one living cell entering another living cell, which usually leads to the death of the internalized cell, specifically through caspase-dependent cell death (emperitosis) or lysosome-dependent cell death (entosis) (Wang et al. 2016). A key factor for differentiating apoptosis from the other RCD modalities is the late permeabilization of plasma membrane, which can be easily monitored by impermeant dyes in cell culture such as propidium iodide.

Regarding mitochondrial role in cell death, it is admitted that they are essential for intrinsic apoptosis, parthanatos and MPT-driven necrosis. In contrast, if mitochondria can contribute (more or less) to extrinsic apoptosis and pyroptotis, these organelles appear to be dispensable for ferroptosis and necroptosis (Galluzzi et al. 2016; Wang et al. 2017). Their role in entosis remain to be explored. Therefore, these latter types of RCD will not be discussed here.

In this chapter, we will review the recent knowledge regarding the role of mitochondria in the regulation of some forms of RCD and discusses how they can influence cell death with an emphasis on apoptosis (Elmore 2007).

2 Mitochondrial Reactive Oxygen Species and Mitochondrial Dysfunction

The redox homeostasis reflects the balance between the amount of ROS produced, the amount of antioxidant enzymes (i.e. superoxide dismutase 1 (SOD1), superoxide dismutase 2 (SOD2), and catalase (CA)) and of pro/anti-oxidant molecules (e.g. GSH, vitamin E...) (for review: (Halliwell and Gutteridge 1999)). More precisely, ROS are unstable molecules such as the anion superoxide ($O_2^{\bullet -}$), hydrogen peroxide (H_2O_2), and hydroxyl radical and are produced in all mammalian cells by mitochondria and plasma membrane-bound enzymes, NADPH oxidase (NOX) (Halliwell and Gutteridge 1999). Within mitochondria, ROS are produced at the level of respiratory chain by the leakage of few electrons acting directly on oxygen producing $O_2^{\bullet -}$. Then, anion superoxide can be detoxified by superoxide dismutase (SOD) to generate H_2O_2, which can be transformed to O_2 by catalase. Locally, ROS can interact with and modify proteins, lipids and nucleic acids, affecting dramatically the function of these macromolecules (Halliwell and Gutteridge 1999).

Oxidative stress implies the notion of cellular damage impacting the cell fate and potentially leading to cell death tissues damages and ultimately human pathologies (for reviews: (Lenaz 2001; Le Bras et al. 2005; Lin and Beal 2006)). In general, independently of the molecular species, ROS stimulate MMP and cell death in many type of cells based on the balance between anion $O_2^{\bullet-}$ and H_2O_2 (Clement and Pervaiz 2001). Although not all targets or oxidative modifications are known, it has been shown that mitochondrial proteins from the OM such as the voltage dependent anion channel (VDAC) (Madesh and Hajnoczky 2001; Tomasello et al. 2009), from the IM such as adenine nucleotide translocase (ANT) (Costantini et al. 2000; Vieira et al. 2001) and proteins from the BCL-2 family (see below) can be modified by ROS. Importantly, *in vivo*, ROS are associated with numerous acute or chronic pathologies exhibiting excessive of impaired cell death (e.g. cancer, cardiac and brain ischemia/reperfusion, myocardial infarction and neurodegenerative diseases) and aging compatible with a causative link between ROS and cell loss (Murphy et al. 2011). Unfortunately, antioxidants are not totally efficient to protect from cell death *in vivo* and in humans, suggesting that (1) they are not the essential and/or the sole trigger of cell death or (2) they are not efficiently targeted to mitochondria. Alternatively, it is also plausible that the short period of life of various ROS and their local action render them difficult to block (Halliwell 1994).

Another mitochondrial source of ROS are the monoamine oxidases, MAO, which are encoded by two isoforms. These enzymes are located in the OM, facing the cytosol and generating H_2O_2 from O_2. The role of this enzyme has been extensively studied in the brain and cardiovascular system (Kaludercic et al. 2011). Thus, MAO activate various pathophysiological signaling cascades including mitochondrial apoptosis through increase of BAX, decrease of BCL-2, opening of PTP, sphingosine kinase inhibition and ceramide production (Bianchi et al. 2005; Pchejetski et al. 2007). H_2O_2 produced by MOA can also affect stress pathways in the cytosol (e.g. p38; JNK) and in the nucleus (e.g. p53), which can lead to cell death, but it is beyond the scope of this chapter (Balaban et al. 2005).

In summary, in a general manner, mitochondria can influence cell death via their roles in ROS production and detoxification, which can activate or inhibit specific lethal signaling pathways. More specifically, depending on the pathological condition, ROS are considered as major determinants in cell death, triggering and amplifying the tissues/organs damages through a mechanism named ROS-induced ROS release (Zorov et al. 2000).

3 Mitochondrial Calcium as a Trigger of Cell Death

In several physiopathological conditions such as ischemia/reperfusion, together with massive ROS production, there is also a concomitant rise in intra-mitochondrial Ca^{2+} coupled with a Ca^{2+} depletion in other cell compartments. Notably, decreased Ca^{2+} concentration was measured in the vicinity of the IP3R receptor in the endoplasmic reticulum (Duchen 2000a, b; Mattson and Chan 2003; Pozzan et al. 1994).

This Ca^{2+} entry into the mitochondrial matrix is mediated by VDAC in the OM (Gincel et al. 2001; Deniaud et al. 2007) and the mitochondrial Ca^{2+} uniport (MCU) in the IM (Baughman et al. 2011; De Stefani et al. 2011). At low levels, Ca^{2+} is believed to activate mitochondrial metabolic enzymes such as dehydrogenases in the Krebs cycle and thus, metabolism. At higher concentrations, Ca^{2+} promotes a rapid collapse in $\Delta\Psi m$. This decrease can be attributed to the opening of the PTP or the arrest of respiration. If sustained below a certain threshold, this $\Delta\Psi m$ decrease leads to cell death (Duchen 2000a, b). In neurons, it has been proposed that the combination of nitric oxide (NO) and high intra-mitochondrial Ca^{2+} cause the collapse of the $\Delta\Psi m$, through an effect on the PTP leading ultimately to cell death. Since it is very difficult technically and quantitatively to study spatiotemporal relationships between ROS and Ca^{2+} *in vivo*, at least *in vitro*, a synergistic and/or co-dependent effect of ROS and Ca^{2+} can be pharmacologically demonstrated. For example, this has been shown by the treatment of isolated mitochondria from rat liver with Ca^{2+} and 5-aminolevulinic acid as a source of ROS and EGTA as a Ca^{2+} chelator (Hermes-Lima et al. 1992; Kanno et al. 2004).

The flux of Ca^{2+} is regulated by its transport through the MCU in the IM (Baughman et al. 2011; De Stefani et al. 2011). This 40 kDa-channel that has been identified by bioinformatics, was found to be expressed in the IM and allows the import of Ca^{2+}. Mutations within the channel impaired the activity of the channel in planar lipid bilayers. $MCU^{-/-}$ mice lack rapid Ca^{2+} accumulation and exhibit a marked exercise impairment. As expected, the mitochondria isolated from these mice failed to open PTP, whereas cells or tissues from mice lacking MCU are still susceptible to cell death induction (Pan et al. 2013).

In summary, Ca^{2+} appears as a master regulator of mitochondrial metabolism as well as mitochondria-dependent cell death (Mattson and Chan 2003).

4 BCL-2 Family Members

Discovered in 1985 (Tsujimoto et al. 1985), BCL-2 has first been described as an onco-protein and demonstrated to inhibit cell death by interaction with multiple proteins (Kroemer and Reed 2000). BCL-2 is the prototype of BLC2 family proteins and shares some sequence homology with bacterial colicin and thus exhibit channel activity in vitro.

BCL-2 proteins family is composed of pro- and anti-apoptotic proteins located in several intracellular compartments including the mitochondrion (e.g. BAX, BAD, BID, BCL-XL, MCL-1) (Cory and Adams 2002). Notably, mitochondria are equipped with a set of constitutive proteins belonging to the BCL-2 family, that, at least in certain conditions, have been shown to function as potent inhibitors of MMP induction and Cyt *c* release (Kroemer et al. 2007). In parallel, pro-apoptotic members of BCL-2 family such as BAX and BID can also translocate and associate with mitochondrial membranes to form supramolecular openings in association with lipids (Harada et al. 1999; Wei et al. 2001; Kuwana et al. 2002). A large body of evidence shows that mitochondria can integrate many pro-apoptotic stimuli, such as ROS and Ca^{2+} and that the protein BCL-2 can block many pro-apoptotic signals at the level of mitochondria led to the definition that mitochondria behave as central integrator of cell death (Brenner and Kroemer 2000). This view is still currently admitted taking in account that multiple lethal pathways co-exist and that all pathways do not involve BCL-2 family members (Galluzzi et al. 2015, 2016).

5 Mitochondrial Intermembrane Space Proteins

The discovery that Cyt *c* release into the cytosol can induce apoptosis had a major impact on the cell death research (Liu et al. 1996). Rapidly, the Cyt *c* release was frequently assimilated to the opening of the Pandora's box and become a hallmark of apoptosis induction (Zamzami and Kroemer 2001). Indeed, this discovery raised numerous questions: is the Cyt *c* release a fast process? Is it concomitant in all cells in a population? Is it accompanied by other mitochondrial proteins? Does it involve membrane rupture or specific pores? Is it observed in all cell death modalities or just in apoptosis? To address these questions, studies based on proteomics analysis of cell free systems have been undertaken, and led to the identification of numerous (if not all) intermembrane space proteins that are released in response to pro-apoptotic signals (Bossy-Wetzel and Green 1999; Patterson et al. 2000). Thus, in view of the scope of the topic, we will focus on Cyt *c* release in apoptosis and two other proteins, namely Smac/Diablo and Omi/HtrA2, for which there is no doubt about their role in the lethal process.

Cyt *c* is located into the mitochondrial intermembrane space where it participates in the electron transport chain (Neupert 1997). Through its heme group, Cyt *c* accepts electrons from the bc1 complex and transfers them to the complex IV. Upon initiation of apoptosis by various endogenous factors such as cardiolipin oxidation, calcium overload, ROS production, caspase activation, BAX or tBID translocation, Cyt *c* can be released into the cytosol, activates pro-caspase 9 in caspase 9 and binds apoptotic protease activating factor-1 (Apaf-1), to form the apoptosome in the presence of ATP and initiates the late steps of cell dismantling (Eskes et al. 1998; Luo et al. 1998; Mattson and Chan 2003; Orrenius and Zhivotovsky 2005; Breckenridge et al. 2003).

Smac/Diablo has been identified as the second protein promoting caspase activation in the Cyt *c*/Apaf-1/caspase-9 axis (Du et al. 2000). In fact, Smac promotes caspase-9 activation by binding to inhibitor of apoptosis proteins, IAPs. As Cyt *c*, Smac resides in mitochondria but is released into the cytosol when cells undergo apoptotic cell death. Interestingly, Smac expression level tunes cell sensitivity to apoptotic induction and Smac mimetics are currently developed as new class of promising anti-cancer agents (Bai et al. 2014).

Proteomic analysis of supernatants obtained from isolated mitochondria that have been exposed to recombinant tBID, a pro-apoptotic BCL-2 member, revealed the presence of a serine protease Omi, also called HtrA2 (Van Loo et al. 2002;

Hegde et al. 2002). This release was prevented in mitochondria derived from BCL-2-transgenic mice. The release of Omi under apoptotic conditions was further evidenced *in vivo* in livers from mice injected with agonistic anti-Fas antibodies and was prevented in livers from BCL-2 transgenic mice (Van Loo et al. 2002). Omi release also occurs in apoptotic dying but not in necrotic dying fibrosarcoma L929 cells, treated with anti-Fas antibodies and TNF, respectively. Omi exerts its activity through a reaper XIAP interaction motif located at its N-terminus (Martins et al. 2002).

6 Mitophagy

6.1 Molecular Mechanisms

Mitophagy is a conserved, mitochondria-specific autophagic clearance process, which regulates the number of mitochondria and maintains quality control within eukaryotic cells (Kim et al. 2007). Mitophagy also contributes to cellular survival by removing damaged mitochondria, eliminating the apoptogenic signals and excessive ROS. Mitophagy is activated by stresses such as hypoxia, nutrient deprivation, DNA damage, inflammation and mitochondrial membrane depolarization (Kim et al. 2007). Several mitophagy mechanisms have been identified, including PTEN-induced putative kinase 1 (PINK1), Parkin, BNIP3L as shown in Fig. 1.

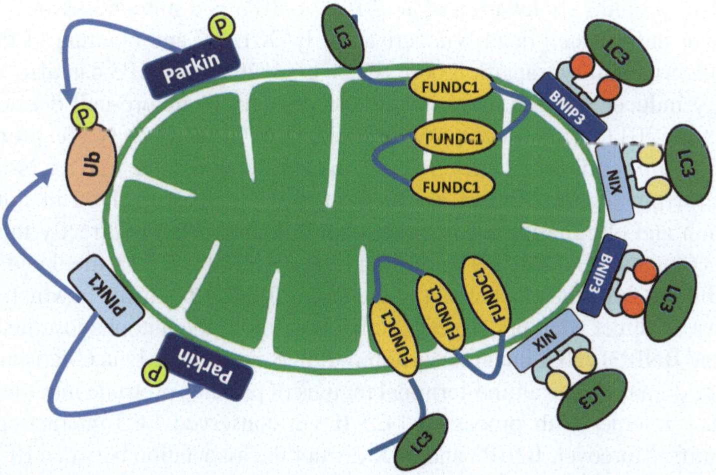

Fig. 1 Mitochondrial membrane depolarization leads to PINK1 kinase accumulation at the outer mitochondrial membrane. PINK1 phosphorylates ubiquitin (Ub) and Parkin (within its ubiquitin-like domain) and this results in recruitment of Parkin ubiquitin ligase activity. BNIP3/NIX and FUNDC1 are induced by hypoxia and interact with LC3 through conserved LC3-interacting region motifs in their respective N termini

PTEN-induced putative kinase1 (PINK1) and ubiquitin E3 ligase Parkin play important roles in the signaling pathways which regulate mitochondrial network homeostasis and quality control, including mitophagy. PINK1 is a serine/threonine-protein kinase, whose activity is critical for the regulation of several cellular processes (Aerts et al. 2015). Mutation in PINK1 is one of the major causes of Parkinson's disease (PD) (Morais et al. 2009). It is admitted that damaged mitochondria are characterized by a loss of $\Delta\Psi m$, which can be induced by depolarizing agents such as carbonyl cyanide m-chlorophenyl hydrazone (CCCP) (Rodriguez-Enriquez et al. 2004). Upon mitochondrial depolarization, PINK1 accumulates rapidly on mitochondrial surface, removes mitochondria by mediating mitophagy. These activities require the recruitment and phosphorylation of multiple substrates including Parkin, which is another key protein involved in mitophagy modulation (Koyano et al. 2014). Parkin, which is an E3 Ubiquitin ligase which functions in early Parkinson's disease, is considered to be a cytosolic protein that is translocated to the mitochondria to participate in mitochondrial quality control mechanisms. The defects in Drosophila PINK1 null mutants can be rescued by expression of Parkin indicating that PINK1 and Parkin are both in the same pathway and Parkin acts downstream of PINK1 (Clark et al. 2006; Yang et al. 2006). When mitochondria are damaged, Parkin is phosphorylated by PINK1 and consequently, it ubiquitinates proteins, which reside on OM, targeting the mitochondria for degradation (Lazarou et al. 2015).

BNIP3 and BNIP3L/NIX are atypical Bcl2-homology 3 (BH3)-only proteins localized on the OMM. NIX is identified as an autophagy receptor which mediates mitochondrial elimination during the erythrocyte development (Novak et al. 2010). Both NIX and BNIP3 contain a highly conserved LC3-interacting region (LIR) motif, which mediates mitophagy through the direct binding with LC3 and GABARAP proteins (Schwarten et al. 2014). BNIP3 is a mitochondrial BH3-only protein that induces cell death via activating BAX/BAK and opening of the mitochondrial permeability transition pore (Chen et al. 2010). BNIP3 can also act as an autophagy inducer through mechanisms independent of its pro-cell death activity (Rikka et al. 2011). It has been shown to serve as an autophagy receptor for the binding of mitochondria to LC3-II on the autophagosome via its N-terminal LC3-interacting region (Zhu et al. 2013). BNIP3 is also involved in mitophagy modulation and elimination of dysfunctional mitochondria, by directly interacting with LC3 (Hanna et al. 2012). The mechanism of BNIP3 and NIX mediated mitophagy is different from the mechanism of the Parkin/PINK1 pathway, in which the proteins serve as direct adaptors targeting mitochondria to the autophagosome. Briefly, BNIP3 and BNIP3L/NIX integrate as homo-dimers into the OM via C-terminal transmembrane domains. The amino-terminal regions of proteins protrude into the cytosol, where they interact with processed LC3-II via conserved LC3-interacting region (LIR) motifs. Moreover, BNIP3 and NIX disrupt the association between BCL-2 and Beclin1 activating the autophagic machinery (Bellot et al. 2009). Therefore, BNIP3 and NIX can increase autophagosome formation and enhance mitophagy through Beclin1 activation.

Similar to BNIP3 and NIX, the mitochondrial outer membrane protein FUNDC1 acts as an autophagy receptor promoting mitochondria removal specifically in response to hypoxia (Liu et al. 2012). Similar to BNIP3 and NIX, FUNDC1 is a hypoxia-inducible mitochondrial adaptor that binds to LC3 directly through a LIR motif. Src and CK2 kinases can phosphorylate the LIR motif of FUNDC1 to prevent mitophagy under normal conditions (Liu et al. 2014). On the other hand, mitochondrial depolarization and hypoxia trigger PGAM5-dependent dephosphorylation of FUNDC1, thereby inducing mitophagy (Chen et al. 2014). Under conditions of hypoxia, ULK1 kinase, a key component of the autophagy pre-initiation complex, translocates to mitochondria, where it phosphorylates FUNDC1 to enhance its interaction with FUNDC1 with LC3 to promote mitophagy (Wu et al. 2014).

The opening of PTP has been considered to be responsible for the mitophagy of depolarized mitochondria in mammalian cells. Mitochondria become permeable to all solutes up to a molecular mass of about 1500 Da after the onset of PTP opening, which can lead to mitochondrial depolarization. Mitochondrial depolarization is a main event for the activation of mitophagy and degradation of dysfunctional mitochondria. Furthermore, it was reported that the change of intracellular pH is a regulator of mitochondrial quality control and also triggers mitophagy. Low FCCP concentration induces a complete loss of mitochondrial membrane potential, but doesn't trigger mitophagy. However, high concentrations of FCCP can change intracellular pH, which induces mitophagy (Berezhnov et al. 2016). Other physiological mitophagy inducers, like hypoxia and starvation also induce cytosolic mild acidification (Burchell et al. 2013; Dechant et al. 2010).

In summary, mitophagy acts as a critical role for maintaining proper cellular functions and mitochondrial population quality by eliminating dysfunctional mitochondria.

6.2 *Mitophagy, Cell Death or Cytoprotection*

The crosstalk of mitophagy activation promoting cell death or cytoprotection has been intensively studied in recent years. Actually mitophagy was first found to be involved in cell death stimulation, but accumulated evidences show that mitophagy is cytoprotective in cells under stress and reduces the tendency of cells to undergo apoptosis, as is well known that mitophagy can maintain mitochondrial quality and regulate mitochondrial homeostasis. Briefly, mitophagy can be described as a double-edged sword: it appears as a cytoprotective mechanism when it is activated by mild physiological stimulus, meanwhile plays a lethal role when over-activated by severe pathological stress such as ischemia (Wong and Cuervo 2010). Examples where mitophagy contributes to cell death processes are presented below.

Purkinje cells are a class of specialized neurons in the cerebellum, and are among the most metabolically active of all neurons. It was reported that degeneration of Purkinje cell neurons in Purkinje cell degeneration mice is characterized by activation of the autophagy and mitophagy pathway. Increased or aberrant mitophagy can contributes to the Purkinje cell degeneration in pcd mice (Chakrabarti et al. 2009).

RCAN1-1L protein, which can be induced by multiple stresses, has both protective and harmful effects. RCAN-1L is identified as a regulator of mitophagy and induction of RCAN-1L can cause dramatic degradation of mitochondria. Overexpression of RCAN1-1L promotes mitophagy and lately promotes cell death (Ermak et al. 2012).

BNIP3 is a member of a unique subfamily of death-inducing mitochondrial proteins. This gene is an important mitophagy modulator as mentioned above, that has also been associated with cell death promotion. BNIP3 gene silencing achieved significant neuroprotection by rescuing neurons from ischemia/hypoxia (I/H)-induced apoptosis and excessive mitophagy has been detected in BNIP3 knockout mice (Shi et al. 2014). In an apparent contradiction, mitophagy induced by BNIP3 can act as a cytoprotective process. For instance, BNIP3 can exhibit dual functionality, with pro-survival activity via LIR-dependent autophagy interaction, induces mitophagic elimination of unhealthy mitochondria prior to dysfunction (i.e. depolarization and release of cytochrome *c*), consequently reduce the cell apoptotic potential (Zhu et al. 2013). Thus, BNIP3 promotes cell survival or cell death, depending on the context and the intensity of the stimulus.

As previously said, it is more and more accepted that mitophagy also participates in cytoprotection by maintaining mitochondrial function and cellular homeostasis.

The attenuated measles virus of the Edmonston strain lineage B (MV-Edm), which is promising oncolytic single-stranded RNA virus, induced SQSTM1-mediated mitophagy, which is crucial for clearance of damaged mitochondria before cytochrome c is released. Thus MV-Edm controls cytochrome c release and subsequently prevents downstream caspase activation by activation of mitophagy (Xia et al. 2014). Staurosporin, a common inducer of apoptosis, which is often used in mechanistic studies of dopaminergic cell death, induces both mitophagy and autophagy, these two pathways exerting a significant neuroprotective effect, rather than a contribution to autophagic cell death (Ha et al. 2014). Moreover, ROS are the by-products of cellular metabolism and are primarily generated by mitochondria. In cases where the formation of ROS overwhelms its antioxidant capacity, the excess ROS might damage mitochondria and lead the depolarization of the mitochondrial membrane. Retigeric acid B (RAB), a pentacyclic triterpenic acid induces ROS production and inhibits the PI3K/Akt/mTOR pathway. These events strongly stimulate mitophagy, which then, attenuates RAB-triggered cell death (Liu et al. 2013).

In addition, mitochondrial dynamics are integrated with mitophagy and cell death. Mitochondrial fission and fusion are regulated by several different GTPases. Mitofusins 1 and 2 (MFN1 and MFN2) regulate fusion of the outer mitochondrial membrane, whereas optic atrophy protein 1 (OPA1) promotes fusion of the inner membrane (Cipolat et al. 2004). Mitochondrial fission is regulated by dynamin-related protein 1 (DRP1) and fission protein 1 (FIS1) (Yoon et al. 2003). Mitochondrial dynamics are also closely integrated with mitophagy. Mitophagy is attenuated in cells when reduced mitochondrial fission is reduced, indicating that fission is a prerequisite for mitophagy activation. For instance, DRP1-deficient MEFs have significantly reduced Parkin-mediated mitophagy compared with wild-type MEFs (Tanaka et al. 2010). Recent studies have connected the PINK1/Parkin mitophagy pathway with mitochondrial dynamics. MFN1 and MFN2 are substrates for Parkin, although their ubiquitination does not serve as a signal for mitophagy (Gegg et al. 2010). Altogether, several evidences highlights the fact that mitochondrial dynamics influence mitophagic process.

7 Conclusion and Perspectives

Mitochondria are central organelles of cell death process. This view has benefited of the development of *in vitro* methodological approaches to study their structure and their function (Belzacq-Casagrande et al. 2008). Given the importance of cell death in development, adulthood and diseases, an active research is still needed to discover all the mechanistic aspects controlled by mitochondria such as the metabolism flexibility. There is no doubt that novel effectors and putative therapeutic targets remain to discover.

Acknowledgments C.B. is funded by ANR (ANR-13-ISV1-0001-01) and the Investment for the Future program ANR-11-IDEX-0003-01 within the LABEX ANR-10-LABX-0033. D.L. is funded by scholarship from Chine Scientific Council (CSC). This work was supported by funds from Agence Nationale de la Recherche (ANR-10-IBHU-0001, ANR-10-LABX33 and ANR-11-IDEX-003-01), Electricité de France, Fondation Gustave Roussy, Institut National du Cancer (INCA 9414), Cancéropôle Ile de France, NATIXIS, SIDACTION and the French National Agency for Research on AIDS and viral Hepatitis (ANRSH) (to J-L.P.).

References

Aerts L, De Strooper B, Morais VA (2015) PINK1 activation-turning on a promiscuous kinase. Biochem Soc Trans 43:280–286

Bai L, Smith DC, Wang S (2014) Small-molecule SMAC mimetics as new cancer therapeutics. Pharmacol Ther 144:82–95

Balaban RS, Nemoto S, Finkel T (2005) Mitochondria, oxidants, and aging. Cell 120:483–495

Baughman JM, Perocchi F, Girgis HS, Plovanich M, Belcher-Timme CA, Sancak Y, Bao XR, Strittmatter L, Goldberger O, Bogorad RL (2011) Integrative genomics identifies MCU as an essential component of the mitochondrial calcium uniporter. Nature 476:341–345

Bellot G, Garcia-Medina R, Gounon P, Chiche J, Roux D, Pouyssegur J, Mazure NM (2009) Hypoxia-induced autophagy is mediated through hypoxia-inducible factor induction of BNIP3 and BNIP3L via their BH3 domains. Mol Cell Biol 29:2570–2581

Belzacq-Casagrande A-S, Martel C, Pertuiset C, Borgne-Sanchez A, Jacotot E, Brenner C (2008) Pharmacological screening and enzymatic assays for apoptosis. Front Biosci 14:3550–3562

Berezhnov AV, Soutar MP, Fedotova EI, Frolova MS, Plun-Favreau H, Zinchenko VP, Abramov AY (2016) Intracellular pH modulates autophagy and mitophagy. J Biol Chem 291:8701–8708

Bianchi P, Kunduzova O, Masini E, Cambon C, Bani D, Raimondi L, Seguelas M-H, Nistri S, Colucci W, Leducq N (2005) Oxidative stress by monoamine oxidase mediates receptor-independent cardiomyocyte apoptosis by serotonin and postischemic myocardial injury. Circulation 112:3297–3305

Bossy-Wetzel E, Green DR (1999) Caspases induce cytochrome c release from mitochondria by activating cytosolic factors. J Biol Chem 274:17484–17490

Breckenridge DG, Stojanovic M, Marcellus RC, Shore GC (2003) Caspase cleavage product of BAP31 induces mitochondrial fission through endoplasmic reticulum calcium signals, enhancing cytochrome c release to the cytosol. J Cell Biol 160:1115–1127

Brenner C, Kroemer G (2000) Apoptosis mitochondria-the death signal integrators. Science 289:1150–1151

Burchell VS, Nelson DE, Sanchez-Martinez A, Delgado-Camprubi M, Ivatt RM, Pogson JH, Randle SJ, Wray S, Lewis PA, Houlden H, Abramov AY, Hardy J, Wood NW, Whitworth AJ, Laman H, Plun-Favreau H (2013) The Parkinson's disease-linked proteins Fbxo7 and Parkin interact to mediate mitophagy. Nat Neurosci 16:1257–1265

Chakrabarti L, Eng J, Ivanov N, Garden GA, La Spada AR (2009) Autophagy activation and enhanced mitophagy characterize the Purkinje cells of pcd mice prior to neuronal death. Mol Brain 2:24

Chen Y, Lewis W, Diwan A, Cheng EH, Matkovich SJ, Dorn GW II (2010) Dual autonomous mitochondrial cell death pathways are activated by nix/BNip3L and induce cardiomyopathy. Proc Natl Acad Sci U S A 107:9035–9042

Chen G, Han Z, Feng D, Chen Y, Chen L, Wu H, Huang L, Zhou C, Cai X, Fu C, Duan L, Wang X, Liu L, Liu X, Shen Y, Zhu Y, Chen Q (2014) A regulatory signaling loop comprising the PGAM5 phosphatase and CK2 controls receptor-mediated mitophagy. Mol Cell 54:362–377

Cipolat S, Martins De Brito O, Dal Zilio B, Scorrano L (2004) OPA1 requires mitofusin 1 to promote mitochondrial fusion. Proc Natl Acad Sci U S A 101:15927–15932

Clark IE, Dodson MW, Jiang C, Cao JH, Huh JR, Seol JH, Yoo SJ, Hay BA, Guo M (2006) Drosophila pink1 is required for mitochondrial function and interacts genetically with parkin. Nature 441:1162–1166

Clement MV, Pervaiz S (2001) Intracellular superoxide and hydrogen peroxide concentrations: a critical balance that determines survival or death. Redox Rep 6:211–214

Cory S, Adams JM (2002) The Bcl2 family: regulators of the cellular life-or-death switch. Nat Rev Cancer 2:647–656

Costantini P, Belzacq AS, Vieira HL, Larochette N, De Pablo MA, Zamzami N, Susin SA, Brenner C, Kroemer G (2000) Oxidation of a critical thiol residue of the adenine nucleotide translocator enforces Bcl-2-independent permeability transition pore opening and apoptosis. Oncogene 19:307–314

De Stefani D, Raffaello A, Teardo E, Szabò I, Rizzuto R (2011) A forty-kilodalton protein of the inner membrane is the mitochondrial calcium uniporter. Nature 476:336–340

Dechant R, Binda M, Lee SS, Pelet S, Winderickx J, Peter M (2010) Cytosolic pH is a second messenger for glucose and regulates the PKA pathway through V-ATPase. EMBO J 29:2515–2526

Deniaud A, Rossi C, Berquand A, Homand J, Campagna S, Knoll W, Brenner C, Chopineau J (2007) Voltage-dependent anion channel transports calcium ions through biomimetic membranes. Langmuir 23:3898–3905

Di Lisa F, Canton M, Menabò R, Kaludercic N, Bernardi P (2007) Mitochondria and cardioprotection. Heart Fail Rev 12:249–260

Du C, Fang M, Li Y, Li L, Wang X (2000) Smac, a mitochondrial protein that promotes cytochrome c-dependent caspase activation by eliminating IAP inhibition. Cell 102:33–42

Duchen M (2000a) Mitochondria and Ca2+ in cell physiology and pathophysiology. Cell Calcium 28:339–348

Duchen MR (2000b) Mitochondria and calcium: from cell signalling to cell death. J Physiol 529:57–68

Elmore S (2007) Apoptosis: a review of programmed cell death. Toxicol Pathol 35:495–516

Ermak G, Sojitra S, Yin F, Cadenas E, Cuervo AM, Davies KJ (2012) Chronic expression of RCAN1-1L protein induces mitochondrial autophagy and metabolic shift from oxidative phosphorylation to glycolysis in neuronal cells. J Biol Chem 287:14088–14098

Eskes R, Antonsson B, Osen-Sand A, Montessuit S, Richter C, Sadoul R, Mazzei G, Nichols A, Martinou J-C (1998) Bax-induced cytochrome C release from mitochondria is independent of the permeability transition pore but highly dependent on Mg2+ ions. J Cell Biol 143:217–224

Galluzzi L, Bravo-San Pedro J, Vitale I, Aaronson S, Abrams J, Adam D, Alnemri E, Altucci L, Andrews D, Annicchiarico-Petruzzelli M (2015) Essential versus accessory aspects of cell death: recommendations of the NCCD 2015. Cell Death Differ 22:58

Galluzzi L, Kepp O, Kroemer G (2016) Mitochondrial regulation of cell death: a phylogenetically conserved control. Microbial Cell 3:101–108

Garrido C, Galluzzi L, Brunet M, Puig PE, Didelot C, Kroemer G (2006) Mechanisms of cytochrome c release from mitochondria. Cell Death Differ 13:1423–1433

Gegg ME, Cooper JM, Chau KY, Rojo M, Schapira AH, Taanman JW (2010) Mitofusin 1 and mitofusin 2 are ubiquitinated in a PINK1/parkin-dependent manner upon induction of mitophagy. Hum Mol Genet 19:4861–4870

Gincel D, Hilal Z, Shoshan-Barmatz V (2001) Calcium binding and translocation by the voltage-dependent anion channel: a possible regulatory mechanism in mitochondrial function. Biochem J 358:147–155

Ha JY, Kim JS, Kim SE, Son JH (2014) Simultaneous activation of mitophagy and autophagy by staurosporine protects against dopaminergic neuronal cell death. Neurosci Lett 561:101–106

Halliwell B (1994) Free radicals, antioxidants, and human disease: curiosity, cause, or consequence? Lancet 344:721–724

Halliwell B, Gutteridge JMC (1999) Free radicals in biology and medicine. Clarendon Press, Oxford, England

Hanna RA, Quinsay MN, Orogo AM, Giang K, Rikka S, Gustafsson AB (2012) Microtubule-associated protein 1 light chain 3 (LC3) interacts with Bnip3 protein to selectively remove endoplasmic reticulum and mitochondria via autophagy. J Biol Chem 287:19094–19104

Harada H, Becknell B, Wilm M, Mann M, Huang LJ-S, Taylor SS, Scott JD, Korsmeyer SJ (1999) Phosphorylation and inactivation of BAD by mitochondria-anchored protein kinase a. Mol Cell 3:413–422

Hegde R, Srinivasula SM, Zhang Z, Wassell R, Mukattash R, Cilenti L, Dubois G, Lazebnik Y, Zervos AS, Fernandes-Alnemri T (2002) Identification of Omi/HtrA2 as a mitochondrial apoptotic serine protease that disrupts inhibitor of apoptosis protein-caspase interaction. J Biol Chem 277:432–438

Hermes-Lima M, Castilho RF, Valle VG, Bechara EJ, Vercesi AE (1992) Calcium-dependent mitochondrial oxidative damage promoted by 5-aminolevulinic acid. Biochim Biophys Acta 1180:201–206

Ichas F, Jouaville LS, Mazat J-P (1997) Mitochondria are excitable organelles capable of generating and conveying electrical and calcium signals. Cell 89:1145–1153

Juhaszova M, Zorov DB, Kim S-H, Pepe S, Fu Q, Fishbein KW, Ziman BD, Wang S, Ytrehus K, Antos CL (2004) Glycogen synthase kinase-3β mediates convergence of protection signaling to inhibit the mitochondrial permeability transition pore. J Clin Investig 113:1535

Kaludercic N, Carpi A, Menabò R, Di Lisa F, Paolocci N (2011) Monoamine oxidases (MAO) in the pathogenesis of heart failure and ischemia/reperfusion injury. Biochim Biophys Acta 1813:1323–1332

Kanno T, Sato EF, Muranaka S, Fujita H, Fujiwara T, Utsumi T, Inoue M, Utsumi K (2004) Oxidative stress underlies the mechanism for Ca2+-induced permeability transition of mitochondria. Free Radic Res 38:27–35

Kim I, Rodriguez-Enriquez S, Lemasters JJ (2007) Selective degradation of mitochondria by mitophagy. Arch Biochem Biophys 462:245–253

Koyano F, Okatsu K, Kosako H, Tamura Y, Go E, Kimura M, Kimura Y, Tsuchiya H, Yoshihara H, Hirokawa T, Endo T, Fon EA, Trempe JF, Saeki Y, Tanaka K, Matsuda N (2014) Ubiquitin is phosphorylated by PINK1 to activate parkin. Nature 510:162–166

Kroemer G, Reed JC (2000) Mitochondrial control of cell death. Nat Med 6:513–519

Kroemer G, Galluzzi L, Brenner C (2007) Mitochondrial membrane permeabilization in cell death. Physiol Rev 87:99–163

Kuwana T, Mackey MR, Perkins G, Ellisman MH, Latterich M, Schneiter R, Green DR, Newmeyer DD (2002) Bid, Bax, and lipids cooperate to form supramolecular openings in the outer mitochondrial membrane. Cell 111:331–342

Lazarou M, Sliter DA, Kane LA, Sarraf SA, Wang C, Burman JL, Sideris DP, Fogel AI, Youle RJ (2015) The ubiquitin kinase PINK1 recruits autophagy receptors to induce mitophagy. Nature 524:309–314

Le Bras M, Clement MV, Pervaiz S, Brenner C (2005) Reactive oxygen species and the mitochondrial signaling pathway of cell death. Histol Histopathol 20:205–220

Lenaz G (2001) The mitochondrial production of reactive oxygen species: mechanisms and implications in human pathology. IUBMB Life 52:159–164

Lewis CA, Parker SJ, Fiske BP, Mccloskey D, Gui DY, Green CR, Vokes NI, Feist AM, Vander Heiden MG, Metallo CM (2014) Tracing compartmentalized NADPH metabolism in the cytosol and mitochondria of mammalian cells. Mol Cell 55:253–263

Lin MT, Beal MF (2006) Mitochondrial dysfunction and oxidative stress in neurodegenerative diseases. Nature 443:787–795

Liu X, Kim CN, Yang J, Jemmerson R, Wang X (1996) Induction of apoptotic program in cell-free extracts: requirement for dATP and cytochrome c. Cell 86:147–157

Liu L, Feng D, Chen G, Chen M, Zheng Q, Song P, Ma Q, Zhu C, Wang R, Qi W, Huang L, Xue P, Li B, Wang X, Jin H, Wang J, Yang F, Liu P, Zhu Y, Sui S, Chen Q (2012) Mitochondrial outer-membrane protein FUNDC1 mediates hypoxia-induced mitophagy in mammalian cells. Nat Cell Biol 14:177–185

Liu YQ, Ji Y, Li XZ, Tian KL, Young CY, Lou HX, Yuan HQ (2013) Retigeric acid B-induced mitophagy by oxidative stress attenuates cell death against prostate cancer cells in vitro. Acta Pharmacol Sin 34:1183–1191

Liu L, Sakakibara K, Chen Q, Okamoto K (2014) Receptor-mediated mitophagy in yeast and mammalian systems. Cell Res 24:787–795

Luo X, Budihardjo I, Zou H, Slaughter C, Wang X (1998) Bid, a Bcl2 interacting protein, mediates cytochrome c release from mitochondria in response to activation of cell surface death receptors. Cell 94:481–490

Madesh M, Hajnoczky G (2001) VDAC-dependent permeabilization of the outer mitochondrial membrane by superoxide induces rapid and massive cytochrome c release. J Cell Biol 155:1003–1015

Martel C, Allouche M, Esposti D, Fanelli E, Boursier C, Henry C, Chopineau J, Calamita G, Kroemer G, Lemoine A (2012) GSK3-mediated VDAC phosphorylation controls outer mitochondrial membrane permeability during lipid accumulation. Hepatology 57:93–102

Martinou JC, Green DR (2001) Breaking the mitochondrial barrier. Nat Rev Mol Cell Biol 2:63–67

Martins LM, Iaccarino I, Tenev T, Gschmeissner S, Totty NF, Lemoine NR, Savopoulos J, Gray CW, Creasy CL, Dingwall C (2002) The serine protease Omi/HtrA2 regulates apoptosis by binding XIAP through a reaper-like motif. J Biol Chem 277:439–444

Martins I, Raza SQ, Voisin L, Dakhli H, Law F, Allouch A, Thoreau M, Brenner C, Deutsch E, Perfettini J-L (2017) Entosis: the emerging face of non-cell-autonomous type IV programmed death. Biomed J 40(3):133–140

Marzo I, Brenner C, Zamzami N, Jurgensmeier JM, Susin SA, Vieira HL, Prevost MC, Xie Z, Matsuyama S, Reed JC, Kroemer G (1998) Bax and adenine nucleotide translocator cooperate in the mitochondrial control of apoptosis. Science 281:2027–2031

Mattson MP, Chan SL (2003) Calcium orchestrates apoptosis. Nat Cell Biol 5:1041–1043

Morais VA, Verstreken P, Roethig A, Smet J, Snellinx A, Vanbrabant M, Haddad D, Frezza C, Mandemakers W, Vogt-Weisenhorn D, Van Coster R, Wurst W, Scorrano L, De Strooper B (2009) Parkinson's disease mutations in PINK1 result in decreased complex I activity and deficient synaptic function. EMBO Mol Med 1:99–111

Murphy MP, Holmgren A, Larsson N-G, Halliwell B, Chang CJ, Kalyanaraman B, Rhee SG, Thornalley PJ, Partridge L, Gems D (2011) Unraveling the biological roles of reactive oxygen species. Cell Metab 13:361–366

Neupert W (1997) Protein import into mitochondria. Annu Rev Biochem 66:863–917

Novak I, Kirkin V, Mcewan DG, Zhang J, Wild P, Rozenknop A, Rogov V, Lohr F, Popovic D, Occhipinti A, Reichert AS, Terzic J, Dotsch V, Ney PA, Dikic I (2010) Nix is a selective autophagy receptor for mitochondrial clearance. EMBO Rep 11:45–51

Orrenius S, Zhivotovsky B (2005) Cardiolipin oxidation sets cytochrome c free. Nat Chem Biol 1:188–189

Pan X, Liu J, Nguyen T, Liu C, Sun J, Teng Y, Fergusson MM, Rovira II, Allen M, Springer DA (2013) The physiological role of mitochondrial calcium revealed by mice lacking the mitochondrial calcium uniporter. Nat Cell Biol 15:1464–1472

Patterson S, Spahr C, Daugas E, Susin S, Irinopoulou T, Koehler C, Kroemer G (2000) Mass spectrometric identification of proteins released from mitochondria undergoing permeability transition. Cell Death Differ 7:137

Pchejetski D, Kunduzova O, Dayon A, Calise D, Seguelas M-H, Leducq N, Seif I, Parini A, Cuvillier O (2007) Oxidative stress-dependent sphingosine kinase-1 inhibition mediates monoamine oxidase A-associated cardiac cell apoptosis. Circ Res 100:41–49

Petit PX, Goubern M, Diolez P, Susin SA, Zamzami N, Kroemer G (1998) Disruption of the outer mitochondrial membrane as a result of large amplitude swelling: the impact of irreversible permeability transition. FEBS Lett 426:111–116

Pozzan T, Rizzuto R, Volpe P, Meldolesi J (1994) Molecular and cellular physiology of intracellular calcium stores. Physiol Rev 74:595–637

Reed JC (2000) Mechanisms of apoptosis. Am J Pathol 157:1415–1430

Rikka S, Quinsay MN, Thomas RL, Kubli DA, Zhang X, Murphy AN, Gustafsson AB (2011) Bnip3 impairs mitochondrial bioenergetics and stimulates mitochondrial turnover. Cell Death Differ 18:721–731

Rodriguez-Enriquez S, He L, Lemasters JJ (2004) Role of mitochondrial permeability transition pores in mitochondrial autophagy. Int J Biochem Cell Biol 36:2463–2472

Schwarten M, Mohrlüder J, Ma P, Stoldt M, Thielmann Y, Stangler T, Hersch N, Hoffmann B, Merkel R, Willbold D (2014) Nix directly binds to GABARAP: a possible crosstalk between apoptosis and autophagy. Autophagy 5:690–698

Shi RY, Zhu SH, Li V, Gibson SB, Xu XS, Kong JM (2014) BNIP3 interacting with LC3 triggers excessive mitophagy in delayed neuronal death in stroke. CNS Neurosci Ther 20:1045–1055

Susin SA, Zamzami N, Castedo M, Hirsch T, Marchetti P, Macho A, Daugas E, Geuskens M, Kroemer G (1996) Bcl-2 inhibits the mitochondrial release of an apoptogenic protease. J Exp Med 184:1331–1341

Tanaka A, Cleland MM, Xu S, Narendra DP, Suen DF, Karbowski M, Youle RJ (2010) Proteasome and p97 mediate mitophagy and degradation of mitofusins induced by Parkin. J Cell Biol 191:1367–1380

Tomasello F, Messina A, Lartigue L, Schembri L, Medina C, Reina S, Thoraval D, Crouzet M, Ichas F, De Pinto V (2009) Outer membrane VDAC1 controls permeability transition of the inner mitochondrial membrane in cellulo during stress-induced apoptosis. Cell Res 19:1363

Tsujimoto Y, Cossman J, Jaffe E, Croce CM (1985) Involvement of the bcl-2 gene in human follicular lymphoma. Science 228:1440–1444

Van Loo G, Van Gurp M, Depuydt B, Srinivasula SM, Rodriguez I, Alnemri ES, Gevaert K, Vandekerckhove J, Declercq W, Vandenabeele P (2002) The serine protease Omi/HtrA2 is released from mitochondria during apoptosis. Omi interacts with caspase-inhibitor XIAP and induces enhanced caspase activity. Cell Death Differ 9:20–26

Vaseva AV, Marchenko ND, Ji K, Tsirka SE, Holzmann S, Moll UM (2012) p53 opens the mitochondrial permeability transition pore to trigger necrosis. Cell 149:1536–1548

Vieira H, Belzacq A-S, Haouzi D, Bernassola F, Cohen I, Jacotot E, Ferri KF, El Hamel C, Bartle LM, Melino G (2001) The adenine nucleotide translocator: a target of nitric oxide, peroxynitrite, and 4-hydroxynonenal. Oncogene 20:4305–4316

Wang S, He M, Li L, Liang Z, Zou Z, Tao A (2016) Cell-in-cell death is not restricted by caspase-3 deficiency in MCF-7 cells. J Breast Cancer 19:231–241

Wang Z, Figueiredo-Pereira C, Oudot C, Vieira H, Brenner C (2017) Mitochondrion: a common organelle for distinct cell deaths? Int Rev Cell Mol Biol 331:245–287

Wei MC, Zong W-X, Cheng EH-Y, Lindsten T, Panoutsakopoulou V, Ross AJ, Roth KA, Macgregor GR, Thompson CB, Korsmeyer SJ (2001) Proapoptotic BAX and BAK: a requisite gateway to mitochondrial dysfunction and death. Science 292:727–730

Wong E, Cuervo AM (2010) Autophagy gone awry in neurodegenerative diseases. Nat Neurosci 13:805–811

Wu W, Tian W, Hu Z, Chen G, Huang L, Li W, Zhang X, Xue P, Zhou C, Liu L, Zhu Y, Zhang X, Li L, Zhang L, Sui S, Zhao B, Feng D (2014) ULK1 translocates to mitochondria and phosphorylates FUNDC1 to regulate mitophagy. EMBO Rep 15:566–575

Xia M, Meng G, Jiang A, Chen A, Dahlhaus M, Gonzalez P, Beltinger C, Wei J (2014) Mitophagy switches cell death from apoptosis to necrosis in NSCLC cells treated with oncolytic measles virus. Oncotarget 5:3907

Yang Y, Gehrke S, Imai Y, Huang Z, Ouyang Y, Wang JW, Yang L, Beal MF, Vogel H, Lu B (2006) Mitochondrial pathology and muscle and dopaminergic neuron degeneration caused by inactivation of drosophila Pink1 is rescued by Parkin. Proc Natl Acad Sci U S A 103:10793–10798

Yoon Y, Krueger EW, Oswald BJ, Mcniven MA (2003) The mitochondrial protein hFis1 regulates mitochondrial fission in mammalian cells through an interaction with the dynamin-like protein DLP1. Mol Cell Biol 23:5409–5420

Zamzami N, Kroemer G (2001) The mitochondrion in apoptosis: how Pandora's box opens. Nat Rev Mol Cell Biol 2:67–71

Zamzami N, Marchetti P, Castedo M, Zanin C, Vayssière J-L, Petit PX, Kroemer G (1995) Reduction in mitochondrial potential constitutes an early irreversible step of programmed lymphocyte death in vivo. J Exp Med 181:1661–1672

Zhu Y, Massen S, Terenzio M, Lang V, Chen-Lindner S, Eils R, Novak I, Dikic I, Hamacher-Brady A, Brady NR (2013) Modulation of serines 17 and 24 in the LC3-interacting region of Bnip3 determines pro-survival mitophagy versus apoptosis. J Biol Chem 288:1099–1113

Zorov DB, Filburn CR, Klotz L-O, Zweier JL, Sollott SJ (2000) Reactive oxygen species (ROS-induced) ROS release. J Exp Med 192:1001–1014

Mitochondria in Liver Diseases

Ignazio Grattagliano, Ornella de Bari, Domenica Di Palo, Fabrizio Montecucco, Federico Carbone, Paulo J. Oliveira, David Q.-H. Wang, and Piero Portincasa

Abstract Mitochondria are not only the main energy source in the hepatocytes but also play a major role in the cell redox homeostasis and maintain normal liver function including signalling pathways and the metabolism of exogenous substances. These roles assign mitochondria a gateway function in protecting hepatocyte from injury since unbalanced mitochondrial function unequivocally affects cell survival by actively causing the onset and perpetuation of liver diseases.

I. Grattagliano
Clinica Medica "A. Murri", Department of Biomedical Sciences and Human Oncology, Policlinico Hospital, University of Bari Medical School, Bari, Italy

Italian College of General Practitioners, Florence, Italy

O. de Bari · D. Di Palo
Clinica Medica "A. Murri", Department of Biomedical Sciences and Human Oncology, Policlinico Hospital, University of Bari Medical School, Bari, Italy

F. Montecucco
Department of Internal Medicine, First Clinic of Internal Medicine, University of Genoa, Genoa, Italy

Ospedale Policlinico San Martino, Genoa, Italy

Centre of Excellence for Biomedical Research (CEBR), University of Genoa, Genoa, Italy

F. Carbone
Department of Internal Medicine, First Clinic of Internal Medicine, University of Genoa, Genoa, Italy

P. J. Oliveira
CNC, Center for Neuroscience and Cell Biology, University of Coimbra, Cantanhede, Portugal

D. Q.-H. Wang
Division of Gastroenterology and Liver Diseases, Department of Medicine, Marion Bessin Liver Research Center, Albert Einstein College of Medicine, Bronx, NY, USA

P. Portincasa (✉)
Clinica Medica "A. Murri", Department of Biomedical Sciences and Human Oncology, Policlinico Hospital, University of Bari Medical School, Bari, Italy
e-mail: piero.portincasa@uniba.it

P. J. Oliveira (ed.), *Mitochondrial Biology and Experimental Therapeutics*,
https://doi.org/10.1007/978-3-319-73344-9_7

Abnormal mitochondrial function is reported to be involved in a variety of liver diseases including drug-induced liver injury, alcoholic liver disease, non-alcoholic fatty liver disease, viral hepatitis, primary and secondary cholestasis, hemochromatosis, and Wilson's disease. These changes lead to the impairment of the electron transport chain and/or oxidative phosphorylation, which induces a decrease in oxidative degradation of many exogenous and endogenous substrates and ATP synthesis, and in general, reduces hepatocyte tolerance towards potentially damaging insults. Structural changes accompany functional impairment of mitochondria, resulting in swelling and formation of aggregates and inclusions within the mitochondrial matrix. In chronic liver diseases, adequate mitochondrial function is maintained by mitochondrial proliferation and/or by an increased activity of critical enzymes. The assessment of mitochondrial functions *in vivo* is a useful tool in patients with liver diseases for diagnostic and prognostic purposes in patients with liver diseases and for the evaluation of therapeutic interventions.

Keywords Alcohol · Cholestasis · Fatty liver · Hemochromatosis · Hepatitis C virus · Nitrosative stress · Non-alcoholic fatty liver disease · Oxidative stress · Primary biliary cirrhosis · Wilson's disease

Abbreviations

ALD	Alcoholic liver disease
AMA	Antimitochondrial antibodies
ANG	Angiotensin
ATP	Adenosine triphosphate
BDL	Bile duct ligation
ER	Endoplasmic reticulum
Fe/S	Iron-sulfur
GSH	Glutathione
HFE	Hemochromatosis gene
iNOs	Inducible nitric oxide synthase
KGDH	α-Ketoglutarate dehydrogenase
KICA	α-Ketoisocaproic acid
LCFA	Long chain fatty acids
MAM	Mitochondria-associated membrane
MPT	Mitochondrial permeability transition
mtDNA	Mitochondrial DNA
NAFL	Non-alcoholic fatty liver
NAFLD	Non-alcoholic fatty liver disease
NASH	Non-alcoholic steatohepatitis
NO	Nitric oxide
PBC	Primary biliary cirrhosis
PDH	Pyruvate dehydrogenase

PSH	Protein sulfhydryls
PSSG	Protein mixed disulfides
ROS	Reactive oxygen species
SAME	S-Adenosyl-L-methionine
SOD	Superoxide dismutase
TG	Triglycerides
WD	Wilson's disease

1 Introduction

Mitochondria are the main energy source in the hepatocytes and play a key role in extensive redox metabolism and function in the liver of healthy subjects and patients with acute and chronic liver diseases. Unbalanced mitochondrial function plays a determining role in the onset and progression of several liver diseases (Krahenbuhl 1993), including fatty degeneration of hepatocytes (Grattagliano et al. 2003, 2008). Mitochondria represent the gateway where signals leading to cell injury or governing cell protection converge (Kass 2006; Susin et al. 1998). The assessment of mitochondrial function is of great importance in experimental studies and in clinical setting to interpret the fine processes of energy metabolism and the complex intracellular network which include the relationships among intracellular organelles, microsomes and nucleus.

The present chapter will address the role of mitochondria in liver diseases. We also discuss the in vivo assessment of mitochondrial function in diagnosis and therapy.

2 Mitochondria Contribute to the Progression of Liver Disease

Mitochondria are actively involved in metabolic pathways and signalling networks (Kass and Price 2008). In addition, mitochondria can be directly or indirectly damaged as the consequence of the activation of intracellular stress cascades or death receptor pathways. Production of reactive oxygen species (ROS), glutathione (GSH) depletion and protein oxidative alkylation lead to mitochondrial dysfunction and are critical events in the pathogenesis of chronic liver diseases.

The molecular and cellular mechanisms leading to mitochondria unbalance can involve alterations in the biochemical reactions involved in the Krebs cycle, which occurs in the mitochondrial matrix, and in mitochondrial respiratory chain and electron transfer, a phenomenon located in the inner mitochondrial membrane. These two chains of reactions are critical for the production of energy under the form of adenosine triphosphate (ATP) and are closely dependent on the integrity of the mitochondrial membrane (Guerrieri et al. 2000). ATP availability is the main conditioner of cell death regulation. For example, the apoptotic signalling requires ATP for its initiation and completion while ATP is required for the process of necroptosis (Feoktistova and Leverkus 2015). Described mechanisms are involved in the activation of apoptotic and necrotic processes. Mitochondrial apoptosis is associated with alterations in the membrane distribution of cardiolipin and phosphatidylcholine, as well as to increased probability of the mitochondrial permeability transition (MPT) and homing of pro-apoptotic proteins (Kagan et al. 2009). MPT occurs upon clustering of proteins and opening of multiple pores distributed on contact sites between the inner and outer mitochondrial membranes, and this step is associated with the loss of proton gradient through the inner mitochondrial membrane with inhibition of ATP synthesis. The increased inner membrane permeability and potential rupture of the outer mitochondrial membrane are associated with release of cytochrome c and other pro-apoptotic factors into the cytosolic compartment, disruption of nuclear chromatin, and activation of Ca^{2+}-dependent hydrolases (Haouzi et al. 2000; Malhi and Gores 2008a; Feldmann et al. 2000; Pessayre et al. 1999a). Indeed, cytochrome c, which binds to a cytoplasmic scaffold (apaf-1), forms a complex in the presence of ATP called "apoptosome", a potent activator of the signalling pro-caspase 9. This apoptotic signalling is further amplified, followed by the activation of the executioner caspase 3 and the interaction with pro-caspases 6, 7, and 2 (Slee et al. 2001). Apoptotic death is featured by cytoplasmic and nuclear condensation and fragmentation without loss of membrane integrity. Apoptotic cell fragments are then removed by phagocytosis but trigger inflammatory reactions.

If the mitochondrial damage becomes irreversible, and ATP production is significantly decreased, necroptotic cell death is favoured over apoptotic cell death. When the initial injury is so severe that MPT occurs very quickly, mitochondrial ATP depletion becomes severe and it does not allow the activation of the apoptotic pathway. In this context, necroptosis occurs with cell swelling and lysis and inflammatory responses, release of cytokines and amplification of the initial injury through sensitization of surrounding hepatocytes. This step favours and perpetuates collateral damages. Differentiation between apoptotic and necrotic hepatocyte death does not always follow a clear separation, rather it is a *continuum*, with overlapping phenomena often observed (Papucci et al. 2004; Formigli et al. 2000), more recent data suggests that necroptosis is a key pathogenic event in non-alcoholic steatohepatitis (Afonso et al. 2015) and cholestasis (Afonso et al. 2016).

A large number of toxic compounds as well as some liver diseases affect the integrity of mitochondrial membranes structures, resulting in functional impairment and decreased energy production (ATP). ROS promote oxidative modifications of lipid and protein components of the mitochondrial membrane and of mitochondrial DNA (mtDNA). Oxidative changes of mitochondrial structural components induce a series of morphological alterations, namely fluidity changes, appearance of hydrophilic changes within hydrophobic regions as well as progressive loss of lipid-lipid and lipid-protein interactions, changes in membrane permeability, and inactivation of carrier and receptor systems. Oxidative damages represent the most frequent damaging process in many chronic liver diseases and may result in mtDNA alterations, stimulation of apoptotic pathways, and increased propensity to necroptosis due to multi-level failure in the synthesis of ATP (Jadeja et al. 2017). Mitochondrial health likely depends on several factors including integrity of mtDNA, membrane lipids, lipoprotein trafficking, pro- and antioxidants balance, and metabolic demands (Caldwell et al. 2004).

Nitrosative stress is another damaging mechanism which involves the local production of nitric oxide (NO) derivatives and their binding to proteins and thiols. These reactions cause enzyme inactivation, and carriers and conformational changes. In mitochondria, NO is knew exerts a controller activity on respiration and biogenesis; changes of local and global NO production and levels may influence mitochondrial function (Grattagliano et al. 2007). ROS and NO disrupt mitochondrial function through post-translational modifications of the mitochondrial proteome. Protein alteration may contribute to mitochondrial dysfunction in some forms of liver diseases. Additional factors include proteomic defects in the assembly of mitochondrial multi-protein complexes as well as the resolution of highly hydrophobic proteins of the inner membrane (Mantena et al. 2007).

All such phenomena are described in alcoholic liver disease, non-alcoholic fatty liver disease, hepatitis C virus infection, hemochromatosis, Wilson's disease, and chronic cholestasis. Drug-induced liver disease will not be focused in this chapter, as it is described in this book by Fromenty et al., as well as in other excellent reviews by Grattagliano et al. (2004b).

3 Mitochondria Changes in Liver Diseases

3.1 Non-alcoholic Fatty Liver Disease (NAFLD)

NAFLD is a condition which encompasses a wide histological spectrum ranging from simple steatosis, i.e. non-alcoholic fatty liver (NAFL) without evidence of inflammation, to the necro inflammatory form of non-alcoholic steatohepatitis (NASH) (histologically indistinguishable from alcoholic steatohepatitis). Due to the increasing epidemics of obesity and metabolic syndrome worldwide and especially in western countries, NAFLD is currently deemed as the most frequent manifestation of liver disease in humans (Vanni et al. 2015). NAFLD patients are at risk of progression of liver disease, and possible steps include liver fibrosis, cirrhosis, and even hepatocellular carcinoma (Ludwig et al. 1980; Chalasani et al. 2012; Burt et al. 1998; Harrison 2015) with or without an underlying cirrhosis (Torres and Harrison 2012; Pocha et al. 2015; Bugianesi et al. 2002). Whether NAFL and NASH are two distinct entities with separate prognostic values is still a matter of debate (Fielding and Angulo 2014; Browning et al. 2004; Wong et al. 2010; Poonawala et al. 2000), the evolution requires several multifactorial "hits" driving from simple steatosis to fibrosis (Singh et al. 2015) (sometime reversible (Wong et al. 2010; Hamaguchi et al. 2010; Argo et al. 2009)).

NAFLD typically occurs in patients not drinking or drinking alcohol in small amounts and often reflects the liver features of the metabolic syndrome, based on several concomitant abnormalities which include visceral obesity, insulin resistance

Table 1 Factors associated with NAFLD progression

• Advanced age (Argo et al. 2009; Ratziu et al. 2000; Angulo et al. 1999; Noureddin et al. 2013)
• Diabetes mellitus (Hossain et al. 2009)
• Hypertransaminasemia (in one study $\geq 2 \times$ ULN) (Ratziu et al. 2000; Noureddin et al. 2013; Hossain et al. 2009; Francque et al. 2012)
• Liver histology showing ballooning degeneration plus Mallory hyaline or fibrosis (Matteoni et al. 1999; Singh et al. 2015)
• *BMI* ≥ 28 kg/m^2 (Ratziu et al. 2000)
• Increased visceral adiposity (waist circumference, BMI, triglycerides, and high-density lipoprotein level) (Petta et al. 2012)

BMI body mass index, *ULN* upper limit of normal

or type 2 diabetes mellitus, arterial hypertension, and dyslipidemia (Pessayre et al. 1999b). Of note, NAFLD also occurs in Asian countries where body mass index is lower and insulin resistance is generally absent, compared to the West (Loomba and Sanyal 2013; Farrell et al. 2013). The role of genomics in NAFLD by GWAS is also contributing to the understanding of genetic influence across different populations with NAFLD (Anstee and Day 2015; Speliotes et al. 2011; Krawczyk et al. 2013). Factors associated with NAFLD progression are listed in Table 1. Notably, coffee consumption appears to protect against the risk of progression (Molloy et al. 2012; Wijarnpreecha et al. 2017).

The pathogenesis of NAFLD is complex: in the most common form, i.e. obesity-associated NAFLD, the early event is the expansion of visceral adipose tissue (containing triglycerides, TG). This step parallels the onset of insulin resistance. TG are fat molecules synthetized from one molecule of glycerol linked to three long-chain fatty acids (LCFA). In a healthy state, appropriate amounts of TG are stored within the adipocytes throughout the body as large droplets. Thus, LCFA are normally stored into the adipocytes as TG during the postprandial period and released intermittently by lipolysis according to the metabolic needs. Both hepatocytes (Ge et al. 2011, 2010a) and cardiomyocytes (Ge et al. 2010b) also host smaller lipid droplets. Upon expansion of visceral adiposity, such as in obesity, a crucial step is the disposal of LCFA which are released into the splanchnic circulation by the adipocytes. Under insulin resistance conditions (a very early step in NAFLD development), adipocyte lipases are de-repressed and LCFA start flowing in excess into the liver, striated muscle and cardiomyocytes, which are stored as TG (Pessayre and Fromenty 2005; Calamita and Portincasa 2007). The incorporation of LCFA follows a mathematical algorithm based on both saturable (i.e. protein-mediated transport of LCFA anions, $\approx$95%) and unsaturable (i.e. passive trans-membrane diffusion of protonated LCFA) function of the unbound LCFA (Berk and Verna 2016). Few proteins have been identified as LCFA transporters (Table 2). Of note, the estimated velocity for the intake of LCFA, according to the Vmax value in the developed algorithm, increases exponentially with the body size, across the three groups of non-obese, obese, and super-obese patients (Greenway 2015). In adipocytes of patients undergoing bariatric surgery, the Vmax has been shown to be markedly increased in both the omental (Ge et al. 2016; Petrescu et al. 2005) and subcutaneous (Ge et al. 2016) fat and these

Table 2 Protein families being involved in the transport of long chain fatty acids (LCFA)

Name	Abbreviation	References
Plasma membrane fatty acid binding protein	FABPpm	Bradbury and Berk (2003), Stremmel et al. (1985), Berk et al. (1990), Bradbury et al. (2011)
Fatty acid translocase	FAT, or CD36	Abumrad et al. (1993)
Fatty acid transporting polypeptide family	FATP	Stahl et al. (2001), Schaffer and Lodish (1994), Digel et al. (2009)
Caveolin-1		Pohl et al. (2002), Trigatti et al. (1999)

Additional transporters might be involved (Kampf and Kleinfeld 2007; Kampf et al. 2007)

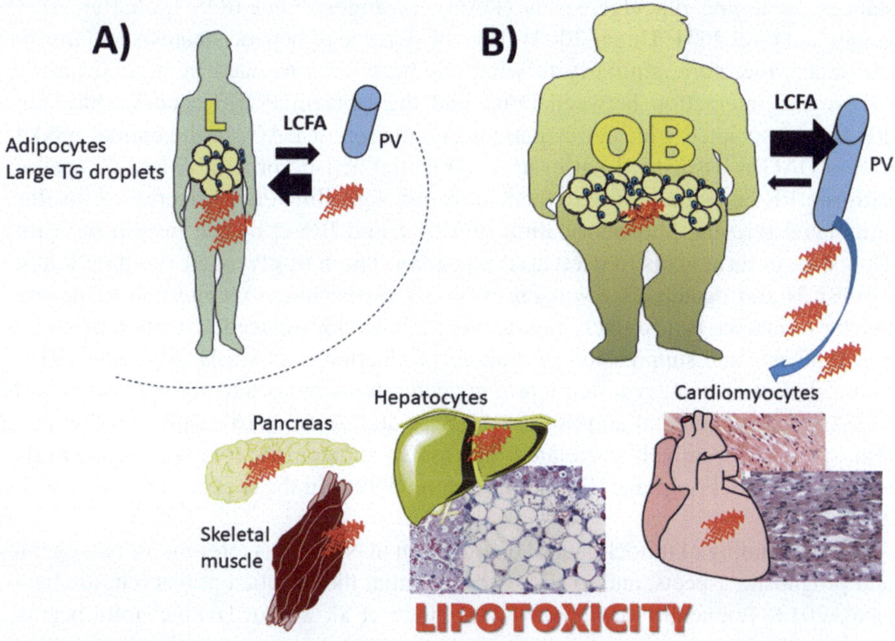

Fig. 1 Schematic view illustrating the distinct regulatory function of adipocytes in lean and obese subjects. (**a**) In the lean (L) subject, uptake of long-chain fatty acids (LCFA) by adipocytes is normally up-regulated and LCFA undergo partitioning and buffering in adipose tissues. This condition serves as a "protective function", because LCFA are packed into large droplets that are composed mainly of triglycerides (TG) and do not undergo oxidative processes. (**b**) Under conditions of obesity (OB), the expansion of visceral adiposity will not prevent the effect of lipases and the excess flow of LCFA via the portal vein (PV) into non-adipose cells, i.e. cardiac myocytes, hepatocytes, pancreatic β-cells and skeletal muscle cells. At this stage lipotoxic consequences occur due to excessive LCFA and TG accumulation and production of ceramides, diacyl-glycerols (DAG), reactive oxygen species (ROS) and cholesterol (Berk and Verna 2016)

findings parallel the up-regulation and the down-regulation of the uptake of LCFA into adipocytes occurring before the onset of obesity (Berk et al. 1997) or before weight loss (Fan et al. 2003), respectively. The potential protective role of adipose tissue in lean subjects with its buffering function of LCFA and TG are depicted in Fig. 1. By contrast, the ectopic accumulation of LCFA and TG is amplified by excess caloric intake (with respect to energy expenditure), hyperinsulinemia and up-regulation of lipogenic enzymes, as well as inhibition of LCFA oxidation.

Other key events include increased intracellular oxidative stress, hepatocellular apoptosis and injury, production of cytokines and inflammation, and activation of stellate cells and fibrosis (Jou et al. 2008; Malhi and Gores 2008b; Mantena et al. 2008; McCullough 2006; Charlton 2007). Following a transient status of hyperleptinemia due to increase of the insulin counter-regulatory hormone leptin, a phenotype involving leptin resistance is initiated (Myers et al. 2008) followed by intracellular increase of the LCFA pool, activation of non-oxidative pathways (Listenberger et al. 2001), and ultimately accumulation of several lipotoxic substances, including diacylglycerols (DAG), ceramides, and ROS (Schaffer 2003; Unger and Orci 2001; Unger 2003). The coexistence of hepatic steatosis and insulin resistance, therefore, similarly to what has been demonstrated in striated muscle (where the interaction between DAG and the isoform PCKθ occurs) (Shulman 2014), is associated with intrahepatic accumulation of DAG. In the animal model, excess DAG is associated with activation of the hepatic protein kinase C, epsilon isoform (PKCε) and, in turn, defects in insulin signalling due to decreased insulin-stimulated tyrosine phosphorylation of IRS-1 and IRS-2 by the insulin receptor. This step, in turn, leads to decreased phosphorylation of glycogen synthase kinase 3 (GSK3) and decreased glycogen synthesis in the liver. A condition of hepatic insulin resistance is ultimately interfering with insulin-induced activation of glycogen synthesis and suppression of glucose production in NAFLD (Shulman 2014). Some studies also suggest that leptin concentrations are positively correlated with LCFA Vmax in omental adipose tissue (Berk and Verna 2016), while another protein, spexin is negatively correlated, playing an antagonistic role with leptin in the regulation of LCFA uptake (Shimomura et al. 1999; Mirabeau et al. 2007; Walewski et al. 2014).

Understanding of mitochondrial dysfunction in NAFLD is relevant for pathogenic and prognostic aspects, and to addresses potential therapeutic options (Grattagliano et al. 2013; Loguercio et al. 2012; Vecchione et al. 2016). Distinct intra-hepatic mechanisms link excessive incorporation of LCFA to mitochondrial dysfunction, insulin resistance and following steps resulting in decreased glycogen synthesis and increased gluconeogenesis (Fig. 2). The link between increased DAG and PKCε activation is essential in this respect. LCFA are oxidized at both mitochondrial and extramitochondrial sites in the hepatocytes. A consequence of excessive incorporation of LCFA is the increase of ROS which mediates hepatocellular and mitochondrial injury (Diogo et al. 2011; Grattagliano et al. 2011, 2013).

The complex scenario where pathogenic mechanisms lead to the sequence NAFLD-NASH-Cirrhosis-HCC are illustrated in Figs. 3 and 4.

One pathway that limits excessive fat accumulation in the liver is the increased mitochondrial oxidation of LCFA which is associated with an impaired respiration (Fromenty and Pessayre 1995). Fatty degeneration exposes hepatocytes to a higher risk of oxidative damage, although a number of adaptive metabolic mechanisms have been described (Grattagliano et al. 2003; Yang et al. 2000). Such mechanisms include expression of intracellular sensors and signalling molecules for lipid metab-

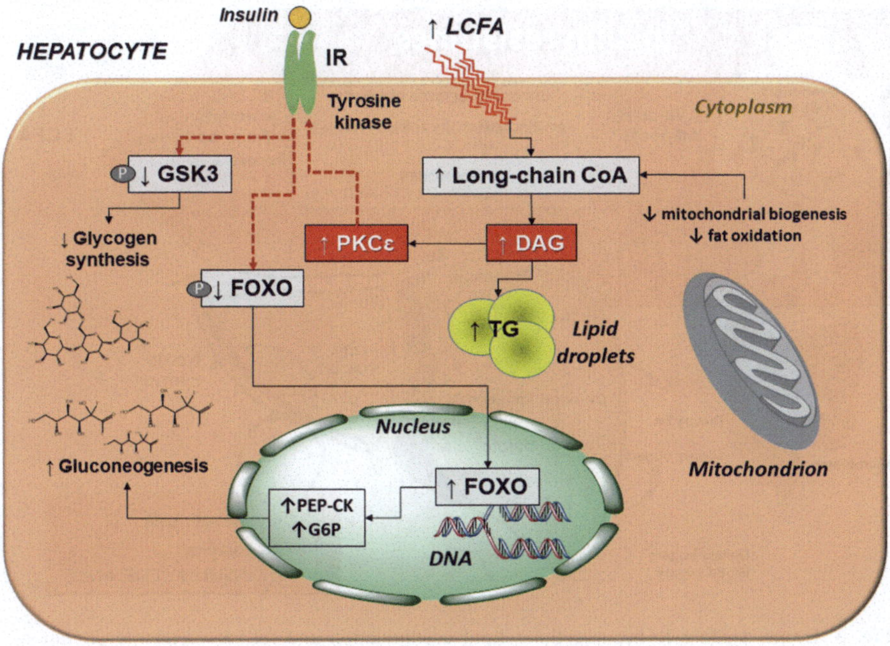

Fig. 2 Events on the basis of lipid-induced insulin resistance in the liver. Increased influx of LCFA in the liver leads to increased DAG and activation of the epsilon isoform of protein kinase C (PKCε). Transient increase in hepatocellular DAG occurs when rates of DAG synthesis (due to fatty acid re-esterification and de novo lipogenesis) are greater than rates of mitochondrial long-chain CoA (fat) oxidation, rates of DAG incorporation into neutral lipid (TAG), or both. DAG activates PKCε that binds and inhibits the insulin receptor tyrosine kinase. This step leads to decreased phosphorylation of GSK3 and decreased glycogen synthesis in the liver. A further step is the decreased phosphorylation of forkhead box subgroup O (FOXO) and decreased insulin suppression of hepatic gluconeogenesis via increased FOXO translocation to the nucleus, and increased gene transcription of the gluconeogenic enzymes like PEP-CK and G6P). Adapted from Shulman, NEJM (2014). Abbreviations: *DAG* diacylglyceride, *FOXO* forkhead box subgroup O, *PEP-CK* phosphoenolpyruvate carboxykinase, *TG* triglycerides

olism and oxidative stress pathways (Merriman et al. 2006; Sanyal et al. 2001). Impairment of these systems may have important pathogenic roles in NAFLD progression, including disturbing ATP synthesis (Cortez-Pinto et al. 1999). mtDNA levels, protein expression and activity of respiratory complexes are also decreased in liver mitochondria (Haque and Sanyal 2002; Pérez-Carreras et al. 2003), with evidence pointing out to an oxidative stress mechanism The same effects were observed in intact HepG2 cells incubated with saturated fatty acids, creating a model for NASH. In this model, not only mitochondrial function was depressed, as a general inhibition of mtDNA gene expression was observed, together with accelerated degradation of respiratory chain subunits (Garcia-Ruiz et al. 2015). In fatty livers, ROS formation is increased at the mitochondrial respiratory chain level and determines oxidation of unsaturated lipids (Grattagliano et al. 2003, 2008; Yang et al.

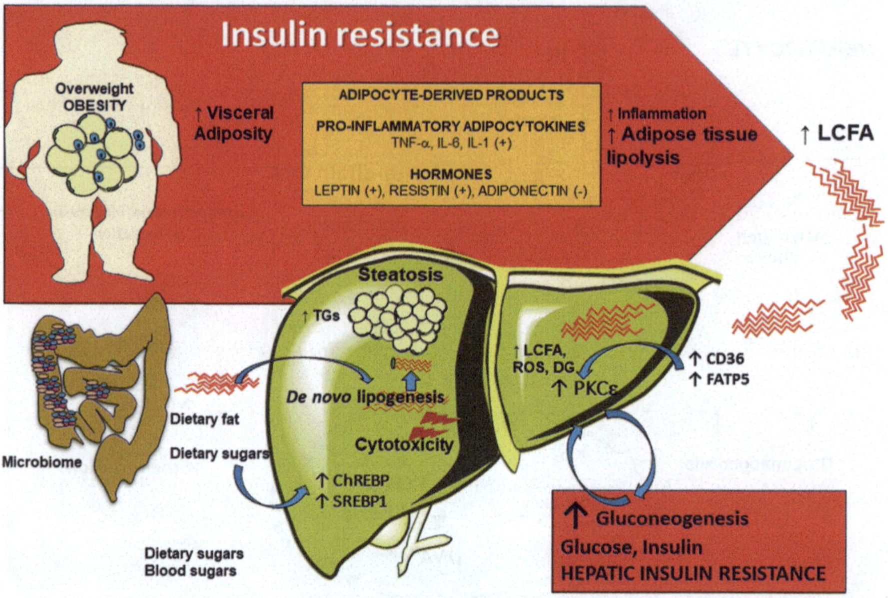

Fig. 3 Events leading to hepatocellular lipid accumulation and further metabolic alterations, resulting in disease progression. Overweight and obesity are followed by early expansion of visceral adipose tissue, local macrophagic inflammation and upregulation of adipocytokines and adipocyte-derived hormones, responsible for the onset of insulin resistance. This step is associated with increased lipolysis and massive flow of circulatory long-chain fatty acids (LCFA) such as the liver (Fabbrini et al. 2010; Bonfrate et al. 2014). Following this pro-inflammatory "obesogenic" scenario, the two LCFA transporters (class B scavenger receptor CD36 and the fatty acid transport protein 5, FATP5 = bile acyl-CoA synthetase) are upregulated and facilitate the FFA uptake into the liver (Zhou et al. 2008; Mitsuyoshi et al. 2009). The intrahepatic events include increased synthesis of triglycerides, accumulation of diacylglycerol (DG) intermediates, reactive oxygen species (ROS), activation of the protein kinase Cε (PKCε) (Samuel et al. 2007) and deficient hepatic insulin signalling. A further step is the increased gluconeogenesis and perpetuation of the hyperglycemic-insulin resistance status. Liver steatosis is enhanced by *de novo* lipogenesis (DNL) from dietary and/or serum sugars. Insulin and glucose (Postic and Girard 2008) upregulate two enzymes: the carbohydrate-responsive element-binding protein (ChREBP) and sterol regulatory element-binding protein 1 (SREBP1) with increase of intrahepatic lipid supply by about 25% (Donnelly et al. 2005; Lambert et al. 2014). Additional pathways (i.e. via the endoplasmic reticulum stress) might render enzymatic activation (i.e. SREBP1) independent of insulin-dependent pathways (Kammoun et al. 2009). DNL is also responsible for the synthesis of long-chain saturated FAs which are cytotoxic to hepatocytes. The toxicity of long chain fatty acids is limited by their conversion to TGs. Dietary fat is also responsible for the increased hepatic uptake and accumulation of LCFA. The intestinal microbiota contributes to the pathogenesis of NAFLD in many respects: dysbiosis (Schnabl and Brenner 2014), enhanced extraction of dietary nutrients (Wu et al. 2011), increased intestinal permeability (Miele et al. 2009), translocation of bacteria and/or bacterial products (endotoxins and endogenous ethanol) into the portal circulation (Krawczyk et al. 2010), and intestinal degradation of choline which is beneficial to lipid homeostasis (Bedogni et al. 2005; Altomare et al. 1998; Grattagliano et al. 2000, 2003, 2008, 2013; Petrosillo et al. 2007). The events increase enormously the intrahepatic content of fatty acids, TGs, cytotoxic diacylglycerols and long-chain FA. In NAFLD, the expanded DNL and dietary fat influx account for about 40% of intrahepatic lipids, as compared to adipose tissue lipolysis (60%) (Donnelly et al. 2005).

2000). The activity of complex I of the respiratory chain is also reduced (−35%) in mitochondria from fatty livers and is associated with changes in state 3 respiration (Petrosillo et al. 2007); hydrogen peroxide generation and oxidized cardiolipin are significantly increased (Petrosillo et al. 2007; Grattagliano et al. 2008). ROS affect the mitochondrial complex I activity by oxidizing cardiolipin which is required for the function of this enzyme complex (Paradies et al. 2002).

Oxidation, glutathionylation and nitrosylation of mitochondrial proteins occur as a response to oxidative stress and result in post-translational modification of proteins by carbonyl and disulfide formation or by thiol nitrogen exchange. All such alterations contribute to a further block of the electron flow in the respiratory chain resulting in subsequent generation of ROS. This vicious circle involves ROS-mediated antioxidant depletion, and the deficient capacity of mitochondria to inactivate ROS (Grattagliano et al. 2003). Ultimately, protein and lipid oxidation, and cytokine production are increased. Hepatocytes react to fat deposition with an early increase of GSH and thioredoxin to prevent lipid and protein oxidation (Grattagliano et al. 2008). Also, increases of protein mixed disulfides (PSSG), nitrates and nitrosothiols are consistent with both pro-oxidant protein modifications and increased NO synthesis. A critical role for mitochondrial GSH in the development of non-alcoholic steatohepatitis (NASH) was in fact proposed (Garcia-Ruiz and Fernandez-Checa 2006). GSH depletion sensitizes hepatocytes to inflammatory cytokines and TNF-alpha. Mitochondrial GSH content declines more rapidly than cytosolic GSH, suggesting mitochondria as specific early target for oxidative changes (Grattagliano et al. 2008).

Other pathogenic factors including NO may be important for the progression of liver steatosis and appearance of fibrosis. A crucial role in the disease progression has been recently assigned to thioredoxin, a redox active protein with several biological activities including regulation of PSH/PSSG ratio. Thioredoxin is actively involved in the regulation of NO activity via cleavage of nitrosothiols (Nikitovic and Holmgren 1996; Stoyanovsky et al. 2005) which are formed by conjugation of NO with free thiols and oppose dangerous reactions such as peroxynitrite formation. Nitrosothiols also act as intracellular messengers that control mitochondrial functions (Arnelle and Stamler 1995; Grattagliano et al. 2004a). Major alterations of thioredoxin levels have been observed with ongoing liver steatosis and have been associated with PSSG and nitrosothiols formation (García-Ruiz et al. 2006). Increased peroxynitrite formation is associated with a variety of interactions, including protein nitration and generation of nitrotyrosine (Sanyal et al. 2001).

Fig. 3 (continued) Abbreviations: *ChREBP* carbohydrate-responsive element-binding protein, *DNL de novo* lipogenesis, *LCFA* long-chain fatty acids, *IL* interleukin, Fig. 3 (continued) *FATP5* fatty acid transport protein 5 (bile acyl-CoA synthetase), *PKCε* protein kinase Cε, *SREBP* sterol regulatory element-binding protein 1, *TG* triglicerydes, *TNF* tumor necroptosis factor. Symbols: ↑, increased. Adapted from Portincasa P, Wang DQH. Nonalcoholic fatty liver and gallstone disease. In: Wang DQH, Portincasa P, eds. Gallstones. Recent advances in epidemiology, pathogenesis, diagnosis and management. New York: Nova Science Publisher Inc., 2017:387–414 (Portincasa and Wang 2017)

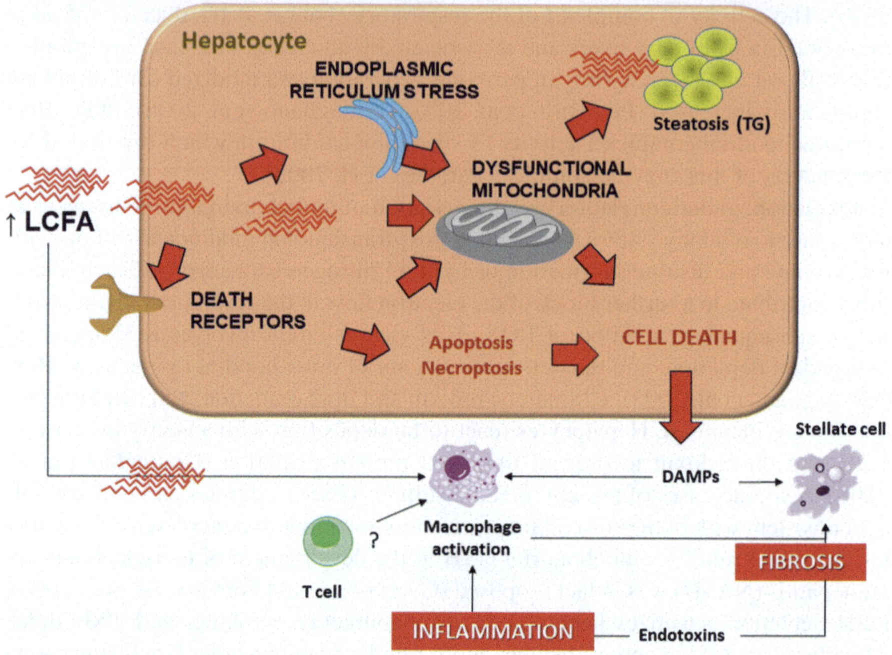

Fig. 4 Events leading to hepatocyte cell death, inflammation and fibrosis in NAFLD-NASH. The damage related to excess of influx of LCFAs in the liver occur via extrinsic upregulation of death receptors (DR5, FAS, TRAIL-R2, TNFRSF1A, see legend) (Cazanave et al. 2011; Inokuchi-Shimizu et al. 2014; Gautheron et al. 2014), intrinsic endoplasmic reticulum stress (involving Caspase-2 cleavage of BH3-interacting domain death agonist (BID)), and mitochondrial dysfunction involving ↑ FFA β-oxidation, uncoupling, generation of free oxygen radicals, activation of Jun N-Terminal Kinase (JNK), increased oxidative stress and lipid peroxidation, decreased antioxidant defences, and impaired ATP production Principal toxic pathways are illustrated and include production of ROS, activation of JNK (Koliaki et al. 2015), caspase-2 pathway activation (Machado et al. 2015; Johnson et al. 2013; Upton et al. 2008), and increased β-oxidation. Hepatocyte death will be associated with production and release of extracellular DAMPs. Subsequent mechanisms linking hepatocyte death with inflammation and fibrosis include the toxic effects of FFAs, intestinal endotoxins and DAMPs on the TLRs (Roh and Seki 2013) of recruited macrophages (sequence: inflammasomes, pro-inflammatory cytokines, chemokines and inflammation (Wehr et al. 2014; Leroux et al. 2012)), and stellate cells (sequence: myofibroblasts, collagen production, fibrosis) (Brunt et al. 2015; Wallace et al. 2015; Seki et al. 2007). Stellate cells are also sensitive to the ongoing oxidative stress, and free cholesterol accumulation (Tomita et al. 2014). A role for T cells in the progression of liver disease is also emerging (Wehr et al. 2014). Abbreviations: *DAMPs* damage-associated molecular patterns, *DR5* death receptor 5, *LCFA* long chain fatty acids, *ROS* reactive oxygen species, *TLRs* Toll-like receptors, *TNFRSF1A = TNFR1* tumour necroptosis factor receptor superfamily member 1A. Adapted from Portincasa P, Wang DQH. Non-alcoholic fatty liver and gallstone disease. In: Wang DQH, Portincasa P, eds. Gallstones. Recent advances in epidemiology, pathogenesis, diagnosis and management. New York: Nova Science Publisher Inc., 2017:387–414 (Portincasa and Wang 2017)

Additional mechanisms of mitochondrial damage include increased production of angiotensin II which is associated with impaired mitochondrial β-oxidation, and oxidative stress. Animals with elevated endogenous angiotensin II levels exhibit mitochondrial alterations with reduced β-oxidation and consequent decreased mitochondrial palmitate oxidation, decreased enzymatic activities, and expression of mitochondrial proteins, including cytochrome c, cytochrome c oxidase subunit 1, and mitochondrial transcription factor A (Tfam). These abnormalities are substantially improved by administration of angiotensin II receptor blockers or superoxide dismutase/catalase mimetic treatment (Wei et al. 2009). All the described biochemical changes play a role in the reduced tolerance of fatty livers to oxidative stress. In addition, fatty livers show a lower catalytic β-F1 subunit of the F0F1-ATP synthase by about 35%. Under starvation, mitochondrial oxidative injury is exacerbated to a greater extent in fatty livers. In the steatotic liver, fasting induces a further decrease of the ATP levels which is accompanied by a 70% fall of the catalytic β-F1 subunit. These changes may account for the observed reduction in the synthesis of ATP which were initially described (Fernández-Checa et al. 1998). In follow-up observations, oxidative mitochondrial damage, occurring during reperfusion after warm (Grattagliano et al. 1999; Nardo et al. 2000) or cold (Minor et al. 2000) ischemia, strongly contributed to the deterioration of hepatic energy metabolism observed after transplantation of fatty livers (Miki et al. 1998) associated with impairment of ketogenesis and glucose oxidation (Minor et al. 2000). The recovery time after reperfusion was shown to be markedly prolonged in steatotic livers (Trevisani et al. 1996) in association with an increased mitochondrial ROS formation (Hensley et al. 2000). Liver steatosis is also favoured by prolonged intake of drugs such as amiodarone or valproate (Berson et al. 1998). These molecules accumulate in mitochondria and induce inhibition of fatty acid oxidation and electron transfer chain (Berson et al. 1998). In support of the key role of mitochondria in NAFLD, recent data indicated the critical role of the mitochondrial pyruvate carrier MPC, a heterologous complex made of MPC1 and MPC2 proteins in the inner mitochondrial membrane (Colca et al. 2017). The complex is required for the entry of pyruvate that is synthetized in the cytosol, in the mitochondrial matrix, where it will be further metabolized. MPC might become therefore the target for treatment of several metabolic and inflammatory diseases, namely diabetes, (Colca et al. 2013; Mccommis et al. 2015; Chen et al. 2012) and even NASH (Mccommis et al. 2016).

Although the above described observations suggests that fatty livers have compromised mitochondrial function, several evidences show that the initial stages of the NAFLD condition are associated with an increase in mitochondrial mass, with or without increased mitochondrial fatty acid oxidation, a step acting as an adaptation response to excessive accumulation in the liver. This was elegantly demonstrated in human liver biopsies by Koliaki et al. (2015). This paper showed that when compared with isolated mitochondria from lean individuals, their counterparts from obese humans with or without NAFL had 4.3- to 5.0-fold higher maximal respiration rates, despite similar mitochondrial content. In opposition to this, and despite the fact that NASH patients featured higher hepatic mitochondrial mass, a 31–40% lower maximal respiration associated with greater hepatic insulin resistance, mitochondrial uncoupling, and leaking activity was described.

It seems that the initial adaptation to a fatty-rich environment contributes to enhance fatty acid oxidation, which means that hypothetically, accelerating the rate of fatty acid burn by mitochondria could be an effective therapeutic strategy. In fact, a liver-targeted mitochondrial protonophore was described to promote a mild depolarization of the inner membrane, contributing to increase electron transfer and fatty acid oxidation. In rodent models for NAFLD and T2DM, this approach appears to reverse hypertriglyceridemia, hepatic steatosis, insulin resistance, and hyperglycemia (Perry et al. 2013). Another approach involved Adenovirus-mediated liver expression of a malonyl-CoA-insensitive CPT1A (CPT1mt) in a high fat/high sugar animal model, with the ultimate objective of accelerating mitochondrial fatty acid beta-oxidation. This approach was able to reverse insulin resistance and glucose intolerance, although not affecting steatosis (Monsenego et al. 2012).

The important role of mitochondria in the context of NAFLD and its progression to NASH is well demonstrated by two recent findings. One of them demonstrates that mtDNA, a pro-inflammatory molecule per se (Zhang et al. 2016; Boyapati et al. 2017), when released from fatty liver hepatocytes, causes liver inflammation by TRL-9 activation (Garcia-Martinez et al. 2016), which can be an important component of the transition between NAFLD and NASH.

In another interesting development, it was recently demonstrated that when mitochondria isolated from hepatoma cells were injected into a rodent model for fatty liver, the phenotype was improved. Since exogenous mitochondria were tagged with green-fluorescence protein (GFP), it was possible to demonstrate accumulation in mouse liver, lung, brain, muscle, and kidney. How mitochondria entered the different cells and were able to maintain the integrity and restore metabolic activity, was not explained.

In conclusion, NAFLD encompasses mitochondrial alterations with an early adaptive protective response but also at an increased predisposition towards pro-oxidant insults (Pessayre and Fromenty 2005), which later results in a dysfunctional phenotype which accelerates the way towards necroptosis and inflammation.

3.2 Alcoholic Liver Disease (ALD)

Excessive alcohol (ethanol) consumption is associated with chronic liver injury (fatty liver, histologically indistinguishable from NAFLD) and potential progression towards fibrosis and cirrhosis. In most experimental animal models of chronic ethanol intoxication, liver steatosis occurs with inflammation, necroptosis and fibrosis, without cirrhosis (Tsukamoto et al. 1990). Liver cirrhosis often develops when ethanol is combined with other toxins (Hall et al. 1991) but in primates liver cirrhosis may represent the final evolution step of alcoholic liver disease (Lieber et al. 1989).

Mechanisms of ethanol toxicity include the participation of subcellular organelles. Abnormalities of mitochondria are common in humans and animals exposed to chronic ethanol toxicity. Morphological changes include megamitochondria (which can even be larger than the nucleus), decreased number of cristae, and appearance of crystalline inclusions. Mitochondria become pleomorphic, increase in size and show disorganization of cristae (Chang 1987); disrupted membranes disappear by forming different shapes or show a U shape (Yan et al. 2007). Taken together, these observations suggest that alcohol is a direct toxicant for mitochondria, and that mitochondrial function is impaired at an early stage (Quintanilla and Tampier 1989), causing progressive disruption of the liver function.

Ethanol is first metabolized to acetaldehyde in the cytosol of hepatocytes by the enzyme alcohol dehydrogenase (ADH). Acetaldehyde is subsequently metabolized to acetate in mitochondria as catalyzed by acetaldehyde dehydrogenase (ALDH). Ethanol-intoxication determines a damage to electron transport chain and decreases ATP synthesis. Acetaldehyde accumulates in mitochondria and causes adduct formation with structural and functional proteins with consequent oxidation (Vendemiale et al. 1998; Wieland and Lauterburg 1995). By using enzyme inhibitors for alcohol dehydrogenase and aldehyde oxidase, it was possible to conclude that acetaldehyde accumulation, but not its metabolites or ethanol itself are responsible for some of the effects related with alcohol-induced mitochondrial toxicity (Vendemiale et al. 1998) and that toxicity is partly associated with oxidative compartmentalization changes of GSH (Fernandez-Checa et al. 1991). The main step in ethanol-induced liver injury is thus represented by the generation of acetaldehyde and free radicals at ethanol metabolic sites, leading to mitochondrial alterations: destruction of components of mitochondrial membranes, consumption of antioxidant molecules, formation of acetaldehyde-protein adducts (Lauterburg and Bilzer 1988), alteration of antigenic and functional properties (Israel 1989). Mitochondrial GSH depletion has a key role in the development of alcoholic liver disease (Hirano et al. 1992): mitochondrial GSH pool is lowered in ethanol-fed animals, as a consequence of the suppression of cytosolic *de novo* synthesis of GSH (Lauterburg et al. 1984), and the impairment of GSH influx from cytosol into mitochondria (Fernandez-Checa et al. 1991). The end result is that the liver of chronic alcoholics becomes more vulnerable to the pro-oxidant effects of hepatotoxic drugs which are detoxified by the GSH-dependent system (e.g. acetaminophen).

A critical reduction of oxidative phosphorylation in liver mitochondria has been described in ethanol-fed rats. This is the consequence of the reduced activities of some subunits of the electron transport chain, decreased ATP synthesis, decreased content of cytochromes, and altered phospholipid and fatty acid membrane composition (Krahenbuhl 1993; Spach et al. 1982; Thayer and Rubin 1980). Mitochondrial respiratory activity is impaired and superoxide radical generation is increased in animals fed chronically with ethanol (Ribiere et al. 1994; Ma et al. 2017). The reduced ATP synthase activity is dependent on the ethanol-induced defect of the F0 subunit of this enzyme (Montgomery et al. 1987), a likely consequence of decreased protein synthesis (Coleman and Cunningham 1990). The same mechanism is invoked in the reduced activity of complexes I and IV of the respiratory chain. Alterations determine the appearance of a lower mitochondrial transmembrane potential, an increased mitochondrial mass, and an increased intracellular

Ca^{2+} concentration with an altered regulation of the MTP pore (Yan et al. 2007). In the presence of hypoxia, the damage greater and occurs as a consequence of decreased mitochondrial and glycolytic activities (Young et al. 2006) or following the increased expression of inducible nitric oxide synthase (iNOs) and enhanced sensitivity of mitochondria to NO (Venkatraman et al. 2004).

Chronic ethanol exposure also leads to changes in membrane lipid composition and redox state in dependence of the dietary lipid content (Schilling and Reitz 1980) and results into changes in membrane permeability and fluidity which impacts both membrane and protein activity. Changes of fatty acid compositions in mitochondrial membranes and decreased antioxidant protection are major causes of lipid peroxidation products (conjugated dienes, malondialdehyde, 4-hydroxynonenal (Vendemiale et al. 1998; Kamimura et al. 1992).

Additional injurious mechanisms promoted by chronic ethanol exposure include decreased mitochondrial translation, depressed respiratory complex levels and mitochondrial respiration rates, and decreased protein synthesis. An increased dissociation of mitoribosomes resulting from decreased sedimentation rates, a larger hydrodynamic volume, an increased levels of non-associated subunits, and changes in the levels of specific ribosomal proteins (Cahill and Sykora 2008) were also previously described. Moreover, S-adenosyl-L-methionine (SAME) which is required for the assembly and subsequent stability of mitoribosomes is depleted by chronic ethanol feeding (Sykora et al. 2009). Other studies revealed that ethanol mediates losses in cytochrome c oxidase subunits, complex IV activity, and up-regulates the mitochondrial stress chaperone prohibitin. SAME administration, by contrast, preserves hepatic SAME levels and prevents several defects of mitochondrial genome and proteome that contribute to the deficiency of bioenergetics as seen in the liver after alcohol consumption (Bailey et al. 2006).

Although the vast majority of studies indicate that chronic alcohol feeding results in progressive disruption of mitochondrial function, a recent study highlighted that the negative phenotype is only a tip of a bigger iceberg. In fact, oral and intra-gastric alcohol feeding to rats increased mitochondrial respiration using glycerol-3-phosphate (which delivers electrons from cytoplasmic NADH to mitochondria) and octanoate (a substrate for beta-oxidation), despite inhibiting Complex II-sustained respiration. Furthermore, chronic ethanol administration increased expression of mitochondrial glycerol phosphate dehydrogenase-2 (GPD2), transcription factor A (TFAM), a protein related with mitochondrial biogenesis, and increased mitochondrial NAD+-NADH and NADP+-NADPH levels in the liver (Han et al. 2017). This picture suggests a similar progression to that of non-alcoholic fatty liver disease, in which a progressive liver mitochondrial remodeling occurs, ultimately leading to a situation of irreversible bioenergetic colapse which pushes the liver to necroptosis and inflammation. This means that targeting oxidative stress and mitochondria in both liver pathologies may prove beneficial. In fact, inhibition of NADPH oxidase activity by means of a NOX4 inhibitor in a rodent model for alcoholic liver disease was able to prevent oxidative stress and mitochondria alterations in the liver, including induction of apoptosis (Sun et al. 2017).

3.3 *Hepatitis C Virus (HCV) Infection*

Liver tissue from patients infected by HCV shows the morphological changes of mitochondria and oxidative stress (Kageyama et al. 2000; Barbaro et al. 1999) with a clear relationship between HCV infection and mitochondrial dysfunction as a cause (Okuda et al. 2002). This may depend on the primary location of HCV core protein in mitochondria other than in the cytoplasm and endoplasmic reticulum (Tsutsumi et al. 2009). NS4A expression causes mitochondrial damage (Nomura-Takigawa et al. 2006) since it seems to directly interfere with mitochondrial function (Campbell et al. 2009). HCV protein expression produces specific inhibition of complex I activity, depression of mitochondrial membrane potential and oxidative phosphorylation coupling efficiency, increased production of ROS and NO species, and mitochondrial Ca^{2+} overload. These events lead to alterations in the bioenergetic balance and nitro-oxidative stress (Piccoli et al. 2007). Among mitochondrial proteins with consistently different expressions, prohibitin, a protein chaperon, is up-regulated not only in core-expressing cells but also in full-genomic replicon cells and livers of core-gene transgenic mice. It is also clear that the interaction of prohibitin with mitochondrial DNA-encoded subunits of cytochrome c oxidase is disturbed by the core protein, resulting in a significant decrease in cytochrome c oxidase activity. This may ultimately lead to an impaired function of the mitochondrial respiratory chain and subsequently to oxidative stress. This complex relationship among subcellular organelles is strongly supported by the observation that proteins from HCV are documented to traffic sequentially from the endoplasmic reticulum into mitochondria, probably through mitochondria-associated membrane (MAM) compartment, a contact site that enables the direct transfer of membrane bound lipids and the generation of high Ca^{2+} microdomains for mitochondrial signalling and responses to cellular stress. HCV core protein is associated with Ca^{2+} regulation and apoptotic signals and directly increases mitochondrial Ca^{2+} uptake via a primary effect on the uniporter. This allows mitochondria to sequester Ca^{2+} and increases ROS production with a higher possibility for MPT induction (Li et al. 2007). It is generally accepted that trafficking of viral proteins to the MAM allows viruses to manipulate a variety of fundamental cellular processes converging at the MAM (i.e. Ca^{2+} signalling, lipid synthesis and transfer, bioenergetics, metabolic flow, and apoptosis). Cause of their distinct topologies and targeted MAM subdomains, mitochondrial trafficking of HCV proteins involve alternative pathways and distinct targeting signals (Williamson and Colberg-Poley 2009). Among damaging mechanisms, an increased hepatic iron content in HCV-infected patients is responsible for the increased ROS generation and lipid peroxidation (Barbaro et al. 2001). These alterations lead to the formation of megamitochondria in these patients, especially in those infected by genotype 1b (Barbaro et al. 2001). The appearance of megamitochondria is likely linked to the fact that core protein alters the signal transduction pathways by binding to the cytoplasmic domains of some receptors that induce a signalling cascade promoting MPT (Matsumoto et al. 1997; Zhu et al. 1998). The inhibition of ROS production by the mitochondrial electron transport chain suggests this as a possible explanation for the inhibitory effect exerted by HCV core protein on mitochondrial function (Okuda et al. 2002). Accordingly, HCV-infected hepatocytes die after activation of caspase 3, nuclear translocation of activated caspase 3, and cleavage of the DNA repair enzyme poly(ADP-ribose)-polymerase, finally resulting in apoptosis. Moreover, HCV infection activates Bax, a pro-apoptotic member of the Bcl-2 family, as revealed by its conformational change and its increased accumulation on mitochondrial membranes, leading to increased mitochondrial outer membrane permeability and release of pro-apoptotic factors (Deng et al. 2008). This state of chronic high susceptibility to oxidative stress determines a reduction in mitochondrial metabolic process which favours the appearance of fatty degeneration of hepatocytes by inhibiting mitochondrial β-oxidation and oxidative damages to mitochondrial DNA.

3.4 Hemochromatosis

Hemochromatosis, a hereditary iron overload disorder, presents with a common phenotype characterised by normal erythropoiesis, increased transferrin saturation and serum ferritin levels, excess liver iron deposition (Janssen and Swinkels 2009). Several other gene defects have been detected since the discovery of the hemochromatosis gene (HFE) and its main mutations (C282Y and H63D). The main biochemical defect is related to a low production of the hepatic peptide hormone hepcidin, which is the central regulator of iron homeostasis (Lee and Beutler 2009).

Iron works as a cofactor for enzymes involved in many metabolic processes, including oxidative phosphorylation, but it can also be harmful especially for the liver, where iron overload causes oxidative stress, mitochondrial damage, stimulation of hepatocyte proliferation (Isom et al. 2009). Iron is transformed into its biologically available form in mitochondria by the iron-sulfur (Fe/S) cluster and heme synthesis pathways (Levi and Rovida 2009). Within mitochondria, iron is complexed in ferritin molecules, which have a restricted tissue distribution and appears to protect mitochondria from iron toxicity and oxidative damage (Arosio et al. 2009). Iron is an important bio-catalyst of oxidation-reduction reactions in the cell, and becomes dangerous when the fraction of redox-active metal ions exceeds that sequestered in specialized proteins or cellular compartments (Corradini et al. 2004).

Changes in mitochondrial membrane composition and function have been observed in rats with chronic iron overload (Britton et al. 1991; Bacon et al. 1989, 1985). The decrease in state 3 respiration rate observed in mitochondria of HFE (−/−) mouse correlates well with the total liver iron content. Lipid peroxidation is the proposed major mechanism that explains mitochondrial iron toxicity (Britton et al. 1990), as a consequence of increased production but also of a reduced degradation of peroxidative products (Jouihan et al. 2008). Altered fatty acid composition of membrane phospholipids with increased amount of saturated fatty acids has been described in rats with iron overload (Pietrangelo et al. 1990). Also a decreased activity of ferrocytochrome C reductase of the respiratory complex II, III and IV has been reported (Bacon et al. 1986). In conclusion, mitochondrial imbalance is a consequence of an oxidized alteration of lipid environment of the electron transport chain (Zhang et al. 2009).

3.5 *Wilson's Disease*

Similar to the iron, the copper is an essential transition metal ion. Its redox reactivity, whilst essential cofactor for over a dozen mitochondrial enzymes, represents a source of harmful ROS (Prohaska and Gybina 2004). Copper is sequestered by protein chaperones and moved across membranes by protein carriers. Mitochondrial-specific ceruloplasmin and metallothionein exert such a function in that organelle (Mehta et al. 2006).

Structural changes in hepatocellular mitochondria are characteristic of Wilson's disease (WD), an autosomal-recessive disorder, which is caused by mutations in a P-type ATPase and is associated with excess copper deposition in the liver (Stromeyer and Ishak 1980). Different types of morphological alterations have been described and include megamitochondria, matrix swelling and granular inclusion.

Mitochondrial dysfunction in copper disorders is associated with lipid accumulation as a consequence of a defect in lipid metabolism related to mitochondrial impairment. The latter is related neither to the extent of steatosis nor to the hepatic concentration of copper and suggests a genetic defect (Sternlieb 1992), although the extent of oxidative stress in mitochondria is dependent on the amount of copper accumulated in that organelle (Zischka and Lichtmannegger 2014). Mitochondrial DNA is oxidatively damaged with consequent multiple deletions and point mutations (Mansouri et al. 1997). Animal studies with dietary copper overload have shown that the increased lipid peroxidation products is likely to be partially prevented by vitamin E administration (Sokol et al. 1990). There is also evidence of severe mitochondrial dysfunction in the liver of patients with WD in which respiratory enzyme activities are decreased: complex I by 62%, complex II + III by 52%, complex IV by 33%, and aconitase by 71% (Gu et al. 2000). Interestingly, copper ions have an inhibitory effect on cell respiration only at 500 μM concentration and after 48 h incubation but produces a significant uncoupling effect at lower concentrations (Belyaeva et al. 2008). Copper induces an early and sharp increase of intracellular production of ROS which, in turn, inhibits pyruvate dehydrogenase (PDH) and α-ketoglutarate dehydrogenase (KGDH) (Sheline and Choi 2004).

The redox imbalance linked with copper accumulation is responsible for the oxidation of GSH to GSSG resulting in the generation of O_2^-. Down-regulation of Cu-Zn superoxide dismutase (SOD) consequent to degradation of this enzyme causes decreased dismutation of O_2^- and elevated levels of O_2^- in the periportal region of the liver. Decreased functioning of MnSOD activity and reduction in mitochondrial thiol/disulphide ratio and generation of O_2^- are much higher in the mitochondria from periportal region with involvement of mitochondria as evidenced from ATP depletion, collapse of mitochondrial membrane potential and induction of MPT (Roy et al. 2009). In fact, incubation of hepatocytes with copper (10 μM) results in a strong stimulation of ROS formation and a significant increase of both apoptotic and necrotic processes and occurrence of the MPT (Krumschnabel et al. 2005).

3.6 Chronic Cholestatic Disorders

A number of biochemical abnormalities including toxic bile salt accumulation and oxidative mitochondrial changes are implicated in the onset and evolution of cholestatic liver disease (Krähenbühl et al. 1995). The contribution of mitochondria to liver injury in primary biliary cholangitis (PBC) is documented by the presence of antimitochondrial antibodies (AMAs) in the serum of over 95% of patients with PBC as a reaction with components of the M2 antigen identified as the 2-oxoacid dehydrogenase complex (Feuchtinger et al. 2009). The progression of hepatic alterations is linked to the exposure of hepatocytes to hydrophobic bile salts, which is in turn associated with time- and concentration-dependent increased generation of ROS and appearance of MPT (Yerushalmi et al. 2001). These bile compounds decrease state 3 respiration and increase state 4 respiration. The concentration-dependent stimulation of state 4 by the above-mentioned bile salts is associated with increased inner membrane permeability of mitochondria (Rolo et al. 2000). This effect has also been reported with other experimental models of cholestasis (Rehman et al. 2008; Rolo et al. 2002a, b).

Cholestatic livers induced by bile duct ligation (BDL) exhibit a decreased oxygen consumption per mitochondrial mass, indicating a reduced oxidative metabolism (Krähenbühl et al. 1992a, b). Since complex I and III are limiting steps in hepatic fatty acid metabolism, these alterations likely depend on oxidative lipid and protein changes and have metabolic significance (Krahenbuhl and Brass 1991). Increased concentration of lipid peroxidation products and accumulation of oxidized proteins have been observed in mitochondria isolated from rats with BDL (Sokol et al. 1991; Grattagliano et al. 2007), confirming the hypothesis of oxidative mitochondrial damages in chronic cholestasis. These processes are associated with a decrease in the content of mitochondrial GSH and ubiquinones 9 and 10. The activity of complex II and III of the electron transport chain is also oxidatively decreased in BDL rats (Krähenbühl et al. 1995). In mitochondria, PSH is decreased and negatively correlates with PSSG (Portincasa et al. 2007). The transcriptional regulation of mitochondrial biogenesis is also imbalanced at ashort time after few hours of complete bile duct obstruction, resulting in mitochondrial dysfunction and consequent cholestatic liver injury via activation of the intrinsic apoptotic pathway (Tiao et al. 2009).

Morphometric and biochemical analyses have shown that the mitochondrial content *per* hepatocyte is increased in rats with BDL, suggesting mitochondrial proliferation as a mechanism to maintain liver mitochondrial function. NO is a known determining factor for mitochondrial proliferation (Carreras et al. 2004; Nisoli et al. 2003);

in fact, high hepatic levels of nitrosothiols during cholestasis may provide beneficial effects by supporting cell survival and mitochondriogenesis, although at this point it is not known if amplification of an already damaged mitochondrial population is beneficial for the cell. However, with the progression of cholestasis, the decreased availability of thioredoxin together with increased production of NO and with GSH depletion (Chen et al. 2005) may result in increased protein nitrosation (Ottesen et al. 2001) and PSH oxidation.

Patients with prolonged extra-hepatic obstructive cholestasis display oxidative alterations of liver proteins (Vendemiale et al. 2002). Changes depend on the retention of toxic products, which contribute to oxidative alterations. Antioxidants may effectively reduce liver injury and MPT stimulation caused by toxic bile salts (Yerushalmi et al. 2001). The mechanism by which hydrophilic bile salts provide protection on mitochondrial electron transport chain, damaged by hydrophobic bile salts accumulation, appears to be related to a decreased incorporation of toxic bile salts into mitochondrial membranes (Krähenbühl et al. 1994).

4 Assessment of Liver Mitochondrial Function *In Vivo*

A number of tests have been suggested to assess hepatic mitochondrial function *in vivo*. Considering the important role of mitochondria in chronic liver disease and the difficult assessment of liver function with current routine laboratory tests, these tools may indeed serve several purposes in supporting diagnosis and prognosis of chronic liver disease. For a complete review on this topic, see Bonfrate et al. (2015).

4.1 Respiratory Chain Activity

The mitochondrial respiratory chain activity can be assessed *in vivo* by measuring the ratio acetoacetate/β-hydroxybutyrate or pyruvate/lactate, which is related to the NAD/NADH ratio in the arterial blood. A reduced activity of the respiratory chain is associated with a decrease in NAD/NADH ratio, which consequently results in decreased acetoacetate/β-hydroxybutyrate (mitochondrial redox state) and pyruvate/lactate (cytosolic redox state) ratios. Patients with chronic liver disease may have a decreased ratio of the above reported parameters (Yamaguchi et al. 1992; Iwata et al. 1991).

4.2 Alpha-Ketoacid Dehydrogenase

Specific mitochondrial metabolic pathways can be assessed *in vivo* by performing breath test using substrates delivering labelled CO_2 during mitochondrial metabolism. The most interesting tools include α-ketoisocaproic acid (KICA), benzoic acid, octanoic acid, and methionine. KICA breath test assesses the activity of branched-chain α-ketoacid dehydrogenase, an enzyme sited in the mitochondrial matrix (Michaletz et al. 1989). This method is particularly reliable to assess mitochondrial function in patients with alcoholic liver disease or after administration of mitochondrial toxic drugs (Lauterburg et al. 1993, 1995). Chronic alcoholics have an impaired KICA decarboxylation to a greater extent than conventional quantitative tests of liver function such as galactose elimination capacity or the aminopyrine clearance. This specific enzymatic alteration is likely related to the ethanol-induced redox shift (NADH/NAD ratio). More recently, KICA breath test helps to distinguish patients with different histological stages of NAFLD (Portincasa et al. 2006).

4.3 *Octanoic Acid*

Mitochondrial β-oxidation can be assessed by the use of fatty acids breath test (Bates 1990). There are a few reports with oral [^{13}C]-octanoate that have shown how the rate of mitochondrial metabolism of octanoate is increased in patients with NAFLD without advanced disease (Miele et al. 2003).

4.4 *Benzoic Acid*

Benzoic acid undergoes hepatic conversion to hippuric acid after mitochondrial activation to form a CoA-derivative (Gatley and Sherratt 1977). Decreased availability of ATP and/or CoA results in a reduced renal excretion of hippurate after administration of benzoate. This test has a good reproducibility in patients with organic acidurias (Roe et al. 1983) and in experimental liver cirrhosis (Krähenbühl et al. 2000).

4.5 *Urea Production*

Quantification of urea production is a method to assess liver function in cirrhotic patients (Hamberg et al. 1992; Bianchi et al. 1991). The principle is based on the assessment of urea synthesis after infusion of amino acids as substrates, resulting in plasma accumulation of α-amino nitrogen. Urea synthesis increases in parallel to plasma α-amino nitrogen, thus representing the "functional hepatic nitrogen clearance" (Hamberg et al. 1992). The results correlate with the galactose elimination capacity, and reflects hepatocellular mass.

4.6 [^{31}P] Nuclear Magnetic Resonance Spectroscopy

This technique allows to assess changes in hepatic phosphate metabolism of cirrhotic patients after the administration of fructose (Dufour et al. 1992) and offers the possibility to monitor oxidative phosphorylation under different metabolic conditions (Dagnelie et al. 1992).

5 Conclusions

Mitochondria play a key role in mediating hepatocyte protection and injury. Unbalanced mitochondrial function unequivocally affects hepatocyte survival by actively contributing to the onset and perpetuation of liver diseases. Abnormal mitochondrial function is reported to participate in a variety of liver diseases including drug-induced liver injury, alcohol-induced liver disease, non-alcoholic fatty liver disease, viral hepatitis, primary and secondary cholestasis, hemochromatosis, and Wilson's disease. How the pathophysiology of each condition contributes and/or is caused by mitochondrial remodelling and further impairment is still an open avenue for research. The assessment of mitochondrial functions *in vivo* is a useful tool in patients with liver diseases for diagnostic and prognostic purposes in patients with liver diseases, and for evaluating the efficacy of therapeutic interventions.

Acknowledgements The present chapter is written in the context of the project FOIE GRAS, which has received funding from the European Union's Horizon 2020 Research and Innovation programme under the Marie Sklodowska-Curie Grant Agreement No. 722619.
Conflicts of Interest We declare that we have no conflicts of interest.

References

Abumrad NA, El-Maghrabi MR, Amri EZ, Lopez E, Grimaldi PA (1993) Cloning of a rat adipocyte membrane protein implicated in binding or transport of long-chain fatty acids that is induced during preadipocyte differentiation. Homology with human CD36. J Biol Chem 268:17665–17668

Afonso MB, Rodrigues PM, Carvalho T, Caridade M, Borralho P, Cortez-Pinto H, Castro RE, Rodrigues CM (2015) Necroptosis is a key pathogenic event in human and experimental murine models of non-alcoholic steatohepatitis. Clin Sci (Lond) 129:721–739

Afonso MB, Rodrigues PM, Simao AL, Ofengeim D, Carvalho T, Amaral JD, Gaspar MM, Cortez-Pinto H, Castro RE, Yuan J, Rodrigues CM (2016) Activation of necroptosis in human and experimental cholestasis. Cell Death Dis 7:e2390

Altomare E, Grattagliano I, Didonna D, Gentile A, Vendemiale G (1998) Gastric and intestinal ethanol toxicity in the rat. Effect on glutathione level and role of alcohol and acetaldehyde metabolisms. Ital J Gastroenterol Hepatol 30:82–90

Angulo P, Keach JC, Batts KP, Lindor KD (1999) Independent predictors of liver fibrosis in patients with nonalcoholic steatohepatitis. Hepatology 30:1356–1362

Anstee QM, Day CP (2015) The genetics of nonalcoholic fatty liver disease: spotlight on PNPla3 and TM6SF2. Semin Liver Dis 35:270–290

Argo CK, Northup PG, Al-Osaimi AM, Caldwell SH (2009) Systematic review of risk factors for fibrosis progression in non-alcoholic steatohepatitis. J Hepatol 51:371–379

Arnelle DR, Stamler JS (1995) NO+, NO., and NO− donation by S-nitrosothiols: implications for regulation of physiological functions by S-nitrosylation and acceleration of disulfide formation. Arch Biochem Biophys 318:279–285

Arosio P, Ingrassia R, Cavadini P (2009) Ferritins: a family of molecules for iron storage, antioxidation and more. Biochim Biophys Acta 1790:589–599

Bacon BR, Park CH, Brittenham GM, Tavill AS (1985) Hepatic mitochondrial oxidative metabolism in rats with chronic dietary iron overload. Hepatology 5:789–797

Bacon BR, O'Neill R, Park CH (1986) Iron-induced peroxidative injury to isolated rat hepatic mitochondria. J Free Radic Biol Med 2:339–347

Bacon BR, Britton RS, O'Neill R (1989) Effects of vitamin E deficiency on hepatic mitochondrial lipid peroxidation and oxidative metabolism in rats with chronic dietary iron overload. Hepatology 9:398–404

Bailey SM, Robinson G, Pinner A, Chamlee L, Ulasova E, Pompilius M, Page GP, Chhieng D, Jhala N, Landar A (2006) S-Adenosylmethionine prevents chronic alcohol-induced mitochondrial dysfunction in the rat liver. Am J Physiol Gastrointest Liver Physiol 291:G857–G867

Barbaro G, Di Lorenzo G, Asti A, Ribersani M, Belloni G, Grisorio B, Filice G, Barbarini G (1999) Hepatocellular mitochondrial alterations in patients with chronic hepatitis C: ultrastructural and biochemical findings. Am J Gastroenterol 94:2198–2205

Barbaro G, Di Lorenzo G, Belloni G, Asti A, Pellicelli A, Barbarini G (2001) Hepatic iron storage and megamitochondria formation in patients with chronic hepatitis C related to the hepatitis C virus genotype. Dig Liver Dis 33:81–83

Bates C (1990) Liberation of CO2 from [C] adipic acid and [C] octanoic acid by adult rats during riboflavin deficiency and its reversal. Br J Nutr 63:553–562

Bedogni G, Miglioli L, Masutti F, Tiribelli C, Marchesini G, Bellentani S (2005) Prevalence of and risk factors for nonalcoholic fatty liver disease: the Dionysos nutrition and liver study. Hepatology 42:44–52

Belyaeva EA, Dymkowska D, Więckowski MR, Wojtczak L (2008) Mitochondria as an important target in heavy metal toxicity in rat hepatoma AS-30D cells. Toxicol Appl Pharmacol 231:34–42

Berk PD, Verna EC (2016) Nonalcoholic fatty liver disease. Clin Liver Dis 20:245–262

Berk PD, Wada H, Horio Y, Potter BJ, Sorrentino D, Zhou SL, Isola LM, Stump D, Kiang CL, Thung S (1990) Plasma membrane fatty acid-binding protein and mitochondrial glutamic-oxaloacetic transaminase of rat liver are related. Proc Natl Acad Sci U S A 87:3484–3488

Berk PD, Zhou SL, Kiang CL, Stump D, Bradbury M, Isola LM (1997) Uptake of long chain free fatty acids is selectively up-regulated in adipocytes of Zucker rats with genetic obesity and non-insulin-dependent diabetes mellitus. J Biol Chem 272:8830–8835

Berson A, De Beco V, Lettéron P, Robin MA, Moreau C, El Kahwaji J, Verthier N, Feldmann G, Fromenty B, Pessayre D (1998) Steatohepatitis-inducing drugs cause mitochondrial dysfunction and lipid peroxidation in rat hepatocytes. Gastroenterology 114:764–774

Bianchi G, Marchesini G, Vilstrup H, Fabbri A, De Mitri MS, Zoli M, Pisiél E (1991) Hepatic amino-nitrogen clearance to urea-nitrogen in control subjects and in patients with cirrhosis: a simplified method. Hepatology 13:460–466

Bonfrate L, Wang DQ, Garruti G, Portincasa P (2014) Obesity and the risk and prognosis of gallstone disease and pancreatitis. Best Pract Res Clin Gastroenterol 28:623–635

Bonfrate L, Grattagliano I, Palasciano G, Portincasa P (2015) Dynamic carbon 13 breath tests for the study of liver function and gastric emptying. Gastroenterol Rep (Oxf) 3:12–21

Boyapati RK, Tamborska A, Dorward DA, Ho GT (2017) Advances in the understanding of mitochondrial DNA as a pathogenic factor in inflammatory diseases. F1000Res 6:169

Bradbury MW, Berk PD (2003) Cellular uptake of long chain free fatty acids: the structure and function of plasma membrane fatty acid binding protein. Adv Mol Cell Biol 33:47–80

Bradbury MW, Stump D, Guarnieri F, Berk PD (2011) Molecular modeling and functional confirmation of a predicted fatty acid binding site of mitochondrial aspartate aminotransferase. J Mol Biol 412:412–422

Britton RS, O'Neill R, Bacon BR (1990) Hepatic mitochondrial malondialdehyde metabolism in rats with chronic iron overload. Hepatology 11:93–97

Britton RS, O'Neill R, Bacon BR (1991) Chronic dietary iron overload in rats results in impaired calcium sequestration by hepatic mitochondria and microsomes 1, 2, and 3. Gastroenterology 101:806–811

Browning JD, Kumar KS, Saboorian MH, Thiele DL (2004) Ethnic differences in the prevalence of cryptogenic cirrhosis. Am J Gastroenterol 99:292–298

Brunt EM, Wong VW, Nobili V, Day CP, Sookoian S, Maher JJ, Bugianesi E, Sirlin CB, Neuschwander-Tetri BA, Rinella ME (2015) Nonalcoholic fatty liver disease. Nat Rev Dis Primers 1:15080

Bugianesi E, Leone N, Vanni E, Marchesini G, Brunello F, Carucci P, Musso A, De Paolis P, Capussotti L, Salizzoni M (2002) Expanding the natural history of nonalcoholic steatohepatitis: from cryptogenic cirrhosis to hepatocellular carcinoma. Gastroenterology 123:134–140

Burt AD, Mutton A, Day CP (1998) Diagnosis and interpretation of steatosis and steatohepatitis. SeminDiagnPathol 15:246–258

Cahill A, Sykora P (2008) Alcoholic liver disease and the mitochondrial ribosome: methods of analysis. Methods Mol Biol 447:381–394

Calamita G, Portincasa P (2007) Present and future therapeutic strategies in non-alcoholic fatty liver disease. Expert Opin Ther Targets 11:1231–1249

Caldwell SH, Chang CY, Nakamoto RK, Krugner-Higby L (2004) Mitochondria in nonalcoholic fatty liver disease. Clin Liver Dis 8:595–617

Campbell RV, Yang Y, Wang T, Rachamallu A, Li Y, Watowich SJ, Weinman SA (2009) Effects of hepatitis C core protein on mitochondrial electron transport and production of reactive oxygen species. Methods Enzymol 456:363–380

Carreras MC, Converso DP, Lorenti AS, Barbich M, Levisman DM, Jaitovich A, Antico Arciuch VG, Galli S, Poderoso JJ (2004) Mitochondrial nitric oxide synthase drives redox signals for proliferation and quiescence in rat liver development. Hepatology 40:157–166

Cazanave SC, Mott JL, Bronk SF, Werneburg NW, Fingas CD, Meng XW, Finnberg N, El-Deiry WS, Kaufmann SH, Gores GJ (2011) Death receptor 5 signaling promotes hepatocyte lipoapoptosis. J Biol Chem 286:39336–39348

Chalasani N, Younossi Z, Lavine JE, Diehl AM, Brunt EM, Cusi K, Charlton M, Sanyal AJ (2012) The diagnosis and management of non-alcoholic fatty liver disease: practice guideline by the American Gastroenterological Association, American Association for the Study of Liver Diseases, and American College of Gastroenterology. Gastroenterology 142:1592–1609

Chang ES (1987) Ultrastructural morphogenesis of mitochondria in alcoholic liver. Pathol Int 37:213–224

Charlton M (2007) Noninvasive indices of fibrosis in NAFLD: starting to think about a three-hit (at least) phenomenon. Am J Gastroenterol 102:409–411

Chen T, Pearce LL, Peterson J, Stoyanovsky D, Billiar TR (2005) Glutathione depletion renders rat hepatocytes sensitive to nitric oxide donor–mediated toxicity. Hepatology 42:598–607

Chen Z, Vigueira PA, Chambers KT, Hall AM, Mitra MS, Qi N, Mcdonald WG, Colca JR, Kletzien RF, Finck BN (2012) Insulin resistance and metabolic derangements in obese mice are ameliorated by a novel peroxisome proliferator-activated receptor γ-sparing thiazolidinedione. J Biol Chem 287:23537–23548

Colca JR, Vanderlugt JT, Adams WJ, Shashlo A, Mcdonald WG, Liang J, Zhou R, Orloff DG (2013) Clinical proof-of-concept study with MSDC-0160, a prototype mTOT-modulating insulin sensitizer. Clin Pharmacol Ther 93:352–359

Colca JR, Mcdonald WG, Mccommis KS, Finck BN (2017) Treating fatty liver disease by modulating mitochondrial pyruvate metabolism. Hepatol Commun 1:193–197

Coleman WB, Cunningham CC (1990) Effects of chronic ethanol consumption on the synthesis of polypeptides encoded by the hepatic mitochondrial genome. Biochim Biophys Acta 1019:142–150

Corradini E, Ferrara F, Pietrangelo A (2004) Iron and the liver. Pediatr Endocrinol Rev 2:245–248
Cortez-Pinto H, Chatham J, Chacko V, Arnold C, Rashid A, Diehl AM (1999) Alterations in liver ATP homeostasis in human nonalcoholic steatohepatitis: a pilot study. JAMA 282:1659–1664
Dagnelie PC, Menon DK, Cox IJ, Bell JD, Sargentoni J, Coutts GA, Urenjak J, Iles RA (1992) Effect of L-alanine infusion on 31P nuclear magnetic resonance spectra of normal human liver: towards biochemical pathology in vivo. Clin Sci 83:183–190
Deng L, Adachi T, Kitayama K, Bungyoku Y, Kitazawa S, Ishido S, Shoji I, Hotta H (2008) Hepatitis C virus infection induces apoptosis through a Bax-triggered, mitochondrion-mediated, caspase 3-dependent pathway. J Virol 82:10375–10385
Digel M, Ehehalt R, Stremmel W, Fullekrug J (2009) Acyl-CoA synthetases: fatty acid uptake and metabolic channeling. Mol Cell Biochem 326:23–28
Diogo CV, Grattagliano I, Oliveira PJ, Bonfrate L, Portincasa P (2011) Re-wiring the circuit: mitochondria as a pharmacological target in liver disease. Curr Med Chem 18:5448–5465
Donnelly KL, Smith CI, Schwarzenberg SJ, Jessurun J, Boldt MD, Parks EJ (2005) Sources of fatty acids stored in liver and secreted via lipoproteins in patients with nonalcoholic fatty liver disease. J Clin Invest 115:1343–1351
Dufour JF, Stoupis C, Lazeyras F, Vock P, Terrier F, Reichen J (1992) Alterations in hepatic fructose metabolism in cirrhotic patients demonstrated by dynamic 31phosphorus spectroscopy. Hepatology 15:835–842
Fabbrini E, Sullivan S, Klein S (2010) Obesity and nonalcoholic fatty liver disease: biochemical, metabolic, and clinical implications. Hepatology 51:679–689
Fan X, Bradbury MW, Berk PD (2003) Leptin and insulin modulate nutrient partitioning and weight loss in ob/ob mice through regulation of long-chain fatty acid uptake by adipocytes. J Nutr 133:2707–2715
Farrell GC, Wong VW, Chitturi S (2013) NAFLD in Asia—as common and important as in the West. Nat Rev Gastroenterol Hepatol 10:307–318
Feldmann G, Haouzi D, Moreau A, Durand-Schneider AM, Bringuier A, Berson A, Mansouri A, Fau D, Pessayre D (2000) Opening of the mitochondrial permeability transition pore causes matrix expansion and outer membrane rupture in Fas-mediated hepatic apoptosis in mice. Hepatology 31:674–683
Feoktistova M, Leverkus M (2015) Programmed necrosis and necroptosis signalling. FEBS J 282:19–31
Fernandez-Checa J, García-Ruiz C, Ookhtens M, Kaplowitz N (1991) Impaired uptake of glutathione by hepatic mitochondria from chronic ethanol-fed rats. Tracer kinetic studies in vitro and in vivo and susceptibility to oxidant stress. J Clin Investig 87:397
Fernández-Checa JC, Kaplowitz N, García-Ruiz C, Colell A (1998) Mitochondrial glutathione: importance and transport. Semin Liver Dis 18:389–401
Feuchtinger M, Christ S, Preuss B, Dengjel J, Duman S, Stevanovic S, Klein R (2009) Detection of novel non-M2-related antimitochondrial antibodies in patients with anti-M2 negative primary biliary cirrhosis. Gut 58:983–989
Fielding CM, Angulo P (2014) Hepatic steatosis and steatohepatitis: are they really two distinct entities? Curr Hepatol Rep 13:151–158
Formigli L, Papucci L, Tani A, Schiavone N, Tempestini A, Orlandini G, Capaccioli S, Zecchi Orlandini S (2000) Aponecrosis: morphological and biochemical exploration of a syncretic process of cell death sharing apoptosis and necrosis. J Cell Physiol 182:41–49
Francque SM, Verrijken A, Mertens I, Hubens G, Van Marck E, Pelckmans P, Michielsen P, Van Gaal L (2012) Noninvasive assessment of nonalcoholic fatty liver disease in obese or overweight patients. Clin Gastroenterol Hepatol 10:1162–8; quiz e87
Fromenty B, Pessayre D (1995) Inhibition of mitochondrial beta-oxidation as a mechanism of hepatotoxicity. Pharmacol Ther 67:101–154
Garcia-Martinez I, Santoro N, Chen Y, Hoque R, Ouyang X, Caprio S, Shlomchik MJ, Coffman RL, Candia A, Mehal WZ (2016) Hepatocyte mitochondrial DNA drives nonalcoholic steatohepatitis by activation of TLR9. J Clin Invest 126:859–864

Garcia-Ruiz C, Fernandez-Checa JC (2006) Mitochondrial glutathione: hepatocellular survival–death switch. J Gastroenterol Hepatol 21:S3–S6
García-Ruiz I, Rodríguez-Juan C, Díaz-Sanjuan T, Del Hoyo P, Colina F, Muñoz-Yagüe T, Solís-Herruzo JA (2006) Uric acid and anti-TNF antibody improve mitochondrial dysfunction in ob/ob mice. Hepatology 44:581–591
Garcia-Ruiz I, Solis-Munoz P, Fernandez-Moreira D, Munoz-Yague T, Solis-Herruzo JA (2015) In vitro treatment of HepG2 cells with saturated fatty acids reproduces mitochondrial dysfunction found in nonalcoholic steatohepatitis. Dis Model Mech 8:183–191
Gatley SJ, Sherratt H (1977) The synthesis of hippurate from benzoate and glycine by rat liver mitochondria. Submitochondrial localization and kinetics. Biochem J 166:39–47
Gautheron J, Vucur M, Reisinger F, Cardenas DV, Roderburg C, Koppe C, Kreggenwinkel K, Schneider AT, Bartneck M, Neumann UP (2014) A positive feedback loop between RIP3 and JNK controls non-alcoholic steatohepatitis. EMBO Mol Med 6:1062–1074
Ge F, Lobdell H IV, Zhou S, Hu C, Berk PD (2010a) Digital analysis of hepatic sections in mice accurately quantitates triglycerides and selected properties of lipid droplets. Exp Biol Med 235:1282–1286
Ge F, Zhou S, Hu C, Lobdell H, Berk PD (2010b) Insulin-and leptin-regulated fatty acid uptake plays a key causal role in hepatic steatosis in mice with intact leptin signaling but not in ob/ob or db/db mice. Am J Physiol Gastrointest Liver Physiol 299:G855–G866
Ge F, Hu C, Hyodo E, Arai K, Zhou S, Lobdell H IV, Walewski JL, Homma S, Berk PD (2011) Cardiomyocyte triglyceride accumulation and reduced ventricular function in mice with obesity reflect increased long chain fatty acid uptake and de novo fatty acid synthesis. J Obes. https://doi.org/10.1155/2012/205648
Ge F, Walewski JL, Torghabeh MH, Lobdell HT, Hu C, Zhou S, Dakin G, Pomp A, Bessler M, Schrope B, Ude-Welcome A, Inabnet WB, Feng T, Carras-Terzian E, Anglade D, Ebel FE, Berk PD (2016) Facilitated long chain fatty acid uptake by adipocytes remains upregulated relative to BMI for more than a year after major bariatric surgical weight loss. Obesity (Silver Spring) 24:113–122
Grattagliano I, Vendemiale G, Lauterburg BH (1999) Reperfusion injury of the liver: role of mitochondria and protection by glutathione ester. J Surg Res 86:2–8
Grattagliano I, Vendemiale G, Caraceni P, Domenicali M, Nardo B, Cavallari A, Trevisani F, Bernardi M, Altomare E (2000) Starvation impairs antioxidant defense in fatty livers of rats fed a choline-deficient diet. J Nutr 130:2131–2136
Grattagliano I, Caraceni P, Portincasa P, Domenicali M, Palmieri VO, Trevisani F, Bernardi M, Palasciano G (2003) Adaptation of subcellular glutathione detoxification system to stress conditions in choline-deficient diet induced rat fatty liver. Cell BiolToxicol 19:355–366
Grattagliano I, Portincasa P, Cocco T, Moschetta A, Di Paola M, Palmieri VO, Palasciano G (2004a) Effect of dietary restriction and N-acetylcysteine supplementation on intestinal mucosa and liver mitochondrial redox status and function in aged rats. ExpGerontol 39:1323–1332
Grattagliano I, Russmann S, Palmieri VO, Juni P, Bihl F, Portincasa P, Palasciano G, Lauterburg BH (2004b) Low membrane protein sulfhydrils but not G6PD deficiency predict ribavirin-induced hemolysis in hepatitis C. Hepatology 39:1248–1255
Grattagliano I, Portincasa P, Palmieri VO, Palasciano G (2007) Mutual changes of thioredoxin and nitrosothiols during biliary cirrhosis: results from humans and cholestatic rats. Hepatology 45:331–339
Grattagliano I, Caraceni P, Calamita G, Ferri D, Gargano I, Palasciano G, Portincasa P (2008) Severe liver steatosis correlates with nitrosative and oxidative stress in rats. Eur J Clin Investig 38:523–530
Grattagliano I, Russmann S, Diogo C, Bonfrate L, Oliveira PJ, Wang DQ, Portincasa P (2011) Mitochondria in chronic liver disease. CurrDrug Targets 12:879–893
Grattagliano I, Diogo CV, Mastrodonato M, De Bari O, Persichella M, Wang DQ, Liquori A, Ferri D, Carratu MR, Oliveira PJ, Portincasa P (2013) A silybin-phospholipids complex counter-

acts rat fatty liver degeneration and mitochondrial oxidative changes. World J Gastroenterol 19:3007–3017
Greenway F (2015) Physiological adaptations to weight loss and factors favouring weight regain. Int J Obes 39:1188–1196
Gu M, Cooper J, Butler P, Walker A, Mistry P, Dooley J, Schapira A (2000) Oxidative-phosphorylation defects in liver of patients with Wilson's disease. Lancet 356:469–474
Guerrieri F, Nicoletti C, Adorisio E, Caraccio G, Leonetti P, Zanotti F, Cantatore P (2000) Correlation between decreased expression of mitochondrial F0F1-ATP synthase and low regenerating capability of the liver after partial hepatectomy in hypothyroid rats. J Bioenerg Biomembr 32:183–191
Hall PD, Plummer JL, Ilsley AH, Cousins MJ (1991) Hepatic fibrosis and cirrhosis after chronic administration of alcohol and "low-dose" carbon tetrachloride vapor in the rat. Hepatology 13:815–819
Hamaguchi E, Takamura T, Sakurai M, Mizukoshi E, Zen Y, Takeshita Y, Kurita S, Arai K, Yamashita T, Sasaki M, Nakanuma Y, Kaneko S (2010) Histological course of nonalcoholic fatty liver disease in Japanese patients: tight glycemic control, rather than weight reduction, ameliorates liver fibrosis. Diabetes Care 33:284–286
Hamberg O, Nielsen K, Vilstrup H (1992) Effects of an increase in protein intake on hepatic efficacy for urea synthesis in healthy subjects and in patients with cirrhosis. J Hepatol 14:237–243
Han D, Johnson HS, Rao MP, Martin G, Sancheti H, Silkwood KH, Decker CW, Nguyen KT, Casian JG, Cadenas E, Kaplowitz N (2017) Mitochondrial remodeling in the liver following chronic alcohol feeding to rats. Free Radic Biol Med 102:100–110
Haouzi D, Lekéhal M, Moreau A, Moulis C, Feldmann G, Robin MA, Letteron P, Fau D, Pessayre D (2000) Cytochrome P450-generated reactive metabolites cause mitochondrial permeability transition, caspase activation, and apoptosis in rat hepatocytes. Hepatology 32:303–311
Haque M, Sanyal AJ (2002) The metabolic abnormalities associated with non-alcoholic fatty liver disease. Best Pract Res Clin Gastroenterol 16:709–731
Harrison SA (2015) Nonalcoholic fatty liver disease and fibrosis progression: the good, the bad, and the unknown. Clin Gastroenterol Hepatol 13:655–657
Hensley K, Kotake Y, Sang H, Pye QN, Wallis GL, Kolker LM, Tabatabaie T, Stewart CA, Konishi Y, Nakae D (2000) Dietary choline restriction causes complex I dysfunction and increased H2O2 generation in liver mitochondria. Carcinogenesis 21:983–989
Hirano A, Kaplowitz N, Tsukamoto H, Kamimura S, Fernandez-Checa JC (1992) Hepatic mitochondrial glutathione depletion and progression of experimental alcoholic liver disease in rats. Hepatology 16:1423–1427
Hossain N, Afendy A, Stepanova M, Nader F, Srishord M, Rafiq N, Goodman Z, Younossi Z (2009) Independent predictors of fibrosis in patients with nonalcoholic fatty liver disease. Clin Gastroenterol Hepatol 7:1224–9, 1229 e1-2
Inokuchi-Shimizu S, Park EJ, Roh YS, Yang L, Zhang B, Song J, Liang S, Pimienta M, Taniguchi K, Wu X, Asahina K, Lagakos W, Mackey MR, Akira S, Ellisman MH, Sears DD, Olefsky JM, Karin M, Brenner DA, Seki E (2014) TAK1-mediated autophagy and fatty acid oxidation prevent hepatosteatosis and tumorigenesis. J Clin Invest 124:3566–3578
Isom HC, Mcdevitt EI, Moon MS (2009) Elevated hepatic iron: a confounding factor in chronic hepatitis C. Biochim Biophys Acta 1790:650–662
Israel Y (1989) Covalent binding of acetaldehyde to liver tubulin: a step in the right direction. Hepatology 9:161–162
Iwata S, Ozawa K, Shimahara Y, Mori K, Kobayashi N, Kumada K, Yamaoka Y (1991) Diurnal fluctuations of arterial ketone body ratio in normal subjects and patients with liver dysfunction. Gastroenterology 100:1371–1378
Jadeja RN, Devkar RV, Nammi S (2017) Oxidative stress in liver diseases: pathogenesis, prevention, and therapeutics. Oxidative Med Cell Longev 2017:8341286
Janssen M, Swinkels D (2009) Hereditary haemochromatosis. Best Pract Res Clin Gastroenterol 23:171–183

Johnson ES, Lindblom KR, Robeson A, Stevens RD, Ilkayeva OR, Newgard CB, Kornbluth S, Andersen JL (2013) Metabolomic profiling reveals a role for caspase-2 in lipoapoptosis. J Biol Chem 288:14463–14475

Jou J, Choi SS, Diehl AM (2008) Mechanisms of disease progression in nonalcoholic fatty liver disease. Semin Liver Dis 28:370–379

Jouihan HA, Cobine PA, Cooksey RC, Hoagland EA, Boudina S, Abel ED, Winge DR, McClain DA (2008) Iron-mediated inhibition of mitochondrial manganese uptake mediates mitochondrial dysfunction in a mouse model of hemochromatosis. Mol Med 14:98

Kagan VE, Bayır HA, Belikova NA, Kapralov O, Tyurina YY, Tyurin VA, Jiang J, Stoyanovsky DA, Wipf P, Kochanek PM (2009) Cytochrome c/cardiolipin relations in mitochondria: a kiss of death. Free Radic Biol Med 46:1439–1453

Kageyama F, Kobayashi Y, Kawasaki T, Toyokuni S, Uchida K, Nakamura H (2000) Successful interferon therapy reverses enhanced hepatic iron accumulation and lipid peroxidation in chronic hepatitis C. Am J Gastroenterol 95:1041–1050

Kamimura S, Gaal K, Britton RS, Bacon BR, Triadafilopoulos G, Tsukamoto H (1992) Increased 4-hydroxynonenal levels in experimental alcoholic liver disease: association of lipid peroxidation with liver fibrogenesis. Hepatology 16:448–453

Kammoun HL, Chabanon H, Hainault I, Luquet S, Magnan C, Koike T, Ferre P, Foufelle F (2009) GRP78 expression inhibits insulin and ER stress-induced SREBP-1c activation and reduces hepatic steatosis in mice. J Clin Invest 119:1201–1215

Kampf JP, Kleinfeld AM (2007) Is membrane transport of FFA mediated by lipid, protein, or both? An unknown protein mediates free fatty acid transport across the adipocyte plasma membrane. Physiology (Bethesda) 22:7–14

Kampf JP, Parmley D, Kleinfeld AM (2007) Free fatty acid transport across adipocytes is mediated by an unknown membrane protein pump. Am J Physiol Endocrinol Metab 293:E1207–E1214

Kass GE (2006) Mitochondrial involvement in drug-induced hepatic injury. Chem Biol Interact 163:145–159

Kass GE, Price SC (2008) Role of mitochondria in drug-induced cholestatic injury. Clin Liver Dis 12:27–51, vii

Koliaki C, Szendroedi J, Kaul K, Jelenik T, Nowotny P, Jankowiak F, Herder C, Carstensen M, Krausch M, Knoefel WT, Schlensak M, Roden M (2015) Adaptation of hepatic mitochondrial function in humans with non-alcoholic fatty liver is lost in steatohepatitis. Cell Metab 21:739–746

Krahenbuhl S (1993) Alterations in mitochondrial function and morphology in chronic liver disease: pathogenesis and potential for therapeutic intervention. PharmacolTher 60:1–38

Krahenbuhl S, Brass EP (1991) Fuel homeostasis and carnitine metabolism in rats with secondary biliary cirrhosis. Hepatology 14:927–934

Krähenbühl S, Krähenbühl-Glauser S, Stucki J, Gehr P, Reichen J (1992a) Stereological and functional analysis of liver mitochondria from rats with secondary biliary cirrhosis: impaired mitochondrial metabolism and increased mitochondrial content per hepatocyte. Hepatology 15:1167–1172

Krähenbühl S, Stucki J, Reichen J (1992b) Reduced activity of the electron transport chain in liver mitochondria isolated from rats with secondary biliary cirrhosis. Hepatology 15:1160–1166

Krähenbühl S, Fischer S, Talos C, Reichen PJ (1994) Ursodeoxycholate protects oxidative mitochondrial metabolism from bile acid toxicity: dose-response study in isolated rat liver mitochondria. Hepatology 20:1595–1601

Krähenbühl S, Talos C, Lauterburg BH, Reichen J (1995) Reduced antioxidative capacity in liver mitochondria from bile duct ligated rats. Hepatology 22:607–612

Krähenbühl L, Ledermann M, Lang C, Krähenbühl S (2000) Relationship between hepatic mitochondrial functions in vivo and in vitro in rats with carbon tetrachloride-induced liver cirrhosis. J Hepatol 33:216–223

Krawczyk M, Bonfrate L, Portincasa P (2010) Nonalcoholic fatty liver disease. Best Pract Res Clin Gastroenterol 24:695–708

Krawczyk M, Portincasa P, Lammert F (2013) PNPLA3-associated steatohepatitis: toward a gene-based classification of fatty liver disease. Semin Liver Dis 33:369–379

Krumschnabel G, Manzl C, Berger C, Hofer B (2005) Oxidative stress, mitochondrial permeability transition, and cell death in Cu-exposed trout hepatocytes. Toxicol Appl Pharmacol 209:62–73

Lambert JE, Ramos-Roman MA, Browning JD, Parks EJ (2014) Increased de novo lipogenesis is a distinct characteristic of individuals with nonalcoholic fatty liver disease. Gastroenterology 146:726–735

Lauterburg BH, Bilzer M (1988) Mechanisms of acetaldehyde hepatotoxicity. J Hepatol 7:384–390

Lauterburg BH, Davies S, Mitchell J (1984) Ethanol suppresses hepatic glutathione synthesis in rats in vivo. J Pharmacol Exp Ther 230:7–11

Lauterburg BH, Liang D, Schwarzenbach FA, Breen KJ (1993) Mitochondrial dysfunction in alcoholic patients as assessed by breath analysis. Hepatology 17:418–422

Lauterburg BH, Grattagliano I, GmüR R, Stalder M, Hildebrand P (1995) Noninvasive assessment of the effect of xenobiotics on mitochondrial function in human beings: studies with acetylsalicylic acid and ethanol with the use of the carbon 13-labeled ketoisocaproate breath test. J Lab Clin Med 125:378–383

Lee PL, Beutler E (2009) Regulation of hepcidin and iron-overload disease. Annu Rev Pathol Mech Dis 4:489–515

Leroux A, Ferrere G, Godie V, Cailleux F, Renoud M-L, Gaudin F, Naveau S, Prévot S, Makhzami S, Perlemuter G (2012) Toxic lipids stored by Kupffer cells correlates with their pro-inflammatory phenotype at an early stage of steatohepatitis. J Hepatol 57:141–149

Levi S, Rovida E (2009) The role of iron in mitochondrial function. Biochim Biophys Acta 1790:629–636

Li Y, Boehning DF, Qian T, Popov VL, Weinman SA (2007) Hepatitis C virus core protein increases mitochondrial ROS production by stimulation of Ca2+ uniporter activity. FASEB J 21:2474–2485

Lieber C, Baraona E, Hernandez-Munoz R, Kubota S, Sato N, Kawano S, Matsumura T, Inatomi N (1989) Impaired oxygen utilization. A new mechanism for the hepatotoxicity of ethanol in sub-human primates. J Clin Investig 83:1682

Listenberger LL, Ory DS, Schaffer JE (2001) Palmitate-induced apoptosis can occur through a ceramide-independent pathway. J Biol Chem 276:14890–14895

Loguercio C, Andreone P, Brisc C, Brisc MC, Bugianesi E, Chiaramonte M, Cursaro C, Danila M, de Sio I, Floreani A, Freni MA, Grieco A, Groppo M, Lazzari R, Lobello S, Lorefice E, Margotti M, Miele L, Milani S, Okolicsanyi L, Palasciano G, Portincasa P, Saltarelli P, Smedile A, Somalvico F, Spadaro A, Sporea I, Sorrentino P, Vecchione R, Tuccillo C, del Vecchio Blanco C, Federico A (2012) Silybin combined with phosphatidylcholine and vitamin E in patients with nonalcoholic fatty liver disease: a randomized controlled trial. Free Radic Biol Med 52:1658–1665

Loomba R, Sanyal AJ (2013) The global NAFLD epidemic. Nat Rev Gastroenterol Hepatol 10:686–690

Ludwig J, Viggiano TR, McGill DB, Oh BJ (1980) Nonalcoholic steatohepatitis: Mayo Clinic experiences with a hitherto unnamed disease. Mayo Clin Proc 55:434–438

Ma L, Dong JX, Wu C, Li XY, Chen J, Zhang H, Liu Y (2017) Spectroscopic, polarographic, and microcalorimetric studies on mitochondrial dysfunction induced by ethanol. J Membr Biol 250:195–204

Machado MV, Michelotti GA, Pereira Tde A, Boursier J, Kruger L, Swiderska-Syn M, Karaca G, Xie G, Guy CD, Bohinc B, Lindblom KR, Johnson E, Kornbluth S, Diehl AM (2015) Reduced lipoapoptosis, hedgehog pathway activation and fibrosis in caspase-2 deficient mice with non-alcoholic steatohepatitis. Gut 64:1148–1157

Malhi H, Gores GJ (2008a) Cellular and molecular mechanisms of liver injury. Gastroenterology 134:1641–1654

Malhi H, Gores GJ (2008b) Molecular mechanisms of lipotoxicity in nonalcoholic fatty liver disease. Semin Liver Dis 28:360–369

Mansouri A, Gaou I, Fromenty B, Berson A, Letteron P, Degott C, Erlinger S, Pessayre D (1997) Premature oxidative aging of hepatic mitochondrial DNA in Wilson's disease. Gastroenterology 113:599–605

Mantena SK, King AL, Andringa KK, Landar A, Darley-Usmar V, Bailey SM (2007) Novel interactions of mitochondria and reactive oxygen/nitrogen species in alcohol mediated liver disease. World J Gastroenterol 13:4967

Mantena SK, King AL, Andringa KK, Eccleston HB, Bailey SM (2008) Mitochondrial dysfunction and oxidative stress in the pathogenesis of alcohol- and obesity-induced fatty liver diseases. Free Radic Biol Med 44:1259–1272

Matsumoto M, Hsieh T-Y, Zhu N, Vanarsdale T, Hwang SB, Jeng K-S, Gorbalenya AE, Lo S-Y, Ou J-H, Ware CF (1997) Hepatitis C virus core protein interacts with the cytoplasmic tail of lymphotoxin-beta receptor. J Virol 71:1301–1309

Matteoni CA, Younossi ZM, Gramlich T, Boparai N, Liu YC, McCullough AJ (1999) Nonalcoholic fatty liver disease: a spectrum of clinical and pathological severity. Gastroenterology 116:1413–1419

Mccommis KS, Chen Z, Fu X, McDonald WG, Colca JR, Kletzien RF, Burgess SC, Finck BN (2015) Loss of mitochondrial pyruvate carrier 2 in the liver leads to defects in gluconeogenesis and compensation via pyruvate-alanine cycling. Cell Metab 22:682–694

Mccommis K, Hodges W, Brunt E, Nalbantoglu I, McDonald W, Holley C, Fujiwara H, Schaffer J, Colca J, Finck B (2016) Targeting the mitochondrial pyruvate carrier attenuates fibrosis in a mouse model of nonalcoholic steatohepatitis. Hepatology (Baltimore, Md) 65(5):1543–1556

McCullough AJ (2006) Pathophysiology of nonalcoholic steatohepatitis. J Clin Gastroenterol 40(Suppl 1):S17–S29

Mehta R, Templeton DM, O'Brien PJ (2006) Mitochondrial involvement in genetically determined transition metal toxicity: II. Copper toxicity. Chem Biol Interact 163:77–85

Merriman RB, Aouizerat BE, Bass NM (2006) Genetic influences in nonalcoholic fatty liver disease. J Clin Gastroenterol 40:S30–S33

Michaletz PA, Cap L, Alpert E, Lauterburg BH (1989) Assessment of mitochondrial function in vivo with a breath test utilizing α-ketoisocaproic acid. Hepatology 10:829–832

Miele L, Grieco A, Armuzzi A, Candelli M, Forgione A, Gasbarrini A, Gasbarrini G (2003) Hepatic mitochondrial beta-oxidation in patients with nonalcoholic steatohepatitis assessed by 13C-octanoate breath test. Am J Gastroenterol 98:2335–2336

Miele L, Valenza V, La Torre G, Montalto M, Cammarota G, Ricci R, Masciana R, Forgione A, Gabrieli ML, Perotti G, Vecchio FM, Rapaccini G, Gasbarrini G, Day CP, Grieco A (2009) Increased intestinal permeability and tight junction alterations in nonalcoholic fatty liver disease. Hepatology 49:1877–1887

Miki C, Iriyama K, Mirza DF, Mayer AD, Buckels JA, Suzuki H, McMaster P (1998) Postperfusion energy metabolism of steatotic graft and its relation to early graft viability following liver transplantation. Dig Dis Sci 43:74–79

Minor T, Akbar S, Tolba R, Dombrowski F (2000) Cold preservation of fatty liver grafts: prevention of functional and ultrastructural impairments by venous oxygen persufflation. J Hepatol 32:105–111

Mirabeau O, Perlas E, Severini C, Audero E, Gascuel O, Possenti R, Birney E, Rosenthal N, Gross C (2007) Identification of novel peptide hormones in the human proteome by hidden Markov model screening. Genome Res 17:320–327

Mitsuyoshi H, Yasui K, Harano Y, Endo M, Tsuji K, Minami M, Itoh Y, Okanoue T, Yoshikawa T (2009) Analysis of hepatic genes involved in the metabolism of fatty acids and iron in nonalcoholic fatty liver disease. Hepatol Res 39:366–373

Molloy JW, Calcagno CJ, Williams CD, Jones FJ, Torres DM, Harrison SA (2012) Association of coffee and caffeine consumption with fatty liver disease, nonalcoholic steatohepatitis, and degree of hepatic fibrosis. Hepatology 55:429–436

Monsenego J, Mansouri A, Akkaoui M, Lenoir V, Esnous C, Fauveau V, Tavernier V, Girard J, Prip-Buus C (2012) Enhancing liver mitochondrial fatty acid oxidation capacity in obese mice improves insulin sensitivity independently of hepatic steatosis. J Hepatol 56:632–639

Montgomery RI, Coleman WB, Eble KS, Cunningham CC (1987) Ethanol-elicited alterations in the oligomycin sensitivity and structural stability of the mitochondrial F0. F1 ATPase. J Biol Chem 262:13285–13289

Myers MG, Cowley MA, Munzberg H (2008) Mechanisms of leptin action and leptin resistance. Annu Rev Physiol 70:537–556

Nardo B, Grattagliano I, Domenicali M, Caraceni P, Catena F, Santoni B, Turi P, Cavallari G, Dall'agata M, Trevisani F (2000) Mitochondrial oxidative injury in rat fatty livers exposed to warm ischemia-reperfusion. Transplant Proc 32:51

Nikitovic D, Holmgren A (1996) S-Nitrosoglutathione is cleaved by the thioredoxin system with liberation of glutathione and redox regulating nitric oxide. J Biol Chem 271:19180–19185

Nisoli E, Clementi E, Paolucci C, Cozzi V, Tonello C, Sciorati C, Bracale R, Valerio A, Francolini M, Moncada S (2003) Mitochondrial biogenesis in mammals: the role of endogenous nitric oxide. Science 299:896–899

Nomura-Takigawa Y, Nagano-Fujii M, Deng L, Kitazawa S, Ishido S, Sada K, Hotta H (2006) Non-structural protein 4A of Hepatitis C virus accumulates on mitochondria and renders the cells prone to undergoing mitochondria-mediated apoptosis. J Gen Virol 87:1935–1945

Noureddin M, Yates KP, Vaughn IA, Neuschwander-Tetri BA, Sanyal AJ, McCullough A, Merriman R, Hameed B, Doo E, Kleiner DE, Behling C, Loomba R, Nash CRN (2013) Clinical and histological determinants of nonalcoholic steatohepatitis and advanced fibrosis in elderly patients. Hepatology 58:1644–1654

Okuda M, Li K, Beard MR, Showalter LA, Scholle F, Lemon SM, Weinman SA (2002) Mitochondrial injury, oxidative stress, and antioxidant gene expression are induced by hepatitis C virus core protein. Gastroenterology 122:366–375

Ottesen LH, Harry D, Frost M, Davies S, Khan K, Halliwell B, Moore K (2001) Increased formation of S-nitrothiols and nitrotyrosine in cirrhotic rats during endotoxemia. Free Radic Biol Med 31:790–798

Papucci L, Formigli L, Schiavone N, Tani A, Donnini M, Lapucci A, Perna F, Tempestini A, Witort E, Morganti M (2004) Apoptosis shifts to necrosis via intermediate types of cell death by a mechanism depending on c-myc and bcl-2 expression. Cell Tissue Res 316:197–209

Paradies G, Petrosillo G, Pistolese M, Ruggiero FM (2002) Reactive oxygen species affect mitochondrial electron transport complex I activity through oxidative cardiolipin damage. Gene 286:135–141

Pérez-Carreras M, Del Hoyo P, Martín MA, Rubio JC, Martín A, Castellano G, Colina F, Arenas J, Solis-Herruzo JA (2003) Defective hepatic mitochondrial respiratory chain in patients with nonalcoholic steatohepatitis. Hepatology 38:999–1007

Perry RJ, Kim T, Zhang X-M, Lee H-Y, Pesta D, Popov VB, Zhang D, Rahimi Y, Jurczak MJ, Cline GW (2013) Reversal of hypertriglyceridemia, fatty liver disease, and insulin resistance by a liver-targeted mitochondrial uncoupler. Cell Metab 18:740–748

Pessayre D, Fromenty B (2005) NASH: a mitochondrial disease. J Hepatol 42:928–940

Pessayre D, Haouzi D, Fau D, Robin MA, Mansouri A, Berson A (1999a) Withdrawal of life support, altruistic suicide, fratricidal killing and euthanasia by lymphocytes: different forms of drug-induced hepatic apoptosis. J Hepatol 31:760–770

Pessayre D, Mansouri A, Haouzi D, Fromenty B (1999b) Hepatotoxicity due to mitochondrial dysfunction. Cell Biol Toxicol 15:367–373

Petrescu O, Fan X, Gentileschi P, Hossain S, Bradbury M, Gagner M, Berk PD (2005) Long-chain fatty acid uptake is upregulated in omental adipocytes from patients undergoing bariatric surgery for obesity. Int J Obes 29:196–203

Petrosillo G, Portincasa P, Grattagliano I, Casanova G, Matera M, Ruggiero FM, Ferri D, Paradies G (2007) Mitochondrial dysfunction in rat with nonalcoholic fatty liver Involvement of complex I, reactive oxygen species and cardiolipin. BiochimBiophysActa 1767:1260–1267

Petta S, Amato MC, Di Marco V, Camma C, Pizzolanti G, Barcellona MR, Cabibi D, Galluzzo A, Sinagra D, Giordano C, Craxi A (2012) Visceral adiposity index is associated with significant fibrosis in patients with non-alcoholic fatty liver disease. Aliment Pharmacol Ther 35:238–247

Piccoli C, Scrima R, Quarato G, D'Aprile A, Ripoli M, Lecce L, Boffoli D, Moradpour D, Capitanio N (2007) Hepatitis C virus protein expression causes calcium-mediated mitochondrial bioenergetic dysfunction and nitro-oxidative stress. Hepatology 46:58–65

Pietrangelo A, Grandi R, Tripodi A, Tomasi A, Ceccarelli D, Ventura E, Masini A (1990) Lipid composition and fluidity of liver mitochondria, microsomes and plasma membrane of rats with chronic dietary iron overload. Biochem Pharmacol 39:123–128

Pocha C, Kolly P, Dufour JF (2015) Nonalcoholic fatty liver disease-related hepatocellular carcinoma: a problem of growing magnitude. Semin Liver Dis 35:304–317

Pohl J, Ring A, Stremmel W (2002) Uptake of long-chain fatty acids in HepG2 cells involves caveolae: analysis of a novel pathway. J Lipid Res 43:1390–1399

Poonawala A, Nair SP, Thuluvath PJ (2000) Prevalence of obesity and diabetes in patients with cryptogenic cirrhosis: a case-control study. Hepatology 32:689–692

Portincasa P, Wang DQH (2017) Nonalcoholic fatty liver and gallstone disease. In: Wang DQH, Portincasa P (eds) Gallstones. Recent advances in epidemiology, pathogenesis, diagnosis and management. Nova Science Publishers, Inc, New York

Portincasa P, Grattagliano I, Lauterburg BH, Palmieri VO, Palasciano G, Stellaard F (2006) Liver breath tests non-invasively predict higher stages of non-alcoholic steatohepatitis. Clin Sci (Lond) 111:135–143

Portincasa P, Grattagliano I, Testini M, Caruso ML, Wang DQ, Moschetta A, Calamita G, Vacca M, Valentini AM, Renna G, Lissidini G, Palasciano G (2007) Parallel intestinal and liver injury during early cholestasis in the rat: modulation by bile salts and antioxidants. Free RadicBiolMed 42:1381–1391

Postic C, Girard J (2008) Contribution of de novo fatty acid synthesis to hepatic steatosis and insulin resistance: lessons from genetically engineered mice. J Clin Invest 118:829–838

Prohaska JR, Gybina AA (2004) Intracellular copper transport in mammals. J Nutr 134:1003–1006

Quintanilla ME, Tampier L (1989) Sensitivity of liver mitochondrial functions to various levels of ethanol intake in the rat. Alcohol Clin Exp Res 13:280–283

Ratziu V, Giral P, Charlotte F, Bruckert E, Thibault V, Theodorou I, Khalil L, Turpin G, Opolon P, Poynard T (2000) Liver fibrosis in overweight patients. Gastroenterology 118:1117–1123

Rehman H, Ramshesh VK, Theruvath TP, Kim I, Currin RT, Giri S, Lemasters JJ, Zhong Z (2008) NIM811 (N-methyl-4-isoleucine cyclosporine), a mitochondrial permeability transition inhibitor, attenuates cholestatic liver injury but not fibrosis in mice. J Pharmacol Exp Ther 327:699–706

Ribiere C, Hininger I, Saffar-Boccara C, Sabourault D, Nordmann R (1994) Mitochondrial respiratory activity and superoxide radical generation in the liver, brain and heart after chronic ethanol intake. Biochem Pharmacol 47:1827–1833

Roe C, Hoppel CL, Stacey T, Chalmers R, Tracey B, Millington D (1983) Metabolic response to carnitine in methylmalonic aciduria. An effective strategy for elimination of propionyl groups. Arch Dis Child 58:916–920

Roh YS, Seki E (2013) Toll-like receptors in alcoholic liver disease, non-alcoholic steatohepatitis and carcinogenesis. J Gastroenterol Hepatol 28:38–42

Rolo AP, Oliveira PJ, Moreno AJ, Palmeira CM (2000) Bile acids affect liver mitochondrial bioenergetics: possible relevance for cholestasis therapy. Toxicol Sci 57:177–185

Rolo AP, Oliveira PJ, SeiçA R, Santos MS, Moreno AJ, Palmeira CM (2002a) Disruption of mitochondrial calcium homeostasis after chronic α-naphthylisothiocyanate administration: relevance for cholestasis. J Investig Med 50:193–200

Rolo AP, Oliveira PJ, Seiça R, Santos MS, Moreno AJ, Palmeira CM (2002b) Improved efficiency of hepatic mitochondrial function in rats with cholestasis induced by an acute dose of α-naphthylisothiocyanate. Toxicol Appl Pharmacol 182:20–26

Roy DN, Mandal S, Sen G, Biswas T (2009) Superoxide anion mediated mitochondrial dysfunction leads to hepatocyte apoptosis preferentially in the periportal region during copper toxicity in rats. Chem Biol Interact 182:136–147
Samuel VT, Liu Z-X, Wang A, Beddow SA, Geisler JG, Kahn M, Zhang X-M, Monia BP, Bhanot S, Shulman GI (2007) Inhibition of protein kinase Cε prevents hepatic insulin resistance in nonalcoholic fatty liver disease. J Clin Invest 117:739–745
Sanyal AJ, Campbell-Sargent C, Mirshahi F, Rizzo WB, Contos MJ, Sterling RK, Luketic VA, Shiffman ML, Clore JN (2001) Nonalcoholic steatohepatitis: association of insulin resistance and mitochondrial abnormalities. Gastroenterology 120:1183–1192
Schaffer JE (2003) Lipotoxicity: when tissues overeat. Curr Opin Lipidol 14:281–287
Schaffer JE, Lodish HF (1994) Expression cloning and characterization of a novel adipocyte long chain fatty acid transport protein. Cell 79:427–436
Schilling RJ, Reitz RC (1980) A mechanism for ethanol-induced damage to liver mitochondrial structure and function. Biochim Biophys Acta 603:266–277
Schnabl B, Brenner DA (2014) Interactions between the intestinal microbiome and liver diseases. Gastroenterology 146:1513–1524
Seki E, De Minicis S, Österreicher CH, Kluwe J, Osawa Y, Brenner DA, Schwabe RF (2007) TLR4 enhances TGF-β signaling and hepatic fibrosis. Nat Med 13:1324–1332
Sheline CT, Choi DW (2004) Cu2+ toxicity inhibition of mitochondrial dehydrogenases in vitro and in vivo. Ann Neurol 55:645–653
Shimomura I, Hammer RE, Ikemoto S, Brown MS, Goldstein JL (1999) Leptin reverses insulin resistance and diabetes mellitus in mice with congenital lipodystrophy. Nature 401:73–76
Shulman GI (2014) Ectopic fat in insulin resistance, dyslipidemia, and cardiometabolic disease. N Engl J Med 371:1131–1141
Singh S, Allen AM, Wang Z, Prokop LJ, Murad MH, Loomba R (2015) Fibrosis progression in nonalcoholic fatty liver vs nonalcoholic steatohepatitis: a systematic review and meta-analysis of paired-biopsy studies. Clin Gastroenterol Hepatol 13:643–54 e1–9; quiz e39–40
Slee EA, Adrain C, Martin SJ (2001) Executioner caspase-3,-6, and-7 perform distinct, non-redundant roles during the demolition phase of apoptosis. J Biol Chem 276:7320–7326
Sokol RJ, Deverbaux M, Mierau GW, Hambidge KM, Shikes RH (1990) Oxidant injury to hepatic mitochondrial lipids in rats with dietary copper overload: modification by vitamin E deficiency. Gastroenterology 99:1061–1071
Sokol RJ, Devereaux M, Khandwala RA (1991) Effect of dietary lipid and vitamin E on mitochondrial lipid peroxidation and hepatic injury in the bile duct-ligated rat. J Lipid Res 32:1349–1357
Spach PI, Bottenus RE, Cunningham CC (1982) Control of adenine nucleotide metabolism in hepatic mitochondria from rats with ethanol-induced fatty liver. Biochem J 202:445–452
Speliotes EK, Yerges-Armstrong LM, Wu J, Hernaez R, Kim LJ, Palmer CD, Gudnason V, Eiriksdottir G, Garcia ME, Launer LJ, Nalls MA, Clark JM, Mitchell BD, Shuldiner AR, Butler JL, Tomas M, Hoffmann U, Hwang SJ, Massaro JM, O'Donnell CJ, Sahani DV, Salomaa V, Schadt EE, Schwartz SM, Siscovick DS, Voight BF, Carr JJ, Feitosa MF, Harris TB, Fox CS, Smith AV, Kao WH, Hirschhorn JN, Borecki IB (2011) Genome-wide association analysis identifies variants associated with nonalcoholic fatty liver disease that have distinct effects on metabolic traits. PLoS Genet 7:e1001324
Stahl A, Gimeno RE, Tartaglia LA, Lodish HF (2001) Fatty acid transport proteins: a current view of a growing family. Trends Endocrinol Metab 12:266–273
Sternlieb I (1992) Fraternal concordance of types of abnormal hepatocellular mitochondria in Wilson's disease. Hepatology 16:728–732
Stoyanovsky DA, Tyurina YY, Tyurin VA, Anand D, Mandavia DN, Gius D, Ivanova J, Pitt B, Billiar TR, Kagan VE (2005) Thioredoxin and lipoic acid catalyze the denitrosation of low molecular weight and protein S-nitrosothiols. J Am Chem Soc 127:15815–15823
Stremmel W, Strohmeyer G, Borchard F, Kochwa S, Berk PD (1985) Isolation and partial characterization of a fatty acid binding protein in rat liver plasma membranes. Proc Natl Acad Sci U S A 82:4–8

Stromeyer FW, Ishak KG (1980) Histology of the liver in Wilson's disease: a study of 34 cases. Am J Clin Pathol 73:12–24

Sun Q, Zhang W, Zhong W, Sun X, Zhou Z (2017) Pharmacological inhibition of NOX4 ameliorates alcohol-induced liver injury in mice through improving oxidative stress and mitochondrial function. Biochim Biophys Acta 1861:2912–2921

Susin SA, Zamzami N, Kroemer G (1998) Mitochondria as regulators of apoptosis: doubt no more. Biochim Biophys Acta 1366:151–165

Sykora P, Kharbanda KK, Crumm SE, Cahill A (2009) S-Adenosyl-L-methionine co-administration prevents the ethanol-elicited dissociation of hepatic mitochondrial ribosomes in male rats. Alcohol Clin Exp Res 33:1–9

Thayer WS, Rubin E (1980) Effects of chronic ethanol consumption on the respiratory chain of rat liver submitochondrial particles. Adv Exp Med Biol 132:385–392

Tiao M-M, Lin T-K, Liou C-W, Wang P-W, Chen J-B, Kuo F-Y, Huang C-C, Chou Y-M, Chuang J-H (2009) Early transcriptional deregulation of hepatic mitochondrial biogenesis and its consequent effects on murine cholestatic liver injury. Apoptosis 14:890–899

Tomita K, Teratani T, Suzuki T, Shimizu M, Sato H, Narimatsu K, Okada Y, Kurihara C, Irie R, Yokoyama H, Shimamura K, Usui S, Ebinuma H, Saito H, Watanabe C, Komoto S, Kawaguchi A, Nagao S, Sugiyama K, Hokari R, Kanai T, Miura S, Hibi T (2014) Free cholesterol accumulation in hepatic stellate cells: mechanism of liver fibrosis aggravation in nonalcoholic steatohepatitis in mice. Hepatology 59:154–169

Torres DM, Harrison SA (2012) Nonalcoholic steatohepatitis and noncirrhotic hepatocellular carcinoma: fertile soil. Semin Liver Dis 32:30–38

Trevisani F, Colantoni A, Caraceni P, Van Thiel DH (1996) The use of donor fatty liver for liver transplantation: a challenge or a quagmire? J Hepatol 24:114–121

Trigatti BL, Anderson RG, Gerber GE (1999) Identification of caveolin-1 as a fatty acid binding protein. Biochem Biophys Res Commun 255:34–39

Tsukamoto H, Gaal K, French SW (1990) Insights into the pathogenesis of alcoholic liver necrosis and fibrosis: status report. Hepatology 12:599–608

Tsutsumi T, Matsuda M, Aizaki H, Moriya K, Miyoshi H, Fujie H, Shintani Y, Yotsuyanagi H, Miyamura T, Suzuki T (2009) Proteomics analysis of mitochondrial proteins reveals overexpression of a mitochondrial protein chaperon, prohibitin, in cells expressing hepatitis C virus core protein. Hepatology 50:378–386

Unger RH (2003) Minireview: weapons of lean body mass destruction: the role of ectopic lipids in the metabolic syndrome. Endocrinology 144:5159–5165

Unger RH, Orci L (2001) Diseases of liporegulation: new perspective on obesity and related disorders. FASEB J 15:312–321

Upton J-P, Austgen K, Nishino M, Coakley KM, Hagen A, Han D, Papa FR, Oakes SA (2008) Caspase-2 cleavage of BID is a critical apoptotic signal downstream of endoplasmic reticulum stress. Mol Cell Biol 28:3943–3951

Vanni E, Marengo A, Mezzabotta L, Bugianesi E (2015) Systemic complications of nonalcoholic fatty liver disease: when the liver is not an innocent bystander. Semin Liver Dis 35:236–249

Vecchione G, Grasselli E, Voci A, Baldini F, Grattagliano I, Wang DQ, Portincasa P, Vergani L (2016) Silybin counteracts lipid excess and oxidative stress in cultured steatotic hepatic cells. World J Gastroenterol 22:6016–6026

Vendemiale G, Grattagliano I, Signorile A, Altomare E (1998) Ethanol-induced changes of intracellular thiol compartmentation and protein redox status in the rat liver: effect of tauroursodeoxycholate. J Hepatol 28:46–53

Vendemiale G, Grattagliano I, Lupo L, Memeo V, Altomare E (2002) Hepatic oxidative alterations in patients with extra-hepatic cholestasis. Effect of surgical drainage. J Hepatol 37:601–605

Venkatraman A, Shiva S, Wigley A, Ulasova E, Chhieng D, Bailey SM, Darley-Usmar VM (2004) The role of iNOS in alcohol-dependent hepatotoxicity and mitochondrial dysfunction in mice. Hepatology 40:565–573

Walewski JL, Ge F, Lobdell HT, Levin N, Schwartz GJ, Vasselli JR, Pomp A, Dakin G, Berk PD (2014) Spexin is a novel human peptide that reduces adipocyte uptake of long chain fatty

acids and causes weight loss in rodents with diet-induced obesity. Obesity (Silver Spring) 22:1643–1652
Wallace MC, Friedman SL, Mann DA (2015) Emerging and disease-specific mechanisms of hepatic stellate cell activation. Semin Liver Dis 35:107–118
Wehr A, Baeck C, Ulmer F, Gassler N, Hittatiya K, Luedde T, Neumann UP, Trautwein C, Tacke F (2014) Pharmacological inhibition of the chemokine CXCL16 diminishes liver macrophage infiltration and steatohepatitis in chronic hepatic injury. PLoS One 9:e112327
Wei Y, Clark SE, Thyfault JP, Uptergrove GM, Li W, Whaley-Connell AT, Ferrario CM, Sowers JR, Ibdah JA (2009) Oxidative stress-mediated mitochondrial dysfunction contributes to angiotensin II-induced nonalcoholic fatty liver disease in transgenic Ren2 rats. Am J Pathol 174:1329–1337
Wieland P, Lauterburg BH (1995) Oxidation of mitochondrial proteins and DNA following administration of ethanol. Biochem Biophys Res Commun 213:815–819
Wijarnpreecha K, Thongprayoon C, Ungprasert P (2017) Coffee consumption and risk of nonalcoholic fatty liver disease: a systematic review and meta-analysis. Eur J Gastroenterol Hepatol 29:e8–e12
Williamson CD, Colberg-Poley AM (2009) Access of viral proteins to mitochondria via mitochondria-associated membranes. Rev Med Virol 19:147–164
Wong VW, Wong GL, Choi PC, Chan AW, Li MK, Chan HY, Chim AM, Yu J, Sung JJ, Chan HL (2010) Disease progression of non-alcoholic fatty liver disease: a prospective study with paired liver biopsies at 3 years. Gut 59:969–974
Wu GD, Chen J, Hoffmann C, Bittinger K, Chen YY, Keilbaugh SA, Bewtra M, Knights D, Walters WA, Knight R, Sinha R, Gilroy E, Gupta K, Baldassano R, Nessel L, Li H, Bushman FD, Lewis JD (2011) Linking long-term dietary patterns with gut microbial enterotypes. Science 334:105–108
Yamaguchi T, Shimahara Y, Takada Y, Ino K, Mori K, Kobayashi N, Yamaoka Y, Ozawa K (1992) Evaluation of ketogenesis in seriously reduced hepatic mitochondrial redox state an analysis of survivors and non-survivors in critically ill hepatectomized patients. Scand J Gastroenterol 27:472–478
Yan M, Zhu P, Liu H-M, Zhang H-T, Liu L (2007) Ethanol induced mitochondria injury and permeability transition pore opening: role of mitochondria in alcoholic liver disease. World J Gastroenterol 13:2352
Yang S, Zhu H, Li Y, Lin H, Gabrielson K, Trush MA, Diehl AM (2000) Mitochondrial adaptations to obesity-related oxidant stress. Arch Biochem Biophys 378:259–268
Yerushalmi B, Dahl R, Devereaux MW, Gumpricht E, Sokol RJ (2001) Bile acid-induced rat hepatocyte apoptosis is inhibited by antioxidants and blockers of the mitochondrial permeability transition. Hepatology 33:616–626
Young TA, Bailey SM, Van Horn CG, Cunningham CC (2006) Chronic ethanol consumption decreases mitochondrial and glycolytic production of ATP in liver. Alcohol Alcohol 41:254–260
Zhang Y, Yang M, Park JH, Singelyn J, Ma H, Sailor MJ, Ruoslahti E, Ozkan M, Ozkan C (2009) A surface-charge study on cellular-uptake behavior of F3-peptide-conjugated iron oxide nanoparticles. Small 5:1990–1996
Zhang L, Deng S, Zhao S, Ai Y, Zhang L, Pan P, Su X, Tan H, Wu D (2016) Intra-peritoneal administration of mitochondrial DNA provokes acute lung injury and systemic inflammation via Toll-like receptor 9. Int J Mol Sci 17:E1425
Zhou J, Febbraio M, Wada T, Zhai Y, Kuruba R, He J, Lee JH, Khadem S, Ren S, Li S, Silverstein RL, Xie W (2008) Hepatic fatty acid transporter Cd36 is a common target of LXR, PXR, and PPARgamma in promoting steatosis. Gastroenterology 134:556–567
Zhu N, Khoshnan A, Schneider R, Matsumoto M, Dennert G, Ware C, Lai MM (1998) Hepatitis C virus core protein binds to the cytoplasmic domain of tumor necrosis factor (TNF) receptor 1 and enhances TNF-induced apoptosis. J Virol 72:3691–3697
Zischka H, Lichtmannegger J (2014) Pathological mitochondrial copper overload in livers of Wilson's disease patients and related animal models. Ann N Y Acad Sci 1315:6–15

Mitochondria in Liver Regeneration: Energy Metabolism and Posthepatectomy Liver Dysfunction

Henrique Alexandrino, Anabela Rolo, José Guilherme Tralhão, Francisco Castro e Sousa, and Carlos Palmeira

Abstract Liver resection is the only curative therapy for most patients with hepatobiliary malignancies. Resection of a substantial amount of liver mass is only possible due to the liver's remarkable capacity to regenerate. However, when this process is hampered Posthepatectomy Liver Failure (PHLF) ensues, resulting in increased postoperative mortality and morbidity. Being a highly energy dependent chain of events, liver regeneration is influenced by the energy status of the main parenchymal cell—the hepatocyte. Mitochondria are the powerhouses of eukaryote cells and key players in cell death. As such, they play a major role in the organ's response to major resection. Although this subject has been the focus of investigation in the past, recent findings have led to a renewal of interest on disordered bioenergetics in liver surgery. In this Chapter we will scrutinize the experimental and clinical evidence supporting a major role for mitochondria in the liver's response to resection, as well as the relevance of mitochondrial derangement in the pathophysiology of PHLF. Furthermore, as a significant proportion of patients undergoing hepatectomy have chronic liver diseases, namely cirrhosis, biliary obstruction, steatosis or sinusoidal obstruction syndrome, which are at increased risk of PHLF, we will recapitulate the disordered bioenergetics in the pathophysiology of these conditions. Finally, we will elaborate on the evidence for a definitive role of boosting energetic status of the liver parenchyma in improving the clinical results of hepatectomy.

Keywords Mitochondria · Bioenergetics · Oxidative phosphorylation · Hepatectomy · Liver regeneration

---

H. Alexandrino · J. G. Tralhão · F. Castro e Sousa
Serviço de Cirurgia A dos Hospitais da Universidade de Coimbra, Centro Hospitalar e Universitário de Coimbra, Coimbra, Portugal

Clínica Universitária de Cirurgia III, Faculdade de Medicina, Universidade de Coimbra, Coimbra, Portugal

A. Rolo · C. Palmeira (✉)
Departamento de Ciências da Vida, Faculdade de Ciências e Tecnologia, Universidade de Coimbra, Coimbra, Portugal

Centro de Neurociências e Biologia Celular, Universidade de Coimbra, Coimbra, Portugal
e-mail: palmeira@ci.uc.pt

P. J. Oliveira (ed.), *Mitochondrial Biology and Experimental Therapeutics*,
https://doi.org/10.1007/978-3-319-73344-9_8

1 Introduction

Liver resection is the only curative therapy for most patients with hepatobiliary malignancies (Agrawal and Belghiti 2011). Resection of a substantial amount of liver mass is only possible due to the liver's remarkable capacity to regenerate (Clavien et al. 2007). However, when this process is hampered Posthepatectomy Liver Failure (PHLF) ensues, resulting in increased postoperative mortality and morbidity (Golse et al. 2013).

Being a highly energy dependent chain of events, liver regeneration is influenced by the energy status of the main parenchymal cell—the hepatocyte. Mitochondria are the powerhouses of eukaryote cells and key players in cell death. As such, they play a major role in the organ's response to major resection. Although this subject has been the focus of investigation in the past, recent findings have led to a renewal of interest on disordered bioenergetics in liver surgery.

In this Chapter we will scrutinize the experimental and clinical evidence supporting a major role for mitochondria in the liver's response to resection, as well as the relevance of mitochondrial derangement in the pathophysiology of PHLF. Furthermore, as a significant proportion of patients undergoing hepatectomy have chronic liver diseases, namely cirrhosis, biliary obstruction, steatosis or sinusoidal obstruction syndrome, which are at increased risk of PHLF, we will recapitulate the disordered bioenergetics in the pathophysiology of these conditions. Finally, we will elaborate on the evidence for a definitive role of boosting energetic status of the liver parenchyma in improving the clinical results of hepatectomy.

2 Liver Regeneration: A Highly Energetic Cellular Process

Liver regeneration is a highly energy dependent process. After hepatectomy liver cells undergo cell cycle progression, DNA replication and protein synthesis, processes that require a large amount of energy (Minuk 2003), mostly derived from β-oxidation of fatty acids (Nakatani et al. 1982). Although the increase in adenosine triphosphate (ATP) synthesis precedes the peak in deoxyribonucleic acid (DNA) synthesis (Yamaoka et al. 1974; Ozawa et al. 1982), the net result is a significant drop in liver ATP, more pronounced the more extended the resection of parenchyma, as more energy is required for the diverse anabolic reactions of cell growth and division. In fact, as early as 30 s after hepatectomy there is a 50% decrease in liver ATP stores in rodents (Crumm et al. 2008). The decrease in ATP reaches a nadir at 48 h but is expected to fully recover by the fifth day. Likewise, there is a decrease in ATP to phosphate (ATP/Pi) ratio, that significantly correlates with postoperative liver function and markers of cellular proliferation (Corbin et al. 2002). Although the evidence for this is mostly experimental, Mann et al. (2002a), using ^{31}Phosphorus Magnetic Resonance Spectroscopy, demonstrated that hepatectomy in human subjects was also associated with an early fall in ATP/Pi ratio.

In order to sustain the increased requirements of ATP to fuel liver regeneration, hepatocytes undergo a series of metabolic adaptations. In the first hours after hepatectomy there is an increase in mitochondrial DNA and RNA in the remnant liver (Koyama et al. 1998), augmented expression of several enzymes involved in electron transport and β-oxidation of fatty acids (FA's), such as cytochrome c oxidase and carnitine *O*-palmitoyltransferase (Sun et al. 2007). Ultimately this leads to an overall overexpression of the energy-producing enzymatic machinery in the first 2–4 days after hepatectomy (Nagino et al. 1989). In fact, of the 87 different proteins expressed during liver regeneration in a rodent model of 50% hepatectomy, 25 are mitochondrial proteins, especially involved in carbohydrate and lipid metabolism (Cao et al. 2009). The importance of mitochondrial biogenesis in liver regeneration is demonstrated by the role played by mitochondrial topoisomerase 1, a key enzyme for the replication of the circular 13-gene mitochondrial DNA. In a model of toxic-induced liver injury, knock-out mice for this enzyme presented with decreased mitochondrial DNA, lower activities of ETC complexes I and IV and impaired hepatocyte replication (Khiati et al. 2015).

These adaptations are paramount for the increased metabolic demand of liver regeneration, as the recovery in energy status is preceded by an enhanced liver oxygen consumption, which in turn is followed by the peak in DNA replication (Yoshioka et al. 1998). As a proof of concept of the importance of hepatic energy status in liver regeneration, Satoh et al., in an elegant experiment, used knock-in mice expressing creatine kinase (CK) in liver cells. CK is normally expressed in skeletal and cardiac muscle, and in brain tissue, but not in the liver. It constitutes an alternate source for adenosine triphosphate (ATP) production by the transfer of a high-energy phosphate from creatine phosphate to ADP. In this experiment, mice with liver expression of CK were fed either a high-creatine diet or a control diet. After 70% hepatectomy, creatine-fed animals had higher hepatic ATP synthesis and displayed increased bromodeoxyuridine incorporation and liver weight gain, versus CK-positive controls with normal diet (Satoh et al. 1996).

Since the availability of ATP is ultimately dependent upon the proper function of mitochondria, it is expected that enhanced mitochondrial function would improve liver regeneration. In fact, this is well illustrated by an experimental work that used N-methyl-4-isoleucine cyclosporine (NIM811), an inhibitor of the mitochondrial membrane permeability transition (MPT). In the setting of extended hepatectomy NIM811 inhibited the MPT, preserved liver energy status and hepatocellular function and improved survival (Rehman et al. 2012).

Having summarized the biological link of energy availability to cell proliferation in liver regeneration, we will now discuss the evidence for a definite role of bioenergetics dysfunction in the pathophysiology of PHLF.

3 Posthepatectomy Liver Failure: Evidence for Disturbed Bioenergetics in Pathophysiology

Clinical success of hepatectomy depends on the liver's unique ability to regenerate. After major hepatectomy, the remnant liver must replace lost hepatocyte mass, produce an acute-phase response and still carry the burden of maintaining acceptable hepatocellular function for whole body homeostasis. After the surgical loss of liver tissue, regeneration does not rely on the proliferation of a progenitor cell population, but on the replication of normally quiescent hepatocytes, cells that already have high metabolic demands (Fausto et al. 2006). The enormous amount of energy for these processes is supplied by the oxidative phosphorylation of fatty acids by mitochondria under aerobic conditions (Fig. 1) (Nakatani et al. 1982; Mann et al. 2002a, b). However, if the energy demand exceeds the supply, liver regeneration can be severely hampered resulting in PHLF and even death.

Disturbed bioenergetics is an important factor in several acute and chronic liver diseases (McGill et al. 2012; Auger et al. 2015). Mitochondrial dysfunction is characterized by decreased cellular ATP stores, compromised cellular processes and viability. Also, uncoupling of oxidative phosphorylation leads to increased production of reactive oxygen species (ROS), which can severely damage both mitochondrial DNA and inner membrane lipids, leading to decreased mitochondrial

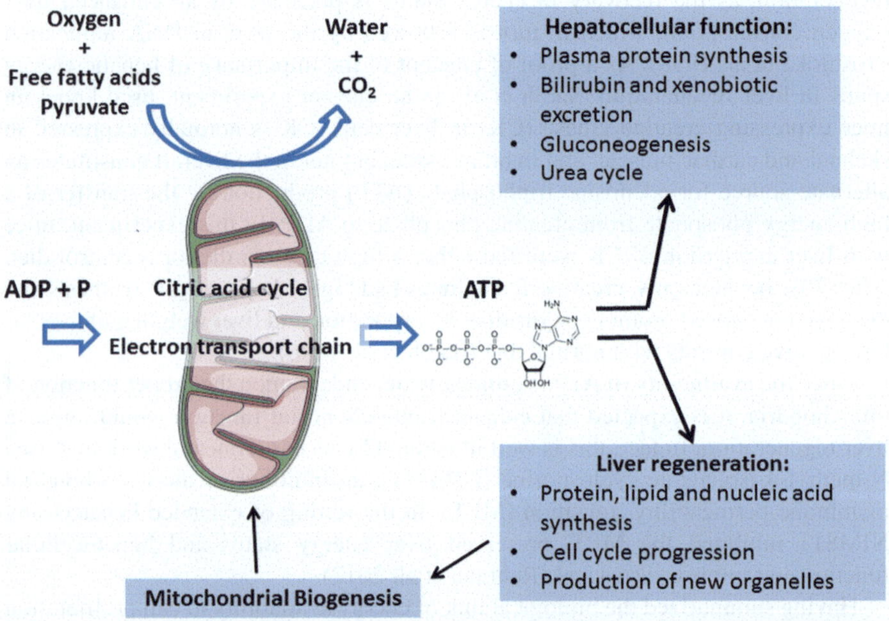

Fig. 1 Metabolic demands on mitochondria after hepatectomy. After hepatectomy mitochondria, through two oxygen-dependent enzymatic chains, the tricarboxylic acid cycle and the electron transport chain, supply hepatocytes with increasing amounts of adenosine triphosphate (ATP) needed to fuel biosynthesis of cell components and progression through cell cycle. In the meantime, hepatocellular function is dependent upon a constant ATP supply. In ideal conditions, hepatocyte function will not be severely hampered and should return to normal as soon as possible. In the clinical setting good posthepatectomy outcome is confirmed by a return to normal values of arterial lactate (reflecting gluconeogenesis), prothrombin time (reflecting adequate protein synthesis) and serum bilirubin (reflecting conjugation and excretion of bilirubin, as well as other endo- and xenobiotics). All these liver functions are endergonic, i.e. energy-dependent. Interestingly, mitochondrial biogenesis is, in itself, a needed step for adequate hepatocyte replication

biogenesis and further uncoupling, respectively. Furthermore, mitochondrial membrane permeabilization releases cytochrome c in the cytoplasm and activates the caspase-mediated pathway of apoptosis, further compromising liver function (Fig. 2). Finally, the delicate equilibrium of the cellular mitochondrial pool is maintained by the balance of formation of new and degradation of damaged mitochondria, by biogenesis and mitophagy, respectively, and if disturbed can further compromise the energetic status of the cell.

Clinical evidence for a role of bioenergetics derangement in posthepatectomy liver dysfunction is mostly indirect. The Arterial Ketone Body Ratio (AKBR), acetoacetate to 3-hydroxybutyrate ratio, is a marker of mitochondrial redox state. When there is impairment of the mitochondrial respiratory chain, NADH cannot be recycled to its reduced form (NAD+). In these conditions, higher quantities of 3-hydroxybutyrate are formed, decreasing the AKBR. Ukikusa et al. (1981) studied the AKBR in a rabbit model of 70% hepatectomy and concluded that the early drop

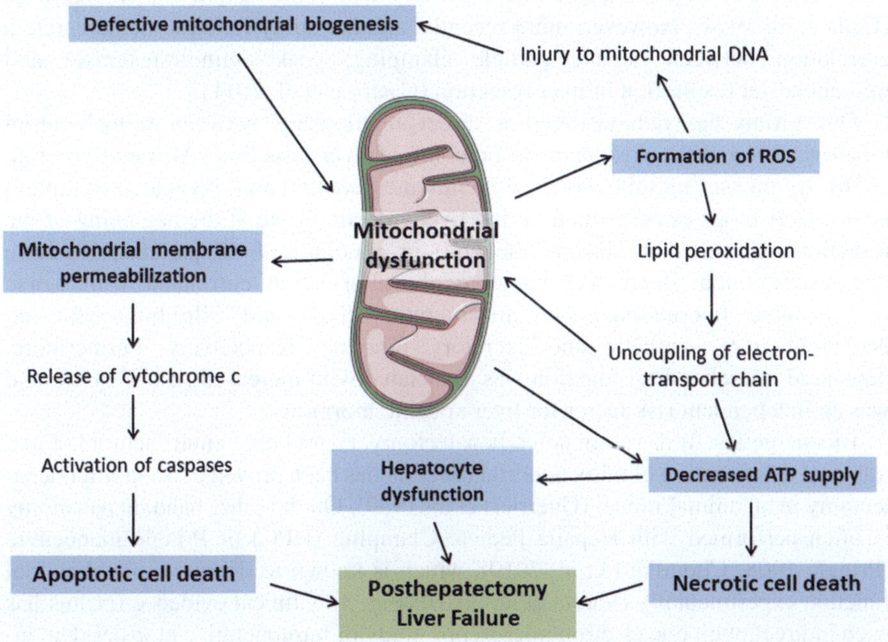

Fig. 2 A look into the mechanisms of mitochondrial dysfunction in Posthepatectomy Liver Failure. Uncoupling of oxidative phosphorylation causes increased production of reactive oxygen species (ROS) and decreased ATP production. ROS cause direct damage to mitochondrial DNA, further compromising mitochondrial biogenesis, while peroxidation of mitochondrial membrane phospholipids decreases energy efficiency. Ultimately, decreased ATP supply will hamper hepatocyte function and, if severe, lead to necrotic cell death through the loss of osmotic gradient because of failure of ATP-dependent ionic pumps. Finally, mitochondrial membrane permeabilization releases cytochrome c in the cytoplasm and activates the caspase-mediated pathway of apoptosis. The net result is a decrease in functioning hepatocyte mass and a derangement in hepatocellular function, clinically recognized as Posthepatectomy Liver Failure

in AKBR was associated with a decrease in energy charge of the remnant liver. These findings were corroborated in the clinical setting, with an AKBR under 0.4 after hepatectomy associated with decreased survival (Yamaguchi et al. 1992).

As the energy requirements for liver regeneration and function depend upon an efficient oxidative phosphorylation, the hepatic venous oxygen saturation could reflect the oxygen supply and demand of the liver parenchyma, and thus its energy charge (Yoshioka et al. 1998). In the clinical setting a sustained decrease in intraoperative hepatic venous haemoglobin oxygen saturation is significantly correlated with peak postoperative aminotransferases, risk of PHLF and death (Kainuma et al. 1992).

Correlation of mitochondrial derangement with clinical variables has been previously described. Patients with more pronounced changes in mitochondrial cytochrome activity in liver biopsies were more likely to present with previous liver dysfunction, and also more prone to postoperative morbidity and mortality (Ozawa et al. 1973). Later, the same author, with others, correlated the AKBR and postoperative complications after hepatectomy, with redox activity in liver samples (Ueda et al. 1994). However, more recently, another study failed to demonstrate a correlation between hepatic pedicle clamping, peak aminotransferases and mitochondrial respiration in liver resection (Castro e et al. 2011).

Our group has demonstrated a direct relationship between mitochondrial bioenergetics and the postoperative outcome of liver resection (Alexandrino et al. 2016). By measuring mitochondrial membrane potential and oxygen consumption in two liver biopsies performed during liver resection (one at the beginning of the resection and the other just at the end) in a cohort of 30 patients, we have demonstrated that depressed oxidative phosphorylation correlated with worse postoperative international normalized ratio (INR) and bilirubin, reflecting decreased liver synthetic and excretory function, respectively. Furthermore, depressed mitochondrial function was associated with increased risk of PHLF and was an independent risk factor for liver-specific morbidity.

Bioenergetics dysfunction after hepatectomy is probably multifactorial. First, transient deterioration of mitochondrial function has been proved to occur after hepatectomy in an animal model (Guerrieri et al. 1999). On the other hand, hepatectomy is often performed with Hepatic Pedicle Clamping (HPC) or Pringle manoeuvre (Pringle 1908; Chouillard et al. 2010), which is known to decrease mitochondrial function experimentally (Varela et al. 2010) (Fig. 3). Clinical evidence for this has been indirect, with one electron microscopy study of intraoperative biopsies demonstrating that hepatectomy with HPC can cause mitochondrial swelling (Wilasrusmee et al. 2004). We sought to investigate on this matter and have found a significant correlation between longer HPC time and worse mitochondrial depolarization and lag phase (Alexandrino et al. 2016).

Moreover, other mechanisms could be at play, including the hemodynamic changes after major hepatectomy. Since portal blood flow is dependent upon the splanchnic bed, after extended hepatectomy the reduced liver mass is exposed to an increased portal vein pressure (PVP). This causes increase in shear stress and induces nitric oxide production, which is recognized as an important stimulus to

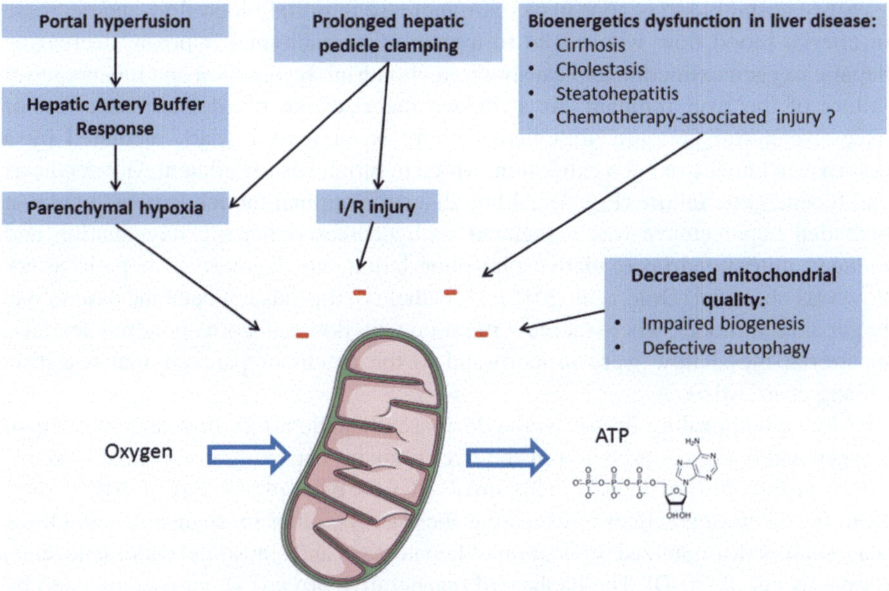

Fig. 3 Pathophysiology of bioenergetics dysfunction in clinical liver surgery. Energetic efficiency can be compromised after hepatectomy and several different mechanisms are at play. After extended hepatectomy, a state of portal hyperperfusion of the liver remnant ensues. This is known as the "small-for-size" or "small-for-flow" syndrome. Portal hyperperfusion causes a reduction in hepatic arterial blood flow and oxygenation of the liver parenchyma, in a process mediated by the Hepatic Artery Buffer Response (HABR). Hepatic pedicle clamping, if used and prolonged in time, also contributes to parenchymal hypoxia and causes ischemia-reperfusion injury (IRI), which is known to cause mitochondrial dysfunction. Also, there is evidence for disturbed bioenergetics in several liver diseases, namely cirrhosis, biliary obstruction and steatohepatitis, making hepatectomy more risky in these settings. Finally, defective mitochondrial biogenesis and autophagy can decrease quality of liver mitochondria, decreasing energy yield and increasing production of reactive oxygen species (see text for details)

liver regeneration (Schoen et al. 2001). Paradoxically, excessive portal pressure can also be deleterious and contribute to PHLF, both in non-cirrhotic and cirrhotic patients, especially when PVP exceeds 20 mmHg (Allard et al. 2013). The clinical presentation of disturbed synthetic function, ascites and increased risk of sepsis recapitulates the "small-for-size" syndrome (SFSS) of the transplant setting, occurring with the transplantation of a reduced size liver graft (Demetris et al. 2006).

However, shear stress is not the only mechanism thought to be involved. There is a physiologic mechanisms regulating total liver flow, the Hepatic Arterial Buffer Response (HABR), consisting of a reciprocal regulation of the hepatic artery flow by the portal venous inflow. When portal blood flow decreases, arterial dilation of the intrahepatic arterial bed compensates, maintaining constant hepatic blood flow. Conversely, portal hyperperfusion leads to a reduction in hepatic arterial blood flow. The HABR is probably mediated by the washout of adenosine in the space of Mall by the increased portal flow, causing arterial vasoconstriction (Lautt 2007).

Since portal blood is poor in oxygen, increase in portal blood flow and decrease in arterial blood flow would lead to a state of parenchymal hypoxia, decreasing hepatic oxygen extraction and causing mitochondrial dysfunction and bioenergetics failure of the liver remnant. As such, we theorize that the deleterious effect of excessive portal pressure after hepatectomy is, at least in part, mediated by a decrease in hepatic oxygen extraction, which in turn leads to deficient ATP synthesis and bioenergetic failure (Fig. 3). Although a small animal model demonstrated that extended hepatectomy was associated with decreased hepatic oxygenation and reduced mitochondrial oxidative phosphorylation, no decrease in hepatic artery flow was observed (Dold et al. 2015). Nonetheless, this has not been the case in one larger animal model, where increase in portal vein flow and corresponding decrease in arterial blood flow were proportional to the extent of parenchymal resection (Xiang et al. 2016).

Although appealing in theory, the decreased hepatic artery flow and subsequent oxygen and energy deprivation of the liver remnant occurring in the small-for-size setting is probably not the only factor involved in the pathophysiology of PHLF. Apart from the deleterious effect of excessive shear stress, other mechanisms could be at play, such as disorganized replication of hepatocytes and sinusoidal endothelial cells (Ninomiya et al. 2010). The haphazard regenerative process is not accompanied by an efficient vascular network, potentially leading, early in the process, to an impaired oxygen supply to the rapidly dividing and metabolically active liver cells.

Bioenergetics dysfunction could potentially be even more relevant in patients with pre-existing mitochondrial impairment, such as chronic liver disease. In the following section we will briefly review the experimental and clinical evidence for impaired bioenergetics in the pathophysiology of PHLF in the setting of abnormal liver parenchyma.

4 Hepatectomy in Chronic Liver Injury: Mitochondrial Dysfunction as a Key Factor in the Pathophysiology of Posthepatectomy Liver Failure

Mitochondrial dysfunction has been recently linked to the pathogenesis of many acute and chronic liver diseases (Auger et al. 2015). And while liver resection is ideally performed in patients with intact liver function, some candidates for hepatectomy present with chronic liver injury, such as steatosis, cirrhosis, chemotherapy-induced liver injury and biliary obstruction. In these conditions there is some degree of hepatic mitochondrial dysfunction, possibly aggravated in the postresection period (Fig. 3). Thus it is no surprise that hepatectomy is fraught with an increased incidence of postoperative liver failure, mandating a correct preoperative stratification of surgical risk. In this section we will elaborate on the evidence for disturbed mitochondrial bioenergetics in the pathophysiology of PHLF in the setting of diseased liver parenchyma.

4.1 Steatosis and Steatohepatitis

Steatosis and steatohepatitis are increasingly recognized as a significant public health problems, associated with the obesity and diabetes epidemics (Ratziu et al. 2010). There are two distinct forms of non-alcoholic fatty liver disease (NAFLD): a more benign form, simple steatosis; and a more progressive, severe form—non-alcoholic steatohepatitis (NASH) often leading to cirrhosis, liver failure and hepatocellular carcinoma. In fact, in developed countries NASH has supplanted other liver diseases as the main risk factor for hepatocellular carcinoma (Mittal et al. 2015).

Mitochondrial dysfunction is a hallmark of NAFLD and this link has been extensively reviewed elsewhere (Grattagliano et al. 2012; Rolo et al. 2012).

Steatosis *per se* has been associated with impaired liver regeneration in experimental models. In an animal model of choline/methionine deficient diet, Veteläinen et al. describe impaired liver regeneration after 70% hepatectomy, both in mild and in severe steatosis (Veteläinen et al. 2007a, b). However, these results were not replicated in western-diet models of steatosis (Sydor et al. 2012; Garnol et al. 2016). Nonetheless, since liver resection is often performed with inflow occlusion, fatty livers are more susceptible to IRI due to microcirculatory disturbances, Kupffer cell dysfunction, increased leukocyte adhesiveness, ATP depletion and mitochondrial failure (Varela et al. 2011). Clinically, steatosis is a known risk factor for worse outcomes after hepatectomy (de Meijer et al. 2010). This, together with the role of mitochondrial dysfunction in fatty liver disease mandates a quest for understanding the mechanisms of bioenergetics dysfunction in steatosis, thus enhancing the safety of liver resection.

4.2 Cirrhosis

Cirrhosis is the end stage of infectious, immune, toxic or metabolic insults to the liver. It is characterized by decreased hepatocellular function and portal hypertension (Schuppan and Afdhal 2008). Hepatocellular carcinoma is a frequent complication and resection is one of the few curative treatments (Torzilli et al. 2013). However, liver resection in cirrhosis is only possible in patients with preserved liver function and even then at the cost of a high risk of PHLF (Cescon et al. 2009).

The decreased regenerative response of cirrhotic livers has been previously noticed (Hashimoto and Watanabe 2005). Among the contributing mechanisms, bioenergetics failure definitely plays a key role. Yang et al. (2004) used a rodent model of chemically-induced cirrhosis to study the mitochondrial respiratory function and antioxidant capacity after hepatectomy. After inducing cirrhosis with intra-peritoneal thioacetamide in Wistar rats, the authors performed 70% partial hepatectomy. After hepatectomy cirrhotic animals displayed decreased state 3 respiration, decreased activities of NADH-cytochrome c reductase and mitochondrial glutathione peroxidase, as well as decreased levels of mitochondrial glutathione, when compared with non-cirrhotic controls. Nishikawa et al. (2014), using a rat model of carbon tetrachloride and phenobarbital-induced cirrhosis, examined energy metabolism in isolated hepatocytes and concluded that in compensated cirrhosis maximal mitochondrial respiration is compromised and that ATP production was maintained by an increase in the glycolytic pathway. As cirrhosis progresses to decompensated form, failure of glycolytic pathway also occurs, leading to energetic failure. The authors confirmed these findings in gene-expression profile of human liver biopsy samples, as patients with compensated cirrhosis (Child A and B) had decreased expression of cytochrome oxidase 1 and 2 genes, with increased expression of genes for glycolytic enzymes. These enzymes were under-expressed in Child C cirrhosis, underlying the energetic failure that occurs in this premorbid state of decompensated liver disease. An interesting conclusion of this study is that mitochondrial function is hampered in early stages of cirrhosis, with glycolysis taking over most of the energy production. As the glycolytic pathway is much less energy-efficient than oxidative phosphorylation, this could explain the decreased adaptation of compensated cirrhotic livers to resection and other acute stressors, such as infection (Arvaniti et al. 2010); as failure to increase ATP production would result in acute decompensation of liver function.

Deficient energetic recovery of the cirrhotic liver after hepatectomy has been demonstrated by Mann et al. (2001). Using ^{31}Phosphorus Magnetic Resonance Spectroscopy, the authors studied nine cirrhotic patients undergoing hepatectomy and compared with nine other patients with normal liver parenchyma. Cirrhotic patients experienced a sustained decrease in ATP to phosphate ratio and delayed regeneration versus patients with normal liver parenchyma.

Given the evidence for disordered energetics in cirrhosis, interventions aimed at improving energy status could be clinically relevant in prevention of hepatocellular dysfunction after liver resection. This will be further explored in Sect. 5.

4.3 Chemotherapy-Induced Liver Injury

While hepatectomy is the gold standard of care for patients with colorectal cancer liver metastases, preoperative chemotherapy is increasingly used (Folprecht et al. 2005; Nordlinger et al. 2008). The most commonly used drugs are 5-fluorouracil (5-FU), irinotecan and oxaliplatin. The cytotoxic agent 5-FU is a thymidylate synthase inhibitor and has been used for over 40 years in systemic therapy for advanced colorectal cancer. Irinotecan, or CPT-11, is a prodrug. It suffers transformation by carboxylesterase into SN-38, a potent inhibitor of topoisomerase-I. Oxaliplatin, a platinum derivative, forms DNA adducts and is directly cytotoxic (Lentz et al. 2005).

However, chemotherapy is associated with hepatocellular injury and increased morbidity and even mortality after hepatectomy (Karoui et al. 2006; Vauthey et al. 2006). Although chemotherapy *per se* produces a decline in hepatocellular function (as measured by indocyanine green retention rate), independently of histologically-perceptible injury (Takamoto et al. 2010), chemotherapy-associated liver injury

(CALI) usually occurs in distinct histologic patterns: steatosis and steatohepatitis (CASH); and sinusoidal obstruction syndrome (SOS).

Steatosis and steatohepatitis (SH) are usually associated with irinotecan-based chemotherapy and one study demonstrated an increase in morbidity and even mortality after liver resection (Vauthey et al. 2006). Although the precise mechanism of hepatoxicity of irinotecan is unknown, it is thought to involve mitochondrial dysfunction (Labbe et al. 2008). Vauthey et al. (2006) reported a higher incidence of SH in patients treated with irinotecan-based chemotherapy, in particular with Body Mass Index over 25 kg/m^2. Patients with SH were more likely to suffer from posthepatectomy liver failure and had higher postoperative mortality. As the typical histologic injury of irinotecan-associated liver injury is steatohepatitis, we refer to the previous section (Sect. 4.1).

Sinusoidal obstruction syndrome (SOS) is another typical histologic pattern, and is usually associated with oxaliplatin chemotherapy (Rubbia-Brandt et al. 2010). It is characterized by severe sinusoidal dilation, sinusoidal endothelial cell necrosis and sloughing, causing extravasation of blood into the space of Disse and decreased sinusoidal flow (Fan and Crawford 2014). SOS is an independent risk factor for postoperative morbidity and usually courses with increased portal pressure, increased intraoperative bleeding, decreased tolerance to hepatic pedicle clamping and lower regenerative capacity (Aloia et al. 2006; Nakano et al. 2008; Narita et al. 2012a, b).

Apart from DNA-directed effects, there is evidence to support that oxaliplatin is directly toxic to mitochondria, inhibiting oxidative phosphorylation in isolated liver mitochondria (Rosen et al. 1992; Tabassum et al. 2015) and inducing mitochondrial-dependent apoptosis in cell cultures (Gourdier et al. 2004).

However, the lack of a valid animal model of oxaliplatin-induced toxicity has been a major drawback. Up until recently the only existing model of posthepatectomy liver dysfunction in toxic-induced SOS was the monocrotaline model (Schiffer et al. 2009). The work of Robinson et al. (2013) has shed new light into the pathophysiology of SOS. The authors validated an animal model of oxaliplatin-induced SOS in C57Bl/6 mice and found increase in oxidative stress in the injured parenchyma.

Although there is still no definitive link between mitochondrial toxicity and the development of SOS, we postulate that mitochondrial homeostasis could be severely hampered in chemotherapy-associated liver injury, thus leading to an increased susceptibility to postoperative liver dysfunction. Further investigation into this field is of paramount importance.

4.4 Chronic Biliary Obstruction

Extended hepatectomy with *en bloc* bile duct resection is potentially curative for hilar cholangiocarcinoma (Nagino et al. 2013). However, liver resection under biliary obstruction is fraught with an extremely high incidence of postoperative liver dysfunction, with significant mortality. Thus, preoperative biliary decompression is recommended to decrease the risk of PHLF and improve results in patients with high malignant biliary obstruction (Iacono et al. 2013). Chronic cholestasis induces liver dysfunction through several mechanisms, including mitochondrial toxicity (Palmeira and Rolo 2004). Conversely, biliary decompression results in an improvement in energy status of the liver. This is supported by clinical and experimental evidence which we will now summarily review.

Bile acids are directly toxic to mitochondria, causing a significant reduction of membrane potential, state 3 respiration and Respiratory Control Ratio, as well as increased susceptibility to Mitochondrial Permeability Transition (MPT) (Rolo et al. 2000; Schulz et al. 2013). Mitochondrial biogenesis is also reduced by chronic biliary obstruction. In a rat model of bile duct ligation (BDL) there was a decrease in respiratory control ratio (RCR) and membrane potential, decreased expression of mitochondrial transcription factor A (TFAM) and depletion of mitochondrial DNA (Arduini et al. 2011). Surprisingly, in another animal model of segmental biliary obstruction, mitochondrial dysfunction was demonstrated not only in the obstructed lobe but also in the non-obstructed one (Kanai et al. 1992).

Partial reversal of physiologic derangements by biliary decompression was demonstrated in an experimental model. Krähenbühl et al. (1998) explored mitochondrial homeostasis in a rat model of BDL followed by Roux-en-Y anastomosis 4 weeks later. The authors demonstrated that oxidative phosphorylation and β-oxidation of fatty acids were inhibited during cholestasis. After relief of obstruction there was an improvement in activities of complexes I and III of the mitochondrial electron transport chain, whereas the activities of complexes II and IV remained impaired. Furthermore, β-oxidation of fatty acids also remained depressed. In the clinical setting, ^{31}Phosphorus magnetic resonance spectroscopy was used to study liver energy status in patients with malignant biliary obstruction (Mann et al. 2002a, b) and after endoscopic or percutaneous biliary drainage, the ATP/Pi ratio, a measure of energy status, significantly increased in 1 week time, reflecting the beneficial effect on the entire liver parenchyma.

In conclusion, part of the biologic rationale for preoperative biliary decompression in the management of jaundiced patients undergoing major hepatectomy and bile duct resection can be related to improvement in mitochondrial function. If proved, further therapies aimed at mitochondrial bioenergetics could lead to an enhanced energetic status of the remnant liver, thereby decreasing the inherently high risk of postoperative liver dysfunction.

5 Improving Liver Regeneration: Hepatocyte Energetics as a Potential Target for Prevention of Postoperative Liver Failure

While in the previous sections we recapitulated the evidence for an important role of mitochondrial function in liver recovery following resection, much more appealing is the possibility of actually preventing PHLF by targeting mitochondrial homeostasis. Plausibly, energetic conditioning of liver cells could aid in the

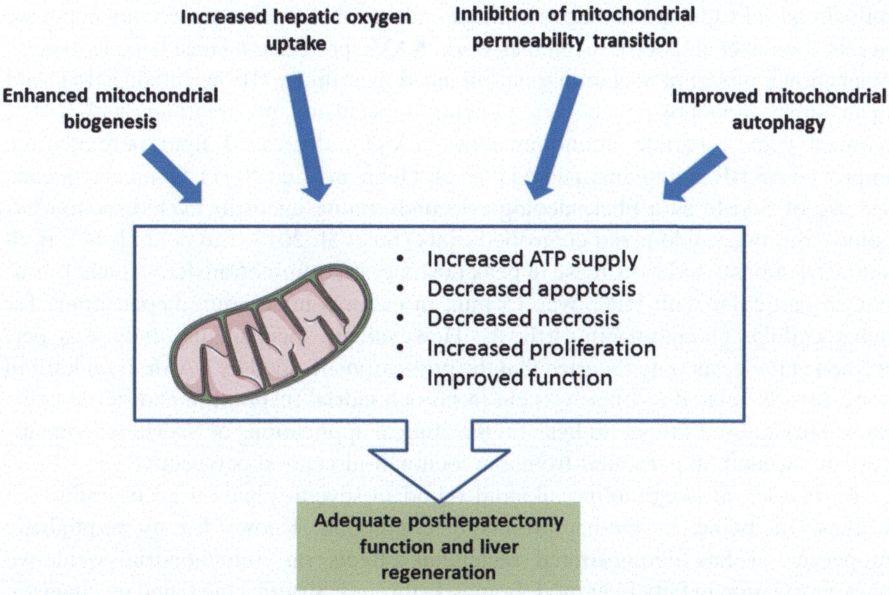

Fig. 4 Mitochondrial-centered perspective on posthepatectomy outcome. Adequate recovery of liver parenchyma after resection depends upon a proper cellular energy homeostasis. Prevention of Posthepatectomy Liver Failure can be achieved by increasing hepatic oxygen uptake, such as intraoperative hemodynamic modulation, prevention of postoperative portal hyperperfusion or hyperbaric oxygen therapy. Enhanced mitochondrial biogenesis is currently the focus of research in pharmacological preconditioning. Improved mitochondrial autophagy and inhibition of the mitochondrial permeability transition are novel and yet clinically unexplored pathways that could potentially improve energy homeostasis in regenerating hepatocytes, enhancing ATP supply and decreasing production of ROS (see text for further details)

maintenance of adequate liver function and in the recovery of functioning liver mass after major hepatectomy. Several possible pathways could be explored, either isolated or in combination, include: improvement in mitochondrial pool by enhanced biogenesis and more efficient mitophagy; more efficient oxidative phosphorylation through increased parenchymal oxygenation; and decrease in mitochondrial permeability transition (Fig. 4).

Because of the considerable overlapping of these pathways, we will now dwell on several therapeutic strategies, presenting them according to different methods: (1) Pharmacological conditioning; (2) Strategies aiming at improving hepatocyte oxygenation; and (3) Stem cell therapies.

5.1 *Pharmacological Therapy*

Mitochondrial-based therapies have been the focus of recent research. S-adenosyl-L-methionine (SAMe) is a methyl donor and a precursor of glutathione, improving mitochondrial respiration and oxidative phosphorylation, while decreasing mitochondrial ROS and damage to mitochondrial DNA in chronic ethanol exposure in rats (Bailey et al. 2006). In another work, SAMe protected against hepatic toxicity in an animal model of acetaminophen-induced liver injury (Brown et al. 2014). And in an animal model of 70% hepatic ischemia-reperfusion, pre-treatment with SAMe reduced peak aspartate aminotransferase (AST), decreased lipid peroxidation, improved AKBR and maintained ATP levels (Jeon and Lee 2001). Clinical evidence for use of SAMe as a pharmacological conditioning agent in liver resection has come from two randomized controlled trials (Su et al. 2013; Liu et al. 2014). Both studies demonstrated a decrease in peak postoperative aminotransferases and bilirubin, in particular with HPC over 15 min, in patients undergoing hepatectomy for hepatocellular carcinoma in cirrhosis. However, no mechanistic study was performed and we can only theorize that the protection afforded by SAMe is, at least in part, directly related to improvement in mitochondrial respiration and in oxidative phosphorylation. Further studies into the clinical applications of SAMe in liver surgery are needed, in particular from a mitochondrial-centred perspective.

Berberine, an isoquinoline alkaloid found in several plants used in traditional Chinese medicine, is another promising compound. Known for its antidiabetic properties, it has demonstrated beneficial effects on mitochondrial oxidative phosphorylation in fatty liver models, mostly through Sirtuin 3 mediated mechanism, possibly involving enhanced mitochondrial biogenesis, among other actions (Teodoro et al. 2013). At least one clinical trial has proved its safety and efficacy in NAFLD (Yan et al. 2015), suggesting that berberine could be a valid alternative for energetic conditioning of liver cells before hepatectomy. Further studies are needed, however.

One of the most exciting candidate molecules for energetic conditioning of liver cells is the Augmenter of Liver Regeneration (ALR). Having been first isolated and purified by Labrecque et al. (1987), ALR is also known as Hepatic Stimulator Substance or Hepatopoietin, for its strong mitogenic effect, far surpassing other known hepatic mitogens such as Epidermal growth factor (EGF) and Transforming growth factor alpha (TGF-α). ALR is secreted in liver cells and acts on Kupffer cells and in autocrine and paracrine fashion on hepatocytes. It is also found in the intracellular compartment, namely in mitochondria, where it regulates oxidative phosphorylation and is essential for mitochondrial biogenesis and ATP synthesis (Gandhi 2012). The importance of ALR in liver regeneration seems to be dependent upon its effect on mitochondrial biogenesis, uphill of other two co-factors, TFAM and Peroxisome proliferator-activated receptor gamma coactivator 1-alpha (PGC-1α) (Han et al. 2015). In one experimental study, transfection of isolated hepatocytes with antisense oligonucleotide against ALR brought about energetic failure and cell death by apoptosis and necrosis (Thirunavukkarasu et al. 2008), while transfection of ALR via adenovirus in rodent models of hepatic ischemia reperfusion and toxic injuries demonstrated improved oxidative phosphorylation, increased ATP production, decreased mitochondrial permeability transition, reduced cell death and improved survival (Li et al. 2011; Jiang et al. 2013). Interestingly, delivery of ALR is not limited to gene transfection with intraperitoneal injection also effective in improving mitochondrial bioenergetics and biogenesis (Tang et al. 2015). This ease of administration, not requiring ALR integration into the host genome, could open the possibility of expanding ALR therapy into the clinical arena for the treatment of acute liver failure of several etiologies, including PHLF.

As previously discussed, another potential therapeutic target is inhibition of the MPT, with an ongoing search for small molecule inhibitors and future advances are expected in the near future (Šileikytė and Forte 2016).

Finally, induction of autophagy is yet another possible mechanism of hepatic energetic conditioning and a plausible therapeutic avenue to pursue with pharmacologic agents (Madrigal-Matute and Cuervo 2016). By the selective removal of damaged mitochondria, enhanced mitophagy is thought to increase the energetic efficiency of cells and decrease oxidative stress, thus improving liver regeneration and survival in a rodent model of 90% hepatectomy (Lin et al. 2015).

5.2 Improving Hepatocyte Oxygenation

Intraoperative increase of hepatic oxygen delivery and extraction could potentially prevent postoperative hepatocellular dysfunction by increasing the energy supply to the rapidly dividing and metabolically active liver parenchymal cells (Kainuma et al. 1992; Meguro et al. 2013). Since the use of inflow occlusion to minimize bleeding during liver resection causes inevitable liver ischemia, diverse clamping techniques and strategies have been developed to minimize the impact on the liver parenchyma. This subject has been extensively detailed elsewhere (Hoekstra et al. 2012; Richardson et al. 2012). We will shed some light on novel or less explored approaches to increase parenchymal oxygenation. Several strategies aiming at this goal have been pursued, some of them with clinically relevant results.

Intraoperative hyperdynamic circulatory pharmacological manipulation with β-adrenergic agonists is an attractive option, as it increases hepatic arterial flow, portal vein oxygen saturation and hepatic oxygen delivery, with subsequent increase in liver lactate uptake (Nonami et al. 1991). In an extremely interesting prospective randomized trial with 30 compensated cirrhotic patients undergoing hepatectomy for HCC, low dose β-adrenergic agent dobutamine decreased portal vein resistance, increased hepatic blood flow and improved hepatic oxygen delivery, consumption and extraction. In the dobutamine-treated group this resulted in increased lactate clearance and decreased peak serum AST and peak postoperative bilirubin (Taurà et al. 2010). Although underpowered to prove a decrease in morbidity with the pharmacologic intervention, this study is fascinating as it conclusively demonstrates that enhancement of hepatic aerobic metabolism improves postoperative liver function.

Modulation of portal flow is another option, since extended hepatectomies result in portal overflow and subsequent arterial vasoconstriction, mediated by the Hepatic Arterial Buffer Response (Lautt 2007) (see Sect. 3). This relative liver hypoxia during the post-resection acute increase in metabolic demands could theoretically contribute to early liver remnant bioenergetics failure. Thus, reversal of excessive portal flow has been explored in order to prevent postoperative liver dysfunction. Several different animal models have explored surgical measures to decrease splanchnic blood flow, namely jejunal resection, splenectomy, splenic artery ligation and portal vein banding. Kawano et al. (2006) performing jejunectomy and 70% hepatectomy in dogs, reported improved hepatocellular function and decreased necrosis and apoptosis. In a rodent model of 90% hepatectomy, splenectomy has been demonstrated to decrease portal vein flow, improve regeneration and decrease postoperative liver failure (Glanemann et al. 2005); this effect is linked to an increase in hepatic arterial blood flow and hepatic oxygenation and is likely mediated by the HABR (Eipel et al. 2010; Zhuang et al. 2012). Decreased mitochondrial swelling has also been described after splenectomy as a portal flow modulation strategy after 80% hepatectomy (Di Domenico et al. 2011). Splenic artery ligation has also been tested in experimental models, with encouraging results (Ito et al. 2007). Portal vein banding is another promising method and one preclinical study in porcine model has reported improved postoperative liver function, however without demonstrating increased parenchymal oxygenation or improved hepatocellular metabolism (Bucur et al. 2017). These studies have been followed by clinical attempts to modulate portal blood flow as a strategy to improve liver regeneration, in the setting of partial liver transplantation (Taniguchi et al. 2007) and extended hepatectomy (Famularo et al. 2015).

Given that the rationale for the aforementioned strategies is the increase in liver oxygen uptake, hyperbaric oxygen therapy (HBO) would seem a reasonable approach. Although there is both preclinical and clinical evidence to support its use, it has been largely unexplored. One experimental study has demonstrated improvement in liver regeneration and ATP content after 90% hepatectomy with 2 atm. 80% O_2 therapy (Nagamine et al. 2004). Mitochondrial function was assessed in one animal model of 70% hepatectomy and although there was a decrease in RCR in the HBO-treated group, there was also an in improvement in liver regeneration rate, DNA content and proliferation index (Tolentino et al. 2006). There is undoubtedly a beneficial effect in liver bioenergetics, as HBO can prevent deterioration of RCR, maintain mitochondrial membrane potential and sustain ATP synthesis in isolated mitochondria in experimental conditions of cold ischemia reperfusion injury (Sgarbi et al. 2011). The clinical evidence for the use of HBO in liver surgery has come mostly from case reports demonstrating its anecdotal use in acute liver failure (Ponikvar et al. 1998), including one case of PHLF (Asanuma et al. 2003). However, one prospective clinical trial have reported encouraging results on its use after hepatectomy. Cirrhotic patients with hepatocellular carcinoma after hepatectomy complicated with major intraoperative blood loss were randomized to either HBO or standard management (Ueno et al. 2011). HBO-treated patients experienced lower postoperative arterial lactate and bilirubin, higher hepatic vein oxygen saturation and a trend towards improved hepatocellular function. However, in this study the definition of PHLF was not standardized so no conclusions can be drawn on this respect.

5.3 *Stem Cell Therapy*

Cell therapy has emerged as an important approach for end-stage liver disease, with both clinical and experimental studies reporting on the beneficial impact of mesenchymal, embryonic and induced pluripotent stem cells on liver regeneration after toxic and surgical insults (Forbes et al. 2015).

In part, the effect of stem-cell therapy could be mediated by an improvement in mitochondrial function, as there is an increase in mitochondrial number (Kuai et al. 2006), decrease in mitochondrial-mediated apoptosis (Cai et al. 2015) and improved mitochondrial respiration (El'chaninov et al. 2015). In fact, bone marrow mononuclear cells transplanted through the jugular vein 14 day after BDL in Wistar rats resulted in improved mitochondrial respiration (state 3 and RCR), decreased lipid peroxidation and decreased fibrogenesis (de Andrade et al. 2015). Tautenhahn et al. (2015) studied the effect of mesenchymal stem cells on an animal model of PHLF. They found that the improvement in liver function, decreased apoptosis, increased regeneration and overall better metabolic profile occurred very early after therapy and were independent of stem cells' engraftment. As such, an interesting phenomenon could be taking place, as stem cells could potentially transfer mitochondria to liver cells, thus improving their metabolic capacity. The transfer of mitochondria is thought to occur through intercellular nanotubes (Vallabhaneni et al. 2012) and direct intrasplenic injection of isolated mitochondria has also proved beneficial in experimental models of liver ischemia-reperfusion injury (Lin et al. 2013). Although stem cell therapy for PHLF has been mostly experimental, some clinical trials have addressed the possibility of using CD133+ stem cells in the prevention of PHLF, in particular as adjuvant to portal vein embolization (Esch et al. 2012). As further clinical studies are underway in cell therapy for liver disease, it would be interesting to evaluate the effect on mitochondrial function, in particular in the setting of acute liver failure and PHLF.

6 Conclusions and Future Prospects

Although the subject of bioenergetics in liver regeneration and clinical liver surgery has been explored in the past, recent clinical and experimental evidence has rekindled the interest in this field. With the increased prevalence of chronic liver disease and the expanding indications for hepatectomy, novel strategies to improve the results of liver surgery are paramount. Albeit a wondrous phenomenon, liver regeneration is not inexhaustible and some patients still die from liver dysfunction in spite of optimal pre-, intra- and postoperative management. As liver physiology is evermore so taken to its limits, a profound knowledge of the factors involved in liver regeneration is desperately needed, including the energy source that fuels it. Mitochondria, the main energy suppliers of eukaryote cells and key players in cell death, find themselves at the heart of this process, and in the spotlight of biomedical research.

As the possibility of direct interventions on liver energy homeostasis proves useful in improving clinical outcomes, a fascinating new field of mitochondrial-directed therapies emerges into the clinical arena. Novel approaches such as pharmacological induction of mitochondrial biogenesis and mitophagy, inhibition of the permeability transition, portal flow modulation or stem cell therapy could ultimately result in an improvement of cellular energetics and substantially decrease the morbidity and mortality of liver resection.

In conclusion, by becoming able to improve energetic conditioning of the liver parenchyma we will safely expand the limits of hepatic surgery; potentially benefiting patients nowadays deemed inoperable due to a predicted insufficient liver remnant.

References

Agrawal S, Belghiti J (2011) Oncologic resection for malignant tumors of the liver. Ann Surg 253(4):656–665. https://doi.org/10.1097/SLA.0b013e3181fc08ca

Alexandrino H, Varela AT, Teodoro JS, Martins MA, Rolo AP, Tralhão JG, Palmeira CM, Castro E, Sousa F (2016) Mitochondrial bioenergetics and posthepatectomy liver dysfunction. Eur J Clin Investig 46(7):627–635. https://doi.org/10.1111/eci.12639

Allard M-A, Adam R, Bucur P-O, Termos S, Cunha AS, Bismuth H, Castaing D, Vibert E (2013) Posthepatectomy portal vein pressure predicts liver failure and mortality after major liver resection on noncirrhotic liver. Ann Surg 258(5):822–829. https://doi.org/10.1097/SLA.0b013e3182a64b38

Aloia T, Sebagh M, Plasse M, Karam V, Lévi F, Giacchetti S, Azoulay D, Bismuth H, Castaing D, Adam R (2006) Liver histology and surgical outcomes after preoperative chemotherapy with fluorouracil plus oxaliplatin in colorectal cancer liver metastases. J Clin Oncol 24(31):4983–4990. https://doi.org/10.1200/JCO.2006.05.8156.

de Andrade DC, de Carvalho SN, Pinheiro D, Thole AA, Moura AS, de Carvalho L, Cortez EAC (2015) Bone marrow mononuclear cell transplantation improves mitochondrial bioenergetics in the liver of cholestatic rats. Exp Cell Res 336(1):15–22. https://doi.org/10.1016/j.yexcr.2015.05.002

Arduini A, Serviddio G, Escobar J, Tormos AM, Bellanti F, Viña J, Monsalve M, Sastre J (2011) Mitochondrial biogenesis fails in secondary biliary cirrhosis in rats leading to mitochondrial DNA depletion and deletions. Am J Physiol Gastrointest Liver Physiol 301(1):G119–G127. https://doi.org/10.1152/ajpgi.00253.2010

Arvaniti V, D'Amico G, Fede G, Manousou P, Tsochatzis E, Pleguezuelo M, Burroughs AK (2010) Infections in patients with cirrhosis increase mortality four-fold and should be used in determining prognosis. Gastroenterology 139(4):1246–1256, 1256–1255. https://doi.org/10.1053/j.gastro.2010.06.019

Asanuma Y, Sato T, Yasui O, Kurokawa T, Koyama K (2003) Treatment for postoperative liver failure after major hepatectomy under hepatic total vascular exclusion. J Artif Organs 6(2):152–156

Auger C, Alhasawi A, Contavadoo M, Appanna VD (2015) Dysfunctional mitochondrial bioenergetics and the pathogenesis of hepatic disorders. Front Cell Dev Biol 3:40. https://doi.org/10.3389/fcell.2015.00040.

Bailey SM, Robinson G, Pinner A, Chamlee L, Ulasova E, Pompilius M, Page GP, Chhieng D, Jhala N, Landar A, Kharbanda KK, Ballinger S, Darley-Usmar V (2006) S-Adenosylmethionine prevents chronic alcohol-induced mitochondrial dysfunction in the rat liver. Am J Physiol Gastrointest Liver Physiol 291(5):G857–G867. https://doi.org/10.1152/ajpgi.00044.2006

Brown JM, Kuhlman C, Terneus MV, Labenski MT, Lamyaithong AB, Ball JG, Lau SS, Valentovic MA (2014) S-Adenosyl-l-methionine protection of acetaminophen mediated oxidative stress

and identification of hepatic 4-hydroxynonenal protein adducts by mass spectrometry. Toxicol Appl Pharmacol 281(2):174–184. https://doi.org/10.1016/j.taap.2014.08.027
Bucur PO, Bekheit M, Audebert C, Othman A, Hammad S, Sebagh M, Allard M-A, Decante B, Friebel A, Miquelestorena-Standley E, Drasdo D, Hengstler JG, Vignon-Clementel IE, Vibert E (2017) Modulating portal hemodynamics with vascular ring allows efficient regeneration after partial hepatectomy in a porcine model. Ann Surg. https://doi.org/10.1097/SLA.0000000000002146
Cai Y, Zou Z, Liu L, Chen S, Chen Y, Lin Z, Shi K, Xu L, Chen Y (2015) Bone marrow-derived mesenchymal stem cells inhibits hepatocyte apoptosis after acute liver injury. Int J Clin Exp Pathol 8(1):107–116
Cao H, Yu J, Xu W, Jia X, Yang J, Pan Q, Zhang Q, Sheng G, Li J, Pan X, Wang Y, Li L (2009) Proteomic analysis of regenerating mouse liver following 50% partial hepatectomy. Proteome Sci 7:48. https://doi.org/10.1186/1477-5956-7-48
Castro e Silva O, Mente ED, Sankarankutty AK, Souza ME, Gomes MC, Picinato MA, Fina CF, Lopes JR Jr (2011) Biochemical liver function after partial hepatic resection with or without partial hepatic vascular exclusion. Acta Cir Bras 26(Suppl 2):120–124
Cescon M, Cucchetti A, Grazi GL, Ferrero A, Viganò L, Ercolani G, Zanello M, Ravaioli M, Capussotti L, Pinna AD (2009) Indication of the extent of hepatectomy for hepatocellular carcinoma on cirrhosis by a simple algorithm based on preoperative variables. Arch Surg 144(1):57–63; discussion 63. https://doi.org/10.1001/archsurg.2008.522
Chouillard EK, Gumbs AA, Cherqui D (2010) Vascular clamping in liver surgery: physiology, indications and techniques. Ann Surg Innov Res 4:2. https://doi.org/10.1186/1750-1164-4-2.
Clavien P-A, Petrowsky H, DeOliveira ML, Graf R (2007) Strategies for safer liver surgery and partial liver transplantation. N Engl J Med 356(15):1545–1559. https://doi.org/10.1056/NEJMra065156.
Corbin IR, Buist R, Volotovskyy V, Peeling J, Zhang M, Minuk GY (2002) Regenerative activity and liver function following partial hepatectomy in the rat using (31)P-MR spectroscopy. Hepatology (Baltimore, Md) 36(2):345–353. https://doi.org/10.1053/jhep.2002.34742.
Crumm S, Cofan M, Juskeviciute E, Hoek JB (2008) Adenine nucleotide changes in the remnant liver: an early signal for regeneration after partial hepatectomy. Hepatology (Baltimore, Md.) 48(3):898–908. https://doi.org/10.1002/hep.22421.
Demetris AJ, Kelly DM, Eghtesad B, Fontes P, Wallis Marsh J, Tom K, Tan HP, Shaw-Stiffel T, Boig L, Novelli P, Planinsic R, Fung JJ, Marcos A (2006) Pathophysiologic observations and histopathologic recognition of the portal hyperperfusion or small-for-size syndrome. Am J Surg Pathol 30(8):986–993
Di Domenico S, Santori G, Traverso N, Balbis E, Furfaro A, Grillo F, Gentile R, Bocca B, Gelli M, Andorno E, Dahame A, Cottalasso D, Valente U (2011) Early effects of portal flow modulation after extended liver resection in rat. Dig Liver Dis 43(10):814–822. https://doi.org/10.1016/j.dld.2011.05.018.
Dold S, Richter S, Kollmar O, Von Heesen M, Scheuer C, Laschke MW, Vollmar B, Schilling MK, Menger MD (2015) Portal hyperperfusion after extended hepatectomy does not induce a hepatic arterial buffer response (HABR) but impairs mitochondrial redox state and hepatocellular oxygenation. PLoS One 10(11):e0141877. https://doi.org/10.1371/journal.pone.0141877
Eipel C, Abshagen K, Ritter J, Cantré D, Menger MD, Vollmar B (2010) Splenectomy improves survival by increasing arterial blood supply in a rat model of reduced-size liver. Transpl Int 23(10):998–1007. https://doi.org/10.1111/j.1432-2277.2010.01079.x.
El'chaninov AV, Volodina MA, Arutyunyan IV, Makarov AV, Tarasova NV, Kananykhina EY, Usman NY, Marei MV, Vysokikh MY, Glinkina VV, Bol'shakova GB, Fatkhudinov TH, Sukhikh GT (2015) Effect of multipotent stromal cells on the function of cell mitochondria in regenerating liver. Bull Exp Biol Med 158(4):566–572. https://doi.org/10.1007/s10517-015-2808-x
Esch JSA, Schmelzle M, Fürst G, Robson SC, Krieg A, Duhme C, Tustas RY, Alexander A, Klein HM, Topp SA, Bode JG, Häussinger D, Eisenberger CF, Knoefel WT (2012) Infusion of CD133+ bone marrow-derived stem cells after selective portal vein embolization enhances

functional hepatic reserves after extended right hepatectomy. Ann Surg 255(1):79–85. https://doi.org/10.1097/SLA.0b013e31823d7d08.

Famularo S, Nefotyou K, Fotiadis N, Khan N, Foxton M, Khan AZ (2015) Small-for-size liver syndrome: a case series with a proposal for management based on portal flow modulation. J Gastrointest Cancer 46(2):185–189. https://doi.org/10.1007/s12029-015-9701-8

Fan CQ, Crawford JM (2014) Sinusoidal obstruction syndrome (hepatic veno-occlusive disease). J Clin Exp Hepatol 4(4):332–346. https://doi.org/10.1016/j.jceh.2014.10.002

Fausto N, Campbell JS, Riehle KJ (2006) Liver regeneration. Hepatology (Baltimore, Md.) 43(2 Suppl 1):S45–S53. https://doi.org/10.1002/hep.20969

Folprecht G, Grothey A, Alberts S, Raab H-R, Köhne C-H (2005) Neoadjuvant treatment of unresectable colorectal liver metastases: correlation between tumour response and resection rates. Ann Oncol 16(8):1311–1319. https://doi.org/10.1093/annonc/mdi246

Forbes SJ, Gupta S, Dhawan A (2015) Cell therapy for liver disease: from liver transplantation to cell factory. J Hepatol 62(S1):S157–S169. https://doi.org/10.1016/j.jhep.2015.02.040

Gandhi CR (2012) Augmenter of liver regeneration. Fibrogenesis Tissue Repair 5(1):10. https://doi.org/10.1186/1755-1536-5-10

Garnol T, Kučera O, Staňková P, Lotková H, Červinková Z (2016) Does simple steatosis affect liver regeneration after partial hepatectomy in rats? Acta Medica 59(2):35–42. https://doi.org/10.14712/18059694.2016.87

Glanemann M, Eipel C, Nussler AK, Vollmar B, Neuhaus P (2005) Hyperperfusion syndrome in small-for-size livers. Eur Surg Res 37(6):335–341. https://doi.org/10.1159/000090333

Golse N, Bucur PO, Adam R, Castaing D, Sa Cunha A, Vibert E (2013) New paradigms in post-hepatectomy liver failure. J Gastrointest Surg 17(3):593–605. https://doi.org/10.1007/s11605-012-2048-6.

Gourdier I, Crabbe L, Andreau K, Pau B, Kroemer G (2004) Oxaliplatin-induced mitochondrial apoptotic response of colon carcinoma cells does not require nuclear DNA. Oncogene 23:7449–7457. https://doi.org/10.1038/sj.onc.1208047

Grattagliano I, de Bari O, Bernardo TC, Oliveira PJ, Wang DQ-H, Portincasa P (2012) Role of mitochondria in nonalcoholic fatty liver disease—from origin to propagation. Clin Biochem 45(9):610–618. https://doi.org/10.1016/j.clinbiochem.2012.03.024

Guerrieri F, Vendemiale G, Grattagliano I, Cocco T, Pellecchia G, Altomare E (1999) Mitochondrial oxidative alterations following partial hepatectomy. Free Radic Biol Med 26(1–2):34–41

Han L-H, Dong L-Y, Yu H, Sun G-Y, Wu Y, Gao J, Thasler W, An W (2015) Deceleration of liver regeneration by knockdown of augmenter of liver regeneration gene is associated with impairment of mitochondrial DNA synthesis in mice. Am J Physiol Gastrointest Liver Physiol 309(2):112–122. https://doi.org/10.1152/ajpgi.00435.2014

Hashimoto M, Watanabe G (2005) Functional restoration of cirrhotic liver after partial hepatectomy in the rat. Hepato-Gastroenterology 52(63):897–902

Hoekstra LT, van Trigt JD, Reiniers MJ, Busch OR, Gouma DJ, van Gulik TM (2012) Vascular occlusion or not during liver resection: the continuing story. Dig Surg 29(1):35–42. https://doi.org/10.1159/000335724

Iacono C, Ruzzenente A, Campagnaro T, Bortolasi L, Valdegamberi A, Guglielmi A (2013) Role of preoperative biliary drainage in jaundiced patients who are candidates for pancreatoduodenectomy or hepatic resection: highlights and drawbacks. Ann Surg 257(2):191–204. https://doi.org/10.1097/SLA.0b013e31826f4b0e

Ito K, Ozasa H, Noda Y, Koike Y, Arii S, Horikawa S (2007) Splenic artery ligation improves remnant liver function in partially hepatectomized rats with ischemia/reperfusion injury. Liver Int 27(3):400–407. https://doi.org/10.1111/j.1478-3231.2006.01432.x

Jeon BR, Lee SM (2001) S-Adenosylmethionine protects post-ischemic mitochondrial injury in rat liver. J Hepatol 34(3):395–401

Jiang S, Li W, An W (2013) Adenoviral gene transfer of hepatic stimulator substance confers resistance against hepatic ischemia – reperfusion injury by improving mitochondrial function. Hum Gene Ther 24(4):443–456. https://doi.org/10.1089/hum.2012.219

Kainuma M, Nakashima K, Sakuma I, Kawase M, Komatsu T, Shimada Y, Nimura Y, Nonami T (1992) Hepatic venous hemoglobin oxygen saturation predicts liver dysfunction after hepatectomy. Anesthesiology 76(3):379–386

Kanai M, Tanaka M, Nimura Y, Nagino M, Katoh T, Ozawa T (1992) Mitochondrial dysfunction in the non-obstructed lobe of rat liver after selective biliary obstruction. Hepato-Gastroenterology 39(5):385–391

Karoui M, Penna C, Amin-Hashem M, Mitry E, Benoist S, Franc B, Rougier P, Nordlinger B (2006) Influence of preoperative chemotherapy on the risk of major hepatectomy for colorectal liver metastases. Ann Surg 243(1):1–7. https://doi.org/10.1097/01.sla.0000193603.26265.c3

Kawano Y, Akimaru K, Takubo K, Matsumoto K, Yoshida H, Mamada Y, Taniai N, Tajiri T (2006) Jejunectomy can reduce excessively elevated portal pressure after major hepatectomy in beagle dogs. J Surg Res 130(1):24–33. https://doi.org/10.1016/j.jss.2005.08.019.

Khiati S, Baechler SA, Factor VM, Zhang H, Huang S-YN, Dalla Rosa I, Sourbier C, Neckers L, Thorgeirsson SS, Pommier Y (2015) Lack of mitochondrial topoisomerase I (TOP1mt) impairs liver regeneration. Proc Natl Acad Sci U S A 112(36):11282–11287. https://doi.org/10.1073/pnas.1511016112

Koyama H, Kurokawa T, Nonami T, Nakao A, Sugiyama S (1998) Increases in the mitochondrial DNA replication and transcription in the remnant liver of rats. Biochem Biophys Res Commun 861(243):858–861

Krähenbühl L, Schäfer M, Krähenbühl S (1998) Reversibility of hepatic mitochondrial damage in rats with long-term cholestasis. J Hepatol 28(6):1000–1007

Kuai XL, Cong XQ, Du ZW, Bian YH, Xiao SD (2006) Treatment of surgically induced acute liver failure by transplantation of HNF4-overexpressing embryonic stem cells. Chin J Dig Dis 7(2):109–116. https://doi.org/10.1111/j.1443-9573.2006.00253.x

Labbe G, Pessayre D, Fromenty B (2008) Drug-induced liver injury through mitochondrial dysfunction: mechanisms and detection during preclinical safety studies. Fundam Clin Pharmacol 22(4):335–353. https://doi.org/10.1111/j.1472-8206.2008.00608.x

LaBrecque DR, Steele G, Fogerty S, Wilson M, Barton J (1987) Purification and physical-chemical characterization of hepatic stimulator substance. Hepatology 7(1):100–106

Lautt WW (2007) Regulatory processes interacting to maintain hepatic blood flow constancy: vascular compliance, hepatic arterial buffer response, hepatorenal reflex, liver regeneration, escape from vasoconstriction. Hepatol Res 37(11):891–903. https://doi.org/10.1111/j.1872-034X.2007.00148.x.

Lentz F, Tran A, Rey E, Pons G, Tréluyer J-M (2005) Pharmacogenomics of fluorouracil, irinotecan, and oxaliplatin in hepatic metastases of colorectal cancer: clinical implications. Am J Pharmacogenomics 5(1):21–33

Li S, Tang Z, Yu H, Li W, Jiang Y, Wang Y, An W (2011) Administration of naked plasmid encoding hepatic stimulator substance by hydrodynamic tail vein injection protects mice from hepatic failure by suppressing the mitochondrial permeability transition. J Pharmacol Exp Ther 338(3):750–757. https://doi.org/10.1124/jpet.111.181305

Lin H-C, Liu S-Y, Lai H-S, Lai I-R (2013) Isolated mitochondria infusion mitigates ischemia-reperfusion injury of the liver in rats. Shock (Augusta, Ga) 39(3):304–310. https://doi.org/10.1097/SHK.0b013e318283035f.

Lin C-W, Chen Y-S, Lin C-C, Chen Y-J, Lo G-H, Lee P-H, Kuo P-L, Dai C-Y, Huang J-F, Chung W-L, Yu M-L (2015) Amiodarone as an autophagy promoter reduces liver injury and enhances liver regeneration and survival in mice after partial hepatectomy. Sci Rep 5:15807. https://doi.org/10.1038/srep15807.

Liu G, Wang W, Jia W, Xu G, Ma J, Ge Y, Yu J, Sun Q, Meng F (2014) Protective effect of S-adenosylmethionine on hepatic ischemia-reperfusion injury during hepatectomy in HCC patients with chronic HBV infection. World J Surg Oncol 12:27. https://doi.org/10.1186/1477-7819-12-27

Madrigal-Matute J, Cuervo AM (2016) Regulation of liver metabolism by autophagy. Gastroenterology 150(2):328–339. https://doi.org/10.1053/j.gastro.2015.09.042

Mann DV, Lam WW, Hjelm NM, So NM, Yeung DK, Metreweli C, Lau WY (2001) Human liver regeneration: hepatic energy economy is less efficient when the organ is diseased. Hepatology 34(3):557–565. https://doi.org/10.1053/jhep.2001.27012

Mann DV, Lam WWM, Hjelm NM, So NMC, Yeung DKW, Metreweli C, Lau WY (2002a) Metabolic control patterns in acute phase and regenerating human liver determined in vivo by 31-phosphorus magnetic resonance spectroscopy. Ann Surg 235(3):408–416

Mann DV, Lam WWM, Magnus Hjelm N, So NMC, Yeung DKW, Metreweli C, Lau WY (2002b) Biliary drainage for obstructive jaundice enhances hepatic energy status in humans: a 31-phosphorus magnetic resonance spectroscopy study. Gut 50(1):118–122

McGill MR, Sharpe MR, Williams CD, Taha M, Curry SC, Jaeschke H (2012) The mechanism underlying acetaminophen-induced hepatotoxicity in humans and mice involves mitochondrial damage and nuclear DNA fragmentation. J Clin Invest 122(4):1574–1583. https://doi.org/10.1172/JCI59755.

Meguro M, Mizuguchi T, Kawamoto M, Nakamura Y, Ota S, Kukita K, Ishii M, Tatsumi H, Hirata K (2013) Continuous monitoring of central venous oxygen saturation predicts postoperative liver dysfunction after liver resection. Surgery 154(2):351–362. https://doi.org/10.1016/j.surg.2013.04.039

de Meijer VE, Kalish BT, Puder M, Ijzermans JNM (2010) Systematic review and meta-analysis of steatosis as a risk factor in major hepatic resection. Br J Surg 97(9):1331–1339. https://doi.org/10.1002/bjs.7194.

Minuk GY (2003) Hepatic regeneration: If it ain't broke, don't fix it. Can J Gastroenterol 17(7):418–424. https://doi.org/10.1038/424709a.

Mittal S, El-Serag HB, Sada YH, Kanwal F, Duan Z, Temple S, May SB, Kramer JR, Richardson PA, Davila JA (2015) Hepatocellular carcinoma in the absence of cirrhosis in US veterans is associated with non-alcoholic fatty liver disease. Clin Gastroenterol Hepatol 14:124–131. https://doi.org/10.1016/j.cgh.2015.07.019.

Nagamine K, Kubota T, Togo S, Nagashima Y, Mori M, Shimada H (2004) Beneficial effect of hyperbaric oxygen therapy on liver regeneration after 90% hepatectomy in rats. Eur Surg Res 36(6):350–356. https://doi.org/10.1159/000081643

Nagino M, Tanaka M, Nishikimi M, Nimura Y, Kubota H, Kanai M (1989) Stimulated rat liver mitochondrial biogenesis after partial hepatectomy. Cancer Res 49(17):4913–4918

Nagino M, Ebata T, Yokoyama Y, Igami T, Sugawara G, Takahashi Y, Nimura Y (2013) Evolution of surgical treatment for perihilar cholangiocarcinoma. Ann Surg 258(1):129–140. https://doi.org/10.1097/SLA.0b013e3182708b57

Nakano H, Oussoultzoglou E, Rosso E, Casnedi S, Chenard-Neu M-P, Dufour P, Bachellier P, Jaeck D (2008) Sinusoidal injury increases morbidity after major hepatectomy in patients with colorectal liver metastases receiving preoperative chemotherapy. Ann Surg 247(1):118–124. https://doi.org/10.1097/SLA.0b013e31815774de

Nakatani T, Yasuda K, Ozawa K, Kawashima S, Tobe T (1982) Effects of (+)-octanoylcarnitine on deoxyribonucleic acid synthesis in regenerating rabbit liver. Clin Sci (London, England: 1979) 62(3):295–297

Narita M, Oussoultzoglou E, Fuchshuber P, Chenard M-P, Rosso E, Yamamoto K, Jaeck D, Bachellier P (2012a) Prolonged portal triad clamping increases postoperative sepsis after major hepatectomy in patients with sinusoidal obstruction syndrome and/or steatohepatitis. World J Surg 36(8):1848–1857. https://doi.org/10.1007/s00268-012-1565-5

Narita M, Oussoultzoglou E, Fuchshuber P, Pessaux P, Chenard M-P, Rosso E, Nobili C, Jaeck D, Bachellier P (2012b) What is a safe future liver remnant size in patients undergoing major hepatectomy for colorectal liver metastases and treated by intensive preoperative chemotherapy? Ann Surg Oncol 19(8):2526–2538. https://doi.org/10.1245/s10434-012-2274-x

Ninomiya M, Shirabe K, Terashi T, Ijichi H, Yonemura Y, Harada N, Soejima Y, Taketomi A, Shimada M, Maehara Y (2010) Deceleration of regenerative response improves the outcome of rat with massive hepatectomy. Am J Transplant 10(7):1580–1587. https://doi.org/10.1111/j.1600-6143.2010.03150.x

Nishikawa T, Bellance N, Damm A, Bing H, Zhu Z, Handa K, Yovchev MI, Sehgal V, Moss TJ, Oertel M, Ram PT, Pipinos II, Soto-Gutierrez A, Fox IJ, Nagrath D (2014) A switch in the source of ATP production and a loss in capacity to perform glycolysis are hallmarks of hepatocyte failure in advance liver disease. J Hepatol 60(6):1203–1211. https://doi.org/10.1016/j.jhep.2014.02.014

Nonami T, Asahi K, Harada A, Nakao A, Takagi H (1991) Effect of hyperdynamic circulatory support on hepatic hemodynamics, oxygen supply and demand after massive hepatectomy. Surgery 109(3 Pt 1):277–283

Nordlinger B, Sorbye H, Glimelius B, Poston GJ, Schlag PM, Rougier P, Bechstein WO, Primrose JN, Walpole ET, Finch-Jones M, Jaeck D, Mirza D, Parks RW, Collette L, Praet M, Bethe U, Van Cutsem E, Scheithauer W, Gruenberger T (2008) Perioperative chemotherapy with FOLFOX4 and surgery versus surgery alone for resectable liver metastases from colorectal cancer (EORTC Intergroup trial 40983): a randomised controlled trial. Lancet 371(9617):1007–1016. https://doi.org/10.1016/S0140-6736(08)60455-9

Ozawa K, Kitamura O, Yamaoka Y, Mizukami T, Kamano T (1973) Quantitative analysis of respiratory enzymes of mitochondria isolated from liver tissue of patients. J Lab Clin Med 81(3):379–392

Ozawa K, Kamiyama Y, Kimura K, Ukikusa M, Kono Y, Yamato T, Shimahara Y, Nakatani T, Asano M, Irie R, Kawashima S, Uchida K, Ohtoshi M, Aoyama H, Hirai F, Yasuda K, Tobe T (1982) The effects of heterologous liver cross-hemodialysis on adenylate energy charge of the remnant liver after major hepatic resection. Artif Organs 6(4):447–452

Palmeira CM, Rolo AP (2004) Mitochondrially-mediated toxicity of bile acids. Toxicology 203(1–3):1–15. https://doi.org/10.1016/j.tox.2004.06.001

Ponikvar R, Buturović J, Cizman M, Mekjavić I, Kandus A, Premru V, Urbancic A, Zakotnik B, Bren A, Ivanovich P (1998) Hyperbaric oxygenation, plasma exchange, and hemodialysis for treatment of acute liver failure in a 3-year-old child. Artif Organs 22(11):952–957

Pringle JH (1908) Notes on the arrest of hepatic hemorrhage due to trauma. Ann Surg 4(4):541–549

Ratziu V, Bellentani S, Cortez-Pinto H, Day C, Marchesini G (2010) A position statement on NAFLD/NASH based on the EASL 2009 special conference. J Hepatol 53(2):372–384. https://doi.org/10.1016/j.jhep.2010.04.008

Rehman H, Sun J, Shi Y, Ramshesh VK, Liu Q, Currin RT, Lemasters JJ, Zhong Z (2012) NIM811 prevents mitochondrial dysfunction, attenuates liver injury, and stimulates liver regeneration after massive hepatectomy. Transplantation 91(4):406–412. https://doi.org/10.1097/TP.0b013e318204bdb2.NIM811

Richardson AJ, Laurence JM, Lam VWT (2012) Portal triad clamping versus other methods of vascular control in liver resection: a systematic review and meta-analysis. HBP (Oxford) 14(6):355–364. https://doi.org/10.1111/j.1477-2574.2012.00466.x

Robinson SM, Mann J, Vasilaki A, Mathers J, Burt AD, Oakley F, White SA, Mann DA (2013) Pathogenesis of FOLFOX induced sinusoidal obstruction syndrome in a murine chemotherapy model. J Hepatol 59(2):318–326. https://doi.org/10.1016/j.jhep.2013.04.014

Rolo AP, Oliveira PJ, Moreno JM, Palmeira CM (2000) Bile acids affect liver mitochondrial bioenergetics: possible relevance for cholestasis therapy. Toxicol Sci 57(1):177–185

Rolo AP, Teodoro JS, Palmeira CM (2012) Role of oxidative stress in the pathogenesis of nonalcoholic steatohepatitis. Free Radic Biol Med 52(1):59–69. https://doi.org/10.1016/j.freeradbiomed.2011.10.003

Rosen M, Figliomeni M, Simpkins H (1992) The interaction of platinum antitumour drugs with mouse liver mitochondria. Int J Exp Pathol 73(1):61–74

Rubbia-Brandt L, Lauwers GY, Wang H, Majno PE, Tanabe K, Zhu AX, Brezault C, Soubrane O, Abdalla EK, Vauthey J-N, Mentha G, Terris B (2010) Sinusoidal obstruction syndrome and nodular regenerative hyperplasia are frequent oxaliplatin-associated liver lesions and partially prevented by bevacizumab in patients with hepatic colorectal metastasis. Histopathology 56(4):430–439. https://doi.org/10.1111/j.1365-2559.2010.03511.x

Satoh S, Tanaka A, Hatano E, Inomoto T, Iwata S, Kitai T, Shinohara H, Tsunekawa S, Chance B, Yamaoka Y (1996) Energy metabolism and regeneration in transgenic mouse liver expressing creatine kinase after major hepatectomy. Gastroenterology 110(4):1166–1174

Schiffer E, Frossard J-L, Rubbia-Brandt L, Mentha G, Pastor CM (2009) Hepatic regeneration is decreased in a rat model of sinusoidal obstruction syndrome. J Surg Oncol 99(7):439–446. https://doi.org/10.1002/jso.21276

Schoen JM, Wang HH, Minuk GY, Lautt WW (2001) Shear stress-induced nitric oxide release triggers the liver regeneration cascade. Nitric Oxide 5(5):453–464. https://doi.org/10.1006/niox.2001.0373.

Schulz S, Schmitt S, Wimmer R, Aichler M, Eisenhofer S, Lichtmannegger J, Eberhagen C, Artmann R, Tookos F, Walch A, Krappmann D, Brenner C, Rust C, Zischka H (2013) Progressive stages of mitochondrial destruction caused by cell toxic bile salts. Biochim Biophys Acta 81(9):2121–2133. https://doi.org/10.1016/j.bbamem.2013.05.007

Schuppan D, Afdhal NH (2008) Liver cirrhosis. Lancet 371(9615):838–851. https://doi.org/10.1016/S0140-6736(08)60383-9

Sgarbi G, Giannone F, Casalena GA, Baracca A, Baldassare M, Longobardi P, Caraceni P, Derenzini M, Lenaz G, Trerè D, Solaini G (2011) Hyperoxia fully protects mitochondria of explanted livers. J Bioenerg Biomembr 43(6):673–682. https://doi.org/10.1007/s10863-011-9390-3

Šileikytė J, Forte M (2016) Shutting down the pore: the search for small molecule inhibitors of the mitochondrial permeability transition. Biochim Biophys Acta 1857(8):1197–1202. https://doi.org/10.1016/j.bbabio.2016.02.016

Su Z-R, Cui Z-L, Ma J-L, Li J-S, Ge Y, Yu J-H, Pan J-H, Xu G-L, Jia W-DJ (2013) Beneficial effects of S-adenosyl-L-methionine on post-hepatectomy residual liver function: a prospective, randomized, controlled clinical trial. Hepato-Gastroenterology 60(125):1136–1141. https://doi.org/10.5754/hge11072

Sun Y, Deng X, Li W, Yan Y, Wei H, Jiang Y, He F (2007) Liver proteome analysis of adaptive response in rat immediately after partial hepatectomy. Proteomics 7(23):4398–4407. https://doi.org/10.1002/pmic.200600913

Sydor S, Gu Y, Schlattjan M, Bechmann LP, Rauen U, Best J, Paul A, Baba HA, Sowa J, Gerken G, Canbay A (2012) Steatosis does not impair liver regeneration after partial hepatectomy. Lab Investig 93(1):20–30. https://doi.org/10.1038/labinvest.2012.142

Tabassum H, Waseem M, Parvez S, Qureshi MI (2015) Oxaliplatin-induced oxidative stress provokes toxicity in isolated rat liver mitochondria. Arch Med Res 6(8):597–603. https://doi.org/10.1016/j.arcmed.2015.10.002.

Takamoto T, Hashimoto T, Sano K, Maruyama Y, Inoue K, Ogata S, Takemura T, Kokudo N, Makuuchi M (2010) Recovery of liver function after the cessation of preoperative chemotherapy for colorectal liver metastasis. Ann Surg Oncol 17(10):2747–2755. https://doi.org/10.1245/s10434-010-1074-4

Tang C, Lin H, Wu Q, Zhang Y, Bie P, Yang J (2015) Recombinant human augmenter of liver regeneration protects hepatocyte mitochondrial DNA in rats with obstructive jaundice. J Surg Res 196(1):90–101. https://doi.org/10.1016/j.jss.2015.02.060

Taniguchi M, Shimamura T, Suzuki T, Yamashita K, Oura T, Watanabe M, Kamiyama T, Matsushita M, Furukawa H, Todo S (2007) Transient portacaval shunt for a small-for-size graft in living donor liver transplantation. Liver Transpl 13(6):932–934. https://doi.org/10.1002/lt.21080.

Taurà P, Fuster J, Mercadal J, Martinez-Palli G, Fondevila C, Blasi A, Balust J, Garcia-Valdecasas J-C (2010) The use of beta-adrenergic drugs improves hepatic oxygen metabolism in cirrhotic patients undergoing liver resection. J Hepatol 52(3):340–347. https://doi.org/10.1016/j.jhep.2009.12.008.

Tautenhahn HM, Bruckner S, Baumann S, Winkler S, Otto W, von Bergen M, Bartels M, Christ B (2015) Attenuation of postoperative acute liver failure by mesenchymal stem cell treatment due to metabolic implications. Ann Surg 263(3):546–556. https://doi.org/10.1097/SLA.0000000000001155.

Teodoro JS, Duarte FV, Gomes AP, Varela AT, Peixoto FM, Rolo AP, Palmeira CM (2013) Berberine reverts hepatic mitochondrial dysfunction in high-fat fed rats: a possible role for SirT3 activation. Mitochondrion 13(6):637–646. https://doi.org/10.1016/j.mito.2013.09.002

Thirunavukkarasu C, Wang LF, Harvey SAK, Watkins SC, Chaillet JR, Prelich J, Starzl TE, Gandhi CR (2008) Augmenter of liver regeneration: an important intracellular survival factor for hepatocytes. J Hepatol 48(4):578–588. https://doi.org/10.1016/j.jhep.2007.12.010

Tolentino EC, Castro e Silva O, Zucoloto S, Souza MEJ, Gomes MCJ, Sankarankutty AK, Oliveira GR, Feres O (2006) Effect of hyperbaric oxygen on liver regeneration in a rat model. Transplant Proc 38(6):1947–1952. https://doi.org/10.1016/j.transproceed.2006.06.066

Torzilli G, Belghiti J, Capussotti L, Vauthey J, Choti MA, Santibanes E, De Morenghi E, Makuuchi M (2013) A snapshot of the effective indications and results of surgery for hepatocellular carcinoma in tertiary referral centers: is it adherent to the EASL/AASLD recommendations? Ann Surg 257(5):929–937. https://doi.org/10.1097/SLA.0b013e31828329b8

Ueda J, Mori K, Sakai Y, Tanaka A, Katayama T, Maki A, Shimahara Y, Yamaoka Y, Ozawa K (1994) Noninvasive evaluation of cytochrome c oxidase activity of the liver. Its prognostic value for hepatic resection. Arch Surg 129(3):303–308

Ueno S, Sakoda M, Kurahara H, Iino S, Minami K, Ando K, Mataki Y, Maemura K, Ishigami S, Shinchi H, Natsugoe S (2011) Safety and efficacy of early postoperative hyperbaric oxygen therapy with restriction of transfusions in patients with HCC who have undergone partial hepatectomy. Langenbeck's Arch Surg 396(1):99–106. https://doi.org/10.1007/s00423-010-0725-z.

Ukikusa M, Ozawa K, Shimahara Y, Asano M, Nakatani T, Tobe T (1981) Changes in blood ketone body ratio: their significance after major hepatic resection. Arch Surg (Chicago, Ill: 1960) 116(6):781–785

Vallabhaneni KC, Haller H, Dumler I (2012) Vascular smooth muscle cells initiate proliferation of mesenchymal stem cells by mitochondrial transfer via tunneling nanotubes. Stem Cells Dev 21(17):3104–3113. https://doi.org/10.1089/scd.2011.0691

Varela AT, Simões AM, Teodoro JS, Duarte FV, Gomes AP, Palmeira CM, Rolo AP (2010) Indirubin-3′-oxime prevents hepatic I/R damage by inhibiting GSK-3beta and mitochondrial permeability transition. Mitochondrion 10(5):456–463. https://doi.org/10.1016/j.mito.2010.04.006.

Varela AT, Rolo AP, Palmeira CM (2011) Fatty liver and ischemia/reperfusion: are there drugs able to mitigate injury? Curr Med Chem 18(32):4987–5002

Vauthey J-N, Pawlik TM, Ribero D, Wu T-T, Zorzi D, Hoff PM, Xiong HQ, Eng C, Lauwers GY, Mino-Kenudson M, Risio M, Muratore A, Capussotti L, Curley SA, Abdalla EK (2006) Chemotherapy regimen predicts steatohepatitis and an increase in 90-day mortality after surgery for hepatic colorectal metastases. J Clin Oncol 24(13):2065–2072. https://doi.org/10.1200/JCO.2005.05.3074.

Veteläinen R, Bennink RJ, van Vliet AK, van Gulik TM (2007a) Mild steatosis impairs functional recovery after liver resection in an experimental model. Br J Surg 94(8):1002–1008. https://doi.org/10.1002/bjs.5672.

Veteläinen R, van Vliet AK, van Gulik TM (2007b) Severe steatosis increases hepatocellular injury and impairs liver regeneration in a rat model of partial hepatectomy. Ann Surg 245(1):44–50. https://doi.org/10.1097/01.sla.0000225253.84501.0e

Wilasrusmee C, Siritheptawee S, Kanchanapanjapon S, Sopon P, Vanichanon C, Limpthong W, Pongchailerks P, Lertsithichai P, Wilasrusmee S, Kittur DS (2004) Ultrastructural changes in cirrhotic and noncirrhotic patients due to hepatectomy. J Hepato-Biliary-Pancreat Surg 11(4):266–271. https://doi.org/10.1007/s00534-004-0902-y

Xiang L, Huang L, Wang X, Zhao Y, Liu Y, Tan J (2016) How much portal vein flow is too much for liver remnant in a stable porcine model? Transplant Proc 48(1):234–241. https://doi.org/10.1016/j.transproceed.2015.12.026

Yamaguchi T, Shimahara Y, Takada Y, Ino K, Mori K, Kobayashi N, Yamaoka Y, Ozawa K (1992) Evaluation of ketogenesis in seriously reduced hepatic mitochondrial redox state. An analysis of survivors and non-survivors in critically ill hepatectomized patients. Scand J Gastroenterol 27(6):472–478

Yamaoka Y, Ohsawa T, Takasan H, Ozawa K (1974) Energy requirement in regenerative and atrophic processes of the liver in man and other mammals. Surg Gynecol Obstet 139(2):234–240

Yan H-M, Xia M-F, Wang Y, Chang X-X, Yao X-Z, Rao S-X, Zeng M-S, Tu Y-F, Feng R, Jia W-P, Liu J, Deng W, Jiang J-D, Gao X (2015) Efficacy of berberine in patients with non-alcoholic fatty liver disease. PLoS One 10(8):e0134172. https://doi.org/10.1371/journal.pone.0134172

Yang S, Tan TMC, Wee A, Leow CK (2004) Mitochondrial respiratory function and antioxidant capacity in normal and cirrhotic livers following partial hepatectomy. Cell Mol Life Sci 61(2):220–229. https://doi.org/10.1007/s00018-003-3357-4

Yoshioka S, Miyazaki M, Shimizu H, Ito H, Nakagawa K, Ambiru S, Nakajima N, Fukuda Y (1998) Hepatic venous hemoglobin oxygen saturation predicts regenerative status of remnant liver after partial hepatectomy in rats. Hepatology (Baltimore, Md) 27(5):1349–1353. https://doi.org/10.1002/hep.510270522.

Zhuang ZG, Xu JR, Qian LJ, Xia Q, Chi JC (2012) Computed tomography perfusion study of hemodynamic changes and portal hyperperfusion in a rabbit model of small-for-size liver. Hepatobiliary Pancreat Dis Int 11(1):74–80. https://doi.org/10.1016/S1499-3872(11)60128-8

Mitochondria Remodeling in Cancer

Kenneth J. Dornfeld and Andrew J. Skildum

Abstract Cancer cells are characterized by altered metabolism and uncontrolled proliferation. It is not surprising that mitochondria, which serve important roles in fuel oxidation (catabolism) and synthesizing biosynthetic precursors (anabolism), are altered in cancer cells. Although most cancer cells have increased uptake of glucose and upregulation of glycolysis, mitochondria remain essential for carcinogenesis. However, cancer cell mitochondria have significant differences that distinguish them from those of normal cells. Mitochondria of cancer cells often have decreased electron transport capacity, depressed ATP generation, and increased production of reactive oxygen species. These alterations in metabolism promote nutrient uptake, support increased proliferation and afford resistance to apoptotic cell death. The specific metabolic processes in cancer provide opportunities to selectively target them pharmacologically to improve outcomes for cancer patients. In cancer diagnosis, tumor metabolism can be inferred from gene expression patterns and mutations, rates of nutrient uptake, and from direct measurement of metabolites. In treatment, these processes can be specifically targeted. Alterations in metabolism, mitochondrial DNA, the electron transport chain, and tricarboxylic acid cycle enzymes in cancer are described. The exploitation of these differences for diagnosis and treatment are discussed, with a review of select agents currently in clinical use or under active investigation.

Keywords Cancer · Chemotherapy · Warburg · Anabolism · Tricarboxylic acid cycle · Glycolysis

K. J. Dornfeld
Department of Biomedical Sciences, University of Minnesota Medical School, Duluth Campus, Duluth, MN, USA

Department of Radiation Oncology, Essentia Health, Duluth, MN, USA
e-mail: Kenneth.Dornfeld@essentiahealth.org

A. J. Skildum (✉)
Department of Biomedical Sciences, University of Minnesota Medical School, Duluth Campus, Duluth, MN, USA
e-mail: askildum@d.umn.edu

P. J. Oliveira (ed.), *Mitochondrial Biology and Experimental Therapeutics*,
https://doi.org/10.1007/978-3-319-73344-9_9

1 Introduction

Cancer refers to the collection of diseases characterized by uncontrolled cellular proliferation. Cellular proliferation is controlled by extrinsic signals such as mitogenic hormones and intrinsic signals such as DNA integrity. Significant investigation into tumor suppressors and proto-oncogenes that function as cell cycle checkpoints and whose function is disrupted in cancer has led to novel targeted therapies that spare patients the off target toxicities associated with traditional broadly cytotoxic chemotherapeutic drugs (e.g. Druker et al. 1996; Baselga et al. 1998). Similar opportunities may exist in targeting the unique mitochondrial metabolism of cancer cells relative to non-transformed cells.

Mitochondria are organelles within cells that serve important anabolic, catabolic, and signal transduction functions. In catabolism, the primary function of mitochondria is the oxidation of carbons from pyruvate and other tricarboxylic acid (TCA) cycle intermediates to carbon dioxide, and using the oxidation of reduced NADH to drive proton transport across the inner mitochondrial membrane and ATP synthesis. The anabolic functions of mitochondria are more diverse, and include the generation of citrate to support fatty acid synthesis, the synthesis of succinyl CoA used for heme synthesis, and the conversion of gluconeogenic precursors to malate. Mitochondria are also important organelles in intracellular signaling; cytochrome C, calcium and reactive oxygen species are all released from the mitochondria and serve as second messengers in signal transduction cascades. For example, the release of cytochrome C from the mitochondria to the cytoplasm causes the activation of caspases that lead to apoptosis (Kluck et al. 1997).

All cells, including embryonic tissues, terminally differentiated adult cells, and malignant cancer cells, have requirements for ATP generation, anabolic reactions to support growth and turnover of cellular biomass, and response to internal and external growth control signals. In cancer cells, these needs are tilted towards increased anabolism and insensitivity to growth inhibitory and apoptotic signals, and this altered metabolism is now appreciated as an important hallmark that distinguishes cancer from normal cells (Hanahan and Weinberg 2011). Early hypothesis of this deranged cancer metabolism proposed that cancer developed due to an injury to respiration (Warburg 1956). However, despite their elevated glucose uptake and reliance on glycolysis, mitochondrial function is required for the development of cancer (Weinberg et al. 2010). Cancer is not a disease of insufficient energy; conversely it is a disease of excessive anabolic reactions (Zielinski et al. 2017). Altered mitochondrial function supports these needs and may offer appealing diagnostic and therapeutic targets for cancer treatment. The major mitochondrial alterations and their consequences are summarized in Fig. 1.

The following sections review and summarize known metabolic differences between normal and cancer cells with an emphasis on mitochondrial metabolism. We will provide examples of the differences in metabolism between treatment sensitive tumors and more aggressive, treatment refractory and metastatic cancer. We conclude with a discussion of the clinical utility of existing or proposed mitochondrially targeted anti-cancer drugs.

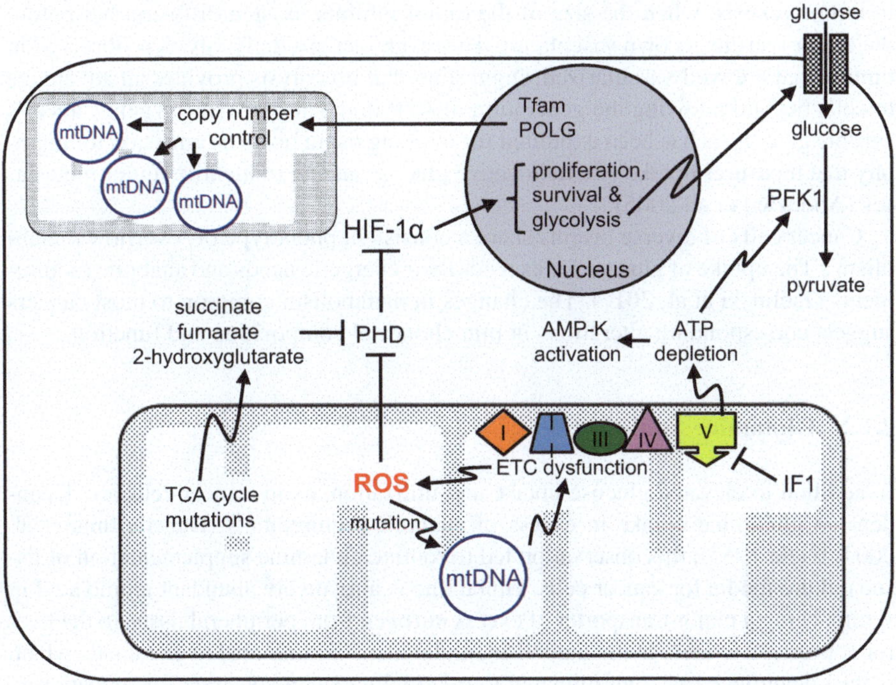

Fig. 1 Mitochondrial alterations that contribute to cancer. Mutations in nuclear and mitochondrial encoded electron transport chain (ETC) proteins increase reactive oxygen species (ROS). ETC complexes produce ROS that oxidizes mitochondrial DNA (mtDNA), causing more mutations. ETC dysfunction causes ATP depletion; many tumor cells increase expression of inhibitory factor-1 (IF1), further depleting ATP. Phosphofructokinase-1 (PFK1), the rate limiting step in glycolysis and glucose uptake, is allosterically inhibited by ATP, and ATP depletion in cancer cells may account for their elevated glucose uptake. In the cytoplasm, ROS inhibits prolyl hydroxylase (PHD), stabilizing hypoxia inducible factor 1α (HIF1α) and allowing it to transactivate genes that promote glycolysis, growth and survival under hypoxic conditions. Mutations in tricarboxylic acid (TCA) enzymes produce elevated succinate, fumarate, and the oncometabolite 2-hydroxyglutarate, that act as inhibitors of prolyl hydroxylase (PHD) and other α-ketoglutarate dependent dioxygenases. In some cancer cells, mitochondrial dysfunction results in a compensatory increase in mitochondrial biogenesis and mtDNA copy number, mediated by DNA polymerase γ (POLG) and mitochondrial transcription factor A (Tfam)

2 Mitochondria in Cancer

2.1 *Changes in Metabolism with Carcinogenesis*

2.1.1 Elevated Glucose Uptake

Cancer cell metabolism is characterized by increased uptake of glucose and increased flow through glycolysis. Reliance on glycolysis, which is not dependent on the availability of oxygen, may provide a growth advantage to tumor cells early in tumorigenesis, when the size of the tumor inhibits oxygen diffusion but before the tumor recruits its own vasculature. However, increased glycolysis is observed in tumors that are well vascularized, suggesting that glycolysis provides an advantage to cells beyond allowing the generation of ATP in the absence of oxygen. Cancer's aerobic glycolysis has been exploited for imaging using positron emission tomography that measures uptake of a radioactive glucose analog to identify tumor metastases (Abouzied et al. 2005).

Cancer cells of diverse origins share a consistent phenotype of 'overflow metabolism': The uptake of glucose far exceeds their energetic needs and anabolic requirements (Zielinski et al. 2017). The changes in metabolism common to most cancers suggest corresponding alterations in mitochondrial composition and function.

2.1.2 Glutamine Addiction

In addition to elevated glucose uptake and utilization, many cancer cells are dependent on glutamine uptake in excess of anabolic requirements (Deberardinis et al. 2007; Eagle 1955); this observation led to routine glutamine supplementation of tissue culture media for cancer cells. Glutamine is a relatively abundant amino acid in serum, being a major transporter of excess nitrogen from peripheral tissue to the liver for conversion to urea. In its catabolism, glutamine is deaminated to glutamate, which is then deaminated or transaminated to α-ketoglutarate, a substrate for oxidation to carbon dioxide by tricarboxylic acid cycle enzymes. Thus glutamine catabolism in the mitochondria can contribute to the generation of reducing equivalents used in the electron transport chain in addition to biosynthetic precursors. Glutamine utilization is facilitated by gene expression patterns driven by oncogenes (Wise et al. 2008; Shroff et al. 2015; Gao et al. 2009; Qie et al. 2014). Glutamine catabolism may contribute to cancer cell proliferation by providing an alternative way of generating reducing equivalents, sparing glucose for biosynthetic reactions that build new cell mass. Glutamine may also promote cancer development through generation of the oncometabolite 2-hydroxyglutarate (Terunuma et al. 2014), a product of α-ketoglutarate metabolism through mutant isocitrate dehydrogenase isoforms.

2.1.3 Elevated Mitochondrial Reactive Oxygen Species

Mitochondria are the major source of reactive oxygen species within normal and cancer cells. Superoxide is generated through single electron transfer to molecular oxygen through 'leakage' from the electron transport chain (Cadenas et al. 1977), and this superoxide can then be converted to hydrogen peroxide through dismutation and to hydroxyl radical by non-enzymatic reactions. Because of accumulated mutations in mitochondrial and nuclear encoded electron transport chain proteins, cancer cells have increased ROS relative to normal cells. While they play important physiological roles, ROS can also damage nucleic acids, membrane lipids, and proteins. The accumulation of ROS driven mutations may promote cancer up to a point, until a lethal threshold of damage is reached and the cell dies by apoptosis or necrosis. At sublethal levels, ROS promotes cell survival, proliferation, and the glycolytic phenotype of cancer through stabilization of hypoxia inducible factor-1α (HIF-1α) (Ma et al. 2013; Park et al. 2003; Mansfield et al. 2005; Jung et al. 2008). Thus for cancer development and progression, mitochondrial ROS must be maintained within a range that is high enough to promote the survival response without being so great that the cell is no longer viable (Sabharwal and Schumacker 2014).

2.2 *Mitochondrial DNA and Cancer*

Changes in the mtDNA abundance or mutations in the mtDNA genotype could underlie the altered metabolism of cancer. Mitochondria have a small circular genome encoding tRNAs, rRNAs, and 13 essential subunits of the electron transport chain. While most of the mitochondrial proteins are encoded by nuclear DNA and synthesized in cytoplasm, the contribution of mitochondrial DNA (mtDNA) encoded proteins is essential to mitochondrial function. Each mitochondrion has multiple copies of mtDNA, and the number of copies of mtDNA is dynamic relative to the fixed number of nuclear chromosomes. mtDNA is especially susceptible to oxidative damage because of its proximity to the electron transport chain, its lack of association with histones, and the mitochondria's lack of the full array of DNA repair systems found in the nucleus; this deficiency in mtDNA repair may be exacerbated in cancer cells (Clayton et al. 1974; Mambo et al. 2002; Croteau et al. 1999).

2.2.1 mtDNA Copy Number

Cells from energetic tissues such as muscle have a higher ratio of mtDNA to nuclear DNA than less active cells such as adipocytes, and the mitochondrial DNA copy number is a surrogate marker of the respiratory capacity of a tissue (Annex and Williams 1990). Decreased mtDNA copy number, along with increased expression of mitochondrial transcripts, is observed upon viral oncogene transformation and in several cancer derived cell lines (Torroni et al. 1990). The progression from normal tissue to hyperplasia to cancer is well described in colorectal cancer (Fearon and Vogelstein 1990), and many studies have examined mtDNA copy number in this system. mtDNA copy number is largely controlled by the mitochondrial transcription factor Tfam (Ekstrand et al. 2004), a nuclear encoded gene that binds mtDNA and regulates its transcription and regulation. The expression and activity of DNA polymerase gamma, the mitochondria specific DNA polymerase, also regulates mtDNA copy number (Lee et al. 2015; Kaguni 2004).

Disruption of the APC tumor suppressor is an early event in colorectal cancer tumorigenesis, and mice with disruption of APC alleles are a useful model of multistage colorectal cancer development (Yamada and Mori 2007). Mice heterozygous for loss of function mutations in Tfam ($Tfam^{+/-}$) have depleted mtDNA and increased oxidative damage to mtDNA, reflecting metabolic stress. Significantly, when these mice were bred with mice heterozygous for APC mutations ($APC^{+/MIN}$), offspring bearing both mutations had significantly increased tumor burden, and this phenotype was rescued by mitochondrial targeted antioxidant (Woo et al. 2012). These results suggest that mtDNA depletion creates metabolic stress that can act to compound the effects of tumor suppressor loss. Interestingly, Tfam has been shown to physically interact with the tumor suppressor p53 to form a complex at damaged mtDNA (Yoshida et al. 2003; Wong et al. 2009), and p53 positively regulates transcription of the Tfam promoter (Wen et al. 2016). Loss of p53 is common in colorectal cancer and many others, and its loss may allow mtDNA depletion, the accumulation of damaged mtDNA and increased oxidative stress (Chen et al. 2006).

Decreased polymerase gamma function may also lead to decreased mtDNA copy number and contribute to carcinogenesis. Increased methylation of the polymerase gamma alpha subunit (POLG) gene results in reduced expression and decreased mtDNA copy number (Kelly et al. 2012). POLG mutations are associated with decreased mtDNA copy number in colorectal cancer (Linkowska et al. 2015). POLG mutations are common in breast cancer, and POLG mutations found in breast cancer both decrease the mtDNA and increase the tumorigenic potential of cells (Singh et al. 2009, 2015).

Clinical studies of mtDNA copy number in cancer have not produced consistent results. A study of Chinese colorectal cancer patients found decreased mtDNA in tumor tissue relative to adjacent normal tissue, and an association between low tumor mtDNA and lymph node metastases and decreased overall survival (Cui et al. 2013). However a larger study of Dutch patients found no difference between mtDNA in normal and tumor tissue, nor any correlation with pathological grade or lymph node metastases (Van Osch et al. 2015). Interestingly, this study found that the middle quintiles of tumor mtDNA copy number had significantly worse overall survival than patients whose tumors had either the highest or the lowest mtDNA copy number, suggesting that tumor cells are able to thrive within a threshold of mitochondrial oxidative stress, with enough to activate cell survival pathways but without overwhelming the antioxidant capacity of the tumor cell and causing irreparable damage. In a Canadian population, a majority of patients' colorectal cancer had decreased mtDNA relative to normal tissue, but no association between mtDNA copy number and overall survival or disease free survival was observed (Mohideen et al. 2015). The mtDNA copy number in circulating blood cells is also associated with the risk of developing colorectal cancer, with middle quartiles of mtDNA copy number having reduced colorectal cancer incidence compared with subjects with higher or lower mtDNA. This association is only found with carcinoma; there is no association between circulating mtDNA and adenoma development (Thyagarajan et al. 2016, 2012). This result suggests that germline factors controlling mtDNA copy number may underlie susceptibility to cancer, and mtDNA may be an easily interrogated biomarker that predicts cancer risk.

In breast cancer, a low mtDNA copy number in the primary tumor predicted worse 10-year disease-free survival, but was not associated with pathological subtype or epithelial mesenchymal transition in a Dutch patient population (Weerts et al. 2016). Similarly, a Taiwanese study showed no correlation between mtDNA content and pathological features of the tumor, however this study observed that patients with increased mtDNA in their primary tumors had worse disease free survival after treatment with anthracycline chemotherapy (Hsu et al. 2010). More studies of mtDNA copy number in different tumor models, along with correlations with the molecular and pathological features of the tumor, are clearly warranted to resolve discrepancies and to unlock the predictive value of mtDNA copy number.

mtDNA copy number has the potential to be a ‘master switch’ in cancer cells, with decreased mtDNA copy number early in tumorigenesis providing decreased electron transport chain capacity that is running at a higher rate and generating more ROS, causing the accumulation of more damage to the mitochondrial and nuclear genomes. Cancer cells’ mtDNA copy number is tightly regulated; cancer cells depleted of mtDNA restore it to normal levels during tumorigenesis, either through mitochondrial biogenesis or acquisition of host mtDNA (Lee et al. 2016). Increased mtDNA copy number, observed in treatment refractory cancer cells (Walter et al. 2015), may be a compensation for mitochondrial dysfunction and may restore or partially restore metabolic flexibility for cells to reduce NADH either through fermentation or respiration, and decoupling redox state with ATP synthesis. This could be advantageous for cancer cells by providing reducing power for anabolic reactions that produce new cell mass, while avoiding the buildup of ATP, which is inhibitory to nutrient uptake through allosteric inhibition of phosphofructokinase-1. Consistent with this idea, post-translational modification of phosphofructokinase-1 by proteolytic cleavage resulting in insensitivity to allosteric inhibition by ATP is a common feature of many cancer cell lines (Smerc et al. 2011).

2.2.2 mtDNA Mutations

Many features of mtDNA make it more susceptible to accumulating mutations than nuclear DNA. Mitochondria are the primary source of oxidative stress within cells, thus mtDNA is in close proximity to a major mutagen. mtDNA is not packaged in histones and is more accessible to oxidative damage, and because mtDNA does not contain introns, that damage is more likely to have functional consequences. Finally, the mitochondria lack the full complement of DNA repair enzymes found in the nucleus. While many mutations in mtDNA are described in cancer, the consensus is that mtDNA mutations are not overtly oncogenic but can modify the potential of other carcinogenic insults to the cell.

Vogelstein provided an early survey of mtDNA mutations in human colorectal cancer; he found mtDNA mutations in ~70% of tumors. Sanger sequencing revealed that mutant tumor mtDNA were primarily T→C and G→A conversions, consistent with oxidative damage, and that mtDNA were homoplasmic in normal tissue, tumor, and cell lines derived from tumors; however the sequencing techniques at the time may have overlooked rare mutations maintained in heteroplasmy. mtDNA harboring mutations replaced wild type mtDNA in cell fusion experiments. Mutations in the coding sequences of NADH dehydrogenase, cytochrome C oxidase, and cytochrome B suggest that mitochondria with mutant mtDNA produce more oxidative stress and provide a growth advantage or additional mutational burden on the nuclear genome (Polyak et al. 1998). A more recent survey of mitochondrial mutations in multiple cancer types confirmed these observations and found that mutations that result in amino acid substitutions in proteins tended to undergo positive selection towards homoplasmy in cancer, while nonsense mutations that caused premature stop codons were maintained in heteroplasmy (Ju et al. 2014). This result implies that electron transport chain function is required for cancer development and progression, but that decreased electron transport chain capacity or fidelity is a cancer driver.

Advances in sequencing technology have yielded more insight into the spectrum of mutations observed in cancer. In contrast to the fixed number of chromosomes in the nucleus, the study of mutations in mitochondria are complicated by the dynamic copy number of mtDNA leading to heteroplasmy, the co-existence of wild type and mutant mtDNA within a cell. mtDNA mutations may need to accumulate to a threshold concentration within cells to cause any pathological effects. Indeed many mtDNA mutations associated with cancer rapidly achieve homoplasmy, completely replacing wild type mtDNA, when introduced to cells. This suggests that a selection process favors cancer associated mtDNA mutations, although it is possible that random assortment of mtDNA during cell division can produce homoplasmy in the absence of a selective advantage (Coller et al. 2001).

2.3 Changes in Electron Transport Chain Complexes in Cancer

2.3.1 Complex I

Complex I of the electron transport chain catalyzes the oxidation of NADH, with transfer of electrons to ubiquinone. Some of the energy from this redox reaction is used to pump protons across the inner mitochondrial matrix. The oxidation of NADH by complex I is critical to replenish the NAD^+ pool and allowing flux through the tricarboxylic acid cycle, and the proton pumping contributes to the electrochemical gradient across the inner mitochondrial membrane that powers ATP synthase. Thus loss of complex I activity impinges on the intracellular metabolite pool and cellular energetics. Complex I in humans consists of 45 proteins, 7 of which are encoded on the mitochondrial genome. Deficiency of NADH dehydrogenase activity can result from inherited or acquired mutations in either the nuclear or mtDNA (Mimaki et al. 2012). Genes encoding complex I subunits are the largest component of the mitochondrial genome, and thus changes to complex I is the most likely consequence of mtDNA mutation.

Defects in complex I activity are observed in oncocytic tumors. The oncocytic phenotype describes the swollen and abnormally abundant mitochondria which fill the cytoplasm. Oncocytic tumors frequently originate from follicular cells of the thyroid gland, though they can appear in other tissues, such as the kidney, as well. Oncocytic tumors are usually benign adenomas, but oncocytic carcinomas occur rarely. Oncocytic tumors have decreased complex I activity, and mutations in mitochondrially encoded complex I subunits are frequent (Gasparre et al. 2007), with a compensatory response of increased mitochondrial biogenesis (Baris et al. 2004). The decrease in complex I activity has the effect of decreasing the activity of α-ketoglutarate dehydrogenase and increasing the ratio of α-ketoglutarate to succinate. α-ketoglutarate acts as a substrate for many dioxygenases including the prolyl hydroxylase that marks HIF-1α for ubiquitination and destruction (Porcelli et al. 2010). HIF-1α stabilization promotes glycolysis and resistance to apoptotic cell death (Carmeliet et al. 1998), and its inactivation in oncocytic tumors may account for their slow growth and benign character. Transmitochondrial hybrid experiments have shown that complex I mutations in mtDNA can modify the character of tumor cells (Iommarini et al. 2014; Cruz-Bermudez et al. 2015; Ishikawa et al. 2008), but there are no reports that the mtDNA complex I mutations found in oncocytic tumors are sufficient by themselves to transform a normal cell into an immortal cancer cell.

NADH dehydrogenase activity modulates the tumorigenic properties of other cancer types as well. Mutations in mitochondrially encoded complex I genes are associated with breast cancer (Czarnecka et al. 2010; Parrella et al. 2001), and complex I activity regulates breast cancer cell phenotype. Decreasing complex I activity increases cell migration and invasion, while increasing its activity pharmacologically decreases tumorigenesis and metastases (Santidrian et al. 2013). This phenomena is the opposite of in oncocytic tumors, wherein increased complex I activity correlates with a more invasive phenotype.

2.3.2 Complex II

Complex II, succinate dehydrogenase, catalyzes the oxidation of succinate to fumarate, providing an alternative source of electrons to ubiquinone and linking the electron transport chain to the tricarboxylic acid cycle. The protein subunits of complex II are all encoded by nuclear DNA, distinguishing it from the other complexes of the electron transport chain, which all have essential subunits encoded by mtDNA (Rutter et al. 2010).

Mutations in the nuclear encoded subunits of complex II cause familial paraganglioma/phaeochromocytoma syndrome (Thompson 2009). Paragangliomas are tumors originating from paraganglia. Paraganglia in the adrenal medulla are responsible for catecholamine secretion, while peripheral paraganglia serve to detect dissolved oxygen and carbon dioxide in the blood. Paraganglioma refers to any tumor of paraganglia origin; phaeochromocytoma refers specifically to tumors within the adrenal medulla. The tumors are usually benign, but cause morbidity by compressing nearby structures or secreting excess catecholamines (Chetty 2010).

While most tumor suppressors serve in direct gatekeeper roles for the cell cycle and DNA integrity, succinate dehydrogenase serves as a metabolic tumor suppressor that indirectly impacts cell survival decisions. In cells with complex II deficiency, diminishment of succinate dehydrogenase activity increases the mitochondrial pool of succinate, which can exit the mitochondria and act as an inhibitor of the prolyl hydroxylase that signals HIF-1α degradation (Hewitson et al. 2007). This increases the expression of HIF-1α regulated genes that promote cell survival, angiogenesis, and glycolysis (Pollard et al. 2006, 2005; Selak et al. 2005). The accumulation of succinate also inhibits other a-ketoglutarate dependent dioxygenases such as 5-methylcytosine hydroxylases, altering nuclear DNA methylation patterns and providing additional transcriptional derangement that promotes growth in cancer cells (Xiao et al. 2012).

2.3.3 Complex III

Complex III forms the cytochrome b-c1 complex, whose function in the electron transport chain is to oxidize coenzyme Q and transfer electrons to cytochrome C. The electrons pass through multiple heme iron and iron sulfur centers, and the electron transfer provides the energy to pump four protons across the inner mitochondrial membrane (Xia et al. 2013). The cytochrome B peptide is encoded in mtDNA, while the other ten subunits and additional assembly factors are encoded on nuclear DNA (Schon et al. 2012). Defects in complex III activity attributed to specific mutations in nuclear encoded subunits are described in breast, ovarian and other cancers.

Rieske iron complexes are proteins that function in redox reactions including in bacterial photosynthesis and in the mitochondria in the cytochrome b-c1 complex. In the cytochrome b-c1 complex, nuclear encoded Rieske iron complexes oxidize coenzyme Q and transfer electrons to the complex III core. The 19q12 chromosomal region is amplified in numerous cancer types (Sait et al. 2002; Leung et al. 2006; Lin et al. 2000; Natrajan et al. 2012; Noske et al. 2017) and contains the gene for cyclin E, an S phase cyclin and potential driver of deregulated cell cycle progression. The gene for the complex III Rieske iron complex, UQCRFS1, is also located on chromosome 19q12 (Pennacchio et al. 1995), and may provide additional metabolic support for the transformed phenotype. Decreasing Rieske expression decreases migration and invasion of breast, prostate and bladder cancer cells, and complex III inhibitors increase the anti-proliferation activities of the antiestrogen tamoxifen in breast cancer cells (Wang et al. 2007; Theodossiou et al. 2012; Owens et al. 2011). Blocking complex III pharmacologically generates oxidative stress in glioblastoma multiforme cells, resulting in DNA damage and cell cycle phase arrest (Bhattacharya et al. 2014). These data suggest that coordinated increases in metabolic capacity support deregulation of growth control, resulting in increased cancer cell proliferation and invasive potential.

Complex III is a source of superoxide generation within mitochondria, however unlike complex I, which generates reactive oxygen species (ROS) in the mitochondrial matrix, complex III produces superoxide in the intermembrane space (Bleier and Drose 2013; Muller et al. 2004; Waypa et al. 2013; Drose and Brandt 2012; Han et al. 2001). ROS produced by complex III plays a key role in stabilizing HIF-1α independently of oxidative phosphorylation (Guzy et al. 2005; Brunelle et al. 2005). Thus complex III plays a role both in energy metabolism and signaling to promote survival and proliferation of cancer cells.

2.3.4 Complex IV

Complex IV is the final electron carrier in the electron transport chain and the rate limiting step for electron transfer. Complex IV, or cytochrome c oxidase, catalyzes the oxidation of cytochrome C and the reduction of oxygen to water. Three of cytochrome oxidase's 13 subunits are encoded in mtDNA, with the remaining 10 subunits, along with a host of assembly factors, encoded by nuclear DNA (Kadenbach and Huttemann 2015).

Mutations and altered expression of complex IV subunits have been reported in numerous types of cancers. In colorectal cancer, non-small cell lung cancer, and glioblastoma multiforme, increased expression of mitochondrial and nuclear encoded subunits is observed, and increased activity of cytochrome c oxidase is associated with advanced pathological grade and poor clinical outcomes (Wallace et al. 2016; Griguer et al. 2013; Chen et al. 2012). Consistent with a role in promoting cancer growth, inhibitors of complex IV decrease cancer proliferation in glioma cells and in a glioblastoma xenograft model (Oliva et al. 2016; Owens et al. 2011). Similar results with an inhibitor of cytochrome c oxidase were attributed to HIF degradation in endometrial and ovarian cancer cell lines (Kim et al. 2015).

While most alterations in complex IV in cancer focus on increased expression and activity, there are exceptions. Mutations resulting in protein loss or decreased activity of complex IV have also been described in colorectal cancer (Namslauer et al. 2011), and in esophageal cancers loss of cytochrome c oxidase promotes increased glycolysis and invasion (Dong et al. 2015; Srinivasan et al. 2016). If the main role of complex IV in supporting cancer cell proliferation is to generate ROS and stabilize HIF, potentially this could be accomplished by increasing the expression and activity of wild type complexes and increasing the basal rate of electron 'leak'. However, increased ROS could also be generated from complex IV mutations that result in decreased activity but more leak.

2.3.5 Complex V

Complex V uses the power of the proton concentration gradient across the inner mitochondrial membrane to catalyze the phosphorylation of ADP to ATP. The complex has two functional parts: The F0 component spans the mitochondrial membrane and forms the proton pore, while the F1 component projects into the mitochondrial matrix and adds inorganic phosphate to ADP to form ATP. Two peptides that form part of the F0 component, mATP6 and mATP8, are encoded on mtDNA. The other complex V subunits and assembly factors are encoded on nuclear DNA. Functional complex V exists as monomers, dimers and oligomers within the mitochondria (Devenish et al. 2008; Ruhle and Leister 2015). Loss of function mutations in mATP6 cause severe seizure disorders characterized by lactic acidosis, such as Leigh syndrome (Moslemi et al. 2005; Sgarbi et al. 2006). Complex V physically interacts with the tumor suppressor p53 in the mitochondria (Bergeaud et al. 2013), thus complex V represents a link between control of growth arrest and metabolism that could be corrupted in cancer cells and exploited in cancer therapy.

Overexpression of mATP6 is observed upon fibroblast transformation and in cancer derived cell lines (Torroni et al. 1990). mATP6 mutations are frequently observed in cancer as well, including in breast cancer (Thapa et al. 2016; Ghaffarpour et al. 2014), lung cancer (Choi et al. 2011), osteosarcoma (Guo et al. 2013), and bladder cancer (Guney et al. 2012). While most reports of mtDNA mutations in cancer have not tested the functional consequences of the mutations, characterization of a handful of ATP6 mutations in cancer showed some were phenotypically silent while others resulted in decreased ATP production (Niedzwiecka et al. 2016b), suggesting that compromised complex V activity can contribute to tumorigenesis but is not a requirement for transformation.

Complex V is regulated by binding to inhibitory factor 1 (IF1), a protein that binds and inhibits mitochondrial ATP synthase by promoting dimer formation (Garcia-Bermudez and Cuezva 2016; Garcia et al. 2006). IF1 induction is implicated in metabolic reprogramming associated with chemical carcinogenesis (Hardonniere et al. 2017). IF1 is expressed at high levels in glioma, breast cancer, colon cancer and lung cancers relative to normal tissue, and overexpression of IF1 causes a compensatory increase in glycolysis (Wu et al. 2015; Sánchez-Aragó et al. 2013; Sanchez-Cenizo et al. 2010). Interestingly, while high expression of IF1 is associated with poor clinical outcomes in many tumor types, in breast cancer cells IF1 overexpression decreases migration and invasion (Garcia-Ledo et al. 2017). In addition to forcing cells to rely on glycolysis for their energy needs, IF1 inhibition of mitochondrial ATP synthase causes hyperpolarization of mitochondria and increased superoxide production, increasing cell proliferation and survival through HIF-1α stabilization (Formentini et al. 2012; Esparza-Molto et al. 2017).

In addition to maintaining ATP homeostasis, complex V plays a major role in apoptosis. In response to high levels of reactive oxygen species (ROS), mitochondrial ATP synthase dimers form calcium dependent permeability transition pores that depolarize mitochondria and allow release of calcium and other small molecular weight substances to the cytoplasm (Bernardi et al. 2015; Giorgio et al. 2013). Mitochondrial permeability transition pore activation is reported to result in Bax localization to the mitochondrial outer membrane and cytochrome c release (Matsuyama et al. 1998; Chiara et al. 2012; Narita et al. 1998; De Giorgi et al. 2002), leading to effector caspase activation and programmed cell death. Resistance to apoptosis is a hallmark of cancer (Hanahan and Weinberg 2011), and cancer cells subvert the normal formation of the permeability transition pore to evade apoptosis (Rasola and Bernardi 2014). Further study of complex V's dual role in energy production and cell death decisions is warranted to identify vulnerabilities that can be exploited to improve cancer therapies.

2.4 Tricarboxylic Acid Cycle Enzymes and Cancer

The tricarboxylic acid (TCA) cycle refers to a set of mitochondrial enzyme complexes with both catabolic and anabolic role. In its basic anabolic role, the TCA cycle functions to oxidize two carbons in acetyl CoA to carbon dioxide, transferring electrons to NADH and FAD(2H) for use in ATP generation through the electron transport chain. The TCA cycle's diverse anabolic functions include converting carbohydrate derived carbon to citrate for fatty acid synthesis, generating malate from gluconeogenic precursors, and serving as a source of succinyl CoA for heme synthesis. The derangement of these nuclear encoded enzymes through mutation or aberrant expression can promote both the altered metabolism and resistance to apoptosis that are hallmarks of cancer.

2.4.1 Fumarate Hydratase

In a similar manner to succinate dehydrogenase (complex II), fumarate hydratase also functions as a tumor suppressor. Fumarate hydratase is a nuclear encoded tricarboxylic acid cycle enzyme that catalyzes that conversion of fumarate to malate, the step immediately after succinate dehydrogenase in the TCA cycle. The energy change of this reaction is very low and the reaction is freely reversible with changes in substrate and product concentration (Gajewski et al. 1985). Fumarate hydratase and succinate dehydrogenase have similar effects on the intracellular succinate and α-ketoglutarate pools, and loss of fumarate hydratase through inherited or acquired mutations causes increased resistance to apoptosis through HIF-1α stabilization and inhibition of other dioxygenases as described above for succinate dehydrogenase mutations.

However, emerging evidence indicates that fumarate hydratase deficiency has distinct features from succinate dehydrogenase deficiency. Fumarate itself acts as a competitive inhibitor of HIF prolyl hydroxylase (Isaacs et al. 2005; Hewitson et al. 2007). Fumarate hydratase is produced both with and without a mitochondrial localization signal via alternative transcriptional initiation (Dik et al. 2016), and an intriguing role for the enzyme in nuclear DNA repair has been described. In response to DNA damage, fumarate hydratase physically interacts with histones near double strand breaks and converts malate to fumarate. The nuclear fumarate in turn inhibits histone demethylases at the site of DNA damage, facilitating non-homologous end joining to repair the double stranded break (Jiang et al. 2015; Yogev et al. 2010). This additional function of fumarate hydratase may explain the different spectrum of tumors associated with fumarate hydratase mutations compared with succinate dehydrogenase mutations. While there is overlap between tumor types associated with fumarate hydratase and succinate dehydrogenase mutations, inherited or acquired loss of fumarate hydratase is predominantly associated with uterine leiomyoma and renal cell carcinoma, while succinate dehydrogenase mutations are associated primarily with paragangliomas (Smit et al. 2011; Mann et al. 2015; Ricketts et al. 2012; Harrison et al. 2016).

2.4.2 Isocitrate Dehydrogenase

Isocitrate dehydrogenase catalyzes the oxidative decarboxylation of isocitrate to α-ketoglutarate. The mitochondrial isoform IDH3 uses NAD^+ as an electron acceptor and is essentially irreversible, while IDH2 uses $NADP^+$ as a redox partner and is reversible. A cytosolic isoform, IDH1, also uses $NADP^+$. IDH1 and IDH2 were first identified as tumor suppressors in gliomas (Yan et al. 2009; Hartmann et al. 2009; Sonoda et al. 2009) and later observed in leukemias (Mardis et al. 2009; Andersson et al. 2011). For most tumor suppressors, multiple tumor causing mutations in each gene have been identified, implying that any mutation that compromises their function will promote cancer. However, the tumor associated IDH mutations were confined to specific catalytic arginines. The tumor causing IDH mutations were shown to be rare neomorphic or gain of function mutations, with the mutant enzymes reducing α-ketoglutarate to 2-hydroxyglutarate and oxidizing NADPH (Dang et al. 2009; Gross et al. 2010; Ward et al. 2010). 2-hydroxyglutarate treatment alone can transform human cells (Koivunen et al. 2012; Losman et al. 2013), thus it is deemed an oncometabolite. The 2-hydroxyglutarate may act as a competitor for the α-ketoglutarate dependent prolyl hydroxylase that targets hypoxia inducible factor for degradation, leading to its stabilization and promotion of survival and growth pathways (Zhao et al. 2009); it also may contribute to cancer cell survival and proliferation through remodeling transcriptional patterns through modification of histone and cytosine methylation (Lu et al. 2012; Chowdhury et al. 2011; Xu et al. 2011).

3 Therapies Exploiting Mitochondrial Differences in Cancer

3.1 Exploiting Metabolic Differences Between Cancer Cell and Normal Cells

The preceding sections outlined how alterations in mitochondrial metabolism optimize cancer cell formation and growth. Differences between normal and cancer mitochondria and metabolism have been known for decades. Unfortunately, very few treatments in current use are based on these differences. The ideal cancer treatment is completely toxic to cancer cells and not toxic to normal cells. Agents capable of delivering selective toxicity should be based on molecular differences between cancer and normal cells. Increased glycolysis and depressed or faulty respiration are hallmarks of cancer. Developing treatments based on these differences has been difficult due to complex and interconnected biochemical pathways and an incomplete understanding of the primary metabolic alteration of cancer. The principle metabolic alteration driving cancer growth remains unknown. Defective oxidative phosphorylation was thought to be the primary defect in cancer metabolism for many years. Increased glycolysis was felt to produce ATP not synthesized from defective cancer mitochondria (Seyfried and Shelton 2010). More recently, increased glycolysis is interpreted as a means to produce more anabolic substrates to support biosynthesis needed for growth and division (Vander Heiden et al. 2009). Central carbon pathways including glycolysis and TCA produce both energy (ATP) and biomass precursors. Manipulating the activity of these pathways will alter production of both ATP and synthetic precursors. Determining if energy or biomass is the key product driving cancer growth is difficulty to know, but is critically important to develop targeted cancer treatments. Despite this challenge, many metabolic therapies have been developed and more are under investigation. Preclinical and clinical research has attempted to exploit metabolic differences between normal and cancer cells despite a precise understanding of these differences.

The following discussion highlights some of the many cancer treatments based on mitochondrial metabolism; agents are summarized in Table 1. The list of agents is not complete; a thorough review of all agents with metabolic activity is beyond the scope of this chapter. We discuss representative agents working through one of the known alterations in cancer: Increased glycolysis for anabolic reactions and inefficient respiration producing less ATP and more ROS. Agents targeting additional mitochondrial functions are reviewed by Boukalova et al. in Chap. 27 of this book. The agents are categorized based on their function in anabolism, ATP or ROS metabolism. These categories are artificial, and alternative categorizations have been proposed (Neuzil et al. 2013). Categorization of agents is limited by the overlapping metabolic functions of the mitochondria and their interconnections with other metabolic compartments. The categorization below is meant to provide an approximate mechanistic framework for discussion.

As described above, altering pathway activity changes not only the carbon products of the pathway but also reducing equivalent and ATP products. The categories are based on current understanding of the predominant mechanism of cancer cytotoxicity. The metabolic consequences of DNA damage are considered given the ubiquitous use of DNA damaging agents as cancer treatments. Optimal response to DNA damage includes cell cycle arrest, which itself requires metabolic

Table 1 Mitochondrial targeted anti-cancer therapies and their proposed mechanisms of action

Drug	Proposed mechanism	Clinical use
3-Bromopyruvate	Enzyme Alkylation, Inhibits glycolysis and glutathione	Yes
Metformin	Complex I inhibition	Yes
Tamoxifen	Complex I inhibition	Yes
CB-839	Glutaminase inhibition, blocking anaplerosis	Yes
Lonidamine	Hexokinase II and Complex II inhibition	Yes
2-Deoxyglucose	Glycolysis and Pentose Phosphate Pathway inhibition	Yes
Azidothymidine	Mitochondrial DNA depletion	Yes
Tamoxifen	Complex I inhibition	Yes
Low calorie/ ketogenic diet	Numerous, key mechanism unknown	Yes
Bedaquilinie	Inhibits ATP synthesis	Yes
Atovaquone	Inhibits ATP synthesis	Yes
Dichloroacetate	Activates Pyruvate Dehydrogenase and Mitochondrial metabolism, Increases ROS production	
Arsenic trioxide	Increase ROS production	Yes
Genipin	Inhibits Uncoupling protein 2	Yes, as herbal medicine

adjustment to slow glycolysis and increase mitochondrial function. Differences in metabolism are already being exploited clinically, as an underappreciated mechanism of DNA damaging agents. The new but rapidly expanding field of metabolomics may provide more precise and targetable definition of the metabolic changes in cancer metabolism.

3.2 Cancer Cells Have Increased Anabolic Activity Supported by TCA Cycle Intermediates: Agents to Slow Anabolism

If excess and irrepressible growth defines cancer, then blocking key anabolic pathways should slow cancer progression. Mitochondria play a key role in synthesis of several important compounds necessary for growth including amino acids, nucleotides, heme, lipids and others from TCA cycle intermediates. Antimetabolites are a foundation of cancer chemotherapy. Examples include methotrexate and fluorouracil which block DNA metabolism. Agents directed toward disrupting cancer specific glucose and mitochondrial metabolism are discussed below. Clinical experience with these agents will be described where available.

3.2.1 3-Bromopyruvate

3-Bromopyruvate (3BP) is a pyruvate analog with alkylating activity. It is taken up by monocarboxylate transporters (MCT) which are preferentially abundant in many cancer cells where MCT facilitates lactate expulsion from the cell. 3BP has several sites of action (Lis et al. 2016). It can inhibit glycolysis via covalent modification (pyruvalation or alkylation) of key metabolic enzymes. 3PB inhibits hexokinase II (HKII) and glyceraldehyde 3-phosphate dehydrogenase. 3BP interaction with HKII may interfere with its association with the mitochondrial voltage dependent anion channel leading to apoptosis (Chen et al. 2009). 3BP has been used in animals and humans with promising results (Ko et al. 2004). Initial studies in rats showed both significant activity against xenografted liver tumors and excellent drug tolerance. 3BP augments the toxicity of more standard cytotoxic cancer treatments (Ihrlund et al. 2008). By blocking glycolysis, it inhibits production of biosynthetic precursors derived from glycolytic intermediates. 3BP can also alkylate glutathione, leading to increased ROS (Niedzwiecka et al. 2016a). Promising results in cancer cell lines showing 3BP can inhibit glycolysis and disrupt oxidative defense led to *in vivo* trials. One case report describes the use of 3BP in a young patient with fibrolamellar hepatocarcinoma (Ko et al. 2012). The patient had an initial abrupt response and transient control for several months. 3BP treatment did cause peripheral neuropathy, but overall was well tolerated. Other clinical experience may be associated with greater toxicity. News reports from a center in Germany using 3BP for cancer treatments describe treatment related deaths in three patients shortly after initiating 3BP therapy (Feldwisch-Drentrup 2016). Currently, no studies registered with clinicaltrials.gov are using 3BP.

3.2.2 Metformin

Metformin lowers blood sugar and is used by millions of patients to treat diabetes. Large retrospective data suggests diabetic patients taking metformin have lower incidence of cancer (Decensi et al. 2010) and lower recurrence rates after cancer treatment (Coyle et al. 2016). Metformin has many effects, including lowering blood glucose, anti-inflammatory effects and decreased mTOR pathway activity. How metformin inhibits cancer is currently under investigation. One potential explanation is biosynthesis inhibition by decreasing TCA cycle and complex I activity. Griss et al. (2015) have shown metformin inhibits fatty acid production from TCA cycle derived citrate. Metformin may decrease cancer cell growth via disruption in $NADH/NAD^+$ homeostasis (Gui et al. 2016). Complex I inhibition by metformin may be useful as a cancer treatment since blocking complex I would lead to an increase in the $NADH/NAD^+$ ratio, inhibiting the TCA cycle and decreasing the production of biosynthetic intermediates such as α-ketoglutarate and oxaloacetic acid. These two compounds provide the carbon backbone for glutamate and aspartate respectively. Nucleotide, protein, fatty acid and many other compounds are derived from these TCA cycle intermediates. Metformin increases the effectiveness of more traditional chemotherapy agents such as paclitaxel. Rocha et al. (2011) found metformin decreased cancer cell and xenograft tumor growth in combination with paclitaxel. However, retrospective clinic data suggests metformin may not improve outcome in all tumor types for uses. For example, diabetic patients using metformin did not have significantly better outcomes than other patients undergoing treatment for esophageal cancer (Spierings et al. 2015). Hundreds of cancer trials using metformin in a variety of applications are now underway. Data from prospective clinical trials is needed to fully appreciate the potential benefit of metformin.

3.2.3 Glutamine Metabolism Inhibitors

Glutaminolysis refers to the use of glutamine as an anaplerotic substrate to replenish the TCA cycle. Many cancers show high rates of glutaminolysis (Zielinski et al. 2017). Increased glutamine consumption by cancer may provide the means to synthesize a number of compounds, including deoxynucleotide triphosphates (dNTPs), other amino acids and fatty acids (via citrate). Glutamine serves as the major source of TCA activity for some cancers. Agents to inhibit the conversion of glutamine to glutamate to α-ketoglutarate have shown selective toxicity against cancer in vitro and in vivo (reviewed by Altman et al. 2016). Currently, clinical trials are examining the safety of CB-839, an allosteric inhibitor of glutamate synthase GLS2, an enzyme acting to convert glutamine to glutamate. Inhibiting glutamate synthase would block the production α-ketoglutarate derived from glutamate and therefore block the ability of glutamine to serve as an anaplerotic substrate. Preliminary data for this agent presented as a poster at the 2016 San Antonio Breast Cancer Conference by DeMichele et al. (2016) show CB-839 is well tolerated and potentially active doses can safely be achieved. Glutaminase inhibition appears synergistically toxic when combined with other agents such as erlotinib and other more traditional cancer treatments in xenografted animals (Momcilovic et al. 2017). Several clinical trials are now underway evaluating CB-839 in a variety of cancers either as a single agent or combined with other anti-cancer treatments.

3.2.4 Lonidamine

Lonidamine was developed as a male fertility drug and subsequently was found to have anticancer activity. Indeed, this drug appears to have several metabolic targets including mitochondrial bound HKII. Lonidamine also inhibits entry of pyruvate into mitochondria and lactate export out of the cell. It may also interfere with electron transport at complex II (Nath et al. 2016). Which of these activities is most important in reducing cancer growth is not clear. Several studies have examined lonidamine in cancer patients. Overall, lonidamine adds a slight increase in tumor response to chemotherapy and radiation therapy but does not provide a substantial improvement in overall survival (Cervantes-Madrid et al. 2015).

3.3 Cancer Cells Have Less Efficient Oxidative Phosphorylation: Agents that Increase ATP Demand

Cancer mitochondria frequently generate more reactive oxygen species (ROS) and less ATP. If cancer cells have limited ATP production capacity, then treatments requiring elevated rates of ATP synthesis should be particularly toxic to cancer.

3.3.1 3-Bromopyruvate

In addition to the activities ascribed to 3BP in the section above, 3BP also depletes ATP in cancer cells. Davidescu et al. (2015) showed 3BP treatment of glioblastoma cells in culture depletes cellular ATP levels resulting in cell death. This is an example of the broad effects of antimetabolic therapies that can make narrow categorizations difficult.

3.3.2 2-Deoxyglucose

2-Deoxyglucose (2DG) is a glucose analog that can be phosphorylated by hexose kinase but cannot be further metabolized. It inhibits glucose utilization in both glycolysis and pentose phosphate pathways. Anti cancer effects of 2DG have been known for several decades (Laszlo et al. 1960). Given the importance of glucose metabolism in many cell functions, the critical action of 2DG is still investigated and potentially different in different cell and cancer types. 2DG can inhibit glycolysis leading to lower ATP levels and decreased biosynthetic precursor production. Pentose phosphate pathway inhibition likewise blocks ribose and subsequent nucleotide production as well as NADPH synthesis used for anabolic reactions and antioxidant enzyme activity. Indeed, several lines of evidence suggest 2DG creates substantial toxicity by blocking NADPH production and increasing susceptibility to oxidative stress (Lin et al. 2003; Simons et al. 2009). Given the central role of glucose in cancer metabolism, 2DG was investigated in clinical trials dating back to the 1950s (Landau et al. 1958). As a single agent, 2DG was able to produce some transient cancer responses. These initial studies also established the safety of 2DG. Subsequent in vitro and in vivo data investigated combining 2DG with other therapies. 2DG augments the cytotoxicity of many cancer treatments including ionizing radiation and various chemotherapy drugs. For example, radiation combined with 2DG augments glioblastoma response and is currently under investigation in a phase III clinical trial (Dwarakanath et al. 2009). 2DG is chemically very similar to the commonly used positron emission tomography (PET) imaging agent 2-F18-Fluorodeoxyglucose (FDG). FDG PET imaging may be useful in defining which patients have tumors with high glucose affinity and may be more susceptible to 2DG (Simons et al. 2007).

3.3.3 POLG Inhibition

Mitochondrial DNA codes for 13 proteins used in electron transport and several rRNA and tRNAs used for mitochondrial protein production. Respiration is not possible without mtDNA. As noted previously in this chapter, mitochondrial DNA polymerase and mtDNA copy number are associated with cancer formation and progression, and mtDNA mutations are associated with human cancers. Although much more work is required for a complete understanding of mtDNA dynamics and cancer, a significant portion of cancers show mtDNA depletion relative to their normal cell counterparts. mtDNA is thought to be more susceptible to DNA damage because it is not packaged in protective chromatin, is physically located next to ROS production and has more limited DNA repair mechanisms available. For all these reasons, mtDNA in cancer may be preferentially susceptible to DNA damaging agents. Azidothymidine or AZT, a nucleotide analog inhibitor of retroviral reverse transcriptase, has significant activity against mtDNA through inhibition of POLG. For example, mtDNA depletion and other mitochondrial effects may be responsible for fatigue, myopathy, and lactic acidosis related to AZT treatment (Scruggs and Dirks Naylor 2008). AZT has been used during treatment for patients with both HIV infection and cancer and appears safe. AZT is part of an effective T-cell lymphoma regimen (Bazarbachi et al. 2010). AZT enhances toxicity of cisplatin and fluorodeoxyuridine chemotherapy and radiation treatments in cultured cancer cells (Chen et al. 2010; Mattson et al. 2009). At least some of the enhanced toxicity from AZT is related to increased oxidative stress. AZT is currently under investigation in several clinical trials mainly in HIV related lymphomas.

3.3.4 Tamoxifen

Tamoxifen is commonly used as adjuvant treatment for breast cancers that express estrogen receptors. Tamoxifen is a partial agonist and partial antagonist of estrogen receptor in different tissues. It has substantial benefit with significant decrease in risk of recurrence and improved survival (Goss et al. 2016). Tamoxifen also reduces the activity of complex I. Tamoxifen leads to ATP depletion independent of its interaction with the estrogen receptor (Moreira et al. 2006). More recent studies confirm the ability of tamoxifen to deplete ATP. Tamoxifen has broad metabolic effects mediated in part through mTOR inhibition (Daurio et al. 2016).

3.3.5 Dietary Restriction

Dietary manipulations limiting the number and type of caloric intake can improve both the safety and efficacy of cancer treatment (Vernieri et al. 2016). Starvation both inhibits cancer growth and sensitizes cancer to chemotherapy and radiation therapy in both cultured cells and xenograft tumor models (Lee et al. 2012). Although the mechanism is not completely understood, restrictive diets likely alter the range of metabolic options open to either normal or cancer cells reducing the ability to generate ATP. Caloric restriction increases markers for DNA damage and apoptosis (Lee et al. 2012). Other mechanisms may include alterations in circulating hormone levels. Several clinical studies are currently underway examining the potential benefits of calorie restriction or ketogenic diets during cancer treatment. Although promising in animals, dietary restrictions may be difficult for patients to tolerate during cancer treatment (Zahra et al. 2017).

3.3.6 Bedaquiline and Atovaquone

Bedaquiline and atovaquone are currently used as antibiotics, and they are known to block electron transport and decrease ATP production (Fiorillo et al. 2016a, b). They appear to have anticancer properties *in vitro* and *in vivo*. Atovaquone also improves the efficacy of radiation, perhaps by reducing hypoxia (Ashton et al. 2016). At least one clinical trial is currently evaluating atovaquone in cancer patients. The safety of these agents is already established since they have been used for several years as antibiotics. Hopefully their proven safety will allow rapid investigation in oncology.

3.4 Cancer Cells Have Impaired Oxidative Phosphorylation: Agents that Increase ROS

Cancer mitochondria frequently produce ROS at elevated rates, promoting cell survival and generating additional mutations that drive cancer. Agents that incite additional ROS production may overwhelm ROS defense mechanisms, or drastically alter activity in ROS signaling pathways in cancer cells resulting in cytotoxicity.

3.4.1 Dichloroacetic Acid

Dichloroacetic acid (DCA) is a structural analog of pyruvate. It has known activity in blocking an inhibitor of pyruvate dehydrogenase (pyruvate dehydrogenase kinase), with the net effect of increasing PDH activity (Whitehouse et al. 1974). PDH converts pyruvate to acetyl CoA in mitochondria, providing more substrate for the TCA cycle and thereby increasing glucose oxidation. DCA is an oral agent previously developed to treat metabolic disorders. It has been used for several years to treat lactic acidosis resulting from inherited mitochondrial disorders. Bonnet et al. (2007) showed DCA enhances mitochondrial activity with a subsequent increase in tumor apoptosis, increased cancer cell reactive oxygen species production and decreased cancer growth rates both *in vitro* and *in vivo* in animal studies. DCA has been described as well tolerated by patients in clinical use both in metabolic disorders and in cancer. DCA has shown clinical activity against lymphoma, glioblastoma and metastatic colon cancer (Michelakis et al. 2010; Khan et al. 2016). Indeed, DCA increases mitochondrial activity, ROS production, apoptosis and tumor response in glioblastoma patients. Michelakis further showed DCA levels sufficient to inhibit pyruvate dehydrogenase kinase could be reached in patients with tolerable side effects. Toxicity was limited to peripheral neuropathy (Michelakis et al. 2010). DCA is currently under investigation in lung, breast and brain cancers (clinicaltrials.gov).

3.4.2 Arsenic Trioxide

Arsenic trioxide (ATO) creates a number of cellular effects, including increased cellular oxidative stress (Flora 2011). ATO is used currently to treat acute promyelocytic leukemia (APL). ATO enhances ROS production in human hepatocellular cells (Alarifi et al. 2013). It also activates apoptosis in a number of cancer cell types *in vitro*. Kumar et al. (2014) showed ATO induces oxidative stress, DNA damage and apoptosis in leukemia cells *in vitro*. ATO also overcomes tamoxifen resistance in cultured breast cancer cells (Martinez-Outschoorn et al. 2011). It has been used in clinical trials and appears to synergize with conventional chemotherapy agents as well as agents with metabolic targets. Work done in leukemias and small cell lung cancer. ATO is currently under investigation in a number of clinical cancer trials, mostly in combination with other agents.

3.4.3 Pentose Phosphate Pathway Inhibition

Approximately 10% of glucose entering a cell is metabolized through the pentose phosphate pathway, which reduces NADPH and generates five carbon sugars for nucleotide synthesis. The pentose phosphate pathway is an attractive pharmacological target for cancer treatment because of cancer cells' high requirements for nucleotides and their high ROS production. Pentose phosphate pathway inhibitors 2-deoxyglucose, dihydroepiandrosterone, and 6-aminonicotinamide show some activity in vitro. Inhibiting thioredoxin may further augment cancer toxicity by altering redox metabolism (Li et al. 2015).

3.5 Uncoupling Inhibitors

Cancer cells may unlink ATP production from NADH oxidation, allowing mitochondria to make synthetic precursors without inhibition from excess ATP production. Uncoupling agents dissipate the mitochondrial proton gradient generated by electron transport that supplies the energy for ATP synthesis. Chemicals such as dinitrophenol can uncouple electron transport from ATP synthesis. Endogenous cell proteins also have this ability. At least four uncoupling proteins have been identified in mammals. Uncoupling protein 1 is important in cold acclimation and increases heat production from brown fat. Metabolic remodeling is associated with uncoupling protein 2 (UCP2) (Samudio et al. 2009). Many cancers over express UCP2, and overexpression is associated with chemotherapy resistance (Baffy et al. 2011). Inhibitors of UCP2 such as genipin may have some activity and are currently under investigation as potential cancer treatments (Mailloux et al. 2010).

3.6 DNA Damaging Agents: Established Cancer Treatments that Already Exploit Mitochondrial Differences

Cancer research over the past several decades has used many different approaches to find new treatments, ranging from open-ended large scale chemical screening to find molecules toxic to cancer cell lines to selectively engineered molecules based on detailed biochemical information on oncogenic proteins. If metabolism is truly a hallmark of cancer, with altered metabolism as an obligate step in cancer formation, progression and mortality, why have more metabolic treatments not been identified?

DNA damage is at least part of the mechanism of action for most standard cancer treatments. Ionizing radiation, cisplatin, alkylating agents, doxorubicin and many others work at least in part by damaging DNA. The metabolic demands of DNA damage and repair are not entirely known, but likely are quite distinct from the demands of rapid growth and division characteristic of cancer. Perhaps the DNA damaging agents in current use as cancer treatments also have significant metabolic activities. The metabolic demands of DNA damage based cancer therapy and the specific metabolic alterations in cancer must both be understood to improve the efficacy and selectivity of treatment. In this section, we suggest the metabolic alterations that define cancer may already be targeted clinically as an unappreciated component of toxicity for DNA damaging agents.

Gene expression changes following DNA damage in yeast show changes in hundreds of genes, many involved in mitochondrial activity and cellular metabolism (Spellman et al. 2000; Jelinsky and Samson 1999). Likewise, screening genome wide collections of mutants for sensitivity to DNA damaging agents such as doxorubicin identify not only DNA repair functions, but also mutants defective in various metabolic activities (Xia et al. 2007; Westmoreland et al. 2009; Hoepfner et al. 2014). While these studies can identify which genes and pathways are active in responding to DNA damage, large scale screens alone cannot define the mechanism of action or provide sufficient detail to explain how metabolic changes protect cells from DNA damage.

Biochemical studies examining metabolic response to DNA damage in yeast and cancer cells suggest glycolysis slows and PPP activity increases (Kitanovic and Wolfl 2006; Kitanovic et al. 2009). Studies in mammalian cells suggest p53 coordinates a decrease in glycolysis and an increase in PPP following DNA damage (Franklin et al. 2016). Mitochondrial responses are harder to characterize and may fluctuate over time. Studies in yeast suggest a complex mitochondrial response to DNA damage. One early response is an increase in dNTP synthesis (Chabes et al. 2003) leading to an increase in mitochondria number (Taylor et al. 2005). Indeed, increased mitochondria copy number appears to improve survival. However, prolonged DNA damage results in decreased mitochondrial activity (Simpson-Lavy et al. 2015). The precise mechanism explaining how mitochondrial activity ameliorates DNA damage toxicity is still not known but actively investigated.

Optimal response to DNA damage includes cell cycle arrest to allow for efficient repair. Cells that continue to grow and divide with damaged DNA will produce inviable or highly mutant progeny. Metabolic changes must occur to allow cells to arrest after experiencing DNA damage. One early and critical participant in DNA damage response is ATM. ATM signaling, perhaps through sirtuin enzyme activation, results in decreased glycolysis and increased PPP activity (Cosentino et al. 2011). Slowing glycolysis may provide the most efficient mechanism to slow production of biosynthetic precursors necessary for growth and division. Slowing glycolysis may be the most efficient means to achieve cell cycle arrest. Biosynthesis cannot be stopped completely or the cell will not be able to produce enzymes and substrates necessary to mitigate stress and effect repair. Mitochondrial activity may temporarily increase to produce damage response and repair precursors from TCA intermediates (Dornfeld et al. 2015), then decrease to facilitate complete cell cycle arrest. If the DNA damaging agent is still present after cell cycle arrest, almost all anabolic reactions should decrease to allow a stable and prolonged arrest. Decreasing mitochondrial activity during this period would allow arrest. Therefore, mitochondrial activity may both increase and decrease in the presence of DNA damage, depending on the time examined. Conflicting observations showing mitochondrial activity both increasing and decreasing after DNA damage may both be correct. Mitochondrial activity may increase initially to allow synthesis of repair intermediates (dNTPs and amino acids through TCA cycle intermediates), then decrease to achieve full cell cycle arrest.

The metabolic derangements of cancer are optimized for growth and may not allow cells to undergo the series of metabolic events is necessary for optimal survival following DNA damage. Cancer metabolism has been described as less flexible than normal cell metabolism (Olson et al. 2016). The lack of flexibility may be useful for optimizing already existing treatments. Mutations in growth promoting genes may severely restrict the ability of cancer cells to slow glycolysis as a means to achieve cell cycle arrest. For example, the myc oncogene is frequently expressed in cancers and its expression is associated with increased glycolysis (Sabnis et al. 2017).

In addition, a survey of breast cancer specimens found an association between mutant p53 and upregulation of glycolytic enzymes (Harami-Papp et al. 2016). Death by mitotic catastrophe is common after treatments such as radiotherapy and is consistent with inability to slow metabolism and arrest cell cycle progression. Activating mitochondrial metabolism to produce biosynthetic precursors from TCA intermediates will also produce mitochondrial NADH. Increasing NADH re-oxidation through complex I may result in a greater flux through defective and toxic and inefficient electron transport leading to elevated ROS and cytotoxicity. In normal cells, greater ETC flux increases ATP to provide energy for repair and also to act as an allosteric inhibitor of anabolic pathways, contributing to cell cycle arrest (Mensonides et al. 2013). In cancer, increased ETC flux may be toxic due to increased ROS production (Panieri and Santoro 2016). Uncoupling NADH oxidation from ATP production may provide cancer cells an opportunity to regenerate NAD^+ allowing for ongoing TCA activity while generating less ATP and ROS. Uncoupling dissipates the proton gradient normally providing the energy for ATP production. Indeed, uncoupling proteins have been described and appear to play a significant role in metabolic remodeling (Samudio et al. 2009). Increased production of UCP2, an uncoupling protein, is associated with chemoresistance, consistent with increased TCA cycle activity promoting growth through dNTP and fatty acid synthesis without producing the potential growth inhibiting allosteric inhibitor ATP (Samudio et al. 2009). Cell cycle arrest following DNA damage is important to allow for faithful repair. Inappropriate progression through the cell cycle with continued growth and division may allow ongoing cell proliferation and "survival" but produce progeny with high mutation rates. This is indeed what is seen with cancer. The mitochondrial alterations allowing for increased growth may also produce toxic and/or inhibitory compounds and increasing mitochondrial activity to respond to DNA damage may further increase toxic compounds.

If increasing mitochondrial activity is helpful to normal cells and toxic to cancer cells, enhancing the mitochondrial response to DNA damage may further separate the effects of DNA damage in normal and cancer cells. For example, supplementing anaplerotic substrates during and after DNA damage may widen the therapeutic window of DNA damaging agents. As described above, starvation prior to a number of DNA damage based cancer treatments leads to both protection in normal cells and extra toxicity in cancer cells (Lee et al. 2012). Mitochondrial amplification in response to starvation may contribute to the extra protection for normal cells and toxicity in cancer cells since starvation can induce mitochondrial activity. Along these lines, inducing mitochondrial biogenesis with troglitazone enhances doxorubicin toxicity in breast cancer cells (Skildum et al. 2011). Combining mitochondrial amplification with gentle electron transport blockade may further lead to excess ROS and toxicity in cancer.

Metabolic treatments against cancer must be performed with a good understanding of the metabolic condition and abilities of the cancer and the metabolic demands of the treatment. Combining DNA damage with metabolic treatment requires knowledge regarding the ability of cancer, relative to normal cells, to meet the metabolic needs of DNA damage. Metabolic treatments that are active against cancer as single treatments may be more or less effective when combined with DNA damaging agents. More work is required to understand the metabolic limitations of cancer, the metabolic demands of DNA damage and the diagnostic tools to assess baseline and response in metabolism in individual patients.

One final thought: If the above series of events is accurate, cells respond to damage by decreasing glycolysis and inducing mitochondrial function. Growth is slower in cells using mitochondrial metabolism without the abundant source of synthetic precursors supplied by glycolysis. If mitochondrial metabolism is a fundamental strategy to reduce growth and division in damaged cells, then overcoming the features of mitochondrial metabolism responsible for slow growth would be a necessary step in carcinogenesis. As argued above, ATP production may slow growth during mitochondrial metabolism by acting as an allosteric inhibitor of biosynthetic pathways. In this view, mitochondrial alterations in cancer would allow rapid growth by decreasing ATP production, allowing TCA and other pathways to produce biosynthetic precursors at higher rates. The primary mitochondrial metabolic defect in cancer may therefore be decreasing ATP production to facilitate anabolic pathways.

3.7 *Selecting the Right Intervention: Tumor Metabolism In Vivo*

Mitochondrial aberrations in cancer allow increased growth and division and avoid normal arrest or apoptosis signals. Mutations or defects in many pathways can achieve these goals, creating many potential mitochondrial and metabolic therapeutic targets. Many of the agents listed above have broad activities affecting many pathways. Improving cancer treatment will depend on defining which metabolic derangements are present in an individual's cancer, then treating with agents specifi cally toxic to that defect. While great strides have been made in genotyping cancer, determining the metabolic signature of a patients cancer is still more an aspiration than reality. Positron emission tomography (PET) scanning with the glucose analog ^{18}F-2 deoxyglucose can identify cancers with relatively low or high glucose consumption. Indeed, the degree of FDG uptake is correlated with aggressiveness for many cancers (Liu et al. 2016; Omloo et al. 2011). However, as noted above with 2DG, glucose consumption involves many pathways. Therefore high FDG avidity can identify a cancer with abnormal glucose metabolism, but cannot specify which of many pathways is important for that cancer.

Magnetic resonance spectroscopy (MRS) is a non-invasive technique to measure relative abundance of metabolites in tissues. MRS can be performed with MRI machines already in widespread use for clinical imaging. MRS is limited by sensitivity and therefore is currently used to measure chemicals with high tissue concentrations. Various isotopes such as ^{13}C can produce stronger signals in MRS detection.

Therefore, labeled compounds can be used to track metabolism. One recent study shows the exciting potential use of the oxidized compound dehydroascorbic acid as a measure of oxidative stress (Timm et al. 2017). This group showed the reduction of ^{13}C dehydroascorbic acid correlated with the ability to generate intracellular reduced glutathione. This technique may be useful in identifying tumors experiencing oxidative stress, and potentially more susceptible to oxidative stress based cancer treatments. MRS has shown promise in differentiating cancer form normal tissue and refinements may further improve metabolomics characterization of individual cancers (Bezabeh et al. 2014).

Metabolomics is a relatively new but very promising field. Chromatographic and spectroscopic techniques with increasing sensitivity and resolution are now able to analyze human tumor samples (Budczies and Denkert 2016). Altered levels of metabolites will suggest which pathways are altered for an individual's cancer and allow more selective and directed cancer treatment. Altered metabolism in cancer produces unusual metabolites. Recent study and appreciation of these metabolites has generated a new field of investigation into oncometabolites (Dang and Su 2017; Sullivan et al. 2016). One particular oncometabolite is 2-hydroxyglutarate, which accumulates to higher concentration in tumors with isocitrate dehydrogenase (IDH) mutations. IDH mutations are well characterized in a subset of glioma and other cancers and are associated with a different natural history. One particularly exciting example of the potential of metabolomic profiling characterized breast cancer genetic and metabolic alterations (Terunuma et al. 2014). This group showed 2-hydroxyglutarate accumulated to higher concentrations in tumors with an activated MYC pathway. High levels of 2-hydroxyglutarate correlated with a poor prognosis. Ongoing research in this area may identify more signature molecules or oncometabolites that specify defective or altered metabolic pathways in individual patients.

4 Summary

Metabolic and mitochondrial changes are common and likely necessary features of cancer. These changes promote autonomous growth and division while escaping normal energetic and allosteric growth inhibition. Energy and biomass building precursors can be derived from nutrients such as glucose via a broad and interconnected set of cytoplasmic and mitochondrial pathways. Complex interactions between biochemical pathways provide cancer cells with many opportunities to increase growth and have stymied metabolic based therapies. Currently available metabolically active agents likely have many effects, limiting their safety and usefulness. Higher resolution definition of metabolic defects in cancer through metabolomics and other techniques will refine our understanding of the metabolic cancer phenotype and suggest more effective targeted treatment.

References

Abouzied MM, Crawford ES, Nabi HA (2005) 18F-FDG imaging: pitfalls and artifacts. J Nucl Med Technol 33:145–155. quiz 162–163

Alarifi S, Ali D, Alkahtani S, Siddiqui MA, Ali BA (2013) Arsenic trioxide-mediated oxidative stress and genotoxicity in human hepatocellular carcinoma cells. Onco Targets Ther 6:75–84

Altman BJ, Stine ZE, Dang CV (2016) From Krebs to clinic: glutamine metabolism to cancer therapy. Nat Rev Cancer 16:619–634

Andersson AK, Miller DW, Lynch JA, Lemoff AS, Cai Z, Pounds SB, Radtke I, Yan B, Schuetz JD, Rubnitz JE, Ribeiro RC, Raimondi SC, Zhang J, Mullighan CG, Shurtleff SA, Schulman BA, Downing JR (2011) IDH1 and IDH2 mutations in pediatric acute leukemia. Leukemia 25:1570–1577

Annex BH, Williams RS (1990) Mitochondrial DNA structure and expression in specialized subtypes of mammalian striated muscle. Mol Cell Biol 10:5671–5678

Ashton TM, Fokas E, Kunz-Schughart LA, Folkes LK, Anbalagan S, Huether M, Kelly CJ, Pirovano G, Buffa FM, Hammond EM, Stratford M, Muschel RJ, Higgins GS, McKenna WG (2016) The anti-malarial atovaquone increases radiosensitivity by alleviating tumour hypoxia. Nat Commun 7:12308

Baffy G, Derdak Z, Robson SC (2011) Mitochondrial recoupling: a novel therapeutic strategy for cancer? Br J Cancer 105:469–474

Baris O, Savagner F, Nasser V, Loriod B, Granjeaud S, Guyetant S, Franc B, Rodien P, Rohmer V, Bertucci F, Birnbaum D, Malthiery Y, Reynier P, Houlgatte R (2004) Transcriptional profiling reveals coordinated up-regulation of oxidative metabolism genes in thyroid oncocytic tumors. J Clin Endocrinol Metab 89:994–1005

Baselga J, Norton L, Albanell J, Kim YM, Mendelsohn J (1998) Recombinant humanized anti-HER2 antibody (Herceptin) enhances the antitumor activity of paclitaxel and doxorubicin against HER2/neu overexpressing human breast cancer xenografts. Cancer Res 58:2825–2831

Bazarbachi A, Plumelle Y, Carlos Ramos J, Tortevoye P, Otrock Z, Taylor G, Gessain A, Harrington W, Panelatti G, Hermine O (2010) Meta-analysis on the use of zidovudine and interferon-alfa in adult T-cell leukemia/lymphoma showing improved survival in the leukemic subtypes. J Clin Oncol 28:4177–4183

Bergeaud M, Mathieu L, Guillaume A, Moll UM, Mignotte B, Le Floch N, Vayssiere JL, Rincheval V (2013) Mitochondrial p53 mediates a transcription-independent regulation of cell respiration and interacts with the mitochondrial F(1)F0-ATP synthase. Cell Cycle 12:2781–2793

Bernardi P, Rasola A, Forte M, Lippe G (2015) The mitochondrial permeability transition pore: channel formation by F-ATP synthase, integration in signal transduction, and role in pathophysiology. Physiol Rev 95:1111–1155

Bezabeh T, Ijare OB, Nikulin AE, Somorjai RL, Smith IC (2014) MRS-based metabolomics in cancer research. Magn Reson Insights 7:1–14

Bhattacharya K, Bag AK, Tripathi R, Samanta SK, Pal BC, Shaha C, Mandal C (2014) Mahanine, a novel mitochondrial complex-III inhibitor induces G0/G1 arrest through redox alteration-mediated DNA damage response and regresses glioblastoma multiforme. Am J Cancer Res 4:629–647

Bleier L, Drose S (2013) Superoxide generation by complex III: from mechanistic rationales to functional consequences. Biochim Biophys Acta 1827:1320–1331

Bonnet S, Archer SL, Allalunis-Turner J, Haromy A, Beaulieu C, Thompson R, Lee CT, Lopaschuk GD, Puttagunta L, Bonnet S, Harry G, Hashimoto K, Porter CJ, Andrade MA, Thebaud B, Michelakis ED (2007) A mitochondria-K+ channel axis is suppressed in cancer and its normalization promotes apoptosis and inhibits cancer growth. Cancer Cell 11:37–51

Brunelle JK, Bell EL, Quesada NM, Vercauteren K, Tiranti V, Zeviani M, Scarpulla RC, Chandel NS (2005) Oxygen sensing requires mitochondrial ROS but not oxidative phosphorylation. Cell Metab 1:409–414

Budczies J, Denkert C (2016) Tissue-based metabolomics to analyze the breast cancer metabolome. Recent Results Cancer Res 207:157–175

Cadenas E, Boveris A, Ragan CI, Stoppani AO (1977) Production of superoxide radicals and hydrogen peroxide by NADH-ubiquinone reductase and ubiquinol-cytochrome c reductase from beef-heart mitochondria. Arch Biochem Biophys 180:248–257

Carmeliet P, Dor Y, Herbert JM, Fukumura D, Brusselmans K, Dewerchin M, Neeman M, Bono F, Abramovitch R, Maxwell P, Koch CJ, Ratcliffe P, Moons L, Jain RK, Collen D, Keshert E (1998) Role of HIF-1alpha in hypoxia-mediated apoptosis, cell proliferation and tumour angiogenesis. Nature 394:485–490

Cervantes-Madrid D, Romero Y, Duenas-Gonzalez A (2015) Reviving lonidamine and 6-diazo-5-oxo-L-norleucine to be used in combination for metabolic cancer therapy. Biomed Res Int 2015:690492

Chabes A, Georgieva B, Domkin V, Zhao X, Rothstein R, Thelander L (2003) Survival of DNA damage in yeast directly depends on increased dNTP levels allowed by relaxed feedback inhibition of ribonucleotide reductase. Cell 112:391–401

Chen D, Yu Z, Zhu Z, Lopez CD (2006) The p53 pathway promotes efficient mitochondrial DNA base excision repair in colorectal cancer cells. Cancer Res 66:3485–3494

Chen Z, Zhang H, Lu W, Huang P (2009) Role of mitochondria-associated hexokinase II in cancer cell death induced by 3-bromopyruvate. Biochim Biophys Acta 1787:553–560

Chen CM, Johnson M, Smith BJ, Dornfeld K (2010) Azidothymidine enhances fluorodeoxyuridine-mediated radiosensitization. Int J Radiat Oncol Biol Phys 76:905–913

Chen WL, Kuo KT, Chou TY, Chen CL, Wang CH, Wei YH, Wang LS (2012) The role of cytochrome c oxidase subunit Va in non-small cell lung carcinoma cells: association with migration, invasion and prediction of distant metastasis. BMC Cancer 12:273

Chetty R (2010) Familial paraganglioma syndromes. J Clin Pathol 63:488–491

Chiara F, Gambalunga A, Sciacovelli M, Nicolli A, Ronconi L, Fregona D, Bernardi P, Rasola A, Trevisan A (2012) Chemotherapeutic induction of mitochondrial oxidative stress activates GSK-3alpha/beta and Bax, leading to permeability transition pore opening and tumor cell death. Cell Death Dis 3:e444

Choi SJ, Kim SH, Kang HY, Lee J, Bhak JH, Sohn I, Jung SH, Choi YS, Kim HK, Han J, Huh N, Lee G, Kim BC, Kim J (2011) Mutational hotspots in the mitochondrial genome of lung cancer. Biochem Biophys Res Commun 407:23–27

Chowdhury R, Yeoh KK, Tian YM, Hillringhaus L, Bagg EA, Rose NR, Leung IK, Li XS, Woon EC, Yang M, McDonough MA, King ON, Clifton IJ, Klose RJ, Claridge TD, Ratcliffe PJ, Schofield CJ, Kawamura A (2011) The oncometabolite 2-hydroxyglutarate inhibits histone lysine demethylases. EMBO Rep 12:463–469

Clayton DA, Doda JN, Friedberg EC (1974) The absence of a pyrimidine dimer repair mechanism in mammalian mitochondria. Proc Natl Acad Sci U S A 71:2777–2781

Coller HA, Khrapko K, Bodyak ND, Nekhaeva E, Herrero-Jimenez P, Thilly WG (2001) High frequency of homoplasmic mitochondrial DNA mutations in human tumors can be explained without selection. Nat Genet 28:147–150

Cosentino C, Grieco D, Costanzo V (2011) ATM activates the pentose phosphate pathway promoting anti-oxidant defence and DNA repair. EMBO J 30:546–555

Coyle C, Cafferty FH, Vale C, Langley RE (2016) Metformin as an adjuvant treatment for cancer: a systematic review and meta-analysis. Ann Oncol 27:2184–2195

Croteau DL, Stierum RH, Bohr VA (1999) Mitochondrial DNA repair pathways. Mutat Res 434:137–148

Cruz-Bermudez A, Vallejo CG, Vicente-Blanco RJ, Gallardo ME, Fernandez-Moreno MA, Quintanilla M, Garesse R (2015) Enhanced tumorigenicity by mitochondrial DNA mild mutations. Oncotarget 6:13628–13643

Cui H, Huang P, Wang Z, Zhang Y, Zhang Z, Xu W, Wang X, Han Y, Guo X (2013) Association of decreased mitochondrial DNA content with the progression of colorectal cancer. BMC Cancer 13:110

Czarnecka AM, Krawczyk T, Zdrozny M, Lubinski J, Arnold RS, Kukwa W, Scinska A, Golik P, Bartnik E, Petros JA (2010) Mitochondrial NADH-dehydrogenase subunit 3 (ND3) polymorphism (A10398G) and sporadic breast cancer in Poland. Breast Cancer Res Treat 121:511–518
Dang L, Su SM (2017) Isocitrate dehydrogenase mutation and (R)-2-hydroxyglutarate: from basic discovery to therapeutics development. Annu Rev Biochem 86:305–331
Dang L, White DW, Gross S, Bennett BD, Bittinger MA, Driggers EM, Fantin VR, Jang HG, Jin S, Keenan MC, Marks KM, Prins RM, Ward PS, Yen KE, Liau LM, Rabinowitz JD, Cantley LC, Thompson CB, Vander Heiden MG, Su SM (2009) Cancer-associated IDH1 mutations produce 2-hydroxyglutarate. Nature 462:739–744
Daurio NA, Tuttle SW, Worth AJ, Song EY, Davis JM, Snyder NW, Blair IA, Koumenis C (2016) AMPK activation and metabolic reprogramming by tamoxifen through estrogen receptor-independent mechanisms suggests new uses for this therapeutic modality in cancer treatment. Cancer Res 76:3295–3306
Davidescu M, Macchioni L, Scaramozzino G, Cristina Marchetti M, Migliorati G, Vitale R, Corcelli A, Roberti R, Castigli E, Corazzi L (2015) The energy blockers bromopyruvate and lonidamine lead GL15 glioblastoma cells to death by different p53-dependent routes. Sci Rep 5:14343
De Giorgi F, Lartigue L, Bauer MK, Schubert A, Grimm S, Hanson GT, Remington SJ, Youle RJ, Ichas F (2002) The permeability transition pore signals apoptosis by directing Bax translocation and multimerization. FASEB J 16:607–609
Deberardinis RJ, Mancuso A, Daikhin E, Nissim I, Yudkoff M, Wehrli S, Thompson CB (2007) Beyond aerobic glycolysis: transformed cells can engage in glutamine metabolism that exceeds the requirement for protein and nucleotide synthesis. Proc Natl Acad Sci U S A 104:19345–19350
Decensi A, Puntoni M, Goodwin P, Cazzaniga M, Gennari A, Bonanni B, Gandini S (2010) Metformin and cancer risk in diabetic patients: a systematic review and meta-analysis. Cancer Prev Res (Phila) 3:1451–1461
DeMichele AM, Harding JJ, Telli ML, Munster P, McKay RR, Illiopoulos O, Whiting S, Orford KW, Bennett MK, Mier JW, Owonikoko TK, Patel MR, Kalinsky K, Carvajal RD, Infante JR, Merit-Bernstam F (2016) Phase 1 study of CB-839, a small molecule inhibitor of glutaminase (GLS), in combination with paclitaxel (Pac) in patients with triple negative breast cancer (TNBC). Abstracts of the 2016 San Antonio Breast Cancer Symposium, P6-11-05
Devenish RJ, Prescott M, Rodgers AJ (2008) The structure and function of mitochondrial F1F0-ATP synthases. Int Rev Cell Mol Biol 267:1–58
Dik E, Naamati A, Asraf H, Lehming N, Pines O (2016) Human fumarate hydratase is dual localized by an alternative transcription initiation mechanism. Traffic 17:720–732
Dong DW, Srinivasan S, Guha M, Avadhani NG (2015) Defects in cytochrome c oxidase expression induce a metabolic shift to glycolysis and carcinogenesis. Genom Data 6:99–107
Dornfeld K, Madden M, Skildum A, Wallace KB (2015) Aspartate facilitates mitochondrial function, growth arrest and survival during doxorubicin exposure. Cell Cycle 14:3282–3291
Drose S, Brandt U (2012) Molecular mechanisms of superoxide production by the mitochondrial respiratory chain. Adv Exp Med Biol 748:145–169
Druker BJ, Tamura S, Buchdunger E, Ohno S, Segal GM, Fanning S, Zimmermann J, Lydon NB (1996) Effects of a selective inhibitor of the Abl tyrosine kinase on the growth of Bcr-Abl positive cells. Nat Med 2:561–566
Dwarakanath BS, Singh D, Banerji AK, Sarin R, Venkataramana NK, Jalali R, Vishwanath PN, Mohanti BK, Tripathi RP, Kalia VK, Jain V (2009) Clinical studies for improving radiotherapy with 2-deoxy-D-glucose: present status and future prospects. J Cancer Res Ther 5(Suppl 1):S21–S26
Eagle H (1955) Nutrition needs of mammalian cells in tissue culture. Science 122:501–514
Ekstrand MI, Falkenberg M, Rantanen A, Park CB, Gaspari M, Hultenby K, Rustin P, Gustafsson CM, Larsson NG (2004) Mitochondrial transcription factor A regulates mtDNA copy number in mammals. Hum Mol Genet 13:935–944

Esparza-Molto PB, Nuevo-Tapioles C, Cuezva JM (2017) Regulation of the H+-ATP synthase by IF1: a role in mitohormesis. Cell Mol Life Sci 74:2151–2166
Fearon ER, Vogelstein B (1990) A genetic model for colorectal tumorigenesis. Cell 61:759–767
Feldwisch-Drentrup H (2016) Candidate cancer drug suspected after death of three patients at an alternative medicine clinic. ScienceInsider 358
Fiorillo M, Lamb R, Tanowitz HB, Cappello AR, Martinez-Outschoorn UE, Sotgia F, Lisanti MP (2016a) Bedaquiline, an FDA-approved antibiotic, inhibits mitochondrial function and potently blocks the proliferative expansion of stem-like cancer cells (CSCs). Aging (Albany NY) 8:1593–1607
Fiorillo M, Lamb R, Tanowitz HB, Mutti L, Krstic-Demonacos M, Cappello AR, Martinez-Outschoorn UE, Sotgia F, Lisanti MP (2016b) Repurposing atovaquone: targeting mitochondrial complex III and OXPHOS to eradicate cancer stem cells. Oncotarget 7:34084–34099
Flora SJ (2011) Arsenic-induced oxidative stress and its reversibility. Free Radic Biol Med 51:257–281
Formentini L, Sanchez-Arago M, Sanchez-Cenizo L, Cuezva JM (2012) The mitochondrial ATPase inhibitory factor 1 triggers a ROS-mediated retrograde prosurvival and proliferative response. Mol Cell 45:731–742
Franklin DA, He Y, Leslie PL, Tikunov AP, Fenger N, Macdonald JM, Zhang Y (2016) p53 coordinates DNA repair with nucleotide synthesis by suppressing PFKFB3 expression and promoting the pentose phosphate pathway. Sci Rep 6:38067
Gajewski E, Goldberg RN, Steckler DK (1985) Thermodynamics of the conversion of fumarate to L-(−)-malate. Biophys Chem 22:187–195
Gao P, Tchernyshyov I, Chang TC, Lee YS, Kita K, Ochi T, Zeller KI, De Marzo AM, Van Eyk JE, Mendell JT, Dang CV (2009) c-Myc suppression of miR-23a/b enhances mitochondrial glutaminase expression and glutamine metabolism. Nature 458:762–765
Garcia JJ, Morales-Rios E, Cortes-Hernandez P, Rodriguez-Zavala JS (2006) The inhibitor protein (IF1) promotes dimerization of the mitochondrial F1F0-ATP synthase. Biochemistry 45:12695–12703
Garcia-Bermudez J, Cuezva JM (2016) The ATPase inhibitory factor 1 (IF1): a master regulator of energy metabolism and of cell survival. Biochim Biophys Acta 1857:1167–1182
Garcia-Ledo L, Nuevo-Tapioles C, Cuevas-Martin C, Martinez-Reyes I, Soldevilla B, Gonzalez-Llorente L, Cuezva JM (2017) Overexpression of the ATPase inhibitory factor 1 favors a non-metastatic phenotype in breast cancer. Front Oncol 7:69
Gasch AP, Spellman PT, Kao CM, Carmel-Harel O, Eisen MB, Storz G, Botstein D, Brown PO (2000) Genomic expression programs in the response of yeast cells to environmental changes. Mol Biol Cell 11:4241–4257
Gasparre G, Porcelli AM, Bonora E, Pennisi LF, Toller M, Iommarini L, Ghelli A, Moretti M, Betts CM, Martinelli GN, Ceroni AR, Curcio F, Carelli V, Rugolo M, Tallini G, Romeo G (2007) Disruptive mitochondrial DNA mutations in complex I subunits are markers of oncocytic phenotype in thyroid tumors. Proc Natl Acad Sci U S A 104:9001–9006
Ghaffarpour M, Mahdian R, Fereidooni F, Kamalidehghan B, Moazami N, Houshmand M (2014) The mitochondrial ATPase6 gene is more susceptible to mutation than the ATPase8 gene in breast cancer patients. Cancer Cell Int 14:21
Giorgio V, Von Stockum S, Antoniel M, Fabbro A, Fogolari F, Forte M, Glick GD, Petronilli V, Zoratti M, Szabo I, Lippe G, Bernardi P (2013) Dimers of mitochondrial ATP synthase form the permeability transition pore. Proc Natl Acad Sci U S A 110:5887–5892
Goss PE, Ingle JN, Pritchard KI, Robert NJ, Muss H, Gralow J, Gelmon K, Whelan T, Strasser-Weippl K, Rubin S, Sturtz K, Wolff AC, Winer E, Hudis C, Stopeck A, Beck JT, Kaur JS, Whelan K, Tu D, Parulekar WR (2016) Extending aromatase-inhibitor adjuvant therapy to 10 years. N Engl J Med 375:209–219
Griguer CE, Cantor AB, Fathallah-Shaykh HM, Gillespie GY, Gordon AS, Markert JM, Radovanovic I, Clement-Schatlo V, Shannon CN, Oliva CR (2013) Prognostic relevance of cytochrome C oxidase in primary glioblastoma multiforme. PLoS One 8:e61035

Griss T, Vincent EE, Egnatchik R, Chen J, Ma EH, Faubert B, Viollet B, Deberardinis RJ, Jones RG (2015) Metformin antagonizes cancer cell proliferation by suppressing mitochondrial-dependent biosynthesis. PLoS Biol 13:e1002309

Gross S, Cairns RA, Minden MD, Driggers EM, Bittinger MA, Jang HG, Sasaki M, Jin S, Schenkein DP, Su SM, Dang L, Fantin VR, Mak TW (2010) Cancer-associated metabolite 2-hydroxyglutarate accumulates in acute myelogenous leukemia with isocitrate dehydrogenase 1 and 2 mutations. J Exp Med 207:339–344

Gui DY, Sullivan LB, Luengo A, Hosios AM, Bush LN, Gitego N, Davidson SM, Freinkman E, Thomas CJ, Vander Heiden MG (2016) Environment dictates dependence on mitochondrial complex I for NAD+ and aspartate production and determines cancer cell sensitivity to metformin. Cell Metab 24:716–727

Guney AI, Ergec DS, Tavukcu HH, Koc G, Kirac D, Ulucan K, Javadova D, Turkeri L (2012) Detection of mitochondrial DNA mutations in nonmuscle invasive bladder cancer. Genet Test Mol Biomarkers 16:672–678

Guo XG, Liu CT, Dai H, Guo QN (2013) Mutations in the mitochondrial ATPase6 gene are frequent in human osteosarcoma. Exp Mol Pathol 94:285–288

Guzy RD, Hoyos B, Robin E, Chen H, Liu L, Mansfield KD, Simon MC, Hammerling U, Schumacker PT (2005) Mitochondrial complex III is required for hypoxia-induced ROS production and cellular oxygen sensing. Cell Metab 1:401–408

Han D, Williams E, Cadenas E (2001) Mitochondrial respiratory chain-dependent generation of superoxide anion and its release into the intermembrane space. Biochem J 353:411–416

Hanahan D, Weinberg RA (2011) Hallmarks of cancer: the next generation. Cell 144:646–674

Harami-Papp H, Pongor LS, Munkacsy G, Horvath G, Nagy AM, Ambrus A, Hauser P, Szabo A, Tretter L, Gyorffy B (2016) TP53 mutation hits energy metabolism and increases glycolysis in breast cancer. Oncotarget 7:67183–67195

Hardonniere K, Fernier M, Gallais I, Mograbi B, Podechard N, Le Ferrec E, Grova N, Appenzeller B, Burel A, Chevanne M, Sergent O, Huc L, Bortoli S, Lagadic-Gossmann D (2017) Role for the ATPase inhibitory factor 1 in the environmental carcinogen-induced Warburg phenotype. Sci Rep 7:195

Harrison WJ, Andrici J, Maclean F, Madadi-Ghahan R, Farzin M, Sioson L, Toon CW, Clarkson A, Watson N, Pickett J, Field M, Crook A, Tucker K, Goodwin A, Anderson L, Srinivasan B, Grossmann P, Martinek P, Ondic O, Hes O, Trpkov K, Clifton-Bligh RJ, Dwight T, Gill AJ (2016) Fumarate hydratase-deficient uterine leiomyomas occur in both the syndromic and sporadic settings. Am J Surg Pathol 40:599–607

Hartmann C, Meyer J, Balss J, Capper D, Mueller W, Christians A, Felsberg J, Wolter M, Mawrin C, Wick W, Weller M, Herold-Mende C, Unterberg A, Jeuken JW, Wesseling P, Reifenberger G, Von Deimling A (2009) Type and frequency of IDH1 and IDH2 mutations are related to astrocytic and oligodendroglial differentiation and age: a study of 1,010 diffuse gliomas. Acta Neuropathol 118:469–474

Hewitson KS, Lienard BM, McDonough MA, Clifton IJ, Butler D, Soares AS, Oldham NJ, McNeill LA, Schofield CJ (2007) Structural and mechanistic studies on the inhibition of the hypoxia-inducible transcription factor hydroxylases by tricarboxylic acid cycle intermediates. J Biol Chem 282:3293–3301

Hoepfner D, Helliwell SB, Sadlish H, Schuierer S, Filipuzzi I, Brachat S, Bhullar B, Plikat U, Abraham Y, Altorfer M, Aust T, Baeriswyl L, Cerino R, Chang L, Estoppey D, Eichenberger J, Frederiksen M, Hartmann N, Hohendahl A, Knapp B, Krastel P, Melin N, Nigsch F, Oakeley EJ, Petitjean V, Petersen F, Riedl R, Schmitt EK, Staedtler F, Studer C, Tallarico JA, Wetzel S, Fishman MC, Porter JA, Movva NR (2014) High-resolution chemical dissection of a model eukaryote reveals targets, pathways and gene functions. Microbiol Res 169:107–120

Hsu CW, Yin PH, Lee HC, Chi CW, Tseng LM (2010) Mitochondrial DNA content as a potential marker to predict response to anthracycline in breast cancer patients. Breast J 16:264–270

Ihrlund LS, Hernlund E, Khan O, Shoshan MC (2008) 3-Bromopyruvate as inhibitor of tumour cell energy metabolism and chemopotentiator of platinum drugs. Mol Oncol 2:94–101

Iommarini L, Kurelac I, Capristo M, Calvaruso MA, Giorgio V, Bergamini C, Ghelli A, Nanni P, De Giovanni C, Carelli V, Fato R, Lollini PL, Rugolo M, Gasparre G, Porcelli AM (2014) Different mtDNA mutations modify tumor progression in dependence of the degree of respiratory complex I impairment. Hum Mol Genet 23:1453–1466

Isaacs JS, Jung YJ, Mole DR, Lee S, Torres-Cabala C, Chung YL, Merino M, Trepel J, Zbar B, Toro J, Ratcliffe PJ, Linehan WM, Neckers L (2005) HIF overexpression correlates with biallelic loss of fumarate hydratase in renal cancer: novel role of fumarate in regulation of HIF stability. Cancer Cell 8:143–153

Ishikawa K, Takenaga K, Akimoto M, Koshikawa N, Yamaguchi A, Imanishi H, Nakada K, Honma Y, Hayashi J (2008) ROS-generating mitochondrial DNA mutations can regulate tumor cell metastasis. Science 320:661–664

Jelinsky SA, Samson LD (1999) Global response of Saccharomyces cerevisiae to an alkylating agent. Proc Natl Acad Sci U S A 96:1486–1491

Jiang Y, Qian X, Shen J, Wang Y, Li X, Liu R, Xia Y, Chen Q, Peng G, Lin SY, Lu Z (2015) Local generation of fumarate promotes DNA repair through inhibition of histone H3 demethylation. Nat Cell Biol 17:1158–1168

Ju YS, Alexandrov LB, Gerstung M, Martincorena I, Nik-Zainal S, Ramakrishna M, Davies HR, Papaemmanuil E, Gundem G, Shlien A, Bolli N, Behjati S, Tarpey PS, Nangalia J, Massie CE, Butler AP, Teague JW, Vassiliou GS, Green AR, Du MQ, Unnikrishnan A, Pimanda JE, Teh BT, Munshi N, Greaves M, Vyas P, El-Naggar AK, Santarius T, Collins VP, Grundy R, Taylor JA, Hayes DN, Malkin D, ICGC Breast Cancer Group; ICGC Chronic Myeloid Disorders Group; ICGC Prostate Cancer Group, Foster CS, Warren AY, Whitaker HC, Brewer D, Eeles R, Cooper C, Neal D, Visakorpi T, Isaacs WB, Bova GS, Flanagan AM, Futreal PA, Lynch AG, Chinnery PF, McDermott U, Stratton MR, Campbell PJ (2014) Origins and functional consequences of somatic mitochondrial DNA mutations in human cancer. Elife 3

Jung SN, Yang WK, Kim J, Kim HS, Kim EJ, Yun H, Park H, Kim SS, Choe W, Kang I, Ha J (2008) Reactive oxygen species stabilize hypoxia-inducible factor-1 alpha protein and stimulate transcriptional activity via AMP-activated protein kinase in DU145 human prostate cancer cells. Carcinogenesis 29:713–721

Kadenbach B, Huttemann M (2015) The subunit composition and function of mammalian cytochrome c oxidase. Mitochondrion 24:64–76

Kaguni LS (2004) DNA polymerase gamma, the mitochondrial replicase. Annu Rev Biochem 73:293–320

Kelly RD, Mahmud A, McKenzie M, Trounce IA, St John JC (2012) Mitochondrial DNA copy number is regulated in a tissue specific manner by DNA methylation of the nuclear-encoded DNA polymerase gamma A. Nucleic Acids Res 40:10124–10138

Khan A, Andrews D, Blackburn AC (2016) Long-term stabilization of stage 4 colon cancer using sodium dichloroacetate therapy. World J Clin Cases 4:336–343

Kim KK, Abelman S, Yano N, Ribeiro JR, Singh RK, Tipping M, Moore RG (2015) Tetrathiomolybdate inhibits mitochondrial complex IV and mediates degradation of hypoxia-inducible factor-1alpha in cancer cells. Sci Rep 5:14296

Kitanovic A, Wolfl S (2006) Fructose-1,6-bisphosphatase mediates cellular responses to DNA damage and aging in Saccharomyces cerevisiae. Mutat Res 594:135–147

Kitanovic A, Walther T, Loret MO, Holzwarth J, Kitanovic I, Bonowski F, Van Bui N, Francois JM, Wolfl S (2009) Metabolic response to MMS-mediated DNA damage in Saccharomyces cerevisiae is dependent on the glucose concentration in the medium. FEMS Yeast Res 9:535–551

Kluck RM, Bossy-Wetzel E, Green DR, Newmeyer DD (1997) The release of cytochrome c from mitochondria: a primary site for Bcl-2 regulation of apoptosis. Science 275:1132–1136

Ko YH, Smith BL, Wang Y, Pomper MG, Rini DA, Torbenson MS, Hullihen J, Pedersen PL (2004) Advanced cancers: eradication in all cases using 3-bromopyruvate therapy to deplete ATP. Biochem Biophys Res Commun 324:269–275

Ko YH, Verhoeven HA, Lee MJ, Corbin DJ, Vogl TJ, Pedersen PL (2012) A translational study "case report" on the small molecule "energy blocker" 3-bromopyruvate (3BP) as a potent anticancer agent: from bench side to bedside. J Bioenerg Biomembr 44:163–170

Koivunen P, Lee S, Duncan CG, Lopez G, Lu G, Ramkissoon S, Losman JA, Joensuu P, Bergmann U, Gross S, Travins J, Weiss S, Looper R, Ligon KL, Verhaak RG, Yan H, Kaelin WG Jr (2012) Transformation by the (R)-enantiomer of 2-hydroxyglutarate linked to EGLN activation. Nature 483:484–488

Kumar S, Yedjou CG, Tchounwou PB (2014) Arsenic trioxide induces oxidative stress, DNA damage, and mitochondrial pathway of apoptosis in human leukemia (HL-60) cells. J Exp Clin Cancer Res 33:42

Landau BR, Laszlo J, Stengle J, Burk D (1958) Certain metabolic and pharmacologic effects in cancer patients given infusions of 2-deoxy-D-glucose. J Natl Cancer Inst 21:485–494

Laszlo J, Humphreys SR, Goldin A (1960) Effects of glucose analogues (2-deoxy-D-glucose, 2-deoxy-D-galactose) on experimental tumors. J Natl Cancer Inst 24:267–281

Lee C, Raffaghello L, Brandhorst S, Safdie FM, Bianchi G, Martin-Montalvo A, Pistoia V, Wei M, Hwang S, Merlino A, Emionite L, De Cabo R, Longo VD (2012) Fasting cycles retard growth of tumors and sensitize a range of cancer cell types to chemotherapy. Sci Transl Med 4:124ra27

Lee W, Johnson J, Gough DJ, Donoghue J, Cagnone GL, Vaghjiani V, Brown KA, Johns TG, St John JC (2015) Mitochondrial DNA copy number is regulated by DNA methylation and demethylation of POLGA in stem and cancer cells and their differentiated progeny. Cell Death Dis 6:e1664

Lee WT, Cain JE, Cuddihy A, Johnson J, Dickinson A, Yeung KY, Kumar B, Johns TG, Watkins DN, Spencer A, St John JC (2016) Mitochondrial DNA plasticity is an essential inducer of tumorigenesis. Cell Death Discov 2:16016

Leung SY, Ho C, Tu IP, Li R, So S, Chu KM, Yuen ST, Chen X (2006) Comprehensive analysis of 19q12 amplicon in human gastric cancers. Mod Pathol 19:854–863

Li L, Fath MA, Scarbrough PM, Watson WH, Spitz DR (2015) Combined inhibition of glycolysis, the pentose cycle, and thioredoxin metabolism selectively increases cytotoxicity and oxidative stress in human breast and prostate cancer. Redox Biol 4:127–135

Lin L, Prescott MS, Zhu Z, Singh P, Chun SY, Kuick RD, Hanash SM, Orringer MB, Glover TW, Beer DG (2000) Identification and characterization of a 19q12 amplicon in esophageal adenocarcinomas reveals cyclin E as the best candidate gene for this amplicon. Cancer Res 60:7021–7027

Lin X, Zhang F, Bradbury CM, Kaushal A, Li L, Spitz DR, Aft RL, Gius D (2003) 2-Deoxy-D-glucose-induced cytotoxicity and radiosensitization in tumor cells is mediated via disruptions in thiol metabolism. Cancer Res 63:3413–3417

Linkowska K, Jawien A, Marszalek A, Malyarchuk BA, Tonska K, Bartnik E, Skonieczna K, Grzybowski T (2015) Mitochondrial DNA polymerase gamma mutations and their implications in mtDNA alterations in colorectal cancer. Ann Hum Genet. https://doi.org/10.1111/ahg.12111

Lis P, Dylag M, Niedzwiecka K, Ko YH, Pedersen PL, Goffeau A, Ulaszewski S (2016) The HK2 dependent "Warburg effect" and mitochondrial oxidative phosphorylation in cancer: targets for effective therapy with 3-bromopyruvate. Molecules 21

Liu J, Dong M, Sun X, Li W, Xing L, Yu J (2016) Prognostic value of 18F-FDG PET/CT in surgical non-small cell lung cancer: a meta-analysis. PLoS One 11:e0146195

Losman JA, Looper RE, Koivunen P, Lee S, Schneider RK, McMahon C, Cowley GS, Root DE, Ebert BL, Kaelin WG Jr (2013) (R)-2-hydroxyglutarate is sufficient to promote leukemogenesis and its effects are reversible. Science 339:1621–1625

Lu C, Ward PS, Kapoor GS, Rohle D, Turcan S, Abdel-Wahab O, Edwards CR, Khanin R, Figueroa ME, Melnick A, Wellen KE, O'rourke DM, Berger SL, Chan TA, Levine RL, Mellinghoff IK, Thompson CB (2012) IDH mutation impairs histone demethylation and results in a block to cell differentiation. Nature 483:474–478

Ma J, Zhang Q, Chen S, Fang B, Yang Q, Chen C, Miele L, Sarkar FH, Xia J, Wang Z (2013) Mitochondrial dysfunction promotes breast cancer cell migration and invasion through HIF1alpha accumulation via increased production of reactive oxygen species. PLoS One 8:e69485

Mailloux RJ, Adjeitey CN, Harper ME (2010) Genipin-induced inhibition of uncoupling protein-2 sensitizes drug-resistant cancer cells to cytotoxic agents. PLoS One 5:e13289

Mambo E, Nyaga SG, Bohr VA, Evans MK (2002) Defective repair of 8-hydroxyguanine in mitochondria of MCF-7 and MDA-MB-468 human breast cancer cell lines. Cancer Res 62:1349–1355

Mann ML, Ezzati M, Tarnawa ED, Carr BR (2015) Fumarate hydratase mutation in a young woman with uterine leiomyomas and a family history of renal cell cancer. Obstet Gynecol 126:90–92

Mansfield KD, Guzy RD, Pan Y, Young RM, Cash TP, Schumacker PT, Simon MC (2005) Mitochondrial dysfunction resulting from loss of cytochrome c impairs cellular oxygen sensing and hypoxic HIF-alpha activation. Cell Metab 1:393–399

Mardis ER, Ding L, Dooling DJ, Larson DE, Mclellan MD, Chen K, Koboldt DC, Fulton RS, Delehaunty KD, McGrath SD, Fulton LA, Locke DP, Magrini VJ, Abbott RM, Vickery TL, Reed JS, Robinson JS, Wylie T, Smith SM, Carmichael L, Eldred JM, Harris CC, Walker J, Peck JB, Du F, Dukes AF, Sanderson GE, Brummett AM, Clark E, McMichael JF, Meyer RJ, Schindler JK, Pohl CS, Wallis JW, Shi X, Lin L, Schmidt H, Tang Y, Haipek C, Wiechert ME, Ivy JV, Kalicki J, Elliott G, Ries RE, Payton JE, Westervelt P, Tomasson MH, Watson MA, Baty J, Heath S, Shannon WD, Nagarajan R, Link DC, Walter MJ, Graubert TA, Dipersio JF, Wilson RK, Ley TJ (2009) Recurring mutations found by sequencing an acute myeloid leukemia genome. N Engl J Med 361:1058–1066

Martinez-Outschoorn UE, Goldberg A, Lin Z, Ko YH, Flomenberg N, Wang C, Pavlides S, Pestell RG, Howell A, Sotgia F, Lisanti MP (2011) Anti-estrogen resistance in breast cancer is induced by the tumor microenvironment and can be overcome by inhibiting mitochondrial function in epithelial cancer cells. Cancer Biol Ther 12:924–938

Matsuyama S, Xu Q, Velours J, Reed JC (1998) The Mitochondrial F0F1-ATPase proton pump is required for function of the proapoptotic protein Bax in yeast and mammalian cells. Mol Cell 1:327–336

Mattson DM, Ahmad I, Dayal D, Parsons AD, Aykin-Burns N, Li L, Orcutt KP, Spitz DR, Dornfeld KJ, Simons AL (2009) Cisplatin combined with zidovudine enhances cytotoxicity and oxidative stress in human head and neck cancer cells via a thiol-dependent mechanism. Free Radic Biol Med 46:232–237

Mensonides FI, Bakker BM, Cremazy F, Messiha HL, Mendes P, Boogerd FC, Westerhoff HV (2013) A new regulatory principle for in vivo biochemistry: pleiotropic low affinity regulation by the adenine nucleotides—illustrated for the glycolytic enzymes of Saccharomyces cerevisiae. FEBS Lett 587:2860–2867

Michelakis ED, Sutendra G, Dromparis P, Webster L, Haromy A, Niven E, Maguire C, Gammer TL, Mackey JR, Fulton D, Abdulkarim B, Mcmurtry MS, Petruk KC (2010) Metabolic modulation of glioblastoma with dichloroacetate. Sci Transl Med 2:31ra34

Mimaki M, Wang X, McKenzie M, Thorburn DR, Ryan MT (2012) Understanding mitochondrial complex I assembly in health and disease. Biochim Biophys Acta 1817:851–862

Mohideen AM, Dicks E, Parfrey P, Green R, Savas S (2015) Mitochondrial DNA polymorphisms, its copy number change and outcome in colorectal cancer. BMC Res Notes 8:272

Momcilovic M, Bailey ST, Lee JT, Fishbein MC, Magyar C, Braas D, Graeber T, Jackson NJ, Czernin J, Emberley E, Gross M, Janes J, Mackinnon A, Pan A, Rodriguez M, Works M, Zhang W, Parlati F, Demo S, Garon E, Krysan K, Walser TC, Dubinett SM, Sadeghi S, Christofk HR, Shackelford DB (2017) Targeted inhibition of EGFR and glutaminase induces metabolic crisis in EGFR mutant lung cancer. Cell Rep 18:601–610

Moreira PI, Custodio J, Moreno A, Oliveira CR, Santos MS (2006) Tamoxifen and estradiol interact with the flavin mononucleotide site of complex I leading to mitochondrial failure. J Biol Chem 281:10143–10152

Moslemi AR, Darin N, Tulinius M, Oldfors A, Holme E (2005) Two new mutations in the MTATP6 gene associated with Leigh syndrome. Neuropediatrics 36:314–318

Muller FL, Liu Y, Van Remmen H (2004) Complex III releases superoxide to both sides of the inner mitochondrial membrane. J Biol Chem 279:49064–49073

Namslauer I, Dietz MS, Brzezinski P (2011) Functional effects of mutations in cytochrome c oxidase related to prostate cancer. Biochim Biophys Acta 1807:1336–1341

Narita M, Shimizu S, Ito T, Chittenden T, Lutz RJ, Matsuda H, Tsujimoto Y (1998) Bax interacts with the permeability transition pore to induce permeability transition and cytochrome c release in isolated mitochondria. Proc Natl Acad Sci U S A 95:14681–14686

Nath K, Guo L, Nancolas B, Nelson DS, Shestov AA, Lee SC, Roman J, Zhou R, Leeper DB, Halestrap AP, Blair IA, Glickson JD (2016) Mechanism of antineoplastic activity of lonidamine. Biochim Biophys Acta 1866:151–162

Natrajan R, Mackay A, Wilkerson PM, Lambros MB, Wetterskog D, Arnedos M, Shiu KK, Geyer FC, Langerod A, Kreike B, Reyal F, Horlings HM, Van De Vijver MJ, Palacios J, Weigelt B, Reis-Filho JS (2012) Functional characterization of the 19q12 amplicon in grade III breast cancers. Breast Cancer Res 14:R53

Neuzil J, Dong LF, Rohlena J, Truksa J, Ralph SJ (2013) Classification of mitocans, anti-cancer drugs acting on mitochondria. Mitochondrion 13:199–208

Niedzwiecka K, Dylag M, Augustyniak D, Majkowska-Skrobek G, Cal-Bakowska M, Ko YH, Pedersen PL, Goffeau A, Ulaszewski S (2016a) Glutathione may have implications in the design of 3-bromopyruvate treatment protocols for both fungal and algal infections as well as multiple myeloma. Oncotarget 7:65614–65626

Niedzwiecka K, Kabala AM, Lasserre JP, Tribouillard-Tanvier D, Golik P, Dautant A, Di Rago JP, Kucharczyk R (2016b) Yeast models of mutations in the mitochondrial ATP6 gene found in human cancer cells. Mitochondrion 29:7–17

Noske A, Brandt S, Valtcheva N, Wagner U, Zhong Q, Bellini E, Fink D, Obermann EC, Moch H, Wild PJ (2017) Detection of CCNE1/URI (19q12) amplification by in situ hybridisation is common in high grade and type II endometrial cancer. Oncotarget 8:14794–14805

Oliva CR, Markert T, Ross LJ, White EL, Rasmussen L, Zhang W, Everts M, Moellering DR, Bailey SM, Suto MJ, Griguer CE (2016) Identification of small molecule inhibitors of human cytochrome c oxidase that target chemoresistant glioma cells. J Biol Chem 291:24188–24199

Olson KA, Schell JC, Rutter J (2016) Pyruvate and metabolic flexibility: illuminating a path toward selective cancer therapies. Trends Biochem Sci 41:219–230

Omloo JM, Van Heijl M, Hoekstra OS, Van Berge Henegouwen MI, Van Lanschot JJ, Sloof GW (2011) FDG-PET parameters as prognostic factor in esophageal cancer patients: a review. Ann Surg Oncol 18:3338–3352

Owens KM, Kulawiec M, Desouki MM, Vanniarajan A, Singh KK (2011) Impaired OXPHOS complex III in breast cancer. PLoS One 6:e23846

Panieri E, Santoro MM (2016) ROS homeostasis and metabolism: a dangerous liason in cancer cells. Cell Death Dis 7:e2253

Park JH, Kim TY, Jong HS, Kim TY, Chun YS, Park JW, Lee CT, Jung HC, Kim NK, Bang YJ (2003) Gastric epithelial reactive oxygen species prevent normoxic degradation of hypoxia-inducible factor-1alpha in gastric cancer cells. Clin Cancer Res 9:433–440

Parrella P, Xiao Y, Fliss M, Sanchez-Cespedes M, Mazzarelli P, Rinaldi M, Nicol T, Gabrielson E, Cuomo C, Cohen D, Pandit S, Spencer M, Rabitti C, Fazio VM, Sidransky D (2001) Detection of mitochondrial DNA mutations in primary breast cancer and fine-needle aspirates. Cancer Res 61:7623–7626

Pennacchio LA, Bergmann A, Fukushima A, Okubo K, Salemi A, Lennon GG (1995) Structure, sequence and location of the UQCRFS1 gene for the human Rieske Fe-S protein. Gene 155:207–211

Pollard PJ, Briere JJ, Alam NA, Barwell J, Barclay E, Wortham NC, Hunt T, Mitchell M, Olpin S, Moat SJ, Hargreaves IP, Heales SJ, Chung YL, Griffiths JR, Dalgleish A, McGrath JA, Gleeson MJ, Hodgson SV, Poulsom R, Rustin P, Tomlinson IP (2005) Accumulation of Krebs cycle intermediates and over-expression of HIF1alpha in tumours which result from germline FH and SDH mutations. Hum Mol Genet 14:2231–2239

Pollard PJ, El-Bahrawy M, Poulsom R, Elia G, Killick P, Kelly G, Hunt T, Jeffery R, Seedhar P, Barwell J, Latif F, Gleeson MJ, Hodgson SV, Stamp GW, Tomlinson IP, Maher ER (2006) Expression of HIF-1alpha, HIF-2alpha (EPAS1), and their target genes in paraganglioma and pheochromocytoma with VHL and SDH mutations. J Clin Endocrinol Metab 91:4593–4598

Polyak K, Li Y, Zhu H, Lengauer C, Willson JK, Markowitz SD, Trush MA, Kinzler KW, Vogelstein B (1998) Somatic mutations of the mitochondrial genome in human colorectal tumours. Nat Genet 20:291–293

Porcelli AM, Ghelli A, Ceccarelli C, Lang M, Cenacchi G, Capristo M, Pennisi LF, Morra I, Ciccarelli E, Melcarne A, Bartoletti-Stella A, Salfi N, Tallini G, Martinuzzi A, Carelli V, Attimonelli M, Rugolo M, Romeo G, Gasparre G (2010) The genetic and metabolic signature of oncocytic transformation implicates HIF1alpha destabilization. Hum Mol Genet 19:1019–1032

Qie S, Chu C, Li W, Wang C, Sang N (2014) ErbB2 activation upregulates glutaminase 1 expression which promotes breast cancer cell proliferation. J Cell Biochem 115:498–509

Rasola A, Bernardi P (2014) The mitochondrial permeability transition pore and its adaptive responses in tumor cells. Cell Calcium 56:437–445

Ricketts CJ, Shuch B, Vocke CD, Metwalli AR, Bratslavsky G, Middelton L, Yang Y, Wei MH, Pautler SE, Peterson J, Stolle CA, Zbar B, Merino MJ, Schmidt LS, Pinto PA, Srinivasan R, Pacak K, Linehan WM (2012) Succinate dehydrogenase kidney cancer: an aggressive example of the Warburg effect in cancer. J Urol 188:2063–2071

Rocha GZ, Dias MM, Ropelle ER, Osorio-Costa F, Rossato FA, Vercesi AE, Saad MJ, Carvalheira JB (2011) Metformin amplifies chemotherapy-induced AMPK activation and antitumoral growth. Clin Cancer Res 17:3993–4005

Ruhle T, Leister D (2015) Assembly of F1F0-ATP synthases. Biochim Biophys Acta 1847:849–860

Rutter J, Winge DR, Schiffman JD (2010) Succinate dehydrogenase - assembly, regulation and role in human disease. Mitochondrion 10:393–401

Sabharwal SS, Schumacker PT (2014) Mitochondrial ROS in cancer: initiators, amplifiers or an Achilles' heel? Nat Rev Cancer 14:709–721

Sabnis HS, Somasagara RR, Bunting KD (2017) Targeting MYC dependence by metabolic inhibitors in cancer. Genes (Basel) 8

Sait SN, Qadir MU, Conroy JM, Matsui S, Nowak NJ, Baer MR (2002) Double minute chromosomes in acute myeloid leukemia and myelodysplastic syndrome: identification of new amplification regions by fluorescence in situ hybridization and spectral karyotyping. Genes Chromosomes Cancer 34:42–47

Samudio I, Fiegl M, Andreeff M (2009) Mitochondrial uncoupling and the Warburg effect: molecular basis for the reprogramming of cancer cell metabolism. Cancer Res 69:2163–2166

Sánchez-Aragó M, Formentini L, Martínez-Reyes I, García-Bermudez J, Santacatterina F, Sánchez-Cenizo L, Willers IM, Aldea M, Nájera L, Juarránz A, López EC, Clofent J, Navarro C, Espinosa E, Cuezva JM (2013) Expression, regulation and clinical relevance of the ATPase inhibitory factor 1 in human cancers. Oncogenesis 22:e46

Sanchez-Cenizo L, Formentini L, Aldea M, Ortega AD, Garcia-Huerta P, Sanchez-Arago M, Cuezva JM (2010) Up-regulation of the ATPase inhibitory factor 1 (IF1) of the mitochondrial H+-ATP synthase in human tumors mediates the metabolic shift of cancer cells to a Warburg phenotype. J Biol Chem 285:25308–25313

Santidrian AF, Matsuno-Yagi A, Ritland M, Seo BB, Leboeuf SE, Gay LJ, Yagi T, Felding-Habermann B (2013) Mitochondrial complex I activity and NAD+/NADH balance regulate breast cancer progression. J Clin Invest 123:1068–1081

Schon EA, Dimauro S, Hirano M (2012) Human mitochondrial DNA: roles of inherited and somatic mutations. Nat Rev Genet 13:878–890

Scruggs ER, Dirks Naylor AJ (2008) Mechanisms of zidovudine-induced mitochondrial toxicity and myopathy. Pharmacology 82:83–88

Selak MA, Armour SM, Mackenzie ED, Boulahbel H, Watson DG, Mansfield KD, Pan Y, Simon MC, Thompson CB, Gottlieb E (2005) Succinate links TCA cycle dysfunction to oncogenesis by inhibiting HIF-alpha prolyl hydroxylase. Cancer Cell 7:77–85

Seyfried TN, Shelton LM (2010) Cancer as a metabolic disease. Nutr Metab (Lond) 7:7

Sgarbi G, Baracca A, Lenaz G, Valentino LM, Carelli V, Solaini G (2006) Inefficient coupling between proton transport and ATP synthesis may be the pathogenic mechanism for NARP and Leigh syndrome resulting from the T8993G mutation in mtDNA. Biochem J 395:493–500

Shroff EH, Eberlin LS, Dang VM, Gouw AM, Gabay M, Adam SJ, Bellovin DI, Tran PT, Philbrick WM, Garcia-Ocana A, Casey SC, Li Y, Dang CV, Zare RN, Felsher DW (2015) MYC oncogene overexpression drives renal cell carcinoma in a mouse model through glutamine metabolism. Proc Natl Acad Sci U S A 112:6539–6544

Simons AL, Fath MA, Mattson DM, Smith BJ, Walsh SA, Graham MM, Hichwa RD, Buatti JM, Dornfeld K, Spitz DR (2007) Enhanced response of human head and neck cancer xenograft tumors to cisplatin combined with 2-deoxy-D-glucose correlates with increased 18F-FDG uptake as determined by PET imaging. Int J Radiat Oncol Biol Phys 15:1222–1230

Simons AL, Mattson DM, Dornfeld K, Spitz DR (2009) Glucose deprivation-induced metabolic oxidative stress and cancer therapy. J Cancer Res Ther 5(Suppl 1):S2–S6

Simpson-Lavy KJ, Bronstein A, Kupiec M, Johnston M (2015) Cross-talk between carbon metabolism and the DNA damage response in S. cerevisiae. Cell Rep 12:1865–1875

Singh KK, Ayyasamy V, Owens KM, Koul MS, Vujcic M (2009) Mutations in mitochondrial DNA polymerase-gamma promote breast tumorigenesis. J Hum Genet 54:516–524

Singh B, Owens KM, Bajpai P, Desouki MM, Srinivasasainagendra V, Tiwari HK, Singh KK (2015) Mitochondrial DNA polymerase POLG1 disease mutations and germline variants promote tumorigenic properties. PLoS One 10:e0139846

Skildum A, Dornfeld K, Wallace K (2011) Mitochondrial amplification selectively increases doxorubicin sensitivity in breast cancer cells with acquired antiestrogen resistance. Breast Cancer Res Treat 129:785–797

Smerc A, Sodja E, Legisa M (2011) Posttranslational modification of 6-phosphofructo-1-kinase as an important feature of cancer metabolism. PLoS One 6:e19645

Smit DL, Mensenkamp AR, Badeloe S, Breuning MH, Simon ME, Van Spaendonck KY, Aalfs CM, Post JG, Shanley S, Krapels IP, Hoefsloot LH, Van Moorselaar RJ, Starink TM, Bayley JP, Frank J, Van Steensel MA, Menko FH (2011) Hereditary leiomyomatosis and renal cell cancer in families referred for fumarate hydratase germline mutation analysis. Clin Genet 79:49–59

Sonoda Y, Kumabe T, Nakamura T, Saito R, Kanamori M, Yamashita Y, Suzuki H, Tominaga T (2009) Analysis of IDH1 and IDH2 mutations in Japanese glioma patients. Cancer Sci 100:1996–1998

Spierings LE, Lagarde SM, Van Oijen MG, Gisbertz SS, Wilmink JW, Hulshof MC, Meijer SL, Anderegg MC, Van Berge Henegouwen MI, Van Laarhoven HW (2015) Metformin use during treatment of potentially curable esophageal cancer patients is not associated with better outcomes. Ann Surg Oncol 22(Suppl 3):S766–S771

Srinivasan S, Guha M, Dong DW, Whelan KA, Ruthel G, Uchikado Y, Natsugoe S, Nakagawa H, Avadhani NG (2016) Disruption of cytochrome c oxidase function induces the Warburg effect and metabolic reprogramming. Oncogene 35:1585–1595

Sullivan LB, Gui DY, Vander Heiden MG (2016) Altered metabolite levels in cancer: implications for tumour biology and cancer therapy. Nat Rev Cancer 16:680–693

Taylor SD, Zhang H, Eaton JS, Rodeheffer MS, Lebedeva MA, O'Rourke TW, Siede W, Shadel GS (2005) The conserved Mec1/Rad53 nuclear checkpoint pathway regulates mitochondrial DNA copy number in Saccharomyces cerevisiae. Mol Biol Cell 16:3010–3018

Terunuma A, Putluri N, Mishra P, Mathe EA, Dorsey TH, Yi M, Wallace TA, Issaq HJ, Zhou M, Killian JK, Stevenson HS, Karoly ED, Chan K, Samanta S, Prieto D, Hsu TY, Kurley SJ, Putluri V, Sonavane R, Edelman DC, Wulff J, Starks AM, Yang Y, Kittles RA, Yfantis HG, Lee DH, Ioffe OB, Schiff R, Stephens RM, Meltzer PS, Veenstra TD, Westbrook TF, Sreekumar A, Ambs S (2014) MYC-driven accumulation of 2-hydroxyglutarate is associated with breast cancer prognosis. J Clin Invest 124:398–412

Thapa S, Lalrohlui F, Ghatak S, Zohmingthanga J, Lallawmzuali D, Pautu JL, Senthil Kumar N (2016) Mitochondrial complex I and V gene polymorphisms associated with breast cancer in mizo-mongloid population. Breast Cancer 23:607–616

Theodossiou TA, Yannakopoulou K, Aggelidou C, Hothersall JS (2012) Tamoxifen subcellular localization; observation of cell-specific cytotoxicity enhancement by inhibition of mitochondrial ETC complexes I and III. Photochem Photobiol 88:1016–1022

Thompson CB (2009) Metabolic enzymes as oncogenes or tumor suppressors. N Engl J Med 360:813–815

Thyagarajan B, Wang R, Barcelo H, Koh WP, Yuan JM (2012) Mitochondrial copy number is associated with colorectal cancer risk. Cancer Epidemiol Biomark Prev 21:1574–1581

Thyagarajan B, Guan W, Fedirko V, Barcelo H, Tu H, Gross M, Goodman M, Bostick RM (2016) No association between mitochondrial DNA copy number and colorectal adenomas. Mol Carcinog 55:1290–1296

Timm KN, Hu DE, Williams M, Wright AJ, Kettunen MI, Kennedy BW, Larkin TJ, Dzien P, Marco-Rius I, Bohndiek SE, Brindle KM (2017) Assessing oxidative stress in tumors by measuring the rate of hyperpolarized [1-13C]dehydroascorbic acid reduction using 13C magnetic resonance spectroscopy. J Biol Chem 292:1737–1748

Torroni A, Stepien G, Hodge JA, Wallace DC (1990) Neoplastic transformation is associated with coordinate induction of nuclear and cytoplasmic oxidative phosphorylation genes. J Biol Chem 265:20589–20593

Van Osch FH, Voets AM, Schouten LJ, Gottschalk RW, Simons CC, Van Engeland M, Lentjes MH, Van Den Brandt PA, Smeets HJ, Weijenberg MP (2015) Mitochondrial DNA copy number in colorectal cancer: between tissue comparisons, clinicopathological characteristics and survival. Carcinogenesis 36:1502–1510

Vander Heiden MG, Cantley LC, Thompson CB (2009) Understanding the Warburg effect: the metabolic requirements of cell proliferation. Science 324:1029–1033

Vernieri C, Casola S, Foiani M, Pietrantonio F, De Braud F, Longo V (2016) Targeting cancer metabolism: dietary and pharmacologic interventions. Cancer Discov 6:1315–1333

Wallace L, Mehrabi S, Bacanamwo M, Yao X, Aikhionbare FO (2016) Expression of mitochondrial genes MT-ND1, MT-ND6, MT-CYB, MT-COI, MT-ATP6, and 12S/MT-RNR1 in colorectal adenopolyps. Tumour Biol 37:12465–12475

Walter W, Thomalla J, Bruhn J, Fagan DH, Zehowski C, Yee D, Skildum A (2015) Altered regulation of PDK4 expression promotes antiestrogen resistance in human breast cancer cells. Springerplus 4:689

Wang F, Zhang R, Xia T, Hsu E, Cai Y, Gu Z, Hankinson O (2007) Inhibitory effects of nitric oxide on invasion of human cancer cells. Cancer Lett 257:274–282

Warburg O (1956) On the origin of cancer cells. Science 123:309–314

Ward PS, Patel J, Wise DR, Abdel-Wahab O, Bennett BD, Coller HA, Cross JR, Fantin VR, Hedvat CV, Perl AE, Rabinowitz JD, Carroll M, Su SM, Sharp KA, Levine RL, Thompson CB (2010) The common feature of leukemia-associated IDH1 and IDH2 mutations is a neomorphic enzyme activity converting alpha-ketoglutarate to 2-hydroxyglutarate. Cancer Cell 17:225–234

Waypa GB, Marks JD, Guzy RD, Mungai PT, Schriewer JM, Dokic D, Ball MK, Schumacker PT (2013) Superoxide generated at mitochondrial complex III triggers acute responses to hypoxia in the pulmonary circulation. Am J Respir Crit Care Med 187:424–432

Weerts MJ, Sieuwerts AM, Smid M, Look MP, Foekens JA, Sleijfer S, Martens JW (2016) Mitochondrial DNA content in breast cancer: impact on in vitro and in vivo phenotype and patient prognosis. Oncotarget 7:29166–29176

Weinberg F, Hamanaka R, Wheaton WW, Weinberg S, Joseph J, Lopez M, Kalyanaraman B, Mutlu GM, Budinger GR, Chandel NS (2010) Mitochondrial metabolism and ROS generation are essential for Kras-mediated tumorigenicity. Proc Natl Acad Sci U S A 107:8788–8793

Wen S, Gao J, Zhang L, Zhou H, Fang D, Feng S (2016) p53 increase mitochondrial copy number via up-regulation of mitochondrial transcription factor A in colorectal cancer. Oncotarget 7:75981–75995

Westmoreland TJ, Wickramasekara SM, Guo AY, Selim AL, Winsor TS, Greenleaf AL, Blackwell KL, Olson JA Jr, Marks JR, Bennett CB (2009) Comparative genome-wide screening identifies a conserved doxorubicin repair network that is diploid specific in Saccharomyces cerevisiae. PLoS One 4:e5830

Whitehouse S, Cooper RH, Randle PJ (1974) Mechanism of activation of pyruvate dehydrogenase by dichloroacetate and other halogenated carboxylic acids. Biochem J 141:761–774

Wise DR, Deberardinis RJ, Mancuso A, Sayed N, Zhang XY, Pfeiffer HK, Nissim I, Daikhin E, Yudkoff M, McMahon SB, Thompson CB (2008) Myc regulates a transcriptional program that stimulates mitochondrial glutaminolysis and leads to glutamine addiction. Proc Natl Acad Sci U S A 105:18782–18787

Wong TS, Rajagopalan S, Freund SM, Rutherford TJ, Andreeva A, Townsley FM, Petrovich M, Fersht AR (2009) Biophysical characterizations of human mitochondrial transcription factor A and its binding to tumor suppressor p53. Nucleic Acids Res 37:6765–6783

Woo DK, Green PD, Santos JH, D'Souza AD, Walther Z, Martin WD, Christian BE, Chandel NS, Shadel GS (2012) Mitochondrial genome instability and ROS enhance intestinal tumorigenesis in APC(Min/+) mice. Am J Pathol 180:24–31

Wu J, Shan Q, Li P, Wu Y, Xie J, Wang X (2015) ATPase inhibitory factor 1 is a potential prognostic marker for the migration and invasion of glioma. Oncol Lett 10:2075–2080

Xia L, Jaafar L, Cashikar A, Flores-Rozas H (2007) Identification of genes required for protection from doxorubicin by a genome-wide screen in Saccharomyces cerevisiae. Cancer Res 67:11411–11418

Xia D, Esser L, Tang WK, Zhou F, Zhou Y, Yu L, Yu CA (2013) Structural analysis of cytochrome bc1 complexes: implications to the mechanism of function. Biochim Biophys Acta 1827:1278–1294

Xiao M, Yang H, Xu W, Ma S, Lin H, Zhu H, Liu L, Liu Y, Yang C, Xu Y, Zhao S, Ye D, Xiong Y, Guan KL (2012) Inhibition of alpha-KG-dependent histone and DNA demethylases by fumarate and succinate that are accumulated in mutations of FH and SDH tumor suppressors. Genes Dev 26:1326–1338

Xu W, Yang H, Liu Y, Yang Y, Wang P, Kim SH, Ito S, Yang C, Wang P, Xiao MT, Liu LX, Jiang WQ, Liu J, Zhang JY, Wang B, Frye S, Zhang Y, Xu YH, Lei QY, Guan KL, Zhao SM, Xiong Y (2011) Oncometabolite 2-hydroxyglutarate is a competitive inhibitor of alpha-ketoglutarate-dependent dioxygenases. Cancer Cell 19:17–30

Yamada Y, Mori H (2007) Multistep carcinogenesis of the colon in Apc(Min/+) mouse. Cancer Sci 98:6–10

Yan H, Parsons DW, Jin G, Mclendon R, Rasheed BA, Yuan W, Kos I, Batinic-Haberle I, Jones S, Riggins GJ, Friedman H, Friedman A, Reardon D, Herndon J, Kinzler KW, Velculescu VE, Vogelstein B, Bigner DD (2009) IDH1 and IDH2 mutations in gliomas. N Engl J Med 360:765–773

Yogev O, Yogev O, Singer E, Shaulian E, Goldberg M, Fox TD, Pines O (2010) Fumarase: a mitochondrial metabolic enzyme and a cytosolic/nuclear component of the DNA damage response. PLoS Biol 8:e1000328

Yoshida Y, Izumi H, Torigoe T, Ishiguchi H, Itoh H, Kang D, Kohno K (2003) P53 physically interacts with mitochondrial transcription factor A and differentially regulates binding to damaged DNA. Cancer Res 63:3729–3734

Zahra A, Fath MA, Opat E, Mapuskar KA, Bhatia SK, Ma DC III, Rodman SN, Snyders TP, Chenard CA, Eichenberger-Gilmore JM, Bodeker KL, Ahmann L, Smith BJ, Vollstedt SA, Brown HA, Hejleh TA, Clamon GH, Berg DJ, Szweda LI, Spitz DR, Buatti JM, Allen BG (2017) Consuming a ketogenic diet while receiving radiation and chemotherapy for locally advanced lung cancer and pancreatic cancer: the University of Iowa experience of two phase 1 clinical trials. Radiat Res 187(6):743–754

Zhao S, Lin Y, Xu W, Jiang W, Zha Z, Wang P, Yu W, Li Z, Gong L, Peng Y, Ding J, Lei Q, Guan KL, Xiong Y (2009) Glioma-derived mutations in IDH1 dominantly inhibit IDH1 catalytic activity and induce HIF-1alpha. Science 324:261–265

Zielinski DC, Jamshidi N, Corbett AJ, Bordbar A, Thomas A, Palsson BO (2017) Systems biology analysis of drivers underlying hallmarks of cancer cell metabolism. Sci Rep 7:41241

Mitochondria and Kidney Disease

Kenneth E. McMartin

Abstract The kidney maintains body homeostasis by regulating the excretion and/or conservation of nutrients such as glucose and amino acids, of electrolytes such as sodium and potassium, and of xenobiotics such as drugs or toxic compounds. Reabsorption and secretion of these compounds utilize transport pathways that rely on the energy produced by mitochondrial oxidative phosphorylation, which generates ATP. Because of the high demand for ATP, renal tubule cells contain a large number of mitochondria, particularly the proximal tubule cells, where most of the transporter pathways are located. Damage to the kidney is generally classified as either chronic kidney disease (CKD) or acute kidney injury (AKI), where CKD is a slow loss of kidney function over an extended period of time, while AKI is a sudden loss of kidney function in hours to days. Both CKD and AKI are associated with damage to tubular cells, which is likely to result from a loss of mitochondrial function. Some drugs and toxic compounds, such as cisplatin, gentamycin, tenofovir, heavy metals and contrast media, are well known to induce mitochondrial damage leading to cell death and eventually to AKI. This chapter will however focus on two less known compounds, ethylene glycol and diethylene glycol that both induce AKI and can result in complete loss of kidney function. Neither compound itself is nephrotoxic, rather they are metabolized to oxalic acid and diglycolic acid, respectively, which in turn damage proximal tubule cells. Both oxalate, in its crystalline form calcium oxalate monohydrate, and diglycolate produce mitochondrial dysfunction, either by direct inhibition of mitochondrial respiration or by increasing reactive oxygen species production.

Keywords Proximal tubular necrosis · Acute kidney injury · Ethylene glycol · Diethylene glycol · Oxalate crystals · Diglycolic acid

K. E. McMartin (✉)
Department of Pharmacology, Toxicology, and Neuroscience, Louisiana State University Health Sciences Center, Shreveport, LA, USA
e-mail: kmcmar@lsuhsc.edu

P. J. Oliveira (ed.), *Mitochondrial Biology and Experimental Therapeutics*,
https://doi.org/10.1007/978-3-319-73344-9_10

1 Introduction to the Kidney

The kidney functions as a major organ in terms of maintaining body homeostasis through its ability to regulate the excretion and/or conservation of nutrients such as glucose and amino acids, of electrolytes such as sodium and potassium, and of xenobiotics such as drugs or toxic compounds. Glomerular filtration allows for removal of non-protein bound substances from the blood (see Fig. 1 for structural components of the nephron, the functional unit of the kidney). Most filtered substances can then be reabsorbed by tubular cells, especially in the proximal tubule, primarily through active transport or facilitated diffusion pathways. Many protein-bound substances can be taken up from the blood by specific transporters in tubular cells and then be secreted from the cells into the lumen of the nephron. Although reabsorption can recover some of these compounds, many are eventually excreted in the urine. Hence body homeostasis of nutrients, electrolytes and xenobiotics is mostly controlled by activity of the reabsorption and secretory pathways. Most of these transport pathways ultimately rely on the energy demanded by Na-K-ATPase, hence on mitochondrial oxidative phosphorylation to produce ATP as the energy source (Soltoff 1986). Hence, renal tubule cells contain a large number of mitochondria (Duann et al. 2016). Energy demands vary among the segments of the nephron such that different tubular cells contain varying amounts of mitochondria, although proximal tubular cells are thought to have the highest density of mitochondria (Hall et al. 2009).

2 Kidney Damage and Mitochondrial Role

Damage to the kidney is generally classified as either chronic kidney disease (CKD) or acute kidney injury (AKI). CKD is described by a slow loss of kidney function over an extended period of time, while AKI is a sudden loss of kidney function in hours to days. Both CKD and AKI are manifested by an accumulation of nitrogen metabolites in the blood (blood urea nitrogen, BUN, or serum creatinine), although other biomarkers such as urinary excretion of KIM-1 have been touted as earlier indicators of a loss of function (Vaidya et al. 2010). AKI primarily results from direct injury to tubular cells from diseases such as ischemia or sepsis or from exposure to nephrotoxic agents. CKD occurs primarily in conjunction with other diseases such as hypertension, diabetes, and genetic diseases of mitochondrial dysfunction (Che et al. 2014). Progression of CKD is related to a decrease in mitochondrial biogenesis, loss of mitochondrial membrane potential, increase in reactive oxygen species (ROS) and release of inflammatory mediators. CKD is ultimately characterized by proteinuria and microalbuminuria. Interestingly, AKI and CKD can be risk factors for the other – patients surviving an episode of AKI may suffer from early onset of CKD later in life (Chevalier 2016).

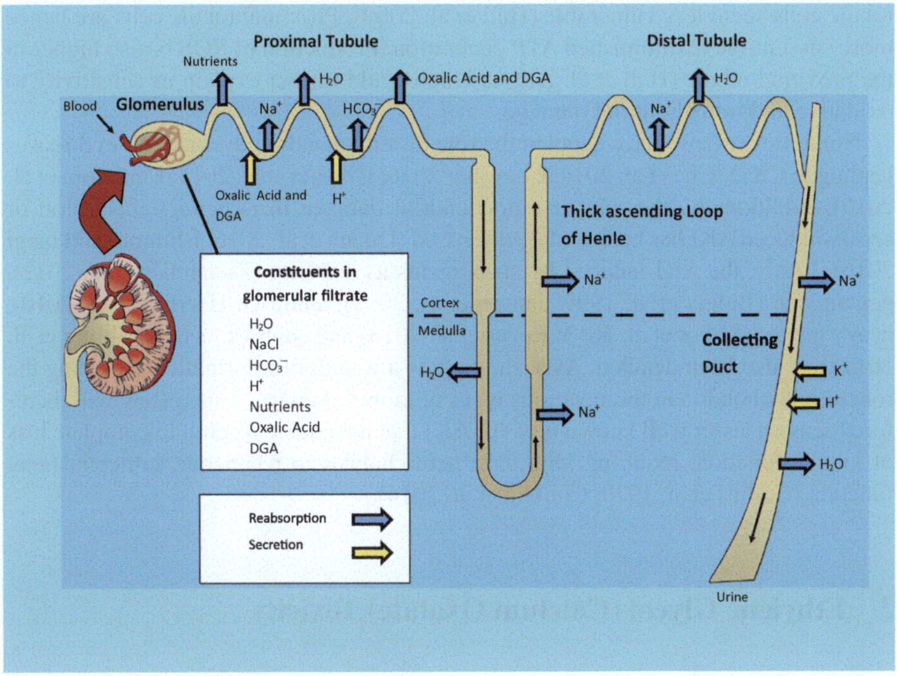

Fig. 1 Diagram of the nephron, the functional unit of the kidney. Blood flows through the capillaries of the glomerulus, where it is filtered into Bowman's capsule and then into the proximal tubule lumen. Major constituents of the glomerular filtrate include electrolytes, such as Na^+, CL^-, H^+ and HCO_3^-, nutrients such as glucose and amino acids, as well as certain drugs and toxicants like oxalic acid and diglycolic acid (DGA). One of the primary functions of the proximal tubule, requiring high energy expenditure by the cells, is reabsorption of 2/3 of the filtered load of Na^+ and water. Other reabsorptive processes take place here as well, including uptake of oxalic acid and DGA. The proximal tubule cell also requires substantial energy to support the secretion (uptake from the blood and passage across the cell into the lumen) of many organic anions (such as oxalic acid and DGA) and organic cations. Distal portions of the nephron are involved in further reabsorption of Na^+ and water, control of acid-base balance, as well as control of potassium homeostasis. Under normal circumstances the nephron functions to retain almost all of the electrolytes and nutrients that are filtered at the glomerulus. The high energy requirement for these processes is a major reason for the susceptibility of the kidney to mitochondrial toxicants like oxalate and DGA

AKI results from a multiplicity of factors in addition to tubular cell injury, including ROS accumulation and inflammatory responses. When tubular cells are injured, they undergo both apoptotic and necrotic cell death, which ultimately results in loss of brush border, loss of cell polarity, release of tubular cell debris into the lumen and luminal obstruction. AKI is primarily associated with damage to proximal tubular cells, which are especially vulnerable to loss of mitochondrial function. As such, mitochondria play a key role in the pathogenesis of AKI (Ishimoto and Inagi 2016). Studies have shown that inhibition of mitochondrial respiration quickly breaks down the mitochondrial membrane potential in the proximal tubule, while the distal tubule cells seem less vulnerable (Hall et al. 2009). Proximal tubule cells are hence more susceptible to diminished ATP generation. Production of ROS is also higher in the proximal tubule (Hall et al. 2009), which could further explain its sensitivity to oxidative and mitochondrial damage.

Numerous reviews have detailed the role of mitochondrial dysfunction in diseases leading to CKD (Che et al. 2014; Chevalier 2016; Granata et al. 2015; Lindblom et al. 2015). Additionally, the role of mitochondrial damage in ischemic-reperfusion or sepsis-induced AKI has been widely discussed (Duann et al. 2016; Ishimoto and Inagi 2016). Lastly, the AKI induced by such agents as cisplatin (Zsengeller et al. 2012), gentamycin (Baliga et al. 1999; Jansen et al. 2016), tenofovir (Herlitz et al. 2010), heavy metals (Miller et al. 2013; Reyes et al. 2013) and contrast media (Fähling et al. 2017) has also been detailed. As such, there is a wealth of information regarding the role of mitochondria in these various types of kidney damage. This review will therefore focus on lesser well known types of AKI that nevertheless result in complete loss of kidney function, requiring long term hemodialysis to regenerate sufficient renal function (Collins et al. 1970; Conklin et al. 2014).

3 Ethylene Glycol (Calcium Oxalate) Toxicity

Ingestion of ethylene glycol (EG) solutions, primarily automobile antifreeze, can lead to severe morbidity (especially AKI) and mortality. In the United States, roughly about 6000 people per year are reported for EG ingestion, with about 20 of these resulting in fatality (Mowry et al. 2016). Beyond human intoxications, EG ingestion also frequently occurs in companion animals such as dogs and cats, with similar morbidity and mortality. Clinical indicators of EG poisoning include severe metabolic acidosis, neurological deficits, cardiopulmonary distress and severe AKI. The AKI is manifested primarily as widespread necrosis, albeit most heavily in the proximal tubules, accompanied by deposition of crystals. EG itself is relatively nontoxic and its toxic syndrome results from the accumulation of various metabolites. EG is metabolized in the liver by alcohol dehydrogenase to glycolaldehyde, which is then quickly converted to glycolic acid by hepatic aldehyde dehydrogenase (Fig. 2). When sufficient EG is ingested to overcome a metabolic threshold, glycolic acid accumulates and produces the metabolic acidosis (Marshall 1982; Jacobsen et al. 1984). Glycolate can be further metabolized to glyoxylic acid. Under subthreshold conditions, glyoxylate is converted by alanine-glyoxylate aminotransferase (AGT) into glycine and thence to CO_2 as the ultimate product. In the genetic disease, primary hyperoxaluria (PH) type 1 as well as in EG poisoning, AGT is either dysfunctional or becomes saturated and then glyoxylate is metabolized to oxalic acid (Hoppe et al. 2009). Oxalate is an excellent calcium chelator, although the resulting calcium oxalate is very insoluble and precipitates as calcium oxalate crystals in tissues such as in the kidney or is excreted in the urine. Two types of oxalate crystals include calcium oxalate monohydrate (COM) and dihydrate (COD) (Jacobsen et al. 1988). Although COD crystals are observed in the urine, only the COM form has been found in kidney tissue.

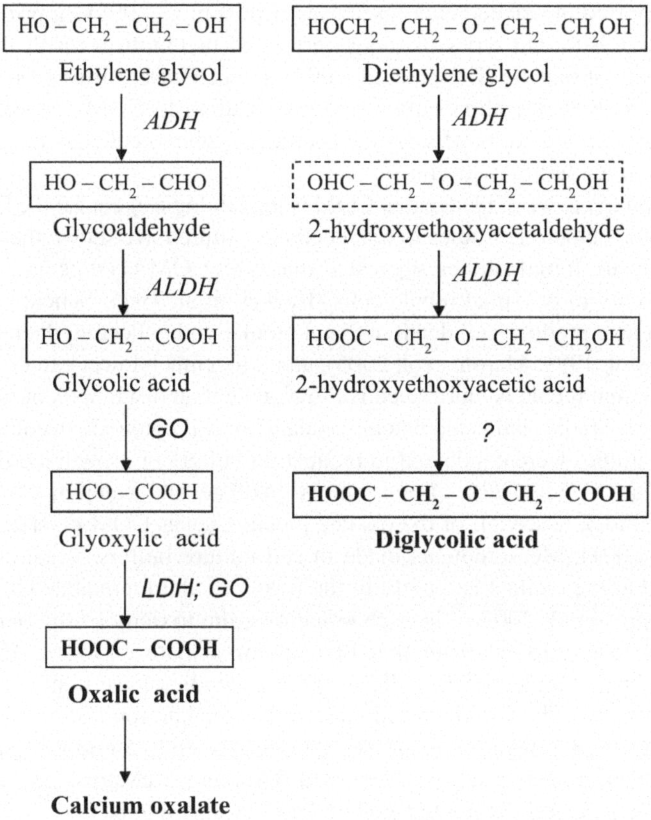

Fig. 2 Metabolism of ethylene glycol and diethylene glycol to their ultimate nephrotoxic metabolites (oxalic acid and diglycolic acid). Only major metabolites are shown—not shown are possible branch points that feed into other pathways (for example, glyoxylic acid can be converted into glycine, which can be further metabolized). Solid boxes indicate confirmed metabolites; dashed boxes for theoretical intermediates. The key enzymes are *ADH* alcohol dehydrogenase, *ALDH* aldehyde dehydrogenase, *GO* glycolate oxidase, *LDH* lactate dehydrogenase, *?* unknown activity

3.1 Role of Oxalate Accumulation in the Renal Toxicity of EG

Recent studies have confirmed that accumulation of COM crystals in the kidney is closely linked with the occurrence of proximal tubular necrosis. Necrotic damage following EG intoxication is only seen microscopically in association with COM deposition (Cruzan et al. 2004; Pomara et al. 2008). Furthermore, animal studies with chronic EG administration have demonstrated a strong dose-response relationship between the severity of the nephropathy and the total kidney oxalate concentration, suggesting causality (Corley et al. 2008). Other situations in which high levels of oxalate are introduced into the body have reported evidence of renal damage. Ingestion of oxalate itself is rare, but AKI associated with renal COM accumulation has been reported in isolated overdose cases (Konta et al. 1998). In primary hyperoxaluria, three different types of inborn errors of metabolism result in markedly elevated levels of oxalate (Hoppe et al. 2009). In such cases, urine oxalate concentrations reach above the supersaturation limit, leading to COM crystal formation, COM accumulation in kidney tissue and eventually tubular cell damage progressing to end-stage renal failure with time.

In addition to *in vivo* evidence that COM is the etiologic agent for the EG-induced renal necrosis, numerous studies in renal cells in culture have shown that COM produces cell death. Initial studies suggested that both COM crystals and oxalate ion produce toxicity to proximal tubule cells (Hackett et al. 1995; Scheid et al. 1995). Oxalate can also produce cell death in distal tubular and collecting duct cells in culture (Khan et al. 1999; Maroni et al. 2005), albeit to a much lower extent, suggesting that the proximal tubule is more sensitive to oxalate than distal parts of the nephron. Although early studies indicated that the oxalate ion was cytotoxic in cultured kidney cells, these studies were conducted in incubation buffers or growth media that contained calcium. Under such circumstances (oxalate and calcium concentrations above 1 mmol/L), nearly all of the oxalate precipitates as COM crystals (Belliveau and Griffin 2001). Measurements made in cell culture buffers containing calcium confirmed that the oxalate is mostly in the form of calcium oxalate (Maroni et al. 2005; Schepers et al. 2005a). In such situations, the toxicity of the oxalate ion is not reliably distinguished from that of calcium oxalate or COM. However, in human proximal tubule (HPT) cells in culture, studies have confirmed that the toxic component is the COM crystal and not the oxalate ion (Guo and McMartin 2005). For example, in solutions of oxalate ion only up to 5 mmol/L (with no calcium in buffer), no cell death was recorded (Guo and McMartin 2005; McMartin 2009). Similarly, COM but not the oxalate ion was shown to be toxic to renal tubule cell lines (LLC-PK1 and MDCK-II) (Schepers et al. 2005a, b).

3.2 Cellular Uptake of Oxalate and Oxalate Crystals

Oxalate crystals could affect mitochondrial function by releasing various mediators as discussed below or by being taken up into the cytoplasm of cells and directly interacting with mitochondria. COM crystals are known to adhere to PT cell membranes and are also readily internalized (Lieske et al. 1994; Verkoelen et al. 1999; Schepers et al. 2003). In fact the degree of COM internalization has been shown to correlate with the degree of cell death (Hovda et al. 2010), suggesting that intracellular uptake is important in the cytotoxicity. COM crystals appear to be taken up by endocytosis, with cytoplasmic appearance within 30–60 min (Lieske et al. 1994; Meimaridou et al. 2005). Evidence has also shown that renal tubular epithelial cells bind and take up COM crystals *in vivo* (Khan 1995). The fate of intracellular crystals includes persistence in the cytoplasm (Lieske et al. 1994), slow dissolution within a week (Lieske et al. 1997), or transport from the cell into the interstitium (Vervaet et al. 2009), where they are known to induce inflammation (De Water et al. 1999).

Oxalate ion can be taken up by proximal tubule cells and could theoretically form calcium oxalate inside the cell. Oxalate ion can reach the lumen of the proximal tubule, either by filtration at the glomerulus or by basolateral uptake into the cell followed by secretion across the apical membrane (Fig. 1). In kidney cell lines cultured on inserts that allow for polarized transport, oxalate ion is transported across the apical membrane (reabsorption) and across the basolateral membrane (secretion) (Ebisuno et al. 1994; Koul et al. 1994). The net transmembrane flux (into the opposite compartment) follows the secretory direction (Koul et al. 1994), which mirrors the movement of the oxalate ion *in vivo* (Verkoelen and Romijn 1996), where there is net secretory clearance into the urine.

3.3 Mitochondrial Effects of Oxalate and Oxalate Crystals

A number of studies have examined possible cell death mechanisms, which are centered on mitochondrial effects, either through generation of second messengers (lipid mediators) that elicit toxicity, production of oxidative stress or direct mitochondrial dysfunction. Oxalate-induced cell death has been associated with increased lipid peroxidation, accumulation of ROS and decreased anti-oxidant enzyme activity both *in vivo* (Huang et al. 2002; Thamilselvan et al. 1997) and in cultured kidney cells (Scheid et al. 1996; Thamilselvan et al. 2000). The increase in ROS is thought to be of mitochondrial origin, through leakage via Complex I or III of the electron transport chain (Khand et al. 2002; Cao et al. 2004). The effects of COM on cellular ROS can be reversed by co-treatment with anti-oxidant proteins (catalase and superoxide dismutase) (Thamilselvan et al. 2000; Khand et al. 2002) or with vitamin E (Thamilselvan et al. 2003). Oxalate treatment of isolated rat kidney mitochondria increases ROS and decreases protein anti-oxidants (Cao et al. 2004). The increase in ROS in kidney cells is mirrored by treatment of cells with arachidonic acid and lyso-phosphatidycholine, (Kohjimoto et al. 1999; Cao et al. 2004), lipid signaling molecules that are known to be released by oxalate-induced activation of phospholipase A_2 (Kohjimoto et al. 1999). These studies suggest that oxalate may indirectly alter mitochondrial ROS production via generation of these signaling molecules. However, COM has also been shown to directly increase mitochondrial superoxide concentrations (Meimaridou et al. 2005). COM can decrease kidney cell content of glutathione, the major non-protein antioxidant in cells (Khand et al. 2002; Rashed et al. 2004). Thus, COM could produce cell death via an oxidative stress response, either by direct mitochondrial ROS production or by an indirect response to lipid signaling molecules.

Studies in mitochondria isolated from rat kidneys or livers have shown that COM or oxalate can directly affect mitochondrial function, not related to any changes in ROS (Bachmann and Golberg 1971). COM directly decreases State 3 mitochondrial respiration in rat renal mitochondria, without affecting State 4 respiration or the ADP/oxygen ratio (McMartin and Wallace 2005). COM is more potent than the oxalate ion in inhibiting mitochondrial respiration (McMartin and Wallace 2005;

Bachmann and Golberg 1971). The decrease in State 3 respiration is not blocked by cyclosporine A, an inhibitor of the mitochondrial permeability transition (MPT), suggesting that the MPT is not involved in the decrease in respiration. The lack of an effect on State 4 respiration and on the ADP/O ratio suggests that COM is not an uncoupler nor does it affect the phosphorylation per se, rather that it affects the electron transport chain.

Concentrations of COM that are known to decrease respiration are also able to induce the MPT, as noted by the COM-induced mitochondrial swelling that is blocked by cyclosporine A (McMartin and Wallace 2005). The lack of an effect of cyclosporine A on COM's reduction in State 3 respiration suggests that it occurs independently from the MPT. However, the decrease in State 3 respiration could be the means by which COM opens the permeability transition pore. The MPT, which is a sudden increase in inner mitochondrial membrane permeability that allows release of matrix components >1500 Da, results from opening of the mitochondrial permeability transition pore. The open pore also allows influx of solutes leading to mitochondrial swelling and to an inability to maintain the pH gradient needed for ATP synthesis (Lindblom et al. 2015). Hence, induction of the MPT leads to ATP reduction, suggesting that these direct effects of COM on mitochondrial functions are the means by which it produces cell death. Oxalate has been shown to interact with the inner mitochondrial membrane (Strzelecki et al. 1989) and to decrease the mitochondrial membrane potential in HPT cells (Cao et al. 2004; Hovda et al. 2010), effects that are consistent with induction of the MPT in isolated mitochondria. Mitochondrial dysfunction has also been observed *in vivo* in rats given EG (0.75%) in their drinking water for 28 days (Veena et al. 2008). This protocol is known to induce hyperoxaluria within 7–14 days and COM accumulation in kidney tissue in 28 days (Green et al. 2005; Li and McMartin 2009). In these rats (Veena et al. 2008), there was a decrease in activity of the respiratory chain complex enzyme activities and an increase in mitochondrial swelling, suggesting induction of the MPT *in vivo*.

The various studies indicate that the renal damage (AKI) produced by EG is a result of the accumulation of COM crystals inside proximal tubular cells. COM crystals can produce cytotoxicity to tubular cells by inducing an oxidative stress response or directly by producing mitochondrial damage leading to a loss in ATP production.

3.4 Oxalate Crystal Toxicity in Other Diseases

Kidney stones occur in about 12% of males and 5% of females in the developed world and recurrence of stones is a common aspect of the disease (Curhan 1999). Most kidney stones are comprised of calcium oxalate and stone formation can be linked with mildly elevated levels of oxalate in the blood and urine (Ramello et al. 2000). Pathogenesis of oxalate stones is similar to that producing crystal aggregation in EG toxicity, including attachment of COM to tubular cells followed by growth of COM crystals until a "stone" is shaped. Differences from EG poisoning include that kidney stone disease is associated with much lower levels of "elevated" oxalate than those in EG poisoning, that there is no overt tubular cell damage (no AKI), and that the interaction of COM occurs not in the proximal tubule but in the distal tubule or collecting duct/papilla (Evan et al. 2005).

In primary hyperoxaluria, hepatic enzymes that usually convert glyoxylate to other products are missing or structurally disformed so as to lack function. As such, glyoxylate is instead converted to oxalate with the end result being very high concentrations of oxalate in blood, urine and tissues (Milliner et al. 2001), similar to those produced by EG ingestion. As a result, COM is readily deposited in kidney tubules and leads to tubular cell damage and eventually end-stage renal disease.

4 Diethylene Glycol (Diglycolic Acid) Toxicity

Diethylene glycol (DEG) poisoning represents a world-wide health challenge because it has produced repeated mass epidemic poisonings from its mistaken use as solvent for liquid medications. From 1995 to 2009, there have been at least nine highly fatal epidemics of renal failure, including those in Haiti (O'Brien et al. 1998), Panama (Sosa et al. 2014) and Nigeria (Akuse et al. 2012). The number of deaths in these epidemics is nearly 800, with more occurring in isolated cases of accidental or suicidal ingestion (Marraffa et al. 2008). Although epidemic poisonings are not likely in the US due to safety regulations, DEG is found in easily accessible consumer products such as anti-freeze solutions, brake fluids and canned heating fluids (Marraffa et al. 2008; Schier et al. 2011), which can be mistakenly or intentionally ingested. In the most recent National Poison Data System report, there were over 800 exposures in the US to brake fluids (containing 10% or more DEG) in one year, with 272 being treated in a health care facility, implying significant exposures requiring some treatment (Mowry et al. 2016).

DEG toxicity in humans targets a number of organ systems including the kidney, liver and nervous system (Rollins et al. 2002; Alfred et al. 2005), although the hallmark of poisoning is AKI. Human autopsies (Ferrari and Giannuzzi 2005) show severe renal cellular swelling, tubular dilation and necrosis with a moderate hepatotoxicity. DEG toxicity in animals mirrors that in humans, including an initial metabolic acidosis, followed by increased BUN by 48 h and eventually anuria (Freundt and Weis 1989; Besenhofer et al. 2010). Pathologic studies show enlarged kidneys with tubular cell necrosis and less severe damage in the liver (Hebert et al. 1978; Besenhofer et al. 2010. Like EG, DEG itself is minimally toxic, such that the toxicity results from the accumulation of one or more metabolites. As shown in Fig. 2, DEG is metabolized by alcohol dehydrogenase (Herold et al. 1989) to 2-hydroxyethoxyacetaldehyde, then to 2-hydroxyethoxyacetate (HEAA) and further to diglycolic acid (DGA), with minimal formation of ethylene glycol (Besenhofer et al. 2011). Early metabolic studies with nontoxic doses of DEG in animals (Mathews et al. 1991; Wiener and Richardson 1989) showed that HEAA and unchanged DEG are the major urinary metabolites. Studies of DEG metabolism at highly toxic doses in rats confirm that HEAA and DEG are the major urinary and blood metabolites of DEG, but that small amounts of DGA (1/50th of the amount of HEAA) are also found in urine and blood (Besenhofer et al. 2010, 2011). Small amounts of EG have also been reported in body fluids after highly toxic doses of DEG, but no increases in glycolate or oxalate are observed (Besenhofer et al. 2011).

4.1 Role of Diglycolate Accumulation in the Renal Toxicity of DEG

Because HEAA has appeared to be the major metabolite of DEG in the urine and blood, it has often been suggested to be the nephrotoxic metabolite. However, a number of recent studies have established that DGA is the metabolite of DEG that is responsible for the renal and hepatic damage produced by DEG. In the rat studies by Besenhofer (Besenhofer et al. 2010, 2011), rats were treated with toxic doses of DEG and with and without fomepizole, a potent inhibitor of alcohol dehydrogenase (Li and Theorell 1969). In fomepizole-treated animals, no HEAA or DGA accumulates in the blood or the urine, indicating a complete blockage of metabolism of DEG. Furthermore, there is no renal or hepatic damage observed in the fomepizole-treated rats, demonstrating that metabolism of DEG is required for the production of the organ toxicity. In these studies, no increase in liver or kidney concentrations of DGA or of HEAA are seen at low, nontoxic doses of DEG nor after the administration of fomepizole. However, both metabolites are observed in the liver and kidney at the toxic doses of DEG. Interestingly, the amount of DGA in both tissues is substantial, roughly 100-fold higher than what is observed in the blood, suggesting a remarkable accumulation of DGA into the kidney tissue. The key finding, that DGA is a major metabolite of DEG in terms of kidney retention, conflicts with the fact that its blood and urine concentrations are so low compared to HEAA.

During a CDC-conducted investigation of the outbreak of DEG poisonings that occurred in Panama from a contaminated cough syrup, serum and urine specimens were analyzed for concentrations of DEG, HEAA and DGA (Schier et al. 2013). In this study, cases are defined as persons presenting with acute renal failure, while age- and sex-matched controls are randomly selected from other hospitalized patients. In both the serum and urine samples from cases and controls, DGA is the only compound whose presence is significantly associated with case status, indicating that it is the only metabolite whose levels are increased in the late stages of DEG poisoning. DGA levels in these humans in the serum and urine of the cases are 0.30 and 0.21 mmol/L (both are not detectable in controls), respectively, which compare to those in the DEG-treated rats of 0.04 and 4 mmol/L (Besenhofer et al. 2010, 2011).

Although the renal accumulation of DGA is a novel finding that suggests a role in the kidney toxicity, the *in vivo* studies can't truly distinguish if HEAA or DGA is the likely toxicant (since both are present). To make this distinction, the toxicity of DEG metabolites has been studied in normal human kidney (HPT) cells in culture.

The results (Landry et al. 2011) indicate that: (1) neither DEG nor HEAA produces any toxicity, even at 100 mmol/L; (2) DGA increases ethidium homodimer uptake and lactate dehydrogenase release in a concentration-dependent manner (significant at >25 mmol/L), indicating development of a necrotic toxicity; (3) neither HEAA nor DGA produces apoptosis (no caspase-3 activation and flow cytometry shows an increase in both PI and annexin V staining, hence necrosis). The renal histopathology from DEG is a severe cortical necrosis localized to the PT cells (Ferrari and Giannuzzi 2005; Besenhofer et al. 2010), hence these findings that DGA produces a necrotic cell death rather than apoptosis is consistent with the *in vivo* findings.

The importance of kidney DGA accumulation to DEG toxicity has been confirmed in rats gavaged with doses of DEG (2, 5 and 10 g/kg) to examine the dose-response of metabolism and toxicity (Landry et al. 2015). These studies show a very steep dose-response, with a marked threshold between 5 and 10 g/kg of DEG. At the lower doses of 2 and 5 g/kg, there is no renal toxicity, yet severe AKI at 10 g/kg (increased serum creatinine to >7 mg/dL and increased BUN to >125 mg/dL with histopathology showing marked vacuolization and tubular necrosis only at the high dose). Also there is no hepatotoxicity at 2 and 5 g/kg, yet moderate injury at 10 g/kg (ALT > 150 U/L, AST > 500 U/L, with severe hepatic glycogen depletion only at the high dose). In close correlation with the toxicity results, there is no DGA in the kidneys at 2 and 5 g/kg, yet marked elevation of kidney DGA at 10 g/kg (equivalent to 10–15 mmol/L). These studies confirm that DEG-induced kidney and liver toxicities depict a threshold dose response and that toxicity is observed only when tissue DGA levels are elevated, asserting its necessary role in DEG-induced toxicity. These studies have also shown the relevance of the toxic DGA concentrations in *in vitro* studies. In the HPT cells (Landry et al. 2011), DGA produces toxicity at concentrations >25 mmol/L, which are somewhat higher, but still in the range of the kidney DGA levels in DEG-treated rats (10–15 mmol/L). It is also important to note that the DGA levels reported in rat kidneys represent levels measured in whole kidney tissue, not in just the proximal tubule cells, such that DGA levels are likely higher in proximal tubule cells *in vivo*.

Direct administration of DGA to rats has verified that DGA itself can produce AKI and moderate hepatotoxicity. Rats have been gavaged with single doses of DGA (0, 100 and 300 mg/kg) and followed afterwards for 48 h (Robinson et al. 2017). At the high dose only, DGA increases serum creatinine and BUN and produces the typical AKI histopathology of tubular vacuolization and necrotic degeneration of proximal tubular cells. DGA also increases the urinary excretion of KIM-1, an early biomarker of drug-induced kidney injury (Bonventre 2014). KIM-1 increases only after the high dose of DGA, starting at 36 h after treatment. DGA also produces a moderate hepatotoxicity, with increases in AST and widespread microvesicular fatty change in heptotocytes (without obvious necrosis), but again only at the high dose. These changes correlate with the concentrations of DGA measured in the kidney and liver tissue, where increases equivalent to about 6 mmol/L are observed only at the doses of DGA that produce toxicity (300 mg/kg). A 28 day study using daily gavage doses of DGA from 0.3 to 300 mg/kg has reported similar results (Sprando et al. 2017). In dose groups up to 100 mg/kg, minimal to no effects are reported even after 28 days; at 300 mg/kg, all rats were euthanized by 5 days due to overwhelming toxicity. Histopathologic effects at 300 mg/kg include renal tubule vacuolar degeneration and necrosis, microvesicular centrilobular vacuolization in the liver and damage to a variety of other organs (thymus, spleen, stomach, intestines). Serum obtained on day 5 in the rats given 300 mg/kg show marked increases in markers of kidney and liver damage (creatinine, BUN and AST). The kidney and liver toxicity seen with DGA is virtually identical to that seen in DEG overdoses. Taken together, these two studies demonstrate unequivocally that DGA is the metabolite of DEG that produces the AKI (and liver damage also).

4.2 Mitochondrial Effects of Diglycolic Acid

In renal PT cells, the major source for energizing the key transport processes, such as reabsorption of two thirds of the filtered sodium, is mitochondrial metabolism, and ATP levels are maintained almost entirely by oxidative phosphorylation (Nowak et al. 2008). Mitochondrial respiration represents 90% of overall cellular O_2 consumption, particularly in the kidney (Lash and Jones 1996). In HPT cells, DGA markedly depletes ATP levels, occurring prior to the increase in LDH release (necrotic cell death) (Landry et al. 2011) and also reduces cellular O_2 consumption (Landry et al. 2013). These data suggest that DGA can reduce mitochondrial oxidative phosphorylation, which eventually leads to renal cell death.

A number of studies have attempted to examine the mechanism by which DGA produces mitochondrial dysfunction. Mitochondria are known to be the primary source of intracellular ROS, through leakage from Complex I or III of the electron transport chain (Khand et al. 2002). DGA produces a time- and concentration-dependent increase in ROS in HPT cells (Landry et al. 2013). Treatment with the antioxidant Trolox (100 μmol/L) completely prevents the increase in ROS and also the cell death at DGA concentrations <25 mmol/L but not the cell death at 50 mmol/L. These data suggest that an increase in ROS may play a role in the cytotoxicity of DGA at low levels, but at high DGA concentrations, toxicity is likely a direct inhibition of mitochondrial metabolism. Further evidence against a major mechanistic role of ROS comes from studies in HPT cells (Conrad et al. 2016). In these cells, DGA produces a dissipation of the mitochondrial membrane potential ($\Delta\psi m$), as measured by relative changes in JC-1 fluorescence. A decrease in $\Delta\psi m$ reflects changes in inner membrane fluidity or conformational changes in pore forming proteins that can lead to the membrane permeability transition (MPT) (Colell et al. 2003). Co-treatment with Trolox (100 μmol/L) does not prevent the DGA-induced changes in mitochondrial membrane potential, indicating that an increase in ROS is not responsible for these mitochondrial effects of DGA. Instead DGA appears to directly alter $\Delta\psi m$.

DGA has a structure (as a 4-carbon dicarboxylic acid) that is remarkably similar to metabolic dicarboxylic acids such as succinate that are important in intermediary metabolism and the Kreb's cycle (Fig. 3). As such, one hypothesis for a direct effect

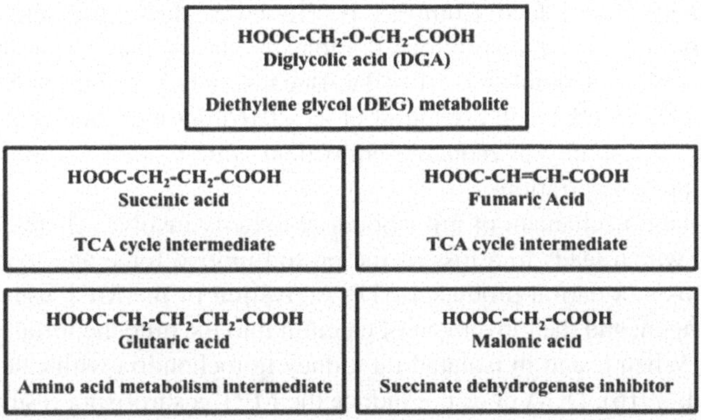

Fig. 3 Molecular mimicry of diglycolic acid. Figure compares structure of diglycolic acid with those of dicarboxylate compounds, including common tricarboxylic acid (TCA) cycle intermediates and a known succinate dehydrogenase inhibitor

of DGA on mitochondrial function would be that it is acting as a molecular mimic, either interfering with the Kreb's cycle or the electron transport chain complexes. DGA inhibits succinate dehydrogenase (SDH) activity from bovine heart mitochondria and from HPT cell mitochondria, but does not inhibit other TCA cycle enzymes such as fumarase, malate dehydrogenase or α-ketoglutarate dehydrogenase (Landry et al. 2013). Kinetic analysis confirms that DGA acts as a competitive inhibitor of bovine heart SDH, with a $K_I = 6.8$ mmol/L, but with a lower affinity than succinate for SDH (measured K_M for succinate = 0.4 mmol/L). Thus, DGA acts similarly to malonate (Fig. 3), another structural analog for succinate, which competitively inhibits SDH and ultimately leads to cell death (Pardee and Potter 1949). This inhibition of SDH might be sufficient to turn off oxidative phosphorylation in the proximal tubule cells. Under such circumstances, these cells would be expected to enhance their glycolytic capacity to preserve ATP production (Dickman and Mandel 1990). However, HPT cells treated with DGA do not show an increase in glycolysis as measured by lactate production (Landry et al. 2013). The inhibition of oxidative phosphorylation without glycolytic compensation could explain the dramatic effects of DGA on HPT cell ATP levels.

Additional studies have examined direct effects of DGA on mitochondrial functions using isolated rat kidney mitochondria. DGA decreases State 3 respiration (ADP-induced) with either succinate or glutamate/malate as energizing substrates (Conrad et al. 2016). DGA does not affect State 4 respiration from either glutamate/malate or succinate nor the ADP/O ratio in isolated mitochondria. These results indicate that DGA is operating as an inhibitor of mitochondrial respiration (State 3) and not through an uncoupling mechanism. Interestingly, the effects of DGA on glutamate/malate respiration occur at very low DGA concentrations (significant by 6 mmol/L), while those on succinate respiration occur only at 100 mmol/L. These results suggest that DGA may be having a potent inhibitory effect on Complex I activity, since glutamate/malate would feed into that complex, while succinate would feed Complex II. However, direct measurement of the enzyme activities of the respiratory complexes shows that DGA inhibits only Complex II and not Complex I, III or IV. Because DGA decreases NADH levels relative to NAD in HPT cells (Conrad et al. 2016), another interpretation is that DGA inhibits the supply of reducing equivalents into Complex I thus decreasing glutamate/malate respiration.

An important mechanism of mitochondrial toxicity involves chemical induction of the MPT, which leads to a loss of the proton-motive force and to inhibition of oxidative phosphorylation-produced ATP. Activation of the MPT usually involves an inducing agent and the movement of calcium into the mitochondrial matrix (Kim et al. 2003). When tested in isolated rat kidney mitochondria with calcium present (Conrad et al. 2016), DGA does not induce the MPT, a surprising result because of its inhibitory effects on $\Delta\psi m$ in HPT cells as noted above. Instead, DGA inhibits the calcium-induced mitochondrial swelling in a concentration-dependent manner. A potential explanation for this inhibition is that DGA chelates calcium outside of the mitochondria (Montekaitis and Martell 1984), thus decreasing the ability of calcium to induce the MPT. When compared to the classic calcium chelator, DGA is equimolar potent as EGTA in its ability to chelate calcium in solution (Conrad et al. 2016).

The fact that DGA can chelate calcium may also explain the differential concentration-related effect on glutamate-induced and succinate-induced State 3 respiration noted above. Glutamate-induced State 3 respiration is regulated by the external concentration of calcium (Gellerich et al. 2009), while succinate-induced respiration is independent of calcium. Extra-mitochondrial calcium is known to control the glutamate/aspartate antiporter (aralar), which is needed to provide glutamate for the Krebs cycle and to transport NADH into the mitochondrial matrix (Satrustegui et al. 2007). Both glutamate and NADH are needed for Complex I respiration, such that DGA could inhibit this respiration by chelating the calcium needed to support aralar activity. The concentration of DGA that chelates calcium (5 mmol/L) is similar to the concentration of DGA that inhibits glutamate-supported respiration (6 mmol.L) and to that which inhibits the calcium-induced mitochondrial swelling.

The AKI that is produced by DEG appears to result from the accumulation of the metabolite DGA in renal proximal tubule cells. Once inside the cells, DGA produces an inhibition of oxidative phosphorylation that results in a loss of the ATP production that is necessary to energize the various transport and metabolic functions of the proximal tubule, eventually leading to the necrotic injury characteristic of AKI. DGA appears to produce the mitochondrial dysfunction by directly inhibiting SDH activity, hence Complex II. Another contributing factor may be its ability to chelate calcium, which in turn decreases the availability of substrates and reducing equivalents that feed into Complex I, thus minimizing Complex I activity also. It is interesting that two calcium chelators, DGA and oxalic acid, both produce mitochondrial dysfunction that leads to kidney cell death and ultimately to AKI. One difference is that calcium oxalate is very insoluble and crystal accumulation results, while the calcium-DGA complex is apparently very soluble as no crystals are observed when calcium and DGA are present (Besenhofer et al. 2011).

References

Akuse RM, Eke FU, Ademola AD et al (2012) Diagnosing renal failure due to diethylene glycol in children in a resource-constrained setting. Pediatr Nephrol 27:1021–1028

Alfred S, Coleman P, Harris D, Wigmore T, Stachowski E, Graudins A (2005) Delayed neurologic sequelae resulting from epidemic diethylene glycol poisoning. Clin Toxicol 43:155–159

Bachmann E, Golberg L (1971) Reappraisal of the toxicology of ethylene glycol. III. Mitochondrial effects. Food Cosmet Toxicol 9:39–55

Baliga R, Ueda N, Walker PD, Shah SV (1999) Oxidant mechanisms in toxic acute renal failure. Drug Metab Rev 31:971–997

Belliveau J, Griffin H (2001) The solubility of calcium oxalate in tissue culture media. Anal Biochem 291:69–73

Besenhofer LM, Adegboyega PA, Bartels M, Filary MJ, Perala AW, McLaren MC, McMartin KE (2010) Inhibition of metabolism of diethylene glycol prevents target organ toxicity in rats. Toxicol Sci 117:25–35

Besenhofer LM, McLaren MC, Latimer B, Bartels M, Filary MJ, Perala AW, McMartin KE (2011) Role of tissue metabolite accumulation in the renal toxicity of diethylene glycol. Toxicol Sci 123:374–383

Bonventre JV (2014) Kidney injury molecule-1: a translational journey. Trans Am Clin Climatol Assoc 125:293–299

Cao LC, Honeyman TW, Cooney R, Kennington L, Scheid CR, Jonassen JA (2004) Mitochondrial dysfunction is a primary event in renal cell oxalate toxicity. Kidney Int 66:1890–1900

Che R, Yuan Y, Huang S, Zhang A (2014) Mitochondrial dysfunction in the pathophysiology of renal diseases. Am J Physiol Renal Physiol 306:F367–F378

Chevalier RL (2016) The proximal tubule is the primary target of injury and progression of kidney disease: role of the glomerulotubular junction. Am J Physiol Renal Physiol 311:F145–F161

Colell A, Garcia-Ruiz C, Lluis JM, Coll O, Mari M, Fernandez-Checa JC (2003) Cholesterol impairs the adenine nucleotide translocator-mediated mitochondrial permeability transition through altered membrane fluidity. J Biol Chem 278:33928–33935

Collins JM, Hennes DM, Holzgang CR et al (1970) Recovery after prolonged oliguria due to ethylene glycol intoxication. Arch Intern Med 125:1059–1062

Conklin L, Sejvar JJ, Kieszak S, Sabogal R, Sanchez C, Flanders D, Tulloch F, Victoria G, Rodriguez G, Sosa N, McGeehin MA, Schier JG (2014) Long-term renal and neurologic outcomes among survivors of diethylene glycol poisoning. JAMA Intern Med 174(6):912–917. https://doi.org/10.1001/jamainternmed.2014.344

Conrad T, Landry GM, Aw TY, Nichols R, McMartin KE (2016) Diglycolic acid, the toxic metabolite of diethylene glycol, chelates calcium and produces renal mitochondrial dysfunction in vitro. Clin Toxicol 54:501–511

Corley RA, Wilson DM, Hard GC, Stebbins KE, Bartels MJ, Soelberg JJ, Dryzga MD, Gingell R, McMartin KE, Snellings WM (2008) Dosimetry considerations in the enhanced sensitivity of male Wistar rats to chronic ethylene glycol-induced nephrotoxicity. Toxicol Appl Pharmacol 228:165–178

Cruzan C, Corley RA, Hard GC, Mertens JJWM, McMartin KE, Snellings WM, Gingell R, Deyo JA (2004) Subchronic toxicity of ethylene glycol in Wistar and F344 rats related to metabolism and clearance of metabolites. Toxicol Sci 81:502–501

Curhan GC (1999) Epidemiologic evidence for the role of oxalate in idiopathic nephrolithiasis. J Endourol 13:629–631

De Water R, Noordermeer C, van der Kwast TH, Nizze H, Boeve ER, Kok DJ, Schroder FH (1999) Calcium oxalate nephrolithiasis: effect of renal crystal deposition on the cellular composition of the renal interstitium. Am J Kidney Dis 33:761–771

Dickman K, Mandel L (1990) Differential effects of respiratory inhibitors on glycolysis in proximal tubules. Am J Physiol 258:F1608–F1615

Duann P, Lianos EA, Ma J, Lin P-H (2016) Autophagy, innate immunity and tissue repair in acute kidney injury. Int J Mol Sci 17:662. https://doi.org/10.3390/ijms17050662

Ebisuno S, Koul H, Menon M, Scheid C (1994) Oxalate transport in a line of porcine renal epithelial cells - LLC-PK1 cells. J Urol 152:237–242

Evan AP, Coe FL, Lingeman JE, Worcester E (2005) Insights on the pathology of kidney stone formation. Urol Res 33:383–389

Fähling M, Seeliger E, Patzak A, Persson PB (2017) Understanding and preventing contrast-induced acute kidney injury. Nat Rev Nephrol 13:169–180

Ferrari LA, Giannuzzi L (2005) Clinical parameters, postmortem analysis and estimation of lethal dose in victims of a massive intoxication with diethylene glycol. Forensic Sci Int 153:45–51

Freundt KJ, Weis N (1989) Transient renal impairment in rats after oral exposure to diethylene glycol. J Appl Toxicol 9:317–321

Gellerich FN, Gizatullina Z, Arandarcikaite O et al (2009) Extramitochondrial Ca^{2+} in the nanomolar range regulates glutamate-dependent oxidative phosphorylation on demand. PLoS One 4:e8181

Granata S, Gassa AD, Tomei P, Lupo A, Zaza G (2015) Mitochondria: a new therapeutic target in chronic kidney disease. Nutr Metab 12:49. https://doi.org/10.1186/s12986-015-0044-z.

Green ML, Hatch M, Freel RW (2005) Ethylene glycol induces hyperoxaluria without metabolic acidosis in rats. Am J Physiol Renal Physiol 289:F536–F543

Guo C, McMartin KE (2005) The cytotoxicity of oxalate, metabolite of ethylene glycol, is due to calcium oxalate monohydrate formation. Toxicology 208:347–355

Hackett RL, Shevock PN, Khan SR (1995) Alteration in MDCK and LLCPK1 cells exposed to oxalate and calcium oxalate monohydrate crystals. Scanning Microsc 9:587–596

Hall AM, Unwin RJ, Parker N, Duchen MR (2009) Multiphoton imaging reveals differences in mitochondrial function between nephron segments. J Am Soc Nephrol 20:1293–1302

Hebert JL, Fabre M, Auzepy P, Paillas J (1978) Acute experimental poisoning by diethylene glycol: acid-base balance and histological data in male rats. Toxicol Eur Res 1:289–294

Herlitz LC, Mohan S, Stokes MB, Radhakrishnan J, D'Agati VD, Markowitz GS (2010) Tenofovir nephrotoxicity: acute tubular necrosis with distinctive clinical, pathological, and mitochondrial abnormalities. Kidney Int 78:1171–1177

Herold DA, Keil K, Bruns DE (1989) Oxidation of polyethylene glycols by alcohol dehydrogenase. Biochem Pharmacol 38:73–76

Hoppe B, Beck BB, Milliner D (2009) The primary hyperoxalurias. Kidney Int 75:1264–1271

Hovda KE, Guo C, Austin R, McMartin KE (2010) Renal toxicity of ethylene glycol results from internalization of calcium oxalate crystals by proximal tubule cells. Toxicol Lett 192:365–372

Huang HS, Ma MC, Chen J, Chen CF (2002) Changes in the oxidant-antioxidant balance in the kidney of rats with nephrolithiasis induced by ethylene glycol. J Urol 167:2584–2593

Ishimoto Y, Inagi R (2016) Mitochondria: a therapeutic target in acute kidney injury. Nephrol Dial Transplant 31:1062–1069

Jacobsen D, Ovrebo S, Ostborg J, Sejersted OM (1984) Glycolate causes the acidosis in ethylene glycol poisoning and is effectively removed by hemodialysis. Acta Med Scand 216:409–416

Jacobsen D, Hewlett TP, Webb R, Brown ST, Ordinario AT, McMartin KE (1988) Ethylene glycol intoxication. Evaluation of kinetics and crystalluria. Am J Med 84:145–152

Jansen D, Peters E, Heemskerk S, Koster-Kamphuis L, Bouw MP, Roelofs HM, van Oeveren W, van Heijst AF, Pickkers P (2016) Tubular injury biomarkers to detect gentamicin-induced acute kidney injury in the neonatal intensive care unit. Am J Perinatol 33:180–187

Khan SR (1995) Calcium oxalate crystal interaction with renal tubular epithelium, mechanism of crystal adhesion and its impact on stone development. Urol Res 23:71–79

Khan SR, Byer KJ, Thamilvan S, Hackett RL, McCormack WT, Benson NA, Vaughn KL, Erdos GW (1999) Crystal-cell interaction and apoptosis in oxalate-associated injury of renal epithelial cells. J Am Soc Nephrol 10:S457–S463

Khand FD, Gordge MP, Robertson WG, Noronha-Dutra AA, Hothersall JS (2002) Mitochondrial superoxide production during oxalate-mediated oxidative stress in renal epithelial cells. Free Radic Biol Med 32:1339–1350

Kim J-S, He L, Lemasters JJ (2003) Mitochondrial permeability transition: a common pathway to necrosis and apoptosis. Biochem Biophys Res Commun 304:463–470

Kohjimoto Y, Kennington L, Scheid CR, Honeyman TW (1999) Role of phospholipase A2 in the cytotoxic effects of oxalate in cultured renal epithelial cells. Kidney Int 56:1432–1441

Konta T, Yamaoka M, Tanida H, Matsunaga T, Tomoike H (1998) Acute renal failure due to oxalate ingestion. Intern Med 37:762–765

Koul H, Ebisuno S, Renzulli L, Yanagawa M, Menon M, Scheid C (1994) Polarized distribution of oxalate transport systems in LLC-PK1 cells, a line of renal epithelial cells. Am J Phys 266:F266–F274

Landry GL, Martin S, McMartin KE (2011) Diglycolic acid is the nephrotoxic metabolite in diethylene glycol poisoning inducing necrosis in human proximal tubule cells in vitro. Toxicol Sci 124:35–44

Landry GM, Dunning CL, Conrad T, Hitt MJ, McMartin KE (2013) Diglycolic acid inhibits succinate dehydrogenase activity in human proximal tubule cells leading to mitochondrial dysfunction and cell death. Toxicol Lett 221:176–184

Landry GM, Dunning CL, Abreo F, Latimer B, Orchard E, McMartin KE (2015) Diethylene glycol-induced toxicities show marked threshold dose response in rats. Toxicol Appl Pharmacol 282:244–251

Lash LH, Jones DP (1996) Mitochondrial toxicity in renal injury. In: Lash LH, Zalups RK (eds) Methods in toxicology. CRC Press, Boca Raton, pp 299–330

Li Y, McMartin KE (2009) Strain differences in urinary factors that promote calcium oxalate crystal formation in kidney in ethylene glycol treated rats. Am J Physiol Renal Physiol 296:1080–1087

Li TK, Theorell H (1969) Human liver alcohol dehydrogenase: inhibition by pyrazole and pyrazole analogs. Acta Chem Scand A 23:892–902

Lieske JC, Swift H, Martin T, Patterson B, Toback FG (1994) Renal epithelial cells rapidly bind and internalize calcium oxalate monohydrate crystals. Proc Natl Acad Sci 91:6987–6991

Lieske JC, Norris R, Swift H, Toback FG (1997) Adhesion, internalization and metabolism of calcium oxalate monohydrate crystals by renal epithelial cells. Kidney Int 52:1291–1301

Lindblom R, Higgins G, Coughlan M, de Haan JB (2015) Targeting mitochondria- and reactive oxygen species-driven pathogenesis in diabetic nephropathy. Rev Diabet Stud 12:134–156

Maroni PD, Koul S, Chandhoke PS, Meacham RB, Koul HK (2005) Oxalate toxicity in culture mouse inner medullary collecting duct cells. J Urol 174:757–760

Marraffa JM, Holland MG, Stork CM, Hoy CD, Hodgman MJ (2008) Diethylene glycol: widely used solvent presents serious poisoning potential. J Emerg Med 35:401–406

Marshall TC (1982) Dose-dependent disposition of ethylene glycol in the rat after intravenous administration. J Toxicol Environ Health 10:397–409

Mathews JM, Parker MK, Mathews HB (1991) Metabolism and disposition of diethylene glycol in rat and dog. Drug Metab Dispos 19:1066–1070

McMartin KE (2009) Are calcium oxalate crystals involved in the mechanism of acute renal failure in ethylene glycol poisoning? Clin Toxicol 47:859–869

McMartin KE, Wallace KB (2005) Calcium oxalate monohydrate, a metabolite of ethylene glycol, is toxic for rat renal mitochondrial function. Toxicol Sci 84:195–200

Meimaridou E, Jacobson J, Seddon AM, Noronha-Dutra AA, Robertson WG, Hothersall JS (2005) Crystal and microparticle effects on MDCK cell superoxide production: oxalate-specific mitochondrial membrane potential changes. Free Radic Biol Med 38:1553–1564

Miller S, Pallan S, Gangji AS, Lukic D, Clase CM (2013) Mercury-associated nephrotic syndrome: a case report and systematic review of the literature. Am J Kidney Dis 62:135–138

Milliner DS, Wilson DM, Smith LH (2001) Phenotypic expression of primary hyperoxalurias: comparative features of types I and II. Kidney Int 59:31–36

Montekaitis RJ, Martell AE (1984) New multidentate ligands. XXV. The coordination chemistry of divalent metal ions with diglycolic acid, carboxymethyltartronic acid and ditartronic acid. J Coord Chem 13:265–271

Mowry JB, Spyker DA, Brooks DE, Zimmerman A, Schauben JL (2016) 2015 Annual Report of the American Association of Poison Control Centers' National Poison Data System (NPDS): 33rd Annual Report. Clin Toxicol 54:924–1109

Nowak G, Clifton GL, Bakajsova D (2008) Succinate ameliorates energy deficits and prevents dysfunction of complex I in injured renal proximal tubular cells. J Pharmacol Exp Ther 324:1155–1162

O'Brien KL, Selanikio JD, Hecdivert C et al (1998) Epidemic of pediatric deaths from acute renal failure caused by diethylene glycol poisoning. JAMA 279:1175–1180

Pardee AB, Potter VR (1949) Malonate inhibition of oxidations in the Krebs tricarboxylic acid cycle. J Biol Chem 178:241–250

Pomara C, Fiore C, D'Errico S, Riezzo I, Fineschi V (2008) Calcium oxalate crystals in acute ethylene glycol poisoning: a confocal laser scanning microscope study in a fatal case. Clin Toxicol 46:322–324

Ramello A, Vitale C, Marangella M (2000) Epidemiology of nephrolithiasis. J Nephrol 13(Suppl 3):S45–S50

Rashed T, Menon M, Thamilselvan S (2004) Molecular mechanism of oxalate-induced free radical production and glutathione redox imbalance in renal epithelial cells: effect of antioxidants. Am J Nephrol 24:557–568

Reyes JL, Molina-Jijón E, Rodríguez-Muñoz R, Bautista-García P, Debray-García Y, Namorado Mdel C (2013) Tight junction proteins and oxidative stress in heavy metals-induced nephrotoxicity. Biomed Res Int 2013:730789. https://doi.org/10.1155/2013/730789

Robinson CN, Latimer B, Abreo F, Broussard K, McMartin KE (2017) In-vivo evidence of nephrotoxicity and altered hepatic function in rats following administration of diglycolic acid, a metabolite of diethylene glycol. Clin Toxicol 55:96–105

Rollins YD, Filley CM, McNutt JT, Chalal S, Kleinschmidt-DeMasters BK (2002) Fulminant ascending paralysis as a delayed sequela of diethylene glycol (Sterno) ingestion. Neurology 59:1460–1463

Satrustegui J, Pardo B, del Arco A (2007) Mitochondrial transporters as novel targets for intracellular calcium signaling. Physiol Rev 87:29–67

Scheid CR, Koul HK, Kennington L, Hill WA (1995) Oxalate-induced damage to renal tubular cells. Scanning Microsc 9:1097–1107

Scheid C, Koul H, Hill WA, Luber-Narod J, Kennington L (1996) Oxalate toxicity in LLC-PK1 cells: role of free radicals. Kidney Int 49:413–419

Schepers MSJ, Duim RAJ, Asselman M, Romijn JC, Schroder FH, Verkoelen CF (2003) Internalization of calcium oxalate crystals by renal tubular cells: a nephron segment-specific process? Kidney Int 64:493–500

Schepers MSJ, Van Ballegooijen ES, Bangma CH, Verkoelen CF (2005a) Crystals cause acute necrotic death in renal proximal tubule cells, but not in collecting tubule cells. Kidney Int 68:1543–1553

Schepers MSJ, Van Ballegooijen ES, Bangma CH, Verkoelen CF (2005b) Oxalate is toxic to renal tubule cells only at supraphysiologic concentrations. Kidney Int 68:1660–1669

Schier JG, Barr DB, Li Z, Wolkin AF, Baker SE, Lewis LS, McGeehin MA (2011) Diethylene glycol in health products sold over-the-counter and imported from Asian countries. J Med Toxicol 7:33–38

Schier JG, Hunt DR, Perala A, McMartin KE, Bartels MJ, Lewis L, McGeehin MA, Flanders WD (2013) Characterizing concentrations of diethylene glycol and suspected metabolites in human serum, urine, and cerebrospinal fluid samples from the Panama DEG mass poisoning. Clin Toxicol 51:923–929

Soltoff SP (1986) ATP and the regulation of renal cell function. Annu Rev Physiol 48:9–31

Sosa NR, Rodriguez GM, Schier JG et al (2014) Clinical, laboratory, diagnostic, and histopathologic features of diethylene glycol poisoning – Panama, 2006. Ann Emerg Med 64:38–47

Sprando RL, Mossoba ME, Black T et al (2017) 28-day repeated dose response study of diglycolic acid: renal and hepatic effects. Food Chem Toxicol 106(Pt A):558–567. https://doi.org/10.1016/j.fct.2017.03.047

Strzelecki T, McGraw BR, Scheid CR, Menon M (1989) Effect of oxalate on function of kidney mitochondria. J Urol 141:423–427

Thamilselvan S, Hackett RL, Khan SR (1997) Lipid peroxidation in ethylene glycol induced hyperoxaluria and calcium oxalate nephrolithiasis. J Urol 157:1059–1063
Thamilselvan S, Byer KJ, Hackett RL, Khan SR (2000) Free radical scavengers, catalase and superoxide dismutase provide protection from oxalate-associated injury to LLC-PK1 and MDCK cells. J Urol 164:224–229
Thamilselvan S, Khan SR, Menon M (2003) Oxalate and calcium oxalate mediated free radical toxicity in renal epithelial cells: effect of antioxidants. Urol Res 31:3–9
Vaidya VS, Ozer JS, Dieterle F, Collings FB, Ramirez V, Troth S, Muniappa N, Thudium D, Gerhold D, Holder DJ, Bobadilla NA, Marrer E, Perentes E, Cordier A, Vonderscher J, Maurer G, Goering PL, Sistare FD, Bonventre JV (2010) Kidney injury molecule-1 outperforms traditional biomarkers of kidney injury in preclinical biomarker qualification studies. Nat Biotechnol 28:478–485
Veena CK, Josephine A, Preetha SP, Rajesh NG, Varalakshmi P (2008) Mitochondrial dysfunction in an animal model of hyperoxaluria: a prophylactic approach with fucoidan. Eur J Pharmacol 579:330–336
Verkoelen CF, Romijn JC (1996) Oxalate transport and calcium oxalate renal stone disease. Urol Res 24:183–191
Verkoelen CF, van der Boom BG, Kok DJ, Houtsmuller AB, Visser P, Schroder FH, Romijn JC (1999) Cell type-specific acquired protection from crystal adherence by renal tubule cells in culture. Kidney Int 55:1426–1433
Vervaet BA, Verhulst A, Dauwe SE, De Broe ME, D'Haese PC (2009) An active renal crystal clearance mechanism in rat and man. Kidney Int 75:41–51
Wiener HL, Richardson KE (1989) Metabolism of diethylene glycol in male rats. Biochem Pharmacol 38:539–541
Zsengeller ZK, Ellezian L, Brown D et al (2012) Cisplatin nephrotoxicity involves mitochondrial injury with impaired tubular mitochondrial enzyme activity. J Histochem Cytochem 60:521–529

Role of Mitochondria in Neurodegenerative Diseases: The Dark Side of the "Energy Factory"

Sónia C. Correia and Paula I. Moreira

Abstract Neurodegenerative disease is an umbrella term for a range of pathological conditions primarily characterized by the progressive dysfunction and loss of selective neuronal populations. Bereft of cure and effective disease-modifying therapies, why and how this selective neuronal loss occurs in neurodegenerative diseases remain as the most intriguing and still unsolved questions in the field. Despite this limited knowledge regarding the trigger(s) underlying the different neurodegenerative phenotypes, during the last decades, mitochondrial dysfunction emerged as a common pathological feature being considered a "convergence point" for neurodegeneration. This is not surprising taking into account that neurons are post-mitotic cells with a complex architecture, long lifespan, and energetic requirements that fluctuate in time and space making them particularly reliant on a functional and dynamic mitochondrial network. Within this scenario, the present chapter provides an overview on the role of mitochondrial pathobiology in Alzheimer, Parkinson and Huntington diseases, the most prevalent neurodegenerative diseases. A more comprehensive view on the fundamental role of mitochondrial (mal)function during the pathological course of the abovementioned diseases may offer a new therapeutic window of opportunity to tackle the neurodegenerative phenotypes by bolstering mitochondrial health.

Keywords Alzheimer disease · Huntington disease · Mitochondrial bioenergetics and dynamics · Neurodegeneration · Parkinson disease

---

S. C. Correia (✉)
CNC – Center for Neuroscience and Cell Biology, University of Coimbra, Coimbra, Portugal

Institute for Interdisciplinary Research, University of Coimbra, Coimbra, Portugal
e-mail: sonia.correia@cnc.uc.pt

P. I. Moreira (✉)
CNC – Center for Neuroscience and Cell Biology, University of Coimbra, Coimbra, Portugal

Laboratory of Physiology, Faculty of Medicine, University of Coimbra, Coimbra, Portugal
e-mail: venta@ci.uc.pt; pimoreira@fmed.uc.pt

P. J. Oliveira (ed.), *Mitochondrial Biology and Experimental Therapeutics*,
https://doi.org/10.1007/978-3-319-73344-9_11

1 Introduction

Everyone ages, but why and how only some individuals acquire age-related neurodegenerative diseases remains an enigma for the scientific community. By definition, neurodegenerative diseases are incurable and devastating neurological disorders characterized by the progressive loss of the structure and function of specific subsets of neurons in the nervous system, culminating in motor, behavioral and cognitive disturbances, and eventually premature death (Dugger and Dickson 2017; Jellinger 2010). The pathogenic mechanisms leading to the onset and progression of neurodegenerative diseases are considered multifactorial caused by genetic, environmental and endogenous factors related to aging (Jellinger 2009; Skovronsky et al. 2006). With a worldwide population that is aging at a rapid pace, the prevalence of neurodegenerative diseases is now reaching alarming proportions emerging as one of the major global epidemics of the twenty-first century (Bales 2004; Kieburtz 2016). Currently, there is no cure or effective disease-modifying therapies able to slow the rate of neurodegeneration or halt the pathological course associated with the most common neurodegenerative diseases, namely Alzheimer (AD), Parkinson (PD) and Huntington diseases (HD). In this sense, neurodegenerative diseases represent a major frontier in medical research, a major challenge for the pharmaceutical industry and a burden for the society, being of tremendous importance to demystify the initial steps underlying the onset of these diseases in order to validate novel biomarkers for the early diagnosis and to design new therapeutic interventions aimed to forestall the neurodegenerative processes before reaching a "point of no return".

Despite the broad range of pathological phenotypes and different vulnerability of neuronal populations, neurodegenerative diseases share several common features including mitochondrial dysfunction (Carvalho et al. 2015; Correia et al. 2012b; Grimm and Eckert 2017; Rodolfo et al. 2017; Sheng and Cai 2012). Typically, connoted as the "powerhouse of the cell" and the "gatekeepers of life and death", it is now recognized that the strict regulation of mitochondrial structure, function and turnover represents an immutable control node for the maintenance of neuronal homeostasis. Within this scenario, the present chapter is aimed to summarize the current views on how mitochondria pathophysiology is involved in the onset and progression of AD, PD and HD, putting a special focus on the dynamic properties of these fascinating organelles.

2 What Is the Physiological Relevance of Mitochondria for Neuronal Populations?

Recent advances in mitochondrial biology posit that the neuronal fate is, in part, dictated by the befitting interplay between mitochondrial bioenergetics and dynamics (Kann and Kovacs 2007; Sheng 2014; Sheng and Cai 2012). Therefore, in order to understand the role of mitochondria in the context of neurodegenerative diseases, it is important to firstly address the physiological relevance of these fascinating organelles for normal neuronal functioning and integrity.

First and foremost, the primary role of mitochondria in neurons is to provide energy via production of adenosine triphosphate (ATP), which is accomplished through the Krebs cycle and oxidative phosphorylation system (OXPHOS). Owing to their limited glycolytic capacity, mitochondria function as "power stations" in neurons to support vital energy-consuming processes, such as axonal transport of macromolecules and organelles, ion pumps activity and synaptic transmission and plasticity (Sheng 2014; Sheng and Cai 2012). However, it is important to mention that mitochondrial electron transport chain (ETC) (specially complexes I and III) is also a major source of reactive oxygen species (ROS) (Finkel and Holbrook 2000). ROS are considered a double-edged sword for neurons. At low to moderate levels, ROS act as signaling molecules playing a vital role in normal neuronal functioning. However, exacerbated ROS levels may directly "attack" mitochondrial components triggering mitochondrial and, subsequently, neuronal oxidative damage (Kim et al. 2015).

Second, mitochondria are highly sophisticated and "tunable" buffers that shape the spatial and temporal patterns of calcium (Ca^{2+}) fluxes in excitable cells as neurons (Kwon et al. 2016). By acting as short-term reservoirs of Ca^{2+}, mitochondria partake in synaptic vesicles mobilization and recycling and neurotransmitter release and thus regulate neuronal communication (Vos et al. 2010). Furthermore, mitochondria can cluster in defined neuronal sub-domains forming a mitochondrial "firewall" to avert the spread of Ca^{2+} signals and consequent excitotoxicity (Celsi et al. 2009).

Lastly, neurons are post-mitotic, long-lived, and highly polarized cells composed by subcellular compartments with distinct metabolic requirements and a non-uniform mitochondrial distribution (Hollenbeck and Saxton 2005). During neuronal lifetime, mitochondrial network is continually reshaped and renewed by the action of four major dynamic processes: mitochondrial fusion-fission, biogenesis, transport and selective degradation by autophagy (mitophagy) (Seo et al. 2010) (Fig. 1). Functional mitochondria are generated *de novo* from pre-existing organelles in the soma through mitochondrial biogenesis, a complex phenomenon coordinated by the peroxisome proliferator-activated receptor gamma (PPARγ) co-activator (PGC-1α), and transported along the microtubule tracks to structures with high metabolic requirements, such as synaptic terminals. The mitochondrial movement in the anterograde direction is mediated by a class of motor proteins, the kinesins. Once in synaptic terminals, mitochondria sustain synaptic transmission by buffering Ca^{2+} and providing ATP. When damaged or no longer able to fuse with healthy mitochondria, mitochondria are retro-transported *via* dynein motor proteins to the soma in order to fuse with lysosomes and to be degraded by mitophagy. Thus, mitophagy represents an important node of mitochondrial quality control in neurons due to their limited regenerative capacity (Lin and Sheng 2015; Sheng 2014; Sheng and Cai 2012; Correia et al. 2016). During transitions between mitochondrial biogenesis and mitophagy, mitochondria experience successive cycles of fusion and fission. The balance of these opposing mechanisms allows the generation of a heterogeneous mitochondrial population and the proper neuronal functioning. Mitochondrial fusion events are crucial for the exchange of contents between mitochondria and enable damaged mitochondria to acquire components from healthy mitochondria,

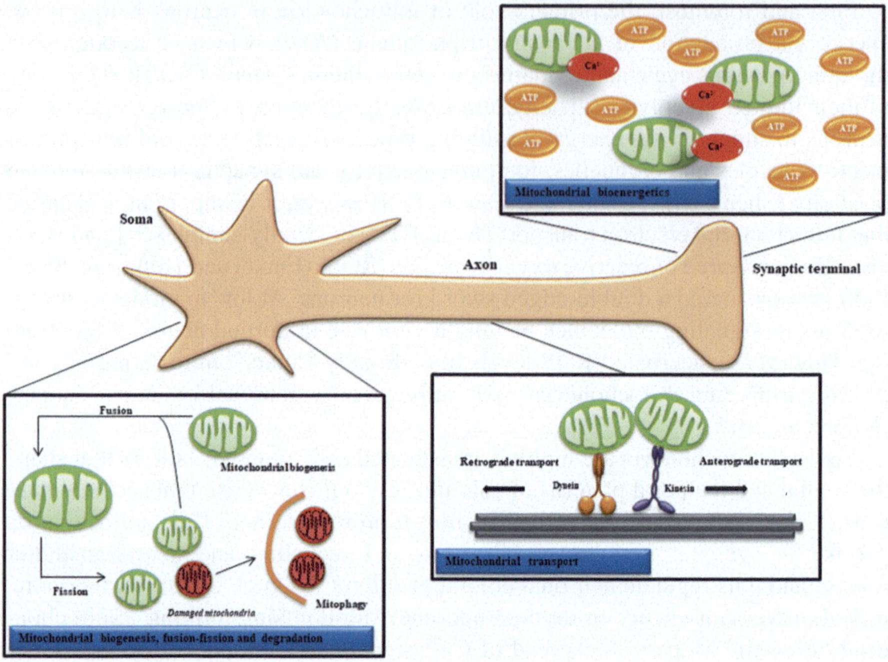

Fig. 1 Mitochondrial bioenergetics and dynamics in neuronal physiology. During neuronal lifetime, mitochondrial network is continually reshaped and renewed in order to maintain a healthy mitochondrial pool. Functional mitochondria are generated *de novo* in the soma by mitochondrial biogenesis transported in the anterograde direction to structures with high metabolic requirements, where these organelles sustain synaptic function by providing ATP and buffering Ca^{2+}. When damaged, mitochondria are retro-transported back to the soma where these organelles fuse with lysosomes and are degraded by mitophagy. In-between, mitochondria experience successive cycles of fusion and fission that allows the maintenance of a heterogeneous mitochondrial population

whereas mitochondrial fission has been recognized a pre-requisite for mitophagy and essential for the bidirectional movement of mitochondria along axons (Sheng and Cai 2012; Westermann 2010).

Overall, given the importance of all these "mitocheckpoints" in the maintenance of a healthy mitochondrial pool in neurons, even subtle disturbances in mitochondrial biogenesis, fusion-fission, transport and turnover are likely to far reaching functional consequences contributing to neurodegenerative diseases.

3 Mitochondrial Dysfunction in Alzheimer Disease

Firstly described by Alois Alzheimer in 1907, AD is currently recognized as the most prevalent age-related neurodegenerative disease affecting more than 30 million people worldwide (Correia et al. 2012a; Querfurth and LaFerla 2010). Existing in both familial and sporadic forms, the clinical symptoms of this devastating neurodegenerative disease comprise the progressive deterioration of memory and other cognitive domains due to the collapse of synaptic connectivity and neural network, culminating in the premature death of the individual typically 3–9 years after diagnosis (Correia et al. 2012a; Querfurth and LaFerla 2010). However, cognitive and behavioral symptoms constitute only the "tip of the iceberg", since the disruption of brain structure and function and consequent neuronal loss precede the clinical signs of the disease by 20–30 years (Braak and Braak 1995). The gold standard for the definitive diagnosis of AD relies on the *postmortem* identification of two distinctive brain lesions: the extracellular deposition of amyloid-β (Aβ) peptide in senile plaques and the intracellular accumulation of neurofibrillary tangles (NFTs) mainly composed of hyperphosphorylated tau protein (Correia et al. 2012a; Querfurth and LaFerla 2010).

One of the most ferocious debates regarding AD etiology concerns the real role of Aβ in the pathological cascade of this neurodegenerative disease. For many years, Aβ was recognized as the pathological trigger of the deleterious events that occur in AD (Karran et al. 2011). However, past approaches focused on blocking Aβ pathology have failed repeatedly (Herrup 2015), leading to the emergence of alternative hypotheses to explain the basis of this neurodegenerative disease. Among them is the "mitochondrial cascade hypothesis" (Swerdlow et al. 2014; Swerdlow and Khan 2004). According to this hypothesis: (1) inheritance determines mitochondrial baseline function and robustness; (2) mitochondrial robustness determines how mitochondria change with age; and (3) when mitochondrial alterations reach a threshold, AD histopathology and symptoms ensue. In line with this hypothesis, this sub-chapter highlights the most relevant mitochondrial abnormalities that contribute to the pathological course of AD.

3.1 *Mitochondrial Bioenergetics in Alzheimer Disease*

Over the past decades, compelling evidence demonstrates that altered mitochondrial metabolism and oxidative damage precedes the occurrence of both Aβ and tau pathologies in AD (Chou et al. 2011; Nunomura et al. 2001; Resende et al. 2008; Yao et al. 2009). The first incontestable clue for the involvement of metabolic defects in AD derived from positron-emission tomography (PET) imaging studies using the tracer 2-[^{18}F] fluoro-2-deoxy-D-glucose, in which reductions in cerebral glucose transport and utilization were detected in brain areas affected by the disease (Friedland et al. 1989; Jagust et al. 1991). Additionally, this glucose hypometabolism was reported to be accompanied by a drastic reduction in the expression of glucose transporter-1 (GLUT-1) and -3 (GLUT-3), especially in the cerebral cortex. Notably, Shah and collaborators (Shah et al. 2012) found that the reduction in GLUT-1 and GLUT-3 expression aggravates AD pathology by facilitating tau hyperphosphorylation and Aβ generation and compromising Aβ clearance. Concerning glucose metabolism, the activity of pyruvate, isocitrate and α-ketoglutarate dehydrogenases, three key enzymes involved in Krebs cycle, was also shown to be impaired in *post-mortem* brain tissue and fibroblasts from AD subjects (Bubber et al. 2005; Mastrogiacomo et al. 1993). Notably, this impairment in the activity of Krebs cycle enzymes is correlated with the clinical state of AD, pinpointing for a causal relation between the progressive deterioration of mitochondrial functioning with the severity of the disease (Bubber et al. 2005). However, the reduction in cytochrome c oxidase (COX) activity is the most consistent and well-characterized mitochondrial abnormality documented in AD (Mutisya et al. 1994). Searching for the origin of COX loss of activity in AD, it has been proposed that mitochondrial components are highly susceptible to oxidative modification and are altered by exposure to a range of pro-oxidants (Tretter and Adam-Vizi 2000). As a matter of fact, oxidative modification of mitochondrial proteins, including COX, have been demonstrated in the AD brain by elevated levels of protein carbonyl and nitration of tyrosine residues (Sultana and Butterfield 2013). Importantly, impaired COX activity, in turn, can potentiate the generation of mitochondrial-derived ROS, fostering the idea that defective mitochondrial bioenergetics and oxidative stress are coupled in a vicious circle (Mutisya et al. 1994).

Being conceivable that mitochondrial bioenergetics deficits occur early and contribute to the progression of sporadic AD, one captivating dilemma is whether or not the dysfunction of these organelles participates in the familial forms of the disease. As the major driving force underlying familial AD, Aβ was shown to directly perturb mitochondrial function. Lessons from clinical and experimental studies revealed a progressive accumulation of Aβ monomers and oligomers within mitochondria in AD (Caspersen et al. 2005; Crouch et al.2005; Devi et al. 2006; Manczak et al. 2006). Hansson and collaborators (Hansson Petersen et al. 2008) found that Aβ is imported into mitochondria via the translocase of the outer membrane (TOM) import machinery and then transported through the inner mitochondrial membrane by the translocase of the inner membrane (TIM). Once inside mitochondria, Aβ binds to heme groups, which constitute critical redox centers of COX, and interact with the mitochondrial protein Aβ-binding alcohol dehydrogenase (ABAD), culminating in hampered COX activity, ROS overproduction and ATP depletion (Atamna 2006; Atamna and Frey 2004). Furthermore, Aβ was also shown to disturb COX activity by blocking TOM40 and TIM23 from importing COX subunits into the mitochondria (Devi et al. 2006). Using the triple transgenic mouse model of AD to dissect the close interrelationship between these organelles and Aβ and tau pathologies, Rhein and collaborators (Rhein et al. 2009) found that disturbances in mitochondrial metabolism are due to a convergence of Aβ and tau on mitochondria, accelerating defects in respiratory capacity and a main impairment in COX activity. Additionally, an age-related oxidative stress was found to further exacerbate the disruption of mitochondrial metabolism, finally culminating in neuronal loss (Rhein et al. 2009). Collectively, these findings consolidate the idea that a synergistic effect of Aβ and tau augments the pathological deterioration of mitochondria in familial AD.

3.2 Mitochondrial Dynamics in Alzheimer Disease

Since the pioneering work developed by Hirai and collaborators (Hirai et al. 2001), mitochondrial morphological changes manifested by reduced size and broken internal membrane cristae have been extensively reported in *postmortem* brain tissue from AD subjects and experimental models of the disease (Baloyannis 2006; Hirai et al. 2001; Gao et al. 2017). More recently, it was also found that AD cybrids possess an altered mitochondrial morphology as denoted by the presence of fragmented, misshaped, and bleb-like mitochondria (Gan et al. 2014). Taking into account the dynamic nature of mitochondria, it is alluring to anticipate that the accumulation of mitochondria with a defective morphology in AD may result from excessive mitochondrial fission, inefficient mitophagy, and faulty retrograde transport or from the catastrophic combination of all these anomalies.

Regarding mitochondria fusion-fission events a shift towards fission has been reported in AD pathology. Reduced levels of the mitochondrial fusion proteins optic atrophy protein-1 (OPA-1) and mitofusins were observed in hippocampal tissue from AD subjects (Wang et al. 2009). Furthermore, the combination of increased translocation of the mitochondrial fission protein dynamin-like protein 1 (DRP-1) to mitochondria and reduced mitofusin-2 levels was also shown to contribute to mitochondrial network fragmentation in AD cybrids (Gan et al. 2014). In a step further, it was also found that the AD brain displays increased levels of DRP-1 S-nitrosylation (Cho et al. 2009). By modulating the GTPase activity of DRP-1, this post-translation modification favors the occurrence of mitochondrial fission and hampers mitochondrial bioenergetics, thus leading to synaptic damage and neuronal demise (Cho et al. 2009).

Having in mind that mitochondrial fission is required for mitophagy and mitochondrial biogenesis, a major question invades our mind: are defective mitochondria being properly degraded and replenished in AD? As a sign of faulty mitochondrial turnover, reduced mitochondrial content was observed in susceptible hippocampal neurons from AD subjects (Baloyannis 2006; Hirai et al. 2001). Recent breakthroughs reported a drastic reduction in PGC-1α protein levels and mitochondrial DNA (mtDNA)-to-nuclear DNA (nDNA) ratio in human *postmortem* brain tissue from AD subjects and experimental models of the disease, suggesting that the refurbishment of an "old" mitochondrial pool via mitochondrial biogenesis is compromised (Qin et al. 2009; Sheng et al. 2012). On the flipside, elevated levels of mitochondrial components, namely COX and lipoic acid, were also detected within autophagosomes in human *postmortem* brain tissue from AD subjects, suggesting an increase in the rate of mitochondrial degradation by autophagy (Moreira et al. 2007a, b). It was proposed that during the initial phase of AD pathology the induction of mitophagy may constitute a compensatory mechanism to "escape" from a neuronal energetic crisis (Shaerzadeh et al. 2014). However, with the pathological progression of the disease the recruitment of the autophagic machinery to mitochondria fails, leading to an abnormal accumulation dysfunctional mitochondria (Ye et al. 2015). In line with these findings, two challenging studies suggest a similar mitophagy failure in sporadic and familial AD, however through different mechanisms (Martin-Maestro et al. 2016, 2017). Briefly, in sporadic cases, the deficit is due to insufficient labeling of mitochondria to be degraded by mitophagy while in familial cases mitochondria are correctly tagged for recycling but a deficiency in the degradation phase of autophagy seems to occur (Martin-Maestro et al. 2016, 2017).

Mitochondrial trafficking is another dynamic property of mitochondria that is compromised in AD pathology as denoted by mitochondrial overcrowding, incorrect localization, and impaired mitochondrial function in metabolic demanding structures (Correia et al. 2012a; Trimmer and Borland 2005). Aβ and tau have been proposed as the major obstacles to the normal mitochondrial trafficking in AD. For instance, oligomeric Aβ was shown to promote an abnormal axonal transport of mitochondria in cultured hippocampal neurons (Wang et al. 2010). Particularly, it was found that oligomeric Aβ reduces mitochondrial motility (the percentage of motile mitochondria over total mitochondria) in both anterograde and retrograde directions (Du et al. 2010). However, oligomeric Aβ seems to induce a more drastic effect on anterograde movement of mitochondria than on their retrograde movement (Calkins et al. 2011; Calkins and Reddy 2011; Du et al. 2010). Being a microtubule-associated protein, the involvement of tau in hampered mitochondrial trafficking in AD is also expected. Tau overexpression in differentiated N2a cells was shown to impair axonal traffic and reduce the number of mitochondria present in neurites (Ebneth et al. 1998). The observed mitochondrial mislocalization is caused by the binding of tau to microtubules tracks, which slows down the anterograde transport (Ebneth et al. 1998). By acting in overlapping and synergistic ways, it has also been proposed that Aβ and tau interact causing mitochondrial anomalies. Indeed, low concentrations of Aβ together with pathologically modified tau protein detrimentally affect mitochondrial transport and function (Quintanilla et al. 2012). Particularly, the presence of truncated tau protein and Aβ decreased the anterograde transport of mitochondria resulting in an increase in the stationary mitochondrial population at the soma (Quintanilla et al. 2012). Moreover, the genetic ablation of tau protein averts the inhibition of mitochondrial transport promoted by oligomeric Aβ, suggesting that Aβ hampers the axonal transport of mitochondria in a tau-dependent manner (Vossel et al. 2010). Finally, taking into account that mitochondrial travelling along the axons is an ATP-dependent process, mitochondrial bioenergetics deficits are also viewed as obstacles for the normal mitochondrial movement in AD (Correia et al. 2016).

4 Mitochondrial Dysfunction in Parkinson Disease

With over ten million people being affected worldwide, PD is the second most common neurodegenerative disease (Muangpaisan et al. 2011). Currently, the clinical diagnosis of PD is based on the manifestation of a classic tetrad of motor symptoms: low-frequency resting tremor, rigidity, bradykinesia, and, in later stages of the disease, postural instability (Berardelli et al. 2013). However, PD is more than a movement disorder being increasingly recognized that a wide range of non-motor symptoms (i.e. autonomic, sleep, and neuropsychiatric disturbances) may develop insidiously in the prodromal phase of the disease preceding the onset of the traditional motor features and the clinical diagnosis by several decades (Goldman and Postuma 2014; Schapira et al. 2017). From a neuropathological point of view, PD-related phenotype results mainly from the progressive and massive loss of midbrain dopaminergic neurons in the *substantia nigra pars compacta* (SNpc) and the concomitant depletion of dopamine content in the striatum (Forno 1996; Kovari et al. 2009). Notably, this specific pattern of neurodegeneration is accompanied by the presence of intracellular inclusions primarily composed of misfolded α-synuclein in the form of Lewy bodies (Forno 1996; Spillantini and Goedert 2000).

Notwithstanding the lack of a clear and unifying pathogenic mechanism underlying the selective vulnerability of dopaminergic neurons in PD pathology, the natural history of both sporadic and familial PD seems to be intimately related with mitochondrial dysfunction (Bose and Beal 2016; Franco-Iborra et al. 2016; Subramaniam and Chesselet 2013). The next sub-chapter is aimed to summarize the current knowledge derived from epidemiological studies and studies performed in experimental models of parkinsonism, that show that mitochondrial dysfunction occurs early and acts as causative event in the pathogenesis of both sporadic and familial forms of the disease.

4.1 Mitochondrial Bioenergetics in Parkinson Disease

A major breakthrough linking mitochondrial pathobiology to PD came from specific cases of idiopathic PD during the 1980's, when several drug addicts accidently injected an heroin analog, the 1-methyl-4-phenyl-1,2,3,6-tetrahydropyridine (MPTP). Within days, these subjects developed parkinsonism and the *postmortem* analysis revealed significant lesions of dopaminergic neurons in the SNpc (Langston et al. 1983). MPP^+ (1-methyl-4-phenyl-1,2,3,6-tetrahydropyridium ion) is the active metabolite of MPTP that concentrates within the dopaminergic neurons and specifically inhibits the activity of mitochondrial complex I (Ali et al. 1994). Shortly after the observation of Parkinsonism caused by MPTP administration, Schapira and collaborators (Schapira et al. 1990) reported a drastic reduction in the activity of the mitochondrial complex I in the SNpc of PD subjects. Subsequent studies consistently showed a mild impairment in the activity of mitochondrial respiratory complex I in frontal cortex (Parker et al. 2008), platelets (Parker et al. 1989; Krige et al. 1992; Haas et al. 1995; Blandini et al. 1998), lymphocytes (Barroso et al. 1993; Yoshino et al. 1992), and skeletal muscle tissue (Taylor et al. 1994; Penn et al. 1995) from PD subjects. Besides MPTP, several other specific inhibitors of the mitochondrial complex I, such as rotenone and 6-hydroxydopamine (6-OHDA), were shown to cause degeneration of the nigral dopaminergic neurons and recapitulate the PD pathological phenotype in *in vivo* models, further emphasizing the critical involvement of mitochondrial bioenergetics deficits in the etiology of PD (Betarbet et al. 2000; Gash et al. 2008; Sherer et al. 2007). But, why SNc appears to be more susceptible to complex I activity impairment than other brain regions? There is no clear-cut answer to this question, but a possible explanation attributes a major role to oxidative stress. Indeed, dopaminergic neurons are particularly prone to oxidative stress due to dopamine metabolism and auto-oxidation combined with increased iron content, decreased antioxidant defenses and mitochondrial complex I inhibition-induced ROS overproduction in the SNpc. All these events can trigger neuronal degeneration and loss by exceeding the oxidative capacity of dopamine-containing neurons in SNpc (Chinta and Andersen 2008).

While the nature of mitochondrial complex I deficiency in PD remains puzzling, a causative role has been attributed to mtDNA anomalies. Indeed, cybrids containing mtDNA from PD subjects present a significant impairment in complex I activity associated with increased oxidative stress levels (Veech et al. 2000; Swerdlow et al. 1996) suggesting mtDNA encoded defects in PD. Additionally, several mtDNA mutations that affect complex I activity and OXPHOS efficiency are associated to an increased risk for developing PD (e.g. mutations in mtDNA-encoded complex I genes) (Autere et al. 2004). Corroborating the notion that mtDNA deletions may contribute to neurodegeneration in PD, increased levels of mtDNA deletions have been observed in the striatum and SNpc neurons of PD subjects (Yao and Wood 2009). Importantly, Swerdlow and collaborators (Swerdlow et al. 1998) reported maternally inherited mutations in mtDNA in one family with PD.

4.2 Mitochondrial Dynamics in Parkinson Disease

Taking into account that mitochondrial bioenergetics and dynamics are closely intertwined, it is expected that alterations in mitochondrial morphology, positioning and turnover also take part in the global picture underlying PD pathology. So far, less is known about the mitochondrial morphological changes occurring in the PD brain however, excessive mitochondrial fragmentation has been persistently reported in peripheral cells from PD subjects (Haylett et al. 2016; Mortiboys et al. 2010; Papkovskaia et al. 2012; Su and Qi 2013). Importantly, the direct link between mitochondrial dynamics and PD pathology is reinforced by the identification of several proteins associated with the familial forms of the disease, namely α-synuclein, parkin, DJ-1, PTEN-induced kinase (PINK1), leucine-rich repeat kinase 2 (LRRK2), and vacuolar protein sorting 35 (VPS35), which are known to be physically associated with mitochondria or mitochondria-associated endoplasmic reticulum membranes (MAM), which are capable of modulating the integrity and dynamic properties of mitochondria (Clark et al. 2006; Guardia-Laguarta et al. 2014; Narendra et al. 2008; Wang et al. 2012a, b, 2016). In the last years, a flurry of experimental studies in genetic models of PD extended our knowledge regarding the mechanistic basis that leads to mitochondrial fragmentation in this neurodegenerative disease. For instance, mutant α-synuclein was shown to mislocalize with mitochondria triggering mitochondrial bioenergetics deficits, oxidative damage, and consequent mitochondrial fragmentation (Hsu et al. 2000; Sherer et al. 2002; Xie and Chung 2012). Genetic studies in *Drosophila* demonstrated that the PINK1-Parkin axis regulates mitochondrial dynamics, the loss-of-function mutations in Parkin and PINK1 genes triggering mitochondrial fragmentation and/or inhibiting mitochondrial elongation by upregulating DRP-1 and/or downregulating mitofusins and OPA-1 function (Deng et al. 2008; Poole et al. 2008). Using genetic models expressing mutant forms (R1441C or G2019S) of LRRK2, the most common cause of autosomal-dominant PD, Wang and collaborators (Wang et al. 2012b) also found that the toxic gain-of-function of LRRK2 promotes mitochondrial fragmentation by increasing the activity of DRP-1 through a direct physical interaction with this critical mediator of mitochondrial fission. Accordingly, alterations in mitochondrial morphology accompanied by changes in the levels of mitochondrial fission proteins DRP-1 and mitochondrial fission protein 1 (Fis1) were observed in a knock-in mice harboring the LRRK2 mutation G2019S (Yue et al. 2015). In *in vitro* and *in vivo* models of MPTP-induced PD, it was also observed that the S-nitrosylation of Parkin reduces its ability as a suppressor of DRP-1 expression, leading to an upregulation of DRP-1 and concomitant mitochondrial fragmentation (Zhang et al. 2016). Recent findings also revealed that mutant forms of VPS35 also promote mitochondrial fragmentation by enhancing the clearance of inactive mitochondrial DRP-1 complexes (Wang et al. 2016). Importantly, the inhibition of mitochondrial fission *per se* is sufficient to alleviate mitochondrial dysfunction in these experimental models of PD, fostering the fundamental role of mitochondrial dynamics in PD pathology and its therapeutic potential (Gao et al. 2017).

As abovementioned, mitochondrial fusion-fission is required for the normal trafficking and distribution of mitochondria along the neurons. Mitochondrial trafficking defects have been observed in genetic and neurotoxin-induced models of PD as denoted by mitochondrial mislocalization and massive accumulation of these organelles in the perinuclear region (Wang et al. 2011a, 2012b). Notably, suppression of mitochondrial fragmentation was shown to avert defective mitochondrial transport in these PD models (Wang et al. 2011a, 2012b), being tempting to speculate that hampered mitochondrial movement in PD is a downstream consequence of unbalance mitochondrial fusion-fission equilibrium. Given the central role of α-synuclein in PD pathology, a recent study also documented early defects in anterograde-to-retrograde mitochondrial flux in human-derived neurons overexpressing α-synuclein, indicating a regulatory role of α-synuclein on mitochondrial axonal transport (Pozo Devoto et al. 2017). Mechanistically, the defective mitochondrial trafficking may arise from the interaction of α-synuclein with the molecular motor machinery, particularly with a complex containing kinesin-1 (Utton et al. 2005).

The most groundbreaking finding pertains to the involvement of PINK-Parkin axis in the regulation of mitochondrial network homeostasis and quality control, including mitophagy (Durcan and Fon 2015; McWilliams and Muqit 2017). Under steady state conditions, PINK1 is constitutively imported into mitochondria, cleaved and degraded *via* the N-end rule pathway. Upon the loss of mitochondrial membrane potential, PINK1 acts as a bioenergetic sensor that recruits the E3 ubiquitin ligase Parkin from the cytosol to the damaged mitochondria. Subsequently, active Parkin ubiquitinates a myriad of substrates that resides in the outer mitochondrial membrane by elongating pre-existing ubiquitin chains or ubiquitinating these substrates *de novo* (Wild and Dikic 2010; Kane et al. 2014; McWilliams and Muqit 2017; Yamano et al. 2016). These polyubiquitin chains serve as a distinct molecular code required for the selective recruitment of autophagy machinery that drives the clearance of the damaged organelle and ultimately, the completion of mitophagy (Wild and Dikic 2010; Kane et al. 2014; McWilliams and Muqit 2017; Yamano et al. 2016). Mutations in either Parkin or PINK1 were shown to alter mitochondrial turnover leading to an abnormal accumulation of damaged mitochondria. Particularly, Geisler and collaborators (Geisler et al. 2010) demonstrated that PINK1 mutations compromise the selective degradation of depolarized mitochondria, mainly due to the decreased physical binding activity of PD-linked PINK1 mutations to Parkin. In this sense, PINK1 mutations abrogate autophagy of impaired mitochondria upstream of Parkin. In addition to compromised PINK1 kinase activity, reduced binding of PINK1 to Parkin causes the failure of Parkin mitochondrial translocation, resulting in the accumulation of damaged mitochondria, and, consequently, contributing to the pathogenesis of PD (Geisler et al. 2010). In a step further, recent advances also cast light on the complex relationship between mitophagy and mitochondrial transport in PD, raising the hypothesis that the impaired clearance of damaged mitochondria is, in part, due to the inappropriate trafficking of these organelles (Choong and Mochizuki 2017). As a matter of fact, PINK1 and Parkin are known to target Miro for phosphorylation and degradation (Wang et al. 2011b). Miro is an outer mitochondrial membrane protein that anchors mitochondria to microtubule motors and is removed to stop mitochondrial motility as an early step in mitophagy. In this sense, Hsieh and collaborators (Hsieh et al. 2016) recently reported that both sporadic and familial forms of PD are characterized by an impaired Miro removal and mitochondrial arrest, which delays the initiation of the mitophagic process (Hsieh et al. 2016).

After the clearance of damaged mitochondria, neurons reestablished the mitochondrial pool through the generation of new organelles by mitochondrial biogenesis (Zhu et al. 2013). Therefore, a major question should arise at this point: is mitochondrial biogenesis affected in PD? The first hint regarding a potential role for compromised mitochondrial biogenesis derived from a study reporting a lower mitochondrial mass in dopaminergic neurons of the SNpc, compared to dopaminergic neurons located in the adjacent ventral tegmental area, or non-dopaminergic cells in the midbrain (Liang et al. 2007). Indeed, genes responsive to the master regulator of mitochondrial biogenesis PGC-1α are underexpressed in PD subjects (Zheng et al. 2010). It was also found that PGC-1α null nigral neurons are more prone to degenerate following overexpression of human α-synuclein, whereas PGC-1α overexpression restores mitochondrial morphology, oxidative stress detoxification and basal respiration (Ciron et al. 2015). The creation of the conditional knockout "MitoPark" mice also revealed that the ablation of mitochondrial transcription factor A (Tfam) gene in dopamine transporter-positive nigral neurons resulted in a robust reduction in mtDNA expression and mitochondrial bioenergetics deficits, which lead to a parkinsonism phenotype with adult onset and characterized by slowly progressive impairment of motor function accompanied by the formation of intraneuronal inclusions and loss of dopaminergic neurons (Ekstrand et al. 2007). Furthermore, it was found that Parkin loss-of-function leads to the accumulation of Parkin interacting substrate (PARIS) that, in turn, represses PGC-1α contributing to the reduction of mitochondrial mass and respiration and progressive loss of dopaminergic neurons in a PARIS-dependent manner (Shin et al. 2011; Stevens et al. 2015).

5 Mitochondrial Dysfunction in Huntington Disease

HD is an autosomal dominant neurodegenerative disease clinically characterized by motor dysfunction, cognitive decline and psychiatric disturbances. At the molecular level, HD is caused by the expansion of an unstable trinucleotide CAG repeat in the gene encoding Huntingtin (*Htt*), resulting in an extended polyglutamine (polyQ) tract in the N-terminus of the Htt protein, thereafter designated by mutant Htt (mHtt) (Bates 2005; Gil and Rego 2008). In unaffected subjects, the Htt gene contains 6-35 CAG repeats. When polyQ residues exceed 40 CAG repeats, mHTT gets unfolded and is cleaved by proteases forming short peptides that have a high propensity to aggregate (Bates 2005; Gil and Rego 2008). Despite the ubiquitous expression of mHtt, it selectively targets GABAergic striatal medium spiny neurons (MSNs), and also leads to cerebral cortex degeneration in early- to-mild stages of HD (Reiner et al. 1988). Among several targets, mHtt affects mitochondrial homeostasis by physically interacting with these organelles or by disturbing the transcriptional regulation of genes involved in mitochondrial biogenesis and function (Oliveira 2010; Guedes-Dias et al. 2016; Naia et al. 2017). In this sense, mitochondrial dysfunction has been considered an integrant feature of HD pathology (Bossy-Wetzel et al. 2008; Oliveira 2010; Guedes-Dias et al. 2016; Naia et al. 2017). Due to the existing discrepancies regarding whether mitochondrial dysfunction is a primary cause of early onset or is a late-stage consequence of mHtt toxicity (Polyzos and McMurray 2017), this sub-chapter is aimed to clarify this "chicken-and-egg dilemma".

5.1 Mitochondrial Bioenergetics in Huntington Disease

The existence of a diminished glucose metabolism in brain regions affected by HD in both pre-symptomatic and symptomatic stages represents the first sign that early metabolic deficits are involved in the pathological course of this neurodegenerative disease (Kuhl et al. 1982; Kuwert et al. 1989; Koroshetz et al. 1997). Mechanistically, it was recently demonstrated that oxidative damage causes decreased expression of GLUT-3, resulting in inhibited glucose uptake and accumulation of lactate (Covarrubias-Pinto et al. 2015). The mitochondrial involvement in HD gained strength when Beal and collaborators (Beal et al. 1993) demonstrated for the first time that the systemic administration of mitochondrial complex II inhibitors (3-nitropropionic acid and malonate) produces selective striatal lesions, recapitulating several histologic and neurochemical features of HD. Supporting the critical role of mitochondrial complex II in HD, it was found that the expression of mHtt is associated with the loss of complex II components, the 30-kDa iron-sulfur (Ip) subunit and the 70-kDa FAD (Fp) (Damiano et al. 2013). However, considering that HD is a monogenic disorder it is questionable if this mitochondrial bioenergetics failure is a trigger of HD or a detrimental consequence of the toxic gain-of-function from mHtt. *Postmortem* studies performed with advanced (grade 3 and 4) HD subjects showed a drastic reduction in the activity of mitochondrial complexes II and III and a moderate decrease in the activity of mitochondrial complex IV in both caudate or putamen (Browne et al. 1997; Gu et al. 1996). However, similar studies performed with pre-symptomatic and early (grade 1) HD subjects revealed no impairment in striatal mitochondrial complexes I–IV (Guidetti et al. 2001), relegating respiratory chain impairment as a secondary event in HD. In this perspective, mHtt was shown to have direct inhibitory effects on respiratory complex II (Panov et al. 2002). Cumulative data indicates that soluble N-terminal mHtt interacts with both outer and inner mitochondrial membranes impairing the import of mitochondrial proteins, which seems to precede the overall mitochondrial dysfunction (Oliveira 2010; Polyzos and McMurray 2017).

Alongside with the impaired mitochondrial respiratory chain activity, disturbances on mitochondrial Ca^{2+} handling and metabolism have been described during the early and late stages of HD pathology as a direct consequence of mitochondrial localization of mHtt (Naia et al. 2017). Previous data from clinical and experimental studies revealed that HD mitochondria have an increased propensity to depolarize and are more susceptible to Ca^{2+} overload and mitochondrial permeability transition pore (mPTP) opening, which reflect their poor Ca^{2+} buffering capacity (Panov et al. 2002; Choo et al. 2004). In this sense, the use of cyclosporine A, a classical mPTP inhibitor, was shown to correct this lower mitochondrial Ca^{2+}buffering capacity in HD.

5.2 *Mitochondrial Dynamics in Huntington Disease*

Abnormal mitochondrial ultrastructure was also reported in HD, raising a possible involvement of mitochondrial dynamics defects in this neurodegenerative disease (Squitieri et al. 2006). In fact, mHtt was to shown to shift mitochondrial network from an elongated to a small round punctuated phenotype due to an unbalance in mitochondrial fusion-fission machinery. Furthermore, this alteration in mitochondrial morphology is accompanied by a defective trafficking of mitochondria in HD neurons (Chang et al. 2006; Trushina et al. 2004). A subsequent study also demonstrated that the exposure of rat cortical neurons to the HD mimetic 3-NP causes mitochondrial fragmentation and condensation (Liot et al. 2009). But, what are the mechanisms that trigger these alterations in the dynamic properties of mitochondria in HD pathology? Concerning mitochondrial morphology, the HD brain exhibits altered expression or post-translational modifications of key proteins involved in mitochondrial fusion-fission, such as DRP-1 and mitofusins (Kim et al. 2010; Shirendeb et al. 2012; Guedes-Dias et al. 2015). For instance, increased DRP-1 S-nitrosylation was observed in the striatum of HD subjects and rodent models of the disease, which is correlated with excessive mitochondrial fragmentation followed by the loss of dendritic spines and massive synaptic damage (Haun et al. 2013). Moreover, mHtt was recently reported to directly interact with the mitochondrial fission protein DRP-1, increasing its GTPase activity (Song et al. 2011; Shirendeb et al. 2012). On the flipside, the inactivation of DRP-1 or the blockage of DRP-1 and Fis1 interaction alleviates mitochondrial dynamic abnormalities in neurons expressing mHtt, strongly suggesting the direct involvement of mHtt in mitochondrial fusion-fission imbalance (Song et al. 2011; Shirendeb et al. 2012).

But, what happens to mitochondrial biogenesis, transport and degradation in HD? As mentioned above, HD pathology is associated with transcriptional deregulation. Particularly, it has been suggested that the mitochondrial dysfunction that characterizes this neurodegenerative disease is due, in part, to an abnormal functioning of PGC-1α (Guedes-Dias et al. 2016). As a matter of fact, *postmortem* studies revealed a progressive reduction in brain levels of PGC-1α mRNA and protein with the severity of HD (Cui et al. 2006; Weydt et al. 2006). Notably, a significant decrease in PGC-1α mRNA was also reported in the caudate nucleus of asymptomatic HD subjects indicating that faulty mitochondrial biogenesis occurs early in HD pathology (Cui et al. 2006). In the same line, *in vivo* studies revealed that R6/2 transgenic mice of HD present a similar decrease in PGC-1α mRNA levels in both striatum and brain cortex however, a significant reduction in mtDNA-to-nDNA was detected only the striatum (Hering et al. 2015), emphasizing the increased susceptibility of this brain region in HD. Mechanistically, mHtt was proposed to repress PGC-1α gene transcription due to its association with the promoter and by interfering with the CREB/TAF4-dependent transcriptional pathway, critical for the regulation of PGC-1α gene expression (Cui et al. 2006).

Another important aspect concerns with the fact that mHtt is prone to aggregate, which may physically act as roadblocks for mitochondrial movement. In this sense, Chang and collaborators (Chang et al. 2006) observed that the expression of mHtt in cortical neurons leads to a drastic reduction in mitochondrial trafficking. The authors observed that those defects were particularly prominent in neuronal regions rich in aggregates, being the degree of mitochondrial motility impairment correlated with the size of mHtt aggregates (Chang et al. 2006). However, mHtt aggregates may also impair movement indirectly, namely by sequestering wild-type Htt, which is essential for fast axonal trafficking (Trushina et al. 2004). However, defective mitochondrial motility was also observed in HD striatal neurons in the absence of aggregates (Trushina et al. 2004), suggesting alternative mechanisms whereby mHtt could affect mitochondrial transport. Particularly, specific N-terminal fragments of mHtt (produced before aggregates formation) were shown to preferentially associate with mitochondria, in an age-dependent manner, directly affecting mitochondrial traffic (Orr et al. 2008). mHtt has also been reported to selectively sequester and inactivate motor proteins such as kinesin and dynactin or disrupt the association of motor proteins with microtubules by interacting with huntingtin-associated protein 1 (HAP1) (Gauthier et al. 2004; Trushina et al. 2004).

Finally, mHtt also affects one of the most important mitochondrial quality control mechanisms, mitophagy. Several studies suggest that the removal of defective mitochondria is compromised in HD. Impaired mitophagy was recently reported in flies expressing neuronal Htt-ex1p-Q93 and in HdhQ111 striatal cells lines, this impairment being associated with decreased targeting of ubiquitinated mitochondria to autophagosomes (Khalil et al. 2015). Data from neuronal HD models, however, failed to identify impaired mitochondrial loading, but rather pointed to the accumulation of undigested mitochondria due to disturbances in the autophagosomal transport, which is required for efficient fusion with lysosomes (Wong and Holzbaur 2014). It was observed that Htt and HAP1 form a scaffolding complex in a subpopulation of LC3-positive neuronal autophagosomes that enhances their dynein-dynactin-driven transport towards the soma (Wong and Holzbaur 2014). However, the polyQ expansion in mHtt enhances its affinity for HAP1 (Li et al. 1995). Consequently, mHtt disrupts the normal Htt-HAP1 association, impairing autophagosomal transport and limiting the clearance of cargo, such as damaged mitochondria (Wong and Holzbaur 2014).

6 Final Remarks and Future Directions

While the quest for the cure is still an utopia in field of neurodegenerative diseases, it is undeniable that mitochondria pathobiology is an important contributing factor for neurodegenerative events that occurs during the pathological course of these diseases. Recent advances highlighted the mechanisms underlying mitochondrial dysfunction in AD, PD and HD, with defects on both mitochondrial bioenergetics and dynamics being transversal to these three diseases. Exhibiting mitochondrial ultrastructural alterations, mislocalization and reduced mass, these neurodegenerative diseases are, in part, the result of a complex interplay that encompasses robust changes in the transport and turnover of mitochondria, contributing to a mitoenergetic crisis, synaptic starvation and ultimately, neuronal degeneration and loss. Within this scenario, mitochondria have emerged, during the last decades, as an attractive therapeutic target to tackle these neurodegenerative diseases in a timely manner. Natural and mitochondria-targeted antioxidants have been extensively used to counteract symptomatic and neuropathological features associated with AD, PD and HD (Moreira et al. 2010; Reddy and Reddy 2011; Federico et al. 2012). Despite the positive outcomes in preclinical models, both classes of antioxidants failed to show efficacy in clinical trials. Why this happens? One possible reason is related with the vital role of ROS as signaling molecules in neurons; antioxidants may neutralize ROS-mediated survival mechanisms. Another important reason relates with the stage of the disease of individuals enrolled in most clinical trials. In fact, most clinical trials enrolled individuals with moderate/late stages of the neurodegenerative disease (Persson et al. 2014), stages that are associated with the loss of a large percentage of susceptible neuronal populations. Future trials must focus on prodromal or asymptomatic stages of the diseases, where hope resides.

Scientific efforts also targeted the dynamic properties of mitochondria to counteract neurodegeneration in *in vitro* and *in vivo* models of AD, PD and HD; the inhibition of mitochondrial fission emerging as a promising strategy. Finally, lifestyle interventions, such as physical exercise and caloric restriction, have gained attention in the last decades as a possible alternatives to fight neurodegenerative diseases, in part, by bolstering mitochondrial health (Moreira et al. 2010; Reddy and Reddy 2011; Federico et al. 2012). Future studies aimed to unveil the temporal alterations that occur during the asymptomatic and prodromal stages of AD, PD and HD and the development of novel pharmacological interventions designed to multi-target different aspects of mitochondrial biology may hold the "holy grail" for neurodegenerative disease therapy.

References

Ali SF, David SN, Newport GD, Cadet JL, Slikker W Jr (1994) MPTP-induced oxidative stress and neurotoxicity are age-dependent: evidence from measures of reactive oxygen species and striatal dopamine levels. Synapse 18(1):27–34. https://doi.org/10.1002/syn.890180105

Atamna H (2006) Heme binding to Amyloid-beta peptide: mechanistic role in Alzheimer's disease. J Alzheimers Disease 10(2–3):255–266

Atamna H, Frey WH 2nd (2004) A role for heme in Alzheimer's disease: heme binds amyloid beta and has altered metabolism. Proc Natl Acad Sci U S A 101(30):11153–11158. https://doi.org/10.1073/pnas.0404349101

Autere J, Moilanen JS, Finnila S, Soininen H, Mannermaa A, Hartikainen P, Hallikainen M, Majamaa K (2004) Mitochondrial DNA polymorphisms as risk factors for Parkinson's disease and Parkinson's disease dementia. Hum Genet 115(1):29–35. https://doi.org/10.1007/s00439-004-1123-9

Bales KR (2004) Neurodegenerative disease research in the 21st century. Drug Discov Today 9(13):553–556. https://doi.org/10.1016/S1359-6446(04)03120-4

Baloyannis SJ (2006) Mitochondrial alterations in Alzheimer's disease. J Alzheimers Disease 9(2):119–126

Barroso N, Campos Y, Huertas R, Esteban J, Molina JA, Alonso A, Gutierrez-Rivas E, Arenas J (1993) Respiratory chain enzyme activities in lymphocytes from untreated patients with Parkinson disease. Clin Chem 39(4):667–669

Bates GP (2005) History of genetic disease: the molecular genetics of Huntington disease - a history. Nat Rev Genet 6(10):766–773

Beal MF, Brouillet E, Jenkins BG, Ferrante RJ, Kowall NW, Miller JM, Storey E, Srivastava R, Rosen BR, Hyman BT (1993) Neurochemical and histologic characterization of striatal excitotoxic lesions produced by the mitochondrial toxin 3-nitropropionic acid. J Neurosci 13(10):4181–4192

Berardelli A, Wenning GK, Antonini A, Berg D, Bloem BR, Bonifati V, Brooks D, Burn DJ, Colosimo C, Fanciulli A, Ferreira J, Gasser T, Grandas F, Kanovsky P, Kostic V, Kulisevsky J, Oertel W, Poewe W, Reese JP, Relja M, Ruzicka E, Schrag A, Seppi K, Taba P, Vidailhet M (2013) EFNS/MDS-ES/ENS [corrected] recommendations for the diagnosis of Parkinson's disease. Eur J Neurol 20(1):16–34. https://doi.org/10.1111/ene.12022

Betarbet R, Sherer TB, MacKenzie G, Garcia-Osuna M, Panov AV, Greenamyre JT (2000) Chronic systemic pesticide exposure reproduces features of Parkinson's disease. Nat Neurosci 3(12):1301–1306

Blandini F, Nappi G, Greenamyre JT (1998) Quantitative study of mitochondrial complex I in platelets of parkinsonian patients. Mov Disord 13(1):11–15

Bose A, Beal MF (2016) Mitochondrial dysfunction in Parkinson's disease. J Neurochem 139(Suppl 1):216–231. https://doi.org/10.1111/jnc.13731

Bossy-Wetzel E, Petrilli A, Knott AB (2008) Mutant huntingtin and mitochondrial dysfunction. Trends Neurosci 31(12):609–616

Braak H, Braak E (1995) Staging of Alzheimer's disease-related neurofibrillary changes. Neurobiol Aging 16(3):271–278. discussion 278-284

Browne SE, Bowling AC, MacGarvey U, Baik MJ, Berger SC, Muqit MM, Bird ED, Beal MF (1997) Oxidative damage and metabolic dysfunction in Huntington's disease: selective vulnerability of the basal ganglia. Ann Neurol 41(5):646–653. https://doi.org/10.1002/ana.410410514

Bubber P, Haroutunian V, Fisch G, Blass JP, Gibson GE (2005) Mitochondrial abnormalities in Alzheimer brain: mechanistic implications. Ann Neurol 57(5):695–703. https://doi.org/10.1002/ana.20474

Calkins MJ, Reddy PH (2011) Amyloid beta impairs mitochondrial anterograde transport and degenerates synapses in Alzheimer's disease neurons. Biochim Biophys Acta 1812(4):507–513. https://doi.org/10.1016/j.bbadis.2011.01.007

Calkins MJ, Manczak M, Mao P, Shirendeb U, Reddy PH (2011) Impaired mitochondrial biogenesis, defective axonal transport of mitochondria, abnormal mitochondrial dynamics and synaptic degeneration in a mouse model of Alzheimer's disease. Hum Mol Genet 20(23):4515–4529. https://doi.org/10.1093/hmg/ddr381

Carvalho C, Correia SC, Cardoso S, Placido AI, Candeias E, Duarte AI, Moreira PI (2015) The role of mitochondrial disturbances in Alzheimer, Parkinson and Huntington diseases. Expert Rev Neurother 15(8):867–884. https://doi.org/10.1586/14737175.2015.1058160

Caspersen C, Wang N, Yao J, Sosunov A, Chen X, Lustbader JW, Xu HW, Stern D, McKhann G, Yan SD (2005) Mitochondrial Abeta: a potential focal point for neuronal metabolic dysfunction in Alzheimer's disease. FASEB J 19(14):2040–2041. https://doi.org/10.1096/fj.05-3735fje

Celsi F, Pizzo P, Brini M, Leo S, Fotino C, Pinton P, Rizzuto R (2009) Mitochondria, calcium and cell death: a deadly triad in neurodegeneration. Biochim Biophys Acta 1787(5):335–344. https://doi.org/10.1016/j.bbabio.2009.02.021

Chang DT, Rintoul GL, Pandipati S, Reynolds IJ (2006) Mutant huntingtin aggregates impair mitochondrial movement and trafficking in cortical neurons. Neurobiol Dis 22(2):388–400. https://doi.org/10.1016/j.nbd.2005.12.007

Chinta SJ, Andersen JK (2008) Redox imbalance in Parkinson's disease. Biochim Biophys Acta 1780(11):1362–1367

Cho DH, Nakamura T, Fang J, Cieplak P, Godzik A, Gu Z, Lipton SA (2009) S-nitrosylation of Drp1 mediates beta-amyloid-related mitochondrial fission and neuronal injury. Science 324(5923):102–105. https://doi.org/10.1126/science.1171091

Choo YS, Johnson GV, MacDonald M, Detloff PJ, Lesort M (2004) Mutant huntingtin directly increases susceptibility of mitochondria to the calcium-induced permeability transition and cytochrome c release. Hum Mol Genet 13(14):1407–1420. https://doi.org/10.1093/hmg/ddh162

Choong CJ, Mochizuki H (2017) Inappropriate trafficking of damaged mitochondria in Parkinson's disease. Stem Cell Investig 4:17. https://doi.org/10.21037/sci.2017.02.07

Chou JL, Shenoy DV, Thomas N, Choudhary PK, Laferla FM, Goodman SR, Breen GA (2011) Early dysregulation of the mitochondrial proteome in a mouse model of Alzheimer's disease. J Proteome 74(4):466–479. https://doi.org/10.1016/j.jprot.2010.12.012

Ciron C, Zheng L, Bobela W, Knott GW, Leone TC, Kelly DP, Schneider BL (2015) PGC-1alpha activity in nigral dopamine neurons determines vulnerability to alpha-synuclein. Acta Neuropathol Commun 3:16. https://doi.org/10.1186/s40478-015-0200-8

Clark IE, Dodson MW, Jiang C, Cao JH, Huh JR, Seol JH, Yoo SJ, Hay BA, Guo M (2006) Drosophila pink1 is required for mitochondrial function and interacts genetically with parkin. Nature 441(7097):1162–1166. https://doi.org/10.1038/nature04779

Correia SC, Santos RX, Cardoso S, Carvalho C, Candeias E, Duarte AI, Placido AI, Santos MS, Moreira PI (2012a) Alzheimer disease as a vascular disorder: where do mitochondria fit? Exp Gerontol 47(11):878–886. https://doi.org/10.1016/j.exger.2012.07.006

Correia SC, Santos RX, Perry G, Zhu X, Moreira PI, Smith MA (2012b) Mitochondrial importance in Alzheimer's, Huntington's and Parkinson's diseases. Adv Exp Med Biol 724:205–221. https://doi.org/10.1007/978-1-4614-0653-2_16

Correia SC, Perry G, Moreira PI (2016) Mitochondrial traffic jams in Alzheimer's disease - pinpointing the roadblocks. Biochim Biophys Acta 1862(10):1909–1917. https://doi.org/10.1016/j.bbadis.2016.07.010

Covarrubias-Pinto A, Moll P, Solis-Maldonado M, Acuna AI, Riveros A, Miro MP, Papic E, Beltran FA, Cepeda C, Concha II, Brauchi S, Castro MA (2015) Beyond the redox imbalance: oxidative stress contributes to an impaired GLUT3 modulation in Huntington's disease. Free Radic Biol Med 89:1085–1096. https://doi.org/10.1016/j.freeradbiomed.2015.09.024

Crouch PJ, Blake R, Duce JA, Ciccotosto GD, Li QX, Barnham KJ, Curtain CC, Cherny RA, Cappai R, Dyrks T, Masters CL, Trounce IA (2005) Copper-dependent inhibition of human cytochrome c oxidase by a dimeric conformer of amyloid-beta1-42. J Neurosci 25(3):672–679. https://doi.org/10.1523/JNEUROSCI.4276-04.2005

Cui L, Jeong H, Borovecki F, Parkhurst CN, Tanese N, Krainc D (2006) Transcriptional repression of PGC-1alpha by mutant huntingtin leads to mitochondrial dysfunction and neurodegeneration. Cell 127(1):59–69. https://doi.org/10.1016/j.cell.2006.09.015

Damiano M, Diguet E, Malgorn C, D'Aurelio M, Galvan L, Petit F, Benhaim L, Guillermier M, Houitte D, Dufour N, Hantraye P, Canals JM, Alberch J, Delzescaux T, Deglon N, Beal MF, Brouillet E (2013) A role of mitochondrial complex II defects in genetic models of Huntington's disease expressing N-terminal fragments of mutant huntingtin. Hum Mol Genet 22(19):3869–3882. https://doi.org/10.1093/hmg/ddt242

Deng H, Dodson MW, Huang H, Guo M (2008) The Parkinson's disease genes pink1 and parkin promote mitochondrial fission and/or inhibit fusion in Drosophila. Proc Natl Acad Sci U S A 105(38):14503–14508. https://doi.org/10.1073/pnas.0803998105

Devi L, Prabhu BM, Galati DF, Avadhani NG, Anandatheerthavarada HK (2006) Accumulation of amyloid precursor protein in the mitochondrial import channels of human Alzheimer's disease brain is associated with mitochondrial dysfunction. J Neurosci 26(35):9057–9068. https://doi.org/10.1523/JNEUROSCI.1469-06.2006

Du H, Guo L, Yan S, Sosunov AA, McKhann GM, Yan SS (2010) Early deficits in synaptic mitochondria in an Alzheimer's disease mouse model. Proc Natl Acad Sci U S A 107(43):18670–18675. https://doi.org/10.1073/pnas.1006586107

Dugger BN, Dickson DW (2017) Pathology of neurodegenerative diseases. Cold Spring Harb Perspect Biol 9(7). https://doi.org/10.1101/cshperspect.a028035

Durcan TM, Fon EA (2015) The three 'P's of mitophagy: PARKIN, PINK1, and post-translational modifications. Genes Dev 29(10):989–999. https://doi.org/10.1101/gad.262758.115

Ebneth A, Godemann R, Stamer K, Illenberger S, Trinczek B, Mandelkow E (1998) Overexpression of tau protein inhibits kinesin-dependent trafficking of vesicles, mitochondria, and endoplasmic reticulum: implications for Alzheimer's disease. J Cell Biol 143(3):777–794

Ekstrand MI, Terzioglu M, Galter D, Zhu S, Hofstetter C, Lindqvist E, Thams S, Bergstrand A, Hansson FS, Trifunovic A, Hoffer B, Cullheim S, Mohammed AH, Olson L, Larsson NG (2007) Progressive parkinsonism in mice with respiratory-chain-deficient dopamine neurons. Proc Natl Acad Sci U S A 104(4):1325–1330

Federico A, Cardaioli E, Da Pozzo P, Formichi P, Gallus GN, Radi E (2012) Mitochondria, oxidative stress and neurodegeneration. J Neurol Sci 322(1–2):254–262. https://doi.org/10.1016/j.jns.2012.05.030

Finkel T, Holbrook NJ (2000) Oxidants, oxidative stress and the biology of ageing. Nature 408(6809):239–247. https://doi.org/10.1038/35041687

Forno LS (1996) Neuropathology of Parkinson's disease. J Neuropathol Exp Neurol 55(3):259–272
Franco-Iborra S, Vila M, Perier C (2016) The Parkinson disease mitochondrial hypothesis: where are we at? Neuroscientist 22(3):266–277. https://doi.org/10.1177/1073858415574600
Friedland RP, Jagust WJ, Huesman RH, Koss E, Knittel B, Mathis CA, Ober BA, Mazoyer BM, Budinger TF (1989) Regional cerebral glucose transport and utilization in Alzheimer's disease. Neurology 39(11):1427–1434
Gan X, Huang S, Wu L, Wang Y, Hu G, Li G, Zhang H, Yu H, Swerdlow RH, Chen JX, Yan SS (2014) Inhibition of ERK-DLP1 signaling and mitochondrial division alleviates mitochondrial dysfunction in Alzheimer's disease cybrid cell. Biochim Biophys Acta 1842(2):220–231. https://doi.org/10.1016/j.bbadis.2013.11.009
Gao J, Wang L, Liu J, Xie F, Su B, Wang X (2017) Abnormalities of mitochondrial dynamics in neurodegenerative diseases. Antioxidants (Basel) 6(2). https://doi.org/10.3390/antiox6020025
Gash DM, Rutland K, Hudson NL, Sullivan PG, Bing G, Cass WA, Pandya JD, Liu M, Choi DY, Hunter RL, Gerhardt GA, Smith CD, Slevin JT, Prince TS (2008) Trichloroethylene: Parkinsonism and complex 1 mitochondrial neurotoxicity. Ann Neurol 63(2):184–192
Gauthier LR, Charrin BC, Borrell-Pages M, Dompierre JP, Rangone H, Cordelieres FP, De Mey J, MacDonald ME, Lessmann V, Humbert S, Saudou F (2004) Huntingtin controls neurotrophic support and survival of neurons by enhancing BDNF vesicular transport along microtubules. Cell 118(1):127–138. https://doi.org/10.1016/j.cell.2004.06.018
Geisler S, Holmstrom KM, Treis A, Skujat D, Weber SS, Fiesel FC, Kahle PJ, Springer W (2010) The PINK1/Parkin-mediated mitophagy is compromised by PD-associated mutations. Autophagy 6(7):871–878
Gil JM, Rego AC (2008) Mechanisms of neurodegeneration in Huntington's disease. Eur J Neurosci 27(11):2803–2820. https://doi.org/10.1111/j.1460-9568.2008.06310.x
Goldman JG, Postuma R (2014) Premotor and nonmotor features of Parkinson's disease. Curr Opin Neurol 27(4):434–441. https://doi.org/10.1097/WCO.0000000000000112
Grimm A, Eckert A (2017) Brain aging and neurodegeneration: from a mitochondrial point of view. J Neurochem 143(4):418–431. https://doi.org/10.1111/jnc.14037
Gu M, Gash MT, Mann VM, Javoy-Agid F, Cooper JM, Schapira AH (1996) Mitochondrial defect in Huntington's disease caudate nucleus. Ann Neurol 39(3):385–389. https://doi.org/10.1002/ana.410390317
Guardia-Laguarta C, Area-Gomez E, Rub C, Liu Y, Magrane J, Becker D, Voos W, Schon EA, Przedborski S (2014) alpha-Synuclein is localized to mitochondria-associated ER membranes. J Neurosci 34(1):249–259. https://doi.org/10.1523/JNEUROSCI.2507-13.2014
Guedes-Dias P, de Proenca J, Soares TR, Leitao-Rocha A, Pinho BR, Duchen MR, Oliveira JM (2015) HDAC6 inhibition induces mitochondrial fusion, autophagic flux and reduces diffuse mutant huntingtin in striatal neurons. Biochim Biophys Acta 1852(11):2484–2493. https://doi.org/10.1016/j.bbadis.2015.08.012
Guedes-Dias P, Pinho BR, Soares TR, de Proenca J, Duchen MR, Oliveira JM (2016) Mitochondrial dynamics and quality control in Huntington's disease. Neurobiol Dis 90:51–57. https://doi.org/10.1016/j.nbd.2015.09.008
Guidetti P, Charles V, Chen EY, Reddy PH, Kordower JH, Whetsell WO Jr, Schwarcz R, Tagle DA (2001) Early degenerative changes in transgenic mice expressing mutant huntingtin involve dendritic abnormalities but no impairment of mitochondrial energy production. Exp Neurol 169(2):340–350. https://doi.org/10.1006/exnr.2000.7626
Haas RH, Nasirian F, Nakano K, Ward D, Pay M, Hill R, Shults CW (1995) Low platelet mitochondrial complex I and complex II/III activity in early untreated Parkinson's disease. Ann Neurol 37(6):714–722
Haun F, Nakamura T, Shiu AD, Cho DH, Tsunemi T, Holland EA, La Spada AR, Lipton SA (2013) S-nitrosylation of dynamin-related protein 1 mediates mutant huntingtin-induced mitochondrial fragmentation and neuronal injury in Huntington's disease. Antioxid Redox Signal 19(11):1173–1184. https://doi.org/10.1089/ars.2012.4928
Hansson Petersen CA, Alikhani N, Behbahani H, Wiehager B, Pavlov PF, Alafuzoff I, Leinonen V, Ito A, Winblad B, Glaser E, Ankarcrona M (2008) The amyloid beta-peptide is imported into

mitochondria via the TOM import machinery and localized to mitochondrial cristae. Proc Natl Acad Sci U S A 105(35):13145–13150. https://doi.org/10.1073/pnas.0806192105

Haylett W, Swart C, van der Westhuizen F, van Dyk H, van der Merwe L, van der Merwe C, Loos B, Carr J, Kinnear C, Bardien S (2016) Altered mitochondrial respiration and other features of mitochondrial function in parkin-mutant fibroblasts from Parkinson'disease patients. Parkinsons Dis 2016:1819209. https://doi.org/10.1155/2016/1819209

Hering T, Birth N, Taanman JW, Orth M (2015) Selective striatal mtDNA depletion in end-stage Huntington's disease R6/2 mice. Exp Neurol 266:22–29. https://doi.org/10.1016/j.expneurol.2015.02.004

Herrup K (2015) The case for rejecting the amyloid cascade hypothesis. Nat Neurosci 18(6): 794–799. https://doi.org/10.1038/nn.4017

Hirai K, Aliev G, Nunomura A, Fujioka H, Russell RL, Atwood CS, Johnson AB, Kress Y, Vinters HV, Tabaton M, Shimohama S, Cash AD, Siedlak SL, Harris PL, Jones PK, Petersen RB, Perry G, Smith MA (2001) Mitochondrial abnormalities in Alzheimer's disease. J Neurosci 21(9):3017–3023

Hollenbeck PJ, Saxton WM (2005) The axonal transport of mitochondria. J Cell Sci 118(Pt 23):5411–5419. https://doi.org/10.1242/jcs.02745

Hsieh CH, Shaltouki A, Gonzalez AE, Bettencourt da Cruz A, Burbulla LF, St Lawrence E, Schule B, Krainc D, Palmer TD, Wang X (2016) Functional impairment in miro degradation and mitophagy is a shared feature in familial and sporadic Parkinson's disease. Cell Stem Cell 19(6):709–724. https://doi.org/10.1016/j.stem.2016.08.002

Hsu LJ, Sagara Y, Arroyo A, Rockenstein E, Sisk A, Mallory M, Wong J, Takenouchi T, Hashimoto M, Masliah E (2000) Alpha-synuclein promotes mitochondrial deficit and oxidative stress. Am J Pathol 157(2):401–410

Jagust WJ, Seab JP, Huesman RH, Valk PE, Mathis CA, Reed BR, Coxson PG, Budinger TF (1991) Diminished glucose transport in Alzheimer's disease: dynamic PET studies. J Cereb Blood Flow Metab 11(2):323–330. https://doi.org/10.1038/jcbfm.1991.65

Jellinger KA (2009) Recent advances in our understanding of neurodegeneration. J Neural Transm 116(9):1111–1162. https://doi.org/10.1007/s00702-009-0240-y

Jellinger KA (2010) Basic mechanisms of neurodegeneration: a critical update. J Cell Mol Med 14(3):457–487. https://doi.org/10.1111/j.1582-4934.2010.01010.x

Kane LA, Lazarou M, Fogel AI, Li Y, Yamano K, Sarraf SA, Banerjee S, Youle RJ (2014) PINK1 phosphorylates ubiquitin to activate Parkin E3 ubiquitin ligase activity. J Cell Biol 205(2):143–153. https://doi.org/10.1083/jcb.201402104

Kann O, Kovacs R (2007) Mitochondria and neuronal activity. Am J Physiol Cell Physiol 292(2):C641–C657. https://doi.org/10.1152/ajpcell.00222.2006

Karran E, Mercken M, De Strooper B (2011) The amyloid cascade hypothesis for Alzheimer's disease: an appraisal for the development of therapeutics. Nat Rev Drug Discov 10(9):698–712. https://doi.org/10.1038/nrd3505

Khalil B, El Fissi N, Aouane A, Cabirol-Pol MJ, Rival T, Lievens JC (2015) PINK1-induced mitophagy promotes neuroprotection in Huntington's disease. Cell Death Dis 6:e1617. https://doi.org/10.1038/cddis.2014.581

Kieburtz K (2016) Treating neurodegenerative disease before illness: a challenge for the 21st century. Lancet Neurol 15(6):540–541. https://doi.org/10.1016/S1474-4422(16)30001-1

Kim J, Moody JP, Edgerly CK, Bordiuk OL, Cormier K, Smith K, Beal MF, Ferrante RJ (2010) Mitochondrial loss, dysfunction and altered dynamics in Huntington's disease. Hum Mol Genet 19(20):3919–3935. https://doi.org/10.1093/hmg/ddq306

Kim GH, Kim JE, Rhie SJ, Yoon S (2015) The role of oxidative stress in neurodegenerative diseases. Exp Neurobiol 24(4):325–340. https://doi.org/10.5607/en.2015.24.4.325

Koroshetz WJ, Jenkins BG, Rosen BR, Beal MF (1997) Energy metabolism defects in Huntington's disease and effects of coenzyme Q10. Ann Neurol 41(2):160–165. https://doi.org/10.1002/ana.410410206

Kovari E, Horvath J, Bouras C (2009) Neuropathology of Lewy body disorders. Brain Res Bull 80(4–5):203–210. https://doi.org/10.1016/j.brainresbull.2009.06.018

Krige D, Carroll MT, Cooper JM, Marsden CD, Schapira AH (1992) Platelet mitochondrial function in Parkinson's disease. The Royal Kings and Queens Parkinson Disease Research Group. Ann Neurol 32(6):782–788

Kuhl DE, Phelps ME, Markham CH, Metter EJ, Riege WH, Winter J (1982) Cerebral metabolism and atrophy in Huntington's disease determined by 18FDG and computed tomographic scan. Ann Neurol 12(5):425–434. https://doi.org/10.1002/ana.410120504

Kuwert T, Lange HW, Langen KJ, Herzog H, Aulich A, Feinendegen LE (1989) Cerebral glucose consumption measured by PET in patients with and without psychiatric symptoms of Huntington's disease. Psychiatry Res 29(3):361–362

Kwon SK, Hirabayashi Y, Polleux F (2016) Organelle-specific sensors for monitoring Ca^{2+} dynamics in neurons. Front Synaptic Neurosci 8:29. https://doi.org/10.3389/fnsyn.2016.00029

Langston JW, Ballard P, Tetrud JW, Irwin I (1983) Chronic Parkinsonism in humans due to a product of meperidine-analog synthesis. Science 219(4587):979–980

Li XJ, Li SH, Sharp AH, Nucifora FC Jr, Schilling G, Lanahan A, Worley P, Snyder SH, Ross CA (1995) A huntingtin-associated protein enriched in brain with implications for pathology. Nature 378(6555):398–402. https://doi.org/10.1038/378398a0

Liang CL, Wang TT, Luby-Phelps K, German DC (2007) Mitochondria mass is low in mouse substantia nigra dopamine neurons: implications for Parkinson's disease. Exp Neurol 203(2):370–380. https://doi.org/10.1016/j.expneurol.2006.08.015

Lin MY, Sheng ZH (2015) Regulation of mitochondrial transport in neurons. Exp Cell Res 334(1):35–44. https://doi.org/10.1016/j.yexcr.2015.01.004

Liot G, Bossy B, Lubitz S, Kushnareva Y, Sejbuk N, Bossy-Wetzel E (2009) Complex II inhibition by 3-NP causes mitochondrial fragmentation and neuronal cell death via an NMDA- and ROS-dependent pathway. Cell Death Differ 16(6):899–909

Manczak M, Anekonda TS, Henson E, Park BS, Quinn J, Reddy PH (2006) Mitochondria are a direct site of A beta accumulation in Alzheimer's disease neurons: implications for free radical generation and oxidative damage in disease progression. Hum Mol Genet 15(9):1437–1449. https://doi.org/10.1093/hmg/ddl066

Martin-Maestro P, Gargini R, Perry G, Avila J, Garcia-Escudero V (2016) PARK2 enhancement is able to compensate mitophagy alterations found in sporadic Alzheimer's disease. Hum Mol Genet 25(4):792–806. https://doi.org/10.1093/hmg/ddv616

Martin-Maestro P, Gargini R, AS A, Garcia E, Anton LC, Noggle S, Arancio O, Avila J, Garcia-Escudero V (2017) Mitophagy failure in fibroblasts and iPSC-derived neurons of Alzheimer's disease-associated presenilin 1 mutation. Front Mol Neurosci 10:291. https://doi.org/10.3389/fnmol.2017.00291

Mastrogiacomo F, Bergeron C, Kish SJ (1993) Brain alpha-ketoglutarate dehydrogenase complex activity in Alzheimer's disease. J Neurochem 61(6):2007–2014

McWilliams TG, Muqit MM (2017) PINK1 and Parkin: emerging themes in mitochondrial homeostasis. Curr Opin Cell Biol 45:83–91. https://doi.org/10.1016/j.ceb.2017.03.013

Moreira PI, Siedlak SL, Wang X, Santos MS, Oliveira CR, Tabaton M, Nunomura A, Szweda LI, Aliev G, Smith MA, Zhu X, Perry G (2007a) Autophagocytosis of mitochondria is prominent in Alzheimer disease. J Neuropathol Exp Neurol 66(6):525–532. https://doi.org/10.1097/01.jnen.0000240476.73532.b0

Moreira PI, Siedlak SL, Wang X, Santos MS, Oliveira CR, Tabaton M, Nunomura A, Szweda LI, Aliev G, Smith MA, Zhu X, Perry G (2007b) Increased autophagic degradation of mitochondria in Alzheimer disease. Autophagy 3(6):614–615

Moreira PI, Zhu X, Wang X, Lee HG, Nunomura A, Petersen RB, Perry G, Smith MA (2010) Mitochondria: a therapeutic target in neurodegeneration. Biochim Biophys Acta 1802(1):212–220. https://doi.org/10.1016/j.bbadis.2009.10.007

Mortiboys H, Johansen KK, Aasly JO, Bandmann O (2010) Mitochondrial impairment in patients with Parkinson disease with the G2019S mutation in LRRK2. Neurology 75(22):2017–2020. https://doi.org/10.1212/WNL.0b013e3181ff9685

Muangpaisan W, Mathews A, Hori H, Seidel D (2011) A systematic review of the worldwide prevalence and incidence of Parkinson's disease. J Med Assoc Thai 94(6):749–755

Mutisya EM, Bowling AC, Beal MF (1994) Cortical cytochrome oxidase activity is reduced in Alzheimer's disease. J Neurochem 63(6):2179–2184

Naia L, Ferreira IL, Ferreiro E, Rego AC (2017) Mitochondrial Ca^{2+} handling in Huntington's and Alzheimer's diseases - role of ER-mitochondria crosstalk. Biochem Biophys Res Commun 483(4):1069–1077. https://doi.org/10.1016/j.bbrc.2016.07.122

Narendra D, Tanaka A, Suen DF, Youle RJ (2008) Parkin is recruited selectively to impaired mitochondria and promotes their autophagy. J Cell Biol 183(5):795–803. https://doi.org/10.1083/jcb.200809125

Nunomura A, Perry G, Aliev G, Hirai K, Takeda A, Balraj EK, Jones PK, Ghanbari H, Wataya T, Shimohama S, Chiba S, Atwood CS, Petersen RB, Smith MA (2001) Oxidative damage is the earliest event in Alzheimer disease. J Neuropathol Exp Neurol 60(8):759–767

Oliveira JM (2010) Nature and cause of mitochondrial dysfunction in Huntington's disease: focusing on huntingtin and the striatum. J Neurochem 114(1):1–12. https://doi.org/10.1111/j.1471-4159.2010.06741.x

Orr AL, Li S, Wang CE, Li H, Wang J, Rong J, Xu X, Mastroberardino PG, Greenamyre JT, Li XJ (2008) N-terminal mutant huntingtin associates with mitochondria and impairs mitochondrial trafficking. J Neurosci 28(11):2783–2792. https://doi.org/10.1523/JNEUROSCI.0106-08.2008

Panov AV, Gutekunst CA, Leavitt BR, Hayden MR, Burke JR, Strittmatter WJ, Greenamyre JT (2002) Early mitochondrial calcium defects in Huntington's disease are a direct effect of polyglutamines. Nat Neurosci 5(8):731–736. https://doi.org/10.1038/nn884

Papkovskaia TD, Chau KY, Inesta-Vaquera F, Papkovsky DB, Healy DG, Nishio K, Staddon J, Duchen MR, Hardy J, Schapira AH, Cooper JM (2012) G2019S leucine-rich repeat kinase 2 causes uncoupling protein-mediated mitochondrial depolarization. Hum Mol Genet 21(19):4201–4213. https://doi.org/10.1093/hmg/dds244

Parker WD Jr, Boyson SJ, Parks JK (1989) Abnormalities of the electron transport chain in idiopathic Parkinson's disease. Ann Neurol 26(6):719–723

Parker WD Jr, Parks JK, Swerdlow RH (2008) Complex I deficiency in Parkinson's disease frontal cortex. Brain Res 1189:215–218

Penn AM, Roberts T, Hodder J, Allen PS, Zhu G, Martin WR (1995) Generalized mitochondrial dysfunction in Parkinson's disease detected by magnetic resonance spectroscopy of muscle. Neurology 45(11):2097–2099

Persson T, Popescu BO, Cedazo-Minguez A (2014) Oxidative stress in Alzheimer's disease: why did antioxidant therapy fail? Oxidative Med Cell Longev 2014:427318. https://doi.org/10.1155/2014/427318

Polyzos AA, McMurray CT (2017) The chicken or the egg: mitochondrial dysfunction as a cause or consequence of toxicity in Huntington's disease. Mech Ageing Dev 161(Pt A):181–197. https://doi.org/10.1016/j.mad.2016.09.003

Poole AC, Thomas RE, Andrews LA, McBride HM, Whitworth AJ, Pallanck LJ (2008) The PINK1/Parkin pathway regulates mitochondrial morphology. Proc Natl Acad Sci U S A 105(5):1638–1643. https://doi.org/10.1073/pnas.0709336105

Pozo Devoto VM, Dimopoulos N, Alloatti M, Pardi MB, Saez TM, Otero MG, Cromberg LE, Marin-Burgin A, Scassa ME, Stokin GB, Schinder AF, Sevlever G, Falzone TL (2017) alphaSynuclein control of mitochondrial homeostasis in human-derived neurons is disrupted by mutations associated with Parkinson's disease. Sci Rep 7(1):5042. https://doi.org/10.1038/s41598-017-05334-9

Qin W, Haroutunian V, Katsel P, Cardozo CP, Ho L, Buxbaum JD, Pasinetti GM (2009) PGC-1alpha expression decreases in the Alzheimer disease brain as a function of dementia. Arch Neurol 66(3):352–361. https://doi.org/10.1001/archneurol.2008.588

Querfurth HW, LaFerla FM (2010) Alzheimer's disease. N Engl J Med 362(4):329–344. https://doi.org/10.1056/NEJMra0909142

Quintanilla RA, Dolan PJ, Jin YN, Johnson GV (2012) Truncated tau and Abeta cooperatively impair mitochondria in primary neurons. Neurobiol Aging 33(3):619.e625–619.e635. https://doi.org/10.1016/j.neurobiolaging.2011.02.007

Reddy PH, Reddy TP (2011) Mitochondria as a therapeutic target for aging and neurodegenerative diseases. Curr Alzheimer Res 8(4):393–409

Reiner A, Albin RL, Anderson KD, D'Amato CJ, Penney JB, Young AB (1988) Differential loss of striatal projection neurons in Huntington disease. Proc Natl Acad Sci U S A 85(15): 5733–5737

Resende R, Moreira PI, Proenca T, Deshpande A, Busciglio J, Pereira C, Oliveira CR (2008) Brain oxidative stress in a triple-transgenic mouse model of Alzheimer disease. Free Radic Biol Med 44(12):2051–2057. https://doi.org/10.1016/j.freeradbiomed.2008.03.012

Rhein V, Song X, Wiesner A, Ittner LM, Baysang G, Meier F, Ozmen L, Bluethmann H, Drose S, Brandt U, Savaskan E, Czech C, Gotz J, Eckert A (2009) Amyloid-beta and tau synergistically impair the oxidative phosphorylation system in triple transgenic Alzheimer's disease mice. Proc Natl Acad Sci U S A 106(47):20057–20062. https://doi.org/10.1073/pnas.0905529106

Rodolfo C, Campello S, Cecconi F (2017) Mitophagy in neurodegenerative diseases. Neurochem Int. https://doi.org/10.1016/j.neuint.2017.08.004

Schapira AH, Cooper JM, Dexter D, Clark JB, Jenner P, Marsden CD (1990) Mitochondrial complex I deficiency in Parkinson's disease. J Neurochem 54(3):823–827

Schapira AHV, Chaudhuri KR, Jenner P (2017) Non-motor features of Parkinson disease. Nat Rev Neurosci 18(7):435–450. https://doi.org/10.1038/nrn.2017.62

Seo AY, Joseph AM, Dutta D, Hwang JC, Aris JP, Leeuwenburgh C (2010) New insights into the role of mitochondria in aging: mitochondrial dynamics and more. J Cell Sci 123(Pt 15):2533–2542. https://doi.org/10.1242/jcs.070490

Shaerzadeh F, Motamedi F, Minai-Tehrani D, Khodagholi F (2014) Monitoring of neuronal loss in the hippocampus of Abeta-injected rat: autophagy, mitophagy, and mitochondrial biogenesis stand against apoptosis. NeuroMolecular Med 16(1):175–190. https://doi.org/10.1007/s12017-013-8272-8

Shah K, Desilva S, Abbruscato T (2012) The role of glucose transporters in brain disease: diabetes and Alzheimer's disease. Int J Mol Sci 13(10):12629–12655. https://doi.org/10.3390/ijms131012629

Sheng ZH (2014) Mitochondrial trafficking and anchoring in neurons: new insight and implications. J Cell Biol 204(7):1087–1098. https://doi.org/10.1083/jcb.201312123

Sheng ZH, Cai Q (2012) Mitochondrial transport in neurons: impact on synaptic homeostasis and neurodegeneration. Nat Rev Neurosci 13(2):77–93. https://doi.org/10.1038/nrn3156

Sheng B, Wang X, Su B, Lee HG, Casadesus G, Perry G, Zhu X (2012) Impaired mitochondrial biogenesis contributes to mitochondrial dysfunction in Alzheimer's disease. J Neurochem 120(3):419–429. https://doi.org/10.1111/j.1471-4159.2011.07581.x

Sherer TB, Betarbet R, Stout AK, Lund S, Baptista M, Panov AV, Cookson MR, Greenamyre JT (2002) An in vitro model of Parkinson's disease: linking mitochondrial impairment to altered alpha-synuclein metabolism and oxidative damage. J Neurosci 22(16):7006–7015. doi:20026721

Sherer TB, Richardson JR, Testa CM, Seo BB, Panov AV, Yagi T, Matsuno-Yagi A, Miller GW, Greenamyre JT (2007) Mechanism of toxicity of pesticides acting at complex I: relevance to environmental etiologies of Parkinson's disease. J Neurochem 100(6):1469–1479

Shin JH, Ko HS, Kang H, Lee Y, Lee YI, Pletinkova O, Troconso JC, Dawson VL, Dawson TM (2011) PARIS (ZNF746) repression of PGC-1alpha contributes to neurodegeneration in Parkinson's disease. Cell 144(5):689–702. https://doi.org/10.1016/j.cell.2011.02.010

Shirendeb UP, Calkins MJ, Manczak M, Anekonda V, Dufour B, McBride JL, Mao P, Reddy PH (2012) Mutant huntingtin's interaction with mitochondrial protein Drp1 impairs mitochondrial biogenesis and causes defective axonal transport and synaptic degeneration in Huntington's disease. Hum Mol Genet 21(2):406–420. https://doi.org/10.1093/hmg/ddr475

Skovronsky DM, Lee VM, Trojanowski JQ (2006) Neurodegenerative diseases: new concepts of pathogenesis and their therapeutic implications. Annu Rev Pathol 1:151–170. https://doi.org/10.1146/annurev.pathol.1.110304.100113

Song W, Chen J, Petrilli A, Liot G, Klinglmayr E, Zhou Y, Poquiz P, Tjong J, Pouladi MA, Hayden MR, Masliah E, Ellisman M, Rouiller I, Schwarzenbacher R, Bossy B, Perkins G, Bossy-Wetzel

E (2011) Mutant huntingtin binds the mitochondrial fission GTPase dynamin-related protein-1 and increases its enzymatic activity. Nat Med 17(3):377–382. https://doi.org/10.1038/nm.2313

Spillantini MG, Goedert M (2000) The alpha-synucleinopathies: Parkinson's disease, dementia with Lewy bodies, and multiple system atrophy. Ann N Y Acad Sci 920:16–27

Squitieri F, Cannella M, Sgarbi G, Maglione V, Falleni A, Lenzi P, Baracca A, Cislaghi G, Saft C, Ragona G, Russo MA, Thompson LM, Solaini G, Fornai F (2006) Severe ultrastructural mitochondrial changes in lymphoblasts homozygous for Huntington disease mutation. Mech Ageing Dev 127(2):217–220

Stevens DA, Lee Y, Kang HC, Lee BD, Lee YI, Bower A, Jiang H, Kang SU, Andrabi SA, Dawson VL, Shin JH, Dawson TM (2015) Parkin loss leads to PARIS-dependent declines in mitochondrial mass and respiration. Proc Natl Acad Sci U S A 112(37):11696–11701. https://doi.org/10.1073/pnas.1500624112

Su YC, Qi X (2013) Inhibition of excessive mitochondrial fission reduced aberrant autophagy and neuronal damage caused by LRRK2 G2019S mutation. Hum Mol Genet 22(22):4545–4561. https://doi.org/10.1093/hmg/ddt301

Subramaniam SR, Chesselet MF (2013) Mitochondrial dysfunction and oxidative stress in Parkinson's disease. Prog Neurobiol 106–107:17–32. https://doi.org/10.1016/j.pneurobio.2013.04.004

Sultana R, Butterfield DA (2013) Oxidative modification of brain proteins in Alzheimer's disease: perspective on future studies based on results of redox proteomics studies. J Alzheimers Dis 33(Suppl 1):S243–S251. https://doi.org/10.3233/JAD-2012-129018

Swerdlow RH, Khan SM (2004) A "mitochondrial cascade hypothesis" for sporadic Alzheimer's disease. Med Hypotheses 63(1):8–20. https://doi.org/10.1016/j.mehy.2003.12.045

Swerdlow RH, Parks JK, Miller SW, Tuttle JB, Trimmer PA, Sheehan JP, Bennett JP Jr, Davis RE, Parker WD Jr (1996) Origin and functional consequences of the complex I defect in Parkinson's disease. Ann Neurol 40(4):663–671. https://doi.org/10.1002/ana.410400417

Swerdlow RH, Parks JK, Davis JN 2nd, Cassarino DS, Trimmer PA, Currie LJ, Dougherty J, Bridges WS, Bennett JP Jr, Wooten GF, Parker WD (1998) Matrilineal inheritance of complex I dysfunction in a multigenerational Parkinson's disease family. Ann Neurol 44(6):873–881

Swerdlow RH, Burns JM, Khan SM (2014) The Alzheimer's disease mitochondrial cascade hypothesis: progress and perspectives. Biochim Biophys Acta 1842(8):1219–1231. https://doi.org/10.1016/j.bbadis.2013.09.010

Taylor DJ, Krige D, Barnes PR, Kemp GJ, Carroll MT, Mann VM, Cooper JM, Marsden CD, Schapira AH (1994) A 31P magnetic resonance spectroscopy study of mitochondrial function in skeletal muscle of patients with Parkinson's disease. J Neurol Sci 125(1):77–81

Tretter L, Adam-Vizi V (2000) Inhibition of Krebs cycle enzymes by hydrogen peroxide: a key role of [alpha]-ketoglutarate dehydrogenase in limiting NADH production under oxidative stress. J Neurosci 20(24):8972–8979

Trimmer PA, Borland MK (2005) Differentiated Alzheimer's disease transmitochondrial cybrid cell lines exhibit reduced organelle movement. Antioxid Redox Signal 7(9–10):1101–1109. https://doi.org/10.1089/ars.2005.7.1101

Trushina E, Dyer RB, Badger JD 2nd, Ure D, Eide L, Tran DD, Vrieze BT, Legendre-Guillemin V, McPherson PS, Mandavilli BS, Van Houten B, Zeitlin S, McNiven M, Aebersold R, Hayden M, Parisi JE, Seeberg E, Dragatsis I, Doyle K, Bender A, Chacko C, McMurray CT (2004) Mutant huntingtin impairs axonal trafficking in mammalian neurons in vivo and in vitro. Mol Cell Biol 24(18):8195–8209. https://doi.org/10.1128/MCB.24.18.8195-8209.2004

Utton MA, Noble WJ, Hill JE, Anderton BH, Hanger DP (2005) Molecular motors implicated in the axonal transport of tau and alpha-synuclein. J Cell Sci 118(Pt 20):4645–4654. https://doi.org/10.1242/jcs.02558

Veech GA, Dennis J, Keeney PM, Fall CP, Swerdlow RH, Parker WD Jr, Bennett JP Jr (2000) Disrupted mitochondrial electron transport function increases expression of anti-apoptotic bcl-2 and bcl-X(L) proteins in SH-SY5Y neuroblastoma and in Parkinson disease cybrid cells through oxidative stress. J Neurosci Res 61(6):693–700

Vos M, Lauwers E, Verstreken P (2010) Synaptic mitochondria in synaptic transmission and organization of vesicle pools in health and disease. Front Synaptic Neurosci 2:139. https://doi.org/10.3389/fnsyn.2010.00139

Vossel KA, Zhang K, Brodbeck J, Daub AC, Sharma P, Finkbeiner S, Cui B, Mucke L (2010) Tau reduction prevents Abeta-induced defects in axonal transport. Science 330(6001):198. https://doi.org/10.1126/science.1194653

Wang X, Su B, Lee HG, Li X, Perry G, Smith MA, Zhu X (2009) Impaired balance of mitochondrial fission and fusion in Alzheimer's disease. J Neurosci 29(28):9090–9103. https://doi.org/10.1523/JNEUROSCI.1357-09.2009

Wang X, Perry G, Smith MA, Zhu X (2010) Amyloid-beta-derived diffusible ligands cause impaired axonal transport of mitochondria in neurons. Neurodegener Dis 7(1–3):56–59. https://doi.org/10.1159/000283484

Wang X, Su B, Liu W, He X, Gao Y, Castellani RJ, Perry G, Smith MA, Zhu X (2011a) DLP1-dependent mitochondrial fragmentation mediates 1-methyl-4-phenylpyridinium toxicity in neurons: implications for Parkinson's disease. Aging Cell 10(5):807–823. https://doi.org/10.1111/j.1474-9726.2011.00721.x

Wang X, Winter D, Ashrafi G, Schlehe J, Wong YL, Selkoe D, Rice S, Steen J, LaVoie MJ, Schwarz TL (2011b) PINK1 and Parkin target Miro for phosphorylation and degradation to arrest mitochondrial motility. Cell 147(4):893–906. https://doi.org/10.1016/j.cell.2011.10.018

Wang X, Petrie TG, Liu Y, Liu J, Fujioka H, Zhu X (2012a) Parkinson's disease-associated DJ-1 mutations impair mitochondrial dynamics and cause mitochondrial dysfunction. J Neurochem 121(5):830–839. https://doi.org/10.1111/j.1471-4159.2012.07734.x

Wang X, Yan MH, Fujioka H, Liu J, Wilson-Delfosse A, Chen SG, Perry G, Casadesus G, Zhu X (2012b) LRRK2 regulates mitochondrial dynamics and function through direct interaction with DLP1. Hum Mol Genet 21(9):1931–1944. https://doi.org/10.1093/hmg/dds003

Wang W, Wang X, Fujioka H, Hoppel C, Whone AL, Caldwell MA, Cullen PJ, Liu J, Zhu X (2016) Parkinson's disease-associated mutant VPS35 causes mitochondrial dysfunction by recycling DLP1 complexes. Nat Med 22(1):54–63. https://doi.org/10.1038/nm.3983

Westermann B (2010) Mitochondrial fusion and fission in cell life and death. Nat Rev Mol Cell Biol 11(12):872–884. https://doi.org/10.1038/nrm3013

Weydt P, Pineda VV, Torrence AE, Libby RT, Satterfield TF, Lazarowski ER, Gilbert ML, Morton GJ, Bammler TK, Strand AD, Cui L, Beyer RP, Easley CN, Smith AC, Krainc D, Luquet S, Sweet IR, Schwartz MW, La Spada AR (2006) Thermoregulatory and metabolic defects in Huntington's disease transgenic mice implicate PGC-1alpha in Huntington's disease neurodegeneration. Cell Metab 4(5):349–362. https://doi.org/10.1016/j.cmet.2006.10.004

Wild P, Dikic I (2010) Mitochondria get a Parkin' ticket. Nat Cell Biol 12(2):104–106. https://doi.org/10.1038/ncb0210-104

Wong YC, Holzbaur EL (2014) The regulation of autophagosome dynamics by huntingtin and HAP1 is disrupted by expression of mutant huntingtin, leading to defective cargo degradation. J Neurosci 34(4):1293–1305. https://doi.org/10.1523/JNEUROSCI.1870-13.2014

Xie W, Chung KK (2012) Alpha-synuclein impairs normal dynamics of mitochondria in cell and animal models of Parkinson's disease. J Neurochem 122(2):404–414. https://doi.org/10.1111/j.1471-4159.2012.07769.x

Yamano K, Matsuda N, Tanaka K (2016) The ubiquitin signal and autophagy: an orchestrated dance leading to mitochondrial degradation. EMBO Rep 17(3):300–316. https://doi.org/10.15252/embr.201541486

Yao Z, Wood NW (2009) Cell death pathways in Parkinson's disease: role of mitochondria. Antioxid Redox Signal 11(9):2135–2149. https://doi.org/10.1089/ARS.2009.2624

Yao J, Irwin RW, Zhao L, Nilsen J, Hamilton RT, Brinton RD (2009) Mitochondrial bioenergetic deficit precedes Alzheimer's pathology in female mouse model of Alzheimer's disease. Proc Natl Acad Sci U S A 106(34):14670–14675. https://doi.org/10.1073/pnas.0903563106

Ye X, Sun X, Starovoytov V, Cai Q (2015) Parkin-mediated mitophagy in mutant hAPP neurons and Alzheimer's disease patient brains. Hum Mol Genet 24(10):2938–2951. https://doi.org/10.1093/hmg/ddv056

Yoshino H, Nakagawa-Hattori Y, Kondo T, Mizuno Y (1992) Mitochondrial complex I and II activities of lymphocytes and platelets in Parkinson's disease. J Neural Transm Park Dis Dement Sect 4(1):27–34

Yue M, Hinkle KM, Davies P, Trushina E, Fiesel FC, Christenson TA, Schroeder AS, Zhang L, Bowles E, Behrouz B, Lincoln SJ, Beevers JE, Milnerwood AJ, Kurti A, McLean PJ, Fryer JD, Springer W, Dickson DW, Farrer MJ, Melrose HL (2015) Progressive dopaminergic alterations and mitochondrial abnormalities in LRRK2 G2019S knock-in mice. Neurobiol Dis 78:172–195. https://doi.org/10.1016/j.nbd.2015.02.031

Zhang Z, Liu L, Jiang X, Zhai S, Xing D (2016) The essential role of Drp1 and its regulation by S-nitrosylation of parkin in dopaminergic neurodegeneration: implications for Parkinson's disease. Antioxid Redox Signal 25(11):609–622. https://doi.org/10.1089/ars.2016.6634

Zheng B, Liao Z, Locascio JJ, Lesniak KA, Roderick SS, Watt ML, Eklund AC, Zhang-James Y, Kim PD, Hauser MA, Grunblatt E, Moran LB, Mandel SA, Riederer P, Miller RM, Federoff HJ, Wullner U, Papapetropoulos S, Youdim MB, Cantuti-Castelvetri I, Young AB, Vance JM, Davis RL, Hedreen JC, Adler CH, Beach TG, Graeber MB, Middleton FA, Rochet JC, Scherzer CR, Global PDGEC (2010) PGC-1alpha, a potential therapeutic target for early intervention in Parkinson's disease. Sci Transl Med 2(52):52ra73. https://doi.org/10.1126/scitranslmed.3001059

Zhu J, Wang KZ, Chu CT (2013) After the banquet: mitochondrial biogenesis, mitophagy, and cell survival. Autophagy 9(11):1663–1676. https://doi.org/10.4161/auto.24135

Mitochondrial Disruption in Cardiovascular Diseases

Edward J. Lesnefsky, Qun Chen, Bernard Tandler, and Charles L. Hoppel

Abstract Mitochondrial dysfunction is a critical factor in the initiation and progression of heart failure. Impairment of mitochondrial metabolism decreases energy production and leads to enhanced oxidative and calcium mediated injury to the myocardium. Mitochondrial dysfunction leads to progressive contractile dysfunction, ultimately to cardiomyocyte death. Myocardial ischemia and infarction, as well as pressure overload, are external causes of heart failure associated with impaired mitochondrial function. Genetic defects in both nuclear-encoded and mitochondrial DNA-encoded mitochondrial proteins lead to cardiomyopathy and eventual heart failure, highlighting those mitochondrial functions required to support normal cardiac physiology. Maladaptive neurohumoral response(s) to decreased cardiac contractile function activates cardiomyocyte signaling that results in

---

E. J. Lesnefsky
Medical Service, McGuire Veterans Affairs Medical Center, Richmond, VA, USA

Division of Cardiology, Department of Medicine, Pauley Heart Center, Virginia Commonwealth University, Richmond, VA, USA
e-mail: edward.lesnefsky@va.gov; edward.lesnefsky@vcuhealth.org

Q. Chen
Division of Cardiology, Department of Medicine, Pauley Heart Center, Virginia Commonwealth University, Richmond, VA, USA
e-mail: qun.chen@vcuhealth.org

B. Tandler
Department of Biological Sciences, Case Western Reserve University, School of Dental Medicine, Cleveland, OH, USA
e-mail: bernard.tandler@case.edu

C. L. Hoppel (✉)
Department of Pharmacology, Case Western Reserve University, School of Medicine, Cleveland, OH, USA

Department of Medicine, Case Western Reserve University, School of Medicine, Cleveland, OH, USA

Center for Mitochondrial Disease, Case Western Reserve University, School of Medicine, Cleveland, OH, USA
e-mail: charles.hoppel@case.edu

P. J. Oliveira (ed.), *Mitochondrial Biology and Experimental Therapeutics*,
https://doi.org/10.1007/978-3-319-73344-9_12

impaired cardiac metabolism. For these reasons, cardiac metabolism in general and mitochondrial function remain novel therapeutic targets. Future treatment approaches include modulation of cardiac substrate selection, biochemical bypass of metabolic defects in mitochondria, and ultimately the manipulation of mitochondrial dynamics to foster the repair or removal of dysfunctional mitochondria, with accompanying biogenesis of fully-functioning mitochondria.

Keywords Oxidative phosphorylation · Respirasomes · Electron transport chain · Reactive oxygen species · Cardiolipin · Fatty acid oxidation · Ubiquinol:cytochrome *c* oxidoreductase

1 Introduction

As a basis for our discussion of cardiac mitochondria, we begin with a simplified overview of the structure of cardiac muscle, highlighting only those features that have relevance to our story. Mammalian hearts are made up mainly of fibers of indeterminate length, some of which are branched. These fibers consist of a series of individual cardiomyocytes that are linked end-to-end by intercalated discs to form a functional syncytium. Each of the constituent cardiac myocytes contains at least one nucleus, sometimes several that are centrally located. The contractile apparatus resembles that of skeletal muscle in that it consists of stacked sarcomeres, which lend the hearts cells a striated appearance. These striations are named after their birefringent appearance, with the isotropic band abbreviated the “I band” and the anisotropic band labeled the “A band”. Conveniently, the dark band contains the letter “A” and the light band the letter “I”, making it a simple matter to apply the correct terminology. The structure of the banding, which is readily apparent in the light microscope, is based on the disposition of actin and myosin filaments. The I bands consist of parallel actin filaments while the “A” bands consist of a mixture of parallel actin and myosin filaments. It is the sliding of the actin filaments relative to the myosin filaments that leads to contraction (Goldstein et al. 1991; Lunkenheimer et al. 2006).

The engines for driving the interaction of thick and thin filaments are the mitochondria, which provide the necessary, critically important ATP. These organelles, which are quite abundant, are distributed throughout the cardiomyocytes. Irrespective of their position within these cells, these organelles appear in transmission electron microscopy to have the same structure in terms of cristae architecture (Hoppel et al. 2009; Riva et al. 2006; Riva et al. 2005). However, scanning electron microscopy has revealed that there are differences in these structures depending on the intracellular site where the mitochondria are situated. Most (77%) mitochondria that lie subjacent to the sarcolemma (which we call subsarcolemmal = SSM) have lamelliform cristae, whereas most (55%) mitochondria that are situated among the myofibrils (which we call interfibrillar = IFM), have tubular cristae (Riva et al. 2006, Riva et al. 2005). These structural differences are consonant with the functional differences that exist between the two populations, i.e., SSM and IFM. That is to say, the IFM operate at a higher rate of oxidation than do the SSM (Palmer et al. 1977).

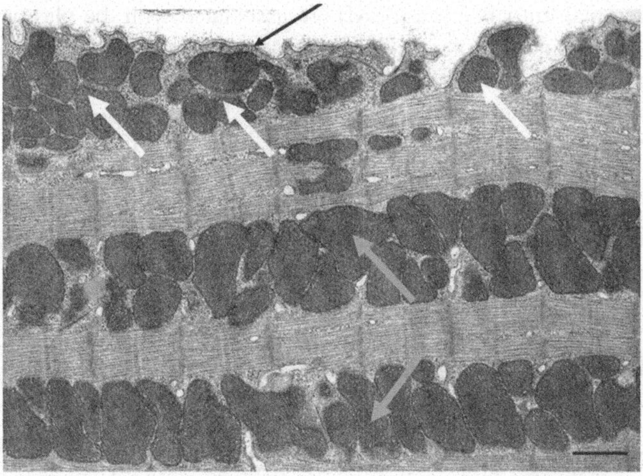

Fig. 1 Elecltron micrograph (EM) of a portion of a mouse cardiomyocyte illustrating the two type of mitochondria, identified by their intracellular location. Subsarcolemmal mitochondria (SSM) (white arrows) lie immediately subjacent to the sarcolemma (black arrow). Interfibrillar mitochondria (IFM) (gray arrows) are situated between the myofibrils. Note that with this type of microscopy (transmission EM), there is no obvious morphological differences between the two populations of these organelles. Bar = 1 µm. This figure is from a spotlight review in Cardiovascular Research (2010) 88, 40–50 2010, Rosca and Hoppel, Mitochondria in heart failure, and reproduced according to the guidelines of Oxford University Press

The sarcoplasmic reticulum (SR), a form of endoplasmic reticulum, consists of numerous ribbon-like, interconnected strands that pervade the myocytes. Mitochondria are closely apposed to the SR and essentially interact via shared membrane domains, the mitochondria-associated membranes (MAM) (Paillard et al. 2013). MAM contribute importantly to phospholipid trafficking, to calcium homeostasis between SR and mitochondria, and to mitochondrial dynamics. MAM are increasingly appreciated for their contribution to cardiac pathophysiologic states, including ischemia and reperfusion (Paillard et al. 2013; Hughes et al. 2014), with their potential role in heart failure remaining a promising area for future investigation.

The two populations of mitochondria to which we have alluded are relatively easily separated (Palmer et al. 1977). Disruption of the sarcolemma permits the SSM to spill out, whereas the IFM remain entrapped, but structurally intact, between the myofibrils. A second disruption of the myofibrils by a protease frees the IFM (Palmer et al. 1977). The first studies of cardiac mitochondria were based largely on the total population of mitochondria derived by exposure of the heart to proteases, which liberated all of the mitochondria, a mélange of SSM and IFM (Lesnefsky and Hoppel 2006). As the Polytron gained in popularity, many of these studies became based on SSM, which were freed by the easy disruption of the sarcolemma by this instrumentation; the IFM were largely ignored (Czerski et al. 2003; Lesnefsky and Hoppel 2006). Our approach to mitochondrial disruption in cardiovascular disease has been to make use of the separated populations of mitochondria in heart afflictions (Fig. 1).

Subsarcolemmal mitochondria (SSM) are associated with the plasma membrane and share lipid rafts and caveolae with the plasma membrane (Fridolfsson et al. 2012). These shared membrane regions have the potential for direct signaling to the mitochondria from plasma membrane receptors. Interfibrillar mitochondria (IFM) are associated with T-tubules via the protein, junctophilin (Jiang et al. 2016). IFM exhibit greater oxidative capacity in part due to their greater content of cytochromes (Palmer et al. 1977). There is a differing response of the subpopulations to superimposed cardiac disease and comorbid conditions. IFM are preferentially affected in aging (Fannin et al. 1999) and type I diabetes (Dabkowski et al. 2009). SSM show a more rapid onset of damage during ischemia (Lesnefsky et al. 1997), perhaps in part because of their decreased tolerance to calcium overload (Palmer et al. 1985). IFM demonstrate a greater involvement in cardiomyopathy, both genetic and acquired. The differences in relative susceptibility raise issues of environment versus innate difference in components, or perhaps the relative ability to remove or repair mitochondria that are present in different regions of the cardiomyocyte. There is a paucity of differences in protein composition between the two populations (Fig. 2).

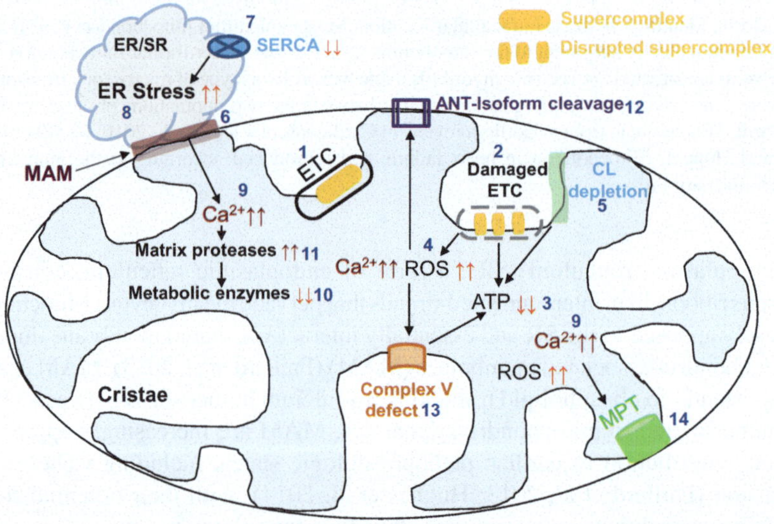

Fig. 2 Mitochondrial sites in the potential development of heart failure. Defects of the electron transport chain (ETC, 1) include damage within individual respiratory complexes and damage to or disruption of supercomplexes (2). These defects result in heart failure through reduction of bioenergy production (3) and augmentation of ROS generation (4). A decrease in cardiolipin content or an alteration in cardiolipin composition may contribute to the disruption of supercomplexes (5). Mitochondria are closely connected to ER (endoplasmic reticulum) via MAM (mitochondria associated membrane, 6). Dysfunction of SERCA (7) triggers ER stress (8) that leads to calcium over load within mitochondria (9). Calcium overload impairs energy metabolism through degradation of metabolic enzymes (10) via activation of proteases, including calpains, within mitochondria (11). Cleavage of subunits of ANT (12) or complex V (13) potentially sensitizes mitochondrial permeability transition pore (MPTP) opening (14), predisposing to myocyte death. The ensuing loss of cardiac myocytes favors progressive ventricular dysfunction and the progression of the heart failure state

Recent reviews have delineated the role of mitochondria in aging (Lesnefsky et al. 2016; Lesnefsky and Hoppel 2006); ischemia-reperfusion injury (Lesnefsky et al. 2017; Morel et al. 2012); permeability transition mediated injury (Weiss et al. 2003; Halestrap 2009); role in the entrance of canonical transcription factors that enter mitochondrial to regulate mitochondrial metabolism (Szczepanek et al. 2012a; Szczepanek et al. 2012b). This review focuses on mitochondria disruption in the evolution of congestive HF and their maladaptation to this condition.

2 Role of Mitochondria in Congestive Heart Failure

2.1 Substrate Selection

The failing heart loses metabolic flexibility as HF progresses. The normal myocardium is able to process multiple classes of substrates including fatty acids, glucose, and lactate for eventual mitochondrial oxidation (Lesnefsky et al. 2001b). Glucose oxidation requires insulin-driven uptake across the plasma membrane via GLUT4 (Abel 2004) and glycolytic generation of pyruvate, which is imported into mitochondria followed by metabolism to acetyl CoA and NADH by the pyruvate dehydrogenase complex (Folmes et al. 2010; Sugden and Holness 2006). Lactate uptake occurs via the monocarboxylate transporter with oxidation to pyruvate by lactate dehydrogenase in the cytosol (Baba and Sharma 1971; Dym et al. 2000). Acetyl-CoA enters the tricarboxylic acid cycle; NADH is a substrate for electron transport chain (ETC) complex I. Fatty acids are taken up by the CD36 plasma membrane transporter followed by the formation of fatty acyl-CoA esters by long-chain fatty acid CoA synthetase (Van Der Vusse et al. 2002). Long chain acyl-CoAs are converted to acylcarnitines by carnitine palmitoyltransferase-I (CPT-I) with activation of mitochondrial uptake by the carnitine-acyl-carnitine transporter (Kerner and Hoppel 2000).

Substrate selection is tightly controlled and is determined by substrate availability and regulation in response to cardiomyocyte signaling activation and pathologic stress (Lesnefsky et al. 2001b). This regulation in large part occurs at the level of pyruvate dehydrogenase complex and CPT-I in response to the content of metabolic intermediates and regulation by signaling cascades (Lesnefsky et al. 2016; Lesnefsky et al. 2001b; Fillmore et al. 2014; Das et al. 2014). In moderate congestive HF, this regulation is maintained with a preserved balance of carbohydrate and fatty acid oxidation (Chandler et al. 2004). CPT-I activity was unaltered in the setting of moderate congestive HF (Chandler et al. 2004; Panchal et al. 1998). The *in vivo* rates of free fatty acid and glucose oxidation remained unaltered (Chandler et al. 2004).

Severe HF states result in a loss of metabolic flexibility (Dolinsky et al. 2016; Goldenthal 2016; Fillmore et al. 2014). Fatty acid oxidation relative to glucose oxidation is increased. Furthermore, glucose utilization by glycolysis is uncoupled from glucose oxidation (Goldenthal 2016, Fillmore et al. 2014). A contributing factor to the dysregulation of glucose metabolism is impaired insulin-mediated signaling in the failing heart (Goldenthal 2016). The failing human heart as well as a variety of models of experimental heart failure exhibit a decreased content of high energy phosphates. The impaired energy production occurs as a consequence of defects in substrate utilization and in mitochondrial oxidative function (Neubauer 2007), although the specific metabolic defects differ according to the model and potentially to the etiology of HF. Angiotensin II appears to contribute to the lack of metabolic flexibility by downregulating PDH-complex activity (Mori et al. 2013).

2.2 Mitochondrial Dysfunction

Cardiac health indisputably is based largely on unimpaired mitochondrial function. The following discussion on heart failure—a leading cause of death—is related to mitochondrial disruption. Samples from patients with heart failure undergoing heart transplant surgery had defects in the mitochondrial phosphorylation system, complex I respiration, and fatty acid oxidation (Park et al. 2016). The last effect also was observed in patients early in the development of HF (Lemieux et al. 2011). In patients with advanced HF, similar mitochondrial defects were present; in addition, the mitochondrial content of the heart was decreased. Reduction of oxidative phosphorylation in end-stage human heart failure was noted by a number of different investigators (Park et al. 2016; Lemieux et al. 2011; Sabbah et al. 1992; Sharov et al. 2000).Two studies found that normal rates of oxidative phosphorylation were preserved in failing human hearts (Cordero-Reyes et al. 2014; Hamilton 2013)—metabolic defects that precede mitochondria involvement as well as decreased numbers of these organelles were thought to be the cause of the defect.

In human HF, the results are less consistent than in animal studies. In permeabilized human cardiac muscle fibers, oxidative phosphorylation is decreased more in ischemia-induced HF than in non-ischemia-induced HF—both are decreased compared to healthy hearts (Park et al. 2016). Citrate synthase activity progressively declined in non-ischemic HF as well as in ischemia-induced HF compared to healthy hearts. After normalization with citrate synthase, there were no significant differences in oxidative phosphorylation among healthy, non-ischemic, and ischemic HF (Park et al. 2016). Moreover, mitochondrial oxidative phosphorylation was not impaired in freshly isolated cardiac mitochondria from failing human hearts (Cordero-Reyes et al. 2014). In that study, results from non-ischemic-induced HF patients were combined with those from ischemia-induced HF patients. In another study, oxidative phosphorylation was decreased in permeabilized myocardial fibers using complex I or fatty acid substrates in both mild human HF and chronic HF (Lemieux et al. 2011). Mitochondrial content also was decreased in failing heart. This loss of mitochondrial density contributes to decreases in oxidative metabolism. In support of the contribution of mitochondrial dysfunction to the severe HF state, oxidative metabolism improves in human hearts following periods of unloading of the failing left ventricle after periods of treatment with mechanical left ventricular assist devices (Sharma and Kass 2014).

With respect to the presence of electron transport defects, the activity of complexes III, IV and V was decreased in mitochondria isolated from transplant hearts of patients with congestive HF stage NYHA IV (Buchwald et al. 1990). In contrast, Scheubel et al. showed a decrease only in complex I activity (Scheubel et al. 2002). Jarreta et al. showed a 35% decrease in complex III activity in both idiopathic dilated and ischemic cardiomyopathy (Jarreta et al. 2000).

Using animal models of HF, heart tissue, permeabilized fibers, or isolated mitochondria were employed. In rats with pressure overload-induced HF, Sanbe et al. (Sanbe et al. 1995) reported a decrease in oxygen consumption with glutamate as substrate in skinned myocardial fibers, a finding consistent with a decrease in high energy phosphates (ATP, creatine-phosphate). Impaired mitochondrial respiration assessed in skinned myocardial fibers also was reported by Sharov et al. in coronary microembolization-induced HF dogs (Sharov et al. 1998). In that model, the mitochondria were smaller in size than normal (Sabbah et al. 1992). Acute HF was accompanied by marked decreases in both state 3 respiratory rates and in the respiratory control index of heart mitochondria isolated from dogs (Takaki et al. 1995). Although the above studies clearly indicate a mitochondrial involvement in the development and/or progression of HF, no attempt has been made to distinguish between the two different mitochondrial populations, i.e., SSM and IFM, to identify the site of the lesion or the mechanism of the defect. In a genetic model of cardiomyopathy in the Syrian hamster, we reported that only the heart IFM are affected (Hoppel et al. 1982).

As noted in ischemia-induced HF in dogs, Sharov et al. (Sharov et al. 1998) reported that permeabilized fibers exhibited decreased oxidative phosphorylation, a finding that was extended to isolated mitochondria from the identical model (Rosca et al. 2008). Use of these two populations of the isolated organelles permitted the identification of the mechanism of this defect. The key findings were a 50% decrease in integrated mitochondrial function measured as oxidative phosphorylation, but there was no defect in the ETC (Rosca et al. 2008). The reason for this discrepancy turned out to be that mitochondrial supercomplexes (complex I, dimeric complex III, and 1–4 copies of complex IV) were reduced in number, which in turn leads to a reduction in oxidative phosphorylation (Rosca et al. 2008).Ide et al. reported a 50% decrease in complex I activity in submitochondrial particles (Ide et al. 1999) in canine pacing-induced HF. In contrast to the above findings, no decrease in complex I activity was found by Marin-Garcia et al. in the same animal model using frozen whole tissue homogenates (Marin-Garcia et al. 2001), but significant decreases were present in complex III and complex V activities. Reduced complex V activity also was reported in a naturally occurring canine model of idiopathic dilated cardiomyopathy (Mccutcheon et al. 1992). Using the two populations of isolated mitochondria from left ventricle from hearts from dogs with pacing-induced HF, Marin-Garcia reported decreased activity of complexes I, III, and V (Marin-Garcia et al. 2009).

Rodent models of HF also have been used to study defects in mitochondrial metabolism and to supplement the larger animal translational models in identifying an electron transport chain dysfunction in the failing heart. In post-infarction-induced HF in the rat, oxidative phosphorylation was decreased with complex I substrates. Complex I activity also was decreased in cardiac mitochondria isolated from the post-infarction rat. The protein content of complex I subunits was lessened in rat heart mitochondria following left anterior descending artery ligation (Knowlton et al. 2014). Exercise improved cardiac function concomitant with improvement in complex I activity post-infarction (Kraljevic et al. 2013). These findings are in line with the development of cardiomyopathy with reduced systolic function in humans with Leigh's syndrome with a complex I defect (Cameron et al. 2015; Loeffen et al. 2000). They support the conclusion that complex I defect is involved in HF following myocardial infarction.

Defective electron transport also is present in left ventricular pressure overload-induced HF in Sprague Dawley rats. Transverse aortic constriction (TAC) induces myocardial hypertrophy in two weeks. Mild HF appears at 6–10 weeks with an unchanged ejection fraction. Heart failure with systolic dysfunction occurs after 20 weeks TAC (Schwarzer et al. 2014). Oxidative phosphorylation is increased in isolated mitochondria after 3 and 6 weeks of TAC. These results indicate that mitochondrial function is initially enhanced to meet increased energy requirements of the concentrically-hypertrophied heart as a component of the initial compensatory response during the early pressure-overload period. Following 20 weeks pressure overload, oxidative phosphorylation is impaired in both SSM and IFM isolated from the rat heart (Schwarzer et al. 2013). The degree of damage is more severe in IFM than in SSM. These results indicate that pressure-overload leads to a mitochondrial defect in late stages of HF. Inhibition of complex I by genetic inactivation of the Ndufs4 gene exacerbates HF development after pressure overload by increasing opening of the MPTP (Karamanlidis et al. 2013).

Cardiolipin is a phospholipid unique to mitochondria, consisting of two phosphatidic acids connected by glycerol. Cardiolipin is synthesized within mitochondria and undergoes final remodeling by the matrix enzyme, tafazzin, a cardiolipin acyltransferase that leads in the heart to specific, defined acyl group composition of cardiolipin, consisting of four linoleic acid (C18:2), residues (Sparagna and Lesnefsky 2009). These residues are susceptible to oxidative damage, with oxidatively-altered cardiolipin undergoing remodeling by phospholipase A2 (Schlame 2013) removing an acyl group to generate monolysocardiolipin (Sparagna and Lesnefsky 2009; Zachman et al. 2010). The latter compound is modified with acyl-groups by alternative acyl-transferases, including ALCAT1 (Liu et al. 2012; Li et al. 2012). This secondary remodeling leads to the replacement of acyl groups other than linoleic acid, such as docosahexaenoic acid, C22:6, which is observed in heart failure in rodents and humans, as well as in the diabetic heart (Chicco et al. 2008; Sparagna et al. 2007; Cole et al. 2016). The monolysocardiolipin, and perhaps the altered molecular species of cardiolipin, serve as signals of mitochondrial dyshomeostasis. The altered molecular species are suspected of fostering metabolic impairment by serving as ineffective cofactors for electron transport complexes, especially for cytochrome oxidase, as well as by not supporting the assembly of supercomplexes. Altered cardiolipin species favor the binding to mitochondria of peptides that signal and initiate programmed cell death (Kagan et al. 2005; Korytowski et al. 2011).

In addition to the complex I downregulation discussed above, other genetic mouse models that lead to cardiomyopathy point to additional relevant mitochondria-based mechanisms in the development of heart failure. The tafazzin knockdown mouse, which recapitulates the cardiolipin defect in Barth's syndrome, a human disease, develops cardiomyopathy (Phoon et al. 2012). Inappropriate activation of the c-myc gene that leads to myocyte hyperplasia results in cardiomyopathy associated with a complex III defect (Lee et al. 2009). Leigh's syndrome with a genetic complex IV defect leads to a cardiomyopathy in humans (Bugiani et al. 2005). In the normal heart, the key role of mitochondrial dynamics in maintaining mitochondrial quality control is supported by the occurrence of cardiomyopathy when pertinent proteins are eliminated. Genetic deletion of parkin, that then impairs the removal of dysfunctional mitochondria, leads to a cardiomyopathy (Kubli et al. 2013). Knockout of mitofusin-2, which impairs mitochondrial fusion, leads to cardiac hypertrophy and contractile dysfunction (Papanicolaou et al. 2011).

2.3 *Reactive Oxygen Species Generation*

The ETC is the main source of ROS generation in cardiac myocytes. Complexes I and III are key sites of ROS production in the ETC (Chen et al. 2008; Chen et al. 2003). Recent studies show complex II also produces ROS (Chen and Zweier 2014; Ockaili et al. 2001). Most of the ROS generated from complex I are released into the matrix and dissipated by mitochondrial antioxidants (Chen et al. 2003). In contrast, ROS generated at the complex III Qo center are mainly released into the intermembrane space and, through VDAC, into the cytosol (Gille and Nohl 2001; Han et al. 2003; St-Pierre et al. 2002). ROS generated at the complex III Qi center also is released into the matrix (Shi et al. 2013).

There are two routes to generate ROS production from complex I: a forward electron flow (complex I to complex III) and a reverse electron flow (complex II to complex I). Blockade of electron flow in the terminal portion of complex I (Chen et al. 2003) or blockade at distal sites in the electron transport chain including cytochrome oxidase increases forward electron flow-mediated ROS generation from complex I (Chen and Lesnefsky 2006). Induction of reverse electron flow from complex II to complex I using succinate as a substrate also markedly increases ROS production from complex I (Chouchani et al. 2014; Miwa and Brand 2003; Ross et al. 2013; Murphy 2016). Since a mitochondrial membrane potential is required for reverse electron flow to occur (Chouchani et al. 2014; Ross et al. 2013; Murphy 2016), depolarization of the inner membrane potential can effectively eliminate reverse flow-mediated ROS generation. Inhibition of complex I at the terminal quinol binding site ("rotenone site") decreases the reverse flow-mediated ROS generation from complex I (Chen et al. 2008).

Increased ROS generation occurs in HF. In the TAC-induced HF model in the rat, ROS generation from complex I begins to increase after 6 weeks TAC (Schwarzer et al. 2014). ROS generation from complex I remains elevated in mitochondria isolated following 20 weeks TAC. ROS generation from complex III is increased after 6 weeks TAC. However, ROS generation from complex III is not elevated in heart mitochondria following 20 weeks TAC. Administration of the ROS scavenger decreases ROS generation following 6 weeks TAC. Treatment with the uncoupling agent, dinitrophenol, also reduces ROS production at 6 weeks TAC, indicating that reverse flow-mediated ROS generation probably contributes to HF development (Schwarzer et al. 2014). In failing human hearts, ROS generation is increased (Park et al. 2016). Ischemia-induced HF produces more ROS compared to non-ischemia-induced HF, suggesting that ROS generation may contribute in a major way to development of ischemia-mediated HF.

2.4 Ischemia-Reperfusion Injury and Mitochondria

Myocardial ischemia leads to progressive damage to the ETC (Lesnefsky et al. 2001b). Ischemia initially decreases complex I activity with evidence of injury primarily to the FMN-containing NADH dehydrogenase portion of the complex and to the iron-sulfur centers that participate in electron flow, the latter immediately upstream of the quinone binding site.(Chen et al. 2008; Ohnishi et al. 2005) An additional mechanism of decreased complex I activity in ischemia is the transition of complex I to an inactive form from an active one (Babot and Galkin 2013; Galkin et al. 2009). Ischemia favors transition of complex I from active to the inactive form, a potentially reversible change (Gorenkova et al. 2013). The inactive form of complex I arises as a consequence of a change in tertiary structure of the complex in the presence of excess NADH and ubiquinol that result in exposure of a redox-active cysteine on the ND3 subunit (Galkin and Moncada 2007). Irreversible oxidation of this vulnerable sulfhydryl group leads to persistent deactivation of the complex. Although persistent deactivation of complex I impairs energy generation, attenuation of the abrupt reactivation of complex I from the inactive to active form at the onset of reperfusion can blunt mitochondria-driven cardiac injury, as discussed below (Drose et al. 2016).

Complex III is damaged by ischemia as a consequence of the functional deactivation of the Rieske iron-sulfur protein subunit, leading to loss of the redox active iron-sulfur catalytic center contained within the subunit (Lesnefsky et al. 2001a). Complex IV activity decreases during ischemia, essentially as a result of a decrease in the content of cardiolipin (Lesnefsky et al. 2001c; Lesnefsky et al. 2004b), a requisite component of the inner membrane complex IV environment that is required for optimal complex IV activity. In addition, post-translational modification via phosphorylation of complex IV subunits that decreases activity also has been described (Fang et al. 2007; Huttemann et al. 2012). The role of cardiolipin in optimizing complex IV activity probably occurs via the cardiolipin-mediated stabilization of supercomplexes that contain complex IV (Mileykovskaya and Dowhan 2014), although alteration of supercomplexes by ischemia is less-well described than the alterations described above in HF.

The bulk of damage to the ETC occurs during ischemia, as clearly shown by studies in rabbit and rat hearts. Reperfusion does not result in additional decreases in oxidative phosphorylation nor in a further decrease in ETC complex enzyme activities (Lesnefsky et al. 2004b). Others have shown modest injury to complex I (Veitch et al. 1992) during reperfusion (Mccully et al. 2004; Murphy and Steenbergen 2008). As a consequence of ischemic damage to electron transport, reperfusion mitochondria damaged by ischemia can wreak biochemical havoc within the cardiac myocyte. Reoxygenation with abrupt reactivation of complex I augments the production of ROS by both forward (Chen et al. 2006b; Chen et al. 2008; Lesnefsky et al. 2016; Lesnefsky et al. 2017) and reverse (Chouchani et al. 2014; Ross et al. 2013) electron transport. Complex I and complex III are the main sites that generate ROS during reperfusion (Ambrosio et al. 1993; Chen et al. 2003; Chen et al. 2007), complex III via forward electron flow from complexes I and II and complex I via both forward and reverse electron transport.

The key contribution of mitochondrial-driven cardiomyocyte injury during early reperfusion was shown by reperfusing the heart that contains mitochondria that are devoid of ischemic damage (Chen et al. 2006b). The blockade of electron transport during ischemia at the quinone binding site of complex I with rotenone (Lesnefsky et al. 2004a) or amobarbital (Chen et al. 2006a) mitigates damage to the ETC during ischemia, supporting the hypothesis that the predominant source of ischemic damage to the ETC during ischemia is the ETC itself. In other words, in order to protect the mitochondria, their function must be reversibly, transiently inhibited. Use of this approach to protect the ETC against ischemic damage allows reperfusion of the heart in the presence of mitochondria with preserved ETC function (Chen et al. 2006b). If the mitochondria are essentially normal, will myocyte injury still occur during reperfusion? When the ETC is reversibly inhibited during ischemia to prevent damage to the ETC (Chen et al. 2006a; Aldakkak et al. 2008; Chen et al. 2006b), ETC function not only remains normal during reperfusion (Chen et al. 2006b; Chen et al. 2007), but cardiac injury is dramatically reduced (Chen et al. 2006b). These protected mitochondria produce less ROS during reperfusion and exhibit a decreased susceptibility to mitochondrial permeability transition (Chen et al. 2006b; Chen et al. 2012a), exhibit a pro-survival mitochondrial phenotype with increased content of pro-survival peptide bcl-2 (Chen and Lesnefsky 2011), and release less cytochrome *c* to activate programmed cell death (Chen et al. 2006b). In sum, when mitochondria are devoid of ischemic damage at the onset of reperfusion (Chen et al. 2006a; Aldakkak et al. 2008), cardiac injury after reperfusion is dramatically reduced (Chen et al. 2006b). Blockade of electron flow at complex I (Chen et al. 2003; Nadtochiy et al. 2007) is critical to prevent flow into more distal sites of the ETC, especially complex III. Blockade of distal complex I also decreases damage due to reverse electron flow that may occur at the onset of reperfusion (Chouchani et al. 2014, Ross et al. 2013).

Although the ETC cannot be immediately repaired or replaced, can the ischemia-damaged ETC be successfully modulated to reduce cardiomyocyte injury during early reperfusion? The transient and reversible blockade of the proximal ETC at reperfusion decreases cardiomyocyte injury (Ambrosio et al. 1993; Chen et al.

2012b; Stewart et al. 2009), although the extent of protection is less than that provided by intervention before ischemia (Stewart et al. 2009). Early reperfusion represents a significant therapeutic window for interventions to attenuate ETC-dependent cardiac injury, even though ischemia-mediated mitochondrial damage has already occurred. Although direct inhibitors of complex I provide important mechanistic insight, they probably will not be useful in clinical settings. Potential translational agents to transiently and reversibly block complex I during early reperfusion include ranolazine (Aldakkak et al. 2011), Mito-SNO (Brown and Borutaite 2007), 3-mercaptopropionylglycine (Nadtochiy et al. 2007), nitrite (Shiva et al. 2007), and high dose metformin (Mohsin et al. 2016). Even persistence during early reperfusion of intracellular acidification from ischemia might be helpful (Xu et al. 2014).

Reperfusion leads to the activation of both canonical and novel ("neopathologic") signaling sequences that are effectors of mitochondrial-driven cell death. Permeation of mitochondria via the onset of mitochondrial permeability transition or targeted permeabilization of the outer membrane by the peptide, bax, leads to the loss of key proteins located within the intermembrane space. Cytochrome *c* (Green and Reed 1998) and apoptosis-inducing factor (Chen et al. 2011, Yu et al. 2002) are peptides that contribute to optimal ETC function, but their release activates cell death programs, either caspase-dependent (cytochrome *c*) (Borutaite and Brown 2003; Mccully et al. 2004) or independent (Yu et al. 2002). Oxidative stress predisposes to the onset of permeability transition opening driven by calcium-overload during reperfusion. Therapeutic modulation of electron transport through complex I decreases susceptibility to permeability transition pore opening (Chen et al. 2012a). The intimate interaction of oxidative and calcium-mediated injury via permeability transition pore opening will continue to evolve as a critical concept relevant to cardiac injury during early reperfusion.

Calcium overload and oxidative stress of early reperfusion drive injury by novel signaling systems. Calpains are Ca^{2+}-dependent cysteine proteases localized to mitochondria (Ozaki et al. 2007), in both the intermembrane space and matrix (Chen and Lesnefsky 2015); these proteases respond to mitochondrial calcium overload injury. In order for release of apoptosis-inducing factor from mitochondria to happen, the protein must be detached from the inner membrane via cleavage by mitochondrial-localized μ-calpain with release occurring when the outer membrane is breached (Chen et al. 2011; Ozaki et al. 2007). Activation of matrix-localized calpains damages complex I, pyruvate dehydrogenase, and may increase the susceptibility to permeability transition (Thompson et al. 2016). Mitochondrial calpains constitute a calcium-responsive mitochondrial-based signaling system to activate caspase-independent cell death programs.

In addition to direct oxidative modification of mitochondrial constituents, increased ROS production during reperfusion contributes to injury by triggering pathological mitochondrial-driven signaling cascades, especially apoptosis-signaling kinase 1 (ASK1), an upstream regulator of JNK/p38 pathways (Tobiume et al. 2001). In mitochondria, ASK1 is bound to thioredoxin 2 (Huang et al. 2015). ROS production oxidizes thiols in thioredoxin 2 leading to release and subsequent activation of ASK1 to initiate caspase-dependent cell death programs (Huang et al. 2015). Blockade of ASK-1 at reperfusion with a cell-permeable, small molecule inhibitor reduced infarct size and improved contractile recovery (Toldo et al. 2012), in part via protection of mitochondria and by decreasing the susceptibility to permeability transition pore opening (Aluri et al. 2012). Thus, ASK1 is a cellular redox sensor that is activated by excessive ROS, supporting a critical role for ROS-mediated signaling in oxidant-mediated cardiac injury. ASK-1 sensing for oxidative stress complements the mitochondrial calpain-AIF system for detection of calcium stress that exceeds homeostatic capability.

In addition to activation of ROS-driven signaling for cell death, oxidative injury from the ETC alters cytochrome *c* to acquire a new function as a peroxidase of the key inner membrane phospholipid, cardiolipin (Kagan et al. 2009; Kagan et al. 2005). Methionine-81 is one of six ligands that coordinate the heme iron in cytochrome *c*. When the sulfhydryl of methionine-81 is oxidized, the heme center is opened, allowing the redox-active iron to catalyze H_2O_2-mediated oxidation of cardiolipin to produce unstable lipid peroxidizes. The appearance of peroxidized cardiolipin and its degradation products favors the insertion of pro-apoptotic bax into the outer membrane (Korytowski et al. 2011), leading to its permeation, intermembrane protein release (Kagan et al. 2009; Kagan et al. 2004), and the activation of cell death pathways, as discussed above. Formation of neoperoxidase cytochrome *c* via oxidative modification occurs in multiple situations of oxidative stress (Belikova et al. 2006; Kagan et al. 2005) including ischemia-reperfusion (Aluri et al. 2014).

Ischemia and early reperfusion leads to a mitochondrial phenotype that favors both direct chemical toxicity and signaling-mediated cardiomyocyte injury. In sum, ischemia-induced damage to mitochondria leads to mitochondrial-driven injury during reperfusion via: first, direct oxidative injury from the ETC (Chen et al. 2007); second, the formation of outer membrane peptide phenotypes that activate programmed cell death (Gottlieb et al. 1994; Chen et al. 2011); third, activation of existing calcium-mediated and oxidant-induced mitochondrial based signaling (Hausenloy and Yellon 2007; Murphy and Steenbergen 2008; Chen et al. 2011); and fourth, the formation of new pathologic mitochondrial-based signaling for cell death (Hausenloy and Yellon 2003; Weiss et al. 2003; Kagan et al. 2005; Toldo et al. 2012). These integrative responses signal the myocyte that a critical mass of mitochondria has been damaged. The cellular response involves activation of repair and compensatory systems at the transcriptional (Chiong et al. 2011; Singh and Kang 2011), post-transcriptional (Kukreja et al. 2011), and post-translational levels (Murphy and Steenbergen 2011; Finichiu et al. 2015). Unfortunately, the persistent damaged mitochondria that perpetuate myocyte injury apparently exceed repair and removal and favors activation of cell death programs (Dorn 2nd 2015). Thus, ETC-dependent, ischemia-induced processes have the potential to extend much farther into prolonged periods of reperfusion,thus contributing to the genesis of myocyte loss, myocyte dysfunction leading to impaired ventricular function, and the genesis of cardiomyopathy and HF.

3 Calcium Transporter Defects and Calcium Overload in CHF?

Calcium is a key factor to activate metabolic enzymes within the mitochondrial matrix. Transient calcium overload within mitochondria is critical to enhance ATP production when the demand for ATP is markedly increased. Calcium also activates several TCA cycle enzymes and F_1F_0 ATP synthase to stimulate ATP production (Hamilton 2013). Additionally, prolonged mitochondrial calcium overload causes mitochondrial damage in pathological conditions, including HF (Santulli et al. 2015).

Import of calcium into mitochondria occurs through the mitochondrial calcium uniporter (MCU), with the mitochondrial inner membrane potential being the driving force (Pan et al. 2011). The maintenance of mitochondrial membrane potential favors calcium uptake, whereas depolarization of the inner membrane potential leads to calcium release. Inhibition of the MCU using ruthenium red decreases mitochondrial calcium loading during ischemia and reperfusion (Cao et al. 2006), supporting the idea that activation of the MCU contributes to calcium overloading during ischemia-reperfusion. However, a recent study showed that knockout of the MCU does not decrease cardiac injury during ischemia-reperfusion (Luongo et al. 2015). These results support the finding that activation of the MCU contributes to calcium overloading during ischemia-reperfusion. Interestingly, there is no significant contractile responsiveness during infusion of isoproterenol in MCU knockout mice (Luongo et al. 2015). Metabolic enzymes (dehydrogenase) within the mitochondrial matrix are not activated during isoproterenol infusion in MCU knockout mice. MCU apparently is required to compensate for reduced energy production during stressed condition through modulation of mitochondrial Ca^{2+} and Ca^{2+}-dependent metabolism. The role of MCU activation in calcium overloading during HF needs further clarification.

One of the detrimental effects of mitochondrial calcium overloading is to increase the opening of mitochondrial permeability transition pores (MPTP) (Weiss et al. 2003). The structure of MPTP is detailed in a separate chapter, and is not discussed here. Opening of the MPTP is a final step in induction of cell death in myocytes following ischemia-reperfusion. Modulation of MPTP opening using pharmacological or genetic approaches decreases cell injury in a variety of animal models following *in vivo* or *in vitro* ischemia or reperfusion (Halestrap et al. 2004; Weiss et al. 2003; Cung et al. 2015). Cyclophilin D (cy-D) is a key regulator of MPTP opening. Knockout of cy-D decreases cardiac injury in the heart following ischemia-reperfusion through inhibition of MPTP opening (Baines et al. 2005). The decreased myocyte death should benefit HF after ischemia-reperfusion. Interestingly, knockout of cy-D enhances rather than decreases HF in mice following transaortic constriction (Santulli et al. 2015), suggesting that inhibition of MPTP is not always beneficial in decreasing HF. The rapid onset of HF in cy-D knockout mice following transaortic constriction is due to decreased calcium efflux through MPTP that leads to increased calcium within mitochondria. A persistent elevation of matrix calcium impairs substrate flexibility of hearts by increasing glucose oxidation relative to fatty acid oxidation through enhanced activity of pyruvate dehydrogenase and α-ketoglutarate dehydrogenase (Santulli et al. 2015). Thus, MPTP may play a role in maintaining homeostatic mitochondrial Ca^{2+} levels to match metabolism during conditions of stress (Elrod et al. 2010). However, another study showed that MPTP may not be involved in Ca^{2+} efflux from mitochondria. Modulation of the expression of the c subunit of F_1F_0 ATP synthase, a proposed component of MPTP, does not affect mitochondrial Ca^{2+} efflux in conditions favoring closing of MPTP (De Marchi et al. 2014). Therefore, the role of MPTP in mitochondrial Ca^{2+} efflux, and especially the role of persistent alterations in HF, warrants further study.

One consequence of intracellular calcium over-load is to activate Ca^{2+}-sensitive cysteine proteases, calpains (Ozaki et al. 2007). In cardiomyocytes, calpain1 and calpain 2 are ubiquitously expressed. Calpain 1 has more proteolytic activity than calpain 2 in that the calcium concentration for calpain 1 activation is more easily reached in cardiomyocytes (Nishida et al. 2015). Activation of calpain 1 degrades myofibrillar proteins such as troponin, tropomyosin, myosin, and titin (Tsuji et al. 2001). The detrimental effect of calpain 1 activation mainly is manifested in ischemia-reperfusion mediated cardiac injury. Activation of calpain 1 contributes to ischemia-reperfusion-induced cardiac dysfunction and cell death through proteolysis of a variety of proteins including those in the sarcolemma, sarcoplasmic reticulum, myofibrillar proteins, and regulatory enzymes (Hausenloy et al. 2005; Hausenloy and Yellon 2007; Inserte et al. 2008). In addition, ischemia-reperfusion also activates mitochondrial localized calpain 1, leading to increased cardiac injury by inhibition of mitochondrial oxidative phosphorylation and translocation of apoptosis-inducing factor (AIF) from mitochondria into the cytosol (Chen et al. 2011). Therefore, prevention of calpain activation using pharmacological or genetic methods is a potential therapeutic approach to decrease myocardial reperfusion injury.

Compared to ischemia-reperfusion injury, the role of calpain activation in HF still has a degree of uncertainty. Overexpression of calpastatin (an endogenous calpain inhibitor) leads to decreased calpain activity and dilated cardiomyopathy (Galvez et al. 2007). Calpain 4 is a critical regulatory subunit of calpain 1 and calpain 2. Knockout of calpain 4 eliminates both calpain 1 and calpain 2 activities (Chen and Lesnefsky 2015). Interestingly, HF occurs early in calpain 4 knockout mice following TAC, compared to wild type mice (Taneike et al. 2011). Infusion of isoproterenol for two weeks induces left ventricular dilatation and contractile dysfunction in calpain 4 knockout mice (Taneike et al. 2011). These results suggest that activation of calpain has a beneficial role in delaying HF occurrence following pressure overload. However, administration of a calpain inhibitor following ischemia-reperfusion improves cardiac function through attenuation of adverse post-infarction remodeling (Poncelas et al. 2017). Thus, calpain activation may have different roles in heart failure depending on its etiology. The participation of mitochondrial calpain activation in different HF models clearly requires further study.

Mitochondria are closely connected to endoplasmic reticulum (ER) (Paillard et al. 2013). A dysfunctional ER leads to ER stress that contributes to intracellular and mitochondrial calcium overload. In heart failure patients, expression of GRP78 (an ER chaperon protein) is markedly increased, indicating that ER stress is involved in the pathophysiology of HF (Castillero et al. 2015; Minamino and Kitakaze 2010).

In human failing hearts, expression of the ER calcium-ATPase isoform 3f (SERCA3f) is increased (Kranias and Hajjar 2012). Overexpression of SERCA3f in mice leads to increased GRP78 expression, indicating that SERCA3f is involved in ER stress (Dally et al. 2009). These results suggest that there is a vicious cycle between calcium overload and ER stress.

ER stress results in mitochondrial dysfunction. Thapsigargin is a calcium-ATPase inhibitor commonly used to induce ER stress. In thapsigargin-treated cardiomyocytes, inner mitochondrial membrane potential is depolarized, suggesting that ER stress sensitizes MPTP opening (Zhang and Ren 2011). Thapsigargin-treated mice also exhibit a marked decrease in oxidative phosphorylation compared to vehicle-treated mice. The calcium retention capacity is decreased in mitochondria isolated from thapsigargin-treated mice, supporting the idea that ER stress sensitizes MPTP opening (Chen et al. 2016). These findings provide clear evidence that ER stress is involved in mitochondrial dysfunction.

4 Co-morbid Conditions to Exacerbate Disease-Induced Mitochondrial Damage

4.1 Aging

Aging leads to defects in the ETC that are centered on the IFM (Lesnefsky and Hoppel 2006). IFM exhibit decreases in the activity of complexes III and IV. The complex IV defect is due to an alteration in the intermembrane environment of the complex, possibly due to an alteration of cardiolipin (Fannin et al. 1999). The complex III defect is in the quinol oxidation site of the cytochrome *b* subunit, encoded by mitochondrial DNA (Moghaddas et al. 2003). IFM with these defects in electron transport exhibit increased production of ROS (Sun and Trumpower 2003) and an increased susceptibility to permeability transition pore opening (Hafner et al. 2010), even in the baseline state. Mitochondria are a likely contributor to cardiomyocyte dropout and increased areas of fibrosis present in the aged heart. The metabolic defects present in the aging heart have been the subject of a recent review (Lesnefsky et al. 2016).

Superimposition of aging on animal models of ischemia and reperfusion leads to enhanced injury (Lesnefsky et al. 1996; Lesnefsky et al. 1994). Support for this concept was provided by the finding that when aging-induced defects were attenuated by therapy to enhance mitochondrial DNA transcription and translation, mitochondrial function improved to that seen in the adult heart (Lesnefsky et al. 2006). Subsequent cardiac injury in these elderly hearts was reduced to the extent of the infarction that occurs in the adult heart (Lesnefsky et al. 2006). Thus, age-induced dysfunction lays the groundwork for enhanced injury elicited by various experimental pathologies. Unfortunately, the role of aging in HF induced defects has been all but ignored.

4.2 *Diabetes*

Diabetes leads to defects in mitochondrial function in the heart, the two populations of cardiac mitochondria being affected differently. In type I diabetes, IFM are profoundly affected with decreases in respiration using complex I, II, or III substrates (Dabkowski et al. 2009). In SSM, only complex II respiration was decreased. Diabetes did not lead to defects in complex IV. The production of reactive oxygen species was increased in IFM. In contrast, in a model of type II diabetes, i.e., the db/db mouse, oxidative phosphorylation was decreased in SSM whereas IFM were unaffected (Dabkowski et al. 2010). The activities of complexes I, III, and IV were decreased in SSM (Hollander et al. 2014).

The defects in the non-failing diabetic heart have the potential to subsequently affect HF. In type I diabetic hearts, the defects may be intensified, likely further impairing energy production. In type II diabetic hearts, the HF-induced defects in IFM will combine the defect in SSM to affect global respiration and to further impair energy production. In the db/db mouse, chronic treatment with tadalafil to activate PKG-mediated signaling improves the respiratory defect (Koka et al. 2014). This approach has the potential to attenuate mitochondrial defects in the failing diabetic heart by addressing the diabetes-induced component of dysfunction.

Although few studies have been performed, the baseline diabetic defects combined with the HF-induced defects probably exacerbate myocardial dysfunction and damage. These mitochondrial defects are superimposed upon a loss of metabolic plasticity, with increased fatty acid oxidation and less glucose-carbohydrate-based oxidation. Mitochondrial dysfunction in the failing diabetic heart leads to impaired high-energy production, increased production of reactive oxygen species, and impaired calcium handling with subsequent myocyte cell death and contractile dysfunction (Hollander et al. 2014; Abel 2004). Removal of the defective mitochondria appears to be impaired, potentiating mitochondria-driven tissue injury (Liang and Kobayashi 2016).

5 Mitochondria: Therapeutic Approaches to Modulate Mitochondria and Their Metabolism

Although there is a plethora of cardiac diseases that originate in defects in heart mitochondria, the picture is not as bleak as first appears. A number of potential therapies have been proposed and, although some of these may be beyond our present capabilities, several have been shown to work.

Treatments to overcome mitochondrial defects involve several approaches. Metabolic and pharmacologic therapies appear to have a place in addressing defects in the failing heart. In patients with genetic mitochondrial disease resulting in electron transport defects, the goal has been to attempt to "bypass" the defect with exogenously provided redox active compounds (Dadabayev et al. 2014; Manning and Albarran 2016). This approach has been only modestly successful. Bypass of dysfunctional complex I with a yeast NADH oxidase that does not contribute to the inner membrane electrochemical gradient (and thus does not indirectly contribute to ADP phosphorylation) nonetheless is beneficial in hearts with a complex I defect (Mentzer Jr et al. 2014; Pepe et al. 2014).

The damaged electron transport chain is a key injury mechanism during ischemia and reperfusion, thus strategically timed, reversible blockade of electron transport can mitigate the cardiac injury (Stewart et al. 2009). However, in the failing heart, metabolic defects are persistent, making transient blockade of electron transport unlikely to be of benefit. However, persistent partial blockade that does not impair mitochondrial membrane potential generation yet decreases oxidant production and makes mitochondria less susceptible to permeability transition, may be of benefit (Szczepanek et al. 2011). This phenotype is present in the mitochondria-targeted STAT3 overexpression mouse. Thus, chronic targeting of STAT3 to mitochondria by pharmacologic manipulation may have therapeutic value by enhancing STAT3 interactions with cyclophilin D to decrease susceptibility to permeability transition pore opening (Szczepanek et al. 2012a; Heusch et al. 2011).

Modulation of mitochondrial function by alteration of upstream substrate supply may have therapeutic benefit. This could be accomplished by downregulation of fatty acid oxidation in the failing heart by trimetazadine (Fillmore et al. 2014) or ranolazine (Bhandari and Subramanian 2007). Thus, CPT-I, the regulator of mitochondrial uptake and fatty acid oxidation, remains an attractive, though yet unrealized, therapeutic target.

The most novel therapeutic targets to reduce mitochondrial driven cardiac dysfunction and injury may not be the mitochondria themselves. Endoplasmic reticulum stress is increased in HF (Castillero et al. 2015, Minamino and Kitakaze 2010), favoring mitochondrial calcium overload and electron transport chain dysfunction and damage. Reduction of endoplasmic reticulum stress attenuates damage to mitochondria-associated membranes and protects mitochondria from the deleterious effects of dysfunctional endoplasmic reticulum (Chen et al. 2016; Sciarretta et al. 2013; Toth et al. 2007; Paillard et al. 2013).

The most evocative emerging approach appears to be modulation of PPAR mediated signaling to potentially enhance mitochondrial biogenesis. PPAR-gamma mediated activation with pioglitazone was used to treat neurologic dysfunction in Harlequin mice that exhibit a complex I defect (Benit et al. 2017). Treatment improved neurologic function, although the mitochondrial dysfunction was unaltered (Benit et al. 2017). The underlying mechanism appeared to be a disease-specific modulation of substrate utilization upstream of the dysfunctional mitochondria. Thus, organ dysfunction can improve even though the mitochondria, at least in genetic mitochondrial disease, remain dysfunctional (Chen and Lesnefsky 2017). The impact on the parallel cardiomyopathy remains to be studied. In a murine model of Barth's syndrome, a cardiomyopathy due to decreased cardiolipin content with electron transport problems due to this defect, apparent activation of mitochondrial biogenesis using the PPAR-pan agonist benzafibrate improved cardiac function, although the signature cardiolipin defect worsened (Huang et al. 2017). Highlighting the disease-specific and organ-specific responses to attempted manipulation of mitochondrial substrate utilization, metabolism, and biogenesis, treatment with benzafibrate did not improve the neurologic deficit in the Harlequin mice (Benit et al. 2017). Thus, in select circumstances, organ phenotypes appear to improve with activation of mitochondrial biogenesis, although the mitochondria themselves remain dysfunctional. It appears that in HF the cardiac mitochondria are open to therapeutic manipulations.

References

Abel ED (2004) Glucose transport in the heart. Front Biosci 9:201–215

Aldakkak M, Stowe DF, Chen Q et al (2008) Inhibited mitochondrial respiration by amobarbital during cardiac ischaemia improves redox state and reduces matrix Ca2+ overload and ROS release. Cardiovasc Res 77:406–415

Aldakkak M, Camara AK, Heisner JS et al (2011) Ranolazine reduces Ca2+ overload and oxidative stress and improves mitochondrial integrity to protect against ischemia reperfusion injury in isolated hearts. Pharmacol Res 64:381–392

Aluri HS, Chen Q, Hu Y et al (2012) Apoptosis Signal Regulating Kinase-1 Inhibitor Decreases Mitochondrial Damage during Ischemia-Reperfusion. Circulation 124:A15526. (abstract)

Aluri HS, Simpson DC, Allegood JC et al (2014) Electron flow into cytochrome c coupled with reactive oxygen species from the electron transport chain converts cytochrome c to a cardiolipin peroxidase: role during ischemia-reperfusion. Biochim Biophys Acta 1840:3199–3207

Ambrosio G, Zweier JL, Duilio C et al (1993) Evidence that mitochondrial respiration is a source of potentially toxic oxygen free radicals in intact rabbit hearts subjected to ischemia and reflow. J Biol Chem 268:18532–18541

Baba N, Sharma HM (1971) Histochemistry of lactic dehydrogenase in heart and pectoralis muscles of rat. J Cell Biol 51:621–635

Babot M, Galkin A (2013) Molecular mechanism and physiological role of active-deactive transition of mitochondrial complex I. Biochem Soc Trans 41:1325–1330

Baines CP, Kaiser RA, Purcell NH et al (2005) Loss of cyclophilin D reveals a critical role for mitochondrial permeability transition in cell death. Nature 434:658–662

Belikova NA, Vladimirov YA, Osipov AN et al (2006) Peroxidase activity and structural transitions of cytochrome c bound to cardiolipin-containing membranes. Biochemistry 45:4998–5009

Benit P, Pelhaitre A, Saunier E et al (2017) Paradoxical inhibition of glycolysis by pioglitazone opposes the mitochondriopathy caused by AIF deficiency. EBioMedicine 17:75–87

Bhandari B, Subramanian L (2007) Ranolazine, a partial fatty acid oxidation inhibitor, its potential benefit in angina and other cardiovascular disorders. Recent Pat Cardiovasc Drug Discov 2:35–39

Borutaite V, Brown GC (2003) Mitochondria in apoptosis of ischemic heart. FEBS Lett 541:1–5

Brown GC, Borutaite V (2007) Nitric oxide and mitochondrial respiration in the heart. Cardiovasc Res 75:283–290

Buchwald A, Till H, Unterberg C et al (1990) Alterations of the mitochondrial respiratory chain in human dilated cardiomyopathy. Eur Heart J 11:509–516

Bugiani M, Tiranti V, Farina L et al (2005) Novel mutations in COX15 in a long surviving Leigh syndrome patient with cytochrome c oxidase deficiency. J Med Genet 42:e28

Cameron JM, Mackay N, Feigenbaum A et al (2015) Exome sequencing identifies complex I NDUFV2 mutations as a novel cause of Leigh syndrome. Eur J Paediatr Neurol 19:525–532

Cao CM, Yan WY, Liu J et al (2006) Attenuation of mitochondrial, but not cytosolic, Ca2+ overload reduces myocardial injury induced by ischemia and reperfusion. Acta Pharmacol Sin 27:911–918

Castillero E, Akashi H, Pendrak K et al (2015) Attenuation of the unfolded protein response and endoplasmic reticulum stress after mechanical unloading in dilated cardiomyopathy. Am J Physiol Heart Circ Physiol 309:H459–H470

Chandler MP, Kerner J, Huang H et al (2004) Moderate severity heart failure does not involve a downregulation of myocardial fatty acid oxidation. Am J Physiol Heart Circ Physiol 287:H1538–H1543

Chen Q, Lesnefsky EJ (2006) Depletion of cardiolipin and cytochrome c during ischemia increases hydrogen peroxide production from the electron transport chain. Free Radic Biol Med 40:976–982

Chen Q, Lesnefsky EJ (2011) Blockade of electron transport during ischemia preserves bcl-2 and inhibits opening of the mitochondrial permeability transition pore. FEBS Lett 585:921–926

Chen Q, Lesnefsky EJ (2015) Heart mitochondria and calpain 1: location, function, and targets. Biochim Biophys Acta 1852:2372–2378

Chen Q, Lesnefsky EJ (2017) A new strategy to treat mitochondrial disease without improvement of mitochondrial function? EBioMedicine 18:19–20

Chen YR, Zweier JL (2014) Cardiac mitochondria and reactive oxygen species generation. Circ Res 114:524–537

Chen Q, Vazquez EJ, Moghaddas S et al (2003) Production of reactive oxygen species by mitochondria: central role of complex III. J Biol Chem 278:36027–36031

Chen Q, Hoppel CL, Lesnefsky EJ (2006a) Blockade of electron transport before cardiac ischemia with the reversible inhibitor amobarbital protects rat heart mitochondria. J Pharmacol Exp Ther 316:200–207

Chen Q, Moghaddas S, Hoppel CL et al (2006b) Reversible blockade of electron transport during ischemia protects mitochondria and decreases myocardial injury following reperfusion. J Pharmacol Exp Ther 319:1405–1412

Chen Q, Camara AK, Stowe DF et al (2007) Modulation of electron transport protects cardiac mitochondria and decreases myocardial injury during ischemia and reperfusion. Am J Physiol Cell Physiol 292:C137–C147

Chen Q, Moghaddas S, Hoppel CL et al (2008) Ischemic defects in the electron transport chain increase the production of reactive oxygen species from isolated rat heart mitochondria. Am J Physiol Cell Physiol 294:C460–C466

Chen Q, Paillard M, Gomez L et al (2011) Activation of mitochondrial mu-calpain increases AIF cleavage in cardiac mitochondria during ischemia-reperfusion. Biochem Biophys Res Commun 415:533–538

Chen Q, Paillard M, Gomez L et al (2012a) Postconditioning modulates ischemia-damaged mitochondria during reperfusion. J Cardiovasc Pharmacol 59:101–108

Chen Q, Ross T, Hu Y et al (2012b) Blockade of electron transport at the onset of reperfusion decreases cardiac injury in aged hearts by protecting the inner mitochondrial membrane. J Aging Res 2012:753949

Chen Q, Thompson J, Hu Y et al (2016) Metformin treatment decreases the er stress-mediated mitochondrial injury. Circulation 134:A17009

Chicco AJ, Sparagna GC, Mccune SA et al (2008) Linoleate-rich high-fat diet decreases mortality in hypertensive heart failure rats compared with lard and low-fat diets. Hypertension 52:549–555

Chiong M, Wang ZV, Pedrozo Z et al (2011) Cardiomyocyte death: mechanisms and translational implications. Cell Death Dis 2:e244

Chouchani ET, Pell VR, Gaude E et al (2014) Ischaemic accumulation of succinate controls reperfusion injury through mitochondrial ROS. Nature 515:431–435

Cole LK, Mejia EM, Vandel M et al (2016) Impaired cardiolipin biosynthesis prevents hepatic steatosis and diet-induced obesity. Diabetes 65:3289–3300

Cordero-Reyes AM, Gupte AA, Youker KA et al (2014) Freshly isolated mitochondria from failing human hearts exhibit preserved respiratory function. J Mol Cell Cardiol 68:98–105

Cung TT, Morel O, Cayla G et al (2015) Cyclosporine before PCI in patients with acute myocardial infarction. N Engl J Med 373:1021–1031

Czerski LW, Szweda PA, Szweda LI (2003) Dissociation of cytochrome c from the inner-mitochondrial membrane during cardiac ischemia. J Biol Chem 278:34499–34504

Dabkowski ER, Williamson CL, Bukowski VC et al (2009) Diabetic cardiomyopathy-associated dysfunction in spatially distinct mitochondrial subpopulations. Am J Physiol Heart Circ Physiol 296:H359–H369

Dabkowski ER, Baseler WA, Williamson CL et al (2010) Mitochondrial dysfunction in the type 2 diabetic heart is associated with alterations in spatially distinct mitochondrial proteomes. Am J Physiol Heart Circ Physiol 299:H529–H540

Dadabayev AR, Yin G, Latchoumycandane C et al (2014) Apolipoprotein A1 regulates coenzyme Q10 absorption, mitochondrial function, and infarct size in a mouse model of myocardial infarction. J Nutr 144:1030–1036

Dally S, Monceau V, Corvazier E et al (2009) Compartmentalized expression of three novel sarco/endoplasmic reticulum Ca2+ATPase 3 isoforms including the switch to ER stress, SERCA3f, in non-failing and failing human heart. Cell Calcium 45:144–154

Das A, Durrant D, Koka S et al (2014) Mammalian target of rapamycin (mTOR) inhibition with rapamycin improves cardiac function in type 2 diabetic mice: potential role of attenuated oxidative stress and altered contractile protein expression. J Biol Chem 289:4145–4160

De Marchi E, Bonora M, Giorgi C et al (2014) The mitochondrial permeability transition pore is a dispensable element for mitochondrial calcium efflux. Cell Calcium 56:1–13

Dolinsky VW, Cole LK, Sparagna GC et al (2016) Cardiac mitochondrial energy metabolism in heart failure: role of cardiolipin and sirtuins. Biochim Biophys Acta 1860:1544–1554

Dorn GW 2nd (2015) Mitochondrial dynamism and heart disease: changing shape and shaping change. EMBO Mol Med 7:865–877

Drose S, Stepanova A, Galkin A (2016) Ischemic A/D transition of mitochondrial complex I and its role in ROS generation. Biochim Biophys Acta 1857:946–957

Dym O, Pratt EA, Ho C et al (2000) The crystal structure of D-lactate dehydrogenase, a peripheral membrane respiratory enzyme. Proc Natl Acad Sci U S A 97:9413–9418

Elrod JW, Wong R, Mishra S et al (2010) Cyclophilin D controls mitochondrial pore-dependent Ca(2+) exchange, metabolic flexibility, and propensity for heart failure in mice. J Clin Invest 120:3680–3687

Fang JK, Prabu SK, Sepuri NB et al (2007) Site specific phosphorylation of cytochrome c oxidase subunits I, IVi1 and Vb in rabbit hearts subjected to ischemia/reperfusion. FEBS Lett 581:1302–1310

Fannin SW, Lesnefsky EJ, Slabe TJ et al (1999) Aging selectively decreases oxidative capacity in rat heart interfibrillar mitochondria. Arch Biochem Biophys 372:399–407

Fillmore N, Mori J, Lopaschuk GD (2014) Mitochondrial fatty acid oxidation alterations in heart failure, ischaemic heart disease and diabetic cardiomyopathy. Br J Pharmacol 171:2080–2090

Finichiu PG, Larsen DS, Evans C et al (2015) A mitochondria-targeted derivative of ascorbate: MitoC. Free Radic Biol Med 89:668–678

Folmes CD, Sawicki G, Cadete VJ et al (2010) Novel O-palmitolylated beta-E1 subunit of pyruvate dehydrogenase is phosphorylated during ischemia/reperfusion injury. Proteome Sci 8:38

Fridolfsson HN, Kawaraguchi Y, Ali SS et al (2012) Mitochondria-localized caveolin in adaptation to cellular stress and injury. FASEB J 26:4637–4649

Galkin A, Moncada S (2007) S-nitrosation of mitochondrial complex I depends on its structural conformation. J Biol Chem 282:37448–37453

Galkin A, Abramov AY, Frakich N et al (2009) Lack of oxygen deactivates mitochondrial complex I: implications for ischemic injury? J Biol Chem 284:36055–36061

Galvez AS, Diwan A, Odley AM et al (2007) Cardiomyocyte degeneration with calpain deficiency reveals a critical role in protein homeostasis. Circ Res 100:1071–1078

Gille L, Nohl H (2001) The ubiquinol/bc1 redox couple regulates mitochondrial oxygen radical formation. Arch Biochem Biophys 388:34–38

Goldenthal MJ (2016) Mitochondrial involvement in myocyte death and heart failure. Heart Fail Rev 21:137–155

Goldstein MA, Schroeter JP, Michael LH (1991) Role of the Z band in the mechanical properties of the heart. FASEB J 5:2167–2174

Gorenkova N, Robinson E, Grieve DJ et al (2013) Conformational change of mitochondrial complex I increases ROS sensitivity during ischemia. Antioxid Redox Signal 19:1459–1468

Gottlieb RA, Burleson KO, Kloner RA et al (1994) Reperfusion injury induces apoptosis in rabbit cardiomyocytes. J Clin Invest 94:1621–1628

Green DR, Reed JC (1998) Mitochondria and apoptosis. Science 281:1309–1312

Hafner AV, Dai J, Gomes AP et al (2010) Regulation of the mPTP by SIRT3-mediated deacetylation of CypD at lysine 166 suppresses age-related cardiac hypertrophy. Aging (Albany NY) 2:914–923
Halestrap AP (2009) What is the mitochondrial permeability transition pore? J Mol Cell Cardiol 46:821–831
Halestrap AP, Clarke SJ, Javadov SA (2004) Mitochondrial permeability transition pore opening during myocardial reperfusion--a target for cardioprotection. Cardiovasc Res 61:372–385
Hamilton DJ (2013) Mechanisms of disease: is mitochondrial function altered in heart failure? Methodist Debakey Cardiovasc J 9:44–48
Han D, Antunes F, Canali R et al (2003) Voltage-dependent anion channels control the release of the superoxide anion from mitochondria to cytosol. J Biol Chem 278:5557–5563
Hausenloy DJ, Yellon DM (2003) The mitochondrial permeability transition pore: its fundamental role in mediating cell death during ischaemia and reperfusion. J Mol Cell Cardiol 35:339–341
Hausenloy DJ, Yellon DM (2007) Reperfusion injury salvage kinase signalling: taking a RISK for cardioprotection. Heart Fail Rev 12:217–234
Hausenloy DJ, Tsang A, Yellon DM (2005) The reperfusion injury salvage kinase pathway: a common target for both ischemic preconditioning and postconditioning. Trends Cardiovasc Med 15:69–75
Heusch G, Musiolik J, Gedik N et al (2011) Mitochondrial STAT3 activation and cardioprotection by ischemic postconditioning in pigs with regional myocardial ischemia/reperfusion. Circ Res 109:1302–1308
Hollander JM, Thapa D, Shepherd DL (2014) Physiological and structural differences in spatially distinct subpopulations of cardiac mitochondria: influence of cardiac pathologies. Am J Physiol Heart Circ Physiol 307:H1–14
Hoppel CL, Tandler B, Parland W et al (1982) Hamster cardiomyopathy. A defect in oxidative phosphorylation in the cardiac interfibrillar mitochondria. J Biol Chem 257:1540–1548
Hoppel CL, Tandler B, Fujioka H et al (2009) Dynamic organization of mitochondria in human heart and in myocardial disease. Int J Biochem Cell Biol 41:1949–1956
Huang Q, Zhou HJ, Zhang H et al (2015) Thioredoxin-2 inhibits mitochondrial reactive oxygen species generation and apoptosis stress kinase-1 activity to maintain cardiac function. Circulation 131:1082–1097
Huang Y, Powers C, Moore V et al (2017) The PPAR pan-agonist bezafibrate ameliorates cardiomyopathy in a mouse model of Barth syndrome. Orphanet J Rare Dis 12:49
Hughes BG, Fan X, Cho WJ et al (2014) MMP-2 is localized to the mitochondria-associated membrane of the heart. Am J Physiol Heart Circ Physiol 306:H764–H770
Huttemann M, Lee I, Grossman LI et al (2012) Phosphorylation of mammalian cytochrome c and cytochrome c oxidase in the regulation of cell destiny: respiration, apoptosis, and human disease. Adv Exp Med Biol 748:237–264
Ide T, Tsutsui H, Kinugawa S et al (1999) Mitochondrial electron transport complex I is a potential source of oxygen free radicals in the failing myocardium. Circ Res 85:357–363
Inserte J, Barba I, Hernando V et al (2008) Effect of acidic reperfusion on prolongation of intracellular acidosis and myocardial salvage. Cardiovasc Res 77:782–790
Jarreta D, Orus J, Barrientos A et al (2000) Mitochondrial function in heart muscle from patients with idiopathic dilated cardiomyopathy. Cardiovasc Res 45:860–865
Jiang M, Zhang M, Howren M et al (2016) JPH-2 interacts with Cai-handling proteins and ion channels in dyads: contribution to premature ventricular contraction-induced cardiomyopathy. Heart Rhythm 13:743–752
Kagan VE, Borisenko GG, Tyurina YY et al (2004) Oxidative lipidomics of apoptosis: redox catalytic interactions of cytochrome c with cardiolipin and phosphatidylserine. Free Radic Biol Med 37:1963–1985
Kagan VE, Tyurin VA, Jiang J et al (2005) Cytochrome c acts as a cardiolipin oxygenase required for release of proapoptotic factors. Nat Chem Biol 1:223–232
Kagan VE, Bayir HA, Belikova NA et al (2009) Cytochrome c/cardiolipin relations in mitochondria: a kiss of death. Free Radic Biol Med 46:1439–1453
Karamanlidis G, Lee CF, Garcia-Menendez L et al (2013) Mitochondrial complex I deficiency increases protein acetylation and accelerates heart failure. Cell Metab 18:239–250

Kerner J, Hoppel C (2000) Fatty acid import into mitochondria. Biochim Biophys Acta 1486:1–17

Knowlton AA, Chen L, Malik ZA (2014) Heart failure and mitochondrial dysfunction: the role of mitochondrial fission/fusion abnormalities and new therapeutic strategies. J Cardiovasc Pharmacol 63:196–206

Koka S, Aluri HS, Xi L et al (2014) Chronic inhibition of phosphodiesterase 5 with tadalafil attenuates mitochondrial dysfunction in type 2 diabetic hearts: potential role of NO/SIRT1/PGC-1alpha signaling. Am J Physiol Heart Circ Physiol 306:H1558–H1568

Korytowski W, Basova LV, Pilat A et al (2011) Permeabilization of the mitochondrial outer membrane by Bax/truncated Bid (tBid) proteins as sensitized by cardiolipin hydroperoxide translocation: mechanistic implications for the intrinsic pathway of oxidative apoptosis. J Biol Chem 286:26334–26343

Kraljevic J, Marinovic J, Pravdic D et al (2013) Aerobic interval training attenuates remodelling and mitochondrial dysfunction in the post-infarction failing rat heart. Cardiovasc Res 99:55–64

Kranias EG, Hajjar RJ (2012) Modulation of cardiac contractility by the phospholamban/SERCA2a regulatome. Circ Res 110:1646–1660

Kubli DA, Zhang X, Lee Y et al (2013) Parkin protein deficiency exacerbates cardiac injury and reduces survival following myocardial infarction. J Biol Chem 288:915–926

Kukreja RC, Yin C, Salloum FN (2011) MicroRNAs: new players in cardiac injury and protection. Mol Pharmacol 80:558–564

Lee HG, Chen Q, Wolfram JA et al (2009) Cell cycle re-entry and mitochondrial defects in myc-mediated hypertrophic cardiomyopathy and heart failure. PLoS One 4:e7172

Lemieux H, Semsroth S, Antretter H et al (2011) Mitochondrial respiratory control and early defects of oxidative phosphorylation in the failing human heart. Int J Biochem Cell Biol 43:1729–1738

Lesnefsky EJ, Hoppel CL (2006) Oxidative phosphorylation and aging. Ageing Res Rev 5:402–433

Lesnefsky EJ, Gallo DS, Ye J et al (1994) Aging increases ischemia-reperfusion injury in the isolated, buffer-perfused heart. J Lab Clin Med 124:843–851

Lesnefsky EJ, Lundergan CF, Hodgson JM et al (1996) Increased left ventricular dysfunction in elderly patients despite successful thrombolysis: the GUSTO-I angiographic experience. J Am Coll Cardiol 28:331–337

Lesnefsky EJ, Tandler B, Ye J et al (1997) Myocardial ischemia decreases oxidative phosphorylation through cytochrome oxidase in subsarcolemmal mitochondria. Am J Phys 273:H1544–H1554

Lesnefsky EJ, Gudz TI, Migita CT et al (2001a) Ischemic injury to mitochondrial electron transport in the aging heart: damage to the iron-sulfur protein subunit of electron transport complex III. Arch Biochem Biophys 385:117–128

Lesnefsky EJ, Moghaddas S, Tandler B et al (2001b) Mitochondrial dysfunction in cardiac disease: ischemia-reperfusion, aging, and heart failure. J Mol Cell Cardiol 33:1065–1089

Lesnefsky EJ, Slabe TJ, Stoll MS et al (2001c) Myocardial ischemia selectively depletes cardiolipin in rabbit heart subsarcolemmal mitochondria. Am J Phys 280:H2770–H2778

Lesnefsky EJ, Chen Q, Moghaddas S et al (2004a) Blockade of electron transport during Ischemia protects cardiac mitochondria. J Biol Chem 279:47961–47967

Lesnefsky EJ, Chen Q, Slabe TJ et al (2004b) Ischemia, rather than reperfusion, inhibits respiration through cytochrome oxidase in the isolated, perfused rabbit heart: role of cardiolipin. Am J Physiol Heart Circ Physiol 287:H258–H267

Lesnefsky EJ, He D, Moghaddas S et al (2006) Reversal of mitochondrial defects before ischemia protects the aged heart. FASEB J 20:1543–1545

Lesnefsky EJ, Chen Q, Hoppel CL (2016) Mitochondrial metabolism in aging heart. Circ Res 118:1593–1611

Lesnefsky EJ, Chen Q, Tandler B et al (2017) Mitochondrial dysfunction and myocardial ischemia-reperfusion: implications for novel therapies. Annu Rev Pharmacol Toxicol 57:535–565

Li J, Liu X, Wang H et al (2012) Lysocardiolipin acyltransferase 1 (ALCAT1) controls mitochondrial DNA fidelity and biogenesis through modulation of MFN2 expression. Proc Natl Acad Sci U S A 109:6975–6980

Liang Q, Kobayashi S (2016) Mitochondrial quality control in the diabetic heart. J Mol Cell Cardiol 95:57–69

Liu X, Ye B, Miller S et al (2012) Ablation of ALCAT1 mitigates hypertrophic cardiomyopathy through effects on oxidative stress and mitophagy. Mol Cell Biol 32:4493–4504

Loeffen JL, Smeitink JA, Trijbels JM et al (2000) Isolated complex I deficiency in children: clinical, biochemical and genetic aspects. Hum Mutat 15:123–134

Lunkenheimer PP, Redmann K, Westermann P et al (2006) The myocardium and its fibrous matrix working in concert as a spatially netted mesh: a critical review of the purported tertiary structure of the ventricular mass. Eur J Cardiothorac Surg 29(Suppl 1):S41–S49

Luongo TS, Lambert JP, Yuan A et al (2015) The mitochondrial calcium uniporter matches energetic supply with cardiac workload during stress and modulates permeability transition. Cell Rep 12:23–34

Manning N, Albarran JW (2016) Low-dose intensive insulin therapy in patients with acute coronary syndrome accompanied by left ventricular failure: audit of two UK hospitals. J Clin Nurs 25:3001–3009

Marin-Garcia J, Goldenthal MJ, Moe GW (2001) Abnormal cardiac and skeletal muscle mitochondrial function in pacing-induced cardiac failure. Cardiovasc Res 52:103–110

Marin-Garcia J, Goldenthal MJ, Damle S et al (2009) Regional distribution of mitochondrial dysfunction and apoptotic remodeling in pacing-induced heart failure. J Card Fail 15:700–708

Mccully JD, Wakiyama H, Hsieh YJ et al (2004) Differential contribution of necrosis and apoptosis in myocardial ischemia-reperfusion injury. Am J Physiol Heart Circ Physiol 286:H1923–H1935

Mccutcheon LJ, Cory CR, Nowack L et al (1992) Respiratory chain defect of myocardial mitochondria in idiopathic dilated cardiomyopathy of Doberman pinscher dogs. Can J Physiol Pharmacol 70:1529–1533

Mentzer RM Jr, Wider J, Perry CN et al (2014) Reduction of infarct size by the therapeutic protein TAT-Ndi1 in vivo. J Cardiovasc Pharmacol Ther 19:315–320

Mileykovskaya E, Dowhan W (2014) Cardiolipin-dependent formation of mitochondrial respiratory supercomplexes. Chem Phys Lipids 179:42–48

Minamino T, Kitakaze M (2010) ER stress in cardiovascular disease. J Mol Cell Cardiol 48:1105–1110

Miwa S, Brand MD (2003) Mitochondrial matrix reactive oxygen species production is very sensitive to mild uncoupling. Biochem Soc Trans 31:1300–1301

Moghaddas S, Hoppel CL, Lesnefsky EJ (2003) Aging defect at the Qo site of complex III augments oxyradical production in rat heart interfibrillar mitochondria. Arch Biochem Biophys 414:59–66

Mohsin A, Chen Q, Quan N et al (2016) Cardioprotection during early reperfusion via complex I inhibition by metformin. Circulation 134:A19919. (abstract)

Morel O, Perret T, Delarche N et al (2012) Pharmacological approaches to reperfusion therapy. Cardiovasc Res 94:246–252

Mori J, Alrob OA, Wagg CS et al (2013) ANG II causes insulin resistance and induces cardiac metabolic switch and inefficiency: a critical role of PDK4. Am J Physiol Heart Circ Physiol 304:H1103–H1113

Murphy MP (2016) Understanding and preventing mitochondrial oxidative damage. Biochem Soc Trans 44:1219–1226

Murphy E, Steenbergen C (2008) Mechanisms underlying acute protection from cardiac ischemia-reperfusion injury. Physiol Rev 88:581–609

Murphy E, Steenbergen C (2011) What makes the mitochondria a killer? Can we condition them to be less destructive? Biochim Biophys Acta 1813:1302–1308

Nadtochiy SM, Burwell LS, Brookes PS (2007) Cardioprotection and mitochondrial S-nitrosation: effects of S-nitroso-2-mercaptopropionyl glycine (SNO-MPG) in cardiac ischemia-reperfusion injury. J Mol Cell Cardiol 42:812–825

Neubauer S (2007) The failing heart--an engine out of fuel. N Engl J Med 356:1140–1151

Nishida K, Yamaguchi O, Otsu K (2015) Degradation systems in heart failure. J Mol Cell Cardiol 84:212–222

Ockaili RA, Bhargava P, Kukreja RC (2001) Chemical preconditioning with 3-nitropropionic acid in hearts: role of mitochondrial K(ATP) channel. Am J Physiol Heart Circ Physiol 280:H2406–H2411

Ohnishi ST, Ohnishi T, Muranaka S et al (2005) A possible site of superoxide generation in the complex I segment of rat heart mitochondria. J Bioenerg Biomembr 37:1–15

Ozaki T, Tomita H, Tamai M et al (2007) Characteristics of mitochondrial calpains. J Biochem 142:365–376

Paillard M, Tubbs E, Thiebaut PA et al (2013) Depressing mitochondria-reticulum interactions protects cardiomyocytes from lethal hypoxia-reoxygenation injury. Circulation 128:1555–1565

Palmer JW, Tandler B, Hoppel CL (1977) Biochemical properties of subsarcolemmal and interfibrillar mitochondria isolated from rat cardiac muscle. J Biol Chem 252:8731–8739

Palmer JW, Tandler B, Hoppel CL (1985) Biochemical differences between subsarcolemmal and interfibrillar mitochondria from rat cardiac muscle: effects of procedural manipulations. Arch Biochem Biophys 236:691–702

Pan S, Ryu SY, Sheu SS (2011) Distinctive characteristics and functions of multiple mitochondrial Ca2+ influx mechanisms. Sci China Life Sci 54:763–769

Panchal AR, Stanley WC, Kerner J et al (1998) Beta-receptor blockade decreases carnitine palmitoyl transferase I activity in dogs with heart failure. J Card Fail 4:121–126

Papanicolaou KN, Khairallah RJ, Ngoh GA et al (2011) Mitofusin-2 maintains mitochondrial structure and contributes to stress-induced permeability transition in cardiac myocytes. Mol Cell Biol 31:1309–1328

Park SY, Trinity JD, Gifford JR et al (2016) Mitochondrial function in heart failure: The impact of ischemic and non-ischemic etiology. Int J Cardiol 220:711–717

Pepe S, Mentzer RM Jr, Gottlieb RA (2014) Cell-permeable protein therapy for complex I dysfunction. J Bioenerg Biomembr 46:337–345

Phoon CK, Acehan D, Schlame M et al (2012) Tafazzin knockdown in mice leads to a developmental cardiomyopathy with early diastolic dysfunction preceding myocardial noncompaction. J Am Heart Assoc 1:ah3-e000455

Poncelas M, Inserte J, Aluja D et al (2017) Delayed, oral pharmacological inhibition of calpains attenuates adverse post-infarction remodeling. Cardiovasc Res 113:950–961

Riva A, Tandler B, Loffredo F et al (2005) Structural differences in two biochemically defined populations of cardiac mitochondria. Am J Physiol Heart Circ Physiol 289:H868–H872

Riva A, Tandler B, Lesnefsky EJ et al (2006) Structure of cristae in cardiac mitochondria of aged rat. Mech Ageing Dev 127:917–921

Rosca MG, Vazquez EJ, Kerner J et al (2008) Cardiac mitochondria in heart failure: decrease in respirasomes and oxidative phosphorylation. Cardiovasc Res 80:30–39

Ross T, Szczepanek K, Bowler E et al (2013) Reverse electron flow-mediated ROS generation in ischemia-damaged mitochondria: role of complex I inhibition vs. depolarization of inner mitochondrial membrane. Biochim Biophys Acta 1830:4537–4542

Sabbah HN, Sharov V, Riddle JM et al (1992) Mitochondrial abnormalities in myocardium of dogs with chronic heart failure. J Mol Cell Cardiol 24:1333–1347

Sanbe A, Tanonaka K, Kobayasi R et al (1995) Effects of long-term therapy with ACE inhibitors, captopril, enalapril and trandolapril, on myocardial energy metabolism in rats with heart failure following myocardial infarction. J Mol Cell Cardiol 27:2209–2222

Santulli G, Xie W, Reiken SR et al (2015) Mitochondrial calcium overload is a key determinant in heart failure. Proc Natl Acad Sci U S A 112:11389–11394

Scheubel RJ, Tostlebe M, Simm A et al (2002) Dysfunction of mitochondrial respiratory chain complex I in human failing myocardium is not due to disturbed mitochondrial gene expression. J Am Coll Cardiol 40:2174–2181

Schlame M (2013) Cardiolipin remodeling and the function of tafazzin. Biochim Biophys Acta 1831:582–588

Schwarzer M, Schrepper A, Amorim PA et al (2013) Pressure overload differentially affects respiratory capacity in interfibrillar and subsarcolemmal mitochondria. Am J Physiol Heart Circ Physiol 304:H529–H537

Schwarzer M, Osterholt M, Lunkenbein A et al (2014) Mitochondrial reactive oxygen species production and respiratory complex activity in rats with pressure overload-induced heart failure. J Physiol 592:3767–3782

Sciarretta S, Zhai P, Shao D et al (2013) Activation of NADPH oxidase 4 in the endoplasmic reticulum promotes cardiomyocyte autophagy and survival during energy stress through the

protein kinase RNA-activated-like endoplasmic reticulum kinase/eukaryotic initiation factor 2alpha/activating transcription factor 4 pathway. Circ Res 113:1253–1264
Sharma K, Kass DA (2014) Heart failure with preserved ejection fraction: mechanisms, clinical features, and therapies. Circ Res 115:79–96
Sharov VG, Goussev A, Lesch M et al (1998) Abnormal mitochondrial function in myocardium of dogs with chronic heart failure. J Mol Cell Cardiol 30:1757–1762
Sharov VG, Todor AV, Silverman N et al (2000) Abnormal mitochondrial respiration in failed human myocardium. J Mol Cell Cardiol 32:2361–2367
Shi Y, Pulliam DA, Liu Y et al (2013) Reduced mitochondrial ROS, enhanced antioxidant defense, and distinct age-related changes in oxidative damage in muscles of long-lived Peromyscus leucopus. Am J Physiol Regul Integr Comp Physiol 304:R343–R355
Shiva S, Sack MN, Greer JJ et al (2007) Nitrite augments tolerance to ischemia/reperfusion injury via the modulation of mitochondrial electron transfer. J Exp Med 204:2089–2102
Singh SS, Kang PM (2011) Mechanisms and inhibitors of apoptosis in cardiovascular diseases. Curr Pharm Des 17:1783–1793
Sparagna GC, Lesnefsky EJ (2009) Cardiolipin remodeling in the heart. J Cardiovasc Pharmacol 53:290–301
Sparagna GC, Chicco AJ, Murphy RC et al (2007) Loss of cardiac tetralinoleoyl cardiolipin in human and experimental heart failure. J Lipid Res 48:1559–1570
Stewart S, Lesnefsky EJ, Chen Q (2009) Reversible blockade of electron transport with amobarbital at the onset of reperfusion attenuates cardiac injury. Transl Res 153:224–231
St-Pierre J, Buckingham JA, Roebuck SJ et al (2002) Topology of superoxide production from different sites in the mitochondrial electron transport chain. J Biol Chem 277:44784–44790
Sugden MC, Holness MJ (2006) Mechanisms underlying regulation of the expression and activities of the mammalian pyruvate dehydrogenase kinases. Arch Physiol Biochem 112:139–149
Sun J, Trumpower BL (2003) Superoxide anion generation by the cytochrome bc1 complex. Arch Biochem Biophys 419:198–206
Szczepanek K, Chen Q, Derecka M et al (2011) Mitochondrial-targeted Signal transducer and activator of transcription 3 (STAT3) protects against ischemia-induced changes in the electron transport chain and the generation of reactive oxygen species. J Biol Chem 286:29610–29620
Szczepanek K, Chen Q, Larner AC et al (2012a) Cytoprotection by the modulation of mitochondrial electron transport chain: the emerging role of mitochondrial STAT3. Mitochondrion 12:180–189
Szczepanek K, Lesnefsky EJ, Larner AC (2012b) Multi-tasking: nuclear transcription factors with novel roles in the mitochondria. Trends Cell Biol 22:429–437
Takaki M, Zhao DD, Zhao LY et al (1995) Suppression of myocardial mitochondrial respiratory function in acute failing hearts made by a short-term Ca2+ free, high Ca2+ coronary perfusion. J Mol Cell Cardiol 27:2009–2013
Taneike M, Mizote I, Morita T et al (2011) Calpain protects the heart from hemodynamic stress. J Biol Chem 286:32170–32177
Thompson J, Hu Y, Lesnefsky EJ et al (2016) Activation of mitochondrial calpain and increased cardiac injury: beyond AIF release. Am J Physiol Heart Circ Physiol 310:H376–H384
Tobiume K, Matsuzawa A, Takahashi T et al (2001) ASK1 is required for sustained activations of JNK/p38 MAP kinases and apoptosis. EMBO Rep 2:222–228
Toldo S, Breckenridge DG, Mezzaroma E et al (2012) Inhibition of apoptosis signal-regulating kinase 1 reduces myocardial ischemia-reperfusion injury in the mouse. J Am Heart Assoc 1:e002360
Toth A, Nickson P, Mandl A et al (2007) Endoplasmic reticulum stress as a novel therapeutic target in heart diseases. Cardiovasc Hematol Disord Drug Targets 7:205–218
Tsuji T, Ohga Y, Yoshikawa Y et al (2001) Rat cardiac contractile dysfunction induced by Ca2+ overload: possible link to the proteolysis of alpha-fodrin. Am J Physiol Heart Circ Physiol 281:H1286–H1294
Van Der Vusse GJ, Van Bilsen M, Glatz JF et al (2002) Critical steps in cellular fatty acid uptake and utilization. Mol Cell Biochem 239:9–15

Veitch K, Hombroeckx A, Caucheteux D et al (1992) Global ischaemia induces a biphasic response of the mitochondrial respiratory chain. Anoxic pre-perfusion protects against ischaemic damage. Biochem J 281:709–715

Weiss JN, Korge P, Honda HM et al (2003) Role of the mitochondrial permeability transition in myocardial disease. Circ Res 93:292–301

Xu A, Szczepanek K, Maceyka MW et al (2014) Transient complex I inhibition at the onset of reperfusion by extracellular acidification decreases cardiac injury. Am J Physiol Cell Physiol 306:C1142–C1153

Yu SW, Wang H, Poitras MF et al (2002) Mediation of poly(ADP-ribose) polymerase-1-dependent cell death by apoptosis-inducing factor. Science 297:259–263

Zachman DK, Chicco AJ, Mccune SA et al (2010) The role of calcium-independent phospholipase A2 in cardiolipin remodeling in the spontaneously hypertensive heart failure rat heart. J Lipid Res 51:525–534

Zhang Y, Ren J (2011) Thapsigargin triggers cardiac contractile dysfunction via NADPH oxidase-mediated mitochondrial dysfunction: role of Akt dephosphorylation. Free Radic Biol Med 51:2172–2184

Drug-Induced Mitochondrial Toxicity

Julie Massart, Annie Borgne-Sanchez, and Bernard Fromenty

Abstract Mitochondrial dysfunction can be a major mechanism whereby different drugs can induce adverse effects affecting different tissues such as liver, heart and skeletal muscle. In the most severe cases, drug-induced mitochondrial dysfunction can require a hospitalization, or lead to the death of the patient. Moreover, these adverse effects can lead to the withdrawal of drugs from the market, or earlier during clinical trials. Drugs can induce mitochondrial dysfunction by different mechanisms including inhibition of fatty acid oxidation, impairment of oxidative phosphorylation and respiratory chain activity as well as alteration of the integrity of the mitochondrial membranes. Some drugs also impair mitochondrial function via the production of reactive oxygen species and the generation of reactive metabolites, which can covalently bind to key mitochondrial proteins. The present chapter focuses on different drugs for which enough clinical and experimental evidence indicates the potential role of mitochondrial dysfunction in the pathogenesis of adverse effects such as liver injury, myopathy and cardiotoxicity: acetaminophen, amiodarone, doxorubicin, nucleoside reverse transcriptase inhibitors (e.g. stavudine, zidovudine, didanosine), statins (e.g. atorvastatin, cerivastatin, simvastatin) and valproic acid. Notably, these drugs epitomize the diversity of the mechanisms whereby xenobiotics can induce mitochondrial dysfunction and also the variety of the targeted tissues. Other drugs affecting mitochondrial function by similar mechanisms are discussed more briefly in the present chapter. Because drug-induced mitochondrial dysfunction

J. Massart
Department of Molecular Medicine and Surgery, Karolinska University Hospital, Karolinska Institutet, Stockholm, Sweden

A. Borgne-Sanchez
Hôpital Robert Debré, MITOLOGICS S.A.S., Paris, France

B. Fromenty (✉)
Univ Rennes, INSERM, INRA, Institut NUMECAN (Nutrition Metabolisms and Cancer) UMR_A 1341, UMR_S 1241, Rennes, France

INSERM UMR 1241 (NuMeCan),
2 avenue du Pr Léon Bernard, Rennes 35033 Cedex, France
e-mail: bernard.fromenty@inserm.fr

P. J. Oliveira (ed.), *Mitochondrial Biology and Experimental Therapeutics*,
https://doi.org/10.1007/978-3-319-73344-9_13

and related adverse events are major issues for public health and pharmaceutical companies, mitochondrial liability should be systematically investigated during preclinical safety studies.

Keywords Adverse effect · Drug · Heart · Liver · Mitochondria · Muscle · Myopathy · Respiratory chain · Steatosis · Toxicity · Oxidative stress

Abbreviations

APAP	Acetaminophen
AZT	Zidovudine
CPT1	Carnitine palmitoyltransferase 1
CYP	Cytochrome P450
d4T	Stavudine
ER	Endoplasmic reticulum
FA	Fatty acid
FIAU	Fialuridine
GSH	Glutathione
LCFA	Long-chain fatty acid
MCFA	Medium-chain fatty acid
MnSOD	Manganese superoxide dismutase
MPT	Mitochondrial permeability transition
MRC	Mitochondrial respiratory chain
mtDNA	Mitochondrial DNA
mtFAO	Mitochondrial fatty acid oxidation
NAPQI	N-Acetyl-*p*-benzoquinone imine
NRTI	Nucleoside reverse transcriptase inhibitors
OXPHOS	Oxidative phosphorylation
POLG	DNA polymerase γ
ROS	Reactive oxygen species
SCFA	Short-chain fatty acid
TCA	Tricarboxylic acid
VPA	Valproic acid

1 Introduction

Many drugs are able to induce adverse effects affecting different tissues such as the liver, heart, kidney, lung, skeletal muscles and peripheral nerves (Begriche et al. 2011; Hohenegger 2012; Marrer and Dieterle 2010; Miltenburg and Boogerd 2014; Tocchetti et al. 2013). For instance, it is estimated that more than 350 drugs of our modern pharmacopeia could induce liver injury such as hepatic cytolysis, steatosis and cholestasis (Björnsson and Hoofnagle 2016; Weng et al. 2015). Notably, a single drug can damage several tissues and organs. An interesting example is acetaminophen (APAP), which is able to damage not only liver but also kidney, heart, lung, pancreas and cochleae (Baudouin et al. 1995; Igarashi et al. 2009; McGill et al. 2016; Price et al. 1991; Vliegenthart et al. 2015).

Although most cases of drug-induced adverse effects present a favorable clinical course, some of them can be severe and require a hospitalization, or even can cause the death of some patients. Thus, drug-induced toxicity is an important issue for public health and the wellbeing of the patients. These adverse effects can also be a major concern for the pharmaceutical companies because significant toxicity can lead to the interruption of clinical trials, or the withdrawal of the incriminated drugs from the market (Elangbam 2010; Labbe et al. 2008).

Among the mechanisms whereby drugs can damage different tissues and organs, mitochondrial dysfunction is deemed to play a primary role (Begriche et al. 2011; John and Herzenberg 2009; Miltenburg and Boogerd 2014; Nadanaciva and Will 2011; Pereira et al. 2011). This is because these organelles are mandatory for energy output and the metabolism of different key endogenous substrates such as fatty acids and amino acids (Kastaniotis et al. 2017; Wallace et al. 2010). However, it should be kept in mind that other mechanisms of toxicity can exist including endoplasmic reticulum (ER) stress, lysosomal permeabilization and nuclear DNA damage (Foufelle and Fromenty 2016; Hu et al. 2016; Pessayre et al. 2010). Importantly, some drugs appear to be toxic by way of several mechanisms, for instance by inducing both mitochondrial dysfunction and ER stress (Foufelle and Fromenty 2016).

It is noteworthy that drug-induced mitochondrial dysfunction can be induced by the parental drug and/or by one or several toxic metabolites generated by cellular enzymes such as cytochromes P450 (CYPs) (Begriche et al. 2011; Fromenty and Pessayre 1995; Pessayre et al. 2010). CYPs and other xenobiotic metabolizing enzymes (XMEs) are mainly expressed in the liver but also in other tissues including the gastrointestinal (GI) tract, lung, kidney, heart, brain and white adipose tissue (Dutheil et al. 2009; Ellero et al. 2010; Knights et al. 2013; Ravindranath and Strobel 2013; Thelen and Dressman 2009). Hence, the generation of toxic metabolites can occur in liver and extra-hepatic tissues (Ding and Kaminsky 2003; Gu et al. 2005).

In the present chapter, we first recall the major roles of mitochondria in fuel oxidation, energy homeostasis, reactive oxygen species (ROS) production and cell death. Then, we describe the main biochemical, histological and clinical manifestations arising from the different mechanisms of drug-induced mitochondrial dysfunction. Finally, we provide information on different drugs for which enough data indicate the potential role of mitochondrial dysfunction in the etiology of adverse effects.

2 Structure and Physiological Roles of Mitochondria

Readers are invited to peruse different chapters of this book in order to get thorough information concerning mitochondrial structure and function. Hence, the paragraphs below are just giving the essential information in order to comprehend how drugs can induce mitochondrial dysfunction and the main consequences of such deleterious effect.

2.1 Structure and Main Components of Mitochondria

Mitochondria are intracellular organelles with two membranes (namely the outer and the inner membrane) that surround the matrix (Fig. 1). This compartment contains numerous enzymes involved in different key oxidative pathways such as mitochondrial fatty acid oxidation (mtFAO) and pyruvate oxidation via the tricarboxylic acid (TCA) cycle (Bénit et al. 2014; Houten et al. 2016). The matrix also contains the

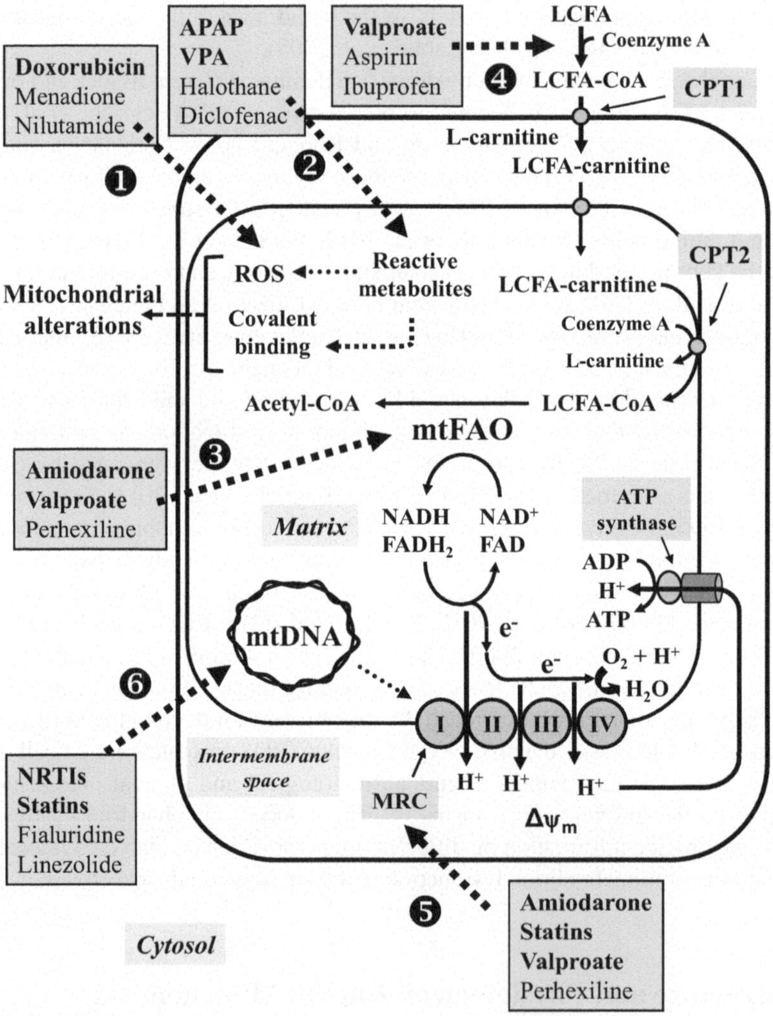

Fig. 1 Metabolism and energy production in mitochondria, and main drug targets. The entry of long-chain fatty acids (LCFAs) within mitochondria requires a specific shuttle system involving four steps. First, LCFAs are activated into LCFA-coenzyme A (acyl-CoA) thioesters by long-chain acyl-CoA synthetases (not shown). Second, the long-chain acyl-CoA thioesters are converted into

mitochondrial genome and all the components (e.g. enzymes and transcription factors) mandatory for its replication, transcription, translation and repair (Taanman 1999; Wallace et al. 2010). Mitochondrial DNA (mtDNA) is a 16.6 kb circular genome present within each mitochondrion in several copies and encoding 13 polypeptides of the mitochondrial respiratory chain (MRC) (Schon and Fromenty 2016; Wallace et al. 2010). These polypeptides are then inserted in the inner membrane within the MRC complexes I, III, IV (cytochrome *c* oxidase) and V (ATP synthase), along with dozens of nuclear DNA-encoded proteins. Permanent mtDNA replication by the DNA polymerase γ allows to keep constant cellular mtDNA levels despite continuous degradation of the most damaged and/or dysfunctional mitochondria. A strong reduction of the mtDNA copy number can lead to OXPHOS deficiency and severe diseases (Schon and Fromenty 2016; Wallace et al. 2010).

2.2 *Oxidation of Substrates*

Mitochondria are able to oxidize many substrates including amino acids, pyruvate and fatty acids (Rui 2014; Wallace et al. 2010). Pyruvate, which is provided by glycolysis, is oxidized by the TCA cycle after its transformation into acetyl-CoA by

Fig. 1 (continued) acyl-carnitine derivatives by carnitine palmitoyltransferase-1 (CPT1) in the outer mitochondrial membrane. Third, the acyl-carnitine derivatives are translocated across the inner mitochondrial membrane into the mitochondrial matrix by carnitine-acylcarnitine translocase. Finally, carnitine palmitoyltransferase-2 (CPT2), located in the inner mitochondrial membrane, transfers the fatty acyl moieties from carnitine back to coenzyme A. LCFA-CoA thioesters are then oxidized into acetyl-CoA molecules via the mitochondrial fatty acid β-oxidation (mtFAO) process. mtFAO and the tricarboxylic acid (TCA) cycle (not shown) generate NADH and $FADH_2$, which transfer their electrons (e^-) to the mitochondrial respiratory chain (MRC), thus regenerating NAD^+ and FAD used for other β-oxidation (or TCA) cycles. Within the MRC, electrons are sequentially transferred to different polypeptide complexes (numbered from I to IV) embedded within the inner membrane. The final transfer of the electrons to oxygen takes place at the level of complex IV, also referred to as cytochrome *c* oxidase. The mitochondrial DNA (mtDNA) encodes 13 polypeptides, which are embedded within complexes I, III, IV and V. The flow of electrons within the MRC is coupled with the extrusion of protons (H^+) from the mitochondrial matrix to the intermembrane space, which creates the mitochondrial transmembrane potential, $\Delta\psi_m$. When energy is needed, these protons re-enter the matrix through ATP synthase (also referred to as complex V), thus liberating energy that is used to phosphorylate ADP into ATP. Drugs can impair mitochondrial function through different mechanisms. (1) By favoring the production of reactive oxygen species (ROS), which subsequently lead to the oxidative alterations of key mitochondrial components such as MRC polypeptides and mtDNA; ROS overproduction can also trigger the opening of the mitochondrial permeability transition (MPT) pores (not shown). (2) By generating toxic reactive metabolites that can favor ROS overproduction and bind covalently to key mitochondrial proteins. (3) By directly altering the mtFAO pathway via the inhibition of β-oxidation enzyme(s) including CPT1. (4) By indirectly impairing mtFAO through the sequestration of the mtFAO cofactor coenzyme A. (5) By directly impairing MRC activity, which can secondarily inhibit mtFAO and the TCA cycle and enhance ROS formation (not shown). (6) By impairing mtDNA replication, or translation. Drugs in bold characters are discussed in detail in this chapter, while drugs in normal characters are only briefly mentioned in the text the mitochondrial enzyme pyruvate dehydrogenase. Succinate dehydrogenase, a FAD-dependent enzyme of the TCA cycle, is one of the five complexes of the MRC and is referred to as complex II. The TCA cycle also includes three different NAD^+-dependent dehydrogenases. Fatty acids (FAs) are oxidized by the β-oxidation process (Fig. 1), a key metabolic pathway mandatory for the preservation of normal energy output, especially during fasting (Houten and Wanders 2010; Spiekerkoetter and Wood 2010). Notably, mitochondria are able to oxidize fatty acids of different lengths into ketone bodies, mainly acetoacetate and β-hydroxybutyrate. Whereas short-chain FAs (SCFAs) and medium-chain FAs (MCFAs) freely enter mitochondria, the entry of long-chain FAs (LCFAs) into mitochondria requires carnitine palmitoyltransferase 1 (CPT1) and 2 (CPT2) (Fromenty and Pessayre 1995; Houten et al. 2016). During their β-oxidation, SCFAs, MCFAs and LCFAs undergo four sequential reactions leading to the release of one acetyl-CoA molecule and a shortened FA, which can be further oxidized by other mtFAO cycles. Two of these reactions are catalyzed by different FAD-dependent and NAD^+-dependent dehydrogenases that have specific activities for SCFAs, MCFAs or LCFAs (Fromenty and Pessayre 1995; Houten et al. 2016).

2.3 *Oxidative Phosphorylation and ATP Production*

Mitochondrial oxidation of substrates such as pyruvate and fatty acids constantly generate $FADH_2$ and NADH. These reduced cofactors subsequently transfer their electrons to the MRC, thus regenerating the NAD^+ and FAD necessary for other cycles of fuel oxidation (Fig. 1) (Begriche et al. 2013; Wallace et al. 2010). The electrons provided by NADH or $FADH_2$ migrate all the way along the respiratory chain, up to cytochrome *c* oxidase (COX), where they safely react with oxygen and protons to form water. Electron transfer across MRC complexes I, III and IV is coupled with the extrusion of protons from the mitochondrial matrix into the intermembrane space of mitochondria, thus creating a large electrochemical potential ($\Delta\psi$) across the inner membrane. When ADP is high, protons reenter the matrix through the F_0 portion of ATP synthase, causing the conversion of ADP into ATP by the F_1 portion of this enzyme. The biochemical process linking substrate oxidation to ATP generation is referred to as oxidative phosphorylation (OXPHOS).

2.4 *Physiological Production of Mitochondrial ROS*

During OXPHOS, a small fraction of electrons going through the MRC complexes I and III can leak from these complexes and react with oxygen to form the superoxide anion radical ($O_2^{\cdot-}$) (Begriche et al. 2013; Wallace et al. 2010). However, this radical is mostly dismutated by manganese superoxide dismutase (MnSOD, also referred to as SOD2) into hydrogen peroxide (H_2O_2), which is more stable compared to the superoxide anion radical. Some enzymes of the β-oxidation pathway could also produce significant amount of hydrogen peroxide within mitochondria (Kakimoto et al. 2015; Seifert et al. 2010). Mitochondrial hydrogen peroxide can be subsequently detoxified into water by peroxiredoxins and glutathione peroxidases, which necessitate reduced glutathione (GSH) as cofactor (Pessayre et al. 2010; Wallace et al. 2010). Hence, mitochondrial GSH plays a key role in order to avoid the excessive accumulation of hydrogen peroxide and subsequent oxidative stress (Mari et al. 2013). There is now mounting evidence that the physiological production of mitochondrial ROS serves as key signaling molecules (Diebold and Chandel 2016; Imhoff and Hansen 2009).

3 Main Consequences of Mitochondrial Dysfunction and Damage

3.1 Mitochondrial Membrane Permeabilization and Cell Death

In some pathophysiological situations, mitochondrial membranes can lose their structural and functional integrity. In the field of toxicology, an important mode of mitochondrial membrane permeabilization is via the opening of the mitochondrial permeability transition (MPT) pores. MPT pore opening is leading to water accumulation in the matrix that causes mitochondrial swelling and the rupture of the outer membrane (Marroquin et al. 2014; Pessayre et al. 2010). Cyclophilin D and mitochondrial ATP synthase c subunit have been shown to be key components of the MPT pores, although other proteins might be involved (Bernardi 2013; Bonora et al. 2015; Pessayre et al. 2010). Notably, the immunosuppressive drug cyclosporin A is able to reduce the opening probability of the MPT pores by specifically binding to cyclophilin D. It is also noteworthy that there are significant tissue-specific differences in MPT pore properties (Berman et al. 2000; Halestrap and Davidson 1990).

MPT pore opening can be triggered or favored by numerous endogenous factors such as calcium, inorganic phosphate and ROS as well as xenobiotics including atractyloside, *tert*-butylhydroperoxide and different drugs (Biasutto et al. 2016; Di Lisa et al. 2007; Izzo et al. 2016). Whatever the triggering factor, MPT pore opening can induce apoptosis or necrosis, depending on the number of damaged mitochondria (Begriche et al. 2011; Pessayre et al. 2010). Finally, it is noteworthy that other modes of mitochondrial membrane permeabilization not involving MPT pores have been described (Begriche et al. 2011; Belosludtsev et al. 2009; Lei et al. 2006). Such mechanisms of mitochondrial membrane permeabilization are thus insensitive to cyclosporin A (Marroquin et al. 2014). Nonetheless, MPT-independent mitochondrial membrane permeabilization can eventually induce cell death (Begriche et al. 2011; Belosludtsev et al. 2009; Pessayre et al. 2010).

3.2 Consequences of MRC Inhibition

MRC inhibition can induce significant ATP depletion, thus leading to cell death (Begriche et al. 2011; Pessayre et al. 2010). In liver, ATP shortage can induce cytolytic hepatitis that encompasses a large spectrum of liver injury of different severity. While the mildest forms of cytolytic hepatitis are characterized by an isolated increase in plasma alanine aminotransferase (ALT) and aspartate aminotransferase (AST), the most severe cases can be associated with fulminant hepatitis and severe hepatic dysfunction.

Another important consequence of MRC inhibition is the secondary impairment of mitochondrial β-oxidation and TCA cycle. Indeed, the lack of NADH and $FADH_2$ oxidation strongly reduces FAD and NAD^+ levels, thus impeding the activity of the different FAD-dependent and NAD^+-dependent dehydrogenases of the β-oxidation and TCA cycle pathways (Fromenty and Pessayre 1995; Massart et al. 2013). Impairment of the TCA cycle can lead to hyperlactatemia and lactic acidosis because the conversion of unmetabolized pyruvate to lactate by lactate dehydrogenase is favored by NADH accumulation (Igoudjil et al. 2006; Margolis et al. 2014). Impairment of mtFAO can induce lipid accumulation, as discussed later on.

Finally, a significant inhibition of MRC can also favor ROS overproduction (Li et al. 2003; Pessayre et al. 2010). Indeed, when the flow of electrons is blocked at some point downstream to complexes I and III, these MRC complexes become overly reduced, thus leading to higher leakage of electrons and superoxide anion overproduction (Begriche et al. 2011; Pessayre et al. 2010). Drug-induced mitochondrial ROS overproduction can favor oxidative stress, lipid peroxidation, and different types of mtDNA oxidative damage (Igoudjil et al. 2006; Lewis et al. 2001; Schon and Fromenty 2016). In some tissues such as lung and liver, ROS overproduction and oxidative stress can favor inflammation and fibrosis (Massart et al. 2013; Pessayre et al. 2010; Yue and Yao 2016).

3.3 Consequences of Mitochondrial β-Oxidation Inhibition

Because mtFAO provides most of the ATP required for cell homeostasis and function, any significant impairment of this metabolic pathway can lead to energy shortage, and cell demise (Begriche et al. 2011; Fromenty and Pessayre 1995). Inhibition of mtFAO also leads to lipid accumulation, which is referred to as steatosis. Although most of the accumulated lipids are triglycerides, other lipid derivatives can be present in a significant amount such as free fatty acids, dicarboxylic acids and acyl-carnitine intermediates (Fromenty and Pessayre 1995; Labbe et al. 2008). Accumulation of free fatty acids and dicarboxylic acids could reinforce drug-induced mitochondrial dysfunction because these derivatives are able to impair key biochemical processes such as OXPHOS and TCA cycle (Begriche et al. 2011; Fromenty and Pessayre 1995). In liver, a severe inhibition of mtFAO can also alter the gluconeogenesis pathway, thus leading to hypoglycemia (Fromenty and Pessayre 1995).

4 Examples of Drugs Inducing Adverse Effects via Mitochondrial Dysfunction

Numerous drugs have been reported to impair mitochondrial function via different mechanisms (Finsterer and Ohnsorge 2013; Fromenty and Pessayre 1995; Massart et al. 2013; Nadanaciva et al. 2012; Porceddu et al. 2012; Varga et al. 2015). Drug-induced mitochondrial dysfunction is often responsible for liver, muscle and heart diseases but other tissues can be damaged such as lung and kidney (Nadanaciva and Will 2011; Porceddu et al. 2012; Sardao et al. 2008; Varga et al. 2015). Because of space limitation, this chapter will only deal with drugs for which different experimental and/or clinical investigations strongly support a major role of primary mitochondrial dysfunction in the occurrence of tissue toxicity. Furthermore, in order to highlight the complexity of the topic, we have selected drugs able to induce mitochondrial dysfunction in different tissues by distinct mechanisms. Other drugs are briefly mentioned because of similar mechanism(s) of toxicity. Figure 1 summarizes the different mitochondrial targets of the drugs mentioned in this chapter.

4.1 Acetaminophen

Acetaminophen (APAP) is a popular drug for the management of pain and fever. Although APAP is usually deemed as a safe drug, APAP intoxication after an overdose can lead to massive hepatocellular necrosis and acute liver failure (Craig et al. 2011; Michaut et al. 2014). APAP overdose has also been reported to induce acute kidney injury, cardiotoxicity, pancreatitis, and ototoxicity (Blakely and McDonald 1995; Jones and Prescott 1997; Kato et al. 2014; McGill et al. 2016).

In tissues such as liver and kidney, APAP is mainly metabolized into the nontoxic glucuronide and sulfate conjugates. However, a small amount of APAP is oxidized by cytochromes P450 2E1 (CYP2E1) and 3A4 to the reactive metabolite N-acetyl-*p*-benzoquinone imine (NAPQI) (Aubert et al. 2012; Michaut et al. 2016). When APAP is taken at the recommended dosage, NAPQI is normally detoxified by cytosolic and mitochondrial GSH. In contrast, after APAP overdose, high levels of NAPQI can induce cell death. Indeed, once GSH is deeply depleted and no longer available for NAPQI detoxification, this reactive metabolite binds to different proteins, especially within the mitochondria (McGill and Jaeschke 2013; Michaut et al. 2014). This is followed by profound MRC impairment and ATP depletion, ROS overproduction, c-jun N-terminal kinase (JNK) activation, and massive cellular necrosis (Das et al. 2010; Michaut et al. 2014; Jaeschke et al. 2012). Among the different components of the MRC, complex II (succinate dehydrogenase) seems to be very sensitive to the inhibitory effect of NAPQI (Burcham and Harman 1991; Lee et al. 2015).

Some data strongly suggest that APAP can also impair the mtFAO pathway. First, APAP can induce hepatic steatosis in intoxicated patients and rodents (Aubert et al. 2012; Biour et al. 2004). Second, APAP intoxication in humans and mice induces the accumulation of circulating acyl-carnitine derivatives (Bhattacharyya et al. 2014;

McGill et al. 2014). Although mtFAO impairment could be secondary to APAP-induced inhibition of MRC, it would be interesting to determine whether APAP could also inactivate one or several enzymes involved in the mtFAO pathway.

Different investigations performed in rodent and human livers consistently showed that CYP2E1 is located not only in the ER compartment but also within mitochondria (Bansal et al. 2013; Knockaert et al. 2011a). The presence of CYP2E1 within mitochondria might play a significant role in APAP-induced mitochondrial dysfunction and liver injury because intra-mitochondrial CYP2E1 is expected to generate NAPQI close to the MRC complexes and other possible key NAPQI mitochondrial targets. In keeping with this assumption, investigations performed in COS-7 cells expressing CYP2E1 only in mitochondria showed that APAP treatment was associated with ROS overproduction, GSH depletion, mitochondrial dysfunction and cytotoxicity (Knockaert et al. 2011b).

In addition to APAP, other drugs such as halothane (Massart et al. 2017), diclofenac (Bort et al. 1999) and valproic acid (Fromenty and Pessayre 1995) could induce mitochondrial dysfunction via the generation of CYP-derived reactive metabolites. Notably, these drugs are also able to induce severe and sometimes fatal hepatotoxicity (Biour et al. 2004; Fromenty and Pessayre 1995; Wang et al. 2013). Valproic acid is discussed below as this drug is able to induce mitochondrial dysfunction via different mechanisms.

4.2 *Amiodarone*

Amiodarone is broad-spectrum antiarrhythmic drug that also presents an antianginal effect. A frequent adverse effect induced by amiodarone is hepatotoxicity, which includes acute hepatic cytolysis, steatosis, steatohepatitis and cirrhosis (Biour et al. 2004; Fromenty and Pessayre 1995; Santangeli et al. 2012; von Vital et al. 2011). Amiodarone can also induce cardiovascular adverse effects such as bradycardia and hypotension, thyroid toxicity (hyper- or hypothyroidism), ocular abnormalities and pulmonary toxicity including bronchiolitis and pulmonary fibrosis (Biour et al. 2004; Fromenty and Pessayre 1995; Santangeli et al. 2012; von Vital et al. 2011).

Numerous studies have shown that mitochondrial dysfunction is a major mechanism of amiodarone-induced toxicity in liver and other tissues such as lung and thyroid (Begriche et al. 2011; Bolt et al. 2001; Di Matola et al. 2000; Fromenty and Pessayre 1995; Nicolescu et al. 2008; Schumacher and Guo 2015). Actually, investigations performed in isolated rodent mitochondria disclosed that this cationic amphiphilic drug presents a dual effect on mitochondrial respiration and OXPHOS depending on its concentration (Bolt et al. 2001; Felser et al. 2013; Fromenty et al. 1990a; Serviddio et al. 2011; Spaniol et al. 2001). Indeed, at low concentrations (<100 μM), amiodarone is able to uncouple OXPHOS and stimulate mitochondrial respiration via a protonophoric effect. However, higher concentrations are leading to intramitochondrial accumulation of the drug that rapidly induces an inhibition of MRC activity, especially at the level of MRC complexes I and II (Fromenty and

Pessayre 1995; Fromenty et al. 1990a; Spaniol et al. 2001). Hence, stimulation of mitochondrial respiration (i.e. OXPHOS uncoupling) is only transient at high amiodarone concentrations (Bolt et al. 2001; Fromenty et al. 1990a; Spaniol et al. 2001). Because both OXPHOS uncoupling and MRC inhibition can impair ATP synthesis, amiodarone-induced energy shortage could occur for low and high intracellular concentrations of this drug (Felser et al. 2013; Fromenty et al. 1993).

Amiodarone is also able to inhibit mtFAO (Felser et al. 2013; Fromenty and Pessayre 1995; Fromenty et al. 1990b; Spaniol et al. 2001). Although this effect could be secondary to MRC inhibition, some investigations also suggested that mtFAO could be directly inhibited by amiodarone, in particular at the level of long-chain acyl-CoA dehydrogenase (LCAD) and CPT1 (Hamdan et al. 2001; Kennedy et al. 1996; Serviddio et al. 2011; Spaniol et al. 2001). The inhibitory effect of amiodarone on both mtFAO and MRC might explain why this drug can cause not only steatosis but also steatohepatitis in some patients (Begriche et al. 2011; Berson et al. 1998; Fromenty and Pessayre 1995). Interestingly, perhexiline and 4,4′-diethylaminoethoxyhexestrol (DEAEH), two cationic amphiphilic drugs withdrawn from the market because of frequent steatohepatitis, were shown to alter mitochondrial function in a similar manner to amiodarone (Berson et al. 1998; Deschamps et al. 1994).

4.3 Doxorubicin

Doxorubicin (also called adriamycin) is a potent chemotherapeutic drug used in the treatment of a wide range of tumors including hematological malignancies as well as breast, prostate, stomach and liver cancers (Carvalho et al. 2009; Tacar et al. 2013). Unfortunately, the use of doxorubicin is limited by the occurrence of a dose-dependent and progressive cardiomyopathy that can be fatal (Sardao et al. 2008; Varga et al. 2015). Moreover, doxorubicin can induce other serious adverse effects such as bone marrow aplasia, liver injury and nephropathy (Biour et al. 2004; Carvalho et al. 2009; Tacar et al. 2013).

There is now strong evidence that mitochondrial dysfunction is a major mechanism involved in doxorubicin-induced cardiotoxicity. Actually, doxorubicin-induced mitochondrial dysfunction is secondary to ROS overproduction and overt oxidative stress. Indeed, this drug undergoes in heart mitochondria a redox-cycling reaction that generates high amounts of superoxide anion radicals (Damiani et al. 2016; Finsterer and Ohnsorge 2013; Lebrecht and Walker 2007; Sardao et al. 2008; Varga et al. 2015). Some experimental investigations also suggest that the direct interaction of doxorubicin with iron could be an additional source of ROS production (Lebrecht et al. 2007a; Saad et al. 2001; Varga et al. 2015). In addition, doxorubicin could favor the mitochondrial accumulation of iron (Ichikawa et al. 2014). Doxorubicin-induced mitochondrial ROS overproduction is leading to direct MRC inhibition but also to indirect alteration of OXPHOS secondary to the occurrence of different types of mtDNA oxidative damage (Lebrecht and Walker 2007; Lebrecht et al. 2007a; Sardao et al. 2008; Varga et al. 2015). Moreover, doxorubicin-induced

ROS overproduction could also trigger MPT pore opening, which secondary leads to mitochondrial dysfunction and cardiomyocyte injury (Chahine et al. 2016; Sardao et al. 2008). Besides MRC alterations, some investigations reported that doxorubicin is able to alter the mtFAO pathway, in particular by inhibiting CPT1 (Hong et al. 2002; Sayed-Ahmed et al. 2000). Finally, it is noteworthy that a previous study suggested that doxorubicin-induced mitochondrial dysfunction could be secondary to p53 activation (Sahin et al. 2011). In this study, p53 activation was shown to induce a profound repression of peroxisome proliferator-activated receptor γ coactivator 1α and 1β (PGC-1α and PGC-1β), which play a major role in mitochondrial biogenesis and function (Sahin et al. 2011).

Doxorubicin is also able to induce oxidative stress and mitochondrial dysfunction in other tissues such as kidney and liver (Kalender et al. 2005; Saad et al. 2001; Serrano et al. 1999; Pereira et al. 2012a). Interestingly, in rats, doxorubicin-induced mitochondrial dysfunction was shown to be less severe in these tissues compared to the heart (Pereira et al. 2012a). Recent investigations also showed in rats that doxorubicin treatment did not induce cytochrome *c* and cardiolipin depletion in liver and kidney whereas the amount of these mitochondrial molecules was significantly reduced in heart (Pereira et al. 2016). This could explain why cardiotoxicity is the main adverse event observed with this anticancer agent. However, it is still unclear why doxorubicin-induced mitochondrial toxicity is significantly greater in heart.

Other drugs can exert their toxicity via a redox-cycling reaction, which can secondarily lead to oxidative stress and mitochondrial dysfunction. Such mechanism of toxicity has been proposed for instance with the antiandrogen nilutamide (Fau et al. 1992) and the vitamin K analogue menadione (He et al. 2015; Thor et al. 1982). In the US, menadione supplements have been banned by the Food and Drug Administration because of their potential toxicity in human use. In contrast, menadione-induced cytotoxicity could be interesting for the treatment of different types of cancer (Badave et al. 2016; He et al. 2015).

4.4 Nucleoside Reverse Transcriptase Inhibitors

The nucleoside reverse transcriptase inhibitors (NRTIs) are the first antiretroviral drugs marketed for the treatment of human immunodeficiency virus (HIV) infection. Notably, these drugs are 2′,3′-dideoxynucleoside analogues in which the hydroxyl group in the 3′ position on the sugar ring is replaced by either an hydrogen atom or another group unable to form a phosphodiester linkage. The lack of a 3′-hydroxyl group is required for the inhibition of HIV-reverse transcriptase activity. NRTIs include zidovudine (AZT), stavudine (d4T), lamivudine (3TC), didanosine (ddI) and abacavir (ABC).

NRTIs can induce numerous adverse effects including hepatotoxicity, renal dysfunction, myopathy, pancreatitis, peripheral neuropathy, bone marrow suppression, lipodystrophy and lactic acidosis (Caron-Debarle et al. 2010; Fromenty and Pessayre

1995; Margolis et al. 2014). NRTI-induced liver injury includes hepatic cytolysis, microvesicular and/or macrovacuolar steatosis, steatohepatitis, cirrhosis and cholestasis (Biour et al. 2004; Massart et al. 2013; Wang et al. 2013).

It is now acknowledged that most of the side effects induced by NRTIs are the consequence of an impairment of mtDNA replication (Fromenty and Pessayre 1995; Gardner et al. 2014; Schon and Fromenty 2016). Indeed, by impairing mtDNA replication, these antiretroviral drugs can induce severe mtDNA depletion and OXPHOS impairment (Igoudjil et al. 2006; Schon and Fromenty 2016). Actually, NRTIs are acting as chain terminators as their incorporation into the growing chain of mtDNA does not allow the addition of endogenous nucleotides by the DNA polymerase γ (Gardner et al. 2014; Igoudjil et al. 2006; Schon and Fromenty 2016). Hence, the lack of a 3′-hydroxyl group is responsible not only for the antiretroviral activity of NRTIs but also explains their mitochondrial toxicity. Interestingly, a mutation (R964C) in the gene encoding the DNA polymerase γ (*POLG*) might favor NRTI-induced mitochondrial toxicity, possibly by enhancing the probability of NRTI incorporation within the mtDNA molecules (Bailey et al. 2009; Yamanaka et al. 2007). *POLG* polymorphisms might also favor NRTI-induced mitochondrial toxicity (Baruffini et al. 2015). Finally, it is noteworthy that some NRTIs such as AZT and d4T could have mitochondrial and metabolic effects through mechanisms unrelated to DNA polymerase γ inhibition and mtDNA depletion (Apostolova et al. 2011; Igoudjil et al. 2006; Stankov et al. 2013).

Inhibition of mtDNA replication has been reported for other drugs including tamoxifen, tacrine and fialuridine (FIAU) (Larosche et al. 2007; Mansouri et al. 2003; Schon and Fromenty 2016). FIAU is a nucleoside analogue that has been developed in the early 90s for the treatment of chronic hepatitis B virus (HBV) infection. However, clinical trials were prematurely interrupted because this antiviral drug induced several cases of severe and sometimes fatal lactic acidosis and hepatic failure (Fromenty and Pessayre 1995; Schon and Fromenty 2016). FIAU hepatotoxicity was sometimes associated with other adverse effects such as pancreatitis, neuropathy and myopathy (McKenzie et al. 1995), suggesting that mitochondrial dysfunction could occur in different tissues. Finally, it is noteworthy that some drugs could cause severe adverse effects by inhibiting translation of mtDNA-encoded mitochondrial proteins (Schon and Fromenty 2016). This has been shown for instance with linezolid, an antibiotic able to induce lactic acidosis, hepatotoxicity, peripheral and optic neuropathy, skeletal myopathy, and renal failure (Ager and Gould 2012; Schon and Fromenty 2016).

4.5 Statins

Statins are lipid-lowering drugs able to reduce blood cholesterol via the inhibition of 3-hydroxy-3-methylglutaryl-CoA (HMG-CoA) reductase. This pharmacological class includes different compounds such as atorvastatin, cerivastatin, fluvastatin, lovastatin, pravastatin and simvastatin. Statins can induce different adverse effects such as gastrointestinal disorders, peripheral neuropathy, hepatotoxicity and myopathy (Biour et al. 2004; Jones et al. 2014; Law and Rudnicka 2006; Sirvent et al. 2008; Wang et al. 2013). Actually, the severity of statin-induced myopathy ranges from asymptomatic serum creatine kinase (CK) elevations to rhabdomyolysis (Thompson et al. 2006; Thompson 2016). Although statin-induced rhabdomyolysis is rather rare, it can be severe and even fatal in some patients (Law and Rudnicka 2006; Sirvent et al. 2008). Cerivastatin was withdrawn from the European and USA markets in 2001 because of a relatively high risk of rhabdomyolysis (Maggini et al. 2004; Will and Dykens 2014). Statin-induced liver injury can also be severe and fatal in some patients (Björnsson 2017). Finally, it is noteworthy that statins significantly increase the risk of new-onset diabetes, although this risk seems to be small (Brault et al. 2014; Thakker et al. 2016).

Several clinical and experimental studies strongly suggest that mitochondrial dysfunction is involved in statin-induced myopathy (Apostolopoulou et al. 2015; Phillips et al. 2002; Seachrist et al. 2005; Sirvent et al. 2008). Notably, statins seem to impair mitochondrial function by different mechanisms that are not mutually exclusive. First, statins are able to directly inhibit the MRC at different levels, although the extent of this inhibition is variable among the different compounds (Kaufmann et al. 2006; Nadanaciva et al. 2007a). In addition, some statins such as cerivastatin and fluvastatin can directly impair the mtFAO pathway, in particular at the CPT1 level (Kaufmann et al. 2006). This could explain the presence of lipid accumulation in muscle biopsies of patients with statin-induced myopathy (Phillips et al. 2002; Thompson et al. 2006). Second, statins could impair mitochondrial function in muscle via a decrease in mtDNA levels (Schick et al. 2007; Stringer et al. 2013). Although the exact mechanism of statins-induced mtDNA depletion in skeletal muscle is currently unknown, some experimental investigations suggested that these drugs could impair mitochondrial biogenesis via ROS overproduction (Bouitbir et al. 2012). A last mechanism might be coenzyme Q depletion in muscle because statin-induced inhibition of HMG-CoA reductase is suspected to reduce the levels of this key MRC component. However, different experimental and clinical investigations do not support a significant role of reduced coenzyme Q levels in statin-induced myopathy (Jones et al. 2014; Marcoff and Thompson 2007). Interestingly, some studies suggested that different inherited mitochondrial diseases could increase the risk of statin-induced muscle disorders (Ghatak et al. 2010; Patel et al. 2015). Finally, it is noteworthy that mitochondrial dysfunction could also be involved in statin-induced liver injury (Nadanaciva et al. 2007b; Porceddu et al. 2012).

4.6 Valproic Acid

Valproic acid (VPA), or 2-propylpentanoic acid, is a broad-spectrum antiepileptic drug used in the treatment of different forms of epilepsy. This drug can induce different types of adverse effects including tremors and convulsions, somnolence, fatigue, thrombocytopenia, pancreatitis, nephrotoxicity, and hepatotoxicity (Fromenty and Pessayre 1995; Nanau and Neuman 2013; Star et al. 2014; Yoshikawa et al. 2002). In most cases, hepatotoxicity consists of an asymptomatic increase in serum transaminase activity that normalizes with dose reduction, or drug discontinuation. A much less frequent hepatotoxic effect is severe, and sometimes fatal, Reye-like syndrome with liver failure, prolonged prothrombin time, hypoglycemia and hyperammonemia (Finsterer 2017; Fromenty and Pessayre 1995; Powell-Jackson et al. 1984). Three major predisposing factors can increase the risk of VPA-induced Reye-like syndrome, namely young age of the patients, polytherapy with other antiepileptic drugs (e.g. phenytoin, phenobarbital and carbamazepine) and underlying genetic diseases affecting mtFAO and MRC activity (Begriche et al. 2011; Fromenty and Pessayre 1995; Li et al. 2015; Star et al. 2014).

VPA is metabolized in the liver by two major pathways. First, this branched-chain fatty acid is activated into VPA-coenzyme A (VPA-CoA), which undergoes complete mitochondrial β-oxidation generating propionyl-CoA and acetyl-CoA (Luis et al. 2011; Silva et al. 2008). Second, VPA can be metabolized via CYP-mediated dehydrogenation leading to the formation of 4-ene-valproate, also referred to as Δ^4-VPA. Inside the mitochondria, Δ^4-VPA is then activated into Δ^4-VPA-CoA that undergoes the mtFAO pathway to form $\Delta^{2,4}$-VPA-CoA (Fromenty and Pessayre 1995; Kassahun et al. 1991).

Numerous *in vitro* and *in vivo* studies have been performed in order to determine the exact mechanism of VPA-induced Reye-like syndrome (reviewed in Baillie 1988; Coulter 1991; Fromenty and Pessayre 1995; Ponchaut and Veitch 1993; Silva et al. 2008). These investigations consistently showed that VPA can inhibit the mtFAO via at least three different mechanisms: (1) sequestration of CoA by VPA and VPA metabolites (Δ^4-VPA and $\Delta^{2,4}$-VPA); (2) depletion of carnitine (another key cofactor necessary for FA activation and oxidation), in particular due to reduced biosynthesis and increased urinary excretion; and (3) direct inhibition of several mtFAO enzymes. More specifically, VPA and/or different VPA metabolites (Δ^4-VPA-CoA and $\Delta^{2,4}$-VPA-CoA) are able to inhibit short-chain acyl-CoA dehydrogenase (SCAD), medium-chain acyl-CoA dehydrogenase (MCAD), CPT1 and the trifunctional enzyme (Aires et al. 2010; Fromenty and Pessayre 1995; Silva et al. 2008). Notably, the CYP-generated VPA metabolites could be potent inhibitors of the mtFAO pathway, and this could explain why CYP inducers such as phenobarbital and phenytoin increase the risk of VPA-induced severe hepatotoxicity (Baillie 1988; Fromenty and Pessayre 1995). Finally, it is noteworthy that different dicarboxylic acids have been detected in the urine of treated patients, thus reflecting VPA-induced inhibition of mitochondrial β-oxidation (Fromenty and Pessayre 1995; Mortensen et al. 1980; Silva et al. 2008).

Other drugs could impair hepatic mtFAO by sequestrating CoA. This has been for instance reported with the nonsteroidal anti-inflammatory drugs aspirin and ibuprofen (Deschamps et al. 1991; Fréneaux et al. 1990). However, some experimental investigations suggested that this mechanism could occur only if endogenous CoA levels are relatively low within the hepatocytes (Deschamps et al. 1991; Fréneaux et al. 1990).

In addition to mtFAO, VPA can impair other mitochondrial metabolic pathways in liver. VPA is able to inhibit ureagenesis and this could explain, at least in part, why VPA can induce hyperammonemia in treated patients (Coude et al. 1983;

Fromenty and Pessayre 1995). Chronic VPA administration in rat was shown to reduce hepatic MRC activity at the level of complex IV (Fromenty and Pessayre 1995; Ponchaut et al. 1992). VPA is also able to directly inhibit mitochondrial respiration stimulated by succinate (Porceddu et al. 2012), but this effect could be secondary to an impairment of succinate transport into mitochondria (Rumbach et al. 1989). Finally, investigations in isolated rat liver mitochondria showed that VPA was able to induce MPT pore opening in condition of calcium pulse (Trost and Lemasters 1996). However, VPA-induced mitochondrial membrane permeabilization could not be observed in isolated mouse liver mitochondria in the absence of this calcium pulse (Porceddu et al. 2012).

It is noteworthy that VPA-induced mitochondrial toxicity has been exclusively studied in experimental investigations performed in order to explain the occurrence of Reye-like syndromes. However, it is unknown whether VPA-induced mitochondrial liability might explain the occurrence of other severe adverse effects such as pancreatitis and nephrotoxicity. Although mitochondrial abnormalities have been observed in the proximal renal tubules of some VPA-treated patients with renal toxicity (Yoshikawa et al. 2002), a direct causal relationship between mitochondrial dysfunction and nephrotoxicity has never been demonstrated. Likewise, it is unknown whether VPA-induced mitochondrial toxicity might be responsible for the ability of this drug to increase the risk of congenital malformations (Jentink et al. 2010; Nanau and Neuman 2013). Although this hypothesis cannot be excluded, some investigations suggested that VPA-induced teratogenic effect could be linked to the inhibition of histone deacetylases (HDAC) (Giavini and Menegola 2014; Ornoy 2009).

5 Conclusion

It is now acknowledged that mitochondrial dysfunction is a key mechanism whereby some drugs can cause different adverse effects and in particular liver injury, myopathy and cardiomyopathy. Importantly, adverse effects secondary to mitochondrial dysfunction can be severe and threaten the life of patients because mitochondria are the main source of energy in most cells. Moreover, these serious adverse events can lead to the premature interruption of clinical trials, or to the withdrawal of drugs from the market (Labbe et al. 2008; Nadanaciva and Will 2011). Hence, a major challenge for the pharmaceutical industry is to detect drug-induced mitochondrial dysfunction during preclinical studies. This can be performed with different *in vitro* methods that allow to determine whether drug candidates are able to impair OXPHOS and MRC activity, or lead to MPT opening (Marroquin et al. 2014; Nadanaciva et al. 2012; Porceddu et al. 2012; Will and Dykens 2014). However, it is important to keep in mind that these methods are not able to detect late-onset mitochondrial dysfunction that are for instance induced by an inhibition of mtDNA replication or mitochondrial protein synthesis. Hence, long-term treatment in relevant experimental models are required to detect such deleterious effects (Gerschenson et al. 2001; Lebrecht et al. 2007b; Setzer et al. 2008). It should also be recalled that the detection of mitochondrial dysfunction can be complicated by interspecies differences, as demonstrated with FIAU (Labbe et al. 2008; Lee et al. 2006).

Another major challenge is to determine the main factors that might predispose to drug-induced mitochondrial dysfunction and related adverse effects. Indeed, drug-induced mitochondrial dysfunction can be silent in most patients but can become severe and life-threatening in a few individuals with different predisposing factors (Begriche et al. 2011; Labbe et al. 2008). Recent investigations suggested that some polymorphisms in the human *POLG* gene could favor mitochondrial dysfunction induced by VPA (Stewart et al. 2010), or some antiretroviral NRTIs (Baruffini et al. 2015), as previously mentioned. Some mutations and polymorphisms (or haplogroups) in the mitochondrial genome are suspected to significantly modulate mitochondrial toxicity for different pharmacological classes including antiretroviral drugs and antibiotics (Del Pozo et al. 2014; Hart et al. 2013; Jing et al. 2015; Pacheu-Grau et al. 2013; Pereira et al. 2012b). The Ala16Val genetic dimorphism of MnSOD could increase the susceptibility to drug-induced mitochondrial dysfunction and hepatotoxicity (Huang et al. 2007; Lucena et al. 2010), by reducing MnSOD importation within mitochondria and thus favoring oxidative stress in these organelles (Degoul et al. 2001; Sutton et al. 2003). Regarding MnSOD, investigations performed in *Sod2*+/− mice could be useful to investigate whether reduced MnSOD activity is able to favor hepatotoxicity induced by some drugs (Ballet 2015; Hsiao et al. 2010; Lee et al. 2013).

Nonalcoholic fatty liver disease (NAFLD) linked to obesity might also favor drug-induced mitochondrial dysfunction and hepatotoxicity (Fromenty 2013; Massart et al. 2017). Indeed, NAFLD is associated with OXPHOS and MRC impairment, in particular when nonalcoholic steatohepatitis (NASH) has developed (Begriche et al. 2013; Koliaki et al. 2015; Sunny et al. 2017). To determine which drugs can pose a significant risk in obese patients is also an important challenge for the future. Hence, animal models such as the *Sod2*$^{+/-}$ and obese *ob/ob* mice could be useful to identify which drugs are able to be particularly hepatotoxic in the presence of underlying mitochondrial dysfunction (Ballet 2015; Buron et al. 2017; Labbe et al. 2008). Finally, it is important to underline that obesity and related metabolic disorders are associated with mitochondrial dysfunction not only in liver but also in skeletal muscle, heart and adipose tissue (Bhatti et al. 2017; Crescenzo et al. 2015; Massart et al. 2016). Thus, it is tempting to speculate that obesity might favor the mitochondrial toxicity of some drugs not only in liver but also in different extra-hepatic tissues. In keeping with this assumption, obese patients seem to have a higher risk of statin-induced hepatotoxicity and myopathy (Marzoa-Rivas et al. 2005) and of stavudine-induced lactic acidosis (Coghlan et al. 2001; Wester et al. 2012).

References

Ager S, Gould K (2012) Clinical update on linezolid in the treatment of Gram-positive bacterial infections. Infect Drug Resist 5:87–102

Aires CC, Ijlst L, Stet F, Prip-Buus C, de Almeida IT, Duran M, Wanders RJ, Silva MF (2010) Inhibition of hepatic carnitine palmitoyl-transferase I (CPT IA) by valproyl-CoA as a possible mechanism of valproate-induced steatosis. Biochem Pharmacol 79:792–799

Apostolopoulou M, Corsini A, Roden M (2015) The role of mitochondria in statin-induced myopathy. Eur J Clin Investig 45:745–754

Apostolova N, Blas-García A, Esplugues JV (2011) Mitochondrial interference by anti-HIV drugs: mechanisms beyond Pol-γ inhibition. Trends Pharmacol Sci 32:715–725

Aubert J, Begriche K, Delannoy M, Morel I, Pajaud J, Ribault C, Lepage S, McGill MR, Lucas-Clerc C, Turlin B, Robin MA, Jaeschke H, Fromenty B (2012) Differences in early acetaminophen hepatotoxicity between obese ob/ob and db/db mice. J Pharmacol Exp Ther 342:676–687

Badave KD, Khan AA, Rane SY (2016) Anticancer vitamin K3 analogues: a review. Anticancer Agents Med Chem 16(8):1017–1030

Bailey CM, Kasiviswanathan R, Copeland WC, Anderson KS (2009) R964C mutation of DNA polymerase γ imparts increased stavudine toxicity by decreasing nucleoside analog discrimination and impairing polymerase activity. Antimicrob Agents Chemother 53:2610–2612

Baillie TA (1988) Metabolic activation of valproic acid and drug-mediated hepatotoxicity. Role of the terminal olefin, 2-n-propyl-4-pentenoic acid. Chem Res Toxicol 1:195–199

Ballet F (2015) Preventing drug-induced liver injury: how useful are animal models? Dig Dis 33:477–485

Bansal S, Anandatheerthavarada HK, Prabu GK, Milne GL, Martin MV, Guengerich FP, Avadhani NG (2013) Human cytochrome P450 2E1 mutations that alter mitochondrial targeting efficiency and susceptibility to ethanol-induced toxicity in cellular models. J Biol Chem 288:12627–12644

Baruffini E, Ferrari J, Dallabona C, Donnini C, Lodi T (2015) Polymorphisms in DNA polymerase γ affect the mtDNA stability and the NRTI-induced mitochondrial toxicity in Saccharomyces cerevisiae. Mitochondrion 20:52–63

Baudouin SV, Howdle P, O'Grady JG, Webster NR (1995) Acute lung injury in fulminant hepatic failure following paracetamol poisoning. Thorax 50:399–402

Begriche K, Massart J, Robin MA, Borgne-Sanchez A, Fromenty B (2011) Drug-induced toxicity on mitochondria and lipid metabolism: mechanistic diversity and deleterious consequences for the liver. J Hepatol 54:773–794

Begriche K, Massart J, Robin MA, Bonnet F, Fromenty B (2013) Mitochondrial adaptations and dysfunctions in nonalcoholic fatty liver disease. Hepatology 58:1497–1507

Belosludtsev KN, Saris NE, Belosludtseva NV, Trudovishnikov AS, Lukyanova LD, Mironova GD (2009) Physiological aspects of the mitochondrial cyclosporin A-insensitive palmitate/ Ca^{2+}-induced pore: tissue specificity, age profile and dependence on the animal's adaptation to hypoxia. J Bioenerg Biomembr 41:395–401

Bénit P, Letouzé E, Rak M, Aubry L, Burnichon N, Favier J, Gimenez-Roqueplo AP, Rustin P (2014) Unsuspected task for an old team: succinate, fumarate and other Krebs cycle acids in metabolic remodeling. Biochim Biophys Acta 1837:1330–1337

Berman SB, Watkins SC, Hastings TG (2000) Quantitative biochemical and ultrastructural comparison of mitochondrial permeability transition in isolated brain and liver mitochondria: evidence for reduced sensitivity of brain mitochondria. Exp Neurol 164:415–425

Bernardi P (2013) The mitochondrial permeability transition pore: a mystery solved? Front Physiol 4:95

Berson A, De Beco V, Lettéron P, Robin MA, Moreau C, El Kahwaji J, Verthier N, Feldmann G, Fromenty B, Pessayre D (1998) Steatohepatitis-inducing drugs cause mitochondrial dysfunction and lipid peroxidation in rat hepatocytes. Gastroenterology 114:764–774

Bhattacharyya S, Yan K, Pence L, Simpson PM, Gill P, Letzig LG, Beger RD, Sullivan JE, Kearns GL, Reed MD, Marshall JD, Van Den Anker JN, James LP (2014) Targeted liquid chromatography-mass spectrometry analysis of serum acylcarnitines in acetaminophen toxicity in children. Biomark Med 8:147–159

Bhatti JS, Bhatti GK, Reddy PH (2017) Mitochondrial dysfunction and oxidative stress in metabolic disorders. A step towards mitochondria based therapeutic strategies. Biochim Biophys Acta 1863(5):1066–1077

Biasutto L, Azzolini M, Szabò I, Zoratti M (2016) The mitochondrial permeability transition pore in AD 2016: an update. Biochim Biophys Acta 1863:2515–2530

Biour M, Ben Salem C, Chazouillères O, Grangé JD, Serfaty L, Poupon R (2004) Drug-induced liver injury; fourteenth updated edition of the bibliographic database of liver injuries and related drugs. Gastroenterol Clin Biol 28:720–759

Björnsson ES (2017) Hepatotoxicity of statins and other lipid-lowering agents. Liver Int 37:173–178

Björnsson ES, Hoofnagle JH (2016) Categorization of drugs implicated in causing liver injury: critical assessment based on published case reports. Hepatology 63:590–603

Blakely P, McDonald BR (1995) Acute renal failure due to acetaminophen ingestion: a case report and review of the literature. J Am Soc Nephrol 6:48–53

Bolt MW, Card JW, Racz WJ, Brien JF, Massey TE (2001) Disruption of mitochondrial function and cellular ATP levels by amiodarone and N-desethylamiodarone in initiation of amiodarone-induced pulmonary cytotoxicity. J Pharmacol Exp Ther 298:1280–1289

Bonora M, Wieckowski MR, Chinopoulos C, Kepp O, Kroemer G, Galluzzi L, Pinton P (2015) Molecular mechanisms of cell death: central implication of ATP synthase in mitochondrial permeability transition. Oncogene 34:1475–1486

Bort R, Ponsoda X, Jover R, Gomez-Lechon MJ, Castell JV (1999) Diclofenac toxicity to hepatocytes: a role for drug metabolism in cell toxicity. J Pharmacol Exp Ther 288:65–72

Bouitbir J, Charles AL, Echaniz-Laguna A, Kindo M, Daussin F, Auwerx J, Piquard F, Geny B, Zoll J (2012) Opposite effects of statins on mitochondria of cardiac and skeletal muscles: a 'mitohormesis' mechanism involving reactive oxygen species and PGC-1. Eur Heart J 33:1397–1407

Brault M, Ray J, Gomez YH, Mantzoros CS, Daskalopoulou SS (2014) Statin treatment and new-onset diabetes: a review of proposed mechanisms. Metabolism 63:735–745

Burcham PC, Harman AW (1991) Acetaminophen toxicity results in site-specific mitochondrial damage in isolated mouse hepatocytes. J Biol Chem 266:5049–5054

Buron N, Porceddu M, Roussel C, Begriche K, Trak-Smayra V, Gicquel T, Fromenty B, Borgne-Sanchez A (2017) Chronic and low exposure to a pharmaceutical cocktail induces mitochondrial dysfunction in liver and hyperglycemia: differential responses between lean and obese mice. Environ Toxicol 32:1375–1389

Caron-Debarle M, Boccara F, Lagathu C, Antoine B, Cervera P, Bastard JP, Vigouroux C, Capeau J (2010) Adipose tissue as a target of HIV-1 antiretroviral drugs. Potential consequences on metabolic regulations. Curr Pharm Des 16:3352–3360

Carvalho C, Santos RX, Cardoso S, Correia S, Oliveira PJ, Santos MS, Moreira PI (2009) Doxorubicin: the good, the bad and the ugly effect. Curr Med Chem 16:3267–3285

Chahine N, Nader M, Duca L, Martiny L, Chahine R (2016) Saffron extracts alleviate cardiomyocytes injury induced by doxorubicin and ischemia-reperfusion in vitro. Drug Chem Toxicol 39:87–96

Coghlan ME, Sommadossi JP, Jhala NC, Many WJ, Saag MS, Johnson VA (2001) Symptomatic lactic acidosis in hospitalized antiretroviral-treated patients with human immunodeficiency virus infection: a report of 12 cases. Clin Infect Dis 33:1914–1921

Coude FX, Grimber G, Parvy P, Rabier D, Petit F (1983) Inhibition of ureagenesis by valproate in rat hepatocytes. Role of N-acetylglutamate and acetyl-CoA. Biochem J 216:233–236

Coulter DL (1991) Carnitine, valproate, and toxicity. J Child Neurol 6:7–14

Craig DG, Bates CM, Davidson JS, Martin KG, Hayes PC, Simpson KJ (2011) Overdose pattern and outcome in paracetamol-induced acute severe hepatotoxicity. Br J Clin Pharmacol 71:273–282

Crescenzo R, Bianco F, Mazzoli A, Giacco A, Liverini G, Iossa S (2015) Mitochondrial efficiency and insulin resistance. Front Physiol 5:512

Damiani RM, Moura DJ, Viau CM, Caceres RA, Henriques JA, Saffi J (2016) Pathways of cardiac toxicity: comparison between chemotherapeutic drugs doxorubicin and mitoxantrone. Arch Toxicol 90:2063–2076

Das J, Ghosh J, Manna P, Sil PC (2010) Taurine protects acetaminophen-induced oxidative damage in mice kidney through APAP urinary excretion and CYP2E1 inactivation. Toxicology 269:24–34

Degoul F, Sutton A, Mansouri A, Cepanec C, Degott C, Fromenty B, Beaugrand M, Valla D, Pessayre D (2001) Homozygosity for alanine in the mitochondrial targeting sequence of superoxide dismutase and risk for severe alcoholic liver disease. Gastroenterology 120:1468–1474

Del Pozo JL, Fernández-Ros N, Sáez E, Herrero JI, Yuste JR, Banales JM (2014) Linezolid-induced lactic acidosis in two liver transplant patients with the mitochondrial DNA A2706G polymorphism. Antimicrob Agents Chemother 58:4227–4229

Deschamps D, Fisch C, Fromenty B, Berson A, Degott C, Pessayre D (1991) Inhibition by salicylic acid of the activation and thus oxidation of long chain fatty acids. Possible role in the development of Reye's syndrome. J Pharmacol Exp Ther 259:894–904

Deschamps D, DeBeco V, Fisch C, Fromenty B, Guillouzo A, Pessayre D (1994) Inhibition by perhexiline of oxidative phosphorylation and the beta-oxidation of fatty acids: possible role in pseudoalcoholic liver lesions. Hepatology 19:948–961

Di Lisa F, Canton M, Menabo R, Kaludercic N, Bernardi P (2007) Mitochondria and cardioprotection. Heart Fail Rev 12:249–260

Di Matola T, D'Ascoli F, Fenzi G, Rossi G, Martino E, Bogazzi F, Vitale M (2000) Amiodarone induces cytochrome c release and apoptosis through an iodine-independent mechanism. J Clin Endocrinol Metab 85:4323–4330

Diebold L, Chandel NS (2016) Mitochondrial ROS regulation of proliferating cells. Free Radic Biol Med 100:86–93

Ding X, Kaminsky LS (2003) Human extrahepatic cytochromes P450: function in xenobiotic metabolism and tissue-selective chemical toxicity in the respiratory and gastrointestinal tracts. Annu Rev Pharmacol Toxicol 43:149–173

Dutheil F, Dauchy S, Diry M, Sazdovitch V, Cloarec O, Mellottée L, Bièche I, Ingelman-Sundberg M, Flinois JP, de Waziers I, Beaune P, Declèves X, Duyckaerts C, Loriot MA (2009) Xenobiotic-metabolizing enzymes and transporters in the normal human brain: regional and cellular mapping as a basis for putative roles in cerebral function. Drug Metab Dispos 37:1528–1538

Elangbam CS (2010) Drug-induced valvulopathy: an update. Toxicol Pathol 38:837–848

Ellero S, Chakhtoura G, Barreau C, Langouët S, Benelli C, Penicaud L, Beaune P, de Waziers I (2010) Xenobiotic-metabolizing cytochromes P450 in human white adipose tissue: expression and induction. Drug Metab Dispos 38:679–686

Fau D, Berson A, Eugene D, Fromenty B, Fisch C, Pessayre D (1992) Mechanism for the hepatotoxicity of the antiandrogen, nilutamide. Evidence suggesting that redox cycling of this nitroaromatic drug leads to oxidative stress in isolated hepatocytes. J Pharmacol Exp Ther 263:69–77

Felser A, Blum K, Lindinger PW, Bouitbir J, Krähenbühl S (2013) Mechanisms of hepatocellular toxicity associated with dronedarone: a comparison to amiodarone. Toxicol Sci 131:480–490

Finsterer J (2017) Toxicity of antiepileptic drugs to mitochondria. Handb Exp Pharmacol 240:473–488

Finsterer J, Ohnsorge P (2013) Influence of mitochondrion-toxic agents on the cardiovascular system. Regul Toxicol Pharmacol 67:434–445

Foufelle F, Fromenty B (2016) Role of endoplasmic reticulum stress in drug-induced toxicity. Pharmacol Res Perspect 4:e00211

Fréneaux E, Fromenty B, Berson A, Labbe G, Degott C, Letteron P, Larrey D, Pessayre D (1990) Stereoselective and nonstereoselective effects of ibuprofen enantiomers on mitochondrial beta-oxidation of fatty acids. J Pharmacol Exp Ther 255:529–535

Fromenty B (2013) Drug-induced liver injury in obesity. J Hepatol 58:824–826

Fromenty B, Pessayre D (1995) Inhibition of mitochondrial beta-oxidation as a mechanism of hepatotoxicity. Pharmacol Ther 67:101–154

Fromenty B, Fisch C, Berson A, Letteron P, Larrey D, Pessayre D (1990a) Dual effect of amiodarone on mitochondrial respiration. Initial protonophoric uncoupling effect followed by inhibition of the respiratory chain at the levels of complex I and complex II. J Pharmacol Exp Ther 255:1377–1384

Fromenty B, Fisch C, Labbe G, Degott C, Deschamps D, Berson A, Letteron P, Pessayre D (1990b) Amiodarone inhibits the mitochondrial β-oxidation of fatty acids and produces microvesicular steatosis of the liver in mice. J Pharmacol Exp Ther 255:1371–1376

Fromenty B, Letteron P, Fisch C, Berson A, Deschamps D, Pessayre D (1993) Evaluation of human blood lymphocytes as a model to study the effects of drugs on human mitochondria. Effects of low concentrations of amiodarone on fatty acid oxidation, ATP levels and cell survival. Biochem Pharmacol 46:421–432

Gardner K, Hall PA, Chinnery PF, Payne BA (2014) HIV treatment and associated mitochondrial pathology: review of 25 years of in vitro, animal, and human studies. Toxicol Pathol 42:811–822

Gerschenson M, Nguyen VT, St Claire MC, Harbaugh SW, Harbaugh JW, Proia LA, Poirier MC (2001) Chronic stavudine exposure induces hepatic mitochondrial toxicity in adult Erythrocebus patas monkeys. J Hum Virol 4:335–342

Ghatak A, Faheem O, Thompson PD (2010) The genetics of statin-induced myopathy. Atherosclerosis 210:337–343

Giavini E, Menegola E (2014) Teratogenic activity of HDAC inhibitors. Curr Pharm Des 20:5438–5442

Gu J, Cui H, Behr M, Zhang L, Zhang QY, Yang W, Hinson JA, Ding X (2005) In vivo mechanisms of tissue-selective drug toxicity: effects of liver-specific knockout of the NADPH-cytochrome P450 reductase gene on acetaminophen toxicity in kidney, lung, and nasal mucosa. Mol Pharmacol 67:623–630

Halestrap AP, Davidson AM (1990) Inhibition of Ca^{2+}-induced large-amplitude swelling of liver and heart mitochondria by cyclosporin is probably caused by the inhibitor binding to mitochondrial-matrix peptidyl-prolyl cis-trans isomerase and preventing it interacting with the adenine nucleotide translocase. Biochem J 268:153–160

Hamdan M, Urien S, Le Louet H, Tillement JP, Morin D (2001) Inhibition of mitochondrial carnitine palmitoyltransferase-1 by a trimetazidine derivative, S-15176. Pharmacol Res 44:99–104

Hart AB, Samuels DC, Hulgan T (2013) The other genome: a systematic review of studies of mitochondrial DNA haplogroups and outcomes of HIV infection and antiretroviral therapy. AIDS Rev 15:213–220

He T, Hatem E, Vernis L, Lei M, Huang ME (2015) PRX1 knockdown potentiates vitamin K3 toxicity in cancer cells: a potential new therapeutic perspective for an old drug. J Exp Clin Cancer Res 34:152

Hohenegger M (2012) Drug induced rhabdomyolysis. Curr Opin Pharmacol 12:335–339

Hong YM, Kim HS, Yoon HR (2002) Serum lipid and fatty acid profiles in adriamycin-treated rats after administration of L-carnitine. Pediatr Res 51:249–255

Houten SM, Wanders RJ (2010) A general introduction to the biochemistry of mitochondrial fatty acid β-oxidation. J Inherit Metab Dis 33:469–477

Houten SM, Violante S, Ventura FV, Wanders RJ (2016) The biochemistry and physiology of mitochondrial fatty acid β-oxidation and its genetic disorders. Annu Rev Physiol 78:23–44

Hsiao CJ, Younis H, Boelsterli UA (2010) Trovafloxacin, a fluoroquinolone antibiotic with hepatotoxic potential, causes mitochondrial peroxynitrite stress in a mouse model of underlying mitochondrial dysfunction. Chem Biol Interact 188:204–213

Hu J, Kholmukhamedov A, Lindsey CC, Beeson CC, Jaeschke H, Lemasters JJ (2016) Translocation of iron from lysosomes to mitochondria during acetaminophen-induced hepatocellular injury: protection by starch-desferal and minocycline. Free Radic Biol Med 97:418–426

Huang YS, Su WJ, Huang YH, Chen CY, Chang FY, Lin HC, Lee SD (2007) Genetic polymorphisms of manganese superoxide dismutase, NAD(P)H:quinone oxidoreductase, glutathione S-transferase M1 and T1, and the susceptibility to drug-induced liver injury. J Hepatol 47:128–134

Ichikawa Y, Ghanefar M, Bayeva M, Wu R, Khechaduri A, Naga Prasad SV, Mutharasan RK, Naik TJ, Ardehali H (2014) Cardiotoxicity of doxorubicin is mediated through mitochondrial iron accumulation. J Clin Invest 124:617–630

Igarashi H, Ito T, Yoshinaga M, Oono T, Sakai H, Takayanagi R (2009) Acetaminophen-induced acute pancreatitis. A case report. JOP 10:550–553

Igoudjil A, Begriche K, Pessayre D, Fromenty B (2006) Mitochondrial, metabolic and genotoxic effects of antiretroviral nucleoside reverse-transcriptase inhibitors. Anti-Infect Agents Med Chem 5:273–292

Imhoff BR, Hansen JM (2009) Extracellular redox status regulates Nrf2 activation through mitochondrial reactive oxygen species. Biochem J 424:491–500

Izzo V, Bravo-San Pedro JM, Sica V, Kroemer G, Galluzzi L (2016) Mitochondrial permeability transition: new findings and persisting uncertainties. Trends Cell Biol 26:655–667

Jaeschke H, McGill MR, Ramachandran A (2012) Oxidant stress, mitochondria, and cell death mechanisms in drug-induced liver injury: lessons learned from acetaminophen hepatotoxicity. Drug Metab Rev 44:88–106

Jentink J, Loane MA, Dolk H, Barisic I, Garne E, Morris JK, de Jong-van den Berg LT (2010) Valproic acid monotherapy in pregnancy and major congenital malformations. N Engl J Med 362:2185–2193

Jing W, Zongjie H, Denggang F, Na H, Bin Z, Aifen Z, Xijiang H, Cong Y, Yunping D, Ring HZ, Ring BZ (2015) Mitochondrial mutations associated with aminoglycoside ototoxicity and hearing loss susceptibility identified by meta-analysis. J Med Genet 52:95–103

John R, Herzenberg AM (2009) Renal toxicity of therapeutic drugs. J Clin Pathol 62:505–515

Jones AL, Prescott LF (1997) Unusual complications of paracetamol poisoning. QJM 90:161–168

Jones JD, Kirsch HL, Wortmann RL, Pillinger MH (2014) The causes of drug-induced muscle toxicity. Curr Opin Rheumatol 26:697–703

Kakimoto PA, Tamaki FK, Cardoso AR, Marana SR, Kowaltowski AJ (2015) H_2O_2 release from the very long chain acyl-CoA dehydrogenase. Redox Biol 4:375–380

Kalender Y, Yel M, Kalender S (2005) Doxorubicin hepatotoxicity and hepatic free radical metabolism in rats. The effects of vitamin E and catechin. Toxicology 209:39–45

Kassahun K, Farrell K, Abbott F (1991) Identification and characterization of the glutathione and N-acetylcysteine conjugates of (E)-2-propyl-2,4-pentadienoic acid, a toxic metabolite of valproic acid, in rats and humans. Drug Metab Dispos 19:525–535

Kastaniotis AJ, Autio KJ, Kerätär JM, Monteuuis G, Mäkelä AM, Nair RR, Pietikäinen LP, Shvetsova A, Chen Z, Hiltunen JK (2017) Mitochondrial fatty acid synthesis, fatty acids and mitochondrial physiology. Biochim Biophys Acta 1862:39–48

Kato H, Fujigaki Y, Inoue R, Asakawa S, Shin S, Shima T, Furunishi J, Higaki M, Tanemoto M, Yamaguchi Y, Hoshimoto K, Uozaki H, Uchida S (2014) Therapeutic dose of acetaminophen as a possible risk factor for acute kidney injury: learning from two healthy young adult cases. Intern Med 53:1531–1534

Kaufmann P, Török M, Zahno A, Waldhauser KM, Brecht K, Krähenbühl S (2006) Toxicity of statins on rat skeletal muscle mitochondria. Cell Mol Life Sci 63:2415–2425

Kennedy JA, Unger SA, Horowitz JD (1996) Inhibition of carnitine palmitoyltransferase-1 in rat heart and liver by perhexiline and amiodarone. Biochem Pharmacol 52:273–280

Knights KM, Rowland A, Miners JO (2013) Renal drug metabolism in humans: the potential for drug-endobiotic interactions involving cytochrome P450 (CYP) and UDP-glucuronosyltransferase (UGT). Br J Clin Pharmacol 76:587–602

Knockaert L, Fromenty B, Robin MA (2011a) Mechanisms of mitochondrial targeting of cytochrome P450 2E1: physiopathological role in liver injury and obesity. FEBS J 278:4252–4260

Knockaert L, Descatoire V, Vadrot N, Fromenty B, Robin MA (2011b) Mitochondrial CYP2E1 is sufficient to mediate oxidative stress and cytotoxicity induced by ethanol and acetaminophen. Toxicol In Vitro 25:475–484

Koliaki C, Szendroedi J, Kaul K, Jelenik T, Nowotny P, Jankowiak F, Herder C, Carstensen M, Krausch M, Knoefel WT, Schlensak M, Roden M (2015) Adaptation of hepatic mitochondrial function in humans with non-alcoholic fatty liver is lost in steatohepatitis. Cell Metab 21:739–746

Labbe G, Pessayre D, Fromenty B (2008) Drug-induced liver injury through mitochondrial dysfunction: mechanisms and detection during preclinical safety studies. Fundam Clin Pharmacol 22:335–353

Larosche I, Lettéron P, Fromenty B, Vadrot N, Abbey-Toby A, Feldmann G, Pessayre D, Mansouri A (2007) Tamoxifen inhibits topoisomerases, depletes mitochondrial DNA, and triggers steatosis in mouse liver. J Pharmacol Exp Ther 321:526–535

Law M, Rudnicka AR (2006) Statin safety: a systematic review. Am J Cardiol 97:52C–60C

Lebrecht D, Walker UA (2007) Role of mtDNA lesions in anthracycline cardiotoxicity. Cardiovasc Toxicol 7:108–113

Lebrecht D, Geist A, Ketelsen UP, Haberstroh J, Setzer B, Walker UA (2007a) Dexrazoxane prevents doxorubicin-induced long-term cardiotoxicity and protects myocardial mitochondria from genetic and functional lesions in rats. Br J Pharmacol 151:771–778

Lebrecht D, Vargas-Infante YA, Setzer B, Kirschner J, Walker UA (2007b) Uridine supplementation antagonizes zalcitabine-induced microvesicular steatohepatitis in mice. Hepatology 45:72–79

Lee EW, Lai Y, Zhang H, Unadkat JD (2006) Identification of the mitochondrial targeting signal of the human equilibrative nucleoside transporter 1 (hENT1): implications for interspecies differences in mitochondrial toxicity of fialuridine. J Biol Chem 281:16700–16706

Lee YH, Goh WW, Ng CK, Raida M, Wong L, Lin Q, Boelsterli UA, Chung MC (2013) Integrative toxicoproteomics implicates impaired mitochondrial glutathione import as an off-target effect of troglitazone. J Proteome Res 12:2933–2945

Lee KK, Imaizumi N, Chamberland SR, Alder NN, Boelsterli UA (2015) Targeting mitochondria with methylene blue protects mice against acetaminophen-induced liver injury. Hepatology 61:326–336

Lei X, Chen Y, Du G, Yu W, Wang X, Qu H, Xia B, He H, Mao J, Zong W, Liao X, Mehrpour M, Hao X, Chen Q (2006) Gossypol induces Bax/Bak-independent activation of apoptosis and cytochrome c release via a conformational change in Bcl-2. FASEB J 20:2147–2149

Lewis W, Copeland WC, Day BJ (2001) Mitochondrial DNA depletion, oxidative stress, and mutation: mechanisms of dysfunction from nucleoside reverse transcriptase inhibitors. Lab Investig 81:777–790

Li N, Ragheb K, Lawler G, Sturgis J, Rajwa B, Melendez JA, Robinson JP (2003) Mitochondrial complex I inhibitor rotenone induces apoptosis through enhancing mitochondrial reactive oxygen species production. J Biol Chem 278:8516–8525

Li S, Guo J, Ying Z, Chen S, Yang L, Chen K, Long Q, Qin D, Pei D, Liu X (2015) Valproic acid-induced hepatotoxicity in Alpers syndrome is associated with mitochondrial permeability transition pore opening-dependent apoptotic sensitivity in an induced pluripotent stem cell model. Hepatology 61:1730–1739

Lucena MI, García-Martín E, Andrade RJ, Martínez C, Stephens C, Ruiz JD, Ulzurrun E, Fernandez MC, Romero-Gomez M, Castiella A, Planas R, Durán JA, De Dios AM, Guarner C, Soriano G, Borraz Y, Agundez JA (2010) Mitochondrial superoxide dismutase and glutathione peroxidase in idiosyncratic drug-induced liver injury. Hepatology 52:303–312

Luis PB, Ruiter JP, Ofman R, Ijlst L, Moedas M, Diogo L, Garcia P, de Almeida IT, Duran M, Wanders RJ, Silva MF (2011) Valproic acid utilizes the isoleucine breakdown pathway for its complete β-oxidation. Biochem Pharmacol 82:1740–1746

Maggini M, Raschetti R, Traversa G, Bianchi C, Caffari B, Da Cas R, Panei P (2004) The cerivastatin withdrawal crisis: a "post-mortem" analysis. Health Policy 69:151–157

Mansouri A, Haouzi D, Descatoire V, Demeilliers C, Sutton A, Vadrot N, Fromenty B, Feldmann G, Pessayre D, Berson A (2003) Tacrine inhibits topoisomerases and DNA synthesis to cause mitochondrial DNA depletion and apoptosis in mouse liver. Hepatology 38:715–725

Marcoff L, Thompson PD (2007) The role of coenzyme Q10 in statin-associated myopathy: a systematic review. J Am Coll Cardiol 49:2231–2237

Margolis AM, Heverling H, Pham PA, Stolbach A (2014) A review of the toxicity of HIV medications. J Med Toxicol 10:26–39

Mari M, Morales A, Colell A, Garcia-Ruiz C, Kaplowitz N, Fernandez-Checa JC (2013) Mitochondrial glutathione: features, regulation and role in disease. Biochim Biophys Acta 1830:3317–3328

Marrer E, Dieterle F (2010) Impact of biomarker development on drug safety assessment. Toxicol Appl Pharmacol 243:167–179

Marroquin L, Swiss R, Will Y (2014) Identifying compounds that induce opening of the mitochondrial permeability transition pore in isolated rat liver mitochondria. Curr Protoc Toxicol 60:25.4.1–25.4.17

Marzoa-Rivas R, Crespo-Leiro MG, Paniagua-Marin MJ, Llinares-García D, Muniz-Garcia J, Aldama-López G, Pinon-Esteban P, Campo-Pérez R, Castro-Beiras A (2005) Safety of statins when response is carefully monitored: a study of 336 heart recipients. Transplant Proc 37:4071–4073

Massart J, Begriche K, Buron N, Porceddu M, Borgne-Sanchez A, Fromenty B (2013) Drug-induced inhibition of mitochondrial fatty acid oxidation and steatosis. Curr Pathobiol Rep 1:147–157

Massart J, Katayama M, Krook A (2016) microManaging glucose and lipid metabolism in skeletal muscle: role of microRNAs. Biochim Biophys Acta 1861:2130–2138

Massart J, Begriche K, Moreau C, Fromenty B (2017) Role of nonalcoholic fatty liver disease as risk factor for drug-induced hepatotoxicity. J Clin Transl Res 3(Suppl 1):212–232

McGill MR, Jaeschke H (2013) Metabolism and disposition of acetaminophen: recent advances in relation to hepatotoxicity and diagnosis. Pharm Res 30:2174–2187

McGill MR, Li F, Sharpe MR, Williams CD, Curry SC, Ma X, Jaeschke H (2014) Circulating acylcarnitines as biomarkers of mitochondrial dysfunction after acetaminophen overdose in mice and humans. Arch Toxicol 88:391–401

McGill MR, Kennon-McGill S, Durham D, Jaeschke H (2016) Hearing, reactive metabolite formation, and oxidative stress in cochleae after a single acute overdose of acetaminophen: an in vivo study. Toxicol Mech Methods 26:104–111

McKenzie R, Fried MW, Sallie R, Conjeevaram H, Di Bisceglie AM, Park Y, Savarese B, Kleiner D, Tsokos M, Luciano C (1995) Hepatic failure and lactic acidosis due to fialuridine (FIAU), an investigational nucleoside analogue for chronic hepatitis B. N Engl J Med 333:1099–1105

Michaut A, Le Guillou D, Moreau C, Bucher S, McGill MR, Martinais S, Gicquel T, Morel I, Robin MA, Jaeschke H, Fromenty B (2016) A cellular model to study drug-induced liver injury in nonalcoholic fatty liver disease: application to acetaminophen. Toxicol Appl Pharmacol 292:40–55

Michaut A, Moreau C, Robin MA, Fromenty B (2014) Acetaminophen-induced liver injury in obesity and nonalcoholic fatty liver disease. Liver Int 34:e171–e179

Miltenburg NC, Boogerd W (2014) Chemotherapy-induced neuropathy: a comprehensive survey. Cancer Treat Rev 40:872–882

Mortensen PB, Gregersen N, Kølvraa S, Christensen E (1980) The occurrence of C6-C10-dicarboxylic acids in urine from patients and rats treated with dipropylacetate. Biochem Med 24:153–161

Nadanaciva S, Will Y (2011) New insights in drug-induced mitochondrial toxicity. Curr Pharm Des 17:2100–2112

Nadanaciva S, Bernal A, Aggeler R, Capaldi R, Will Y (2007a) Target identification of drug induced mitochondrial toxicity using immunocapture based OXPHOS activity assays. Toxicol In Vitro 21:902–911

Nadanaciva S, Dykens JA, Bernal A, Capaldi RA, Will Y (2007b) Mitochondrial impairment by PPAR agonists and statins identified via immunocaptured OXPHOS complex activities and respiration. Toxicol Appl Pharmacol 223:277–287

Nadanaciva S, Rana P, Beeson GC, Chen D, Ferrick DA, Beeson CC, Will Y (2012) Assessment of drug-induced mitochondrial dysfunction via altered cellular respiration and acidification measured in a 96-well platform. J Bioenerg Biomembr 44:421–437

Nanau RM, Neuman MG (2013) Adverse drug reactions induced by valproic acid. Clin Biochem 46:1323–1338

Nicolescu AC, Ji Y, Comeau JL, Hill BC, Takahashi T, Brien JF, Racz WJ, Massey TE (2008) Direct mitochondrial dysfunction precedes reactive oxygen species production in amiodarone-induced toxicity in human peripheral lung epithelial HPL1A cells. Toxicol Appl Pharmacol 227:370–379

Ornoy A (2009) Valproic acid in pregnancy: how much are we endangering the embryo and fetus? Reprod Toxicol 28:1–10

Pacheu-Grau D, Gómez-Durán A, Iglesias E, López-Gallardo E, Montoya J, Ruiz-Pesini E (2013) Mitochondrial antibiograms in personalized medicine. Hum Mol Genet 22:1132–9113

Patel J, Superko HR, Martin SS, Blumenthal RS, Christopher-Stine L (2015) Genetic and immunologic susceptibility to statin-related myopathy. Atherosclerosis 240:260–271

Pereira GC, Silva AM, Diogo CV, Carvalho FS, Monteiro P, Oliveira PJ (2011) Drug-induced cardiac mitochondrial toxicity and protection: from doxorubicin to carvedilol. Curr Pharm Des 17:2113–2129

Pereira GC, Pereira SP, Pereira CV, Lumini JA, Magalhães J, Ascensão A, Santos MS, Moreno AJ, Oliveira PJ (2012a) Mitochondrionopathy phenotype in doxorubicin-treated Wistar rats depends on treatment protocol and is cardiac-specific. PLoS One 7:e38867

Pereira CV, Oliveira PJ, Will Y, Nadanaciva S (2012b) Mitochondrial bioenergetics and drug-induced toxicity in a panel of mouse embryonic fibroblasts with mitochondrial DNA single nucleotide polymorphisms. Toxicol Appl Pharmacol 264:167–181

Pereira GC, Pereira SP, Tavares LC, Carvalho FS, Magalhães-Novais S, Barbosa IA, Santos MS, Bjork J, Moreno AJ, Wallace KB, Oliveira PJ (2016) Cardiac cytochrome c and cardiolipin depletion during anthracycline-induced chronic depression of mitochondrial function. Mitochondrion 30:95–104

Pessayre D, Mansouri A, Berson A, Fromenty B (2010) Mitochondrial involvement in drug-induced liver injury. Handb Exp Pharmacol 196:311–365

Phillips PS, Haas RH, Bannykh S, Hathaway S, Gray NL, Kimura BJ, Vladutiu GD, England JD (2002) Statin-associated myopathy with normal creatine kinase levels. Ann Intern Med 137:581–585

Ponchaut S, Veitch K (1993) Valproate and mitochondria. Biochem Pharmacol 46:199–204

Ponchaut S, van Hoof F, Veitch K (1992) Cytochrome aa3 depletion is the cause of the deficient mitochondrial respiration induced by chronic valproate administration. Biochem Pharmacol 43:644–647

Porceddu M, Buron N, Roussel C, Labbe G, Fromenty B, Borgne-Sanchez A (2012) Prediction of liver injury induced by chemicals in human with a multiparametric assay on isolated mouse liver mitochondria. Toxicol Sci 129:332–345

Powell-Jackson PR, Tredger JM, Williams R (1984) Hepatotoxicity to sodium valproate: a review. Gut 25:673–681

Price LM, Poklis A, Johnson DE (1991) Fatal acetaminophen poisoning with evidence of subendocardial necrosis of the heart. J Forensic Sci 36:930–935

Ravindranath V, Strobel HW (2013) Cytochrome P450-mediated metabolism in brain: functional roles and their implications. Expert Opin Drug Metab Toxicol 9:551–558

Rui L (2014) Energy metabolism in the liver. Compr Physiol 4:177–197

Rumbach L, Cremel G, Marescaux C, Warter JM, Waksman A (1989) Succinate transport inhibition by valproate in rat renal mitochondria. Eur J Pharmacol 164:577–581

Saad SY, Najjar TA, Al-Rikabi AC (2001) The preventive role of deferoxamine against acute doxorubicin-induced cardiac, renal and hepatic toxicity in rats. Pharmacol Res 43:211–218

Sahin E, Colla S, Liesa M, Moslehi J, Müller FL, Guo M, Cooper M, Kotton D, Fabian AJ, Walkey C, Maser RS, Tonon G, Foerster F, Xiong R, Wang YA, Shukla SA, Jaskelioff M, Martin ES, Heffernan TP, Protopopov A, Ivanova E, Mahoney JE, Kost-Alimova M, Perry SR, Bronson R, Liao R, Mulligan R, Shirihai OS, Chin L, DePinho RA (2011) Telomere dysfunction induces metabolic and mitochondrial compromise. Nature 470:359–365

Santangeli P, Di Biase L, Burkhardt JD, Bai R, Mohanty P, Pump A, Natale A (2012) Examining the safety of amiodarone. Expert Opin Drug Saf 11:191–214

Sardao VA, Pereira SL, Oliveira PJ (2008) Drug-induced mitochondrial dysfunction in cardiac and skeletal muscle injury. Expert Opin Drug Saf 7:129–146

Sayed-Ahmed MM, Shouman SA, Rezk BM, Khalifa MH, Osman AM, El-Merzabani MM (2000) Propionyl-L-carnitine as potential protective agent against adriamycin-induced impairment of fatty acid beta-oxidation in isolated heart mitochondria. Pharmacol Res 41:143–150

Schick BA, Laaksonen R, Frohlich JJ, Päivä H, Lehtimäki T, Humphries KH, Côté HC (2007) Decreased skeletal muscle mitochondrial DNA in patients treated with high-dose simvastatin. Clin Pharmacol Ther 81:650–653

Schon E, Fromenty B (2016) Alterations of mitochondrial DNA in liver diseases. In: Kaplowitz N, Han D (eds) Mitochondria in liver disease. Taylor & Francis, New-York, pp 279–309

Schumacher JD, Guo GL (2015) Mechanistic review of drug-induced steatohepatitis. Toxicol Appl Pharmacol 289:40–47

Seachrist JL, Loi CM, Evans MG, Criswell KA, Rothwell CE (2005) Roles of exercise and pharmacokinetics in cerivastatin-induced skeletal muscle toxicity. Toxicol Sci 88:551–561

Seifert EL, Estey C, Xuan JY, Harper ME (2010) Electron transport chain-dependent and -independent mechanisms of mitochondrial H_2O_2 emission during long-chain fatty acid oxidation. J Biol Chem 285:5748–5758

Serrano J, Palmeira CM, Kuehl DW, Wallace KB (1999) Cardioselective and cumulative oxidation of mitochondrial DNA following subchronic doxorubicin administration. Biochim Biophys Acta 1411:201–205

Serviddio G, Bellanti F, Giudetti AM, Gnoni GV, Capitanio N, Tamborra R, Romano AD, Quinto M, Blonda M, Vendemiale G, Altomare E (2011) Mitochondrial oxidative stress and respiratory chain dysfunction account for liver toxicity during amiodarone but not dronedarone administration. Free Radic Biol Med 51:2234–2242

Setzer B, Lebrecht D, Walker UA (2008) Pyrimidine nucleoside depletion sensitizes to the mitochondrial hepatotoxicity of the reverse transcriptase inhibitor stavudine. Am J Pathol 172:681–690

Silva MF, Aires CC, Luis PB, Ruiter JP, IJlst L, Duran M, Wanders RJ, Tavares de Almeida I (2008) Valproic acid metabolism and its effects on mitochondrial fatty acid oxidation: a review. J Inherit Metab Dis 31:205–216

Sirvent P, Mercier J, Lacampagne A (2008) New insights into mechanisms of statin-associated myotoxicity. Curr Opin Pharmacol 8:333–338

Spaniol M, Bracher R, Ha HR, Follath F, Krähenbühl S (2001) Toxicity of amiodarone and amiodarone analogues on isolated rat liver mitochondria. J Hepatol 35:628–636

Spiekerkoetter U, Wood PA (2010) Mitochondrial fatty acid oxidation disorders: pathophysiological studies in mouse models. J Inherit Metab Dis 33:539–546

Stankov MV, Panayotova-Dimitrova D, Leverkus M, Schmidt RE, Behrens GM (2013) Thymidine analogues suppress autophagy and adipogenesis in cultured adipocytes. Antimicrob Agents Chemother 57:543–551

Star K, Edwards IR, Choonara I (2014) Valproic acid and fatalities in children: a review of individual case safety reports in VigiBase. PLoS One 9:e108970

Stewart JD, Horvath R, Baruffini E, Ferrero I, Bulst S, Watkins PB, Fontana RJ, Day CP, Chinnery PF (2010) Polymerase γ gene POLG determines the risk of sodium valproate-induced liver toxicity. Hepatology 52:1791–1796

Stringer HA, Sohi GK, Maguire JA, Côté HC (2013) Decreased skeletal muscle mitochondrial DNA in patients with statin-induced myopathy. J Neurol Sci 325:142–147

Sunny NE, Bril F, Cusi K (2017) Mitochondrial adaptation in nonalcoholic fatty liver disease: novel mechanisms and treatment strategies. Trends Endocrinol Metab 28(4):250–260

Sutton A, Khoury H, Prip-Buus C, Cepanec C, Pessayre D, Degoul F (2003) The Ala16Val genetic dimorphism modulates the import of human manganese superoxide dismutase into rat liver mitochondria. Pharmacogenetics 13:145–157

Taanman JW (1999) The mitochondrial genome: structure, transcription, translation and replication. Biochim Biophys Acta 1410:103–123

Tacar O, Sriamornsak P, Dass CR (2013) Doxorubicin: an update on anticancer molecular action, toxicity and novel drug delivery systems. J Pharm Pharmacol 65:157–170

Thakker D, Nair S, Pagada A, Jamdade V, Malik A (2016) Statin use and the risk of developing diabetes: a network meta-analysis. Pharmacoepidemiol Drug Saf 25:1131–1149

Thelen K, Dressman JB (2009) Cytochrome P450-mediated metabolism in the human gut wall. J Pharm Pharmacol 61:541–558

Thompson PD (2016) What to believe and do about statin-associated adverse effects. JAMA 316:1969–1970

Thompson PD, Clarkson PM, Rosenson RS (2006) An assessment of statin safety by muscle experts. Am J Cardiol 97:69C–76C

Thor H, Smith MT, Hartzell P, Bellomo G, Jewell SA, Orrenius S (1982) The metabolism of menadione (2-methyl-1,4-naphthoquinone) by isolated hepatocytes. A study of the implications of oxidative stress in intact cells. J Biol Chem 257:12419–12425

Tocchetti CG, Gallucci G, Coppola C, Piscopo G, Cipresso C, Maurea C, Giudice A, Iaffaioli RV, Arra C, Maurea N (2013) The emerging issue of cardiac dysfunction induced by antineoplastic angiogenesis inhibitors. Eur J Heart Fail 15:482–489

Trost LC, Lemasters JJ (1996) The mitochondrial permeability transition: a new pathophysiological mechanism for Reye's syndrome and toxic liver injury. J Pharmacol Exp Ther 278:1000–1005

Varga ZV, Ferdinandy P, Liaudet L, Pacher P (2015) Drug-induced mitochondrial dysfunction and cardiotoxicity. Am J Physiol Heart Circ Physiol 309:H1453–H1467

von Vital JM, Karachristos A, Singhal A, Thomas R, Jain A (2011) Acute amiodarone hepatotoxicity after liver transplantation. Transplantation 91:e62–e64

Vliegenthart AD, Shaffer JM, Clarke JI, Peeters LE, Caporali A, Bateman DN, Wood DM, Dargan PI, Craig DG, Moore JK, Thompson AI, Henderson NC, Webb DJ, Sharkey J, Antoine DJ, Park BK, Bailey MA, Lader E, Simpson KJ, Dear JW (2015) Comprehensive microRNA profiling in acetaminophen toxicity identifies novel circulating biomarkers for human liver and kidney injury. Sci Rep 5:15501

Wallace DC, Fan W, Procaccio V (2010) Mitochondrial energetics and therapeutics. Annu Rev Pathol 5:297–348

Wang Y, Lin Z, Liu Z, Harris S, Kelly R, Zhang J, Ge W, Chen M, Borlak J, Tong W (2013) A unifying ontology to integrate histological and clinical observations for drug-induced liver injury. Am J Pathol 182:1180–1187

Weng Z, Wang K, Li H, Shi Q (2015) A comprehensive study of the association between drug hepatotoxicity and daily dose, liver metabolism, and lipophilicity using 975 oral medications. Oncotarget 6:17031–17038

Wester CW, Eden SK, Shepherd BE, Bussmann H, Novitsky V, Samuels DC, Hendrickson SL, Winkler CA, O'Brien SJ, Essex M, D'Aquila RT, DeGruttola V, Marlink RG (2012) Risk factors for symptomatic hyperlactatemia and lactic acidosis among combination antiretroviral therapy-treated adults in Botswana: results from a clinical trial. AIDS Res Hum Retrovir 28:759–765

Will Y, Dykens J (2014) Mitochondrial toxicity assessment in industry—a decade of technology development and insight. Expert Opin Drug Metab Toxicol 10:1061–1067

Yamanaka H, Gatanaga H, Kosalaraksa P, Matsuoka-Aizawa S, Takahashi T, Kimura S, Oka S (2007) Novel mutation of human DNA polymerase gamma associated with mitochondrial toxicity induced by anti-HIV treatment. J Infect Dis 195:1419–1425

Yoshikawa H, Watanabe T, Abe T (2002) Fanconi syndrome caused by sodium valproate: report of three severely disabled childred. Eur J Paediatr Neurol 6:165–167

Yue L, Yao H (2016) Mitochondrial dysfunction in inflammatory responses and cellular senescence: pathogenesis and pharmacological targets for chronic lung diseases. Br J Pharmacol 173:2305–2318

Potential Mechanisms of Mitochondrial DNA Mediated Acquired Mitochondrial Disease

Afshan N. Malik and Hannah S. Rosa

Abstract Mitochondria are cellular organelles which contain mitochondrial DNA (MtDNA) in the form of an extranuclear genome. MtDNA can be present in 100s to thousands to copies per cell in the body depending on the bioenergetic requirements of the host cell. MtDNA encodes subunits of the electron transport chain and therefore is required to produce cellular energy in the form of ATP. However, MtDNA can also act as an inflammatory molecule since it resembles bacterial DNA, resulting in activation of pathways leading to enhanced cytokine production and chronic inflammation. In the current chapter, we suggest that MtDNA mediated mechanisms that cause systemic mitochondrial dysfunction are involved in many common diseases not traditionally recognised as mitochondrial disease, and we suggest that such disorders could be considered as "acquired mitochondrial diseases", distinct from primary and secondary mitochondrial disease. Acquired mitochondrial diseases include cardiovascular and neurodegenerative disease as well as diabetic complications, and are associated with oxidative stress and sterile/chronic inflammation in their pathophysiology. We propose a mechanism of how systemic damage to MtDNA, mediated through oxidative stress, can cause inflammation and bioenergetic deficit. Some recent evidence of the proposed mechanism is provided for diabetic nephropathy, a complication of diabetes, which has not traditionally been regarded as a disease of mitochondrial dysfunction. According to the hypothesis proposed, it may be possible to use MtDNA levels in body fluids or cells to predict risk of acquired diseases of mitochondrial dysfunction, and to design novel therapies targeting the specific MtDNA mediated pathways.

Keywords Mitochondrial DNA · Mitochondrial dysfunction · Acquired disease · Metabolism · Biomarker · MtDNA · Acquired mitochondrial disease · Inflammation · Oxidative stress

A. N. Malik (✉) · H. S. Rosa
Diabetes Research Group, Faculty of Life Sciences and Medicine, King's College London, London, UK
e-mail: afshan.malik@kcl.ac.uk

P. J. Oliveira (ed.), *Mitochondrial Biology and Experimental Therapeutics*,
https://doi.org/10.1007/978-3-319-73344-9_14

1 Introduction

Mitochondria are double-membrane bound organelles found in all human cells, with the exception of erythrocytes. Mitochondria are often described as the "powerhouses of the cell" since they produce the majority of cellular energy via oxidative phosphorylation (OXPHOS), a process involving the transfer of electrons along a series of protein complexes in the inner mitochondrial membrane. The process of electron transport results in the production of some reactive oxygen species (ROS) (Boveris et al. 1972), which play an important role in normal intracellular signalling pathways (D'Autreaux and Toledano 2007). In addition to cellular energy production, mitochondria play many other roles in the cell including homeostasis of cellular calcium, formation of iron-sulphur clusters and regulation of the intrinsic apoptosis pathway. Given their importance in maintaining normal cellular function, it is perhaps no surprise that mitochondria have become a focus of increased attention in terms of their potential impact on the pathogenic mechanisms of human disease.

It is believed that mitochondria originate from the integration of a eubacterium into a host cell, resulting in what we now call eukaryotic cells. Through reductive evolution, the majority of genes were transferred from the eubacterial genome to the nucleus of the host, however a small DNA genome remained within the mitochondrial organelle. The mitochondrial genome (MtDNA) differs from the nuclear genome in terms of size, structure and mode of inheritance. With only 13 proteins encoded by MtDNA, the mitochondrion is under dual control relying on many other proteins encoded by the nuclear genome to function normally. Mutations in these genes can lead to mitochondrial dysfunction, manifesting at a cellular level in various ways, including OXPHOS deficiency, reduced MtDNA copy number or disruption of the mitochondrial network. The clinical symptoms associated with mitochondrial dysfunction fall under the umbrella term of "mitochondrial disease", which encompasses a clinically and genetically diverse group of conditions with systemic effects on the body's cell and organs (Fig. 1). These are broadly categorised as either primary or secondary mitochondrial disease, however there is some overlap regarding the definition of these two main groups. In addition, mitochondrial dysfunction has been postulated to play a key role in many other common diseases, which were not traditionally considered as mitochondrial disease (Wallace 1999). To distinguish these from primary or secondary mitochondrial disease, here we will refer to them as "acquired mitochondrial disease" and they are the main topic of the current review (Table 1).

Primary mitochondrial disease is usually associated with inherited mutations in MtDNA or nuclear DNA which result in impaired OXPHOS (Niyazov et al. 2016). Pathogenic mutations can occur in genes encoding structural subunits of the OXPHOS system, such as the complex I subunits *MT-ND1* in the MtDNA or *NDUFS1* in the nuclear DNA. In addition, there are approximately 1500 nuclear genes encoding proteins indirectly affecting OXPHOS via processes like MtDNA replication, which have been implicated in a number of clinical syndromes. The prevalence of primary adult mitochondrial disease due to nuclear and mitochondrial

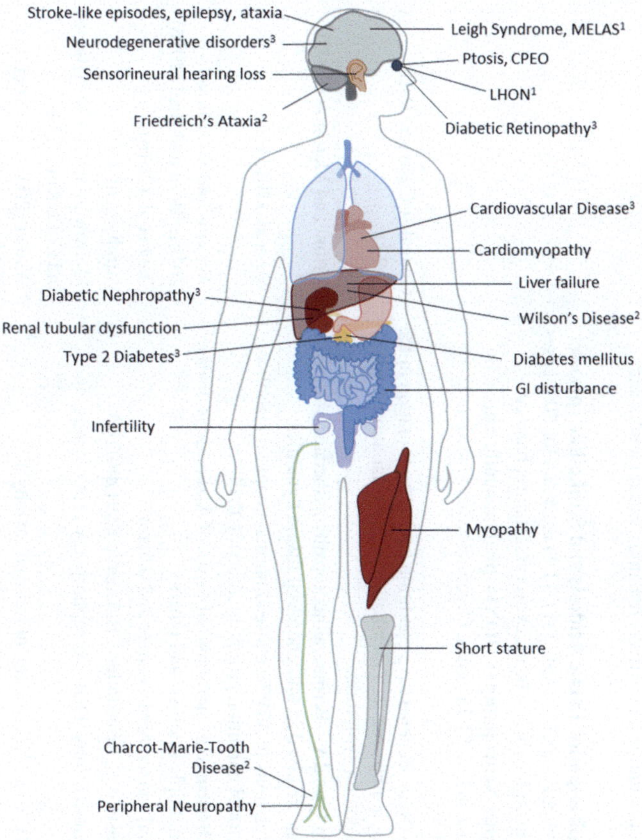

Fig. 1 Systemic impact of mitochondrial disease. Common symptoms of mitochondrial disease are shown alongside examples of primary[1], secondary[2] and acquired[3] mitochondrial diseases. Abbreviations: *MELAS* Mitochondrial Encephalopathy, Lactic Acidosis and Stroke-like episodes, *CPEO* Chronic Progressive Ophthalmoplegia, *LHON* Leber's Hereditary Optic Neuropathy, *GI* gastrointestinal

DNA mutations is estimated at 1 in 4300 (Gorman et al. 2015), although accurately estimating this is challenging due to the genetic and clinical heterogeneity of this group of diseases. Clinically, these conditions can affect any tissue with onset at any age, making diagnosis and management extremely challenging. Furthermore, there is little consistency between the clinical phenotype and the underlying mutation. For example, the m.3243A>G mutation is associated with a severe neurological condition termed MELAS (mitochondrial encephalopathy, lactic acidosis and stroke-like episodes), but can also manifest as a much milder CPEO (chronic external ophthalmoplegia) phenotype. Secondary mitochondrial disease arises due to mutations in "genes encoding neither function nor production of the OXPHOS proteins" (Niyazov et al. 2016). These mutations can be inherited, as in the case of

Table 1 Primary, secondary and acquired mitochondrial disease

Group and definition	Definition	Example: Genes if known: Effects	References
Primary mitochondrial disease	Mostly inherited mutations which result in impaired OXPHOS	**Leigh Syndrome:** Progressive neurological disorder with infantile onset >75 nDNA and MtDNA genes implicated. Affects thalamus, basal ganglia and brain stem.	Gorman et al. (2016)
		Mitochondrial Encephalopathy, Lactic Acidosis and Stroke-like episodes (MELAS): MtDNA mutation m.3243A>G (MT-TL1) accounts for 80% cases. Affect many systems including brain, heart, eyes, ears, stature, balance.	
		Leber Hereditary Optic Neuropathy (LHON)—predominantly caused by three MtDNA mutations in OXPHOS genes. Organ-specific condition associated with vision loss.	
Secondary mitochondrial disease	Mostly inherited mutations which affect mitochondrial function (not in genes encoding OXPHOS)	**Friedreich's Ataxia**—Neurodegenerative condition caused by mutation in the Frataxin gene resulting in impaired iron metabolism. Cerebellar ataxia and cardiomyopathy are common symptoms.	González-Cabo and Palau (2013)
		Charcot-Marie-Tooth type 2A—Peripheral neuropathy associated with mutation in MFN2 gene involved in mitochondrial fusion dynamics.	Cartoni and Martinou (2009)
		Wilson's disease—Mutations in ATP7B cause abnormal copper metabolism, resulting in a multifactorial condition primarily affecting the liver.	Wu et al. (2015)
Acquired mitochondrial disease	Somatic mutations/ deletions in MtDNA caused by environmental factors (e.g., oxidative stress) affecting specific target tissues	**Neurodegenerative Disease** (e.g., non-familial forms of Alzheimer's disease[1], Parkinson's disease[2], multiple sclerosis and others)	[1] Wilkins and Swerdlow (2016); [2]Bose and Beal (2016)
		Cancer (e.g., colorectal cancer, breast cancer and others)	Chen et al. (2016)
		Cardiovascular disease (e.g., heart failure, ischaemic heart disease)	Murphy et al. (2016)
		Diabetes[1] and its complications (Diabetic nephropathy[2], retinopathy, and others)	[1]Morrow et al. (2017); [2]Higgins and Coughlan (2014)
		Drug induced toxicity[1] (e.g., HIV treatment induced toxicity[2])	[1]Walker (2008); [2]Gardner et al. (2014)

Friedreich's ataxia where a mutation in the frataxin gene (*FXN*) leads to dysregulation of iron homeostasis (González-Cabo and Palau 2013). A detailed description of primary and secondary mitochondrial disease is beyond the scope of this chapter and the reader is referred to an excellent recent review by Gorman et al. (2016).

In addition to primary and secondary mitochondrial disease, it has been postulated that mitochondrial dysfunction plays a significant part in the pathogenesis of many common disorders including common metabolic disorders such as diabetes, cardiovascular disease, neurodegenerative diseases and cancers (Schon et al. 2012; Picard et al. 2016). A growing body of evidence implicates changes in MtDNA in cells organs and body fluids in such diseases of acquired mitochondrial dysfunction, including prevalent and chronic health conditions including neurodegenerative disease, obesity, cardiovascular disease, diabetes and its complications, fatty liver disease, and also include complications arising from drug toxicity (reviewed by Malik and Czajka 2013; Malik 2017). Furthermore, emerging evidence indicates that the development of mitochondrial dysfunction can also contribute to chronic inflammation and therefore may also be involved in a number of inflammatory pathways (Boyapati et al. 2017).

In the current chapter, we consider the role of MtDNA in the mechanisms of pathogenesis in acquired mitochondrial disease. Accordingly, we suggest that mitochondrial dysfunction plays a central role in many more diseases than traditionally recognised as mitochondrial disease. We propose that such disorders could be described as putative "acquired mitochondrial diseases" (Table 1) and we propose a mechanism of how systemic damage to MtDNA, mediated through processes such as oxidative stress, and contributing to chronic inflammation and bioenergetic deficit, may be involved in the processes leading to acquired mitochondrial disease. To illustrate how the proposed mechanisms may contribute to common diseases, we review some recent evidence using as an example one of the diseases shown in Table 1, namely diabetic nephropathy. Diabetic nephropathy has not traditionally been regarded as a disease resulting from mitochondrial dysfunction.

2 The Mitochondrial Genome

In the eukaryotic cell, mitochondria are the only organelle, apart from the nucleus, to harbour a DNA genome. The nuclear genome comprises of linear double-stranded DNA molecules, complexed with histone proteins in the form of chromosomes located within the nucleus, whereas the mitochondrial genome is a circular 16,569 bp, double-stranded molecule (Anderson et al. 1981), present within the mitochondrial matrix as a protein-DNA complex of approximately 100 nm length and known as a nucleoid (Table 2). Nucleoids, located at the inner mitochondrial membrane, are estimated to contain between 1.4 and 3 copies of MtDNA (Kukat et al. 2011; Brown et al. 2011) which is complexed with several abundant proteins. The proteins within nucleoids are involved in the mitochondrial replication and transcription machinery, for example the mitochondrial transcription factor A (TFAM),

Table 2 Comparison of Human Mitochondrial and Nuclear DNA Genomes.

	Mitochondrial DNA	Nuclear DNA
Structure	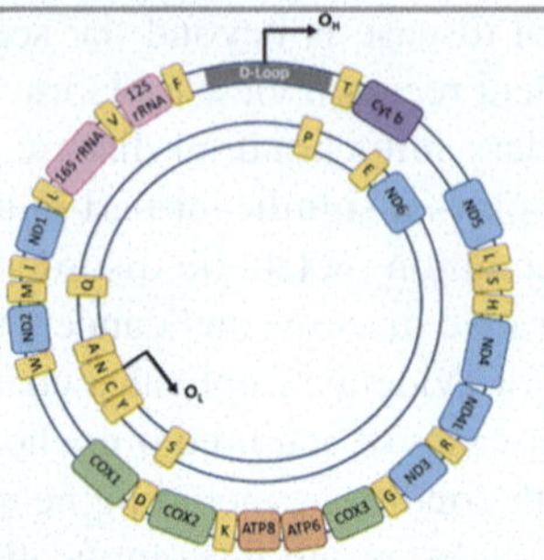	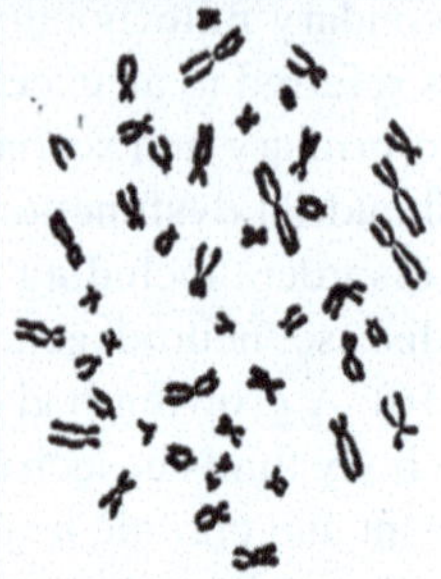
Size (nt)	16,569	~30,000,000,000
Major Function	Information storage DNA replication and transcription, signalling	Information storage DNA replication and transcription
Organisation	Double-stranded circular DNA molecule complexed with TFAM and other proteins to form nucleoids	Double-stranded duplex linear DNA molecules (chromosomes) complexed with histones to form chromatin
Genetic code	Different use of start and stop codons	Universal
Replication	Bidirectional from a single origin of replication	Numerous origins of replication
Transcription	Poly-cistronic mRNAs from 2 promoters	Highly regulated and mostly individual mRNA transcription from thousands of individual promoters
Introns/exons	No introns, very few non-coding regions, contiguous and overlapping	Contain introns and large stretches of non-coding regions
Inheritance	Maternal	Bi-parental
Replication	Independent of the cell cycle	Dependent on the cell cycle
Number of copies per cell	10s to many 1000s–variable and can change in response to physiological stimuli	2 (1 in the case of the sex chromosomes)-fixed
Methylation	low levels of CpG methylation (resembles bacterial DNA)	high levels of CpG methylation
Sequence identity between the two genomes	Contains very few regions which are unique (>90% is duplicated in the nuclear genome)	Contains variable regions known as NUMTs (nuclear mitochondrial DNA segments) which are identical to MtDNA scattered across most chromosomes

mitochondrial DNA helicase (TWINKLE), single-stranded DNA binding protein (mtSSB) and mitochondrial DNA polymerase (Polγ) (Gilkerson et al. 2013).

In total, the 16.5 kb MtDNA contains 37 genes, encoding 22 transfer RNAs (mt-tRNAs), 13 hydrophobic proteins involved in the OXPHOS system and two ribosomal RNAs (mt-rRNAs). Most of these genes are located on the outer, guanine-rich heavy strand, whereas the inner, cysteine-rich light strand holds only 8 tRNA genes and one protein-encoding gene (*MT-ND6*) (Andrews et al. 1999). Unlike nuclear DNA, MtDNA is compact and contains no introns. One or two non-coding base-pairs separate genes, except in the case of *MT-ATP6* and *MT-ATP8*, where there is a region of overlap between these two genes. There is only one non-coding region, known as the D-loop. This region, extending approximately 1 kb in human MtDNA, contains the major control elements for MtDNA transcription and is the site of origin for heavy strand replication (Kasamatsu et al. 1971; Shadel and Clayton 1997).

In terms of genome copy numbers, there is a vast excess of MtDNA relative to nuclear DNA in the human body. With the exception of erythrocytes, human cells contain multiple mitochondria, each carrying MtDNA. The number of mitochondria within a cell, and therefore the number of mtDNA copies, can vary depending upon the cellular energy requirements of the cell (Lightowlers et al. 1997). For example, a cardiac muscle cell may have around 7000 copies of MtDNA (Miller et al. 2003), a white blood cell may have less than a hundred (Selak et al. 2011) whereas oocytes may contain several hundred thousand copies (Shoubridge and Wai 2007). The number of mitochondria can also change in response to the cell's physiological environment (Trinei et al. 2006), consequently, the amount of MtDNA found in different cell types varies depending on the mitochondrial content and bioenergetic requirements of the cell, as well as the physiological status of the cell.

The inheritance of MtDNA follows a non-Mendelian pattern and is instead strictly maternally inherited (Giles et al. 1980), although there is one documented case of paternal inheritance of MtDNA with a deletion in *MT-ND2* (Schwartz and Vissing 2002), yet this has not been reported by others. Paternal MtDNA is thought to be eliminated by selective proteasomal degradation or autophagy post-fertilization, although degradation prior to fertilisation has also been reported (Luo et al. 2013). Mutations in MtDNA are passed from mother to child, however there is great variability in the heteroplasmy levels and phenotypic expression across generations (Taylor and Turnbull 2005).

3 Mitochondrial DNA Content in Different Cells and Organs

Organs and cells in the body have differing levels of mitochondrial content, reflected by their MtDNA content. Mitochondrial content in different cells and organs varies depending on the bioenergetic requirements of the cells, with organs requiring high amounts of energy usually having a higher mitochondrial mass, therefore it follows that with altered mitochondrial mass there will be altered MtDNA content. Currently there is little consensus in the literature on MtDNA content in different cells and organs. We have previously shown that in the mouse, the heart contains the highest MtDNA content followed by the kidney (Fig. 2a, Malik et al. 2016), this observation correlates with earlier proteomic quantification of mitochondrial proteins which found the highest level of mitochondrial cytochrome *C* in the heart followed by the kidney (Fig. 2b, Pagliarini et al. 2008). The observation of high mitochondrial content in heart and

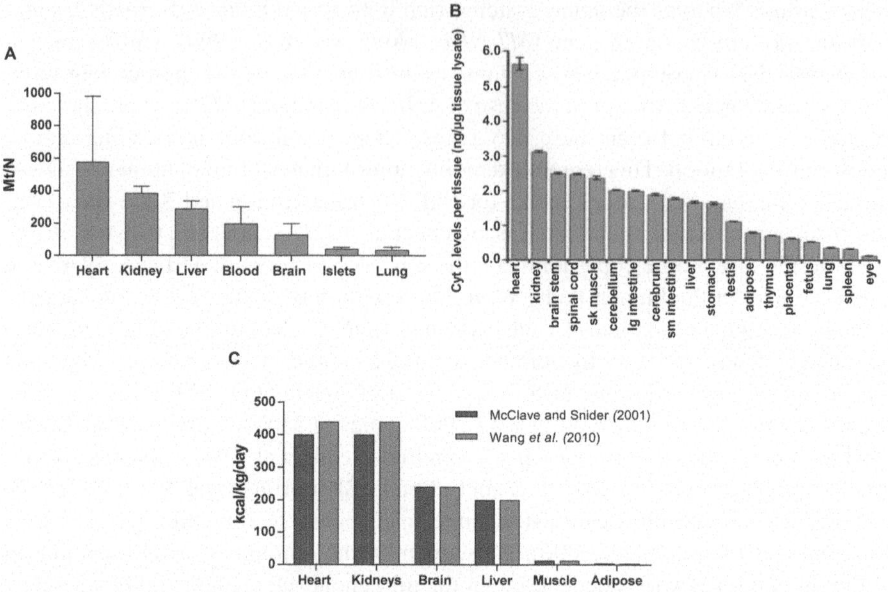

Fig. 2 Mitochondrial content in different tissues. (**a**) Mitochondrial DNA copy number in mouse tissue (Malik et al. 2016). (**b**) Mitochondrial content, represented by cytochrome *c* levels, in mouse tissues (Pagliarini et al. 2008). (**c**) Metabolic rates of human tissues and organs (McClave and Snider 2001; Wang et al. 2010)

kidney is further supported by the assessment of metabolic rate of different organs where two reports of very similar levels with heart and kidney showing the highest metabolic rates (Fig. 2c, McClave and Snider 2001; Wang et al. 2010).

4 Oxidative Stress and Acquired Mitochondrial Dysfunction

The key role of oxidative stress in the chronic disorders we designate as acquired mitochondrial disease has long been recognised and many therapeutic strategies are focused around the correction of redox impairment and scavenging of ROS (Halliwell and Gutteridge 2007). The process of energy production via OXPHOS uses oxygen and metabolic substrates to produce energy in the form of ATP in mitochondria, and can directly result in oxidative stress in the cell, a condition in which the redox balance of the cell has shifted towards a more oxidative state. It is largely accepted that oxidative stress is a common feature in many diseases including diabetic complications, cardiovascular disease, neurodegenerative disease, cancer, renal disease and others.

OXPHOS involves the oxidation of nutrients by oxygen, mediated via the transfer of electrons through protein complexes known as the electron transport chain in the inner mitochondrial membrane. However, not all of the oxygen being used during OXPHOS is utilised to produce ATP, since ~1–3% is converted into reactive oxygen species (ROS) (Halliwell and Gutteridge 2007). Some of the ROS are highly reactive molecules with unpaired electrons and can oxidise and impair the function of cellular DNA, proteins and lipids. In the case of DNA, ROS damage can cause mutations, deletions and oxidation. Since MtDNA is located close to the electron transport chain, it is more likely to incur ROS-induced damage than nuclear DNA.

Under normal physiological conditions, most cells use cellular proteins such as glutathione peroxidise, catalase, and superoxide dismutase (Nordberg and Arnér 2001) as antioxidants to sequester ROS by accepting electrons and becoming oxidised. These endogenous antioxidant proteins are highly abundant in cells and are recycled by donating their electrons to chains of acceptors such as reduced nicotinamide adenine dinucleotide phosphate (NADPH) (Rydstrom 2006). However, in conditions of chronic ROS production, the cell's ROS levels exceed their detoxification and cause a shift in the redox balance resulting in oxidative stress. Free radicals which escape the cells' antioxidant response are highly reactive molecules which oxidise proteins, lipids and DNA molecules within the cell, leading to altered properties and cellular damage. Many common drugs cause mitochondrial oxidative stress (reviewed by Mehta et al. 2008), and many common diseases such as diabetes and its complications, cancer, neurodegenerative disorders as well as ageing have been shown to involve mitochondrial redox impairment (Halliwell and Gutteridge 2007; Wallace 1999; Ying 2008).

Mitochondria are directly affected by oxidative stress, which can impair their functioning at different levels, both by direct damage to MtDNA, mitochondrial lipids and proteins, and also indirectly by altering cellular signalling. Oxidative stress induced MtDNA damage is a central concept in the "free radical theory of ageing", which proposes that oxidative stress and ROS directly lead to damage to MtDNA in the form of single mutations or deletions. Mutations and damage to MtDNA are clearly involved in ageing and several acquired diseases, although damage to MtDNA repair and replication errors are also likely to be involved in MtDNA damage rather than direct ROS induced damage.

Abnormal signalling has been shown to result in an adaptive response through enhanced production of mitochondria (Michel et al. 2012). We have shown that oxidative stress induced by hyperglycemia can cause an increase initial increase in MtDNA; however, in cells with oxidative stress the increased MtDNA is not functional as such cells show decreased viability, decreased cellular respiration, and decreased transcription of MtDNA (Czajka et al. 2015). The effect of the shifted redox balance can also impact on mitophagy, the degradation and removal of damaged mitochondria. Reduced removal results in the accumulation of damaged mitochondria (Kim et al. 2007) as is the case for diabetes where blockage of the electron transport chain at complex III results in accumulation of excess ROS (Giacco and Brownlee 2010; Newsholme et al. 2007). As MtDNA is located close to the source of ROS production, the DNA itself can become damaged resulting in accumulation of deletions and mutations (Croteau and Bohr 1997).

Theoretically, the maternally inherited mitochondrial genome in an individual should be the same in all cell types unless it has been damaged via mutations or deletions either before or after conception, in which case cells may exhibit heteroplasmy in their mitochondria (Taylor and Turnbull 2005). Since most cells contain many copies of MtDNA, ranging from 100s to 1000s depending on their bioenergetic requirements, a small percentage of damaged MtDNA may not have much functional impact, and further MtDNA repair mechanisms and mitophagy may suffice to contain any impact of such damage. However, once a functional threshold is crossed, damage to MtDNA could result in metabolic perturbation in the tissue. Cells need sufficient energy supplies for cellular processes and a fully functional ETC is crucial especially in tissues with high energy requirements such as the heart, kidney and the brain. Damage to MtDNA, which encodes 37 genes, all crucial for mitochondrial function, in the form of acquired mutations, could lead to energy deficit by affecting the function of the ETC (Wallace 1999). The inability to produce sufficient energy to carry out cellular processes is likely to have significant effects on cells, impairing cellular repair and growth and could lead to cell loss.

5 Inflammation and Acquired Mitochondrial Dysfunction

It is increasingly being reported that MtDNA can directly cause inflammation in the body (reviewed by Malik and Czajka 2013). Unlike eukaryotic nuclear DNA which is methylated, MtDNA resembles bacterial DNA in that it is largely unmethylated. Therefore, when MtDNA is "out of place", i.e., not contained within the mitochondria, for example in damaged cells where MtDNA may leak into the cytosol or into circulation, it can be recognised by the immune system as foreign DNA. MtDNA can therefore cause immune responses, including through the activation of the intracellular Toll like receptor 9 (TLR) 9 which normally recognises exogenous bacterial DNA released from pathogen. (Zhang et al. 2010). Activation of TLR9 by MtDNA can cause the activation of an inflammatory cascade via myeloid differentiation primary response gene 88 (MYD88) and nuclear factor kappa-light-chain-enhancer of activated B cells (NF-κB) and results in activation of cytokines such as TNFalpha leading to an inflammatory response which can persist in the absence of any exogenous infection (McCarthy et al. 2015). Indeed, MtDNA-mediated inflammation may be the cause of phenomena described as "sterile inflammation", where an enhanced inflammatory response occurs in the absence of infection (Chen and Nunez 2010).

The diseases listed as "acquired mitochondrial diseases" (Table 1) share the feature of chronic or sterile inflammation since an enhanced inflammatory response ensues in the absence of any detectable pathogens or infection. There is overwhelming evidence showing a causal role of immune-inflammatory responses in cardiovascular disease where the presence of low grade inflammation in patients has been postulated to be causative of and to precede adverse cardiovascular events (Welsh et al. 2017). There are many convincing studies using both animal models and human populations showing that in diabetes inflammation plays a significant role, and increased inflammatory markers are associated with diabetic nephropathy (Donate-Correa et al. 2015) and diabetic retinopathy (Semeraro et al. 2015) complications. Therefore, it is tempting to speculate that inflammation present in these diseases may be mediated by MtDNA.

As well as playing a potential role in chronic inflammatory responses, the release of large amounts of MtDNA can result in acute inflammatory responses with serious consequences. For example, blocking of MtDNA degradation in cardiac cells led to heart failure in a mouse model (Oka et al. 2012). Cell-free MtDNA levels have been shown to strongly correlate with risk of mortality in intensive care units and in patients with sepsis (Nakahira et al. 2014; Yamanouchi et al. 2013).

6 Emerging Evidence of MtDNA Changes in a Large Number of Chronic and Other Human Diseases

MtDNA lends itself to rapid detection and quantification via methods such as qPCR and digital PCR, making it an attractive high-throughput biomarker. MtDNA levels can be measured in human clinical samples and disease associated changes can be detected in populations. There is widespread interest in using MtDNA as a biomarker in human populations and in the last decade, the evidence for reported alterations in MtDNA in body fluids, cells, or of human subjects in correlation with many diseases has grown rapidly and strongly supports the view that alterations in MtDNA levels are associated with many human diseases. MtDNA levels have been measured in numerous types of human samples such as blood cells (e.g., lymphocytes, PBMCs), body fluids (urine, plasma, serum, saliva, cerebrospinal fluid), and in tissues or biopsies (normal and diseased e.g., cancer). Alterations in MtDNA levels have been described in numerous different types of cancers, cardiovascular disease, diabetes and its complications and others. A detailed description of this rapidly growing evidence base is beyond the scope of the current chapter (Reviewed by Malik and Czajka 2013); however, it is clear that many human diseases are associated with measurable changes in MtDNA. Furthermore, there is increasing evidence showing a link between patient drug response and circulating MtDNA levels showing that there is strong potential for the future use of MtDNA in the field of personalised medicine (Malik 2017).

7 Proposed Mechanisms of Mitochondrial DNA-Mediated Damage in Acquired Mitochondrial Disease

We previously proposed a novel hypothesis stating that acquired damage to MtDNA, distinct from inherited genetic mitochondrial disease, can result in mitochondrial dysfunction in a specific pathway involving an early increase in MtDNA levels

(Malik and Czajka 2013; Czajka et al. 2015). According to this hypothesis, changes in MtDNA may impact on cellular health via two distinct but interconnected processes: (a) Bio-energetic deficit in cells caused by mutations in MtDNA leading to a defective electron transport chain resulting in cellular damage, and release of damaged MtDNA into circulation (b) Accumulation of damaged MtDNA in the cytosol (instead of within mitochondria) and of cell-free MtDNA in circulation leading to chronic inflammation. Therefore, this hypothesis predicts an early adaptive increase in MtDNA in response to oxidative stress, and subsequently a decrease in levels of MtDNA as well as increased damage to MtDNA, followed by increased inflammation and reduced bioenergetic capacity (*see* Fig. 3). In order to detect the changes predicted by this hypothesis, we have used experimental model systems and clinical samples for diabetic nephropathy.

Diabetes results in increased risk of numerous other complications which affect major organs, including eyes (retinopathy), heart (diabetic cardiomyopathy), blood vessels (peripheral vascular disease) and brain (dementia) (Fig. 1) and therefore resembles genetic forms of mitochondrial disease. Diabetic nephropathy is a kidney disease which affects approximately 30% of patients with diabetes. The disease, detected using microalbuminuria, the appearance of protein in the urine, develops

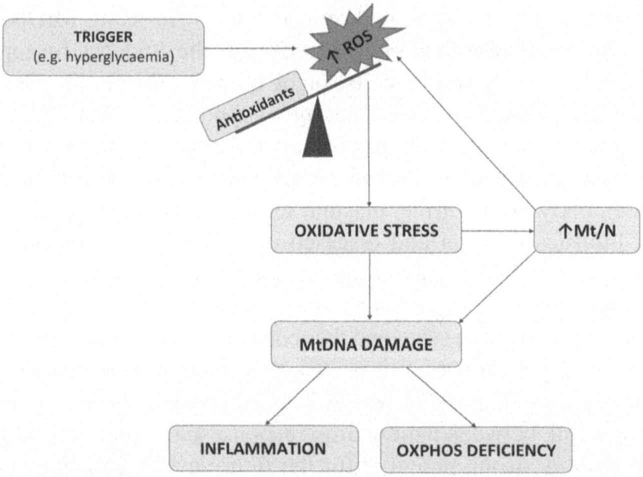

Fig. 3 Proposed mechanism of MtDNA mediated damage in acquired mitochondrial diseases. A trigger such as hyperglycaemia leads to increased intracellular reactive oxygen species (ROS). Under normal conditions, the cell's endogenous antioxidant response recovers the redox balance by scavenging the ROS. However, under conditions of chronic ROS, the cell's antioxidant response is overwhelmed resulting in oxidative stress. Increased intracellular ROS leads to increased mitochondrial biogenesis, an adaptive response, which can be detected by measuring the mitochondrial genome (Mt) to nuclear genome (N) ratio (Mt/N). Over time the oxidative stress would cause damage to mitochondrial membranes, proteins and DNA leading to mitochondrial dysfunction. Accumulation of damaged MtDNA would lead to an inflammatory response as MtDNA is unmethylated and resembles bacterial DNA. According to this hypothesis changes in Mt/N would precede mitochondrial dysfunction. Figure adapted from Malik and Czajka (2013)

over a period of clinical silence and can lead to gradual loss of renal function resulting in end stage renal failure. With the epidemic rise in the incidence of diabetes currently affecting more 350 million people worldwide, more than 100 million people are at risk of DN and approximately 30% of these are likely to progress to end stage renal failure despite therapy. Kidneys are organs requiring large amounts of energy in form of ATP due to the re-absorption processes and although their mass account of less than 1% of total body mass, they use almost 10% of the body's oxygen which is utilized in cellular respiration via OXPHOS, and therefore they are rich in mitochondrial content and MtDNA. We used human primary renal glomerular mesangial cells (HMCs) as well as patient samples to detect the changes predicted by the hypothesis described in Fig. 3 (Malik and Czajka 2013; Czajka and Malik 2016). We showed that when HMCs are grown in hyperglycaemic conditions, the high glucose leads to increased ROS and at the same time we observed more than 200% increase in cellular MtDNA (Fig. 4a), however this MtDNA was damaged (Fig. 4b). Moreover, continued cell culturing in high glucose led to damaged

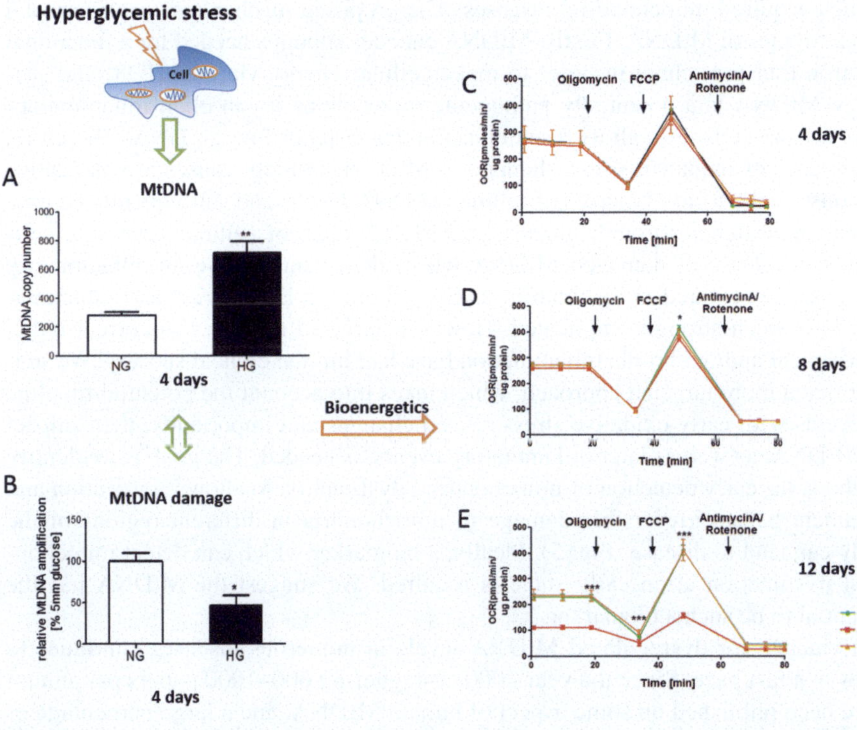

Fig. 4 Changes in cellular MtDNA precede metabolic dysfunction in conditions of oxidative stress. Growth of human mesangial cells (HMCs) in high glucose led to a significant increase in cellular MtDNA within 24 h, (**a**) after 4 days ($n = 12$, $P < 0.001$), however the MtDNA was damaged as illustrated by reduced amplification of an MtDNA 8.6 kb fragment (**b**). Cells showed normal bio-energetic profile at 4 days (**c**), however after 8 days, maximal respiration and reserve capacity were significantly reduced in hyperglycemia cells but unaffected in normoglycmeic cells (**d**, **e**). *See* Czajka et al. (2015) for the full data

cellular respiration in parallel with increased ROS (Fig. 4e). We also found evidence of similar changes in blood samples taken from diabetes patients with and without diabetic kidney disease (*see* Czajka et al. 2015) and in patients with diabetic retinopathy, where circulating MtDNA changes correlated with inflammation (Malik et al. 2015). Specific MtDNA mutations can lead to diabetes and kidney disease in patients with mitochondrial genetic disease (Wallace 1999; Mazzaccara et al. 2012; Seidowsky et al. 2013; D'Aco et al. 2013). These genetic forms of diabetes and kidney disease support our hypothesis as they show that MtDNA damage can cause these specific pathologies.

8 Concluding Remarks

In the current chapter, we have described potential mechanisms by which changes in MtDNA levels may mediate mechanisms of common diseases which we have termed acquired mitochondrial diseases. The proposed mechanisms centre around two attributes of MtDNA. Firstly, MtDNA encodes subunits needed for a functional electron transport chain in order to make cellular energy via OXPHOS, and secondly MtDNA is a potentially pathogenic molecule as it can elicit inflammatory responses in cells through its resemblance to bacterial or foreign DNA. Therefore, in diseases of oxidative stress, damage to MtDNA, initially causes a maladaptive response in the form of increased cellular MtDNA levels, and subsequently cause a bioenergetic deficit through impaired OXPHOS. Lack of cellular energy in cells results in release of damaged MtDNA which then would cause an inflammatory response. In acquired mitochondrial diseases, both oxidative stress and inflammation have been strongly implicated. However, targeted therapies to correct redox impairment and/or chronic inflammation have had limited clinical success. We suggest that a more targeted approach, which takes into account the potential timeline of events with early oxidative stress related changes and importantly, their impact on MtDNA, as well as later inflammatory events, is needed. The goal is to identify methods for early detection of mitochondrial dysfunction to allow intervention and treatment before irreversible damage to mitochondria in different regions of the body can lead to disease (Fig. 5). Ideally, a biomarker which can detect mitochondrial dysfunction at an early stage is required. We suggest the MtDNA has the potential to be such a biomarker.

Evidence for dysregulated MtDNA levels in numerous diseases continues to grow at a fast pace. Since the year 2000, a staggering 600–1000 papers per annum have been published on some aspect of human MtDNA, and a large percentage of these relate to altered MtDNA in chronic disease in human body fluids or cells. Interestingly, our data suggests that early changes in MtDNA, caused by altered redox balance in the cell, may provide a predictive biomarker (Fig. 4), since we found changes in MtDNA were detectable early, and preceded bioenergetic deficit.

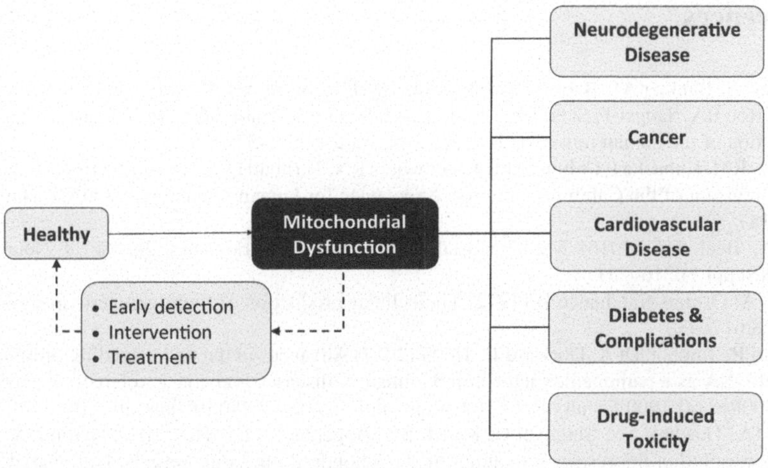

Fig. 5 Early detection of mitochondrial dysfunction to prevent chronic disease. Environmental/lifestyles triggers such as high fat and/or glucose or drugs can result in oxidative stress and altered signalling which in turn damages mitochondria in organs (e.g., kidney, heart, liver), cells (blood cells, adipocytes), and blood vessels, the damage may take decades to manifest itself and cause pathology. Identification of biomarkers for the early detection of the metabolic and bioenergetic changes associated with these pathologies could allow intervention and prevention of irreversible bioenergetic dysfunction

One potential way to identify the stage of events is to use non-invasive methods to measure MtDNA content in human subjects. However, although the wider availability of qPCR as a methodology has led to a substantial increase in publications reporting changes in MtDNA content in human body fluids and tissues, methodological issues remain. Protocols for accurately quantifying MtDNA in small volumes of body fluids remain variable and affected by problems. Such methodology issues have hindered the successful use of MtDNA as a biomarker and led to conflicting and unreproducible findings in many cases. Of particular note in this regard is the presence of nuclear mitochondrial DNA segments (NUMTs) in the nuclear genome which can skew data by co-amplifying nuclear genes when MtDNA levels are being assessed. In addition, assays currently in use seldom distinguish between cell free and cellular MtDNA, the former is of importance as it may be an indicator of inflammation, and the latter is important as it may be an indicator of bioenergetic deficit in the cell. Therefore there is a growing need to accurately measure MtDNA levels in peripheral blood, compartmentalised as PMBCs for cellular and plasma for cell-free, as well as in urine, compartmentalised as urinary pellet for cellular debris and cell free urinary supernatant, or other body fluids, such as saliva, semen, or cerebrospinal fluid. The field of MtDNA may hold the potential for the elucidation of many more diseases than traditionally thought of as mitochondrial genetic disease.

Acknowledgements Thanks to Dr. Anna Czajka for donating Fig. 4. HSR is supported by an EFSD grant.

References

Anderson S, Bankier AT, Barrell BG, De Bruijn MH, Coulson AR, Drouin J, Eperon IC, Nierlich DP, Roe BA, Sanger F, Schreier PH, Smith AJ, Staden R, Young IG (1981) Sequence and organization of the human mitochondrial genome. Nature 290:457–465

Andrews RM, Kubacka I, Chinnery PF, Lightowlers RN, Turnbull DM, Howell N (1999) Reanalysis and revision of the Cambridge reference sequence for human mitochondrial DNA. Nat Genet 23:147

Bose A, Beal MF (2016) Mitochondrial dysfunction in Parkinson's disease. J Neurochem 139(Suppl 1):216–231

Boveris A, Oshino N, Chance B (1972) The cellular production of hydrogen peroxide. Biochem J 128:617–630

Boyapati R, Tamborska A, Dorward D, Ho G (2017) Advances in the understanding of mitochondrial DNA as a pathogenic factor in inflammatory diseases [version 1; referees: 3 approved]. F1000Res 6(F1000 Faculty Rev):169. https://doi.org/10.12688/f1000research.10397.1

Brown TA, Tkachuk AN, Shtengel G, Kopek BG, Bogenhagen DF, Hess HF, Clayton DA (2011) Superresolution fluorescence imaging of mitochondrial nucleoids reveals their spatial range, limits, and membrane interaction. Mol Cell Biol 31:4994–5010

Cartoni R, Martinou J-C (2009) Role of mitofusin 2 mutations in the physiopathology of Charcot–Marie–Tooth disease type 2A. Exp Neurol 218:268–273

Chen GY, Nunez G (2010) Sterile inflammation: sensing and reacting to damage. Nat Rev Immunol 10:826–837

Chen N, Wen S, Sun X, Fang Q, Huang L, Liu S, Li W, Qiu M (2016) Elevated mitochondrial DNA copy number in peripheral blood and tissue predict the opposite outcome of cancer: a meta-analysis. Sci Rep 6:37404

Croteau DL, Bohr VA (1997) Repair of oxidative damage to nuclear and mitochondrial DNA in mammalian cells. J Biol Chem 272:25409–25412

Czajka A, Malik AN (2016) Hyperglycemia induced damage to mitochondrial respiration in renal mesangial and tubular cells: implications for diabetic nephropathy. Redox Biol 10:100–107

Czajka A, Ajaz S, Gnudi L, Parsade CK, Jones P, Reid F, Malik AN (2015) Altered mitochondrial function, mitochondrial DNA and reduced metabolic flexibility in patients with diabetic nephropathy. EBioMedicine 2:499–512

D'aco KE, Manno M, Clarke C, Ganesh J, Meyers KEC, Sondheimer N (2013) Mitochondrial tRNAPhe mutation as a cause of end-stage renal disease in childhood. Pediatr Nephrol 28:515–519

D'Autreaux B, Toledano MB (2007) ROS as signalling molecules: mechanisms that generate specificity in ROS homeostasis. Nat Rev Mol Cell Biol 8:813–824

Donate-Correa J, Martin-Nunez E, Muros-de-Fuentes M, Mora-Fernandez C, Navarro-Gonzalez JF (2015) Inflammatory cytokines in diabetic nephropathy. J Diabetes Res 2015:9

Gardner K, Hall PA, Chinnery PF, Payne BA (2014) HIV treatment and associated mitochondrial pathology: review of 25 years of in vitro, animal, and human studies. Toxicol Pathol 42:811–822

Giacco F, Brownlee M (2010) Oxidative stress and diabetic complications. Circ Res 107:1058–1070

Giles RE, Blanc H, Cann HM, Wallace DC (1980) Maternal inheritance of human mitochondrial DNA. Proc Natl Acad Sci U S A 77:6715–6719

Gilkerson R, Bravo L, Garcia I, Gaytan N, Herrera A, Maldonado A, Quintanilla B (2013) The mitochondrial nucleoid: integrating mitochondrial DNA into cellular homeostasis. Cold Spring Harb Perspect Biol 5:a011080

González-Cabo P, Palau F (2013) Mitochondrial pathophysiology in Friedreich's ataxia. J Neurochem 126:53–64

Gorman GS, Schaefer AM, Ng Y, Gomez N, Blakely EL, Alston CL, Feeney C, Horvath R, Yu-Wai-Man P, Chinnery PF, Taylor RW, Turnbull DM, Mcfarland R (2015) Prevalence of nuclear and mitochondrial DNA mutations related to adult mitochondrial disease. Ann Neurol 77:753–759

Gorman GS, Chinnery PF, Dimauro S, Hirano M, Koga Y, McFarland R, Suomalainen A, Thorburn DR, Zeviani M, Turnbull DM (2016) Mitochondrial diseases. Nat Rev Dis Primers 2:16080

Halliwell B, Gutteridge JMC (2007) Free radicals in biology and medicine. Oxford University Press, Oxford

Higgins GC, Coughlan MT (2014) Mitochondrial dysfunction and mitophagy: the beginning and end to diabetic nephropathy? Br J Pharmacol 171:1917–1942

Kasamatsu H, Robberson DL, Vinograd J (1971) A novel closed-circular mitochondrial DNA with properties of a replicating intermediate. Proc Natl Acad Sci U S A 68:2252–2257

Kim I, Rodriguez-Enriquez S, Lemasters JJ (2007) Selective degradation of mitochondria by mitophagy. Arch Biochem Biophys 462:245–253

Kukat C, Wurm CA, Spahr H, Falkenberg M, Larsson NG, Jakobs S (2011) Super-resolution microscopy reveals that mammalian mitochondrial nucleoids have a uniform size and frequently contain a single copy of mtDNA. Proc Natl Acad Sci U S A 108:13534–13539

Lightowlers RN, Chinnery PF, Turnbull DM, Howell N (1997) Mammalian mitochondrial genetics: heredity, heteroplasmy and disease. Trends Genet 13:450–455

Luo S-M, Schatten H, Sun Q-Y (2013) Sperm mitochondria in reproduction: good or bad and where do they go? J Genet Genomics 40:549–556

Malik AN (2017) Mitochondrial DNA as a potential translational biomarker of mitochondrial dysfunction in drug induced toxicity studies. In: Dykens JA, Wills Y (eds) Drug induced mitochondrial dysfunction: progress towards the clinic. John Wiley & Sons Inc., New Jersey

Malik AN, Czajka A (2013) Is mitochondrial DNA content a potential biomarker of mitochondrial dysfunction? Mitochondrion 13:481–492

Malik AN, Parsade CK, Ajaz S, Crosby-Nwaobi R, Gnudi L, Czajka A, Sivaprasad S (2015) Altered circulating mitochondrial DNA and increased inflammation in patients with diabetic retinopathy. Diabetes Res Clin Pract 110:257–265

Malik AN, Czajka A, Cunningham P (2016) Accurate quantification of mouse mitochondrial DNA without co-amplification of nuclear mitochondrial insertion sequences. Mitochondrion 29:59–64

Mazzaccara C, Iafusco D, Liguori R, Ferrigno M, Galderisi A, Vitale D, Simonelli F, Landolfo P, Prisco F, Masullo M, Sacchetti L (2012) Mitochondrial diabetes in children: seek and you will find it. PLoS One 7:e34956

McCarthy CG, Wenceslau CF, Goulopoulou S, Ogbi S, Baban B, Sullivan JC, Matsumoto T, Webb RC (2015) Circulating mitochondrial DNA and Toll-like receptor 9 are associated with vascular dysfunction in spontaneously hypertensive rats. Cardiovasc Res 107:119–130

McClave SA, Snider HL (2001) Dissecting the energy needs of the body. Curr Opin Clin Nutr Metab Care 4:143–147

Mehta R, Chan K, Lee O, Tafazoli S, O'brien PJ (2008) Drug-associated mitochondrial toxicity. In: Dykens JA, Will Y (eds) Drug induced mitochondrial dysfunction. John Wiley & Sons, Inc., Hoboken, NJ

Michel S, Wanet A, De Pauw A, Rommelaere G, Arnould T, Renard P (2012) Crosstalk between mitochondrial (dys)function and mitochondrial abundance. J Cell Physiol 227:2297–2310

Miller FJ, Rosenfeldt FL, Zhang C, Linnane AW, Nagley P (2003) Precise determination of mitochondrial DNA copy number in human skeletal and cardiac muscle by a PCR-based assay: lack of change of copy number with age. Nucleic Acids Res 31:e61–e61

Morrow RM, Picard M, Derbeneva O, Leipzig J, McManus MJ, Gouspillou G, Barbat-Artigas S, dos Santos C, Hepple RT, Murdock DG, Wallace DC (2017) Mitochondrial energy deficiency leads to hyperproliferation of skeletal muscle mitochondria and enhanced insulin sensitivity. Proc Natl Acad Sci 114:2705–2710

Murphy E, Ardehali H, Balaban RS, Dilisa F, Dorn GW 2nd, Kitsis RN, Otsu K, Ping P, Rizzuto R, Sack MN, Wallace D, Youle RJ (2016) Mitochondrial function, biology, and role in disease: a scientific statement from the American heart association. Circ Res 118:1960–1991

Nakahira K, Kyung S-Y, Rogers AJ, Gazourian L, Youn S, Massaro AF, Quintana C, Osorio JC, Wang Z, Zhao Y, Lawler LA, Christie JD, Meyer NJ, Causland FRM, Waikar SS, Waxman AB, Chung RT, Bueno R, Rosas IO, Fredenburgh LE, Baron RM, Christiani DC, Hunninghake

GM, Choi AMK (2014) Circulating mitochondrial DNA in patients in the ICU as a marker of mortality: derivation and validation. PLoS Med 10:e1001577
Newsholme P, Haber EP, Hirabara SM, Rebelato EL, Procopio J, Morgan D, Oliveira-Emilio HC, Carpinelli AR, Curi R (2007) Diabetes associated cell stress and dysfunction: role of mitochondrial and non-mitochondrial ROS production and activity. J Physiol 583:9–24
Niyazov DM, Kahler SG, Frye RE (2016) Primary mitochondrial disease and secondary mitochondrial dysfunction: importance of distinction for diagnosis and treatment. Mol Syndromol 7:122–137
Nordberg J, Arnér ESJ (2001) Reactive oxygen species, antioxidants, and the mammalian thioredoxin system1 1This review is based on the licentiate thesis "Thioredoxin reductase—interactions with the redox active compounds 1-chloro-2,4-dinitrobenzene and lipoic acid" by Jonas Nordberg, 2001, Karolinska Institute, Stockholm, ISBN 91-631-1064-4. Free Radic Biol Med 31:1287–1312
Oka T, Hikoso S, Yamaguchi O, Taneike M, Takeda T, Tamai T, Oyabu J, Murakawa T, Nakayama H, Nishida K, Akira S, Yamamoto A, Komuro I, Otsu K (2012) Mitochondrial DNA that escapes from autophagy causes inflammation and heart failure. Nature 485:251–255
Pagliarini DJ, Calvo SE, Chang B, Sheth SA, Vafai SB, Ong S-E, Walford GA, Sugiana C, Boneh A, Chen WK, Hill DE, Vidal M, Evans JG, Thorburn DR, Carr SA, Mootha VK (2008) A mitochondrial protein compendium elucidates complex I disease biology. Cell 134:112–123
Picard M, Wallace DC, Burelle Y (2016) The rise of mitochondria in medicine. Mitochondrion 30:105–116
Rydstrom J (2006) Mitochondrial NADPH, transhydrogenase and disease. Biochim Biophys Acta 1757:721–726
Schon EA, Dimauro S, Hirano M (2012) Human mitochondrial DNA: roles of inherited and somatic mutations. Nat Rev Genet 13:878–890
Schwartz M, Vissing J (2002) Paternal inheritance of mitochondrial DNA. N Engl J Med 347:576–580
Seidowsky A, Hoffmann M, Glowacki F, Dhaenens CM, Devaux JP, de Sainte Foy CL, Provot F, Gheerbrant JD, Hummel A, Hazzan M, Dracon M, Dieux-Coeslier A, Copin MC, Noel C, Buob D (2013) Renal involvement in MELAS syndrome–a series of 5 cases and review of the literature. Clin Nephrol 80:456–463
Selak MA, Lyver E, Micklow E, Deutsch EC, Önder Ö, Selamoglu N, Yager C, Knight S, Carroll M, Daldal F, Dancis A, Lynch DR, Sarry J-E (2011) Blood cells from Friedreich ataxia patients harbor frataxin deficiency without a loss of mitochondrial function. Mitochondrion 11:342–350
Semeraro F, Cancarini A, Dell'omo R, Rezzola S, Romano MR, Costagliola C (2015) Diabetic retinopathy: vascular and inflammatory disease. J Diabetes Res 2015:16
Shadel GS, Clayton DA (1997) Mitochondrial DNA maintenance in vertebrates. Annu Rev Biochem 66:409–435
Shoubridge EA, Wai T (2007) Mitochondrial DNA and the mammalian oocyte. Curr Top Dev Biol 77:87–111
Taylor RW, Turnbull DM (2005) Mitochondrial DNA mutations in human disease. Nat Rev Genet 6:389–402
Trinei M, Berniakovich I, Pelicci PG, Giorgio M (2006) Mitochondrial DNA copy number is regulated by cellular proliferation: a role for Ras and p66Shc. Biochim Biophys Acta 1757:624–630
Wallace DC (1999) Mitochondrial diseases in man and mouse. Science 283:1482–1488
Wang Z, Ying Z, Bosy-Westphal A, Zhang J, Schautz B, Later W, Heymsfield SB, Müller MJ (2010) Specific metabolic rates of major organs and tissues across adulthood: evaluation by mechanistic model of resting energy expenditure. Am J Clin Nutr 92:1369–1377
Welsh P, Grassia G, Botha S, Sattar N, Maffia P (2017) Targeting inflammation to reduce cardiovascular disease risk: a realistic clinical prospect? Br J Pharmacol 174(22):3898–3913
Wilkins HM, Swerdlow RH (2016) Relationships between mitochondria and neuroinflammation: implications for Alzheimer's disease. Curr Top Med Chem 16:849–857
Wu F, Wang J, Pu C, Qiao L, Jiang C (2015) Wilson's disease: a comprehensive review of the molecular mechanisms. Int J Mol Sci 16:6419–6431

Walker UA. Lipoatrophy and Other Manifestations of antiretroviral therapeutics. In: J.A. D, Will Y, editors. Drug Induced Mitochondrial Dysfunction Hoboken, NJ, USA; 2008
Yamanouchi S, Kudo D, Yamada M, Miyagawa N, Furukawa H, Kushimoto S (2013) Plasma mitochondrial DNA levels in patients with trauma and severe sepsis: time course and the association with clinical status. J Crit Care 28:1027–1031
Ying W (2008) NAD+/NADH and NADP+/NADPH in cellular functions and cell death: regulation and biological consequences. Antioxid Redox Signal 10:179–206
Zhang Q, Raoof M, Chen Y, Sumi Y, Sursal T, Junger W, Brohi K, Itagaki K, Hauser CJ (2010) Circulating mitochondrial DAMPs cause inflammatory responses to injury. Nature 464:104–107

Part II
Mitochondrial Experimental Therapeutics: From the Bench to Industry and Patient

Part II
Mitochondrial Experimental Therapeutics: From the Bench to Industry and Patient

The Evolution of Mitochondrial Toxicity Assessment in Industry

James Hynes and Yvonne Will

Abstract Safety-related clinical attrition is of particular concern for the pharmaceutical industry due to the significant financial implications associated with late-stage failures. Considerable efforts have therefore been made over the past decade to address this attrition, leading to an industry-wide paradigm shift towards *in vitro* drug discovery toxicology. Traditionally such toxicological assessments were animal-model focused, with *in vitro* assays used largely as investigative tools; however, the last decade has seen significant prospective deployment of *in vitro* toxicology during series selection and lead development. Such *in vitro* assays have been utilised at this early stage, to better inform early decision making and to facilitate iterative structure-activity relationships, with the objective of progressing only the most promising compounds into animal studies and later into humans. Focus on the early deployment of *in vitro* models has also been motivated by the observation that several off-target human organ toxicities, particularly liver, are poorly predicted by conventional animal models. During this period, interest in drug-induced mitochondrial dysfunction as a source of compound attrition was generated by literature reports suggesting mitochondrial impairment played a role in organ toxicities associated with post-market drug withdrawals.

Keywords Isolated mitochondria · High throughput · soluble sensor · Plate reader · Mitochondrial uncouplers · Mitochondrial inhibitors

1 Introduction

Safety-related clinical attrition is of particular concern for the pharmaceutical industry due to the significant financial implications associated with late-stage failures. Considerable efforts have therefore been made over the past decade to

J. Hynes
Luxcel Biosciences, Cork, Ireland

Y. Will (✉)
Pfizer Drug Safety and Development, Groton, CT, USA
e-mail: yvonne.will@pfizer.com

P. J. Oliveira (ed.), *Mitochondrial Biology and Experimental Therapeutics*,
https://doi.org/10.1007/978-3-319-73344-9_15

address this attrition, leading to an industry-wide paradigm shift towards *in vitro* drug discovery toxicology. Traditionally such toxicological assessments were animal-model focused, with *in vitro* assays used largely as investigative tools; however, the last decade has seen significant prospective deployment of *in vitro* toxicology during series selection and lead development. Such *in vitro* assays have been utilised at this early stage, to better inform early decision making and to facilitate iterative SAR, with the objective of progressing only the most promising compounds into animal studies and later into FIH (first in human) (Cook et al. 2014; Hay et al. 2014). Focus on the early deployment of *in vitro* models has also been motivated by the observation that several off-target human organ toxicities, particularly liver, are poorly predicted by conventional animal models (Olson et al. 2000).

During this period, interest in drug-induced mitochondrial dysfunction as a source of compound attrition was generated by literature reports suggesting mitochondrial impairment played a role in organ toxicities associated with post-market drug withdrawals. Specific examples include the withdrawal of troglitazone in 1997 due to severe liver injury an cerivastatin in 2001 due to rhabdomyolysis, with both compounds subsequently shown to deleteriously impact mitochondrial function (Kaufmann et al. 2006; Seachrist et al. 2005; Tirmenstein et al. 2002; Westwood et al. 2005). In addition, mitochondrial liabilities were also being associated with drugs carrying a black box label for hepatic or cardiac toxicity (Dykens et al. 2008a), highlighting again the importance of mitochondrial toxicity when considering off-target effects.

Taken together, such observations suggested an inability of the evolving drug discovery toxicology paradigm to sensitively detect mitochondrial perturbation. Specifically, mitochondrial impairment did not appear to reveal itself easily in conventional preclinical species, likely due in part, to the use of young, drug naïve animals fed a standard diet and lacking genetic diversity which are unsuited to the identification of compounds that may cause sub-lethal reductions of mitochondrial capacity, a problem compounded by the fact that traditional serum biomarkers do not assess mitochondrial impairment, while histopathology only detects late stage mitochondrial toxicity and cannot inform whether observed ultrastructural abnormalities are a cause or a consequence of any observed etiology. Efforts to address this perceived gap included the development of a heterozygous superoxide dismutase 2 (SOD2) gene knockout (Sod2+/−) mouse model and its application to the study of troglitazone (Ong et al. 2007), flutamide (Kashimshetty et al. 2009) and nimesulide (Ong et al. 2006) liver toxicities. Although changes in mitochondrial enzymes were observed in all cases, liver necrosis was only observed with troglitazone, prompting the continued search for more sensitive biomarkers. Analogous weaknesses were believed to exist with conventional *in vitro* assays where late-stage cytotoxicity indicators such as LDH release or cell membrane integrity coupled with immortalised cell lines cultured in high glucose resulted in an insensitivity to mitochondrial toxicity.

The parallel realisations of, (1) the importance of mitochondrial dysfunction in drug-induced toxicity and (2) the weaknesses of deployed technologies in detecting this dysfunction prompted our hypothesis that the development and deployment of suitable *in vitro* mitochondrial toxicity assays would facilitate better mechanistic understanding of specific organ toxicities and more informed decision-making during drug development.

2 Developing the Assays

As outlined in detail elsewhere in this volume, the mitochondrial reticulum plays a central role in a vast array of critical cellular processes. It is therefore no surprise that there are multiple sites through which off-target effects can be mediated and that the implications of such perturbations can be severe. One of the earliest examples identified was the diminished bioenergetic capacity caused by long-term exposure to nucleoside reverse transcriptase inhibitors (NRTIs) which inhibit the off-target DNA polymerase γ, leading to impairment of mtDNA replication and, subsequent depletion of mtDNA-encoded proteins (Brinkman et al. 1999; Lewis and Dalakas 1995). It was later observed that NRTI-mediated mitochondrial toxicity can also occur as a result of direct inhibition of mtDNA-encoded protein synthesis (McKee et al. 2006; Nagiec et al. 2005), and similarly, that some antibacterials which inhibit bacterial protein synthesis can also inhibit mtDNA-encoded protein synthesis due to the structural similarity between the bacterial and mitochondrial ribosome (Böttger et al. 2001). Direct effects on cellular bioenergetics are more common however, often mediated through inhibition of one or more components of the electron transport chain (ETC) or the uncoupling of ADP phosphorylation from ETC activities (Brunmair et al. 2004; Burkhardt et al. 1993; Esposti 1998). Traditionally, analysis of mitochondrial function was performed using a clark-type electrode (Bachmann and Zbinden 1979; Clark 1959), with subsequent technological advancements facilitating detailed bioenergetics investigations (Gnaiger 2008). Also used, albeit less commonly, were single-sample spectroscopic investigations of mitochondrial OXPHOS complex activities (Birch-Machin et al. 1994) and mitochondria permeability transition (Bernardi et al. 1992). These were the first tools used in mechanistic investigation and, during 2005, were validated within Pfizer using troglitazone (a Pfizer attrited drug), cerivastatin and other members of these two drug classes. It quickly became apparent however that these approaches lacked the through-put to facilitate convenient compound screening and dose response generation. What was required was a series of relevant microplate-compatible high-throughput mitochondrial function assays that could be used, in the first instance to underpin mechanistic investigation of drug-induced mitochondrial dysfunction, but which also had the potential to be deployed at an earlier stage in the discovery process. Two discrete but complementary strategies were pursued to achieve this. The first, informed by in-house findings and literature reports implicating ETC complex I inhibition focused on the development of a microplate-based complex I activity assay. Developed in collaboration with Roderick Capaldi and his team at MitoSciences, Eugene, OR (now Abcam) a microplate-based immune-capture complex 1 activity assay was developed using highly purified bovine heart

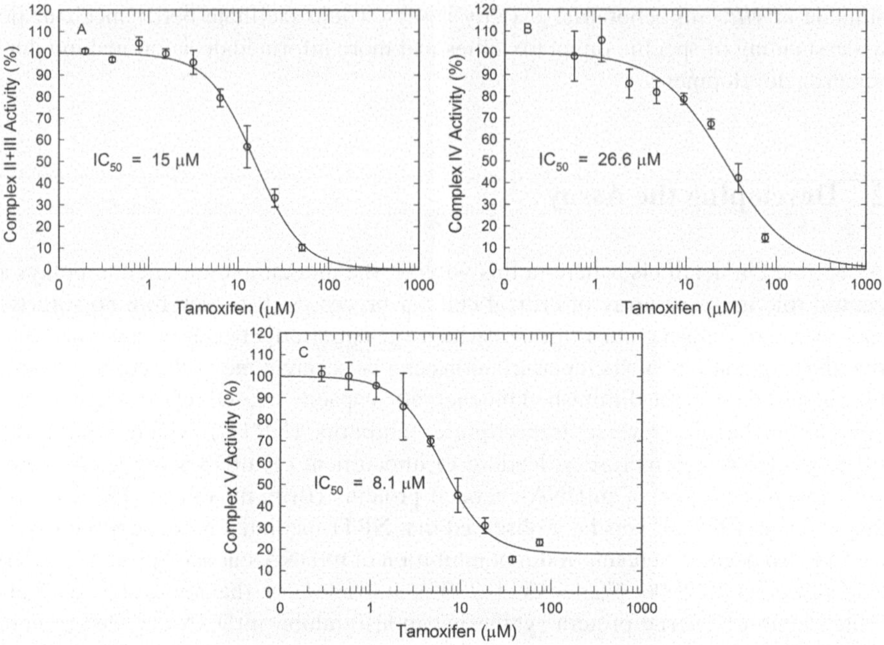

Fig. 1 Effect of tamoxifen on Complex II + III activity (a), Complex IV activity (b) and Complex V activity (c). 100% represents the activity of uninhibited enzyme. Data are expressed as mean ± SD (n = 3). The line is a best fit using a four-parameter logistic equation. Adapted from Nadanaciva et al., Toxicol In Vitro. 2007 Aug;21(5):902–11, with permission

mitochondria and subsequently expanded to include II/III, IV and complex V, the F_1/F_0 ATPase, measured either decreased NADH autofluoresence (later amplified by linking at colorimetric reporter) or cytochrome c oxidation, as appropriate for the individual complex under study (Nadanaciva et al. 2007a) (Fig. 1). These ETC complex activity assays have since been productised as kits containing microplates pre-coated with the relevant capture antibody (http://docs.abcam.com/pdf/kits/MitoTox-guide-Web.pdf). Using these tools to investigate the mechanisms involved in drug-induced mitochondrial dysfunction, it became apparent that compounds often inhibited multiple complexes. For example, troglitazone most potently inhibits complex IV, but also inhibits complexes II/III and complex V (Nadanaciva et al. 2007b). This was in marked contrast to the specificity exhibited by classical ETC modulators such as rotenone, antimycin and oligomycin, and while these observations were highly informative as to the breath of potential targets within oxidative phosphorylation, incorporating them into a reductive risk assessment proved challenging.

The second approach strove to address the throughput limitations associated with traditional polarograhic oxygen consumption measurements by combining the information content of an electrode-based oxygraph measurement with the throughput and convenience of a microplate assay. Harnessing porphyrin-based phosphorescent water-soluble oxygen probes and related oxygen consumption assays

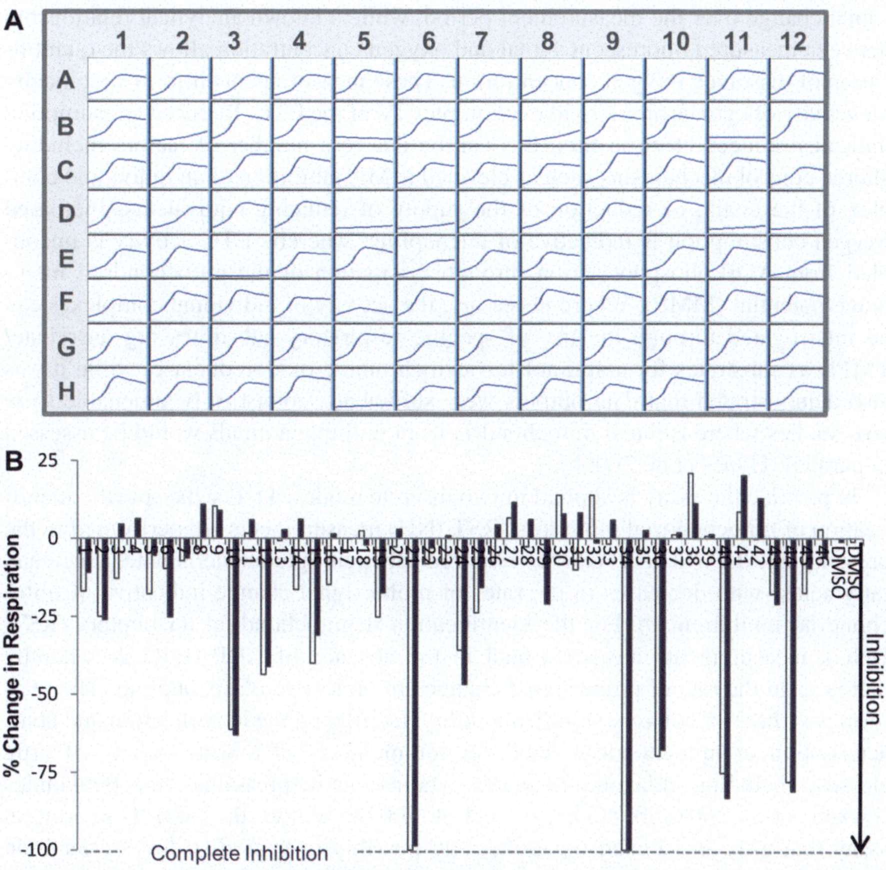

Fig. 2 RST Screening: (**a**) Sample data output; compounds C1 through C46 screened at a single dose using isolated rat liver mitochondria in State as per Figure 2.16.1B. (**b**) Processed data illustrating replicate measurements (solid and empty) of drugs no. 1 to 46. Inhibition represented by a negative value. For this analysis, samples showing >25% change are considered 'responders'. Adapted from Hynes et al., Curr Protoc Toxicol. 2009 Chapter 2:Unit 2.16, with permission

(Hynes et al. 2005, 2006a), dedicated assay protocols were developed and cross-validated against the oxygraph to facilitate robust measurement of drug-induced mitochondrial dysfunction using mitochondria isolated from liver, skeletal muscle and cardiac muscle (Hynes et al. 2006b; Will et al. 2007). Becoming known as known Respiratory Screening Technology (RST), this approach, facilitated specific measurement of ETC activity using conventional fluorescence plate readers on both 96 and 384 well plates (Fig. 2). Probe phosphorescent emission is quenched by molecular oxygen *via* a collisional mechanism such that the oxygen dissolved in the sample partially quenches probe emission. As oxygen is consumed, the concentration of dissolved oxygen in the sample is reduced resulting in an increase in probe signal. Measurement conditions are optimised such that untreated mitochondria give a reliable signal change over the measurement period, while a known analytical relationship between measured fluorescent signal and oxygen concentration allows the quantification of dissolved oxygen concentration. These measurements inform specifically on activity of cytochrome *c* oxidase, complex IV of the ETC. Reduced consumption indicates reduced electron flux; this can be due to a number of factors including altered control mechanisms such as elevated MMP, inhibition of an individual complex of the chain, or reduction of the supply of reducing equivalents. Increased oxygen consumption is indicative of uncoupling, whereby ETC activity is uncoupled from ADP phosphorylation through dissipation of the mitochondrial membrane potential (MMP). Where necessary, the activity of individual complexes can be interrogated through the use of specific respiratory substrates (eg ascorbate/TMPD as substrates for ubiquinol-ferricytochrome-c-oxidoreductase) while it was also demonstrated that data outputs were sufficiently robust to be amenable to *in vivo* studies where isolated mitochondria from multiple animals would be assessed in parallel (Hynes et al. 2006b).

In practice the assay is applied in two discrete modes: 1). For the specific identification of mitochondrial inhibitors (RST-INH) measurements are performed in the presence of ADP (State 3) and the required substrate (glutamate/malate, succinate, fatty acids) with decreases in the rates of probe signal change indicative of mitochondrial inhibition. 2). For the identification if mitochondrial uncouplers (RST-UNC), measurements are performed in the absence of ADP (state 2) whereby increases in the rate of probe signal change are indicative of uncoupling. The platform was therefore of very significant utility, facilitating the identification and characterisation of mitochondrial liabilities for members of a wide variety of drug classes, including thiazolidonediones, statins, antidepressants and biguanides (Dykens et al. 2008a, b; Nadanaciva et al. 2007b; Will et al. 2008). However, an assay that uses an isolated organelle may possibly over-predict direct effects on respiration due the free compound access to the OXPHOS machinery and/or under predict indirect effects or insults amplified by the activity of specific enzymes and transporters. It was therefore felt that cell-based approaches should also be developed and evaluated.

Two strategies were pursued to this end, one focusing on the metabolic modulation to ensure that test cells were sensitive to the downstream effects of mitochondrial perturbation and a second applying a fluorescence-based metabolic analysis similar in principle to that described above, to whole cells. Metabolic modulation was achieved by culturing cells in galactose, the oxidation of which results in an 80% reduction in glycolytic ATP due to the slower reaction kinetics associated with galactose catabolism (Reitzer et al. 1979). This forces cells that are capable of this adaptation to rely on OXPHOS for cellular ATP requirements (Rossignol et al. 2004). Development and evaluation of the galactose model clearly demonstrated that the combined use of late stage cytotoxicity end-points and glucose-grown cell lines were wholly unsuited to the detection of drug-induced mitochondrial toxicity due to their capacity to continue generating sufficient ATP via glycolysis, and also demonstrated that galactose-grown cells were significantly more sensitive to the cytotoxic implications of mitochondrial perturbation (Marroquin et al. 2007) (Fig. 3).

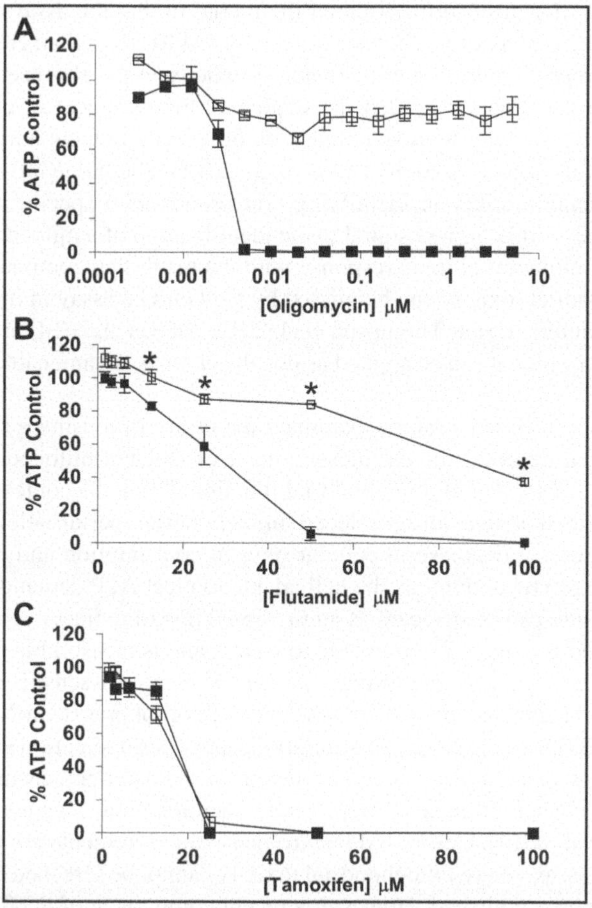

Fig. 3 Dose responses at 24 h for high-glucose–grown (25 mM) (open symbol) and galactose-grown (10 mM) (filled symbol) HepG2 cells treated with oligomycin (**a**), flutamine (**b**) and tamoxifen (**c**). Exposures for time courses were 10, 3.6, 5.0, and 200 μM, respectively. Data are mean ± SD ($n = 3$). Statistically significant differences of the two groups are marked as $*p < 0.05$ Adapted from Marroquin et al., Toxicol Sci. 2007 Jun;97(2):539–47, with permission

Differential sensitivities of glucose and galactose grown cells were therefore used as an identifier of specific drug-induced mitochondrial impairment, with glucose- and galactose-based ATP IC_{50} values compared and an IC_{50} Glucose:IC_{50} Galactose ratio of >3 taken as indicative of mitochondrial toxicity (Marroquin et al. 2007). Mitochondrial liabilities for members of the biguanide family as well as certain antidepressants (nefazodone) have been identified in this manner (Dykens et al. 2008a, b), and the approach has since found widespread use across the pharmaceutical industry with a recent survey conducted by the European MIP-DILI consortium (https://www.imi.europa.eu/content/mip-dili) revealing that 8/8 of the participating companies run the Glucose/Galactose assay regularly for mitochondrial toxicity assessment.

This consortium has also further refined the model in that the assays are run at an early (2 h) time point as well as a later time point (24 h) (Kamalian et al. 2015) with the earlier time point revealing mitochondrial toxicity in the absence of cell injury. Other improvements have included the sequential measurement of cell membrane integrity and ATP to help improve delineation of specific mitochondrial insult from non-specific/multifactorial toxicity (Swiss et al. 2013). The 'Glu/Gal' assay is therefore of considerable utility in identifying compounds with specific mitochondrial liabilities, however it is far less suited to the identification of a mitochondrial liabilities within a multifactorial toxic response with the result that compounds known to cause mitochondrial toxicity can be missed by the Glu/Gal assay in instances where toxicities are multifactorial (Thompson et al. 2012; Aleo et al. 2014) and is limited in the cell types to which it can be applied as not all cell lines are amenable to galactose-based culture.

The second cell-based strategy examined the utility of a using a similar optical approach to that descried for the measurement of isolated mitochondria, thereby facilitating an assessment of mitochondrial function within the context of the whole cell. As described above, drug-induced mitochondrial perturbation in glucose-grown cells causes a decrease in cellular oxygen consumption and a concomitant increase in glycolytic activity as the cell adapts to meet ATP demand. This shift in glycolytic activity can be detected as an increased rate of extracellular acidification due to increased conversion of pyruvate to lactate and is measurable either on plate readers (96 & 384 well plates) through the use of using pH-sensitive fluorophores (e.g. pH-Xtra™) measured alongside O_2-sensitive fluorophores (MitoXpressa®-Xtra) (Hynes et al. 2009, 2013, 2016), or similarly, using instrumentation such as the XF Analyzer and associated plasticware, designed for the parallel measurement of cellular oxygen consumption and extracellular acidification (Ferrick et al. 2008; Nadanaciva et al. 2012). Cell-based respirometric measurements are typically used as a follow up assay where mitochondrial toxicity cannot be screened out and potential risk has to be established. However, a recent comparison of RST, Glu/Gal, and a 384 plate based cell metabolic assessment conduced in dose response across 200 compounds revealed that direct functional measurements (RST and cell-based respirometry) were significantly more sensitive to drug-induced mitochondrial dysfunction, likely due to the inability of the Glu/Gal assay to detect mitochondrial dysfunction when part of a multifactorial off-target response (Hynes et al. 2013). These observations suggest that RST and/or cells based metabolic assessments are an important component in the screening for and characterisation of off-target effects.

While the mitochondrial assays already described are of significant utility in the identification and characterization of perturbed mitochondrial function, they are not ideally suited to the robust identification of inhibited mtDNA transcription due the exposure times required for such inhibition to become apparent. As outlined above, such inhibition is an important off-target effect within antiviral and antibiotics programs due largely to mitochondrial endosymbiotic evolution and the corresponding similarity of mitochondrial and bacterial ribosomes. A suitable assay was therefore developed using an antibody approach where mtDNA- and nuclear-encoded mitochondrial protein levels were measured using high-content imaging (Nadanaciva et al. 2010). Diminished mtDNA-encoded mitochondrial protein observed alongside unchanged nuclear encoded mitochondrial protein was taken as indicative of impairment of mitochondrial transcription or translation, although the assay does not distinguish between impaired transcription versus translation. Due to changes within the Pfizer portfolio, this assay is currently not in use.

3 Learnings from Using the Assays

Between 2008 and 2013 we utilized the above described assays to study a variety of drug classes for their effects on mitochondria, such as the thiazolidinediones, statins, fibrates, biguanides, serotonin/norepinephrine reuptake inhibitors, kinase inhibitors and members of the NSAIDs (Dykens et al. 2008a, b, Nadanaciva et al. 2007a, b, 2013; Will et al. 2008). Taken together, these studies provided a set of key learnings:

1. The rank order of toxicity *in vitro* correlated with the rank order of toxicity observed in the clinic. For example nefazodone, (attrited due to human liver toxicity), troglitazone (attrited due to liver toxicity) and phenformin (attrited for severe lactic acidosis) were more potent mitochondrial toxins than buspirone, pioglitazone and metformin respectively, which are all well tolerated by the majority of patients (Dykens et al. 2008a, b; Nadanaciva et al. 2007b).
2. The compound concentration required to cause detectable toxicity *in vitro* often exceeds the reported therapeutic concentration, however, when interpreting such data, receptor-mediated and nerenstian bioaccumulation should be considered. Examples include likely monocarboxylate transport mediated statin bioaccumulation (Nadanaciva et al. 2007b) and nerstian biguanide bioaccumulation, the latter becoming more evident with the introduction of a 40 min pre-incubation period to allow MMP-driven bioaccumulation to occur (Dykens et al. 2008b).
3. Normalisation to C_{max} (maximum drug plasma concentration in humans) improves overall predictivity. For example in the case of the antidepressants nefazodoen, trazodone, and buspirone (Dykens et al. 2008a) although all three compounds impact mitochondrial function and OXPHOS complex activities, post C_{max} normalisation the margin of safety over efficacy for buspirone and trazodone is 5 and >1000 higher than that of nefazodone respectively.
4. Absence of phase I and phase II enzymes in RST assay or in some cell-models can impact the detection of drug-induced mitochondrial function where that dysfunction is mediated by a compound metabolite. For example, the non-steroidal anti-inflammatory sulindac does not exhibit significant toxicity, whereas it's metabolite sulindac sulfate does exhibit toxicity (Nadanaciva et al. 2012).
5. The glucose-galactose assay has a much lower hit rate than the RST assay which suggests that most toxicity is multifactorial (Hynes et al. 2013).
6. Although toxicity is seen across a wide variety of drug classes, each drug class contains compounds that impair mitochondrial function, but others that do not, or do so with substantially less potency offering the opportunity for appropriate preclinical assessments to better inform compound selection.

7. Many drugs are rather promiscuous due unfavorable physical–chemical properties, such as high lipophilicity (clogP > 3) which frequently yields unspecific binding to many targets, including mitochondrial targets (Will and Dykens 2014). For example, we know today that troglitazone also forms a reactive metabolite (Madsen et al. 2008), and inhibits BSEP (Snow and Moseley 2007) in addition to its here described mitochondrial toxicity. The same is true for nefazodone, which inhibits BSEP (Kostrubsky 2005) and formed a reactive metabolite (Bauman et al. 2008).

4 Implementation Within the Drug Discovery Process

The goal of testing within the drug discovery toxicology paradigm is the appropriate positioning of effective assays with considering for data value, through-put and cost, such that only the most promising compounds are progressed into animal studies and later into FIH. Each pharmaceutical company will differ slightly in which screens/assays they find important to apply for both series and later lead selection, however, early screening is favoured due to the availability of chemical substrate and the opportunity for SAR strategy implementation. Here we outline how cytotoxicity and mitotoxicity assessments integrate within series and lead selection and how output criteria are used to trigger subsequent assay deployment.

At Pfizer, for series selection we first assess cytotoxicity using both THLE and HepG2 cells, measuring ATP depletion after a 72 h dosing period. This implementation is underpinned by observations that binning compounds based on *in vitro* IC_{50} into potent ($IC_{50} < 10\ \mu M$) and non-potent ($IC_{50} > 100\ \mu M$) categories facilitated an approximation of compound safety characteristics, in that the potent category exhibited higher overall severity scores at lower exposures *in vivo* (Benbow et al. 2010). Also supporting this implementation are observations that compounds with a THLE-based $IC_{50} < 50\ \mu M$ were five times more likely to exhibit toxicity findings in an exploratory toxicology study at a $C_{max} < 10\ \mu M$ than if the THLE IC_{50} value was >50 μM (Greene et al. 2010). These measurements are conducted on both THLE and glucose-grown HepG2 due to observations that THLE cells have a relatively higher sensitivity to acidic and neutral compound, whereas HepG2 cells were more sensitive to basic compounds (Shah et al. 2014). The available data therefore suggests that these *in vitro* cytotoxicity assays when combined with specific physical-chemical criteria (cLogP >3 and topological polar surface area <75 Å), are the most effective applicable means to prioritize compounds with a reduced probability of causing adverse events *in vivo*.

As outlined above, the combination of conventional cytotoxicity assays and glucose-grown cell lines are not well suited to the detection of drug-induced mitochondrial dysfunction. A parallel mitochondrial toxicity strategy is therefore implemented to address this deficit. The first assay deployed is a differential sensitivity assessment of glucose- and galactose-grown cells to a 24 h compound treatment. Compounds which exhibit unfavourable physicochemical properties, have a cytotoxicity

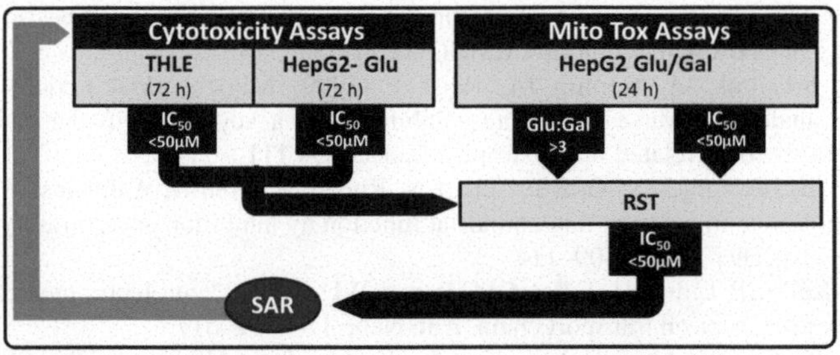

Fig. 4 Screening paradigm for NCEs (new chemical entities) for cytotoxicity and mitochondrial toxicity

(ATP) IC_{50} value of <50 μM, or display a glucose-galactose ratio of >3 are subsequently submitted for testing using the RST assay using both RST-INH and RST-UNC modes to generate IC_{50}/UC_{50} values. Alternatively, to increase throughput, compounds can be initially screened at single dose (100 μM) with hits subsequently submitted to full dose-response analysis. Compounds which flag as mitotox-positive can then undergo SAR strategies, with resultant new compounds resubmitted to cytotoxicity testing and if still positive again to mitochondrial toxicity testing (Fig. 4).

The same screening paradigm is true for lead compounds whereby, if a lead compound is positive in cytotoxicity, it will be submitted to mitochondrial toxicity testing (RST). If the compound remains positive, it will be subjected to further mitochondrial assessment such as in the Seahorse platform, or subjected to specific organ toxicity assessment such as cardiac or liver toxicity (not discussed here). Note that the overall goal would be to advance only compounds which are devoid of cytotoxicity and mitochondrial toxicity.

References

Aleo MD, Luo Y, Swiss R, Bonin PD, Will Y (2014) Human drug-induced liver injury severity is highly associated with dual inhibition of liver mitochondrial function and bile salt export pump. Hepatology 60(3):1015–22

Bachmann E, Zbinden G (1979) Effect of doxorubicin and rubidazone on respiratory function and Ca^{2+} transport in rat heart mitochondria. Toxicol Lett 3:29–34

Bauman JN, Frederick KS, Sawant A, Walsky RL, Cox LM, Obach RS, Kalgutkar AS (2008) Comparison of the bioactivation potential of the antidepressant and hepatotoxin nefazodone with aripiprazole, a structural analog and marketed drug. Drug Metab Dispos 36:1016–1029

Benbow JW, Aubrecht J, Banker MJ, Nettleton D, Aleo MD (2010) Predicting safety toleration of pharmaceutical chemical leads: cytotoxicity correlations to exploratory toxicity studies. Toxicol Lett 197:175–182

Bernardi P, Vassanelli S, Veronese P, Colonna R, Szabó I, Zoratti M (1992) Modulation of the mitochondrial permeability transition pore. Effect of protons and divalent cations. J Biol Chem 267:2934–2939

Birch-Machin MA, Briggs HL, Saborido AA, Bindoff LA, Turnbull DM (1994) An evaluation of the measurement of the activities of complexes I-IV in the respiratory chain of human skeletal muscle mitochondria. Biochem Med Metab Biol 51:35–42

Böttger EC, Springer B, Prammananan T, Kidan Y, Sander P (2001) Structural basis for selectivity and toxicity of ribosomal antibiotics. EMBO Rep 2:318–323

Brinkman K, Smeitink JA, Romijn JA, Reiss P (1999) Mitochondrial toxicity induced by nucleoside-analogue reverse-transcriptase inhibitors is a key factor in the pathogenesis of antiretroviral-therapy-related lipodystrophy. Lancet 354:1112–1115

Brunmair B, Lest A, Staniek K, Gras F, Scharf N, Roden M, Nohl H, Waldhäusl W, Fürnsinn C (2004) Fenofibrate impairs rat mitochondrial function by inhibition of respiratory complex I. J Pharmacol Exp Ther 311(1):109–114

Burkhardt C, Kelly JP, Lim Y-H, Filley CM, Parker WD (1993) Neuroleptic medications inhibit complex I of the electron transport chain. Ann Neurol 33:512–517

Clark JLC (1959). Electrochemical device for chemical analysis. US patent 2,913,386

Cook D, Brown D, Alexander R, March R, Morgan P, Satterthwaite G, Pangalos MN (2014) Lessons learned from the fate of AstraZeneca's drug pipeline: a five-dimensional framework. Nat Rev Drug Discov 13:419–431

Dykens JA, Jamieson JD, Marroquin LD, Nadanaciva S, Xu JJ, Dunn MC, Smith AR, Will Y (2008a) In vitro assessment of mitochondrial dysfunction and cytotoxicity of nefazodone, trazodone, and buspirone. Toxicol Sci 103:335–345

Dykens JA, Jamieson J, Marroquin L, Nadanaciva S, Billis PA, Will Y (2008b) Biguanide-induced mitochondrial dysfunction yields increased lactate production and cytotoxicity of aerobically-poised HepG2 cells and human hepatocytes in vitro. Toxicol Appl Pharmacol 233:203–210

Esposti DM (1998) Inhibitors of NADH-ubiquinone reductase: an overview. Biochim Biophys Acta 1364:222–235

Ferrick D, Wu M, Swift A, Neilson A (2008) Mitochondrial dysfunction assessed quantitatively in real time by measuring the extracellular flux of oxygen and protons. In: Dykens JA, Will Y (eds) Drug-induced mitochondrial dysfunction. Wiley, Hoboken, NJ, pp 373–382

Gnaiger E (2008) Polarographic oxygen sensors, the oxygraph, and high-resolution respirometry to assess mitochondrial function. In: Dykens JA, Will Y (eds) Drug-induced mitochondrial dysfunction. Wiley, Hoboken, NJ, pp 325–352

Greene N, Aleo MD, Louise-May S, Price DA, Will Y (2010) Using an in vitro cytotoxicity assay to aid in compound selection for in vivo safety studies. Bioorg Med Chem Lett 20:5308–5312

Hay M, Thomas DW, Craighead JL, Economides C, Rosenthal J (2014) Clinical development success rates for investigational drugs. Nat Biotechnol 32:40–51

Hynes J, O'Riordan TC, Curtin J, Cotter TG, Papkovsky DB (2005) Fluorescence based oxygen uptake analysis in the study of metabolic responses to apoptosis induction. J Immunol Methods 306:193–201

Hynes J, Hill R, Papkovsky DB (2006a) The use of a fluorescence-based oxygen uptake assay in the analysis of cytotoxicity. Toxicol In Vitro 20:785–792

Hynes J, Marroquin LD, Ogurtsov VI, Christiansen KN, Stevens GJ, Papkovsky DB, Will Y (2006b) Investigation of drug-induced mitochondrial toxicity using fluorescence-based oxygen-sensitive probes. Toxicology 92:186–200

Hynes J, O'Riordan TC, Zhdanov AV, Uray G, Will Y, Papkovsky DB (2009) vitro analysis of cell metabolism using a long-decay pH-sensitive lanthanide probe and extracellular acidification assay. Anal Biochem 390:21–28

Hynes J, Nadanaciva S, Swiss R, Carey C, Kirwan S, Will Y (2013) A high-throughput dual parameter assay for assessing drug-induced mitochondrial dysfunction provides additional predictivity over two established mitochondrial toxicity assays. Toxicol In Vitro 27:560–569

Hynes J, Carey C, Will Y (2016) Fluorescence-based microplate assays for in vitro assessment of mitochondrial toxicity, metabolic perturbation, and cellular oxygenation: fluorescent assay for mitochondrial function. In: Costa LG, Davila JC, Lawrence DA, Reed DJ (eds) Current protocols in toxicology. Wiley, Hoboken, NJ, pp 2.16.1–2.16.30

Kamalian L, Chadwick AE, Bayliss M, French NS, Monshouwer M, Snoeys J, Park BK (2015) The utility of HepG2 cells to identify direct mitochondrial dysfunction in the absence of cell death. Toxicol In Vitro 29:732–740

Kashimshetty R, Desai VG, Kale VM, Lee T, Moland CL, Branham WS, New LS, Chan ECY, Younis H, Boelsterli UA (2009) Underlying mitochondrial dysfunction triggers flutamide-induced oxidative liver injury in a mouse model of idiosyncratic drug toxicity. Toxicol Appl Pharmacol 238:150–159

Kaufmann P, Török M, Zahno A, Waldhauser KM, Brecht K, Krähenbühl S (2006) Toxicity of statins on rat skeletal muscle mitochondria. Cell Mol Life Sci 63:2415–2425

Kostrubsky SE (2005) Inhibition of hepatobiliary transport as a predictive method for clinical hepatotoxicity of nefazodone. Toxicol Sci 90:451–459

Lewis W, Dalakas MC (1995) Mitochondrial toxicity of antiviral drugs. Nat Med 1:417–422

Madsen KG, Grönberg G, Skonberg C, Jurva U, Hansen SH, Olsen J (2008) Electrochemical oxidation of troglitazone: identification and characterization of the major reactive metabolite in liver microsomes. Chem Res Toxicol 21:2035–2041

Marroquin LD, Hynes J, Dykens JA, Jamieson JD, Will Y (2007) Circumventing the crabtree effect: replacing media glucose with galactose increases susceptibility of HepG2 cells to mitochondrial toxicants. Toxicol Sci 97:539–547

McKee EE, Ferguson M, Bentley AT, Marks TA (2006) Inhibition of mammalian mitochondrial protein synthesis by oxazolidinones. Antimicrob Agents Chemother 50:2042–2049

Nadanaciva S, Bernal A, Aggeler R, Capaldi R, Will Y (2007a) Target identification of drug induced mitochondrial toxicity using immunocapture based OXPHOS activity assays. Toxicol In Vitro 21:902–911

Nadanaciva S, Dykens JA, Bernal A, Capaldi RA, Will Y (2007b) Mitochondrial impairment by PPAR agonists and statins identified via immunocaptured OXPHOS complex activities and respiration. Toxicol Appl Pharmacol 223:277–287

Nadanaciva S, Dillman K, Gebhard DF, Shrikhande A, Will Y (2010) High-content screening for compounds that affect mtDNA-encoded protein levels in eukaryotic cells. J Biomol Screen 15:937–948

Nadanaciva S, Rana P, Beeson GC, Chen D, Ferrick DA, Beeson CC, Will Y (2012) Assessment of drug-induced mitochondrial dysfunction via altered cellular respiration and acidification measured in a 96-well platform. J Bioenerg Biomembr 44:421–437

Nadanaciva S, Aleo MD, Strock CJ, Stedman DB, Wang H, Will Y (2013) Toxicity assessments of nonsteroidal anti-inflammatory drugs in isolated mitochondria, rat hepatocytes, and zebrafish show good concordance across chemical classes. Toxicol Appl Pharmacol 272:272–280

Nagiec EE, Wu L, Swaney SM, Chosay JG, Ross DE, Brieland JK, Leach KL (2005) Oxazolidinones inhibit cellular proliferation via inhibition of mitochondrial protein synthesis. Antimicrob Agents Chemother 49:3896–3902

Olson H, Betton G, Robinson D, Thomas K, Monro A, Kolaja G, Lilly P, Sanders J, Sipes G, Bracken W et al (2000) Concordance of the toxicity of pharmaceuticals in humans and in animals. Regul Toxicol Pharmacol 32:56–67

Ong MMK, Wang AS, Leow KY, Khoo YM, Boelsterli UA (2006) Nimesulide-induced hepatic mitochondrial injury in heterozygous Sod2+/− mice. Free Radic Biol Med 40:420–429

Ong MMK, Latchoumycandane C, Boelsterli UA (2007) Troglitazone-induced hepatic necrosis in an animal model of silent genetic mitochondrial abnormalities. Toxicol Sci 97:205–213

Reitzer LJ, Wice BM, Kennell D (1979) Evidence that glutamine, not sugar, is the major energy source for cultured HeLa cells. J Biol Chem 254:2669–2676

Rossignol R, Gilkerson R, Aggeler R, Yamagata K, Remington SJ, Capaldi RA (2004) Energy substrate modulates mitochondrial structure and oxidative capacity in cancer cells. Cancer Res 64:985–993

Seachrist JL, Loi C-M, Evans MG, Criswell KA, Rothwell CE (2005) Roles of exercise and pharmacokinetics in cerivastatin-induced skeletal muscle toxicity. Toxicology 88:551–561

Shah F, Louise-May S, Greene N (2014) Chemotypes sensitivity and predictivity of in vivo outcomes for cytotoxic assays in THLE and HepG2 cell lines. Bioorg Med Chem Lett 24:2753–2757

Snow KL, Moseley RH (2007) Effect of thiazolidinediones on bile acid transport in rat liver. Life Sci 80:732–740

Swiss R, Niles A, Cali JJ, Nadanaciva S, Will Y (2013) Validation of a HTS-amenable assay to detect drug-induced mitochondrial toxicity in the absence and presence of cell death. Toxicol In Vitro 27:1789–1797

Thompson RA, Isin EM, Li Y, Weidolf L, Page K, Wilson I, Swallow S, Middleton B, Stahl S, Foster AJ et al (2012) In vitro approach to assess the potential for risk of idiosyncratic adverse reactions caused by candidate drugs. Chem Res Toxicol 25:1616–1632

Tirmenstein MA, Hu CX, Gales TL, Maleeff BE, Narayanan PK, Kurali E, Hart TK, Thomas HC, Schwartz LW (2002) Effects of troglitazone on HepG2 viability and mitochondrial function. Toxicol Sci 69:131–138

Westwood FR, Bigley A, Randall K, Marsden AM, Scott RC (2005) Statin-induced muscle necrosis in the rat: distribution, development, and fibre selectivity. Toxicol Pathol 33:246–257

Will Y, Dykens J (2014) Mitochondrial toxicity assessment in industry – a decade of technology development and insight. Expert Opin Drug Metab Toxicol 10(8):1061–1067

Will Y, Hynes J, Ogurtsov VI, Papkovsky DB (2007) Analysis of mitochondrial function using phosphorescent oxygen-sensitive probes. Nat Protoc 1:2563–2572

Will Y, Dykens JA, Nadanaciva S, Hirakawa B, Jamieson J, Marroquin LD, Hynes J, Patyna S, Jessen BA (2008) Effect of the multitargeted tyrosine kinase inhibitors imatinib, dasatinib, sunitinib, and sorafenib on mitochondrial function in isolated rat heart mitochondria and H9c2 cells. Toxicol Sci 106:153–161

Targeting Mitochondria: The Road to Mitochondriotropic Antioxidants and Beyond

Ricardo Amorim, Sofia Benfeito, José Teixeira, Fernando Cagide, Paulo J. Oliveira, and Fernanda Borges

Abstract Mitochondria are sub-cellular organelles that play a central role in energy metabolism, being these organelles currently recognized as one important target for new drug discovery programs addressed to find innovative therapeutic solutions for diverse pathologic events, such as cancer, cardiovascular, and neurological diseases. Although attractive, the success of the strategies developed so far has been hampered by several challenges and limitations, and until now the Food and Drug Administration (FDA) has not approved a drug for mitochondrial therapy. Currently, the most effective strategy to deliver drugs specifically to mitochondria is the covalent link of a lipophilic cation, namely triphenylphosphonium (TPP), to a pharmacophore of interest. Within this framework two mitochondriotropic antioxidants (MitoQ and SkQ1) have entered in human clinical trials as a therapeutic solution for oxidative-stress related diseases. In this chapter, the efforts done so far to target small-molecule antioxidants to mitochondria as potential therapeutics or diagnostic tools have been reviewed. Although TPP cation has been the most extensively used mitochondrial-targeting cation, there are still controversies surrounding this approach, namely related with its intrinsic toxicity. Consequently, efforts must be done in finding new cation carriers, and to guarantee that the cargo does indeed access the mitochondrial matrix and does not merely associate with the mitochondrial membranes. Moreover,

R. Amorim · J. Teixeira
CIQUP/Department of Chemistry and Biochemistry, Faculty of Sciences, University of Porto, Porto, Portugal

CNC – Center for Neuroscience and Cell Biology, University of Coimbra, Cantanhede, Portugal

S. Benfeito · F. Cagide · F. Borges (✉)
CIQUP/Department of Chemistry and Biochemistry, Faculty of Sciences, University of Porto, Porto, Portugal
e-mail: fborges@fc.up.pt

P. J. Oliveira
CNC – Center for Neuroscience and Cell Biology, University of Coimbra, Cantanhede, Portugal

P. J. Oliveira (ed.), *Mitochondrial Biology and Experimental Therapeutics*,
https://doi.org/10.1007/978-3-319-73344-9_16

in *vivo* biodistribution, pharmacokinetics and long-term toxic effects studies to provide accurate information about efficacy and toxicity are still an emergent issue to make available the translation from bench to bedside.

Keywords Mitochondrial (dys)function · Triphenylphosphonium cation · Mitochondriotropic antioxidants · Mitochondria-targeted probes/sensors

Abbreviations

2-DG	2-Deoxy-D-glucose
CoQ	Co-enzyme Q
EPR	Electron paramagnetic resonance
ETC	Electron transport chain
FAD	Flavin adenine dinucleotide
GPx	Glutahione peroxidase
GSH	Reduced glutathione
H_2O_2	Hydrogen peroxide
H_2S	Hydrogen sulfide
MIM	Mitochondrial inner membrane
mPTP	Mitochondrial permeability transition pore
mtDNA	Mitochondrial DNA
$ONOO^-$	Peroxynitrite
OXPHOS	Oxidative phosphorylation
Prx3	Peroxiredoxine 3
ROS	Reactive oxygen species
SOD	Superoxide dismutase
TCA	Tricarboxylic acid
TP	Two-photon
TPP or PPh_3	Triphenylphosphonium
$\Delta\Psi m$	Mitochondrial membrane potential

1 Mitochondrial Function and Dysfunction

Mitochondria are sub-cellular organelles present in virtually every mammalian cell. Mitochondria are at the heart of metabolism supplying most of the cell's ATP and housing a myriad of metabolic network processes, including oxidative phosphorylation (OXPHOS), tricarboxylic (TCA) cycle, fatty acid oxidation and amino acid metabolism (Smith et al. 2012). Mitochondria are also responsible for modulating calcium (Ca^{2+}) fluxes throughout the cell, regulation of redox signaling and in apoptotic and necrotic cell death events (Fig. 1) (Smith et al. 2012). Mitochondria have their own DNA (mtDNA), which encodes only 37 genes in human cells, organized as a circular, covalently closed, double-stranded DNA. Mitochondrial DNA is

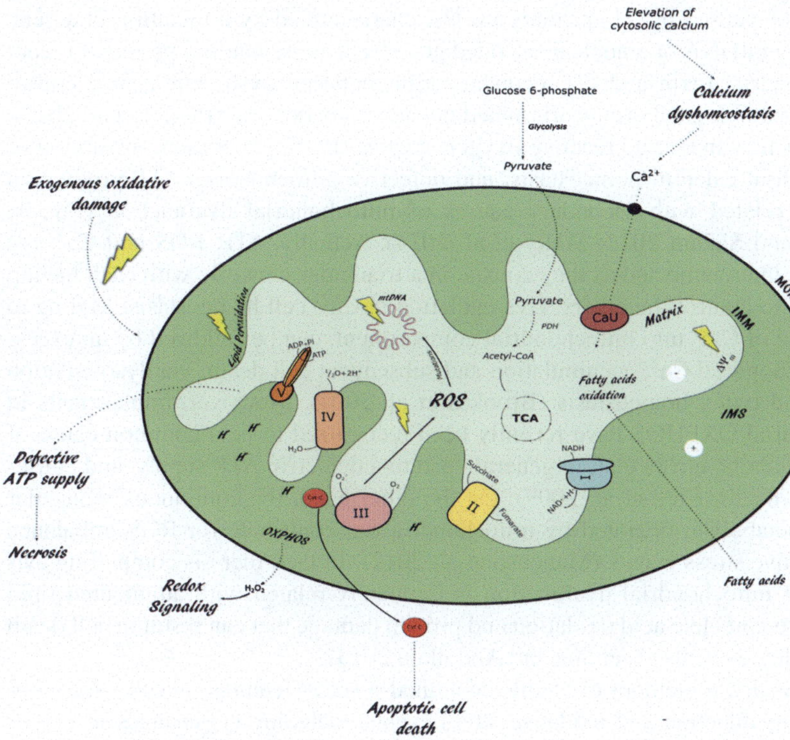

Fig. 1 Mitochondrial (dys)function. Disruption of mitochondrial function can be caused by primary events, such as mutations in mtDNA and/or nDNA. Secondary mitochondrial dysfunction arises from events outside mitochondria. Factors such as increased oxidative stress, disruption of calcium homeostasis and defective mitochondrial ATP synthesis are often related with mitochondrial dysfunction. Oxidative events lead to augmented lipid peroxidation, nucleic acid oxidation and protein damage, which further disrupt mitochondrial function culminating in cell death

replicated by the DNA polymerase gamma complex in a process that require separate transcription and translation machinery (Kanabus et al. 2014). Cellular viability depends on mitochondria metabolic framework and the proper exchange of metabolites, such as amino acids, lipids, prosthetic groups and other intermediates, across the organelle membranes (Figueira et al. 2013). Chief among all mitochondrial metabolic processes, the predominant physiological function of mitochondria is the generation of ATP from the oxidation of nutrients by OXPHOS and the supplying of precursors for the synthesis of macromolecules and post-translational modifications. To participate in energy production, mitochondria use an intricate system that interplays fatty acid oxidation, glycolysis, TCA cycle and OXPHOS (Fig. 1) (Andreux et al. 2013). Abnormality in any of these processes result in mitochondrial dysfunction.

The terms mitochondrial 'function' and 'dysfunction' are widely employed in bioenergetics and cell biology, but a precise definition is rather consensual. Mitochondrial dynamics, including fusion, fission and translocation, are crucial to cellular homeostasis, and intimately linked to apoptosis and to the removal of dysfunctional mitochondria. Mitochondrial dysfunction, which can be linked to disease, can be caused by: (1) primary events, characterized by a mutation in a gene encoded by mtDNA or a nuclear-encoded gene for a mitochondrial protein, or from a mitochondrial toxin and (2) secondary mitochondrial dysfunction, as a consequence of pathological events originated outside mitochondria (Pfeffer et al. 2012). Factors such as increased reactive oxygen species (ROS) generation, disruption of mitochondrial calcium homeostasis, and defective mitochondrial ATP production are often related with secondary causes of mitochondrial dysfunction (Fig. 1) (Alfadda and Sallam 2012; Malty et al. 2014). Actually, ATP, ROS and Ca^{2+} are intimately interconnected as they coexist in a triangular network, with each having the ability to control the others. Dysregulation in Ca^{2+} cell homeostasis, leading to an increase of Ca^{2+} into mitochondrial compartment, can be followed by an excessive mitochondrial Ca^{2+} accumulation and subsequent cell death, via dysregulation in ATP and redox homeostasis (Brookes et al. 2004). Moreover, impairments in mitochondrial OXPHOS have recently been recognized to be a common cause of inborn metabolic errors that, in general, lead to a defective ATP supply, and consequent cell death (Cree et al. 2009). On the other hand, the continuous molecular oxygen metabolism operated by mitochondria can contribute for ROS imbalance and oxidative stress events (Mancuso et al. 2012). In fact, there is compelling evidence that mitochondrial dysfunction is intimately related with augmented lipid peroxidation, nucleic acid oxidation and protein damage that can result in cell death and pathologic events (Saeidnia and Abdollahi 2013).

Additionally, a plethora of studies identified a set of features related with mitochondrial dysfunction and oxidative stress events, including (1) changes in one or more electron transport chain (ETC) activities; (2) decreased expression of Krebs cycle-associated enzymes; (3) defect in manganese superoxide anion ($O_2^{-\bullet}$) dismutase (MnSOD) and/or peroxiredoxin 3 (Prx3) expression; (4) mtDNA deletions, and decrease in mtDNA copy number and mitophagy; (5) decreased levels of co-enzyme C10 (CoQ10) and/or ATP; (6) changes in lactate/pyruvate ratio; and (7) changes in mitochondrial membrane potential ($\Delta\Psi m$) (Pagano et al. 2014).

2 Mitochondria as Pharmacological Targets

The recognition of the mitochondrion as the gatekeeper of cell life and death together with impairments of mitochondrial functions observed in a diversity of pathological states, such cancer, metabolic disorders and age-related diseases, spurred active drug discovery projects.

In mitochondrial-based drug discovery programs, two complementary methodologies have been followed: a bottom-up approach that focuses on individual rare heritable disorders caused by gene mutations or on top-down approach that addresses a defined disorder in which mitochondrial dysfunction may be part of a more complex pathophysiology (Hurko 2013; Kanabus et al. 2014; Suomalainen and Battersby 2017). In this chapter, only the top-down approach centred in the discovery of new molecular-based therapies will be focused.

As mitochondria are key places for life and death decisions, it is not surprising that most of studies have been mostly focused on the mitochondrial targets that regulate mitochondrial biogenesis, ROS production and respiration. Up till now, the drug discovery projects were aimed to find drugs able to restore mitochondrial function as well as regulate mitochondrial ROS production (Sorriento et al. 2014).

To modulate mitochondria dysfunction, a drug must selectively accumulate within mitochondria and bind to one, or more, mitochondrial targets or modulate a process outside of mitochondria that ultimately can modify a mitochondrial dysfunction. Although attractive, the success of this strategy has been hampered by a number of challenges and limitations, namely clinical trials design and implementing, the lack of animal models and the poor druggability and drug-likeness of the drug candidates (Pfeffer et al. 2013). To date, neither of the approaches has resulted in drug approved by Food and Drug Administration (FDA) for a mitochondrial therapy.

3 Mitochondriotropic Antioxidants: The Road to MitoQ and Beyond

Applying a concept pioneered by Skulachev et al. (Liberman and Topaly 1969; Levitskii and Skulachev 1972) to measure mitochondrial membrane potential ($\Delta\Psi m$), during the past decade considerable progress has been made in the development of mitochondria-targeted bioactive molecules. The use of lipophilic cations as carriers is a recognized mitochondriotropic strategy as it can be rapidly and extensively taken up by mitochondria driven by the large $\Delta\Psi m$, which is negative inside (Smith et al. 2011). Generally, the strategy encloses the expenditure of carrier systems suitable to couple to a diversity of bioactive agents that are poorly taken up by mitochondria (e.g. due to the lack of hydrophobicity) allowing their targeting to mitochondria (Murphy and Smith 2007; Smith et al. 2012; Anders 2013).

Most of the studies done so far were focused on triphenylphosphonium (TPP) cation either as a carrier of a variety of agents or as a tool for study and/or diagnose mitochondrial (dys)function. In the both cases the selection of the linker between the cation and the bioactive agent is a crucial step for the success. Currently, this approach is considered to be versatile and robust to target small bioactive molecules, enzymes and probes to mitochondria (Smith et al. 2011). Despite the efficiency of this smart carrier, other lipophilic systems, such as quaternary ammonium salt (choline esters), rhodamine, pyridinium, cyanine, berberine derivatives and peptides, have been used with the same endeavour.

3.1 Triphenylphosphonium Cation as a Versatile Carrier

Triphenylphosphonium (TPP) cation consists of a positively charged phosphorus atom surrounded by three hydrophobic phenyl groups that contribute to an extended hydrophobic surface and charge stabilization. Triphenylphosphonium cations can pass easily through lipid bilayers and mitochondria inner membrane (MIM) due to their extensive hydrophobic surface area and the large out ionic radius of the cation, which can effectively lower the activation energy needed for membrane passage (Smith et al. 2012).

The uptake of TPP and other lipophilic cations by mitochondria is dependent on the ΔΨm and can be estimated by the Nernst equation: 10-fold increase uptake for every ~60 mV can lead to 100–500-fold accumulation (Gerő et al. 2016). Once inside mitochondria, TPP conjugated agents are in general located on the mitochondrial matrix upon surpassing the phospholipid bilayer, maintaining the linker and bioactive molecule positioned within the MIM (Apostolova and Victor 2015). Inside mitochondria, mitochondriotropic agents can elicit beneficial effects in conditions related with oxidative stress and mitochondrial dysfunction by diverse mechanisms, namely by scavenging ROS, for instance preventing and/or minimizing membrane lipid peroxidation damage, or by modulating mitochondrial redox signaling processes (Smith et al. 2012). Targeting mitochondrial ROS can help to regulate altered bioenergetics and other mitochondrial events implied in a range of oxidative stress-induced diseases including metabolic and neurodegenerative diseases, asthma and chronic obstructive pulmonary disease (COPD), cancer, inflammation and ageing-related disorders, such as cardiac dysfunction (Andreux et al. 2013; Agrawal and Mabalirajan 2016; Yang et al. 2016).

Currently, several mitochondria-targeted antioxidants obtained by conjugation to the TPP lipophilic cation have been developed reinforcing the hypothesis that targeting mitochondria is an effective avenue to find a new therapeutic solution (Ross et al. 2005; Wagner et al. 2008).

At the moment, a noteworthy number of studies has been focused on the discovery of mitochondriotropic antioxidants able to prevent and/or minimize mitochondrial oxidative damage (Smith et al. 2012). Within the scope of the chapter, mitochondriotropic antioxidants having TPP cation as carrier, assembled by the type of antioxidant core (endogenous, exogenous and miscellaneous), will be highlighted. In general, they share the same drug design strategy that was depicted on strategy one in Fig. 2.

3.2 Mitochondriotropic Agents Based on Endogenous Antioxidants

MitoE (Fig. 3) was the first synthesized TPP-linked antioxidant, which consists of a TPP cation moiety linked to vitamin E by means of a two-carbon aliphatic chain spacer (Zielonka et al. 2017). MitoE is rapidly taken up by isolated mitochondria and cells protecting them from oxidative damage. MitoE was able to prevent/minimize lipid peroxidation, to reduce H_2O_2-mediated oxidative stress, to maintain proteosomal function in endothelial cells (Dhanasekaran et al. 2004), and to protect cells against H_2O_2-induced caspase activation (Hughes et al. 2005). When administered intravenously in a mice model, it rapidly accumulated in the tissues most

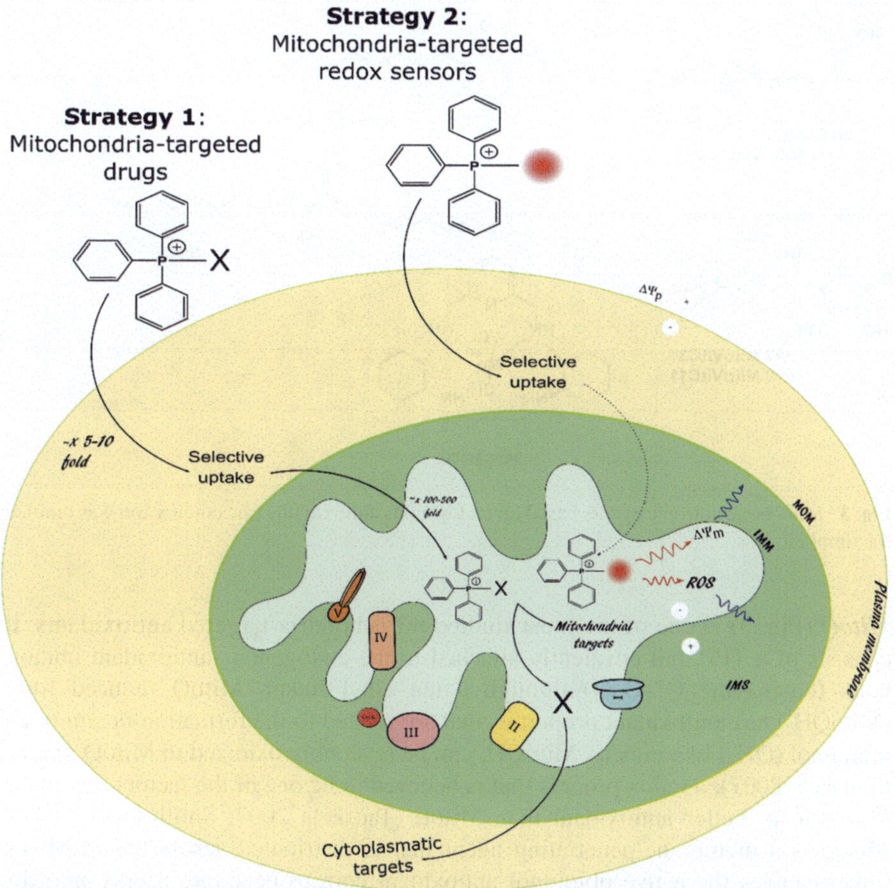

Fig. 2 Strategies to pharmacologically intervene (strategy 1) and evaluate (strategy 2) mitochondrial dysfunction. (1) Targeting small molecules to mitochondria by conjugation to lipophilic cations, such as TPP, lead to the selective uptake of the attached bioactive molecule into the mitochondrial matrix. This type of mitochondriotropic compounds can act directly into the target or can be released from the targeting moiety within mitochondria. (2) Targeting compounds to mitochondria by conjugation to lipophilic cations, such as TPP, lead to the selective accumulation of the attached redox sensor/probe into the mitochondrial matrix

affected by mitochondrial dysfunction and oxidative damage, such as heart, brain, muscle, liver and kidney (Smith et al. 2003).

Based on the attained data, and to optimize MitoE efficacy, several analogues having modifications on length of the linker have been synthesized. The new compounds can accumulate in mitochondria and preliminary biological data demonstrated a greater efficacy than non-targeted compound (vitamin E) in preventing lipid peroxidation, mitochondrial oxidative stress and mtDNA transcription/mitogenesis (Jameson et al. 2015). Mito-Vitamin E derivative with a eleven alkyl linker (Fig. 3) was shown to have antitumor properties, which are related with its capacity to affect the energy metabolism and promoting cell death (Dong et al. 2011).

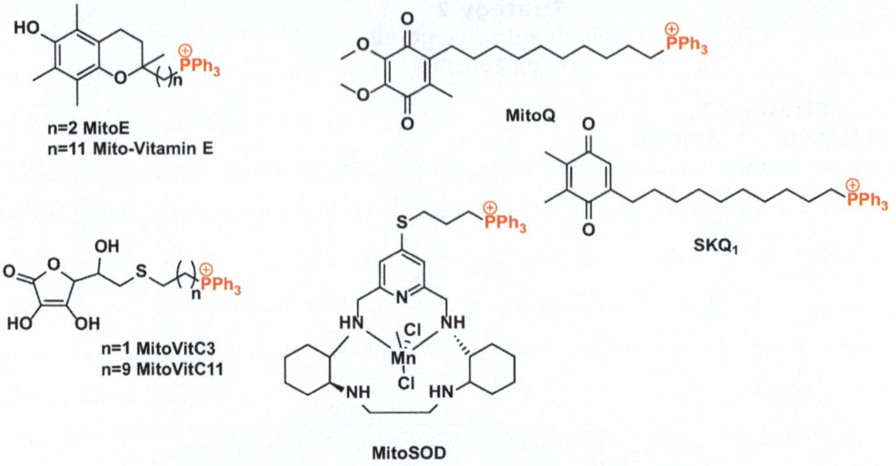

Fig. 3 Mitochondriotropic agents based on endogenous antioxidants (the counter ion was omitted for simplicity)

MitoQ (Fig. 3) is one of the most studied mitochondria-targeted antioxidants. It consists of a TPP unit covalently attached to the endogenous antioxidant ubiquinone (co-enzyme Q, CoQ) through a ten alkyl linker. MitoQ reduced form (MitoQH_2) has antioxidant properties that are related to the formation *in situ* of an ubiquinol (QH_2) like moiety. MitoQH_2 can be reversibly oxidized to MitoQ *in vivo* (Tauskela 2007), a redox process that is believed to be one of the factors that make MitoQ a far better antioxidant than MitoE (Tauskela 2007; Smith et al. 2011). MitoQ is a membrane-penetrating agent that is distributed across the MIM: its linker enables the active ubiquinol antioxidant core to penetrate deeply into the membrane, an orientation that *per se* can justify its antioxidant performance against lipid peroxidation.

MitoQ was found to exert antioxidant protection in a number of animal models linked with oxidative stress related pathological events, namely alcoholic fatty liver and neurodegenerative diseases, ischemia-reperfusion, hypertension, sepsis and kidney damage in type I diabetes (Murphy 2016). The successful MitoQ pre-clinical data stimulated its advance to human clinical trials for several disease conditions such as Parkinson's disease (PD) (NCT00329056), Non-alcoholic Fatty Liver Disease (NAFLD) (NCT01167088) and Hepatitis C (NCT00433108) (**http://clinicaltrials.gov**).

SkQ1 (Fig. 3) is a mitochondrial-targeted antioxidant developed by Skulachev et al. having plastoquinone, a quinone present in chloroplasts, as core moiety, a ten alkyl linker and the carrier TPP (Skulachev et al. 2009). Along the drug discovery process a series of compounds have been synthesized being SkQ1 the most potent derivative (Zielonka et al. 2017). SkQ1H2 (the reduced form of SkQ1) was found to be a much better antioxidant than MitoQH2 and a less active prooxidant drug (Skulachev et al. 2010). The SkQ1 antioxidant efficiency was demonstrated in lipid micelles, artificial membranes, isolated mitochondria and living cells. Data acquired so far showed that SkQ1 can prevent/minimize lipid peroxidation, namely of the mitochondrial lipid cardiolipin, and ROS-induced apoptosis (Antonenko et al. 2008). Extensive animal studies have demonstrated beneficial roles of SkQ1 in many diseases associated with oxidative stress including ageing, stroke, myocardial infarction, sarcopenia (the gradual loss of muscle mass and function), dry eye syndrome, vascular inflammation (Bakeeva et al. 2008). Due to the promising *in vitro* and *in vivo* data, SkQ1 entered human clinical trials for dry eye condition (NCT02121301, **http://clinicaltrials.gov**), being well tolerated with no adverse events. SKQ1 is being tested for other pathologies, such as multiple sclerosis and acute kidney injury. More on this molecule can be read on Chaps. **23** and **28**.

Mito C (Fig. 3) is a mitochondriotropic antioxidant in which vitamin C, one hydrophilic antioxidants found in biological systems, was linked to the TPP moiety via a thioalkyl linker (Mirvish 1986; Finichiu et al. 2015). Along MitoC optimization process aimed to increase the amount of the antioxidant in mitochondria, different carbon aliphatic linkers (the number of methylene groups ranged from 3 to 21) were synthesized and tested. As expected, different data was obtained MitoVitC$_{11}$ is accumulated in response to the $\Delta\Psi m$ in contrastto MitoVitC$_3$ that is likely to be too hydrophilic for mitochondrial uptake. MitoVitC$_{11}$ was described to be the best derivative as it can prevent mitochondrial lipid peroxidation and protect mitochondrial aconitase, one of the most susceptible proteins to $O_2^{\bullet-}$ inactivation (Finichiu et al. 2015). MitoVitC$_{11}$ is oxidized by $O_2^{-\bullet}$, peroxyl radicals and Fe^{3+}, but not by H_2O_2, and within mitochondria is rapidly recycled back to the active ascorbate moiety by the glutathione and thioredoxin systems (Finichiu et al. 2015).

MitoSOD (Fig. 3) is a mitochondria-targeted agent based on M40403, a well-established macrocyclic Mn(II) SOD mimetic system (Salvemini et al. 2001). The core moiety was linked to TPP cation through a thioalkyl linker. MitoSOD is rapidly accumulated within mitochondria (approximately 3000-fold) presenting SOD-like activity (Jin et al. 2014) (Maroz et al. 2008). Kelso and co-workers describe MitoSOD as a selective modulator of mitochondrial redox network able to convert $O_2^{-\bullet}$ to H_2O_2 (Kelso et al. 2012). MitoSOD was also able to revert the rapid and progressive inhibition of aconitase by paraquat, a herbicide that generates $O_2^{-\bullet}$ through redox cycling (Kelso et al. 2012). Stability studies showed that MitoSOD can retain Mn^{2+}, except under acidic conditions, a property that limits its therapeutic potential, unless its administration is carried out by avoiding the passage through the stomach (Mitchell et al. 2012).

3.3 Mitochondriotropic Agents Based on Exogenous Antioxidants

Another approach to develop mitochondriotropic antioxidants is related with the use of dietary antioxidants as core moiety. Dietary antioxidants have been described to reduce oxidative stress by acting primarily as direct ROS scavengers and/or

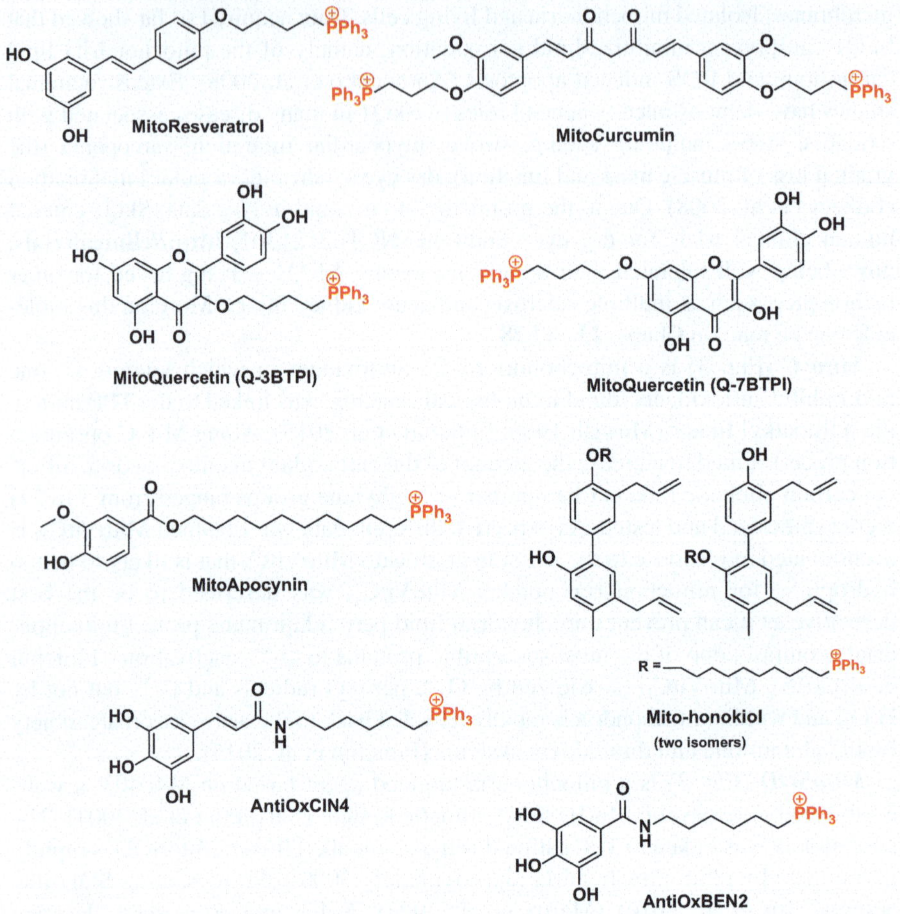

Fig. 4 Mitochondriotropic agents based on exogenous antioxidants (the counter ion was omitted for simplicity)

modulators of ROS-removing and ROS-generating enzymes and secondarily by up-regulating endogenous antioxidant defenses (Obrenovich et al. 2010; Sandoval-Acuña et al. 2014).

MitoResveratrol (Fig. 4) consists of a TPP unit covalently attached to resveratrol through a four alkyl linker Resveratrol has been described as a potent dietary antioxidant mainly due to its ability to scavenger free radicals and homeostasis properties by inducing the concentrations of some components of endogenous antioxidant systems (Biasutto et al. 2008; Smoliga et al. 2011). MitoResveratrol significantly increased the antiproliferative activity of resveratrol against colon tumor cells, via mitochondrial generation of H_2O_2, formed from dismutation of $O_2^{-\bullet}$ acting like a cytotoxic entity rather than a cytoprotective agent (Sassi et al. 2014).

Complementary studies with analogues revealed that the methylation of the hydroxyl groups increased cytotoxic effects of MitoResveratrol (Sassi et al. 2014).

MitoCurcumin (Fig. 4) was synthesized by the conjugation of curcumin antioxidant with two TPP cations through a three carbon aliphatic linkers (Reddy et al. 2014). Curcumin is a lipophilic dietary polyphenol able to prevent oxidative damage, namely by scavenging $O_2^{-\bullet}$ and $^{\bullet}OH$ radicals, by protecting membranes from lipid peroxidation processes and by chelation of redox active transition metals, such as Fe^{2+} and Cu^{+} (Aggarwal and Harikumar 2009; Barzegar and Moosavi-Movahedi 2011). The studies performed so far with MitoCurcumin, and its analogues showed superior antiproliferative effects (cytotoxic effects) when compared with curcumin in various cancer cell lines, but not in non-tumorigenic MCF-10A cells. The mechanism of action responsible for the noteworthy antiproliferative effects was proposed to be the induction of mitochondrial $O_2^{-\bullet}$ generation leading to apoptosis by decreasing the phosphorylation of Akt and STAT3, enhanced ERK phosphorylation, and upregulation of proapoptotic BNIP3 expression (Reddy et al. 2014). In summary, MitoCurcumin increased mitochondrial ROS, decreased the mitochondrial glutathione (GSH) levels and induced caspase-3 activity and strand breaks in the mtDNA. Moreover, MitoCurcumin was also shown to increase BAX to BCL-2 ratio, cytochrome C release into the cytosol and loss of $\Delta\Psi m$ suggesting that it can activate the intrinsic apoptotic pathway of lung cancer cells, which an increase of 25–50-fold antitumour efficacy when compared to curcumin (Jayakumar et al. 2017). MitoCurcumin can be considered a cytotoxic entity rather than a cytoprotective agent.

MitoQuercetins (Fig. 4) are mitochondrial-targeted TPP^{+}- based compounds linked to quercetin by a four carbon aliphatic chain located on a different hydroxyl functions (Q-3BTPI and Q-7BTPI). Quercetin is a flavonoid with a noticeable antioxidant activity due to its free radical scavenger and iron chelating properties (Cheng et al. 2010) and ability to increase the intracellular GSH content (Russo et al. 2012). Beneficial effects on osteoporosis, cancer, pulmonary, cardiovascular and aging-related diseases have been ascribed to quercetin (Miles et al. 2014). MitoQuercetins displayed cytotoxicity on fast-growing cells, probably due to their pro-oxidant properties (Mattarei et al. 2008; Sassi et al. 2012). Despite structural similarities, the two TPP quercetin derivatives have been shown to exert different effects on mitochondria $\Delta\Psi m$ and permeability transition pore (mPTP). While Q-3BTPI was reported to induce mPTP opening, and act as a mild OXPHOS uncoupler, causing mitochondrial depolarization and increasing of oxygen consumption (Biasutto et al. 2010), Q-7BTPI did not stimulate oxygen consumption or induced mPTP (Sassi et al. 2012). Studies in permeabilized rat liver mitochondria showed that Q-3BTPI can cause ATPase inhibition being effective inhibitors of colon tumor cells proliferation (Biasutto et al. 2010). In studies with intact cells, Q-3BTPI induced mitochondrial depolarization and oxidation of the MitoSOX Red probe, which can be explained by *in situ* generation of mitochondrial oxidants (Biasutto et al. 2010; Mattarei et al. 2011). MitoQuercetins are cytotoxic entities rather than cytoprotective agents.

MitoApocynin (Fig. 4) is an antioxidant based on apocynin, a natural compound structurally related with vanillin known to be a NADPH oxidase inhibitor, which showed significant beneficial effects in a preclinical MPTP mouse model of PD (Jin et al. 2014). MitoApocynin was also shown to exhibit neuroprotective effects in cellular and pre-clinical animal models of PD by attenuating oxidative damage and neuroinflammatory processes (Ghosh et al. 2016).

Mito-honokiol (Mito- HNK) (Fig. 4) is a derivative of honokiol linked to a TPP cation that has been developed to increase the accumulation into mitochondria. Honokiol is a natural biphenolic compound present in Magnolia bark extracts that has been reported to promote neurite outgrowth and have neuroprotective effects in rat cortical neurons. In addition, its antitumor effects in several *in vitro* and *in vivo* models of cancer has been demonstrated (Kalyanaraman et al. 2016). Honokiol has also been proposed as an effective antioxidant against lipid peroxidation by interfering with ROS production. The development of mitochondria-targeted honokiol-based agent lead to its successful delivery to mitochondria that inhibits cellular respiration and mitochondrial function at a concentration more than 100-fold lower than honokiol in intact H2030-BrM3 brain metastatic lung cancer cells and in highly invasive DMS-273 small cell lung cancer cells. This preliminary data reveal that Mito-honokiol is a cytotoxic entity rather than a cytoprotective agent.

AntiOxCIN$_4$ (Fig. 4) is a mitochondriotropic antioxidant based on the dietary antioxidant caffeic acid. The hydroxycinnamic acid was linked to lipophilic TPP cation through a six alkyl linker (Teixeira et al. 2012; Teixeira et al. 2017). AntiOxCIN$_4$ did not alter mitochondrial morphology and polarization and showed remarkable antioxidant and iron-chelation properties, preventing iron- and hydrogen peroxide-induced damage either in isolated liver mitochondria and hepatic cells. AntiOxCIN$_4$ can play a role on the maintenance of intracellular GSH homeostasis by increasing its supply (Teixeira et al. 2017). AntiOxCIN$_4$ was proposed as a potential drug candidate for mitochondrial oxidative stress-related diseases or mitochondrial and metabolic disorders involving iron overload (Teixeira et al. 2017).

AntiOxBEN$_2$ (Fig. 4) is a mitochondria-targeted antioxidant based on gallic acid (dietary antioxidant). The hydroxybenzoic acid was conjugated with TPP cation a six-carbon aliphatic chain (Jara et al. 2014; Teixeira et al. 2017). AntiOxBEN$_2$ can accumulate inside mitochondria driven by the $\Delta\Psi m$ achieving intramitochondrial millimolar concentrations and can prevent/minimize lipid peroxidation (Teixeira et al. 2017). AntiOxBEN$_2$ have a safety margin shown in rat cardiomyocytes H9c2, human dermal fibroblasts (HNDF) and human hepatocytes (HepG2) cells and can prevent oxidative stress-induced cytotoxicity in the same type of cells. For the first time it was shown in different cell models that is possible to target a gallic acid derivative to mitochondria without significantly affecting mitochondrial bioenergetics and ATP levels (Teixeira et al. 2017).

3.4 Mitochondriotropic Agents Based on Miscellaneous Antioxidant Moieties

MitoPeroxidase (Fig. 5) is a TPP-based mitochondria-targeted ebselen derivative, a seleno-organic antioxidant that has glutathione peroxidase (GPx)-like activity. The chemical moieties were linked through a four carbon aliphatic chain (Filipovska et al. 2005). Its antioxidant activity requires the presence of GSH or thioredoxin to generate *in situ* its reduced form (selenol). MitoPeroxidase prevented Fe^{2+}/H_2O_2-induced lipid peroxidation in intact mitochondria and decreased 2-deoxy-D-glucose (2DG) or H_2O_2-induced caspase-3-associated apoptosis in RBL-2H3 cells (Nickel et al. 2014).

MitoTEMPOL (Fig. 5) was developed by conjugating TEMPOL (a stable piperidine nitroxide radical) to TPP by means of a five alkyl linker. MitoTEMPOL accumulated inside of energized isolated mitochondria by 1000-fold, when compared with TEMPOL (Trnka et al. 2008). MitoTEMPOL is readily reduced to a hydroxylamine derivative (MitoTEMPOL-H) by ubiquinol (Jin et al. 2014), a process that involve the formation of a semiquinone radical that can rapidly dismutate to ubiquinone (Trnka et al. 2008). MitoTEMPOL has the capacity to detoxify ferrous iron by oxidizing it to ferric form and can act like a SOD mimetic system as it can convert $O2^{\bullet-}$ into water. MitoTEMPOL showed beneficial effects in several *in vitro* settings

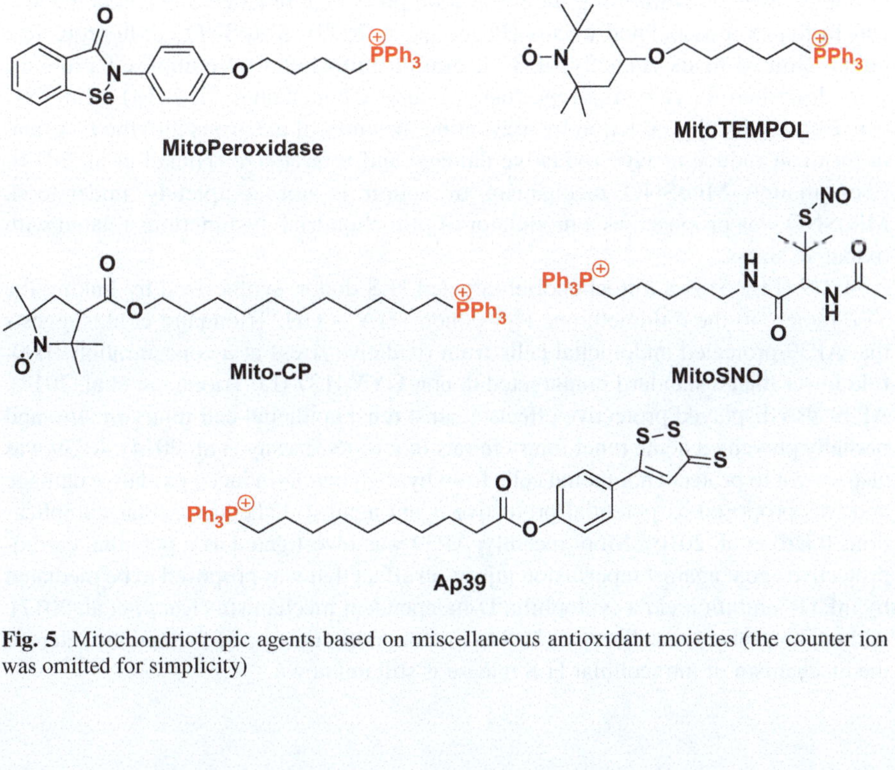

Fig. 5 Mitochondriotropic agents based on miscellaneous antioxidant moieties (the counter ion was omitted for simplicity)

of mitochondrial oxidative stress, for instance in the protection of pancreatic β-cells against oxidative stress (Lim et al. 2011b) and in a model of ischemia-reperfusion, as it can inhibit the ATP depletion mediated by mPTP opening and cell death (Liang et al. 2010). In an animal model of diabetes, MitoTEMPOL was shown to prevent mitochondrial and cytosolic ROS production (Pung et al. 2012).

Mito-CP (Fig. 5) is a mitochondria-targeted antioxidant, aimed to be a SOD mimetic agent, in which a 3-carboxyproxyl radical moiety was linked to TPP throughout an eleven alkyl linker (Dhanasekaran et al. 2005). To better understand Mito-CP SOD-like activity, studies were also performed with an acetamide analogue (Mito-CP-Ac), which do not have the nitroxide moiety thought to be responsible for SOD activity. Both compounds altered mitochondrial and glycolytic functions and inhibited the proliferation of various cancer cells, probably by a change of mitochondrial bioenergetics pathways rather than by dismutating and detoxifying mitochondrial $O_2^{-\bullet}$ (Cheng et al. 2015). Weinberg and co-workers suggested that Mito-CP decreases mitochondrial ROS levels, which was linked to the stimulation of cell proliferation (Weinberg et al. 2010). Interestingly, Mito-CP was proposed to be used as a probe for investigating the role of mitochondrial $O_2^{-\bullet}$ radical in cancer cell proliferation (Zielonka et al. 2017).

MitoSNO (Fig. 5) has a mitochondria-targeted *S*-nitrosothiol core that was linked by a five alkyl linker to TPP cation. MitoSNO is extensively accumulated in energized mitochondria of cells and tissues allowing *in situ* mitochondrial generation of $^{\bullet}$NO. The produced $^{\bullet}$NO can compete with O_2 at cytochrome C oxidase (complex IV) level inhibiting the respiration process, a practice that can be beneficial in hypoxia associated events (Prime et al. 2009). MitoSNO cardioprotective effects, linked to its capacity of S-nitrosation mitochondrial complex I proteins, were described for ischemia-reperfusion injury (Chouchani et al. 2013). MitoSNO can decrease ROS production by preventing the entry of electrons into the ETC and in turn can reduce *in vivo* oxidative damage and infarct size (Nickel et al. 2014). Even though MitoSNO mechanism of action is not completely understood, MitoSNO was proposed as a modulator of mitochondrial dysfunctions related with oxidative stress.

AP39 (Fig. 5) is a mitochondrial-targeted H_2S donor, synthesized by linking the TPP moiety to the dithiolethione H_2S donor, DTA − OH. Trionnaire et al. suggests that AP39 protected endothelial cells from oxidative stress at a concentration 1000-fold lower than a standard nontargeted donor, GYY4137 (Le Trionnaire et al. 2014). AP39 also displayed protective effects against renal epithelial cell injury *in vitro* and partially prevented acute renal injury in rats *in vivo* (Szczesny et al. 2014). AP39 was also shown to protect endothelial cells from hypoglycemia-induced oxidative damage and was proposed as potential protective agent against diabetic vascular complications (Gerő et al. 2016). More recently, AP39 was investigated as a potential cardioprotective agent against reperfusion injury, an effect that was proposed to be mediated by mPTP inhibition *via* a cyclophilin D-independent mechanism (Karwi et al. 2017). Although, a selective delivery of H_2S to mitochondria may be of therapeutic interest, the mechanism of intracellular H_2S release is still unknown.

4 One Step Forward: Development of Mitochondria-Targeted Probes/Sensors

As previously point out, TPP cation has been extensively used as mitochondrial-targeting carrier for a range of small molecules along some time. Noticeably, TPP was recently used for surface modification allowing the effective delivery of a variety of nanosystems to mitochondria (Murphy 2008; Smith et al. 2011; Wojtala et al. 2014; Madak and Neamati 2015).

Although the role of mitochondrion in disease is rather consensual, the evaluation of mitochondrial dysfunction is not always straightforward. This is the reason why the knowledge gathered so far in the development of mitochondriotropic small molecules was translated for the development of diagnostic tools (*e.g.* probes and sensors) suitable to measure the *in vitro* and/or *in vivo* imbalances on mitochondrial function (Strategy two; Fig. 2). In this context, several mitochondria redox probes and sensors have been advanced. This topic has been recently reviewed (Wisnovsky et al. 2016; Yang et al. 2017; Zielonka et al. 2017) and so they will briefly presented in this chapter.

4.1 Mitochondria-Targeted Probes/Sensors for Detection of Reactive Species

NPFR2 (Fig. 6). Inspired by flavin redox chemistry, a coenzyme that is fluorescent in its oxidized form (Visser et al. 1979), a number of reversible ROS sensors have been developed using tricyclic heterocycle isoalloxazine as core (Yamada et al. 2008; Yeow et al. 2014; Kaur et al. 2015b). One of the most known is NpFR2, a mitochondria-targeted probe consisting in a flavine type moiety, a lipophilic TPP cation and a three carbon aliphatic spacer, which exhibits a 115-fold increase in fluorescence when oxidized (Kaur et al. 2015a). NpFR2 can reversibly measure changes in the mitochondrial redox environment. As it responds non-selectively to a range of ROS, some studies were performed in primary mouse bone marrow, thymus and spleen cells in order to evaluate the global redox state (Kaur et al. 2016).

MitoSOX (MitoHEt) (Fig. 6). Based on the performance of hydroethidine (HEt), a chemical reporter often employed as fluorescent sensor for detecting cellular ROS (Wojtala et al. 2014), efforts have been done to target it to mitochondria. Therefore, it was linked to TPP cation throughout a six-carbon aliphatic spacer and MitoSOX (also called MitoHEt) was obtained (Robinson et al. 2006; Robinson et al. 2008; Roelofs et al. 2015). MitoSOX is in fact accumulated in mitochondria but it can disturb its function by inhibiting complex IV (Zielonka et al. 2008). It was found that in the presence of trace metal ions or haem proteins such as cytochrome C, nonspecific oxidation products, similar to those formed with HEt are also obtained (Lippert et al. 2011). So, in this type of studies the fluorescence measurements

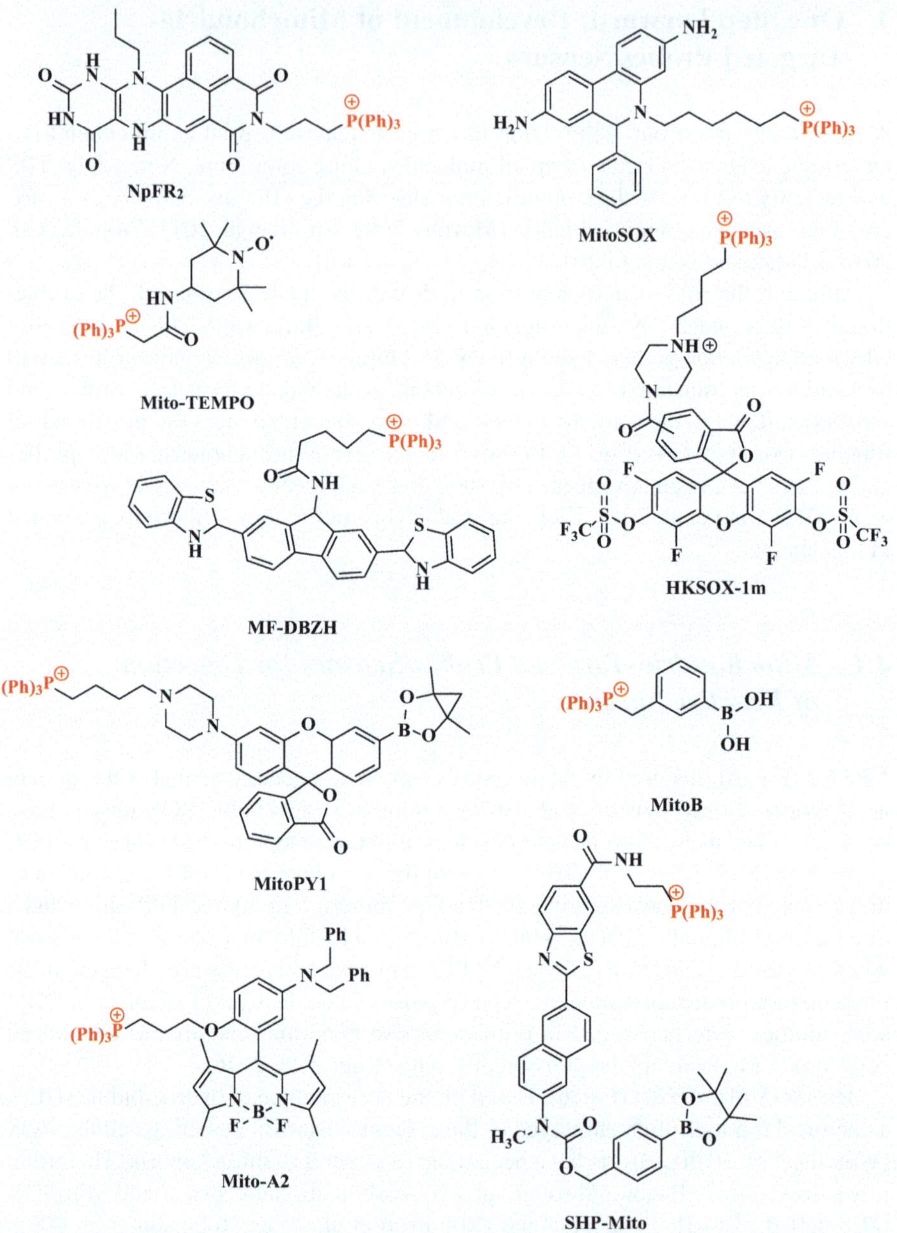

Fig. 6 Mitochondria-targeted probes/sensors for detection of reactive species (the counter ion was omitted for simplicity)

should be followed by HPLC or LC/MS analyses quantitating the amounts of 2-OH-Mito-E$^+$ and Mito-E$^+$ (Kalyanaraman et al. 2014).

Mito-TEMPO (Fig. 6). Electron paramagnetic resonance (EPR) has been used as a tool for monitorization of $O_2^{\bullet-}$ production in cell cultures and tissues (Dikalov et al. 2011). However, due the short-life time of $O_2^{-\bullet}$, indirect determinations, based on the formation of spin-adducts, with spin traps are often used. As the cyclic nitrone TEMPO may be reduced to an electron paramagnetic resonance silent hydroxylamine species it has been used as an EPR spin-trap in a diversity of biological studies.

Due its radical trapping performance, TEMPO was used for detection of mitochondrial $O_2^{-\bullet}$. To accumulate TEMPO in mitochondria, it was conjugated to a TPP cation via an acetamido (-$CH_2C(O)$-NH-) linker (Dikalov et al. 2011). Some studies performed in isolated mitochondria in the presence of ETC inhibitors, such as rotenone or antimycin A, showed that the intensity of the EPR signal is significantly higher, validating Mito-TEMPO accumulation within the organelle. However, some doubts still remained about its selective $O_2^{-\bullet}$ radical trapping capacity, as other *in situ* oxidants can react and change the EPR signal (Dikalov et al. 2011).

HKSOX-1 (Fig. 6), in particular HKSOX-1 m, is a mitochondria-targeted probe for imaging and detection of endogenous $O_2^{-\bullet}$ in cells and in *in vivo* (Hu et al. 2015). It is based on a fluorescein (5-carboxy-2′, 4′, 5′, 7′-tetrafluorofluorescein) fluorophore core, chosen mainly because its low pKa of 3.7 a property that allows a minimum fluorescence quenching even in the most acidic cellular conditions (Hu et al. 2015). Studies performed in differentiated human THP-1 cells reveal that HKSOX-1 can be accumulated by mitochondria and is effective when mitochondrial $O_2^{-\bullet}$ production was stimulated with antimycin A, a complex III inhibitor by a mechanism in which $O_2^{-\bullet}$ can trigger aryl triflate cleavage forming a fluorescent derivative with a 650-fold increment in fluorescence intensity. The probe shows good selectivity toward $O_2^{\bullet-}$ over a broad range of pH. However, during the reaction, $CF_3SO_2OO^\bullet$ radical is formed and can change the redox systems in study (Hu et al. 2015).

MF-DBZH (Fig. 6) is a two-photon (TP) fluorescence imaging mitochondria targeted probe obtained by the conjugation of a dibenzothiazolineflurene fluorophore with TPP cation. Benzothiazoline was used as the receptor due to its good selectivity toward $O_2^{\bullet-}$. MF-DBZH have high selectivity toward mitochondrial $O_2^{\bullet-}$ fluxes, observed by a fluorescence increment (λexc = 483 nm, λemi = 512 nm), responding proportionally to a $O_2^{\bullet-}$ concentration change. Moreover, MF-DBZH has high photostability, insensibility toward pH changes and an excellent biocompatibility (Li et al. 2013). Favourable features of this probe also include convenient cell loading and easy staining in cells. Most importantly, MF-DBZH gives a consistent TP fluorescent signal to changes of $O_2^{\bullet-}$ levels *in vivo*. However, the selectivity toward $O_2^{-\bullet}$ is not reliable, as other one-electron oxidants are able to react with the probe (Li et al. 2013).

MitoPY1 (Fig. 6). Mitochondria peroxy yellow 1 is a mitochondria-targeted fluorescent probe that selectively tracks to the mitochondria and responds to local fluxes of H_2O_2. This probe combines a boronate-masked xanthene fluorophore core and a

TPP cation targeting group. MitoPY1 boronate moiety is converted *in situ* by a H_2O_2-mediated reaction into a phenol moiety by a process that triggers the subsequent opening of the bottom-ring lactone. Along this process, a fully conjugated xanthene fluorophore is formed that has a yellow fluorescence (MitoPY1ox). The boronate-based molecular switch selectively respond to H_2O_2 over competing ROS inside the mitochondria (Dickinson and Chang 2008; Dickinson et al. 2013). MitoPY1 was used in HeLa cells to image H_2O_2 in mitochondria (Dickinson and Chang 2008).

MitoB (Fig. 6) is a mitochondria-targeted probe used to report *in vivo* levels of mitochondrial H_2O_2 (Cochemé et al. 2011; Cochemé et al. 2012). MitoB contains a TPP moiety, that drives its accumulation into mitochondria and an arylboronic moiety that selectively reacts with H_2O_2 producing i*n situ* a phenol product (MitoP or *m*-MitoPhOH). It was proposed as a diagnostic exomarker as when administered to an organism it is oxidized by reactive species into the phenolic product. The exomarker and the precursor probe can be analysed *ex vivo* by LC-MS/MS allowing to identify and quantify the extent of MitoB oxidation, which can be expressed as a ratio of MitoP/MitoB, a parameter that showed the real distribution of the probe in the tissue under study (Cochemé et al. 2011; Logan et al. 2014). The MitoP/MitoB ratio will reflect the H_2O_2 concentration present *in vivo* in the mitochondrial matrix (Logan et al. 2014; Cairns et al. 2015).

SHP-Mito (Fig. 6) is a boronate-based TP fluorescent probe conjugated with TPP cation developed to detect mitochondrial H_2O_2 in living cells and intact tissues. It is a specific, selective and ratiometric TP probe (Masanta et al. 2012). SHP-Mito shows a significant TP cross-section, a marked blue-to-yellow emission color change in response to H_2O_2, a low cytotoxicity, and insensitivity to pH in the biologically relevant pH range (Masanta et al. 2012).

MitoA2 (Fig. 6). This probe is constituted by a boron-dipyrromathene (BODIPY)-based fluorogenic core that is conjugated to TPP cation (Miao et al. 2016). MitoA2 is useful for detection of peroxynitrite ($ONOO^-$), a reactive species that is the product of the diffusion-controlled reaction of nitric oxide with $O_2^{-\bullet}$ radical (Szabó et al. 2007). MitoA2 mechanism is based on a $ONOO^-$-induced oxidation and subsequent nitrosation of its aromatic amine moiety (Miao et al. 2016). It has a quite fast fluorescence off-on response, sensitivity and selectivity over a series of biologically relevant ROS as well as transition metal cations. However, further studies are needed to validate MitoA2 as an imaging tool for the analyze of the distribution and pathophysiological functions of $ONOO^-$ in cells, organelles and tissues.

4.2 Mitochondria-Targeted Probes/Sensors for Detection of Other Species

SSH-Mito (Fig. 7). This probe has a BTDAN (6-(benzo[d]thiazol-20-yl)-2-(N,N-dimethylamino)naphthalene) fluorophore core conjugated with TPP cation and was developed to detect hydrogen sulfide (H_2S), an important mediator of the thiol redox status in cell signaling (Li et al. 2011; Lim et al. 2011b). The wavelength of BTDAN

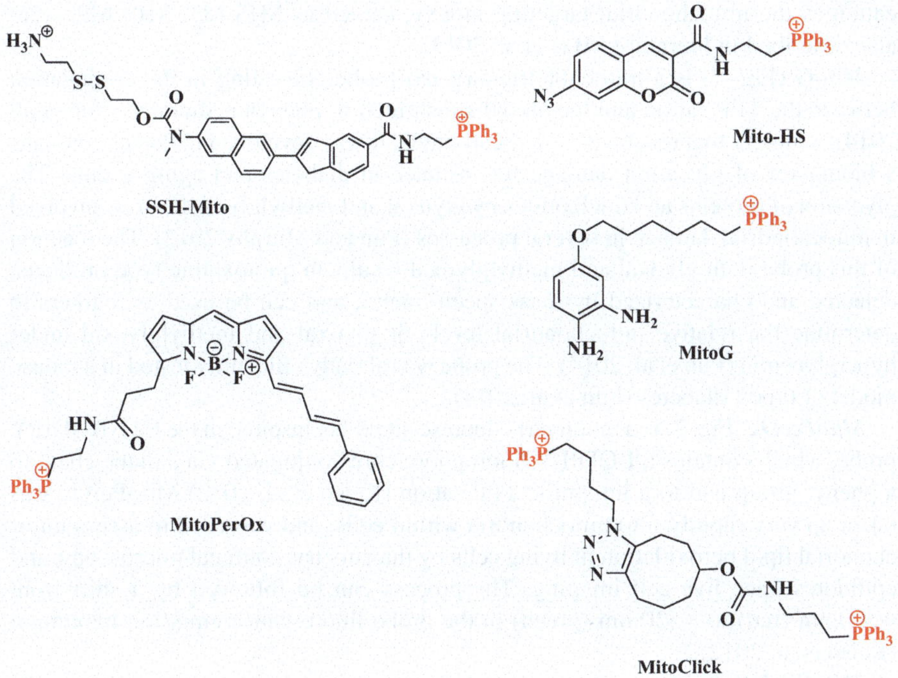

Fig. 7 Mitochondria-targeted probes/sensors for detection of other type of damaging species (the counter ion was omitted for simplicity)

fluorescence is significantly modulated by the amine present on the naphthalene ring, which in SSH-Mito is protected by a carbamate function. SSH-Mito contains a redox-responsive disulfide bond, which upon reduction cause an intramolecular nucleophilic attack of the SH group to the carbamate, and production of a naphthylamine. This reaction is irreversible and the fluorescence of the probe changes from blue to yellow, a feature that can be determined by monitoring the ratio of yellow to blue fluorescence (at 550 nm and 450 nm, respectively). The probe efficiency was demonstrated in cell-based experiences (Kim and Cho 2013) and on rat hippocampal slices to image thiol levels (Lim et al. 2011a).

Mito-HS (Fig. 7) is a bioorthogonal fluorescent probe, having a coumarin azide derivative core conjugated with TPP cation, for detecting mitochondrial H_2S. The mechanism is based on the reduction of the azidyl group to the corresponding amine by H_2S. The sensitivity of Mito-HS toward H_2S was found to be as low as 24.3 nM. The probe showed high selectivity toward H_2S over other competitive ubiquitous entities in living systems. Mito-HS was able to detect H_2S formation in cancer cells without adding external stimulators. Based on the endogenous tracking ability of H_2S, Mito-HS was proposed as a biomarker for distinguishing cancer cells from normal cells (Wu et al. 2016).

It is important to note that other mitochondria-targeted H_2S radiometric probes, based on 4-azidobenzyl carbamate as the H_2S response site and having the TPP cation as the mitochondrial targeting moiety, named as SHS-M1, SHS-M2, were also described in literature. (Bae et al. 2013)

MitoG (Fig. 7) is a mitochondria-targeted probe consisting in the conjugation between the TPP cation and the alkoyl-substituted o- phenylenediamine (Pun et al. 2014), a moiety that react with 1,2-dicarbonyls. It was developed to detect glyoxals, a biomarker of glycation damage for instance in diabetes and aging events. The glycation of proteins and nucleotides by glyoxal and methylglyoxal can be involved in mitochondrial damage in several processes (Pun and Murphy 2012). The reaction of this probe with glyoxal and methylglyoxal results in quinoxaline type products, detected and characterized by mass spectrometry, and can be used as markers to determine the relative mitochondrial levels of glyoxal and methylglyoxal under hyperglycemia (Pun et al. 2014). The probe was already efficiently tested in a mouse model of type I diabetes (Pun et al. 2014).

MitoPerOx (Fig. 7) is a ratiometric fluorescent probe, inspired in the C11-BODIPY probe, which contains a BODIPY fluorophore core conjugated via a diene chain to a phenyl group and to a lipophilic TPP cation (Kelso et al. 2012) MitoPerOx was taken up very rapidly into mitochondria within cells, and was used to assess mitochondrial lipid peroxidation in living cells by fluorimetry, confocal microscopy, and epifluorescence live cell imaging. The process can be followed by a shift from ~590 nm (red) to ~520 nm (green) in the probe fluorescence emission maximum (Kelso et al. 2012).

MitoClick (Fig. 7) is a probe constituted by two systems, one containing an azide and the other a cycloctene, independently conjugated to the TPP cation (MitoAzido and MitoOct) that, when administrated simultaneous, are extensively accumulated in mitochondria. The copper-free click-chemistry-based reaction between azide (present in MitoAzido) and cyclooctyne groups of MitoOct probe lead to the formation *in situ*, in a concentration dependent manner, of a new TPP cation labeled target molecule called MitoClick. The process occurs in the mitochondrial matrix and can be followed qualitatively and quantitatively by mass spectrometry (Logan et al. 2016). This new probe is highly sensitive and can be used for the $\Delta\Psi m$ assessment *in vivo*. This probe opens a new possibility to measure small changes in $\Delta\Psi m$ in physiological and pathological processes (Logan et al. 2016).

5 Concluding Remarks

The knowledge about mitochondria has expanded considerably over the past decade and as a result the impairment of mitochondrial functions has been linked to an increasing number of human illnesses, such as cancer, neurodegenerative diseases, metabolic disorders, and chronic inflammation. Although the wide-ranging impact of mitochondria in so many diseases makes them a promising drug target currently there are not therapies available to specifically treat mitochondrial dysfunctions.

Designing mitochondrial-active molecules that can modulate mitochondrial function is a hot issue that is still in a developmental stage despite major advances in the understanding of the pathophysiology of mitochondrial related diseases. Till now the majority of approaches to target small molecules, in particular antioxidants, to mitochondria have been focused on the use of the lipophilic TPP cation as carrier. Generally, the pharmacophore of interest (cargo) is tethered to TPP cation through a carbon aliphatic linker. As shown along this chapter the performance of the mitochondriotropic agent is dependent of the nature and length of the linker, as it modulates lipophilicity, cellular uptake, site of mitochondrial action (matrix *versus* membrane) and toxicity. The cytoprotective or cytotoxic performance of the new chemical entity is strongly dependent of the length of linker and its position on the bioactive molecule (core). Noteworthy the knowledge has been also applied for the delivery of cytotoxic agents (e.g. Mito-Doxorubicin (Han et al. 2014), MitoK$_3$ (Teixeira et al. 2018), TPP + C10 (Jara et al. 2014), Mito-Metformin (Cheng et al. 2016), Mito-Chlorambucil (Millard et al. 2013)) to improve their selective accumulation in mitochondria.

Based on the gathered knowledge, the TPP carrier has also been used to develop mitochondria targeted redox probes/sensors, imaging agents and nanoparticle platforms, although more data is needed to validate the claims.

Although the TPP cation has been the most extensively used mitochondrial-targeting group, there are still controversies surrounding this approach, namely related with its intrinsic toxicity. Consequently, efforts must be done in finding new cation carriers, and to guarantee that the cargo does indeed access the mitochondrial matrix and does not merely associate with the mitochondrial membranes. Additionally, *in vivo* biodistribution, pharmacokinetics and long-term toxic effects studies to provide accurate information about efficacy and toxicity are still needed. The translation from bench to bedside is still an emergent issue.

Acknowledgements This work was funded by FEDER funds through the Operational Programme Competitiveness Factors—COMPETE and national funds by FCT—Foundation for Science and Technology under research grants PTDC/DTP-FTO/2433/2014, POCI-01-0145-FEDER-016659, POCI-01-0145-FEDER-007440, POCI-01-0145-FEDER-006980, and NORTE-01-0145-FEDER-000028. R. Amorim (PTDC/DTP-FTO/2433/2014), S. Benfeito (SFRH/BD/99189/2013), J. Teixeira (NORTE-01-0145-FEDER-000028) and F. Cagide (NORTE-01-0145-FEDER-000028) grants are supported by the European Regional Development Fund (ERDF) through the COMPETE 2020—Operational Programme for Competitiveness and Internationalisation and Portuguese national funds.

Declaration of Interest The authors declare no competing financial interest. PJO and FB are co-founders of the CNC spin-off company MitoTAG.

References

Aggarwal BB, Harikumar KB (2009) Potential therapeutic effects of curcumin, the anti-inflammatory agent, against neurodegenerative, cardiovascular, pulmonary, metabolic, autoimmune and neoplastic diseases. Int J Biochem Cell Biol 41(1):40–59

Agrawal A, Mabalirajan U (2016) Rejuvenating cellular respiration for optimizing respiratory function: targeting mitochondria. Am J Physiol Lung Cell Mol Physiol 310(2):L103–L113

Alfadda AA, Sallam RM (2012) Reactive oxygen species in health and disease. BioMed Res Int 2012:936486

Anders M (2013) Exploiting endobiotic metabolic pathways to target xenobiotic antioxidants to mitochondria. Mitochondrion 13(5):454–463

Andreux PA et al (2013) Pharmacological approaches to restore mitochondrial function. Nat Rev Drug Discov 12(6):465–483

Antonenko YN et al (2008) Protective effects of mitochondria-targeted antioxidant SkQ in aqueous and lipid membrane environments. J Membr Biol 222(3):141

Apostolova N, Victor VM (2015) Molecular strategies for targeting antioxidants to mitochondria: therapeutic implications. Antioxid Redox Signal 22(8):686–729

Bae SK et al (2013) A ratiometric two-photon fluorescent probe reveals reduction in mitochondrial H2S productiond in Parkinson's disease gene knockout astrocytes. J Am Chem Soc 135(26):9915–9923

Bakeeva L et al (2008) Mitochondria-targeted plastoquinone derivatives as tools to interrupt execution of the aging program. 2. Treatment of some ROS-and age-related diseases (heart arrhythmia, heart infarctions, kidney ischemia, and stroke). Biochem Mosc 73(12):1288–1299

Barzegar A, Moosavi-Movahedi AA (2011) Intracellular ROS protection efficiency and free radical-scavenging activity of curcumin. PLoS One 6(10):e26012

Biasutto L et al (2008) Development of mitochondria-targeted derivatives of resveratrol. Bioorg Med Chem Lett 18(20):5594–5597

Biasutto L et al (2010) Impact of mitochondriotropic quercetin derivatives on mitochondria. Biochimica et Biophysica Acta (BBA)-Bioenergetics 1797(2):189–196

Brookes PS et al (2004) Calcium, ATP, and ROS: a mitochondrial love-hate triangle. Am J Physiol Cell Physiol 287(4):C817–C833

Cairns AG et al (2015) Targeting mitochondria with small molecules: the preparation of MitoB and MitoP as exomarkers of mitochondrial hydrogen peroxide. Methods Mol Biol 1265:25–50. https://doi.org/10.1007/978-1-4939-2288-8_3

Cheng S et al (2010) Quercetin induces tumor-selective apoptosis through downregulation of Mcl-1 and activation of Bax. Clin Cancer Res 16(23):5679–5691

Cheng G et al (2015) Antiproliferative effects of mitochondria-targeted cationic antioxidants and analogs: Role of mitochondrial bioenergetics and energy-sensing mechanism. Cancer Lett 365(1):96–106

Cheng G et al (2016) Mitochondria-targeted analogues of metformin exhibit enhanced antiproliferative and radiosensitizing effects in pancreatic cancer cells. Cancer Res 76(13):3904–3915

Chouchani ET et al (2013) Cardioprotection by S-nitrosation of a cysteine switch on mitochondrial complex I. Nat Med 19(6):753–759

Cochemé HM et al (2011) Measurement of H 2 O 2 within living Drosophila during aging using a ratiometric mass spectrometry probe targeted to the mitochondrial matrix. Cell Metab 13(3):340–350

Cochemé HM et al (2012) Using the mitochondria-targeted ratiometric mass spectrometry probe MitoB to measure H2O2 in living Drosophila. Nat Protoc 7(5):946

Cree L et al (2009) The inheritance of pathogenic mitochondrial DNA mutations. Biochimica et Biophysica Acta (BBA)-Molecular Basis of Disease 1792(12):1097–1102

Dhanasekaran A et al (2004) Supplementation of endothelial cells with mitochondria-targeted antioxidants inhibit peroxide-induced mitochondrial iron uptake, oxidative damage, and apoptosis. J Biol Chem 279(36):37575–37587

Dhanasekaran A et al (2005) Mitochondria superoxide dismutase mimetic inhibits peroxide-induced oxidative damage and apoptosis: role of mitochondrial superoxide. Free Radic Biol Med 39(5):567–583

Dickinson BC, Chang CJ (2008) A targetable fluorescent probe for imaging hydrogen peroxide in the mitochondria of living cells. J Am Chem Soc 130(30):9638–9639

Dickinson BC et al (2013) Preparation and use of MitoPY1 for imaging hydrogen peroxide in mitochondria of live cells. Nat Protoc 8(6):1249–1259

Dikalov SI et al (2011) EPR detection of cellular and mitochondrial superoxide using cyclic hydroxylamines. Free Radic Res 45(4):417–430
Dong L-F et al (2011) Mitochondrial targeting of vitamin E succinate enhances its pro-apoptotic and anti-cancer activity via mitochondrial complex II. J Biol Chem 286(5):3717–3728
Figueira TR et al (2013) Mitochondria as a source of reactive oxygen and nitrogen species: from molecular mechanisms to human health. Antioxid Redox Signal 18(16):2029–2074
Filipovska A et al (2005) Synthesis and characterization of a triphenylphosphonium-conjugated peroxidase mimetic insights into the interaction of ebselen with mitochondria. J Biol Chem 280(25):24113–24126
Finichiu PG et al (2015) A mitochondria-targeted derivative of ascorbate: MitoC. Free Radic Biol Med 89:668–678
Gerő D et al (2016) The novel mitochondria-targeted hydrogen sulfide (H 2 S) donors AP123 and AP39 protect against hyperglycemic injury in microvascular endothelial cells in vitro. Pharmacol Res 113:186–198
Ghosh A et al (2016) Mitoapocynin treatment protects against neuroinflammation and dopaminergic neurodegeneration in a preclinical animal model of Parkinson's disease. J Neuroimmune Pharmacol 11(2):259–278
Han M et al (2014) Mitochondrial delivery of doxorubicin via triphenylphosphine modification for overcoming drug resistance in MDA-MB-435/DOX cells. Mol Pharm 11(8):2640–2649
Hu JJ et al (2015) Fluorescent probe HKSOX-1 for imaging and detection of endogenous superoxide in live cells and in vivo. J Am Chem Soc 137(21):6837–6843
Hughes G et al (2005) Mitochondrial reactive oxygen species regulate the temporal activation of nuclear factor κB to modulate tumour necrosis factor-induced apoptosis: evidence from mitochondria-targeted antioxidants. Biochem J 389(1):83–89
Hurko O (2013) Drug development for rare mitochondrial disorders. Neurotherapeutics 10(2): 286–306
Jameson VJ et al (2015) Synthesis of triphenylphosphonium vitamin E derivatives as mitochondria-targeted antioxidants. Tetrahedron 71(44):8444–8453
Jara JA et al (2014) Antiproliferative and uncoupling effects of delocalized, lipophilic, cationic gallic acid derivatives on cancer cell lines. Validation in vivo in singenic mice. J Med Chem 57(6):2440–2454
Jayakumar S et al (2017) Mitochondrial targeted curcumin exhibits anticancer effects through disruption of mitochondrial redox and modulation of TrxR2 activity. Free Radic Biol Med 113:530–538
Jin H et al (2014) Mitochondria-targeted antioxidants for treatment of Parkinson's disease: preclinical and clinical outcomes. Biochimica et Biophysica Acta (BBA)-Molecular Basis of Disease 1842(8):1282–1294
Kalyanaraman B, Dranka BP, Hardy M, Michalski R, Zielonka J (2014) HPLC-based monitoring of products formed from hydroethidine-based fluorogenic probes–the ultimate approach for intra- and extracellular superoxide detection. Biochim Biophys Acta. 1840(2):739–744. https://doi.org/10.1016/j.bbagen.2013.05.008
Kalyanaraman B et al (2016) Mito-honokiol compounds and methods of synthesis and use thereof, Google Patents
Kanabus M et al (2014) Development of pharmacological strategies for mitochondrial disorders. Br J Pharmacol 171(8):1798–1817
Karwi QG et al (2017) AP39, a mitochondria-targeting hydrogen sulfide (H2S) donor, protects against myocardial reperfusion injury independently of salvage kinase signalling. Br J Pharmacol 174(4):287–301
Kaur A et al (2015a) Mitochondrially targeted redox probe reveals the variations in oxidative capacity of the haematopoietic cells. Org Biomol Chem 13(24):6686–6689
Kaur A et al (2015b) A FRET-based ratiometric redox probe for detecting oxidative stress by confocal microscopy, FLIM and flow cytometry. Chem Commun 51(52):10510–10513
Kaur A et al (2016) Reversible fluorescent probes for biological redox states. Angew Chem Int Ed 55(5):1602–1613

Kelso GF et al (2012) A mitochondria-targeted macrocyclic Mn (II) superoxide dismutase mimetic. Chem Biol 19(10):1237–1246

Kim HM, Cho BR (2013) Mitochondrial-targeted two-photon fluorescent probes for zinc ions, and thiols in living tissues. Oxid Med Cell Longev 2013:323619

Le Trionnaire S et al (2014) The synthesis and functional evaluation of a mitochondria-targeted hydrogen sulfide donor,(10-oxo-10-(4-(3-thioxo-3 H-1, 2-dithiol-5-yl) phenoxy) decyl) triphenylphosphonium bromide (AP39). MedChemComm 5(6):728–736

Levitskii D, Skulachev V (1972) Effects of penetrating synthetic ions on the respiration of mitochondria and submitochondrial particles. Mol Biol 6(3):267

Li L et al (2011) Hydrogen sulfide and cell signaling. Annu Rev Pharmacol Toxicol 51:169–187

Li P et al (2013) Mitochondria-targeted reaction-based two-photon fluorescent probe for imaging of superoxide anion in live cells and in vivo. Anal Chem 85(20):9877–9881

Liang HL et al (2010) SOD1 and MitoTEMPO partially prevent mitochondrial permeability transition pore opening, necrosis, and mitochondrial apoptosis after ATP depletion recovery. Free Radic Biol Med 49(10):1550–1560

Liberman E, Topaly V (1969) Permeability of bimolecular phospholipid membranes for fat-soluble ions. Biofizika 14(3):452

Lim CS et al (2011a) Ratiometric detection of mitochondrial thiols with a two-photon fluorescent probe. J Am Chem Soc 133(29):11132–11135

Lim S et al (2011b) Mitochondria-targeted antioxidants protect pancreatic β-cells against oxidative stress and improve insulin secretion in glucotoxicity and glucolipotoxicity. Cell Physiol Biochem 28(5):873–886

Lippert AR et al (2011) Boronate oxidation as a bioorthogonal reaction approach for studying the chemistry of hydrogen peroxide in living systems. Acc Chem Res 44(9):793–804

Logan A et al (2014) Using exomarkers to assess mitochondrial reactive species in vivo. Biochimica et Biophysica Acta (BBA)-General Subjects 1840(2):923–930

Logan A et al (2016) Assessing the mitochondrial membrane potential in cells and in vivo using targeted click chemistry and mass spectrometry. Cell Metab 23(2):379–385

Madak JT, Neamati N (2015) Membrane permeable lipophilic cations as mitochondrial directing groups. Curr Top Med Chem 15(8):745–766

Malty RH et al (2014) Mitochondrial targets for pharmacological intervention in human disease. J Proteome Res 14(1):5–21

Mancuso M et al (2012) Drugs and mitochondrial diseases: 40 queries and answers. Expert Opin Pharmacother 13(4):527–543

Maroz A et al (2008) Pulse radiolysis investigation on the mechanism of the catalytic action of Mn(II)−pentaazamacrocycle compounds as superoxide dismutase mimetics. Chem A Eur J 112(22):4929–4935

Masanta G et al (2012) A mitochondria-localized two-photon fluorescent probe for ratiometric imaging of hydrogen peroxide in live tissue. Chem Commun 48(29):3518–3520

Mattarei A et al (2008) A mitochondriotropic derivative of quercetin: a strategy to increase the effectiveness of polyphenols. Chembiochem 9(16):2633–2642

Mattarei A et al (2011) Redox properties and cytotoxicity of synthetic isomeric mitochondriotropic derivatives of the natural polyphenol quercetin. Eur J Org Chem 2011(28):5577–5586

Miao J et al (2016) A new class of fast-response and highly selective fluorescent probes for visualizing peroxynitrite in live cells, subcellular organelles, and kidney tissue of diabetic rats. Biomaterials 107:33–43

Miles SL et al (2014) Molecular and physiological actions of quercetin: need for clinical trials to assess its benefits in human disease. Nutr Rev 72(11):720–734

Millard M et al (2013) A selective mitochondrial-targeted chlorambucil with remarkable cytotoxicity in breast and pancreatic cancers. J Med Chem 56(22):9170–9179

Mirvish SS (1986) Effects of vitamins C and E on N-nitroso compound formation, carcinogenesis, and cancer. Cancer 58(S8):1842–1850

Mitchell T et al (2012) Controlling radicals in the powerhouse: development of MitoSOD. Chem Biol 19(10):1217–1218

Murphy MP (2008) Targeting lipophilic cations to mitochondria. Biochimica et Biophysica Acta (BBA)-Bioenergetics 1777(7):1028–1031

Murphy MP (2016) Understanding and preventing mitochondrial oxidative damage. Biochem Soc Trans 44(5):1219–1226

Murphy MP, Smith RA (2007) Targeting antioxidants to mitochondria by conjugation to lipophilic cations. Annu Rev Pharmacol Toxicol 47:629–656

Nickel A et al (2014) Mitochondrial reactive oxygen species production and elimination. J Mol Cell Cardiol 73:26–33

Obrenovich ME et al (2010) The role of polyphenolic antioxidants in health, disease, and aging. Rejuvenation Res 13(6):631–643

Pagano G et al (2014) Oxidative stress and mitochondrial dysfunction across broad-ranging pathologies: toward mitochondria-targeted clinical strategies. Oxid Med Cell Longev 2014:541230

Pfeffer G, Majamaa K, Turnbull DM, Thorburn D, Chinnery PF (2012) Treatment for mitochondrial disorders. Cochrane Database Syst Rev 4:CD004426. https://doi.org/10.1002/14651858.CD004426.pub3

Pfeffer G et al (2013) New treatments for mitochondrial disease—no time to drop our standards. Nat Rev Neurol 9(8):474–481

Prime TA et al (2009) A mitochondria-targeted S-nitrosothiol modulates respiration, nitrosates thiols, and protects against ischemia-reperfusion injury. Proc Natl Acad Sci U S A 106(26):10764–10769

Pun PBL, Murphy MP (2012) Pathological significance of mitochondrial glycation. Int J Cell Biol 2012:843505

Pun PBL et al (2014) A mitochondria-targeted mass spectrometry probe to detect glyoxals: implications for diabetes. Free Radic Biol Med 67:437–450

Pung YF et al (2012) Resolution of mitochondrial oxidative stress rescues coronary collateral growth in Zucker obese fatty rats. Arterioscler Thromb Vasc Biol 32(2):325–334

Reddy CA et al (2014) Mitochondrial-targeted curcuminoids: a strategy to enhance bioavailability and anticancer efficacy of curcumin. PLoS One 9(3):e89351

Robinson KM et al (2006) Selective fluorescent imaging of superoxide in vivo using ethidium-based probes. Proc Natl Acad Sci U S A 103(41):15038–15043

Robinson KM et al (2008) The selective detection of mitochondrial superoxide by live cell imaging. Nat Protoc 3(6):941

Roelofs BA et al (2015) Low micromolar concentrations of the superoxide probe MitoSOX uncouple neural mitochondria and inhibit complex IV. Free Radic Biol Med 86:250–258

Ross M et al (2005) Lipophilic triphenylphosphonium cations as tools in mitochondrial bioenergetics and free radical biology. Biochem Mosc 70(2):222–230

Russo M et al (2012) The flavonoid quercetin in disease prevention and therapy: facts and fancies. Biochem Pharmacol 83(1):6–15

Saeidnia S, Abdollahi M (2013) Toxicological and pharmacological concerns on oxidative stress and related diseases. Toxicol Appl Pharmacol 273(3):442–455

Salvemini D et al (2001) Pharmacological manipulation of the inflammatory cascade by the superoxide dismutase mimetic, M40403. Br J Pharmacol 132(4):815–827

Sandoval-Acuña C et al (2014) Polyphenols and mitochondria: an update on their increasingly emerging ROS-scavenging independent actions. Arch Biochem Biophys 559:75–90

Sassi N et al (2012) Cytotoxicity of a mitochondriotropic quercetin derivative: mechanisms. Biochimica et Biophysica Acta (BBA)-Bioenergetics 1817(7):1095–1106

Sassi N et al (2014) Mitochondria-targeted resveratrol derivatives act as cytotoxic pro-oxidants. Curr Pharm Des 20(2):172–179

Skulachev VP et al (2009) An attempt to prevent senescence: a mitochondrial approach. Biochimica et Biophysica Acta (BBA)-Bioenergetics 1787(5):437–461

Skulachev VP et al (2010) Prevention of cardiolipin oxidation and fatty acid cycling as two antioxidant mechanisms of cationic derivatives of plastoquinone (SkQs). Biochimica et Biophysica Acta (BBA)-Bioenergetics 1797(6):878–889

Smith RA et al (2003) Delivery of bioactive molecules to mitochondria in vivo. Proc Natl Acad Sci U S A 100(9):5407–5412

Smith RA et al (2011) Mitochondria-targeted small molecule therapeutics and probes. Antioxid Redox Signal 15(12):3021–3038

Smith RA et al (2012) Mitochondrial pharmacology. Trends Pharmacol Sci 33(6):341–352

Smoliga JM et al (2011) Resveratrol and health–a comprehensive review of human clinical trials. Mol Nutr Food Res 55(8):1129–1141

Sorriento D et al (2014) Targeting mitochondria as therapeutic strategy for metabolic disorders. Sci World J 2014:604685

Suomalainen A, Battersby BJ (2017) Mitochondrial diseases: the contribution of organelle stress responses to pathology. Nat Rev Mol Cell Biol. https://doi.org/10.1038/nrm.2017.66

Szabó C et al (2007) Peroxynitrite: biochemistry, pathophysiology and development of therapeutics. Nat Rev Drug Discov 6(8):662

Szczesny B et al (2014) AP39, a novel mitochondria-targeted hydrogen sulfide donor, stimulates cellular bioenergetics, exerts cytoprotective effects and protects against the loss of mitochondrial DNA integrity in oxidatively stressed endothelial cells in vitro. Nitric Oxide 41:120–130

Tauskela JS (2007) MitoQ--a mitochondria-targeted antioxidant. IDrugs 10(6):399–412

Teixeira J et al (2012) Rational discovery and development of a mitochondria-targeted antioxidant based on cinnamic acid scaffold. Free Radic Res 46(5):600–611

Teixeira J et al (2017) Development of a mitochondriotropic antioxidant based on caffeic acid: proof of concept on cellular and mitochondrial oxidative stress models. J Med Chem 60(16):7084–7098

Teixeira J et al (2018) Disruption of mitochondrial function as mechanism for anti-cancer activity of a novel mitochondriotropic menadione derivative. Toxicology 393:123–139

Trnka J et al (2008) A mitochondria-targeted nitroxide is reduced to its hydroxylamine by ubiquinol in mitochondria. Free Radic Biol Med 44(7):1406–1419

Visser AJ et al (1979) Fluorescence properties of reduced flavins and flavoproteins. FEBS J 101(1):13–21

Wagner BK et al (2008) Large-scale chemical dissection of mitochondrial function. Nat Biotechnol 26(3):343–351

Weinberg F et al (2010) Mitochondrial metabolism and ROS generation are essential for Kras-mediated tumorigenicity. Proc Natl Acad Sci U S A 107(19):8788–8793

Wisnovsky S et al (2016) Mitochondrial Chemical Biology: new Probes elucidate the secrets of the Powerhouse of the cell. Cell Chem Biol 23(8):917–927

Wojtala A et al (2014) Methods to monitor ROS production by fluorescence microscopy and fluorometry. Methods Enzymol 542:243–262

Wu Z et al (2016) Visualizing hydrogen sulfide in mitochondria and lysosome of living cells and in tumors of living mice with positively charged fluorescent chemosensors. Anal Chem 88(18):9213–9218

Yamada Y et al (2008) BODIPY-based fluorescent redox potential sensors that utilize reversible redox properties of flavin. Chembiochem 9(6):853–856

Yang Y et al (2016) Mitochondria and mitochondrial ROS in cancer: novel targets for anticancer therapy. J Cell Physiol 231(12):2570–2581

Yang K et al (2017) Mitochondrially targeted fluorescent redox sensors. Interface Focus 7(2):20160105

Yeow J et al (2014) A novel flavin derivative reveals the impact of glucose on oxidative stress in adipocytes. Chem Commun 50(60):8181–8184

Zielonka J et al (2008) Cytochrome c-mediated oxidation of hydroethidine and mito-hydroethidine in mitochondria: identification of homo- and heterodimers. Free Radic Biol Med 44(5):835–846

Zielonka J et al (2017) Mitochondria-targeted triphenylphosphonium-based compounds: syntheses, mechanisms of action, and therapeutic and diagnostic applications. Chem Rev 117(15):10043–10120

Recovering Mitochondrial Function in Patients' Fibroblasts

Mariusz R. Wieckowski, Alberto Danese, Giampaolo Morciano, Maciej Pronicki, Jerzy Duszynski, Paolo Pinton, and Agnieszka Karkucinska-Wieckowska

Abstract Despite the fact that majority of studies done using different compounds with antioxidant properties showing pivotal effect on oxidative phosphorylation or glycolytic ATP production, it is still difficult to discuss efficient therapeutic solutions for patients affected by mitochondrial diseases or mitochondrial dysfunction-associated disorders. Since most of the mitochondrial disorders are manifested in tissues or organs that demand high-energy, many experimental studies have described that the pivotal effect of the tested compounds comes from the use of the skin fibroblasts from patients. In this chapter, we have gathered information about these studies and describe the effect of such treatment on mitochondrial function and the attenuation of oxidative stress in patients' fibroblasts.

Keywords Mitochondrial disorders · Patients' fibroblasts · Reactive oxygen species (ROS) · Oxidative stress

M. R. Wieckowski · J. Duszynski
Department of Biochemistry, Nencki Institute of Experimental Biology, Warsaw, Poland
e-mail: m.wieckowski@nencki.gov.pl

A. Danese · P. Pinton
Department of Morphology, Surgery and Experimental Medicine, University of Ferrara, Ferrara, Italy

Maria Cecilia Hospital, GVM Care & Research, E.S: Health Science Foundation, Cotignola, Italy

G. Morciano
Department of Morphology, Surgery and Experimental Medicine, University of Ferrara, Ferrara, Italy

M. Pronicki · A. Karkucinska-Wieckowska (✉)
Department of Pathology, The Children's Memorial Health Institute, Warsaw, Poland
e-mail: a.karkucinska-wieckowska@IPCZD.pl

P. J. Oliveira (ed.), *Mitochondrial Biology and Experimental Therapeutics*,
https://doi.org/10.1007/978-3-319-73344-9_17

1 Introduction

Mitochondria perform central functions in cells, such as buffering the cytosolic calcium concentration, regulating apoptosis through the mitochondrial permeability transition pore (mPTP) (Bonora et al. 2015; Ferrari et al. 2017; Morciano et al. 2015; Bonora et al. 2017), generating reactive oxygen species (ROS) (Morató et al. 2013) and many others. However, their most important role is the production of ATP (Patergnani et al. 2014). Human fibroblasts (e.g., from a skin biopsy) are a valuable and reliable source of biological material for the study of a wide range of diseases (with one limitation—mitochondrial defects must be expressed in fibroblasts), especially those caused by DNA mutations and in instances where it is not always possible to obtain fresh samples for research, such as neurodegenerative disorders (Hirashima et al. 1996). Moreover, the collection of skin biopsies is a much less invasive procedure compared to muscle or liver biopsies. Patients' skin fibroblasts can be used directly or can be used to generate transmitochondrial cybrids or create induced pluripotent stem cells (iPSCs) (Saada 2014). Based on the available literature, fibroblasts obtained from patients suffering from mitochondrial disorders seem to be a good model not only to confirm diagnosis or find the cause of the metabolic defect but also to prove the effectiveness of potential therapies. Different classes of compounds and experimental approaches have been used to improve the mitochondrial function or to decrease mitochondrial dysfunction-related oxidative stress in patients' fibroblasts; we describe some of them in this chapter.

2 Therapeutic Approaches

Supplementation of patients' fibroblasts with compounds that show antioxidant properties should result in the attenuation of mitochondrial respiratory chain dysfunction caused by the intracellular oxidative stress. Very often, oxidative stress is accompanied by mitochondrial dysfunction and the oxidative phosphorylation (OXPHOS) pathology, which usually manifests as an increased level of ROS, as well as the presence of oxidatively damaged proteins, lipid peroxides and DNA (Giorgi C et al. 2010a). Several compounds have been tested in patients' fibroblasts to investigate their positive or negative impact on mitochondrial metabolism and the ROS level. Among them are vitamins, cofactors and classical antioxidants that can enhance the cellular antioxidant capacity to remove ROS as well as improve mitochondrial function. In this group of compounds, we can find vitamin A, B vitamins, including thiamine (B1), riboflavin (B2) and nicotinamide (B3), riboflavin, folic acid and many others. Thanks to the targeting module, a special class of artificial antioxidants can be specifically targeted to the mitochondria. However, it is necessary to mention that mitochondrially-targeted antioxidants, such as ubiquinone (MitoQ) and tocopherol (MitoE), are not useful in all cases of mitochondrial dysfunction. This limitation comes from the fact that these compounds accumulate in the mitochondria due to the high mitochondrial membrane potential. However, in cases of many mitochondrial disorders, decreased mitochondrial membrane potential has been observed. This makes it impossible for the mitochondria to accumulate mitochondria-targeted antioxidants, as is observed in "healthy" fibroblasts (Smith and Murphy 2010). In cases of energy production perturbations, another approach to ameliorate mitochondrial metabolism can be the induction of mitochondrial biogenesis. In animal models of X-linked adrenoleukodystrophy, pioglitazone, a PPAR agonist, has been found to increase mitochondrial mass, decrease DNA oxidative damage, decrease the level of carbonylated proteins and improve bioenergetics parameters (Morató et al. 2013). Similarly, another PPAR agonist, bezafibrate, could improve mitochondrial parameters in the fibroblasts of patients with a complex I deficiency (caused by mutations in the *NDUFS2* gene), however, there was no effect on the fibroblasts of patients with a complex I deficiency, due to mutated complex I assembly factor C20ORF7 (Golubitzky et al. 2011). In the same studies, Golubitzky et al. (2011) has shown that 5-aminoimidazole-4-carboxamide ribonucleotide (AICAR), which is an activator of adenosine monophosphate kinase (AMPK), significantly decreased oxidative stress and increased mitochondrial biogenesis and ATP production in fibroblasts from patients with mutations in the genes encoding NDUFS2 complex I subunit, NDUFA12L and NDUFAF4 complex I assembly factors. Interestingly, such treatment was ineffective in patients' fibroblasts that harbored a mutation in the gene encoding the NDUFS4 subunit of complex I (Golubitzky et al. 2011). Another strategy to abrogate mitochondrial defects is pharmacologically induced metabolic reprograming. The involvement of AMPK, Sirt1 and Sirt3 activation in the metabolic adaptation of human cells harboring mitochondrial DNA mutations induced by resveratrol supplementation has been reviewed by Wu et al. (2014). Moreover, resveratrol can ameliorate the aging process in human primary keratinocytes by preventing dysfunctions in proliferation and decreasing senescence (dependent on AMPK, SIRT1 and FOXO3) (Ido et al. 2015). In addition, this polyphenolic flavonoid improved the phenotypical condition given by Graves' Orbitopathy disease in primary cultured orbital fibroblasts of affected patients, highlighting its potential use in a wide range of disorders (Kim et al. 2015). The positive effect of resveratrol on the elevated oxidative stress of patients' fibroblast is mostly mediated by modulation of antioxidant enzyme levels, including the superoxide dismutases, thioredoxin, glutathione peroxidase-1, heme oxygenase-1 and catalase. It has also been shown that resveratrol treatment was responsible for the increased oxygen consumption and decreased lactate production in moderately OXPHOS-deficient fibroblasts. Moreover, resveratrol has a positive effect on the mitochondrial respiratory capacities in parkin-mutated fibroblasts, which is possibly due to the up-regulation of key regulatory enzymes involved in cellular and mitochondrial metabolism (Ferretta et al. 2014). On the other hand, resveratrol was shown to have harmful effects on patients' fibroblasts (De Paepe et al. 2014; Golubitzky et al. 2011; Lopes Costa et al. 2014). So, the positive or negative effect of resveratrol treatment depends on the type of OXPHOS defect. In general, as mentioned above, the compounds act on multiple sites and modulate mitochondrial metabolism, as well as influence the status of intracellular oxidative stress. A recent review by Koopman et al. provides an overview of the small molecules that are currently being developed for treatment of mitochondrial disease (Koopman et al. 2016). Another experimental approach that can be used to improve cellular metabolism in patients' fibroblasts relies on the influence of proteins involved in calcium homeostasis. This issue will be discussed below.

2.1 Patients' Fibroblasts: Trials for Rescue of Metabolic Defects by Modulation of Calcium Homeostasis

The endoplasmic reticulum (ER) serves as the first calcium store in the cell. Calcium (Ca^{2+}) release occurs through inositol 1,4,5-trisphosphate receptor (IP_3R) channels in the cytosol and thus reaches the mitochondria and other organelles (Sbano et al. 2017). Calcium is essential for cellular bioenergetics regulation (Kaufman and Malhotra 2014), autophagy (Cárdenas and Foskett 2012; Decuypere et al. 2013), ROS production (Singh et al. 2005) and cell death (Danese et al. 2017; Giorgi et al. 2015a, 2015b, 2010; Marchi et al. 2017). Given its crucial involvement in all these physiological contexts, the modulation of calcium homeostasis with pharmacological (or genetic) approaches could be useful to amend the onset of a pathological state. Autophagy is an important response to energetic defects, as well as the lysosomal-dependent elimination of damaged organelles (Marchi et al. 2017). In addition, mitochondrial Ca^{2+} signaling is closely related to the fine regulation of this process (Patergnani et al. 2013; Pinton et al. 2004); thus, pharmacological (and genetic) calcium modulation could be used to regulate autophagy levels in those pathologies where cell bioenergetics properties are impaired. In a study by Granatiero et al., an important increase in the autophagic flux has been observed in fibroblasts carrying the mA13514G mutation of the *MTND5* gene encoding ND5 subunit of the mitochondrial Complex I (Granatiero et al. 2016). Due to a decrease in ER-mitochondria contact sites and defects in the mitochondrial calcium uniporter (MCU) complex, a perturbation of calcium homeostasis translated into reduced mitochondrial calcium uptake in m.A13514G cells and led to atypical MELAS and Leigh syndromes with a late onset and slow progression in patients carrying this mutation (Granatiero et al. 2016). A faster mitochondrial turnover and accelerated autophagy were associated with a milder syndrome phenotype. The authors have shown how the use of MCU activators, such as kaempferol (a plant-derived antioxidant flavonoid) (Vay et al. 2007) and SB202190 (Düzgün et al. 2017), enhanced mitochondrial Ca^{2+} uptake and slowed down the autophagic flux in 13514A4G fibroblasts, giving a phenotype completely comparable to that of control cell lines. These MCU activators restored a normal bioenergetics condition but their prolonged treatment with these compounds decreased cell viability. Indeed, lowering mitochondrial Ca^{2+} could be a compensatory and pro-survival mechanism that allows for a less severe neurodegenerative syndrome. In complex I-deficient fibroblasts from patients carrying a homozygous missense mutation (G364A) in the nuclear *NDUFS7* gene, agonist-induced mitochondrial Ca^{2+} handling and the ensuing stimulation of mitochondrial ATP production are impaired (Visch et al. 2004). Alterations in ATP production were completely restored upon acute treatment with the CGP37157 compound (7-chloro-5-(2-chlorophenyl)-1,5-dihydro-4,1-benzothiazepin-2(3H)-one—a mitochondrial Na^+-Ca^{2+} exchanger inhibitor), which restored the bradykinin-induced mitochondrial Ca^{2+} uptake. This relation can be explained simply by the fact that "repeated" agonist-induced mitochondrial Ca^{2+} uptake leads to an increase in mitochondrial ATP production (Jouaville et al. 1999) and CGP37157 by restoring the bradykinin-induced increase in mitochondrial calcium concentration. This was able to restore the bioenergetics state of the cell. These findings demonstrated that, although the OXPHOS machinery is composed of a defective complex I, modulation of calcium homeostasis can improve mitochondrial ATP synthesis.

Pathological conditions that could benefit from calcium modulation are countless, although, in many cases, no pharmacological treatments have been conducted. An example derives from the triad of Autism Spectrum Disorder (ASD)—Fragile X Syndrome (FXS)—Tourette Syndrome (TS), where a perturbation in IP_3-mediated Ca^{2+} signaling has been reported. Indeed, skin fibroblasts from patients with FXS and TS have significantly decreased Ca^{2+} response compared with control cell lines. This impaired signal is not due to decreased ER calcium content or a reduced expression of IP_3R proteins, but to fewer sites of Ca^{2+} release and a general dysfunction of the IP_3R channel gating (Schmunk et al. 2015, 2017).

Recently, it has been demonstrated that calcium imbalance, ER stress, unfolded protein response (UPR) and oxidative stress are consequences of skin exposure to UV radiations (Farrukh et al. 2014). Indeed, although mammalians own protective systems to overwhelm this damage starting from plasma membrane to the lipids, subcellular organelles and DNA, repeated exposures to UV lower the defenses of the human body. Additional help, such as ROS detoxification and the restoration of calcium homeostasis, may be required. It was shown that glycyrrhizic acid (GA) treatment significantly protects against Ca^{2+} perturbation by lowering ER stress and apoptosis in UV-B treated human skin fibroblasts (Farrukh et al. 2015).

In summary, the findings presented above have remarkable translational relevance that supports the involvement and targeting of Ca^{2+} signaling (and oxidative stress) in cells directly derived from patients. These studies provide important information about the use of fibroblasts from biopsy samples as a functional diagnostic tool and surrogate pharmacological trial.

3 Possible Therapeutic Approaches Carried Out in Fibroblasts Derived from Patients with Different Mitochondrial and Metabolic Abnormalities, as well as Other Disorders Characterized by Oxidative Stress

Fibroblasts derived from patients suffering from mitochondrial disorders have been used repeatedly to investigate the effect of potential pharmacological compounds designed to improve the affected cellular bioenergetics as well as to decrease oxidative stress in these cells. Below, we present examples of therapeutic trials performed with the use of patients' fibroblasts with different abnormalities in the OXPHOS machinery.

3.1 NARP Patients' Fibroblasts

NARP (Neuropathy, Ataxia and *Retinitis Pigmentosa*) and MILS (Maternally Inherited Leigh's Syndrome) are mitochondrial disorders associated with mutations in the *MTATP6* gene encoding subunit a of the mitochondrial ATP synthase. Typical m.T8993G mutations, which are responsible for NARP/MILS, are related to the transversion of thymine to guanine at mtDNA nucleotide 8993, which causes the conversion of a highly conserved leucine to arginine. The clinical phenotype associated with the m.T8993G mutation depends on the heteroplasmy level (White et al. 1999). The NARP phenotype is considered when the mutation load is between 70 and 90%. When it is higher, it may be responsible for fatal infantile encephalopathy MILS. Generally, mutation in the ATPase 6 gene, which encodes a subunit that is a part of the F_1F_0-ATPase c-ring, results in alterations to the mitochondrial ATP production process. It has been found that mitochondrial ATP synthesis can be reduced by 50–70% in cells harboring 100% m.T8993G mutation load (Vazquez-Memije et al. 1996). Detailed characterization of NARP fibroblasts made by Lebiedzinska et al. showed an increased mitochondrial membrane potential, decreased activity of the mitochondrial respiratory chain, reduced NADH/NAD ratio, alterations of mitochondrial calcium homeostasis and an increased level of mitochondrial superoxide and oxidatively damaged proteins (Lebiedzinska et al. 2013). They found that inhibition of p66Shc (an alternatively spliced isoform of the growth factor adaptor that belongs to the ShcA family) phosphorylation at Ser36 by hispidin (inhibitor of PKCβ) results in decreased mitochondrial superoxide anion production, which acts downstream of p66Shc activation and reduces the vicious cycle of ROS production in the studied NARP fibroblasts. Interestingly, in NARP fibroblasts, hispidin treatment increased the level of carbonylated proteins (Lebiedzinska et al. 2013). The other work presented by Mattiazzi et al. showed that mitochondrial dysfunction caused by m.T8993G mutation can be partially reverted by antioxidant treatment. To improve the oxygen consumption and ATP production in primary fibroblasts obtained from a patient harboring a 97% m.T8993G mutation load, 2.5 mM NAC was used (Mattiazzi et al. 2004). The studies of Wojewoda et al. showed that selenite, an inorganic form of selenium, increased the level of antioxidant enzymes, which can explain the decreased level of ROS in NARP cybrids (Wojewoda et al. 2011). Additionally, they found that selenite treated cells had a higher level of mitochondrial respiratory chain subunits, which resulted in higher intracellular ATP levels. An interesting study has been performed by Sgarbi et al. discussing whether α-ketoglutarate and aspartate treatment can have a positive impact on the viability and ATP level of NARP/MILS patients' fibroblasts carrying 2 distinct point mutations, m.T8993G (with severe impact) and m.T8993C (with only mild impact on OXPHOS) (Sgarbi et al. 2009). Interestingly, the protective effect of α-ketoglutarate/aspartate was observed only in fibroblasts harboring the m.T8993G mutation. The treatment had absolutely no effect on the viability of cells with mildly impaired ATP synthase (m.T8993C mutation). Nevertheless, the authors believe that α-ketoglutarate/aspartate dietary supplementation can be considered a potential pharmacological therapeutic approach (Sgarbi et al. 2009).

3.2 LHON Patients' Fibroblasts

Leber Hereditary Optic Neuropathy (LHON) is a primary mtDNA disorder that initially causes a painless and acute unilateral loss of central vision among young adults and later manifests as total bilateral vision loss and blindness. Most of the mtDNA mutations responsible for LHON affect the mitochondrial complex I subunits. Practically, three point mutations, m.G3406A (in the *MTND1* gene), m.G11778A (in the *MTND4* gene) and m.T14484C (in the *MTND6* gene), are responsible for approximately 90% of all LHON cases. Interestingly, among them, the mutation in the gene encoding the *ND4* subunit gene is the most prevalent (60–80%) cause of LHON. As a synthetic analog of CoQ10 in mitochondria, idebenone can act as an electron carrier in the respiratory chain, and it is considered a compound with antioxidant properties. Interestingly, in contrast to CoQ10, idebenone participates in redox reactions outside the mitochondrial compartment (Haefeli et al. 2011). Idebenone in the cytoplasm is reduced by the NAD(P)H quinone oxidoreductase 1 (NQO1) and can be re-oxidized by complex III, which in turn enables bypass of the affected complex I (Haefeli et al. 2011). Additionally, in contrast to CoQ10, idebenone seems to stimulate complex II activity (Gueven et al. 2016). A positive effect of idebenone on fibroblasts derived from LHON patients was observed by Angebault et al. already several years ago (Angebault et al. 2011). They observed that lower activity of complex I in fibroblasts derived from LHON patients after incubation with 10 μM idebenone was increased by approximately 42%. However, idebenone treatment had variable effects on oxygen consumption, indicating that there were not equal benefits from the idebenone treatment (Angebault et al. 2011). Recently, Yu-Wai-Man et al. also evaluated the therapeutic potential of idebenone and other quinone analogues in LHON patients' fibroblasts (Yu-Wai-Man et al. 2017). They found that idebenone treatment partially compensated for the deleterious effect of the m.G11778A mutation. Moreover, idebenone increased ATP production and reduced oxidative stress; however, this effect was observed in only a subgroup of studied patients' fibroblasts. Other quinone analogues tested by this group, like CoQ1, CoQ10 and decylubiquinone, showed variable effects on oxygen consumption and ROS level (Yu-Wai-Man et al. 2017).

3.3 Leigh Syndrome Patients' Fibroblasts

Leigh syndrome (Leigh disease) (OMIM 256000) is an inherited, mitochondrial, neurodegenerative disorder mostly manifested in the central nervous system and is known as subacute necrotic encephalo(mio)pathy. First, symptoms of Leigh syndrome very often apparent in infancy; however, in some cases early symptoms can begin in the teenage or adult years. Leigh syndrome can be caused by mutations in more than 75 different genes (in mitochondrial and nuclear DNA) encoding proteins involved mostly in oxidative phosphorylation. For this reason, we can identify two groups of Leigh syndrome: (a) mitochondrial DNA-associated Leigh syndrome (approximately 20% of cases) caused by mutations in at least 11 mitochondrially-encoded genes, with an estimated incidence of 1 in 100,000 to 1 in 140,000 births, and (b) nuclear DNA-associated Leigh syndrome (approximately 80% of all cases), with an estimated incidence of 1 in 30,000 to 1 in 40,000 people at birth (Lake et al. 2015). The most common causes of Leigh syndrome are mutations in the complex I-encoding genes (mutations in at least 25 genes have been identified). Leigh syndrome is associated also with a defect in complex IV (approximately 15% of cases), with the most frequent mutations in genes encoding SURF1 and SCO2 proteins. The other frequent mutation causing Leigh syndrome affects the *MTATP6* gene-encoding subunit a of mitochondrial ATP synthase (10% of Leigh cases). Leigh syndrome can also be caused by mutations in genes encoding subunits of pyruvate dehydrogenase complex or in genes encoding proteins involved in CoQ10 biosynthesis.

CoQ10, a natural lipid-soluble quinone analogue (known also as ubiquinone), is a component of the mitochondrial respiratory chain involved in electron transport. CoQ10 supplementation partially restores the activities of the mitochondrial respiratory chain enzymes in MELAS fibroblasts and MERRF cybrids (Cotán et al. 2011; De la Mata et al. 2012). Hirano et al. showed that coenzyme Q10 supplementation had a positive effect in fibroblasts with CoQ10 deficiency (Hirano et al. 2012). Interestingly, treatment of fibroblasts from patients with CoQ10 deficiency with coenzyme Q2, a shorter chain analogue of CoQ10, has no effect on mitochondrial parameters (López et al. 2010). Several studies demonstrated that treatment with coenzyme Q and its analogs can be beneficial in Leigh syndrome, as well as in Leigh-like syndrome (Haas 2007; Rahman 2015).

Recently, Kanabus et al. (2016) reported that decanoic acid supplementation of fibroblasts derived from patients with Leigh syndrome associated with nuclear-encoded defects of complex I increases mitochondrial biogenesis (via PPAR-γ receptor) in approximately 50% of studied fibroblast lines. Moreover, decanoic acid increases cellular resistance to oxidative stress by increasing catalase expression (Kanabus et al. 2016).

Treatment of fibroblasts derived from patients with decreased levels of mitochondrial complex I with Trolox, a water-soluble vitamin E derivate, causes a significant decrease in the ROS level (Koopman et al. 2008) and increases complex I level. Koopman et al. speculates that the level of active complex I in the mitochondria is under regulatory control of the cell's oxidative balance (Koopman et al. 2008). Therefore, the antioxidant Trolox can mitigate complex I deficiency. Importantly, the authors claim that such treatment is beneficial to only patients with predominant expression of complex I, rather than an intrinsic catalytic defect in this respiratory chain complex. Moreover, Distelmaier et al. showed that Trolox supplementation also has a positive effect on mitochondrial membrane potential and calcium-stimulated ATP production in complex I-deficient human fibroblasts (Distelmaier et al. 2009).

It has been demonstrated that riboflavin (precursor of flavin mononucleotide—FMN) also had a positive effect on fibroblasts derived from patients with a complex

I defect related to mutation in the *NDUFS2* gene and genes encoding assembly factors (ACAD9 and AIF), as well as in MELAS patients' fibroblasts (Garrido-Maraver et al. 2012; Gerards et al. 2011; Saada 2011). Vitamins such as vitamin K and vitamin C also have positive effects on patients' fibroblasts. Vitamin K (menadione) and vitamin C (ascorbate) are potential electron donors for complexes II and IV, respectively. For this reason, these compounds have positive effects on fibroblasts with an affected complex I and fibroblasts with a CoQ deficiency (Saada 2011). Additionally, ascorbate was found to decrease superoxide levels and reduce manifestation of oxidative stress in QoQ2-deficient fibroblasts (López et al. 2010). In turn, resveratrol, a polyphenol of natural origin present in the skin of fruits, such as grapes, blueberries, raspberries and mulberries, can decrease oxidative stress in complex I-deficient patients' cell lines, as well as restore oxygen consumption in these cells (Lopes Costa et al. 2014; Mathieu et al. 2016). On the other hand, resveratrol can inhibit mitochondrial ATP synthase, which makes possible therapy risky for individuals harboring defects in the ATP synthesis process (Gledhill et al. 2007).

Recent studies by Ehinger et al. using fibroblasts from patients with Leigh syndrome (with a recessive mutation in *NDUFS2* gene) showed that cell membrane permeable succinate prodrugs (diacetoxymethyl succinate; bis-(1-acetoxy-ethyl) succinate and 1-acetoxyethyl acetoxymethyl succinate) access the intracellular space and release succinate, which enables transport of electrons from complex II and ATP production by bypassing the deficiency of complex I (Ehinger et al. 2016).

The positive effect of bezafibrate on mitochondrial parameters has been observed in fibroblasts of patients with complex IV deficiency caused by mutation in the *SCO2* gene. Casarin et al. found that copper (100 μM $CuCl_2$) and 200 μM bezafibrate had no effect on SCO2 fibroblasts when supplemented separately (Casarin et al. 2012). However, when used together, they caused a complete rescue of COX activity in SCO2 cells. Ten years earlier, the same group showed that 100 μM $CuCl_2$ alone can fully restore activity of cytochrome c oxidase not only in fibroblasts but also in myotubes and mioblasts from patients with *SCO2* gene mutation (Salviati et al. 2002).

Menzies et al. studied the effect of a thyroid hormone (3,3′,5-triiodothyronine; T3) on mitochondrial parameters and the status of oxidative stress in two patients with Leigh's syndrome (one harboring a m.G13513A mutation in the *MTND5* gene and the second with a m.T9185C mutation in the *MTATP6* gene) (Menzies et al. 2009). They observed that ROS production in T3-treated patients' fibroblasts was decreased by 40%, accompanied by a 1.3-fold increase in complex IV activity and a 1.6-fold increase in the ATP level; moreover, the level of MnSOD was restored to control levels. The positive effect of T3 treatment was not related to the transcriptional coactivator peroxisome proliferator-activated receptor gamma coactivator 1-alpha (PGC1α) or mitochondrial transcription factor A (TFAM). This is because the level of these proteins was not changed by T3, and the mitochondrial mass was the same before and after T3 treatment of the studied Leigh's syndrome patients' fibroblasts (Menzies et al. 2009).

3.4 Fibroblasts from Patients with Combined Deficiency of Complexes I, III, IV and V with a Normal Complex II Level

The combined deficiency of complexes I, III, IV and V with a normal level of complex II in patients can be caused by (a) deletions in mtDNA and point mutations in mt tRNA-encoding genes (Kemp et al. 2011); (b) mtDNA depletion related to autosomal recessive mutations in nuclear genes involved in mtDNA replication and maintenance (Spinazzola et al. 2009) and (c) mutations in nuclear-encoded components of the mitochondrial translation machinery (Smits et al. 2010). Soiferman et al. presented that ascorbate can reduce the ROS level/production in fibroblasts of patients with defects in elongation factors (EFTs) (Soiferman et al. 2014). Moreover, he found that after ascorbate administration, the activity of complex IV was increased, which was accompanied by a higher level of ATP (Soiferman et al. 2014). In the same work, he showed that another compound with antioxidant properties, N-acetyl cysteine (NAC), which is a precursor of cysteine and glutathione, decreases the ROS level in these fibroblasts as well as in fibroblasts from patients with defects in mitochondrial t-RNA uridylation (TRMU) (Soiferman et al. 2014). Interesting studies have been performed by Wang et al. with the use of Kearns-Sayre patients' fibroblasts (Wang et al. 1996). Kearns-Sayre syndrome, a commonly diagnosed mitochondrial cytopathy, is caused by mitochondrial DNA deletion (removal of a 4977-base pair segment of the mtDNA encoding mitochondrial respiratory chain subunits). They found that in Kearns-Sayre syndrome fibroblasts, azidothymidine and dideoxynucleosides cause a depletion of wild-type mtDNA, while increasing the number of copies of mtDNA with deletions (Wang et al. 1996).

3.5 Fibroblasts from Patients with Complex II Deficiency

Complex II deficiency is an autosomal recessive mitochondrial disorder with a highly variable phenotype that can be caused by mutations in the *SDHA*, *SDHB*, *SDHC*, *SDHD*, or *SDHAF1* and *SDHAF2* genes encoded in the nuclear DNA. Symptoms of mitochondrial complex II deficiency can vary from severe to life-threatening symptoms in infancy to muscle disease beginning in adulthood. It has been described that two inheriting mutations in the *SDHA* gene are associated with myoclonic seizures and Leigh's syndrome. The studies performed by De Paepe et al. showed that resveratrol supplementation had no effect on complex II deficient fibroblasts (derived from patients with complex II activity close to the method detection limit; one patient with an unknown pathogenic mutation and a second one harboring a homozygous c.G622T mutation in the *NFU1* gene-encoding protein involved in the formation of iron-sulfur (Fe-S) clusters) (De Paepe et al. 2014). In fibroblasts from these patients, complex II activity after resveratrol treatment was still negligible (De Paepe et al. 2014). In contrast, fibroblasts from patients harboring the homozygous c.G1663A mutation in the *SDHA* gene and characterized by a higher basal complex II activity (comparing to the above described two other patients with complex II deficiency) when treated with resveratrol showed a significant increase in complex II activity. This indicates that the pivotal effect of resveratrol supplementation depends on the residual complex II activity (De Paepe et al. 2014).

3.6 MERRF Patients' Fibroblasts

Myoclonic epilepsy with ragged red fibers (MERRF) syndrome is a maternally-inherited mitochondrial encephalomyopathy. In the case of MERRF syndrome, the most common symptoms are myoclonus epilepsy, generalized seizures, ataxia and myopathy. With MERRF, four different point mutations are associated; however, the most common one (found in 80–90% of MERRF patients) is the m.A8344G mutation in the $tRNA^{Lys}$ gene of mitochondrial DNA. This mutation is associated with severe defects in mitochondrial protein synthesis, which affects the mitochondrial respiratory chain and ATP synthesis. Skin fibroblasts from patients with MERRF syndrome are characterized by significantly increased ROS production and an increased level of matrix metalloproteinases (MMPs), which can be considered as a progressive marker of neurodegenerative diseases (Wu et al. 2010). The increased level and activity of MMPs may contribute to the cytoskeleton remodeling involved in the weakness and atrophy of muscles commonly seen in MERRF patients (Wu et al. 2010). Interestingly, no significant changes in the antioxidant defense system have been observed. Only the level and activity of SOD2 was increased in MERRF patients' fibroblasts. Increased oxidative stress can be responsible for the oxidative damage of the voltage-dependent anion channel (VDAC) and aconitase in the MERRF fibroblasts (Wu et al. 2010). Fascinating studies with the use of MERRF patients' fibroblasts have been performed by Chang et al. (2013). They demonstrated that, using the cell-penetrating peptide (Pep-1), they could deliver functional mitochondria isolated from "healthy" fibroblasts into the MERRF fibroblasts (peptide-mediated mitochondrial delivery). The MERRF fibroblasts receiving 3 days of treatment with peptide-mediated mitochondrial delivery restored mitochondrial respiratory chain subunits of complexes I, III and IV. This was accompanied by recovery of the mitochondrial membrane potential, ATP synthesis and a decrease in the ROS level and the recovery of the mitochondrial function has been maintained for at least 21 days (Chang et al. 2013). Interestingly, an opposite experiment where healthy cells were treated with mitochondria isolated from MERRF fibroblasts showed that previously healthy recipient cells showed a MERRF phenotype, which was characterized by increased ROS production and MMP activity (Chang et al. 2013; Clauser and Scibak 1990). More about delivering healthy mitochondria as a potential tool in the therapy of mitochondrial disorders can be found in the review by Liu et al. (2014).

3.7 *Fibroblasts from Patients with Propionic Acidemia*

Propionic acidemia (PA) is caused by a deficiency in propionyl-CoA carboxylase (mitochondrial enzyme) and is one of the most frequent organic acidurias in humans (incidence of 1 in 150,000 inhabitants). Propionic acidemia patients during the neonatal period develop different neurological symptoms and movement disorders. Gallego-Villar et al. found that the fibroblasts derived from PA patients have increased ROS levels (Gallego-Villar et al. 2013). Later studies of Gallego-Villar et al. showed that antioxidant treatment successfully decreases high ROS levels in PA patients' fibroblasts, as well as the levels of mitochondrial superoxide dismutase and GPx1 (depending on used compound) (Gallego-Villar et al. 2014). They tested the effect of vitamin E, trolox, tiron, N-acetyl-cysteine (NAC), melatonin, resveratrol and MitoQ on oxidative stress manifestation in PA fibroblasts. The compounds used have different antioxidant actions. Vitamin E and Trolox neutralize lipid-derived radicals. NAC is a precursor of glutathione and can scavenge different types of ROS. Melatonin, which is a direct free radical scavenger, increases the efficiency of the antioxidant defense system. Resveratrol inhibits lipid peroxidation and is a direct ROS scavenger. MitoQ is a mitochondria-targeted antioxidant that can protect against oxidative damage within the mitochondria. Interestingly, they demonstrated that resveratrol, Trolox, Tiron and MitoQ decreased the ROS level in all studied PA-derived fibroblasts (Gallego-Villar et al. 2014). The strongest antioxidant effect was observed for Tiron (50–80%), then for MitoQ (25–30%) and finally for Trolox (15–30%). However, it is necessary to mention that the effect of individual compounds depends on the patient's cell line. The strongest effect was observed for the fibroblasts with the highest ROS level. On the contrary, melatonin, N-acetyl cysteine (NAC) and vitamin E had absolutely no effect on the oxidative stress in these fibroblasts (Gallego-Villar et al. 2014).

3.8 *Fibroblasts from Patients with Friedreich Ataxia*

Friedreich ataxia (FRDA), is the most common recessively inherited ataxia, which is caused by defective expression of frataxin (mitochondrial protein), leading to the progressive loss of neuromuscular function. The decreased level of frataxin is responsible for the accumulation of iron within the mitochondria, increased oxidative stress and decreased activity of iron-sulfur cluster-containing enzymes. Altogether, this causes mitochondrial dysfunction in FRDA patients. An interesting study performed by Jauslin et al. compared the protective effect of mitochondria-targeted and untargeted antioxidants in fibroblasts from FRDA patients (Jauslin et al. 2002). They studied the effect of these compounds on the viability of FRDA patients' fibroblasts treated with an inhibitor of GSH biosynthesis (BSO) and thus showed an artificially-reduced glutathione level. Their experimental approach was based on the fact that BSO treatment leads to a decrease in the GSH level in control cells and FRDA fibroblasts but caused cell death only in FRDA fibroblasts (Jauslin et al. 2002). They found that MitoQ, was approximately 800-fold more effective than idebenone, the coenzyme Q10 untargeted analog, in protecting FRDA fibroblasts against GSH depletion-related cell death. Similarly, mitochondria-targeted MitoVit E was 350-fold more efficient at protecting the cells than Trolox, the water-soluble analog of vitamin E (Jauslin et al. 2003). The other classes of compounds, lipophilic iron chelators, have been investigated by Lim et al. (2008). They investigated the properties of the 2-pyridylcarboxaldehyde isonicotinoyl hydrazone (PCIH) class of chelators as agents rescuing Friedreich's ataxia patients' fibroblasts from H_2O_2-induced cytotoxicity.

3.9 Fibroblasts from Patients with Alzheimer's Disease

Alzheimer's Disease (AD) is a chronic neurodegenerative disease. The most common early symptom of Alzheimer's can be difficulty remembering newly learned information (short-term memory loss), and at later stages of disease progression, symptoms can also include disorientation, problems with language and not managing self-care. Interestingly, most people with Down syndrome develop Alzheimer's disease. The cause of Alzheimer's disease is poorly understood, and AD is often attributed to a variety of causes. It has been found that the apolipoprotein E (APOE) gene is involved in the late-onset form of AD (symptoms become apparent in their mid-60s). APOE ε4 increases a person's risk of developing AD. The accumulation of intracellular aggregates of tau protein in the neurofibrillary tangles and extracellular aggregates of a set of polypeptides called amyloid-β peptides in the senile plaques is the major histological hallmark of AD. Alzheimer's disease is not a mitochondrial disease *per se*; however, there is direct link between AD, mitochondrial dysfunction and oxidative stress. For this reason, oxidative stress and elevated ROS production are used as a target to ameliorate cellular and mitochondrial parameters in AD patients' fibroblasts. Among others, Moreira et al. showed that administration of lipoic acid and N-acetyl cysteine inhibits apoptosis and decreases oxidative stress in fibroblasts from patients with Alzheimer's disease, which suggests that antioxidant therapy based on these compounds may be promising (Moreira et al. 2007).

3.10 Fibroblasts from Patients with Influenza-Associated Encephalopathy

Influenza-associated encephalopathy (IAE) is an acute brain dysfunction that usually occurs at the early stage of infectious diseases caused mainly by influenza virus, human herpes virus-6 (HHV-6) and many other viruses. IAF incidence is highest in infancy and early childhood and occurs more frequently in East Asians than in Caucasians (Kasai et al. 2000). In addition to brain dysfunction, IAE patients also show development of multiple-organ failure. Interestingly, Yao et al. showed that a high number of patients with a disabling or fatal form of IAE have a thermolabile phenotype of compound variants of carnitinepalmitoyltransferase II (CPT II) (Yao et al. 2011). Such patients are characterized by a mitochondrial energy crisis during high fever. It is related to dysfunction of the mitochondrial fatty acid β-oxidation, which is caused by heat-inactivation of carnitine palmitoyltransferase II (CPT II) in patients with the thermolabile phenotype of CPT II (Yao et al. 2011). Treatment of such fibroblasts for 24 h with bezafibrate significantly increased the CPT II activity, increased mitochondrial fatty acid β-oxidation, restored decreased ATP levels and increased the mitochondrial membrane potential in fibroblasts of IAE patients cultured at both 37 °C and 41 °C. The studies of Yamaguchi et al. indicate the possible therapeutic properties of bezafibrate in IAE patients with thermolabile variants of CPT II (Yamaguchi et al. 2012).

4 Conclusion

As presented in this chapter, several different pharmacological treatments of patients' fibroblasts have been performed to find the most potent and appropriate way to ameliorate mitochondrial defect or mitigate oxidative stress. More details about the different strategies and treatments of mitochondrial disorders can be found in an excellent review written by Scarpelli et al. (2014). Moreover, the paper of Voets et al. elegantly shows how fibroblast analysis enables identification of patients who potentially can benefit from the antioxidant therapy (Voets et al. 2012). Similarly, a comprehensive review describing the use of individual patient fibroblasts in the search for personalized treatment has been presented by Saada (2011). Finally, we also recommend a systematic review written by Rai et al. describing pharmacological therapeutics tested using in vitro models (Rai et al. 2015), a review by Rajendran et al. summarizing the impact and involvement of antioxidants in selected human diseases (Rajendran et al. 2014), and a review by Kanabus et al. (2014) and Koopman et al. (2016).

Acknowledgments This work was supported by the Polish National Science Centre grant (UMO-2014/15/B/NZ1/00490) to MRW and AKW and by the Internal Projects of the Children's Memorial Health Institute No S125/2012 for MP and AKW. PP is grateful to Camilla degli Scrovegni for providing continuous support. PP is grateful to Camilla degli Scrovegni for continuous support. PP is supported by the Italian Ministry of Education, University and Research; the Italian Ministry of Health; Telethon (GGP15219/B); the Italian Association for Cancer Research (AIRC: IG-18624); and by local funds from the University of Ferrara).

Conflict of Interest The authors state that there are no conflicts of interest relevant for this publication.

References

Angebault C, Gueguen N, Desquiret-Dumas V, Chevrollier A, Guillet V, Verny C, Cassereau J, Ferre M, Milea D, Amati-Bonneau P, Bonneau D, Procaccio V, Reynier P, Loiseau D (2011) Idebenone increases mitochondrial complex I activity in fibroblasts from LHON patients while producing contradictory effects on respiration. BMC Res Notes 4:557

Bonora M, Wieckowski MR, Chinopoulos C, Kepp O, Kroemer G, Galluzzi L, Pinton P (2015) Molecular mechanisms of cell death: central implication of ATP synthase in mitochondrial permeability transition. Oncogene 34(12):1475–1486

Bonora M, Morganti C, Morciano G, Pedriali G, Lebiedzinska-Arciszewska M, Aquila G, Giorgi C, Rizzo P, Campo G, Ferrari R, Kroemer G, Wieckowski MR, Galluzzi L, Pinton P. (2017) Mitochondrial permeability transition involves dissociation of F1FO ATP synthase dimers and C-ring conformation. EMBO Rep 18(7):1077–1089

Cárdenas C, Foskett JK (2012) Mitochondrial Ca(2+) signals in autophagy. Cell Calcium 52(1):44–51

Casarin A, Giorgi G, Pertegato V, Siviero R, Cerqua C, Doimo M, Basso G, Sacconi S, Cassina M, Rizzuto R, Brosel S, M Davidson M, Dimauro S, Schon EA, Clementi M, Trevisson E, Salviati L (2012) Copper and bezafibrate cooperate to rescue cytochrome c oxidase deficiency in cells of patients with SCO2 mutations. Orphanet J Rare Dis 7:21

Chang JC, Liu KH, Chuang CS, Su HL, Wei YH, Kuo SJ, Liu CS (2013) Treatment of human cells derived from MERRF syndrome by peptide-mediated mitochondrial delivery. Cytotherapy 15(12):1580–1596

Clauser B, Scibak JW (1990) Direct and generalized effects of food satiation in reducing rumination. Res Dev Disabil 11(1):23–36

Cotán D, Cordero MD, Garrido-Maraver J, Oropesa-Ávila M, Rodríguez-Hernández A, Gómez Izquierdo L, De la Mata M, De Miguel M, Lorite JB, Infante ER, Jackson S, Navas P, Sánchez-Alcázar JA (2011) Secondary coenzyme Q10 deficiency triggers mitochondria degradation by mitophagy in MELAS fibroblasts. FASEB J 25(8):2669–2687

Danese A, Patergnani S, Bonora M, Wieckowski MR, Previati M, Giorgi C, Pinton P (2017) Calcium regulates cell death in cancer: Roles of the mitochondria and mitochondria-associated membranes (MAMs). Biochim Biophys Acta 1858(8):615–627. pii: S0005-2728(17)30004-X. https://doi.org/10.1016/j.bbabio.2017.01.003

De la Mata M, Garrido-Maraver J, Cotán D, Cordero MD, Oropesa-Ávila M, Izquierdo LG, De Miguel M, Lorite JB, Infante ER, Ybot P, Jackson S, Sánchez-Alcázar JA (2012) Recovery of MERRF fibroblasts and cybrids pathophysiology by coenzyme Q10. Neurotherapeutics 9(2):446–463

De Paepe B, Vandemeulebroecke K, Smet J, Vanlander A, Seneca S, Lissens W, Van Hove JL, Deschepper E, Briones P, Van Coster R (2014) Effect of resveratrol on cultured skin fibroblasts from patients with oxidative phosphorylation defects. Phytother Res 28(2):312–316

Decuypere JP, Kindt D, Luyten T, Welkenhuyzen K, Missiaen L, De Smedt H, Bultynck G, Parys JB (2013) mTOR-Controlled Autophagy Requires Intracellular Ca^{2+}. Signaling. PLoS One 8(4):e61020

Distelmaier F, Visch HJ, Smeitink JA, Mayatepek E, Koopman WJ, Willems PH (2009) The antioxidant Trolox restores mitochondrial membrane potential and Ca2+ −stimulated ATP production in human complex I deficiency. J Mol Med 87(5):515–522

Düzgün ŞA, Yerlikaya A, Zeren S, Bayhan Z, Okur E, Boyacı İ (2017) Differential effects of p38 MAP kinase inhibitors SB203580 and SB202190 on growth and migration of human MDA-MB-231 cancer cell line. Cytotechnology 69(4):711–724. https://doi.org/10.1007/s10616-017-0079-2

Ehinger JK, Piel S, Ford R, Karlsson M, Sjövall F, Frostner EÅ, Morota S, Taylor RW, Turnbull DM, Cornell C, Moss SJ, Metzsch C, Hansson MJ, Fliri H, Elmér E (2016) Cell-permeable succinate prodrugs bypass mitochondrial complex I deficiency. Nat Commun 7:12317

Farrukh MR, Nissar UA, Afnan Q, Rafiq RA, Sharma L, Amin S, Kaiser P, Sharma PR, Tasduq SA (2014) Oxidative stress mediated Ca(2+) release manifests endoplasmic reticulum stress leading to unfolded protein response in UV-B irradiated human skin cells. J Dermatol Sci 75(1):24–35

Farrukh MR, Nissar UA, Kaiser PJ, Afnan Q, Sharma PR, Bhushan S, Tasduq SA (2015) Glycyrrhizic acid (GA) inhibits reactive oxygen Species mediated photodamage by blocking ER stress and MAPK pathway in UV-B irradiated human skin fibroblasts. J Photochem Photobiol B 148:351–357

Ferrari R, Balla C, Malagù M, Guardigli G, Morciano G, Bertini M, Biscaglia S, Campo G (2017) Reperfusion damage–a story of success, failure, and hope. Circ J 81(2):131–141

Ferretta A, Gaballo A, Tanzarella P, Piccoli C, Capitanio N, Nico B, Annese T, Di Paola M, Dell'aquila C, De Mari M, Ferranini E, Bonifati V, Pacelli C, Cocco T (2014) Effect of resveratrol on mitochondrial function: implications in parkin-associated familiar Parkinson's disease. Biochim Biophys Acta 1842(7):902–915

Gallego-Villar L, Pérez-Cerdá C, Pérez B, Abia D, Ugarte M, Richard E, Desviat LR (2013) Functional characterization of novel genotypes and cellular oxidative stress studies in propionic acidemia. J Inherit Metab Dis 36(5):731–740

Gallego-Villar L, Pérez B, Ugarte M, Desviat LR, Richard E (2014) Antioxidants successfully reduce ROS production in propionic acidemia fibroblasts. Biochem Biophys Res Commun 452(3):457–461

Garrido-Maraver J, Cordero MD, Moñino ID, Pereira-Arenas S, Lechuga-Vieco AV, Cotán D, De la Mata M, Oropesa-Ávila M, De Miguel M, Bautista Lorite J, Rivas Infante E, Alvarez-Dolado M, Navas P, Jackson S, Francisci S, Sánchez-Alcázar JA (2012) Screening of effective pharmacological treatments for MELAS syndrome using yeasts, fibroblasts and cybrid models of the disease. Br J Pharmacol 167(6):1311–1328

Gerards M, van den Bosch BJ, Danhauser K, Serre V, van Weeghel M, Wanders RJ, Nicolaes GA, Sluiter W, Schoonderwoerd K, Scholte HR, Prokisch H, Rötig A, de Coo IF, Smeets HJ (2011) Riboflavin-responsive oxidative phosphorylation complex I deficiency caused by defective ACAD9: new function for an old gene. Brain 134(Pt 1):210–219

Giorgi C, Agnoletto C, Baldini C, Bononi A, Bonora M, Marchi S, Missiroli S, Patergnani S, Poletti F, Rimessi A, Zavan B, Pinton P (2010a) Redox control of protein kinase C: cell- and disease-specific aspects. Antioxid Redox Signal 13(7):1051–85

Giorgi C, Ito K, Lin HK, Santangelo C, Wieckowski MR, Lebiedzinska M, Bononi A, Bonora M, Duszynski J, Bernardi R, Rizzuto R, Tacchetti C, Pinton P, Pandolfi PP (2010b) PML regulates apoptosis at endoplasmic reticulum by modulating calcium release. Science 330(6008):1247–1251

Giorgi C, Bonora M, Sorrentino G, Missiroli S, Poletti F, Suski JM, Galindo Ramirez F, Rizzuto R, Di Virgilio F, Zito E, Pandolfi PP, Wieckowski MR, Mammano F, Del Sal G, Pinton P (2015a) p53 at the endoplasmic reticulum regulates apoptosis in a Ca^{2+}-dependent manner. Proc Natl Acad Sci U S A 112(6):1779–1784

Giorgi C, Missiroli S, Patergnani S, Duszynski J, Wieckowski MR, Pinton P (2015b) Mitochondria-associated membranes: composition, molecular mechanisms, and physiopathological implications. Antioxid Redox Signal 22(12):995-1019

Gledhill JR, Montgomery MG, Leslie AG, Walker JE (2007) Mechanism of inhibition of bovine F1-ATPase by resveratrol and related polyphenols. Proc Natl Acad Sci U S A 104(34):13632–13637

Golubitzky A, Dan P, Weissman S, Link G, Wikstrom JD, Saada A (2011) Screening for active small molecules in mitochondrial complex I deficient patient's fibroblasts, reveals AICAR as the most beneficial compound. PLoS One 6(10):e26883

Granatiero V, Giorgio V, Calì T, Patron M, Brini M, Bernardi P, Tiranti V, Zeviani M, Pallafacchina G, De Stefani D, Rizzuto R (2016) Reduced mitochondrial Ca(2+) transients stimulate autophagy in human fibroblasts carrying the 13514A>G mutation of the ND5 subunit of NADH dehydrogenase. Cell Death Differ 23(2):231–241

Gueven N, Nadikudi M, Daniel A, Chhetri J (2016) Targeting mitochondrial function to treat optic neuropathy. Mitochondrion S1567-7249(16):30115–30115. https://doi.org/10.1016/j.mito.2016.07.013

Haas RH (2007) The evidence basis for coenzyme Q therapy in oxidative phosphorylation disease. Mitochondrion 7(Suppl):S136–S145

Haefeli RH, Erb M, Gemperli AC, Robay D, Courdier Fruh I, Anklin C, Dallmann R, Gueven N (2011) NQO1-dependent redox cycling of idebenone: effects on cellular redox potential and energy levels. PLoS One 6(3):e17963

Hirano M, Garone C, Quinzii CM (2012) CoQ(10) deficiencies and MNGIE: two treatable mitochondrial disorders. Biochim Biophys Acta 1820(5):625–631

Hirashima N, Etcheberrigaray R, Bergamaschi S, Racchi M, Battaini F, Binetti G, Govoni S, Alkon DL (1996) Calcium responses in human fibroblasts: a diagnostic molecular profile for Alzheimer's disease. Neurobiol Aging 17(4):549–555

Ido Y, Duranton A, Lan F, Weikel KA, Breton L, Ruderman NB (2015) Resveratrol prevents oxidative stress-induced senescence and proliferative dysfunction by activating the AMPK-FOXO3 cascade in cultured primary human keratinocytes. PLoS One 10(2):e0115341

Jauslin ML, Wirth T, Meier T, Schoumacher F (2002) A cellular model for Friedreich Ataxia reveals small-molecule glutathione peroxidase mimetics as novel treatment strategy. Hum Mol Genet 11(24):3055–3063

Jauslin ML, Meier T, Smith RA, Murphy MP (2003) Mitochondria-targeted antioxidants protect Friedreich Ataxia fibroblasts from endogenous oxidative stress more effectively than untargeted antioxidants. FASEB J 17(13):1972–1974

Jouaville LS, Pinton P, Bastianutto C, Rutter GA, Rizzuto R (1999) Regulation of mitochondrial ATP synthesis by calcium: evidence for a long-term metabolic priming. Proc Natl Acad Sci U S A 96(24):13807–13812

Kanabus M, Heales SJ, Rahman S (2014) Development of pharmacological strategies for mitochondrial disorders. Br J Pharmacol 171(8):1798–1817

Kanabus M, Fassone E, Hughes SD, Bilooei SF, Rutherford T, Donnell MO, Heales SJ, Rahman S (2016) The pleiotropic effects of decanoic acid treatment on mitochondrial function in fibroblasts from patients with complex I deficient Leigh syndrome. J Inherit Metab Dis 39(3):415–426

Kasai T, Togashi T, Morishima T (2000) Encephalopathy associated with influenza epidemics. Lancet 355(9214):1558–1559

Kaufman RJ, Malhotra JD (2014) Calcium trafficking integrates endoplasmic reticulum function with mitochondrial bioenergetics. Biochim Biophys Acta 1843(10):2233–2239

Kemp JP, Smith PM, Pyle A, Neeve VC, Tuppen HA, Schara U, Talim B, Topaloglu H, Holinski-Feder E, Abicht A, Czermin B, Lochmüller H, McFarland R, Chinnery PF, Chrzanowska-Lightowlers ZM, Lightowlers RN, Taylor RW, Horvath R (2011) Nuclear factors involved in mitochondrial translation cause a subgroup of combined respiratory chain deficiency. Brain 134(Pt 1):183–195

Kim CY, Lee HJ, Chae MK, Byun JW, Lee EJ, Yoon JS (2015) Therapeutic effect of resveratrol on oxidative stress in graves' orbitopathy orbital fibroblasts. Invest Ophthalmol Vis Sci 56(11):6352–6361

Koopman WJ, Verkaart S, van Emst-de Vries SE, Grefte S, Smeitink JA, Nijtmans LG, Willems PH (2008) Mitigation of NADH: ubiquinone oxidoreductase deficiency by chronic Trolox treatment. Biochim Biophys Acta 1777(7–8):853–859

Koopman WJ, Beyrath J, Fung CW, Koene S, Rodenburg RJ, Willems PH, Smeitink JA (2016) Mitochondrial disorders in children: toward development of small-molecule treatment strategies. EMBO Mol Med 8(4):311–327

Lake NJ, Bird MJ, Isohanni P, Paetau A (2015) Leigh syndrome: neuropathology and pathogenesis. J Neuropathol Exp Neurol 74(6):482–492

Lebiedzinska M, Karkucinska-Wieckowska A, Wojtala A, Suski JM, Szabadkai G, Wilczynski G, Wlodarczyk J, Diogo CV, Oliveira PJ, Tauber J, Ježek P, Pronicki M, Duszynski J, Pinton P,

Wieckowski MR (2013) Disrupted ATP synthase activity and mitochondrial hyperpolarisation-dependent oxidative stress is associated with p66Shc phosphorylation in fibroblasts of NARP patients. Int J Biochem Cell Biol 45(1):141–150

Lim CK, Kalinowski DS, Richardson DR (2008) Protection against hydrogen peroxide-mediated cytotoxicity in Friedreich's ataxia fibroblasts using novel iron chelators of the 2-pyridylcarboxaldehyde isonicotinoyl hydrazone class. Mol Pharmacol 74(1):225–235

Liu CS, Chang JC, Kuo SJ, Liu KH, Lin TT, Cheng WL, Chuang SF (2014) Delivering healthy mitochondria for the therapy of mitochondrial diseases and beyond. Int J Biochem Cell Biol 53:141–146

Lopes Costa A, Le Bachelier C, Mathieu L, Rotig A, Boneh A, De Lonlay P, Tarnopolsky MA, Thorburn DR, Bastin J, Djouadi F (2014) Beneficial effects of resveratrol on respiratory chain defects in patients' fibroblasts involve estrogen receptor and estrogen-related receptor alpha signaling. Hum Mol Genet 23(8):2106–2119

López LC, Quinzii CM, Area E, Naini A, Rahman S, Schuelke M, Salviati L, Dimauro S, Hirano M (2010) Treatment of CoQ(10) deficient fibroblasts with ubiquinone, CoQ analogs, and vitamin C: time- and compound-dependent effects. PLoS One 5(7):e11897

Marchi S, Bonora M, Patergnani S, Giorgi C, Pinton P (2017) Methods to assess mitochondrial morphology in mammalian cells mounting autophagic or mitophagic responses. Methods Enzymol 588:171–186

Mathieu L, Costa AL, Le Bachelier C, Slama A, Lebre AS, Taylor RW, Bastin J, Djouadi F (2016) Resveratrol attenuates oxidative stress in mitochondrial Complex I deficiency: Involvement of SIRT3. Free Radic Biol Med 96:190–198

Mattiazzi M, Vijayvergiya C, Gajewski CD, DeVivo DC, Lenaz G, Wiedmann M, Manfredi G (2004) The mtDNA T8993G (NARP) mutation results in an impairment of oxidative phosphorylation that can be improved by antioxidants. Hum Mol Genet 13(8):869–879

Menzies KJ, Robinson BH, Hood DA (2009) Effect of thyroid hormone on mitochondrial properties and oxidative stress in cells from patients with mtDNA defects. Am J Physiol Cell Physiol 296(2):C355–C362

Morató L, Galino J, Ruiz M, Calingasan NY, Starkov AA, Dumont M, Naudí A, Martínez JJ, Aubourg P, Portero-Otín M, Pamplona R, Galea E, Beal MF, Ferrer I, Fourcade S, Pujol A (2013) Pioglitazone halts axonal degeneration in a mouse model of X-linked adrenoleukodystrophy. Brain 136(Pt 8):2432–2443

Morciano G, Giorgi C, Bonora M, Punzetti S, Pavasini R, Wieckowski MR, Campo G, Pinton P (2015) Molecular identity of the mitochondrial permeability transition pore and its role in ischemia-reperfusion injury. J Mol Cell Cardiol 78:142–153

Moreira PI, Harris PL, Zhu X, Santos MS, Oliveira CR, Smith MA, Perry G (2007) Lipoic acid and N-acetyl cysteine decrease mitochondrial-related oxidative stress in Alzheimer disease patient fibroblasts. J Alzheimers Dis 12(2):195–206

Patergnani S, Marchi S, Rimessi A, Bonora M, Giorgi C, Mehta KD, Pinton P (2013) PRKCB/protein kinase C, beta and the mitochondrial axis as key regulators of autophagy. Autophagy 9(9):1367–1385

Patergnani S, Baldassari F, De Marchi E, Karkucinska-Wieckowska A, Wieckowski MR, Pinton P (2014) Methods to monitor and compare mitochondrial and glycolytic ATP production. Methods Enzymol 542:313–332

Pinton P, Leo S, Wieckowski MR, Di Benedetto G, Rizzuto R (2004) Long-term modulation of mitochondrial Ca2+ signals by protein kinase C isozymes. J Cell Biol 165(2):223–32

Rahman S (2015) Emerging aspects of treatment in mitochondrial disorders. J Inherit Metab Dis 38(4):641–653

Rai PK, Russell OM, Lightowlers RN, Turnbull DM (2015) Potential compounds for the treatment of mitochondrial disease. Br Med Bull 116:5–18

Rajendran P, Nandakumar N, Rengarajan T, Palaniswami R, Gnanadhas EN, Lakshminarasaiah U, Gopas J, Nishigaki I (2014) Antioxidants and human diseases. Clin Chim Acta 436:332–347

Saada A (2011) The use of individual patient's fibroblasts in the search for personalized treatment of nuclear encoded OXPHOS diseases. Mol Genet Metab 104(1–2):39–47

Saada A (2014) Mitochondria: mitochondrial OXPHOS (dys) function ex vivo–the use of primary fibroblasts. Int J Biochem Cell Biol 48:60–65

Salviati L, Hernandez-Rosa E, Walker WF, Sacconi S, DiMauro S, Schon EA, Davidson MM (2002) Copper supplementation restores cytochrome c oxidase activity in cultured cells from patients with SCO2 mutations. Biochem J 363(Pt 2):321–327

Sbano L, Bonora M, Marchi S, Baldassari F, Medina DL, Ballabio A, Giorgi C, Pinton P (2017) TFEB-mediated increase in peripheral lysosomes regulates store-operated calcium entry. Sci Rep 7:40797

Scarpelli M, Todeschini A, Rinaldi F, Rota S, Padovani A, Filosto M (2014) Strategies for treating mitochondrial disorders: an update. Mol Genet Metab 113(4):253–260

Schmunk G, Boubion BJ, Smith IF, Parker I, Gargus JJ (2015) Shared functional defect in IP_3R-mediated calcium signaling in diverse monogenic autism syndromes. Transl Psychiatry 5:e643

Schmunk G, Nguyen RL, Ferguson DL, Kumar K, Parker I, Gargus JJ (2017) High-throughput screen detects calcium signaling dysfunction in typical sporadic autism spectrum disorder. Sci Rep 7:40740

Sgarbi G, Casalena GA, Baracca A, Lenaz G, DiMauro S, Solaini G (2009) Human NARP mitochondrial mutation metabolism corrected with alpha-ketoglutarate/aspartate: a potential new therapy. Arch Neurol 66(8):951–957

Singh DK, Kumar D, Siddiqui Z, Basu SK, Kumar V, Rao KV (2005) The strength of receptor signaling is centrally controlled through a cooperative loop between Ca^{2+} and an oxidant signal. Cell 121(2):281–293

Smith RA, Murphy MP (2010) Animal and human studies with the mitochondria-targeted antioxidant MitoQ. Ann N Y Acad Sci 1201:96–103

Smits P, Smeitink J, van den Heuvel L (2010) Mitochondrial translation and beyond: processes implicated in combined oxidative phosphorylation deficiencies. J Biomed Biotechnol 2010:737385

Soiferman D, Ayalon O, Weissman S, Saada A (2014) The effect of small molecules on nuclear-encoded translation diseases. Biochimie 100:184–191

Spinazzola A, Invernizzi F, Carrara F, Lamantea E, Donati A, Dirocco M, Giordano I, Meznaric-Petrusa M, Baruffini E, Ferrero I, Zeviani M (2009) Clinical and molecular features of mitochondrial DNA depletion syndromes. J Inherit Metab Dis 32(2):143–158

Vay L, Hernández-Sanmiguel E, Santo-Domingo J, Lobatón CD, Moreno A, Montero M, Alvarez J (2007) Modulation of Ca(2+) release and Ca(2+) oscillations in HeLa cells and fibroblasts by mitochondrial Ca(2+) uniporter stimulation. J Physiol Lond 580(Pt 1):39–49

Vazquez-Memije ME, Shanske S, Santorelli FM, Kranz-Eble P, Davidson E, DeVivo DC, DiMauro S (1996) Comparative biochemical studies in fibroblasts from patients with different forms of Leigh syndrome. J Inherit Metab Dis 19(1):43–50

Visch HJ, Rutter GA, Koopman WJ, Koenderink JB, Verkaart S, de Groot T, Varadi A, Mitchell KJ, van den Heuvel LP, Smeitink JA, Willems PH (2004) Inhibition of mitochondrial Na^+-Ca^{2+} exchange restores agonist-induced ATP production and Ca^{2+} handling in human complex I deficiency. J Biol Chem 279(39):40328–40336

Voets AM, Lindsey PJ, Vanherle SJ, Timmer ED, Esseling JJ, Koopman WJ, Willems PH, Schoonderwoerd GC, De Groote D, Poll-The BT, de Coo IF, Smeets HJ (2012) Patient-derived fibroblasts indicate oxidative stress status and may justify antioxidant therapy in OXPHOS disorders. Biochim Biophys Acta 1817(11):1971–1978

Wang H, Lemire BD, Cass CE, Weiner JH, Michalak M, Penn AM, Fliegel L (1996) Zidovudine and dideoxynucleosides deplete wild-type mitochondrial DNA levels and increase deleted mitochondrial DNA levels in cultured Kearns-Sayre syndrome fibroblasts. Biochim Biophys Acta 1316(1):51–59

White SL, Collins VR, Wolfe R, Cleary MA, Shanske S, DiMauro S, Dahl HH, Thorburn DR (1999) Genetic counseling and prenatal diagnosis for the mitochondrial DNA mutations at nucleotide 8993. Am J Hum Genet 65(2):474–482

Wojewoda M, Duszyński J, Szczepanowska J (2011) NARP mutation and mtDNA depletion trigger mitochondrial biogenesis which can be modulated by selenite supplementation. Int J Biochem Cell Biol 43(8):1178–1186

Wu SB, Ma YS, Wu YT, Chen YC, Wei YH (2010) Mitochondrial DNA mutation-elicited oxidative stress, oxidative damage, and altered gene expression in cultured cells of patients with MERRF syndrome. Mol Neurobiol 41(2–3):256–266

Wu YT, Wu SB, Wei YH (2014) Metabolic reprogramming of human cells in response to oxidative stress: implications in the pathophysiology and therapy of mitochondrial diseases. Curr Pharm Des 20(35):5510–5526

Yamaguchi S, Li H, Purevsuren J, Yamada K, Furui M, Takahashi T, Mushimoto Y, Kobayashi H, Hasegawa Y, Taketani T, Fukao T, Fukuda S (2012) Bezafibrate can be a new treatment option for mitochondrial fatty acid oxidation disorders: evaluation by in vitro probe acylcarnitine assay. Mol Genet Metab 107(1–2):87–91

Yao M, Yao D, Yamaguchi M, Chida J, Yao D, Kido H (2011) Bezafibrate upregulates carnitine palmitoyltransferase II expression and promotes mitochondrial energy crisis dissipation in fibroblasts of patients with influenza-associated encephalopathy. Mol Genet Metab 104(3):265–272

Yu-Wai-Man P, Soiferman D, Moore DG, Burté F, Saada A (2017) Evaluating the therapeutic potential of idebenone and related quinone analogues in Leber hereditary optic neuropathy. Mitochondrion S1567-7249(17):30012–30010. https://doi.org/10.1016/j.mito.2017.01.004

Targeting Mitochondria with Sweat: Improving Mitochondrial Function with Physical Activity

Jorge Beleza, David Rizo-Roca, António Ascensão, and José Magalhães

Abstract It is well established that physical exercise imposes increased levels of mechanical and metabolic stress to the human organism, altering the homeostasis and stimulating the inherent ability of tissues to structurally and functionally adapt to cope with the inflicted challenges. These adaptations usually result in an increased resistance against the harmful effects characterizing senescence as well as those associated with disease conditions. These include muscle myopathies, cardiac dysfunction induced by ischaemia-reperfusion, obesity, diabetes or toxicants exposure, liver and neurodegenerative diseases. In fact, exercise may directly alter cellular energy status or increase mechanical load in contractile tissues, such as skeletal and cardiac muscles, and may also indirectly induce an endocrine-like effect through the release of distinct molecules by striated muscles, which may exert consequent stimulation in non-contractile tissues, such as brain, liver or adipocytes. Mitochondrial remodelling is among the most important mechanisms targeted by exercise that contribute to the mentioned protective phenotype. The powerful influence of exercise in mitochondrial physiology include favourable changes in bioenergetics and substrate utilization, alterations in redox homeostasis, changes in network dynamics through biogenesis, fusion and fission mechanisms, an important involvement in the control of cellular death mechanisms as well as influence in cell signalling, autophagy-related renewal and quality control processes. The present review analyses the effects of exercise in the modulation of mitochondrial physiology, examining distinct proposed mechanisms targeting mitochondria and potentially responsible tissue boosting and consequent defect rescuing. The emerging role of epigenetic-based contribution to these cross-tolerance effects is also addressed.

Keywords Exercise · Mitochondrial remodelling · Tissue protection · Aging · Disease · Epigenetic

Jorge Beleza and David Rizo-Roca contributed equally to this work.

J. Beleza · D. Rizo-Roca · A. Ascensão · J. Magalhães (✉)
Laboratory of Metabolism and Exercise (LaMetEx), Department of Sport Biology, Faculty of Sport, Research Centre in Physical Activity Health and Leisure (CIAFEL), University of Porto, Porto, Portugal
e-mail: jmaga@fade.up.pt

P. J. Oliveira (ed.), *Mitochondrial Biology and Experimental Therapeutics*,
https://doi.org/10.1007/978-3-319-73344-9_18

1 Introduction

It is well established that physical exercise imposes an increased level of metabolic stress to the human organism, altering the homeostasis during the time exercise is performed, which can be further perpetrated during minutes, hours or days after the completion of exercise. To cope with the physiological challenges imposed by physical exercise, tissues have an inherent ability to structurally and functionally adapt. The most pronounced and direct physiological adaptations can be observed in the effector systems directly related to physical exercise, of which locomotor, cardiovascular and nervous systems are examples. In particular, the physiological adaptations occurring in the skeletal muscle system are well documented (Booth et al. 2015; Lundby and Jacobs 2016). Skeletal muscle is a very specialized tissue with a remarkable plasticity and adaptability to the physiological constrains imposed by contractile activity leading to several structural, functional and metabolic alterations, and ultimately promoting alterations in key molecular pathways that impact tissue phenotype. Besides the direct effect of contractile activity on skeletal muscle itself and in the cardiac tissue, the metabolic and mechanical alterations induced by physical exercise can have pleiotropic repercussions affecting also non-contractile tissues. This endocrine-like effect results, among other reasons, from the release of mediator proteins secreted by muscle cells to the blood stream (the so-called myokines) (Pedersen et al. 2003; Hoffmann and Weigert 2017), including interleukin 6 (IL-6), brain derived neurotrophic factor (BDNF) and irisin, which have specific receptors on the different organs and further activate multiple signalling cascades (Fig. 1). The disruption of homeostasis caused by physical exercise boosts tissue phenotype, i.e., augmenting the metabolic efficiency and/or increasing resistance against harmful stimuli, thus allowing to consider exercise as an outstanding strategy aiming at preserving cellular, tissue and organ viability and able to mitigate distinct stressful conditions (Fiuza-Luces et al. 2013; Pedersen and Saltin 2015). Accordingly, work from our laboratory and from others provide morphologic, functional and molecular evidence of the protective and therapeutic effect of exercise in different contractile and non-contractile tissues, including skeletal and cardiac muscles, brain, liver and adipose tissue, against harmful conditions associated with senescence, drug exposure and pathological states (Ascensão et al. 2005a; Lumini et al. 2008; Santos-Alves et al. 2014; Gonçalves et al. 2014a; Marques-Aleixo et al. 2016; Rocha-Rodrigues et al. 2017). Due to their important roles and interactions with several essential features of cell physiology, special attention to mitochondrial-mediated mechanisms targeted by physical exercise has been devoted (Ji et al. 2016). Among the distinct effects of physical exercise on the organism, the modulation of mitochondrial network seems to play a pivotal role in exercise adaptations. For instance, due to the crucial mitochondrial role in cellular energy production, an increase in energy demand promoted by physical exercise activates a considerable amount of mitochondrial remodelling processes leading to improvements in the mitochondrial network performance, seen as the ability to produce energy more efficiently, to more actively operate in the

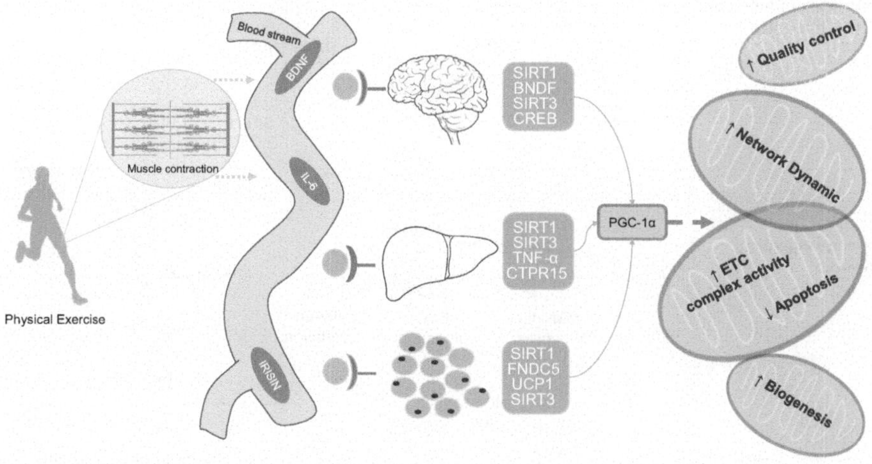

Fig. 1 Examples of how physical exercise-induced skeletal muscle endocrine-like effect may result in influenced mitochondrial phenotype in non-contractile tissues (Brain; Liver and Adipose Tissue). Muscle contractile activity promoted by physical exercise leads to the release of several muscle-derived molecules (myokines) into the blood stream, including brain-derived neurotrophic factor (BDNF), interleukin 6 (IL-6) and FNDC5/irisin. These myokines reach the different non-contractile tissues through tissue-specific receptors activating multiple signalling cascades that modulate tissue phenotype. Some of the most suggested molecules that are modulated in the different tissues as a consequence of myokine interaction include: sirtuin 1 (SIRT1), brain-derived neurotrophic factor (BDNF), sirtuin 3 (SIRT3) and cAMP response-element-binding protein (CREB) in the brain; SIRT1, SIRT3, tumor necrosis factor alpha, (TNF-α) and myonectin (CTPR15) in the liver; and SIRT1, fibronectin type III domain-containing protein 5 (FNDC5), uncoupling protein 1 (UCP1) and SIRT3 in adipose tissue (for references see: Fiuza-Luces et al. 2013; Iizuka et al. 2014; Schnyder and Handschin 2015; Kapilevich et al. 2015; Hoffmann and Weigert 2017). Ultimately, these molecules activate the master regulator of mitochondrial physiology, peroxisome proliferator-activated receptor-γ coactivator-1α (PGC-1α), therefore, activating a considerable amount of mitochondrial remodelling processes leading to improvements in the mitochondrial network performance through positive modulation (up-arrow) of quality control, network dynamic, electron transport chain (ETC) complex activity and biogenesis and downregulation (down-arrow) of apoptotic signalling markers. The mitochondrial remodelling in liver, brain and adipose tissues, further contributes for increasing the resistance of these non-contractile tissues against harmful stimuli, such as those resulting from aging process, toxins exposure and disease conditions

control of cell signalling, death and quality control (Fig. 2). One of the classic adaptations to exercise is the increase in mitochondrial content, being peroxisome proliferator-activated receptor-γ coactivator-1α (PGC-1α) the master regulator of this process of mitochondrial biogenesis (Booth et al. 2015). Also crucial to the adaptive process of mitochondrial turnover and metabolic efficiency is the ability to preserve healthy mitochondria and remove the damaged and non-functional reticulum. Similarly to many other stimuli causing mitochondrial network remodelling, this dynamic process in response to exercise is regulated by fusion, fission and auto(mito) phagic events (Konopka et al. 2014). Altogether, these quality control mechanisms

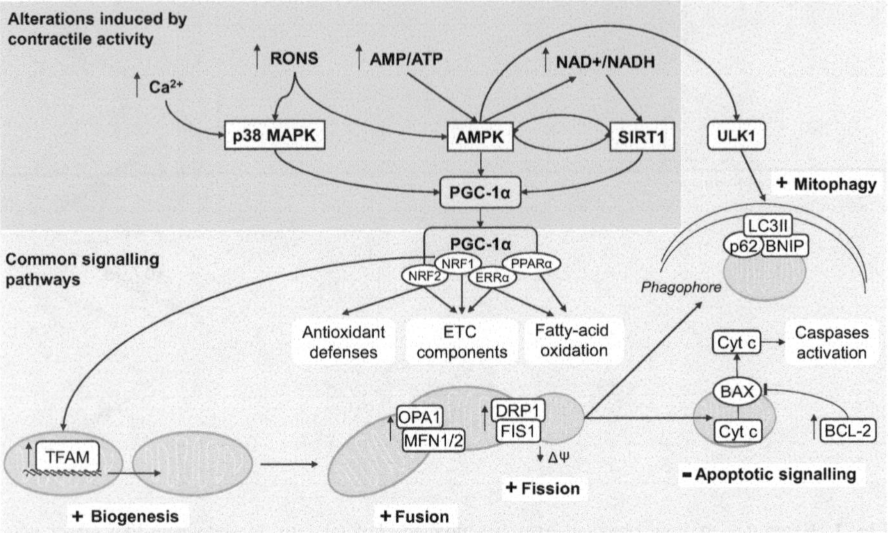

Fig. 2 Main alterations induced by contractile activity in cellular homeostasis, which activate different signalling pathways that converge in the mitochondrial master regulator PGC-1α -1α. After its activation, PGC-1α-1α interacts with multiple transcription factors (such as NRF1 and 2, ERRα and PPARα) which result in a myriad of cellular outcomes, including (1) the formation of new mitochondria, (2) increased mitochondrial dynamics, (3) boosted antioxidant defences, (4) increased content of ETC subunits, (5) enhanced fatty-acid oxidation, and (6) anti-apoptotic adaptations. Furthermore, exercise-induced activation of AMPK leads to the phosphorylation of ULK1, stimulating the mitophagy of damaged mitochondria. Overall, these adaptations boost the tissue oxidative capacity. *RONS* reactive oxygen and nitrogen species, *p38* p38 mitogen-activated protein kinase, *AMPK* AMP-activated protein kinase, *SIRT1* Sirtuin 1, *PGC-1α-1α* peroxisome proliferator-activated receptor-γ coactivator-1α, *NRF1* nuclear respiratory factor 1, *NRF2* nuclear factor erythroid 2-related factor 2, *ERRα* estrogen-related receptor α, *PPARα* peroxisome proliferator-activated receptor α, *TFAM* mitochondrial transcription factor A, *OPA1* optic atrophy 1, *MFN 1 and 2* mitofusin 1 and 2, respectively, *DRP1* dynamin-related protein 1, *FIS1* mitochondrial fission protein 1, *cyt c* cytochrome c, *Δψ* mitochondrial membrane potential

allow the cells to be more metabolically capable and functionally efficient, which are essential requisites to decrease fatigability and to improve endurance capacity in striated muscles. Furthermore, as will be addressed in the next subsections of this chapter, the above-mentioned alterations in non-contractile tissues caused by exercise, which are mainly mediated by circulating cytokines, also translate in mitochondrial remodelling in liver, brain and adipose tissue, further contributing to increase the resistance of these non-contractile tissues against disease-related deleterious consequences (Marques-Aleixo et al. 2016; Gonçalves et al. 2016; Rocha-Rodrigues et al. 2016).

In summary, among the different strategies to modulate mitochondrial physiology, physical exercise is now considered one of the most effective non-pharmacological interventions showing multiple beneficial features to improve several cellular processes (Fiuza-Luces et al. 2013). These include the control of cellular redox status, cellular signalling, calcium homeostasis, cell differentiation and death, and autophagy-related renewal processes.

In the next subsections, we will address how exercise can interfere with physiological and metabolic processes by "shaping and fitting" the mitochondrial network.

2 An Overview of How Physical Exercise Targets Different Features of Mitochondrial Structure, Metabolism and Function

2.1 Mitochondrial Bioenergetics

During exercise, substrate utilization depends on exercise modality, intensity and duration. Thus, the power output determines the ATP demand, while intensity and duration determine the percentage use of each metabolized fuel (lipids or carbohydrates) and its predominant source (hepatic, intramuscular and circulating). Generically, circulating free fatty acids (FFA) (supplied by adipocyte lipolysis) and glucose (either from oral ingestion or hepatic glycogenolysis) are the main substrates during exercise performed at low or moderate intensity (van Loon et al. 2001). On the other hand, during high intensity exercise, the fuel source shifts to a higher glucose contribution both through anaerobic glycolysis and pyruvate oxidation, and a larger contribution of muscle glycogen (van Loon et al. 2001). Hence, exercise can greatly challenge the muscle energy homeostasis. For example, in strenuous aerobic exercise, muscle blood flow and oxygen consumption can increase up to >30-fold, leading to an estimated tricarboxylic acid cycle (TCA) cycle flux increase of 100-fold (Gibala et al. 1998).

It has been extensively described that endurance training induces metabolic adaptations leading to increased content and/or activity of proteins and enzymes involved in glycolysis, glucose transport, TCA cycle, mitochondrial ATP production, and transport and oxidation of fatty acids (Schantz et al. 1983; Holloszy and Coyle 1984; Egan and Zierath 2013), which overall leads to a more efficient oxidative phosphorylation. Furthermore, endurance-trained subjects exhibit, at a given intensity, a relative decrease in circulating glucose and muscle glycogen utilization and greater reliance in fatty acid oxidation (Henriksson 1977; Jansson and Kaijser 1987), which is accompanied by increased content of fatty acid transporters, such as CD36 (Talanian et al. 2010), and increased levels of β-oxidation regulators, including medium-chain and very long chain acyl-CoA dehydrogenases (Horowitz et al. 2000). This decrease in carbohydrate use during submaximal exercise is a consequence of training-induced increased mitochondrial content and function, and allows the organism to spare glycogen; therefore reducing its depletion within the muscle and improving endurance performance and fatigue resistance.

Regardless the acetyl-CoA source, endurance training increases the TCA cycle flux (Befroy et al. 2008), leading to increased production of NADH and succinate, which are oxidized in the respiratory chain. Furthermore, it also induces an increase of the malate-aspartate shuttle (Schantz et al. 1986), which carries reducing equivalents across the mitochondrial membranes. These chronic exercise-based adaptations have the ultimate purpose of improving the abundance of electrons for "fuelling" the electron transport chain (ETC) and prompt cells with a more powerful and efficient aerobic machinery.

2.2 *Mitochondrial Biogenesis*

As stated before, endurance training is associated with an increase of the mitochondrial content (Hood et al. 2016). Tissues respond to perturbations in the energy status, adjusting the mechanisms of ATP production to meet the energy demand. Among these mechanisms, the increase in mitochondrial content and oxidative phosphorylation rate are central to the metabolic plasticity. The expansion of the mitochondrial reticulum occurs through the induction of mitochondrial biogenesis. The first evidence of exercise-induced mitochondrial biogenesis was suggested by John Holloszy in 1967, when he observed a significant increase of cytochrome c content in skeletal muscle of trained rats (Holloszy 1967). Since then, many studies have demonstrated that different modalities of physical exercise can induce mitochondrial biogenesis not only in skeletal muscle and heart, but also in non-contractile tissues, such as brain, liver and adipose tissue (Little et al. 2011). To assess this, several techniques and biomarkers have been developed over the last five decades, either by direct analysis (estimating the organelle volume in relation to total cellular volume) or by indirect evaluation (quantifying the content of mitochondrial proteins, such cytochrome c, or measuring the activity of enzyme markers, of which cytochrome c oxidase and citrate synthase are examples) (Schwerzmann et al. 1989; Larsen et al. 2012).

Accordingly, in the classic papers of Hickson (1981) and Dudley et al. (1982) using a rat model, it was reported that the magnitude of exercise-induced mitochondrial biogenesis was dependent of the intensity, duration and frequency of the exercise sessions, as well as the recruited fibre type. For many years, mitochondrial biogenesis has been associated with endurance training. For instance, Ingjer (1979) reported significantly higher mitochondrial content regardless the fibre type in quadriceps muscle cross-sections of six elite cross-country skiers compared to untrained subjects. Similarly, 6 weeks of endurance training on bicycle ergometers induced a 40% increase in mitochondrial volume/density of *vastus lateralis* (Hoppeler et al. 1985). Increases in mitochondrial enzymes have also been described in response to short-term endurance training programs. For example, 7–10 days of cycle ergometer exercise (2 h/day at 60–70% of VO_{2max}) resulted in a 30% increase of citrate synthase activity, β-hydroxyacyl-CoA dehydrogenase and carnitine acetyltransferase in both men and women (Rodas et al. 2000). However, there is growing evidence that other types of training protocols, namely high-intensity interval training (HIIT) can also result in similar responses regarding mitochondrial biogenesis (Burgomaster et al. 2008; Jacobs and Lundby 2013; MacInnis et al. 2016). In addition, a significant amount of data also suggest that physical exercise increases the mitochondrial content in cardiac tissue (Marques-Aleixo et al. 2015b). As suggested above, physical exercise, namely endurance training can also exert systemic effects in an endocrine-like fashion (Fig. 1), therefore inducing mitochondrial biogenesis even in non-contractile tissues (Little et al. 2011). The first evidence of exercise-induced mitochondrial biogenesis in the adipose tissue of rats after 10 weeks of daily swimming came from Stallknecht in 1991 (Stallknecht et al. 1991), which was further confirmed in more recent successive studies (Sutherland et al. 2009; Rocha-Rodrigues et al. 2017). Intriguingly, the evidence for training-induced mitochondrial network adaptations in liver were more controversial. In fact, while some previous data suggested no alteration or even a decrease in the liver mitochondrial content after voluntary physical activity and endurance training (Terblanche et al. 2001; Matiello et al. 2010), more recent studies have clearly shown the opposite (Haase et al. 2011; Santos-Alves et al. 2015). Similarly and despite some controversy (Herbst et al. 2015), physical exercise also seems to stimulate mitochondrial biogenesis in brain (Boveris and Navarro 2008; Dietrich et al. 2008; Safdar et al. 2011; Marques-Aleixo et al. 2015a), although studies are still scarce.

The signalling pathways that mediate the induction of mitochondrial biogenesis by chronic exercise rely in an orchestrated series of events, involving both nuclear and mitochondrial genomes, which begins immediately after the first bout of exercise (Perry et al. 2010). It involves the upregulation of several transcription factors, such as the nuclear respiratory factor 1 (NRF1), the estrogen-related receptor (ERR) and their co-activator PGC-1α and p53, as well as the incorporation of both mitochondrial and nuclear gene products into the expanding organelle reticulum (Hood et al. 2016). Among the 37 genes encoded by mitochondrial DNA (mtDNA), only 13 encode for proteins of the electron transport chain (Taanman 1999), being the transcription of these genes mediated by the exercise-induced PGC-1α upregulation (Perry et al. 2010; Scarpulla 2011). On the other hand, the upstream regulation of PGC-1α by exercise in the skeletal muscle is modulated by signalling kinases that are activated by events generated during the contractile activity, such as cytosolic Ca^{2+}, mechanical stress, accumulation of AMP and RONS (reactive oxygen and nitrogen species) (Carter et al. 2015). Among several kinases, p38 MAPK (p38 mitogen-activated protein kinase) and AMPK (AMP-activated protein kinase) are particularly relevant. Elevated cytosolic Ca^{2+} results in CaMKII (Ca^{2+}/calmodulin-dependent protein kinase II)-mediated activation of p38 MAPK (Wright et al. 2007), which stimulates upstream transcription factors of *PGC-1α*, such as activating transcription factor 2 and myocyte enhancer factor-2 (Puigserver et al. 2001; Cao et al. 2004). On the other hand, the rise of the AMP/ATP ratio mediated by exercise activates AMPK, which can directly phosphorylate and activate PGC-1α (Jäger et al. 2007). Furthermore, AMPK increases NAD^+ levels leading to the activation of SIRT1 (Sirtuin 1) (Cantó et al. 2009), a NAD-dependent deacetylase that also contributes to the activation of both AMPK and PGC-1α (Ruderman et al. 2010). The sensitivity of cells to such early events explains why mitochondrial biogenesis signalling begins early during the onset of exercise.

The role of PGC-1α in mitochondrial biogenesis is so prominent that it is usually referred as the master regulator of mitochondrial biogenesis and oxidative capacity (Fig. 2). In fact, the overexpression of PGC-1α in transgenic mice leads to fibre type transition from fast glycolytic to slow oxidative fibres (Lin et al. 2002), evidencing the crucial role of this transcriptional co-activator on deep mitochondrial metabolic and morphological changes.

Through the described mechanisms, physical exercise leads to the accumulation of mitochondrial and nuclear encoded proteins required for the assembling of the respiratory chain subunits, as well as other proteins necessary for mitochondrial function (Hood et al. 2016). In addition to an increase in oxidative capacity, concomitant with an increase in ETC complexes content, the formation of new mitochondria facilitates the elongation of the mitochondrial network, which has a positive impact in tissue metabolic capacity and overall bioenergetics. Furthermore, new mitochondria are required to replace the damaged fragments, contributing to the maintenance of a healthy and functional mitochondrial reticulum.

Taken together, growing evidence suggest that physical exercise can exerts a positive influence on mitochondrial network of contractile and some non-contractile tissues, although the underlying mechanisms remain to be completely understood.

2.3 Mitochondrial Network Dynamics and Quality Control

As mentioned in previous sections, when targeted by stimuli, mitochondria are highly dynamic organelles that can change their morphology to cope different energy demands or to activate specific signalling pathways. In fact, it has been suggested that fragmented mitochondria are preferentially found under low energy requirements and low respiratory activity, whereas mitochondrial fusion increases under higher energy demanding conditions leading to mitochondrial reticulum elongation, which allows a better transmission of electrogenic events and distribution of oxygen and metabolites (Westermann 2012).

Accordingly, the mitochondrial network seems to be remodelled under the cellular energy stress imposed by physical exercise. It has been extensively described that endurance training induces an expansion of the mitochondrial reticulum (Carter et al. 2015); therefore optimizing mitochondrial capability to deal with cellular bioenergetics needs. However, despite the considerable amount of work reporting increases in mitochondrial content and function induced by exercise (Hood et al. 2016), few studies devoted specific attention to its dynamics, particularly in human subjects. Cartoni et al. (2005) found increased mRNA levels of *MFN1* and *MFN2* (mitofusin 1 and 2, respectively) in muscle biopsies obtained from trained cyclists 24 h after endurance exercise. On the other hand, Perry et al. (2010) reported progressively increased content of specific proteins associated to dynamic-related signalling processes, namely MFN1, FIS1 (mitochondrial fission 1 protein) and DRP1 (dynamin-related protein 1) after a HIIT program. Similarly, Garnier et al. (2005) described a positive correlation between the training status and the skeletal muscle mRNA levels of *MFN2* and *DRP1* of healthy and diseased humans. These results strongly suggest that different types of exercise can effectively modulate mitochondrial dynamics, increasing both fusion and fission activities. Interestingly, also in the context of physical exercise, both fusion and fission can lead to adequate mitochondrial turnover and result in cellular beneficial adaptations, despite their antagonistic direct implications on mitochondrial network.

In close association with the mechanisms of mitochondrial dynamics, the mitochondrial quality control machinery is also considered a key player in mitochondrial turnover. In fact, among the processes associated with tissue adaptation and survival, mitophagy, which can be activated by increased oxidative stress and damage, encompasses a series of events that ultimately result in the elimination of dysfunctional mitochondria from the network to preserve mitochondrial bioenergetics. In the last few years, some studies, especially performed in rodents, have focused on the effects of exercise in mitochondrial autophagy. Data suggest that both acute and chronic exercise modulate the expression of proteins associated with auto(mito)phagy signalling, such as LC3II, p62 and BNIP3 (Pagano et al. 2014; Vainshtein et al. 2015). Interestingly, Lira et al. (2013) suggested that increased basal mitophagy is necessary for exercise-induced skeletal muscle adaptations, such as mitochondrial biogenesis, and for the improvement of physical performance.

The molecular mechanisms by which exercise modulates the expression of these mitochondrial dynamics and quality control-related proteins have not been completely elucidated. It is clear, however, that PGC-1α, AMPK, RONS and Ca^{2+}signaling play an important role in the exercise-induced mitochondrial dynamics (Fig. 2). For instance, in addition to its well-known role in mitochondrial biogenesis, PGC-1α activation induces the expression of MFN2 (Soriano et al. 2006). Furthermore, AMPK also activates ULK1, a crucial protein for the autophagy induction (Egan et al. 2011). On the other hand, Ca^{2+}-dependent calcineurin dephosphorylates DRP1 at its serine 637, promoting its activation and translocation to the mitochondrial surface (Cribbs and Strack 2007). Thus, it seems that these processes are regulated by the same signalling pathways that are responsible for most of the known exercise-induced skeletal muscle adaptations, suggesting that mitochondrial morphology, dynamics and quality control are an important part of the orchestrated muscle cell response to contractile activity.

In conclusion, there is growing evidence that exercise has a great impact in the mitochondrial dynamics and quality control, not only as epiphenomena, but due to the intrinsic properties of the mitochondrial network, which plays an important role in cellular bioenergetics, signalling and homeostasis. However, reported data and suggestions are limited to mRNA and protein expression. Therefore, extrapolations from these results should be taken with caution, and the application of more advanced techniques is needed to contrast the current hypothesis.

2.4 Involvement of Mitochondria in Cell Fate

In addition to its central role in energy metabolism, mitochondria are key organelles that also participate in distinct but overlapping pathways involved in cellular death, including apoptosis (for references and details see Wang and Youle 2009) and necrosis (Whelan et al. 2012). Interestingly, endurance training has been shown, particularly in rodent models, to modulate the expression of some proteins usually involved in apoptotic signalling pathways towards a more anti-apoptotic phenotype. For example, 8 weeks of endurance training downregulated *Bax* mRNA and upregulated *Bcl-2* mRNA as well as increased the protein content of ARC (apoptosis repressor with caspases recruitment domain protein) in rat *soleus* muscle (Siu et al. 2004). Similarly, increased BCL-2 and decreased AIF (apoptosis inducing factor) content were found in the *soleus* muscle and left ventricle of voluntary wheel-exercised rats (Vainshtein et al. 2011). Furthermore, H_2O_2-induced increase of BAX, cytochrome c and AIF was ablated in the left ventricle of these exercised animals. In accordance, chronic electrostimulation of *tibialis* anterior rat muscle (3 h/day/7 days) did not induce alterations in the BAX/BCL-2 ratio, although increased ARC content and less cytochrome c release were observed under both basal and stressful (exogenous ROS treatment) conditions (Adhihetty et al. 2006).

In addition to the direct modulation of anti- and pro-apoptotic proteins, exercise-induced adaptations of mitochondrial morphology and dynamics also seem to confer a protective phenotype against apoptosis. It is well-known that fragmented mitochondria are more prone to the release of the cytochrome c (Suen et al. 2008) and that mitochondrial fission machinery proteins, such as DRP1 and FIS1 are involved in programmed cell death (Frank et al. 2001). Thus, exercise-induced increased rate of mitochondrial fusion and concomitant network elongation not only contributes to energy-production efficiency, but also confers a more apoptotic-resistant phenotype. In fact, it has been described that stress-induced degradation of MFN2 leads to mitochondrial fragmentation and enhanced apoptotic cell death (Leboucher et al. 2012). In contrast, exercise-induced upregulation of PGC-1α seems to indirectly contribute to mitigate this deleterious phenotype. Adhihetty et al. 2007 reported increased apoptotic susceptibility in denervated skeletal muscle, which was associated to reduced PGC-1α expression and mitochondrial content. These results were further confirmed in a *Pgc-1α* KO rat model (Adhihetty et al. 2009). Additionally, lack or decrease of PGC-1α content was associated with increased RONS production, which also contributes to the activation of the apoptotic signalling cascade (Simon et al. 2000). As will be discussed in the next subsection, physical exercise also boosts cellular antioxidant defences and protects against oxidative stress and damage (Radak et al. 2017), therefore contributing to the aforementioned apoptotic-resistant phenotype.

In summary, physical exercise, particularly endurance training, increases resistance against apoptosis through the direct/indirect regulation of anti- and pro-apoptotic proteins, namely by remodelling the mitochondrial reticulum and by decreasing oxidative stress.

2.5 *Mitochondria as Active Players in the Regulation of Cell Signalling: Dual Role of RONS and Antioxidants*

During the contractile activity, mitochondria not only increase ATP synthesis, but also the production of RONS. For many years, mitochondrial RONS were considered mere harmful by-products of respiration. Surprisingly, Reid et al. (1993) reported that increased doses of antioxidant treatment (superoxide dismutase and catalase) progressively decreased the ability of isolated fibre bundles from rat diaphragm to generate force, which suggested that moderate amounts RONS have a determinant impact on excitation-contraction coupling and are necessary to achieve optimal contractile function. Thereafter, data from several studies reinforced the central involvement of RONS in signalling pathways, including gene transcription, mitochondrial dynamics and activation of apoptosis.

While acute strenuous exercise has been associated with excessive oxidative stress and tissue deleterious consequences, chronic exercise plays an important role in the organism adaptive response to exercise via RONS signalling, including the increase of antioxidant defences. For example, NRF2 (nuclear factor erythroid 2-related factor 2), a master regulator of antioxidant responses, is activated by RONS during physical exercise (Dembele et al. 2009). In fact, NRF2 is consistently shown to be upregulated after exercise in rodents (Gounder et al. 2012) as well as in active humans (Ballmann et al. 2014; Scott et al. 2015). Upon activation, NRF2 regulates the transcription of more than 200 genes, including genes encoding glutathione reductase, glutathione peroxidase and NADPH oxidase. Similarly, RONS-mediated activation of forkhead-box O (FOXO) stimulates the transcription of genes coding for antioxidant enzymes, such as superoxide dismutase-2 (SOD2) and catalase under exercise conditions (Kavazis et al. 2014; Klotz et al. 2015). In addition, factor nuclear kappa B (NFκB) has also been shown to be stimulated by exercise-derived RONS, therefore activating the transcription of antioxidant-related genes and increasing the expression of manganese-dependent superoxide dismutase (MnSOD), inducible nitric oxide synthase and glutathione (Morgan and Liu 2011). Thus, exercise-derived mitochondrial RONS production exhibit a dose-dependent hormetic behaviour, being high doses associated to cellular deleterious effects and adequate amounts related to beneficial and adaptive responses that can protect from following oxidative insults.

In addition to this autoregulatory behaviour, RONS also mediate signalling pathways involved in mitochondrial network remodelling and quality control, such as mitochondrial biogenesis, fission and mitophagy. Accordingly, RONS stimulate the activation and transcription of PGC-1α through p38 MAPK and AMPK (Kang et al. 2009), whereas high-doses of antioxidant supplementation seems to ablate the exercise-induced upregulation of PGC-1α, NRF1 and TFAM (mitochondrial transcription factor A) as well as the concomitant increase in mitochondrial content (Gomez-Cabrera et al. 2008). Similarly, vitamin C and vitamin E supplementation blocked the exercise-dependent upregulation of PGC-1α, SOD1 and SOD2, glutathione peroxidase preventing health-promoting effects of physical exercise in humans (Ristow et al. 2009). As multiple studies demonstrated that exacerbated mitochondrial RONS production positively correlates with mitochondrial fission and fragmentation (Yu et al. 2006; Jendrach et al. 2008), therefore stimulating apoptotic signalling, the exercise-induced protective effects against oxidative stress could also contribute to the anti-apoptotic phenotype in trained subjects.

Altogether, these data show that, within a physiological range, chronic exercise-derived mitochondrial RONS are necessary for the cellular adaptive response against oxidative stress as well as for the regulation of several mitochondrial pathways.

3 Physical Exercise Antagonizes Mitochondriopathies Characterizing Aging and Diseased Conditions

Considering physical exercise as an important stimulus that targets and boosts several mitochondrial features, including those related to morphology, biochemistry and function, it is clearly expected that it may represent a very interesting tool for abrogating the harmful consequences associated with the physiopathology of aging and disease. In fact, assuming that most of the previously referred mechanisms associated with cellular and tissue dysfunction in disease, such as exacerbated inflammatory response, increased oxidative stress and damage, apoptosis, altered auto(mitophagy), mitochondrial dynamics and quality control are also regulated by physical exercise, in this subsection we propose to analyse the cross-tolerance effects of different types of exercise against the mitochondriopathies characterizing aging and some diseased conditions. These include cardiac dysfunction induced by ischaemia-reperfusion (IR), doxorubicin (DOX) treatment and hyperglycaemia, liver diseases such as NAFLD (non-alcoholic fat liver diseases) or induced by anti-inflammatory drugs, neurodegenerative diseases and obesity-induced adiposopathy. We here intended to discuss the potential of physical exercise (preventive and/or therapeutic) against these deleterious effects in several features of mitochondrial physiology, and also to address the potential implications of the epigenetic regulation on exercise-induced mitochondrial adaptations.

3.1 Aging

It is sufficiently clear that senescence causes deterioration of cellular function in all tissues, increasing their susceptibility to dysfunction and disease. Also, among the numerous theories associated with the aetiology and pathophysiology of this process is the increased oxidative stress (Dröge 2002), in which mitochondria have a recognized role as a source and target of the generated RONS. With increasing age, mitochondrial function suffers alterations that predispose them to enhance production of RONS, mainly at the level of the respiratory chain. As known, the interaction of these species with other molecules composing the phospholipidic structure of membranes, proteins and nucleic acids of DNA that lack protective histones, causes damage and consequent dysfunction, phenomena that are further exacerbated by the diminished antioxidant capacity of aged cells. As physical exercise, particularly the chronic forms, is known to decrease the levels of oxidative stress by both decreasing RONS production and boosting antioxidant capacity (through the increase of inducible scavenging molecules) at several levels of tissue and cell organization, including in mitochondria, we will further discuss in brief some examples of how chronic physical exercise mitigates the increased aging-induced organ and tissue dysfunction/damage by targeting mitochondria.

For instance, in addition to the exercise-induced benefits against the incompetent autonomic activity in aged hearts through re-sensitization of β-adrenergic receptors, improvements in the diminished calcium handling and hypertrophy, as well as in the increased fibrosis, exercise also restores/attenuates the compromised mitochondrial function characterizing senescence (Roh et al. 2016). These ameliorations were seen through increased mitochondrial respiration and decreased RONS production (Bo et al. 2014), increased PGC-1α expression and associated signalling pathways (Leick et al. 2010; Bayod et al. 2012; Kang et al. 2013), mitochondrial dynamics and quality control factors (Koltai et al. 2012) and mitochondrial DNA (Safdar et al. 2011) in aged animals submitted to chronic exercise programs. It is proposed that exercise enhances antioxidant defences and restores redox homeostasis in aging tissues, at least in part, via NRF2, activating antioxidant response genes (Gounder et al. 2012). NRF2 is also co-activated by PGC-1α under oxidative stress conditions as PGC-1α induces GPx1 (Glutathione peroxidase 1) and SOD2 in models of neurodegeneration (St-Pierre et al. 2006). It was recently suggested that exercise training in old mice improves brain mitochondrial capacity by targeting electron transport chain function and mitochondrial dynamics without affecting mitochondrial biogenesis (Gusdon et al. 2017).

Age-related sarcopenia and muscle dysfunction in humans were antagonized by 9 weeks of either neuromuscular electrical stimulation or leg press, being the improved muscle function and structure induced by training stimulations associated to increased protein levels of mitochondrial calcium uniporter and of fusion protein OPA1 (optic atrophy 1) (Zampieri et al. 2016). Also, other results indicate that the exercise training-related functional improvements found in aged muscles were associated with changes in mitochondrial proteins involved in redox and calcium homeostasis, bioenergetics, remodelling and dynamics (Figueiredo et al. 2009; Alves et al. 2010; White et al. 2016).

To further unravel evidence for a mitochondrial polymerase γ-independent mtDNA repair pathway mediated by exercise, Safdar et al. (2015) reported that the tumour suppressor protein p53 translocates to mitochondria and facilitates mtDNA mutation repair and mitochondrial biogenesis in response to endurance training of moderate intensity. Corroborating these data, genetic-based approaches showed that exercise failed to prevent mtDNA mutations, to induce mitochondrial biogenesis, to preserve mitochondrial morphology, to reverse sarcopenia, or to mitigate premature mortality in mutator mice with muscle-specific deletion of p53. In addition to previous research from David Hood's laboratory suggesting mitochondrial p53 as an important mediator, although not necessary, for mitochondrial adaptations induced by voluntary activity, including biogenesis (Saleem et al. 2009), and demonstrating that acute endurance exercise induce p53 translocation to mitochondria where it may serve to positively modulate the activity of TFAM (Saleem and Hood 2013), the results from Safdar et al. (2015) established p53 as a novel target for exercise-mediated maintenance of the mtDNA genome and suggest mitochondrially targeted p53 as a therapeutic mechanism counteracting aging and diseases of mitochondrial aetiology. Detailed and comprehensive reviews on the known mechanisms of how exercise mitigates mitochondrial senescence, particularly in skeletal muscle are available (Carter et al. 2015; Joseph et al. 2016).

3.2 Cardiac Dysfunction

The direct important role of mitochondrial physiology abnormalities in heart failure is attested by many studies, justifying a recent consensus report. In fact, experts in the field recently pointed out that, despite the pathophysiology of heart failure-associated diseases is complex, mitochondrial dysfunction seems to be a crucial therapeutic target to improve cardiac function (Brown et al. 2016). The emerging treatments with the potential to improve the function of the failing heart by targeting mitochondria include many pharmacological approaches, being physical exercise probably the most effective non-pharmacological countermeasure. In fact, several studies demonstrated that distinct forms of physical exercise, including acute single pre-conditioning bout, short and long-term endurance training, intermittent exercise and HIIT as well as voluntary physical activity mitigate (by preventing and/or treating) the deleterious effects characterizing cardiac diseases and toxicity, targeting mitochondria in the adaptive process (for references see Ascensão et al. 2007, 2011a, b; Powers et al. 2014a, b). Considering IR injury as a model of cardiac damage, against which exercise has shown to be protective, several mechanisms and related candidate proteins have been described to contribute to the more resistant phenotype induced by exercise rescuing several harmful effects of IR. These include the improvement in mitochondrial antioxidant and anti-apoptotic signalling profile (Ascensão et al. 2005b; Kavazis et al. 2008, 2009; Magalhães et al. 2013), adaptations in proteins known to be involved in the composition and/or sensitizing the mtPTP (mitochondrial permeability transition pore) (Lumini-Oliveira et al. 2011; Magalhães et al. 2014), probably leading to increased capacity of cardiac mitochondria to tolerate calcium (Marcil et al. 2005; Ciminelli et al. 2006) and, despite controversy (Brown et al. 2005), the inhibition of mitochondrial K_{ATP} channels (Quindry et al. 2010). The mentioned positive adaptations observed in trained animals, which correlate with improved haemodynamic endpoints, also translate into boosted mitochondrial respiratory performance (Leichtweis et al. 1997; Ascensão et al. 2006a, b; Lee et al. 2012; Alleman et al. 2016), meaning that, among other performance features, cardiac tissue is more capable to oxidize substrates and to produce ATP. It was also recently reported that regular treadmill exercise inhibits mitochondrial accumulation of cholesterol and oxysterols during myocardial IR in wild-type and *ob/ob* mice, concomitantly improving oxidative phosphorylation (Musman et al. 2016).

The protective mitochondrial phenotype induced by physical exercise against the damage caused by IR events is also evident against DOX exposure (Ascensão et al. 2012b), a very effective anticancer agent against several malignancies, but limited by the dose dependent dilated cardiomyopathy. Data from our group and others revealed the benefits of acute and chronic exercise, as well as of voluntary physical activity, against the deterioration of functional capacity and structure of hearts from animals exposed to acute single bolus and sub-chronic dosages of DOX (Ascensão et al. 2005a, 2006a, 2011b, Chicco et al. 2005, 2006; Marques-Aleixo et al. 2015b). Similarly to IR, these ameliorations in hearts/mitochondria from DOX treated animals were accompanied by favourable adaptations in several molecular markers related to antioxidant capacity, calcium-induced mtPTP opening, mitochondrial biogenesis and dynamics as well as quality control and apoptotic signalling (Kavazis et al. 2010; Dolinsky et al. 2013; Marques-Aleixo et al. 2015b). Actually, exercise-induced cardioprotection could also be associated with an improvement of mitochondrial quality control (Campos et al. 2012; Sun et al. 2013). Exercise increases cardiac mitochondria renewal and remodelling by promoting mitochondrial biogenesis, fusion, and auto(mito)phagy in order to eliminate damaged mitochondria (Sun et al. 2013). Together with redox-related and anti-apoptotic adaptations, the mitochondrial network adjustments provided by chronic physical exercise can be critical in the attenuation or reversion of DOX-induced cardiac and mitochondrial toxicity (Dolinsky et al. 2013; Smuder et al. 2013).

3.3 Liver Diseases-Induced Toxicity

It is likewise known that liver mitochondria are structurally and functionally affected by several deleterious conditions that include obesity and high-fat diet-induced NAFLD as well as by exposure to toxicants, such as anti-inflammatory drugs (Fromenty et al. 2004; Wei et al. 2008; Pessayre et al. 2010) and that increased aerobic fitness and physical exercise represent valuable counteracting tools against the resulting deleterious effects (Rector and Thyfault 2011; Gonçalves et al. 2013; Ascensão et al. 2013). Thyfault et al. (2009) clearly showed that rats selectively bred for low aerobic capacity have reduced hepatic mitochondrial oxidative capacity and increased susceptibility to hepatic steatosis and injury compared to high aerobic capacity animals. Moreover, accumulating evidence agrees that increased voluntary physical activity, but particularly endurance training, is able to positively modify liver tissue and mitochondrial phenotypes from diseased or injured conditions towards more resistant ones. Accordingly, several works from our lab demonstrated that mitochondrial features and endpoints from wild-type animals (Santos-Alves et al. 2015) and rats exposed to high-fat diet induced non-alcoholic steatohepatitis (Gonçalves et al. 2014a, b, 2016) or *in vitro* anti-inflammatory drugs (Ascensão et al. 2012a;

Santos-Alves et al. 2014) were positively modulated by exercise/increased physical activity. These features include histopathological hallmarks, mitochondrial membrane phospholipidic composition, *in vitro* respiratory capacity, transmembrane potential fluctuations, calcium resistance-dependent mtPTP and apoptotic signalling, oxidative stress and damage markers, mitochondrial dynamics comprising biogenesis, fusion and fission, as well as alterations in (auto)mitophagy signalling molecules, closely related to quality control mechanisms.

3.4 *Obesity-Induced Adiposopathy*

While in conditions of obesity and insulin resistance mitochondrial reticulum in adipose tissue is clearly compromised; physical exercise is proposed to increase mitochondrial content (Townsend et al. 2017), thus augmenting the metabolic activity of this tissue and contributing to important energy expenditure and functional health benefits. These physical exercise-related adaptations seem to be based, at least in part, on the lipolysis-dependent increased AMPK-PGC-1α and consequent mitochondrial biogenesis. In fact, exercise stimulates AMPK, PGC-1α and the overexpression of UCP1 (uncoupling protein 1), important hallmark of adipose tissue browning (Boström et al. 2012; Mottillo et al. 2016). Additionally, despite some controversy (Norheim et al. 2014) it has been suggested that exercise prompts the secretion of catecholamines, and several myokines, such as FNDC5, IL-6 or meteorin-like, which seem to be important contributors to mitochondrial-associated "beiging/browning" of white adipose tissue (Stanford et al. 2015; Ost et al. 2016). Accordingly, recent data from our group reported that physical exercise modulated muscle and adipose tissue myokines expression and brown adipose-like phenotype in rats fed a high-fat diet, and improved epididymal white adipose tissue mitochondrial complex subunits as well as markers of mitochondrial biogenesis and dynamics (Rocha-Rodrigues et al. 2016, 2017). Chronic exercise-induced adipose tissue mitochondrial biogenesis is known to occur in a eNOS-dependent manner as exercise increased mitochondrial DNA, mitochondrial enzymes, and expression of NRF1, TFAM and PGC-1α in subcutaneous adipose tissue in wild type, which was not evident in eNOS null animals (Trevellin et al. 2014). These changes occurring in adipose tissue phenotype in response to exercise training may be part of the mechanism by which exercise improves whole-body metabolic health, despite human studies failed to observe increased adipose tissue mitochondrial content and functional capacity to oxidize fat after high-intensity training (Larsen et al. 2015).

3.5 *Neurodegenerative Diseases*

Despite some controversy (Herbst et al. 2015), regular physical activity and exercise may also play an important role in the modulation of brain mitochondrial machinery (Marques-Aleixo et al. 2015a), thus mitigating mitochondrial malfunction characterizing aging and neurodegenerative diseases (Marques-Aleixo et al. 2012; Bernardo et al. 2016). In accordance, cross-tolerance studies have been designed to test physical exercise rescues mitochondrial defects characterizing several pathophysiological conditions as follows. Recently, Gusdon et al. (2017) reported that endurance training increased brain cortex mitochondrial complex I activity and the expression of DRP1 in aged mice, confirming early results pointing out training-induced increased mitochondrial biogenesis signalling markers in most of the brain regions (Steiner et al. 2011). Reinforcing the role of mitochondrial mediating mechanisms and target therapies in neurodegenerative diseases, Cheng et al. (2016) found that neurons lacking the mitochondrial deacetylase SIRT3 were more vulnerable to dysfunction and degeneration in mouse models of epilepsy and Huntington's disease. In contrast, the running wheel-induced hippocampal SIRT3 overexpression modulated mitochondrial protein acetylation and bolster neuronal resistance to oxidative stress and apoptosis. Moreover, long-term endurance training significantly improved the cognitive function and reduced the expression of amyloid β-42 peptide in APP/PS1 transgenic mouse model of Alzheimer Disease (Bo et al. 2014). Concomitantly, improved mitochondrial respiratory function by increasing the complexes I, and IV and ATP synthase activities, whereas attenuated ROS generation and mtDNA oxidative damage were observed in APP/PS1 mice. Furthermore, the impaired mitochondrial antioxidant enzymes and mitochondrial 8-oxoguanine DNA glycosylase-1 (OGG1) activities seen in the transgenic mice were reversed with training. Also, the acetylation levels of mitochondrial OGG1 and MnSOD were markedly suppressed in transgenic mice after exercise training, in parallel with increased content of SIRT3.

In addition to senescence and neurodegenerative diseases-associated mitochondrial defects, exercise has also been effective in counteracting toxicants-induced brain mitochondrial dysfunction and related mechanisms. For instance, brain cortex and cerebellum mitochondrial modifications associated with cumulative sub-chronic DOX administration, including perturbations in mitochondrial oxidative phosphorylation capacity, increased oxidative stress and damage, augmented mPTP opening susceptibility and apoptotic signalling and deregulation of mitochondrial quality control were positively modulated by long-term physical exercise performed before and during treatments (Marques-Aleixo et al. 2016). Moreover, 3,4-methyle nedioxymethamphetamine (MDMA)-induced impairments in cognitive and brain mitochondrial function, namely in acquisition and retention of spatial memory, RONS production, mitochondrial membrane potential, mitochondrial swelling, mitochondrial outer membrane damage, cytochrome c release, and ADP/ATP ratio, were mitigated by physical exercise (Taghizadeh et al. 2016).

4 Is There a Role for Epigenetic Modifications in Exercise-Targeting Mitochondrial Function?

Some of the favourable effects associated with physical exercise are mediated by alterations in gene expression, which are in part orchestrated by epigenetic modifications to DNA (histone modifications and DNA methylation) and by the expression/repression of non-coding miRNAs (McGee et al. 2009; Denham et al. 2014; Jiménez-Chillarón et al. 2015). Through these processes, physical exercise regulates the expression of genes associated to a myriad of cellular events, such as metabolic and inflammatory processes, aging, cancer, and central nervous system degeneration. Reported regulated genes include *PGC-1α*, pyruvate dehydrogenase lipoamide kinase isozyme 4, apoptosis-associated speck-like protein containing a CARD, and BDNF (Pareja-Galeano et al. 2014). It has been proposed that physical exercise regulates the epigenetic landscape indirectly through a metabolic remodelling, involving alterations in the production of TCA cycle intermediates that are involved in epigenetic engineering (Donovan and Miller 2011). Moreover, Barrès et al. (2012) showed that high-intensity acute exercise decreased human skeletal muscle DNA methylation and increased the expression of the respective genes associated to mitochondrial biogenesis, such as *PGC-1α* and *TFAM*.

Whether exercise-induced epigenetic modifications can be transmitted to subsequent generations and cause a persistent modulation of gene expression with consequences at mitochondrial level remains questionable. It is known that physical exercise performed during pregnancy improves long-term health of the offspring in later life through increased mitochondrial fitness in brains and hearts. Actually, Chung et al. (2017) recently reported that maternal exercise up-regulated genes encoding proteins associated with mitochondrial dynamics, namely NRF1, NRF2 and MFN1 expression, and increased ATP production and cytochrome c oxidase activity while decreased H_2O_2 levels on foetal mouse hearts. Suggestions exist that maternal exercise alters mitochondrial enzymatic activity, improves mitochondrial antioxidant defence capacity and markers of mitochondrial biogenesis in brains of the offspring (Park et al. 2013; Marcelino et al. 2013). Moreover, physical exercise performed before pregnancy improves indices of embryos development, which were accompanied by increases in ATP levels, mitochondrial membrane potential and mtDNA copy number, and a decrease in H_2O_2 concentration (Xu et al. 2016). In this study, methylation levels of CpG sites on imprinting control regions of imprinted genes (*Igf2*, *Igf2r*, *Meg 3* and *H19*) in the embryos of exercised groups were normalized when compared to those of the non-exercised groups. Additionally, global DNA and histone methylation (H3K4m2, H3K9m3 and H3K36m) confirmed that physical activity in pre-pregnancy facilitates the maintenance of epigenetic modifications (Xu et al. 2016). Six weeks of voluntary wheel running prior to and during pregnancy prevented maternal high-fat diet-induced skeletal muscle *Pgc-1α* promoter hypermethylation and increased its expression as well as its target genes, an effect that correlated with an improvement of age-associated metabolic dysfunction observed at 9 months of age in the offspring (Laker et al. 2014).

In addition to transcription factors activity, DNA methylation and histone modification, the influence of physical exercise on transcription and translation processes is also related to the expression of miRNAs (Kirby and McCarthy 2013; Russell and Lamon 2015). Despite the recognition that the findings to date, supporting a role for miRNAs in muscle plasticity, remain for the most part correlative, several studies reported that endurance training-modulated animal and human miRNAs (miR-494, miR-696 and miR-761) predicted to target genes that control mitochondrial biogenesis, including *PGC-1α, TFAM and FOXJ3* (Aoi et al. 2010; Yamamoto et al. 2012; Xu et al. 2015).

5 Conclusions

In summary, from the literature consulted, it is clear that increased physical activity and endurance training boost skeletal and cardiac muscles, liver, brain and adipose tissues, increasing its resistance to aging and disease-induced injury by targeting mitochondrial network physiology and related mechanisms. Among the different mechanisms by which physical exercise improves molecular, functional and morphologic mitochondrial features, it is tempted to state that epigenetics has also an important and emergent role.

References

Adhihetty PJ, Ljubicic V, Hood DA (2006) Effect of chronic contractile activity on SS and IMF mitochondrial apoptotic susceptibility in skeletal muscle. AJP Endocrinol Metab 292:E748–E755. https://doi.org/10.1152/ajpendo.00311.2006

Adhihetty PJ, O'Leary MFN, Chabi B et al (2007) Effect of denervation on mitochondrially mediated apoptosis in skeletal muscle. J Appl Physiol 102:1143–1151. https://doi.org/10.1152/japplphysiol.00768.2006

Adhihetty PJ, Uguccioni G, Leick L et al (2009) The role of PGC-1alpha on mitochondrial function and apoptotic susceptibility in muscle. Am J Physiol Cell Physiol 297:C217–C225. https://doi.org/10.1152/ajpcell.00070.2009

Alleman RJ, Tsang AM, Ryan TE et al (2016) Exercise-induced protection against reperfusion arrhythmia involves stabilization of mitochondrial energetics. Am J Physiol Heart Circ Physiol 310:H1360–H1370. https://doi.org/10.1152/ajpheart.00858.2015

Alves RMP, Vitorino R, Figueiredo P et al (2010) Lifelong physical activity modulation of the skeletal muscle mitochondrial proteome in mice. J Gerontol Ser A Biol Sci Med Sci 65A:832–842. https://doi.org/10.1093/gerona/glq081

Aoi W, Naito Y, Mizushima K et al (2010) The microRNA miR-696 regulates PGC-1 in mouse skeletal muscle in response to physical activity. AJP Endocrinol Metab 298:E799–E806. https://doi.org/10.1152/ajpendo.00448.2009

Ascensão A, Magalhães J, Soares J et al (2005a) Endurance training attenuates doxorubicin-induced cardiac oxidative damage in mice. Int J Cardiol 100:451–460. https://doi.org/10.1016/j.ijcard.2004.11.004

Ascensão A, Magalhães J, Soares JMC et al (2005b) Moderate endurance training prevents doxorubicin-induced in vivo mitochondriopathy and reduces the development of cardiac apoptosis. Am J Phys 289:H722–H731. https://doi.org/10.1152/ajpheart.01249.2004

Ascensão A, Ferreira R, Oliveira PJ, Magalhães J (2006a) Effects of endurance training and acute doxorubicin treatment on rat heart mitochondrial alterations induced by in vitro anoxia-reoxygenation. Cardiovasc Toxicol 6:159–172

Ascensão A, Magalhães J, Soares JMC et al (2006b) Endurance training limits the functional alterations of heart rat mitochondria submitted to in vitro anoxia-reoxygenation. Int J Cardiol 109:169–178. https://doi.org/10.1016/j.ijcard.2005.06.003

Ascensão A, Ferreira R, Magalhães J (2007) Exercise-induced cardioprotection—biochemical, morphological and functional evidence in whole tissue and isolated mitochondria. Int J Cardiol 117:16–30. https://doi.org/10.1016/j.ijcard.2006.04.076

Ascensão A, Lumini-Oliveira J, Oliveira P, Magalhaes J (2011a) Mitochondria as a target for exercise-induced cardioprotection. Curr Drug Targets 12:860–871. https://doi.org/10.2174/138945011795529001

Ascensão A, Lumini-Oliveira J, Machado NG et al (2011b) Acute exercise protects against calcium-induced cardiac mitochondrial permeability transition pore opening in doxorubicin-treated rats. Clin Sci (Lond) 120:37–49. https://doi.org/10.1042/CS20100254

Ascensão A, Gonçalves IO, Lumini-Oliveira J et al (2012a) Endurance training and chronic intermittent hypoxia modulate in vitro salicylate-induced hepatic mitochondrial dysfunction. Mitochondrion 12:607–616. https://doi.org/10.1016/j.mito.2012.10.007

Ascensão A, Oliveira PJ, Magalhães J (2012b) Exercise as a beneficial adjunct therapy during doxorubicin treatment-role of mitochondria in cardioprotection. Int J Cardiol 156:4–10. https://doi.org/10.1016/j.ijcard.2011.05.060

Ascensão A, Martins MJ, Santos-Alves E et al (2013) Modulation of hepatic redox status and mitochondrial metabolism by exercise: therapeutic strategy for liver diseases. Mitochondrion 13:862–870. https://doi.org/10.1016/j.mito.2013.07.002

Ballmann C, McGinnis G, Peters B et al (2014) Exercise-induced oxidative stress and hypoxic exercise recovery. Eur J Appl Physiol 114:725–733. https://doi.org/10.1007/s00421-013-2806-5

Barrès R, Yan J, Egan B et al (2012) Acute exercise remodels promoter methylation in human skeletal muscle. Cell Metab 15:405–411. https://doi.org/10.1016/j.cmet.2012.01.001

Bayod S, del Valle J, Lalanza JF et al (2012) Long-term physical exercise induces changes in sirtuin 1 pathway and oxidative parameters in adult rat tissues. Exp Gerontol 47:925–935. https://doi.org/10.1016/j.exger.2012.08.004

Befroy DE, Petersen KF, Dufour S et al (2008) Increased substrate oxidation and mitochondrial uncoupling in skeletal muscle of endurance-trained individuals. Proc Natl Acad Sci U S A 105:16701–16706. https://doi.org/10.1073/pnas.0808889105

Bernardo TC, Marques-Aleixo I, Beleza J et al (2016) Physical exercise and brain mitochondrial fitness: the possible role against Alzheimer's disease. Brain Pathol 26:648–663. https://doi.org/10.1111/bpa.12403

Bo H, Kang W, Jiang N et al (2014) Exercise-induced neuroprotection of hippocampus in APP/PS1 transgenic mice via upregulation of mitochondrial 8-oxoguanine DNA glycosylase. Oxidative Med Cell Longev 2014:834502. https://doi.org/10.1155/2014/834502

Booth FW, Ruegsegger GN, Toedebusch RG, Yan Z (2015) Endurance exercise and the regulation of skeletal muscle metabolism. Prog Mol Biol Transl Sci 135:129–151

Boström P, Wu J, Jedrychowski MP et al (2012) A PGC1-α-dependent myokine that drives brown-fat-like development of white fat and thermogenesis. Nature 481:463–468. https://doi.org/10.1038/nature10777

Boveris A, Navarro A (2008) Systemic and mitochondrial adaptive responses to moderate exercise in rodents. Free Radic Biol Med 44:224–229. https://doi.org/10.1016/j.freeradbiomed.2007.08.015

Brown DA, Chicco AJ, Jew KN et al (2005) Cardioprotection afforded by chronic exercise is mediated by the sarcolemmal, and not the mitochondrial, isoform of the K_{ATP} channel in the rat. J Physiol 569:913–924. https://doi.org/10.1113/jphysiol.2005.095729

Brown DA, Perry JB, Allen ME et al (2016) Expert consensus document: mitochondrial function as a therapeutic target in heart failure. Nat Rev Cardiol 14:238–250. https://doi.org/10.1038/nrcardio.2016.203

Burgomaster KA, Howarth KR, Phillips SM et al (2008) Similar metabolic adaptations during exercise after low volume sprint interval and traditional endurance training in humans. J Physiol 586:151–160. https://doi.org/10.1113/jphysiol.2007.142109

Campos JC, Queliconi BB, Dourado PMM et al (2012) Exercise training restores cardiac protein quality control in heart failure. PLoS One 7:e52764. https://doi.org/10.1371/journal.pone.0052764

Cantó C, Gerhart-Hines Z, Feige JN et al (2009) AMPK regulates energy expenditure by modulating NAD+ metabolism and SIRT1 activity. Nature 458:1056–1060. https://doi.org/10.1038/nature07813

Cao W, Daniel KW, Robidoux J et al (2004) p38 mitogen-activated protein kinase is the central regulator of cyclic AMP-dependent transcription of the brown fat uncoupling protein 1 gene. Mol Cell Biol 24:3057–3067. https://doi.org/10.1128/mcb.24.7.3057-3067.2004

Carter HN, Chen CCW, Hood DA (2015) Mitochondria, muscle health, and exercise with advancing age. Physiology 30:208–223. https://doi.org/10.1152/physiol.00039.2014
Cartoni R, Léger B, Hock MB et al (2005) Mitofusins 1/2 and ERRalpha expression are increased in human skeletal muscle after physical exercise. J Physiol 567:349–358. https://doi.org/10.1113/jphysiol.2005.092031
Cheng A, Yang Y, Zhou Y et al (2016) Mitochondrial SIRT3 mediates adaptive responses of neurons to exercise and metabolic and excitatory challenges. Cell Metab 23:128–142. https://doi.org/10.1016/j.cmet.2015.10.013
Chicco AJ, Schneider CM, Hayward R (2005) Voluntary exercise protects against acute doxorubicin cardiotoxicity in the isolated perfused rat heart. Am J Physiol Regul Integr Comp Physiol 289:R424–R431. https://doi.org/10.1152/ajpregu.00636.2004
Chicco AJ, Hydock DS, Schneider CM, Hayward R (2006) Low-intensity exercise training during doxorubicin treatment protects against cardiotoxicity. J Appl Physiol 100:519–527. https://doi.org/10.1152/japplphysiol.00148.2005
Chung E, Joiner HE, Skelton T et al (2017) Maternal exercise upregulates mitochondrial gene expression and increases enzyme activity of fetal mouse hearts. Physiol Rep 5:e13184. https://doi.org/10.14814/phy2.13184
Ciminelli M, Ascah A, Bourduas K, Burelle Y (2006) Short term training attenuates opening of the mitochondrial permeability transition pore without affecting myocardial function following ischemia-reperfusion. Mol Cell Biochem 291:39–47. https://doi.org/10.1007/s11010-006-9192-9
Cribbs JT, Strack S (2007) Reversible phosphorylation of Drp1 by cyclic AMP-dependent protein kinase and calcineurin regulates mitochondrial fission and cell death. EMBO Rep 8:939–944. https://doi.org/10.1038/sj.embor.7401062
Dembele K, Nguyen KH, Hernandez TA, Nyomba BLG (2009) Effects of ethanol on pancreatic beta-cell death: interaction with glucose and fatty acids. Cell Biol Toxicol 25:141–152. https://doi.org/10.1007/s10565-008-9067-9
Denham J, Marques FZ, O'Brien BJ, Charchar FJ (2014) Exercise: putting action into our epigenome. Sport Med 44:189–209. https://doi.org/10.1007/s40279-013-0114-1
Dietrich MO, Andrews ZB, Horvath TL (2008) Exercise-induced synaptogenesis in the hippocampus is dependent on UCP2-regulated mitochondrial adaptation. J Neurosci 28:10766–10771. https://doi.org/10.1523/JNEUROSCI.2744-08.2008
Dolinsky VW, Rogan KJ, Sung MM et al (2013) Both aerobic exercise and resveratrol supplementation attenuate doxorubicin-induced cardiac injury in mice. AJP Endocrinol Metab 305:E243–E253. https://doi.org/10.1152/ajpendo.00044.2013
Donovan EL, Miller BF (2011) Exercise during pregnancy: developmental origins of disease prevention? Exerc Sport Sci Rev 39:111. https://doi.org/10.1097/JES.0b013e31821f7d78
Dröge W (2002) Free radicals in the physiological control of cell function. Physiol Rev 82:47–95. https://doi.org/10.1152/physrev.00018.2001
Dudley GA, Abraham WM, Terjung RL (1982) Influence of exercise intensity and duration on biochemical adaptations in skeletal muscle. J Appl Physiol 53:844–850
Egan B, Zierath JR (2013) Exercise metabolism and the molecular regulation of skeletal muscle adaptation. Cell Metab 17:162–184. https://doi.org/10.1016/j.cmet.2012.12.012
Egan D, Kim J, Shaw RJ, Guan K-L (2011) The autophagy initiating kinase ULK1 is regulated via opposing phosphorylation by AMPK and mTOR. Autophagy 7:643–644
Figueiredo PA, Powers SK, Ferreira RM et al (2009) Impact of lifelong sedentary behavior on mitochondrial function of mice skeletal muscle. J Gerontol Ser A Biol Sci Med Sci 64A:927–939. https://doi.org/10.1093/gerona/glp066
Fiuza-Luces C, Garatachea N, Berger NA, Lucia A (2013) Exercise is the real polypill. Physiology 28:330–358. https://doi.org/10.1152/physiol.00019.2013
Frank S, Gaume B, Bergmann-Leitner ES et al (2001) The role of dynamin-related protein 1, a mediator of mitochondrial fission, in apoptosis. Dev Cell 1:515–525
Fromenty B, Robin MA, Igoudjil A et al (2004) The ins and outs of mitochondrial dysfunction in NASH. Diabetes Metab 30:121–138

Garnier A, Fortin D, Zoll J et al (2005) Coordinated changes in mitochondrial function and biogenesis in healthy and diseased human skeletal muscle. FASEB J 19:43–52. https://doi.org/10.1096/fj.04-2173com

Gibala MJ, MacLean DA, Graham TE, Saltin B (1998) Tricarboxylic acid cycle intermediate pool size and estimated cycle flux in human muscle during exercise. Am J Phys 275:E235–E242

Gomez-Cabrera M-C, Domenech E, Romagnoli M et al (2008) Oral administration of vitamin C decreases muscle mitochondrial biogenesis and hampers training-induced adaptations in endurance performance. Am J Clin Nutr 87:142–149

Gonçalves IO, Oliveira PJ, Ascensão A, Magalhães J (2013) Exercise as a therapeutic tool to prevent mitochondrial degeneration in nonalcoholic steatohepatitis. Eur J Clin Invest 43(11):1184–1194. https://doi.org/10.1111/eci.12146

Gonçalves IO, Maciel E, Passos E et al (2014a) Exercise alters liver mitochondria phospholipidomic profile and mitochondrial activity in non-alcoholic steatohepatitis. Int J Biochem Cell Biol 54:163–173. https://doi.org/10.1016/j.biocel.2014.07.011

Gonçalves IO, Passos E, Rocha-Rodrigues S et al (2014b) Physical exercise prevents and mitigates non-alcoholic steatohepatitis-induced liver mitochondrial structural and bioenergetics impairments. Mitochondrion 15:40–51. https://doi.org/10.1016/j.mito.2014.03.012

Gonçalves IO, Passos E, Diogo CV et al (2016) Exercise mitigates mitochondrial permeability transition pore and quality control mechanisms alterations in nonalcoholic steatohepatitis. Appl Physiol Nutr Metab 41:298–306. https://doi.org/10.1139/apnm-2015-0470

Gounder SS, Kannan S, Devadoss D et al (2012) Impaired transcriptional activity of Nrf2 in age-related myocardial oxidative stress is reversible by moderate exercise training. PLoS One 7:e45697. https://doi.org/10.1371/journal.pone.0045697

Gusdon AM, Callio J, Distefano G et al (2017) Exercise increases mitochondrial complex I activity and DRP1 expression in the brains of aged mice. Exp Gerontol 90:1–13. https://doi.org/10.1016/j.exger.2017.01.013

Haase TN, Ringholm S, Leick L et al (2011) Role of PGC-1 in exercise and fasting-induced adaptations in mouse liver. AJP Regul Integr Comp Physiol 301:R1501–R1509. https://doi.org/10.1152/ajpregu.00775.2010

Henriksson J (1977) Training induced adaptation of skeletal muscle and metabolism during submaximal exercise. J Physiol 270:661–675

Herbst EAF, Roussakis C, Matravadia S, Holloway GP (2015) Chronic treadmill running does not enhance mitochondrial oxidative capacity in the cortex or striatum. Metabolism 64:1419–1425. https://doi.org/10.1016/j.metabol.2015.07.002

Hickson RC (1981) Skeletal muscle cytochrome c and myoglobin, endurance, and frequency of training. J Appl Physiol 51:746–749

Hoffmann C, Weigert C (2017) Skeletal muscle as an endocrine organ: the role of myokines in exercise adaptations. Cold Spring Harb Perspect Med 7(11):a029793. https://doi.org/10.1101/cshperspect.a029793

Holloszy JO (1967) Biochemical adaptations in muscle. Effects of exercise on mitochondrial oxygen uptake and respiratory enzyme activity in skeletal muscle. J Biol Chem 242:2278–2282

Holloszy JO, Coyle EF (1984) Adaptations of skeletal muscle to endurance exercise and their metabolic consequences. J Appl Physiol 56:831–838

Hood DA, Tryon LD, Carter HN et al (2016) Unravelling the mechanisms regulating muscle mitochondrial biogenesis. Biochem J 473:2295–2314. https://doi.org/10.1042/BCJ20160009

Hoppeler H, Howald H, Conley K et al (1985) Endurance training in humans: aerobic capacity and structure of skeletal muscle. J Appl Physiol 59:320–327

Horowitz JF, Leone TC, Feng W et al (2000) Effect of endurance training on lipid metabolism in women: a potential role for PPARalpha in the metabolic response to training. Am J Physiol Endocrinol Metab 279:E348–E355

Iizuka K, Machida T, Hirafuji M (2014) Skeletal muscle is an endocrine organ. J Pharmacol Sci 125:125–131

Ingjer F (1979) Capillary supply and mitochondrial content of different skeletal muscle fiber types in untrained and endurance-trained men. A histochemical and ultrastructural study. Eur J Appl Physiol Occup Physiol 40:197–209

Jacobs RA, Lundby C (2013) Mitochondria express enhanced quality as well as quantity in association with aerobic fitness across recreationally active individuals up to elite athletes. J Appl Physiol 114:344–350. https://doi.org/10.1152/japplphysiol.01081.2012

Jäger S, Handschin C, St-Pierre J, Spiegelman BM (2007) AMP-activated protein kinase (AMPK) action in skeletal muscle via direct phosphorylation of PGC-1alpha. Proc Natl Acad Sci U S A 104:12017–12022. https://doi.org/10.1073/pnas.0705070104

Jansson E, Kaijser L (1987) Substrate utilization and enzymes in skeletal muscle of extremely endurance-trained men. J Appl Physiol 62:999–1005

Jendrach M, Mai S, Pohl S et al (2008) Short- and long-term alterations of mitochondrial morphology, dynamics and mtDNA after transient oxidative stress. Mitochondrion 8:293–304. https://doi.org/10.1016/j.mito.2008.06.001

Ji LL, Kang C, Zhang Y (2016) Exercise-induced hormesis and skeletal muscle health. Free Radic Biol Med 98:113–122. https://doi.org/10.1016/j.freeradbiomed.2016.02.025

Jiménez-Chillarón JC, Nijland MJ, Ascensão AA et al (2015) Back to the future: transgenerational transmission of xenobiotic-induced epigenetic remodeling. Epigenetics 10:259–273. https://doi.org/10.1080/15592294.2015.1020267

Joseph A-M, Adhihetty PJ, Leeuwenburgh C (2016) Beneficial effects of exercise on age-related mitochondrial dysfunction and oxidative stress in skeletal muscle. J Physiol 594:5105–5123. https://doi.org/10.1113/JP270659

Kang C, O'Moore KM, Dickman JR, Ji LL (2009) Exercise activation of muscle peroxisome proliferator-activated receptor-γ coactivator-1α signaling is redox sensitive. Free Radic Biol Med 47:1394–1400. https://doi.org/10.1016/j.freeradbiomed.2009.08.007

Kang C, Chung E, Diffee G, Ji LL (2013) Exercise training attenuates aging-associated mitochondrial dysfunction in rat skeletal muscle: role of PGC-1α. Exp Gerontol 48:1343–1350. https://doi.org/10.1016/j.exger.2013.08.004

Kapilevich LV, Kironenko TA, Zaharova AN et al (2015) Skeletal muscle as an endocrine organ: role of [Na+]i/[K+]i-mediated excitation-transcription coupling. Genes Dis 2:328–336. https://doi.org/10.1016/j.gendis.2015.10.001

Kavazis AN, McClung JM, Hood DA, Powers SK (2008) Exercise induces a cardiac mitochondrial phenotype that resists apoptotic stimuli. Am J Physiol Heart Circ Physiol 294:H928–H935. https://doi.org/10.1152/ajpheart.01231.2007

Kavazis AN, Alvarez S, Talbert E et al (2009) Exercise training induces a cardioprotective phenotype and alterations in cardiac subsarcolemmal and intermyofibrillar mitochondrial proteins. AJP Heart Circ Physiol 297:H144–H152. https://doi.org/10.1152/ajpheart.01278.2008

Kavazis AN, Smuder AJ, Min K et al (2010) Short-term exercise training protects against doxorubicin-induced cardiac mitochondrial damage independent of HSP72. Am J Physiol Heart Circ Physiol 299:H1515–H1524. https://doi.org/10.1152/ajpheart.00585.2010

Kavazis AN, Smuder AJ, Powers SK (2014) Effects of short-term endurance exercise training on acute doxorubicin-induced FoxO transcription in cardiac and skeletal muscle. J Appl Physiol 117:223–230. https://doi.org/10.1152/japplphysiol.00210.2014

Kirby TJ, McCarthy JJ (2013) MicroRNAs in skeletal muscle biology and exercise adaptation. Free Radic Biol Med 64:95–105. https://doi.org/10.1016/j.freeradbiomed.2013.07.004

Klotz L-O, Sánchez-Ramos C, Prieto-Arroyo I et al (2015) Redox regulation of FoxO transcription factors. Redox Biol 6:51–72. https://doi.org/10.1016/j.redox.2015.06.019

Koltai E, Hart N, Taylor AW et al (2012) Age-associated declines in mitochondrial biogenesis and protein quality control factors are minimized by exercise training. AJP Regul Integr Comp Physiol 303:R127–R134. https://doi.org/10.1152/ajpregu.00337.2011

Konopka AR, Suer MK, Wolff CA, Harber MP (2014) Markers of human skeletal muscle mitochondrial biogenesis and quality control: effects of age and aerobic exercise training. J Gerontol Ser A Biol Sci Med Sci 69:371–378. https://doi.org/10.1093/gerona/glt107

Laker RC, Lillard TS, Okutsu M et al (2014) Exercise prevents maternal high-fat diet-induced hypermethylation of the *Pgc-1α* gene and age-dependent metabolic dysfunction in the offspring. Diabetes 63:1605–1611. https://doi.org/10.2337/db13-1614

Larsen S, Nielsen J, Hansen CN et al (2012) Biomarkers of mitochondrial content in skeletal muscle of healthy young human subjects. J Physiol 590:3349–3360. https://doi.org/10.1113/jphysiol.2012.230185

Larsen S, Danielsen JH, Søndergård SD et al (2015) The effect of high-intensity training on mitochondrial fat oxidation in skeletal muscle and subcutaneous adipose tissue. Scand J Med Sci Sports 25:e59–e69. https://doi.org/10.1111/sms.12252

Leboucher GP, Tsai YC, Yang M et al (2012) Stress-induced phosphorylation and proteasomal degradation of mitofusin 2 facilitates mitochondrial fragmentation and apoptosis. Mol Cell 47:547–557. https://doi.org/10.1016/j.molcel.2012.05.041

Lee Y, Min K, Talbert EE et al (2012) Exercise protects cardiac mitochondria against ischemia-reperfusion injury. Med Sci Sports Exerc 44:397–405. https://doi.org/10.1249/MSS.0b013e318231c037

Leichtweis SB, Leeuwenburgh C, Parmelee DJ et al (1997) Rigorous swim training impairs mitochondrial function in post-ischaemic rat heart. Acta Physiol Scand 160:139–148. https://doi.org/10.1046/j.1365-201X.1997.00138.x

Leick L, Lyngby SS, Wojtasewski JF et al (2010) PGC-1α is required for training-induced prevention of age-associated decline in mitochondrial enzymes in mouse skeletal muscle. Exp Gerontol 45:336–342. https://doi.org/10.1016/j.exger.2010.01.011

Lin J, Wu H, Tarr PT et al (2002) Transcriptional co-activator PGC-1α drives the formation of slow-twitch muscle fibres. Nature 418:797–801. https://doi.org/10.1038/nature00904

Lira VA, Okutsu M, Zhang M et al (2013) Autophagy is required for exercise training-induced skeletal muscle adaptation and improvement of physical performance. FASEB J 27:4184–4193. https://doi.org/10.1096/fj.13-228486

Little JP, Safdar A, Benton CR, Wright DC (2011) Skeletal muscle and beyond: the role of exercise as a mediator of systemic mitochondrial biogenesis. Appl Physiol Nutr Metab 36:598–607. https://doi.org/10.1139/h11-076

van Loon LJ, Greenhaff PL, Constantin-Teodosiu D et al (2001) The effects of increasing exercise intensity on muscle fuel utilisation in humans. J Physiol 536:295–304. https://doi.org/10.1111/j.1469-7793.2001.00295.x

Lumini JA, Magalhães J, Oliveira PJ, Ascensão A (2008) Beneficial effects of exercise on muscle mitochondrial function in diabetes mellitus. Sports Med 38:735–750

Lumini-Oliveira J, Magalhães J, Pereira CV et al (2011) Endurance training reverts heart mitochondrial dysfunction, permeability transition and apoptotic signaling in long-term severe hyperglycemia. Mitochondrion 11:54–63. https://doi.org/10.1016/j.mito.2010.07.005

Lundby C, Jacobs RA (2016) Adaptations of skeletal muscle mitochondria to exercise training. Exp Physiol 101:17–22. https://doi.org/10.1113/EP085319

MacInnis MJ, Zacharewicz E, Martin BJ et al (2016) Superior mitochondrial adaptations in human skeletal muscle after interval compared to continuous single-leg cycling matched for total work. J Physiol. https://doi.org/10.1113/JP272570

Magalhães J, Falcão-Pires I, Gonçalves IO et al (2013) Synergistic impact of endurance training and intermittent hypobaric hypoxia on cardiac function and mitochondrial energetic and signaling. Int J Cardiol 168:5363–5371. https://doi.org/10.1016/j.ijcard.2013.08.001

Magalhães J, Gonçalves IO, Lumini-Oliveira J et al (2014) Modulation of cardiac mitochondrial permeability transition and apoptotic signaling by endurance training and intermittent hypobaric hypoxia. Int J Cardiol 173:40–45. https://doi.org/10.1016/j.ijcard.2014.02.011

Marcelino TB, Longoni A, Kudo KY et al (2013) Evidences that maternal swimming exercise improves antioxidant defenses and induces mitochondrial biogenesis in the brain of young Wistar rats. Neuroscience 246:28–39. https://doi.org/10.1016/j.neuroscience.2013.04.043

Marcil M, Bourduas K, Ascah A, Burelle Y (2005) Exercise training induces respiratory substrate-specific decrease in Ca2+-induced permeability transition pore opening in heart mitochondria. AJP Heart Circ Physiol 290:H1549–H1557. https://doi.org/10.1152/ajpheart.00913.2005

Marques-Aleixo I, Oliveira PJ, Moreira PI et al (2012) Physical exercise as a possible strategy for brain protection: evidence from mitochondrial-mediated mechanisms. Prog Neurobiol 99:149–162. https://doi.org/10.1016/j.pneurobio.2012.08.002

Marques-Aleixo I, Santos-Alves E, Balça MM et al (2015a) Physical exercise improves brain cortex and cerebellum mitochondrial bioenergetics and alters apoptotic, dynamic and auto(mito)phagy markers. Neuroscience 301:480–495. https://doi.org/10.1016/j.neuroscience.2015.06.027

Marques-Aleixo I, Santos-Alves E, Mariani D et al (2015b) Physical exercise prior and during treatment reduces sub-chronic doxorubicin-induced mitochondrial toxicity and oxidative stress. Mitochondrion 20:22–33. https://doi.org/10.1016/j.mito.2014.10.008

Marques-Aleixo I, Santos-Alves E, Balça MM et al (2016) Physical exercise mitigates doxorubicin-induced brain cortex and cerebellum mitochondrial alterations and cellular quality control signaling. Mitochondrion 26:43–57. https://doi.org/10.1016/j.mito.2015.12.002

Matiello R, Fukui RT, Silva ME et al (2010) Differential regulation of PGC-1alpha expression in rat liver and skeletal muscle in response to voluntary running. Nutr Metab (Lond) 7:36. https://doi.org/10.1186/1743-7075-7-36

McGee SL, Fairlie E, Garnham AP, Hargreaves M (2009) Exercise-induced histone modifications in human skeletal muscle. J Physiol 587:5951–5958. https://doi.org/10.1113/jphysiol.2009.181065

Morgan MJ, Liu Z (2011) Crosstalk of reactive oxygen species and NF-κB signaling. Cell Res 21:103–115. https://doi.org/10.1038/cr.2010.178

Mottillo EP, Desjardins EM, Crane JD et al (2016) Lack of adipocyte AMPK exacerbates insulin resistance and hepatic steatosis through Brown and Beige adipose tissue function. Cell Metab 24:118–129. https://doi.org/10.1016/j.cmet.2016.06.006

Musman J, Pons S, Barau C et al (2016) Regular treadmill exercise inhibits mitochondrial accumulation of cholesterol and oxysterols during myocardial ischemia-reperfusion in wild-type and ob/ob mice. Free Radic Biol Med 101:317–324. https://doi.org/10.1016/j.freeradbiomed.2016.10.496

Norheim F, Langleite TM, Hjorth M et al (2014) The effects of acute and chronic exercise on PGC-1α, irisin and browning of subcutaneous adipose tissue in humans. FEBS J 281:739–749. https://doi.org/10.1111/febs.12619

Ost M, Coleman V, Kasch J, Klaus S (2016) Regulation of myokine expression: role of exercise and cellular stress. Free Radic Biol Med 98:78–89. https://doi.org/10.1016/j.freeradbiomed.2016.02.018

Pagano AF, Py G, Bernardi H et al (2014) Autophagy and protein turnover signaling in slow-twitch muscle during exercise. Med Sci Sports Exerc 46:1314–1325. https://doi.org/10.1249/MSS.0000000000000237

Pareja-Galeano H, Sanchis-Gomar F, García-Giménez JL (2014) Physical exercise and epigenetic modulation: elucidating intricate mechanisms. Sport Med 44:429–436. https://doi.org/10.1007/s40279-013-0138-6

Park J-W, Kim M-H, Eo S-J et al (2013) Maternal exercise during pregnancy affects mitochondrial enzymatic activity and biogenesis in offspring brain. Int J Neurosci 123:253–264. https://doi.org/10.3109/00207454.2012.755969

Pedersen BK, Saltin B (2015) Exercise as medicine – evidence for prescribing exercise as therapy in 26 different chronic diseases. Scand J Med Sci Sports 25:1–72. https://doi.org/10.1111/sms.12581

Pedersen BK, Steensberg A, Fischer C et al (2003) Searching for the exercise factor: is IL-6 a candidate? J Muscle Res Cell Motil 24:113–119

Perry CGR, Lally J, Holloway GP et al (2010) Repeated transient mRNA bursts precede increases in transcriptional and mitochondrial proteins during training in human skeletal muscle. J Physiol 588:4795–4810. https://doi.org/10.1113/jphysiol.2010.199448

Pessayre D, Mansouri A, Berson A, Fromenty B (2010) Mitochondrial involvement in drug-induced liver injury. Handb Exp Pharmacol 196:311–365

Powers SK, Smuder AJ, Kavazis AN, Quindry JC (2014a) Mechanisms of exercise-induced cardioprotection. Physiology 29:27–38. https://doi.org/10.1152/physiol.00030.2013

Powers SK, Sollanek KJ, Wiggs MP et al (2014b) Exercise-induced improvements in myocardial antioxidant capacity: the antioxidant players and cardioprotection. Free Radic Res 48:43–51. https://doi.org/10.3109/10715762.2013.825371

Puigserver P, Rhee J, Lin J et al (2001) Cytokine stimulation of energy expenditure through p38 MAP kinase activation of PPARgamma coactivator-1. Mol Cell 8:971–982

Quindry JC, Schreiber L, Hosick P et al (2010) Mitochondrial KATP channel inhibition blunts arrhythmia protection in ischemic exercised hearts. Am J Physiol Heart Circ Physiol 299:H175–H183. https://doi.org/10.1152/ajpheart.01211.2009

Radak Z, Ishihara K, Tekus E et al (2017) Exercise, oxidants, and antioxidants change the shape of the bell-shaped hormesis curve. Redox Biol 12:285–290. https://doi.org/10.1016/j.redox.2017.02.015

Rector RS, Thyfault JP (2011) Does physical inactivity cause nonalcoholic fatty liver disease? J Appl Physiol 111:1828–1835. https://doi.org/10.1152/japplphysiol.00384.2011

Reid MB, Khawli FA, Moody MR (1993) Reactive oxygen in skeletal muscle. III. Contractility of unfatigued muscle. J Appl Physiol 75:1081–1087

Ristow M, Zarse K, Oberbach A et al (2009) Antioxidants prevent health-promoting effects of physical exercise in humans. Proc Natl Acad Sci U S A 106:8665–8670. https://doi.org/10.1073/pnas.0903485106

Rocha-Rodrigues S, Rodríguez A, Gouveia AM et al (2016) Effects of physical exercise on myokines expression and brown adipose-like phenotype modulation in rats fed a high-fat diet. Life Sci 165:100–108. https://doi.org/10.1016/j.lfs.2016.09.023

Rocha-Rodrigues S, Rodríguez A, Becerril S et al (2017) Physical exercise remodels visceral adipose tissue and mitochondrial lipid metabolism in rats fed a high-fat diet. Clin Exp Pharmacol Physiol 44:386–394. https://doi.org/10.1111/1440-1681.12706

Rodas G, Ventura JL, Cadefau JA et al (2000) A short training programme for the rapid improvement of both aerobic and anaerobic metabolism. Eur J Appl Physiol 82:480–486. https://doi.org/10.1007/s004210000223

Roh J, Rhee J, Chaudhari V, Rosenzweig A (2016) The role of exercise in cardiac aging. Circ Res 118:279–295. https://doi.org/10.1161/CIRCRESAHA.115.305250

Ruderman NB, Xu XJ, Nelson L et al (2010) AMPK and SIRT1: a long-standing partnership? Am J Physiol Endocrinol Metab 298:E751–E760. https://doi.org/10.1152/ajpendo.00745.2009

Russell AP, Lamon S (2015) Exercise, skeletal muscle and circulating microRNAs. Prog Mol Biol Transl Sci 135:471–496

Safdar A, Bourgeois JM, Ogborn DI et al (2011) Endurance exercise rescues progeroid aging and induces systemic mitochondrial rejuvenation in mtDNA mutator mice. Proc Natl Acad Sci 108:4135–4140. https://doi.org/10.1073/pnas.1019581108

Safdar A, Khrapko K, Flynn JM et al (2015) Exercise-induced mitochondrial p53 repairs mtDNA mutations in mutator mice. Skelet Muscle 6:7. https://doi.org/10.1186/s13395-016-0075-9

Saleem A, Hood DA (2013) Acute exercise induces tumour suppressor protein p53 translocation to the mitochondria and promotes a p53-Tfam-mitochondrial DNA complex in skeletal muscle. J Physiol 591:3625–3636. https://doi.org/10.1113/jphysiol.2013.252791

Saleem A, Adhihetty PJ, Hood DA (2009) Role of p53 in mitochondrial biogenesis and apoptosis in skeletal muscle. Physiol Genomics 37:58–66. https://doi.org/10.1152/physiolgenomics.90346.2008

Santos-Alves E, Marques-Aleixo I, Coxito P et al (2014) Exercise mitigates diclofenac-induced liver mitochondrial dysfunction. Eur J Clin Investig 44:668–677. https://doi.org/10.1111/eci.12285

Santos-Alves E, Marques-Aleixo I, Rizo-Roca D et al (2015) Exercise modulates liver cellular and mitochondrial proteins related to quality control signaling. Life Sci 135:124–130. https://doi.org/10.1016/j.lfs.2015.06.007

Scarpulla RC (2011) Metabolic control of mitochondrial biogenesis through the PGC-1 family regulatory network. Biochim Biophys Acta 1813:1269–1278. https://doi.org/10.1016/j.bbamcr.2010.09.019

Schantz P, Henriksson J, Jansson E (1983) Adaptation of human skeletal muscle to endurance training of long duration. Clin Physiol 3:141–151

Schantz PG, Sjoberg B, Svedenhag J (1986) Malate-aspartate and alpha-glycerophosphate shuttle enzyme levels in human skeletal muscle: methodological considerations and effect of endurance training. Acta Physiol Scand 128:397–407. https://doi.org/10.1111/j.1748-1716.1986.tb07993.x

Schnyder S, Handschin C (2015) Skeletal muscle as an endocrine organ: PGC-1α, myokines and exercise. Bone 80:115–125. https://doi.org/10.1016/j.bone.2015.02.008

Schwerzmann K, Hoppeler H, Kayar SR, Weibel ER (1989) Oxidative capacity of muscle and mitochondria: correlation of physiological, biochemical, and morphometric characteristics. Proc Natl Acad Sci 86:1583–1587. https://doi.org/10.1073/pnas.86.5.1583

Scott HA, Latham JR, Callister R et al (2015) Acute exercise is associated with reduced exhaled nitric oxide in physically inactive adults with asthma. Ann Allergy Asthma Immunol 114:470–479. https://doi.org/10.1016/j.anai.2015.04.002

Simon H-U, Haj-Yehia A, Levi-Schaffer F (2000) Role of reactive oxygen species (ROS) in apoptosis induction. Apoptosis 5:415–418. https://doi.org/10.1023/A:1009616228304

Siu PM, Bryner RW, Martyn JK, Alway SE (2004) Apoptotic adaptations from exercise training in skeletal and cardiac muscles. FASEB J 18:1150–1152. https://doi.org/10.1096/fj.03-1291fje

Smuder AJ, Kavazis AN, Min K, Powers SK (2013) Doxorubicin-induced markers of myocardial autophagic signaling in sedentary and exercise trained animals. J Appl Physiol 115:176–185. https://doi.org/10.1152/japplphysiol.00924.2012

Soriano FX, Liesa M, Bach D et al (2006) Evidence for a mitochondrial regulatory pathway defined by peroxisome proliferator-activated receptor- coactivator-1 , estrogen-related receptor- , and mitofusin 2. Diabetes 55:1783–1791. https://doi.org/10.2337/db05-0509

Stallknecht B, Vinten J, Ploug T, Galbo H (1991) Increased activities of mitochondrial enzymes in white adipose tissue in trained rats. Am J Phys 261:E410–E414

Stanford KI, Middelbeek RJW, Goodyear LJ (2015) Exercise effects on white adipose tissue: beiging and metabolic adaptations. Diabetes 64:2361–2368. https://doi.org/10.2337/db15-0227

Steiner JL, Murphy EA, McClellan JL et al (2011) Exercise training increases mitochondrial biogenesis in the brain. J Appl Physiol 111:1066–1071. https://doi.org/10.1152/japplphysiol.00343.2011

St-Pierre J, Drori S, Uldry M et al (2006) Suppression of reactive oxygen species and neurodegeneration by the PGC-1 transcriptional coactivators. Cell 127:397–408. https://doi.org/10.1016/j.cell.2006.09.024

Suen D-F, Norris KL, Youle RJ (2008) Mitochondrial dynamics and apoptosis. Genes Dev 22:1577–1590. https://doi.org/10.1101/gad.1658508

Sun M, Huang C, Wang C et al (2013) Ginsenoside Rg3 improves cardiac mitochondrial population quality: mimetic exercise training. Biochem Biophys Res Commun 441:169–174. https://doi.org/10.1016/j.bbrc.2013.10.039

Sutherland LN, Bomhof MR, Capozzi LC et al (2009) Exercise and adrenaline increase PGC-1α mRNA expression in rat adipose tissue. J Physiol 587:1607–1617. https://doi.org/10.1113/jphysiol.2008.165464

Taanman J-W (1999) The mitochondrial genome: structure, transcription, translation and replication. Biochim Biophys Acta Bioenerg 1410:103–123. https://doi.org/10.1016/S0005-2728(98)00161-3

Taghizadeh G, Pourahmad J, Mehdizadeh H et al (2016) Protective effects of physical exercise on MDMA-induced cognitive and mitochondrial impairment. Free Radic Biol Med 99:11–19. https://doi.org/10.1016/j.freeradbiomed.2016.07.018

Talanian JL, Holloway GP, Snook LA et al (2010) Exercise training increases sarcolemmal and mitochondrial fatty acid transport proteins in human skeletal muscle. Am J Physiol Endocrinol Metab 299:E180–E188. https://doi.org/10.1152/ajpendo.00073.2010

Terblanche SE, Gohil K, Packer L et al (2001) The effects of endurance training and exhaustive exercise on mitochondrial enzymes in tissues of the rat (Rattus norvegicus). Comp Biochem Physiol A Mol Integr Physiol 128:889–896

Thyfault JP, Rector RS, Uptergrove GM et al (2009) Rats selectively bred for low aerobic capacity have reduced hepatic mitochondrial oxidative capacity and susceptibility to hepatic steatosis and injury. J Physiol 587:1805–1816. https://doi.org/10.1113/jphysiol.2009.169060

Townsend LK, Knuth CM, Wright DC (2017) Cycling our way to fit fat. Physiol Rep 5:e13247. https://doi.org/10.14814/phy2.13247

Trevellin E, Scorzeto M, Olivieri M et al (2014) Exercise training induces mitochondrial biogenesis and glucose uptake in subcutaneous adipose tissue through eNOS-dependent mechanisms. Diabetes 63:2800–2811. https://doi.org/10.2337/db13-1234

Vainshtein A, Kazak L, Hood DA (2011) Effects of endurance training on apoptotic susceptibility in striated muscle. J Appl Physiol 110:1638–1645. https://doi.org/10.1152/japplphysiol.00020.2011

Vainshtein A, Tryon LD, Pauly M, Hood DA (2015) Role of PGC-1α during acute exercise-induced autophagy and mitophagy in skeletal muscle. Am J Physiol Cell Physiol 308:C710–C719. https://doi.org/10.1152/ajpcell.00380.2014

Wang C, Youle RJ (2009) The role of mitochondria in apoptosis. Annu Rev Genet 43:95–118. https://doi.org/10.1146/annurev-genet-102108-134850

Wei Y, Rector RS, Thyfault JP, Ibdah JA (2008) Nonalcoholic fatty liver disease and mitochondrial dysfunction. World J Gastroenterol 14:193–199

Westermann B (2012) Bioenergetic role of mitochondrial fusion and fission. Biochim Biophys Acta Bioenerg 1817:1833–1838. https://doi.org/10.1016/j.bbabio.2012.02.033

Whelan RS, Konstantinidis K, Wei A-C et al (2012) Bax regulates primary necrosis through mitochondrial dynamics. Proc Natl Acad Sci 109:6566–6571. https://doi.org/10.1073/pnas.1201608109

White Z, Terrill J, White RB et al (2016) Voluntary resistance wheel exercise from mid-life prevents sarcopenia and increases markers of mitochondrial function and autophagy in muscles of old male and female C57BL/6J mice. Skelet Muscle 6:45. https://doi.org/10.1186/s13395-016-0117-3

Wright DC, Geiger PC, Han D-H et al (2007) Calcium induces increases in peroxisome proliferator-activated receptor coactivator-1 and mitochondrial biogenesis by a pathway leading to p38 mitogen-activated protein kinase activation. J Biol Chem 282:18793–18799. https://doi.org/10.1074/jbc.M611252200

Xu Y, Zhao C, Sun X et al (2015) MicroRNA-761 regulates mitochondrial biogenesis in mouse skeletal muscle in response to exercise. Biochem Biophys Res Commun 467:103–108. https://doi.org/10.1016/j.bbrc.2015.09.113

Xu WH, Wu H, Xia WL et al (2016) Physical exercise before pregnancy helps the development of mouse embryos produced in vitro. Mitochondrion. https://doi.org/10.1016/j.mito.2016.12.004

Yamamoto H, Morino K, Nishio Y et al (2012) MicroRNA-494 regulates mitochondrial biogenesis in skeletal muscle through mitochondrial transcription factor A and Forkhead box j3. AJP Endocrinol Metab 303:E1419–E1427. https://doi.org/10.1152/ajpendo.00097.2012

Yu T, Robotham JL, Yoon Y (2006) Increased production of reactive oxygen species in hyperglycemic conditions requires dynamic change of mitochondrial morphology. Proc Natl Acad Sci U S A 103(8):2653. https://doi.org/10.1073/pnas.0511154103

Zampieri S, Mammucari C, Romanello V et al (2016) Physical exercise in aging human skeletal muscle increases mitochondrial calcium uniporter expression levels and affects mitochondria dynamics. Physiol Rep 4(24):e13005. https://doi.org/10.14814/phy2.13005

Modulation of Cell Fate by Tauroursodeoxycholic Acid: All Paths Lead to Mitochondria

Susana Solá, Maria F. Ribeiro, Tânia Genebra, and Cecília M. P. Rodrigues

Abstract Specific endogenous bile acids, such as ursodeoxycholic (UDCA) acid and its taurine conjugated form, tauroursodeoxycholic acid (TUDCA), are potent modulators of cell fate by regulating pathways that involve mitochondria. Curiously, emerging evidence suggests that mitochondrial changes induced by TUDCA result from its influence on mitochondrial redox state, mitochondrial membrane permeabilization, mitochondrial apoptosis and mitophagy. In fact, the pleiotropic cellular function of TUDCA ranges from its direct interaction with mitochondrial membranes to modulation of kinase survival pathways or gene expression that ultimately impact on mitochondria. Further, regulation of the functional endoplasmatic reticulum (ER)-mitochondria unit by this bile acid has also been proven as a key counterpart of mitochondria-targeted TUDCA effect. In this chapter, we summarize mechanisms by which this hydrophilic bile acid affects mitochondria and subsequently cell survival, cell cycle and differentiation. We also discuss the potential therapeutic application of TUDCA in several pathological conditions associated with mitochondrial dysfunction.

Keywords Bile acids · Ursodeoxycholic acid · Tauroursodeoxycholic acid · Cell cycle · Cell death · Mitochondria

1 Introduction

Bile acids, the major components of bile, are produced in the liver through a complex chain of chemical reactions. In humans and most animals, bile acids are produced from cholesterol, being the major pathway of catabolism and elimination of cholesterol from the body. Certain hydrophobic bile acids are cytotoxic, inducing cell death by necrosis and apoptosis, while others, more hydrophilic, are

S. Solá (✉) · M. F. Ribeiro · T. Genebra · C. M. P. Rodrigues
Research Institute for Medicines (iMed.ULisboa), Faculty of Pharmacy, Universidade de Lisboa, Lisbon, Portugal
e-mail: susana.sola@ff.ulisboa.pt

P. J. Oliveira (ed.), *Mitochondrial Biology and Experimental Therapeutics*,
https://doi.org/10.1007/978-3-319-73344-9_19

cytoprotective. Tauroursodeoxycholic acid (TUDCA) is the taurine conjugated derivative of cytoprotective ursodeoxycholic acid (UDCA), a US Food and Drug Administration-approved hydrophilic bile acid for the treatment of certain human diseases (Hohenester et al. 2009). In the human organism, UDCA is immediately conjugated with glycine or taurine, thus forming glycoursodeoxycholic acid (GUDCA) and TUDCA, respectively. However, even though these bile acids have been used in Chinese medicine for more than 3000 years, a thorough and complete molecular characterization of cellular effects has only recently been explored.

Evidence has catapulted mitochondria to the center of TUDCA-mediated cellular actions. In fact, due to its cholesterol-derived structure, TUDCA has been shown to have enormous affinity for cell membranes, such as mitochondrial membranes. Moreover, by stabilizing mitochondria and preserving mitochondria integrity, this endogenous bile acid appears to also act as antioxidant (Sokol et al. 2005; Oveson et al. 2011; Gaspar et al. 2013), cell fate regulator and antiapoptotic agent.

In fact, while all previous studies have revealed a progressive destruction of mitochondrial membrane potential and induction of mitochondrial permeability transition (MPT) by toxic bile acids, the targeting of mitochondria by TUDCA or UDCA has not been shown to promote such devastating events (Schulz et al. 2013).

UDCA and TUDCA have also been described to indirectly regulate mitochondria through other mechanisms including ER-mitochondria axis modulation, activation of kinase signaling pathways, interaction with nuclear steroid receptors (NSR), and transcriptional regulation (Sola et al. 2006a, b; Xavier et al. 2016). These mitochondrial-targeted effects of TUDCA have been described in many cell types, including hepatocytes, fibroblasts, cardiomyocytes, adipocytes, and neural cells. Noteworthy, as an orally bioavailable and central nervous system (CNS) penetrating agent (Keene et al. 2002), the potential therapeutic role of TUDCA has also been extended to neurological disorders associated with mitochondrial dysfunction (Rodrigues et al. 2003a, b; Nunes et al. 2012; Moreira et al. 2017; Rosa et al. 2017). A better understanding of the molecular mechanisms underlying the impact of TUDCA in mitochondria will surely be helpful to design novel pharmacologic targeted strategies. In this review, we discuss the several faces of TUDCA-mediated mitochondrial control that regulates cell fate decision, namely apoptosis, proliferation and differentiation.

2 Inhibition of Mitochondrial Damage and Apoptosis

The initial studies demonstrating a protective role of TUDCA in mitochondrial damage were first performed in hepatic cells. Despite no agreement regarding the most efficient form to reduce mitochondrial toxic effects of lipophilic bile acids, (Krahenbuhl et al. 1994; Neuman et al. 1995), all studies broadly accepted that both TUDCA and unconjugated UDCA positively interfere with mitochondria toward cell survival. This was particularly evident in ethanol-induced cellular damage, when bile acid exposure restored mitochondrial internal organization, including number and length of cristae, when compared with untreated cells (Neuman et al. 1995).

Similarly, the antiapoptotic effect of UDCA was first described in liver tissue in rats fed a diet containing deoxycholic acid (Rodrigues et al. 1998). TUDCA was also effective at partially preventing mitochondrial apoptosis in an *in vitro* model of Huntington diseases (HD). Indeed, TUDCA inhibited mitochondrial depolarization and outer membrane disruption in neuronal cells exposed to 3-nitropropionic acid (3-NP), a toxin that mimics many aspects of HD (Rodrigues et al. 2000). In this cellular model, TUDCA rescued other features of mitochondrial apoptosis, including mitochondrial translocation of the proapoptotic protein Bax, cytochrome *c* release, caspase activation and poly(ADP-ribose) polymerase (PARP) cleavage.

The first demonstration that TUDCA was neuroprotective in vivo came from the study of Keene et al. In this study, rats receiving a combination of 3-NP and TUDCA exhibited sensorimotor and cognitive task performances indistinguishable from control rats (Keene et al. 2001). In a transgenic HD mouse, TUDCA was also shown to prevent the production of reactive oxygen species (ROS) and mitigate mitochondrial insufficiency and apoptosis. Indeed, systemically administered TUDCA inhibited the translocation of proapoptotic protein Bax from cytosol to the mitochondria, reducing striatal neuropathology and the size of ubiquitinated neuronal intranuclear huntingtin inclusions (Keene et al. 2002).

Later, the impact of TUDCA in preventing Bax mitochondrial insertion and release of cytochrome *c* were further dissected in isolated mitochondria. Curiously, TUDCA-repressed Bax mitochondrial translocation relied on its effects on mitochondrial membrane lipid and protein structure. Indeed, the final product of the biosynthesis of bile acids is a molecule with amphiphilic properties, which in turn facilitates interaction with proteins, and insertion into lipid bilayers. Notably, TUDCA reverted profound structural mitochondrial alterations triggered by amyloid β (Aβ), such as modified membrane lipid polarity and disrupted protein mobility (Rodrigues et al. 2001). This, in turn, was shown to promote obvious effects in reducing oxidative injuries in mitochondrial membranes.

Further studies showed that TUDCA administration to ethanol-fed rats prevented tumor necrosis factor α (TNFα)-induced apoptosis in hepatocytes through regulation of mitochondrial membrane fluidity and subsequent normalization of mitochondrial glutathione (GSH) levels (Colell et al. 2001). In agreement, a significant protection by TUDCA against mitochondrial matrix and cristae damage during harvesting and cold storage of human liver has been described (Falasca et al. 2001). Likewise, the protective effect of TUDCA in a well-characterized model of transient focal cerebral ischemia was associated with reduced mitochondrial swelling within the ischemic penumbra, supporting a role for TUDCA in preventing mitochondrial perturbations associated with apoptosis (Rodrigues et al. 2002). This, in turn, had a huge impact on neurological function and reduction in infarct size in rats.

On the other hand, in an ischemia-reperfusion liver injury *in vivo* model, TUDCA antiapoptotic effect was also shown to be partially dependent on regulation of Bax expression and its mitochondrial translocation (Ishigami et al. 2001). In fact, ischemia-reperfusion injury is the main cause of early graft dysfunction after liver transplantation, adding a clinical significance of these studies in human liver transplantation sceneries. Also in the context of liver disorders, TUDCA has been proposed to exert significant efficacy in preventing apoptosis induced by toxic bile acids in hepatic cells. Indeed, it has been revealed that taurodeoxycholic acid (TDCA)-induced apoptosis in HepG2 cells was markedly inhibited by TUDCA through mechanisms dependent on cytochrome *c* release (Xie et al. 2003). In this study TUDCA was also shown to prevent DNA fragmentation and the activation of pro-caspase 9 and 3 by TDCA.

Nevertheless, the fact that TUDCA could indeed physically inhibit mitochondrial perturbation by Bax translocation was only reported using recombinant Bax protein in isolated mitochondrial organelles. Spectroscopic analyses of mitochondria exposed to Bax demonstrated that TUDCA suppresses Bax-induced mitochondrial membrane alterations, including polarity and fluidity of lipids and proteins. This, in turn, was independent of the MPT pore (MPTP) (Rodrigues et al. 2003a, b). Furthermore, the interaction between TUDCA and Bax itself was later revealed by assays using native fluorescence of the tryptophan residues of Bax. In fact, Zhang et al., demonstrated that although recombinant Bax protein does not directly interact with UDCA and DCA, it can potentially interact with the TUDCA (Zhang et al. 2006).

The direct effect of TUDCA in the mitochondria organelle was also explored during cardiac function and under high fat diet-induced obesity. Briefly, by interfering with release of intracellular Ca^{2+}, insulin signaling molecules and mitochondrial permeability, TUDCA could alleviate *in vivo* cardiomyocyte contractile defects induced by high fat diet (Mello-Vieira et al. 2013; Turdi et al. 2013; Sousa et al. 2015).

Finally, it has been recently demonstrated that the neuroprotective role of TUDCA might also be dependent on mitochondrial turnover. In fact, TUDCA could prevent mitochondrial uncoupler carbonyl cyanide m-chlorophenyl hydrazone-induced neuronal death by increasing the mechanisms mediated by parkin mitochondrial translocation and mitophagy (Fonseca et al. 2016). It remains, however, to be determined whether TUDCA-mediated mitophagy and parkin modulation (Rosa et al. 2017) is dependent on its mitochondrial stabilizer role or other upstream cellular effect.

3 Modulation of Mitochondria Through Kinase Signaling Cascades

The impact of TUDCA in mitochondrial fate appears to be dependent on specific intracellular kinase cascades, such as the phosphoinositide 3-kinase (PI3K) signaling pathway. In fact, the activation of this survival pathway often results in repression of mitochondrial apoptotic events. Once activated, PI3K phosphorylates Akt-1 protein kinase B, which subsequently inactivates, the proapoptotic member of B-cell lymphoma 2 (Bcl-2) family, Bad, via phosphorylation at Ser-136 (Datta et al. 1997, 1999). When not phosphorylated, Bad induces apoptosis by inhibiting the antiapoptotic Bcl-2-family members, Bcl-2 and Bcl-xL, and allowing the pro-apoptotic proteins of this family, such as Bak and Bax, to aggregate, trigger MPT and induce release of cytochrome *c* from the mitochondria (Bergmann 2002). The involvement of the PI3K pathway in preventing mitochondrial apoptosis by TUDCA was firstly described in a study with neuronal cells exposed to Aβ peptide. Both Aβ-mediated mitochondrial translocation of Bax and release of mitochondrial cytochrome *c* were abrogated by TUDCA in a PI3K-dependent manner (Sola et al. 2003). This mitochondrial road was also demonstrated using an acute hemorrhagic stroke rat model. Here, TUDCA was shown to activate the Akt-1 protein kinase B survival pathway and induce Bad phosphorylation, which ultimately prevented mitochondrial membrane permeabilization and cytochrome *c* release (Rodrigues et al. 2003a, b). The effect of TUDCA on the PI3K signaling pathway was later corroborated in a study where PI3K inhibition reduced the protective effect of TUDCA during glutamate neurotoxicity (Castro et al. 2004). In fact, it was evident that inhibition of this specific survival pathway abrogated the effect of TUDCA in promoting Bad phosphorylation and its translocation to the cytosol. Importantly, the alterations in Bcl-2 family members and other mitochondrial apoptotic features were no longer prevented.

Other survival signaling pathways may also be involved in the antiapoptotic effect of TUDCA. The involvement of the activation of p38, the mitogen-activated protein kinase (MAPK) and the extracellular signal–regulated kinase (ERK) signaling pathways, has been also described in the protective effect of TUDCA against mitochondrial injury and apoptosis (Schoemaker et al. 2004). In the same line, in an acute kidney injury model, TUDCA was recently shown to inhibit the mitochondrial pathway of apoptosis by upregulating ERK1/2 survival pathways (Gupta et al. 2012). Here, the functional protection against acute kidney injury by TUDCA was supported by less severe histological injury seen in kidneys of TUDCA-treated rats, being the first report of activation of ERK1/2 pathways by TUDCA following cryopreservation injury.

4 Targeting Mitochondria Though ER

ER is in close contact with the mitochondria network (Grimm 2012). Curiously, the inhibitory effects of TUDCA in ER stress have increasingly been elucidated (Xie et al. 2002). TUDCA effect on ER can easily influence the mitochondrial state, turning difficult to precisely identify the main TUDCA targeted-organelle.

Several examples in the literature demonstrate that TUDCA-regulated ER stress can indeed have a huge impact on mitochondrial integrity and cell survival. A strong correlation between ER chaperone properties of TUDCA and mitochondrial dysfunction was firstly reported in a rat model of type 2 diabetes. In this model, the authors showed that myocardium levels of ER chaperones and mitochondrial non-phosphophorylated GSK-3β were significantly higher in diabetic mice than in controls. In addition, and as expected, the authors showed a negative association between MPTP opening threshold and mitochondrial translocation of GSK-3β. In this case, TUDCA was shown to normalize ER stress and mitochondrial integrity of diabetic mice, suggesting that the bile acid may also act on mitochondria through ER-dependent mechanisms (Miki et al. 2009).

Other examples come from evidence of TUDCA mitochondrial protection in studies where this bile acid was used as a chaperone to suppress ER stress signaling. For example, by reducing the activation of two unfolded protein response (UPR) pathways and their targets, including caspase-12, c-Jun N-terminal kinase and CCAAT/enhancer binding proteins (C/EBP) homologous protein-10, TUDCA could prevent mitochondrial damage, mitochondrial voltage-dependent anion channel, cytochrome *c* release, and caspase-9 activation induced by partial hepatectomy under ischemia-reperfusion (Ben Mosbah et al. 2010).

Other link between ER and mitochondrial anomalies was also established in murine cardiomyocyte contractile function. In fact, TUDCA not only reverted several hallmarks of tunicamycin-induced ER stress but also the permeabilization of mitochondrial membrane via inhibition of GSK-3β activation and production of ROS (Zhang et al. 2011). Similar conclusions were *in vivo* under cardiac hypertrophy and myocardial infarction (Mitra et al. 2013). In a high-fat diet-induced obesity scenario, TUDCA was also shown to ameliorate cardiomyocyte contractile and intracellular Ca^{2+} defects via ER and mitochondrial-related mechanisms (Turdi et al. 2013).

Interestingly, the functional ER-mitochondria unit was also shown to be regulated by TUDCA in a pulmonary arterial hypertension model (Dromparis et al. 2013). Again, used as a chaperone to suppress ER stress, this hydrophilic bile acid inhibited not only several ER stress markers, such as the activation of transcription factor 6, but also the decrease in mitochondrial Ca^{2+} and function induced by hypoxia in monocrotaline-induced pulmonary hypertension rat model. Noteworthy, reduction of ER stress by TUDCA has been shown to attenuate the defects induced by mitofusin knockout, providing another evidence for the relevance of ER and mitochondria interplay during the protective properties of this bile acid in modulating cell fate decision (Debattisti et al. 2014).

More recently, other studies have also implicated the repression of ER stress by TUDCA on its inhibitory effect on mitochondrial apoptosis. This has been described for programmed cell death induced by increased fatty acids in nonalcoholic steatohepatitis (NASH) conditions, such as palmitic acid and myristic acid (Martinez et al. 2015), or even during obstructive sleep apnea-associated intermittent hypoxia (Xu et al. 2015).

Interestingly, when trying to understand the toxicity of titanium dioxide nanoparticles (TiO2-NP), it was recently reported that disruption of mitochondria-associated ER membranes and calcium ion balance was prevented by TUDCA (Yu et al. 2015). Of note, GSK-3β phosphorylation on Serine 9 residue has been shown to reduce ER-mitochondrial associations and Ca^{2+} exchange between the two organelles (Paillusson et al. 2016), being also a target of TUDCA regulatory mechanisms. In fact, both GSK-3β activation and ER–mitochondria disruption have been reported for several neurodegenerative disorders (Paillusson et al. 2016).

Once TUDCA protects many of mitochondrial alteration in *in vitro* models of these diseases, it is possible that TUDCA chaperone properties in both ER and mitochondria may somehow preserve the ER-mitochondrial interactions, crucial for the maintenance of cellular function.

Interestingly, Leber's hereditary optic neuropathy is the most common mitochondrial DNA-related disease, involving mutations in mitochondrial complex I. Recently, it has been reported that TUDCA rescues the metabolic signature of affected patients through the involvement of ER (Chao de la Barca et al. 2016). In this study, TUDCA reverted several key molecules of unfolded protein response (UPR), including phosphorylation of protein kinase RNA-like endoplasmic reticulum kinase (PERK) and eukaryotic initiation factor 2 α (eIF2α), as well as the enhanced expression of C/EBP homologous protein and the increase of XBP1 splicing in fibroblasts.

Other example implicating both ER and mitochondria on TUDCA-mediated mechanisms of cellular protection was demonstrated when TUDCA significantly attenuated miltirone effects on ROS production and apoptosis in human leukemia cell lines and primary leukemia cells (Zhou et al. 2016). Miltirone is an antileukemic agent that triggers ER stress and therefore mitochondrial dysfunction through a ROS-dependent destructive effect. Although also inducing Ca^{2+} release from mitochondria and promoting mitochondrial membrane depolarization and cytochrome *c* release, miltirone treatment also mediates ER stress, namely Ca^{2+} dissipation from ER stores, exacerbating its effects on mitochondrial dysfunction. In this study, TUDCA inhibitory effects on ER stress were also shown to counteract in a positive way for the mitochondrial damage abrogation.

5 Nuclear Ways to Target Mitochondria

As a strong antioxidant molecule (Sokol et al. 2005), TUDCA necessarily interferes with mitochondria through the regulation of nuclear factors sensitive to cellular oxidative state. In a rat model of acute hemorrhagic stroke, TUDCA has been shown to be neuroprotector by modulating transcriptional expression of certain Bcl-2 family members via NF-κB nuclear translocation (Rodrigues et al. 2003a, b). In fact, in response to cytokines and ROS, specific kinases, including Akt protein kinase B, phosphorylate IκB and lead to its proteolysis, promoting NF-κB release and nuclear translocation. In the nucleus, NF-κB then regulates several target genes, including proinflammatory cytokines but also antiapoptotic Bcl-2 elements, such as Bcl-2 and Bcl-xL (Chen et al. 2000; Catz and Johnson 2001). Hence, TUDCA can alter activity of transcription factors, such as NF-κB, affecting mitochondria and thus cell survival.

TUDCA also regulates p53 transcription factor. Activated by stimulus such as increased ROS levels, p53 has been extensively documented to induce mitochondrial apoptosis by inducing Bax and inhibiting Bcl-2 via PUMA and Noxa upregulation (Khoo et al. 2014). In this respect, TUDCA impacts on p53-mediated apoptosis in Alzheimer's disease mutant neuroblastoma cells (Ramalho et al. 2006). In this study, TUDCA was shown to prevent p53-induced apoptosis by activating PI3K survival signaling. More importantly, TUDCA reduced the transcriptional changes of Bcl-2 family members Bcl-2 and Bax, regulators of mitochondrial membrane permeability, via p53.

As cholesterol-derived molecules, bile acids may also regulate nuclear gene expression through NSRs. In fact, this represents another road of TUDCA to modulate mitochondria integrity and cell fate. It has been shown that functional modulation of NSR by TUDCA also reduces Aβ peptide-induced mitochondrial apoptosis (Sola et al. 2006a, b). In this work, a fluorescently labeled bile acid was used and detected in both cytoplasm and nucleus of rat cortical neurons. By silencing mineralocorticoid receptor (MR), TUDCA antiapoptotic effect was shown to be MR-dependent. Importantly, under this cellular context the bile acid directly interacted with the MR, promoting MR dissociation from its cytosolic chaperone with consequently nuclear translocation as a steroid-receptor complex. These results identified a direct target and provided novel insights into TUDCA cellular mechanisms to regulate mitochondria during apoptosis.

6 Mitochondria-to-Nucleus Retrograde Signaling

Over the past years, the potential role of TUDCA in the treatment of apoptosis-associated neurological disorders has been actively explored. Thus far, the mitochondrial protective effect of TUDCA has been described in a wide range of *in vitro* and *in vivo* animal models of neurodegenerative disorders, including Huntington's disease, Parkinson's disease, Alzheimer's disease but also acute ischemic and hemorrhagic stroke (Duan et al. 2002; Keene et al. 2002; Rodrigues et al. 2003a, b; Sola et al. 2003; Ved et al. 2005; Mortiboys et al. 2013, 2015). TUDCA has also been tested in clinical trials in amyotrophic lateral sclerosis patients. In fact, these discoveries opened new horizons in the treatment of degenerative diseases, such as Alzheimer's disease. TUDCA inhibited Bax-induced mitochondrial pathway of apoptosis by the vasculotropic E22Q mutant form of Aβ peptide in primary human cerebral endothelial cells (Viana et al. 2009). TUDCA was also shown to attenuate amyloid precursor protein processing and Aβ deposition in familiar mice models of Alzheimer's disease (Nunes et al. 2012).

TUDCA potential in regulating neural cells was recently shown to take part in other cellular contexts, such as proliferation and differentiation processes, which also rely on the apoptotic machinery, mitochondrial integrity and cell survival. In fact, it has been previously demonstrated that TUDCA prevents differentiation-induced mitochondrial alterations, including mitochondrial ROS production, mitochondrial depolarization, cytochrome *c* release and ATP depletion (Xavier et al. 2014). TUDCA also diminished p53 mitochondrial translocation during early-stage differentiation of mouse neural stem cells (NSCs), thus suggesting that this bile acid alone was capable of controlling differentiation-mediated stress events, and hence preventing the need for p53-mediated stress protection. More importantly, it has been demonstrated that mitochondrial stress relieved by TUDCA allows NSCs to re-enter the cell cycle promoting proliferation, while eliciting a differentiation shift from glia to neurons in differentiating NSCs (Xavier et al. 2014). In fact, recent evidence has already indicated that the extent of mitochondrial damage predicts NSC lineage determination, where mitochondrial levels of ROS and ATP in the mitochondria affect cell cycle progression and differentiation of cells. In this regard, it has been shown that TUDCA-regulated mitochondrial alterations and NSC fate were indeed dependent on mitochondrial ROS and ATP modulation (Xavier et al. 2014). Several groups had already supported the idea that distinct mitochondrial retrograde signals control cell cycle checkpoints and that low levels of ROS might function as intracellular messengers of cell cycle arrest (Sauer et al. 2001; Deng et al. 2003). Curiously, it has been demonstrated that mitochondrial dysfunction activates two retrograde signals resulting in cell cycle modulation. The decrease in ATP production is usually followed by downregulation of cyclin E, while increased ROS production induces upregulation of p27 (Owusu-Ansah et al. 2008; Finkel and Hwang 2009). Importantly, this TUDCA-mediated effect in modulating mitochondria, NSC pool and early neurogenesis was also demonstrated in adult rats (Soares et al. 2017).

This example of mitochondria-retrograde signaling regulation could also occur in other TUDCA mechanisms detailed above. In fact, although TUDCA was shown to induce mitochondria alterations through nuclear-mediated mechanisms, this bile acid may also interfere with nuclear events via mitochondria modulation. Interestingly, in an in vitro model of diabetic retinopathy, TUDCA inhibited mitochondria-nuclear translocation of the apoptosis-inducing factor (AIF) (Gaspar et al. 2013). Once in the nucleus, AIF promotes stress-induced cell death, contributing to DNA and nuclear fragmentation as well as chromatin condensation (Sevrioukova 2011). In this study, the authors have demonstrated that by decreasing both protein carbonyl groups and ROS production in the mitochondria, TUDCA was able to maintain nuclear stability and significantly prevent retinal neural cell death induced by high levels of glucose.

7 Conclusion and Future Perspectives

Presenting favorable pharmacokinetics properties, TUDCA is unique in its ability to increase cell survival and inhibit mitochondrial apoptosis. In fact, an increasing number of pre-clinical and clinical studies highlight the potential benefit of this bile acid. Thus far, TUDCA has been proposed as a therapeutic molecule for several diseases associated with mitochondrial dysfunction, including obesity, gastrointestinal disorders, and renal injury (Vang et al. 2014).

TUDCA may act directly on mitochondria, or affect mitochondria through several indirect mechanisms, such as ER, kinase signaling pathways and NSRs (Fig. 1). The road to stabilize mitochondria and affect cell fate, in turn, appears to be highly dependent on the cellular model, context and insult. For example, in human embryonic

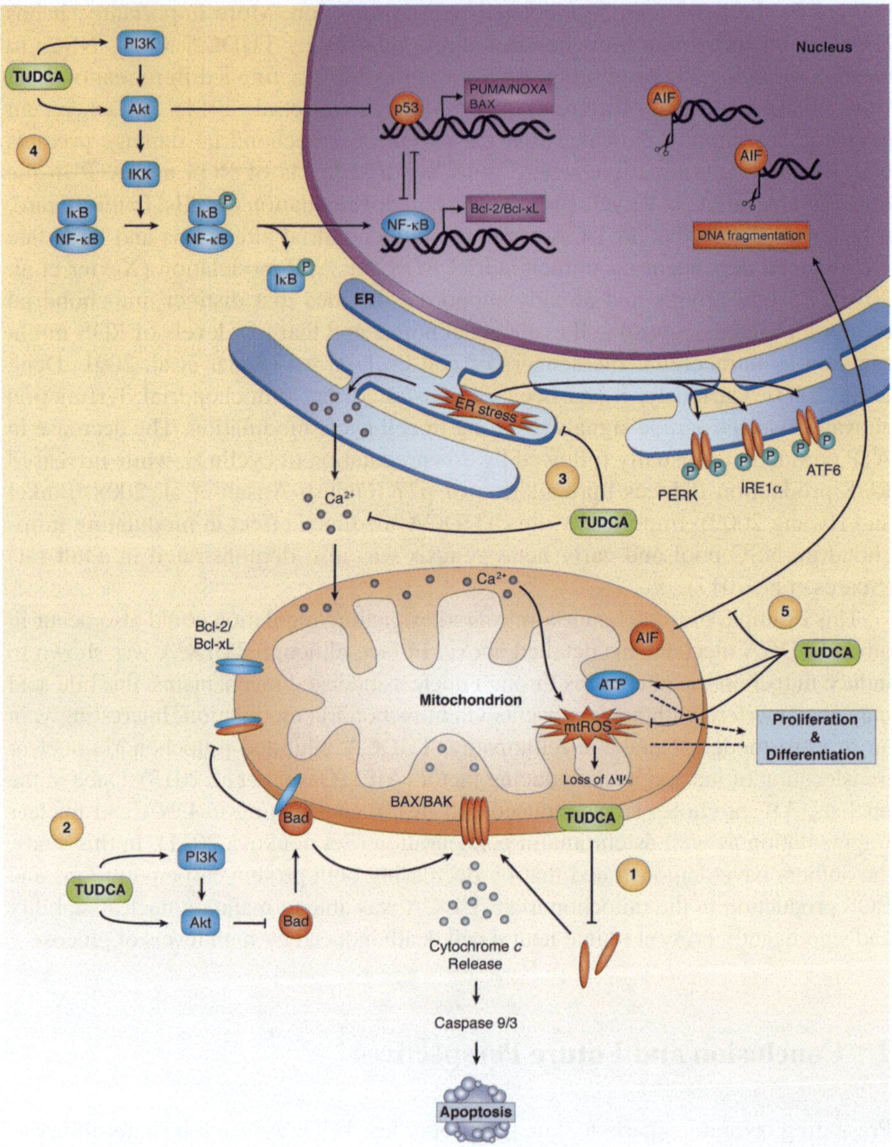

Fig. 1 Mitochondria-targeted roads of TUDCA to regulate cell fate decision. The hydrophilic bile acid TUDCA was shown to regulate cell viability, proliferation and differentiation in different cell types through distinct mitochondria-related pathways depicted in the figure. These include the direct inhibition of mitochondrial membrane permeabilization and damage (1); the activation of survival signaling pathways that ultimately regulate the balance of mitochondrial permeabilization-associated proteins, such as Bad (2); ER stress-mediated events that influence mitochondria state, such as Ca^{2+} release (3); nuclear-mediated mechanisms that regulate expression of mitochondria-related genes (4); and, finally, a mitochondria-to-nucleus retrograde signaling that controls nuclear stability and consequently cell viability (5)

kidney cells, TUDCA- and UDCA-mediated mechanisms of action are not associated with changes in cellular membrane structure (Mello-Vieira et al. 2013; Sousa et al. 2015). Of note, the precise sequence by which many of these TUDCA-triggered mechanisms occur in cells remain to be clarified, as well as how cells integrate all this complex signaling networks to survive, proliferate or differentiate. Importantly, the non-conjugated form of TUDCA is an FDA-approved molecule widely used to treat cholestatic liver diseases and has been used in clinical trials for amyotrophic lateral sclerosis (Parry et al. 2010; Elia et al. 2016), providing a new framework to further exploit the potential use of this bile acid as a therapeutic agent.

Nevertheless, the molecular mechanisms underlying TUDCA neuroprotection are multifaceted and future studies will surely help to clarify whether TUDCA might engage other faces of the mitochondrial machinery critical to influence cell fate.

Acknowledgements We would like to thank all our laboratory colleagues for their support and insightful discussions.

Declaration of Conflicting Interests The authors declared no potential conflicts of interest with respect to the authorship, and/or publication of this article.

References

Ben Mosbah I, Alfany-Fernandez I, Martel C, Zaouali MA, Bintanel-Morcillo M, Rimola A, Rodes J, Brenner C, Rosello-Catafau J, Peralta C (2010) Endoplasmic reticulum stress inhibition protects steatotic and non-steatotic livers in partial hepatectomy under ischemia-reperfusion. Cell Death Dis 1:e52

Bergmann A (2002) Survival signaling goes BAD. Dev Cell 3(5):607–608

Castro RE, Sola S, Ramalho RM, Steer CJ, Rodrigues CM (2004) The bile acid tauroursodeoxycholic acid modulates phosphorylation and translocation of bad via phosphatidylinositol 3-kinase in glutamate-induced apoptosis of rat cortical neurons. J Pharmacol Exp Ther 311(2):845–852

Catz SD, Johnson JL (2001) Transcriptional regulation of bcl-2 by nuclear factor kappa B and its significance in prostate cancer. Oncogene 20(50):7342–7351

Chao de la Barca JM, Simard G, Amati-Bonneau P, Safiedeen Z, Prunier-Mirebeau D, Chupin S, Gadras C, Tessier L, Gueguen N, Chevrollier A, Desquiret-Dumas V, Ferre M, Bris C, Kouassi Nzoughet J, Bocca C, Leruez S, Verny C, Milea D, Bonneau D, Lenaers G, Martinez MC, Procaccio V, Reynier P (2016) The metabolomic signature of Leber's hereditary optic neuropathy reveals endoplasmic reticulum stress. Brain 139(11):2864–2876

Chen C, Edelstein LC, Gelinas C (2000) The Rel/NF-kappaB family directly activates expression of the apoptosis inhibitor Bcl-x(L). Mol Cell Biol 20(8):2687–2695

Colell A, Coll O, Garcia-Ruiz C, Paris R, Tiribelli C, Kaplowitz N, Fernandez-Checa JC (2001) Tauroursodeoxycholic acid protects hepatocytes from ethanol-fed rats against tumor necrosis factor-induced cell death by replenishing mitochondrial glutathione. Hepatology 34(5):964–971

Datta SR, Dudek H, Tao X, Masters S, Fu H, Gotoh Y, Greenberg ME (1997) Akt phosphorylation of BAD couples survival signals to the cell-intrinsic death machinery. Cell 91(2):231–241

Datta SR, Brunet A, Greenberg ME (1999) Cellular survival: a play in three Akts. Genes Dev 13(22):2905–2927

Debattisti V, Pendin D, Ziviani E, Daga A, Scorrano L (2014) Reduction of endoplasmic reticulum stress attenuates the defects caused by Drosophila mitofusin depletion. J Cell Biol 204(3):303–312

Deng X, Gao F, May WS Jr (2003) Bcl2 retards G1/S cell cycle transition by regulating intracellular ROS. Blood 102(9):3179–3185

Dromparis P, Paulin R, Stenson TH, Haromy A, Sutendra G, Michelakis ED (2013) Attenuating endoplasmic reticulum stress as a novel therapeutic strategy in pulmonary hypertension. Circulation 127(1):115–125

Duan WM, Rodrigues CM, Zhao LR, Steer CJ, Low WC (2002) Tauroursodeoxycholic acid improves the survival and function of nigral transplants in a rat model of Parkinson's disease. Cell Transplant 11(3):195–205

Elia AE, Lalli S, Monsurro MR, Sagnelli A, Taiello AC, Reggiori B, La Bella V, Tedeschi G, Albanese A (2016) Tauroursodeoxycholic acid in the treatment of patients with amyotrophic lateral sclerosis. Eur J Neurol 23(1):45–52

Falasca L, Tisone G, Palmieri G, Anselmo A, Di Paolo D, Baiocchi L, Torri E, Orlando G, Casciani CU, Angelico M (2001) Protective role of tauroursodeoxycholate during harvesting and cold storage of human liver: a pilot study in transplant recipients. Transplantation 71(9):1268–1276

Finkel T, Hwang PM (2009) The Krebs cycle meets the cell cycle: mitochondria and the G1-S transition. Proc Natl Acad Sci U S A 106(29):11825–11826

Fonseca I, Gordino G, Moreira S, Nunes MJ, Azevedo C, Gama MJ, Rodrigues E, Rodrigues CM, Castro-Caldas M (2016) Tauroursodeoxycholic acid protects against mitochondrial dysfunction and cell death via mitophagy in human neuroblastoma cells. Mol Neurobiol 54(8):6107–6119

Gaspar JM, Martins A, Cruz R, Rodrigues CM, Ambrosio AF, Santiago AR (2013) Tauroursodeoxycholic acid protects retinal neural cells from cell death induced by prolonged exposure to elevated glucose. Neuroscience 253:380–388

Grimm S (2012) The ER-mitochondria interface: the social network of cell death. Biochim Biophys Acta 1823(2):327–334

Gupta S, Li S, Abedin MJ, Noppakun K, Wang L, Kaur T, Najafian B, Rodrigues CM, Steer CJ (2012) Prevention of acute kidney injury by tauroursodeoxycholic acid in rat and cell culture models. PLoS One 7(11):e48950

Hohenester S, Oude-Elferink RP, Beuers U (2009) Primary biliary cirrhosis. Semin Immunopathol 31(3):283–307

Ishigami F, Naka S, Takeshita K, Kurumi Y, Hanasawa K, Tani T (2001) Bile salt tauroursodeoxycholic acid modulation of Bax translocation to mitochondria protects the liver from warm ischemia-reperfusion injury in the rat. Transplantation 72(11):1803–1807

Keene CD, Rodrigues CM, Eich T, Linehan-Stieers C, Abt A, Kren BT, Steer CJ, Low WC (2001) A bile acid protects against motor and cognitive deficits and reduces striatal degeneration in the 3-nitropropionic acid model of Huntington's disease. Exp Neurol 171(2):351–360

Keene CD, Rodrigues CM, Eich T, Chhabra MS, Steer CJ, Low WC (2002) Tauroursodeoxycholic acid, a bile acid, is neuroprotective in a transgenic animal model of Huntington's disease. Proc Natl Acad Sci U S A 99(16):10671–10676

Khoo KH, Verma CS, Lane DP (2014) Drugging the p53 pathway: understanding the route to clinical efficacy. Nat Rev Drug Discov 13(3):217–236

Krahenbuhl S, Fischer S, Talos C, Reichen J (1994) Ursodeoxycholate protects oxidative mitochondrial metabolism from bile acid toxicity: dose-response study in isolated rat liver mitochondria. Hepatology 20(6):1595–1601

Martinez L, Torres S, Baulies A, Alarcon-Vila C, Elena M, Fabrias G, Casas J, Caballeria J, Fernandez-Checa JC, Garcia-Ruiz C (2015) Myristic acid potentiates palmitic acid-induced lipotoxicity and steatohepatitis associated with lipodystrophy by sustaning de novo ceramide synthesis. Oncotarget 6(39):41479–41496

Mello-Vieira J, Sousa T, Coutinho A, Fedorov A, Lucas SD, Moreira R, Castro RE, Rodrigues CM, Prieto M, Fernandes F (2013) Cytotoxic bile acids, but not cytoprotective species, inhibit the ordering effect of cholesterol in model membranes at physiologically active concentrations. Biochim Biophys Acta 1828(9):2152–2163

Miki T, Miura T, Hotta H, Tanno M, Yano T, Sato T, Terashima Y, Takada A, Ishikawa S, Shimamoto K (2009) Endoplasmic reticulum stress in diabetic hearts abolishes erythropoietin-induced myocardial protection by impairment of phospho-glycogen synthase kinase-3beta-mediated suppression of mitochondrial permeability transition. Diabetes 58(12):2863–2872

Mitra A, Basak T, Datta K, Naskar S, Sengupta S, Sarkar S (2013) Role of alpha-crystallin B as a regulatory switch in modulating cardiomyocyte apoptosis by mitochondria or endoplasmic reticulum during cardiac hypertrophy and myocardial infarction. Cell Death Dis 4:e582

Moreira S, Fonseca I, Nunes MJ, Rosa A, Lemos L, Rodrigues E, Carvalho AN, Outeiro TF, Rodrigues CMP, Gama MJ, Castro-Caldas M (2017) Nrf2 activation by tauroursodeoxycholic acid in experimental models of Parkinson's disease. Exp Neurol 295:77–87

Mortiboys H, Aasly J, Bandmann O (2013) Ursocholanic acid rescues mitochondrial function in common forms of familial Parkinson's disease. Brain 136(Pt 10):3038–3050

Mortiboys H, Furmston R, Bronstad G, Aasly J, Elliott C, Bandmann O (2015) UDCA exerts beneficial effect on mitochondrial dysfunction in LRRK2(G2019S) carriers and in vivo. Neurology 85(10):846–852

Neuman MG, Cameron RG, Shear NH, Bellentani S, Tiribelli C (1995) Effect of tauroursodeoxycholic and ursodeoxycholic acid on ethanol-induced cell injuries in the human Hep G2 cell line. Gastroenterology 109(2):555–563

Nunes AF, Amaral JD, Lo AC, Fonseca MB, Viana RJ, Callaerts-Vegh Z, D'Hooge R, Rodrigues CM (2012) TUDCA, a bile acid, attenuates amyloid precursor protein processing and amyloid-β deposition in APP/PS1 mice. Mol Neurobiol 45(3):440–454

Oveson BC, Iwase T, Hackett SF, Lee SY, Usui S, Sedlak TW, Snyder SH, Campochiaro PA, Sung JU (2011) Constituents of bile, bilirubin and TUDCA, protect against oxidative stress-induced retinal degeneration. J Neurochem 116(1):144–153

Owusu-Ansah E, Yavari A, Mandal S, Banerjee U (2008) Distinct mitochondrial retrograde signals control the G1-S cell cycle checkpoint. Nat Genet 40(3):356–361

Paillusson S, Stoica R, Gomez-Suaga P, Lau DH, Mueller S, Miller T, Miller CC (2016) There's something wrong with my MAM; the ER-mitochondria axis and neurodegenerative diseases. Trends Neurosci 39(3):146–157

Parry GJ, Rodrigues CM, Aranha MM, Hilbert SJ, Davey C, Kelkar P, Low WC, Steer CJ (2010) Safety, tolerability, and cerebrospinal fluid penetration of ursodeoxycholic Acid in patients with amyotrophic lateral sclerosis. Clin Neuropharmacol 33(1):17–21

Ramalho RM, Borralho PM, Castro RE, Sola S, Steer CJ, Rodrigues CM (2006) Tauroursodeoxycholic acid modulates p53-mediated apoptosis in Alzheimer's disease mutant neuroblastoma cells. J Neurochem 98(5):1610–1618

Rodrigues CM, Fan G, Ma X, Kren BT, Steer CJ (1998) A novel role for ursodeoxycholic acid in inhibiting apoptosis by modulating mitochondrial membrane perturbation. J Clin Invest 101(12):2790–2799

Rodrigues CM, Stieers CL, Keene CD, Ma X, Kren BT, Low WC, Steer CJ (2000) Tauroursodeoxycholic acid partially prevents apoptosis induced by 3-nitropropionic acid: evidence for a mitochondrial pathway independent of the permeability transition. J Neurochem 75(6):2368–2379

Rodrigues CM, Sola S, Brito MA, Brondino CD, Brites D, Moura JJ (2001) Amyloid beta-peptide disrupts mitochondrial membrane lipid and protein structure: protective role of tauroursodeoxycholate. Biochem Biophys Res Commun 281(2):468–474

Rodrigues CM, Spellman SR, Sola S, Grande AW, Linehan-Stieers C, Low WC, Steer CJ (2002) Neuroprotection by a bile acid in an acute stroke model in the rat. J Cereb Blood Flow Metab 22(4):463–471

Rodrigues CM, Sola S, Nan Z, Castro RE, Ribeiro PS, Low WC, Steer CJ (2003a) Tauroursodeoxycholic acid reduces apoptosis and protects against neurological injury after acute hemorrhagic stroke in rats. Proc Natl Acad Sci U S A 100(10):6087–6092

Rodrigues CM, Sola S, Sharpe JC, Moura JJ, Steer CJ (2003b) Tauroursodeoxycholic acid prevents Bax-induced membrane perturbation and cytochrome C release in isolated mitochondria. Biochemistry 42(10):3070–3080

Rosa AI, Fonseca I, Nunes MJ, Moreira S, Rodrigues E, Carvalho AN, Rodrigues CMP, Gama MJ, Castro-Caldas M (2017) Novel insights into the antioxidant role of tauroursodeoxycholic acid in experimental models of Parkinson's disease. Biochim Biophys Acta 1863(9):2171–2181

Sauer H, Wartenberg M, Hescheler J (2001) Reactive oxygen species as intracellular messengers during cell growth and differentiation. Cell Physiol Biochem 11(4):173–186

Schoemaker MH, Conde de la Rosa L, Buist-Homan M, Vrenken TE, Havinga R, Poelstra K, Haisma HJ, Jansen PL, Moshage H (2004) Tauroursodeoxycholic acid protects rat hepatocytes from bile acid-induced apoptosis via activation of survival pathways. Hepatology 39(6):1563–1573

Schulz S, Schmitt S, Wimmer R, Aichler M, Eisenhofer S, Lichtmannegger J, Eberhagen C, Artmann R, Tookos F, Walch A, Krappmann D, Brenner C, Rust C, Zischka H (2013) Progressive stages of mitochondrial destruction caused by cell toxic bile salts. Biochim Biophys Acta 1828(9):2121–2133

Sevrioukova IF (2011) Apoptosis-inducing factor: structure, function, and redox regulation. Antioxid Redox Signal 14(12):2545–2579

Soares R, Ribeiro FF, Xapelli S, Genebra T, Ribeiro MF, Sebastião AM, Rodrigues CMP, Solá S (2017) Tauroursodeoxycholic acid enhances mitochondrial biogenesis, neural stem cell pool and early neurogenesis in adult rats. Mol Neurobiol. (in press). https://doi.org/10.1007/s12035-017-0592-5

Sokol RJ, Dahl R, Devereaux MW, Yerushalmi B, Kobak GE, Gumpricht E (2005) Human hepatic mitochondria generate reactive oxygen species and undergo the permeability transition in response to hydrophobic bile acids. J Pediatr Gastroenterol Nutr 41(2):235–243

Sola S, Castro RE, Laires PA, Steer CJ, Rodrigues CM (2003) Tauroursodeoxycholic acid prevents amyloid-beta peptide-induced neuronal death via a phosphatidylinositol 3-kinase-dependent signaling pathway. Mol Med 9(9–12):226–234

Sola S, Amaral JD, Aranha MM, Steer CJ, Rodrigues CM (2006a) Modulation of hepatocyte apoptosis: cross-talk between bile acids and nuclear steroid receptors. Curr Med Chem 13(25):3039–3051

Sola S, Amaral JD, Borralho PM, Ramalho RM, Castro RE, Aranha MM, Steer CJ, Rodrigues CM (2006b) Functional modulation of nuclear steroid receptors by tauroursodeoxycholic acid reduces amyloid {beta}-peptide-induced apoptosis. Mol Endocrinol 10:2292–2303

Sousa T, Castro RE, Pinto SN, Coutinho A, Lucas SD, Moreira R, Rodrigues CM, Prieto M, Fernandes F (2015) Deoxycholic acid modulates cell death signaling through changes in mitochondrial membrane properties. J Lipid Res 56(11):2158–2171

Turdi S, Hu N, Ren J (2013) Tauroursodeoxycholic acid mitigates high fat diet-induced cardiomyocyte contractile and intracellular Ca2+ anomalies. PLoS One 8(5):e63615

Vang S, Longley K, Steer CJ, Low WC (2014) The unexpected uses of Urso- and tauroursodeoxycholic acid in the treatment of non-liver diseases. Glob Adv Health Med 3(3):58–69

Ved R, Saha S, Westlund B, Perier C, Burnam LG, Sluder A, Hoener M, Mp Rodrigues C, Alfonso A, Steer C, Liu L, Przedborski S, Wolozin B (2005) Similar patterns of mitochondrial vulnerability and rescue induced by genetic modification of alpha -synuclein, parkin and DJ-1 in C. elegans. J Biol Chem 280(52):42655–42668

Viana RJ, Nunes AF, Castro RE, Ramalho RM, Meyerson J, Fossati S, Ghiso J, Rostagno A, Rodrigues CM (2009) Tauroursodeoxycholic acid prevents E22Q Alzheimer's Abeta toxicity in human cerebral endothelial cells. Cell Mol Life Sci 66(6):1094–1104

Xavier JM, Morgado AL, Rodrigues CM, Sola S (2014) Tauroursodeoxycholic acid increases neural stem cell pool and neuronal conversion by regulating mitochondria-cell cycle retrograde signaling. Cell Cycle 13(22):3576–3589

Xavier JM, Rodrigues CM, Sola S (2016) Mitochondria: major regulators of neural development. Neuroscientist 22(4):346–358

Xie Q, Khaoustov VI, Chung CC, Sohn J, Krishnan B, Lewis DE, Yoffe B (2002) Effect of tauroursodeoxycholic acid on endoplasmic reticulum stress-induced caspase-12 activation. Hepatology 36(3):592–601

Xie Q, Li GM, Zhou XQ, Liao D, Yu H, Guo Q (2003) Effect of Tauroursodeoxycholic acid on cytochrome C-mediated apoptosis in HepG2 cells. Zhonghua Gan Zang Bing Za Zhi 11(5):298–301

Xu LH, Xie H, Shi ZH, Du LD, Wing YK, Li AM, Ke Y, Yung WH (2015) Critical role of endoplasmic reticulum stress in chronic intermittent hypoxia-induced deficits in synaptic plasticity and long-term memory. Antioxid Redox Signal 23(9):695–710

Yu KN, Chang SH, Park SJ, Lim J, Lee J, Yoon TJ, Kim JS, Cho MH (2015) Titanium dioxide nanoparticles induce endoplasmic reticulum stress-mediated autophagic cell death via mitochondria-associated endoplasmic reticulum membrane disruption in normal lung cells. PLoS One 10(6):e0131208

Zhang W, Steer CJ, Douglas KT, Rodrigues CM (2006) Binding studies of bile acids using the native fluorescence of the tryptophan residue Of Bax protein. Biosci Rep 26(3):245–250

Zhang Y, Xia Z, La Cour KH, Ren J (2011) Activation of Akt rescues endoplasmic reticulum stress-impaired murine cardiac contractile function via glycogen synthase kinase-3beta-mediated suppression of mitochondrial permeation pore opening. Antioxid Redox Signal 15(9):2407–2424

Zhou L, Jiang L, Xu M, Liu Q, Gao N, Li P, Liu EH (2016) Miltirone exhibits antileukemic activity by ROS-mediated endoplasmic reticulum stress and mitochondrial dysfunction pathways. Sci Rep 6:20585

Pharmacological Targeting of the Mitochondrial Permeability Transition Pore for Cardioprotection

Filomena S. G. Silva, Cláudio F. Costa, Ricardo J. Marques, Paulo J. Oliveira, and Gonçalo C. Pereira

Abstract The mitochondrial permeability transition pore (mPTP) has been shown to play an important role in different pathologies, with particular relevance in cardiovascular diseases. Although different compounds and strategies have been shown to have cardioprotection in experimental models for ischaemia reperfusion injury, so far none of them was effective when translated to clinical practice. The poor clinical success is based in part on the fact that the molecular identity of the pore and its regulators is not completely understood. Additionally, most of the molecules tested have low bioavailability, with only a reduced percentage reaching mitochondria, preventing mPTP opening. The mitochondrial-targeting strategies should be a promising solution to increase the selectivity of compounds to the mPTP, reducing also their potential side effects. In this chapter we perform an integrative description of the effects of different mPTP inhibitors, that directly target one of the mPTP components, as well as developed strategies for desensitizing or inhibition of mPTP through indirect pathways. Moreover, strategies that inhibit mPTP opening by modulating mitochondrial calcium uptake, reactive oxygen species, and intracellular pH, as well as protocols of ischaemia and temperature preconditioning are also included. As a new perspective, we emphasize what should be clarified in the future in order to find potential cardioprotective strategies that may be successfully translated to clinical practice.

Keywords Mitochondrial permeability transition pore · Cardioprotective agents · Heart mitochondria · Mitochondria-targeted antioxidants · Oxidative stress · Calcium overload · Drugs

F. S. G. Silva · C. F. Costa · P. J. Oliveira
CNC – Center for Neuroscience and Cell Biology, University of Coimbra, Cantanhede, Portugal

R. J. Marques
Endocrine Oncology Research Group, Department of Surgery, Royal College of Surgeons in Ireland, Dublin, Ireland

G. C. Pereira (✉)
School of Biochemistry, University of Bristol, Bristol, UK
e-mail: g.pereira@bristol.ac.uk

P. J. Oliveira (ed.), *Mitochondrial Biology and Experimental Therapeutics*,
https://doi.org/10.1007/978-3-319-73344-9_20

Abbreviations

6-MeAla-CsA	NMe-Ala-6-cyclosporin A
β-AR	β-Adrenergic receptors
ΔΨm	Mitochondria membrane potential
Akt	Phosphoinositide-dependent serine/threoninekinase
AMI	Acute myocardial infarction
AMPK	AMP-activated protein kinase
ANT	Adenine nucleotide translocator
BAPTA-AM	1,2-bis(2-Aminophenoxy)ethane-N,N,N9,N9-tetraacetic acid-acetoxymethyl ester
Bax	Bcl-2 associated X protein
BKA	Bongkrekic acid
BNP	B-Type natriuretic peptide
CAGB	Coronary artery bypass grafting
CaMKII	Calcium/calmodulin-dependent protein Kinase II
CDZ	4′-Chlorodiazepam
cGMP	Cyclic guanosine monophosphate
CsA	Cyclosporin A
CsA-NP	Nanoparticles poly-lactic/glycolic acid incorporated with CsA
CVDs	Cardiovascular diseases
Cyp-D	Cyclophilin D
Debio-025	D-3-Methyl-Ala-4-ethyl-Val-CsA
DKO	Double knockout
DMA	Dimethylamiloride
DPAG	Dipyruvyl-acetyl-glycerol
Drp-1	Dynamin-related protein-1
EP	Ethyl pyruvate
Epac	Exchange protein activated by cAMP
EPO	Erythropoietin
Erk	Extracellular signal-regulated kinase
ETC	Electron transport chain
GS	Donor S-nitrosoglutathione
GSK3-β	Glycogen synthase kinase 3 β
hANP	Human atrial natriuretic peptide
HK-II	Hexokinase-II
I/R	Ischaemia/reperfusion
IMM	Inner mitochondrial membrane
IP	Ischaemic preconditioning
KO	Knockout
LIF	Leukaemia inhibitory factor
MCU	Mitochondrial calcium uniporter

Mfn-1 and 2	Mitochondrial fusion-1 and 2
MitoQ	Mitoquinone
mK_{ATP}	Mitochondrial ATP-sensitive potassium channels
mNHE	Mitochondrial sodium/proton exchangers
mPTP	Mitochondrial permeability transition
mtCK	Mitochondrial creatine kinase
mtCSA	Mitochondrial-targeted CsA
NADPH	Nicotinamide adenine dinucleotide phosphate
NEM	*N*-Ethylmaleimide
NHE	Sodium/proton exchangers
NIM811	*N*-Methyl-4-isoleucine-CsA
NO	Nitric oxide
OMM	Outer mitochondria membrane
OPA1	Optic atrophy 1
OSCP	Oligomycin sensitivity-conferring protein
PAO	Phenylarsine oxide
PCI	Percutaneous coronary intervention
pHi	Internal pH
Pi3k	Phosphoinositide 3-kinase
PiC	Phosphate carrier protein
PKA	Protein kinase A
PKC	Protein kinase C
PKG	protein kinase G
PTMs	Post-translational modifications
Pyr2	Proline-rich tyrosine kinase 2
RI	Reperfusion injury
RIPC	Remote ischaemic preconditioning
RISK	Reperfusion injury salvage kinase
ROS	Reactive oxygen species
RuR	Ruthenium red
SfA	Sanglifehrin A
sh-RNA	Short-hairpin RNA
SIRT3	Sirtuin 3
SkQs	Plastoquinone
SNO	S-Nitrosylation
SS	Szeto-Schiller
STAT3	Signal transducer and activator of transcription 3
STEMI	ST segment elevation myocardial infarction
TP	Temperature preconditioning
TPP	Triphenylphosphonium (TPP)
TRO40303	3,5-Seco-4-nor-cholestan-5-one oxime-3-ol
TSPO	Translocator protein of 18 kDa
VDAC	Voltage-dependent anion channel

1 Introduction

The mitochondrial permeability transition pore (mPTP) is a megachannel present in the inner mitochondrial membrane (IMM) involved in physiological and pathological conditions which may rapidly switch between fully open, fully closed, and intermediate sub-states (Zoratti and Szabo 1994; Petronilli et al. 1999). Different studies proposed that the transient opening of mPTP occurs under physiological conditions, serving as a rapid mitochondrial efflux mechanism for excessive reactive oxygen species (ROS) and calcium overload (Wang et al. 2008; Kwong and Molkentin 2015), while others authors refute also the indispensable effect of mPTP in mitochondrial calcium homeostasis (De Marchi et al. 2014; Eriksson et al. 1999). Nevertheless, it has been proposed that sustained mPTP opening is a crucial element in many pathological conditions, such as neurodegenerative and cardiovascular diseases (CVDs) (Bernardi et al. 2015; Kwong and Molkentin 2015). In contrast to the transient opening of mPTP, this megachannel exhibits in the fully open state a non-selective conductance of 1–1.3 nS (Kinnally et al. 1989; Petronilli et al. 1989), which involves a sudden efflux of matrix solutes of up to 1.5 kDa in size, disruption of the mitochondrial membrane potential ($\Delta\Psi m$), and an influx of water into the matrix. This in turn, results in mitochondrial swelling (Bernardi et al. 2015; Kwong and Molkentin 2015), a phenomenon which can be measured in isolated mitochondria or in tissue homogenates using electron (Lee et al. 2012; Lim et al. 2002) or atomic force microscopy (Lee et al. 2011), or monitored in intact cells by fluorescence microscopy (Wilson et al. 2005; Leonard et al. 2015). Indeed, prolonged mPTP opening causes a rapid swelling resulting from the osmotic pressure of matrix solutes, rupture of the outer mitochondria membrane (OMM) and collapse of the $\Delta\Psi m$, which culminates in apoptotic cell death if enough ATP is available or necrosis if ATP levels are not sufficient (Halestrap 2009; Kinnally et al. 2011). Although the role of the mPTP has been evidenced in pathophysiological conditions, the molecular identity of the pore forming unit and its regulators is not completely understood.

In the last 30 years, numerous studies have been dedicated to the identification of the molecular structure and regulatory factors of the mPTP (Hurst et al. 2017; Bernardi et al. 2015). Although it was initially proposed that the mPTP was a supramolecular assembly residing at contact sites between the OMM and the IMM (Kottke et al. 1988) this hypothesis was recently questioned with novel evidences pointing for other players as pore components. The initial model proposed the mPTP to comprise the adenine nucleotide translocator (ANT) in the IMM, cyclophilin D (Cyp-D) in the mitochondrial matrix, and the voltage-dependent anion channel (VDAC) in the OMM (Halestrap and Davidson 1990). Notwithstanding, further studies also proposed the mPTP to be formed by other proteins besides the ANT, CyP-D and VDAC, including Bcl-2 associated X protein (Bax), hexokinase-II (HK-II), mitochondrial creatine kinase (mtCK) and translocator protein of 18 kDa (TSPO) (Hurst et al. 2017; Bernardi et al. 2015). However, subsequent mouse genetic knockout (KO) studies put these models into question.

Several studies using ANT (Kokoszka et al. 2004), VDAC (Krauskopf et al. 2006; Baines 2007), CyP-D (Baines et al. 2005; Basso et al. 2005) and TSPO (Sileikyte et al. 2014) KO models have demonstrated that the mPTP is formed in the absence of these proteins, suggesting that these should be interpreted as functional regulators of the pore opening, and not essential pore components as they had been considered so far.

Later, the phosphate carrier protein (PiC) was also proposed as a component of the mPTP pore, based on the assumption of its interaction with CyP-D and ANT (Leung and Halestrap 2008). This hypothesis was also supported by the discovery that PiC overexpression induces mitochondrial dysfunction and apoptosis (Alcala et al. 2008), together with the earlier finding that a nonspecific pore is generated in liposomes by reconstituting the PiC (Schroers et al. 1997). However, results obtained by patch clamp analysis of the reconstituted PiC did not match the electrophysiological mPTP features (Herick et al. 1997), neither the overexpressing or downregulating of PiC support the idea that this protein is an essential component for mPTP formation (Gutierrez-Aguilar et al. 2014; Varanyuwatana and Halestrap 2012), but a regulatory component instead.

Meantime, He and Lemasters (2002) proposed a different model for the mPTP, in which the pore does not form from a specific protein but rather from aggregation of misfolded integral membrane proteins damaged by oxidants. In this model, the mPTP would be blocked by chaperone-like proteins (including CyP-D) and modulated by calcium. Opening of unregulated pores would occur when protein clusters exceed the CyP-D units available (He and Lemasters 2002). The apparent involvement of the ANT and PiC merely reflects the abundance of these proteins in the IMM and their susceptibility to oxidative damage. However, it is not entirely clear how mPTP regulation by voltage, matrix pH, and different ligands of the ANT and PiC would occur in this model (Bernardi 2013; Halestrap and Pasdois 2009).

A novel mechanism for mPTP formation was also proposed based on the involvement of the ATP synthase, which is formed by a catalytic domain F1 and a membrane sector Fo linked by central and peripheral stalks. The F1 domain situated in the mitochondrial matrix consists of five different subunits (α, β, γ, δ and ε), while the Fo domain includes the regular subunits c, a, b, d, F6, oligomycin sensitivity-conferring protein (OSCP), and the accessory subunits e, f, g and A6L (Jonckheere et al. 2012). The ability of ATP synthase to form channels with the features of the mPTP made it an attractive candidate for mediating the mPTP (Bonora et al. 2013; Azarashvili et al. 2014; Chinopoulos and Szabadkai 2014; Giorgio et al. 2017). According to this model, it was initially proposed that the *c* subunit of mitochondrial ATP synthase played a critical role in mPTP phenomena (Bonora et al. 2013; de Macedo et al. 1993; Jonas et al. 2015; Alavian et al. 2014). However, atomistic simulations indicate that the lumen of the *c*-subunit ring is not an aqueous pore, which is a prerequisite for the ion conductivity of the mPTP (Zhou et al. 2017; Chinopoulos 2017). Moreover, the mPTP is still observed in the absence of the subunit *c* (He et al. 2017), suggesting that a *c*-subunit ring is not an essential, but rather a structural component of the mPTP. Since the *c*-subunit on its own is not sufficient to explain the behavior of the mPTP channel (Bernardi et al. 2015) it was proposed that the mPTP regulation only occurs in the presence of a complete ATP synthase monomer whereas calcium and cyclosporin A (CsA), a well-known mPTP inhibitor (Crompton et al. 1988), regulate the mPTP by interaction with OSCP (Alavian et al. 2014). Additionally, it was suggested by Halestrap (2014) that the regulation of the mPTP may occur in association with ANT and PiC, forming a complex called the synthasome, in which each of these molecules may regulate the structure and activity of ATP synthase, modulating also the opening of the mPTP (Jonas et al. 2015). In parallel to the findings above, another hypothesis was proposed to explain mPTP formation based on the channel formed at the interface of two adjacent monomers in the dimer structure (Giorgio et al. 2013). This hypothesis was supported by the fact that gel-excised dimers, but not monomers, of ATP synthase rapidly gave rise to mPTP-like channels in lipid bilayers (Giorgio et al. 2013; Carraro et al. 2014; von Stockum et al. 2015). However, since cells depleted of mitochondrial DNA (Rho0) still display mPTP activity (Masgras et al. 2012), the role of ATP synthase dimers in forming the mPTP was placed into question because subunits *a* and A6L encoded by mtDNA are important for dimerization of ATP synthase. In agreement with this observation, Pinton's lab proposed more recently that the mPTP opening requires the dissociation of ATP synthase dimers. This assumption is supported by the fact that stabilisation of ATP synthase dimers opposes mPTP opening; while the destabilization of the dimers triggers mPTP opening, by a mechanism that requires a correctly folded *c*-subunit rings (Bonora et al. 2017). This is an hypothesis that is also in agreement with the fact that ATP synthase dimers do not exhibit pore-forming features, being the dimerization associated with improved bioenergetic metabolism and cell survival, rather than cell death mechanisms (Daum et al. 2013). In summary, although several studies suggest a regulatory role of the ATP synthase in mPTP opening, further studies are required to discover its structural role on mPTP opening.

Over the years additional proteins have also been identified as having a critical role in mPTP pore opening, including p53, Bcl-2 family proteins, glycogen synthase kinase 3 beta (GSK3-β), sirtuin 3 (SIRT3), protein kinase G (PKG), protein kinase c isoform ε (PKCε) and complement component 1 Q subcomponent-binding protein (C1QBP); however, their regulatory role remains to be clarified (Hurst et al. 2017; Morciano et al. 2015). In summary, despite all the proposed models the mPTP still remains a highly debated and undefined protein complex in part due to the difficulty in distinguishing the major pore-forming subunits from those merely serving a regulatory role. Therefore, the definitive identification of the molecular identity of the pore forming-component is of utmost importance because knowing its identity should facilitate the development of specific inhibitors and therapies for several diseases which have mitochondrial dysfunction and sustained mPTP opening as central etiologies, such as CVDs (Fancelli et al. 2014; Stavrovskaya et al. 2004).

In this chapter, we make an integrative description of the effects of different mPTP inhibitors that directly target structural or regulatory components of the mPTP as well as indirect strategies used to desensitize or inhibit the mPTP.

2 Directly Targeting Mitochondrial Permeability Transition Pore Components

2.1 *Cyclophilin D*

Cyclophilin D, a member of the greater cyclophilin protein family, is a mitochondrial peptidyl-prolyl cis-trans isomerase, that besides being involved in protein folding and having a chaperone function has also been shown to regulate mPTP opening (Elrod and Molkentin 2013). As a regulatory component of the mPTP, it is expected that CyP-D modulation confers remarkable protection against ischaemia/reperfusion (I/R) injury. Indeed, the pharmacological inhibition of CyP-D or deletion of the gene encoding this protein showed a high level of resistance to I/R injury (Baines et al. 2005; Javadov et al. 2009a; Griffiths and Halestrap 1993, 1995), suggesting that mPTP modulation by CyP-D can prevent the development of CVDs. Both cyclosporin A (CsA) and its derivatives which inhibit mPTP by interaction with CyP-D have been showed to have cardioprotective effects in different models (Javadov et al. 2009a). However, it should be stress out that CyP-D-dependent inhibition of the mPTP has some limitation since mPTP opening still occurs in the presence of extreme calcium concentration, even in CyP-D KO mitochondria (Basso et al. 2005). Consequently, this will decrease the utility of CsA-derivatives in the clinic and explains why drug screenings should identify new mPTP inhibitors independent of CyP-D, at least until the molecular identity of the pore is confirmed. This rationale allowed the identification of a novel class of molecules capable of modulating the mPTP (Briston et al. 2016; Fancelli et al. 2014; Roy et al. 2015).

2.1.1 Cyclosporin A

Since the 1990s, different studies have mentioned that CsA inhibits mPTP opening by binding to Cyp-D, reducing ischaemic death in both *ex vivo* and *in vivo* models of cardiac I/R injury (Fig. 1) (Griffiths and Halestrap 1993, 1995; Lim et al. 2007). In fact, pre-clinical assays have demonstrated cardioprotection in *in vivo* mouse and rabbit models of reperfusion injury (RI) (Argaud et al. 2005; Gomez et al. 2007). However, the lack of protection by CsA has also been reported in *in vivo* rat models of I/R (Dow and Kloner 2007; De Paulis et al. 2013) and pig hearts (Karlsson et al. 2010). These discrepancies are likely due to the duration of ischaemia, which determines the severity of damage (Ruiz-Meana et al. 2011), and CsA dosage (Griffiths and Halestrap 1993), being possible that the concentrations of CsA defined as optimal in one model may prove ineffective in others.

Despite this controversy, studies in human atrial tissue have showned that mPTP inhibition by CsA improves contractile function and cell survival against lethal hypoxia-reoxygenation injury in patients undergoing cardiac surgery (Shanmuganathan et al. 2005). The first clinical pilot study to investigate the mPTP

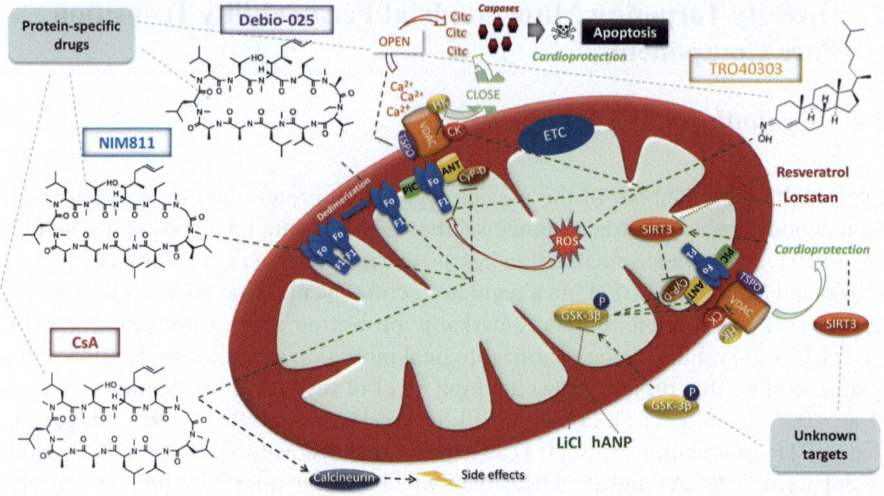

Fig. 1 Different strategies to directly inhibit the mitochondrial permeability transition pore (mPTP) opening by different agents. Cyclosporin A (CsA) inhibits the mPTP opening by binding to cyclophilin D (Cyp-D), reducing the ischaemic death and promoting cardioprotective effects. However, CSA also inhibits calcineurin activity, a mechanism which is responsible for the undesirable effects of this pharmacological approach. The similar cardioprotective effects are also attained with N-methyl-4-isoleucine-CsA (NIM811) and D-3-methyl-Ala-4-ethyl-Val-CsA (Debio-025), two CsA derivatives which also block the mPTP opening by binding to Cyp-D, without effect on calcineurin activity. The cardioprotective effect of 3,5-Seco-4-nor-cholestan-5-one oxime-3-ol (TRO40303), a translocator protein of 18 kDa (TSPO) ligand, is also described. It is mediated by inhibition of mPTP opening which includes its antioxidant activity as well its ability to bind with other possible regulatory components of the pore such as the voltage-dependent anion channel (VDAC) or the newly described pore component, the ATP synthase. Other possible but still unknown targets and mechanisms include the phosphorylation of GSK-3β and the acetylation of sirtuin 3 (SIRT3). The phosphorylation and inactivation of GSK-3β may be modulated by lithium chloride or the human atrial natriuretic peptide (hANP), leading to CyP-D phosphorylation and VDAC de-phosphorylation in addition to the inhibition of CyP-D interaction with the adenine nucleotide tranlocator (ANT). Furthermore, the acetylation of SIRT3 leads to the deacetylation of Cyp-D thereby inhibiting mPTP opening. It is therefore expected that resveratrol and losartan, two compounds that modulate SIRT3 function, may in turn exert cardioprotective effects by inhibition of mPTP opening; however, further studies are needed to confirm these effects. *Cit C* cytochrome c, *CK* creatine kinase, *ETC* electron transport chain, *HK* hexokinase, *PiC* phosphate carrier protein, *ROS* reactive oxygen species

as a therapeutic target in the clinical setting of acute myocardial infarction (AMI) was performed by Piot et al. (2008). In this study, Piot et al. examined the effect of administration of an intravenous bolus of 2.5 mg/Kg CsA in 58 patients with an acute ST segment elevation myocardial infarction (STEMI) immediately before undergoing percutaneous coronary intervention (PCI). The authors showed that patients who received CsA therapy had reduced infarct size by about 40% when compared to placebo control (Piot et al. 2008).

Based on this successful clinical study, a large multicenter, double-blind, randomized phase III clinical trial named "Cyclosporine and Prognosis in Acute Myocardial Infarction Patients" (CIRCUS, #NCT01502774, Hospices Civils de Lyon, France) (Table 1) was performed in 970 patients with an acute STEMI undergoing PCI in order to determine whether CsA (2.5 mg/kg, IV) could improve clinical outcomes after 1 year. Unexpectedly, this study found that CsA did not reduce infarction size neither prevented long-term adverse left ventricular remodeling in comparison with

Table 1 Different strategies of inhibition of mPTP opening for cardioprotective approaches

Strategies	Compounds	Mitochondrial effects	Clinical use and state of application
Direct inhibition by protein-specific drugs	CsA	Interacts with CyP-D, Inhibits mPTP Interacts with CyP-A and inhibits calcineurin	Phase 3 clinical trials for AMI
	NIM811 Debio-025 Sfa	Interacts with CyP-D, Inhibits mPTP Not affect calcineurin activity	Pre-clinical assays Phase 2 clinical trials for hepatitis C treatment
	mtCsA	Interacts with CyP-D, Enhances 12-fold the inhibitory potency mPTP when compared with CsA	Pre-clinical assays
	CsA-NP	Interacts with CyP-D, Inhibits mPTP	Pre-clinical assays
	CZD	TSPO ligands Inhibits mPTP	Pre-clinical assays
	TRO40303		Phase 2 clinical trial for treatment of STEMI
Direct inhibition by unknown targets	LiCl	GSK-3β inhibitor	In clinical use for treatment of bipolar mood disorders
	hANP	GSK-3β inhibitor	Phase 4 clinical trial for treatment of cardiac and renal failure In clinical use for treatment of acute heart failure in Japan
	Resveratrol	Activates SIRT1 and 3	Phases 1/2 and 3 of clinical trials to test its cardiovascular effects Pre-clinical assays
	Losartan	Normalizes the levels of SIRT3	In clinical use for hypertension

(continued)

Table 1 (continued)

Strategies	Compounds	Mitochondrial effects	Clinical use and state of application
Indirect inhibition via HK	Metformin	Activates AMPK Promotes the translocation of HK-II to mitochondria	In clinical use for treatment of diabetes
	LIF	Increases HK-II phosphorylation by Akt. Activates STAT3	Pre-clinical studies
	Morphine	Inhibits GSK-3β Inhibit the HK-II release from mitochondria	Phase 3 of clinical trial for AMI
	Mg^{2+}	Enhances HK-II binding to mitochondria Inhibits mPTP pore opening	Phase 3 of clinical trial for AMI
Indirect inhibition via Drp1	Mdivi-1	Blocks Drp1	Pre-clinical studies
	AIP	Inhibits CaMKII Inhibits mPTP opening	
	KN93		
	Propranolol	β1-AR receptor antagonist Inhibits the phosphorylation of Drp1 at S616, inhibits mPTP opening	In clinical use for hypertension
Indirect inhibition via mitofusins	EPO	Activates Akt, promotes Mfn-1 phosphorylation and its activity, Induces mitochondrial elongation and delays mPTP opening.	Phase 4 of clinical trial for ischaemic cardiomyopathy
	Sevoflurane	Increases Mfn-2 and OPA1 content and decreases in Drp1 content	In clinical use as anesthetic agent
	SRT1720	Activates SIRT1	Pre-clinical studies
Indirect inhibition via mtCK	Creatine	Stabilises the interaction between mtCK and ANT	Pre-clinical studies
	Cyclocreatine		
Indirect inhibition via PKC/PKA	Dipyridamole	Pharmacological activators of PKC through a Gi-protein coupled adenosine receptor-dependent signalling pathway	Phase 4 of clinical trial for I/R injury
	Bradykinin	Activates PI3K and, downstream PKG Results in opening of mKATP and ROS production, that activates PKC	Pilot clinical trials for I/R injury
	8-Br	Activates both PKA and Epac PKA and Epac activate PKC	Pre-clinical studies
	6-Bnz + CPT	6-Bnz is a selective activator of PKA CPT is a selective activator of Epac. PKA and Epac activate PKC	

(continued)

Table 1 (continued)

Strategies	Compounds	Mitochondrial effects	Clinical use and state of application
Indirect inhibition via MCU	BNP	Inhibits MCU, and Ca^{2+} uptake	Phase 1/2 of clinical trial for congestive heart failure
	Doxycycline		Phase 2 of clinical trial for AMI
	Minocycline		Phase 4 of clinical trial for acute ischaemic brain stroke
	RuR		Pre-clinical studies
	RU360		
	PF-431396	Inhibitor of Pyr2. Inhibits a signalling pathway that leads to MCU activation and Ca^{2+} uptake	Pre-clinical studies
	Prazosin	Antagonist of α1-AR. Inhibits a signalling pathway that leads to MCU activation and Ca^{2+} uptake	In clinical use for hypertension
Antioxidants	Carvedilol	Inhibits lipid peroxidation Prevents the depletion of endogenous antioxidants Inhibits mPTP	In clinical use for treatment of hypertension and congestive heart failure
	MitoQ	Inhibits lipid peroxidation	Phase 2 clinical assay for Parkinson's disease and chronic hepatitis C
	SkQs	Inhibits lipid peroxidation ROS scavengers	Phase 2 clinical trial for the treatment of Dry Eye syndrome
	SS-peptides	Inhibits ROS production Inhibits lipid peroxidation Inhibits mPTP opening	Pre-clinical studies
	AntiOxBENs	Inhibits lipid peroxidation ROS scavengers Iron chelators	Pre-clinical studies
Pyruvate derivatives	EP	Antioxidant Anti-apoptotic effects	Phase 2 clinical trial for coronary disease and heart valve disease
	DPAG	–	Pre-clinical studies
NHE inhibition	EMD-87580	Suppresses ROS production Inhibits mPTP opening	Pre-clinical studies Phase 1 clinical trial for patient with Duchenne muscular dystrophy
	AVE-4890	Inhibits mPTP opening via GSK-3β inhibition	Pre-clinical studies
	DMA	Inhibits mPTP	
	BIIB-723	Suppresses ROS production	
	Cariporide	Suppresses ROS production, calcium overload and mPTP opening	Phase 2/3 clinical trials for coronary artery bypass graft Phase 3 clinical trial for MI
	Eniporide	Decreases of calcium overload and ROS production	Phase 2 clinical trial for MI

(continued)

Table 1 (continued)

Strategies	Compounds	Mitochondrial effects	Clinical use and state of application
Ischaemia preconditioning	–	Inhibits mPTP opening Induces mKATP opening Reduces ROS production Activates PKCε pathway Inactivates of GSK-3β and VDAC regulation	Large-scale trials are need or the existent present no overall beneficial effects
Temperature preconditioning	–	Reduces ROS production Inhibits mPTP opening Activates PKCε pathway	Pre-clinical studies

AIP, autocamtide 2-related inhibitory peptide, Akt, phosphoinositide-dependent serine/threoninekinase, AMI, acute myocardial infarction, AMPK, AMP-activated protein kinase, ANT, adenine nucleotide translocator, β1-AR receptor, β-adrenergic receptors, BNP, B-type natriuretic peptide, 6-Bnz, N6-benzoyladenosine-30,50-cyclic monophosphate, acetoxymethyl ester, 8-Br, 8-bromoadenosine-30,50-cyclic monophosphate, acetoxymethyl ester, CaMKII, calcium/calmodulin-dependent protein Kinase II, CPT, 8-(4-chlorophenylthio)-20-O-methyladenosine-30,50-cyclic monophosphate, acetoxymethyl ester, CsA, cyclosporin A, CsA-NP, Nanoparticles poly-lactic/glycolic acid incorporated with CsA, CyP-A, cyclophilin A, CyP-D, cyclophilin D, CZD, 4′-chlorodiazepam, Debio-025, D-3-methyl-Ala-4-ethyl-Val-CsA, DMA, dimethylamiloride, DPAG, dipyruvyl-acetyl-glycerol, Drp1, dynamin-related protein −1, EP, ethyl pyruvate, Epac, exchange protein activated by cAMP, EPO, erythropoietin, GSK-3β, glycogen synthase kinase 3 β, hANP, human atrial natriuretic peptide, HK-II, hexokinase –II, LIF, leukaemia inhibitory factor, I/R, ischaemia/reperfusion, MCU, mitochondrial calcium uniporter, Mfn-1 and 2, mitochondrial fusion-1 and 2, MitoQ, mitoquinone, mK_{ATP}, mitochondrial ATP-sensitive potassium channels, mPTP, mitochondrial permeability transition pore, mtCK, mitochondrial creatine kinase, mtCSA, mitochondrial-targeted CsA, NIM811, N-methyl-4-isoleucine-CsA, OPA1, optic atrophy 1, PKA, protein kinase A, PKC, protein kinase C, Pyr2, proline-rich tyrosine kinase 2, ROS, reactive oxygen species, RuR, ruthenium red, Sfa, sanglifehrin A, SIRT1 and 3, sirtuin 1and 3, SkQs, plastoquinone, STAT3, signal transducer and activator of transcription 3, STEMI, ST segment elevation myocardial infarction, TSPO, translocator protein of 18 kDa, VDAC, voltage-dependent anion channel

placebo (Cung et al. 2015). Furthermore, another randomized multicenter phase III study named "CYCLosporinE A in Reperfused Acute Myocardial Infarction" (CYCLE, #NCT01650662, Mario Negri Institute for Pharmacological Research, Italy) (Ottani et al. 2016) and involving 410 patients with large STEMI showed that a single intravenous CsA bolus just before PCI did not improve clinical outcomes nor left ventricular remodeling up to 6 months. Taken together these clinical trials questioned the cardioprotective effect of CsA in humans, making it extremely important to clarify whether the route, dose and time of administration used are appropriated for CsA to inhibit Cyp-D or whether off-target effects are responsible for the lack of effect.

Despite these negative results, others clinical trials have tested the cardioprotective effect of CsA in cardiac surgery. Hausenloy et al. (2014) demonstrated that a single intravenous bolus of CsA (2.5 mg/kg) administered prior to coronary artery bypass graft surgery could reduce the extent of perioperative myocardial injury since lower postoperative cardiac troponin T serum levels were observed when compared to control. Similarly, it was also demonstrated that CsA protected against RI, promoting a decrease in cardiac troponin I (Chiari et al. 2014). Although these studies emphasize the importance of CsA clinical use for cardioprotection there is still much to unravel and several authors pinpointed two reasons why these compounds may not be ideal agents for cardioprotection. First, CsA provides only modest inhibition of the initial mPTP opening during reperfusion after prolonged ischaemia (Javadov et al. 2003; Javadov and Karmazyn 2007). This is probably due to a very high concentration of calcium and ROS during the first minutes of reperfusion (Griffiths and Halestrap 1995) since it is known that CsA does not inhibit the mPTP under conditions of high calcium overload and oxidative stress (Javadov and Karmazyn 2007). Second, it may be related with the fact that CsA exerts unwanted side-effects by interaction with cyclophilin A and inhibition of calcineurin, which consequently may lead to undesirable effects on heart function (Wilkins and Molkentin 2004) and immunosuppressive activity (Ho et al. 1996). The later is in contrast with different CsA derivatives which do not affect the calcineurin activity (Gomez et al. 2007; Argaud et al. 2005; Javadov et al. 2003) and which will be described in more detail in the next section. Nevertheless, the possible adverse effects of CsA are only associated with chronic treatment, which is distinct from the single-dose protocol required for cardioprotection (Dongworth et al. 2014).

More recently, a new cardioprotective approach was achieved by linking CsA to a triphenylphosphonium cation (TPP^+) in order to take advantage of the charge separation across the mitochondrial membrane and improve accumulation of the drug in the organelle. This mitochondrial-targeted CsA (mtCsA) was shown accumulate in the mitochondrial matrix, to increase 18-fold the binding affinity for CyP-D and to enhance 12-fold the inhibitory potency regarding mPTP opening when compared to CsA in isolated heart mitochondria (Dube et al. 2012). In cardiomyocytes, mtCSA has also showned stonger cytoprotective effect than CsA at lower concentrations (3–15 nM compared with 50–100 nM) (Dube et al. 2012). However, further studies are needed to assess its safety and cardioprotective efficacy *in vivo*. Another recent approach, was based on the development of nanoparticles poly-lactic/glycolic acid to encapsulate and deliver CsA (CsA-NP) to mitochondria. This strategy also showed the ability to increase the cardioprotective effect of CsA against I/R injury *in vivo* by inhibition of mPTP opening (Ikeda et al. 2016). However, the lack of studies in humans places into question the true cardioprotective effect of this strategy, clearly highlighting the need for future human studies.

2.1.2 Cyclosporin A-Derivatives

A number of small-molecule CsA analogs such as *N*-methyl-4-isoleucine-CsA (NIM811) (Argaud et al. 2005; Waldmeier et al. 2002; Hansson et al. 2004), N-Me-Val-4-cyclosporin A (Khaspekov et al. 1999; Nicolli et al. 1996),

NMe-Ala-6- cyclosporine A (6-MeAla-CsA) (Griffiths and Halestrap 1995), and D-3-methyl-Ala-4-ethyl-Val-CsA (Debio-025) (Gomez et al. 2007) have been developed to block mPTP opening through binding to Cyp-D. This effect may in turn decrease mPTP opening without inhibiting calcineurin activity (Gomez et al. 2007; Argaud et al. 2005; Javadov et al. 2003), which is responsible for the immunosuppressive effects of CsA (Ho et al. 1996; Wilkins and Molkentin 2004). As a consequence of mPTP inhibition, treatment with NIM811, 6-MeAla-CsA or Debio-025 reduce cardiac dysfunction induced by I/R injury in rats (Griffiths and Halestrap 1995) as well as infarct size in rabbits (Argaud et al. 2005) and mice (Gomez et al. 2007). Similarly, sanglifehrin A (SfA), an Cyp-D inhibitor with unrelated CsA structure (Clarke et al. 2002; Hausenloy et al. 2003; Lim et al. 2007), also blocked mPTP opening via a conformational change of the pore (Fig. 1) (Clarke et al. 2002) and exerted cardioprotection against *in vivo* (Lim et al. 2007) and *ex vivo* (Hausenloy et al. 2003; Javadov et al. 2003) I/R injury without inhibiting calcineurin activity. Although NIM811 and Debio-025 or SfA can be employed to inhibit mPTP opening (Fig. 1) (Clarke et al. 2002; Griffiths and Halestrap 1995) these compounds have yet to be tested in clinical assays for cardioprotection. Nevertheless, the compounds are currently in clinical trials for hepatitis C treatment, a pathology caused by oxidative stress and mitochondrial dysfunction, providing an interesting cytoprotective model to access the activity of these drugs in humans (NIM-811: #NCT00983060, Novartis Pharmaceuticals, Switzerland; and Debio-0125: #NCT00537407, Debiopharm International SA, Switzerland (Flisiak et al. 2009)) (Table 1).

2.2 *Adenine Nucleotide Translocator*

The direct effect of adenine nucleotides (ADP and ATP) on the mPTP and particularly its regulation by the ADP/ATP ratio has long been observed as powerful inhibitors of mPTP opening by affecting the conformation of the translocase (Kwong and Molkentin 2015; Hurst et al. 2017). Additionally, two inhibitors of ANT, bongkrekic acid (BKA), and atractyloside have been identified as mPTP regulators (Haworth and Hunter 2000; Xu et al. 2001; Halestrap and Davidson 1990). While BKA inhibited calcium-induced mPTP opening in mitochondria isolated from heart and liver (Halestrap and Davidson 1990), atractyloside had opposite effects and sensitized the mPTP to calcium (Haworth and Hunter 2000; Xu et al. 2001). The opposite effect of the inhibitors was explained on the ability of BKA and atractyloside to bind ANT at two distinct sides of the transporter: matrix and cytoplasmic, respectively, and thus inducing "*m*" or "*c*" conformations, respectively (de Macedo et al. 1993; Halestrap and Brenner 2003). Agents stabilising the "*c*" conformation such as high calcium concentration favor mPTP opening, whereas agents which promotes "*m*" conformations are associated with blockage of the mPTP. Oxidative stress is also another factor which induces mPTP pore opening in an ANT conformation-dependent manner (Halestrap et al. 1997b). In fact, ANT represents a major site of oxidative stress and thiol modulation of the mPTP function (Halestrap et al. 1997b; McStay et al. 2002). The cross-linking of two matrix facing cysteine residues (Cys^{56} and Cys^{159}) in the ANT induced by oxidative stress or thiol reagents was described to enhance Cyp-D-ANT interaction and mPTP pore opening (Halestrap and Brenner 2003). A noteworthy tool in the study of ANT-dependent oxidative stress-induced mPTP opening is phenylarsine oxide (PAO), a vicinal thiol reagent which promotes the cross-link between cysteines, stabilising the "*c*" conformation of the ANT and increasing CyP-D binding to the translocase which greatly sensitizes mPTP to calcium (McStay et al. 2002; Halestrap and Brenner 2003). On the contrary, two ubiquinone analogs Ro 68–3400 and UQo induce the "*m*" conformation and reduce the binding between PiC and ANT (Leung et al. 2008), preventing mPTP opening by PAO. Although some compounds have been shown to have cardioprotective effects through inhibition of mPTP opening via interaction with the ANT only a few have been studied. Nevertheless, this strategy should be considered with some caution since the use of the ANT inhibitors may also inhibit catalytic activity of the ADP/ATP carrier and compromise cell viability in energy-demanding tissues.

2.3 *Phosphate Carrier Protein*

The induction of mPTP opening by phosphate (Pi) is well-documented in different studies (Crompton and Costi 1988; Crompton et al. 1988; Roos et al. 1980), suggesting the involvement of the PiC as a component of mPTP pore (Leung et al. 2008; Vaseva et al. 2012). In fact, *N*-ethylmaleimide (NEM), a specific inhibitor of PiC, blocks mPTP opening by covalent modification of Cys^{42} in rat PiC (Leung et al. 2008). However, as NEM also binds to ANT Cys^{57}, inducing its "*m*" conformation (Aquila and Klingenberg 1982), it is expected that the interaction NEM--ANT may also explain the inhibitory effect on the mPTP (McStay et al. 2002). Moreover, ubiquinone analogues (UQo and Ro 68–3400) were also described to inhibit PiC in rat liver mitochondria by inducing a conformation change similar to NEM that favors the close state of the pore (Leung et al. 2008). Despite the protective action, these ubiquinone analogues have been described to cause damage in the heart rather than being cardioprotective (Halestrap et al. 2004). Considerering its role on oxidative phosphorylation it is possible that PiC targetting may also have harmful effects in cellls, similarly to those describe for the ANT. More recently, a new compound of the family of highly potent cinnamic anilide-based inhibitors of the mPTP (Fancelli et al. 2014; Martin et al. 2014) designated by GNX-4975 was described to bind to the interface of ANT-PiC stabilising the interaction between the two proteins. It was proposed that a protein conformational change is required to form the mPTP complex and generate the GNX-4975-binding site. Overall, it suggests a new strategy of mPTP inhibition which involve a synergism between the ANT and PiC but which stills needs to be further explored *in vivo* (Richardson and Halestrap 2016).

2.4 ATP Synthase

The ATP synthase is increasingly believed as a molecular switch between life and death by regulating energy metabolism and sensitizing cells to death (Nesci et al. 2016, 2014). Although a wide variety of ATPase inhibitors have been studied (Hong and Pedersen 2008), including antibiotics, polyphenolic compounds, estrogens, polyketides and organotin compounds among others, compounds which inhibit mPTP pore through a conformational modulation of the ATP synthase have been poorly investigated. However, ATP synthase plays an important role in ischaemia because upon oxygen deprivation it switches from ATP synthesis to ATP hydrolysis in an attempt to maintain a membrane potential. This waste of energy is not only detrimental for the mechanical work of the heart as it can also sensitize the mPTP due to the loss of the adenine nucleotide pool. Therefore, inhibition of ATP synthase activity during ischaemia would be cardioprotective and indeed oligomycin and aurovertin protect the heart during I/R injury (Grover et al. 2004). However, a downside of these drugs is that they inhibit ATP synthase in both forward and reverse mode. This limitation prompted a drug screening by Bristol-Myers for small molecules which inhibit the hydrolase activity with minor effects on the synthase activity. The screening identified a compound, BMS-199264, which was capable of preserving ATP during ischaemia and exerting significant cardioprotection (Grover et al. 2004). More recently, a chemoinformatic screen based on BMS-199264 identified a new drug candidate, BTB06584, which selectively inhibits the reversal of ATP synthase in a IF1-dependent manner (Ivanes et al. 2014). Nonetheless, neither compounds have moved to clinical trials. However, it should be stress out that even though the target of these compounds is the ATP synthase their effect on the mPTP would be indirect and a consequence on the maintenance of the adenine nucleotide pool.

Regarding the ATP synthase role in the mPTP context, a noteworthy study performed by Campo et al. (2016) attempted to assess the *c* subunit levels in a cohort of 158 patients with STEMI treated with PCI. Results showed that elevated *c* subunit serum levels were significantly related to a worse prognosis. Although additional studies are necessary to investigate this correlation, the collected data clearly indicates that patients with higher *c* subunit content are most likely to suffer from hyper-responsive mPTP activity at the moment of reperfusion and thus a higher propensity for RI (Campo et al. 2016). However, further studies are needed to elucidate the molecular composition of the mPTP for correctly designing pharmacological approaches targeting mPTP modules.

2.5 Unknown Direct Target

Post-translational modifications (PTMs) such as phosphorylation (Rasola et al. 2010), acetylation (Bochaton et al. 2015) and nitrosylation (Nguyen et al. 2011) of the regulatory components of the pore have been identified as strategies for mPTP inhibition (Marquez et al. 2016; Rasola and Bernardi 2011). An example is the dynamic network of kinase/phosphatase pathways which have been postulated to promote regulatory effects on the mPTP. Of particular interest is the inhibition of GSK-3β (Miura and Tanno 2010; Gomez et al. 2008; Tong et al. 2002) which now represents an attractive strategy for cardioprotection (Zorov et al. 2009; Miura and Miki 2009). In fact, GSK-3β phosphorylation and thus its inactivation results in mPTP desensitization as a consequence of CyP-D phosphorylation. This renders cells more refractory to cell death (Rasola et al. 2010). In addition to the phosphorylation of CyP-D GSK-3β other mechanisms were also proposed such as the association of phospho-GSK-3β with the ANT (Nishihara et al. 2007) or the suppression of GSK-3β-mediated VDAC phosphorylation which leads to maintenance of HK-II in the OMM (Camara et al. 2010). Both mechanisms prevent the interaction of Cyp-D with ANT leading to the inhibition of mPTP opening (Fig. 1). However, this point raises an interesting question on how OMM events can be transmitted and assimilated by proteins in the IMM and/or in the mitochondrial matrix. A possible explanation is that this "communication" may occur through contact sites between IMM and OMM although further studies are required to unveil these effects.

A different mechanism proposed by Das et al. (2008) suggested that GSK-3β inhibition allows the de-phosphorylation of VDAC which prevents the entry of adenine nucleotides into mitochondria, promoting mPTP opening. Altogether these mechanisms may contribute to mPTP inhibition, suggesting that pharmacological modulation of GSK-3β activity represents an attractive therapeutic strategy against heart failure. *In vitro* studies have shown that pharmacological inhibition of GSK-3β by lithium chloride mimics the cardioprotective effect of ischaemic preconditioning (IP) and post-conditioning. In contrast to this, the constitutively active form of GSK-3β, containing a serine-9-to-alanine mutation which prevents its inhibition by phosphorylation, clearly abolishes its cardioprotection (Juhaszova et al. 2004, 2009). Moreover, the cardioprotection induced by IP and post-conditioning through modulation of the mPTP opening and inhibition of GSK-3β have also been observed in rat, rabbit and mouse hearts treated with different compounds, including bradykinin (Park et al. 2006b), opioids (Obame et al. 2008), erythropoietin (EPO) (Kobayashi et al. 2008), adenosine A1/A2 (Forster et al. 2006) and A3 receptor agonist (Park et al. 2006a), volatile anesthetic agents (isoflurane (Feng et al. 2005), sevoflurane (Onishi et al. 2012)), rosuvastatin (Liu et al. 2017), and resveratrol (Xi et al. 2009), thus representing promisors therapeutic agents against cardiac damage. However, poorly is known about efficacy of these GSK-3β inhibitors for CVDs, being lithium chloride the only GSK-3β inhibitor in clinical use for bipolar mood disorders (Table 1) (Cohen and Goedert 2004).

Another compound that represents a promisor therapeutic agent against cardiac damage is the human atrial natriuretic peptide (hANP). Recombinant hANP, called carperitide, has been used successfully in Japan for the treatment of acute heart failure, showing its safety and effectiveness in the COMPASS trial (Nomura et al. 2008) and by improving the prognosis in the PROTECT trial (Hata et al. 2008). In an earlier clinical study, administration of hANP showed significant reduction in infarct size in AMI patients (Kitakaze et al. 2007). This effect could be in part due to its ability to inactivate GSK3-β and thus to inhibit mPTP (Hong et al. 2012), but to our knowledge this possibility has not been explored. More recently, a phase IV clinical trial was performed with carperitide to investigate the effects of the compound on short- and long-term prognosis in patients with cardiac and renal failure (#NCT00613964, Nara Medical University, Japan). Although the trial has finished no results have been disclosed at the present time (Table 1). Therefore, the role of GSK-3β in the human myocardium and efficient strategies to inhibit GSK-3β for protection of cardiomyocytes warrant further investigation.

Another PTM described to regulate mPTP involves the acetylation/deacetylation of mitochondrial proteins (Hafner et al. 2010). However, whether hyper-acetylation of a single protein or a select group of proteins is the main contributor to the increase in sensitivity of the mPTP remains to be clarified (Karamanlidis et al. 2013). It has been proposed that SIRT3, belonging to the sirtuin family of proteins (SIRT1–7), deacetylates Cyp-D on lysine^{166} thereby inhibiting mPTP opening; however, the full mechanism remains to be elucidated. It was suggested that loss of SIRT3 activity leads to increased activation of the mPTP in response to calcium rise resulting in cardiac mitochondrial dysfunction (Hafner et al. 2010). More recently, it has also been described that ischaemic post-condition attenuates RI via the activation of SIRT3 with subsequent deacetylation of CyP-D and inhibition of mPTP opening (Bochaton et al. 2015). Therefore, strategies that promote SIRT3 expression or its activity may represent pharmacological approaches to prevent cardiac failure.

Although drugs which inhibit mPTP through SIRT3-dependent pathways have not been reported so far, some pharmacological agents which promote cardioprotection by activating SIRT3 have been described (Hu et al. 2016; Koentges et al. 2016). One example is resveratrol which was originally reported to activate SIRT1 and to decrease the risk of CVDs (Zordoky et al. 2015). More recently, it was described that resveratrol can prevent cardiac hypertrophy *via* SIRT3 activation (Fig. 1) (Chen et al. 2015). Resveratrol is being investigated in clinical phase I/II and III trials to evaluate its effects in cardiovascular health in elderly (#NCT01842399, National Institute on Aging, US) and in patients with non-ischaemic cardiomyopathy (#NCT01914081, St. Boniface General Hospital Research Centre, Canada) (Table 1). Another compound that has been proposed to exert protective effects in IR injury through SIRT3 is Losartan, an angiotensin II type I receptor blocker that was introduced as an anti-hypotensor medicine (Fig. 1) (Klishadi et al. 2015). The resulting increase in SIRT3 levels by Losartan promoted the transcription of catalase, manganese superoxide dismutase (MnSOD) and thioredoxin-1 (Klishadi et al. 2015), three antioxidant enzymes which inhibit cardiac dysfunction. Therefore, the modulation of SIRT3 has emerged as a promiser pharmacological intervention for CVDs; however, further studies are needed to clarify the effect of these compounds on clinical trials.

S-nitrosylation (SNO), a covalent attachment of nitric oxide (NO) to a protein thiol group, has also been described to interfere with mPTP (Nguyen et al. 2011) but it remains to be clarified whether SNO really targets mPTP pore directly (Piantadosi 2012). It is known that SNO exerts protective effects by modifying cysteine residues and thereby shielding crucial cysteine residues from irreversible oxidative modification during I/R injury (Sun et al. 2006b), suggesting that SNO may play an important role in cardioprotection. Moreover, it has been shown that over-expression of inducible nitric oxide synthase inhibits mPTP opening during I/R injury (West et al. 2008). However, some controversy exists, for example, while low NO concentrations delay pore opening in response to calcium (Brookes et al. 2000; Leite et al. 2010), higher concentrations (>20 μM) behave as strong oxidants that facilitate pore opening (Piantadosi et al. 2002). Nevertheless, pharmacological modulation of the mPTP with a NO donor S-nitrosoglutathione (GSNO) results in SNO of cysteine^{203} on Cyp-D (Kokoszka et al. 2004) and subsequent inhibition of mPTP opening (Nguyen et al. 2011). Furthermore, other NO donors prevent RI by inhibition of mPTP opening in an ANT-dependent manner (Wang et al. 2005). S-nitrosylation was also described to have a regulatory role in VDAC function. In fact, it was shown that low levels of NO (120–900 nM) afford cardioprotection through inhibition of VDAC and in consequence a delay on mPTP opening (Cheng et al. 2011). Additionally, Kohr et al. (2011) showed that all three VDAC isoforms can be S-nitrosylated, either as a consequence of IP or treatment with GSNO, protecting the heart from I/R injury.

ATP synthase has also been considered as another target for nitrosative modifications, affecting its activity and potentially increasing the occurrence of the mPTP (Kaludercic and Giorgio 2016). In fact, a protective effect of GSNO against I/R injury was reported to occur through SNO of thesubunit of ATPase resulting in decreased ATP hydrolysis and desensitization to the mPTP (Sun et al. 2007). Moreover, SNO has important mitochondrial and non-mitochondrial effects, such as reducing calcium loading during ischaemia and early reperfusion, which may impact mPTP function and thereby reduce I/R injury (Adachi et al. 2004; Manevich et al. 2010; Sun et al. 2006a). Interestingly, evidences show that only about 1% of the mitochondrial proteome is available for SNO which represents a small but significant number of mitochondrial proteins (Prime et al. 2009), suggesting that SNO modulation may represent an important therapeutic strategy in cardioprevention. However, the full mechanism of how SNO regulates/modulates mitochondrial proteins is still unclear and demonstrates that a complete understanding of mitochondrial SNO will improve our insights into cell physiology and pathology, in addition to reveal new avenues for mitochondrial therapy.

2.6 Outer Mitochondrial Membrane Proteins

2.6.1 Voltage-Dependent Anion Channel

Voltage-dependent anion channel, also known as mitochondrial porin, is the most abundant protein in the OMM and mediates the passages of metabolites, nucleotides, and ions into mitochondria, in addition to its role in apoptosis (Camara et al. 2017). Over the years, several models have suggested VDAC to be part of the mPTP (Veenman et al. 2008; Azoulay-Zohar et al. 2004) but the pharmacological modulation of the pore mediated by VDAC has not been considerably exploited yet (Baines et al. 2007). The translocator protein of 18 kDa (Veenman et al. 2008) and HK (Azoulay-Zohar et al. 2004) were described to modulate the mPTP through interaction with VDAC (McCommis and Baines 2012); however, their cardioprotective effects remain to be clarified. The interaction between TSPO and VDAC has also been shown to promote an overproduction of ROS that modulates autophagy and apoptosis (Gatliff et al. 2014; Veenman et al. 2007). It was hypothesized that ROS would act as mediator between TSPO and VDAC such that TSPO ligands would inhibit its activity reducing ROS generation and consequently prevent mPTP pore opening (Veenman et al. 2008). Although poorly understood, different studies have proposed that ROS induce alterations in VDAC and/or ANT which promotes cytochrome *c* release from mitochondria and its consequent induction of apoptotic cascade (Madesh and Hajnoczky 2001; Le Bras et al. 2005). Despite the abovementioned effects it should be kept in mind that the consequences of the TSPO-VDAC interaction have not been described in CVDs. Whether this interaction plays a role during myocardial I/R injury and may be a target for the cardioprotective effect of TSPO ligands requires further investigations.

Hexokinases are also considered to play a role in the regulation of VDAC function. In fact, it has been proposed that HK-II binding to VDAC on the OMM prevents the binding of pro-apoptotic proteins such as Bax resulting in a inhibition of mPTP opening (Mathupala et al. 2006; Pastorino et al. 2002) (described in detail in Sect. 3.1).

2.6.2 Translocator Protein of 18 kDa

Translocator protein of 18 kDa first denominated PBR for "peripheral benzodiazepine receptors" is a transmembrane protein (Anholt et al. 1986) predominantly located in intimate contact sites between OMM and IMM membranes (Culty et al. 1999). TSPO can form a multimeric complex with the VDAC and ANT (McEnery et al. 1992; Garnier et al. 1994). As a consequence of this structural organization, TSPO has been proposed as a regulatory component of the mPTP (Ricchelli et al. 2011). The idea that TSPO was a component of the mPTP was also supported by the fact that TSPO ligands regulated mPTP opening (Azarashvili et al. 2007) and were able to alter the electrophysiological activity of the channel (Kinnally et al. 1993). However, its molecular nature is still a source of debate and several studies questioned the role of TSPO in the molecular structure of the mPTP (Sileikyte et al. 2014).

Several studies have reported the ability of TSPO ligands to improve cardiac functions, making TSPO a potential target for CVDs (Obame et al. 2007; Brown et al. 2008; Surinkaew et al. 2011) even though its role in the cardiac tissue still remains elusive. Some studies with TSPO ligands demonstrated their beneficial effects in different animal models of myocardial infarction by inhibition of mPTP opening after I/R injury (Obame et al. 2007; Xiao et al. 2010). One example involves

4′-chlorodiazepam (CDZ) which protects against myocardial injury by increasing the mitochondrial resistance to calcium-induced mPTP opening (Obame et al. 2007). The inhibitory effect of CDZ on mPTP was also confirmed by Xiao et al. (2010). These authors showed that CDZ improves cardiac functional recovery during reperfusion probably through inhibition of mPTP opening, reduced ROS production and thereby affecting the activities of xanthine oxidase and nicotinamide adenine dinucleotide phosphate (NADPH) oxidase (Xiao et al. 2010); however, some controversy still exists. For example, at higher concentrations CDZ (~100 μM) was not able to limit mPTP opening *in vitro* but rather induced the phenomenon (Obame et al. 2007). Different studies in isolated cardiac mitochondria also showed that other TSPO ligands, such as (1-(2-chlorophenyl-N-methyl-1-methylpropyl)-3-isoquinolinecarboxamide (Li et al. 2007), 7-chloro-5-(4-chlorophenyl)-1,3-dihydro-1-methyl-2H-1,4-benzodiazepin-2-one, and diazepam produce a dose-dependent inhibition of the mPTP pore opening (Chelli et al. 2001). These results suggest that TSPO ligands can activate or inhibit mPTP opening in the same cell type depending on the ligand concentration used (Obame et al. 2007; Xiao et al. 2010).

Another TSPO ligand, 3,5-Seco-4-nor-cholestan-5-one oxime-3-ol (TRO40303) was recently described as having cardioprotective properties by reducing infarct size in a rat model of cardiac I/R injury and through inhibition of mPTP pore opening (Schaller et al. 2010). In fact, TRO40303 reduced mPTP opening in isolated cardiomyocytes without showing any effect on isolated cardiac mitochondria, suggesting that inhibition of mPTP by TSPO ligands is not compatible with a direct effect of the drugs on the channel but should be due to indirect effects (Schaller et al. 2010).

Different mechanisms have been proposed to explain the indirect effect of TSPO ligands on mPTP opening. One theory proposes that TSPO ligands may act as antioxidants, reducing ROS production under conditions of I/R injury (Morin et al. 2016). Others have suggested that the interaction of TSPO ligands with other regulatory components of the pore, such as with VDAC (Veenman et al. 2008; Gatliff et al. 2014) or ATP synthase (Cleary et al. 2007; Stelzer et al. 2010; Giorgio et al. 2013), rather than with TSPO itself explains the modulation of the mPTP; however, further studies are needed to clarify these hypothesis (Fig. 1).

Despite promising preclinical results of TRO40303 and the demonstration of its safety, tolerability and pharmacokinetic parameters in humans in a Phase I trial (EudraCT number: 2010-021453-39) (Le Lamer et al. 2014), a Phase II proof of concept trial (#NCT01374321,Trophos, France (Group MS 2012)) was also performed (Table 1). In this clinical trial, the administration of TRO40303 in patients undergoing revascularization for STEMI failed to show a significant reduction in infarct size and thus did not lend support for the use of TRO40303 as a pharmacological agent to reduce I/R injury in the infarcted myocardium (Atar et al. 2015). In fact, the lack of clinical efficacy of these compounds (Atar et al. 2015) together with the fact that the use of TSPO KO models (Sileikyte et al. 2014) have practical no effect on the pore opening point in question its use as a clinical target for the development of therapies based on mPTP opening.

3 Indirectly Inhibition of Permeability Transition Using mPTP-Independent Targets

3.1 *Hexokinase*

Hexokinase-II plays an important role in glucose metabolism by catalyzing the first step in glycolytic pathway in addition to its effects in cell survival (Robey and Hay 2006). In fact, the anti-apoptotic functions of HK-II in mitochondria have been extensively described in the last years. Although the mechanisms responsible for the anti-apoptotic effect of HK remains poorly understood, it has been proposed that it competes with Bcl-2 family proteins for VDAC binding, influencing the balance of pro-and anti-apoptotic proteins and therefore OMM permeabilization (Pastorino and Hoek 2008). Actually, it has been shown that HK-II binds to VDAC at the OMM preventing in turn the binding of the pro-apoptotic protein Bax to VDAC. Therefore, when HK-II is released from OMM VDAC becomes available for Bax binding, promoting the release of cytochrome *c* and consequent activation of caspases and induction of apoptotic cell death (Mathupala et al. 2006; Pastorino et al. 2002). One of the forms of inducing HK-II translocation from mitochondria to cytosol is through phosphorylation of HK-II binding site on VDAC *via* GSK-3β activation (Pastorino et al. 2005).

The loss of mitochondrial-bound HK-II was first shown by the group of Zuurbier (Gurel et al. 2009; Smeele et al. 2011) followed by Halestrap (Pasdois et al. 2012). It was demonstrated that the accumulation of glucose-6-phosphate and low pH contributed to the loss of mt-HK-II, an event that was associated with loss of cytochrome *c* and membrane potential (Pasdois et al. 2011, 2012) and which could sensitise mitochondria to mPTP. Interestingly, the extent of mt-HK-II loss during ischaemia showed a strong positive correlation with the damage attained to the heart upon reperfusion. Considering that during ischaemic conditions HK-II dissociates from mitochondria, Halestrap et al. (2015) postulated that contact sites of mitochondrial membranes are disrupted in these conditions because HK-II is stabilised at these contact sites in mitochondria under physiological conditions. Since HK-II interference leads to cristae remodelling and destabilization of ATP synthase, with consequent ATP hydrolysis and cytochrome *c* release (Halestrap et al. 2015) it is possible that strategies that inhibit HK-II dissociation can prevent these detrimental events. Therefore, an increase HK-II binding to VDAC is a way of indirectly inhibiting mPTP opening. An example is the use of metformin, an anti-diabetic drug, which increases the translocation of HK-II to mitochondria in diabetic hearts, resulting in cardioprotective effects (Halestrap et al. 2015; Da Silva et al. 2012). However, the mechanism of action underlying the cardioprotective effects of metformin remains unclear. Nederlof et al. (2014), proposed that the activation of AMP-activated protein kinase (AMPK) can lead to HK-II translocation to mitochondria. Considering that AMPK pathway can be activated by increased AMP levels resulting among others from inhibition of complex I (Owen et al. 2000), it is possible that compounds that inhibit complex I, as is the case of metformin (Bhamra et al. 2008;

Halestrap et al. 2015; Calvert et al. 2008; Nederlof et al. 2014) can promote cardioprotective effects through the activation of the AMPK pathway (Fig. 2). However, further studies are needed to clarify these effects and also to explore whether AMPK pathway can be a strategy to inhibit mPTP opening.

Another strategy to increase HK-II binding to mitochondria is by its upregulation using miR-155 which induces HK-II expression through activation of the signal transducer and activator of transcription 3 (STAT3) and inhibition of the miR-143, an activator and a negative regulator of HK-II transcription, respectively (Nederlof et al. 2014; Halestrap et al. 2015; Jiang et al. 2012). However, as several studies suggest that miR-155 promotes inflammation (Caballero-Garrido et al. 2015; Xing et al. 2016; Eisenhardt et al. 2015) further studies evaluating the hypothetical role of miR-155 in cardioprotection are required to clarify its side effects. Since the activation of STAT3 is also associated with cardioprotection, STAT3-directed therapies may also be an attractive alternative (Wu et al. 2017). An example is the case of leukaemia inhibitory factor (LIF) which activates STAT3 (Zouein et al. 2013). In fact, LIF showed an inhibitory effect on calcium- and H_2O_2-induced mPTP in myocytes requiring phosphoinositide-dependent serine/threonine kinase (Akt) phosphorylation. It was demonstrated that LIF increases HK-II phosphorylation via Akt pathway, increasing its binding to mitochondria (Miyamoto et al. 2008) (Table 1). However, further studies are required to understand LIF cardioprotective role.

Phosphorylation and consequent inhibition of GSK-3β is another therapeutic approach since this inhibition can de-phosphorylate VDAC and preserve HK-II binding to mitochondria (Pastorino and Hoek 2008). An example of one compound which promotes cardioprotective effect by promoting the inhibition of GSK-3β with consequent increase of mitochondria-bound HK-II is morphine (Zuurbier et al. 2005; Gross et al. 2004) (Table 1). A phase 3 clinical trial was performed to evaluate the cardioprotective effect of intracoronary injection of morphine during reperfusion in AMI (#NCT01186445, French Cardiology Society, France). The results remain to be disclosed. However, a recent study showed that morphine administration does not appear to have effect on STEMI patients subjected to PCI, which may have results from an incorrect dosage of morphine, exposure to previous morphine treatment (influencing measured parameters), or to an incorrect administration strategy, rather than a lack of effect of morphine, so, further studies are needed to clarify whether morphine can be useful for CVDs (Gwag et al. 2017).

Finally, another approach includes the use of Mg^{2+}, a competitive inhibitor for Ca^{2+} binding sites on the mPTP (Bernardi et al. 1992; Golshani-Hebroni 2016) and a stabilising agent for HK-II binding to mitochondria, both promoting mPTP closure (Golshani-Hebroni 2016; Pastorino and Hoek 2008; Wilson 1982). In this context, a phase 3 clinical trial (code NCT00000610, National Heart, Lung, and Blood Institute, US) was performed in patients with suspected acute myocardial infarction in order to reduce mortality. However, this clinical trial showed that intravenous Mg^{2+} administration in STEMI patients did not reduce the mortality rate (Table 1) (Magnesium in Coronaries Trial 2002). Therefore, further studies are needed to clarify whether drugs that increase Mg^{2+} bioavailability or Mg^{2+} itself can augment the anti-apoptotic effects of HK (Golshani-Hebroni 2016).

Nevertheless, the most efficient strategy to prevent the HK-II dissociation from OMM during ischaemia is by ischaemic preconditioning. It was proposed that IP decreases glycolysis during ischaemia with a reduction of lactic acid accumulation which prevents the pH drop and the HK-II dissociation (Halestrap and Richardson 2015; Pasdois et al. 2012). Based on this assumption, any pharmacological strategy that can prevent the pH drop during ischaemic insult may prevent HK-II dissociation, promoting the mPTP inhibition.

In sum, the different strategies described above such as prevention of pH drop as well as inhibition of GSK-3β or activation of AMPK and STAT3 pathways lead to increase binding of HK-II to mitochondria and may represent different therapeutic strategies to prevent CVDs through inhibition of the mPTP (Fig. 2). These are strategies that should be addressed because at the present time there is no drug approved to counteract this effect. However, despite the advantages of a cardioprotective strategy involving small molecules that promote HK-II-VDAC association serious limitations will coexist. For example, increased HK-II-VDAC association protects tumour cells from mitochondrial outer membrane permeabilization, making them resistant to chemio- and radiotherapy (Galluzzi et al. 2008; Kochel et al. 2017).

3.2 Dynamin-Related Protein-1/Mitofusin

Mitochondrial health is maintained by quality control mechanisms of mitochondrial dynamics (fission and fusion) and mitophagy (Twig et al. 2008; Hall et al. 2016). Alterations of mitochondrial fusion and fission have been described to influence mPTP opening (Cribbs and Strack 2007; Hall et al. 2016). An example of such class of proteins is the dynamin-1-like protein, also known as dynamin-related protein 1 (Drp1), a GTPase found in the cytosol that translocate to mitochondria to mediate mitochondrial fission allowing the maintenance of a reticular network. Adaptor proteins such as Mid49/51 and mitochondrial fission factor facilitate Drp1 translocation while other proteins such as mitochondrial fission-1 will anchor Drp1 to the OMM. Drp1 association with the mPTP was never direct but as with other OMM-bound proteins there are several lines of evidence suggesting that it might regulate mPTP opening. Mitochondria have been reported to undergo fission during I/R (Ong et al. 2010; Disatnik et al. 2013; Gao et al. 2013; Sharp et al. 2014) and genetic ablation (Ong et al. 2010) or transgenic expression, of a dominant negative form of Drp1 (Zepeda et al. 2014) reduces infarct size upon reperfusion. This was associated with lower oxidative stress, decreased sensitivity to mPTP opening, preserved mitochondrial physiology and morphology, and reduced cytochrome *c* release. Moreover, the use of Mdivi-1, a Drp1 blocker, was shown to inhibit mPTP opening and to have beneficial effects against cardiac dysfunction (Cribbs and Strack 2007; Zhang et al. 2017), emerging as a strategy to prevent CVDs (see Table 1). However, further studies are needed to clarify these effects on humans. It is still unclear how Drp1 mediates mPTP opening but considering Drp1 requirement during apoptosis for cristae remodeling and cytochrome *c* release (Prudent et al. 2015) there may also

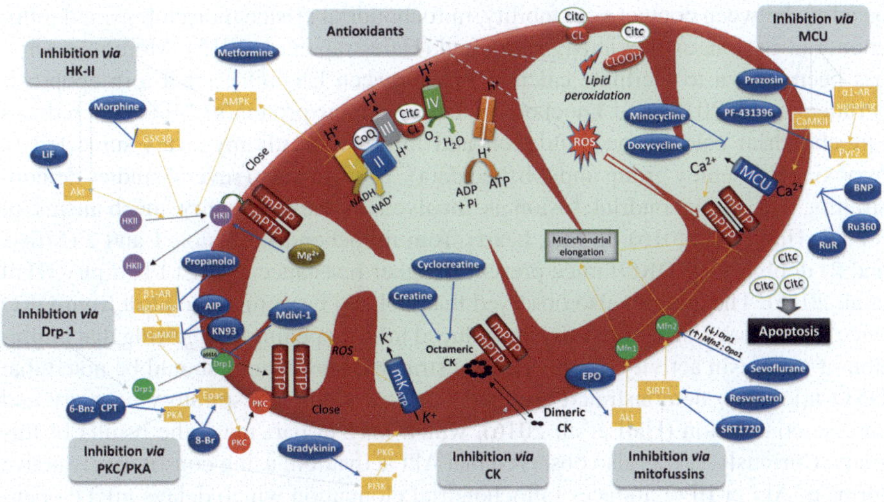

Fig. 2 Different strategies that indirectly inhibit the mitochondrial permeability transition pore (mPTP). Some compounds and signaling pathways described in this figure promote HK-II binding to mitochondria and therefore inhibit the mPTP. These include the activation of AMP-activated protein kinase (AMPK) and phosphoinositide-dependent serine/threoninekinase (Akt) pathways, as well as the inhibition of glycogen synthase kinase 3 β (GSK-3β). Metformin and leukaemia inhibitory factor (LIF) are two other compounds that promote HK-II binding to mitochondria through activation of AMPK and Akt signaling pathways. On the contrary, morphine blocks mPTP opening by inhibition of GSK-3β pathway. The effect of Mg^{2+} is also indicated leading to mPTP closure by increasing HK-II binding to mitochondria. Considering that the phosphorylation of dynamin-related protein-1 (Drp1) at S616 leads to mPTP opening and myocyte death, different strategies to inhibit Drp1 are also referred. These include Drp1 direct inhibition, using Mdivi-1, as well as the inhibition of β-adrenergic receptors (β-AR) or calcium/calmodulin-dependent protein kinase II (CaMKII) using propranolol or autocamtide 2-related inhibitory peptide (AIP) and KN93, respectively. The indirect inhibition of mPTP by modulation of mitofussins is also described by the use of sevoflurane, erythropoietin (EPO), resveratrol or SRT1720. While EPO leads to mitochondrial fusion-1 (Mfn-1) activation in a Akt-dependent manner, sevoflurane shows cardioprotective effects by reducing Mfn-2 and optic atrophy 1 (Opa-1) levels while increasing Drp1 protein expression. Resveratrol or SRT1720 also modulate Mfn-2 in a sirtuin 1 (SIRT1)-dependent manner. The inhibition of mPTP opening by activation of protein kinase C (PKC)/protein kinase A (PKA) pathways is also depicted, either by activation of Phosphoinositide 3-kinase (PI3k) using bradykinin, which consequently leads to mitochondrial ATP-sensitive potassium channels (mK_{ATP}) opening and reactive oxygen species (ROS) production, or by activation of PKA/exchange protein activated by cAMP (Epac) pathway, using 8-bromoadenosine-30,50-cyclic monophosphate, acetoxymethyl ester (8-Br) or N6-benzoyladenosine-30,50-cyclic monophosphate, acetoxymethyl ester (6-Bnz) and 8-(4-chlorophenylthio)-20-O-methyladenosine-30,50-cyclic monophosphate, acetoxymethyl ester (CPT). The stabilisation of the octameric form of creatine kinase using creatine or cyclocreatine of creatine kinase (CK) is other mechanism to inhibit mPTP opening. The inhibition of mitochondrial calcium uniporter (MCU) is another mechanism which blocks mPTP pore, either by the use of direct inhibitors of MCU, as B-type natriuretic peptide (BNP), doxycycline, minocycline, ruthenium red (RuR) or RU360, or by the modulation of the channel. This modulation may be performed using antagonist of α1-AR (Prazosin) or by inhibition of proline-rich tyrosine Kinase 2 (Pyr2), using PF-431396 that leads to phosphorylation and inhibition of MCU. Other strategies to inhibit mPTP opening are also shown in this figure, including the use of different antioxidants, which act as ROS scavengers, inhibit lipid peroxidation and oxidation of protein thiols.

be a link between contact site stability, mitochondrial cristae morphology and mitochondrial fission, as previously suggested (Halestrap et al. 2015). Moreover, Drp1 has been shown to facilitate calcium flux between ER-mitochondria in apoptosis (Prudent et al. 2015) and knockout of Drp1 adaptor protein(s) Mid49/51 reduces mitochondrial calcium accumulation and injury in cardiomyocytes subjected to hypoxia (Hausenloy group unpublished data). There are also several studies demonstrating that mitochondrial fusion is involved in the regulatory mechanisms of mPTP (Hall et al. 2016). In fact, hearts from mitochondrial fusion-1 and 2 (Mfn-1 and 2) double-KO (DKO) mice present a higher resistance against I/R injury (Hall et al. 2016). The authors also observed that isolated heart mitochondria from DKO mice were more resistant to calcium-induced mPTP opening, suggesting that inhibition of mitofusin activity can be a useful strategy. However, it should be noted that DKO mice also demonstrated a deficit in mitochondrial respiratory capacity and cardiac contraction (Hall et al. 2016), which may in part mask the results of this study. Curiously, it was also observed that Akt activation, using constitutively active forms of Akt or EPO, induces mitochondrial elongation which delays mPTP opening and reduces cell death, in a Mfn1-dependent manner (Ong et al. 2015b). It was proposed that Akt promotes Mfn1 phosphorylation and as a result its activation (Ong et al. 2015b). This suggests that EPO or other compounds which activate Akt pathway represent a new therapeutic strategy to prevent CVDs mediated by inhibition of mPTP opening. In fact, EPO has already been used in a clinical trial (code NCT00568542, Charite University, Berlin, Germany) with positive results (Table 1). It was shown that low doses of EPO following PCI increases global ejection fraction (Bergmann et al. 2011).

Another interesting aspect of Drp1 is the regulation of its translocation to and from mitochondria which is complex and involves several types of post-translational modifications (Otera et al. 2013). One of the most studied and possibly more relevant modification is the phosphorylation at sites Ser^{616} and Ser^{637}. The former is mediated by Cdk1/cyclin B activity during mitosis or by protein kinase Cδ (PKCδ) under oxidative stress conditions, such as in ischaemia (Otera et al. 2013). Both promote Drp1 translocation to mitochondria and Drp1-dependent mitochondrial fission. Conversely, phosphorylation at Ser^{637} is catalyzed by the cAMP-dependent protein kinase A (PKA) that becomes active during pre-conditioning and temperature preconditioning and leads to Drp1 inactivation preventing its translocation to mitochondria. Interestingly, dephosphorylation of Ser^{637} occurs under the activity of calcineurin which becomes active during ischaemia-induced calcium overload. Theoretically, inhibition of calcineurin should thus be cardioprotective and in fact previously work has shown positive results (Feng et al. 2011). Recently, Xu et al. (2016) showed that Drp1 can also be phosphorylated by the calcium/calmodulin-dependent protein Kinase II (CaMKII) under conditions of chronic stimulation of β-adrenergic receptors (β-AR). In their study, inhibition of CaMKII using autocamtide 2-related inhibitory peptide or KN93 was shown to have inhibitory effects on mPTP opening (Xu et al. 2016), making these strategies feasible approaches to block the harmful action of Drp1 in I/R injury. Additionally, the β1-AR antagonist propranolol led to a decrease in the phosphorylation of Drp1 in S^{616} (Xu et al. 2016), reinforcing the therapeutic potential of this strategy which is already under clinical use for hypertension (Reiter 2004) (Table 1).

Overall, it seems that Drp1 acts as a hub of multicellular pathways converging in mitochondria to determine its fate. One obvious approach to modulate the mPTP in a Drp1-dependent manner is thus to pharmacologically inhibit/activate particular kinases. The specific cases of PKC and PKA will be explored in more details over the next sections.

The modulation of optic atrophy 1 (OPA1) was also considered as a possible strategy for cardioprevention mainly because a delay in calcium-induced mPTP opening was observed in cardiomyocytes isolated from Opa1$^{+/-}$ mice (Piquereau et al. 2012). Interestingly, although *in vivo* assays using Opa1$^{+/-}$ mice hearts did not show increased predisposition to mPTP opening it showed a higher susceptibility of these hearts to I/R injury instead (Le Page et al. 2016; Piquereau et al. 2012). Therefore, further studies are required to provide additional information about the *in vivo* effects of OPA1 downregulation (Le Page et al. 2016).

Postconditioning of cardiomyocytes with the anesthetic sevoflurane showed cardioprotection by inhibition of mPTP opening (Yu et al. 2015) (Table 1). This cardioprotective mechanism is probably related to the reduction of Mfn-2 and OPA1 levels and increased Drp1 protein expression (Yu et al. 2016), suggesting that modulation of mitochondrial dynamics may be exploited to regulate mPTP opening and promote cardioprotection.

Another proposed strategy involves the regulation of nuclear/cytosolic SIRT1 levels, based on the assumption that during I/R injury SIRT1 levels decrease. In fact, it was proposed that overexpression of SIRT1 or its pharmacological activation using resveratrol or SRT1720 induced a cardioprotective effect, through inhibition of the mPTP. Additionally, the study also showed that the inhibitory effect is lost after silencing of Mfn-2 because Mfn-2 is a substrate of SIRT1. Since SIRT1 and Mfn-2 are involved in autophagy, an important process for the degradation of proteins and organelles damaged under stress conditions, including I/R injury (Biel et al. 2016), it was predicted that defects in autophagy were associated with mitochondrial dysfunction and induction of mPTP (Kim et al. 2008). Overall, it suggests that the modulation of SIRT1 may be another therapeutic strategy for mPTP inhibition and therefore, should be explored in more detail in the future (see Table 1).

In conclusion, modulation of mitochondrial morphology (Fig. 2), using sevoflurane, EPO, resveratrol or SRT1720 has shown cardioprotective effects through inhibition of mPTP opening, although more studies are necessary to understand the beneficial effects on mitochondrial morphology.

3.3 Creatine Kinase

Mitochondrial creatine kinase is an isoenzyme which resides in the intermembrane space as a dimer or as an octamer and is responsible for the transfer of phosphate groups from mitochondrial ATP to cytosolic creatine (Schlattner et al. 2006).

The octamer form of mtCK forms a complex with ANT and VDAC and was suggested to play a role in mPTP regulation (Zorov et al. 2009; Schlattner et al. 2006; Wallimann et al. 2011). Datler et al. (2014) showed that KO of mtCK led to mPTP opening and apoptosis, which was blocked by BKA, an ANT inhibitor. However, the pharmacologic inhibition or downregulation of VDAC or Cyp-D did not affect apoptosis induced by mtCK depletion (Datler et al. 2014). Interestingly, the shift between octameric and dimeric form of mtCK have opposite effects on mPTP opening: while the dimeric form stimulates pore opening by destabilizing the interaction between mtCK and ANT, the octameric form promotes mPTP in a closed state instead (Dolder et al. 2001; Zorov et al. 2009; O'Gorman et al. 1997). Additionally, mtCK contains sulfhydryl groups making it particularly susceptible to oxidative stress (Zorov et al. 2009; Yuan et al. 1992). Thus, during I/R injury oxidative stress promotes a decrease in the octamer/dimer ratio (Soboll et al. 1999; Dolder et al. 2001). Accordingly, chemicals that promote stabilisation of octameric form of mtCK can be a strategy to inhibit mPTP opening. Corroborating this idea, creatine and cyclocreatine are two compounds that stabilise the octameric form and inhibit mPTP opening (Table 1) (Zorov et al. 2009; O'Gorman et al. 1997). The fact that creatine and cyclocreatine inhibit pore opening may also indicate that mtCK works as an energetic sensor to couple the energetic state of the cell and apoptosis (Schlattner et al. 2006).

In summary, stabilisation of the octameric form of mtCK has therapeutic potential to inhibit mPTP opening, as shown in Fig. 2 (Dolder et al. 2001). Although creatine supplements are commonly taken by athletes, data suggests that a normal dosage of 20 g for 5 days and 3–5 g thereafter would be perfectly safe for the general population as well (Francaux and Poortmans 2006). However, excessive supplementation may be harmful for the organs responsible for its metabolism, namely, liver and kidney, and for the brain, interfering with cerebral metabolism (Poortmans and Francaux 2000). The current challenge is to find a pharmacological way to increase the physiological levels of creatine and at the same time to avoid the harmful effects associated with excessive creatine supplementation.

3.4 PKC/PKA

Protein kinase A is a family of kinases activated by increased levels of cAMP and which may act as a regulator of metabolism and mechanisms of vasodilation, among other functions (Yan et al. 2016; Turnham and Scott 2016). Activation of PKA kinase under conditions of increased ROS (Amin et al. 2011; Novalija et al. 2003; Khaliulin et al. 2017) and/or calcium overload (Khaliulin et al. 2017; Bugrim 1999; Dekker et al. 1999) leads to activation of PKC, a family of kinases involved in several physiological functions, including vasoconstriction mechanisms (Rameshrad et al. 2016).

The activation of the PKC pathway has been shown to have cardioprotective effects (Khaliulin et al. 2010; Saurin et al. 2002; Halestrap et al. 2007). An example involves the protective effects of IP (Saurin et al. 2002; Halestrap et al. 2007) (described in Sect. 4.2.1). Another strategy involves the use of adenosine, a pharmacological activator of PKC, which showed a protective effect by reducing mPTP opening (Khaliulin et al. 2010). This molecule binds to G_i-protein-coupled adenosine receptors inducing a signalling pathway that leads to PKC activation (Costa et al. 2008; Yang et al. 2010). Based on this assumption, a phase IV clinical trial with dipyridamole, a molecule that increases endogenous adenosine levels is underway to test whether the activation of the adenosine receptor protects against I/R injury [#NCT00430170, Radboud University] (Table 1). Another strategy to activate PKC pathway was described using endogenous opioids (whose action can be mimicked by morphine) and bradykinin which activate PKC through the phosphoinositide 3-kinase (PI3k)-dependent signalling pathway (Table 1) (Costa et al. 2008; Yang et al. 2010). In this process, PI3k signalling phosphorylates Akt and triggers a signalling cascade that culminates with cGMP production and consequently PKG activation. Once activated, PKG leads to the opening of mitochondrial ATP-sensitive potassium channels (mK_{ATP}) in the IMM. The opening of mK_{ATP} causes alkalinization of the matrix which results from the influx of K^+ which is balanced by the efflux of H^+ ejected by the respiratory chain. The matrix alkalinization, in turn, retards the reduction of Q to QH2 at Complex I of the respiratory chain, increasing the production of superoxide anion, that finally activates PKC, and consequently inhibits mPTP opening (Costa et al. 2008; Yang et al. 2010). Thus, the activation of PI3k signalling pathways should be considered in the future as a cardioprotective strategy to inhibit mPTP opening.

The activation of PKA is also a good strategy to activate PKC. In agreement with this, temperature preconditioning (TP) (described in Sect. 4.2.2) stimulates β-adrenergic response and increases cAMP levels, promoting activation of PKA with consequent activation of PKC and mPTP inhibition. This idea is also supported by the fact that isoproterenol, a β-adrenergic agonist, prevented mPTP opening resulting from I/R injury (Khaliulin et al. 2010). However, it should be noted that β-adrenergic hyperstimulation can cause heart failure, representing a disadvantage of this therapeutic strategy (Khaliulin et al. 2017). More recently, a new strategy was also explored by Khaliulin et al. (2017), showing that the simultaneous activation of both PKA and the exchange protein activated by cAMP (Epac) confers an additive cardioprotection against I/R injury. The authors demonstrated that the cardioprotective effect of N6-benzoyladenosine-30,50-cyclic monophosphate, acetoxymethyl ester (6-Bnz-cAMP-AM, 6-Bnz), a selective activator of PKA signalling can be potentiated using 8-(4-chlorophenylthio)-20-O-methyladenosine-30,50-cyclic monophosphate, acetoxymethyl ester (8-pCPT-20-O-Me-cAMP-AM, CPT), a selective activator of Epac which consequently activates the PKC pathway. Additionally, the activation of both PKA and Epac pathways, using cAMP analogue 8-bromoadenosine-30,50-cyclic monophosphate, acetoxymethyl ester (8-Br-cAMP-AM, 8-Br) demonstrated equivalent cardioprotection against I/R injury (Khaliulin et al. 2017) (see Table 1).

In conclusion, the different strategies of PKC activation and inhibition of mPTP through the use of IP protocols or by promoting the activation of PI3k and PKA signalling pathways represent practicable strategies for cardioprotection that may be explored in the future.

3.5 Mitochondrial Calcium Uniporter

The mitochondrial calcium uniporter (MCU) is a transmembrane protein in the IMM that regulates calcium influx from the cytosol to mitochondria (Rizzuto et al. 2009), playing a significant role in myocardium contractility after β-adrenergic stimulation (Luongo et al. 2015). Calcium influx occurs under conditions where cytosolic calcium rises above a reasonable threshold detected by MICU1/2 in the MCU complex. Initially, the calcium entry is buffered by matrix inorganic phosphate and other proteins but once this mitochondrial buffering capacity is exceeded a massive calcium overload is observed, leading to mPTP opening. Thus, it is no surprise that during I/R the increase in mitochondrial calcium levels culminate in mPTP opening (Griffiths 2009; Delcamp et al. 1998). In line with this it is therefore sensible to assume that inhibition of MCU may then be an obvious strategy to inhibit mPTP opening (Fig. 2). In fact, ruthenium red (RuR), an MCU inhibitor (Ying et al. 1991; Griffiths 2009), was protective in myocytes and in hearts subjected to I/R injury when present in low concentrations (Griffiths 2009; Leperre et al. 1995; Grover et al. 1990). However, in order to decrease mitochondrial calcium levels higher concentrations of RuR are required but unfortunately this precludes its cardioprotection. This suggests that the effect cannot be attributed to calcium influx inhibition (Griffiths 2009) denoting the need of more studies to clarify this issue. Nevertheless, the stronger MCU inhibitor Ruthenium 360 (Ru360), an RuR analogue, was also able to decrease infarct size in hearts subjected to I/R injury by preventing mitochondrial calcium overload (Seidlmayer et al. 2015) and thereby inhibition of the mPTP (Zhang et al. 2006) and by (Table 1).

Minocycline and doxycycline, two tetracycline-derived compounds have been described to confer cardioprotective effects by inhibition of MCU and mPTP opening. In fact, the use of these two tetracycline-derived compounds had the ability to inhibit calcium uptake into mitochondria and mPTP opening after I/R injury in hepatocytes, preventing cell death (Schwartz et al. 2013) (Table 1). The same compounds (minocycline and doxycycline) together with (−)-epicatechin also exerted cardiprotection in hearts subjected to I/R injury (Sloan et al. 2012; Ortiz-Vilchis et al. 2014). In this work, the use of doxycycline was able to inhibit mitochondria swelling, suggesting that this is a good strategy to indirectly inhibit mPTP opening and to promote cardioprotection (Ortiz-Vilchis et al. 2014). Doxycycline and minocycline are both in clinical trials for treatment of acute myocardial infarction (NCT00469261, Careggi Hospital, Italy) and acute ischaemic brain stroke (NCT00930020, Singhealth Foundation, Singapore), respectively. Although the results with minocycline are still to be disclosed, results for the doxycycline trial are very encouraging in part because the drug reduced left ventricular remodelling in patients with acute STEMI and left ventricular dysfunction (Cerisano et al. 2014).

Until the identification of the molecular identity of the MCU the number of MCU inhibitors was scarce and its rate of discovery low. This was recently challenged in a chemical screen of more than 600 compounds performed in order to identify new small molecules that selectively target the MCU. The innovative approach consisted in minimizing false-positive hits that could modulate MCU-mediated Ca^{2+} uptake as a consequence of undesired effects on the electron transport chain (ETC), tricarboxylic acid cycle, mitochondrial membrane integrity, and/or other components of the intracellular Ca^{2+}-signalling networks. To accomplish that authors reconstituted the human MCU and its regulator EMRE in yeast mitochondria and assayed mitochondrial calcium flux in isolated fractions under conditions where the mitochondrial membrane potential is essentially ETC-independent and insensitive to uncouplers. The high-throughput screen identified mitoxantrone, an anti-tumoural compound, as a direct inhibitor of the human MCU both in permeabilized HeLa cells and in yeast mitochondria providing strong evidences for the need to develop MCU modulators (Arduino et al. 2017).

Additional strategies which promote modulation of signalling pathways have been described to confer cardioprotective effects through inhibition of MCU and mPTP opening (Sun et al. 2010; Joiner et al. 2012; Gutierrez-Aguilar et al. 2014) but not in detail. In one of these studies, Sun et al. (2010) showed that treatment of cardiomyocytes with B-type natriuretic peptide (BNP) was an effective strategy to prevent apoptosis during I/R injury through activation of PI3k pathway and subsequent inhibition of the MCU (Sun et al. 2010) (Table 1). The authors concluded that calcium uptake through MCU is partially regulated *via* PI3k pathway (Sun et al. 2010), suggesting that modulation of this target may induce cardioprotection by inhibition of MCU and mPTP opening. As shown in Table 1, a clinical trial investigated the use of BNP for treatment against heart failure (NCT00252187, Horng Chen, Mayo Clinic, US) and concluded that chronic subcutaneous BNP therapy successfully decreased left ventricular remodelling and improved left ventricular filling pressure (Chen et al. 2012).

Modulation of CaMKII has also been identified as a therapeutic target to inhibit MCU. In fact, Joiner et al. (2012) showed that CaMKII promotes mPTP opening by increasing current flow through MCU. The work showed for the first time that targeting CaMKIIN, a potent and specific inhibitor of CaMKII (Singh et al. 2009), to the mitochondria inhibits CaMKII and confers increased resistance against I/R injury (Joiner et al. 2012). The inhibition of CaMKII is thus another possible strategy to promote cardioprotection (Table 1).

A different approach was described by O-Uchi et al. (Gutierrez-Aguilar et al. 2014) showing that signalling through α_1-AR and Proline-rich Tyrosine Kinase 2 (Pyr2) can lead to phosphorylation and activation of MCU, with consequent uptake of calcium and induction of the mPTP (Gutierrez-Aguilar et al. 2014). Therefore, inhibition of this pathway either by the use of PF-431396, a Pyr2 inhibitor, or using prazosin, an α_1-AR antagonist, constitutes a strategy for indirect inhibition of the mPTP (Gutierrez-Aguilar et al. 2014) (Table 1).

In conclusion, considering that an increase in mitochondrial calcium levels leads to mPTP opening and that MCU is considered a critical entry route for calcium, its inhibition is an obvious strategy to prevent mPTP and thereby to induce cardioprotection (see Fig. 2). However, despite several studies have collected data supporting this idea there are not any approved drugs targeting MCU as a mean to provide cardioprotection at the moment.

3.6 *No Specific Target(s)*

3.6.1 Antioxidants

Excessive generation of ROS and mitochondrial calcium overload increases the probability of mPTP opening in different pathological conditions. One example occurs during reperfusion post-cardiac I/R injury; however, the exact mechanism of ROS production under these circumstances is poorly understood. Recently, Murphy has proposed succinate accumulation that occurs during ischaemia to be a hallmark of I/R injury (Chouchani et al. 2014). It was suggested that its fast oxidation upon reoxygenation during reperfusion drives reverse-electron transfer through Complex I, an infamous route in ROS production (Chouchani et al. 2014). However, this theory has been challenged by others (Korge et al. 2017a, b; Andrienko et al. 2017). Contrarily to Murphy, it is suggested that during the initial phase of reperfusion the requirements for reverse-electron flow-induced ROS production are not met. Therefore, the fast oxidation of succinate observed upon reperfusion reflects its use by the ETC after mPTP opening and uncoupling of respiration. The sustained respiration in the absence of antioxidant defenses and full cytochrome *c* pool leads to a significant production of ROS (Korge et al. 2017a, b). Still, others have reported the opposite (Briston et al. 2017).

Notwithstanding, it has been proposed that the mPTP is stimulated by thiol oxidation and depletion of mitochondrial NADPH (Kowaltowski et al. 2001; Castilho et al. 1996) and that interventions which decrease oxidative stress can decrease I/R injury, in part through inhibition of the mPTP (Carreira et al. 2011).

It was previously demonstrated that carvedilol, [1-(9H-carbazol-4-yloxy)-3-[2-(2-methoxyphenoxy) ethylamino]propan-2-ol] a third generation non-selective β-blocker used for treatment of hypertension and congestive heart failure (Chen-Scarabelli et al. 2012; Carreira et al. 2006) affords extra cardioprotective effects due to its antioxidant activity (Sgobbo et al. 2007). Different studies proposed that carvedilol protects mitochondria against oxidative damage by inhibiting lipid peroxidation (Abreu et al. 2000) and also by preventing the depletion of endogenous antioxidants (Table 1) (Silva et al. 2016; Pereira et al. 2011). Previous work from Oliveira lab has demonstrated that carvedilol acts as a weak protonophore (Oliveira et al. 2000a, 2001b) and inhibitor of the high-conductance state of the mPTP due to its antioxidant properties (Oliveira et al. 2001a, 2004b) in isolated cardiac mitochondria. The antioxidant and anti-mPTP activity was referred as important in the protection afforded by carvedilol against the cardiotoxicity of the anti-cancer agent doxorubicin, which also contributes to mPTP induction at lower calcium concentrations (Oliveira et al. 2004a; Santos et al. 2002; Matsui et al. 1999).

Despite its low bioavailability and low distribution in tissues and plasma (Neugebauer et al. 1987), which limits its beneficial effects, it is likely that carvedilol accumulation in the heart may still protect mitochondrial membranes from oxidative injuries and inhibit the mPTP (Silva et al. 2016; Pereira et al. 2011). Interestingly, carvedilol was demonstrated to inhibit Complex I in cardiac mitochondria which was suggested to cause a protective pre-conditioning-like effect (Sgobbo et al. 2007; Oliveira et al. 2000b).

Another intervention that decreases oxidative stress involves the use of the cellular natural antioxidant defenses or mimetics. Numerous natural antioxidants have been emerged over the past years as potential therapeutic agents against oxidative damage which occur in CVDs. Their effects span from ROS neutralization and heavy metal chelation to inhibition of enzymes involved in ROS overproduction as well as gene expression modulation of such enzymes (Benfeito et al. 2013; Silva et al. 2016). All these effects can *per se* afford protection against mPTP opening. In spite of the promising biological effects of antioxidant compounds in cells and animal models none of them has been effective when translated to CVDs patients. The poor bioavailability of antioxidant compounds represents a major challenge and a drawback when moving forward to clinical trials. In fact, antioxidants are widely distributed in the body but only a reduced percentage reach mitochondria, which are one important source of ROS (Oyewole and Birch-Machin 2015). In this sense, the delivery of compounds to mitochondria is a promising therapeutic to mitigate mitochondrial oxidative damage (Murphy and Smith 2000), to overcome the classical limitations of antioxidant compounds, including low bioavailability and insolubility (Apostolova and Victor 2015).

Basically, antioxidants have been fitted with suitably bulky hydrophobic groups, producing hydrophobic molecules providing the entry into hydrophobic biological membranes (Murphy and Smith 2007). Mitochondrial-targeted antioxidants are able to cross the hydrophobic core of the phospholipid bilayer of biological membranes and to accumulate inside mitochondria, which is the only cell compartment negatively charged relative to its environment (Skulachev 2007). This accumulation inside mitochondria is driven by the $\Delta\psi m$ (Apostolova and Victor 2015; Murphy and Smith 2000), where the negative $\Delta\psi m$ present in mitochondria (-150 to -180 mV) in relation to the plasma membrane potential (-30 to -60 mV) supports the targeting of lipophilic cations and their tenfold accumulation in the cytoplasm and 10,000-fold within mitochondria (Apostolova and Victor 2015; Smith et al. 2012; Murphy and Smith 2007).

One possible carrier is the TPP molecule, composed by a positively charged phosphorus atom surrounded by three aromatic groups, conferring an hydrophobic surface despite its positive charge (Gruber et al. 2013). One particular mitochondria-targeted compound that has received greater attention is mitoquinone (MitoQ), which links a TPP^+ moiety to ubiquinone through a 10-carbon aliphatic chain [102, 110]. Mitoquinone is reduced by the mitochondrial respiratory chain to its active ubiquinol form, which prevents lipid peroxidation and mitochondrial damage (Kelso et al. 2001). Considering that cardiolipin peroxidation leads to dissociation of cytochrome *c* and induction of mPTP opening (Szeto 2006), it is also possible that the inhibition of cardiolipin peroxidation may confer protection *per se* against mPTP opening. Still, there is little evidence that MitoQ prevents mPTP induction (Hansson et al. 2015), So far, no mitochondrial-directed antioxidants was approved by the FDA and only one successful phase II clinical assay of oral MitoQ tablets was reported (Table 1) (Snow et al. 2010). Although data from different works and clinical trials suggest that MitoQ is one of the most potent mitochondrial protectors available at the moment there are no trials demonstrating its clinical efficacy in the prevention of CVDs. Obviously, this raises questions about its cardioprotective ability.

Another possible cardioprotective approach is the use of a new class of mitochondria-targeted molecules. Recent examples include modified dietary antioxidants inspired in a natural polyphenol scaffold—hydroxybenzoic acid (AntiOxBENs). Although very few works have evaluated the cardioprotective effects of these novel antioxidants, the molecules accumulate inside mitochondria and prevent lipid peroxidation (Table 1) with lower toxicity compared to MitoQ (Teixeira et al. 2017). This suggests that AntiOxBENs can prevent cardiac oxidative damage and therefore inhibit mPTP opening. However, further studies are required to test the cardio-potential effects of AntiOxBENs.

Another mitochondrial-directed antioxidant construct are the cationic derivatives of plastoquinone (SkQs), developed by Skulachev et al., which take advantage of the quinone present in the ETC of chloroplasts instead of using the ubiquinone antioxidant moiety of MitoQ (Gruber et al. 2013; Skulachev et al. 2009). SkQ antioxidant activity involves the direct quenching of ROS and inhibition of cardiolipin peroxidation (Skulachev et al. 2010), two effects that possibly afford protection against mPTP opening. However, further studies are needed to confirm these effects and its use as a cardioprotective agent.

Additionally, SS (Szeto-Schiller) peptides are antioxidants with structural motif centers on alternating aromatic residues and basic amino acids (aromatic-cationic peptides), and represent a novel approach for targeted delivery of antioxidants to the IMM (Szeto 2008; Szeto and Schiller 2011). These peptides are predominantly targeted to the IMM despite their cationic nature and only 20% reach the mitochondrial matrix *via* $\Delta\psi$m-driven mechanisms. Thus, contrary to MitoQ which caused toxicity at relatively low concentrations (>10 μM), the uptake of SS peptides was not self-limiting and they did not cause mitochondrial depolarisation even at 1 mM (Zhao et al. 2004). SS peptides were shown to scavenge hydrogen peroxide and peroxynitrite and inhibit lipid peroxidation. By reducing mitochondrial ROS, these peptides inhibit mPTP and cytochrome *c* release, thus preventing oxidant-induced cell death (Szeto 2006), a mechanism that may have an important outcome in the prevention of I/R injury. However, further studies are needed to clarify the beneficial effect of antioxidant therapies in the prevention of I/R injury. In fact, it should also be noted that ROS production may also occur as a consequence and not a cause of I/R injury. Moreover, it still remains to make clear the efficacy of antioxidant-mediated therapies in the prevention of CVDs.

3.6.2 Pyruvate

Pyruvate is a product of glycolysis and a substrate in the TCA cycle, and has been shown to improve cardiac functions during I/R injury (reviewed in (Mallet 2000; Mallet et al. 2005)). It has been observed that pyruvate can protect hearts against I/R and anoxia/reoxygenation injury (Bunger et al. 1989; Cavallini et al. 1990; DeBoer et al. 1993). The cardioprotective effect of pyruvate has been attributed to beneficial metabolic alterations (Bunger et al. 1989; Cavallini et al. 1990) and its ability to decrease free radical production (Constantopoulos and Barranger 1984; DeBoer et al. 1993). Still, the full cellular mechanisms of pyruvate action has remained unclear.

In an attempt to explain the cardioprotection afforded by pyruvate, Kerr et al. (1999) evaluated mPTP inhibition in rat hearts subjected to I/R. It was concluded that pyruvate has no direct effect on the mPTP pore but instead its inhibition occurred through an indirect process. Probably due to several pyruvate characteristics unrelated to mPTP, for instance, ROS scavenger, respiration substrate, and inducer of acidosis (Kerr et al. 1999). In fact, pyruvate was described to significantly lower pHi at the end of ischaemia and during the reperfusion phase, a renowned mechanism to maintain mPTP closed and prevent I/R injury (Halestrap et al. 1997a). Additionally, the energetic properties of pyruvate should also be considered as a factor in mPTP inhibition because pyruvate is an excellent respiration substrate which may help to maintain the mitochondrial membrane potential at values that inhibit the pore (Bernardi 1992). Lastly, pyruvate is known to act as an efficient free radical scavenger which contributes to its cardioprotective profile (Dobsak et al. 1999; Constantopoulos and Barranger 1984; DeBoer et al. 1993).

Despite its beneficial effects supplementation of pyruvate at physiological concentrations (0.2 mM) lead only to modest improvements in cardiac performance (Bunger et al. 1989), suggesting that any beneficial effects of pyruvate may be restricted to supra-physiological concentrations (~5–10 mM) (Mallet 2000). Although administration of supra-physiological concentrations of pyruvate has yielded encouraging results in heart failure patients (Hermann et al. 1999) its clinical utility has been hampered due to its instability in aqueous solution and low solubility in water and saline solutions. To overcome these limitations electroneutral and chemically stable pyruvate derivatives were developed. Ethyl pyruvate (EP) is very soluble in calcium solutions and forms a dimolecular complex with the enolate anion which is further stabilised by calcium cations. EP has greater stability than pyruvate and may act as a pyruvate precursor to neutralize H_2O_2 in a direct, nonenzymatic reaction centered on the covalent bond between the carbonyl and carboxyl groups of its pyruvate moiety although at slower rates than its precursor (Mallet et al. 2005). Additionally, EP has shown anti-inflammatory activity and provided myocardial protection against I/R damage in an *in vivo* rat model (Jang et al. 2010). More recently, it was also proposed that the myocardial protective effect of EP could be related to its anti-apoptotic actions (Shim et al. 2016). However, further studies are needed to clarify this effect on mPTP opening. Despite these evidences, only one randomized clinical trial of EP is presently available (NCT00107666, Critical Therapeutics Inc., US) and aims to evaluate the safety and efficacy of EP treatment in patients undergoing cardiopulmonary bypass. Unfortunately, the trial has stopped in Phase II due to concerns related to the stability of the proprietary EP formulation questioning the clinical benefit of EP treatment (Abarbanell 2010).

Dipyruvyl-acetyl-glycerol (DPAG) ester is another pyruvate derivative. An intravenous infusion of DPAG in pig hearts subjected to I/R injury increased arterial pyruvate concentration and reduced myocardial infarct size following myocardial ischaemia (Stanley et al. 2003). Overall, it suggests that pyruvate analogues are a feasible strategy to prevent CVDs although the precise mechanism is still to be uncovered.

4 General Approaches to Inhibit mPTP

4.1 Modulation of pH

During ischaemia cardiomyocytes are submitted to ionic dysregulation eventually leading to intracellular acidosis. Additionally, the drop in pH is accelerated by accumulation of lactic acid due to increased glycolysis occurring as a consequence of the lack of oxygen and ATP shortage. In order to restore pHi the sodium/proton exchangers (NHE) are activated resulting in the rise of sodium levels and ultimately in calcium overload. Nonetheless, a low pHi is preserved throughout ischaemia maintaining the mPTP closed (Halestrap 1991; Bernardi et al. 1992; Halestrap and Pasdois 2009; Kerr et al. 1999). Although damage is inflicted during prolonged ischaemia most of the injury is imposed during reperfusion when pH is restored and the mPTP is no longer inhibited (Kerr et al. 1999). This is in part the concept of post-conditioning where the heart is subjected to short cycles of ischaemia and reperfusion at the immediate onset of reperfusion allowing the maintenance of a low pHi (Boengler et al. 2011). Additionally, early reperfusion at low pH also confers cardioprotection (Kitakaze et al. 1997). Keeping in mind the modulatory role of pHi on mPTP activation, NHE exchangers were considered as promising therapeutic targets with several molecules being evaluated.

4.1.1 Na/H Exchanger

Several families of ion channels are involved in proton exchange across the plasma membrane and hence in pHi regulation. Sodium/proton exchangers contribute to the regulation of both proton and sodium in different compartments and its inhibition has shown cardioprotection in I/R injury (Karmazyn 1988, 1999; Rohmann et al. 1995; Javadov et al. 2008; Hurtado and Pierce 2000). Ten NHE (1–10) isoforms have been described with a diverse distribution among various tissues as well as cellular localizations (Alvarez and Villa-Abrille 2013; Kemp et al. 2008; Odunewu-Aderibigbe and Fliegel 2014). The integral membrane protein NHE1 isoform was the first to be reported (Sardet et al. 1989) and although ubiquitously expressed it is extensively studied and abundant in the heart (Alvarez and Villa-Abrille 2013;

Odunewu-Aderibigbe and Fliegel 2014). Inhibition of NHE1 by specific compounds such as EMD-87580, AVE-4890, dimethylamiloride (DMA) or cariporide has shown cardioprotective effects by inhibiting mPTP opening (Javadov et al. 2008; Javadov et al. 2005; Prendes et al. 2008; Garciarena et al. 2008).

Garciarena et al. (2008) showed that cariporide prevents mitochondrial swelling trough mPTP inhibition similarly to the effects observed with the pore inhibitor CsA. Additionally, AVE-4890 and DMA were also reported to block mPTP opening (Javadov et al. 2008; Prendes et al. 2008), with the effect of AVE-4890 being mediated through the GSK-3β pathway (Javadov et al. 2009b). Moreover, NHE blockers such as cariporide, BIIB-723 and EMD-87580 have shown cardioprotection in a mitochondrial-dependent manner by preventing ROS production despite not having direct scavenger activity (Fig. 3) (Fantinelli et al. 2006; Garciarena et al. 2008). The reduction of diazoxide-induced O_2^- production observed in the presence of cariporide suggests a direct effect on mitochondria (Garciarena et al. 2008) even if mK_{ATP}-independent actions of diazoxide have also been reported (Hanley et al. 2002). The cellular effects of cariporide span from reduction of oxidative stress, lower cytosolic sodium and calcium accumulation, lower mitochondria calcium overload, and maintenance of low pHi during reperfusion (Hartmann and Decking 1999; Teshima et al. 2003) which consequently prevents mitochondrial swelling, implying an effect on the mPTP (Fig. 3) (Ruiz-Meana et al. 2003). In a different work it was demonstrated that NHE inhibition with eniporide at an alkaline pH decreased mitochondrial calcium overload, ROS production and improved heart recovery after reperfusion (Table 1) (Aldakkak et al. 2008).

Unfortunately, the evaluation of NHE inhibitors cariporide and eniporide in clinical trials (GUARDIAN, Expedition, ESCAMI) has not unconditionally confirmed the utility of NHE blockade (Theroux et al. 2000; Mentzer et al. 2008; Zeymer et al. 2001) as it was predicted in pre-clinical studies.

The assessment of eniporide as an adjunct therapy for early reperfusion in acute myocardial infarction was performed in a phase I clinical trial (ESCAMI). In this study, eriporide reduced infarct size from 10% to 23% in a dose dependent manner but in a phase II trial the benefits were not confirmed (Zeymer et al. 2001). Regarding cariporide, after an encouraging small pilot study showed reduced markers of myocardial injury and improved left ventricular dysfunction (Rupprecht et al. 2000) a new trial was established. The GUARDIAN (GUARD during ischemia against necrosis) trial showed a non-significant 10% risk reduction even though cariporide significantly induced cardioprotection in patients undertaking coronary artery bypass grafting (CAGB) surgery (Theroux et al. 2000; Boyce et al. 2003). More recently a phase 3 clinical trial (EXPEDITION) was performed to confirm whether cariporide could reduce the risk of death and myocardial infarction in patients undergoing CAGB (Table 1). The study revealed a reduction of the risk of death and myocardial infarction at different endpoints (5 days, 30 days and 6 months) even though a non-significant mortality increased at 6 months (Mentzer et al. 2008). Despite the drawbacks and the unwanted side effects this trial proved that it is

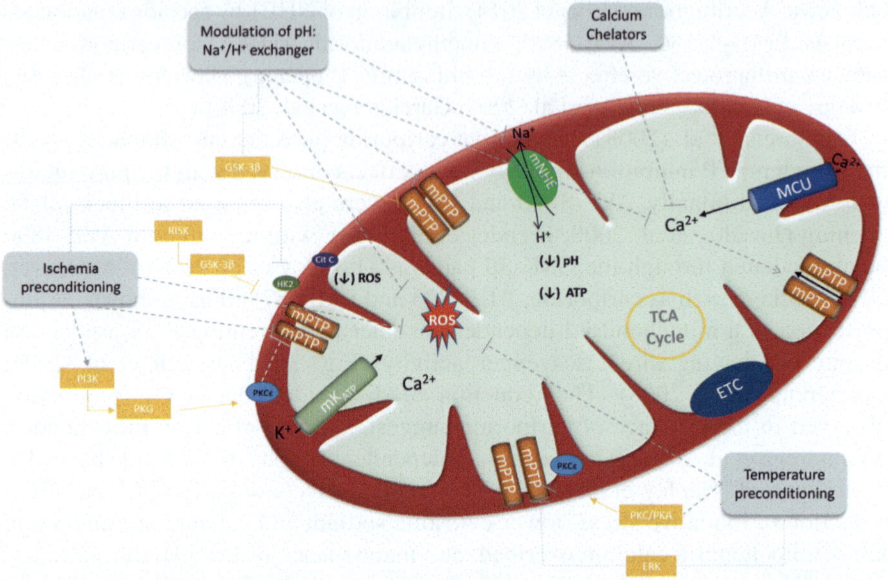

Fig. 3 Different strategies for cardioprotection targeting mitochondria. Schematic illustration of calcium chelators, modulation of pH:Na^+/H^+ exchanger, ischaemia preconditioning, and temperature preconditioning approaches. Calcium chelators reduce calcium levels contributing to the inhibition of mitochondrial permeability transition pore (mPTP) opening. pH:Na^+/H^+ exchanger effects are mediated by reducing Ca^{2+} levels and radical oxygen species (ROS) production, mPTP inhibition that can eventually be mediated by glycogen synthase kinase 3 beta (GSK-3β) or blocking of the mitochondrial sodium/proton exchanger (mNHE). The ischaemia preconditioning (IP) can affect mPTP opening through ROS reduction, hexokinase 2 (HK-II) maintenance at outer mitochondrial membrane, while mPTP inhibition by IP can be also mediated by reperfusion injury kinase (RISK) pathway through GSK-3β or even by PKCε translocation to mitochondria leading to mitochondrial ATP-sensitive potassium channels (mK_{ATP}) activation. The temperature preconditioning reduces ROS levels and inhibits mPTP opening dependent on PKA/PKC or ERK pathway. *Cit C* cytochrome *c*, *PKG* protein kinase G, *ETC* electron transport chain, *TCA* tricarboxylic acid cycle

possible to achieve cardioprotection in humans by pharmacological inhibition of NHE. The beneficial effects observed still hold the promise that this approach has a potential therapeutic usefulness for different heart disease conditions.

An interestingly work evaluated the NHE role in cardioprotection in a drug-free setup (Villa-Abrille et al. 2011). The authors used short-hairpin RNA (sh-RNA) NHE1 to reduce mNHE1 expression in mitochondria and observed a decreased sensitivity to mPTP opening. As expected, cariporide also prevented mitochondrial swelling in sh-RNA control group but lacked any effect if low levels of mNHE1 were present (Villa-Abrille et al. 2011). These results show the effectiveness of targeting NHE with specific drugs or by new genetic approaches in order to promote cardioprotection.

4.2 Drug-Free Methods

4.2.1 Ischaemia Pre-, Post- and Remote-Conditioning

A renowned strategy conferring cardioprotection is to subject hearts to IP. Ischaemia pre-conditioning was initially reported by Murry et al. (1986) which showed that a brief exposure of hearts to sequential cycles of ischaemia and reperfusion protects the myocardium from a further prolonged ischaemic insult. Other protocols such as calcium-preconditioning also improve cell viability, attenuate apoptosis and mitochondria calcium accumulation in myocytes similar to CsA treatment, suggesting a possible role of the mPTP (Xu et al. 2001). Modulation of calcium levels appeared to be an obvious event downstream of IP but more evidences are required to confirm this idea (Ong et al. 2015c). Notwithstanding, hearts subjected to IP are reported to be more resistant to calcium-induced mPTP opening (Clarke et al. 2008) and show lower levels of sodium and calcium during reperfusion (Steenbergen et al. 1993). A possible mechanism through which IP might protect mitochondria from calcium overload during I/R injury is by activation of mK_{ATP}. This hypothesis is still poorly understood and does not gather consensus (Hausenloy et al. 2002; Murata et al. 2001; Wang et al. 2001; Hanley and Daut 2005). Nevertheless, several reports have reinforced IP in cardioprotection through inhibition of mPTP opening during reperfusion (Fig. 3) (Hausenloy et al. 2002; Javadov et al. 2003; Argaud et al. 2004).

The sodium/proton exchangers activity has been reported to be inhibited by preconditioning which may minimize the damage during reperfusion by preventing the increase of sodium and calcium levels and consequently delaying pH recover (Xiao and Allen 1999, 2000). However, the initial hypothesis for the reduced intracellular acidosis was thought to be a consequence of glycogen depletion during IP (Asimakis et al. 1992; Wolfe et al. 1993). Still, the mechanism of pH modulation is poorly understood and more evidences are necessary to confirm it. On the contrary, oxidative stress is a notable factor which production occurs mainly upon re-oxygenation and sensitizes mPTP opening (Kim et al. 2006). In accordance with this hearts submitted to IP demonstrated attenuated oxidative stress (Tosaki et al. 1994) as well as less oxidation of mitochondrial proteins (Clarke et al. 2008) following reperfusion, supporting a correlation between preconditioning and oxidative stress.

Although no direct link has been established between IP and mPTP inhibition the reduction of oxidative stress appears to be a rationale connection. For example, hypoxic preconditioning has previously delayed mPTP opening from local oxidative stress induction (Hausenloy et al. 2004b). However, the underlying mechanism by which IP confers mPTP inhibition in a ROS-related fashion remains unclear. In this regard a different model suggests that IP maintains OMM integrity and reduces cytochrome *c* release during ischaemia, thus contributing to attenuate oxidative stress through the antioxidant capacity of cytochrome *c* (Pasdois et al. 2011, 2012).

The activation of several signaling pathways has been implicated in IP protection due to the release of autocoids such as adenosine, bradykinin or opioids. The binding of this molecules to the correspondent receptors activates several signaling cascades that exert long-term effects on mitochondria (Perricone and Vander Heide 2014; Ong et al. 2015a). In fact, the release and binding of adenosine to its receptor during IP has shown cardioprotection (Liu et al. 1991) in part due to up-regulation of PI3k and activation of downstream targets such as Akt and PKC (isoform PKCε) (Ytrehus et al. 1994; Tong et al. 2000; Mocanu et al. 2002). It is proposed that the cGMP-dependent protein kinase, PKG (previous described) activates mitochondrial PKCε and that the latter stimulates mK_{ATP} opening and ROS formation activating a second pool of PKCε by a feedback loop that inhibits the mPTP (Fig. 3) (Costa and Garlid 2008; Costa et al. 2005, 2006). Accordingly, it was described that PKCε is imported to mitochondria via heat shock protein 90 (Budas et al. 2010). How exactly PKCε is activated and able to block mPTP remains to be elucidated.

Another player(s) in the IP mechanism is the reperfusion injury salvage kinase (RISK) pathway that includes PI3K, Akt, and extracellular signal-regulated kinase (Erk) which become activate during early reperfusion (Hausenloy et al. 2004a, 2005b; Hausenloy and Yellon 2004). The Akt and Erk pathways appear to be key players for IP cardioprotection since mPTP inhibition can only occur if the kinases are active (Hausenloy et al. 2004a, b, 2005a; Tong et al. 2000; Mocanu et al. 2002; Davidson et al. 2006). Moreover, downstream targets of Akt and Erk such as PKG, GSK-3β and HK-II also seem to be essential for protection. Glycogen synthase kinase 3 β inactivation by phosphorylation occurs after IP and confers protection and the effect can also be achieved by pharmacologic inhibition of GSK-3β (Tong et al. 2002). The GSK-3β-mediated cardioprotection is thought to be due to inhibition of the mPTP (Juhaszova et al. 2004) through VDAC regulation (Das et al. 2008). Regarding HK-II, the RISK pathway participates in its phosphorylation promoting its binding to mitochondria. If the mechanism is disturbed IP protection is lost (Roberts et al. 2013; Smeele et al. 2011; Gurel et al. 2009).

Given the specificity of the protocol involving IP, its implementation is difficult in humans. However, the several small trials that have been conducted so far were systematically reviewed by a meta-analysis and although some protection exist authors state that a large-scale trial is required for a better judgment (Walsh et al. 2008). Nevertheless, the IP concept has evolved to a new type of preconditioning designated remote ischaemic preconditioning (RIPC), whereas a nonvital tissue is submitted to ischaemic conditioning to yield remote protection to the heart (Przyklenk et al. 1993). This approach will undoubtedly allow a better translation to clinical settings. Despite this breakthrough the identification of the mechanisms are more complex and some evidence suggest PI3K/Akt/GSK-3β activation (Li et al. 2011) and modulation of mPTP opening (Turrell et al. 2014). The mechanisms of RIPC actions and clinical implications were recently and extensively reviewed elsewhere (Hausenloy and Yellon 2016). Overall it is fair to say that despite the extensive research, the clinical outcomes are still scarce since the large-scale trials show no beneficial effects (Hausenloy and Yellon 2016; Meybohm et al. 2015).

In the last decades, countless efforts have created new approaches to promote cardioprotection and a deeper knowledge on the possible mechanisms behind them. Notwithstanding, the advances made until now have not yet generated a broad consensus and some hypothesis and gaps are being questioned. Nevertheless, the evidence gathered is believed to provide new therapeutically options in the near future. Capable to mitigate ischaemia and that will hopefully translate into effective clinical improvements.

4.2.2 Temperature Preconditioning

The search for new methods to prevent heart from ischaemic damage together with evidences showing benefits from hypothermia in surgical procedures had prompted researcher to address TP as a possible venue to promote cardioprotection. Khaliulin et al. (2007) first reported the benefits of the use of TP by applying short cycles of hypothermic perfusion (26 °C) during normothermic perfusion and observed a similar protection to IP treatment. Although cardioprotection is temperature-dependent, the best effects are seen at 26 °C (Khaliulin et al. 2011). From the parameters assessed, TP had similar or improved results on hemodynamic function, ventricular arrhythmias or lactate dehydrogenase release compared to IP. The study also presented evidences of a decrease in ROS production and mPTP inhibition. Accordingly, some of the signaling pathways that were previously described to mediate IP protection were also involved, namely increased PKCε activation/translocation (Khaliulin et al. 2007). Subsequent work suggested that PKA activation and downstream PKC activation are involved in TP signaling (Khaliulin et al. 2010). Finally, it was reported that ROS release during hypothermic episodes induces ROS-dependent Erk activation and is associated with a delay in mPTP opening, although using a different protocol (Fig. 3) (Bhagatte et al. 2012). As with other strategies, more data are required to fully understand the mechanisms and potential of this application.

5 Conclusion

The mitochondrial permeability transition has been shown to play an important role in different pathologies, with particular relevance in CVDs (Bernardi et al. 2015; Kwong and Molkentin 2015). Taking into account the importance of this event, different strategies for mPTP inhibition have been tested as potential cardioprotective approaches (Fancelli et al. 2014; Stavrovskaya et al. 2004). Despite numerous compounds and strategies have demonstrated cardioprotective effects in experimental models (Argaud et al. 2005; Gomez et al. 2007; Schaller et al. 2010) their clinical applications are still modest. The main reason that can explain this poor clinical success is based on the fact that the true molecular composition of this pore is unknown. This limitation precludes the discovery of true inhibitors of the pore by *in silico* or orthogonal genetic approaches. To date, the majority of compounds used to inhibit the mPTP target modulatory proteins rather than structural proteins of the pore, an exception would be those targeting Cyp-D. The new generation of mPTP inhibitors (Briston et al. 2016; Fancelli et al. 2014; Roy et al. 2015) which mechanism of action is CyP-D-independent show some promise; however, until the mPTP identity is solved they are merely another inhibitor of modulatory mPTP proteins. Moreover, despite their potency in *in vitro* assays, most of these compounds fail to move to the next stage of drug discovery—preclinical and clinical development. This is in part explained by the fact that it has been extremely difficult to evaluate and confirm their effect in intact cells. In fact, over the past years several pharmaceutical companies have discontinued projects on the development of new mPTP inhibitors because no reliable and robust *in cell* assay is available (GC Pereira, personal communication). Therefore, the second reason to explain the lack of success resides on the requirement to develop a specific and reliable assay to evaluate pore opening in cells and thus confirm drug's effects on mPTP and discard any side effects. To date, the majority of works rely on the measurement of membrane potential as a surrogate of pore opening. However, this will eventually lead to false-positives as mitochondrial depolarisation can occur independently of the mPTP. The other technique widely in use is based on the quenching of calcein fluorescence by cobalt (Petronilli et al. 1998). However, in addition to all the problems associated with cobalt toxicity and calcein compartmentalization problems (Jones et al. 2002; Lemasters and Nieminen 1999) caution should also be taken when interpreting the results because depending on the stimulus used to trigger the mPTP it could lead to false-negative hits (GC Pereira, unpublished results; Panel et al. 2017). In our opinion, in order for new methods to be specific and reliable they should evaluate the integrity of the IMM which becomes compromised upon the mPTP. This could be achieved by developing small molecules that accumulate in the cytoplasm and cannot permeate the mitochondria under physiological conditions. Considering their small size (<1.5 kDa) they would permeate through the mPTP after pore opening and exert their action on the matrix. A possible pair would be the expression of the bright luciferase NanoLuc (Promega) and the use of one of their inhibitors (Walker et al. 2017); however, to our knowledge these compounds either permeate the cell or remain in the extracellular compartment. It would be of great interest to develop similar molecules with the ability to enter the cell but not the endomembrane systems. Alternatively, one could use bipartite protein systems such as the split-GFP (Cabantous et al. 2005) or NanoBiT (split-NanoLuc, Dixon et al. 2016). In this case, the large fragments would be expressed in the mitochondrial matrix while the small fragments (16aa for split-GFP and 11aa for NanoBiT; 1.5–1.8 kDa) would remain in the cytosol. Generation of a luminescent or fluorescent signal would occur upon mPTP opening. Although this approach is attractive a possible limitation is the expression and stability of such small peptides in the cell.

The third reason to explain the lack of success of the different approaches is based on the fact that most of compost tested have a decreased bioavailability and only a reduced amount reach to the mitochondria and thus are not adequate to prevent mPTP pore opening.

In this chapter, we examined some compounds and approaches to inhibit mPTP pore opening which have been described in preclinical studies and clinical trials for cardioprovention, namely the development of strategies of desensitizing or inhibition of mPTP by modulation of mPTP-dependent and independent targets, as well as other approaches based on the modulation of calcium, ROS and pHi. While in the past the reduction of calcium and ROS or the inhibition of Cyp-D were the main approaches to block mPTP opening, nowadays other proposals have emerged. Some examples are the different strategies of inhibition of mPTP to promote binding of HK-II to mitochondria, to favour the stabilisation of octameric form of mtCK or mechanisms that modulate mitochondrial morphology. Strategies to inhibit mPTP pore opening by promoting PTMs in regulatory components of pore, and different protocols of ischaemia and temperature preconditioning were also other proposals described in this chapter. Although all the above mentioned approaches have been developed none have been effective when translated to a clinical setting. The success in the discovery of new cardioprotective strategies should be based on the search of new approaches with low toxicity and high clinical efficacy.

Altogether, we make an integrative description of different strategies of cardioprotection mediated by inhibition of mPTP opening and focus on the key issues that researchers should consider and clarify in the future in order to successfully identify strategies to be translated to clinic.

Acknowledgments Work in the authors laboratory is funded by FEDER funds through the Operational Programme Competitiveness Factors -COMPETE and national funds by FCT—Foundation for Science and Technology under research grants PTDC/DTP-FTO/2433/2014, POCI-01-0145-FEDER-016659, and POCI-01-0145-FEDER-007440. Filomena Silva is recipient of a Post-Doctoral Fellowship from the Foundation for Science and Technology, SFRH/BPD/122648/2016.

References

Abarbanell AM (2010) Structure and lipophilicity—the keys to understanding the function of pyruvate derivatives for ischemia/reperfusion? J Surg Res 164(1):72–73. https://doi.org/10.1016/j.jss.2009.12.007

Abreu RM, Santos DJ, Moreno AJ (2000) Effects of carvedilol and its analog BM-910228 on mitochondrial function and oxidative stress. J Pharmacol Exp Ther 295(3):1022–1030

Adachi T, Weisbrod RM, Pimentel DR, Ying J, Sharov VS, Schoneich C, Cohen RA (2004) S-Glutathiolation by peroxynitrite activates SERCA during arterial relaxation by nitric oxide. Nat Med 10(11):1200–1207. https://doi.org/10.1038/nm1119

Alavian KN, Beutner G, Lazrove E, Sacchetti S, Park HA, Licznerski P, Li H, Nabili P, Hockensmith K, Graham M, Porter GA Jr, Jonas EA (2014) An uncoupling channel within the c-subunit ring of the F1FO ATP synthase is the mitochondrial permeability transition pore. Proc Natl Acad Sci U S A 111(29):10580–10585. https://doi.org/10.1073/pnas.1401591111

Alcala S, Klee M, Fernandez J, Fleischer A, Pimentel-Muinos FX (2008) A high-throughput screening for mammalian cell death effectors identifies the mitochondrial phosphate carrier as a regulator of cytochrome c release. Oncogene 27(1):44–54. https://doi.org/10.1038/sj.onc.1210600

Aldakkak M, Stowe DF, Heisner JS, Spence M, Camara AK (2008) Enhanced Na+/H+ exchange during ischemia and reperfusion impairs mitochondrial bioenergetics and myocardial function. J Cardiovasc Pharmacol 52(3):236–244. https://doi.org/10.1097/FJC.0b013e3181831337

Alvarez BV, Villa-Abrille MC (2013) Mitochondrial NHE1: a newly identified target to prevent heart disease. Front Physiol 4:152. https://doi.org/10.3389/fphys.2013.00152

Amin P, Singh M, Singh K (2011) Beta-adrenergic receptor-stimulated cardiac myocyte apoptosis: role of beta1 integrins. J Signal Transduct 2011:179057. https://doi.org/10.1155/2011/179057

Andrienko TN, Pasdois P, Pereira GC, Ovens MJ, Halestrap AP (2017) The role of succinate and ROS in reperfusion injury – a critical appraisal. J Mol Cell Cardiol 110:1–14. https://doi.org/10.1016/j.yjmcc.2017.06.016

Anholt RR, Pedersen PL, De Souza EB, Snyder SH (1986) The peripheral-type benzodiazepine receptor. Localization to the mitochondrial outer membrane. J Biol Chem 261(2):576–583

Apostolova N, Victor VM (2015) Molecular strategies for targeting antioxidants to mitochondria: therapeutic implications. Antioxid Redox Signal 22(8):686–729. https://doi.org/10.1089/ars.2014.5952

Aquila H, Klingenberg M (1982) The reactivity of -SH groups in the ADP/ATP carrier isolated from beef heart mitochondria. Eur J Biochem 122(1):141–145

Arduino DM, Wettmarshausen J, Vais H, Navas-Navarro P, Cheng Y, Leimpek A, Ma Z, Delrio-Lorenzo A, Giordano A, Garcia-Perez C, Medard G, Kuster B, Garcia-Sancho J, Mokranjac D, Foskett JK, Alonso MT, Perocchi F (2017) Systematic identification of MCU modulators by orthogonal interspecies chemical screening. Mol Cell 67(4):711–723. e717. https://doi.org/10.1016/j.molcel.2017.07.019

Argaud L, Gateau-Roesch O, Chalabreysse L, Gomez L, Loufouat J, Thivolet-Bejui F, Robert D, Ovize M (2004) Preconditioning delays Ca2+-induced mitochondrial permeability transition. Cardiovasc Res 61(1):115–122

Argaud L, Gateau-Roesch O, Muntean D, Chalabreysse L, Loufouat J, Robert D, Ovize M (2005) Specific inhibition of the mitochondrial permeability transition prevents lethal reperfusion injury. J Mol Cell Cardiol 38(2):367–374. https://doi.org/10.1016/j.yjmcc.2004.12.001

Asimakis GK, Inners-McBride K, Medellin G, Conti VR (1992) Ischemic preconditioning attenuates acidosis and postischemic dysfunction in isolated rat heart. Am J Phys 263(3 Pt 2):H887–H894

Atar D, Arheden H, Berdeaux A, Bonnet JL, Carlsson M, Clemmensen P, Cuvier V, Danchin N, Dubois-Rande JL, Engblom H, Erlinge D, Firat H, Halvorsen S, Hansen HS, Hauke W, Heiberg E, Koul S, Larsen AI, Le Corvoisier P, Nordrehaug JE, Paganelli F, Pruss RM, Rousseau H, Schaller S, Sonou G, Tuseth V, Veys J, Vicaut E, Jensen SE (2015) Effect of intravenous TRO40303 as an adjunct to primary percutaneous coronary intervention for acute ST-elevation myocardial infarction: MITOCARE study results. Eur Heart J 36(2):112–119. https://doi.org/10.1093/eurheartj/ehu331

Azarashvili T, Grachev D, Krestinina O, Evtodienko Y, Yurkov I, Papadopoulos V, Reiser G (2007) The peripheral-type benzodiazepine receptor is involved in control of Ca2+-induced permeability transition pore opening in rat brain mitochondria. Cell Calcium 42(1):27–39. https://doi.org/10.1016/j.ceca.2006.11.004

Azarashvili T, Odinokova I, Bakunts A, Ternovsky V, Krestinina O, Tyynela J, Saris NE (2014) Potential role of subunit c of F0F1-ATPase and subunit c of storage body in the mitochondrial permeability transition. Effect of the phosphorylation status of subunit c on pore opening. Cell Calcium 55(2):69–77. https://doi.org/10.1016/j.ceca.2013.12.002

Azoulay-Zohar H, Israelson A, Abu-Hamad S, Shoshan-Barmatz V (2004) In self-defence: hexokinase promotes voltage-dependent anion channel closure and prevents mitochondria-mediated apoptotic cell death. Biochem J 377(Pt 2):347–355. https://doi.org/10.1042/BJ20031465

Baines CP (2007) The mitochondrial permeability transition pore as a target of cardioprotective signaling. Am J Physiol Heart Circ Physiol 293(2):H903–H904. https://doi.org/10.1152/ajpheart.00575.2007

Baines CP, Kaiser RA, Purcell NH, Blair NS, Osinska H, Hambleton MA, Brunskill EW, Sayen MR, Gottlieb RA, Dorn GW, Robbins J, Molkentin JD (2005) Loss of cyclophilin D reveals a critical role for mitochondrial permeability transition in cell death. Nature 434(7033):658–662. https://doi.org/10.1038/nature03434

Baines CP, Kaiser RA, Sheiko T, Craigen WJ, Molkentin JD (2007) Voltage-dependent anion channels are dispensable for mitochondrial-dependent cell death. Nat Cell Biol 9(5):550–555. https://doi.org/10.1038/ncb1575

Basso E, Fante L, Fowlkes J, Petronilli V, Forte MA, Bernardi P (2005) Properties of the permeability transition pore in mitochondria devoid of cyclophilin D. J Biol Chem 280(19):18558–18561. https://doi.org/10.1074/jbc.C500089200

Benfeito S, Oliveira C, Soares P, Fernandes C, Silva T, Teixeira J, Borges F (2013) Antioxidant therapy: still in search of the 'magic bullet'. Mitochondrion 13(5):427–435. https://doi.org/10.1016/j.mito.2012.12.002

Bergmann MW, Haufe S, von Knobelsdorff-Brenkenhoff F, Mehling H, Wassmuth R, Munch I, Busjahn A, Schulz-Menger J, Jordan J, Luft FC, Dietz R (2011) A pilot study of chronic, low-dose epoetin-{beta} following percutaneous coronary intervention suggests safety, feasibility, and efficacy in patients with symptomatic ischaemic heart failure. Eur J Heart Fail 13(5):560–568. https://doi.org/10.1093/eurjhf/hfr002

Bernardi P (1992) Modulation of the mitochondrial cyclosporin A-sensitive permeability transition pore by the proton electrochemical gradient. Evidence that the pore can be opened by membrane depolarization. J Biol Chem 267(13):8834–8839

Bernardi P (2013) The mitochondrial permeability transition pore: a mystery solved? Front Physiol 4:95. https://doi.org/10.3389/fphys.2013.00095

Bernardi P, Vassanelli S, Veronese P, Colonna R, Szabo I, Zoratti M (1992) Modulation of the mitochondrial permeability transition pore. Effect of protons and divalent cations. J Biol Chem 267(5):2934–2939

Bernardi P, Rasola A, Forte M, Lippe G (2015) The mitochondrial permeability transition pore: channel formation by F-ATP synthase, integration in signal transduction, and role in pathophysiology. Physiol Rev 95(4):1111–1155. https://doi.org/10.1152/physrev.00001.2015

Bhagatte Y, Lodwick D, Storey N (2012) Mitochondrial ROS production and subsequent ERK phosphorylation are necessary for temperature preconditioning of isolated ventricular myocytes. Cell Death Dis 3:e345. https://doi.org/10.1038/cddis.2012.84

Bhamra GS, Hausenloy DJ, Davidson SM, Carr RD, Paiva M, Wynne AM, Mocanu MM, Yellon DM (2008) Metformin protects the ischemic heart by the Akt-mediated inhibition of mitochondrial permeability transition pore opening. Basic Res Cardiol 103(3):274–284. https://doi.org/10.1007/s00395-007-0691-y

Biel TG, Lee S, Flores-Toro JA, Dean JW, Go KL, Lee MH, Law BK, Law ME, Dunn WA Jr, Zendejas I, Behrns KE, Kim JS (2016) Sirtuin 1 suppresses mitochondrial dysfunction of ischemic mouse livers in a mitofusin 2-dependent manner. Cell Death Differ 23(2):279–290. https://doi.org/10.1038/cdd.2015.96

Bochaton T, Crola-Da-Silva C, Pillot B, Villedieu C, Ferreras L, Alam MR, Thibault H, Strina M, Gharib A, Ovize M, Baetz D (2015) Inhibition of myocardial reperfusion injury by ischemic postconditioning requires sirtuin 3-mediated deacetylation of cyclophilin D. J Mol Cell Cardiol 84:61–69. https://doi.org/10.1016/j.yjmcc.2015.03.017

Boengler K, Heusch G, Schulz R (2011) Mitochondria in postconditioning. Antioxid Redox Signal 14(5):863–880. https://doi.org/10.1089/ars.2010.3309

Bonora M, Bononi A, De Marchi E, Giorgi C, Lebiedzinska M, Marchi S, Patergnani S, Rimessi A, Suski JM, Wojtala A, Wieckowski MR, Kroemer G, Galluzzi L, Pinton P (2013) Role of the c subunit of the FO ATP synthase in mitochondrial permeability transition. Cell Cycle 12(4):674–683. https://doi.org/10.4161/cc.23599

Bonora M, Morganti C, Morciano G, Pedriali G, Lebiedzinska-Arciszewska M, Aquila G, Giorgi C, Rizzo P, Campo G, Ferrari R, Kroemer G, Wieckowski MR, Galluzzi L, Pinton P (2017) Mitochondrial permeability transition involves dissociation of F1FO ATP synthase dimers and C-ring conformation. EMBO Rep 18(7):1077–1089. https://doi.org/10.15252/embr.201643602

Boyce SW, Bartels C, Bolli R, Chaitman B, Chen JC, Chi E, Jessel A, Kereiakes D, Knight J, Thulin L, Theroux P, Investigators GDIANS (2003) Impact of sodium-hydrogen exchange inhibition by cariporide on death or myocardial infarction in high-risk CABG surgery patients: results of the CABG surgery cohort of the GUARDIAN study. J Thorac Cardiovasc Surg 126(2):420–427

Briston T, Lewis S, Koglin M, Mistry K, Shen Y, Hartopp N, Katsumata R, Fukumoto H, Duchen MR, Szabadkai G, Staddon JM, Roberts M, Powney B (2016) Identification of ER-000444793, a cyclophilin D-independent inhibitor of mitochondrial permeability transition, using a high-throughput screen in cryopreserved mitochondria. Sci Rep 6:37798. https://doi.org/10.1038/srep37798

Briston T, Roberts M, Lewis S, Powney B, Staddon MJ, Szabadkai G (2017) Mitochondrial permeability transition pore: sensitivity to opening and mechanistic dependence on substrate availability. Sci Rep 7(1):10492. https://doi.org/10.1038/s41598-017-10673-8

Brookes PS, Salinas EP, Darley-Usmar K, Eiserich JP, Freeman BA, Darley-Usmar VM, Anderson PG (2000) Concentration-dependent effects of nitric oxide on mitochondrial permeability transition and cytochrome c release. J Biol Chem 275(27):20474–20479. https://doi.org/10.1074/jbc.M001077200

Brown DA, Aon MA, Akar FG, Liu T, Sorarrain N, O'Rourke B (2008) Effects of 4′-chlorodiazepam on cellular excitation-contraction coupling and ischaemia-reperfusion injury in rabbit heart. Cardiovasc Res 79(1):141–149. https://doi.org/10.1093/cvr/cvn053

Budas GR, Churchill EN, Disatnik MH, Sun L, Mochly-Rosen D (2010) Mitochondrial import of PKCepsilon is mediated by HSP90: a role in cardioprotection from ischaemia and reperfusion injury. Cardiovasc Res 88(1):83–92. https://doi.org/10.1093/cvr/cvq154

Bugrim AE (1999) Regulation of Ca2+ release by cAMP-dependent protein kinase. A mechanism for agonist-specific calcium signaling? Cell Calcium 25(3):219–226. https://doi.org/10.1054/ceca.1999.0027

Bunger R, Mallet RT, Hartman DA (1989) Pyruvate-enhanced phosphorylation potential and inotropism in normoxic and postischemic isolated working heart. Near-complete prevention of reperfusion contractile failure. Eur J Biochem 180(1):221–233

Caballero-Garrido E, Pena-Philippides JC, Lordkipanidze T, Bragin D, Yang Y, Erhardt EB, Roitbak T (2015) In vivo inhibition of miR-155 promotes recovery after experimental mouse stroke. J Neurosci 35(36):12446–12464. https://doi.org/10.1523/JNEUROSCI.1641-15.2015

Cabantous S, Terwilliger TC, Waldo GS (2005) Protein tagging and detection with engineered self-assembling fragments of green fluorescent protein. Nat Biotechnol 23(1):102–107. https://doi.org/10.1038/nbt1044

Calvert JW, Gundewar S, Jha S, Greer JJ, Bestermann WH, Tian R, Lefer DJ (2008) Acute metformin therapy confers cardioprotection against myocardial infarction via AMPK-eNOS-mediated signaling. Diabetes 57(3):696–705. https://doi.org/10.2337/db07-1098

Camara AK, Lesnefsky EJ, Stowe DF (2010) Potential therapeutic benefits of strategies directed to mitochondria. Antioxid Redox Signal 13(3):279–347. https://doi.org/10.1089/ars.2009.2788

Camara AKS, Zhou Y, Wen PC, Tajkhorshid E, Kwok WM (2017) Mitochondrial VDAC1: a key gatekeeper as potential therapeutic target. Front Physiol 8:460. https://doi.org/10.3389/fphys.2017.00460

Campo G, Morciano G, Pavasini R, Bonora M, Sbano L, Biscaglia S, Bovolenta M, Pinotti M, Punzetti S, Rizzo P, Aquila G, Giorgi C, Ferrari R, Pinton P (2016) Fo ATP synthase C subunit serum levels in patients with ST-segment elevation myocardial infarction: preliminary findings. Int J Cardiol 221:993–997. https://doi.org/10.1016/j.ijcard.2016.07.125

Carraro M, Giorgio V, Sileikyte J, Sartori G, Forte M, Lippe G, Zoratti M, Szabo I, Bernardi P (2014) Channel formation by yeast F-ATP synthase and the role of dimerization in the mitochondrial permeability transition. J Biol Chem 289(23):15980–15985. https://doi.org/10.1074/jbc.C114.559633

Carreira RS, Monteiro P, Gon Alves LM, Providencia LA (2006) Carvedilol: just another Beta-blocker or a powerful cardioprotector? Cardiovasc Hematol Disord Drug Targets 6(4):257–266

Carreira RS, Lee P, Gottlieb RA (2011) Mitochondrial therapeutics for cardioprotection. Curr Pharm Des 17(20):2017–2035

Castilho RF, Kowaltowski AJ, Vercesi AE (1996) The irreversibility of inner mitochondrial membrane permeabilization by Ca2+ plus prooxidants is determined by the extent of membrane protein thiol cross-linking. J Bioenerg Biomembr 28(6):523–529

Cavallini L, Valente M, Rigobello MP (1990) The protective action of pyruvate on recovery of ischemic rat heart: comparison with other oxidizable substrates. J Mol Cell Cardiol 22(2):143–154

Cerisano G, Buonamici P, Valenti R, Sciagra R, Raspanti S, Santini A, Carrabba N, Dovellini EV, Romito R, Pupi A, Colonna P, Antoniucci D (2014) Early short-term doxycycline therapy in patients with acute myocardial infarction and left ventricular dysfunction to prevent the omi-

nous progression to adverse remodelling: the TIPTOP trial. Eur Heart J 35(3):184–191. https://doi.org/10.1093/eurheartj/eht420
Chelli B, Falleni A, Salvetti F, Gremigni V, Lucacchini A, Martini C (2001) Peripheral-type benzodiazepine receptor ligands: mitochondrial permeability transition induction in rat cardiac tissue. Biochem Pharmacol 61(6):695–705
Chen HH, Glockner JF, Schirger JA, Cataliotti A, Redfield MM, Burnett JC Jr (2012) Novel protein therapeutics for systolic heart failure: chronic subcutaneous B-type natriuretic peptide. J Am Coll Cardiol 60(22):2305–2312. https://doi.org/10.1016/j.jacc.2012.07.056
Chen T, Li J, Liu J, Li N, Wang S, Liu H, Zeng M, Zhang Y, Bu P (2015) Activation of SIRT3 by resveratrol ameliorates cardiac fibrosis and improves cardiac function via the TGF-beta/Smad3 pathway. Am J Physiol Heart Circ Physiol 308(5):H424–H434. https://doi.org/10.1152/ajpheart.00454.2014
Cheng Q, Sedlic F, Pravdic D, Bosnjak ZJ, Kwok WM (2011) Biphasic effect of nitric oxide on the cardiac voltage-dependent anion channel. FEBS Lett 585(2):328–334. https://doi.org/10.1016/j.febslet.2010.12.008
Chen-Scarabelli C, Saravolatz L Jr, Murad Y, Shieh WS, Qureshi W, Di Rezze J, Abrencillo R, Gardin T, Gidwani UK, Saravolatz L, Faggian G, Scarabelli TM (2012) A critical review of the use of carvedilol in ischemic heart disease. Am J Cardiovasc Drugs 12(6):391–401. https://doi.org/10.2165/11636090-000000000-00000
Chiari P, Angoulvant D, Mewton N, Desebbe O, Obadia JF, Robin J, Farhat F, Jegaden O, Bastien O, Lehot JJ, Ovize M (2014) Cyclosporine protects the heart during aortic valve surgery. Anesthesiology 121(2):232–238. https://doi.org/10.1097/ALN.0000000000000331
Chinopoulos C (2017) Mitochondrial permeability transition pore: back to the drawing board. Neurochem Int. https://doi.org/10.1016/j.neuint.2017.06.010
Chinopoulos C, Szabadkai G (2014) What makes you can also break you, part III: mitochondrial permeability transition pore formation by an uncoupling channel within the C-subunit ring of the F1FO ATP synthase? Front Oncol 4:235. https://doi.org/10.3389/fonc.2014.00235
Chouchani ET, Pell VR, Gaude E, Aksentijevic D, Sundier SY, Robb EL, Logan A, Nadtochiy SM, Ord ENJ, Smith AC, Eyassu F, Shirley R, Hu CH, Dare AJ, James AM, Rogatti S, Hartley RC, Eaton S, Costa ASH, Brookes PS, Davidson SM, Duchen MR, Saeb-Parsy K, Shattock MJ, Robinson AJ, Work LM, Frezza C, Krieg T, Murphy MP (2014) Ischaemic accumulation of succinate controls reperfusion injury through mitochondrial ROS. Nature 515(7527):431–435. https://doi.org/10.1038/nature13909
Clarke SJ, McStay GP, Halestrap AP (2002) Sanglifehrin A acts as a potent inhibitor of the mitochondrial permeability transition and reperfusion injury of the heart by binding to cyclophilin-D at a different site from cyclosporin A. J Biol Chem 277(38):34793–34799. https://doi.org/10.1074/jbc.M202191200
Clarke SJ, Khaliulin I, Das M, Parker JE, Heesom KJ, Halestrap AP (2008) Inhibition of mitochondrial permeability transition pore opening by ischemic preconditioning is probably mediated by reduction of oxidative stress rather than mitochondrial protein phosphorylation. Circ Res 102(9):1082–1090. https://doi.org/10.1161/CIRCRESAHA.107.167072
Cleary J, Johnson KM, Opipari AW Jr, Glick GD (2007) Inhibition of the mitochondrial F1F0-ATPase by ligands of the peripheral benzodiazepine receptor. Bioorg Med Chem Lett 17(6):1667–1670. https://doi.org/10.1016/j.bmcl.2006.12.102
Cohen P, Goedert M (2004) GSK3 inhibitors: development and therapeutic potential. Nat Rev Drug Discov 3(6):479–487. https://doi.org/10.1038/nrd1415
Constantopoulos G, Barranger JA (1984) Nonenzymatic decarboxylation of pyruvate. Anal Biochem 139(2):353–358
Costa AD, Garlid KD (2008) Intramitochondrial signaling: interactions among mitoKATP, PKCepsilon, ROS, and MPT. Am J Physiol Heart Circ Physiol 295(2):H874–H882. https://doi.org/10.1152/ajpheart.01189.2007
Costa AD, Garlid KD, West IC, Lincoln TM, Downey JM, Cohen MV, Critz SD (2005) Protein kinase G transmits the cardioprotective signal from cytosol to mitochondria. Circ Res 97(4):329–336. https://doi.org/10.1161/01.RES.0000178451.08719.5b

Costa AD, Jakob R, Costa CL, Andrukhiv K, West IC, Garlid KD (2006) The mechanism by which the mitochondrial ATP-sensitive K+ channel opening and H2O2 inhibit the mitochondrial permeability transition. J Biol Chem 281(30):20801–20808. https://doi.org/10.1074/jbc.M600959200

Costa AD, Pierre SV, Cohen MV, Downey JM, Garlid KD (2008) cGMP signalling in pre- and post-conditioning: the role of mitochondria. Cardiovasc Res 77(2):344–352. https://doi.org/10.1093/cvr/cvm050

Cribbs JT, Strack S (2007) Reversible phosphorylation of Drp1 by cyclic AMP-dependent protein kinase and calcineurin regulates mitochondrial fission and cell death. EMBO Rep 8(10):939–944. https://doi.org/10.1038/sj.embor.7401062

Crompton M, Costi A (1988) Kinetic evidence for a heart mitochondrial pore activated by Ca2+, inorganic phosphate and oxidative stress. A potential mechanism for mitochondrial dysfunction during cellular Ca2+ overload. Eur J Biochem 178(2):489–501

Crompton M, Ellinger H, Costi A (1988) Inhibition by cyclosporin A of a Ca2+-dependent pore in heart mitochondria activated by inorganic phosphate and oxidative stress. Biochem J 255(1):357–360

Culty M, Li H, Boujrad N, Amri H, Vidic B, Bernassau JM, Reversat JL, Papadopoulos V (1999) In vitro studies on the role of the peripheral-type benzodiazepine receptor in steroidogenesis. J Steroid Biochem Mol Biol 69(1–6):123–130

Cung TT, Morel O, Cayla G, Rioufol G, Garcia-Dorado D, Angoulvant D, Bonnefoy-Cudraz E, Guerin P, Elbaz M, Delarche N, Coste P, Vanzetto G, Metge M, Aupetit JF, Jouve B, Motreff P, Tron C, Labeque JN, Steg PG, Cottin Y, Range G, Clerc J, Claeys MJ, Coussement P, Prunier F, Moulin F, Roth O, Belle L, Dubois P, Barragan P, Gilard M, Piot C, Colin P, De Poli F, Morice MC, Ider O, Dubois-Rande JL, Unterseeh T, Le Breton H, Beard T, Blanchard D, Grollier G, Malquarti V, Staat P, Sudre A, Elmer E, Hansson MJ, Bergerot C, Boussaha I, Jossan C, Derumeaux G, Mewton N, Ovize M (2015) Cyclosporine before PCI in patients with acute myocardial infarction. N Engl J Med 373(11):1021–1031. https://doi.org/10.1056/NEJMoa1505489

Da Silva D, Ausina P, Alencar EM, Coelho WS, Zancan P, Sola-Penna M (2012) Metformin reverses hexokinase and phosphofructokinase downregulation and intracellular distribution in the heart of diabetic mice. IUBMB Life 64(9):766–774. https://doi.org/10.1002/iub.1063

Das S, Wong R, Rajapakse N, Murphy E, Steenbergen C (2008) Glycogen synthase kinase 3 inhibition slows mitochondrial adenine nucleotide transport and regulates voltage-dependent anion channel phosphorylation. Circ Res 103(9):983–991. https://doi.org/10.1161/CIRCRESAHA.108.178970

Datler C, Pazarentzos E, Mahul-Mellier AL, Chaisaklert W, Hwang MS, Osborne F, Grimm S (2014) CKMT1 regulates the mitochondrial permeability transition pore in a process that provides evidence for alternative forms of the complex. J Cell Sci 127(Pt 8):1816–1828. https://doi.org/10.1242/jcs.140467

Daum B, Walter A, Horst A, Osiewacz HD, Kuhlbrandt W (2013) Age-dependent dissociation of ATP synthase dimers and loss of inner-membrane cristae in mitochondria. Proc Natl Acad Sci U S A 110(38):15301–15306. https://doi.org/10.1073/pnas.1305462110

Davidson SM, Hausenloy D, Duchen MR, Yellon DM (2006) Signalling via the reperfusion injury signalling kinase (RISK) pathway links closure of the mitochondrial permeability transition pore to cardioprotection. Int J Biochem Cell Biol 38(3):414–419. https://doi.org/10.1016/j.biocel.2005.09.017

de Macedo DV, Nepomuceno ME, Pereira-da-Silva L (1993) Involvement of the ADP/ATP carrier in permeabilization processes of the inner mitochondrial membrane. Eur J Biochem 215(3):595–600

De Marchi E, Bonora M, Giorgi C, Pinton P (2014) The mitochondrial permeability transition pore is a dispensable element for mitochondrial calcium efflux. Cell Calcium 56(1):1–13. https://doi.org/10.1016/j.ceca.2014.03.004

De Paulis D, Chiari P, Teixeira G, Couture-Lepetit E, Abrial M, Argaud L, Gharib A, Ovize M (2013) Cyclosporine A at reperfusion fails to reduce infarct size in the in vivo rat heart. Basic Res Cardiol 108(5):379. https://doi.org/10.1007/s00395-013-0379-4

DeBoer LW, Bekx PA, Han L, Steinke L (1993) Pyruvate enhances recovery of rat hearts after ischemia and reperfusion by preventing free radical generation. Am J Phys 265(5 Pt 2):H1571–H1576

Dekker LR, Coronel R, VanBavel E, Spaan JA, Opthof T (1999) Intracellular Ca2+ and delay of ischemia-induced electrical uncoupling in preconditioned rabbit ventricular myocardium. Cardiovasc Res 44(1):101–112

Delcamp TJ, Dales C, Ralenkotter L, Cole PS, Hadley RW (1998) Intramitochondrial [Ca2+] and membrane potential in ventricular myocytes exposed to anoxia-reoxygenation. Am J Phys 275(2 Pt 2):H484–H494

Disatnik MH, Ferreira JC, Campos JC, Gomes KS, Dourado PM, Qi X, Mochly-Rosen D (2013) Acute inhibition of excessive mitochondrial fission after myocardial infarction prevents long-term cardiac dysfunction. J Am Heart Assoc 2(5):e000461. https://doi.org/10.1161/JAHA.113.000461

Dixon AS, Schwinn MK, Hall MP, Zimmerman K, Otto P, Lubben TH, Butler BL, Binkowski BF, Machleidt T, Kirkland TA, Wood MG, Eggers CT, Encell LP, Wood KV (2016) NanoLuc complementation reporter optimized for accurate measurement of protein interactions in cells. ACS Chem Biol 11(2):400–408. https://doi.org/10.1021/acschembio.5b00753

Dobsak P, Courderot-Masuyer C, Zeller M, Vergely C, Laubriet A, Assem M, Eicher JC, Teyssier JR, Wolf JE, Rochette L (1999) Antioxidative properties of pyruvate and protection of the ischemic rat heart during cardioplegia. J Cardiovasc Pharmacol 34(5):651–659

Dolder M, Wendt S, Wallimann T (2001) Mitochondrial creatine kinase in contact sites: interaction with porin and adenine nucleotide translocase, role in permeability transition and sensitivity to oxidative damage. Biol Signals Recept 10(1–2):93–111

Dongworth RK, Hall AR, Burke N, Hausenloy DJ (2014) Targeting mitochondria for cardioprotection: examining the benefit for patients. Futur Cardiol 10(2):255–272. https://doi.org/10.2217/fca.14.6

Dow J, Kloner RA (2007) Postconditioning does not reduce myocardial infarct size in an in vivo regional ischemia rodent model. J Cardiovasc Pharmacol Ther 12(2):153–163. https://doi.org/10.1177/1074248407300897

Dube H, Selwood D, Malouitre S, Capano M, Simone MI, Crompton M (2012) A mitochondrial-targeted cyclosporin A with high binding affinity for cyclophilin D yields improved cytoprotection of cardiomyocytes. Biochem J 441(3):901–907. https://doi.org/10.1042/BJ20111301

Eisenhardt SU, Weiss JB, Smolka C, Maxeiner J, Pankratz F, Bemtgen X, Kustermann M, Thiele JR, Schmidt Y, Bjoern Stark G, Moser M, Bode C, Grundmann S (2015) MicroRNA-155 aggravates ischemia-reperfusion injury by modulation of inflammatory cell recruitment and the respiratory oxidative burst. Basic Res Cardiol 110(3):32. https://doi.org/10.1007/s00395-015-0490-9

Elrod JW, Molkentin JD (2013) Physiologic functions of cyclophilin D and the mitochondrial permeability transition pore. Circ J 77(5):1111–1122

Eriksson O, Pollesello P, Geimonen E (1999) Regulation of total mitochondrial Ca2+ in perfused liver is independent of the permeability transition pore. Am J Phys 276(6 Pt 1):C1297–C1302

Fancelli D, Abate A, Amici R, Bernardi P, Ballarini M, Cappa A, Carenzi G, Colombo A, Contursi C, Di Lisa F, Dondio G, Gagliardi S, Milanesi E, Minucci S, Pain G, Pelicci PG, Saccani A, Storto M, Thaler F, Varasi M, Villa M, Plyte S (2014) Cinnamic anilides as new mitochondrial permeability transition pore inhibitors endowed with ischemia-reperfusion injury protective effect in vivo. J Med Chem 57(12):5333–5347. https://doi.org/10.1021/jm500547c

Fantinelli JC, Cingolani HE, Mosca SM (2006) Na+/H+ exchanger inhibition at the onset of reperfusion decreases myocardial infarct size: role of reactive oxygen species. Cardiovasc Pathol 15(4):179–184. https://doi.org/10.1016/j.carpath.2006.04.005

Feng J, Lucchinetti E, Ahuja P, Pasch T, Perriard JC, Zaugg M (2005) Isoflurane postconditioning prevents opening of the mitochondrial permeability transition pore through inhibition of glycogen synthase kinase 3beta. Anesthesiology 103(5):987–995

Feng X, Li J, Liu J, Jin M, Liu X, Du H, Zhang L, Sun Z, Li X (2011) Protective effect of FK506 on myocardial ischemia/reperfusion injury by suppression of CaN and ASK1 signaling circuitry. Cardiovasc Toxicol 11(1):18–27. https://doi.org/10.1007/s12012-010-9095-6

Flisiak R, Feinman SV, Jablkowski M, Horban A, Kryczka W, Pawlowska M, Heathcote JE, Mazzella G, Vandelli C, Nicolas-Metral V, Grosgurin P, Liz JS, Scalfaro P, Porchet H, Crabbe R (2009) The cyclophilin inhibitor Debio 025 combined with PEG IFNalpha2a significantly reduces viral load in treatment-naive hepatitis C patients. Hepatology 49(5):1460–1468. https://doi.org/10.1002/hep.22835

Forster K, Paul I, Solenkova N, Staudt A, Cohen MV, Downey JM, Felix SB, Krieg T (2006) NECA at reperfusion limits infarction and inhibits formation of the mitochondrial permeability transition pore by activating p70S6 kinase. Basic Res Cardiol 101(4):319–326. https://doi.org/10.1007/s00395-006-0593-4

Francaux M, Poortmans JR (2006) Side effects of creatine supplementation in athletes. Int J Sports Physiol Perform 1(4):311–323

Galluzzi L, Kepp O, Tajeddine N, Kroemer G (2008) Disruption of the hexokinase-VDAC complex for tumor therapy. Oncogene 27(34):4633–4635. https://doi.org/10.1038/onc.2008.114

Gao D, Zhang L, Dhillon R, Hong TT, Shaw RM, Zhu J (2013) Dynasore protects mitochondria and improves cardiac lusitropy in Langendorff perfused mouse heart. PLoS One 8(4):e60967. https://doi.org/10.1371/journal.pone.0060967

Garciarena CD, Caldiz CI, Correa MV, Schinella GR, Mosca SM, Chiappe de Cingolani GE, Cingolani HE, Ennis IL (2008) Na+/H+ exchanger-1 inhibitors decrease myocardial superoxide production via direct mitochondrial action. J Appl Physiol (1985) 105(6):1706–1713. https://doi.org/10.1152/japplphysiol.90616.2008

Garnier M, Dimchev AB, Boujrad N, Price JM, Musto NA, Papadopoulos V (1994) In vitro reconstitution of a functional peripheral-type benzodiazepine receptor from mouse Leydig tumor cells. Mol Pharmacol 45(2):201–211

Gatliff J, East D, Crosby J, Abeti R, Harvey R, Craigen W, Parker P, Campanella M (2014) TSPO interacts with VDAC1 and triggers a ROS-mediated inhibition of mitochondrial quality control. Autophagy 10(12):2279–2296. https://doi.org/10.4161/15548627.2014.991665

Giorgio V, von Stockum S, Antoniel M, Fabbro A, Fogolari F, Forte M, Glick GD, Petronilli V, Zoratti M, Szabo I, Lippe G, Bernardi P (2013) Dimers of mitochondrial ATP synthase form the permeability transition pore. Proc Natl Acad Sci U S A 110(15):5887–5892. https://doi.org/10.1073/pnas.1217823110

Giorgio V, Burchell V, Schiavone M, Bassot C, Minervini G, Petronilli V, Argenton F, Forte M, Tosatto S, Lippe G, Bernardi P (2017) Ca2+ binding to F-ATP synthase beta subunit triggers the mitochondrial permeability transition. EMBO Rep 18(7):1065–1076. https://doi.org/10.15252/embr.201643354

Golshani-Hebroni S (2016) Mg(++) requirement for MtHK binding, and Mg(++) stabilization of mitochondrial membranes via activation of MtHK & MtCK and promotion of mitochondrial permeability transition pore closure: a hypothesis on mechanisms underlying Mg(++)'s antioxidant and cytoprotective effects. Gene 581(1):1–13. https://doi.org/10.1016/j.gene.2015.12.046

Gomez L, Thibault H, Gharib A, Dumont JM, Vuagniaux G, Scalfaro P, Derumeaux G, Ovize M (2007) Inhibition of mitochondrial permeability transition improves functional recovery and reduces mortality following acute myocardial infarction in mice. Am J Physiol Heart Circ Physiol 293(3):H1654–H1661. https://doi.org/10.1152/ajpheart.01378.2006

Gomez L, Paillard M, Thibault H, Derumeaux G, Ovize M (2008) Inhibition of GSK3beta by postconditioning is required to prevent opening of the mitochondrial permeability transition pore during reperfusion. Circulation 117(21):2761–2768. https://doi.org/10.1161/CIRCULATIONAHA.107.755066

Griffiths EJ (2009) Mitochondrial calcium transport in the heart: physiological and pathological roles. J Mol Cell Cardiol 46(6):789–803. https://doi.org/10.1016/j.yjmcc.2009.03.001

Griffiths EJ, Halestrap AP (1993) Protection by cyclosporin A of ischemia/reperfusion-induced damage in isolated rat hearts. J Mol Cell Cardiol 25(12):1461–1469. https://doi.org/10.1006/jmcc.1993.1162

Griffiths EJ, Halestrap AP (1995) Mitochondrial non-specific pores remain closed during cardiac ischaemia, but open upon reperfusion. Biochem J 307(Pt 1):93–98

Gross ER, Hsu AK, Gross GJ (2004) Opioid-induced cardioprotection occurs via glycogen synthase kinase beta inhibition during reperfusion in intact rat hearts. Circ Res 94(7):960–966. https://doi.org/10.1161/01.RES.0000122392.33172.09

Group MS (2012) Rationale and design of the 'MITOCARE' study: a phase II, multicenter, randomized, double-blind, placebo-controlled study to assess the safety and efficacy of TRO40303 for the reduction of reperfusion injury in patients undergoing percutaneous coronary intervention for acute myocardial infarction. Cardiology 123(4):201–207. https://doi.org/10.1159/000342981

Grover GJ, Dzwonczyk S, Sleph PG (1990) Ruthenium red improves postischemic contractile function in isolated rat hearts. J Cardiovasc Pharmacol 16(5):783–789

Grover GJ, Atwal KS, Sleph PG, Wang FL, Monshizadegan H, Monticello T, Green DW (2004) Excessive ATP hydrolysis in ischemic myocardium by mitochondrial F1F0-ATPase: effect of selective pharmacological inhibition of mitochondrial ATPase hydrolase activity. Am J Phys Heart Circ Phys 287(4):H1747–H1755. https://doi.org/10.1152/ajpheart.01019.2003

Gruber J, Fong S, Chen CB, Yoong S, Pastorin G, Schaffer S, Cheah I, Halliwell B (2013) Mitochondria-targeted antioxidants and metabolic modulators as pharmacological interventions to slow ageing. Biotechnol Adv 31(5):563–592. https://doi.org/10.1016/j.biotechadv.2012.09.005

Gurel E, Smeele KM, Eerbeek O, Koeman A, Demirci C, Hollmann MW, Zuurbier CJ (2009) Ischemic preconditioning affects hexokinase activity and HKII in different subcellular compartments throughout cardiac ischemia-reperfusion. J Appl Physiol (1985) 106(6):1909–1916. https://doi.org/10.1152/japplphysiol.90537.2008

Gutierrez-Aguilar M, Douglas DL, Gibson AK, Domeier TL, Molkentin JD, Baines CP (2014) Genetic manipulation of the cardiac mitochondrial phosphate carrier does not affect permeability transition. J Mol Cell Cardiol 72:316–325. https://doi.org/10.1016/j.yjmcc.2014.04.008

Gwag HB, Kim EK, Park TK, Lee JM, Yang JH, Song YB, Choi JH, Choi SH, Lee SH, Chang SA, Park SJ, Lee SC, Park SW, Jang WJ, Lee M, Chun WJ, Oh JH, Park YH, Choe YH, Gwon HC, Hahn JY (2017) Cardioprotective effects of intracoronary morphine in ST-segment elevation myocardial infarction patients undergoing primary percutaneous coronary intervention: a prospective, randomized trial. J Am Heart Assoc 6(4). https://doi.org/10.1161/JAHA.116.005426

Hafner AV, Dai J, Gomes AP, Xiao CY, Palmeira CM, Rosenzweig A, Sinclair DA (2010) Regulation of the mPTP by SIRT3-mediated deacetylation of CypD at lysine 166 suppresses age-related cardiac hypertrophy. Aging (Albany NY) 2(12):914–923. https://doi.org/10.18632/aging.100252

Halestrap AP (1991) Calcium-dependent opening of a non-specific pore in the mitochondrial inner membrane is inhibited at pH values below 7. Implications for the protective effect of low pH against chemical and hypoxic cell damage. Biochem J 278(Pt 3):715–719

Halestrap AP (2009) What is the mitochondrial permeability transition pore? J Mol Cell Cardiol 46(6):821–831. https://doi.org/10.1016/j.yjmcc.2009.02.021

Halestrap AP (2014) The C ring of the F1Fo ATP synthase forms the mitochondrial permeability transition pore: a critical appraisal. Front Oncol 4:234. https://doi.org/10.3389/fonc.2014.00234

Halestrap AP, Brenner C (2003) The adenine nucleotide translocase: a central component of the mitochondrial permeability transition pore and key player in cell death. Curr Med Chem 10(16):1507–1525

Halestrap AP, Davidson AM (1990) Inhibition of Ca2(+)-induced large-amplitude swelling of liver and heart mitochondria by cyclosporin is probably caused by the inhibitor binding to mitochondrial-matrix peptidyl-prolyl cis-trans isomerase and preventing it interacting with the adenine nucleotide translocase. Biochem J 268(1):153–160

Halestrap AP, Pasdois P (2009) The role of the mitochondrial permeability transition pore in heart disease. Biochim Biophys Acta 1787(11):1402–1415. https://doi.org/10.1016/j.bbabio.2008.12.017

Halestrap AP, Richardson AP (2015) The mitochondrial permeability transition: a current perspective on its identity and role in ischaemia/reperfusion injury. J Mol Cell Cardiol 78:129–141. https://doi.org/10.1016/j.yjmcc.2014.08.018

Halestrap AP, Connern CP, Griffiths EJ, Kerr PM (1997a) Cyclosporin A binding to mitochondrial cyclophilin inhibits the permeability transition pore and protects hearts from ischaemia/reperfusion injury. Mol Cell Biochem 174(1–2):167–172

Halestrap AP, Woodfield KY, Connern CP (1997b) Oxidative stress, thiol reagents, and membrane potential modulate the mitochondrial permeability transition by affecting nucleotide binding to the adenine nucleotide translocase. J Biol Chem 272(6):3346–3354

Halestrap AP, Clarke SJ, Javadov SA (2004) Mitochondrial permeability transition pore opening during myocardial reperfusion—a target for cardioprotection. Cardiovasc Res 61(3):372–385. https://doi.org/10.1016/S0008-6363(03)00533-9

Halestrap AP, Clarke SJ, Khaliulin I (2007) The role of mitochondria in protection of the heart by preconditioning. Biochim Biophys Acta 1767(8):1007–1031. https://doi.org/10.1016/j.bbabio.2007.05.008

Halestrap AP, Pereira GC, Pasdois P (2015) The role of hexokinase in cardioprotection – mechanism and potential for translation. Br J Pharmacol 172(8):2085–2100. https://doi.org/10.1111/bph.12899

Hall AR, Burke N, Dongworth RK, Kalkhoran SB, Dyson A, Vicencio JM, Dorn GW II, Yellon DM, Hausenloy DJ (2016) Hearts deficient in both Mfn1 and Mfn2 are protected against acute myocardial infarction. Cell Death Dis 7:e2238. https://doi.org/10.1038/cddis.2016.139

Hanley PJ, Daut J (2005) K(ATP) channels and preconditioning: a re-examination of the role of mitochondrial K(ATP) channels and an overview of alternative mechanisms. J Mol Cell Cardiol 39(1):17–50. https://doi.org/10.1016/j.yjmcc.2005.04.002

Hanley PJ, Mickel M, Loffler M, Brandt U, Daut J (2002) K(ATP) channel-independent targets of diazoxide and 5-hydroxydecanoate in the heart. J Physiol 542(Pt 3):735–741

Hansson MJ, Mattiasson G, Mansson R, Karlsson J, Keep MF, Waldmeier P, Ruegg UT, Dumont JM, Besseghir K, Elmer E (2004) The nonimmunosuppressive cyclosporin analogs NIM811 and UNIL025 display nanomolar potencies on permeability transition in brain-derived mitochondria. J Bioenerg Biomembr 36(4):407–413. https://doi.org/10.1023/B:JOBB.0000041776.31885.45

Hansson MJ, Llwyd O, Morin D, de Paulis D, Arnoux T, Gouarne C, Koul S, Engblom H, Bordet T, Tissier R, Arheden H, Erlinge D, Halestrap AP, Berdeaux A, Pruss RM, Schaller S (2015) Differences in the profile of protection afforded by TRO40303 and mild hypothermia in models of cardiac ischemia/reperfusion injury. Eur J Pharmacol 760:7–19. https://doi.org/10.1016/j.ejphar.2015.04.009

Hartmann M, Decking UK (1999) Blocking Na(+)-H+ exchange by cariporide reduces Na(+)-overload in ischemia and is cardioprotective. J Mol Cell Cardiol 31(11):1985–1995. https://doi.org/10.1006/jmcc.1999.1029

Hata N, Seino Y, Tsutamoto T, Hiramitsu S, Kaneko N, Yoshikawa T, Yokoyama H, Tanaka K, Mizuno K, Nejima J, Kinoshita M (2008) Effects of carperitide on the long-term prognosis of patients with acute decompensated chronic heart failure: the PROTECT multicenter randomized controlled study. Circ J 72(11):1787–1793

Hausenloy DJ, Yellon DM (2004) New directions for protecting the heart against ischaemia-reperfusion injury: targeting the Reperfusion Injury Salvage Kinase (RISK)-pathway. Cardiovasc Res 61(3):448–460. https://doi.org/10.1016/j.cardiores.2003.09.024

Hausenloy DJ, Yellon DM (2016) Ischaemic conditioning and reperfusion injury. Nat Rev Cardiol 13(4):193–209. https://doi.org/10.1038/nrcardio.2016.5

Hausenloy DJ, Maddock HL, Baxter GF, Yellon DM (2002) Inhibiting mitochondrial permeability transition pore opening: a new paradigm for myocardial preconditioning? Cardiovasc Res 55(3):534–543

Hausenloy DJ, Duchen MR, Yellon DM (2003) Inhibiting mitochondrial permeability transition pore opening at reperfusion protects against ischaemia-reperfusion injury. Cardiovasc Res 60(3):617–625

Hausenloy DJ, Mocanu MM, Yellon DM (2004a) Cross-talk between the survival kinases during early reperfusion: its contribution to ischemic preconditioning. Cardiovasc Res 63(2):305–312. https://doi.org/10.1016/j.cardiores.2004.04.011

Hausenloy DJ, Yellon DM, Mani-Babu S, Duchen MR (2004b) Preconditioning protects by inhibiting the mitochondrial permeability transition. Am J Physiol Heart Circ Physiol 287(2):H841–H849. https://doi.org/10.1152/ajpheart.00678.2003

Hausenloy DJ, Tsang A, Mocanu MM, Yellon DM (2005a) Ischemic preconditioning protects by activating prosurvival kinases at reperfusion. Am J Physiol Heart Circ Physiol 288(2):H971–H976. https://doi.org/10.1152/ajpheart.00374.2004

Hausenloy DJ, Tsang A, Yellon DM (2005b) The reperfusion injury salvage kinase pathway: a common target for both ischemic preconditioning and postconditioning. Trends Cardiovasc Med 15(2):69–75. https://doi.org/10.1016/j.tcm.2005.03.001

Hausenloy D, Kunst G, Boston-Griffiths E, Kolvekar S, Chaubey S, John L, Desai J, Yellon D (2014) The effect of cyclosporin-A on peri-operative myocardial injury in adult patients undergoing coronary artery bypass graft surgery: a randomised controlled clinical trial. Heart 100(7):544–549. https://doi.org/10.1136/heartjnl-2013-304845

Haworth RA, Hunter DR (2000) Control of the mitochondrial permeability transition pore by high-affinity ADP binding at the ADP/ATP translocase in permeabilized mitochondria. J Bioenerg Biomembr 32(1):91–96

He L, Lemasters JJ (2002) Regulated and unregulated mitochondrial permeability transition pores: a new paradigm of pore structure and function? FEBS Lett 512(1–3):1–7

He J, Ford HC, Carroll J, Ding S, Fearnley IM, Walker JE (2017) Persistence of the mitochondrial permeability transition in the absence of subunit c of human ATP synthase. Proc Natl Acad Sci U S A 114(13):3409–3414. https://doi.org/10.1073/pnas.1702357114

Herick K, Kramer R, Luhring H (1997) Patch clamp investigation into the phosphate carrier from Saccharomyces cerevisiae mitochondria. Biochim Biophys Acta 1321(3):207–220

Hermann HP, Pieske B, Schwarzmuller E, Keul J, Just H, Hasenfuss G (1999) Haemodynamic effects of intracoronary pyruvate in patients with congestive heart failure: an open study. Lancet 353(9161):1321–1323

Ho S, Clipstone N, Timmermann L, Northrop J, Graef I, Fiorentino D, Nourse J, Crabtree GR (1996) The mechanism of action of cyclosporin A and FK506. Clin Immunol Immunopathol 80(3 Pt 2):S40–S45

Hong S, Pedersen PL (2008) ATP synthase and the actions of inhibitors utilized to study its roles in human health, disease, and other scientific areas. Microbiol Mol Biol Rev 72(4):590–641. https://doi.org/10.1128/MMBR.00016-08

Hong L, Xi J, Zhang Y, Tian W, Xu J, Cui X, Xu Z (2012) Atrial natriuretic peptide prevents the mitochondrial permeability transition pore opening by inactivating glycogen synthase kinase 3beta via PKG and PI3K in cardiac H9c2 cells. Eur J Pharmacol 695(1–3):13–19. https://doi.org/10.1016/j.ejphar.2012.07.053

Hu DX, Liu XB, Song WC, Wang JA (2016) Roles of SIRT3 in heart failure: from bench to bedside. J Zhejiang Univ Sci B 17(11):821–830. https://doi.org/10.1631/jzus.B1600253

Hurst S, Hoek J, Sheu SS (2017) Mitochondrial Ca2+ and regulation of the permeability transition pore. J Bioenerg Biomembr 49(1):27–47. https://doi.org/10.1007/s10863-016-9672-x

Hurtado C, Pierce GN (2000) Inhibition of Na(+)/H(+) exchange at the beginning of reperfusion is cardioprotective in isolated, beating adult cardiomyocytes. J Mol Cell Cardiol 32(10):1897–1907. https://doi.org/10.1006/jmcc.2000.1222

Ikeda G, Matoba T, Nakano Y, Nagaoka K, Ishikita A, Nakano K, Funamoto D, Sunagawa K, Egashira K (2016) Nanoparticle-mediated targeting of cyclosporine A enhances cardioprotection against ischemia-reperfusion injury through inhibition of mitochondrial permeability transition pore opening. Sci Rep 6:20467. https://doi.org/10.1038/srep20467

Ivanes F, Faccenda D, Gatliff J, Ahmed AA, Cocco S, Cheng CH, Allan E, Russell C, Duchen MR, Campanella M (2014) The compound BTB06584 is an IF1-dependent selective inhibitor of the mitochondrial F1 Fo-ATPase. Br J Pharmacol 171(18):4193–4206. https://doi.org/10.1111/bph.12638

Jang IS, Park MY, Shin IW, Sohn JT, Lee HK, Chung YK (2010) Ethyl pyruvate has anti-inflammatory and delayed myocardial protective effects after regional ischemia/reperfusion injury. Yonsei Med J 51(6):838–844. https://doi.org/10.3349/ymj.2010.51.6.838

Javadov S, Karmazyn M (2007) Mitochondrial permeability transition pore opening as an endpoint to initiate cell death and as a putative target for cardioprotection. Cell Physiol Biochem 20(1–4):1–22. https://doi.org/10.1159/000103747

Javadov SA, Clarke S, Das M, Griffiths EJ, Lim KH, Halestrap AP (2003) Ischaemic preconditioning inhibits opening of mitochondrial permeability transition pores in the reperfused rat heart. J Physiol 549(Pt 2):513–524. https://doi.org/10.1113/jphysiol.2003.034231

Javadov S, Huang C, Kirshenbaum L, Karmazyn M (2005) NHE-1 inhibition improves impaired mitochondrial permeability transition and respiratory function during postinfarction remodelling in the rat. J Mol Cell Cardiol 38(1):135–143. https://doi.org/10.1016/j.yjmcc.2004.10.007

Javadov S, Choi A, Rajapurohitam V, Zeidan A, Basnakian AG, Karmazyn M (2008) NHE-1 inhibition-induced cardioprotection against ischaemia/reperfusion is associated with attenuation of the mitochondrial permeability transition. Cardiovasc Res 77(2):416–424. https://doi.org/10.1093/cvr/cvm039

Javadov S, Karmazyn M, Escobales N (2009a) Mitochondrial permeability transition pore opening as a promising therapeutic target in cardiac diseases. J Pharmacol Exp Ther 330(3):670–678. https://doi.org/10.1124/jpet.109.153213

Javadov S, Rajapurohitam V, Kilic A, Zeidan A, Choi A, Karmazyn M (2009b) Anti-hypertrophic effect of NHE-1 inhibition involves GSK-3beta-dependent attenuation of mitochondrial dysfunction. J Mol Cell Cardiol 46(6):998–1007. https://doi.org/10.1016/j.yjmcc.2008.12.023

Jiang S, Zhang LF, Zhang HW, Hu S, Lu MH, Liang S, Li B, Li Y, Li D, Wang ED, Liu MF (2012) A novel miR-155/miR-143 cascade controls glycolysis by regulating hexokinase 2 in breast cancer cells. EMBO J 31(8):1985–1998. https://doi.org/10.1038/emboj.2012.45

Joiner ML, Koval OM, Li J, He BJ, Allamargot C, Gao Z, Luczak ED, Hall DD, Fink BD, Chen B, Yang J, Moore SA, Scholz TD, Strack S, Mohler PJ, Sivitz WI, Song LS, Anderson ME (2012) CaMKII determines mitochondrial stress responses in heart. Nature 491(7423):269–273. https://doi.org/10.1038/nature11444

Jonas EA, Porter GA Jr, Beutner G, Mnatsakanyan N, Alavian KN (2015) Cell death disguised: the mitochondrial permeability transition pore as the c-subunit of the F(1)F(O) ATP synthase. Pharmacol Res 99:382–392. https://doi.org/10.1016/j.phrs.2015.04.013

Jonckheere AI, Smeitink JA, Rodenburg RJ (2012) Mitochondrial ATP synthase: architecture, function and pathology. J Inherit Metab Dis 35(2):211–225. https://doi.org/10.1007/s10545-011-9382-9

Jones RA, Smail A, Wilson MR (2002) Detecting mitochondrial permeability transition by confocal imaging of intact cells pinocytically loaded with calcein. Eur J Biochem 269(16):3990–3997

Juhaszova M, Zorov DB, Kim SH, Pepe S, Fu Q, Fishbein KW, Ziman BD, Wang S, Ytrehus K, Antos CL, Olson EN, Sollott SJ (2004) Glycogen synthase kinase-3beta mediates convergence of protection signaling to inhibit the mitochondrial permeability transition pore. J Clin Invest 113(11):1535–1549. https://doi.org/10.1172/JCI19906

Juhaszova M, Zorov DB, Yaniv Y, Nuss HB, Wang S, Sollott SJ (2009) Role of glycogen synthase kinase-3beta in cardioprotection. Circ Res 104(11):1240–1252. https://doi.org/10.1161/CIRCRESAHA.109.197996

Kaludercic N, Giorgio V (2016) The dual function of reactive oxygen/nitrogen species in bioenergetics and cell death: the role of ATP synthase. Oxidative Med Cell Longev 2016:3869610. https://doi.org/10.1155/2016/3869610

Karamanlidis G, Lee CF, Garcia-Menendez L, Kolwicz SC Jr, Suthammarak W, Gong G, Sedensky MM, Morgan PG, Wang W, Tian R (2013) Mitochondrial complex I deficiency increases protein acetylation and accelerates heart failure. Cell Metab 18(2):239–250. https://doi.org/10.1016/j.cmet.2013.07.002

Karlsson LO, Zhou AX, Larsson E, Astrom-Olsson K, Mansson C, Akyurek LM, Grip L (2010) Cyclosporine does not reduce myocardial infarct size in a porcine ischemia-reperfusion model. J Cardiovasc Pharmacol Ther 15(2):182–189. https://doi.org/10.1177/1074248410362074

Karmazyn M (1988) Amiloride enhances postischemic ventricular recovery: possible role of Na+-H+ exchange. Am J Phys 255(3 Pt 2):H608–H615

Karmazyn M (1999) Mechanisms of protection of the ischemic and reperfused myocardium by sodium-hydrogen exchange inhibition. J Thromb Thrombolysis 8(1):33–38

Kelso GF, Porteous CM, Coulter CV, Hughes G, Porteous WK, Ledgerwood EC, Smith RA, Murphy MP (2001) Selective targeting of a redox-active ubiquinone to mitochondria within cells: antioxidant and antiapoptotic properties. J Biol Chem 276(7):4588–4596. https://doi.org/10.1074/jbc.M009093200

Kemp G, Young H, Fliegel L (2008) Structure and function of the human Na+/H+ exchanger isoform 1. Channels (Austin) 2(5):329–336

Kerr PM, Suleiman MS, Halestrap AP (1999) Reversal of permeability transition during recovery of hearts from ischemia and its enhancement by pyruvate. Am J Phys 276(2 Pt 2):H496–H502

Khaliulin I, Clarke SJ, Lin H, Parker J, Suleiman MS, Halestrap AP (2007) Temperature preconditioning of isolated rat hearts—a potent cardioprotective mechanism involving a reduction in oxidative stress and inhibition of the mitochondrial permeability transition pore. J Physiol 581(Pt 3):1147–1161. https://doi.org/10.1113/jphysiol.2007.130369

Khaliulin I, Parker JE, Halestrap AP (2010) Consecutive pharmacological activation of PKA and PKC mimics the potent cardioprotection of temperature preconditioning. Cardiovasc Res 88(2):324–333. https://doi.org/10.1093/cvr/cvq190

Khaliulin I, Halestrap AP, Suleiman MS (2011) Temperature preconditioning is optimal at 26 degrees C and confers additional protection to hypothermic cardioplegic ischemic arrest. Exp Biol Med (Maywood) 236(6):736–745. https://doi.org/10.1258/ebm.2011.010357

Khaliulin I, Bond M, James AF, Dyar Z, Amini R, Johnson JL, Suleiman MS (2017) Functional and cardioprotective effects of simultaneous and individual activation of protein kinase A and Epac. Br J Pharmacol 174(6):438–453. https://doi.org/10.1111/bph.13709

Khaspekov L, Friberg H, Halestrap A, Viktorov I, Wieloch T (1999) Cyclosporin A and its nonimmunosuppressive analogue N-Me-Val-4-cyclosporin A mitigate glucose/oxygen deprivation-induced damage to rat cultured hippocampal neurons. Eur J Neurosci 11(9):3194–3198

Kim JS, Jin Y, Lemasters JJ (2006) Reactive oxygen species, but not Ca2+ overloading, trigger pH- and mitochondrial permeability transition-dependent death of adult rat myocytes after ischemia-reperfusion. Am J Physiol Heart Circ Physiol 290(5):H2024–H2034. https://doi.org/10.1152/ajpheart.00683.2005

Kim JS, Nitta T, Mohuczy D, O'Malley KA, Moldawer LL, Dunn WA Jr, Behrns KE (2008) Impaired autophagy: a mechanism of mitochondrial dysfunction in anoxic rat hepatocytes. Hepatology 47(5):1725–1736. https://doi.org/10.1002/hep.22187

Kinnally KW, Campo ML, Tedeschi H (1989) Mitochondrial channel activity studied by patch-clamping mitoplasts. J Bioenerg Biomembr 21(4):497–506

Kinnally KW, Zorov DB, Antonenko YN, Snyder SH, McEnery MW, Tedeschi H (1993) Mitochondrial benzodiazepine receptor linked to inner membrane ion channels by nanomolar actions of ligands. Proc Natl Acad Sci U S A 90(4):1374–1378

Kinnally KW, Peixoto PM, Ryu SY, Dejean LM (2011) Is mPTP the gatekeeper for necrosis, apoptosis, or both? Biochim Biophys Acta 1813(4):616–622. https://doi.org/10.1016/j.bbamcr.2010.09.013

Kitakaze M, Takashima S, Funaya H, Minamino T, Node K, Shinozaki Y, Mori H, Hori M (1997) Temporary acidosis during reperfusion limits myocardial infarct size in dogs. Am J Phys 272(5 Pt 2):H2071–H2078

Kitakaze M, Asakura M, Kim J, Shintani Y, Asanuma H, Hamasaki T, Seguchi O, Myoishi M, Minamino T, Ohara T, Nagai Y, Nanto S, Watanabe K, Fukuzawa S, Hirayama A, Nakamura N, Kimura K, Fujii K, Ishihara M, Saito Y, Tomoike H, Kitamura S, investigators JW (2007) Human atrial natriuretic peptide and nicorandil as adjuncts to reperfusion treatment for acute myocardial infarction (J-WIND): two randomised trials. Lancet 370(9597):1483–1493. https://doi.org/10.1016/S0140-6736(07)61634-1

Klishadi MS, Zarei F, Hejazian SH, Moradi A, Hemati M, Safari F (2015) Losartan protects the heart against ischemia reperfusion injury: sirtuin3 involvement. J Pharm Pharm Sci 18(1):112–123

Kobayashi H, Miura T, Ishida H, Miki T, Tanno M, Yano T, Sato T, Hotta H, Shimamoto K (2008) Limitation of infarct size by erythropoietin is associated with translocation of Akt to the mitochondria after reperfusion. Clin Exp Pharmacol Physiol 35(7):812–819. https://doi.org/10.1111/j.1440-1681.2008.04925.x

Kochel K, Tomczyk MD, Simoes RF, Fraczek T, Sobon A, Oliveira PJ, Walczak KZ, Koceva-Chyla A (2017) Evaluation of biological properties of 3,3′,4,4′-benzophenonetetracarboxylic dianhydride derivatives and their ability to inhibit hexokinase activity. Bioorg Med Chem Lett 27(3):427–431. https://doi.org/10.1016/j.bmcl.2016.12.055

Koentges C, Bode C, Bugger H (2016) SIRT3 in cardiac physiology and disease. Front Cardiovasc Med 3:38. https://doi.org/10.3389/fcvm.2016.00038

Kohr MJ, Aponte AM, Sun J, Wang G, Murphy E, Gucek M, Steenbergen C (2011) Characterization of potential S-nitrosylation sites in the myocardium. Am J Physiol Heart Circ Physiol 300(4):H1327–H1335. https://doi.org/10.1152/ajpheart.00997.2010

Kokoszka JE, Waymire KG, Levy SE, Sligh JE, Cai J, Jones DP, MacGregor GR, Wallace DC (2004) The ADP/ATP translocator is not essential for the mitochondrial permeability transition pore. Nature 427(6973):461–465. https://doi.org/10.1038/nature02229

Korge P, Calmettes G, John SA, Weiss JN (2017a) Reactive oxygen species production induced by pore opening in cardiac mitochondria: the role of complex III. J Biol Chem 292(24):9882–9895. https://doi.org/10.1074/jbc.M116.768317

Korge P, John SA, Calmettes G, Weiss JN (2017b) Reactive oxygen species production induced by pore opening in cardiac mitochondria: the role of complex II. J Biol Chem 292(24):9896–9905. https://doi.org/10.1074/jbc.M116.768325

Kottke M, Adam V, Riesinger I, Bremm G, Bosch W, Brdiczka D, Sandri G, Panfili E (1988) Mitochondrial boundary membrane contact sites in brain: points of hexokinase and creatine kinase location, and control of Ca2+ transport. Biochim Biophys Acta 935(1):87–102

Kowaltowski AJ, Castilho RF, Vercesi AE (2001) Mitochondrial permeability transition and oxidative stress. FEBS Lett 495(1–2):12–15

Krauskopf A, Eriksson O, Craigen WJ, Forte MA, Bernardi P (2006) Properties of the permeability transition in VDAC1(−/−) mitochondria. Biochim Biophys Acta 1757(5–6):590–595. https://doi.org/10.1016/j.bbabio.2006.02.007

Kwong JQ, Molkentin JD (2015) Physiological and pathological roles of the mitochondrial permeability transition pore in the heart. Cell Metab 21(2):206–214. https://doi.org/10.1016/j.cmet.2014.12.001

Le Bras M, Clement MV, Pervaiz S, Brenner C (2005) Reactive oxygen species and the mitochondrial signaling pathway of cell death. Histol Histopathol 20(1):205–219. https://doi.org/10.14670/HH-20.205

Le Lamer S, Paradis S, Rahmouni H, Chaimbault C, Michaud M, Culcasi M, Afxantidis J, Latreille M, Berna P, Berdeaux A, Pietri S, Morin D, Donazzolo Y, Abitbol JL, Pruss RM, Schaller S (2014) Translation of TRO40303 from myocardial infarction models to demonstration of safety and tolerance in a randomized phase I trial. J Transl Med 12:38. https://doi.org/10.1186/1479-5876-12-38

Le Page S, Niro M, Fauconnier J, Cellier L, Tamareille S, Gharib A, Chevrollier A, Loufrani L, Grenier C, Kamel R, Sarzi E, Lacampagne A, Ovize M, Henrion D, Reynier P, Lenaers G, Mirebeau-Prunier D, Prunier F (2016) Increase in cardiac ischemia-reperfusion injuries in Opa1+/− mouse model. PLoS One 11(10):e0164066. https://doi.org/10.1371/journal.pone.0164066

Lee GJ, Chae SJ, Jeong JH, Lee SR, Ha SJ, Pak YK, Kim W, Park HK (2011) Characterization of mitochondria isolated from normal and ischemic hearts in rats utilizing atomic force microscopy. Micron 42(3):299–304. https://doi.org/10.1016/j.micron.2010.09.002

Lee HL, Chen CL, Yeh ST, Zweier JL, Chen YR (2012) Biphasic modulation of the mitochondrial electron transport chain in myocardial ischemia and reperfusion. Am J Phys Heart Circ Phys 302(7):H1410–H1422. https://doi.org/10.1152/ajpheart.00731.2011

Leite AC, Oliveira HC, Utino FL, Garcia R, Alberici LC, Fernandes MP, Castilho RF, Vercesi AE (2010) Mitochondria generated nitric oxide protects against permeability transition via formation of membrane protein S-nitrosothiols. Biochim Biophys Acta 1797(6–7):1210–1216. https://doi.org/10.1016/j.bbabio.2010.01.034

Lemasters JJ, Nieminen AL (1999) Negative contrast imaging of mitochondria by confocal microscopy. Biophys J 77(3):1747–1750. https://doi.org/10.1016/S0006-3495(99)77020-7

Leonard AP, Cameron RB, Speiser JL, Wolf BJ, Peterson YK, Schnellmann RG, Beeson CC, Rohrer B (2015) Quantitative analysis of mitochondrial morphology and membrane potential in living cells using high-content imaging, machine learning, and morphological binning. Biochim Biophys Acta 1853(2):348–360. https://doi.org/10.1016/j.bbamcr.2014.11.002

Leperre A, Millart H, Prevost A, Trenque T, Kantelip JP, Keppler BK (1995) Compared effects of ruthenium red and cis [Ru(NH3)4Cl2]Cl on the isolated ischaemic-reperfused rat heart. Fundam Clin Pharmacol 9(6):545–553

Leung AW, Halestrap AP (2008) Recent progress in elucidating the molecular mechanism of the mitochondrial permeability transition pore. Biochim Biophys Acta 1777(7–8):946–952. https://doi.org/10.1016/j.bbabio.2008.03.009

Leung AW, Varanyuwatana P, Halestrap AP (2008) The mitochondrial phosphate carrier interacts with cyclophilin D and may play a key role in the permeability transition. J Biol Chem 283(39):26312–26323. https://doi.org/10.1074/jbc.M805235200

Li J, Wang J, Zeng Y (2007) Peripheral benzodiazepine receptor ligand, PK11195 induces mitochondria cytochrome c release and dissipation of mitochondria potential via induction of mitochondria permeability transition. Eur J Pharmacol 560(2–3):117–122. https://doi.org/10.1016/j.ejphar.2006.12.027

Li J, Xuan W, Yan R, Tropak MB, Jean-St-Michel E, Liang W, Gladstone R, Backx PH, Kharbanda RK, Redington AN (2011) Remote preconditioning provides potent cardioprotection via PI3K/Akt activation and is associated with nuclear accumulation of beta-catenin. Clin Sci (Lond) 120(10):451–462. https://doi.org/10.1042/CS20100466

Lim KH, Javadov SA, Das M, Clarke SJ, Suleiman MS, Halestrap AP (2002) The effects of ischaemic preconditioning, diazoxide and 5-hydroxydecanoate on rat heart mitochondrial volume and respiration. J Physiol 545(Pt 3):961–974

Lim SY, Davidson SM, Hausenloy DJ, Yellon DM (2007) Preconditioning and postconditioning: the essential role of the mitochondrial permeability transition pore. Cardiovasc Res 75(3):530–535. https://doi.org/10.1016/j.cardiores.2007.04.022

Liu GS, Thornton J, Van Winkle DM, Stanley AW, Olsson RA, Downey JM (1991) Protection against infarction afforded by preconditioning is mediated by A1 adenosine receptors in rabbit heart. Circulation 84(1):350–356

Liu CW, Yang F, Cheng SZ, Liu Y, Wan LH, Cong HL (2017) Rosuvastatin postconditioning protects isolated hearts against ischemia-reperfusion injury: the role of radical oxygen species, PI3K-Akt-GSK-3beta pathway, and mitochondrial permeability transition pore. Cardiovasc Ther 35(1):3–9. https://doi.org/10.1111/1755-5922.12225

Luongo TS, Lambert JP, Yuan A, Zhang X, Gross P, Song J, Shanmughapriya S, Gao E, Jain M, Houser SR, Koch WJ, Cheung JY, Madesh M, Elrod JW (2015) The mitochondrial calcium uniporter matches energetic supply with cardiac workload during stress and modulates permeability transition. Cell Rep 12(1):23–34. https://doi.org/10.1016/j.celrep.2015.06.017

Madesh M, Hajnoczky G (2001) VDAC-dependent permeabilization of the outer mitochondrial membrane by superoxide induces rapid and massive cytochrome c release. J Cell Biol 155(6):1003–1015. https://doi.org/10.1083/jcb.200105057

Magnesium in Coronaries Trial I (2002) Early administration of intravenous magnesium to high-risk patients with acute myocardial infarction in the magnesium in coronaries (MAGIC) trial: a randomised controlled trial. Lancet 360(9341):1189–1196

Mallet RT (2000) Pyruvate: metabolic protector of cardiac performance. Proc Soc Exp Biol Med 223(2):136–148

Mallet RT, Sun J, Knott EM, Sharma AB, Olivencia-Yurvati AH (2005) Metabolic cardioprotection by pyruvate: recent progress. Exp Biol Med (Maywood) 230(7):435–443

Manevich Y, Townsend DM, Hutchens S, Tew KD (2010) Diazeniumdiolate mediated nitrosative stress alters nitric oxide homeostasis through intracellular calcium and S-glutathionylation of nitric oxide synthetase. PLoS One 5(11):e14151. https://doi.org/10.1371/journal.pone.0014151

Marquez J, Lee SR, Kim N, Han J (2016) Post-translational modifications of cardiac mitochondrial proteins in cardiovascular disease: not lost in translation. Korean Circ J 46(1):1–12. https://doi.org/10.4070/kcj.2016.46.1.1

Martin LJ, Fancelli D, Wong M, Niedzwiecki M, Ballarini M, Plyte S, Chang Q (2014) GNX-4728, a novel small molecule drug inhibitor of mitochondrial permeability transition, is therapeutic in a mouse model of amyotrophic lateral sclerosis. Front Cell Neurosci 8:433. https://doi.org/10.3389/fncel.2014.00433

Masgras I, Rasola A, Bernardi P (2012) Induction of the permeability transition pore in cells depleted of mitochondrial DNA. Biochim Biophys Acta 1817(10):1860–1866. https://doi.org/10.1016/j.bbabio.2012.02.022

Mathupala SP, Ko YH, Pedersen PL (2006) Hexokinase II: cancer's double-edged sword acting as both facilitator and gatekeeper of malignancy when bound to mitochondria. Oncogene 25(34):4777–4786. https://doi.org/10.1038/sj.onc.1209603

Matsui H, Morishima I, Numaguchi Y, Toki Y, Okumura K, Hayakawa T (1999) Protective effects of carvedilol against doxorubicin-induced cardiomyopathy in rats. Life Sci 65(12):1265–1274

McCommis KS, Baines CP (2012) The role of VDAC in cell death: friend or foe? Biochim Biophys Acta 1818(6):1444–1450. https://doi.org/10.1016/j.bbamem.2011.10.025

McEnery MW, Snowman AM, Trifiletti RR, Snyder SH (1992) Isolation of the mitochondrial benzodiazepine receptor: association with the voltage-dependent anion channel and the adenine nucleotide carrier. Proc Natl Acad Sci U S A 89(8):3170–3174

McStay GP, Clarke SJ, Halestrap AP (2002) Role of critical thiol groups on the matrix surface of the adenine nucleotide translocase in the mechanism of the mitochondrial permeability transition pore. Biochem J 367(Pt 2):541–548. https://doi.org/10.1042/BJ20011672

Mentzer RM Jr, Bartels C, Bolli R, Boyce S, Buckberg GD, Chaitman B, Haverich A, Knight J, Menasche P, Myers ML, Nicolau J, Simoons M, Thulin L, Weisel RD, Investigators ES (2008) Sodium-hydrogen exchange inhibition by cariporide to reduce the risk of ischemic cardiac events in patients undergoing coronary artery bypass grafting: results of the EXPEDITION study. Ann Thorac Surg 85(4):1261–1270. https://doi.org/10.1016/j.athoracsur.2007.10.054

Meybohm P, Bein B, Brosteanu O, Cremer J, Gruenewald M, Stoppe C, Coburn M, Schaelte G, Boning A, Niemann B, Roesner J, Kletzin F, Strouhal U, Reyher C, Laufenberg-Feldmann R, Ferner M, Brandes IF, Bauer M, Stehr SN, Kortgen A, Wittmann M, Baumgarten G, Meyer-Treschan T, Kienbaum P, Heringlake M, Schon J, Sander M, Treskatsch S, Smul T, Wolwender E, Schilling T, Fuernau G, Hasenclever D, Zacharowski K, Collaborators RIS (2015) A multicenter trial of remote ischemic preconditioning for heart surgery. N Engl J Med 373(15):1397–1407. https://doi.org/10.1056/NEJMoa1413579

Miura T, Miki T (2009) GSK-3beta, a therapeutic target for cardiomyocyte protection. Circ J 73(7):1184–1192

Miura T, Tanno M (2010) Mitochondria and GSK-3beta in cardioprotection against ischemia/reperfusion injury. Cardiovasc Drugs Ther 24(3):255–263. https://doi.org/10.1007/s10557-010-6234-z

Miyamoto S, Murphy AN, Brown JH (2008) Akt mediates mitochondrial protection in cardiomyocytes through phosphorylation of mitochondrial hexokinase-II. Cell Death Differ 15(3):521–529. https://doi.org/10.1038/sj.cdd.4402285

Mocanu MM, Bell RM, Yellon DM (2002) PI3 kinase and not p42/p44 appears to be implicated in the protection conferred by ischemic preconditioning. J Mol Cell Cardiol 34(6):661–668. https://doi.org/10.1006/jmcc.2002.2006

Morciano G, Giorgi C, Bonora M, Punzetti S, Pavasini R, Wieckowski MR, Campo G, Pinton P (2015) Molecular identity of the mitochondrial permeability transition pore and its role in ischemia-reperfusion injury. J Mol Cell Cardiol 78:142–153. https://doi.org/10.1016/j.yjmcc.2014.08.015

Morin D, Musman J, Pons S, Berdeaux A, Ghaleh B (2016) Mitochondrial translocator protein (TSPO): from physiology to cardioprotection. Biochem Pharmacol 105:1–13. https://doi.org/10.1016/j.bcp.2015.12.003

Murata M, Akao M, O'Rourke B, Marban E (2001) Mitochondrial ATP-sensitive potassium channels attenuate matrix Ca(2+) overload during simulated ischemia and reperfusion: possible mechanism of cardioprotection. Circ Res 89(10):891–898

Murphy MP, Smith RA (2000) Drug delivery to mitochondria: the key to mitochondrial medicine. Adv Drug Deliv Rev 41(2):235–250

Murphy MP, Smith RA (2007) Targeting antioxidants to mitochondria by conjugation to lipophilic cations. Annu Rev Pharmacol Toxicol 47:629–656. https://doi.org/10.1146/annurev.pharmtox.47.120505.105110

Murry CE, Jennings RB, Reimer KA (1986) Preconditioning with ischemia: a delay of lethal cell injury in ischemic myocardium. Circulation 74(5):1124–1136

Nederlof R, Eerbeek O, Hollmann MW, Southworth R, Zuurbier CJ (2014) Targeting hexokinase II to mitochondria to modulate energy metabolism and reduce ischaemia-reperfusion injury in heart. Br J Pharmacol 171(8):2067–2079. https://doi.org/10.1111/bph.12363

Nesci S, Ventrella V, Trombetti F, Pirini M, Pagliarani A (2014) Thiol oxidation of mitochondrial F0-c subunits: a way to switch off antimicrobial drug targets of the mitochondrial ATP synthase. Med Hypotheses 83(2):160–165. https://doi.org/10.1016/j.mehy.2014.05.004

Nesci S, Trombetti F, Ventrella V, Pagliarani A (2016) The c-ring of the F1FO-ATP synthase: facts and perspectives. J Membr Biol 249(1–2):11–21. https://doi.org/10.1007/s00232-015-9860-3

Neugebauer G, Akpan W, von Mollendorff E, Neubert P, Reiff K (1987) Pharmacokinetics and disposition of carvedilol in humans. J Cardiovasc Pharmacol 10(Suppl 11):S85–S88

Nguyen TT, Stevens MV, Kohr M, Steenbergen C, Sack MN, Murphy E (2011) Cysteine 203 of cyclophilin D is critical for cyclophilin D activation of the mitochondrial permeability transition pore. J Biol Chem 286(46):40184–40192. https://doi.org/10.1074/jbc.M111.243469

Nicolli A, Basso E, Petronilli V, Wenger RM, Bernardi P (1996) Interactions of cyclophilin with the mitochondrial inner membrane and regulation of the permeability transition pore, and cyclosporin A-sensitive channel. J Biol Chem 271(4):2185–2192

Nishihara M, Miura T, Miki T, Tanno M, Yano T, Naitoh K, Ohori K, Hotta H, Terashima Y, Shimamoto K (2007) Modulation of the mitochondrial permeability transition pore complex in GSK-3beta-mediated myocardial protection. J Mol Cell Cardiol 43(5):564–570. https://doi.org/10.1016/j.yjmcc.2007.08.010

Nomura F, Kurobe N, Mori Y, Hikita A, Kawai M, Suwa M, Okutani Y (2008) Multicenter prospective investigation on efficacy and safety of carperitide as a first-line drug for acute heart failure syndrome with preserved blood pressure: COMPASS: carperitide effects observed through monitoring dyspnea in acute decompensated heart failure study. Circ J 72(11):1777–1786

Novalija E, Kevin LG, Camara AK, Bosnjak ZJ, Kampine JP, Stowe DF (2003) Reactive oxygen species precede the epsilon isoform of protein kinase C in the anesthetic preconditioning signaling cascade. Anesthesiology 99(2):421–428

Obame FN, Zini R, Souktani R, Berdeaux A, Morin D (2007) Peripheral benzodiazepine receptor-induced myocardial protection is mediated by inhibition of mitochondrial membrane permeabilization. J Pharmacol Exp Ther 323(1):336–345. https://doi.org/10.1124/jpet.107.124255

Obame FN, Plin-Mercier C, Assaly R, Zini R, Dubois-Rande JL, Berdeaux A, Morin D (2008) Cardioprotective effect of morphine and a blocker of glycogen synthase kinase 3 beta, SB216763 [3-(2,4-dichlorophenyl)-4(1-methyl-1H-indol-3-yl)-1H-pyrrole-2,5-dione], via inhibition of the mitochondrial permeability transition pore. J Pharmacol Exp Ther 326(1):252–258. https://doi.org/10.1124/jpet.108.138008

Odunewu-Aderibigbe A, Fliegel L (2014) The Na(+) /H(+) exchanger and pH regulation in the heart. IUBMB Life 66(10):679–685. https://doi.org/10.1002/iub.1323

O'Gorman E, Beutner G, Dolder M, Koretsky AP, Brdiczka D, Wallimann T (1997) The role of creatine kinase in inhibition of mitochondrial permeability transition. FEBS Lett 414(2):253–257

Oliveira PJ, Marques MP, Batista de Carvalho LA, Moreno AJ (2000a) Effects of carvedilol on isolated heart mitochondria: evidence for a protonophoretic mechanism. Biochem Biophys Res Commun 276(1):82–87. https://doi.org/10.1006/bbrc.2000.3374

Oliveira PJ, Santos DJ, Moreno AJ (2000b) Carvedilol inhibits the exogenous NADH dehydrogenase in rat heart mitochondria. Arch Biochem Biophys 374(2):279–285. https://doi.org/10.1006/abbi.1999.1624

Oliveira PJ, Coxito PM, Rolo AP, Santos DL, Palmeira CM, Moreno AJ (2001a) Inhibitory effect of carvedilol in the high-conductance state of the mitochondrial permeability transition pore. Eur J Pharmacol 412(3):231–237

Oliveira PJ, Rolo AP, Sardao VA, Coxito PM, Palmeira CM, Moreno AJ (2001b) Carvedilol in heart mitochondria: protonophore or opener of the mitochondrial K(ATP) channels? Life Sci 69(2):123–132

Oliveira PJ, Bjork JA, Santos MS, Leino RL, Froberg MK, Moreno AJ, Wallace KB (2004a) Carvedilol-mediated antioxidant protection against doxorubicin-induced cardiac mitochondrial toxicity. Toxicol Appl Pharmacol 200(2):159–168. https://doi.org/10.1016/j.taap.2004.04.005

Oliveira PJ, Esteves T, Rolo AP, Palmeira CM, Moreno AJ (2004b) Carvedilol inhibits the mitochondrial permeability transition by an antioxidant mechanism. Cardiovasc Toxicol 4(1):11–20

Ong SB, Subrayan S, Lim SY, Yellon DM, Davidson SM, Hausenloy DJ (2010) Inhibiting mitochondrial fission protects the heart against ischemia/reperfusion injury. Circulation 121(18):2012–2022. https://doi.org/10.1161/CIRCULATIONAHA.109.906610

Ong SB, Dongworth RK, Cabrera-Fuentes HA, Hausenloy DJ (2015a) Role of the MPTP in conditioning the heart – translatability and mechanism. Br J Pharmacol 172(8):2074–2084. https://doi.org/10.1111/bph.13013

Ong SB, Hall AR, Dongworth RK, Kalkhoran S, Pyakurel A, Scorrano L, Hausenloy DJ (2015b) Akt protects the heart against ischaemia-reperfusion injury by modulating mitochondrial morphology. Thromb Haemost 113(3):513–521. https://doi.org/10.1160/TH14-07-0592

Ong SB, Samangouei P, Kalkhoran SB, Hausenloy DJ (2015c) The mitochondrial permeability transition pore and its role in myocardial ischemia reperfusion injury. J Mol Cell Cardiol 78:23–34. https://doi.org/10.1016/j.yjmcc.2014.11.005

Onishi A, Miyamae M, Kaneda K, Kotani J, Figueredo VM (2012) Direct evidence for inhibition of mitochondrial permeability transition pore opening by sevoflurane preconditioning in cardiomyocytes: comparison with cyclosporine A. Eur J Pharmacol 675(1–3):40–46. https://doi.org/10.1016/j.ejphar.2011.11.040

Ortiz-Vilchis P, Yamazaki KG, Rubio-Gayosso I, Ramirez-Sanchez I, Calzada C, Romero-Perez D, Ortiz A, Meaney E, Taub P, Villarreal F, Ceballos G (2014) Co-administration of the flavanol (−)-epicatechin with doxycycline synergistically reduces infarct size in a model of ischemia reperfusion injury by inhibition of mitochondrial swelling. Eur J Pharmacol 744:76–82. https://doi.org/10.1016/j.ejphar.2014.09.042

Otera H, Ishihara N, Mihara K (2013) New insights into the function and regulation of mitochondrial fission. Biochim Biophys Acta 1833(5):1256–1268. https://doi.org/10.1016/j.bbamcr.2013.02.002

Ottani F, Latini R, Staszewsky L, La Vecchia L, Locuratolo N, Sicuro M, Masson S, Barlera S, Milani V, Lombardi M, Costalunga A, Mollichelli N, Santarelli A, De Cesare N, Sganzerla P, Boi A, Maggioni AP, Limbruno U, Investigators C (2016) Cyclosporine A in reperfused myocardial infarction: the multicenter, controlled, open-label CYCLE trial. J Am Coll Cardiol 67(4):365–374. https://doi.org/10.1016/j.jacc.2015.10.081

Owen MR, Doran E, Halestrap AP (2000) Evidence that metformin exerts its anti-diabetic effects through inhibition of complex 1 of the mitochondrial respiratory chain. Biochem J 348(Pt 3):607–614

Oyewole AO, Birch-Machin MA (2015) Mitochondria-targeted antioxidants. FASEB J 29(12):4766–4771. https://doi.org/10.1096/fj.15-275404

Panel M, Ghaleh B, Morin D (2017) Ca2+ ionophores are not suitable for inducing mPTP opening in murine isolated adult cardiac myocytes. Sci Rep 7(1):4283. https://doi.org/10.1038/s41598-017-04618-4

Park SS, Zhao H, Jang Y, Mueller RA, Xu Z (2006a) N6-(3-iodobenzyl)-adenosine-5'-N-methylcarboxamide confers cardioprotection at reperfusion by inhibiting mitochondrial permeability transition pore opening via glycogen synthase kinase 3 beta. J Pharmacol Exp Ther 318(1):124–131. https://doi.org/10.1124/jpet.106.101477

Park SS, Zhao H, Mueller RA, Xu Z (2006b) Bradykinin prevents reperfusion injury by targeting mitochondrial permeability transition pore through glycogen synthase kinase 3beta. J Mol Cell Cardiol 40(5):708–716. https://doi.org/10.1016/j.yjmcc.2006.01.024

Pasdois P, Parker JE, Griffiths EJ, Halestrap AP (2011) The role of oxidized cytochrome c in regulating mitochondrial reactive oxygen species production and its perturbation in ischaemia. Biochem J 436(2):493–505. https://doi.org/10.1042/BJ20101957

Pasdois P, Parker JE, Halestrap AP (2012) Extent of mitochondrial hexokinase II dissociation during ischemia correlates with mitochondrial cytochrome c release, reactive oxygen species production, and infarct size on reperfusion. J Am Heart Assoc 2(1):e005645. https://doi.org/10.1161/JAHA.112.005645

Pastorino JG, Hoek JB (2008) Regulation of hexokinase binding to VDAC. J Bioenerg Biomembr 40(3):171–182. https://doi.org/10.1007/s10863-008-9148-8

Pastorino JG, Shulga N, Hoek JB (2002) Mitochondrial binding of hexokinase II inhibits Bax-induced cytochrome c release and apoptosis. J Biol Chem 277(9):7610–7618. https://doi.org/10.1074/jbc.M109950200

Pastorino JG, Hoek JB, Shulga N (2005) Activation of glycogen synthase kinase 3beta disrupts the binding of hexokinase II to mitochondria by phosphorylating voltage-dependent anion channel and potentiates chemotherapy-induced cytotoxicity. Cancer Res 65(22):10545–10554. https://doi.org/10.1158/0008-5472.CAN-05-1925

Pereira GC, Silva AM, Diogo CV, Carvalho FS, Monteiro P, Oliveira PJ (2011) Drug-induced cardiac mitochondrial toxicity and protection: from doxorubicin to carvedilol. Curr Pharm Des 17(20):2113–2129

Perricone AJ, Vander Heide RS (2014) Novel therapeutic strategies for ischemic heart disease. Pharmacol Res 89:36–45. https://doi.org/10.1016/j.phrs.2014.08.004

Petronilli V, Szabo I, Zoratti M (1989) The inner mitochondrial membrane contains ion-conducting channels similar to those found in bacteria. FEBS Lett 259(1):137–143

Petronilli V, Miotto G, Canton M, Colonna R, Bernardi P, Di Lisa F (1998) Imaging the mitochondrial permeability transition pore in intact cells. Biofactors 8(3–4):263–272

Petronilli V, Miotto G, Canton M, Brini M, Colonna R, Bernardi P, Di Lisa F (1999) Transient and long-lasting openings of the mitochondrial permeability transition pore can be monitored directly in intact cells by changes in mitochondrial calcein fluorescence. Biophys J 76(2):725–734. https://doi.org/10.1016/S0006-3495(99)77239-5

Piantadosi CA (2012) Regulation of mitochondrial processes by protein S-nitrosylation. Biochim Biophys Acta 1820(6):712–721. https://doi.org/10.1016/j.bbagen.2011.03.008

Piantadosi CA, Tatro LG, Whorton AR (2002) Nitric oxide and differential effects of ATP on mitochondrial permeability transition. Nitric Oxide 6(1):45–60. https://doi.org/10.1006/niox.2001.0368

Piot C, Croisille P, Staat P, Thibault H, Rioufol G, Mewton N, Elbelghiti R, Cung TT, Bonnefoy E, Angoulvant D, Macia C, Raczka F, Sportouch C, Gahide G, Finet G, Andre-Fouet X, Revel D, Kirkorian G, Monassier JP, Derumeaux G, Ovize M (2008) Effect of cyclosporine on reperfusion injury in acute myocardial infarction. N Engl J Med 359(5):473–481. https://doi.org/10.1056/NEJMoa071142

Piquereau J, Caffin F, Novotova M, Prola A, Garnier A, Mateo P, Fortin D, Huynh le H, Nicolas V, Alavi MV, Brenner C, Ventura-Clapier R, Veksler V, Joubert F (2012) Down-regulation of OPA1 alters mouse mitochondrial morphology, PTP function, and cardiac adaptation to pressure overload. Cardiovasc Res 94(3):408–417. https://doi.org/10.1093/cvr/cvs117

Poortmans JR, Francaux M (2000) Adverse effects of creatine supplementation: fact or fiction? Sports Med 30(3):155–170

Prendes MG, Torresin E, Gonzalez M, Fernandez MA, Perazzo JC, Savino EA, Varela A (2008) Protection of ischaemic-reperfused rat heart by dimethylamiloride is associated with inhibition of mitochondrial permeability transition. Clin Exp Pharmacol Physiol 35(2):201–206. https://doi.org/10.1111/j.1440-1681.2007.04800.x

Prime TA, Blaikie FH, Evans C, Nadtochiy SM, James AM, Dahm CC, Vitturi DA, Patel RP, Hiley CR, Abakumova I, Requejo R, Chouchani ET, Hurd TR, Garvey JF, Taylor CT, Brookes PS, Smith RA, Murphy MP (2009) A mitochondria-targeted S-nitrosothiol modulates respiration, nitrosates thiols, and protects against ischemia-reperfusion injury. Proc Natl Acad Sci U S A 106(26):10764–10769. https://doi.org/10.1073/pnas.0903250106

Prudent J, Zunino R, Sugiura A, Mattie S, Shore GC, McBride HM (2015) MAPL SUMOylation of Drp1 stabilizes an ER/mitochondrial platform required for cell death. Mol Cell 59(6):941–955. https://doi.org/10.1016/j.molcel.2015.08.001

Przyklenk K, Bauer B, Ovize M, Kloner RA, Whittaker P (1993) Regional ischemic 'preconditioning' protects remote virgin myocardium from subsequent sustained coronary occlusion. Circulation 87(3):893–899

Rameshrad M, Babaei H, Azarmi Y, Fouladia DF (2016) Rat aorta as a pharmacological tool for in vitro and in vivo studies. Life Sci 145:190–204. https://doi.org/10.1016/j.lfs.2015.12.043

Rasola A, Bernardi P (2011) Mitochondrial permeability transition in Ca(2+)-dependent apoptosis and necrosis. Cell Calcium 50(3):222–233. https://doi.org/10.1016/j.ceca.2011.04.007

Rasola A, Sciacovelli M, Chiara F, Pantic B, Brusilow WS, Bernardi P (2010) Activation of mitochondrial ERK protects cancer cells from death through inhibition of the permeability transition. Proc Natl Acad Sci U S A 107(2):726–731. https://doi.org/10.1073/pnas.0912742107

Reiter MJ (2004) Cardiovascular drug class specificity: beta-blockers. Prog Cardiovasc Dis 47(1):11–33

Ricchelli F, Sileikyte J, Bernardi P (2011) Shedding light on the mitochondrial permeability transition. Biochim Biophys Acta 1807(5):482–490. https://doi.org/10.1016/j.bbabio.2011.02.012

Richardson AP, Halestrap AP (2016) Quantification of active mitochondrial permeability transition pores using GNX-4975 inhibitor titrations provides insights into molecular identity. Biochem J 473(9):1129–1140. https://doi.org/10.1042/BCJ20160070

Rizzuto R, Marchi S, Bonora M, Aguiari P, Bononi A, De Stefani D, Giorgi C, Leo S, Rimessi A, Siviero R, Zecchini E, Pinton P (2009) Ca(2+) transfer from the ER to mitochondria: when, how and why. Biochim Biophys Acta 1787(11):1342–1351. https://doi.org/10.1016/j.bbabio.2009.03.015

Roberts DJ, Tan-Sah VP, Smith JM, Miyamoto S (2013) Akt phosphorylates HK-II at Thr-473 and increases mitochondrial HK-II association to protect cardiomyocytes. J Biol Chem 288(33):23798–23806. https://doi.org/10.1074/jbc.M113.482026

Robey RB, Hay N (2006) Mitochondrial hexokinases, novel mediators of the antiapoptotic effects of growth factors and Akt. Oncogene 25(34):4683–4696. https://doi.org/10.1038/sj.onc.1209595

Rohmann S, Weygandt H, Minck KO (1995) Preischaemic as well as postischaemic application of a Na+/H+ exchange inhibitor reduces infarct size in pigs. Cardiovasc Res 30(6):945–951

Roos I, Crompton M, Carafoli E (1980) The role of inorganic phosphate in the release of Ca2+ from rat-liver mitochondria. Eur J Biochem 110(2):319–325

Roy S, Sileikyte J, Schiavone M, Neuenswander B, Argenton F, Aube J, Hedrick MP, Chung TD, Forte MA, Bernardi P, Schoenen FJ (2015) Discovery, synthesis, and optimization of diarylisoxazole-3-carboxamides as potent inhibitors of the mitochondrial permeability transition pore. ChemMedChem 10(10):1655–1671. https://doi.org/10.1002/cmdc.201500284

Ruiz-Meana M, Garcia-Dorado D, Pina P, Inserte J, Agullo L, Soler-Soler J (2003) Cariporide preserves mitochondrial proton gradient and delays ATP depletion in cardiomyocytes during ischemic conditions. Am J Physiol Heart Circ Physiol 285(3):H999–1006. https://doi.org/10.1152/ajpheart.00035.2003

Ruiz-Meana M, Inserte J, Fernandez-Sanz C, Hernando V, Miro-Casas E, Barba I, Garcia-Dorado D (2011) The role of mitochondrial permeability transition in reperfusion-induced cardiomyocyte death depends on the duration of ischemia. Basic Res Cardiol 106(6):1259–1268. https://doi.org/10.1007/s00395-011-0225-5

Rupprecht HJ, vom Dahl J, Terres W, Seyfarth KM, Richardt G, Schultheibeta HP, Buerke M, Sheehan FH, Drexler H (2000) Cardioprotective effects of the Na(+)/H(+) exchange inhibitor cariporide in patients with acute anterior myocardial infarction undergoing direct PTCA. Circulation 101(25):2902–2908

Santos DL, Moreno AJ, Leino RL, Froberg MK, Wallace KB (2002) Carvedilol protects against doxorubicin-induced mitochondrial cardiomyopathy. Toxicol Appl Pharmacol 185(3):218–227

Sardet C, Franchi A, Pouyssegur J (1989) Molecular cloning, primary structure, and expression of the human growth factor-activatable Na+/H+ antiporter. Cell 56(2):271–280

Saurin AT, Pennington DJ, Raat NJ, Latchman DS, Owen MJ, Marber MS (2002) Targeted disruption of the protein kinase C epsilon gene abolishes the infarct size reduction that follows ischaemic preconditioning of isolated buffer-perfused mouse hearts. Cardiovasc Res 55(3):672–680

Schaller S, Paradis S, Ngoh GA, Assaly R, Buisson B, Drouot C, Ostuni MA, Lacapere JJ, Bassissi F, Bordet T, Berdeaux A, Jones SP, Morin D, Pruss RM (2010) TRO40303, a new cardioprotective compound, inhibits mitochondrial permeability transition. J Pharmacol Exp Ther 333(3):696–706. https://doi.org/10.1124/jpet.110.167486

Schlattner U, Tokarska-Schlattner M, Wallimann T (2006) Mitochondrial creatine kinase in human health and disease. Biochim Biophys Acta 1762(2):164–180. https://doi.org/10.1016/j.bbadis.2005.09.004

Schroers A, Kramer R, Wohlrab H (1997) The reversible antiport-uniport conversion of the phosphate carrier from yeast mitochondria depends on the presence of a single cysteine. J Biol Chem 272(16):10558–10564

Schwartz J, Holmuhamedov E, Zhang X, Lovelace GL, Smith CD, Lemasters JJ (2013) Minocycline and doxycycline, but not other tetracycline-derived compounds, protect liver cells from chemical hypoxia and ischemia/reperfusion injury by inhibition of the mitochondrial calcium uniporter. Toxicol Appl Pharmacol 273(1):172–179. https://doi.org/10.1016/j.taap.2013.08.027

Seidlmayer LK, Juettner VV, Kettlewell S, Pavlov EV, Blatter LA, Dedkova EN (2015) Distinct mPTP activation mechanisms in ischaemia-reperfusion: contributions of Ca2+, ROS, pH, and inorganic polyphosphate. Cardiovasc Res 106(2):237–248. https://doi.org/10.1093/cvr/cvv097

Sgobbo P, Pacelli C, Grattagliano I, Villani G, Cocco T (2007) Carvedilol inhibits mitochondrial complex I and induces resistance to H2O2-mediated oxidative insult in H9C2 myocardial cells. Biochim Biophys Acta 1767(3):222–232. https://doi.org/10.1016/j.bbabio.2007.01.023

Shanmuganathan S, Hausenloy DJ, Duchen MR, Yellon DM (2005) Mitochondrial permeability transition pore as a target for cardioprotection in the human heart. Am J Physiol Heart Circ Physiol 289(1):H237–H242. https://doi.org/10.1152/ajpheart.01192.2004

Sharp WW, Fang YH, Han M, Zhang HJ, Hong Z, Banathy A, Morrow E, Ryan JJ, Archer SL (2014) Dynamin related protein 1 (Drp1)-mediated diastolic dysfunction in myocardial ischemia-reperfusion injury: therapeutic benefits of Drp1 inhibition to reduce mitochondrial fission. FASEB J 28(1):316–326. https://doi.org/10.1096/fj.12-226225

Shim HS, Park JH, Kim YA, Han JY, Park M, Song YG, Kim JS, Shin IW (2016) Ethyl pyruvate has anti-apoptotic and myocardial protective effects after regional ischaemia/reperfusion myocardial damage in an in vivo rat model. Singapore Med J. https://doi.org/10.11622/smedj.2016190

Sileikyte J, Blachly-Dyson E, Sewell R, Carpi A, Menabo R, Di Lisa F, Ricchelli F, Bernardi P, Forte M (2014) Regulation of the mitochondrial permeability transition pore by the outer membrane does not involve the peripheral benzodiazepine receptor (translocator protein of 18 kDa (TSPO)). J Biol Chem 289(20):13769–13781. https://doi.org/10.1074/jbc.M114.549634

Silva FS, Simoes RF, Couto R, Oliveira PJ (2016) Targeting mitochondria in cardiovascular diseases. Curr Pharm Des 22(37):5698–5717

Singh MV, Kapoun A, Higgins L, Kutschke W, Thurman JM, Zhang R, Singh M, Yang J, Guan X, Lowe JS, Weiss RM, Zimmermann K, Yull FE, Blackwell TS, Mohler PJ, Anderson ME (2009) Ca2+/calmodulin-dependent kinase II triggers cell membrane injury by inducing complement factor B gene expression in the mouse heart. J Clin Invest 119(4):986–996. https://doi.org/10.1172/JCI35814

Skulachev VP (2007) A biochemical approach to the problem of aging: "megaproject" on membrane-penetrating ions. The first results and prospects. Biochemistry (Mosc) 72(12):1385–1396

Skulachev VP, Anisimov VN, Antonenko YN, Bakeeva LE, Chernyak BV, Erichev VP, Filenko OF, Kalinina NI, Kapelko VI, Kolosova NG, Kopnin BP, Korshunova GA, Lichinitser MR, Obukhova LA, Pasyukova EG, Pisarenko OI, Roginsky VA, Ruuge EK, Senin II, Severina II, Skulachev MV, Spivak IM, Tashlitsky VN, Tkachuk VA, Vyssokikh MY, Yaguzhinsky LS, Zorov DB (2009) An attempt to prevent senescence: a mitochondrial approach. Biochim Biophys Acta 1787(5):437–461. https://doi.org/10.1016/j.bbabio.2008.12.008

Skulachev VP, Antonenko YN, Cherepanov DA, Chernyak BV, Izyumov DS, Khailova LS, Klishin SS, Korshunova GA, Lyamzaev KG, Pletjushkina OY, Roginsky VA, Rokitskaya TI, Severin FF, Severina II, Simonyan RA, Skulachev MV, Sumbatyan NV, Sukhanova EI, Tashlitsky VN, Trendeleva TA, Vyssokikh MY, Zvyagilskaya RA (2010) Prevention of cardiolipin oxidation and fatty acid cycling as two antioxidant mechanisms of cationic derivatives of plastoquinone (SkQs). Biochim Biophys Acta 1797(6–7):878–889. https://doi.org/10.1016/j.bbabio.2010.03.015

Sloan RC, Moukdar F, Frasier CR, Patel HD, Bostian PA, Lust RM, Brown DA (2012) Mitochondrial permeability transition in the diabetic heart: contributions of thiol redox state and mitochondrial calcium to augmented reperfusion injury. J Mol Cell Cardiol 52(5):1009–1018. https://doi.org/10.1016/j.yjmcc.2012.02.009

Smeele KM, Southworth R, Wu R, Xie C, Nederlof R, Warley A, Nelson JK, van Horssen P, van den Wijngaard JP, Heikkinen S, Laakso M, Koeman A, Siebes M, Eerbeek O, Akar FG, Ardehali H, Hollmann MW, Zuurbier CJ (2011) Disruption of hexokinase II-mitochondrial binding blocks ischemic preconditioning and causes rapid cardiac necrosis. Circ Res 108(10):1165–1169. https://doi.org/10.1161/CIRCRESAHA.111.244962

Smith RA, Hartley RC, Cocheme HM, Murphy MP (2012) Mitochondrial pharmacology. Trends Pharmacol Sci 33(6):341–352. https://doi.org/10.1016/j.tips.2012.03.010

Snow BJ, Rolfe FL, Lockhart MM, Frampton CM, O'Sullivan JD, Fung V, Smith RA, Murphy MP, Taylor KM, Protect Study G (2010) A double-blind, placebo-controlled study to assess the mitochondria-targeted antioxidant MitoQ as a disease-modifying therapy in Parkinson's disease. Mov Disord 25(11):1670–1674. https://doi.org/10.1002/mds.23148

Soboll S, Brdiczka D, Jahnke D, Schmidt A, Schlattner U, Wendt S, Wyss M, Wallimann T (1999) Octamer-dimer transitions of mitochondrial creatine kinase in heart disease. J Mol Cell Cardiol 31(4):857–866. https://doi.org/10.1006/jmcc.1998.0925

Stanley WC, Kivilo KM, Panchal AR, Hallowell PH, Bomont C, Kasumov T, Brunengraber H (2003) Post-ischemic treatment with dipyruvyl-acetyl-glycerol decreases myocardial infarct size in the pig. Cardiovasc Drugs Ther 17(3):209–216

Stavrovskaya IG, Narayanan MV, Zhang W, Krasnikov BF, Heemskerk J, Young SS, Blass JP, Brown AM, Beal MF, Friedlander RM, Kristal BS (2004) Clinically approved heterocyclics act on a mitochondrial target and reduce stroke-induced pathology. J Exp Med 200(2):211–222. https://doi.org/10.1084/jem.20032053

Steenbergen C, Perlman ME, London RE, Murphy E (1993) Mechanism of preconditioning. Ionic alterations. Circ Res 72(1):112–125

Stelzer AC, Frazee RW, Van Huis C, Cleary J, Opipari AW Jr, Glick GD, Al-Hashimi HM (2010) NMR studies of an immunomodulatory benzodiazepine binding to its molecular target on the mitochondrial F(1)F(0)-ATPase. Biopolymers 93(1):85–92. https://doi.org/10.1002/bip.21306

Sun J, Picht E, Ginsburg KS, Bers DM, Steenbergen C, Murphy E (2006a) Hypercontractile female hearts exhibit increased S-nitrosylation of the L-type Ca2+ channel alpha1 subunit and reduced ischemia/reperfusion injury. Circ Res 98(3):403–411. https://doi.org/10.1161/01.RES.0000202707.79018.0a

Sun J, Steenbergen C, Murphy E (2006b) S-Nitrosylation: NO-related redox signaling to protect against oxidative stress. Antioxid Redox Signal 8(9–10):1693–1705. https://doi.org/10.1089/ars.2006.8.1693

Sun J, Morgan M, Shen RF, Steenbergen C, Murphy E (2007) Preconditioning results in S-nitrosylation of proteins involved in regulation of mitochondrial energetics and calcium transport. Circ Res 101(11):1155–1163. https://doi.org/10.1161/CIRCRESAHA.107.155879

Sun Y, Deng T, Lu N, Yan M, Zheng X (2010) B-type natriuretic peptide protects cardiomyocytes at reperfusion via mitochondrial calcium uniporter. Biomed Pharmacother 64(3):170–176. https://doi.org/10.1016/j.biopha.2009.09.024

Surinkaew S, Chattipakorn S, Chattipakorn N (2011) Roles of mitochondrial benzodiazepine receptor in the heart. Can J Cardiol 27(2):262 e263–262 e213. https://doi.org/10.1016/j.cjca.2010.12.023

Szeto HH (2006) Cell-permeable, mitochondrial-targeted, peptide antioxidants. AAPS J 8(2):E277–E283. https://doi.org/10.1208/aapsj080232

Szeto HH (2008) Mitochondria-targeted cytoprotective peptides for ischemia-reperfusion injury. Antioxid Redox Signal 10(3):601–619. https://doi.org/10.1089/ars.2007.1892

Szeto HH, Schiller PW (2011) Novel therapies targeting inner mitochondrial membrane—from discovery to clinical development. Pharm Res 28(11):2669–2679. https://doi.org/10.1007/s11095-011-0476-8

Teixeira J, Oliveira C, Ricardo A, Cagide F, Garrido J, Ribeiro JA, Pereira CM, Silva AF, Andrade PB, Oliveira PJ, Borges F (2017) Development of hydroxybenzoic-based platforms as a solution to deliver dietary antioxidants to mitochondria. Sci Rep 7(1):6842

Teshima Y, Akao M, Jones SP, Marban E (2003) Cariporide (HOE642), a selective Na+-H+ exchange inhibitor, inhibits the mitochondrial death pathway. Circulation 108(18):2275–2281. https://doi.org/10.1161/01.CIR.0000093277.20968.C7

Theroux P, Chaitman BR, Danchin N, Erhardt L, Meinertz T, Schroeder JS, Tognoni G, White HD, Willerson JT, Jessel A (2000) Inhibition of the sodium-hydrogen exchanger with cariporide to prevent myocardial infarction in high-risk ischemic situations. Main results of the GUARDIAN trial. Guard during ischemia against necrosis (GUARDIAN) Investigators. Circulation 102(25):3032–3038

Tong H, Chen W, Steenbergen C, Murphy E (2000) Ischemic preconditioning activates phosphatidylinositol-3-kinase upstream of protein kinase C. Circ Res 87(4):309–315

Tong H, Imahashi K, Steenbergen C, Murphy E (2002) Phosphorylation of glycogen synthase kinase-3beta during preconditioning through a phosphatidylinositol-3-kinase-dependent pathway is cardioprotective. Circ Res 90(4):377–379

Tosaki A, Cordis GA, Szerdahelyi P, Engelman RM, Das DK (1994) Effects of preconditioning on reperfusion arrhythmias, myocardial functions, formation of free radicals, and ion shifts in isolated ischemic/reperfused rat hearts. J Cardiovasc Pharmacol 23(3):365–373

Turnham RE, Scott JD (2016) Protein kinase A catalytic subunit isoform PRKACA; history, function and physiology. Gene 577(2):101–108. https://doi.org/10.1016/j.gene.2015.11.052

Turrell HE, Thaitirarot C, Crumbie H, Rodrigo G (2014) Remote ischemic preconditioning of cardiomyocytes inhibits the mitochondrial permeability transition pore independently of reduced calcium-loading or sarcKATP channel activation. Physiol Rep 2(11). https://doi.org/10.14814/phy2.12231

Twig G, Elorza A, Molina AJ, Mohamed H, Wikstrom JD, Walzer G, Stiles L, Haigh SE, Katz S, Las G, Alroy J, Wu M, Py BF, Yuan J, Deeney JT, Corkey BE, Shirihai OS (2008) Fission and selective fusion govern mitochondrial segregation and elimination by autophagy. EMBO J 27(2):433–446. https://doi.org/10.1038/sj.emboj.7601963

Varanyuwatana P, Halestrap AP (2012) The roles of phosphate and the phosphate carrier in the mitochondrial permeability transition pore. Mitochondrion 12(1):120–125. https://doi.org/10.1016/j.mito.2011.04.006

Vaseva AV, Marchenko ND, Ji K, Tsirka SE, Holzmann S, Moll UM (2012) p53 opens the mitochondrial permeability transition pore to trigger necrosis. Cell 149(7):1536–1548. https://doi.org/10.1016/j.cell.2012.05.014

Veenman L, Papadopoulos V, Gavish M (2007) Channel-like functions of the 18-kDa translocator protein (TSPO): regulation of apoptosis and steroidogenesis as part of the host-defense response. Curr Pharm Des 13(23):2385–2405

Veenman L, Shandalov Y, Gavish M (2008) VDAC activation by the 18 kDa translocator protein (TSPO), implications for apoptosis. J Bioenerg Biomembr 40(3):199–205. https://doi.org/10.1007/s10863-008-9142-1

Villa-Abrille MC, Cingolani E, Cingolani HE, Alvarez BV (2011) Silencing of cardiac mitochondrial NHE1 prevents mitochondrial permeability transition pore opening. Am J Physiol Heart Circ Physiol 300(4):H1237–H1251. https://doi.org/10.1152/ajpheart.00840.2010

von Stockum S, Giorgio V, Trevisan E, Lippe G, Glick GD, Forte MA, Da-Re C, Checchetto V, Mazzotta G, Costa R, Szabo I, Bernardi P (2015) F-ATPase of Drosophila melanogaster forms 53-picosiemen (53-pS) channels responsible for mitochondrial Ca2+-induced Ca2+ release. J Biol Chem 290(8):4537–4544. https://doi.org/10.1074/jbc.C114.629766

Waldmeier PC, Feldtrauer JJ, Qian T, Lemasters JJ (2002) Inhibition of the mitochondrial permeability transition by the nonimmunosuppressive cyclosporin derivative NIM811. Mol Pharmacol 62(1):22–29

Walker JR, Hall MP, Zimprich CA, Robers MB, Duellman SJ, Machleidt T, Rodriguez J, Zhou W (2017) Highly potent cell-permeable and impermeable NanoLuc luciferase inhibitors. ACS Chem Biol 12(4):1028–1037. https://doi.org/10.1021/acschembio.6b01129

Wallimann T, Tokarska-Schlattner M, Schlattner U (2011) The creatine kinase system and pleiotropic effects of creatine. Amino Acids 40(5):1271–1296. https://doi.org/10.1007/s00726-011-0877-3

Walsh SR, Tang TY, Kullar P, Jenkins DP, Dutka DP, Gaunt ME (2008) Ischaemic preconditioning during cardiac surgery: systematic review and meta-analysis of perioperative outcomes in randomised clinical trials. Eur J Cardiothorac Surg 34(5):985–994. https://doi.org/10.1016/j.ejcts.2008.07.062

Wang L, Cherednichenko G, Hernandez L, Halow J, Camacho SA, Figueredo V, Schaefer S (2001) Preconditioning limits mitochondrial Ca(2+) during ischemia in rat hearts: role of K(ATP) channels. Am J Physiol Heart Circ Physiol 280(5):H2321–H2328

Wang G, Liem DA, Vondriska TM, Honda HM, Korge P, Pantaleon DM, Qiao X, Wang Y, Weiss JN, Ping P (2005) Nitric oxide donors protect murine myocardium against infarction via modulation of mitochondrial permeability transition. Am J Physiol Heart Circ Physiol 288(3):H1290–H1295. https://doi.org/10.1152/ajpheart.00796.2004

Wang W, Fang H, Groom L, Cheng A, Zhang W, Liu J, Wang X, Li K, Han P, Zheng M, Yin J, Wang W, Mattson MP, Kao JP, Lakatta EG, Sheu SS, Ouyang K, Chen J, Dirksen RT, Cheng H (2008) Superoxide flashes in single mitochondria. Cell 134(2):279–290. https://doi.org/10.1016/j.cell.2008.06.017

West MB, Rokosh G, Obal D, Velayutham M, Xuan YT, Hill BG, Keith RJ, Schrader J, Guo Y, Conklin DJ, Prabhu SD, Zweier JL, Bolli R, Bhatnagar A (2008) Cardiac myocyte-specific expression of inducible nitric oxide synthase protects against ischemia/reperfusion injury by preventing mitochondrial permeability transition. Circulation 118(19):1970–1978. https://doi.org/10.1161/CIRCULATIONAHA.108.791533

Wilkins BJ, Molkentin JD (2004) Calcium-calcineurin signaling in the regulation of cardiac hypertrophy. Biochem Biophys Res Commun 322(4):1178–1191. https://doi.org/10.1016/j.bbrc.2004.07.121

Wilson JE (1982) "Ambiquitous" behavior of brain hexokinase: rapid and reversible interaction of hexokinase with the outer mitochondrial membrane. Biophys J 37(1):18–19. https://doi.org/10.1016/S0006-3495(82)84577-3

Wilson JD, Bigelow CE, Calkins DJ, Foster TH (2005) Light scattering from intact cells reports oxidative-stress-induced mitochondrial swelling. Biophys J 88(4):2929–2938. https://doi.org/10.1529/biophysj.104.054528

Wolfe CL, Sievers RE, Visseren FL, Donnelly TJ (1993) Loss of myocardial protection after preconditioning correlates with the time course of glycogen recovery within the preconditioned segment. Circulation 87(3):881–892

Wu J, Yu J, Xie P, Maimaitili Y, Wang J, Yang L, Ma H, Zhang X, Yang Y, Zheng H (2017) Sevoflurane postconditioning protects the myocardium against ischemia/reperfusion injury via activation of the JAK2-STAT3 pathway. Peer J 5:e3196. https://doi.org/10.7717/peerj.3196

Xi J, Wang H, Mueller RA, Norfleet EA, Xu Z (2009) Mechanism for resveratrol-induced cardioprotection against reperfusion injury involves glycogen synthase kinase 3beta and mitochondrial permeability transition pore. Eur J Pharmacol 604(1–3):111–116. https://doi.org/10.1016/j.ejphar.2008.12.024

Xiao XH, Allen DG (1999) Role of Na(+)/H(+) exchanger during ischemia and preconditioning in the isolated rat heart. Circ Res 85(8):723–730

Xiao XH, Allen DG (2000) Activity of the Na(+)/H(+) exchanger is critical to reperfusion damage and preconditioning in the isolated rat heart. Cardiovasc Res 48(2):244–253

Xiao J, Liang D, Zhang H, Liu Y, Li F, Chen YH (2010) 4′-Chlorodiazepam, a translocator protein (18 kDa) antagonist, improves cardiac functional recovery during postischemia reperfusion in rats. Exp Biol Med (Maywood) 235(4):478–486. https://doi.org/10.1258/ebm.2009.009291

Xing G, Luo Z, Zhong C, Pan X, Xu X (2016) Influence of miR-155 on cell apoptosis in rats with ischemic stroke: role of the ras homolog enriched in brain (Rheb)/mTOR pathway. Med Sci Monit 22:5141–5153. https://doi.org/10.12659/msm.898980

Xu M, Wang Y, Hirai K, Ayub A, Ashraf M (2001) Calcium preconditioning inhibits mitochondrial permeability transition and apoptosis. Am J Physiol Heart Circ Physiol 280(2):H899–H908

Xu S, Wang P, Zhang H, Gong G, Gutierrez Cortes N, Zhu W, Yoon Y, Tian R, Wang W (2016) CaMKII induces permeability transition through Drp1 phosphorylation during chronic beta-AR stimulation. Nat Commun 7:13189. https://doi.org/10.1038/ncomms13189

Yan K, Gao LN, Cui YL, Zhang Y, Zhou X (2016) The cyclic AMP signaling pathway: exploring targets for successful drug discovery (review). Mol Med Rep 13(5):3715–3723. https://doi.org/10.3892/mmr.2016.5005

Yang X, Cohen MV, Downey JM (2010) Mechanism of cardioprotection by early ischemic preconditioning. Cardiovasc Drugs Ther 24(3):225–234. https://doi.org/10.1007/s10557-010-6236-x

Ying WL, Emerson J, Clarke MJ, Sanadi DR (1991) Inhibition of mitochondrial calcium ion transport by an oxo-bridged dinuclear ruthenium ammine complex. Biochemistry 30(20):4949–4952

Ytrehus K, Liu Y, Downey JM (1994) Preconditioning protects ischemic rabbit heart by protein kinase C activation. Am J Phys 266(3 Pt 2):H1145–H1152

Yu P, Zhang J, Yu S, Luo Z, Hua F, Yuan L, Zhou Z, Liu Q, Du X, Chen S, Zhang L, Xu G (2015) Protective effect of sevoflurane postconditioning against cardiac ischemia/reperfusion injury via ameliorating mitochondrial impairment, oxidative stress and rescuing autophagic clearance. PLoS One 10(8):e0134666. https://doi.org/10.1371/journal.pone.0134666

Yu J, Wu J, Xie P, Maimaitili Y, Wang J, Xia Z, Gao F, Zhang X, Zheng H (2016) Sevoflurane postconditioning attenuates cardiomyocyte hypoxia/reoxygenation injury via restoring mitochondrial morphology. Peer J 4:e2659. https://doi.org/10.7717/peerj.2659

Yuan G, Kaneko M, Masuda H, Hon RB, Kobayashi A, Yamazaki N (1992) Decrease in heart mitochondrial creatine kinase activity due to oxygen free radicals. Biochim Biophys Acta 1140(1):78–84

Zepeda R, Kuzmicic J, Parra V, Troncoso R, Pennanen C, Riquelme JA, Pedrozo Z, Chiong M, Sanchez G, Lavandero S (2014) Drp1 loss-of-function reduces cardiomyocyte oxygen dependence protecting the heart from ischemia-reperfusion injury. J Cardiovasc Pharmacol 63(6):477–487. https://doi.org/10.1097/FJC.0000000000000071

Zeymer U, Suryapranata H, Monassier JP, Opolski G, Davies J, Rasmanis G, Linssen G, Tebbe U, Schroder R, Tiemann R, Machnig T, Neuhaus KL, Investigators E (2001) The Na(+)/H(+) exchange inhibitor eniporide as an adjunct to early reperfusion therapy for acute myocardial infarction. Results of the evaluation of the safety and cardioprotective effects of eniporide in acute myocardial infarction (ESCAMI) trial. J Am Coll Cardiol 38(6):1644–1650

Zhang SZ, Gao Q, Cao CM, Bruce IC, Xia Q (2006) Involvement of the mitochondrial calcium uniporter in cardioprotection by ischemic preconditioning. Life Sci 78(7):738–745. https://doi.org/10.1016/j.lfs.2005.05.076

Zhang H, Wang P, Bisetto S, Yoon Y, Chen Q, Sheu SS, Wang W (2017) A novel fission-independent role of dynamin-related protein 1 in cardiac mitochondrial respiration. Cardiovasc Res 113(2):160–170. https://doi.org/10.1093/cvr/cvw212

Zhao K, Zhao GM, Wu D, Soong Y, Birk AV, Schiller PW, Szeto HH (2004) Cell-permeable peptide antioxidants targeted to inner mitochondrial membrane inhibit mitochondrial swelling, oxidative cell death, and reperfusion injury. J Biol Chem 279(33):34682–34690. https://doi.org/10.1074/jbc.M402999200

Zhou W, Marinelli F, Nief C, Faraldo-Gomez JD (2017) Atomistic simulations indicate the c-subunit ring of the F1Fo ATP synthase is not the mitochondrial permeability transition pore. Elife 6. https://doi.org/10.7554/eLife.23781

Zoratti M, Szabo I (1994) Electrophysiology of the inner mitochondrial membrane. J Bioenerg Biomembr 26(5):543–553

Zordoky BN, Robertson IM, Dyck JR (2015) Preclinical and clinical evidence for the role of resveratrol in the treatment of cardiovascular diseases. Biochim Biophys Acta 1852(6):1155–1177. https://doi.org/10.1016/j.bbadis.2014.10.016

Zorov DB, Juhaszova M, Yaniv Y, Nuss HB, Wang S, Sollott SJ (2009) Regulation and pharmacology of the mitochondrial permeability transition pore. Cardiovasc Res 83(2):213–225. https://doi.org/10.1093/cvr/cvp151

Zouein FA, Kurdi M, Booz GW (2013) LIF and the heart: just another brick in the wall? Eur Cytokine Netw 24(1):11–19. https://doi.org/10.1684/ecn.2013.0335

Zuurbier CJ, Eerbeek O, Meijer AJ (2005) Ischemic preconditioning, insulin, and morphine all cause hexokinase redistribution. Am J Physiol Heart Circ Physiol 289(1):H496–H499. https://doi.org/10.1152/ajpheart.01182.2004

Targeting the Mitochondrial Genome Through a Nanocarrier and the Regulation of Mitochondrial Gene Expression

Yuma Yamada and Hideyoshi Harashima

Abstract Mitochondria are attractive organelles that have the potential to contribute greatly to medical therapy. This organnelle is responsible for a variety of essential functions including ATP production and the regulation of apoptosis, and they have their own genome, mitochondrial DNA (mtDNA). It has recently become evident that a variety of human diseases are associated with mitochondrial dysfunctions caused by mutations and defects in mtDNA. Therefore, the ability to successfully target the mitochondrial genome and to regulate mitochondrial gene expression would contribute to mitochondrial gene therapy for various human diseases. To achieve such an innovative objective, it will be necessary to deliver various cargoes to mitochondria in living cells. However, only a limited number of approaches are available for accomplishing this. In this chapter, we discuss problems associated with mitochondrial delivery systems and mitochondrial gene expression, and propose a strategy for overcoming these problems based our current efforts. To this end, we highlight a MITO-Porter, a mitochondrial delivery system, and show some examples of the regulation of mitochondrial gene expression including mitochondrial RNA knockdown and mitochondrial transgene expression.

Keywords Mitochondria · Mitochondrial drug delivery · MITO-Porter · Mitochondrial genome · RNA knockdown · Transgene expression · Mitochondrial gene therapy

1 Current Status of Mitochondrial DDS Development

The mitochondrion possesses a two membrane structure consisting of an outer membrane and an inner membrane. Drugs with molecular weights less than 5 kDa can pass through the outer membrane, but they cannot penetrate through the inner membrane which has tight membrane structure. To achieve mitochondrial gene

Y. Yamada · H. Harashima (✉)
Laboratory for Molecular Design of Pharmaceutics, Faculty of Pharmaceutical Sciences, Hokkaido University, Sapporo, Japan
e-mail: harasima@pharm.hokudai.ac.jp

P. J. Oliveira (ed.), *Mitochondrial Biology and Experimental Therapeutics*,
https://doi.org/10.1007/978-3-319-73344-9_21

therapy, it will be necessary to deliver functional agents into the mitochondrial matrix, where the control system for mitochondrial gene expression is located, by overcoming the barrier that is comprised of the cellular membrane as well as the mitochondrial double membranes. During this past decade, numerous developments regarding mitochondrial drug delivery systems (DDS) have been reported (Fig. 1) (Zhang et al. 2011; Biswas and Torchilin 2014; Sato et al. 2017). Many researchers have reported on the use of mitochondrial DDS for low molecular weight drugs, while reports dealing with targeting the mitochondrial DNA for delivery are limited. The use of a mitochondrial targeting signal peptide (MTS) makes it possible to selectively deliver oligo DNA to isolated mitochondria (Flierl et al. 2003). However, to be effective, the MTS and the associated cargo must first be taken up by cells, because the MTS conjugate itself cannot be internalized through the cellular membrane. Moreover, MTS cannot deliver macromolecules, such as plasmid DNA (pDNA) and mitochondrial DNA (mtDNA), to mitochondria because mitochondrial delivery using MTS is severely limited by the size of the cargo (Endo et al. 1995). Weissig and coworkers reported on the development of DQAsomes, which are mitochondriotropic and cationic vesicles designed for mitochondrial-targeted DNA delivery (D'Souza et al. 2003). They reported that DQAsomes specifically release pDNA, proximal to mitochondria in living cells, although the delivery of cargoes into the interior of the mitochondria has not been validated.

2 A MITO-Porter System for the Delivery of Macromolecules to Mitochondria

We recently proposed a new concept for mitochondrial delivery, a MITO-Porter, a liposome-based nanocarrier system that introduces macromolecular cargos into mitochondria via membrane fusion (Fig. 1c) (Yamada et al. 2008). The MITO-Porter, which is modified with a cell penetrating peptide, octaarginine (R8), is first internalized into the cytosol via macropinocytosis. The MITO-Porter then binds to mitochondria via electrostatic interactions between the cationic R8 and negatively charged mitochondria. Finally, the cargoes are delivered to mitochondria *via* membrane fusion. Mitochondria actively repeat fusion and fission to share bio macromolecules in living cells. Therefore, we hypothesized that a strategy involving membrane fusion using a MITO-Porter system would permit a cargo to be delivered to mitochondria, independent of the size and physical properties of the particle.

We first screened for mitochondrial fusogenic liposomes using isolated rat liver mitochondria by varying the lipid composition of a panel of liposomes. As a result, among more than 100 different liposomes, we succeeded in identifying liposomes with a high mitochondrial fusogenic activity, and refer to such liposomes as MITO-Porters (Fig. 2a). The successful MITO-Porter had the following lipid composition: 1,2-dioleoyl-sn-glycero-3-phosphatidylethanolamine (DOPE)/sphingomyelin (SM)/stearylated R8 (STR-R8) [9:2:1, molar ratio] or DOPE/phosphatidic acid (PA)/STR-R8 [9:2:1, molar ratio] (Yamada and Harashima 2014; Yamada et al. 2008). To evaluate

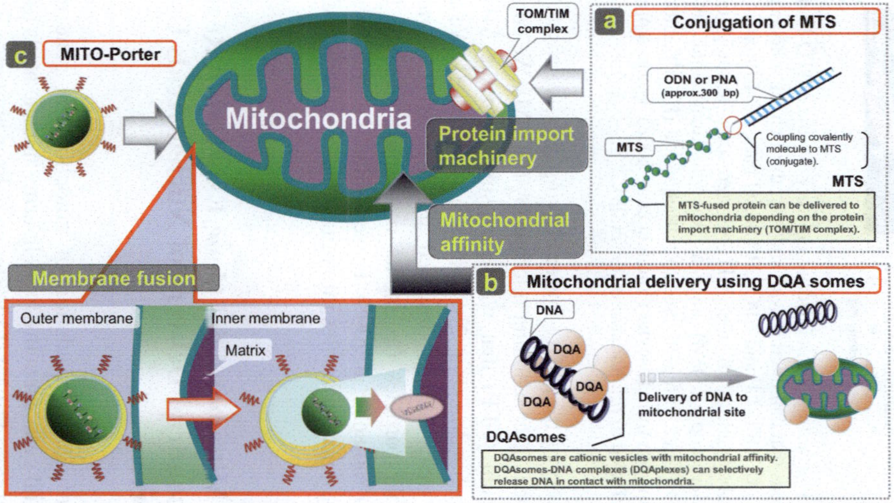

Fig. 1 Current strategies for mitochondria-targeted DNA delivery. Schematic image of mitochondria-targeted DNA delivery by MTS (**a**), DQAsomes (**b**) and MITO-Porter (**c**) are shown. This figure is used with permission from John Wiley and Sons (Sato et al. 2017)

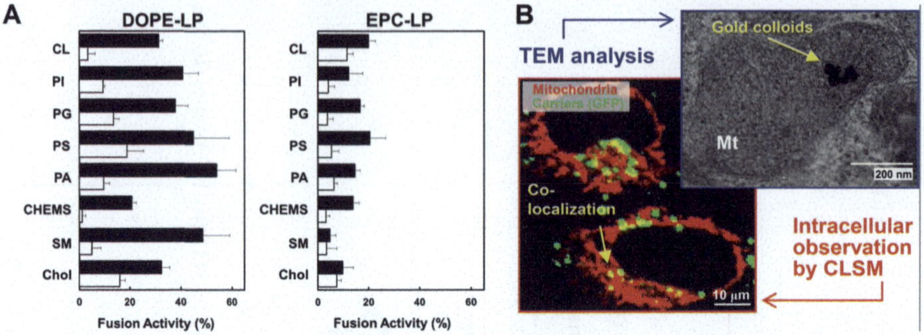

Fig. 2 Experimental data regarding mitochondrial delivery by the MITO-Porter. Screening of fusogenic lipid compositions that can interact with and pass through the mitochondrial membrane (**a**). Fusion activities (%) of DOPE-LP and EPC-LP. Closed bars, R8-LP; open bars, unmodified liposome. Intracellular observations of the MITO-Porter after staining mitochondria and transmission electron microscopy (TEM) analysis showing that Gold colloids that were encapsulated in the MITO-Porter were delivered to mitochondria (**b**). Mt., mitochondria. These figures are used with permission from Elsevier (Yamada et al. 2008). For interpreting the colors in this figure, the reader is referred to ref. Yamada et al. (2008).

the mitochondrial delivery by the MTIO-Porter in living cells, we observed intracellular trafficking using confocal laser scanning microscopy (CLSM). We confirmed that the green fluorescent protein (GFP) that was encapsulated in the MITO-Porter was co-localized with mitochondria (Fig. 2b) (Yamada et al. 2008). We also performed electron microscopy analyses, and observed that the MITO-Porter successfully delivered colloidal gold particles (as model macromolecules) to the interior of mitochondria (Fig. 2b) (Yamada et al. 2008). These results confirm that the MITO-Porter is capable of delivering macromolecules into mitochondria of living cells.

3 Validation of Mitochondrial Gene Expression Regulation by the Mitochondrial Delivery of Nucleic Acids Using the MITO-Porter

In our research regarding the regulation of mitochondrial gene expression, we attempted mitochondrial gene silencing by delivering nucleic acids to mitochondria. In this experiment, we used the MITO-Porter to deliver antisense RNA targeted to endogenous mitochondrial mRNA. We chose an mRNA that codes for COX 2 as a target. COX 2 is an endogenous mitochondrial protein related to the maintenance of mitochondrial membrane potential. If the knockdown of the COX 2 mRNA was successful, the level of expression of the target mitochondrial protein would be decreased, followed by a decrease in mitochondrial membrane potential.

Figure 3A provides information on the quantification of mRNA-levels coding COX2 after transfection by the MITO-Porter. As a result, the mRNA levels of COX2 were decreased as the result of the mitochondrial delivery of antisense RNA (Furukawa et al. 2015) (Fig. 3A). We also quantified the levels of expression of the

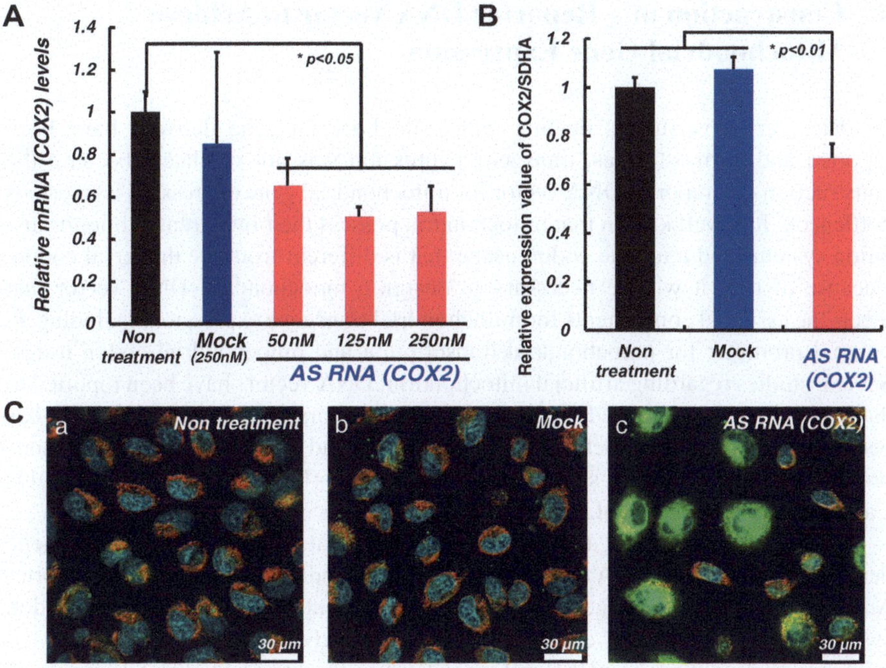

Fig. 3 Summary of experimental results for regulating mitochondrial gene expression levels using MITO-Porter system. (**A**) Quantification of target mRNA levels was performed by qRT-PCR. Bars indicates the mean with SEM (n = 3–4). *Significant differences were calculated by one-way ANOVA ($p < 0.05$). COX, cytochrome c oxidase subunit II. (**B**) Evaluation of relative COX 2/SDHA protein expression levels in mitochondrial protein knockdown using immunostaining. Bars are the mean with SEM (n = 64). *Significant differences were calculated by one-way ANOVA ($p < 0.01$). (**C**) Visualization of mitochondrial membrane potential using JC-1. After 48 h transfection of MITO-Porter system (Non treatment (**a**), Mock (**b**), AS RNA (COX 2) (**c**)), HeLa cells were incubated with JC-1 and observed by CLSM. These figures are reproduced from ref. Furukawa et al. (2015). For interpreting the colors in this figure, the reader is referred to ref. Furukawa et al. (2015)

target protein using Immunostaining, and confirmed that the use of antisense RNA resulted in a significant decrease in the level of expression of the target protein (Furukawa et al. 2015) (Fig. 3B).

We then evaluated mitochondrial membrane potential using JC-1. JC-1 emits a red fluorescence in mitochondria when the mitochondria have a high membrane potential, while the fluorescence changes to green in the cytosol when the potential is low. As a result, in the case of non-treated and mock treated cells, mitochondria containing red fluorescent material were observed. On the other hand, we observed a green color in the cytosol as result of the mitochondrial delivery of antisense RNA (Fig. 3C) (please refer to ref. Furukawa et al. (2015) for an interpretation of the color in this figure). These results indicate that the MITO-Porter system is able to regulate mitochondrial gene expression and mitochondrial functions.

4 Construction of a Reporter DNA Vector to Achieve Mitochondrial Gene Expression

To date, very few studies dealing with mitochondrial gene delivery have been reported and, in most cases, transgene expression was not validated, because the construction of a reporter DNA vector for mitochondrial gene expression is a serious bottleneck. It is well known that mitochondria possess their own transcription/translation systems and a unique codon usage that is different from the universal codon. Because of this, it will be necessary to design a mitochondrial DNA vector that meets the essential components for mitochondrial transgene expression, including an optimal promoter for mitochondrial transcription and mitochondrial codon usage. Several studies regarding artificial mitochondrial DNA vectors have been reported to date, and most have involved the design of DNA vectors that contain a gene that has been optimized for the mitochondrial codon system and a mitochondrial endogenous promotor such as a heavy strand mtDNA promoter (HSP) (Lyrawati et al. 2011; Cardoso et al. 2015; Yu et al. 2012).

We recently constructed a pHSP-mtLuc (CGG) that possesses HSP, a *Nd 4* gene derived from mouse mtDNA and an artificial mitochondrial genome with the reporter *NanoLuc (Nluc) luciferase* gene that records adjustments to the mitochondrial codon system (Fig. 4) (Yasuzaki et al. 2015). In this study, we examined the *in vivo* mitochondrial delivery of pHSP-mtLuc (CGG) to the liver and skeletal muscle of mice *via* hydrodynamic injection, and confirmed that mitochondrial transgene expression had occurred. Hydrodynamic injection, in which a large volume of naked pDNA is rapidly injected, has frequently been used as an efficient method for the *in vivo* nuclear delivery of naked pDNA in a wide variety of basic and translational studies (Suda and Liu 2007). We also succeeded in *in vivo* mitochondrial gene delivery *via* hydrodynamic injection (Yasuzaki et al. 2013). Moreover, we showed that the hydrodynamic injection of pHSP-mtLuc (CGG) resulted in the expression of

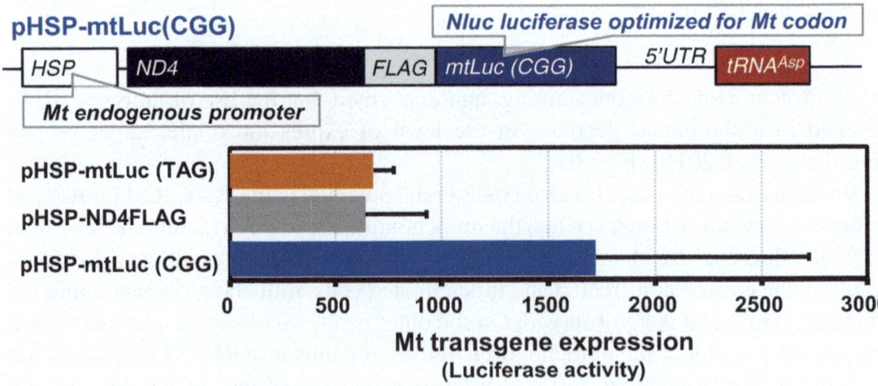

Fig. 4 Structure of pHSP-mtLuc (CGG) and evaluation of the mitochondrial transgene expression. This figure is reproduced with permission from ACS Publications (Yasuzaki et al. 2015)

the mitochondrial Nluc luciferase protein in the liver and skeletal muscle, while no mitochondrial transgene expression was observed in the case of control pDNA which contains no promoter (pHSP-ND4FLAG) or "TAG" codon for mitochondrial/universal translational arrest (pHSP-mtLuc (TAG)) (Fig. 4) (Yasuzaki et al. 2015). We are currently attempting to control mitochondrial gene expression by integrating a MITO-Porter system and a pHSP-mtLuc (CGG) design.

5 Perspective

The focus of the material presented herein was on our current efforts regarding mitochondrial delivery and mitochondrial gene expression regulation. We are currently conducting trials to establish a viable mitochondrial therapeutic strategy. Our ultimate goal is to achieve successful mitochondrial gene therapy. To reach this ultimate goal, we plan to validate the utility of mitochondrial gene therapeutic strategy using mitochondrial DDS, in order to target cells in patients with mitochondrial diseases. It is our hope that these optimized delivery systems will provide effective therapies for the many patients who currently suffer from various mitochondrial diseases.

Acknowledgements This work was supported, in part, by the Scientific Research (B) from the Ministry of Education, Culture, Sports, Science and Technology of Japanese Government (MEXT). We also thank Dr. Milton Feather for his helpful advice in writing the manuscript.

References

Biswas S, Torchilin VP (2014) Nanopreparations for organelle-specific delivery in cancer. Adv Drug Deliv Rev 66:26–41. https://doi.org/10.1016/j.addr.2013.11.004

Cardoso AM, Morais CM, Cruz AR, Cardoso AL, Silva SG, do Vale ML, Marques EF, Pedroso de Lima MC, Jurado AS (2015) Gemini surfactants mediate efficient mitochondrial gene delivery and expression. Mol Pharm 12(3):716–730. https://doi.org/10.1021/mp5005349

D'Souza GG, Rammohan R, Cheng SM, Torchilin VP, Weissig V (2003) DQAsome-mediated delivery of plasmid DNA toward mitochondria in living cells. J Control Release 92(1–2):189–197

Endo T, Nakayama Y, Nakai M (1995) Avidin fusion protein as a tool to generate a stable translocation intermediate spanning the mitochondrial membranes. J Biochem (Tokyo) 118(4):753–759

Flierl A, Jackson C, Cottrell B, Murdock D, Seibel P, Wallace DC (2003) Targeted delivery of DNA to the mitochondrial compartment via import sequence-conjugated peptide nucleic acid. Mol Ther 7(4):550–557

Furukawa R, Yamada Y, Kawamura E, Harashima H (2015) Mitochondrial delivery of antisense RNA by MITO-porter results in mitochondrial RNA knockdown, and has a functional impact on mitochondria. Biomaterials 57:107–115. https://doi.org/10.1016/j.biomaterials.2015.04.022

Lyrawati D, Trounson A, Cram D (2011) Expression of GFP in the mitochondrial compartment using DQAsome-mediated delivery of an artificial mini-mitochondrial genome. Pharm Res 28(11):2848–2862. https://doi.org/10.1007/s11095-011-0544-0

Sato Y, Sakurai Y, Kajimoto K, Nakamura T, Yamada Y, Akita H, Harashima H (2017) Innovative technologies in nanomedicines: from passive targeting to active targeting/from controlled pharmacokinetics to controlled intracellular pharmacokinetics. Macromol Biosci 17(1). https://doi.org/10.1002/mabi.201600179

Suda T, Liu D (2007) Hydrodynamic gene delivery: its principles and applications. Mol Ther 15(12):2063–2069. https://doi.org/10.1038/sj.mt.6300314

Yamada Y, Harashima H (2014) A method for screening mitochondrial fusogenic envelopes for use in mitochondrial drug delivery. Methods Mol Biol 1141:57–66. https://doi.org/10.1007/978-1-4939-0363-4_2

Yamada Y, Akita H, Kamiya H, Kogure K, Yamamoto T, Shinohara Y, Yamashita K, Kobayashi H, Kikuchi H, Harashima H (2008) MITO-porter: a liposome-based carrier system for delivery of macromolecules into mitochondria via membrane fusion. Biochim Biophys Acta 1778(2):-423–432. https://doi.org/10.1016/j.bbamem.2007.11.002

Yasuzaki Y, Yamada Y, Kanefuji T, Harashima H (2013) Localization of exogenous DNA to mitochondria in skeletal muscle following hydrodynamic limb vein injection. J Control Release 172(3):805–811. https://doi.org/10.1016/j.jconrel.2013.09.029

Yasuzaki Y, Yamada Y, Ishikawa T, Harashima H (2015) Validation of mitochondrial gene delivery in liver and skeletal muscle via hydrodynamic injection using an artificial mitochondrial reporter DNA vector. Mol Pharm 12(12):4311–4320. https://doi.org/10.1021/acs.molpharmaceut.5b00511

Yu H, Koilkonda RD, Chou TH, Porciatti V, Ozdemir SS, Chiodo V, Boye SL, Boye SE, Hauswirth WW, Lewin AS, Guy J (2012) Gene delivery to mitochondria by targeting modified adenoassociated virus suppresses Leber's hereditary optic neuropathy in a mouse model. Proc Natl Acad Sci U S A 109(20):E1238–E1247. https://doi.org/10.1073/pnas.1119577109

Zhang E, Zhang C, Su Y, Cheng T, Shi C (2011) Newly developed strategies for multifunctional mitochondria-targeted agents in cancer therapy. Drug Discov Today 16(3–4):140–146. https://doi.org/10.1016/j.drudis.2010.12.006. S1359-6446(10)00844-5 [pii]

Mitochondrial Dysfunction and Chronic Disease: Treatment with Membrane Lipid Replacement and Other Natural Supplements

Garth L. Nicolson, Gonzalo Ferreira, Robert Settineri, Rita R. Ellithorpe, Paul Breeding, and Michael E. Ash

Abstract Mitochondria provide most of our cellular energy needs; therefore, mitochondrial dysfunction can cause fatigue and other symptoms that are commonly found in every chronic and many acute conditions. Reductions in mitochondrial function occur when there is loss of maintenance of inner mitochondrial membrane trans-membrane potential, modifications in the electron transport chain, damage to mitochondrial DNA, altered mitochondrial transcription, and reductions in the transport of critical substrates and metabolites into mitochondria. These events can result in reduced efficiency of oxidative phosphorylation and reductions in ATP production. Several components of mitochondria require routine replacement, and this can be facilitated with dietary changes and the use of natural supplements. Clinical trials have shown the utility of using oral mitochondrial replacement supplements, such as replacement glycerolphospholipids, L-carnitine, alpha-Lipoic acid, coenzyme Q10, NADH, pyrroloquinoline quinone and other mitochondrial supplements to improve mitochondrial function. Membrane Lipid Replacement supplements with or without other mitochondrial supplements can significantly

G. L. Nicolson (✉)
Department of Molecular Pathology, The Institute for Molecular Medicine, S. Laguna Beach, CA, USA
e-mail: gnicolson@immed.org

G. Ferreira
Ion Channels Laboratory, Department of Biophysics, Universidad de la Republica, Montevideo, Uruguay

R. Settineri
Sierra Productions Research, Irvine, CA, USA

R. R. Ellithorpe
Tustin Longevity Center, Tustin, CA, USA

P. Breeding
Blue Hole Wellness, Kimberley, TX, USA

M. E. Ash
Clinical Education, Newton Abbot, Devon, UK

P. J. Oliveira (ed.), *Mitochondrial Biology and Experimental Therapeutics*,
https://doi.org/10.1007/978-3-319-73344-9_22

diminish fatigue and other mitochondria-associated symptoms found in aging, cancer and chronic diseases.

Keywords Membranes · Phospholipids · Coenzyme Q10 · L-Carnitine · Alpha-Lipoic acid · Fatigue · Mitochondrial function · Metabolite transport · Chronic diseases · Cancer · Aging

1 Introduction

Mitochondria are thought to perform multiple functions beyond their energy production, including the regulation of cellular communication, nuclear gene expression, synaptic transmission, inflammatory responses, and complex metabolic pathways (Picard et al. 2016). Mitochondria are absolutely required for health, and if mitochondria become dysfunctional, a number of adverse events ensue that result in illness (Piecznik and Neustadt 2007). Mitochondrial dysfunction is most often characterized by loss of efficiency in the electron transport chain (ETC) and resulting decreased synthesis of high-energy molecules, such as ATP (Aw and Jones 1989; Smeitink et al. 2006). The dysfunctinal state of mitochondria is so comonly found that it is a characteristic of aging and essentially all chronic and some acute diseases (Green et al. 2011; Nicolson 2014; Nicolson and Ash 2014, 2017; Picard et al. 2016; Piecznik and Neustadt 2007; Reddy 2008; Reddy and Reddy 2011; Smeitink et al. 2006).

In terms of chronic, non-genetic diseases, the most obvious illnesses that are typified by mitochondrial dysfunction are the fatiguing illnesses, such as chronic fatigue, chronic fatigue syndrome, myalgic encephalomyelitis, fibromyalgia and Gulf War illness (Agadjanyan et al. 2003; Booth et al. 2012; Breeding et al. 2012; Cordero et al. 2010; DiMauro and Rustin 2009; Myhill et al. 2009; Morris and Maes 2014; Nicolson 2014; Nicolson and Nicolson 1996; Nicolson et al. 2003; Norheim et al. 2011; Park et al. 2000). In addition to the fatiguing illnesses, this list also includes: (1) cardiovascular diseases, such as atherosclerosis and other heart and vascular conditions (Limongelli et al. 2012; Nicolson 2007; Rabinovich and Vilaro 2010; Victor et al. 2009); (2) metabolic diseases, such as metabolic syndrome and type 2 diabetes (Joseph et al. 2012; Ma et al. 2012; Nicolson 2007); (3) neurodegenerative diseases, such as Alzheimer's disease, Parkinson's disease, Huntington's disease, amyotrophic lateral sclerosis and Freidriech ataxia (Ghafourifar et al. 2008; Karbowski and Neutzner 2012; Mao and Reddy 2010; Reddy 2008; Reddy and Reddy 2011); (4) autoimmune diseases, such as systemic lupus erythematosus, multiple sclerosis and type 1 diabetes (Fernandez and Perl 2009; Fernandez and Perl 2009; Maiese et al. 2007; Mao and Reddy 2010); (5) neurobehavioral and psychiatric diseases, such as schizophrenia, bipolar and mood disorders and autism spectrum disorders (Konradi et al. 2004; Marazziti et al. 2012; Palmieri and Peerscio 2010; Prince et al. 2000; Rossignol and Frye 2012; Stork and

Table 1 Some natural ingredients and cofactors that have been used or suggested to treat mitochondrial dysfuction[a]

Category	Some examples
Minerals	Magnesium, calcium, phosphate
Vitamins	Vitamins C, D and E, thiamine, riboflavin
Metabolites	Creatine, pyruvate
Cofactors	CoQ10, NADH, alpha-lipoic acid, nicotinic acid
Lipids	Membrane glycerolphospholipids, unsaturated fatty acids
Transporters	L-Carnitine, membrane phospholipids
Antioxidants	CoQ10, alpha-lipoic acid, NADH, glutathione, resveratrol
Enzyme inhibitors	Alpha-lipoic acid, dichloroacetate
Biogenesis stimulants	Pyrroloquinoline quinone
Herbs	Curcimin, schisandrin

[a]Modified from Kerr (2010) and Nicolson (2014)

Renshaw 2005); (6) gastrointestinal disorders (Breeding et al. 2012; Chitkara et al. 2003; Di Donato 2009); (7) chronic infections (Anand and Tikoo 2013; Ashida et al. 2011); (8) cancers (Gabridge 1987; Nicolson 2010; Sotgia et al. 2011; Wallace 2005); and (9) sytemic environmantal contaminantion (Leung et al. 2013; Mayer et al. 2013).

Genetic alterations in mitochondrial genes also contribute to mitochondrial dysfunction. This has been reviewed elsewhere (Kerr 2010; Holt 2010; Picard et al. 2016; Wallace and Fan 2010) and will not be part of this review. Here we will concentrate on non-genetic or acquired mechanisms that could explain mitochondrial dysfuction and its treatment with natural supplements. We will also discuss the use of Membrane Lipid Replacement (MLR) strategies using glycerolphospholipid supplements (Nicolson 2010, 2014, 2016; Nicolson and Ellithorpe 2006; Nicolson and Ash 2014, 2017; Nicolson et al. 2016) and combinations of natural supplements that include antioxidants, vitamins, minerals, enzyme cofactors, metabolites, transporters, and other natural products (Table 1) (Nicolson 2014; Nicolson et al. 2012a, b).

2 Mitochondrial Membrane Dysfunction

There are several reasons for mitochondria failing to provide enough necessary high energy molecules for cell and tissue functions. For example, there could be an inadequate number of mitochondria, damaged mitochondrial membranes, a problem in providing the necessary substrates for mitochondrial function, mitochondial genetic damage or defects in the ETC or ATP synthesis machinery of the mitochondrial inner membrane (MIM) (Nicolson 2014; Nicolson and Ash 2014, 2017).

Inside cells the numbers, locations and functional status of mitochondria can be changed by several processes, such as the fusion of partially dysfunctional mitochondria in order to provide enough undamaged components to improve overall function, the generation of entirely new mitochondria from existing mitochondria by fission, and the removal and degradation of dysfunctional mitochondria in a process called mitophagy (Mishra and Chan 2016; Twig and Shirihi 2011). Mitophagy and the fission and fusion of mitochondria are controlled by complex cellular processes that sense the deterioration of mitochondria. For example, some of mitochondrial properties that characterize deterioration of mitochondria are the depolarization of the MIM and the activation of certain transcription pathways (Lee et al. 2012; Priault et al. 2005).

The MIM is especially sensitive to environmental and cellular changes, and the ability of the MIM to produce high-energy molecules like ATP is directly related to the capacity of the ETC to convert the energy of metabolites to NADH and transfer electrons from the ETC to molecular oxygen while simultaneously pumping protons from the mitochondrial matrix across the MIM to the intermembrane space. This process creates trans-membrane proton (Δp) and electrochemical gradients ($\Delta\Psi_m$) across the MIM that are critical for mitochondrial function (Nicholls 2010; Rich and Marechal 2010). The trans-membrane potential created by the proton gradient can then use ATP synthase to flow protons back across the MIM, using the energy from this process to synthesize ATP (Divakaruni and Brand 2011; Nicholls 2010). Since MIM trans-membrane potential is directly related to ETC function through reversible phosphorylation, mitochondrial dysfunction can be monitored by the loss of MIM trans-membrane potential. In higher organisms, the normal MIM trans-membrane potential associated with an efficient energy production is approximately -120 mV (Hüttemann et al. 2008).

An interesting and important consequence of the transfer of electrons through the ETC is the production of Reactive Oxygen Species (ROS), highly reactive free radicals that are produced as a byproduct of oxidative phosphorylation. Although these oxidant molecules can be produced by other events inside cells, significant sources of ROS and related Reactive Nitrogen Species (RNS) are the mitochondria. Low concentrations of ROS are essential as secondary cellular messengers, for example, passing messages between the mitochondria and the nucleus (Reczek and Chandel 2015). However, high concentrations of ROS and RNS can react with and damage cellular DNA, lipids and proteins (Richter et al. 1998; Spiteller 2010; Spector and Yorek 1985; Stadtman 2002).

The appropriate balance of ROS and RNS concentrations within cells is essential in order to maintain cellular homeostasis—thus there are natural mechanisms inside cells to neutralize excess ROS/RNS. Among the mechanisms that control ROS/RNS concentrations, the most important are dismutase enzymes and antioxidants that are used to neutralize exces amounts of ROS/RNS (Duchen and Szabadkai 2010; Nicholls 2010). In addition to the conrolled flow of protons across the MIM that allows ATP production, the leakage of protons back across the MIM can also be facilitated by inducing uncoupling proteins that allow protons to flow back across the proton gradient (Nicholls 2010; Rich and Marechal 2010). This reduces the production of ATP while still consuming excess oxygen (Duchen and Szabadkai 2010). This process has been described as a controlled proton leak, where excess oxygen is consumed in the process (Divakaruni and Brand 2011). The resulting excess ROS that is produced as a byproduct of this process, however, can damage mitochondrial as well as other cellular membrane lipids, proteins and DNA (Divakaruni and Brand 2011; Richter et al. 1998; Wei and Lee 2002). Since membrane phospholipids maintain membrane matrix structure and associate and stabilize critical membrane proteins, phospholipid modification by events such as peroxidation is an important cause of mitochondrial dysfunction, which in turn is linked to fatigue and aging-associated diseases (Ademowo et al. 2017).

Within the mitochondria the most sensitive substrates of ROS and RNS oxidation are the MIM lipids, especially the MIM cardiolipin. Cardiolipin is a specialized phospholipid that is highly integrated into the ETC. It is bound by essential proteins in the MIM, and is absolutely required for mitochondrial ETC function (Spector and Yorek 1985). Alterations in cardiolipid have profound consequences for the function and activity of the ETC (Chicco and Sparagna 2007; Houtkooper and Vaz 2008). Although the exact molecular mechanisms that yield peroxidized lipids, such as the cardiolipins, are still poorly understood, the formation of 4-hydroxynoneal reactive lipids seems to be a common intermediate pathway for many of the events leading to lipid peroxidation (Xiao et al. 2017). ROS and RNS damage to MIM cardiolipin and other membrane phospholipids can result in increased proton and ion leakage back across the MIM and partial loss of the proton/electrochemical gradient. Since cardiolipin provides stability for the cytochrome/enzyme complexes in the MIM, its damage by ROS/RNS results in loss of ETC function and can result in initiation of mitophagy (Houtkooper and Vaz 2008).

Cellular antioxidant defenses neutralize excess oxidants, and this usually maintains ROS/RNS levels at concentrations that prevent excess damage to cellular molecules (Barber and Harris 1994; Sun 1990). Some of the most important endogenous cellular antioxidant defenses are glutathione peroxidase, catalase and superoxide dismutase, among others (Fridovich 1995; Jagetia et al. 2003; Sun 1990). These antioxidant defense enzymes are essential in preventing excess damage to cells and initiating cell death programs.

There are also low molecular weight antioxidants that can affect ROS/RNS levels, and some of these can be provided by dietary supplementation (Aeschbach et al. 1994; Jagetia et al. 2003; Schwartz 1996). Dietary antioxidants have been used as natural preventive agents to shift the levels of oxidative molecules to physiological levels that can be maintained in redox balance by the natural cellular antioxidant system (Prasad et al. 2001). However, dietary antioxidants alone cannot provide replacement molecules for damaged cellular and mitochondrial components.

Finally, calcium levels inside mitochondria are also related to the normal function of mitochondria (Marchi et al. 2017). The balance of calcium in mitochondria is maintained through a variety of ion channels and transporters in mitochondrial membranes. For example, mitochondrial permeability transition pore (mPTP), calcium uniporter, mitochondrial BK channels, among other mitochondrial membrane components, maintain intra-mitochondria calcium levels (Pérez and Quintanilla

2017; Satrustegui et al. 2007). The function of these channels and transporters is influenced and dependent on the lipid composition of the mitochondrial membrane (Lee 2004; Szabo et al. 2004).

3 Fatigue and Mitochondrial Dysfunction

Fatigue is a multidimensional sensation that is perceived to be a lack of overall energy, an inability to perform even simple tasks without exertion, and prolonged recovery required after physical activity. Although the etiologic mechanisms that cause fatigue are not well understood, fatigue is a hallmark symptom of mitochondrial dysfunction (Filler et al. 2014). Mild fatigue can be caused by different conditions, including depression and other psychological conditions, but moderate to severe fatigue involves cellular energy systems, such as those provided by mitochondria (Kroenke et al. 1988; Nicolson 2010; Nicolson and Settineri 2011). At the tissue and cellular level moderate to severe fatigue is related to loss of mitochondrial ETC function and reduced production of ATP (Booth et al. 2012; Myhill et al. 2009; Nicolson 2007, 2010; Nicolson and Ash 2014, 2017).

There are different types of fatigue, including short-term fatigue, often associated with physical exertion, and long-lasting or chronic fatigue. Intractable fatigue lasting more than 6 months that is not reversed by sleep is generally considered to be chronic fatigue, and this is the most common complaint of patients seeking general medical care (Kroenke et al. 1988; Morrison 1980). Fatigue and chronic fatigue are important secondary symptoms in many clinical conditions, and they often occur early in the progression of disease (Kroenke et al. 1988).

Mitochondrial dysfunction is directly related to excess fatigue that is not reversed by rest. During aging and chronic diseases oxidative ROS/RNS damage to mitochondrial membranes and the ETC decreases mitochondrial function, which is perceived as fatigue (Huang and Manton 2004; Logan and Wong 2001), and when this persists for some time it is perceived as chronic fatigue. Chronic fatigue often presents with additional symptoms, among them musculoskeletal pain, sleep disturbance, cognition and memory problems, headaches, digestive symptoms and other complaints. These chronic fatigue-associated symptoms have been collectively described as chronic fatigue syndrome (CFS) (Fukuda et al. 1994). CFS pateints also show evidence of oxidative damage, such as oxidative damage to DNA and lipids (Logan and Wong 2001; Manuel y Keenoy et al. 2001), oxidized blood markers (Richards et al. 2000) and oxidatively damaged membrane lipids, including sphingolipids and glycosphingolipids (Fulle et al. 2000). CFS patients also have continuously elevated levels of peroxynitrite due to excess nitric oxide. RNS such as peroxynitrite can cause lipid peroxidation and loss of mitochondrial function as well as changes in cytokine levels that exert a positive feedback on nitric oxide production (Pall 2000).

4 Membrane Lipid Replacement and Mitochondrial Dysfunction

The dietary replacement of cellular membrane glycerolphospholipids using natural whole foods is a difficult way to provide enough lipid replacement molecules on a daily basis (Nicolson and Ash 2014, 2017). To more efficiently remove and replace damaged, oxidized, membrane phospholipids in mitochondria and other cellular organelles specific dietary supplements have been used. For example, Membrane Lipid Replacement (MLR) with antioxidants has proved very effective in replacing damaged, oxidized cellular lipids and restoring function to organelles such as mitochondria (recently reviewed in Nicolson 2016; Nicolson et al. 2016; Nicolson and Ash 2014, 2017). To various degrees antioxidant supplements can reduce ROS/RNS levels and prevent some membrane phospholipid oxidation (Prasad et al. 2001; Schwartz 1996), but antioxidants alone cannot repair the damage done to cellular membranes, and in particular, to their mitochondria ETC and other critical membrane systems (Nicolson and Ash 2014, 2017).

The MLR use of oral membrane glycerolphospholipids with unsaturated fatty acids in doses ranging from 2–4 g per day along with antioxidants has proven safe and effective for the natural medicine treatment of certain conditions, such as fatiguing illnesses (reviewed in Nicolson 2016; Nicolson and Ash 2014, 2017; Nicolson et al. 2016). MLR results in the actual replacement of damaged membrane phospholipids with undamaged (unoxidized) phospholipids with unsaturated fatty acids to ensure proper functioning of cellular and intraellular membranes. The reason that MLR has been so effective is likely due to the natural process of lipid uptake and transport, which can translocate and transport high concentrations of dietary phospholipids as lipid granules, liposomes, droplets, vesciles and lipoproteins in a 'bulk flow' gradient process that disseminates the MLR lipids to every cell and tissue and returns damaged, oxidized phospholipids for excretion by the same 'bulk flow' process (Nicolson and Ash 2017).

In clinical situations, oral MLR phospholipids can increase mitochondrial ETC function and decrease fatigue in CFS, fibromyalgia, and other fatiguing conditions, including natural aging (Table 2). For example, a membrane phospholipid-vitamin mixture (Propax™ with NTFactor®) was utilized by Ellithorpe et al. (2003) in a study on aging patients with severe chronic fatigue and was found to reduce fatigue by 40.5% within 8 weeks. Also, a cross-over study was initiated to study the effects of MLR phospholipids on fatigue and mitochondrial function in patients with moderate to severe chronic fatigue (Agadjanyan et al. 2003). Oral administration of NTFactor® for 12 weeks resulted in a 35.5% reduction in fatigue and 26.8% increase in mitochondrial function. Switching these patients to placebo without their knowledge resulted in slow increases in fatigue along with decreases in mitochondrial function towards control levels. Similar levels of fatigue reduction were observed in CFS and fibromyalgia patients given oral MLR phospholipids for 8 weeks (Nicolson and Ellithorpe 2006). The timing of fatigue reduction may depend on the oral MLR supplement used and whether it is combined with other supplements. For example,

Table 2 Some examples of oral membrane glycerolphospholipid supplementation and effects on fatigue in chronic illnesses[a]

Subjects/patients	n	Av. age	Time on MLR	Fatigue Scale fatigue reduction (%)	References
Chronic fatigue	34	50.3	8 weeks	40.5**	Ellithorpe et al. (2003)
Aging, chronic fatigue	20	68.9	12 weeks	35.5*	Agadjanyan et al. (2003)
Chronic fatigue syndrome (and/or fibromyalgia syndrome)	15	44.8	8 weeks	43.1*	Nicolson and Ellithorpe (2006)
Aging, fatigue	67	57.3	1 week	36.8*	Nicolson et al. (2010)
Chronic illnesses	58	55.0	8 weeks	30.7*	Nicolson et al. (2012a)
Chronic fatigue syndrome	30	56.2	8 weeks	34.3*	Nicolson et al. (2012a)
Lyme disease	18	52.4	8 weeks	26.7*	Nicolson et al. (2012b)
Gulf War illnesses	16	42.2	8 weeks	34.6*	Nicolson et al. (2012a)

**$p < 0.0001$, *$p < 0.001$ compared to no supplement
[a]Modified from Nicolson (2016)

using a higher dose formulation of NTFactor® and adding vitamins, minerals and other supplements in patients with moderate chronic fatigue resulted in a 36.8% reduction in fatigue within 1 week (Nicolson et al. 2010).

In multiple studies in animals and humans the use of MLR supplements has proven to be incredibly safe, and there was absolutely no evidence of any type of toxicity or adverse events (reviewed in Nicolson and Ash 2014, 2017). Up to 45 g of MLR phospholipids have been administered per day without any evidence of toxic effects (Polinsky et al. 1980). In fact, high doses of MLR phospholipids have had the opposite effect—they actually reduced the adverse symptoms caused by drugs and other treatments (reviewed in Nicolson 2010; Nicolson and Ash 2017).

5 Other Natural Supplements and Mitochondrial Dysfunction

A number of natural dietary supplements have been used to reduce non-psychological fatigue and increase mitochondrial function (DiMauro and Rustin 2009; Nicolson 2014; Nicolson and Settineri 2011). Some of these supplements include vitamins, minerals, antioxidants, metabolites, enzyme inhibitors and cofactors, mitochondrial transporters, herbs and membrane phospholipids (Table 1). As interventions for the treatment and management of patients' symptoms, such as fatigue, several dietary supplements have been utilized. However, for the most part they have not been considered effective (Chambers et al. 2006). Some of the most promising supplements beyond MLR are listed in Table 1 and discussed below.

5.1 Coenzyme Q10

The mitochondrial cofactor coenzyme Q10 (CoQ10) or ubiquinone is a key component of the mitochondrial ETC and one of the most widely used natural supplements that target mitochondrial dysfunction (Mancuso et al. 2010; Orsucci et al. 2011; Potgieter et al. 2013; Rich and Marechal 2010). CoQ10 is also a strong antioxidant in its reduced form, and it can affect the expression of certain genes involved in cell signaling, metabolism and transport (Groneberg et al. 2005; Littarru and Tiano 2010). The main use of CoQ10 as a natural supplement has been to improve the efficiency of electron transfer between the various complexes of the ETC (Littarru and Tiano 2010; Mancuso et al. 2010).

An important use of CoQ10 suppelments has been to increase the levels of cellular CoQ10 in various conditions that typically show reduced CoQ10 levels (Mancuso et al. 2010; Orsucci et al. 2011). For example, analysis of patient samples shows reduced levels of CoQ10 in neuromuscular and neurodegenerative diseases, such as Parkinson's disease, Huntington's disease, amyotrophic lateral sclerosis, Friedreich's ataxia and other disorders (Mancuso et al. 2009, 2010).

CoQ10 supplements have been used to reduce symptoms and progression in neurodegenerative diseases (Littarru and Tiano 2010; Mancuso et al. 2009; Yang et al. 2010). In a randomized, placebo-controlled clinical trial on 98 Alzheimer's patients Galasko et al. (2012) used a mixture of CoQ10, vitamins C and E and alpha-Lipoic acid in the test arm. When the trial was completed, patients in the test arm showed significant reductions in oxidative stress markers compared to placebo, but did not show significant changes in cerebrospinal fluid markers related to beta-amyloid or tau pathology. Parkinson disease patients generally showed increased oxidized-to-total CoQ10 ratios as well as significant increases in markers of oxidative damage in the cerebrospinal fluid (Isobe et al. 2010). This situation can be partially reversed with CoQ10 administration. In patients with early Huntington's disease CoQ10 administration for 30 months slowed the usual decline in total functional capacity but the differences were not statistically significance (Kieburtz and Huntington Study Group 2001).

Many of the neurodegenerative disease studies with CoQ10 supplementation did not reach clinical significance. For example, a randomized, placebo-controlled, multi-center phase II trial with amyotrophic lateral sclerosis patients given CoQ10 did not show differences in functional decline over a 9 month period (Thompson et al. 2009).

Other uses of CoQ10 supplementation include physical performance, hypertnesion and coronary heart disease. Mortensen et al. (2014) and Rosenfeldt et al. (2003) have reviewed the use of CoQ10 supplements for improving physical exercise, hypertension and heart failure. In 6 of 11 of the published studies modest improvements in exercise capacity were found in the subjects given dietary CoQ10. In eight publications on the effects of CoQ10 on hypertension there was a mean decrease in systolic blood pressure (16 mmHg) and diastolic pressure (10 mmHg). In nine randomized trials on the use of CoQ10 in heart failure patients there were non-significant trends towards increased injection fraction and reduced mortality. For example, Rosenfeldt et al. (2003) utilized a randomized, placebo-controlled trial to study the effects of oral CoQ10 in 35 patients with heart failure. They found in the test arm, but not in the control arm, that patients had significant improvements in symptoms and a trend towards improvements in mean exercise times. In a randomized, placebo-controlled, multi-center trial on chronic heart failure Mortensen et al. (2014) found that although there were no significant differences in short-term endpoints, the long-term use of CoQ10 resulted in lower cardiovascular mortality (9% versus 16%), all-cause mortality (10% versus 18%) and hosptial stays in the test arm of the study.

The effects of administration of oral CoQ10 during physical exercise have also been examined in a placebo-controlled, blinded, cross-over trial. Healthy subjects received CoQ10 or placebo for 8 days, and their performance was then evaluated at fixed workloads on a bicycle ergometer with a rest period in-between (Mizuno et al. 2008). While on CoQ10 the subjects were able to achieve higher work outputs. They also had less fatigue, and their need for more recovery time was reduced compared to the placebo arm.

To improve the delivery of CoQ10 to mitochondria hydrophobic carriers have been used. For example, the products MitoQ and SkQ are derivatives of quinone covalently conjugated to lipophilic cations to imrove the delivery of CoQ10 to mitochondria. By using hydrophobic derivatives of CoQ10 to increase the transfer of CoQ10 to mitochondrial MIM cardiolipin peroxidation by ROS could be largely prevented (Feniouk and Skulachev 2017).

5.2 Alpha-Lipoic Acid

Alpha-Lipoic acid (1,2-dithiolane-3-pentanoic acid) is an anti-inflammatory agent, potent antioxidant, transition metal ion chelator and redox transcription regulator (Marczurek et al. 2008; Shay et al. 2009; Smith et al. 2004). Alpha-Lipoic acid is also an important cofactor in mitochondrial alpha-ketoacid dehydrogenases, and thus it is a critical cofactor in mitochondrial oxidative decarboxylation reactions (Shay et al. 2009). In terms of its clinical use alpha-Lipoic acid has been utilized as an oral supplement in the treatment of complications associated with diabetes mellitus, where it has been shown to reduce various diabetic-associated neuropathies, inflammation and vascular health (Chambers et al. 2006; Marczurek et al. 2008). Glucose uptake and metabolism are also affected by alpha-Lipoic acid (Estrada et al. 1997).

In chronic illnesses and during natural aging certain sphingolipids, especially ceramides, and in particular short-chain ceramides, accumulate in mitochondria. This occurs due to hydrolysis of sphingomyelin by sphingomyelinase. If accumulation occurs over a certain threshold, eventually the excess ceramides can reduce ETC activity (Di Paola et al. 2000; Gudz et al. 1997). Ceramide accumulation in mitochondria is especially damaging to mitochondria in cardiac tissue. Thus feeding aging rodents with dietary alpha-Lipoic acid has been used to lower ceramide levels in vascular endothelial cells of cardiac muscle, and this has resulted in a restoration of mitochondrial glutathione levels—the result was increased ETC function (Monette et al. 2011).

There are certain uses of alpha-Lipoic acid that have proven to be effective. For example, in diabetes patients alpha-Lipoic acid has been used to reduce diabetic complications, such as sensorimotor polyneuropathies (Ziegler et al. 2004). In a 4-year blinded study that used oral alpha-Lipoic acid supplementation in diabetic patients neuropathic impairments improved significantly but not attributes of nerve conduction. Alpha-Lipoic acid was found to be safe for long-term use in diabetic patients (Ziegler et al. 2004).

When given as an oral supplement, alpha-Lipoic acid rarely accumulates in tissues above micromolar levels. Because of this, Shay et al. (2009) have argued that alpha-Lipoic acid is unlikely to be directly involved as an important primary cellular antioxidant. On the other hand, alpha-Lipoic acid can increase cellular glutathione levels by regulating glutathione synthesis and thus reducing cellular oxidative stress (Yoshida et al. 1995). In addition, alpha-Lipoic acid can have effects on the regulation of the nuclear transcription factor NF-κB. Because of its effects on transcription factor NF-κB, alpha-Lipoic acid may cause widespread transcriptional effects (Goraca et al. 2011). Alpha-Lipoic acid also has transition metal chelation properties and can remove excess copper, iron and other metals from cells. These metal can accumulate in chronic diseases, such as hemochromatosis, end-stage renal failure, Alzheimer's and Parkinson's diseases, and thus alpha-Lipoic acid has been promoted as a potential therapeutic agent in heavy metal poisoning (Smith et al. 2004). Finally there are reports that alpha-Lipoic acid also improves cognitive function along with its positive effects on mitochondrial function, suggesting a link between oxidative damage to mitochondria and congnition (Head et al. 2009).

Alpha-Lipoic acid supplementation has been promoted as a useful supplement for fatigue, although it has not been used as a sole supplement in clinical trials on chronic fatigue and CFS (Shay et al. 2009). However, due to its widespread use as a safe oral supplement (usually at 200–600 mg/day) alpha-Lipoic acid has been added to many mitochondrial support supplement mixtures to support mitochondrial function and reduce oxidative stress (Goraca et al. 2011; Nicolson et al. 2012a, b; Ziegler et al. 2011).

5.3 L-Carnitine

L-Carnitine (3-hydroxy-4-*N*-trimethylaminobutyrate), a naturally occurring fatty acid transporter, is directly involved in the transport of fatty acids into the mitochondrial matrix and in the removal of excess acyl groups. L-Carnitine functions in beta-oxidation and is important in the regulation of coenzyme A homeostatasis (Marcovina et al. 2012; Reuter and Evans 2012). Due to its mode of action L-carnitine must be maintained within a relatively narrow concentration range. Thus dietary supplementation of L-carnitine is important in maintaining its concentration within cells (Reuter and Evans 2012). When L-carnitine falls below essential concentrations, mitochondrial function deteriorates. For example, L-carnitine deficiency disorders are associated with reduced mitochondrial function, and this is further associated with insulin resistance and coronary artery disease (Koves et al. 2008; Newgard et al. 2009; Shah et al. 2009).

It has been known for some time that physicial performance is dependent on the use of L-carnitine. Supplements containing L-carnitine have been extensively by athlethes to improve performance (Spriet et al. 2008). The rationale has been that increasing the reliance on fat as the principle substrate for energy production during extreme exercise reduces the need for carbohydrates and delays the depletion of carbohydrate stores. The use of fat instead of sugars can increase overall energy production and reduce exercise-induced fatigue. L-Carnitine plays an important role in increasing the transport of fatty acids into mitochondria for their use as substrates. However, supplementation with oral L-carnitine for several weeks prior to extreme exercise did not increase skeletal muscle carnitine content, and thus during extreme exercise it is unlikely that L-carnitine supplementation alters muscle metabolism (Brass 2000; Wachter et al. 2002).

Oral L-carnitine has been used in clinical disorders that are characterized by low L-carnitine concentrations or impaired fatty acid oxidation, such as diabetes, sepsis, renal disease and cardiomyopathy (Yoshida et al. 1995). In patients with congestive heart failure L-carnitine supplementation has been used to improve physical status. For example, in a small study of 18 patients with congestive heart failure compared to 12 placebo controls propionyl-L-carnitine supplementation improved peak heart rate (12%), exercise capacity (21%) and peak oxygen consumption (45%) in the treatment group (Anand et al. 1998). Other uses for L-carnitine supplementation include alcoholism, hepatic encephalopathy, coronary heart diseases, Peyronie's disease, cerebral ischemia and infertility (Anon 2010).

L-Carnitine has also been used as an anti-aging suppement. It is known that the rate of mitochondrial oxidative phosphorylation naturally declines during aging. In animals acetyl-L-carnitine was found to reverse age-related decreases in L-carnitine levels and increase fatty acid metabolism. L-Carnitine supplementation also reversed age-related declines in cellular glutathione levels while improving mitochondrial complex IV activity (Brass 2000). Although L-carnitine dietary supplementation at doses up to 2 g per day has been reported to be safe and potentially useful in increasing mitochondrial function in the aged, multiple clinical trials that show its effectiveness in age-related chronic illnesses are lacking (Anon 2010). An exception to this was a randomized, controlled clinical trial using 70 centenarians who were treated with L-carnitine for 6 months. Before the trial participants were found to have muscle weakness, decreasing mental health, impaired mobility and poor endurance. By the end of the trial the treated group showed significant improvements in physical fatigue, mental fatigue and fatigue severity. The participants also showed reductions in total fat mass, increased muscle mass and increased physical and cognitive activity (Malaguarnera et al. 2007).

5.4 Reduced Nicotinamide Adenine Dinucleotide (NADH)

NADH is a cellular redox cofactor in over 200 redox reactions, and it is also a substrate for many enzymes (Penberthy 2009; Kirkland 2009). NAD/NADH also performs a balancing act between the nucleus and mitochondria over the control of cellular energy homeostasis (Canto et al. 2015). There is a universal requirement for NAD/NADH, and its deficiency results in pellagra, which is characterized by dermatitis, diarrhea, dementia and eventually death (Penberthy 2009). In the mitochondria NADH delivers electrons from metabolite hydrolysis to the ETC, but in its reduced form NADH can also act as a strong antioxidant. The usual dietary supplementation has been the use of NADH precursors, such as niacin, nicotinic acid or nicotinamide, but recently microcarriers have been used to stabilize oral NADH so that it can be directly ingested in small doses in the gastrointestinal system. This turns out to be more effective than using large oral doses of NADH, which are susceptible to oxidation and degradation and are generally considered ineffective (Kirkland 2009).

In many diseases, such as neurodegenerative diseases, oxidative damage is extensive, and various mitochondrial antioxidants have been used to treat disease and delay progression (Aw and Jones 1989; Kerr 2010; Pieczenik and Neustadt 2007; Reddy 2008; Smeitink et al. 2006). Use of oral NADH in unstabilized or stablized form has resulted in mixed results. In Alzheimer's disease Birkmayer (1996) used stabilized oral NADH to improve cognitive functioning and dementia. However, in another clinical trial there were no improvements in cognition or dementia using oral NADH (Rainer et al. 2000). In a controlled trial using 26 Alzheimer's patients who were supplemented with stabilized NADH or placebo Demarin et al. (2004) found significantly better performance scores in the test arm compared to the placebo arm in verbal fluency and visual construction, with a trend toward increased performance on abstract verbal reasoning. In this trial there was no evidence of better performance in attention, memory or on scores of dementia severity (Demarin et al. 2004).

In order to bypass the problems with oral NADH supplementation Birkmayer et al. (1993) use stabilized oral NADH and IV NADH to reduce the symptoms of Parkinson's disease. In an open label trial using over 800 Parkinson disease patients the effects of IV and oral NADH was studied. Birkmayer et al. (1993) found that

19.3% of patients exhibited 30–50% improvements in disability, while 58.8% had moderate improvements (10–30%), and 21.8% did not respond ($p < 0.01$). They found that younger patients with a shorter history of Parkinson's disease had a better chance of responding and displaying significant disability improvements than older patients and patients with a longer duration of disease. No difference was found between the oral NADH and the IV NADH (Birkmayer et al. (1993). However, when Dizdar et al. (1994) repeated this study with a similar trial, they found no statistically significant improvements in Parkinson's disease symptom scores.

Stabilized, oral NADH has also been used to reduce symptoms in CFS patients and in patients with chronic fatigue. For example, CFS patients were given stabilized, microencapsulated, oral NADH or placebo for 4 weeks in a blinded cross-over trial (Forsyth et al. 1999). In this clinical study 8/26 (30.7%) patients responded to the NADH compared with 2 of 26 (8%) in the placebo arm ($p < 0.05$). Thus these results were not considered significant by Colquhoun and Senn (2000). In another clinical trial using CFS patients oral, stabilized NADH was compared to psychological/nutritional therapy. The stabilized NADH reduced fatigue in the first 4 months of a 12 month trial. However, after the first 4 months, symptom scores were similar in the NADH and psychological/nutritional arms of the trial (Santaella et al. 2004). In an open label study stabilized NADH was given orally for 2 months to CFS patients (Alegre et al. 2010). They found decreases in anxiety and maximum heart rate after a stress test, but no differences were found in the functional impact of fatigue, quality of life, sleep quality, exercise capacity, or functional reserve (Alegre et al. 2010). The clinical studies described above suggest mixed results with stabilized NADH. Some patients responded to the oral, stabilized NADH, but others did not.

6 Combination Natural Supplements and Mitochondrial Dysfunction

There are a few examples of using combinations of MLR supplements to reduce the effects of mitochondrial dysfunction. Ellithorpe et al. (2003) used a membrane phospholipid-vitamin mixture (Propax™ with NTFactor®, various vitamins and minerals) to treat aging patients with severe chronic fatigue. They found reductions in overall fatigue of 40.5% within 8 weeks. Examination of the subcategories of fatigue in this study indicated that the combination MLR supplement reduced all subcategories of fatigue, such as the behavioral/severity, affective/meaning, sensory, and cognitive/mood dimensions of fatigue. This same oral supplement was used to reduce cancer-associated fatigue and the fatigue-effects of cancer therapy (Colodny et al. 2001). For example, Propax™ reduced chemotherapy-induced fatigue, nausea, vomiting, malaise, diarrhea, headaches and other side effects of cancer therapy. Eighty-one percent of the patients on chemotherapy that used the combination MLR supplement experienced an overall improvement in quality of life parameters. In a subsequent double-blind, placebo-controlled cross-over trial 36 patients on chemotherapy plus Propax™ showed fewer adverse effects, resulting in improvements in fatigue, nausea, diarrhea, impaired taste, constipation, insomnia and other quality of life indicators (Colodny et al. 2001).

In a study using long-term intractable fatigue patients with a variety of diagnoses a MLR supplement containing membrane glycerolphospholipids, CoQ10, microencapsulated NADH, alpha-Lipoic acid, L-carnitine, alpha-Ketoglutaric acid and other nutrients (ATP Fuel®) was to treat fatigue and mitochondrial dysfunction (Nicolson et al. 2012a). The 58 participants in this trial had moderate to severe intractable fatigue for an average of more than 17 years and had seen an average of more than 15 practitioners without resolution of their fatigue. These patients had tried unsuccessfully an average of over 35 drugs and supplements to reduce their fatigue without success.

The chronic illness patients in the ATP Fuel® trial took the combination oral MLR supplement for 8 weeks, and fatigue was determined at several intermediate end points during and at the end of the trial. After 8 weeks of supplement the mean fatigue scores improved by 30.7% (t-test, $p < 0.0001$ and Wilcoxon signed-rank, $p < 0.0001$) (Nicolson et al. 2012a). The fatigue scores were further dissected into four fatigue subcategories (Behavior/Severity subcategory, which deals with completing tasks, socializing, engaging in sexual activity and other activities, and intensity or degree of fatigue; Affective/Meaning subcategory, which determines fatigue/tiredness is pleasant/unpleasant, whether the patient is agreeable/disagreeable, protective/destructive, or feels normal/abnormal; Sensory subcategory, which determines whether the patient is strong/weak, awake/sleepy, refreshed/tired, or energetic/unenergetic; and Cognitive/Mood subcategory, which assesses whether a patient feels relaxed/tense, exhilarated/depressed, able/unable to concentrate, remember, and think clearly). All of the subcategories showed significant reductions by the end of the trial ($p < 0.0001$) (Nicolson et al. 2012a).

Regression analysis was used to determine if the downward trends in fatigue over time during the ATP Fuel® trial were consistent, occurred with a high degree of confidence, and could predict further reductions in fatigue. The regression analyses of overall fatigue and each of the subcategories of fatigue indicated significant and consistent downward trends in the fatigue data. The regression R^2 values for the various subgroups varied from 0.950 to 0.980. Regression analysis of the overall fatigue yielded a R^2 of 0.960. This indicated that there was a high level of confidence and reproducibility in the downward trends in all fatigue data. The combination MLR supplement was safe, and there were no safety issues that came up during the trial (Nicolson et al. 2012a). Separately Lyme patients were examined to see if they responded with decreases in fatigue while on ATP Fuel® (Nicolson et al. 2012b). Similar to the study on intractable fatigue patients, the Lyme disease patients also showed dignificant reductions in overall fatigue and in all fatigue subcategories during an 8-week open label trial using ATP Fuel® (Nicolson et al. 2012b).

7 Other Natural Supplements That Have Not Been Extensively Tested

One natural supplement that has stimulated attention because of its effects on mitochondria is pyrroloquinoline quinone (PQQ). PQQ is a redox cofactor that displays anti-oxidant, anti-inflammatory and neuroprotective effects by modulating mitochondrial lipid and energy metabolism (Lee et al. 2014; Tao et al. 2007; Yang et al. 2014; Zhang et al. 2012). It also acts as a growth factor or even vitamin in laboratory animals (Bauerly et al. 2011; Killgore et al. 1989; Rucker et al. 2009; Steinberg et al. 1994), although it appears that PQQ does not meet the criteria for a vitamin (Felton and Anthony 2005). PQQ modifies the quantity and function of mitochondria in mice, such as promoting the spontaneous generation of new mitochondria in aging cells—thus it appears to be a stimulator of mitochondrial biogenesis (Chowandisai et al. 2010; Stites et al. 2006).

PQQ occurs normally in human tissues, and some (100–400 ng/day) is thought to be synthesized in humans (Harris et al. 2013). It is also found in various foods at different concentrations (Kumazawa et al. 1995). PQQ is thought to induce mitochondrial biogenesis through a cell signaling mechanism mediated through activation of specific nuclear and mitochondrial transcription pathways (Chowandisai et al. 2010).

Although PQQ is now available as a commercial supplement, there are few clinical studies in the literature on its effectiveness in promoting mitochondrial function in humans. Using PQQ as a dietary supplement does appear to be safe. Harris et al. (2013) used a placebo-controlled cross-over study to examine the effects of PQQ on pathways of inflammation, lipid and carbohydrate metabolism and urinary metabolites that are related to oxidative damage. After 76 h they found some changes in their measurements on markers of inflammation (C-reactive protein, interleukin-6), other blood markers (cholesterol, glucose, high- and low-density lipoproteins, triglycerides and other markers) and urinary metablolites by ^{1}H-nuclear magnetic resonance. Although the standard clinical markers were not altered by PQQ supplementation, there were significant reductions in C-reactive protein, interleukin-6 and urinary methylated amines consistent with reductions in inflammation as well as changes in antioxidant potential and enhancement in mitochondria-related functions. Additional studies will be need to confirm these findings and expand the effects to include other clinical aspects of mitochondrial function.

8 Final Comment

Oral mitochondrial supplements, such as the use of all-natural membrane glycerol phospholipids with or without other supplements and nutrients, can increase mitochondrial function and reduce fatigue and other symptoms and improve quality of life in patients with chronic illnesses. Future studies will concentrate on the use of

MLR supplements to improve clinical symptoms in a variety of chronic illnesses. For example, based on preliminary studies we have initiated a new double-blinded, placebo-conrolled, cross-over clinical trial using fibromyalgia patients to assess the use of NTFactor Lipids® in reducing pain, fatigue, gastrointestinal and other symptoms. Future studies will need to carefully document the effectiveness of natural supplements using randomized, controlled clinical trials.

References

Ademowo OS, Dias HKI, Burton DGA, Griffiths HR (2017) Lipid (per) oxidation in mitochondria: an emerging target in the ageing process? Biogerontology 18:1–21

Aeschbach R, Loliger J, Scott BC, Murcia A, Butler J, Halliwell B, Anuoma OI (1994) Antioxidant actions of thymol, carvacrol, 6-gingerol, zingerone and hydroxytyrosol. Food Chem Toxicol 32:31–36

Agadjanyan M, Vasilevko V, Ghochikyan A, Berns P, Kesslak P, Settineri, Nicolson GL (2003) Nutritional supplement (NTFactor) restores mitochondrial function and reduces moderately severe fatigue in aged subjects. J Chronic Fatigue Syndr 11(3):23–26

Alegre J, Rosés JM, Javierre C, Ruiz-Baqués A, Segundo MJ, de Sevilla TF (2010) Nicotinamide adenine dinucleotide (NADH) in patients with chronic fatigue syndrome. Rev Clin España 210(6):284–288

Anand SK, Tikoo SK (2013) Viruses and modulators of mitochondrial functions. Adv Virol 2013:738794

Anand I, Chandrashekhan Y, De Giuli F, Pasini E, Mazzoletti A, Confortini R, Ferrari R (1998) Acute and chronic effects of propionyl-L-carnitine on the hemodynamics, exercise capacity and hormones in patients with congestive heart failure. Cardiovasc Drugs Ther 12:291–299

Anon (2010) Acetyl-L-carnitine monograph. Altern Med Rev 15(1):76–83

Ashida H, Mimuro H, Ogawa M, Kobayashi T, Sanada T, Kim M, Sasakawa C (2011) Cell death and infection: a double-edged sword for host and pathogen survival. J Cell Biol 195:931–942

Aw TY, Jones DP (1989) Nutrient supply and mitochondrial function. Annu Rev Nutr 9:229–251

Barber DA, Harris SR (1994) Oxygen free radicals and antioxidants: a review. Am Pharmacol 34:26–35

Bauerly K, Harris C, Chowanadisai W, Graham J, Havel PJ, Tchaparian E, Satre M, Karliner JS, Rucker RB (2011) Altering pyrrologuinoline quinone nutritional status modulates mitochondrial, lipid and energy metabolism in rats. PLoS One 6(7):e21779

Birkmayer JG (1996) Coenzyme nicotinamide adenine dinucleotide: new therapeutic approach for improving dementia of the Alzheimer type. Ann Clin Lab Sci 26:1–9

Birkmayer JG, Vrecko C, Volc D, Birkmayer W (1993) Nicotinamide adenine dinucleotide—a new therapeutic approach to Parkinson's disease. Comparison of oral and parenteral application. Acta Neurol Scand Suppl 146:32–35

Booth NE, Myhill S, McLaren-Howard J (2012) Mitochondrial dysfunction and the pathophysiology of myalgic encephalomyelitis/chronic fatigue syndrome (ME/CFS). Int J Clin Exp Med 5:208–220

Brass EP (2000) Supplemental carnitine and exercise. Am J Clin Nutr 72:618S–623S

Breeding PC, Russell NC, Nicolson GL (2012) An integrative model of chronically activated immune-hormonal pathways important in the generation of fibromyalgia. Brit J Med Practit 5(3):a524

Canto C, Menzies K, Auwerx J (2015) NAD+ metabolism and the control of energy homeostasis—a balancing act between mitochondria and the nucleus. Cell Metab 22:31–53

Chambers D, Bagnall A-M, Hempel S, Forbes C (2006) Interventions for the treatment, management and rehabilitation of patients with chronic fatigue syndrome/myalgic encepthalomyelitis: an updated systematic review. J R Soc Med 99:506–520
Chicco AJ, Sparagna GC (2007) Role of cardiolipin alterations in mitochondrial dysfunction and disease. Am J Physiol Cell Physiol 292:C33–C44
Chitkara DK, Nurko S, Shoffner JM, Buie T, Flores A (2003) Abnormalities in gastrointestinal motility are associated with diseases of oxidative phosphorylation in children. Am J Gastroenterol 98:871–877
Chowandisai W, Bauerly KA, tchaparian E, Wong A, Cortopassi GA, Rucker RB (2010) Biogenesis through cAMP response element-binding protein phosphorylation and increased PGC-1alpha expression. J Biol Chem 285:142–152
Colodny L, Lynch K, Farber C, Papish R, Phillips K, Sanchez M, Cooper K, Pickus, Palmer O, Percy TB, Faroqui M, Block JB (2001) Results of a study to evaluate the use of Propax to reduce adverse effects of chemotherapy. J Am Nutraceut Assoc 3(1):17–25
Colquhoun D, Senn S (2000) Is NADH effective in the treatment of chronic fatigue syndrome? Ann Allergy Asthma Immunol 84:639–640
Cordero MD, de Miguel M, Carmona-Lopez I, Bonal P, Campa F, Moreno-Fernandez AM (2010) Oxidative stress and mitochondrial dysfunction in fibromyalgia. Neuro Endocrinol Lett 31(2):169–173
Demarin V, Podobnik SS, Storga-Tomic D, Kay G (2004) Treatment of Alzheimer's disease with stabilized oral nicotinamide adenine dinucleotide: a randomized, double-blind study. Drugs Exp Clin Res 30:27–33
Di Donato S (2009) Multisystem manifestations of mitochondrial disorders. J Neurol 256:693–710
Di Paola M, Cocco T, Lorusso M (2000) Ceramide interaction with the respiratory chain of heart mitochondria. Biochemist 39:6660–6668
DiMauro S, Rustin P (2009) A critical approach to the therapy of mitochondrial respiratory chain and oxidative phosphorylation diseases. Biochim Biophys Acta 1792:1159–1167
Divakaruni AS, Brand MD (2011) The regulation and physiology of mitochondrial proton leak. Physiology 26:192–205
Dizdar N, Kagedal B, Lindvall B (1994) Treatment of Parkinson's disease with NADH. Acta Neurol Scand 90:345–347
Duchen MR, Szabadkai G (2010) Roles of mitochondria in human disease. Essays Biochem 47:115–137
Ellithorpe RR, Settineri R, Nicolson GL (2003) Reduction of fatigue by use of a dietary supplement containing glycophospholipids. J Am Nutraceut Assoc 6(1):23–28
Estrada DE, Ewart HS, Tsakiridis T, Volchuk A, Ramtal T, Tritschler H, Klip A (1997) Stimulation by alpha-lipoic acid of glucose transport activity in skeletal muscle of lean and obese Zucker rats. Life Sci 61:805–812
Felton LM, Anthony C (2005) Role of PQQ as a mammalian enzyme cofactor? Nature 433(7025):E10. discussion E11–12
Feniouk BA, Skulachev VP (2017) Cellular and molecular mechanisms of action of mitochondria-targeted antioxidants. Curr Aging Sci 10:41–48
Fernandez D, Perl A (2009) Metabolic control of T cell activation and death in SLE. Autoimmun Rev 8:184–189
Filler K, Lyon D, Bennett J, McCain N, Elswick R, Lukkahatai N, Saligan LN (2014) Association of mitochondrial dysfunction and fatigue: a review of the literature. BBA Clin 1:12–23
Forsyth LM, Preuss HG, MacDowell AL, Chiazze L Jr, Birkmayer GD, Bellanti JA (1999) Therapeutic effects of oral NADH on the symptoms of patients with chronic fatigue syndrome. Ann Allergy Asthma Immunol 82:185–191
Fridovich I (1995) Superoxide radical and superoxide dismutases. Annu Rev Biochem 64:97–112
Fukuda K, Straus S, Hickie I, Sharpe M, Drobbins J, Komaroff A (1994) The chronic fatigue syndrome: a comprehensive approach to its definition and study. Ann Intern Med 121:953–959

Fulle S, Mecocci P, Fano G, Vecchiet I, Racciotti D, Cherubini A, Pizzigallo E, Vecchiet L, Senin U, Beal MF (2000) Specific oxidative alterations in vastus lateralis muscle of patients with the diagnosis of chronic fatigue syndrome. Free Radic Biol Med 29:1252–1259

Gabridge MG (1987) Metabolic consequences of Mycoplasma pneumoniae infection. Isr J Med Sci 23:574–579

Galasko DR, Peskind E, Clark CM, Quinn JF, Ringman JM, Jicha GA, Cotman C, Cotrell B, Montine TJ, Thomas RG, Aisen P (2012) Antioxidants for Alzheimer's disease: a randomized clinical trial with cerebrospinal fluid biomarker measures. Arch Neurol 69:836–841

Ghafourifar P, Mousavizadeh K, Parihar MS, Nazarewicz RR, Parihar A, Zenebe WJ (2008) Mitochondria in multiple sclerosis. Front Biosci 13:3116–3126

Goraca A, Huk-Kolega H, Piechota A, Kleniewska P, Ciejka E, Skibska B (2011) Lipoic acid—biological activity and therapeutic potential. Pharmacol Rep 63:849–858

Green DR, Galluzzi L, Kroemer G (2011) Mitochondria and the autophagy-inflammation-cell death axis in organismal aging. Science 333:1109–1112

Groneberg DA, Kindermann B, Althammer M, Klapper M, Vormann J, Littarru GP, Doring F (2005) Coenzyme Q10 affects expression of genes involved in cell signaling, metabolism and transport in CaCo-2 cells. Int J Biochem Cell Biol 37:1208–1218

Gudz TI, Tserng KY, Hoppel CL (1997) Direct inhibition of mitochondrial respiratory chain complex III by cell-permeable ceramide. J Biol Chem 272:24154–24158

Harris CB, Chowanadisai W, Mishchuk DO, Satre MA, Slupsky CM, Rucker RB (2013) Dietary pyrroloquinoline quinone (PQQ) alters indicators of inflammation and mitochondrial-related metabolism in human subjects. J Nutr Biochem 24:2076–2084

Head E, Nukala VN, Fenoglio KA, Muggenburg BA, Cotman CW, Sullivan PG (2009) Effects of age dietary and behavioral enrichment on brain mitochondria in a canine model of human aging. Exp Neurol 220:171–176

Holt IJ (2010) Zen and the art of mitochondrial DNA maintenance. Trends Genet 26:103–109

Houtkooper RH, Vaz FM (2008) Cardiolipin, the heart of mitochondrial mechanisms. Cell Mol Life Sci 65:2493–2506

Huang H, Manton KG (2004) The role of oxidative damage in mitochondria during aging: a review. Front Biosci 9:1100–1117

Hüttemann M, Lee I, Pecinova A, Pecina P, Przyklenk K, Doan JW (2008) Regulation of oxidative phosphorylation, the mitochondrial membrane potential, and their role in human disease. J Bioenerg Biomembr 40:445–456

Isobe C, Abe T, Terayama Y (2010) Levels of reduced and oxidized coenzyme Q10 and 8-hydroxy-2′-deoxyguanosine in the cerebrospinal fluid of patients with living Parkinson's disease demonstrates that mitochondrial oxidative damage contributes to the neurodegenerative process. Neurosci Lett 220:41–48

Jagetia GC, Rajanikant GK, Rao SK, Shrinath Baliga M (2003) Alteration in the glutathione, glutathione peroxidase, superoxide dismutase and lipid peroxidation by ascorbic acid in the skin of mice exposed to fractionated gamma radiation. Clin Chim Acta 332:111–121

Joseph A-M, Joanisse DR, Baillot RG, Hood DA (2012) Mitochondrial dysregulation in the pathogenesis of diabetes: potential for mitochondrial biogenesis-mediated interventions. Exp Diabetes Res 2012:1–16. https://doi.org/10.1155/2012/642038

Karbowski M, Neutzner A (2012) Neurodegeneration as a consequence of failed mitochondrial maintenance. Acta Neuropathol 123:157–171

Kerr DS (2010) Treatment of mitochondrial electron transport chain disorders: a review of clinical trials over the past decade. Mol Genet Metab 99:246–255

Kieburtz K, Huntington Study Group (2001) A randomized, placebo-controlled trial of coenzyme Q10 and remacemide in Huntington's disease. Neurology 57:397–404

Killgore J, Smidt C, Duich L, Romero-Chapman N, Tinker D, Reiser K, Melko M, Hyde D, Rucker RB (1989) Nutritional importance of pyrroloquinoline quinone. Science 245:850–852

Kirkland JB (2009) Niacin status, NAD distribution and ADP-ribose metabolism. Curr Pharm Des 15:3–11

Konradi C, Eaton M, MacDonald ML, Walsh J, Benes FM, Heckers S (2004) Molecular evidence for mitochondrial dysfunction in bipolar disorder. Arch Gen Psychiatry 61:300–308
Koves TR, Ussher JR, Noland RC, Slentz D, Mosedale M, Ilkayeva O, Bain J, Stevens R, Dyck JR, Newgard CB, Lopaschuk GD, Muoio DM (2008) Mitochondrial overload and incomplete fatty acid oxidation contribute to skeletal muscle insulin resistance. Cell Metab 7(1):45–56
Kroenke K, Wood DR, Mangelsdorff AD, Meier NJ, Powell JB (1988) Chronic fatigue in primary care. Prevalence, patient characteristics, and outcome. JAMA 260:929–934
Kumazawa T, Sato K, Seno H, Ishii A, Suzuki O (1995) Levels of pyrroloquinoline quinone in various foods. Biochem J 307:331–333
Lee AG (2004) How lipids affect the activities of integral membrane proteins. Biochim Biophys Acta 1666:62–87
Lee J, Giordano S, Zhang J (2012) Autophagy, mitochondria and oxidative stress: cross-talk and rexox signaling. Biochem J 441:523–540
Lee Y, Lee S-R, Choi SS, Yeo H-G, Chang K-T, Lee HJ (2014) Therapeutically targeting neuroinflammation and microglia after acute ishemic stroke. Biomed Res Int 2014:297241
Leung MCK, Rooney JP, Ryde IR, Bernal AJ, Bess AS, Crocker TL, Ji AQ, Meyer JN (2013) Effects of early life exposure to ultraviolet C radiation on mitochondrial DNA content, transcription, ATP production, and oxygen consumption in developing Caenorhabditis elegans. BMC Pharmacol Toxicol 14:9
Limongelli G, Msarone D, D'Alessandro R, Elliott PM (2012) Mitochondrial diseases and the heart: an overview of molecular basis, diagnosis, treatment and clinical course. Futur Cardiol 8:2–18
Littarru GP (2014) The effect of coenzyme Q10 on morbidity and mortality in chronic heart failure: results from Q-SYMBIO, a randomized, double-blind trial. JACC Heart Fail 2:641–649
Littarru GP, Tiano L (2010) Clinical aspects of coenzyme Q10: an update. Curr Opin Clin Nutr Metab Care 26:250–254
Logan AC, Wong C (2001) Chronic fatigue syndrome: oxidative stress and dietary modifications. Altern Med Rev 6:450–459
Ma ZA, Zhao Z, Turk J (2012) Mitochondrial dysfunction and β-cell failure in type 2 diabetes mellitus. Exp Diabetes Res 2012:1–11. https://doi.org/10.1155/2012/703538
Maiese K, Morhan SD, Chong ZZ (2007) Oxidative stress biology and cell injury during type 1 and type 2 diabetes mellitus. Curr Neurovasc Res 4:63–71
Malaguarnera M, Cammalleri L, Gargante MP, Vacante M, Colonna V, Motta M (2007) L-Carnitine treatment reduces severity of physician and mental fatigue and increases cognitive functions in centenarians: a randomized and controlled clinical trial. Am J Clin Nutr 86:1738–1744
Mancuso M, Orsucci D, Calsolaro V, Choub A, Siciliano G (2009) Coenzyme Q and neurological diseases. Pharmaceuticals (Basal) 2:134–149
Mancuso M, Orsucci D, Volpi L, Calsolaro V, Siciliano G (2010) Coenzyme Q10 in neuromuscular and neurodegenerative disorders. Curr Drug Targets 11:111–121
Manuel y Keenoy B, Moorkens G, Vertommen J, De Leeuw I (2001) Antioxidant status and lipoprotein peroxidation in chronic fatigue syndrome. Life Sci 68:2037–2049
Mao P, Reddy PH (2010) Is multiple sclerosis a mitochondrial disease? Biochim Biophys Acta 1802:66–79
Marazziti D, Baroni S, Piccheti M, Landi P, Silvestri S, Vatteroni E, Dell'Osso MC (2012) Psychiatric disorders and mitochondrial dysfunctions. Eur Rev Med Pharmacol Sci 16:270–275
Marchi S, Patergnani S, Missiroli S, Morciano G, Rimessi A, Wieckowski MR, Pinton P (2017) Mitochondrial and endoplasmic reticulum calcium homeostasis and cell death. Cell Calcium. https://doi.org/10.1016/j.ceca.2017.05.003
Marcovina SM, Sirtori C, Peracino A, Gheorghiade M, Borum P, Remuzzi G, Ardehali H (2012) Translating the basic knowledge of mitochondrial functions to metabolic therapy: role of L-carnitine. Translat Res 2012. https://doi.org/10.1016/j.trsl2012.10.006. pii: S1931-5244(12)00367-2

Marczurek A, Hager K, Kenklies M, Sharman M, Martins R, Engel J, Carlson DA, Munch G (2008) Lipoic acid as an anti-inflammatory and neuroprotective treatment for Alzheimer's disease. Adv Drug Deliv Rev 60:1463–1470

Mayer JN, Leung MCK, Rooney JP, Sendoel A et al (2013) Mitochondria as the target of environmental toxicants. Toxicol Sci 134:1–17

Mishra P, Chan DC (2016) Metabolic regulation of mitochondrial dynamics. J Cell Biol 212:379–387

Mizuno K, Tanaka M, Nozaki S, Mizuma H, Ataka S, Tahara T, Sugino T, Shirai T, Kajimoto Y, Kuratsune H, Kajimoto O, Watanabe Y (2008) Antifatigue effects of coenzyme Q10 during physical fatigue. Nutrition 24:293–299

Monette JS, Gomez LA, Moreau RF, Dunn KC, Butler JA, Finlay LA, Michels AJ, Shay KP, Smith EJ, Hagen TM (2011) (R)-alpha-Lipoic acid treatment restores ceramide balance in aging rat cardiac muscle. Pharmacol Res 63:23–29

Morris G, Maes M (2014) Mitochondrial dysfunction in myalgic encephalomyelitis/chronic fatigue syndrome explained by activated immuno-inflammatory, oxidative and nitrosative stress pathways. Metab Brain Dis 29:19–36

Morrison JD (1980) Fatigue as a presenting complaint in family practice. J Fam Pract 10:795–801

Mortensen SA, Rosenfeldt F, Kumar A, Dolliner P, Filipiak KJ, Pella D, Alehagen U, Steurer G, Littarru GP (2014) The effect of coenzyme Q10 on morbidity and mortality in chronic heart failure: results from Q-SYMBIO, a randomized, double-blind trial. JACC Heart Fail 2: 641–649.

Myhill S, Booth NE, McLaren-Howard J (2009) Chronic fatigue syndrome and mitochondrial dysfunction. Int J Clin Exp Med 2:1–16

Newgard CB, An J, Bain JR, Meuhlbauer MJ, Stevens RD, Lien LF, Haqq AM, Shah SH, Arlotto M, Slentz CA, Rochon J, Gallup D, Ilkayeva O, Wenner BR, Yancy WS Jr, Eisenson H, Musante G, Survit RS, Millington DS, Butler MD, Svetkey LP (2009) A branched-chain amino acid-related metabolic signature that differentiates obese and lean humans and contributes to insulin resistance. Cell Metab 9:311–326

Nicholls DG (2010) Mitochondrial ion circuits. Essays Biochem 47:25–35

Nicolson GL (2003) Lipid replacement as an adjunct to therapy for chronic fatigue, anti-aging and restoration of mitochondrial function. J Am Nutraceut Assoc 6(3):22–28

Nicolson GL (2007) Metabolic syndrome and mitochondrial function: molecular replacement and antioxidant supplements to prevent membrane oxidation and restore mitochondrial function. J Cell Biochem 100:1352–1369

Nicolson GL (2010) Lipid replacement therapy: a nutraceutical approach for reducing cancer-associated fatigue and the adverse effects of cancer therapy while restoring mitochondrial function. Cancer Metastasis Rev 29:543–552

Nicolson GL (2014) Mitochondrial dysfunction and chronic disease: treatment with natural supplements. Alt Ther Health Med 20(Suppl. 1):18–25

Nicolson GL (2016) Membrane lipid replacement: clinical Studies using a natural medicine approach to restoring membrane function and improving health. Intern J Clin Med 7(2):133–143

Nicolson GL, Ash ME (2014) Lipid replacement therapy: a natural medicine approach to replacing damaged phospholipids in cellular membranes and organelles and restoring function. Biochim Biophys Acta 1838:1657–1679

Nicolson GL, Ash ME (2017) Membrane lipid replacement for chronic illnesses, aging and cancer using oral glycerolphospholipid formulations with fructooligosaccharides to restore phospholipid function in cellular membranes, organelles, cells and tissues. Biochim Biophys Acta. https://doi.org/10.1016/j.bbamem.2017.04.013

Nicolson GL, Ellithorpe RR (2006) Lipid replacement and antioxidant nutritional therapy for restoring mitochondrial function and reducing fatigue in chronic fatigue syndrome and other fatiguing illnesses. J Chronic Fatigue Syndr 13(1):57–68

Nicolson GL, Nicolson L (1996) Chronic fatigue illnesses and operation desert storm. J Occup Environ Med 38:14–16

Nicolson GL, Settineri R (2011) Lipid replacement therapy: a functional food approach with new formulations for reducing cellular oxidative damage, cancer-associated fatigue and the adverse effects of cancer therapy. Funct Food Health Dis 1(4):135–160

Nicolson GL, Berns P, Nasralla M, Haier J, Nicolson NL, Nass M (2003) Gulf War illnesses: chemical, radiological and biological exposures resulting in chronic fatiguing illnesses can be identified and treated. J Chronic Fatigue Syndr 11(1):135–154

Nicolson GL, Ellithorpe RR, Ayson-Mitchell C, Jacques B, Settineri R (2010) Lipid replacement therapy with a glycophospholipid-antioxidant-vitamin formulation significantly reduces fatigue within one week. J Am Nutraceut Assoc 13(1):11–15

Nicolson GL, Settineri R, Ellithorpe E (2012a) Lipid replacement therapy with a glycophospholipid formulation with NADH and CoQ10 significantly reduces fatigue in intractable chronic fatiguing illnesses and chronic Lyme disease. Int J Clin Med 3(3):163–170

Nicolson GL, Settineri R, Ellithorpe E (2012b) Glycophospholipid formulation with NADH and CoQ10 significantly reduces intractable fatigue in Western blot-positive chronic Lyme disease patients: preliminary report. Funct Food Health Dis 2(3):35–47

Nicolson GL, Rosenblatt S, Ferreira de Mattos G, Settineri R, Breeding PC, Ellithorpe RR and Ash ME (2016) Clinical uses of Membrane Lipid Replacement supplements in restoring membrane function and reducing fatigue in chronic diseases and cancer. Discoveries 4(1):e54.

Norheim KB, Jonsson G, Omdal R (2011) Biological mechanisms of chronic fatigue. Rheumatology 50(6):1009–1018

Orsucci D, Mancuso M, Ienco EC, LoGerfo A, Siciliano G (2011) Targeting mitochondrial dysfunction and neurodegeneration by means of coenzyme Q10 and its analogues. Curr Med Chem 18:4053–4064

Pall ML (2000) Elevated, sustained peroxynitrite levels as the cause of chronic fatigue syndrome. Med Hypotheses 54:115–125

Palmieri L, Peerscio AM (2010) Mitochondrial dysfunction in autism spectrum disorders: cause or effect? Biochim Biophys Acta 1797:1130–1137

Park JH, Niermann KJ, Olsen N (2000) Evidence for metabolic abnormalities in the muscles of patients with fibromyalgia. Curr Rheumatol Rep 2(2):131–140

Penberthy WT (2009) Nicotinamide adenine dinucleotide biology and disease. Curr Pharm Des 15:1–2

Pérez MJ, Quintanilla RA (2017) Development or disease: duality of the mitochondrial permeability transition pore. Dev Biol 426:1–7

Picard M, Wallace DC, Burelle Y (2016) The rise of mitochondria in medicine. Mitochondrion 30:105–116

Pieczník SR, Neustadt J (2007) Mitochondrial dysfunction and molecular pathways of disease. Exp Mol Pathol 83:84–92

Polinsky AJ, Ebert M, Cain ED, Ludlow C, Bassich CJ (1980) Cholinergic treatment in Tourette syndrome. N Engl J Med 302:1310–1311

Potgieter M, Pretorius E, Pepper MS (2013) Primary and secondary coenzyme Q10 deficiency: the role of therapeutic supplementation. Nutr Rev 71:180–188

Prasad KN, Cole WC, Kumar B, Prasad KC (2001) Scientific rationale for using high-dose multiple micronutrients as an adjunct to standard and experimental cancer therapies. J Am Coll Nutr 20(5 Suppl 1):450S–453S

Priault M, Salin B, Schaeffer J, Vallette FM, di Rago JP, Martinou JC (2005) Impairing the bioenergetic status and the biogenesis of mitochondria triggers mitophagy in yeast. Cell Death Differ 12:1613–1621

Prince JA, Harro J, Blennow K, Gottfries CG, Putamen OL (2000) Mitochondrial energy metabolism is highly correlated to emotional and intellectual impairment in schizophrenics. Neuropsychopharmacology 22:284–292

Rabinovich RA, Vilaro J (2010) Structural and functional changes of peripheral muscles in chronic obstructive pulmonary disease patients. Curr Opin Pulm Med 16(2):123–133

Rainer M, Kraxberger E, Haushofer M, Mucke HA, Jellinger KA (2000) No evidence for cognitive improvement from oral nicotiamide adenine dinucleotide (NADH) in dementia. J Neural Transm 107:1475–1481

Reczek CR, Chandel NS (2015) ROS-dependent signal transduction. Curr Opin Cell Biol 33:8–13

Reddy PH (2008) Mitochondrial medicine for aging and neurodegenerative diseases. Neuromolecular Med 10(4):291–315

Reddy PH, Reddy TP (2011) Mitochondria as a therapeutic target for aging and neurodegenerative diseases. Curr Alzheimer Res 8:393–409

Reuter SE, Evans AM (2012) Carnitine and acylcarnitines: pharmacokinetic, pharmacological and clinical aspects. Clin Pharmacokinet 51:553–572

Rich PR, Marechal A (2010) The mitochondrial respiratory chain. Essays Biochem 47:1–23

Richards RS, Roberts TK, McGregor NR, Dunstan RH, Butt HL (2000) Blood parameters indicative of oxidative stress are associated with symptom expression in chronic fatigue syndrome. Redox Rep 5:35–41

Richter C, Par JW, Ames B (1998) Normal oxidative damage to mitochondrial and nuclear DNA is extensive. Proc Natl Acad Sci U S A 85:6465–6467

Rosenfeldt F, Hilton D, Pepe S, Krum H (2003) Systematic review of effect of coenzyme Q10 in physical exercise, hypertension and heart failure. Biofactors 18:91–100

Rossignol DA, Frye RE (2012) Mitochondrial dysfunction in autism spectrum disorders: a systematic review and meta-analysis. Mol Psychiatry 17:290–314

Rucker R, Chowanadisai W, Nakano M (2009) Potential physiological importance of pyrroloquinoline quinone. Altern Med Rev 14(3):179–183

Santaella ML, Font I, Disdier OM (2004) Comparison of oral nicotinamide adenine dinucleotide (NADH) versus conventional therapy for chronic fatigue syndrome. P R Health Sci J 23(2):89–93

Satrustegui J, Pardo B, Del Arco A (2007) Mitochondrial transporters as novel targets for intracellular calcium signaling. Physiol Rev 87:29–67

Schwartz JL (1996) The dual roles of nutrients as antioxidants and prooxidants: their effects on tumor cell growth. J Nutr 126:1221S–1227S

Shah SH, Hauser ER, Bain JR, Muehlbauer MJ, Haynes C, Stevens RD, Wenner BR, Dowdy ZE, Granger CB, Ginsburg GS, Newgard CB, Kraus WE (2009) High heritability of metabolomic profiles in families burdened with premature cardiovascular disease. Mol Syst Biol 5:258. https://doi.org/10.1038/msb.2009.11

Shay KP, Moreau RF, Smith EJ, Smith AR, Hagen TM (2009) Alpha-lipoic acid as a dietary supplement: molecular mechanisms and therapeutic potential. Biochim Biophys Acta 1790:1149–1160

Smeitink JA, Zeviani M, Turnbull DM, Jacobs HT (2006) Mitochondrial medicine: a metabolic perspective on the pathology of oxidative phosphorylation disorders. Cell Metabol 3:9–13

Smith AR, Shenvi SV, Widlansky M, Suh JH, Hagen TM (2004) Lipoic acid as a potential therapy for chronic diseases associated with oxidative stress. Curr Med Chem 11:1135–1146

Sotgia F, Martinez-Outschoorn UE, Lisanti MP (2011) Mitochondrial oxidative stress drives tumor progression and metastasis: should we use antioxidants as a key component of cancer treatment and prevention? BMC Med 9:62–67

Spector AA, Yorek MA (1985) Membrane lipid composition and cellular function. J Lipid Res 26:101–105

Spiteller G (2010) Is lipid peroxidation of polyunsaturated acids the only source of free radicals that induce aging and age-related diseases? Rejuvenation Res 13:91–103

Spriet LL, Perrry CGR, Talanian JL (2008) Legal pre-event nutritional supplements to assist energy metabolism. Essays Biochem 44:27–43

Stadtman E (2002) Introduction to serial reviews on oxidatively modified proteins in aging and disease. Free Radic Biol Med 32:789

Steinberg FM, Gershwin ME, Rucker RB (1994) Dietary pyrroloquinoline quinone: growth and immune response in BALB/c mice. J Nutr 124:744–753

Stites T, Storms D, Bauerly K, Mah J, Harris C, Fascetti A, Rogers Q, Tchaparian E, Satre M, Rucker RB (2006) Pyrroloquinoline quinone modulates mitochondrial quantity and function in mice. J Nutr 136:390–396

Stork C, Renshaw PF (2005) Mitochondrial dysfunction in bipolar disorder: evidence from magnetic resonance spectroscopy research. Mol Psychiatry 10:900–919

Sun Y (1990) Free radicals, antioxidant enzymes and carcinogenesis. Free Radic Biol Med 8:583–599

Szabo I, Adams C, Gulbins E (2004) Ion channels and membrane rafts in apoptosis. Pflugers Arch 448:304–312

Tao R, Karliner JS, Simonis U, Zheng J, Zhang J, Honbo N, Alano CC (2007) Pyrroloquinoline quinone preserves mitochondrial function and prevents oxidative injury in adult rat cardia myocytes. Biochem Biophys Res Commun 363:257–262

Thompson KP, Levy G, Buchsbaum R, Shefner J, Krivickas LS, Katz J, Rollins Y, Barohn RJ, Jackson CE, Tiryaki E, Lomen-Hoerth C, Armon C, Tandan R, Rudnicki SA, Rezania K, Sufit R, Pestronk A, Novella SP, Heiman-Patterson T, Kasarskis EJ, Pioro EP, Montes J, Arbing R, Vecchio D, Barsdorf A, Mitsumoto H, Levin B (2009) Phase II trial of CoQ10 for ALS finds insufficient evidence to justify phase III. Ann Neurol 66:235–244

Twig G, Shirihi OS (2011) The interplay between mitochondrial dynamics and mitophagy. Antioxid Redox Signal 14:1939–1951

Victor VM, Apostolova N, Herance R, Hernandez-Mijares A, Rocha M (2009) Oxidative stress and mitochondrial dysfunction in atherosclerosis: mitochondria-targeted antioxidants as potential therapy. Curr Med Chem 16:4654–4667

Wachter S, Vogt M, Kreis R, Boesch C, Bigler P, Hoppeler H, Krahenbuhl S (2002) Long term administration of L-carnitine to huans: effects on skeletal mscle carnitine content and physical performance. Clin Chim Acta 318:51–61

Wallace DC (2005) A mitochondrial paradigm of metabolic and degenerative diseases, aging, and cancer: a dawn for evolutionary medicine. Annu Rev Genet 39:359–407

Wallace DC, Fan W (2010) Energetics, epigenetics, mitochondrial genetics. Mitochondrion 10:12–31

Wei YH, Lee HC (2002) Oxidative stress, mitochondrial DNA mutation and impairment of antioxidant enzymes in aging. Exp Biol Med 227(9):671–682

Xiao M, Zhong H, Xia L, Tao Y, Yin H (2017) Pathophysiology of mitochondrial lipid oxidation: role of 4-hydroxynonenal (4-HNE) and other bioactive lipids in mitochondria. Free Radic Biol Med 111:316–327

Yang X, Dai G, Li G, Yang ES (2010) Coenzyme Q10 reduces beta-amyloid plaque in an APP/PSI transgenic mouse model of Alzheimer's disease. J Mol Neurosci 41:110–113

Yang C, Yu L, Kong L, Ma R, Zhang J, Zhu Q, Zhu J, Hao D (2014) Pyrroloquinoline quinone (PQQ) inhibits lipopolysaccharide-induced inflammation in part via down-regulation of NF-κB and p38/JNK activation in microglial and attenuates microglia activation in lipopolysaccharide tratmet mice. PLoS One 9(10):e109502

Yoshida K, Hirokawa J, Tagami S, Kawakami Y, Urata Y, Kondo T (1995) Weakened cellular scavenging activity against oxidative stress in diabetes mellitus: regulation of glutathione synthesis and efflux. Diabetologia 38(2):201–210

Zhang L, Liu J, Cheng C, Yuan Y, Yu B, Shen A, Yan M (2012) The neuroprotective effect of pyrroloquinoline quinone on traumatic brain injury. J Neurotrauma 29:851–864

Ziegler D, Nowak H, Kempler P, Vargha P, Low PA (2004) Treatment of symptomatic diabetic polyneuropathy with the antioxidant alpha-lipoic acid: a meta-analysis. Diabet Med 21:114–121

Ziegler D, Low PA, Litchy WJ, Boulton AJM, Vink AI, Freeman R, Samigullin R, Tritschler H, Munzel U, Maus J, Schute K, Dyck PJ (2011) Efficacy and safety of antioxidant treatment with alpha-Lipoic acid over 4 years in diabetic polyneuropathy. Diabetes Care 34:2054–2060

Studies on Mitochondria Directed Plastoquinones

Boris A. Feniouk and Vladimir P. Skulachev

Abstract Mitochondria-targeted cationic plastoquinone derivatives (SkQs, e.g. SkQ1, SkQR1) and their analogs lacking plastoquinol moiety (C_{12}TPP, C_{12}R1) can pass through bilayer phospholipid membrane. Their cationic forms are accumulated in isolated mitochondria or in mitochondria of living cells, driven by membrane potential. These compounds were extensively tested in model lipid membranes, isolated mitochondria and in living human cells in culture. It was found that SkQs are antioxidants that quench reactive oxygen species (ROS) in mitochondria, and mild uncouplers that facilitate transmembrane proton transport by fatty acids. Both properties result in efficient prevention of oxidative stress and protection of mitochondria and cells from damage by ROS, making SkQs a promising drug candidate against pathologies caused by excess mitochondrial ROS generation. Recent discovery of SkQ1 antibacterial activity at concentrations not toxic to human cells opens a perspective for development of new antibiotics. In this chapter, we summarize recent in vitro experiments with mitochondria-targeted plastoquinones.

Keywords Mitochondria-targeted cationic plastoquinone derivatives · SkQ1 · Mitochondria · Oxidative stress · Model lipid membranes

Abbreviations

MitoQ	10-(6′-Ubiquinonyl) decyltriphenylphosphonium
MPC	membranophilic penetrating cation
MPI	membranophilic penetrating ion
PQ	Plastoquinone

B. A. Feniouk · V. P. Skulachev (✉)
Faculty of Bioengineering and Bioinformatics, Lomonosov Moscow State University, Moscow, Russia

A.N. Belozersky Institute of Physico-Chemical Biology, Lomonosov Moscow State University, Moscow, Russia
e-mail: skulach@genebee.msu.ru

P. J. Oliveira (ed.), *Mitochondrial Biology and Experimental Therapeutics*,
https://doi.org/10.1007/978-3-319-73344-9_23

ROS	Reactive oxygen species
SkQ	Cationic derivative of plastoquinone or methyl plastoquinone
SkQ1	10-(6′-Plastoquinonyl) decyltriphenylphosphonium
SkQ2M	10-(6′-Plastoquinonyl) decylmethylcarnitine
SkQ3	10-(6′-Methylplastoquinonyl) decyltriphenylphosphonium
SkQ4	10-(6′-Plastoquinonyl) decyltributylammonium
SkQ5	5-(6′-Plastoquinonyl) amyltriphenylphosphonium
SkQR1	10(Plastoquinonyl) decylrhodamine 19
TPB	Tetraphenylborate
TPP	Tetraphenylphosphonium
$\Delta\psi$	Transmembrane electric potential difference

1 Introduction

For a long time, mitochondria were regarded merely as "power plants" of eukaryotic cells that convert the energy derived from respiration into ATP. However, in the last decades it was found that these organelles play a key role in several other biological processes, including autophagy, programmed cell death, inflammation, reactive oxygen species (ROS) signaling, and aging (Feniouk and Skulachev 2017; Ott et al. 2007; Green et al. 2011; Lopez-Armada et al. 2013; Kasahara and Scorrano 2014; Li et al. 2013). If produced in excess amounts, ROS generated by mitochondria (mROS) can cause oxidative stress that results in lipids oxidation, can damage proteins and DNA and trigger cell death.

The idea to use membranophilic penetrating ions (MPIs) as "locomotives" that can drag payload, e.g. antioxidants, into mitochondria was first introduced by our group in 1970 (Severin et al. 1970). Unlike ions with charge localized in a relatively small volume (e.g. Na^+, H_3O^+, Cl^-, ATP^{4-}), MPIs have the charge delocalized over relatively large hydrophobic groups (Fig. 1) and can easily pass through lipid bilayers.

If electrical potential difference ($\Delta\psi$) is present between two aqueous compartments separated by a lipid membrane, MPIs will distribute according to the Nernst equation. This means that at 37 °C the concentration of a positively charged MPI in the negatively charged compartment will exceed its concentration in the positively charged one by approximately one order of magnitude for 60 mV $\Delta\psi$ (Fig. 2).

MPIs quickly became a universal invaluable tool for mitochondrial research. In fact, they could be used to determine the direction of electron and proton transport and the magnitude of $\Delta\psi$ generated by the respiratory enzymes and H^+-ATP synthase. In isolated mitochondria at $\Delta\psi$ of ~180 mV (interior negative) that is present under normal physiological conditions cationic MPIs accumulate in mitochondria in 1000-fold excess. When applied to whole cells, the relative concentration of membranophilic penetrating cations (MPCs) increases even further because of the $\Delta\psi$ at the cellular outer membrane (about 60 mV, negative inside).

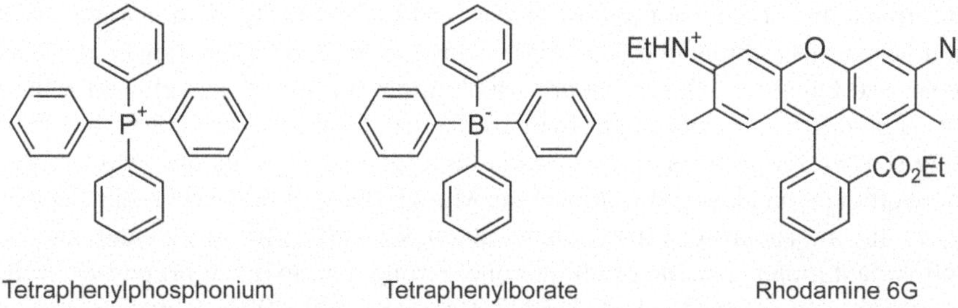

Fig. 1 Examples of membranophilic penetrating ions

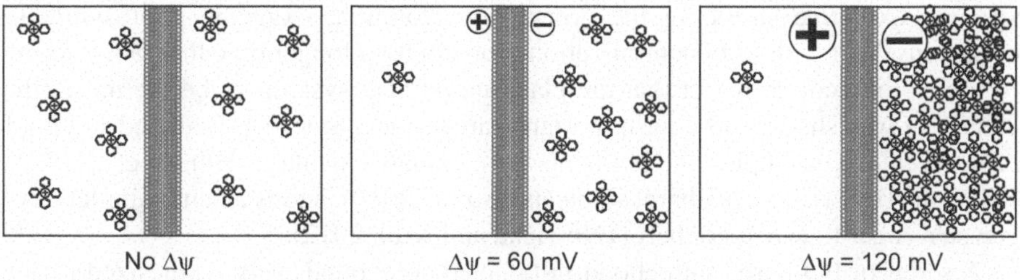

Fig. 2 Distribution of a positively charged MPI (e.g. tetraphenylphosphonium) driven by $\Delta\psi$ between two aqueous compartments separated by a lipid membrane

The first compounds electrophoretically targeted to mitochondria were designed by Murphy and co-workers. MPCs were conjugated to antioxidants, such as thiobutyl (Burns et al. 1995), vitamin E (Smith et al. 1999), and ubiquinone (Kelso et al. 2001), yielding a tool to prevent excessive mROS production. As explained above, they could be applied at very low concentrations, and therefore adverse side-effects are negligible. Noteworthy, higher $\Delta\psi$ increases MPCs accumulation in mitochondria, and the most dangerous "over-energized" organelles that produce excess mROS (Korshunov et al. 1997; Votyakova and Reynolds 2001; Miwa and Brand 2003) receive the highest dose of the antioxidant. Moreover, MPCs can act as mild uncouplers, mediating transmembrane H^+ transport via fatty acid cycling (Severin et al. 2010) and lowering excessively high $\Delta\psi$, thereby also decreasing the mROS production that was found to sharply increase if $\Delta\psi$ exceeds a certain threshold (Korshunov et al. 1997; Votyakova and Reynolds 2001; Miwa and Brand 2003; Skulachev et al. 2010).

The most promising among the first mitochondria-targeted compounds proved to be decyltriphenylphosphonium conjugated with ubiquinone (MitoQ) (Kelso et al. 2001). It was a rechargeable antioxidant: after being oxidized, it could be effectively converted back to the reduced (i.e. active) form by the mitochondrial respiratory chain.

Subsequent experiments demonstrated that MitoQ protects mitochondrial lipids from oxidation by OH radical (·OH·) in vitro, increases the survival of ROS-hypersensitive cells much more efficiently than CoQ or α-tocopherol, and prolongs lifespan of cultivated fibroblasts at high O_2 concentration (Kelso et al. 2001, 2002; Murphy and Smith 2007; James et al. 2005; Saretzki et al. 2003; Jauslin et al. 2003).

Unfortunately, at concentrations above ~500 nM MitoQ demonstrates pro-oxidant properties (Antonenko et al. 2008; James et al. 2005; O'malley et al. 2006; Doughan and Dikalov 2007), a feature which might limit its practical use for therapy of pathologies associated with oxidative stress (Skulachev et al. 2011). This is why we started a search for mitochondria-targeted rechargeable antioxidants that are more efficient and less pro-oxidant than MitoQ. Plastoquinone (PQ), an electron carrier in the chloroplast electron transfer chain, seemed a promising candidate for the antioxidant moiety instead of ubiquinone. Stronger antioxidant properties of the former are not surprising (Kruk et al. 1997; Roginsky et al. 2003). Chloroplasts that generate oxygen, in contrast to oxygen-consuming mitochondria, produce more ROS (Gill and Tuteja 2010; Foyer and Noctor 2003) and are exposed to a much stronger oxidative stress. In the process of evolution, oxygenic photosynthetic organisms dealt with ROS neutralization problem for a longer time than other forms of life, so one could expect that the plant antioxidant systems to be the most efficient. PQ was shown to be an important part of these systems. Reduced PQ pool protects chloroplast lipids and pigments from oxidative damage (Hundal et al. 1995; Kruk and Trebst 2008; Mubarakshina and Ivanov 2010), and is a better singlet oxygen scavenger than α-tocopherol (Nowicka and Kruk 2012).

A series of PQ-based mitochondria-targeted compounds was synthesized in our group (Fig. 3). These compounds were named SkQs, where Sk stands for the membranophilic penetrating cation ("Skulachev ion", a term introduced by Green (1974)), and Q stands for quinone. The physico-chemical and biological properties of these new compounds were studied in the framework of the «SkQ-Project»—an effort to develop drugs that might prevent or decrease the mROS-induced oxidative damage and thereby alleviate related pathologies and slow aging.

Our compounds differed by the nature of the penetrating cation, i.e. triphenylphosphonium (SkQ1, SkQ3, SkQ5), Rhodamine 19 (SkQR1), Rhodamine B (SkQR4), or natural cations methylcarnitine (SkQ2M), berberine (SkQB), and palmatine (SkQP). Demethylplastoquinone derivatives (SkQT1) were also found to be very promising mitochondria-targeted drugs of the quinone series (Severina et al. 2013). A saturated hydrocarbon linker between the quinone and cation moieties usually contained ten C-atoms, but only five in SkQ5. Plastoquinonyl decyltributylammonium (SkQ4) was used as a non-penetrating analog of SkQ1, and C_{12}TPP and C_{12}R1 were used as compounds containing a penetrating cation and a linker, but no PQ. Details on the chemical synthesis for these compounds can be found in Korshunova et al. (2017).

It was experimentally confirmed that PQ-MPC conjugates penetrate planar phospholipid membranes, are electrophoretically accumulated in mitochondria, but are almost undetectable in other cellular compartments, and can be reduced by the respiratory chain (Antonenko et al. 2008). Subsequent studies indicating that PQ-MPC conjugates are efficient in suppressing or preventing the mROS-dependent oxidative stress, inflammatory response and age-related pathological events are already covered in several reviews (Feniouk and Skulachev 2017, 2018; Skulachev

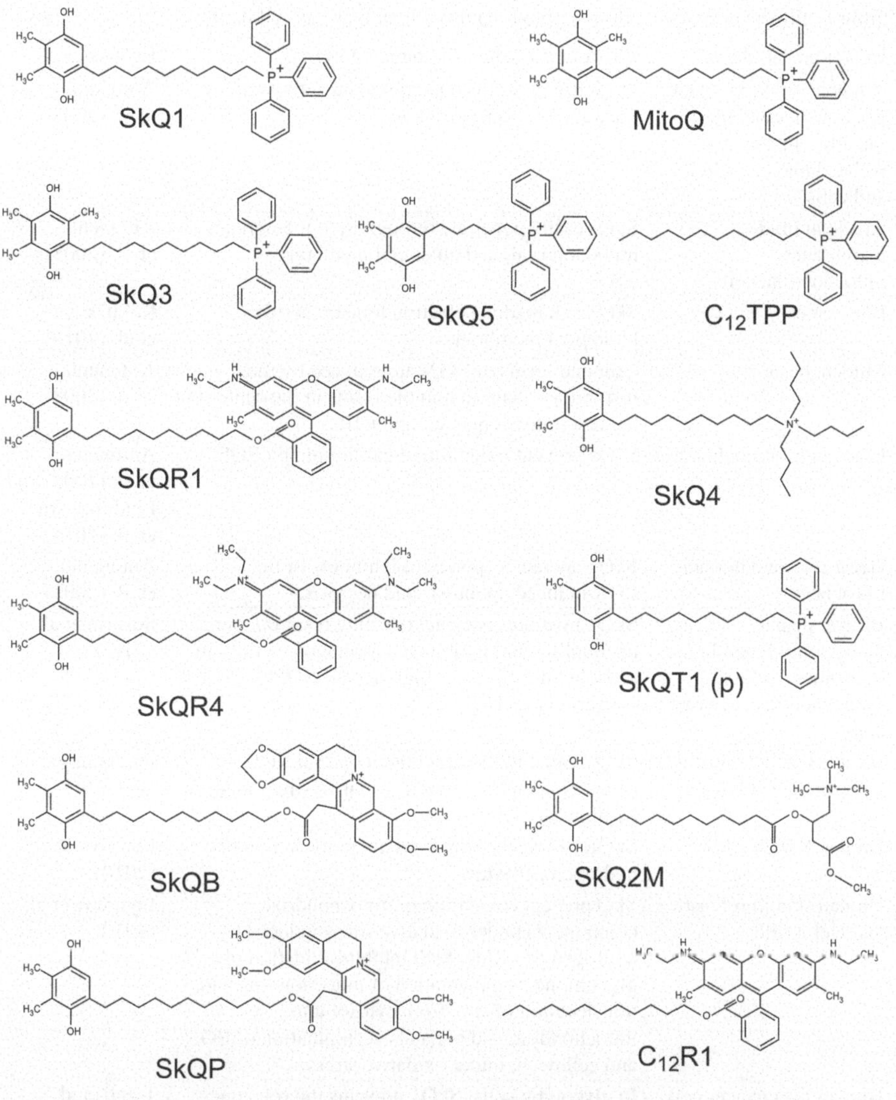

Fig. 3 Formulae of mitochondria-targeted compounds used in our studies

et al. 2011; Lukashev et al. 2014). In the Table 1 below we provide summary of recent advances in this field, with special accent on experimental works done on cellular and sub-cellular level.

A quite new promising direction of research for PQ-based mitochondria-targeted compounds is design of antibacterial drugs. Recently it was shown that SkQ1 is toxic to both Gram-positive and Gram-negative bacteria in sub-micromolar concentrations that were not dangerous for eukaryotic cells (Nazarov et al. 2017).

Table 1 Effects of mitochondria-targeted PQ on cellular and sub-cellular level

Experimental object	Effect of mitochondria-targeted PQ	Reference
Aqueous solution, black lipid membrane, lipid micelles, liposomes, isolated mitochondria, and cells	SkQs has pronounced antioxidant activity at low (nanomolar) concentrations	Antonenko et al. (2008)
Artificial lipid membranes, mitochondria, cells	SkQs easily penetrate through lipid membranes, mitochondrial, and outer cell membranes	Antonenko et al. (2008)
Liposomes	SkQ1 can mediate electron transfer across biological membranes	Rokitskaya et al. (2016)
Mitochondria	Quinone forms of SkQs are reduced by the respiratory chain to quinole forms in the center i of the respiratory chain complex III	Antonenko et al. (2008)
Liposomes, mitochondria	SkQs prevent oxidation of cardiolipin by ROS	Antonenko et al. (2008) and Lokhmatikov et al. (2014)
HeLa cells and human fibroblasts	SkQs operate as powerful inhibitors of the ROS-induced apoptosis and necrosis.	Antonenko et al. (2008)
Bilayer planar phospholipid membrane, liposomes, isolated mitochondria, and yeast cells	SkQ1 mediates fatty acid cycling resulting in mild uncoupling that inhibits the formation of ROS in mitochondrial state 4 (high $\Delta\psi$; all ADP is converted to ATP)	Severin et al. (2010)
Murine C2C12 muscle cells, mice on high-fat diet	SkQ treatment reduces mitochondrial ROS production and prevented oxidative stress induced by high-fat diet in mice	Paglialunga et al. (2012)
Human K562 cells	SkQR1 protects cells from apoptotic action of hydrogen peroxide	Fetisova et al. (2010)
Human skin fibroblasts and HeLa cells	SkQ prevents oxidation of mitochondrial components under oxidative stress induced by hydrogen peroxide. SkQ inhibits oxidation of glutathione, fragmentation of mitochondria, and translocation of Bax from cytosol into mitochondria, and prevents accumulation of ROS and cell death under oxidative stress	Izyumov et al. (2010)
Human liver cancer cell line Hep G2	In glycolytic cells, SkQ1 prevents the rotenone-induced ROS generation, but not basal generation without rotenone	Jezek et al. (2017)
Human neutrophils	Scavenging of mROS with SkQ1 slightly accelerates spontaneous apoptosis and significantly stimulates apoptosis of fMLP-activated neutrophils	Vorobjeva et al. (2017)
Cervical cancer SiHa and Ca-Ski cells	SkQ1 induces the formation of β-cytoplasmic actin stress fibers and circumferential rings, initiates reversal of epithelial-to-mesenchymal transition and suppresses proliferation	Shagieva et al. (2017)

(continued)

Table 1 (continued)

Experimental object	Effect of mitochondria-targeted PQ	Reference
Human endothelial cell line EAhy926	As little as 0.2 nM SkQR1 prevents TNF-induced apoptosis, while SkQ analogs lacking the antioxidant moiety have no such effect, suggesting that mitochondria-targeted antioxidants might be promising vasoprotector drug candidates	Galkin et al. (2014)
Endothelial cells	SkQR1 inhibited TNF-induces increase in endothelial permeability, as well as caspase-3 activation, β-catenin cleavage, and MMP9-dependent VE-cadherin shedding	Galkin et al. (2016)
K562 erythroleukemia cells	SkQR1 protects against nuclear DNA damage induced by gamma radiation and prevents the early accumulation of phosphorylated histone H2AX (γ-H2AX) an indicator of DNA double-strand break formation, as well as the radiation-induced increase in chromosomal aberrations	Fetisova et al. (2015)
Human rhabdomyosarcoma cell culture	SkQ1, SkQR1 and SkQT1 strongly suppress the growth of human rhabdomyosarcoma cells	Severina et al. (2013)
Human colon carcinoma cell line RKO and in diploid human dermal fibroblasts	SkQ1 and SkQR1 decreases the oxidative stress induced by hydrogen peroxide or by serum deprivation	Galimov et al. (2014)
Human erythrocytes	SkQ1 protects erythrocytes from oxidative hemolysis induced by a lipophilic free radical initiator 2,2′-azobis(2,4-dimethylvaleronitrile) (AMVN) and a water-soluble free radical initiator 2,2′-azobis(2-methylpropionamidine) dihydrochloride (AAPH)	Omarova and Antonenko (2014)
Planar and liposomal lipid membranes	SkQBerb, SkQPalm, and their analogs lacking the plastoquinone moiety (C10Berb and C10Palm) facilitate the protonophoretic effect of free fatty acids	Pustovidko et al. (2013)
Planar and liposomal lipid membranes, isolated mitochondria, yeast cells	$C_{12}R1$ and SkQR1 demonstrates protonophorous activity in artificial membranes and decreases $\Delta\psi$ and stimulates respiration of rat liver mitochondria in State 4. Also, $C_{12}R1$ partially stimulates respiration of yeast cells without suppression of growth	Antonenko et al. (2011)
Yeast cells (*D. magnusii* and *Y. lipolytica*)	SkQ1 prevents oxidative stress and related mitochondrial fragmentation and yeast cell death	Rogov et al. (2017)
Bacteria	SkQ1 has strong antibacterial activity towards Gram-positive *Bacillus subtilis*, *Mycobacterium* sp. and *Staphylococcus aureus* and Gram-negative *Photobacterium phosphoreum* and *Rhodobacter sphaeroides* in submicromolar and micromolar concentrations. *Escherichia coli* mutants lacking AcrAB-TolC have similar SkQ1 sensitivity, as *B. subtilis*, while the wild-type *E. coli* is much less sensitive	Nazarov et al. (2017)

2 Conclusions

PQ-based mitochondria-targeted compounds are promising drug candidates against a broad spectrum of pathologies related to oxidative stress and mitochondrial dysfunction, including chronic inflammation, SIRS, ischemia-reperfusion injury, and neurodegenerative diseases. One of them is already being implemented as a drug (eye drops); the results of the clinical trials (described in detail in Feniouk and Skulachev (2018)) indicated it can be safely administered to the patients and is efficient against dry eye syndrome and cataract.

Besides antioxidant and anti-inflammatory applications, these compounds might be used in future as bactericidal agents.

Acknowledgements This work was supported by the Russian Science Foundation (Project No. 14-50-00029 (B.A.F.) and Project No. 14–24-00107 (V.P.S.)).

Conflict of Interest V.P.S. is a board member of Mitotech LLC, a biotech company which owns rights for compounds of SkQ type.

References

Antonenko YN, Avetisyan AV, Bakeeva LE, Chernyak BV, Chertkov VA, Domnina LV, Ivanova OY, Izyumov DS, Khailova LS, Klishin SS, Korshunova GA, Lyamzaev KG, Muntyan MS, Nepryakhina OK, Pashkovskaya AA, Pletjushkina OY, Pustovidko AV, Roginsky VA, Rokitskaya TI, Ruuge EK, Saprunova VB, Severina II, Simonyan RA, Skulachev IV, Skulachev MV, Sumbatyan NV, Sviryaeva IV, Tashlitsky VN, Vassiliev JM, Vyssokikh MY, Yaguzhinsky LS, Zamyatnin AA Jr, Skulachev VP (2008) Mitochondria-targeted plastoquinone derivatives as tools to interrupt execution of the aging program. 1. Cationic plastoquinone derivatives: synthesis and in vitro studies. Biochemistry (Mosc) 73:1273–1287

Antonenko YN, Avetisyan AV, Cherepanov DA, Knorre DA, Korshunova GA, Markova OV, Ojovan SM, Perevoshchikova IV, Pustovidko AV, Rokitskaya TI, Severina II, Simonyan RA, Smirnova EA, Sobko AA, Sumbatyan NV, Severin FF, Skulachev VP (2011) Derivatives of rhodamine 19 as mild mitochondria-targeted cationic uncouplers. J Biol Chem 286:17831–17840

Burns RJ, Smith RA, Murphy MP (1995) Synthesis and characterization of thiobutyltriphenylphosphonium bromide, a novel thiol reagent targeted to the mitochondrial matrix. Arch Biochem Biophys 322:60–68

Doughan AK, Dikalov SI (2007) Mitochondrial redox cycling of mitoquinone leads to superoxide production and cellular apoptosis. Antioxid Redox Signal 9:1825–1836

Feniouk BA, Skulachev VP (2017) Cellular and molecular mechanisms of action of mitochondria-targeted antioxidants. Curr Aging Sci 10:41–48

Feniouk BA, Skulachev VP (2018) Studies on mitochondria directed plastoquinones. In: Oliveira PJ (ed) Mitochondrial biology and experimental therapeutics. Springer, New York

Fetisova EK, Avetisian AV, Iziumov DS, Korotetskaia MV, Tashlitskii VN, Skulachev VP, Cherniak BV (2010) Multidrug resistance p-glycoprotein inhibits the antiapoptotic action of mitochondria-targeted antioxidant SkQR1. Tsitologiia 52:1031–1040

Fetisova EK, Antoschina MM, Cherepanynets VD, Izumov DS, Kireev II, Kireev RI, Lyamzaev KG, Riabchenko NI, Chernyak BV, Skulachev VP (2015) Radioprotective effects of mitochondria-targeted antioxidant SkQR1. Radiat Res 183:64–71

Foyer CH, Noctor G (2003) Redox sensing and signalling associated with reactive oxygen in chloroplasts, peroxisomes and mitochondria. Physiol Plant 119:355–364

Galimov ER, Chernyak BV, Sidorenko AS, Tereshkova AV, Chumakov PM (2014) Prooxidant properties of p66shc are mediated by mitochondria in human cells. PLoS One 9:e86521

Galkin II, Pletjushkina OY, Zinovkin RA, Zakharova VV, Birjukov IS, Chernyak BV, Popova EN (2014) Mitochondria-targeted antioxidants prevent TNFalpha-induced endothelial cell damage. Biochemistry (Mosc) 79:124–130

Galkin II, Pletjushkina OY, Zinovkin RA, Zakharova VV, Chernyak BV, Popova EN (2016) Mitochondria-targeted antioxidant SkQR1 reduces TNF-induced endothelial permeability in vitro. Biochemistry (Mosc) 81:1188–1197

Gill SS, Tuteja N (2010) Reactive oxygen species and antioxidant machinery in abiotic stress tolerance in crop plants. Plant Physiol Biochem 48:909–930

Green DE (1974) The electromechanochemical model for energy coupling in mitochondria. Biochim Biophys Acta 346:27–78

Green DR, Galluzzi L, Kroemer G (2011) Mitochondria and the autophagy-inflammation-cell death axis in organismal aging. Science 333:1109–1112

Hundal T, Forsmark-Andree P, Ernster L, Andersson B (1995) Antioxidant activity of reduced plastoquinone in chloroplast thylakoid membranes. Arch Biochem Biophys 324:117–122

Izyumov DS, Domnina LV, Nepryakhina OK, Avetisyan AV, Golyshev SA, Ivanova OY, Korotetskaya MV, Lyamzaev KG, Pletjushkina OY, Popova EN, Chernyak BV (2010) Mitochondria as source of reactive oxygen species under oxidative stress. Study with novel mitochondria-targeted antioxidants—the "Skulachev-ion" derivatives. Biochemistry (Mosc) 75:123–129

James AM, Cocheme HM, Smith RA, Murphy MP (2005) Interactions of mitochondria-targeted and untargeted ubiquinones with the mitochondrial respiratory chain and reactive oxygen species. Implications for the use of exogenous ubiquinones as therapies and experimental tools. J Biol Chem 280:21295–21312

Jauslin ML, Meier T, Smith RAJ, Murphy MP (2003) Mitochondria-targeted antioxidants protect Friedreich Ataxia fibroblasts from endogenous oxidative stress more effectively than untargeted antioxidants. FASEB J 17:1972–1974

Jezek J, Engstova H, Jezek P (2017) Antioxidant mechanism of mitochondria-targeted plastoquinone SkQ1 is suppressed in aglycemic HepG2 cells dependent on oxidative phosphorylation. Biochim Biophys Acta 1858:750–762

Kasahara A, Scorrano L (2014) Mitochondria: from cell death executioners to regulators of cell differentiation. Trends Cell Biol 24:761–770

Kelso GF, Porteous CM, Coulter CV, Hughes G, Porteous WK, Ledgerwood EC, Smith RA, Murphy MP (2001) Selective targeting of a redox-active ubiquinone to mitochondria within cells: antioxidant and antiapoptotic properties. J Biol Chem 276:4588–4596

Kelso GF, Porteous CM, Hughes G, Ledgerwood EC, Gane AM, Smith RAJ, Murphy MP (2002) Prevention of mitochondrial oxidative damage using targeted antioxidants. Ann N Y Acad Sci 959:263–274

Korshunov SS, Skulachev VP, Starkov AA (1997) High protonic potential actuates a mechanism of production of reactive oxygen species in mitochondria. FEBS Lett 416:15–18

Korshunova GA, Shishkina AV, Skulachev MV (2017) Design, synthesis and some aspects of biological activity of mitochondria-targeted antioxidants. Biochemistry (Mosc) 82(7):760–777

Kruk J, Trebst A (2008) Plastoquinol as a singlet oxygen scavenger in photosystem II. BBA-Bioenergetics 1777:154–162

Kruk J, Jemiolarzeminska M, Strzalka K (1997) Plastoquinol and alpha-tocopherol quinol are more active than ubiquinol and alpha-tocopherol in inhibition of lipid peroxidation. Chem Phys Lipids 87:73–80

Li L, Chen YQ, Gibson SB (2013) Starvation-induced autophagy is regulated by mitochondrial reactive oxygen species leading to AMPK activation. Cell Signal 25:50–65

Lokhmatikov AV, Voskoboynikova NE, Cherepanov DA, Sumbatyan NV, Korshunova GA, Skulachev MV, Steinhoff HJ, Skulachev VP, Mulkidjanian AY (2014) Prevention of peroxidation of cardiolipin liposomes by quinol-based antioxidants. Biochemistry (Mosc) 79:1081–1100

Lopez-Armada MJ, Riveiro-Naveira RR, Vaamonde-Garcia C, Valcarcel-Ares MN (2013) Mitochondrial dysfunction and the inflammatory response. Mitochondrion 13:106–118
Lukashev AN, Skulachev MV, Ostapenko V, Savchenko AY, Pavshintsev VV, Skulachev VP (2014) Advances in development of rechargeable mitochondrial antioxidants. Prog Mol Biol Transl Sci 127:251–265
Miwa S, Brand MD (2003) Mitochondrial matrix reactive oxygen species production is very sensitive to mild uncoupling. Biochem Soc Trans 31:1300–1301
Mubarakshina MM, Ivanov BN (2010) The production and scavenging of reactive oxygen species in the plastoquinone pool of chloroplast thylakoid membranes. Physiol Plant 140:103–110
Murphy MP, Smith RAJ (2007) Targeting antioxidants to mitochondria by conjugation to lipophilic cations. Annu Rev Pharmacol Toxicol 47:629–656
Nazarov PA, Osterman IA, Tokarchuk AV, Karakozova MV, Korshunova GA, Lyamzaev KG, Skulachev MV, Kotova EA, Skulachev VP, Antonenko YN (2017) Mitochondria-targeted antioxidants as highly effective antibiotics. Sci Rep 7:1394
Nowicka B, Kruk J (2012) Plastoquinol is more active than alpha-tocopherol in singlet oxygen scavenging during high light stress of Chlamydomonas reinhardtii. BBA-Bioenergetics 1817:389–394
O'malley Y, Fink BD, Ross NC, Prisinzano TE, Sivitz WI (2006) Reactive oxygen and targeted antioxidant administration in endothelial cell mitochondria. J Biol Chem 281:39766–39775
Omarova EO, Antonenko YN (2014) Inhibition of oxidative hemolysis in erythrocytes by mitochondria-targeted antioxidants of SkQ series. Biochemistry (Mosc) 79:139–145
Ott M, Gogvadze V, Orrenius S, Zhivotovsky B (2007) Mitochondria, oxidative stress and cell death. Apoptosis 12:913–922
Paglialunga S, van Bree B, Bosma M, Valdecantos MP, Amengual-Cladera E, Jorgensen JA, van Beurden D, den Hartog GJM, Ouwens DM, Briede JJ, Schrauwen P, Hoeks J (2012) Targeting of mitochondrial reactive oxygen species production does not avert lipid-induced insulin resistance in muscle tissue from mice. Diabetologia 55:2759–2768
Pustovidko AV, Rokitskaya TI, Severina II, Simonyan RA, Trendeleva TA, Lyamzaev KG, Antonenko YN, Rogov AG, Zvyagilskaya RA, Skulachev VP, Chernyak BV (2013) Derivatives of the cationic plant alkaloids berberine and palmatine amplify protonophorous activity of fatty acids in model membranes and mitochondria. Mitochondrion 13:520–525
Roginsky V, Barsukova T, Loshadkin D, Pliss E (2003) Substituted p-hydroquinones as inhibitors of lipid peroxidation. Chem Phys Lipids 125:49–58
Rogov AG, Ovchenkova AP, Goleva TN, Kireev II, Zvyagilskaya RA (2017) New yeast models for studying mitochondrial morphology as affected by oxidative stress and other factors. Anal Biochem. https://doi.org/10.1016/j.ab.2017.04.003
Rokitskaya TI, Murphy MP, Skulachev VP, Antonenko YN (2016) Ubiquinol and plastoquinol triphenylphosphonium conjugates can carry electrons through phospholipid membranes. Bioelectrochemistry 111:23–30
Saretzki G, Murphy MP, von Zglinicki T (2003) MitoQ counteracts telomere shortening and elongates lifespan of fibroblasts under mild oxidative stress. Aging Cell 2:141–143
Severin SE, Skulachev VP, Yaguzhinsky LS (1970) Possible role of carnitine in transport of fatty acids through mitochondrial membrane. Biokhimiya 35:1250–1252
Severin FF, Severina II, Antonenko YN, Rokitskaya TI, Cherepanov DA, Mokhova EN, Vyssokikh MY, Pustovidko AV, Markova OV, Yaguzhinsky LS, Korshunova GA, Sumbatyan NV, Skulachev MV, Skulachev VP (2010) Penetrating cation/fatty acid anion pair as a mitochondria-targeted protonophore. Proc Natl Acad Sci U S A 107:663–668
Severina II, Severin FF, Korshunova GA, Sumbatyan NV, Ilyasova TM, Simonyan RA, Rogov AG, Trendeleva TA, Zvyagilskaya RA, Dugina VB, Domnina LV, Fetisova EK, Lyamzaev KG, Vyssokikh MY, Chernyak BV, Skulachev MV, Skulachev VP, Sadovnichii VA (2013) In search of novel highly active mitochondria-targeted antioxidants: thymoquinone and its cationic derivatives. FEBS Lett 587:2018–2024

Shagieva G, Domnina L, Makarevich O, Chernyak B, Skulachev V, Dugina V (2017) Depletion of mitochondrial reactive oxygen species downregulates epithelial-to-mesenchymal transition in cervical cancer cells. Oncotarget 8:4901–4913

Skulachev VP, Antonenko YN, Cherepanov DA, Chernyak BV, Izyumov DS, Khailova LS, Klishin SS, Korshunova GA, Lyamzaev KG, Pletjushkina OY, Roginsky VA, Rokitskaya TI, Severin FF, Severina II, Simonyan RA, Skulachev MV, Sumbatyan NV, Sukhanova EI, Tashlitsky VN, Trendeleva TA, Vyssokikh MY, Zvyagilskaya RA (2010) Prevention of cardiolipin oxidation and fatty acid cycling as two antioxidant mechanisms of cationic derivatives of plastoquinone (SkQs). Biochim Biophys Acta 1797:878–889

Skulachev MV, Antonenko YN, Anisimov VN, Chernyak BV, Cherepanov DA, Chistyakov VA, Egorov MV, Kolosova NG, Korshunova GA, Lyamzaev KG, Plotnikov EY, Roginsky VA, Savchenko AY, Severina II, Severin FF, Shkurat TP, Tashlitsky VN, Shidlovsky KM, Vyssokikh MY, Zamyatnin AA Jr, Zorov DB, Skulachev VP (2011) Mitochondrial-targeted plastoquinone derivatives. Effect on senescence and acute age-related pathologies. Curr Drug Targets 12:800–826

Smith RA, Porteous CM, Coulter CV, Murphy MP (1999) Selective targeting of an antioxidant to mitochondria. Eur J Biochem 263:709–716

Vorobjeva N, Prikhodko A, Galkin I, Pletjushkina O, Zinovkin R, Sud'ina G, Chernyak B, Pinegin B (2017) Mitochondrial reactive oxygen species are involved in chemoattractant-induced oxidative burst and degranulation of human neutrophils in vitro. Eur J Cell Biol 96:254–265

Votyakova TV, Reynolds IJ (2001) DeltaPsi(m)-dependent and -independent production of reactive oxygen species by rat brain mitochondria. J Neurochem 79:266–277

Targeted Mitochondrial Genome Elimination

Sandra R. Bacman, Claudia V. Pereira, and Carlos T. Moraes

Abstract Mitochondrial diseases are a heterogeneous group of pathologies, presenting mitochondrial oxidative phosphorylation (OXPHOS) deficiency and subsequent deterioration of affected tissues, usually the ones with high energy demand. Nuclear and mitochondrial DNA (mtDNA) mutations have been associated with mitochondrial disorders, and only palliative therapies are available, with no approach available to correct the underlining genetic problem. Our and other groups, taking advantage of mtDNA heteroplasmy, were able to develop gene therapy strategies to target the mtDNA, inducing the increase of the wild-type haplotype. The gene editing tools used by our group and others such as specific endonucleases targeted to the mitochondria, and more recently mitoTALENs and mitochondrial Zinc-Finger Nucleases were shown to effectively shift mtDNA heteroplasmy. This change towards an increase in wild-type mtDNA molecules occurred when the nuclease was specifically designed to cleave DNA regions harboring point mutations or the common deletion. In this chapter we will focus on strategies to target specific mtDNA haplotypes to shift heteroplasmy *in vitro* and *in vivo*. A brief overview on heteroplasmic animal models and viral delivery will be introduced, as these will be critical for future translation of this approach into human gene therapy.

Keywords mtDNA · DNA editing · TALEN · Heteroplasmy · Gene therapy · Mitochondrial diseases

Mitochondrial diseases are a heterogeneous group of pathologies, presenting mitochondrial oxidative phosphorylation (OXPHOS) deficiency and subsequent deterioration of affected tissues, usually the ones with high energy demand. Nuclear and mitochondrial DNA mutations have been associated with mitochondrial disorders, and only palliative therapies are available, with no approach available to correct the underlining genetic problem. Mitochondrial DNA (mtDNA) mutations are usually heteroplasmic, where co-existence of wild-type and mutations vary in ratios (Stefano and Kream 2016). If a mutation is pathogenic, the cell can usually tolerate some levels of this variant, until a biochemical threshold is exceeded and a defect in

S. R. Bacman · C. V. Pereira · C. T. Moraes (✉)
Miller School of Medicine, University of Miami, Miami, FL 33136, USA
e-mail: CMoraes@med.miami.edu

P. J. Oliveira (ed.), *Mitochondrial Biology and Experimental Therapeutics*,
https://doi.org/10.1007/978-3-319-73344-9_24

the respiratory chain is triggered. Typically, this threshold level is >80% for a deleterious phenotype, suggesting that most mtDNA mutations are "recessive" (Durham et al. 2007).

Our and other groups, taking advantage of mtDNA heteroplasmy, were able to develop gene therapy strategies to target the mtDNA, inducing the increase of the wild-type haplotype. The gene editing tools used by our group and others such as specific endonucleases targeted to the mitochondria, and more recently mitoTALENs and mitochondrial Zinc-Finger Nucleases were shown to effectively shift mtDNA heteroplasmy. This change towards an increase in wild-type mtDNA molecules occurred when the nuclease was specifically designed to cleave DNA regions harboring point mutations or the common deletion. In this chapter we will focus on strategies to target specific mtDNA haplotypes to shift heteroplasmy *in vitro* and *in vivo*. A brief overview on heteroplasmic animal models and viral delivery will also be introduced, as these will be critical for future translation of this approach into human gene therapy.

1 Mitochondrial DNA: Heteroplasmy and the Bottleneck Theory

The human mitochondrial genome is a small double-stranded circular genome composed of 16,569 bp which encodes: the large and small ribosomal RNAs (rRNAs, 12S and 16S) genes, 22 transfer RNAs (tRNA) and 13 protein-encoding genes that are part of the OXPHOS complexes (Andrews et al. 1999). These subunits, together with nuclear-coded ones, are responsible for mitochondrial ATP production (Vafai and Mootha 2012). Aerobic tissues depend on OXPHOS to survive, therefore the overall process is highly regulated and dependent on intermediary metabolites (e.g. Krebs cycle) and regulatory ions (e.g. calcium) (Pozzan and Rizzuto 2000). Because most of the OXPHOS complex subunits are actually encoded by the nuclear DNA, synthetized in the cytosol and imported to mitochondria, there is a requirement for constant communication and high coordination between these genomes. As mentioned above, there are only a small set of proteins encoded by the mtDNA, including: seven subunits of Complex I (ND1-ND6), one subunit of Complex III (cytochrome b), three subunits of Complex IV (COXI-III) and two subunits of Complex V (ATPase 6 and 8) (Anderson et al. 1981). On the other hand, there are hundreds of proteins required to assemble the OXPHOS complexes that are encoded by the nuclear genome. Altogether, a plethora of nuclear and mitochondrial subunits, nuclear transcription factors, replication and translation machineries, antioxidant defenses are involved in mtDNA maintenance and cell survival. Therefore, the optimal functioning of the cell is greatly dependent on how fit mitochondria are and how efficiently these two genomes can coordinate to respond to energy demands.

The mtDNA is located within the matrix and is packaged into nucleoids, a DNA/protein complex which contains one or two mDNA molecules (Akhmedov and Marin-Garcia 2015; Gilkerson et al. 2008). A large number of proteins has been found in nucleoids, including factors required to replicate the mtDNA (Iborra et al. 2004; Chen and Butow 2005; Malka et al. 2006). The number of mitochondria per cell differs from one cell type to another but each mammalian cell can contain approximately one thousand copies of mtDNA (Miller et al. 2003). The mtDNA copy number per cell also varies among different tissues due to the tissue-specific regulation of enzymes involved in mtDNA replication, including polymerase γ (Pol γ), which is responsible for the replication of the mtDNA. In addition, mtDNA replication is not linked to the cell division; therefore competition between mtDNA molecules within the same cell may favor those that have a replication or survival advantage, even if such variants reduce the overall cell fitness. The proximity of the mtDNA to the OXPHOS enzymes is believed to make it prone to oxidative damage, and consequently to mutations (Bandy and Davison 1990; Shokolenko et al. 2009). Besides, in general, the repair mechanisms of mtDNA are less efficient than in the nucleus and the exposure to reactive oxygen species (ROS) or replication errors contribute to mtDNA sequence variations within the same cell. Mutations in the mtDNA are also believed to contribute to aging phenotypes such as blindness, deafness, cardiac and neurological diseases, endocrine and renal dysfunction, and some studies also reported their involvement in cancer (Lu et al. 2009).

To secure the maintenance of healthy mitochondrial populations inside a cell, the dynamic mechanisms of fission (division) and fusion (re-joining) are constantly occurring (Scott and Youle 2010). These are believed to help to compensate for the presence of low levels of mutant mtDNA as well as assist on organelle movement (Twig et al. 2008). Moreover, it is well established that the mammalian mitochondrial genome is maternally inherited (Hutchison et al. 1974; Giles et al. 1980). The importance of this occurrence is that pathogenic mutations in an affected mother can be transmitted to all her children, but only the females will be able to transmit the disease to their offspring. A possible explanation for maternal inheritance lies in the significant downregulation of the mtDNA copy number during spermatogenesis ("dilution effect") and ubiquitination of the sperm mitochondria after fertilization (Chinnery et al. 2000; Sutovsky 2003). Recent studies based on a mathematical model suggested that the selection against heteroplasmy may explain the uniparental mitochondrial inheritance (Christie et al. 2015).

Another important concept that is involved in the complexity of mitochondrial disease inheritance is the mitotic segregation, due to a random drift in the percentage of mutant mtDNA between the parental and daughter cells, which may change quickly. This means that the proportion of the mutant mtDNA passed onto daughter cells during cell division may not be identical to that of the parental cell. In addition, the replication of the WT and mutant mtDNA may also vary according to the energetic demands of the cell, which leads to differential replication (Wallace and Fan 2010). During female germline development, there is a significant reduction in the number of mtDNA molecules in the oocytes followed by re-amplification to a final number of more than 100,000 copies. This "genetic bottleneck" may account for the generational shifts in mtDNA mutation loads (Jenuth et al. 1996; Wai et al. 2008). This phenomenon, first described in Holstein cows, is also observed in mice as well as in several human pedigrees (Olivo et al. 1983; Larsson et al. 1992; Howell et al. 1992; Chinnery et al. 2000). Mutations that escape the "bottleneck" can be filtered out by purifying selection mechanisms (Stewart et al. 2008b; Fan et al. 2008). Studies performed in heteroplasmic mice have shown that the heteroplasmy levels within individual oocytes were actually determined before birth (Freyer et al. 2012). Recently, another study tried to address the same fundamental question, whether mtDNA mutations are or not selected in human oocytes. In this study, 60 mitochondrial DNA genomes from 17 sets of oocytes, first and second polar bodies (PBs) and blood from nine women (between 38 and 43 years old) were analyzed by sequencing. The results suggested the existence of a filtering mechanism for disadvantageous mtDNA variants, which may occur between the expulsion of the first and second polar bodies (PBs) (De Fanti et al. 2017). Possible explanations for the occurrence could be justified by the absence of mtDNA replication during early embryogenesis (Ebert et al. 1988) which could also lead to very low mtDNA amounts in the primordial germline cells (PGCs). It is possible that this is a prerequisite to selectively filter out deleterious mtDNA mutations, since only when the mtDNA load is small the differences in organelle fitness can manifest quickly and be selected for (Wai et al. 2010).

With the advance in sequencing techniques, many mtDNA segregation studies have been reported. The largest *in silico* analysis to date performed in 577 mother-child pairs revealed that different mtDNA mutations segregate at different rates, although they did not find any evidence of selection during transmission (Wilson et al. 2016). Ultimately, interpretations of next-generation sequencing data require caution as the data can have the interference of pseudogene sequences which are known to be present in the nuclear DNA, termed nuclear mitochondrial DNA sequences (NUMTs) (Woischnik and Moraes 2002).

Mitochondrial diseases can be classified in two mayor categories; from the genetic standpoint, depending whether it is the nuclear or the mitochondrial DNA carrying the pathogenic mutation (Viscomi et al. 2015). Nuclear mutations have been found in genes directly or indirectly related to the respiratory chain with proteins involved in maintenance and replication, structural subunits of the OXPHOS complexes, assembly and translations factors and proteins involved in dynamics as fission/fusion, among others (Koopman et al. 2012). MtDNA mutations include point mutations, deletions and large-scale rearrangements (Gorman et al. 2015). In any case, they disrupt OXPHOS and lead to a range of genetic diseases that have mostly no treatments (only palliative ones). Mitochondrial DNA mutations can be classified as heteroplasmic or homoplasmic. Pathogenic mutations in mtDNA can affect the mitochondrial tRNA, rRNA or protein coding genes. In non-dividing (postmitotic) heteroplasmic cells, mtDNA is continuously being turned over, leading to the possibility that mutant molecules might be replicated preferentially over the wild-type molecule (Stewart and Chinnery 2015). MtDNA molecules with a pathogenic deletion are smaller than wild-type mtDNA molecules, which were suggested to lead to a selective advantage favoring smaller deleted molecules, which would accumulate over time (Diaz et al. 2002). Clinical and genetic heterogeneity of the mtDNA mutations can lead to a wide variety of clinical phenotypes (Schon et al. 2012), but generally mtDNA point mutations can affect multiple individuals in the same family, whereas mtDNA deletions are rarely inherited and are never homoplasmic (Chinnery et al. 2004). Homoplasmic mtDNA point mutations usually cause a relatively mild biochemical defect that typically affects only one organ or tissue (for example, Leber's Hereditary Optic Neuropathy or LHON) (McFarland et al. 2002). Heteroplasmic mutations can affect multiple organs, such as the brain, spinal cord, muscle, peripheral nerves, heart and endocrine organs (Stewart and Chinnery 2015) and higher mutation loads are associated with strong biochemical phenotypes (Sciacco et al. 1994). In general, mtDNA deletions cause diseases at lower mutations loads than point mutations, 60% vs 80% (Moraes et al. 1995; Russell and Turnbull 2014). The threshold can vary among members of the same family, where small but significant differences in mutation load can lead to a healthy life or devastating diseases (Hao et al. 1995; Bacman et al. 2014). One of the consequences of heteroplasmy and threshold effects is that by removing or reducing a small percentage of the mutant load below a critical threshold, it is possible to ameliorate OXPHOS dysfunction, as the increased wild-type mtDNA pool can compensate for the mutant allele.

2 Animal Models to Study Heteroplasmy

Several reasons contribute to the scarcity of heteroplasmic animal models that could recapitulate human mitochondrial diseases generated by mtDNA mutations. There are no available techniques to genetically modify mammalian mtDNA, so cell biology approaches have been used to transfer organelles between cells (Bacman and Moraes 2007). One of the main challenges for genetic engineering is the difficulty to trespass the two double mitochondrial lipid membranes, which would facilitate the proper mtDNA genome targeting; secondly, since there are hundreds to thousands of copies of mtDNA in a cell, the efficiency of mitochondrial transfection or to modify the DNA is lower (McGregor et al. 2001). At last, trying to introduce mtDNA molecules that confer a disadvantage (such as decreased respiratory capacity) are usually selectively eliminated through the germline (Fan et al. 2008; Pinkert and Trounce 2002) accounting for one of the main barriers to build heteroplasmic mice. Therefore, until recently, the attempts to generate mouse models of mitochondrial DNA mutations have been focused on introducing pre-existing mtDNA mutations from cell lines or somatic tissues into mice. Currently, there are only a few models available that can be used to explore the mtDNA heteroplasmy and segregation mechanisms in mammals and mimic certain aspects of human mitochondrial pathologies induced by mtDNA mutations. In addition, mouse models of OXPHOS dysfunction are also scarce due to embryonic lethality, so the majority of the studies have been carried in conditional knockout (cKO) mice, including: NDUFS6 (Ke et al. 2012), NDUFA5 (Peralta et al. 2014) complex I subunits and cytochrome c oxidase related models such as COX10, Surf1, COX15, SCO1, SCO2 and LRPRC (Tiranti et al. 1998; Diaz et al. 2005; Zhu et al. 1998; Valnot et al. 2000; Papadopoulou et al. 1999; Xu et al. 2004) which are assembly factors. Cytochrome c (cyt c) models have also been characterized, such as the cytochrome c testis (cyt c_T) isoform KO animals and the double KO model in which both somatic and the testis isoforms were disrupted. Other models for the study of proteins involved in the interaction and mtDNA stability have also been studied such as the TFAM KO models (Larsson et al. 1998; Wang et al. 1999; Li et al. 2000), a protein involved in mtDNA maintenance; Polymerase gamma (Pol γ) (Trifunovic et al. 2004; Hance et al. 2005), ANT1 (adenine nucleotide translocator isoform 1 (Graham et al. 1997; Esposito et al. 1999), Twinkle (mitochondrial helicase required for mtDNA replication (Tyynismaa et al. 2005), MTERF3 (mitochondrial termination factor 3, that promotes transcription termination, (Park et al. 2007), among others. A complete KO of NDUFS4 was found to be viable despite dying early (Kruse et al. 2008; Quintana et al. 2010, 2012a; Sterky et al. 2012; Karamanlidis et al. 2013; Leong et al. 2012). For a comprehensive review on animal models of mitochondrial diseases please refer to recent reviews (Torraco et al. 2015, 2009; Iommarini et al. 2015; Peralta et al. 2015; Ruzzenente et al. 2016).

This section will focus on the currently available heteroplasmic mtDNA mouse models. In 1978, the first attempt was developed to establish a method which would allow the introduction of specific mitochondrial-encoded mutant genes into mice (Watanabe et al. 1978). Watanabe and others used mouse teratocarcinoma stem cells as vehicles for the cytoplasmic markers because of their capacity to differentiate after injection into blastocysts. Cybrids were then created by fusing mouse mutagenized melanoma cytoplasts carrying a 16S rRNA mutation which confers chloramphenicol (CAP^R) resistance, with the teratocarcinoma stem cells. After 16 weeks, the cells were still CAP^R resistant even in the absence of the selective agent. The selected and stable cybrid clones were then microinjected into blastocysts of another inbred mouse strain and then transferred to foster mothers. Mosaic mice were obtained which comprised both cybrid and blastocyst-derived cells in various tissues (Watanabe et al. 1978). Unfortunately, the analysis of the first generation (F1) did not clearly demonstrate transmission of the mtDNA mutation. Many years later, the technique was amended by the introduction of a step to remove the endogenous mitochondria of the embryonic stem cells by pre-treating the cells with rhodamine 6-G, a potent mitochondrial toxicant, prior to the introduction of foreign mitochondria (Sligh et al. 2000). The cybrids were injected into B6 blastocyst and resulting chimeric mice showed congenital cataracts, retinal abnormalities and optic nerve hamartomas. The female chimeras transmitted the mutation to the next generation and about 50% of the CAP^R was detected in the pups. Furthermore, their heteroplasmic and homoplasmic progeny were more severely affected. This study was proof of principle that transmitochondrial mice harboring pathogenic mutations can transmit the mutation to the next generation, nevertheless some of the pups died as embryos and the newborn pups succumbed 11 days after birth. Furthermore, the phenotype of the mice was characterized by myopathy, growth retardation and cardiomyopathy, which was consistent with skeletal muscle and heart degeneration and abnormal mitochondria (Sligh et al. 2000). In contrast to these mice, the transmitochondrial mice produced by Inoue et al. (2000) were generated by the fusion of enucleated cytoplasts with fertilized pro-nuclear state embryos. After 24–48 h, the embryos were implanted in a pseudo-pregnant mother. The deleted mtDNA (4.7 kb deletion) with a breaking point from positions nt 7759–12,454 including six tRNA and seven structural genes was transmitted with success to the progeny, but the amount of the mutation found in the analyzed tissues was less than 100%, likely related to the loss of eggs carrying the highest amount of the deletion. This deletion in the so called "mito-miceΔ" is similar to the common deletion, which induces Kearns-Sayre syndrome. The analysis of the mtDNA showed the presence of wild-type mtDNA, deleted mtDNA and partially duplicated mtDNA, which could be derived by recombination of the wild-type with the deleted mtDNA during embryogenesis. The duplicated mtDNA molecules can be transmitted and can cause mitochondrial disease. The threshold level for COX negative fibers was higher than 80%, after analysis of single muscle fibers from F1-F3 mice. The most affected organs were the kidney, muscle and heart. The mito-miceΔ showed low body weight, lactic acidosis, and systemic ischemia, auriculoventricular block, hearing loss, renal loss and male infertility, which mimics mitochondrial disease. The mice with more than 70% of deletion only lived for 6 months and died from renal failure, which was the organ that showed the highest percentage of COX negative cells. On the other hand, mice with 30–50% of deletion were healthy at birth and died at 1.5 years when the mtDNA deletion load became higher than 75%. The same authors generated the transmitochondrial "mito-mice" carrying a pathogenic missense T6589C (Val 412 Ala) mutation in a mitochondrially encoded cytochrome c oxidase I (COI) gene (Kasahara et al. 2006) and a mito-miceND6 which has a pathogenic missense A13997G (Pro 25 Leu) mutation in the mitochondrial NADH dehydrogenase 6 (ND6) gene (Yokota et al. 2010). Both COI and ND6 mutant mice had mild mitochondrial respiratory defects when compared to the mito-miceΔ which showed a vast phenotype characterized by a defective respiration when the deletion loads were high. The mutant mice generated were homoplasmic for each mutation, and the COI mice showed 20–30% loss of COX activity whereas the ND6 mice showed 20–30% decreased complex I and III activities, in various tissues. Two years later a study was conducted in the aged ND6 mice showing late-onset and age-related disorders could not be justified by respiration defects but to increased ROS production, which was suppressed by antioxidants administration (Hashizume et al. 2012). In 2008, Fan et al. (2008) generated transmitochondrial mice by the fusion of cytoplast carrying the ND6 and the COI mutations with ES cells devoid of mitochondria. The authors compared the transmission rate of these two mtDNA mutations with differential pathogenicity. The ND6 mutation/insertion (13885insC) completely abolished complex I activity and the COI missense mutation only decreased 50% of complex IV activity, showing a milder phenotype. It was shown that the amount of heteroplasmy of the ND6 decreased in the female germline and the mutation was completely lost after four generations. The COI mutation was maintained overtime inducing myopathy and cardiac hypertrophy. This study provided evidence that the elimination of mtDNA mutations may occur before ovulation, when the number of proto-oocytes is reduced. Another hypothesis is that the elimination of the mutated mtDNA occurs during the rapid proliferation of mtDNA in the germline through the loss of segregating units (Cree et al. 2008). The current view on this topic is that the female germline progressively and selectively eliminates the potentially harmful mtDNA that may affect fecundity (Stewart et al. 2008a).

Heteroplasmic mice have as well been generated by cytoplasmic fusion between two normal inbred mouse strains (NZB and BALB), yielding mice with different levels of heteroplasmy (Jenuth et al. 1996). In this case, the cytoplasts derived from either BALB or NZB were electrofused to a one-cell embryo of the other type and then transplanted to a pseudomother at the two cells embryo stage. These mice were asymptomatic but directional selection for one mtDNA haplotype in some tissues was observed (Battersby et al. 2003) and likely mediated by tissue-specific factors (Bayona-Bafaluy et al. 2005). With aging, NZB segregated preferentially in liver and kidney whereas BALB in spleen and blood. Besides, proliferating hepatocytes *in vitro* accumulated BALB mtDNA instead of NZB, by a still unknown mechanism of selection for a certain haplotype in cells and tissues. More recently, transmitochondrial mice carrying a $tRNA^{Lys}$ gene encoded mtDNA mutation were generated by introducing the G7731A mutant mtDNA into mouse ES cells and F0 chimeric mice were obtained (Shimizu et al. 2014). The F1 generation named mito-mice-$tRNA^{Lys7731}$ was generated and maternal inheritance and random segregation occurred in the succeeding generations. Mice with low, intermediate and high mutation levels were generated. The authors found that the higher mutant mice showed short body length and muscle weakness. Furthermore, biochemical analysis revealed respiration defects in skeletal muscle and kidney; therefore the authors concluded that the accumulation of this tRNA mutation is responsible for the respiration defects and phenotype. On the other hand, other metabolic parameters usually described in mitochondrial disease patients were not observed in these mice, which could be related with the threshold effect of disease manifestation for this particular mutation. In addition, mice with more than 85% of mutation could not be obtained in this study. Interestingly, these heteroplasmic mice segregated stochastically in a similar manner as the BALB/NZB mice, and as those with or without point mutations. The high variation in the levels of heteroplasmy of the pups may reflect "bottleneck effects" with or without decrease in the mtDNA copy number during female germline transmission of heteroplasmic mtDNA (Shimizu et al. 2014). In 2015, another report from the same group showed that an orthologous mutation to the MELAS A3302G mutation, the A2748G was responsible for mitochondrial respiration defects in mouse ES cells. Thus, the study could be useful for the future generation of another transmitochondrial mouse model, as a potential model for MELAS disease (Shimizu et al. 2015).

Unfortunately, a clear disadvantage of the cytoplasmic fusion strategy is that it is very time-consuming and the choice of mutations is restricted to those existing in cell lines or somatic tissues. Furthermore, some of the introduced mutations cannot be propagated in mice (Fan et al. 2008) because of strong purifying germline selection (Stewart et al. 2008a). Many mice models have been created by modifying nuclear genes involved in mitochondrial disease, such as the mtDNA mutator mouse (PolgAMUT/MUT). The homozygous knockin mutator mice express a proofreading-deficient Polg with much reduced 30–50 exonucleolytic activity (Ross et al. 2013); (Trifunovic et al. 2004). These mice have extensive somatic and germline mutagenesis of mtDNA and transmit mutant mtDNA via germline (Stewart et al. 2008a; Ross et al. 2013). Mouse lines derived by breeding PolgAMUT/MUT females have been valuable tools to study purifying selection (Stewart et al. 2008a, b). It was observed that tRNA gene mutations are better tolerated and mainly undergo purifying selection during embryonic development (Freyer et al. 2012). Unfortunately, the breeding of PolgAMUT/MUT mice results in mouse lines that have many linked mutations in the same mtDNA molecule, which difficults genotype-phenotype correlations. These models are therefore of limited use when it comes to understanding the pathophysiological effects of single specific mtDNA mutations. To overcome these limitations, a technically simple and screening approach to create mice with pathogenic mtDNA mutations has been described by Kauppila et al. (2016). Briefly, heterozygous mtDNA-mutator (PolgA+/MUT) females were bred to establish mouse lines with a wild-type nuclear background that each carries a very limited number of mtDNA mutations. In a second step, the founder individual of each line was sacrificed after having established a maternal mouse line and colonic crypts were analyzed by enzyme histochemistry to detect mosaic cytochrome c oxidase (COX) deficiency. After, laser-capture dissection and mtDNA sequencing were performed on single colonic crypts in order to identify the pathogenic mtDNA mutation and establish its pathogenicity. Finally, identified mouse lines, where the founder mouse harbors a specific heteroplasmic pathogenic mtDNA mutation are bred as maternal lines and extensively characterized. As a proof of principle, the creation of a mouse model for mitochondrial disease caused by an heteroplasmic C5024T mutation in the $tRNA^{Ala}$ (mt-Ta) gene of mtDNA was shown to recapitulate important aspects of human mitochondrial disease (Kauppila et al. 2016). The importance of creating mouse models to better understand the mechanisms of inherited mitochondrial disease and segregation of pathogenic mutations, in order to further advance in targeted and effective mitochondrial-DNA therapies cannot be overemphasized. In conclusion, in the past few years important advances have been made and more suitable models are now available to address the fundamental questions of mtDNA segregation, heteroplasmy and mitochondrial biology.

3 Mitochondrial DNA Genome Editing

Although great progress has been made in understanding the fundamental pathogenic processes which cause mitochondrial disease, effective treatments are not available (Russell and Turnbull 2014; Viscomi et al. 2015). Therefore, gene therapy is an attractive therapeutic strategy for the treatment of mitochondrial diseases. Here we described the available exploratory tools for gene therapy in mitochondrial diseases, focusing on the ones that can directly alter the levels of mutant mtDNA.

3.1 Targeting Mitochondria with Specific Endonucleases

Because heteroplasmy levels define tissue involvement (Stewart and Chinnery 2015), in the last 15 years our group and others have attempted to shift heteroplasmy as a potentially strategy to treat mtDNA disorders. The first study using restriction endonucleases (RE) to change heteroplasmy *in vitro* was developed in our laboratory. The *Pst*I-RE was targeted to mitochondria and showed the ability to change heteroplasmy in a predicted way. The *Pst*I-RE could differentiate between mouse and rat mtDNA in xenomitochondrial cybrids. These cells were created by fusing mouse LM(TK) cells (which contain two sites for *Pst*I) and rat NRK cells (which contain none) (Dey et al. 2000). These experiments provided proof of principle that restriction endonucleases are viable tools for mtDNA heteroplasmy manipulation (Srivastava and Moraes 2001).

The practicality of using mitochondrially targeted restriction endonucleases to modulate mtDNA heteroplasmy is limited by the presence of a bona fide mutation-created restriction site. However, a mitochondrial disease caused by a T → G transversion in the mitochondrial ATP6 gene (at mtDNA position 8399) creates a unique *Sma*I/*Xma*I site that could be targeted. This mutation has been associated with neuropathy, ataxia and retinitis pigmentosa (NARP) and maternal inherited leigh syndrome (MILS) (Holt et al. 1990; Tatuch et al. 1992, 1994). In most patients, a milder NARP harbors 60–90% of the T8399G mutation (Tatuch et al. 1992; Tsao et al. 2001), while higher levels show a severe MILS phenotype (>90%). Using a mitochondria-targeted *Sma*I restriction endonuclease a reduction in the level of the mutation in heteroplasmic cybrids cells was successfully achieved (Tanaka et al. 2002). Further studies were done using the *Sma*I isoschizomer *Xma*I RE expressed from an adenovirus system. Mitochondria-targeted *Xma*I reduced the mutant mtDNA levels in a time and dose-dependent manner, restoring some of the phenotypes such as oxygen consumption and ATP levels that were decreased in the original cybrids carrying the mtDNA mutation, and decreasing lactic acid production. These results showed that specific endonucleases can be useful for the treatment of NARP and MILS (Alexeyev et al. 2008).

In our laboratory, we took advantage of a well described heteroplasmic mouse model that contains two haplotypes of mtDNA, NZB and BALB (Jenuth et al. 1997), using an inducible system *ApaL*I-RE (Bayona-Bafaluy et al. 2005). The goal was to decrease specific BALB mtDNA haplotype by expressing a mito-ApaLI in the mitochondria of cells that carry two mtDNA haplotypes (BALB/NZB) as there is one *ApaL*I site in the BALB and none in the NZB. This system showed to be extremely rapid and efficient, as the NZB haplotype was already significantly increased after 2 h of inducing the expression of the mito*ApaL*I (Bayona-Bafaluy et al. 2005) in cultured hepatocytes derived from the heteroplasmy mice.

Our group also showed the applicability *in vivo* in both skeletal muscle and brain in NZB/BALB heteroplasmic mice using either a recombinant adenovirus vector (rAd5) or AAV vector (rAAV1,2). rAd5-mito*ApaL*I transduction resulted in mtDNA heteroplasmy shift with significant higher proportion of NZB haplotype. In the brain, injections of rAd5-mito*ApaL*I in the right cerebral hemisphere of NZB/BALB heteroplasmic mice showed efficient transduction of rAd5 in mostly neurons but also astrocytes and other glial cells, and transduced areas expressing mito*ApaL*I showed a shift in mtDNA heteroplasmy (Bayona-Bafaluy et al. 2005).

We also delivered another mitochondria-targeted RE, the *Sca*I-RE that recognizes multiple restriction sites in both haplotypes, creating a system of a "differential multiple cleavage-site" in the same heteroplasmic mouse model described above. There are five *Sca*I sites in the NZB mtDNA and three in BALB mtDNA. In this case expression was obtained in both liver, after intravenous injection, and in skeletal muscle, after intramuscular injection using recombinant adenovirus as a vehicle. MtDNA depletion was an undesirable side effect probably due to the numerous RE-sites targeted in this approach, but heteroplasmy shift was detected (Bacman et al. 2007).

To broaden this approach, we took advantage of the recombinant AAV serotypes to target mito-RE to different tissues (Bacman et al. 2010). In this case, the mitochondrial-targeted *ApaL*I was able to increase the proportion of NZB mtDNA in targeted tissues. This was observed in the heart after systemic injections of a cardiotropic adeno-associated virus type-6 (AAV6) or in liver, using the hepatotropic recombinant adenovirus type-5 (Ad5). Cardiac expression induced a significant shift in mtDNA heteroplasmy that persisted for 12 weeks after injection, demonstrating that a single injection can induce long term heteroplasmy shift (Bacman et al. 2010). Neither mitochondrial DNA depletion, nor cytochrome c oxidase deficiency was observed probably due to *ApaL*I-RE having only a single restriction site in the mouse NZB/BALB (Bacman et al. 2010). We also tested AAV transduction in newborns, to study the applicability of these treatments in human neonates that are widely described to be afflicted with a mitochondrial disease at early ages (Uusimaa et al. 2000). Using recombinant AAV9 carrying the mito*ApaL*I, we delivered via intraperitoneal (IP) or temporal vein (TV) injection to neonates (Bacman et al. 2012) and a single systemic injection of rAAV9[mito*ApaL*I] induced shifts in mtDNA heteroplasmy in all skeletal muscles tissues and heart which persisted for at least 6 months. This long-lasting expression was in agreement with previous reports showing that robust transduction could be achieved in skeletal muscle after delivery of AAV9 vectors (Inagaki et al. 2006; Ghosh et al. 2007).

These groups of experiments contributed to advance the knowledge on the efficacy of the usage of mitochondrial-targeted restriction endonucleases to change mtDNA heteroplasmy for specific mtDNA mutations, but also revealed its applicability *in vivo* and the benefit that a unique injection, when using the adequate viral vector, it appears to promote a long-lasting change in heteroplasmy. This approach requires a highly specific nuclease that recognizes one haplotype but not the other(s).

3.2 *Targeting Mitochondria DNA Heteroplasmy with DNA Editing Enzymes*

The limitation of using restriction endonucleases to target specific mutations, can be overcome by the recent developments of gene editing nucleases that has broadened the spectrum of mtDNA mutations that can be targeted (Gaj et al. 2013). Our group and others have investigated the use of endonucleases with modular DNA recognition domains, which can be designed to bind many and almost any predetermined DNA sequence. Various gene editing nucleases are actually available to modify nuclear genes, such as bacterial *Streptococcus pyogene* Clustered Regularly Interspaced Short Palindromic Repeats, CRISPR-associated protein-9 nuclease (Cas9) (Cong et al. 2013). Other available tools include Zinc-Finger nucleases (ZFNs), by fusing a zinc finger DNA-binding domain to a DNA-cleavage domain (Urnov et al. 2010; Ramirez et al. 2012) and Transcription activator-like effectors nucleases (TALENs) that are secreted by Xanthomonas bacteria via their type III secretion system (Boch and Bonas 2010; Bogdanove and Voytas 2011; Scharenberg et al. 2013) and meganucleases (Stoddard 2011), among others. ZFNs, TALENs and meganucleases achieve specific DNA binding via protein-DNA interactions, whereas Cas9 is targeted to specific DNA sequences by a short RNA guide molecule that base-pairs directly with the target DNA and by protein-DNA interactions (Cox et al. 2015).

Both TALENs and ZFNs share a common basic structure, a pair of customized designed monomers that bind the region of interest and enable the dimerization of non-specific *Fok*I domains inducing efficient and precise double-strand cleavage (DSBs) in the targeted DNA (Fig. 1). In the nucleus, such break stimulates cellular DNA repair mechanisms, non-homologous end joining (NHEJ) and homology-directed repair (HDR) (Wyman and Kanaar 2006), (Bacman et al. 2014). Mitochondrial DNA double strand breaks lead to rapid degradation of molecules (Bayona-Bafaluy et al. 2005). The recognition site varies: traditionally, 9–18 bp per ZFN monomer, 14–20 bp per TALEN monomer, 22 bp (20-bp guide sequence plus 2-bp protospacer adjacent motif (PAM) for Cas9 and between 14 and 40 bp for meganucleases (Cox et al. 2015). ZFNs, TALENs and CRISPR systems have been successfully approached in nuclear gene editing, although targeting to mitochondria offers different challenges. Nuclear genes can be targeted at different positions of the genome to avoid off-target sites with similar sequences. Although this is a less important problem in the small mtDNA, the elimination of mutant mtDNA requires the specific recognition of as few as a single base. Furthermore, the import of such enzymes to mitochondria is critical. This is more difficult for ZFNs which have a strong nuclear tropism (Urnov et al. 2005; Bacman et al. 2014). Some advances have been done to reduce off-target effects and improve target specificity of CRISPR. However, as discussed below, CRISPR/Cas9 has not been convincingly shown to work in mitochondria, therefore more studies are needed to confirm its feasibility in editing mitochondrial DNA.

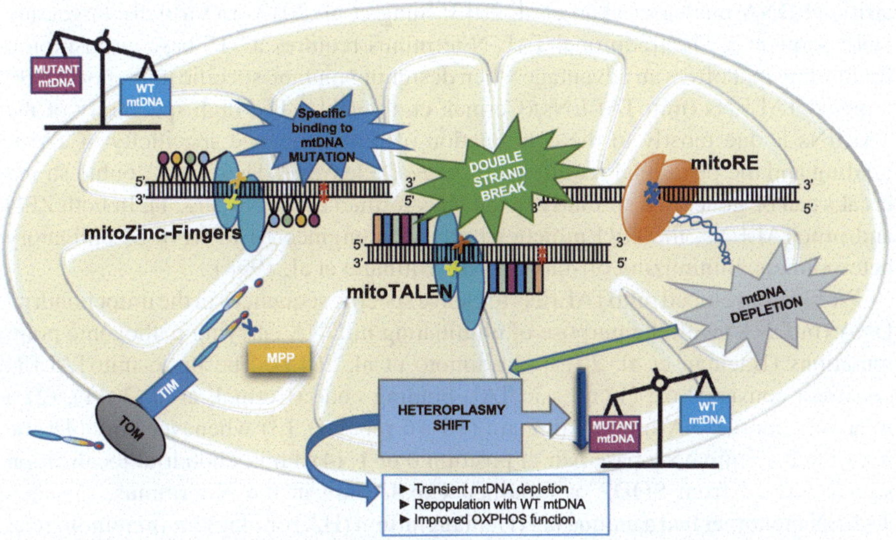

Fig. 1 Nucleases used to modify mtDNA heteroplasmy. Cells with high levels of mutant mtDNA and low levels of wild-type mtDNA have energy deficits and are phenotypically altered. Specific DNA nucleases, such as ZFN, TALEN and Restriction Endonucleases can be modified to localize to mitochondria by the addition of a mitochondrial targeting sequence (MTS). This MTS is removed by metal proteases upon import (MPP). After specific binding to the mutant mtDNA, a double strand break leads to subsequent degradation of the DNA. This causes a transient depletion, until the residual WT mtDNA replicates to make up for the reduced levels. This change in heteroplasmy improves the biochemical defect and the functional phenotypes

3.2.1 TALENs as Genetic Tools to Alter mtDNA Heteroplasmy

TALE-containing proteins were initially isolated from *Xanthomonas* bacteria (Boch and Bonas 2010), as positive transcription factors activating expression of downstream genes, which may contribute to bacterial colonization, symptom development, or pathogen dissemination (Boch and Bonas 2010). They recognize specific DNA sequences through the Repeat Variable Di-residue motif (RVD), which commonly recognize 15–20 bases (Valton et al. 2014; Bogdanove and Voytas 2011; Scharenberg et al. 2013) and its length is determined to avoid off-target cleavage (Hockemeyer et al. 2011; Valton et al. 2014; Cox et al. 2015). Each repeat is identical, and the polymorphism among the repeats are almost exclusively localized to the pair of residues at positions 12 and 13 (so-called RVDs) with the four most common being HD, NG, NI, and NN accounting for the binding to each one of the four nucleotides C, T, A, and G respectively, providing the basis to the engineering of novel specificities (Moscou and Bogdanove 2009; Boch et al. 2009; Bogdanove and Voytas 2011).When attached to a dimeric nuclease domain such as *Fok*I they are denominated TALENs and form heterodimers that can be used for DNA cleavage and editing (Cermak et al. 2011; Pan et al. 2013). Their modular DNA recognition code makes it an ideal system for the customized design of artificial DNA nucleases (Pan et al. 2013; Sung et al. 2013) to virtually target any DNA sequence. The traditional TAL N-terminus requires a "T" base at position 0 for binding that offers an advantage when designing mutant-specific mitochondrially targeted TALENs (mitoTALENs)(Cermak et al. 2011). The high specificity of the TALENs is due mostly to the combination of both sequence specificity of TALE binding and the positional requirements of *FokI* cleavage. Therefore, double strand breaks can be generated similarly to the ones formed by ZFNs (Fig. 1). In both ZFN and mitoTALENs, the *Fok*I moieties have been engineered to work as obligatory heterodimers, minimizing off-target DSB (Bitinaite et al. 1998).

We have developed mitoTALENs to cleave specific sequences in the mitochondrial DNA (mtDNA) with the purpose of eliminating mtDNA carrying pathogenic point mutations (Bacman et al. 2013; Hashimoto et al. 2015). The basic mitoTALEN construct consisted of: (1) a basic TAL-binding code (Cermak et al. 2011), (2) a relatively short DNA-binding domain (10–16 repeats), (3) whenever possible, the use of a discriminatory base pair at position 0 or 1, (4) a mitochondrial localization signal derived from SOD2 or modified Cox8A/Su9 in the N-terminus, (5) each TALEN monomer had a unique tag (Hemagglutinin (HA) or Flag) for immunological detection (6) each TALEN monomer had GFP or mCherry in the C-terminus as a useful tool for selecting transfected cells in a cell sorter equipment (7) inclusion of 3′-UTR untranslated region from a known nuclear gene coding for a mitochondrial protein (ATP5B or SOD2 mRNA). These sequences appear to help localize mRNA to ribosomes contacting mitochondria (8) both mitoTALEN and the fluorescence marker genes were expressed by a single promoter (CMV) using a recoded picoviral 2A-like sequence (T2A) (Szymczak et al. 2004) that allows the simultaneous transcription of different proteins from a unique promoter (Bacman et al. 2014).

We have successfully targeted the mtDNA "common deletion" (m.8483_13459del4977) present in approximately 30% of patients with mtDNA deletions (Schon et al. 1989) and also in normal aging tissues (Corral-Debrinski et al. 1992). The "common deletion"-mitoTALEN (Δ5-mitoTALEN) co-localized to mitochondria, and was tested in human heteroplasmic cybrids for the common deletion by transfecting two monomer-containing plasmids. When the mitoTALENs are close enough to allow *Fok*I dimerization and subsequent cleavage of the mtDNA, which is the case of the deleted Δ5-mitochondria DNA, the mitoTALEN effectively reduced the mtDNA deletion load and changed mtDNA heteroplasmy to a predominance of wild-type mtDNA. We then measured the levels of the different mtDNA species and found that the change in heteroplasmy was primarily caused by a reduction in the absolute levels of deletion-mutant mtDNA with the trend toward a reduction in the total mtDNA levels after 2 days, compensated by an increase in wild-type mtDNA levels at 14 days (Bacman et al. 2013). We were also able to target more challenging mtDNA point mutations that require the specific recognition of one base to be discriminated from the non-targeted DNA (wild-type DNA), such as the mtDNA mutation m.14459G > A in the MT-ND6 (Jun et al. 1996), that causes the Leber's hereditary optic neuropathy plus dystonia (Jun et al. 1996). In this case, one monomer binds the wild-type sequence adjacent to the mutation (or wild-type strand) and the cleavage is dictated by the binding of the recognition sequence where the m.14459A is present (mutant strand). Transfection of heteroplasmic cells showed a significant increase in the wild-type mtDNA that persisted for 14 days. In addition, complex I activity which was defective in human osteosarcoma heteroplasmic cells carrying the point mutation in the MT-ND6 was improved after transfection with the mitoTALENs (Bacman et al. 2013). When the levels of mutation were very high (more than 90%), such as in cybrids harboring the m.14459G > A in MT-ND6 mutation, a transient decrease in the levels of total mtDNA 2 days after transfection was observed, but these levels were back to normal when cells were analyzed at 14 days (Bacman et al. 2013).

We were able to differentiate other targets with single nucleotide differences such as the m.8344A > G $tRNA^{Lys}$ gene mutation associated with myoclonic epilepsy with ragged red fibers (MERRF) syndrome (Shoffner et al. 1990; Berkovic et al. 1991), and the m.13513G > A ND5mutation associated with MELAS/Leigh syndrome (Chol et al. 2003; Shanske et al. 2008). In both cases, the designed mitoTALENs co-localized to mitochondria and were able to promote a robust change in mtDNA heteroplasmy in both mtDNA point mutations while improving the OXPHOS function (Hashimoto et al. 2015).

Although mitoTALENs are a feasible tool for gene therapy in patients because of the large size of the constructs, packaging into viral vectors is a concern. Nevertheless, we were able to reduce the size of the TALE-DNA binding domain, downsizing the original m.8344A > G mitoTALEN sense and antisense monomers, which originally contained 15.5 and 9.5 RVDs, to 10.5 and 7.5 RVDs, respectively. These shorter mitoTALENs were still very effective in recognizing a single base difference and shifting heteroplasmy (Hashimoto et al. 2015).

3.2.2 ZFNs as Genetic Tools to Alter mtDNA Heteroplasmy

Zinc finger nucleases (ZNFs) are chimeric enzymes with a modular Cys2-His2-zinc finger protein (ZFP). The DNA-binding zinc-finger domains (ZF) recognize 3–4 bp that account for a total recognition of a unique 9–18 bp DNA sequence (Urnov et al. 2010; Cox et al. 2015). Fusing a particular ZFP to a *Fok*I nuclease domain, results in double strand DNA cleavage (Wu et al. 2007). Similarly to TALENs, ZFNs are developed as heterodimers, recognizing two adjacent stretches of DNA (Pan et al. 2013) and to achieve dimerization, pairs of ZFNs are used to bind adjacent sequences of the double strand DNA to cleave it (Smith et al. 2000), providing sequence specificity to target the DNA (Kim et al. 1996). ZFNs technology has been successfully applied to nuclear genome editing in many organisms including human cell lines (Carroll 2008) to correct specific genes via homologous recombination after achieving double-strand breaks of nuclear DNA (Urnov et al. 2005). The incorporation of nuclear export sequences into ZFN monomers between an MLS and the DNA-binding domain was used to build ZFNs targeted to mitochondria to overcome the innate tropism for the nucleus (Minczuk et al. 2006), besides using a pair of heterodimeric domains to reduce homodimer activity (Doyon et al. 2011; Gammage et al. 2014). ZFNs' technology has been expended to target pathogenic mitochondrial genomes in cells harboring the 8993T > G-NARP/Leigh's syndrome mutation and the "common deletion" (m.8483_13459del4977) (Minczuk et al. 2006, 2008; Gammage et al. 2014). Mitochondrial localization was nicely achieved, and the shift in heteroplasmy decreasing the common-deletion mtDNA to WT was associated with improved oxygen consumption rates and increased respiratory subunits of complexes I-IV (Gammage et al. 2014). A controlled system allowing lower expression levels of the mitoZFNs, resulted in higher heteroplasmy shift with less mtDNA depletion and better mitochondrial function (Gammage et al. 2016).

3.2.3 Can CRISPR/Cas9 Be Used to Alter mtDNA Heteroplasmy?

Recent advances in genome engineering technologies based on the CRISPR-associated RNA-guided endonuclease Cas9 are being developed. Cas9 can be guided to specific locations within complex genomes by a short RNA search string. Using genome engineering technologies based on the CRISPR-associated RNA-guided endonuclease, Cas9 DNA sequences within the endogenous genome can be easily edited or modulated in virtually any organism of choice. It has been shown that Cas9 can be guided to specific locations within complex genomes by a short RNA search string (Hsu et al. 2014). Related to the mitochondrial genome, cleavage of the specific mtDNA Cox1 and Cox3 in HEK-293 cells by FLAG-Cas9 together with gRNA targeted to mitochondria has been reported (Jo et al. 2015). Also with a mitochondria-targeted Cas9 (mitoCas9) together with gRNA targeting mitochondrial CoxI, and CoxIII, they reported specific cleavage of mtDNA, leading to decreased copy number, mitochondrial membrane potential disruption and cell growth inhibition. In our laboratory, we were not able to observe mtDNA cleavage using a similar approach. Cas9 was imported into mitochondria, but no change in heteroplasmy was observed for two different mtDNA mutations (unpublished). Even though we used different forms of gRNA, which included RNA structures previously described as import structures (Wang et al. 2012), we are doubtful that the gRNA was efficiently imported into mammalian mitochondria (unpublished observations). A more comprehensive discussion on this issue was recently published (Gammage et al. 2017).

3.3 Applications of Specific Nucleases to Reduce the Levels of mtDNA Mutations in the Germline

Reddy et al. (2015) using the NZB/BALB heteroplasmic mouse model (Jenuth et al. 1997) selectively prevented germline transmission of specific mtDNA haplotypes using either mitochondria-targeted restriction endonucleases or mitoTALENs. Using mito*ApalL* (Bayona-Bafaluy et al. 2005; Bacman et al. 2010) mRNA injection in oocytes of the heteroplasmic mice and in one cell embryos, they were able to specifically reduce BALB mtDNA (*ApaLI* has one restriction site in the BALB and none in the NZB haplotype) in MII oocytes and blastocysts resulting in a decrease of mtDNA levels proportional to the BALB mtDNA levels. With this approach, a selective clearance of a specific mtDNA in the offspring of heteroplasmic mothers was achieved, and F1 showed normal developed animals, with normal mtDNA levels. When mitoTALENs RNA injections in the oocytes against the NZB haplotype were used, a specific decrease of the NZB haplotype was achieved 48 h after. When the authors generated artificial mammalian oocytes, by fusing patient's cells harboring the LHON/NARP mutation m.14459G > A (Taylor and Turnbull 2005) to mouse MII oocytes, and injected with mRNA encoding the LHON mitoTALEN described before (Bacman et al. 2013), RFLP analysis 48 h after mRNA injection demonstrated the specific reduction of LHON mtDNA in fused oocytes (Reddy et al. 2015).

4 *In Vivo* Delivery of Gene-Editing Tools to Mitochondria: Challenges and Perspectives

As mentioned above, delivery of gene editing tools into mitochondria poses some challenges. The relatively impermeable mitochondrial membrane and the fact that there are thousands of copies of mtDNA in a cell can be disadvantageous. On the other hand, the mitochondrial genome is small and may be easier to be specifically targeted when compared to the genomic DNA. There are some studies reporting effective gene therapy with adeno-associated-virus (AAV) for nuclear genes of mitochondrial disorders. AAV vectors are the most widely used vectors for nuclear gene therapy in humans because they are known to remain episomic therefore reducing the risk of insertional mutagenesis (Chandler and Venditti 2016). In addition, an expanding number of natural and also engineered serotypes targeting different tissues have been described allowing a vast and selective tropism. There are nine different serotypes of AAV (1–9) currently available which differ from each other primarily on their capsid protein sequence (Watanabe et al. 1978). Variations in the capside confer different cell and tissue affinities and play an important role in the definition of the expression onset and in the levels of transgene expression. The main restrictions of their usage are their limited packing capacity, no more than 4.7 kb should be included between ITRs and the handicap to target more tissues at the same time, which is fundamental in multisystem disorders such as mitochondrial diseases. AAV serotype 2 (AAV2) was the first used for gene transfer. A few examples of the application of *in vivo* viral delivery of AAV vectors will be described for mitochondrial disease models of nuclear-encoded genes. The AAV2/8 serotype was successfully applied to express a wild-type form of mitochondrial sulfide di-oxygenase (SDO) Ethe1 in the liver of Ethe1 KO mice, a model for ethylmalonic encephalopathy (EE), a very severe mitochondrial disease. The virus administration in mice fully rescued enzyme activity and prolonged life span (Di Meo et al. 2012).

Mutations in TYMP (thymidine phosphorylase) have been related with mitochondrial neurogastrointestinal encephalopathy (MNGIE), characterized by mtDNA depletion, and formation of mtDNA deletions and point mutations. The use of AAV2/8-hc*TYMP* which has a liver-specific promoter has been shown to be an effective tool to reduce the systemic toxic levels of thymidine found in the systemic circulation (Torres-Torronteras et al. 2014). Another example is the recombinant adeno-associated viral vector (rAAV2) carrying the mouse Ant1 cDNA that was used to transduce muscle cells and muscle from Ant1 mutant mice, which have a mitochondrial myopathy. AAV-ANT1 injection in heart and muscle resulted in long-term and stable expression of the Ant1 transgene reversing the histopathological changes associated with the mitochondrial myopathy. This approach had the potential to provide symptomatic relief for the ophthalmoplegia and ptosis resulting from paralysis of the extraocular eye muscles caused by mutations in the Ant1 gene (Flierl et al. 2005).

Furthermore, the intra-vitreal administration of an AAV2/2_AIF1 vector has been shown to counteract the retinal and optic nerve complex I deficiency responsible for glial and microglial cell activation, retinal ganglion loss and optic atrophy in a mouse model of a complex I deficiency of a severely reduced AIF expression (Bouaita et al. 2012). In addition, the MPV17 KO mice phenotype was rescued by AAV viral expression of human MPV17 cDNA with a liver specific promoter (Bottani et al. 2014). However, is still unclear whether the tissue-specific expression of the gene will correct the brain disorder. Another study showed that the loss of *Ndufs4* function in the vestibular nucleus (VN) contributed to the motor phenotype and breathing abnormalities of NDUFS4 KO mice (AAV-VN-KO). The viral rescue with a AAV1 construct encoding NDUFS4-IRES-GFP was shown to delay the clinical progression of the disease and significantly increased the life span of the AAV-VN-VR mice when compared with the KO mice (Quintana et al. 2012b). However, the prolonged life span of the AAV-VN-VR mice revealed that other brain regions (i.e. striatum) where developing gliosis at a slower rate. An alternative to target mitochondria is to harness established techniques for nuclear genes therapy to "allotopically" express mtDNA encoded proteins. The mtDNA m.11778G:A mutation in the NADH subunit 4 (ND4 of complex I) has been implicated in 50% of the cases of LHON patients. Several *in vitro* reports focused in the allotopic expression of complex I subunits and this mutation has been the main focus of these type of studies, to date. *In vitro* and pre-clinical studies have shown some success by using this approach. Currently, in spite of all the concerns raised by other studies showing that allotopically-expressed mitochondrial proteins are not effectively imported and assembled into OXPHOS complexes (Oca-Cossio et al. 2003; McKenzie et al. 2003; Perales-Clemente et al. 2011; Figueroa-Martinez et al. 2011), ongoing clinical trials to examine intraocular AAV-ND4 for LHON are under way in the USA (Lam et al. 2010; Feuer et al. 2016), China (Wan et al. 2016) and Europe (https://clinicaltrials.gov/show/NCT02064569).

Previously, it has been demonstrated by our group and others that is possible to shift heteroplasmy *in vivo* by using mitochondrial-targeted endonucleases by either using different rAAV serotypes or adenovirus. Therefore, we believe it is possible to apply an identical strategy to deliver mitoTALENs to shift heteroplasmy in heteroplasmic mice. The big size of the mitoTALENs constructs as well as mitochondrial targeted zinc-fingers may difficult their packing into viral systems nevertheless our group and others are currently attempting to test the efficacy of this approach *in vivo*.

Investigators have been successful by using AAV/ZFNs in animal models, such as the model of hemophilia (Li et al. 2011). In regards to TALENs, their repetitive characteristics can induce unwanted recombination events when packaged to viral systems such as lentivirus (Holkers et al. 2013). Considering this, proper delivery vectors such as high-capacity adenovirus (HCAdVs), cationic polymers, and cell penetrating peptides (CPPs)-TAT may be used as alternatives and in future clinical trials. Actually, one vector made of cationic polymers (TurboFect) has been used to carry TALENs targeting the human papillomavirus (HPV) directly to the cervix of transgenic mice (Hu et al. 2015). After therapy, the tumor size was reduced and no off-target effects were observed (Hu et al. 2015). In spite of the undeniable challenges, the cases of success of *in vivo* delivery of gene-editing tools to the nuclear genome, give us hope that soon these techniques can also be successfully applied to mtDNA diseases.

References

Akhmedov AT, Marin-Garcia J (2015) Mitochondrial DNA maintenance: an appraisal. Mol Cell Biochem 409(1–2):283–305. https://doi.org/10.1007/s11010-015-2532-x

Alexeyev MF, Venediktova N, Pastukh V, Shokolenko I, Bonilla G, Wilson GL (2008) Selective elimination of mutant mitochondrial genomes as therapeutic strategy for the treatment of NARP and MILS syndromes. Gene Ther 15(7):516–523. https://doi.org/10.1038/sj.gt.2008.11

Anderson S, Bankier AT, Barrell BG, de Bruijn MH, Coulson AR, Drouin J, Eperon IC, Nierlich DP, Roe BA, Sanger F, Schreier PH, Smith AJ, Staden R, Young IG (1981) Sequence and organization of the human mitochondrial genome. Nature 290(5806):457–465

Andrews RM, Kubacka I, Chinnery PF, Lightowlers RN, Turnbull DM, Howell N (1999) Reanalysis and revision of the Cambridge reference sequence for human mitochondrial DNA. Nat Genet 23(2):147. https://doi.org/10.1038/13779

Bacman SR, Moraes CT (2007) Transmitochondrial technology in animal cells. Methods Cell Biol 80:503–524. https://doi.org/10.1016/S0091-679X(06)80025-7

Bacman SR, Williams SL, Hernandez D, Moraes CT (2007) Modulating mtDNA heteroplasmy by mitochondria-targeted restriction endonucleases in a 'differential multiple cleavage-site' model. Gene Ther 14(18):1309–1318. https://doi.org/10.1038/sj.gt.3302981

Bacman SR, Williams SL, Garcia S, Moraes CT (2010) Organ-specific shifts in mtDNA heteroplasmy following systemic delivery of a mitochondria-targeted restriction endonuclease. Gene Ther 17(6):713–720. https://doi.org/10.1038/gt.2010.25

Bacman SR, Williams SL, Duan D, Moraes CT (2012) Manipulation of mtDNA heteroplasmy in all striated muscles of newborn mice by AAV9-mediated delivery of a mitochondria-targeted restriction endonuclease. Gene Ther 19(11):1101–1106. https://doi.org/10.1038/gt.2011.196

Bacman SR, Williams SL, Pinto M, Peralta S, Moraes CT (2013) Specific elimination of mutant mitochondrial genomes in patient-derived cells by mitoTALENs. Nat Med 19(9):1111–1113. https://doi.org/10.1038/nm.3261

Bacman SR, Williams SL, Pinto M, Moraes CT (2014) The use of mitochondria-targeted endonucleases to manipulate mtDNA. Methods Enzymol 547:373–397. https://doi.org/10.1016/B978-0-12-801415-8.00018-7

Bandy B, Davison AJ (1990) Mitochondrial mutations may increase oxidative stress: implications for carcinogenesis and aging? Free Radic Biol Med 8(6):523–539

Battersby BJ, Loredo-Osti JC, Shoubridge EA (2003) Nuclear genetic control of mitochondrial DNA segregation. Nat Genet 33(2):183–186. https://doi.org/10.1038/ng1073

Bayona-Bafaluy MP, Blits B, Battersby BJ, Shoubridge EA, Moraes CT (2005) Rapid directional shift of mitochondrial DNA heteroplasmy in animal tissues by a mitochondrially targeted restriction endonuclease. Proc Natl Acad Sci U S A 102(40):14392–14397. https://doi.org/10.1073/pnas.0502896102

Berkovic SF, Shoubridge EA, Andermann F, Andermann E, Carpenter S, Karpati G (1991) Clinical spectrum of mitochondrial DNA mutation at base pair 8344. Lancet 338(8764):457

Bitinaite J, Wah DA, Aggarwal AK, Schildkraut I (1998) FokI dimerization is required for DNA cleavage. Proc Natl Acad Sci U S A 95(18):10570–10575

Boch J, Bonas U (2010) Xanthomonas AvrBs3 family-type III effectors: discovery and function. Annu Rev Phytopathol 48:419–436. https://doi.org/10.1146/annurev-phyto-080508-081936

Boch J, Scholze H, Schornack S, Landgraf A, Hahn S, Kay S, Lahaye T, Nickstadt A, Bonas U (2009) Breaking the code of DNA binding specificity of TAL-type III effectors. Science 326(5959):1509–1512. https://doi.org/10.1126/science.1178811

Bogdanove AJ, Voytas DF (2011) TAL effectors: customizable proteins for DNA targeting. Science 333(6051):1843–1846. https://doi.org/10.1126/science.1204094

Bottani E, Giordano C, Civiletto G, Di Meo I, Auricchio A, Ciusani E, Marchet S, Lamperti C, d'Amati G, Viscomi C, Zeviani M (2014) AAV-mediated liver-specific MPV17 expression restores mtDNA levels and prevents diet-induced liver failure. Mol Ther 22(1):10–17. https://doi.org/10.1038/mt.2013.230

Bouaita A, Augustin S, Lechauve C, Cwerman-Thibault H, Benit P, Simonutti M, Paques M, Rustin P, Sahel JA, Corral-Debrinski M (2012) Downregulation of apoptosis-inducing factor in Harlequin mice induces progressive and severe optic atrophy which is durably prevented by AAV2-AIF1 gene therapy. Brain 135(Pt 1):35–52. https://doi.org/10.1093/brain/awr290

Carroll D (2008) Progress and prospects: zinc-finger nucleases as gene therapy agents. Gene Ther 15(22):1463–1468. https://doi.org/10.1038/gt.2008.145

Cermak T, Doyle EL, Christian M, Wang L, Zhang Y, Schmidt C, Baller JA, Somia NV, Bogdanove AJ, Voytas DF (2011) Efficient design and assembly of custom TALEN and other TAL effector-based constructs for DNA targeting. Nucleic Acids Res 39(12):e82. https://doi.org/10.1093/nar/gkr218

Chandler RJ, Venditti CP (2016) Gene therapy for metabolic diseases. Transl Sci Rare Dis 1(1):73–89. https://doi.org/10.3233/TRD-160007

Chen XJ, Butow RA (2005) The organization and inheritance of the mitochondrial genome. Nat Rev Genet 6(11):815–825. https://doi.org/10.1038/nrg1708

Chinnery PF, Thorburn DR, Samuels DC, White SL, Dahl HM, Turnbull DM, Lightowlers RN, Howell N (2000) The inheritance of mitochondrial DNA heteroplasmy: random drift, selection or both? Trends Genet 16(11):500–505

Chinnery PF, DiMauro S, Shanske S, Schon EA, Zeviani M, Mariotti C, Carrara F, Lombes A, Laforet P, Ogier H, Jaksch M, Lochmuller H, Horvath R, Deschauer M, Thorburn DR, Bindoff LA, Poulton J, Taylor RW, Matthews JN, Turnbull DM (2004) Risk of developing a mitochondrial DNA deletion disorder. Lancet 364(9434):592–596. https://doi.org/10.1016/S0140-6736(04)16851-7

Chol M, Lebon S, Benit P, Chretien D, de Lonlay P, Goldenberg A, Odent S, Hertz-Pannier L, Vincent-Delorme C, Cormier-Daire V, Rustin P, Rotig A, Munnich A (2003) The mitochondrial DNA G13513A MELAS mutation in the NADH dehydrogenase 5 gene is a frequent cause of Leigh-like syndrome with isolated complex I deficiency. J Med Genet 40(3):188–191

Christie JR, Schaerf TM, Beekman M (2015) Selection against heteroplasmy explains the evolution of uniparental inheritance of mitochondria. PLoS Genet 11(4):e1005112. https://doi.org/10.1371/journal.pgen.1005112

Cong L, Ran FA, Cox D, Lin S, Barretto R, Habib N, Hsu PD, Wu X, Jiang W, Marraffini LA, Zhang F (2013) Multiplex genome engineering using CRISPR/Cas systems. Science 339(6121):819–823. https://doi.org/10.1126/science.1231143

Corral-Debrinski M, Horton T, Lott MT, Shoffner JM, Beal MF, Wallace DC (1992) Mitochondrial DNA deletions in human brain: regional variability and increase with advanced age. Nat Genet 2(4):324–329. https://doi.org/10.1038/ng1292-324

Cox DB, Platt RJ, Zhang F (2015) Therapeutic genome editing: prospects and challenges. Nat Med 21(2):121–131. https://doi.org/10.1038/nm.3793

Cree LM, Samuels DC, de Sousa Lopes SC, Rajasimha HK, Wonnapinij P, Mann JR, Dahl HH, Chinnery PF (2008) A reduction of mitochondrial DNA molecules during embryogenesis explains the rapid segregation of genotypes. Nat Genet 40(2):249–254. https://doi.org/10.1038/ng.2007.63

De Fanti S, Vicario S, Lang M, Simone D, Magli C, Luiselli D, Gianaroli L, Romeo G (2017) Intra-individual purifying selection on mitochondrial DNA variants during human oogenesis. Hum Reprod 32(5):1100–1107. https://doi.org/10.1093/humrep/dex051

Dey R, Barrientos A, Moraes CT (2000) Functional constraints of nuclear-mitochondrial DNA interactions in xenomitochondrial rodent cell lines. J Biol Chem 275(40):31520–31527. https://doi.org/10.1074/jbc.M004053200

Di Meo I, Auricchio A, Lamperti C, Burlina A, Viscomi C, Zeviani M (2012) Effective AAV-mediated gene therapy in a mouse model of ethylmalonic encephalopathy. EMBO Mol Med 4(9):1008–1014. https://doi.org/10.1002/emmm.201201433

Diaz F, Bayona-Bafaluy MP, Rana M, Mora M, Hao H, Moraes CT (2002) Human mitochondrial DNA with large deletions repopulates organelles faster than full-length genomes under relaxed copy number control. Nucleic Acids Res 30(21):4626–4633

Diaz F, Thomas CK, Garcia S, Hernandez D, Moraes CT (2005) Mice lacking COX10 in skeletal muscle recapitulate the phenotype of progressive mitochondrial myopathies associated with cytochrome c oxidase deficiency. Hum Mol Genet 14(18):2737–2748. https://doi.org/10.1093/hmg/ddi307

Doyon Y, Vo TD, Mendel MC, Greenberg SG, Wang J, Xia DF, Miller JC, Urnov FD, Gregory PD, Holmes MC (2011) Enhancing zinc-finger-nuclease activity with improved obligate heterodimeric architectures. Nat Methods 8(1):74–79. https://doi.org/10.1038/nmeth.1539

Durham SE, Samuels DC, Cree LM, Chinnery PF (2007) Normal levels of wild-type mitochondrial DNA maintain cytochrome c oxidase activity for two pathogenic mitochondrial DNA mutations but not for m.3243A→G. Am J Hum Genet 81(1):189–195. https://doi.org/10.1086/518901

Ebert KM, Liem H, Hecht NB (1988) Mitochondrial DNA in the mouse preimplantation embryo. J Reprod Fertil 82(1):145–149

Esposito LA, Melov S, Panov A, Cottrell BA, Wallace DC (1999) Mitochondrial disease in mouse results in increased oxidative stress. Proc Natl Acad Sci U S A 96(9):4820–4825

Fan W, Waymire KG, Narula N, Li P, Rocher C, Coskun PE, Vannan MA, Narula J, Macgregor GR, Wallace DC (2008) A mouse model of mitochondrial disease reveals germline selection against severe mtDNA mutations. Science 319(5865):958–962. https://doi.org/10.1126/science.1147786

Feuer WJ, Schiffman JC, Davis JL, Porciatti V, Gonzalez P, Koilkonda RD, Yuan H, Lalwani A, Lam BL, Guy J (2016) Gene therapy for leber hereditary optic neuropathy: initial results. Ophthalmology 123(3):558–570. https://doi.org/10.1016/j.ophtha.2015.10.025

Figueroa-Martinez F, Vazquez-Acevedo M, Cortes-Hernandez P, Garcia-Trejo JJ, Davidson E, King MP, Gonzalez-Halphen D (2011) What limits the allotopic expression of nucleus-encoded mitochondrial genes? The case of the chimeric Cox3 and Atp6 genes. Mitochondrion 11(1):147–154. https://doi.org/10.1016/j.mito.2010.09.003

Flierl A, Chen Y, Coskun PE, Samulski RJ, Wallace DC (2005) Adeno-associated virus-mediated gene transfer of the heart/muscle adenine nucleotide translocator (ANT) in mouse. Gene Ther 12(7):570–578. https://doi.org/10.1038/sj.gt.3302443

Freyer C, Cree LM, Mourier A, Stewart JB, Koolmeister C, Milenkovic D, Wai T, Floros VI, Hagstrom E, Chatzidaki EE, Wiesner RJ, Samuels DC, Larsson NG, Chinnery PF (2012) Variation in germline mtDNA heteroplasmy is determined prenatally but modified during subsequent transmission. Nat Genet 44(11):1282–1285. https://doi.org/10.1038/ng.2427

Gaj T, Gersbach CA, Barbas CF 3rd (2013) ZFN, TALEN, and CRISPR/Cas-based methods for genome engineering. Trends Biotechnol 31(7):397–405. https://doi.org/10.1016/j.tibtech.2013.04.004

Gammage PA, Moraes CT, Minczuk M (2017) Mitochondrial genome engineering: the revolution may not be CRISPR-Ized. Trends Genet pii:S0168-9525(17)30191-9. https://doi.org/10.1016/j.tig.2017.11.001

Gammage PA, Rorbach J, Vincent AI, Rebar EJ, Minczuk M (2014) Mitochondrially targeted ZFNs for selective degradation of pathogenic mitochondrial genomes bearing large-scale deletions or point mutations. EMBO Mol Med 6(4):458–466. https://doi.org/10.1002/emmm.201303672

Gammage PA, Van Haute L, Minczuk M (2016) Engineered mtZFNs for manipulation of human mitochondrial DNA heteroplasmy. Methods Mol Biol 1351:145–162. https://doi.org/10.1007/978-1-4939-3040-1_11

Ghosh A, Yue Y, Long C, Bostick B, Duan D (2007) Efficient whole-body transduction with trans-splicing adeno-associated viral vectors. Mol Ther 15(6):1220. https://doi.org/10.1038/sj.mt.6300153

Giles RE, Blanc H, Cann HM, Wallace DC (1980) Maternal inheritance of human mitochondrial DNA. Proc Natl Acad Sci U S A 77(11):6715–6719

Gilkerson RW, Schon EA, Hernandez E, Davidson MM (2008) Mitochondrial nucleoids maintain genetic autonomy but allow for functional complementation. J Cell Biol 181(7):1117–1128. https://doi.org/10.1083/jcb.200712101

Gorman GS, Schaefer AM, Ng Y, Gomez N, Blakely EL, Alston CL, Feeney C, Horvath R, Yu-Wai-Man P, Chinnery PF, Taylor RW, Turnbull DM, McFarland R (2015) Prevalence of nuclear and mitochondrial DNA mutations related to adult mitochondrial disease. Ann Neurol 77(5):753–759. https://doi.org/10.1002/ana.24362

Graham BH, Waymire KG, Cottrell B, Trounce IA, MacGregor GR, Wallace DC (1997) A mouse model for mitochondrial myopathy and cardiomyopathy resulting from a deficiency in the heart/muscle isoform of the adenine nucleotide translocator. Nat Genet 16(3):226–234. https://doi.org/10.1038/ng0797-226

Hance N, Ekstrand MI, Trifunovic A (2005) Mitochondrial DNA polymerase gamma is essential for mammalian embryogenesis. Hum Mol Genet 14(13):1775–1783. https://doi.org/10.1093/hmg/ddi184

Hao H, Bonilla E, Manfredi G, DiMauro S, Moraes CT (1995) Segregation patterns of a novel mutation in the mitochondrial tRNA glutamic acid gene associated with myopathy and diabetes mellitus. Am J Hum Genet 56(5):1017–1025

Hashimoto M, Bacman SR, Peralta S, Falk MJ, Chomyn A, Chan DC, Williams SL, Moraes CT (2015) MitoTALEN: a general approach to reduce mutant mtDNA loads and restore oxidative phosphorylation function in mitochondrial diseases. Mol Ther 23(10):1592–1599. https://doi.org/10.1038/mt.2015.126

Hashizume O, Shimizu A, Yokota M, Sugiyama A, Nakada K, Miyoshi H, Itami M, Ohira M, Nagase H, Takenaga K, Hayashi J (2012) Specific mitochondrial DNA mutation in mice regulates diabetes and lymphoma development. Proc Natl Acad Sci U S A 109(26):10528–10533. https://doi.org/10.1073/pnas.1202367109

Hockemeyer D, Wang H, Kiani S, Lai CS, Gao Q, Cassady JP, Cost GJ, Zhang L, Santiago Y, Miller JC, Zeitler B, Cherone JM, Meng X, Hinkley SJ, Rebar EJ, Gregory PD, Urnov FD, Jaenisch R (2011) Genetic engineering of human pluripotent cells using TALE nucleases. Nat Biotechnol 29(8):731–734. https://doi.org/10.1038/nbt.1927

Holkers M, Maggio I, Liu J, Janssen JM, Miselli F, Mussolino C, Recchia A, Cathomen T, Goncalves MA (2013) Differential integrity of TALE nuclease genes following adenoviral and lentiviral vector gene transfer into human cells. Nucleic Acids Res 41(5):e63. https://doi.org/10.1093/nar/gks1446

Holt IJ, Harding AE, Petty RK, Morgan-Hughes JA (1990) A new mitochondrial disease associated with mitochondrial DNA heteroplasmy. Am J Hum Genet 46(3):428–433

Howell N, Halvorson S, Kubacka I, McCullough DA, Bindoff LA, Turnbull DM (1992) Mitochondrial gene segregation in mammals: is the bottleneck always narrow? Hum Genet 90(1–2):117–120

Hsu PD, Lander ES, Zhang F (2014) Development and applications of CRISPR-Cas9 for genome engineering. Cell 157(6):1262–1278. https://doi.org/10.1016/j.cell.2014.05.010

Hu Z, Ding W, Zhu D, Yu L, Jiang X, Wang X, Zhang C, Wang L, Ji T, Liu D, He D, Xia X, Zhu T, Wei J, Wu P, Wang C, Xi L, Gao Q, Chen G, Liu R, Li K, Li S, Wang S, Zhou J, Ma D, Wang H (2015) TALEN-mediated targeting of HPV oncogenes ameliorates HPV-related cervical malignancy. J Clin Invest 125(1):425–436. https://doi.org/10.1172/JCI78206

Hutchison CA 3rd, Newbold JE, Potter SS, Edgell MH (1974) Maternal inheritance of mammalian mitochondrial DNA. Nature 251(5475):536–538

Iborra FJ, Kimura H, Cook PR (2004) The functional organization of mitochondrial genomes in human cells. BMC Biol 2:9. https://doi.org/10.1186/1741-7007-2-9

Inagaki K, Fuess S, Storm TA, Gibson GA, McTiernan CF, Kay MA, Nakai H (2006) Robust systemic transduction with AAV9 vectors in mice: efficient global cardiac gene transfer superior to that of AAV8. Mol Ther 14(1):45–53. https://doi.org/10.1016/j.ymthe.2006.03.014

Inoue K, Nakada K, Ogura A, Isobe K, Goto Y, Nonaka I, Hayashi JI (2000) Generation of mice with mitochondrial dysfunction by introducing mouse mtDNA carrying a deletion into zygotes. Nat Genet 26(2):176–181. https://doi.org/10.1038/82826

Iommarini L, Peralta S, Torraco A, Diaz F (2015) Mitochondrial diseases part II: mouse models of OXPHOS deficiencies caused by defects in regulatory factors and other components required for mitochondrial function. Mitochondrion 22:96–118. https://doi.org/10.1016/j.mito.2015.01.008

Jenuth JP, Peterson AC, Fu K, Shoubridge EA (1996) Random genetic drift in the female germline explains the rapid segregation of mammalian mitochondrial DNA. Nat Genet 14(2):146–151. https://doi.org/10.1038/ng1096-146

Jenuth JP, Peterson AC, Shoubridge EA (1997) Tissue-specific selection for different mtDNA genotypes in heteroplasmic mice. Nat Genet 16(1):93–95. https://doi.org/10.1038/ng0597-93

Jo A, Ham S, Lee GH, Lee YI, Kim S, Lee YS, Shin JH, Lee Y (2015) Efficient mitochondrial genome editing by CRISPR/Cas9. Biomed Res Int 2015:305716. https://doi.org/10.1155/2015/305716

Jun AS, Trounce IA, Brown MD, Shoffner JM, Wallace DC (1996) Use of transmitochondrial cybrids to assign a complex I defect to the mitochondrial DNA-encoded NADH dehydrogenase subunit 6 gene mutation at nucleotide pair 14459 that causes Leber hereditary optic neuropathy and dystonia. Mol Cell Biol 16(3):771–777

Karamanlidis G, Lee CF, Garcia-Menendez L, Kolwicz SC Jr, Suthammarak W, Gong G, Sedensky MM, Morgan PG, Wang W, Tian R (2013) Mitochondrial complex I deficiency increases protein acetylation and accelerates heart failure. Cell Metab 18(2):239–250. https://doi.org/10.1016/j.cmet.2013.07.002

Kasahara A, Ishikawa K, Yamaoka M, Ito M, Watanabe N, Akimoto M, Sato A, Nakada K, Endo H, Suda Y, Aizawa S, Hayashi J (2006) Generation of trans-mitochondrial mice carrying homoplasmic mtDNAs with a missense mutation in a structural gene using ES cells. Hum Mol Genet 15(6):871–881. https://doi.org/10.1093/hmg/ddl005

Kauppila JH, Baines HL, Bratic A, Simard ML, Freyer C, Mourier A, Stamp C, Filograna R, Larsson NG, Greaves LC, Stewart JB (2016) A phenotype-driven approach to generate mouse models with pathogenic mtDNA mutations causing mitochondrial disease. Cell Rep 16(11):2980–2990. https://doi.org/10.1016/j.celrep.2016.08.037

Ke BX, Pepe S, Grubb DR, Komen JC, Laskowski A, Rodda FA, Hardman BM, Pitt JJ, Ryan MT, Lazarou M, Koleff J, Cheung MM, Smolich JJ, Thorburn DR (2012) Tissue-specific splicing of an Ndufs6 gene-trap insertion generates a mitochondrial complex I deficiency-specific cardiomyopathy. Proc Natl Acad Sci U S A 109(16):6165–6170. https://doi.org/10.1073/pnas.1113987109

Kim YG, Cha J, Chandrasegaran S (1996) Hybrid restriction enzymes: zinc finger fusions to Fok I cleavage domain. Proc Natl Acad Sci U S A 93(3):1156–1160

Koopman WJ, Willems PH, Smeitink JA (2012) Monogenic mitochondrial disorders. N Engl J Med 366(12):1132–1141. https://doi.org/10.1056/NEJMra1012478

Kruse SE, Watt WC, Marcinek DJ, Kapur RP, Schenkman KA, Palmiter RD (2008) Mice with mitochondrial complex I deficiency develop a fatal encephalomyopathy. Cell Metab 7(4):312–320. https://doi.org/10.1016/j.cmet.2008.02.004

Lam BL, Feuer WJ, Abukhalil F, Porciatti V, Hauswirth WW, Guy J (2010) Leber hereditary optic neuropathy gene therapy clinical trial recruitment: year 1. Arch Ophthalmol 128(9):1129–1135. https://doi.org/10.1001/archophthalmol.2010.201

Larsson NG, Tulinius MH, Holme E, Oldfors A, Andersen O, Wahlstrom J, Aasly J (1992) Segregation and manifestations of the mtDNA tRNA(Lys) A→G(8344) mutation of myoclonus epilepsy and ragged-red fibers (MERRF) syndrome. Am J Hum Genet 51(6):1201–1212

Larsson NG, Wang J, Wilhelmsson H, Oldfors A, Rustin P, Lewandoski M, Barsh GS, Clayton DA (1998) Mitochondrial transcription factor A is necessary for mtDNA maintenance and embryogenesis in mice. Nat Genet 18(3):231–236. https://doi.org/10.1038/ng0398-231

Leong DW, Komen JC, Hewitt CA, Arnaud E, McKenzie M, Phipson B, Bahlo M, Laskowski A, Kinkel SA, Davey GM, Heath WR, Voss AK, Zahedi RP, Pitt JJ, Chrast R, Sickmann A, Ryan MT, Smyth GK, Thorburn DR, Scott HS (2012) Proteomic and metabolomic analyses of mitochondrial complex I-deficient mouse model generated by spontaneous B2 short interspersed nuclear element (SINE) insertion into NADH dehydrogenase (ubiquinone) Fe-S protein 4 (Ndufs4) gene. J Biol Chem 287(24):20652–20663. https://doi.org/10.1074/jbc.M111.327601

Li H, Wang J, Wilhelmsson H, Hansson A, Thoren P, Duffy J, Rustin P, Larsson NG (2000) Genetic modification of survival in tissue-specific knockout mice with mitochondrial cardiomyopathy. Proc Natl Acad Sci U S A 97(7):3467–3472

Li H, Haurigot V, Doyon Y, Li T, Wong SY, Bhagwat AS, Malani N, Anguela XM, Sharma R, Ivanciu L, Murphy SL, Finn JD, Khazi FR, Zhou S, Paschon DE, Rebar EJ, Bushman FD, Gregory PD, Holmes MC, High KA (2011) In vivo genome editing restores haemostasis in a mouse model of haemophilia. Nature 475(7355):217–221. https://doi.org/10.1038/nature10177

Lu J, Sharma LK, Bai Y (2009) Implications of mitochondrial DNA mutations and mitochondrial dysfunction in tumorigenesis. Cell Res 19(7):802–815. https://doi.org/10.1038/cr.2009.69

Malka F, Lombes A, Rojo M (2006) Organization, dynamics and transmission of mitochondrial DNA: focus on vertebrate nucleoids. Biochim Biophys Acta 1763(5–6):463–472. https://doi.org/10.1016/j.bbamcr.2006.04.001

McFarland R, Clark KM, Morris AA, Taylor RW, Macphail S, Lightowlers RN, Turnbull DM (2002) Multiple neonatal deaths due to a homoplasmic mitochondrial DNA mutation. Nat Genet 30(2):145–146. https://doi.org/10.1038/ng819

McGregor A, Temperley R, Chrzanowska-Lightowlers ZM, Lightowlers RN (2001) Absence of expression from RNA internalised into electroporated mammalian mitochondria. Mol Gen Genomics 265(4):721–729

McKenzie M, Chiotis M, Pinkert CA, Trounce IA (2003) Functional respiratory chain analyses in murid xenomitochondrial cybrids expose coevolutionary constraints of cytochrome b and nuclear subunits of complex III. Mol Biol Evol 20(7):1117–1124. https://doi.org/10.1093/molbev/msg132

Miller FJ, Rosenfeldt FL, Zhang C, Linnane AW, Nagley P (2003) Precise determination of mitochondrial DNA copy number in human skeletal and cardiac muscle by a PCR-based assay: lack of change of copy number with age. Nucleic Acids Res 31(11):e61

Minczuk M, Papworth MA, Kolasinska P, Murphy MP, Klug A (2006) Sequence-specific modification of mitochondrial DNA using a chimeric zinc finger methylase. Proc Natl Acad Sci U S A 103(52):19689–19694. https://doi.org/10.1073/pnas.0609502103

Minczuk M, Papworth MA, Miller JC, Murphy MP, Klug A (2008) Development of a single-chain, quasi-dimeric zinc-finger nuclease for the selective degradation of mutated human mitochondrial DNA. Nucleic Acids Res 36(12):3926–3938. https://doi.org/10.1093/nar/gkn313

Moraes CT, Sciacco M, Ricci E, Tengan CH, Hao H, Bonilla E, Schon EA, DiMauro S (1995) Phenotype-genotype correlations in skeletal muscle of patients with mtDNA deletions. Muscle Nerve Suppl 3:S150–S153

Moscou MJ, Bogdanove AJ (2009) A simple cipher governs DNA recognition by TAL effectors. Science 326(5959):1501. https://doi.org/10.1126/science.1178817

Oca-Cossio J, Kenyon L, Hao H, Moraes CT (2003) Limitations of allotopic expression of mitochondrial genes in mammalian cells. Genetics 165(2):707–720

Olivo PD, Van de Walle MJ, Laipis PJ, Hauswirth WW (1983) Nucleotide sequence evidence for rapid genotypic shifts in the bovine mitochondrial DNA D-loop. Nature 306(5941):400–402

Pan Y, Xiao L, Li AS, Zhang X, Sirois P, Zhang J, Li K (2013) Biological and biomedical applications of engineered nucleases. Mol Biotechnol 55(1):54–62. https://doi.org/10.1007/s12033-012-9613-9

Papadopoulou LC, Sue CM, Davidson MM, Tanji K, Nishino I, Sadlock JE, Krishna S, Walker W, Selby J, Glerum DM, Coster RV, Lyon G, Scalais E, Lebel R, Kaplan P, Shanske S, De Vivo DC, Bonilla E, Hirano M, DiMauro S, Schon EA (1999) Fatal infantile cardioencephalomyopathy with COX deficiency and mutations in SCO2, a COX assembly gene. Nat Genet 23(3):333–337. https://doi.org/10.1038/15513

Park CB, Asin-Cayuela J, Camara Y, Shi Y, Pellegrini M, Gaspari M, Wibom R, Hultenby K, Erdjument-Bromage H, Tempst P, Falkenberg M, Gustafsson CM, Larsson NG (2007) MTERF3 is a negative regulator of mammalian mtDNA transcription. Cell 130(2):273–285. https://doi.org/10.1016/j.cell.2007.05.046

Perales-Clemente E, Fernandez-Silva P, Acin-Perez R, Perez-Martos A, Enriquez JA (2011) Allotopic expression of mitochondrial-encoded genes in mammals: achieved goal, undemonstrated mechanism or impossible task? Nucleic Acids Res 39(1):225–234. https://doi.org/10.1093/nar/gkq769

Peralta S, Torraco A, Wenz T, Garcia S, Diaz F, Moraes CT (2014) Partial complex I deficiency due to the CNS conditional ablation of Ndufa5 results in a mild chronic encephalopathy but no increase in oxidative damage. Hum Mol Genet 23(6):1399–1412. https://doi.org/10.1093/hmg/ddt526

Peralta S, Torraco A, Iommarini L, Diaz F (2015) Mitochondrial diseases part III: therapeutic interventions in mouse models of OXPHOS deficiencies. Mitochondrion 23:71–80. https://doi.org/10.1016/j.mito.2015.01.007

Pinkert CA, Trounce IA (2002) Production of transmitochondrial mice. Methods 26(4):348–357. https://doi.org/10.1016/S1046-2023(02)00041-5

Pozzan T, Rizzuto R (2000) High tide of calcium in mitochondria. Nat Cell Biol 2(2):E25–E27. https://doi.org/10.1038/35000095

Quintana A, Kruse SE, Kapur RP, Sanz E, Palmiter RD (2010) Complex I deficiency due to loss of Ndufs4 in the brain results in progressive encephalopathy resembling Leigh syndrome. Proc Natl Acad Sci U S A 107(24):10996–11001. https://doi.org/10.1073/pnas.1006214107

Quintana A, Morgan PG, Kruse SE, Palmiter RD, Sedensky MM (2012a) Altered anesthetic sensitivity of mice lacking Ndufs4, a subunit of mitochondrial complex I. PLoS One 7(8):e42904. https://doi.org/10.1371/journal.pone.0042904

Quintana A, Zanella S, Koch H, Kruse SE, Lee D, Ramirez JM, Palmiter RD (2012b) Fatal breathing dysfunction in a mouse model of Leigh syndrome. J Clin Invest 122(7):2359–2368. https://doi.org/10.1172/JCI62923

Ramirez CL, Certo MT, Mussolino C, Goodwin MJ, Cradick TJ, McCaffrey AP, Cathomen T, Scharenberg AM, Joung JK (2012) Engineered zinc finger nickases induce homology-directed repair with reduced mutagenic effects. Nucleic Acids Res 40(12):5560–5568. https://doi.org/10.1093/nar/gks179

Reddy P, Ocampo A, Suzuki K, Luo J, Bacman SR, Williams SL, Sugawara A, Okamura D, Tsunekawa Y, Wu J, Lam D, Xiong X, Montserrat N, Esteban CR, Liu GH, Sancho-Martinez I, Manau D, Civico S, Cardellach F, Del Mar O'CM, Campistol J, Zhao H, Campistol JM, Moraes CT, Izpisua Belmonte JC (2015) Selective elimination of mitochondrial mutations in the germline by genome editing. Cell 161(3):459–469. https://doi.org/10.1016/j.cell.2015.03.051

Ross JM, Stewart JB, Hagstrom E, Brene S, Mourier A, Coppotelli G, Freyer C, Lagouge M, Hoffer BJ, Olson L, Larsson NG (2013) Germline mitochondrial DNA mutations aggravate ageing and can impair brain development. Nature 501(7467):412–415. https://doi.org/10.1038/nature12474

Russell O, Turnbull D (2014) Mitochondrial DNA disease-molecular insights and potential routes to a cure. Exp Cell Res 325(1):38–43. https://doi.org/10.1016/j.yexcr.2014.03.012

Ruzzenente B, Rotig A, Metodiev MD (2016) Mouse models for mitochondrial diseases. Hum Mol Genet 25(R2):R115–R122. https://doi.org/10.1093/hmg/ddw176

Scharenberg AM, Duchateau P, Smith J (2013) Genome engineering with TAL-effector nucleases and alternative modular nuclease technologies. Curr Gene Ther 13(4):291–303

Schon EA, Rizzuto R, Moraes CT, Nakase H, Zeviani M, DiMauro S (1989) A direct repeat is a hotspot for large-scale deletion of human mitochondrial DNA. Science 244(4902):346–349

Schon EA, DiMauro S, Hirano M (2012) Human mitochondrial DNA: roles of inherited and somatic mutations. Nat Rev Genet 13(12):878–890. https://doi.org/10.1038/nrg3275

Sciacco M, Bonilla E, Schon EA, DiMauro S, Moraes CT (1994) Distribution of wild-type and common deletion forms of mtDNA in normal and respiration-deficient muscle fibers from patients with mitochondrial myopathy. Hum Mol Genet 3(1):13–19

Scott I, Youle RJ (2010) Mitochondrial fission and fusion. Essays Biochem 47:85–98. https://doi.org/10.1042/bse0470085

Shanske S, Coku J, Lu J, Ganesh J, Krishna S, Tanji K, Bonilla E, Naini AB, Hirano M, DiMauro S (2008) The G13513A mutation in the ND5 gene of mitochondrial DNA as a common cause of MELAS or Leigh syndrome: evidence from 12 cases. Arch Neurol 65(3):368–372. https://doi.org/10.1001/archneurol.2007.67

Shimizu A, Mito T, Hayashi C, Ogasawara E, Koba R, Negishi I, Takenaga K, Nakada K, Hayashi J (2014) Transmitochondrial mice as models for primary prevention of diseases caused by mutation in the tRNA(Lys) gene. Proc Natl Acad Sci U S A 111(8):3104–3109. https://doi.org/10.1073/pnas.1318109111

Shimizu A, Enoki S, Ishikawa K, Mito T, Obata K, Nagashima R, Yonekawa H, Nakada K, Hayashi J (2015) Mouse somatic mutation orthologous to MELAS A3302G mutation in the mitochondrial tRNA(Leu(UUR)) gene confers respiration defects. Biochem Biophys Res Commun 467(4):1097–1102. https://doi.org/10.1016/j.bbrc.2015.09.072

Shoffner JM, Lott MT, Lezza AM, Seibel P, Ballinger SW, Wallace DC (1990) Myoclonic epilepsy and ragged-red fiber disease (MERRF) is associated with a mitochondrial DNA tRNA(Lys) mutation. Cell 61(6):931–937

Shokolenko I, Venediktova N, Bochkareva A, Wilson GL, Alexeyev MF (2009) Oxidative stress induces degradation of mitochondrial DNA. Nucleic Acids Res 37(8):2539–2548. https://doi.org/10.1093/nar/gkp100

Sligh JE, Levy SE, Waymire KG, Allard P, Dillehay DL, Nusinowitz S, Heckenlively JR, MacGregor GR, Wallace DC (2000) Maternal germ-line transmission of mutant mtDNAs from embryonic stem cell-derived chimeric mice. Proc Natl Acad Sci U S A 97(26):14461–14466. https://doi.org/10.1073/pnas.250491597

Smith J, Bibikova M, Whitby FG, Reddy AR, Chandrasegaran S, Carroll D (2000) Requirements for double-strand cleavage by chimeric restriction enzymes with zinc finger DNA-recognition domains. Nucleic Acids Res 28(17):3361–3369

Srivastava S, Moraes CT (2001) Manipulating mitochondrial DNA heteroplasmy by a mitochondrially targeted restriction endonuclease. Hum Mol Genet 10(26):3093–3099

Stefano GB, Kream RM (2016) Mitochondrial DNA heteroplasmy in human health and disease. Biomed Rep 4(3):259–262. https://doi.org/10.3892/br.2016.590

Sterky FH, Hoffman AF, Milenkovic D, Bao B, Paganelli A, Edgar D, Wibom R, Lupica CR, Olson L, Larsson NG (2012) Altered dopamine metabolism and increased vulnerability to MPTP in mice with partial deficiency of mitochondrial complex I in dopamine neurons. Hum Mol Genet 21(5):1078–1089. https://doi.org/10.1093/hmg/ddr537

Stewart JB, Chinnery PF (2015) The dynamics of mitochondrial DNA heteroplasmy: implications for human health and disease. Nat Rev Genet 16(9):530–542. https://doi.org/10.1038/nrg3966

Stewart JB, Freyer C, Elson JL, Larsson NG (2008a) Purifying selection of mtDNA and its implications for understanding evolution and mitochondrial disease. Nat Rev Genet 9(9):657–662. https://doi.org/10.1038/nrg2396

Stewart JB, Freyer C, Elson JL, Wredenberg A, Cansu Z, Trifunovic A, Larsson NG (2008b) Strong purifying selection in transmission of mammalian mitochondrial DNA. PLoS Biol 6(1):e10. https://doi.org/10.1371/journal.pbio.0060010

Stoddard BL (2011) Homing endonucleases: from microbial genetic invaders to reagents for targeted DNA modification. Structure 19(1):7–15. https://doi.org/10.1016/j.str.2010.12.003

Sung YH, Baek IJ, Kim DH, Jeon J, Lee J, Lee K, Jeong D, Kim JS, Lee HW (2013) Knockout mice created by TALEN-mediated gene targeting. Nat Biotechnol 31(1):23–24. https://doi.org/10.1038/nbt.2477

Sutovsky P (2003) Ubiquitin-dependent proteolysis in mammalian spermatogenesis, fertilization, and sperm quality control: killing three birds with one stone. Microsc Res Tech 61(1):88–102. https://doi.org/10.1002/jemt.10319

Szymczak AL, Workman CJ, Wang Y, Vignali KM, Dilioglou S, Vanin EF, Vignali DA (2004) Correction of multi-gene deficiency in vivo using a single 'self-cleaving' 2A peptide-based retroviral vector. Nat Biotechnol 22(5):589–594. https://doi.org/10.1038/nbt957

Tanaka M, Borgeld HJ, Zhang J, Muramatsu S, Gong JS, Yoneda M, Maruyama W, Naoi M, Ibi T, Sahashi K, Shamoto M, Fuku N, Kurata M, Yamada Y, Nishizawa K, Akao Y, Ohishi N, Miyabayashi S, Umemoto H, Muramatsu T, Furukawa K, Kikuchi A, Nakano I, Ozawa K, Yagi K (2002) Gene therapy for mitochondrial disease by delivering restriction endonuclease SmaI into mitochondria. J Biomed Sci 9(6 Pt 1):534–541. doi:64726

Tatuch Y, Christodoulou J, Feigenbaum A, Clarke JT, Wherret J, Smith C, Rudd N, Petrova-Benedict R, Robinson BH (1992) Heteroplasmic mtDNA mutation (T–G) at 8993 can cause Leigh disease when the percentage of abnormal mtDNA is high. Am J Hum Genet 50(4):852–858

Tatuch Y, Pagon RA, Vlcek B, Roberts R, Korson M, Robinson BH (1994) The 8993 mtDNA mutation: heteroplasmy and clinical presentation in three families. Eur J Hum Genet 2(1):35–43

Taylor RW, Turnbull DM (2005) Mitochondrial DNA mutations in human disease. Nat Rev Genet 6(5):389–402. https://doi.org/10.1038/nrg1606

Tiranti V, Hoertnagel K, Carrozzo R, Galimberti C, Munaro M, Granatiero M, Zelante L, Gasparini P, Marzella R, Rocchi M, Bayona-Bafaluy MP, Enriquez JA, Uziel G, Bertini E, Dionisi-Vici C, Franco B, Meitinger T, Zeviani M (1998) Mutations of SURF-1 in Leigh disease associated with cytochrome c oxidase deficiency. Am J Hum Genet 63(6):1609–1621. https://doi.org/10.1086/302150

Torraco A, Diaz F, Vempati UD, Moraes CT (2009) Mouse models of oxidative phosphorylation defects: powerful tools to study the pathobiology of mitochondrial diseases. Biochim Biophys Acta 1793(1):171–180. https://doi.org/10.1016/j.bbamcr.2008.06.003

Torraco A, Peralta S, Iommarini L, Diaz F (2015) Mitochondrial diseases part I: mouse models of OXPHOS deficiencies caused by defects in respiratory complex subunits or assembly factors. Mitochondrion 21:76–91. https://doi.org/10.1016/j.mito.2015.01.009

Torres-Torronteras J, Viscomi C, Cabrera-Perez R, Camara Y, Di Meo I, Barquinero J, Auricchio A, Pizzorno G, Hirano M, Zeviani M, Marti R (2014) Gene therapy using a liver-targeted AAV vector restores nucleoside and nucleotide homeostasis in a murine model of MNGIE. Mol Ther 22(5):901–907. https://doi.org/10.1038/mt.2014.6

Trifunovic A, Wredenberg A, Falkenberg M, Spelbrink JN, Rovio AT, Bruder CE, Bohlooly YM, Gidlof S, Oldfors A, Wibom R, Tornell J, Jacobs HT, Larsson NG (2004) Premature ageing in mice expressing defective mitochondrial DNA polymerase. Nature 429(6990):417–423. https://doi.org/10.1038/nature02517

Tsao CY, Mendell JR, Bartholomew D (2001) High mitochondrial DNA T8993G mutation (<90%) without typical features of Leigh's and NARP syndromes. J Child Neurol 16(7):533–535. https://doi.org/10.1177/088307380101600716

Twig G, Hyde B, Shirihai OS (2008) Mitochondrial fusion, fission and autophagy as a quality control axis: the bioenergetic view. Biochim Biophys Acta 1777(9):1092–1097. https://doi.org/10.1016/j.bbabio.2008.05.001

Tyynismaa H, Mjosund KP, Wanrooij S, Lappalainen I, Ylikallio E, Jalanko A, Spelbrink JN, Paetau A, Suomalainen A (2005) Mutant mitochondrial helicase Twinkle causes multiple mtDNA deletions and a late-onset mitochondrial disease in mice. Proc Natl Acad Sci U S A 102(49):17687–17692. https://doi.org/10.1073/pnas.0505551102

Urnov FD, Miller JC, Lee YL, Beausejour CM, Rock JM, Augustus S, Jamieson AC, Porteus MH, Gregory PD, Holmes MC (2005) Highly efficient endogenous human gene correction using designed zinc-finger nucleases. Nature 435(7042):646–651. https://doi.org/10.1038/nature03556

Urnov FD, Rebar EJ, Holmes MC, Zhang HS, Gregory PD (2010) Genome editing with engineered zinc finger nucleases. Nat Rev Genet 11(9):636–646. https://doi.org/10.1038/nrg2842

Uusimaa J, Remes AM, Rantala H, Vainionpaa L, Herva R, Vuopala K, Nuutinen M, Majamaa K, Hassinen IE (2000) Childhood encephalopathies and myopathies: a prospective study in a defined population to assess the frequency of mitochondrial disorders. Pediatrics 105(3 Pt 1):598–603

Vafai SB, Mootha VK (2012) Mitochondrial disorders as windows into an ancient organelle. Nature 491(7424):374–383. https://doi.org/10.1038/nature11707

Valnot I, Osmond S, Gigarel N, Mehaye B, Amiel J, Cormier-Daire V, Munnich A, Bonnefont JP, Rustin P, Rotig A (2000) Mutations of the SCO1 gene in mitochondrial cytochrome c oxidase deficiency with neonatal-onset hepatic failure and encephalopathy. Am J Hum Genet 67(5):1104–1109. https://doi.org/10.1016/S0002-9297(07)62940-1

Valton J, Cabaniols JP, Galetto R, Delacote F, Duhamel M, Paris S, Blanchard DA, Lebuhotel C, Thomas S, Moriceau S, Demirdjian R, Letort G, Jacquet A, Gariboldi A, Rolland S, Daboussi F, Juillerat A, Bertonati C, Duclert A, Duchateau P (2014) Efficient strategies for TALEN-mediated genome editing in mammalian cell lines. Methods 69(2):151–170. https://doi.org/10.1016/j.ymeth.2014.06.013

Viscomi C, Bottani E, Zeviani M (2015) Emerging concepts in the therapy of mitochondrial disease. Biochim Biophys Acta 1847(6–7):544–557. https://doi.org/10.1016/j.bbabio.2015.03.001

Wai T, Teoli D, Shoubridge EA (2008) The mitochondrial DNA genetic bottleneck results from replication of a subpopulation of genomes. Nat Genet 40(12):1484–1488. https://doi.org/10.1038/ng.258

Wai T, Ao A, Zhang X, Cyr D, Dufort D, Shoubridge EA (2010) The role of mitochondrial DNA copy number in mammalian fertility. Biol Reprod 83(1):52–62. https://doi.org/10.1095/biolreprod.109.080887

Wallace DC, Fan W (2010) Energetics, epigenetics, mitochondrial genetics. Mitochondrion 10(1):12–31. https://doi.org/10.1016/j.mito.2009.09.006

Wan X, Pei H, Zhao MJ, Yang S, Hu WK, He H, Ma SQ, Zhang G, Dong XY, Chen C, Wang DW, Li B (2016) Efficacy and safety of rAAV2-ND4 treatment for Leber's hereditary optic neuropathy. Sci Rep 6:21587. https://doi.org/10.1038/srep21587

Wang J, Wilhelmsson H, Graff C, Li H, Oldfors A, Rustin P, Bruning JC, Kahn CR, Clayton DA, Barsh GS, Thoren P, Larsson NG (1999) Dilated cardiomyopathy and atrioventricular conduction blocks induced by heart-specific inactivation of mitochondrial DNA gene expression. Nat Genet 21(1):133–137. https://doi.org/10.1038/5089

Wang G, Shimada E, Zhang J, Hong JS, Smith GM, Teitell MA, Koehler CM (2012) Correcting human mitochondrial mutations with targeted RNA import. Proc Natl Acad Sci U S A 109(13):4840–4845. https://doi.org/10.1073/pnas.1116792109

Watanabe T, Dewey MJ, Mintz B (1978) Teratocarcinoma cells as vehicles for introducing specific mutant mitochondrial genes into mice. Proc Natl Acad Sci U S A 75(10):5113–5117

Wilson IJ, Carling PJ, Alston CL, Floros VI, Pyle A, Hudson G, Sallevelt SC, Lamperti C, Carelli V, Bindoff LA, Samuels DC, Wonnapinij P, Zeviani M, Taylor RW, Smeets HJ, Horvath R, Chinnery PF (2016) Mitochondrial DNA sequence characteristics modulate the size of the genetic bottleneck. Hum Mol Genet 25(5):1031–1041. https://doi.org/10.1093/hmg/ddv626

Woischnik M, Moraes CT (2002) Pattern of organization of human mitochondrial pseudogenes in the nuclear genome. Genome Res 12(6):885–893. https://doi.org/10.1101/gr.227202

Wu J, Kandavelou K, Chandrasegaran S (2007) Custom-designed zinc finger nucleases: what is next? Cell Mol Life Sci 64(22):2933–2944. https://doi.org/10.1007/s00018-007-7206-8

Wyman C, Kanaar R (2006) DNA double-strand break repair: all's well that ends well. Annu Rev Genet 40:363–383. https://doi.org/10.1146/annurev.genet.40.110405.090451

Xu F, Morin C, Mitchell G, Ackerley C, Robinson BH (2004) The role of the LRPPRC (leucine-rich pentatricopeptide repeat cassette) gene in cytochrome oxidase assembly: mutation causes lowered levels of COX (cytochrome c oxidase) I and COX III mRNA. Biochem J 382(Pt 1):331–336. https://doi.org/10.1042/BJ20040469

Yokota M, Shitara H, Hashizume O, Ishikawa K, Nakada K, Ishii R, Taya C, Takenaga K, Yonekawa H, Hayashi J (2010) Generation of trans-mitochondrial mito-mice by the introduction of a pathogenic G13997A mtDNA from highly metastatic lung carcinoma cells. FEBS Lett 584(18):3943–3948. https://doi.org/10.1016/j.febslet.2010.07.048

Zhu Z, Yao J, Johns T, Fu K, De Bie I, Macmillan C, Cuthbert AP, Newbold RF, Wang J, Chevrette M, Brown GK, Brown RM, Shoubridge EA (1998) SURF1, encoding a factor involved in the biogenesis of cytochrome c oxidase, is mutated in Leigh syndrome. Nat Genet 20(4):337–343. https://doi.org/10.1038/3804

Targeting Therapeutic Nucleic Acids into Mitochondria: A Long Challenge

Frédérique Weber-Lotfi and André Dietrich

Abstract Mitochondria resulted from an endosymbiosis event and subsequently kept their own genome. In the course of evolution, the mitochondrial DNA shrunk down but it still encodes essential components of the oxidative phosphorylation chain. Point mutations and deletions in the human mitochondrial DNA cause severe incurable neurodegenerative diseases and accumulate during aging. Rearrangements in the plant mitochondrial genome contribute to evolution and agronomical traits. Development of human mitochondrial gene therapy strategies or plant mitochondrial biotechnologies suffer from the inability of conventional methodologies to genetically transform mitochondria. The importance of these issues led to the development of a large set of alternative strategies aiming to target DNA or RNA into mitochondria, mainly in mammalian cells. A first group relied on natural RNA uptake pathways of mitochondria, using tRNA derivatives, tRNA mimics, 5S rRNA, and stem-loop structures of RNase P and RNase MRP RNAs as import shuttles, or taking a special RNA import complex as a carrier. Other strategies took advantage of the regular protein uptake pathway of mitochondria to design a series of DNA or RNA-binding plaforms driven to the organelles by mitochondrial targeting peptides. In a third category of approaches, elaborate DNA-binding lipophilic vesicles were rendered mitochondriotropic and served as carriers for organelle targeting. Finally, atypical protocols like hydrodynamic vein injection and magnetofection were adapted for the challenge. A number of these methodologies were claimed to be successful on the basis of functional or genetic observations, but there is still little consensus in the field.

Keywords Genetic transformation · Mitochondria · Nanocarriers · Neurodegenerative disease · RNA and DNA trafficking · Transfection

F. Weber-Lotfi · A. Dietrich (✉)
Université de Strasbourg, CNRS, IBMP UPR 2357, Strasbourg, France
e-mail: andre.dietrich@ibmp-cnrs.unistra.fr

P. J. Oliveira (ed.), *Mitochondrial Biology and Experimental Therapeutics*,
https://doi.org/10.1007/978-3-319-73344-9_25

1 Introduction

Several models are put forward for the origin and formation of eukaryotic cells, but mitochondria are still considered to result from a single endosymbiosis of an α-proteobacterium-like organism into some proto-eukaryotic cell (Baum and Baum 2014; Martin et al. 2015). In the course of evolution, the genome of the endosymbiont rapidly shrunk to a small set of genes, but was maintained as the mitochondrial DNA (mtDNA). Acquisition of mitochondria increased the cellular energy production potential, but at the cost of a parallel increase in the production of mutagenic reactive oxygen species (ROS). Differences then occurred in the evolution of the mtDNA, with different adaptations for coping with the mutation burden within the different organisms (Otten and Smeets 2015). In mammals, the mtDNA shrunk down to a 16.5 kb genome that encodes 13 proteins, all essential for building the oxidative phosphorylation (OXPHOS) chain, 2 ribosomal RNAs (rRNAs) and 22 transfer RNAs (tRNAs) (Legros et al. 2004). It is prone to point mutations or large deletions. Plant mitochondrial genomes are larger, commonly in the range of 200–700 kb in angiosperms, and encode in addition proteins involved in the biogenesis of OXPHOS complexes, as well as some ribosomal proteins. Plant mtDNAs mainly undergo rearrangements that generate new configurations. Initially substoichiometric, these new mitotypes can become predominent upon segregation and shifting (Gualberto and Newton 2017).

Point mutations and deletions in the human mtDNA cause severe neurodegenerative diseases that are currently incurable and await the development of gene therapy strategies (Alston et al. 2017; Scarpelli et al. 2017). Hundreds of pathogenic mutations have been identified in protein, rRNA or tRNA genes (http://www.mitomap.org). Mutations in the mtDNA also accumulate during aging, a process that has long been attributed to a vicious circle of genotoxic oxidative stress, but now seems to come from replication errors and failure of the repair mechanisms (Kauppila and Stewart 2015; Payne and Chinnery 2015; Pinto and Moraes 2015). Rearrangements in the plant mitochondrial genome contribute to evolution, are often lethal or generate cytoplasmic male sterility (CMS) (Gualberto and Newton 2017; Touzet and Meyer 2014; Horn et al. 2014; Hu et al. 2014). CMS is a highly valued trait in plant breeding for hybrid generation, but it is available only for a small number of species.

Since many years, progress in all these fields has remained hindered by the persistant failure to genetically transform animal and plant mitochondria *via* conventional methodologies. A paradox of such a situation is that isolated mammalian and plant mitochondria are naturally competent for DNA uptake (Koulintchenko et al. 2003, 2006). Mitochondrial transformation through regular biolistics has been established in yeast and *Chlamydomonas reinhardtii* (Bonnefoy et al. 2007; Zhou et al. 2010), but remained restricted to this couple of unicellular organisms for which selection benefits from powerful genetic tools. The importance of the issue stimulated the development of a wealth of alternative strategies aiming to target nucleic acids, whether DNA or RNA, into mitochondria *in vivo* and these will be the subject of the present Chapter. For obvious reasons, *i.e.* the seek for gene therapy approaches to cure diseases caused by mtDNA mutations, the efforts of the different laboratories were mainly concentrated on mammals. A number of approaches have also been established to model and exploit mitochondrial transfection with isolated organelles from various organisms. These *in organello* studies will not be discussed here. They provided important information on the mechanisms of DNA integration and expression, as well as on RNA editing and splicing (Hinrichsen et al. 2009; Mileshina et al. 2011; Farre et al. 2012).

2 Targeting Strategies Using the Natural RNA Import Pathways

In a number of organisms, some nuclear-encoded RNAs are naturally imported into the mitochondria and function in the organelles (Schneider 2011; Sieber et al. 2011a; Wang et al. 2012a). The targeting and translocation mechanisms are not the same for all RNAs and in all organisms. Nevertheless, molecular strategies have been developed based on these natural import pathways so as to target various RNAs into mitochondria. The tRNA-like structure of the *Turnip yellow mosaic virus* genomic RNA has also been used for such purposes. A synthetic view of the strategies based on the natural RNA import pathways is given in Table 1.

2.1 Transfer RNA Derivatives as Mitochondrial Shuttles

In many organisms, the mtDNA encodes an incomplete set of transfer RNAs (tRNAs). Depending on the species, some of the mitochondrial tRNAs, or even all of them in extreme cases, are encoded by the nuclear genome and imported from the cytosol into the mitochondria (Salinas et al. 2008; Schneider 2011). They often partition between the cytosol and the organelles. In the yeast *Saccharomyces cerevisiae*, the cytosolic $tRNA^{Lys}_{(CUU)}$ partially localizes to mitochondria (Tarassov and Entelis 1992) and is used in particular by the organellar translation in non permissive temperatures. This tRNA turned out to be also importable into isolated human mitochondria (Entelis et al. 2001). Experiments were thus designed to import a functional yeast $tRNA^{Lys}$ into the mitochondria of human cells carrying the A8344G pathogenic mutation, which can cause the MERRF (myoclonic epilepsy with ragged-red fibers) syndrome, in the corresponding mitochondrial $tRNA^{Lys}$ gene. The strategy successfully led to a partial rescue of the mitochondrial functions affected by the mutation (Kolesnikova et al. 2004). Similarly, expression of recombinant tRNAs bearing the identity elements for human mitochondrial leucyl-tRNA synthetase could partially rescue the phenotype caused by the A3243G pathogenic mutation in the $tRNA^{Leu}_{(UUR)}$ gene, which causes the MELAS

Table 1 Synthetic view of the strategies using the natural RNA import pathways to target nucleic acids into mitochondria in whole cells

Nucleic acid targeted	Strategy for cellular expression/uptake	Strategy for mitochondrial uptake	References
tRNA	– Transient transgene expression – Stable transgene expression	Natural tRNA import 	Kolesnikova et al. (2004); Karicheva et al. (2011)
tRNA derivatives	– Stable transgene expression	Natural tRNA import 	Small et al. (1992); Sbicego et al. (1998)
Antigenomic RNA	– Transfection with RNA – Stable transgene expression – Cholesterol conjugation	FD-RNAs: D-arm and F-hairpin of yeast $tRNA^{Lys}$ 	Comte et al. (2013); Tonin et al. (2014); Dovydenko et al. (2016)
Ribozyme	– Stable transgene expression	PKTLS: 3′-end of the TYMV genomic RNA 	Val et al. (2011)

Antigenomic RNA	– Transfection with RNA – Stable transgene expression	Natural 5S rRNA import 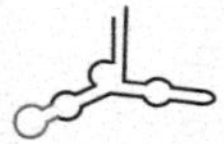	Smirnov et al. (2008); Comte et al. (2013)
Antisense RNA	– Transfection with RNA – Transient transgene expression – Stable transgene expression	Human 5S rRNA MAM domain 	Zelenka et al. (2014)
mRNA	– Transient transgene expression	RP: stem-loop from human RNase P or RNase MRP RNA 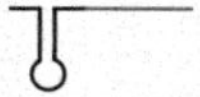	Wang et al. (2012b)
tRNA	– Transient transgene expression	RP + 3′-UTR human *MRPS12* 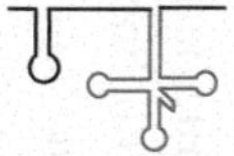	Wang et al. (2012b)
– Antisense DNA – Antisense RNA – Polycistronic coding RNAs	– Incubation of cells with RIC/nucleic acid complex – Muscle injection of RIC/RNA complex	RIC complex and tRNA^{Tyr} D-arm of *L. tropica* 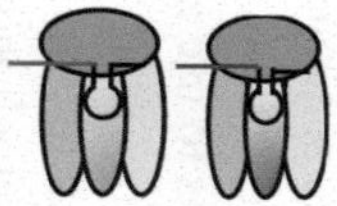	Mukherjee et al. (2008); Mahato et al. (2011); Jash and Adhya (2011); Jash and Adhya (2012); Jash et al. (2012)

In the schemes, shuttle sequences are represented in black, cargo RNAs in green and cargo DNA in red
TYMV Turnip yellow mosaic virus; *RNase MRP* RNase for mitochondrial RNA processing, *RIC* RNA import complex of *Leishmania tropica*

(mitochondrial encephalomyopathy with lactic acidosis and stroke-like episodes) syndrome, in human transmitochondrial cybrid cells (Karicheva et al. 2011).

Further strategies were developed using naturally imported tRNAs as shuttles to introduce cargo RNAs into mitochondria. Modified tRNA genes were introduced into the nuclear genome, expressed and the transcripts were imported into mitochondria. Small et al. (1992) showed that a bean (*Phaseolus vulgaris*) $tRNA^{Leu}$ containing a four nucleotide insertion in the anticodon loop and expressed from a nuclear transgene was imported into the mitochondria of transformed potato (*Solanum tuberosum*) plants. Further, Sbicego et al. (1998) exchanged the natural intron in $tRNA^{Tyr}$ for synthetic sequences and transformed splicing-deficient *Leishmania tarentolae* cell lines with the corresponding gene constructs. In this way, they could obtain the organellar import of modified tRNAs containing up to 38 nucleotides of additional sequence, demonstrating that short sequences can be introduced into *L. tarentolae* mitochondria *in vivo*.

Another strategy was to use as a shuttle derivatives of the *S. cerevisiae* $tRNA^{Lys}_{CUU}$ that is partially imported into mitochondria (Tarassov and Entelis 1992). Extensive analyses led to the characterization of $tRNA^{Lys}$ derivatives that are imported into human mitochondria (Entelis et al. 2001) and allowed to characterize short synthetic RNAs containing two domains of the tRNA, the D-arm and a domain that was called the F-hairpin, joined by a central linker domain. The resulting structures were called FD-RNAs and showed high efficiency of mitochondrial import (Kolesnikova et al. 2010). This design was used to develop an antigenomic strategy aiming to repress the replication of mutated mtDNA carrying a large pathogenic deletion in human cells. For that purpose, the central linker domain of FD-RNAs was replaced by a short sequence spanning over the deletion junction and destined to specifically anneal to the mutated mtDNA. Transient expression of such a construct in KSS (Kearns Sayre Syndrome) cybrids led to a shift of the heteroplasmy towards wild-type mtDNA in the cells, validating the anti-replicative approach *in vivo* (Comte et al. 2013). By introducing as a central linker domain into a further FD-RNA design, a 20-nucleotide sequence complementary to a sequence of the ND5 mitochondrial gene containing a point mutation, Tonin et al. (2014) also obtained a significant decrease of the proportion of mutated mtDNA molecules in transfected human cybrid cells. Towards potential prospects in gene therapy, Dovydenko et al. developed a strategy for carrier-free delivery of such therapeutic RNAs into human cells. A protocol was set up for chemical conjugation of the RNAs with a cholesterol residue through cleavable covalent bonds (Dovydenko et al. 2016). The 5′-end of a 26 nucleotide anti-replicative RNA was conjugated to cholesterol through a covalent hydrazone bond. The conjugates showed efficient carrier-free cellular uptake. The hydrazone bond was stable during the cell transfection procedure, but was rapidly cleaved in the acidic conditions of endosomal compartments. The released RNA was targeted to mitochondria as above through the RNA import pathway. Transfection of primary skin fibroblasts from a patient bearing a mutation in the *ND5* gene at 30% heteroplasmy with the cholesterol-anti-replicative RNA conjugate led to a shift of heteroplasmy in favour of the wild-type mtDNA copies (Dovydenko et al. 2016).

2.2 A tRNA Mimic as a Mitochondrial Shuttle

To import longer RNA sequences into mitochondria in plants, cargo RNAs were linked as 5′-trailors to a 120 nucleotide sequence taken from the 3′-end of the *Turnip yellow mosaic virus* (TYMV) genomic RNA (Val et al. 2011). This sequence, which was called PKTLS, forms a pseudoknot (PK) followed by a tRNA-like structure (TLS) (Matsuda and Dreher 2004). The TYMV TLS is a $tRNA^{Val}$ mimic and was a proper candidate as a mitochondrial shuttle in plant cells because $tRNAs^{Val}$ are imported into mitochondria in all plant species studied so far (Salinas et al. 2008; Sieber et al. 2011a). The pseudoknot was included because it contributes to the binding of the TLS to valyl-tRNA synthetase (Matsuda and Dreher 2004) and recognition by the cognate aminoacyl-tRNA synthetase is needed for mitochondrial import of tRNAs in plants (Dietrich et al. 1996). A construct coding for a *trans*-cleaving hammerhead ribozyme directed against the *atp9* mitochondrial mRNA, fused to the PKTLS shuttle through a linker was introduced into the nuclear genome in *Nicotiana tabacum* cell suspensions and in *Arabidopsis thaliana* plants. The chimeric RNA was expressed from the nuclear transgene and was translocated into the mitochondria of the transformed plant cells and plants. A drastic decrease in the level of the *atp9* target mRNA was observed, demonstrating that the ribozyme was active inside mitochondria (Val et al. 2011). The strategy appears to be suitable to study mitochondrial gene functions or to manipulate organelle gene expression in plants.

2.3 The 5S Ribosomal RNA as a Mitochondrial Shuttle

In mammals, the nuclear-encoded 5S rRNA is imported into the mitochondrial matrix (Magalhaes et al. 1998), although its role in the organelles remains unclear (Smirnov et al. 2011). Functional analysis of the structural elements in human 5S rRNA highlighted three domains and established that parts of the domains α and γ are important for mitochondrial import *in vitro* and *in vivo*. Conversely, domain ß is dispensable for organelle uptake (Smirnov et al. 2008). Starting from these results, strategies were developed to produce mitochondrial shuttles exploiting 5S rRNA trafficking. First, it was shown that 5S rRNA with a mutated ß domain was actually imported twenty times more efficiently *in vivo* into human mitochondria than the wild-type 5S rRNA and such a mutated form was taken as a shuttle. Recombinant 5S rRNAs were designed by exchanging part of the ß domain for short foreign cargo sequences of 12–14 nucleotides. These recombinant forms were expressed in human cells from appropriate constructs and were imported into mitochondria (Smirnov et al. 2008; Comte et al. 2013). When using as cargo sequences in the recombinant 5S rRNA short antigenomic oligoribonucleotides binding specifically to the deletion junction in mutated mtDNA causing KSS, a shift of the mitochondrial heteroplasmy in favor of the wild-type mtDNA was recorded (Comte et al. 2013). A second shuttling system was designed, consisting of a cargo sequence attached to the 3′-end of a modified 5S rRNA γ domain named MAM (Mitochondria-Addressing Domain) (Zelenka et al. 2014; Zelenka and Jezek 2016). A sequence of 21 nucleotides destined to hybridize to a region of the ND5 mitochondrial mRNA or gene was used as a cargo. The chimeric RNA was synthesized and labeled with a fluorescent dye *in vitro*. The probe was introduced into HepG2 human cells and driven to the mitochondrial RNA import pathway upon loading on dequalinium micelles (see also Sect. 4.2). Confocal microscopy analysis supported localization of the probe to the mitochondrial matrix (Zelenka et al. 2014). Alternatively, the sequence coding for the cargo/MAM chimeric RNA was cloned into an expression vector. After transfection of HEK 293 human cells and induced expression of the construct, mitochondrial delivery and enrichment of the resulting transcript was tested by RT-qPCR (Zelenka et al. 2014).

2.4 Stem-Loop Structures of RNase P or RNase MRP RNAs as Mitochondrial Shuttles

Further strategies were established on the basis of the RNase P and RNase MRP RNA moieties (Hernandez-Cid et al. 2012). RNase P is an ubiquitous endoribonuclease involved in 5′-end maturation of tRNA precursors (Klemm et al. 2016). Two types of RNase P enzymes have been characterized, depending on the organism and the cellular compartment: (1) RNA-dependent enzymes that include a catalytic RNA responsible for the activity and (2) enzymes made only of proteins. The yeast mitochondrial RNase P is a ribonucleoprotein (Dang and Martin 1993). The RNA is encoded by the mitochondrial genome and the protein by the nuclear genome. For mammalian cells, the situation is complex. Doersen et al. (1985) reported in early studies that the nuclear-encoded catalytic RNA moiety of RNase P is imported into mitochondria, implying that the mammalian organellar RNase P would be a ribonucleoprotein with a catalytic RNA. Subsequently, the scheme was supported by further data (Puranam and Attardi 2001), but was contradicted when the mammalian mitochondrial RNase P was characterized and shown to be a protein-only enzyme (Holzmann et al. 2008; Walker and Engelke 2008). The role that the nuclear-encoded RNase P RNA moiety might have in mitochondria thus remained a question mark. RNase MRP stands for “RNase for mitochondrial RNA processing” and corresponds to a ribonucleoprotein that is considered in mammals to cleave RNA transcripts complementary to the mtDNA origin of replication, forming RNA primers for mtDNA leading strand replication (Chang and Clayton 1987, 1989). It is also present in the nucleolus. The mitochondrial and nucleolar MRP RNases were further characterized in yeast and shown to contain distinct protein components, but an identical RNA moiety that is involved in catalysis (Lu et al. 2010). The RNA moieties of RNase P and RNase MRP are structurally related (Chamberlain et al. 1998). Data on mitochondrial RNase MRP in mammmals are also puzzling, as the presence of the nuclear-encoded RNA moiety in the organelles has been challenged (Kiss and Filipowicz 1992).

Nevertheless, Wang et al. (2010) used *in vitro* RNA uptake assays with isolated yeast mitochondria to analyze the sequences of the human RNase P RNA moiety that are important for mitochondrial import. A 20-nucleotide stem-loop structure was identified and a similar stem-loop structure was found in the RNase MRP RNA. These stem-loop structures were fused to the 5′-end of the GAPDH mRNA, an RNA that naturally is not imported into the organelles. The resulting fusion transcripts were taken up into isolated yeast mitochondria (Wang et al. 2010). Further experiments were run with the stem-loop structure from the RNA component of the human RNase P. This structure, which was called RP, proved able to direct the mouse *COX2* mRNA into mitochondria in human HeLa cells (Wang et al. 2012b). The RP structure was also sufficient to direct import of tRNA precursors into isolated human mitochondria *in vitro* (Wang et al. 2012b). However, for *in vivo* mitochondrial import of tRNAs, an additional sequence had to be fused to the 3′-end of the RP-containing tRNA precursors, *i.e.* the 3′-UTR of the human mitochondrial ribosomal protein S12 (MRPS12). Precursors of wild-type mitochondrial tRNAs flanked by the 5′ RP structure and the MRPS12 3′-UTR were expressed in cells derived from MERRF and MELAS patients harboring mutations in the mitochondrial $tRNA^{Lys}_{AAA}$ and $tRNA^{Leu}_{AUU}$, respectively. These tRNA defects in MERRF and MELAS cells cause inefficient mitochondrial translation and result in defective cell respiration. Stable nuclear expression of the wild-type mitochondrial tRNA precursors linked to the RP and the MRPS12 3′-UTR led to a respiration increase in MELAS and MERRF cells (Wang et al. 2012b).

2.5 *Exploiting the Leishmania RNA Import Complex for Mitochondrial Shuttling*

In trypanosomatids, the mtDNA does not code for tRNAs and all mitochondrial tRNAs are imported from the cytosol (Schneider and Maréchal-Drouard 2000). In this respect, an RNA import complex (RIC) active in the presence of ATP was characterized in the inner membrane of *Leshmania tropica* mitochondria. This large complex was composed of 11 subunits, among which 8 were nuclear-encoded (Bhattacharyya et al. 2003; Chatterjee et al. 2006; Mukherjee et al. 2007). Several proteins of this complex turned out to be dispensable for tRNA import into mitochondria from *L. tropica* (Mukherjee et al. 2007; Koley and Adhya 2013) or from other species (Paris et al. 2009). Alternatively, the complete RIC complex, a partial complex made of the 8 nuclear-encoded subunits (R8) and a partial complex made of 6 of them (R6) were tested to import RNAs into mitochondria *in vivo*. In initial studies, the RIC complex allowed the import of the human $tRNA^{Lys}$ into isolated human mitochondria *in vitro*. Import was followed by aminoacylation of the tRNA and participation to mitochondrial protein synthesis (Mahata et al. 2005). When the assay was applied to mitochondria from patients with MERRF or KSS syndrome due to mutant $tRNA^{Lys}$ genes, translation efficiency was corrected (Mahata et al. 2005). Mahata et al. then incubated human cultured cells with isolated RIC complex and found that the complex could enter the cells by a caveolin-1-dependent pathway and subsequently support mitochondrial import of endogenous tRNAs (Mahata et al. 2006; Mukherjee et al. 2014). The next step was to deliver exogenous functional RNAs into mitochondria in cells. Previous studies had identified the D-arm hairpin of the *Leishmania* $tRNA^{Tyr}$ as a determinant for mitochondrial import (Goswami et al. 2006). This structure was thus fused with various short RNA sequences, each complementary to the 5′-end region of a mitochondrial mRNA. The fusion RNAs were preincubated with isolated RIC, so as to form a ribonucleoprotein complex that was applied on human HepG2 cell monolayers (Mukherjee et al. 2008). The results showed an efficient import of the chimeric antisense RNAs into most of the mitochondria in the cells and a specific degradation of the target mRNAs in the organelles. Chimeras of single-stranded antisense DNA oligonucleotides covalently attached to the D-arm RNA and loaded on RIC were taken up as well into cellular mitochondria and induced degradation of the target mRNA (Mukherjee et al. 2008). Further experiments were run on the same principle with the partial RIC complex R8 loaded with transcripts combining the $tRNA^{Tyr}$ D-arm with long polycistronic RNAs (pcRNAs) carrying native mtDNA coding sequences. Application of the R8 RIC/pcRNA complexes to human cybrid cells carrying a 1.9 kb mtDNA deletion from a KSS patient led to the translation of the pcRNA-encoded mRNAs and to the rescue of the mitochondria functions (Mahato et al. 2011). A similar strategy was applied to a specially generated HepG2-derived cell line carrying multiple mtDNA deletions, also resulting in mitochondrial rescue (Jash and Adhya 2011). Finally, the R6 RIC complex combined with a pan-genome cocktail of pcRNAs was injected into middle-aged and old rats, stimulating mitochondrial translation and respiratory capacity of skeletal muscles (Jash et al. 2012). Administration to injured muscles induced muscle regeneration (Jash and Adhya 2012) and accelerated satellite cell activation (Jash et al. 2014).

3 Targeting Strategies Using the Natural Protein Import Pathways

The vast majority of the mitochondrial proteins are nuclear-encoded and imported from the cytosol through dedicated translocases (Wiedemann and Pfanner 2017). Sorting, recruitment and mitochondrial translocation of the polypeptides rely on specific targeting sequences, on mechanisms and on channels that have been extensively studied, hence the idea of exploiting these further import pathways that can accommodate macromolecules, so as to target nucleic acids to the organelles. A synthetic view of the strategies based on the protein import pathways is given in Table 2.

Table 2 Synthetic view of the strategies using the natural protein import pathways to target nucleic acids into mitochondria

Nucleic acid targeted	Strategy for cellular expression/uptake	Strategy for mitochondrial uptake	References
– ssDNA – dsDNA	*Only isolated mitochondria*	MTS 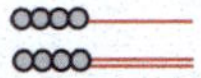	Vestweber and Schatz (1989); Seibel et al. (1999)
Oligodeoxyribonucleotide	Addition of synthetic polycations or membrane permeabilizing toxins	MTS linked to complementary PNA 	Flierl et al. (2003)
Plasmid DNA with a reporter gene	Incubation of rat cells with the complex	MTS linked to PEI 	Lee et al. (2007)
– Human mtDNA – LHON mt DNA	Incubation of human cells with the complex containing the transduction domain	MTS linked to TFAM 	Keeney et al. (2009); Iyer et al. (2012a, b)
Plasmid DNA with a reporter gene	– HK peptide + BP100 peptide – Infiltration of plant leaves with the complex – Incubation of human cells with the complex	MTS 	Chuah et al. (2015); Chuah et al. (2016)
Plasmid DNA with a human wild-type or mutant gene	AAV viral capsid – Transfection – Injection into eyes – Injection into fertilized oocytes	MTS added to the capsid 	Yu et al. (2012a); Yu et al. (2012b); Yu et al. (2015)
– mRNAs – tRNAs	*Only isolated mitochondria*	MTS linked to DHFR 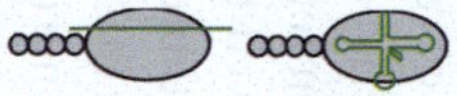	Sieber et al. (2011b)
Guide RNA	Transient transgene expression	MTS linked to a modified Cas9 	Jo et al. (2015)

In the schemes, cargo RNAs are represented in green and cargo DNA in red
MTS mitochondrial targeting sequence, *PNA* peptide nucleic acid, *PEI* polyethylene immine, *TFAM* mitochondrial transcription factor A, *AAV* adeno-associated virus, DHFR dihydrofolate reductase

3.1 A Mitochondrial Targeting Peptide as a Shuttle for DNA Conjugates

Nuclear-encoded proteins destined to be imported into the mitochondrial matrix through the TOM (translocase of the outer membrane) and TIM (translocase of the inner membrane) pathway carry an N-terminal sequence, the so-called mitochondrial targeting sequence (MTS). MTS peptides were thus tentatively used as mitochondrial shuttles for nucleic acids *in vitro*. Vestweber and Schatz first showed that a 24 nucleotide single-stranded or double-stranded DNA oligonucleotide whose 5′-end was covalently linked to the C-terminus of a fusion protein containing a sequence derived from the MTS of the COX4 subunit was imported into isolated *S. cerevisiae* mitochondria (Vestweber and Schatz 1989). Similarly, Seibel et al. (1995) showed translocation of 17 or 322 bp DNA fragments covalently linked to the rat ornithine transcarbamylase MTS into rat (*Rattus norvegicus*) liver mitochondria. In this strategy, the synthetic peptide-DNA conjugates were processed by the mitochondrial proteolytic machinery (Seibel et al. 1999). Follow up of such approaches was hindered by the sensitivity of the peptide-DNA conjugates to hydrolysis.

3.2 Polymers Driven by a Mitochondrial Targeting Peptide as a Shuttle for DNA

Instead of linking covalently the MTS to the DNA, further strategies fused the MTS to a DNA carrier and used the resulting platform to shuttle cargo DNA across the double membrane of mitochondria. First, the platform was based on a peptide nucleic acid (PNA) fragment. PNAs are synthetic polymers made of a peptide bond backbone carrying regular nucleobases. They pair with complementary DNA or RNA with high affinity and the duplex is less sensitive to nucleases than regular nucleic acids. To make the platform, MTS peptides were covalently linked to a PNA fragment complementary to the DNA to be targeted to mitochondria. The cargo DNA was subsequently annealed with the platform. Both in assays with isolated mouse liver mitochondria and in experiments with a mouse cell line the MTS-PNA/cargo DNA complex was recovered in the mitochondrial compartment (Flierl et al. 2003). However, in cell line experiments, cytosolic uptake of the complex needed permeabilization of the cells with synthetic polycations or membrane permeabilizing toxins. In a similar strategy, the MTS was conjugated to a stretch of polyethyleneimine (PEI) through a disulfide bond. PEI is a cationic polymer that binds DNA. The MTS-PEI/DNA complex associated with isolated mitochondria in cell-free assays and localized to mitochondria in living cells (Lee et al. 2007). Interestingly, the MTS-PEI conjugate showed lower cytotoxicity than PEI alone.

3.3 Protofection, a TFAM-Based Carrier for Mitochondrial Shuttling of DNA

To transfect mammalian mitochondria, Khan and Bennet developed another MTS-based approach that they called protofection (Khan and Bennett Jr. 2004). The platform carrying the cargo DNA was designed so as to ensure both cell penetration and mitochondrial delivery. To promote the entry into the cell, the strategy used a short cationic peptide, *i.e.* a protein transduction domain (PTD). Such specific protein domains are found for instance in some transcription factors or in viral proteins. They can permeate cell membranes and deliver proteins or bioactive materials into living cells (Suzuki 2012). The PTD domain followed by the MTS were fused to the N-terminus of the mitochondrial factor TFAM known to be involved in mtDNA packaging in the nucleoids. Thus, the PTD allows the platform to enter the cell, the MTS ensures mitochondrial targeting and the TFAM protein carries the DNA (Khan and Bennett Jr. 2004). Iyer et al. showed that such a platform was rapidly recovered in the mitochondria of cybrid cells carrying a G11778A mutation leading to the LHON (Leber's Hereditary Optic Neuropathy) syndrome (Iyer et al. 2009). *In vivo* treatment of mice with the PTD-MTS-TFAM platform increased complex I-driven respiration in mitochondria from brain and skeletal muscle (Iyer et al. 2009). In view of mitochondrial gene therapy modeling, the platform was loaded with bulk mtDNA derived from commercial human genomic DNA and the complex was used to transfect Parkinson's disease cybrid cells impacted in respiration and containing reduced mtDNA levels (Keeney et al. 2009). After 9–11 weeks, the transfected cells increased their respiration rate and their mtDNA copy number (Keeney et al. 2009). It has to be noted that treatment of the cells with the PTD-MTS-TFAM platform deprived of cargo DNA led to activation of mitochondrial biogenesis through an undefined mechanism, possibly resulting from a direct effect of increased TFAM levels (Iyer et al. 2009; Keeney et al. 2009; Thomas et al. 2011). Treatment of LHON (G11778A mutant mtDNA) and LS (Leigh's syndrome; T8993G mutant mtDNA) cells with healthy bulk mtDNA loaded on the PTD-MTS-TFAM platfom improved mitochondrial respiration in LHON cells and partially rescued ATP synthase function in LS cells (Iyer et al. 2012a). In a reverse approach, the platform was used to introduce a LHON mtDNA (G11778A mutation) into human neural progenitor cells. Prior to transfection, the cells were treated with dideoxycytidine, so as to reduce their endogenous mtDNA level and gene expression without loss of phenotypic markers. Transfection of the cells with the PTD-MTS-TFAM platform loaded with LHON mtDNA resulted in the expression of the pathogenic mtDNA and persistance of the capacity of the cells to differentiate into neurons (Iyer et al. 2012b).

3.4 A Double/Triple Component Targeting Peptide as a Mitochondrial Shuttle for DNA

It has been shown that positively charged peptides composed of lysine (K) and histidine (H) can be used as carriers to transfect siRNAs or plasmids into cells (Leng et al. 2007; Chou et al. 2014). In such peptides, lysine residues allow the condensation of the RNA or DNA, while histidines are necessary for endosomal lysis. On that basis, Chuah et al. (2015) developed a dual-domain peptide by fusing the first twelve amino acids of the MTS of the yeast COX4 subunit to a polycationic copolymer of alternating lysine and histidine residues (nine KH motifs). Plasmid DNA carrying the luciferase marker gene under the control of the mitochondrial COX2 promoter was loaded on this platform as a polyanion. Infiltration of the complexes into *A. thaliana* leaves led to the expression of the marker gene, pointing to mitochondrial targeting of the plasmid (Chuah et al. 2015). Addition of the BP100 (KKLFKKILKYL-amide) cell-penetrating peptide (Eggenberger et al. 2011) to the platform increased transfection levels. In further assays, the platform was loaded with a plasmid carrying the GFP marker gene under the control of the COX2 promoter and the complex was infiltrated into *A. thaliana* leaves. Subsequent confocal microscopy analysis showed co-localization of the expressed GFP fluorescence with mitochondria in leaf epidermal cells (Chuah et al. 2015). It was proposed that the plasmid DNA dissociates from the platform upon contact with the surface of the organelles (as shown for DQAsomes, see Sect. 4.2 below) and is subsequently imported through the natural competence mechanism extensively characterized with isolated mitochondria (Koulintchenko et al. 2003; Weber-Lotfi et al. 2015). The same strategy was developed to transfect human HEK-293 cell cultures, with an additional platform associating 32 residues from the MTS of the human ornithine transcarbamylase to the KH motifs (Chuah et al. 2016). Again, expression of the luciferase and GFP markers was observed upon transfection with plasmids carrying the corresponding genes under the control of the mitochondrial COX2 promoter and GFP fluorescence was shown to co-localize with mitochondria. The peptides and their complexes with DNA were more extensively characterized in terms of structure and interactions. Notably, the viability of the human cells was not affected by the process in the most transfection-effective conditions (Chuah et al. 2016).

3.5 A Modified Viral Capsid as a Mitochondrial Shuttle for DNA

An alternative strategy to deliver DNA to the organelles was to use as a platform an adeno-associated virus (AAV) capsid modified so as to target mitochondria instead of the nucleus (Yu et al. 2012a). For this, the 23 amino acid MTS of the COX8 subunit was added to VP2, one of the three capsid proteins of the AAV. VP2 is not essential for the infectivity of AAV and corresponds to only 3 of 60 subunits in the capsid. A plasmid carrying the cDNA for the wild-type human mtDNA-encoded ND4 subunit under the control of the mitochondrial heavy-strand promoter (HSP) was packaged into AAV virions bearing MTS-VP2 subunits. The ND4 sequence was tagged with a FLAG epitope for specific probing. The virions were used to infect human cybrid cell lines containing 100% mtDNA copies with a G11778A LHON mutation in the *ND4* gene. The construct was recovered in mitochondrial fractions and expressed, leading to wild-type ND4-FLAG protein production and assembly into Complex I. The defective respiration of the LHON cells due to the *ND4* mutation was rescued (Yu et al. 2012a). The strategy was subsequently extended to mice through injection of the MTS-VP2 virions loaded with the *ND4* construct into the eyes (Yu et al. 2012a). Wild-type human ND4-FLAG was expressed in mouse eyes and co-localized with mitochondria. Upon injection into the eyes of mice carrying a R340H *ND4* mutation, early visual loss induced by the mutant ND4 was prevented. The rescue persisted on the long term and prevented optic atrophy, the hallmark of LHON induced by the mutant R340H ND4, although next-generation sequencing showed that the exogenous human *ND4* did not integrate by homologous recombination into the mtDNA of the injected mice and remained episomal (Yu et al. 2012a, 2013).

In reciprocal experiments, MTS-VP2 virions loaded with a construct encoding human G11778A mutant *ND4* under the control of the HSP were injected into the vitreous cavity of normal mice. Expression of the mutated ND4 led to significant loss of visual function, loss of retinal ganglion cells, and optic nerve degeneration, thereby recapitulating the hallmarks of human LHON (Yu et al. 2012b). Finally, to generate transgenic LHON mice, the MTS-VP2 viral platform was used to deliver the human mutant *ND4* DNA into fertilized eggs. Offspring females with higher expression were backcrossed with wild-type males, generating mutant *ND4* mice. The delivered DNA represented up to 20% of the transgenic mouse *ND4*, but did not integrate into the mouse mtDNA. The mutant ND4 protein was expressed and assembled into Complex I. The transgenic mice showed a decrease of the respiratory chain function and the hallmarks of human LHON, with progression to blindness 8 months after birth (Yu et al. 2015). Notably, the process could be reversed by intraocular injection of the MTS-VP2 viral platform loaded with the plasmid expressing wild-type human ND4 (Yu et al. 2015).

3.6 DHFR-Driven Shuttling of RNA to Mitochondria

To deliver RNAs into isolated organelles, Sieber et al. (2011b) developed a strategy that uses a modified dihydrofolate reductase (DHFR) as a platform. This protein is able to interact *in vivo* with its own mRNA (Tai et al. 2002, 2008), but *in vitro* it can bind RNAs in an unspecific manner (Sieber et al. 2011b). To develop a mitochondrial shuttle, the 69 amino acid MTS of the mitochondrial ATP9 subunit of *Neurospora crassa* was fused to the mouse DHFR. The fusion protein was overexpressed in bacteria and introduced into RNA import assays with isolated plant mitochondria (Sieber et al. 2011b). The MTS-DHFR strongly stimulated tRNA translocation into isolated mitochondria, but also allowed import of longer RNAs. The larch $tRNA^{His}$ precursor and the full-length potato *atp9* mRNA were translocated into potato mitochondria in the presence of the fusion protein. The internalized $tRNA^{His}$ precursor was correctly processed *in organello*, while the *atp9* mRNA was edited. Using the same approach, plant $tRNA^{Ala}$ could be imported into isolated yeast and human mitochondria. Finally, transcripts up to 630 nucleotides were successfully imported into isolated rat mitochondria in the presence of MTS-DHFR (Sieber et al. 2011b).

3.7 Cas9-Driven Guide RNA Shuttling to Mitochondria

The CRISPR/Cas9 system has become in few years a major approach for genome engineering (Hsu et al. 2014). It relies on site-specific cleavage of the DNA mediated by the Cas9 (CRISPR-associated protein-9) endonuclease of the microbial adaptive immune system CRISPR ("clustered regularly interspaced short palindromic repeats"). Cas9 cleavage is directed to a specific site by a short guide RNA carrying a 20 nucleotide sequence complementary to the selected region in the DNA. Appropriately adapting the sequence of the guide RNA allows to target virtually any site of interest in any genome, hence the idea of developing the strategy to target the mtDNA. Such a strategy requires to translocate both the Cas9 endonuclease and the specific guide RNA into the mitochondria. Jo et al. (2015) developed a mitoCas9 by combining the endonuclease with the MTS of cytochrome c and designed a guide RNA against the mtDNA COX1 sequence. The mitoCas9 enzyme was expressed in human HEK-293 cell cultures upon transfection with the appropriate gene construct and shown to be imported into the mitochondria. When the cells were transfected with both the mitoCas9 construct and a second gene construct encoding the guide RNA, specific cleavage of the mtDNA at the targeted COX1 locus was observed, together with functional defects of the mitochondria (Jo et al. 2015). Similar results were obtained when targeting a region in the mtDNA COX3 region. This data suggests that either the Cas9 protein can by itself serve as an import platform for guide RNAs, on the same line as the other platforms described so far, or that the endonuclease can drive the guide RNA to one of the mitochondrial RNA import pathways mentioned above (Schneider 2011; Sieber et al. 2011a; Wang et al. 2012a).

4 Targeting Strategies Based on Lipophilic Carriers

Instead of hijacking import pathways based on protein channels, the third set of strategies to target nucleic acids into mitochondria through the cellular membrane and the double mitochondrial membrane was developed on the idea of an interplay of lipophilic carriers with the lipid bilayers of the membranes. A synthetic view of these strategies is given in Table 3.

Table 3 Synthetic view of the strategies using lipophilic carriers to target nucleic acids into mitochondria in whole cells

Nucleic acid targeted	Strategy for cellular uptake	Strategy for mitochondrial uptake	References
Antisense RNA	MITO-Porter	MITO-Porter and D-arm hairpin of the *L. tropica* $tRNA^{Tyr}$ 	Yamada et al. (2015)
Oligonucleotide	Outer envelope of DF-MITO-Porter	Inner envelope of DF-MITO-Porter 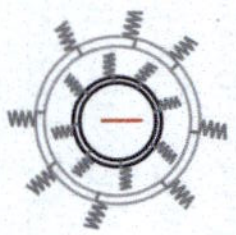	Yamada et al. (2012)
– Fluorescent oligonucleotide and plasmid DNA – Plasmid DNA with a reporter gene – Linear and circular DNA	DQAsome	DQAsome + MTS 	D'Souza et al. (2003, 2005); Lyrawati et al. (2011)
Fluorescent linear plasmid DNA	Cationic liposome	Cationic liposome 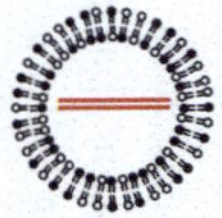	Boddapati et al. (2005); Wagle et al. (2011)

In the schemes, cargo RNAs are represented in green and cargo DNA in red
DF dual-function, *DQA* dequalinium

4.1 MITO-Porters

MITO-Porters are liposome-based nanocarriers used to deliver molecules of various sizes into mitochondria *via* membrane fusion. In 2008, Yamada et al. (2008) developed a MITO-Porter made of liposomes carrying a high density of octaarginine surface modifications, so as to stimulate their entry into cells as intact vesicles. Octaarginine (R8) residues stimulate the internalization into the cell by macropinocytosis, allow the escape of the Mito-Porter from the macropinosomes to the cytosol and subsequently promote binding to mitochondria *via* electrostatic interactions. In a final step, fusion with the mitochondrial membranes promoted by the highly fusogenic lipid composition of the liposomes, including sphingomyelin or phosphatidic acid, leads to the delivery of the MITO-Porter content into mitochondria. In initial experiments, GFP was encapsulated into the MITO-Porter and incubated with isolated rat liver mitochondria. The results of western blot analyses suggested that the MITO-Porter delivered the GFP cargo to the intermembrane space by passing through the outer membrane (Yamada et al. 2008). When living HeLa cells were exposed to GFP-loaded MITO-Porter and observed by confocal microscopy, GFP fluorescence co-localized with mitochondria. In further assays, MITO-Porter loaded with propidium iodide, a fluorescent intercalating agent used to stain nucleic acids, was incubated as before with isolated rat liver mitochondria and with living HeLa cells. In both cases, staining of the mtDNA was observed, supporting MITO-Porter-driven delivery of the propidium iodide cargo into the mitochondria (Yasuzaki et al. 2010).

To further enhance the escape of the carriers from the macropinosomes, Furukawa et al. (2015) added the pH-sensitive membrane fusogenic peptide GALA to the octaarginine modified MITO-Porter, resulting in higher levels of mitochondrial targeting. GALA is a 30 amino acid synthetic peptide long enough to span a lipid bilayer (Li et al. 2004). The GALA/octaarginine MITO-Porter was loaded with an antisense RNA oligonucleotide directed against the mitochondrial *COX2* mRNA and fused to the D-arm hairpin of the *Leishmania* $tRNA^{Tyr}$, *i.e.* the above described mitochondrial import determinant (see Sect. 2.5). The loaded carrier was incubated with HeLa cells and up to 50% knockdown of the *COX2* target RNA was recorded (Furukawa et al. 2015). As a consequence, the COX2 protein was significantly decreased and the mitochondrial membrane potential was depolarized. Further testing of different MITO-Porter lipid combinations led to the selection of lipid fusogenic compositions appropriate for both the outer and the inner mitochondrial membrane. *In vitro* import of a 19 nucleotide model RNA was evaluated with the optimized MITO-Porter and isolated rat mitochondria (Yamada et al. 2015).

Along with the other variants of the original design, a dual function MITO-Porter (DF-MITO-Porter) concept was developed. The DF-MITO-Porter possesses two envelopes. The outer one contains endosome fusogenic lipids and octaarginine, the inner one is composed of mitochondrion fusogenic lipids and octaarginine (Yamada et al. 2011). Altogether, the composition provides the ability to pass through both endosomal and mitochondrial membranes *via* step-wise membrane fusion. For escape from the endosome to the cytosol, the outer envelope of the DF-MITO-Porter fuses with the endosomal membrane, releasing a particle surrounded by the second envelope. The latter subsequently fuses with the mitochondrial membranes to deliver the cargo to the organelles. DF-MITO-Porters were more efficient for mitochondrial delivery of cargos than the original MITO-Porters, as tested in HeLa cells with encapsulated DNase I enzyme (Yamada et al. 2011; Yamada and Harashima 2012). The DF-MITO-Porter was subsequently loaded with a 24 nucleotide, cyanine-labeled DNA oligonucleotide and incubated with HeLa cells. Confocal microscope analysis of the cells showed a co-localization of the fluorescent DNA oligonucleotide with the mitochondria (Yamada et al. 2012). Further ways of optimizing the MITO-Porter functionality were explored by grafting an MTS (Yamada and Harashima 2013) or RNA aptamers corresponding to the above described mitochondrial import determinants, *i.e.* the stem-loop structures of RNase P or RNase MRP RNAs (Sect. 2.4) or the D-arm hairpin of the *Leishmania* $tRNA^{Tyr}$ (Sect. 2.5) (Yamada et al. 2016).

4.2 DQAsomes

DQAsomes are vesicles made of dequalinium, a dicationic amphiphile compound known to accumulate in mitochondria in response to the electrochemical gradient at the mitochondrial membrane. It was shown that dequalinium vesicles can bind and condensate DNA and are able to transfect cells (Weissig et al. 1998). *In vitro*, the DQAsome–DNA complexes ('DQAplexes') released their DNA when they got in contact with cardiolipin-rich liposomes mimicking mitochondrial membranes (Weissig et al. 2000) or with intact mouse liver mitochondria (Weissig et al. 2001). DQAplexes were shown to transfect BT20 human cells and to escape from the endosomes without losing their cargo DNA. They subsequently migrated towards mitochondria where they released at least part of their DNA when contacting the organelles (D'Souza et al. 2003). In further assays, DQAsomes were loaded with a DNA oligonucleotide or a full plasmid, both conjugated to the MTS of the mouse mitochondrial ornithine transcarbamylase (OTC). The hypothesis was that, upon release at the mitochondrial surface the conjugates would be imported through the regular protein import pathway. The DQAsomes drove the fluorescently labeled conjugates into the cells and to the mitochondria, where the DNA kept localized, suggesting that uptake had occurred (D'Souza et al. 2005). Notably, higher levels of mitochondrial co-localization were obtained with linearized DNA than with circular plasmid. In another set of experiments, Lyrawati et al. (2011) constructed a minimal mitochondrial genome designed to express the GFP in the organelles. The construct was encapsulated into DQAsomes and the resulting DQAplexes served to transfect RAW264.7 mouse macrophage cells. Punctuated GFP fluorescence co-localizing with mitochondria was obtained, supporting expression of the construct in the organelles. Expression of the GFP mRNA and protein was assessed by RT-PCR and immunohistochemistry assays (Lyrawati et al. 2011). The artificial construct could be maintained for longer time, but whether this resulted from recombination with the endogenous mtDNA was not established. The approach was extended to a series of further mammalian cell lines (Lyrawati et al. 2011).

4.3 Cationic Liposomes

Conventional liposomes could be rendered mitochondriotropic by surface modification with known mitochondriotropic compounds like stearyl-triphenylphosphonium (STPP) (Boddapati et al. 2005; Weissig et al. 2006; D'Souza et al. 2007). Different improved liposome formulations were explored, incorporating a mitochondrial fraction or mitochondrial lipids for better targeting and the cationic amphiphile lipid 1,2-dioleoyl-3 trimethyl ammonium-propane (DOTAP) to facilitate DNA complexation (Wagle et al. 2011). Like DQAsomes, such liposomes were able to mediate cell uptake of fluorescent DNA and the DNA subsequently co-localized with mitochondria (Wagle et al. 2011). Mitochondriotropic liposomes proved less toxic, especially when containing a mitochondrial fraction.

5 Hydrodynamic Vein Injection

A surprising strategy was the use of hydrodynamic limb vein (HLV) injection to deliver DNA to mitochondria *in vivo*. The HLV injection procedure refers to rapid injection of a large volume of naked DNA into the distal vein of a limb in the anterograde direction (Hagstrom et al. 2004). The method was described for nuclear transgene expression or siRNA strategies in skeletal muscle. It relies on the hypothesis that the hydrodynamic force would induce a transient opening of cellular membranes. Upon HLV injection of a naked plasmid DNA in rats, the exogenous DNA was recovered in a mitochondria-enriched fraction extracted from the rat skeletal muscle. Recovery was resistant to DNase I treatment of the organelles (Yasuzaki et al. 2013). The HLV injection procedure showed no toxicity for mitochondrial functions. Condensing the plasmid DNA by treatment with the polycation protamine resulted in a more effective association with the mitochondria-enriched fraction (Yasuzaki et al. 2014). In further assays, mitochondrial expression constructs carrying an *ND4*-FLAG gene or a recoded luciferase gene under the control of the HSP promoter were delivered into rats by HLV injection and into mice by hydrodynamic tail vein (HTV) injection (Liu et al. 1999; Yasuzaki et al. 2015). Expression of the *ND*-FLAG and luciferase mRNAs, as well as of luciferase protein, was detected, pointing to mitochondrial incorporation of the injected constructs (Yasuzaki et al. 2015).

6 Superparamagnetic Iron Oxide Nanoparticule Directed to Mitochondria

Recently, Kim et al. (2016) proposed a further original strategy aiming to introduce DNA into mitochondria based on magnetofection with a modified superparamagnetic iron oxide nanoparticle (SPION). SPION is a biocompatible nanoparticle for non-invasive transfection under an external magnetic field (Grzeskowiak et al. 2015). It was coated with chitosan-*graft*-PEI (CHI-*g*-PEI, CP), to improve binding and condensation of the DNA, and with PK11195 (PK), an isoquinoline carboxamide that binds selectively to the TSPO mitochondrial 18 kDa translocator protein (Kim et al. 2016). Interaction between PK11195 and TSPO might open mitochondrial permeability transition pores (mPTP) and mPTP components were proposed to be involved in DNA import into isolated mitochondria (Koulintchenko et al. 2003; Weber-Lotfi et al. 2015). PK11195-CHI-*g*-PEI-modified SPION (PK-CP-SPION) was able to condense DNA and protect the loaded DNA against nucleases. Under an external magnetic field, the PK-CP-SPION showed a high efficiency of human cell transfection with a low toxicity (Kim et al. 2016). The nanoparticles specifically associated with mitochondria after entering the cells. It was thus proposed that the PK-CP-SPION holds the prospect to serve as a mitochondria-targeting gene vector with high transfection efficiency, but at the same time association with the mitochondria led to the activation of the apoptosis pathway (Kim et al. 2016).

7 Conclusion

The present chapter illustrates how the variety of strategies that have been explored to target RNA or DNA into mitochondria is large, extending from cell cultures to whole model organisms. Many of these approaches were claimed to be successful on the basis of functional or genetic observations. One might thus wonder why there is still so little consensus in the field and what keeps slowing down common progress towards modeling realistic mitochondrial gene therapy. First, a number of the documented strategies would be strengthened if direct molecular evidence would be brought for the mitochondrial uptake. Second, each of the strategies essentially remained confined to one laboratory, while validation of independently confirmed and generally accepted biotechnological tools is needed (Lightowlers 2011). As a third issue, only a few approaches were tested in animal models and the biocompatibility of the methods and compounds remains to be assessed in a number of cases. This is especially true for nanocarriers in a context of growing public debate and concern about nanotechnologies in medicine (Su et al. 2017). The fourth aspect is the fate of the DNA potentially taken up into the organelles. In the studies that included such analyses, exogenous gene constructs carrying mitochondrial sequences did not recombine with the endogenous mtDNA and remained as independent episomes. How long such episomes will be maintained, how they will segregate between cells or within tissues and whether integration into the mitochondrial genome will be possible are completely open and essential questions that will need further work on mtDNA recombination and segregation mechanisms. The occurrence of mtDNA homologous recombination in mammals is a matter of debate (Elson and Lightowlers 2006), but is supported by different lines of evidence (Chen 2013). Active recombination is likely to occur at least in mitochondria of oxyradical-rich tissues like heart, brain or skeletal muscle (Kajander et al. 2001). Nevertheless, the set of recombination factors of prokaryotic origin that ensure mtDNA dynamics in plant mitochondria (Gualberto and Newton 2017) is not found in mammals. In the absence of integration into the mitochondrial genome, therapeutic DNA remains a drug to be administrated as such. Notably, awaiting strategies for curative gene therapy, disease prevention through mtDNA replacement has emerged, based on the transfer of the nuclear genetic material from the egg of a woman with an mtDNA mutation into a donated healthy enucleated egg (Xu and Shi 2016). Clinical application of such germline modification approaches developed in animal models in turn raises its own ethical, scientific and technical concerns (Eyre-Walker 2017; Marlow 2017; Rulli 2017).

Acknowledgments We acknowledge support of our work by the French Centre National de la Recherche Scientifique (CNRS, UPR2357), the University of Strasbourg, the Agence Nationale de la Recherche (grant numbers ANR-06-MRAR-037-02, ANR-09-BLAN-0240-01) and the Ministère de la Recherche et de l'Enseignement Supérieur (Investissements d'Avenir/Laboratoire d'Excellence MitoCross, grant number ANR-11-LABX-0057_MITOCROSS).

References

Alston CL, Rocha MC, Lax NZ, Turnbull DM, Taylor RW (2017) The genetics and pathology of mitochondrial disease. J Pathol 241:236–250

Baum DA, Baum B (2014) An inside-out origin for the eukaryotic cell. BMC Biol 12:76

Bhattacharyya SN, Chatterjee S, Goswami S, Tripathi G, Dey SN, Adhya S (2003) "Ping-pong" interactions between mitochondrial tRNA import receptors within a multiprotein complex. Mol Cell Biol 23:5217–5224

Boddapati SV, Tongcharoensirikul P, Hanson RN, D'Souza GG, Torchilin VP, Weissig V (2005) Mitochondriotropic liposomes. J Liposome Res 15:49–58

Bonnefoy N, Remacle C, Fox TD (2007) Genetic transformation of *Saccharomyces cerevisiae* and *Chlamydomonas reinhardtii* mitochondria. Methods Cell Biol 80:525–548

Chamberlain JR, Lee Y, Lane WS, Engelke DR (1998) Purification and characterization of the nuclear RNase P holoenzyme complex reveals extensive subunit overlap with RNase MRP. Genes Dev 12:1678–1690

Chang DD, Clayton DA (1987) A mammalian mitochondrial RNA processing activity contains nucleus-encoded RNA. Science 235:1178–1184

Chang DD, Clayton DA (1989) Mouse RNAase MRP RNA is encoded by a nuclear gene and contains a decamer sequence complementary to a conserved region of mitochondrial RNA substrate. Cell 56:131–139

Chatterjee S, Home P, Mukherjee S, Mahata B, Goswami S, Dhar G, Adhya S (2006) An RNA-binding respiratory component mediates import of type II tRNAs into *Leishmania* mitochondria. J Biol Chem 281:25270–25277

Chen XJ (2013) Mechanism of homologous recombination and implications for aging-related deletions in mitochondrial DNA. Microbiol Mol Biol Rev 77:476–496

Chou ST, Hom K, Zhang D, Leng Q, Tricoli LJ, Hustedt JM, Lee A, Shapiro MJ, Seog J, Kahn JD, Mixson AJ (2014) Enhanced silencing and stabilization of siRNA polyplexes by histidine-mediated hydrogen bonds. Biomaterials 35:846–855

Chuah JA, Yoshizumi T, Kodama Y, Numata K (2015) Gene introduction into the mitochondria of *Arabidopsis thaliana* via peptide-based carriers. Sci Rep 5:7751

Chuah JA, Matsugami A, Hayashi F, Numata K (2016) Self-assembled peptide-based system for mitochondrial-targeted gene delivery: functional and structural insights. Biomacromolecules 17:3547–3557

Comte C, Tonin Y, Heckel-Mager AM, Boucheham A, Smirnov A, Aure K, Lombes A, Martin RP, Entelis N, Tarassov I (2013) Mitochondrial targeting of recombinant RNAs modulates the level of a heteroplasmic mutation in human mitochondrial DNA associated with Kearns Sayre syndrome. Nucleic Acids Res 41:418–433

D'Souza GG, Rammohan R, Cheng SM, Torchilin VP, Weissig V (2003) DQAsome-mediated delivery of plasmid DNA toward mitochondria in living cells. J Control Release 92:189–197

D'Souza GG, Boddapati SV, Weissig V (2005) Mitochondrial leader sequence—plasmid DNA conjugates delivered into mammalian cells by DQAsomes co-localize with mitochondria. Mitochondrion 5:352–358

D'Souza GG, Boddapati SV, Weissig V (2007) Gene therapy of the other genome: the challenges of treating mitochondrial DNA defects. Pharm Res 24:228–238

Dang YL, Martin NC (1993) Yeast mitochondrial RNase P. Sequence of the RPM2 gene and demonstration that its product is a protein subunit of the enzyme. J Biol Chem 268:19791–19796

Dietrich A, Maréchal-Drouard L, Carneiro V, Cosset A, Small I (1996) A single base change prevents import of cytosolic tRNA(Ala) into mitochondria in transgenic plants. Plant J 10:913–918

Doersen CJ, Guerrier-Takada C, Altman S, Attardi G (1985) Characterization of an RNase P activity from HeLa cell mitochondria. Comparison with the cytosol RNase P activity. J Biol Chem 260:5942–5949

Dovydenko I, Tarassov I, Venyaminova A, Entelis N (2016) Method of carrier-free delivery of therapeutic RNA importable into human mitochondria: lipophilic conjugates with cleavable bonds. Biomaterials 76:408–417

Eggenberger K, Mink C, Wadhwani P, Ulrich AS, Nick P (2011) Using the peptide BP100 as a cell-penetrating tool for the chemical engineering of actin filaments within living plant cells. Chembiochem 12:132–137

Elson JL, Lightowlers RN (2006) Mitochondrial DNA clonality in the dock: can surveillance swing the case? Trends Genet 22:603–607

Entelis NS, Kolesnikova OA, Dogan S, Martin RP, Tarassov IA (2001) 5 S rRNA and tRNA import into human mitochondria. Comparison of in vitro requirements. J Biol Chem 276:45642–45653

Eyre-Walker A (2017) Mitochondrial replacement therapy: are mito-nuclear interactions likely to be a problem? Genetics 205:1365–1372

Farre JC, Aknin C, Araya A, Castandet B (2012) RNA editing in mitochondrial trans-introns is required for splicing. PLoS One 7:e52644

Flierl A, Jackson C, Cottrell B, Murdock D, Seibel P, Wallace DC (2003) Targeted delivery of DNA to the mitochondrial compartment *via* import sequence-conjugated peptide nucleic acid. Mol Ther 7:550–557

Furukawa R, Yamada Y, Kawamura E, Harashima H (2015) Mitochondrial delivery of antisense RNA by MITO-Porter results in mitochondrial RNA knockdown, and has a functional impact on mitochondria. Biomaterials 57:107–115

Goswami S, Dhar G, Mukherjee S, Mahata B, Chatterjee S, Home P, Adhya S (2006) A bifunctional tRNA import receptor from *Leishmania* mitochondria. Proc Natl Acad Sci U S A 103:8354–8359

Grzeskowiak BF, Sanchez-Antequera Y, Hammerschmid E, Doblinger M, Eberbeck D, Wozniak A, Slomski R, Plank C, Mykhaylyk O (2015) Nanomagnetic activation as a way to control the efficacy of nucleic acid delivery. Pharm Res 32:103–121

Gualberto JM, Newton KJ (2017) Plant mitochondrial genomes: dynamics and mechanisms of mutation. Annu Rev Plant Biol 68:225–252

Hagstrom JE, Hegge J, Zhang G, Noble M, Budker V, Lewis DL, Herweijer H, Wolff JA (2004) A facile nonviral method for delivering genes and siRNAs to skeletal muscle of mammalian limbs. Mol Ther 10:386–398

Hernandez-Cid A, Aguirre-Sampieri S, Diaz-Vilchis A, Torres-Larios A (2012) Ribonucleases P/MRP and the expanding ribonucleoprotein world. IUBMB Life 64:521–528

Hinrichsen I, Bolle N, Paun L, Kempken F (2009) RNA processing in plant mitochondria is independent of transcription. Plant Mol Biol 70:663–668

Holzmann J, Frank P, Loffler E, Bennett KL, Gerner C, Rossmanith W (2008) RNase P without RNA: identification and functional reconstitution of the human mitochondrial tRNA processing enzyme. Cell 135:462–474

Horn R, Gupta KJ, Colombo N (2014) Mitochondrion role in molecular basis of cytoplasmic male sterility. Mitochondrion 19(Pt B):198–205

Hsu PD, Lander ES, Zhang F (2014) Development and applications of CRISPR-Cas9 for genome engineering. Cell 157:1262–1278

Hu J, Huang W, Huang Q, Qin X, Yu C, Wang L, Li S, Zhu R, Zhu Y (2014) Mitochondria and cytoplasmic male sterility in plants. Mitochondrion 19(Pt B):282–288

Iyer S, Thomas RR, Portell FR, Dunham LD, Quigley CK, Bennett JP Jr (2009) Recombinant mitochondrial transcription factor A with N-terminal mitochondrial transduction domain increases respiration and mitochondrial gene expression. Mitochondrion 9:196–203

Iyer S, Bergquist K, Young K, Gnaiger E, Rao RR, Bennett JP Jr (2012a) Mitochondrial gene therapy improves respiration, biogenesis, and transcription in G11778A Leber's hereditary optic neuropathy and T8993G Leigh's syndrome cells. Hum Gene Ther 23:647–657

Iyer S, Xiao E, Alsayegh K, Eroshenko N, Riggs MJ, Bennett JP Jr, Rao RR (2012b) Mitochondrial gene replacement in human pluripotent stem cell-derived neural progenitors. Gene Ther 19:469–475

Jash S, Adhya S (2011) Suppression of reactive oxygen species in cells with multiple mitochondrial DNA deletions by exogenous protein-coding RNAs. Mitochondrion 11:607–614

Jash S, Adhya S (2012) Induction of muscle regeneration by RNA-mediated mitochondrial restoration. FASEB J 26:4187–4197

Jash S, Chowdhury T, Adhya S (2012) Modulation of mitochondrial respiratory capacity by carrier-mediated transfer of RNA *in vivo*. Mitochondrion 12:262–270

Jash S, Dhar G, Ghosh U, Adhya S (2014) Role of the mTORC1 complex in satellite cell activation by RNA-induced mitochondrial restoration: dual control of cyclin D1 through microRNAs. Mol Cell Biol 34:3594–3606

Jo A, Ham S, Lee GH, Lee YI, Kim S, Lee YS, Shin JH, Lee Y (2015) Efficient mitochondrial genome editing by CRISPR/Cas9. Biomed Res Int 2015:305716

Kajander OA, Karhunen PJ, Holt IJ, Jacobs HT (2001) Prominent mitochondrial DNA recombination intermediates in human heart muscle. EMBO Rep 2:1007–1012

Karicheva OZ, Kolesnikova OA, Schirtz T, Vysokikh MY, Mager-Heckel AM, Lombes A, Boucheham A, Krasheninnikov IA, Martin RP, Entelis N, Tarassov I (2011) Correction of the consequences of mitochondrial 3243A>G mutation in the MT-TL1 gene causing the MELAS syndrome by tRNA import into mitochondria. Nucleic Acids Res 39:8173–8186

Kauppila JH, Stewart JB (2015) Mitochondrial DNA: radically free of free-radical driven mutations. Biochim Biophys Acta 1847:1354–1361

Keeney PM, Quigley CK, Dunham LD, Papageorge CM, Iyer S, Thomas RR, Schwarz KM, Trimmer PA, Khan SM, Portell FR, Bergquist KE, Bennett JP Jr (2009) Mitochondrial gene therapy augments mitochondrial physiology in a Parkinson's disease cell model. Hum Gene Ther 20:897–907

Khan SM, Bennett JP Jr (2004) Development of mitochondrial gene replacement therapy. J Bioenerg Biomembr 36:387–393

Kim YK, Zhang M, Lu JJ, Xu F, Chen BA, Xing L, Jiang HL (2016) PK11195-chitosan-graft-polyethylenimine-modified SPION as a mitochondria-targeting gene carrier. J Drug Target 24:457–467

Kiss T, Filipowicz W (1992) Evidence against a mitochondrial location of the 7-2/MRP RNA in mammalian cells. Cell 70:11–16

Klemm BP, Wu N, Chen Y, Liu X, Kaitany KJ, Howard MJ, Fierke CA (2016) The diversity of ribonuclease P: protein and RNA catalysts with analogous biological functions. Biomol Ther 6:E27

Kolesnikova OA, Entelis NS, Jacquin-Becker C, Goltzene F, Chrzanowska-Lightowlers ZM, Lightowlers RN, Martin RP, Tarassov I (2004) Nuclear DNA-encoded tRNAs targeted into mitochondria can rescue a mitochondrial DNA mutation associated with the MERRF syndrome in cultured human cells. Hum Mol Genet 13:2519–2534

Kolesnikova O, Kazakova H, Comte C, Steinberg S, Kamenski P, Martin RP, Tarassov I, Entelis N (2010) Selection of RNA aptamers imported into yeast and human mitochondria. RNA 16:926–941

Koley S, Adhya S (2013) A voltage-gated pore for translocation of tRNA. Biochem Biophys Res Commun 439:23–29

Koulintchenko M, Konstantinov Y, Dietrich A (2003) Plant mitochondria actively import DNA *via* the permeability transition pore complex. EMBO J 22:1245–1254

Koulintchenko M, Temperley RJ, Mason PA, Dietrich A, Lightowlers RN (2006) Natural competence of mammalian mitochondria allows the molecular investigation of mitochondrial gene expression. Hum Mol Genet 15:143–154

Lee M, Choi JS, Choi MJ, Pak YK, Rhee BD, Ko KS (2007) DNA delivery to the mitochondria sites using mitochondrial leader peptide conjugated polyethylenimine. J Drug Target 15:115–122

Legros F, Malka F, Frachon P, Lombes A, Rojo M (2004) Organization and dynamics of human mitochondrial DNA. J Cell Sci 117:2653–2662

Leng Q, Goldgeier L, Zhu J, Cambell P, Ambulos N, Mixson AJ (2007) Histidine-lysine peptides as carriers of nucleic acids. Drug News Perspect 20:77–86

Li W, Nicol F, Szoka FC Jr (2004) GALA: a designed synthetic pH-responsive amphipathic peptide with applications in drug and gene delivery. Adv Drug Deliv Rev 56:967–985

Lightowlers RN (2011) Mitochondrial transformation: time for concerted action. EMBO Rep 12:480–481

Liu F, Song Y, Liu D (1999) Hydrodynamics-based transfection in animals by systemic administration of plasmid DNA. Gene Ther 6:1258–1266

Lu Q, Wierzbicki S, Krasilnikov AS, Schmitt ME (2010) Comparison of mitochondrial and nucleolar RNase MRP reveals identical RNA components with distinct enzymatic activities and protein components. RNA 16:529–537

Lyrawati D, Trounson A, Cram D (2011) Expression of GFP in the mitochondrial compartment using DQAsome mediated delivery of an artificial mini-mitochondrial genome. Pharm Res 28:2848–2862

Magalhaes PJ, Andreu AL, Schon EA (1998) Evidence for the presence of 5S rRNA in mammalian mitochondria. Mol Biol Cell 9:2375–2382

Mahata B, Bhattacharyya SN, Mukherjee S, Adhya S (2005) Correction of translational defects in patient-derived mutant mitochondria by complex-mediated import of a cytoplasmic tRNA. J Biol Chem 280:5141–5144

Mahata B, Mukherjee S, Mishra S, Bandyopadhyay A, Adhya S (2006) Functional delivery of a cytosolic tRNA into mutant mitochondria of human cells. Science 314:471–474

Mahato B, Jash S, Adhya S (2011) RNA-mediated restoration of mitochondrial function in cells harboring a Kearns Sayre syndrome mutation. Mitochondrion 11:564–574

Marlow FL (2017) Mitochondrial matters: mitochondrial bottlenecks, self-assembling structures, and entrapment in the female germline. Stem Cell Res 21:178–186

Martin WF, Garg S, Zimorski V (2015) Endosymbiotic theories for eukaryote origin. Philos Trans R Soc Lond Ser B Biol Sci 370:20140330

Matsuda D, Dreher TW (2004) The tRNA-like structure of *Turnip yellow mosaic virus* RNA is a 3′-translational enhancer. Virology 321:36–46

Mileshina D, Koulintchenko M, Konstantinov Y, Dietrich A (2011) Transfection of plant mitochondria and *in organello* gene integration. Nucleic Acids Res 39:e115

Mukherjee S, Basu S, Home P, Dhar G, Adhya S (2007) Necessary and sufficient factors for the import of transfer RNA into the kinetoplast mitochondrion. EMBO Rep 8:589–595

Mukherjee S, Mahata B, Mahato B, Adhya S (2008) Targeted mRNA degradation by complex-mediated delivery of antisense RNAs to intracellular human mitochondria. Hum Mol Genet 17:1292–1298

Mukherjee J, Mahato B, Adhya S (2014) Vesicular transport of a ribonucleoprotein to mitochondria. Biol Open 3:1083–1091

Otten AB, Smeets HJ (2015) Evolutionary defined role of the mitochondrial DNA in fertility, disease and ageing. Hum Reprod Update 21:671–689

Paris Z, Rubio MA, Lukes J, Alfonzo JD (2009) Mitochondrial tRNA import in *Trypanosoma brucei* is independent of thiolation and the Rieske protein. RNA 15:1398–1406

Payne BA, Chinnery PF (2015) Mitochondrial dysfunction in aging: much progress but many unresolved questions. Biochim Biophys Acta 1847:1347–1353

Pinto M, Moraes CT (2015) Mechanisms linking mtDNA damage and aging. Free Radic Biol Med 85:250–258

Puranam RS, Attardi G (2001) The RNase P associated with HeLa cell mitochondria contains an essential RNA component identical in sequence to that of the nuclear RNase P. Mol Cell Biol 21:548–561

Rulli T (2017) The mitochondrial replacement 'therapy' myth. Bioethics 31:368–374

Salinas T, Duchêne AM, Maréchal-Drouard L (2008) Recent advances in tRNA mitochondrial import. Trends Biochem Sci 33:320–329

Sbicego S, Nabholz CE, Hauser R, Blum B, Schneider A (1998) *In vivo* import of unspliced tRNATyr containing synthetic introns of variable length into mitochondria of *Leishmania tarentolae*. Nucleic Acids Res 26:5251–5255

Scarpelli M, Todeschini A, Volonghi I, Padovani A, Filosto M (2017) Mitochondrial diseases: advances and issues. Appl Clin Genet 10:21–26

Schneider A (2011) Mitochondrial tRNA import and its consequences for mitochondrial translation. Annu Rev Biochem 80:1033–1053

Schneider A, Maréchal-Drouard L (2000) Mitochondrial tRNA import: are there distinct mechanisms? Trends Cell Biol 10:509–513

Seibel P, Trappe J, Villani G, Klopstock T, Papa S, Reichmann H (1995) Transfection of mitochondria: strategy towards a gene therapy of mitochondrial DNA diseases. Nucleic Acids Res 23:10–17

Seibel M, Bachmann C, Schmiedel J, Wilken N, Wilde F, Reichmann H, Isaya G, Seibel P, Pfeiler D (1999) Processing of artificial peptide-DNA-conjugates by the mitochondrial intermediate peptidase (MIP). Biol Chem 380:961–967

Sieber F, Duchêne AM, Maréchal-Drouard L (2011a) Mitochondrial RNA import: from diversity of natural mechanisms to potential applications. Int Rev Cell Mol Biol 287:145–190

Sieber F, Placido A, El Farouk-Ameqrane S, Duchêne AM, Maréchal-Drouard L (2011b) A protein shuttle system to target RNA into mitochondria. Nucleic Acids Res 39:e96

Small I, Maréchal-Drouard L, Masson J, Pelletier G, Cosset A, Weil JH, Dietrich A (1992) *In vivo* import of a normal or mutagenized heterologous transfer RNA into the mitochondria of transgenic plants: towards novel ways of influencing mitochondrial gene expression? EMBO J 11:1291–1296

Smirnov A, Tarassov I, Mager-Heckel AM, Letzelter M, Martin RP, Krasheninnikov IA, Entelis N (2008) Two distinct structural elements of 5S rRNA are needed for its import into human mitochondria. RNA 14:749–759

Smirnov A, Entelis N, Martin RP, Tarassov I (2011) Biological significance of 5S rRNA import into human mitochondria: role of ribosomal protein MRP-L18. Genes Dev 25:1289–1305

Su H, Wang Y, Gu Y, Bowman L, Zhao J, Ding M (2017) Potential applications and human biosafety of nanomaterials used in nanomedicine. J Appl Toxicol. https://doi.org/10.1002/jat.3476

Suzuki Y (2012) Exploring transduction mechanisms of protein transduction domains (PTDs) in living cells utilizing single-quantum dot tracking (SQT) technology. Sensors (Basel) 12:549–572

Tai N, Ding Y, Schmitz JC, Chu E (2002) Identification of critical amino acid residues on human dihydrofolate reductase protein that mediate RNA recognition. Nucleic Acids Res 30:4481–4488

Tai N, Schmitz JC, Chen TM, O'Neill MB, Chu E (2008) Identification of a cis-acting element of human dihydrofolate reductase mRNA. Biochem Biophys Res Commun 369:795–800

Tarassov IA, Entelis NS (1992) Mitochondrially-imported cytoplasmic tRNA(Lys)(CUU) of *Saccharomyces cerevisiae*: *in vivo* and *in vitro* targetting systems. Nucleic Acids Res 20:1277–1281

Thomas RR, Khan SM, Portell FR, Smigrodzki RM, Bennett JP Jr (2011) Recombinant human mitochondrial transcription factor A stimulates mitochondrial biogenesis and ATP synthesis, improves motor function after MPTP, reduces oxidative stress and increases survival after endotoxin. Mitochondrion 11:108–118

Tonin Y, Heckel AM, Vysokikh M, Dovydenko I, Meschaninova M, Rotig A, Munnich A, Venyaminova A, Tarassov I, Entelis N (2014) Modeling of antigenomic therapy of mitochondrial diseases by mitochondrially addressed RNA targeting a pathogenic point mutation in mitochondrial DNA. J Biol Chem 289:13323–13334

Touzet P, Meyer EH (2014) Cytoplasmic male sterility and mitochondrial metabolism in plants. Mitochondrion 19(Pt B):166–171

Val R, Wyszko E, Valentin C, Szymanski M, Cosset A, Alioua M, Dreher TW, Barciszewski J, Dietrich A (2011) Organelle trafficking of chimeric ribozymes and genetic manipulation of mitochondria. Nucleic Acids Res 39:9262–9274

Vestweber D, Schatz G (1989) DNA-protein conjugates can enter mitochondria *via* the protein import pathway. Nature 338:170–172

Wagle MA, Martinville LE, D'Souza GG (2011) The utility of an isolated mitochondrial fraction in the preparation of liposomes for the specific delivery of bioactives to mitochondria in live mammalian cells. Pharm Res 28:2790–2796

Walker SC, Engelke DR (2008) A protein-only RNase P in human mitochondria. Cell 135:412–414

Wang G, Chen HW, Oktay Y, Zhang J, Allen EL, Smith GM, Fan KC, Hong JS, French SW, McCaffery JM, Lightowlers RN, Morse HC 3rd, Koehler CM, Teitell MA (2010) PNPASE regulates RNA import into mitochondria. Cell 142:456–467

Wang G, Shimada E, Koehler CM, Teitell MA (2012a) PNPASE and RNA trafficking into mitochondria. Biochim Biophys Acta 1819:998–1007

Wang G, Shimada E, Zhang J, Hong JS, Smith GM, Teitell MA, Koehler CM (2012b) Correcting human mitochondrial mutations with targeted RNA import. Proc Natl Acad Sci U S A 109:4840–4845

Weber-Lotfi F, Koulintchenko MV, Ibrahim N, Hammann P, Mileshina DV, Konstantinov YM, Dietrich A (2015) Nucleic acid import into mitochondria: new insights into the translocation pathways. Biochim Biophys Acta 1853:3165–3181

Weissig V, Lasch J, Erdos G, Meyer HW, Rowe TC, Hughes J (1998) DQAsomes: a novel potential drug and gene delivery system made from dequalinium. Pharm Res 15:334–337

Weissig V, Lizano C, Torchilin VP (2000) Selective DNA release from DQAsome/DNA complexes at mitochondria-like membranes. Drug Deliv 7:1–5

Weissig V, D'Souza GG, Torchilin VP (2001) DQAsome/DNA complexes release DNA upon contact with isolated mouse liver mitochondria. J Control Release 75:401–408

Weissig V, Boddapati SV, Cheng SM, D'Souza GG (2006) Liposomes and liposome-like vesicles for drug and DNA delivery to mitochondria. J Liposome Res 16:249–264

Wiedemann N, Pfanner N (2017) Mitochondrial machineries for protein import and assembly. Annu Rev Biochem. https://doi.org/10.1146/annurev-biochem-060815-014352

Xu L, Shi R (2016) Weigh and wait: the prospect of mitochondrial gene replacement. Hum Fertil (Camb) 19:222–229

Yamada Y, Harashima H (2012) Delivery of bioactive molecules to the mitochondrial genome using a membrane-fusing, liposome-based carrier, DF-MITO-Porter. Biomaterials 33: 1589–1595

Yamada Y, Harashima H (2013) Enhancement in selective mitochondrial association by direct modification of a mitochondrial targeting signal peptide on a liposomal based nanocarrier. Mitochondrion 13:526–532

Yamada Y, Akita H, Kamiya H, Kogure K, Yamamoto T, Shinohara Y, Yamashita K, Kobayashi H, Kikuchi H, Harashima H (2008) MITO-Porter: a liposome-based carrier system for delivery of macromolecules into mitochondria *via* membrane fusion. Biochim Biophys Acta 1778:423–432

Yamada Y, Furukawa R, Yasuzaki Y, Harashima H (2011) Dual function MITO-Porter, a nano carrier integrating both efficient cytoplasmic delivery and mitochondrial macromolecule delivery. Mol Ther 19:1449–1456

Yamada Y, Kawamura E, Harashima H (2012) Mitochondrial-targeted DNA delivery using a DF-MITO-Porter, an innovative nano carrier with cytoplasmic and mitochondrial fusogenic envelopes. J Nanopart Res 14:1013

Yamada Y, Fukuda Y, Harashima H (2015) An analysis of membrane fusion between mitochondrial double membranes and MITO-Porter, mitochondrial fusogenic vesicles. Mitochondrion 24:50–55

Yamada Y, Furukawa R, Harashima H (2016) A dual-ligand liposomal system composed of a cell-penetrating peptide and a mitochondrial RNA aptamer synergistically facilitates cellular uptake and mitochondrial targeting. J Pharm Sci 105:1705–1713

Yasuzaki Y, Yamada Y, Harashima H (2010) Mitochondrial matrix delivery using MITO-Porter, a liposome-based carrier that specifies fusion with mitochondrial membranes. Biochem Biophys Res Commun 397:181–186

Yasuzaki Y, Yamada Y, Kanefuji T, Harashima H (2013) Localization of exogenous DNA to mitochondria in skeletal muscle following hydrodynamic limb vein injection. J Control Release 172:805–811

Yasuzaki Y, Yamada Y, Fukuda Y, Harashima H (2014) Condensation of plasmid DNA enhances mitochondrial association in skeletal muscle following hydrodynamic limb vein injection. Pharmaceuticals (Basel) 7:881–893

Yasuzaki Y, Yamada Y, Ishikawa T, Harashima H (2015) Validation of mitochondrial gene delivery in liver and skeletal muscle via hydrodynamic injection using an artificial mitochondrial reporter DNA vector. Mol Pharm 12:4311–4320

Yu H, Koilkonda RD, Chou TH, Porciatti V, Ozdemir SS, Chiodo V, Boye SL, Boye SE, Hauswirth WW, Lewin AS, Guy J (2012a) Gene delivery to mitochondria by targeting modified adenoassociated virus suppresses Leber's hereditary optic neuropathy in a mouse model. Proc Natl Acad Sci U S A 109:E1238–E1247

Yu H, Ozdemir SS, Koilkonda RD, Chou TH, Porciatti V, Chiodo V, Boye SL, Hauswirth WW, Lewin AS, Guy J (2012b) Mutant NADH dehydrogenase subunit 4 gene delivery to mitochondria by targeting sequence-modified adeno-associated virus induces visual loss and optic atrophy in mice. Mol Vis 18:1668–1683

Yu H, Mehta A, Wang G, Hauswirth WW, Chiodo V, Boye SL, Guy J (2013) Next-generation sequencing of mitochondrial targeted AAV transfer of human ND4 in mice. Mol Vis 19:1482–1491

Yu H, Koilkonda RD, Chou TH, Porciatti V, Mehta A, Hentall ID, Chiodo VA, Boye SL, Hauswirth WW, Lewin AS, Guy J (2015) Consequences of zygote injection and germline transfer of mutant human mitochondrial DNA in mice. Proc Natl Acad Sci U S A 112:E5689–E5698

Zelenka J, Jezek P (2016) Import of fluorescent RNA into mitochondria of living cells. Methods Mol Biol 1351:175–181

Zelenka J, Alan L, Jaburek M, Jezek P (2014) Import of desired nucleic acid sequences using addressing motif of mitochondrial ribosomal 5S-rRNA for fluorescent *in vivo* hybridization of mitochondrial DNA and RNA. J Bioenerg Biomembr 46:147–156

Zhou J, Liu L, Chen J (2010) Mitochondrial DNA heteroplasmy in *Candida glabrata* after mitochondrial transformation. Eukaryot Cell 9:806–814

Cell-Penetrating Peptides Targeting Mitochondria

Carmine Pasquale Cerrato and Ülo Langel

Abstract Mitochondria are key organelles with essential functions and fundamental roles in cell death and survival signaling. Consequently, they are involved in a wide range of diseases with a great diversity of clinical appearance, which makes them attractive as target for drugs to treat metabolic and degenerative diseases and cancer. Efficient methods for specific intracellular delivery of exogenous compounds, including biochemically active small molecules, imaging agents, peptides, peptide nucleic acids, proteins, RNA, DNA, and nanoparticles, would be beneficial for research and patients. A sustained effort in the last 20 years has been done to exploit cell-penetrating peptides (CPPs) for the delivery of such useful cargoes *in vitro* and *in vivo* because of their biocompatibility, ease of synthesis, and controllable physical chemistry. Here, we discuss the mechanisms by which CPPs can function, the use of this alternative as well as strategies used to target mitochondria and the implications for drug delivery.

Keywords Cell-penetrating peptides · Peptide agents · Uptake pathways · Drug delivery · Mitochondrial targeting

1 Introduction

There are many hurdles to overcome and that separate a therapeutic molecule at the site of administration from the targeted site or an intracellular organelle, which is its final destination. The first barrier is imposed by the permeability of the cellular membrane that limits the access of exogenous molecules to the inner part of the cell.

C. P. Cerrato (✉)
The Arrhenius Laboratories for Natural Sciences, Department of Biochemistry and Biophysics, Stockholm University, Stockholm, Sweden
e-mail: carmine.pasquale.cerrato@dbb.su.se

Ü. Langel
The Arrhenius Laboratories for Natural Sciences, Department of Biochemistry and Biophysics, Stockholm University, Stockholm, Sweden

Laboratory of Molecular Biotechnology, Institute of Technology, University of Tartu, Tartu, Estonia

P. J. Oliveira (ed.), *Mitochondrial Biology and Experimental Therapeutics*,
https://doi.org/10.1007/978-3-319-73344-9_26

Another barrier is represented by the mitochondrial membrane. There are also other challenges that affect intracellular targeting as for example escape from endosomes in case of endosomal uptake as well as lysosomal or cytosolic degradation processes. Furthermore, the mitochondrial membrane potential (ΔΨm) that prevails across the outer (OMM) and inner mitochondrial membranes (IMM) needs to be overcome in order to achieve a selective targeting and accumulation in mitochondria. Several strategies have been approached toward mitochondrial targeting spacing from invasive or biological techniques to non-peptide and peptide based delivery systems as well as nanocarriers. The development of drug delivery technology began in the 1950s with the introduction of the first micro-encapsulated drug particles. Pharmacokinetic researches in the 1960s provided an initial understanding for therapeutic drug monitoring (Levy 1965). At the same time, polymers began to be used to deliver drugs. In the late 1970s were described the conditions for attaching anticancer drugs to polyalkylcyanoacrylate nanoparticles and deliver agents to cells that were capable of endocytic uptake (Couvreur et al. 1979). From the 1980s the so-called "smart" delivery systems have been developed where polymer, hydrogel and biodegradable micro-particles were used to deliver peptides and proteins. In the last decade nanoparticles-based drug delivery technologies have been developed. Since the 2000 nanotechnologies have been intensively supported and significant advances has been made. The goal of drug delivery technology is to deliver the drug to the target site in order to produce formulations clinically useful to treat patients affected by various diseases.

There are demand and need for effective treatments with highly potent and targeting therapeutic peptides, proteins and new drug delivery systems due to the increasing occurrence of life-threatening and serious debilitating diseases. The plasma membrane impermeability required a strategy that facilities entry, essential for successful delivery to the target site. This applies to therapeutic peptides and proteins having an intracellular target. Such a membrane permeation-enhancing strategy may also be applicable for drugs dosed via a non-injectable route of administration (e.g., oral, nasal, pulmonary), where cells organized in a tight epithelium must be traversed in order to reach the systemic circulation and subsequently a target receptor.

The scope of the present chapter will be applications of the CPPs as transport vectors for the delivery of peptides and proteins to mitochondria, and studies within the fields of CPP-mediated delivery across cellular membranes and mitochondrial targeting will be highlighted. In addition, the choice of formulation approach, mechanism of membrane permeation, and limitations in the use of CPPs as delivery vectors will be discussed.

2 Cell-Penetrating Peptides

CPPs comprise a family of delivery vectors, which are able to cross the cellular membranes when present in non-toxic submicromolar concentrations, thereby making them one of the powerful tools used to deliver drugs, proteins, and nucleic acids.

The CPPs are typically between 4 and 30 residues in length and often carry a positive charge, which facilitates electrostatic interactions with negatively charged cell-surface constituents, such as the glycosaminoglycans, as an initial step prior to membrane translocation. In 1988, Frankel and Pabo discovered that the Human Immunodeficiency Virus (HIV) Trans-activator of transcription (Tat) protein possessed the ability to translocate across cellular membranes (Frankel and Pabo 1988), giving the birth to the CPP field. In 1991, Joliot et al. first demonstrated effective cellular uptake of the *Drosophila* antennapedia Homeodomain protein, and a few years later, in 1994, the peptide sequence responsible for membrane traslocation was narrowed down to the third helix of the full-length protein, named as penetratin (Joliot et al. 1991; Derossi et al. 1994). Since the discovery of Tat and penetratin, more than 1000 different peptides—cationic, amphipathic, hydrophobic, and anionic—have been added to the still growing family of CPPs. They are classified based on their origin as naturally derived, synthetic, or chimera sequences. The best-known CPPs are reported in Table 1. However, a big effort is now given to the improvement of the transfection efficacy of already know CPPs more then to the discovery of new CPPs. Small molecule drugs (Lindgren et al. 2006; Shin et al. 2014), oligonucleotides (Ezzat et al. 2011; Cerrato et al. 2015a, b), proteins (Kamei et al. 2013; Morris et al. 2001), liposomes (Ding et al. 2015; Nakamura et al. 2013) and imaging agents (Qian et al. 2015; Jean et al. 2016) are some of the cargo molecules successfully delivered using the CPP-based delivery technology. Various strategies have been used to achieve better efficacy via a better understanding of uptake mechanisms, complex-formation, controlled physicochemical properties, as well as via chemical modifications like changing amino acid stereochemistry, non-coded amino acids, or branch modifications. In Table 1 is reported a selection of best-known CPPs.

Different CPPs, with or without a cargo molecule, different conditions, cell lines or animal models result in different modes of uptake. The direct penetration pathway is energy-independent; It has been shown in early studies that some CPPs were able to enter in the cells at 4 °C (Derossi et al. 1994; Vivès et al. 1997). The CPPs that enter via direct penetration first bind to the cell membrane via electrostatic or hydrophobic interactions and induce membrane destabilization at different levels. The internalization level is usually directly related to the peptide concentration. CPPs are predominantly internalized to cells via energy-dependent mechanisms involving different sorts of endocytosis or direct translocation (Brock 2014; Margus et al. 2013; Cleal et al. 2013; El-Sayed and Harashima 2013), as shown in Fig. 1. All various major mechanism of endocytosis can be involved in the uptake process, including macropinocytosis, clathrin- and caveolae-mediated endocytosis. The different route of uptake can be highly dependent on the properties of the cargo molecules, their relative concentrations, as well as several endocytic routes can be used at the same time, or alternative pathways can compensate for the direct inhibition of another (Lundin et al. 2008). The endosomal entrapment is a consequence of endocytic internalization process for CPPs. This is one of the major key limiting factors in the trafficking and activity of CPPs due to a great extent sequestration in endo-lysosomal compartments. This so-called endosomal entrapment also serves as

Table 1 Selection of best-known cell-penetrating peptides

Sequence	Name (origin)	Type[a]	Ref.
Cataionic			
RKKRRRESRKKRRRES	DPV3	N	de Coupade et al. (2005)
GRPRESGKKRKRKRLKP	DPV6	N	de Coupade et al. (2005)
RKKRRQRRR	HIV-1 pTat(49–57)	N	Vivès et al. (1997)
RRRRNRTRRNRRRVR[b]	FHV coat	N	Futaki et al. (2001) and Nakase et al. (2009)
TRQARRNRRRRWRERQR[b]	HIV-1 Rev	N	Futaki et al. (2001) and Nakase et al. (2009)
PRRRRSSSRPVRRRRRPRVSRRRRRRGGRRRR	Protamine 1	N	Reynolds et al. (2005)
RIKAERKRMRNRIAASKSRKRKLERIAR	Human cJun	N	Futaki et al. (2001) and Nakase et al. (2009)
RQIKIWFQNRRMKWKK	Penetratin	N	Derossi et al. (1994)
RVIRVWFQNKRCKDKK	Islet-1	N	Kilk et al. (2001)
SQIKIWFQNKRAKIKK	Engrailed-2	N	Han et al. (2000)
RQVTIWFQNRRVKEKK	HoxA-13	N	Balayssac et al. (2006)
(R)n[b]; n = 6–12	Polyarginine	S	Futaki et al. (2001) and Mitchell et al. (2000)
Amphipathic			
MVKSKIGSWILVLFVAMWSDVGLCKKRPKP	BPrPp(1–30)	N	Magzoub et al. (2006)
KCFQWQRNMRKVRGPPVSCIKR[b]	hLF peptide(19–40)	N	Duchardt et al. (2009)
TRSSRAGLQWPVGRVHRLLRK	Buforin 2	N	Kobayashi et al. (2000)
YKQCHKKGGKKGSG	Crotamine	N	Kerkis et al. (2004)
NAATATRGRSAASRPTQRPRAPARSASRPRRPVQ	VP22	N	Elliott and Hare (1997)
DPKGDPKGVTVTVTVTVTGKGDPKPD	VT5	N	Oehlke et al. (1997)

(continued)

Table 1 (continued)

Sequence	Name (origin)	Type[a]	Ref.
MVRRFLVTLRIRRACGPPRVRV	ARF(1–22)	N	Johansson et al. (2008)
MVTVLFRRLRIRRACGPPRVRV[b]	M918	N	El-Andaloussi et al. (2007)
LLIILRRRIRKQAHAHSK[b]	pVEC	N	Elmquist et al. (2001)
LSTAADMQGVVTDGMASG	Azurin p18	N	Taylor et al. (2009)
KFHTFPQTAIGVGAP[b]	hCT peptide 18–32	N	Tréhin et al. (2004)
ALWKTLLKKVLKAPKKKRKV	S4 13 -PV rev	C	Mano et al. (2006)
KETWWETWWTEWSQPKKKRKV[c]	Pep-1	C	Morris et al. (2001)
GALFLGFLGAAGSTMGA[c]	MPG	C	Morris et al. (1997)
GWTLNSAGYLLGKINLKALAALAKKIL[b]	Transportan	C	Pooga et al. (1998)
AGYLLGKINLKALAALAKKIL[b]	TP10 (analogue)	C	Soomets et al. (2000)
St-AGYLLGK(ε-TMQ)INLKALAALAK KIL[b]	PepFect 6	S	Andaloussi et al. (2011)
St-AGYLLGKLLOOLAAAALOOLL[b]	PepFect 14	S	Ezzat et al. (2011)
St-AGY(PO3)LLGKTNLKALAALAKKIL[b]	NickFect 1	S	Oskolkov et al. (2011)
δ-(St-AGYLLG)OINLKALAALAKKIL[b]	NickFect 51	S	Arukuusk et al. (2013)
GLWRALWRLLRSLWRLLWRA[c]	CADY	S	Crombez et al. (2009)
RRWWRRWRR	W/R	S	Delaroche et al. (2007)
KLALKALKALKAALKLA[b]	MAP (model amphipathic peptide)	S	Oehlke et al. (1998)
QLALQLALQALQAALQLA	MAP17	S	Scheller et al. (1999)
(PPR)3, (PPR)4, (PPR)5, (PPR)6	(PPR)*n*	S	Daniels and Schepartz (2007)
VRLPPPVRLPPPVRLPPP	SAP	S	Martín et al. (2011)
YTAIAWVKAFIRKLRK[b]	YTA2	S	Lindgren et al. (2006)

(continued)

Table 1 (continued)

Sequence	Name (origin)	Type[a]	Ref.
IAWVKAFIRKLRKGPLG[b]	YTA4	S	Lindgren et al. (2006)
Palmitoyl-SFLLRN	PAR1	S	Covic et al. (2002)
Palmitoyl-KIHKKGMIKS	F2Pal$_{10}$	S	Forsman et al. (2013)
Hydrophobic			
AAVLLPVLLAAP	K-FGF	N	Lin et al. (1995)
VPTLK (PMLKE, VPALR, VSALK, IPALK)	Bip	N	Gomez et al. (2007) and Gomez et al. (2010)
PFVYLI	C105Y	N	Rhee and Davis (2006)
SDLWEMMMVSLACQY	Pep-7	S	Gao et al. (2002)
PLILLRLLRGQF	Pept1	S	Marks et al. (2011)
Anionic			
LKTLTETLKELTKTLTEL	MAP12	S	Oehlke et al. (2002)
VELPPPVELPPPVELPPP	SAP(E)	S	Martín et al. (2011)
TSF d EYWYLL[d]	Stapled p53 peptide MO6	S	Chee et al. (2014)

[a]*N* natural, *C* chimeric, *S* synthetic
[b]C-terminal amide
[c]C-terminal cycteamide
[d]Stample tethering site

one of the key limiting factors in the bioavailability of CPPs (Margus et al. 2013; El-Sayed et al. 2009). Consequently, the activity of CPPs is to a high extent dependent on their ability to escape from the endo-lysosomal compartments.

Targeted drug delivery systems may facilitate a better therapeutic outcome as they are supposed to overcome limitations of conventional drug application such as unfavorable bio-distribution, low bioavailability, lack of water solubility, low therapeutic response despite high dosages, side effects, drug resistance, toxicity and barriers in the body such as the blood brain barrier (Cerrato et al. 2015a, b; Stewart et al. 2008; Dowdy 2017). Various targeting approaches for cancer therapy (Woldetsadik et al. 2017; Shin et al. 2014), neurodegenerative diseases e.g. Alzheimer's disease (Reddy et al. 2017), infectious diseases e.g. tuberculosis (Sparr et al. 2013), autoimmune diseases (Lim et al. 2015; Li et al. 2014) and several other disorders were reported.

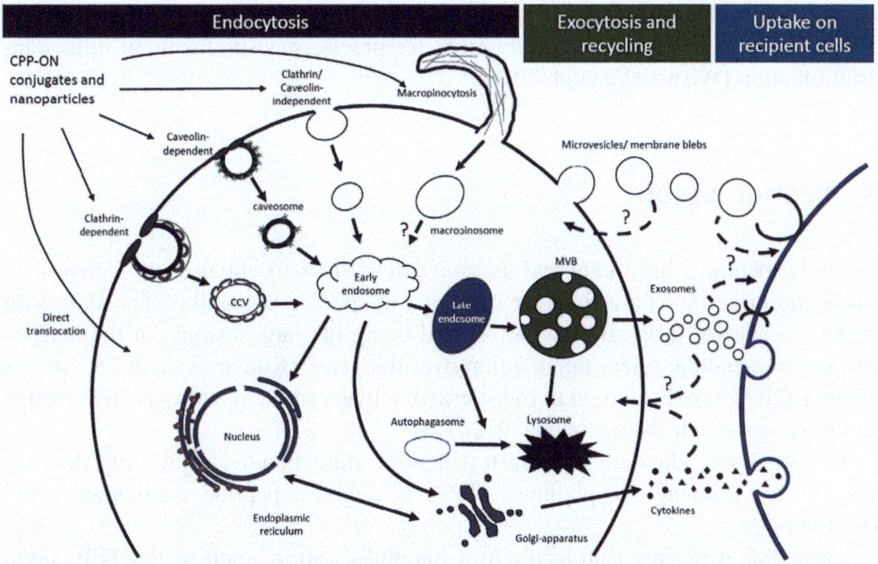

Fig. 1 Uptake and trafficking pathways of CPPs. CPPs use predominantly different sorts of endocytosis to gain access to the interior of cells. In the context of this, all major pathways for CPP uptake have been described, including clathrin- and caveolae-mediated endocytosis, as well as micropinocytosis- Less information is available for other less-defined pathways, such as several clathrin- and caveolae-independent endocytosis mechanisms. Uptake of the cargo molecule is followed by complex intracellular trafficking events towards early/sorting endosomes, late endosomes/microvesicular bodies (MVBs), lysosomes or Golgi network. Typically, endo-/lysososmal maturation is characterized by gradual drop in the pH. Of note, recycling pathways can direct cargo also through late endosomes/MVBs for being released to extracellular milieu via extracellular vesicle (EV) release. In this case, the cargo could become incorporated into exosomes and subsequently be taken up by other cells (re-distribute) or by the same cell (reuptake). In addition to the more dominant endocytic pathways, some membrane active CPPs have also been reported to be taken up by direct translocation over the cellular membrane, which potentially allows them to avoid endocytic pathways altogether. Reproduced from Lehto et al. (2016)

3 Strategies to Target Mitochondria

One approach used to target bioactive molecules to mitochondria is based the on biophysical membrane property of mitochondria. Since the mitochondria-specific phospholipid cardiolipin maintains a strong negative internal potential required for the electron transport chain functioning, cationic molecules are attracted and accumulate preferentially within the negatively charged mitochondrial matrix (Murphy and Smith 2007). Liberman and colleagues were the first to demonstrate that lipophilic cations can pass directly through phospholipid bilayers without requiring a specific uptake mechanism and they can subsequently accumulate into mitochondria due to the membrane potential (Liberman et al. 1969). Wisnovsky et al. have recently highlighted significant advances in mitochondrial chemical biology and the development of new tools and techniques for precise measurements of mitochondrial function (Wisnovsky et al. 2016).

4 Peptide Agents

Recently, peptides have emerged as safer alternatives to enhance the delivery of small and large molecules into the cells and in specific organelles. The skin is an important site for drug administration, and being the largest organ of the body is gaining tremendous scientific attention over the years (Kumar et al. 2015). In this context, CPPs have been used as transdermal delivery of therapeutics is an attractive and novel approach (Nasrollahi et al. 2012).

In Fig. 2 are schematically illustrated some mitochondriotropic drug delivery systems, as triphenylphosphonium (TPP) conjugates, peptide antioxidants, and DQAsomes.

Conjugation of cargo molecule to a lipophilic cation, such as the TPP cation (Fig. 2a) is one of the first used approach to target mitochondria due to the orally bioavailability and the accumulation into the mitochondria driven by the plasma and mitochondrial membrane potential.

Dequalinium (DQA) has been used for long time as topical antimicobrial agent and shows anticarcinoma activity. Weissig and colleagues showed that DQA forms liposome-like aggregates, termed DQAsomes (Fig. 2c). DQAsomes are able to deliver DNA and low-molecular weight molecules to mitochondria (Weissig 2015).

The developed Szeto-Schiller (SS) peptides are the most promising mitochondria-targeted antioxidant peptides. They are small tetra-peptides with the ability to accumulate in the IMM and very potent at reducing intracellular ROS and preventing cell death caused by oxidant (Zhao et al. 2004). They represent an alternative to lipophilic cations to deliver antioxidants to mitochondria (Szeto 2008). CPPs have an inherent ability to traverse biological membrane without causing significant membrane damage (Milletti 2012). It has been previously described the mechanism by which these peptides exert their antioxidant effect. It has been attributed to the presence of the modified tyrosine, specifically the dimethyltyrosine (Dmt), the ability to scavenge ROS leading to the formation of unreactive tyrosyl or dityrosine radicals, which also react with superoxide radicals to form tyrosine hydroperoxide (Zhao et al. 2004). It was demonstrated that SS peptides (Fig. 2b) target mitochondria and they localize in the IMM up to 1000-fold, selectively binding to the cardiolipin by electrostatic and hydrophobic interactions (Szeto 2006; Birk et al. 2013). The advantage of the SS peptides for being uptake not in mitochondrial membrane potential ($\Delta\psi_m$) dependent manner lies in the fact that the delivery to pathological mitochondria with reduced $\Delta\psi_m$ is then not affected. This peptides has been proved to be not toxic at concentration up to 100 μM, to have protective effects against oxidation by H_2O_2, to significantly reduced the intracellular ROS production, to prevent oxidative cell death, and to protect from Ca^{2+}-induced mitochondrial

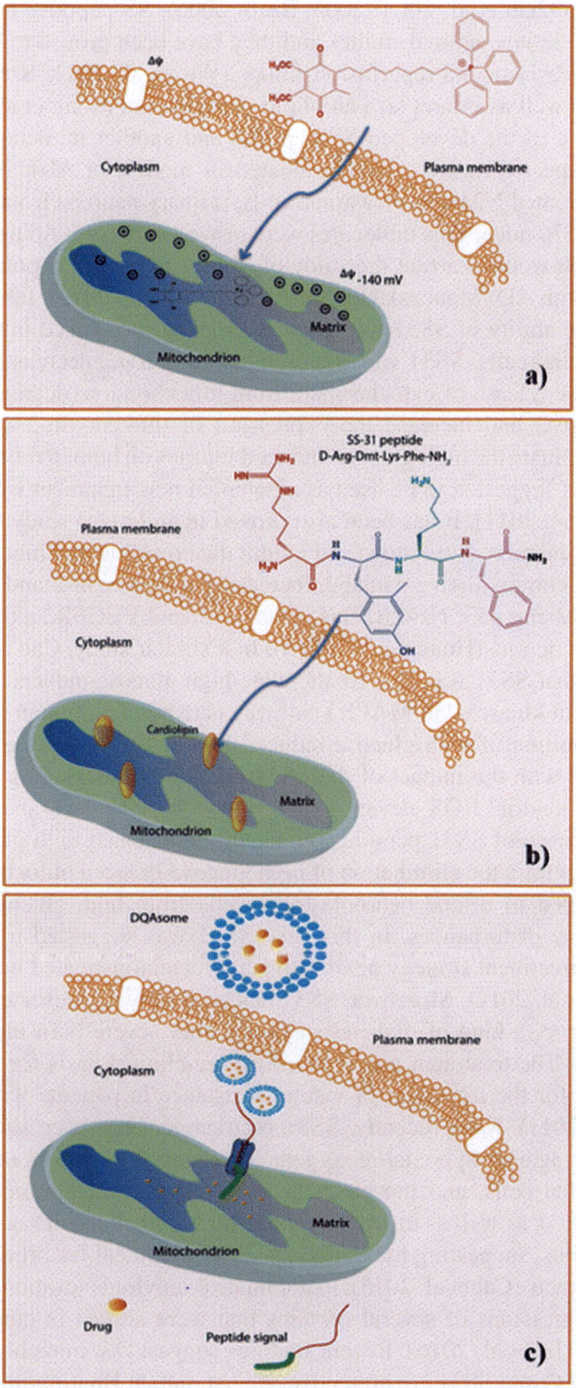

Fig. 2 Schematic illustration of mitochondriotropic drug delivery systems: (**a**) triphenylphosphonium (TPP) conjugates, (**b**) peptide antioxidants, and (**c**) DQAsomes. Reproduced from Guzman-Villanueva and Weissig (2016) with permission

depolarization (Zhao et al. 2004; 2005; Szeto 2006). SS peptides have been used also for several *in vivo* animal studies and they have been proved to be able to significantly reduce ischemia-reperfusion injury (Wu et al. 2002; Szeto 2008; Cho et al. 2007), as well as infarct size and lipid peroxidation (Song et al. 2005). SS31 and MitoQ, one of the developed SS peptides and another mitochondria-targeted antioxidant respectively, were later evaluated in models of Alzheimer's disease, including Aβ-treated N2a neuroblastoma cells, primary neurons from Tg2576 mice and aging Tg2576 mice. This molecules were proved to prevent Aβ toxicity in mitochondria, which would warrant the study of MitoQ and SS31 as potential drugs to treat patients with AD (Manczak et al. 2010; Calkins et al. 2012). It has been investigated also the ability of SS31 to prevent high glucose-induced injury on human retinal endothelial cells. SS31 was found to stabilize $\Delta\psi_m$, decrease ROS production, prevent the release of cytochrome c from mitochondria, decrease the expression of caspase-3 and increase the expression of thioredoxin-2 with a general outcome to attenuate the high glucose-induced injuries on human retinal endothelial cells. SS31 was suggested to be used as a potential new treatment for diabetic retinopathy (Li et al. 2011). It has been also proved in an *in vivo* study that SS31 was able to protect the retinal structures and inhibit the breakdown of inner blood retinal barrier by reducing oxidative damage, increasing thioredoxin-2 and Bcl-2 expression, and decreasing p53, NF-κB, Bax, caspase-3, and VEGFR2 expression in the retinas of diabetic rats (Huang et al. 2013). In a similar study, Cao and colleagues demonstrated that SS31 was able to attenuate high glucose-induced P38 mitogen-activated protein kinase (P38 MAPK) pathway activation in human neuroblastoma cells. The inhibition of high glucose-induced P38 activation by the SS31 peptide was associated with the impact of the SS31 peptide on attenuating high glucose-induced mitochondrial ROS elevation and mitochondrial membrane potential collapse. The addition of SS31 peptide significantly attenuated high glucose-induced apoptosis. Therefore, the elimination of high glucose-induced mitochondrial oxidative stress helped to rescue neuroblastoma cells from high glucose-related P38 MAPK pathway disturbances. In this case SS31 was suggested to be used as a potential new treatment strategy against hyperglycemia-instigated neuronal perturbations (Cao et al. 2012). Moreover, SS31 has been shown to reverse burn-induced insulin resistance, a kind of diabetes of injury after severe burn injury and other major traumas. The treatment provided a firm pre-clinical basis for future clinical trials of SS31 for the treatment of insulin resistance in patients with burn injury (Carter et al. 2011). Most recently, SS31 pretreatment has been shown to have a protective role against hypoxia//reoxygenation-induced apoptosis of human renal tubular epithelial cells, and the mechanism is related to suppression of p66Shc (Zhao et al. 2013), as well as to protect human lens epithelial cells against oxidative damage and, thus, suggesting to be used as a new treatment for preventing the formation of cataracts (Cai et al. 2015), to attenuate β-amyloid elevation and to restore the protein expressions of several proteins that were altered in control AD mice animal model (Jia et al. 2016). Recent findings suggest that mitochondria-targeted molecules MitoQ and SS31 are protective against mutant Huntingtin-induced mitochondrial and synaptic damage in Huntington's disease (HD) neurons, and these

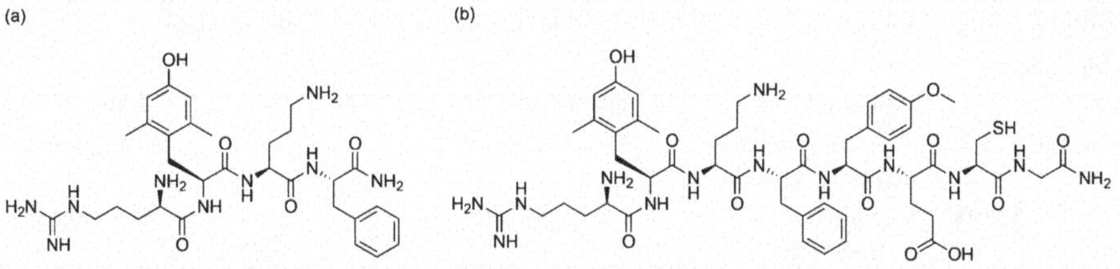

Fig. 3 General structure of mtCPP1 (**a**) and mtgCPP (**b**)

mitochondria-targeted molecules are potential therapeutic molecules for the treatment of HD neurons (Yin et al. 2016). In another recent investigation to understand the protective effects of SS31 against amyloid beta-induced mitochondrial and synaptic toxicities in AD progression, Aβ production and mitochondrial dysfunction were reduced in AD mice model (APP) treated with SS31, and mitochondrial dynamics were maintained and mitochondrial biogenesis and synaptic activity were enhanced. All this effects confirmed that SS31 confer protective effects against mitochondrial and synaptic toxicities in APP transgenic mice and that it may be a promising drug molecule to treat patients with AD (Reddy et al. 2017). SS31 has been furthermore developed and from a small library of modified peptides mtCPP1 (Fig. 3a) emerged having a twofold better effect on preventing ROS formation and rescuing $\Delta\psi_m$ at physiological levels in cells under stress condition (Cerrato et al. 2015a, b). This peptide has been shown to transport into the cells and specifically to the mitochondria different cargo molecules like fluorophore or another peptide. mtCPP1 fused by covalent conjugation to a glutathione analog peptide was showed to have even better antioxidant effects compared to the parent peptides. The fused peptide, called mtgCPP (Fig. 3b) has superior protective effect from oxidative damage compared with mtCPP1 and UPF25 against H_2O_2 insult, preventing ROS formation by two- and threefold, respectively. It has been hypothesized that the fused peptide has a synergistic effect, where the mitochondrial peptide segment is targeting mitochondria, whereas the glutathione analog peptide segment is active in also in the cytosol, resulting in increased scavenging ability (Cerrato and Langel 2017).

SS31 has been developed under the name of Bendavia first and Elamipretide later undergoing preclinical and clinical studies conducted by laboratories and clinicians worldwide. The main goal is to treat both rare mitochondrial diseases and all other major diseases where mitochondrial dysfunctions are involved. A list of completed and ongoing clinical trial for Elamipretide is presented in Table 2. Mitochondrial diseases characterized by genetic defect are categorized as inherited or rare mitochondrial diseases. Another category of mitochondrial diseases is represented by common diseases such as heart failure, renal disease and age-related macular degeneration where the mitochondrial dysfunctions are involved. Preclinical, phase I, and phase II have been conducted for both disease categories, and the compound has been administrated via subcutaneous injection, intravenous infusion or topical ophthalmic delivery resulting in all cases safe.

Table 2 Clinical trails for Elamipretide (Bendavia) peptide listed on clinicaltrial.gov

Identifier number	End point	Phase	Subjects	Condition	Status
NCT01513200	Pharmacodynamics and pharmacokinetics	I	Healthy	–	Completed
NCT01518985	Endothelial dysfunction	I	Healthy	–	Completed
NCT01786915	Safety and pharmacokinetics	I	Healthy	–	Completed
NCT01755858	Pharmacodynamics (renal flow and function)	I/II	Patients	Renal artery obstruction; hypertension, renovascular; ischemia reperfusion injury	Terminated
NCT01754818	Pharmacokinetics and tolerability	I	Healthy	–	Completed
NCT02245620	Safety and efficacy (skeletal function)	II	Patients	Skeletal muscle mitochondrial dysfunction in the elderly	Completed
NCT01572909	Effectiveness, safety and tolerability	II	Patients	Reperfusion injury; STEMI	Completed
NCT01115920	Safety, tolerability and pharmacokinetics	I	Healthy	–	Completed
NCT02388529	Safety, tolerability, pharmacokinetics and efficacy	I	Patients	Heart failure	Withdrawn
NCT02388464	Safety, tolerability and pharmacokinetics	I	Patients	Congestive heart failure	Completed
NCT02436447	Safety and pharmacokinetics	I	Patients	Normal and impaired renal function	Completed
NCT02367014	Safety, tolerability and efficacy	I/II	Patients	Mitochondrial myopathy	Completed
NCT02314299	Safety and efficacy	I	Patients	–	Recruiting
NCT02914665	Pharmacodynamics (cardiac and renal function)	II	Patients	Heart failure	Recruiting
NCT02814097	Safety and efficacy	II	Patients	Chronic heart failure	Recruiting
NCT02805790	Safety, tolerability and efficacy	II	Patients	Primary mitochondrial disease	Active, not recruiting
NCT02848313	Safety, tolerability and efficacy	I	Patients	Age-related macular degeneration	Recruiting
NCT02788747	Effectiveness, safety and tolerability	II	Patients	Heart failure	Recruiting

Adapted from (Guzman-Villanueva and Weissig (2016) with permission

5 Requirements for Mitochondrial Localization of Peptides

Horton and colleagues have been the first to systematically identify critical levels of lipophilicity that allow molecules to access mitochondria and to also better understand the effect of molecular charge on lipophilicity thresholds. They synthesized peptides incorporating several hydrophobic amino acids, including unnatural residues displaying diphenyl, naphtyl, or hexyl functionalities as well as fluorinated, methylated tyrosine were used to enhance chemical diversity (Fig. 4a). It was noted that within the peptides bearing the same net charge, some peptides displayed exclusively mitochondrial localization, others were present in both mitochondria and nuclei, and others were not localized in mitochondria at all (Fig. 4b). Interestingly, they found a striking trend examining the correlation between mitochondrial localization and the lipophilicity of the peptides. It was observed that peptides with a log P value (measured via octanol portioning by a modification of the shake-flask method) higher than −1.7 were localized in mitochondria, while peptides with log P values lower than −2.0 were localized in the nuclei and cytoplasm. Peptides with log P values between −1.7 and −2.0 showed to have a mixed localization into cells. They concluded that there is a critical lipophilicity threshold where mitochondrial entry is permitted, where instead higher level of hydrophobicity would result in decreased compatibility with the mitochondrial membrane precluding the access (Fig. 5). The trends observed with those peptides give insight into the exact requirements that impart mitochondrial localization (Horton et al. 2008).

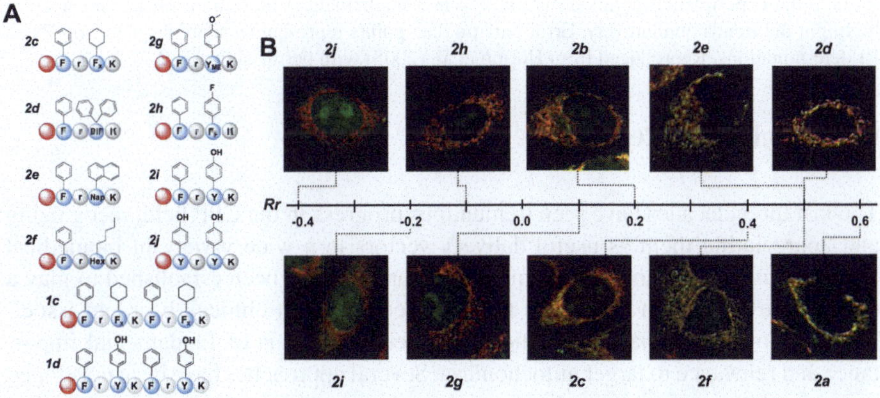

Fig. 4 Evaluation of mitochondrial localization for synthetic peptide conjugates. (**a**) The peptido-conjugate scaffolds tested consisted of four or eight amino acids linked to the fluorophore, thiazole orange. The lysine and arginine residues shown schematically in gray were held constant (along with the fluorophore, thiazole orange), and the residues in red were altered to modulate lipophilicity. The tetramer conjugates possess a charge of +3, while the octamers possess a charge of +5. (**b**) Quantitation of mitochondrial localization via calculation of Rr for +3 peptide conjugates. Reproduced from Horton et al. (2008) with permission

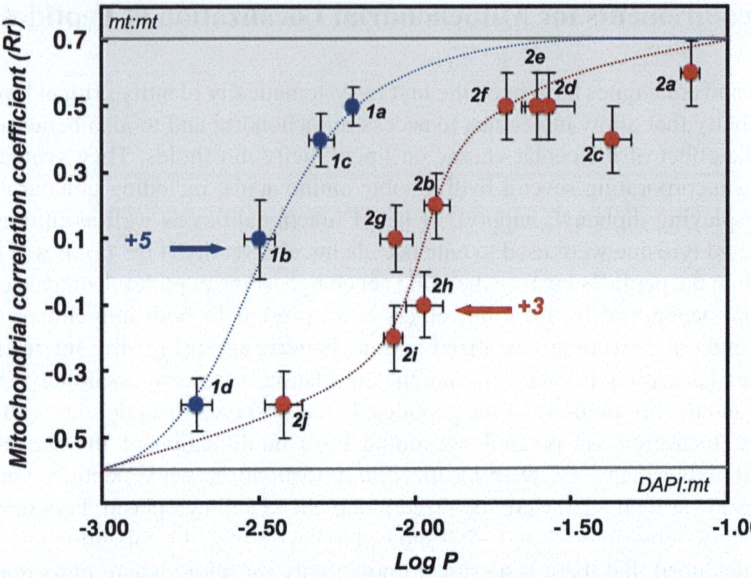

Fig. 5 Correlation between mitochondrial localization and lipophilicity. Rr values are plotted against experimentally determined log P values. Compounds bearing a +3 charge are plotted in red, and those with a +5 charge are plotted in blue. Solid gray lines represent limits for Rr values obtained when two mitochondrial dyes were colocalized (mt:mt) versus when one mitochondrial dye and one nuclear dye were colocalized (DAPI:mt). These control experiments indicate that the maximum positive correlation coefficient that can be obtained with the protocols employed is +0.7, and that the minimum negative correlation coefficient is –0.6. The gray-shaded area on the plot represents the values of Rr reflecting mitochondrial localization. Please note that the lines drawn through the experimental points do not represent a mathematical fit, but are instead intended to highlight the trends qualitatively. Error bars on data points represent standard error for log P and Rr determinations. Reproduced from Horton et al. (2008) with permission

6 Concluding Remarks

The last three decades have seen tremendous progress in the CPP field, recognizing and appreciating them as useful delivery vectors in a wide variety of biomedical applications. In the same timeframe, mitochondria have been established to play a central role for human health due to the large number of clinical disorders associated with mitochondrial dysfunctions. Consequently, it is of fundamental importance and relevance to target mitochondria. Several approaches have been developed and several basic research studies have shown successful results for cancer therapy, cardiac protection and other cardiovascular diseases as well as neuroprotection in Parkinson's and Alzheimer's diseases or for metabolic disorders. Encouraged by new positive clinical trials results and motivate by the advance of research in the area of mitochondrial targeting it is still needed to continue persuing the targeted delivery of different cargos in order to develop useful tools for research and therapeutics.

Acknowledgments The authors have no relevant affiliations or financial involvement with any organization or entity with a financial interest in or financial conflict with the subject matter or materials discussed in the manuscript. This includes employment, consultancies, honoraria, stock ownership or options, expert testimony, grants or patents received or pending, or royalties. No writing assistance was utilized in the production of this manuscript. This work was supported by the Swedish Research Council for Natural Sciences (621-2011-5902), the Swedish Research Council for Medical Research (K2012-66X-21148-04-5), and the Swedish Cancer Foundation (CAN 2014/259).

References

Andaloussi SE, Lehto T, Mäger I, Rosenthal-Aizman K, Oprea II, Simonson OE, Sork H, Ezzat K, Copolovici DM, Kurrikoff K, Viola JR, Zaghloul EM, Sillard R, Johansson HJ, Said Hassane F, Guterstam P, Suhorutšenko J, Moreno PM, Oskolkov N, Hälldin J, Tedeba U, Metspalu A, Lebleu B, Lehtiö J, Smith CI, Langel U (2011) Design of a peptide-based vector, PepFect6, for efficient delivery of siRNA in cell culture and systemically in vivo. Nucleic Acids Res 39(9):3972–3987

Arukuusk P, Pärnaste L, Oskolkov N, Copolovici DM, Margus H, Padari K, Möll K, Maslovskaja J, Tegova R, Kivi G, Tover A, Pooga M, Ustav M, Langel U (2013) New generation of efficient peptide-based vectors, NickFects, for the delivery of nucleic acids. Biochim Biophys Acta 1828(5):1365–1373

Balayssac S, Burlina F, Convert O, Bolbach G, Chassaing G, Lequin O (2006) Comparison of penetratin and other homeodomain-derived cell-penetrating peptides: a interaction in a membrane-mimicking environment and cellular uptake efficiency. Biochemistry 45(5):1408–1420

Birk AV et al (2013) The mitochondrial-targeted compound SS-31 re-energizes ischemic mitochondria by interacting with cardiolipin. J Am Soc Nephrol 24(8):1250–1261

Brock R (2014) The uptake of arginine-rich cell-penetrating peptides: putting the puzzle together. Bioconjug Chem 25:863–868

Cai M et al (2015) Mitochondria-targeted antioxidant peptide SS31 protects cultured human lens epithelial cells against oxidative stress. Curr Eye Res 40(8):822–829

Calkins MJ, Manczak M, Reddy PH (2012) Mitochondria-targeted antioxidant SS31 prevents amyloid beta-induced mitochondrial abnormalities and synaptic degeneration in Alzheimer's disease. Pharmaceuticals (Basel, Switzerland) 5(10):1103–1119

Cao M et al (2012) Mitochondria-targeted antioxidant attenuates high glucose-induced P38 MAPK pathway activation in human neuroblastoma cells. Mol Med Rep 5(4):929–934

Carter EA et al (2011) Evaluation of the antioxidant peptide SS31 for treatment of burn-induced insulin resistance. Int J Mol Med 28(4):589–594

Cerrato CP, Langel Ü (2017) Effect of a fusion peptide by covalent conjugation of a mitochondrial cell-penetrating peptide and a glutathione analog peptide. Mol Ther Methods Clin Dev 5:221–231

Cerrato CP, Pirisinu M et al (2015a) Novel cell-penetrating peptide targeting mitochondria. FASEB J 29(11):4589–4599

Cerrato CP, Veiman K-L, Langel Ü (2015b) Advances in peptide delivery. In: Kruger HG, Albericio F (eds) Advances in the discovery and development of peptide therapeutics. Future Science, Future Science Group, London, pp 160–171

Chee SM, Wongsantichon J, Soo Tng Q, Robinson R, Joseph TL, Verma C, Lane DP, Brown CJ, Ghadessy FJ (2014) Structure of a stapled peptide antagonist bound to nutlin-resistant Mdm2. PLoS One 9(8):e104914

Cho S et al (2007) A novel cell-permeable antioxidant peptide, SS31, attenuates ischemic brain injury by down-regulating CD36. J Biol Chem 282(7):4634–4642

Cleal K et al (2013) Endocytosis, intracellular traffic and fate of cell penetrating peptide based conjugates and nanoparticles. Curr Pharm Des 19(16):2878–2894
de Coupade C, Fittipaldi A, Chagnas V, Michel M, Carlier S, Tasciotti E, Darmon A, Ravel D, Kearsey J, Giacca M, Cailler F (2005) Novel human-derived cell-penetrating peptides for specific subcellular delivery of therapeutic biomolecules. Biochem J 390(2):407–418
Couvreur P et al (1979) Adsorption of antineoplastic drugs to polyalkylcyanoacrylate nanoparticles and their release in calf serum. J Pharm Sci 68(12):1521–1524
Covic L, Misra M, Badar J, Singh C, Kuliopulos A (2002) Pepducin-based intervention of thrombin-receptor signaling and systemic platelet activation. Nat Med 8(10):1161–1165
Crombez L, Aldrian-Herrada G, Konate K, Nguyen QN, McMaster GK, Brasseur R, Heitz F, Divita G (2009) A new potent secondary amphipathic cell-penetrating peptide for siRNA delivery into mammalian cells. Mol Ther 17:95–103
Daniels DS, Schepartz A (2007) Intrinsically cell-permeable miniature proteins based on a minimal cationic PPII motif. J Am Chem Soc 129(47):14578–14579
Delaroche D, Aussedat B, Aubry S, Chassaing G, Burlina F, Clodic G, Bolbach G, Lavielle S, Sagan S (2007) Tracking a new cell-penetrating (W/R) nonapeptide, through an enzyme-stable mass spectrometry reporter tag. Anal Chem 79(5):1932–1938
Derossi D et al (1994) The third helix of the Antennapedia homeodomain translocates through biological membranes. J Biol Chem 269:10444–10450
Ding Y et al (2015) An efficient PEGylated liposomal nanocarrier containing cell-penetrating peptide and pH- sensitive hydrazone bond for enhancing tumor- targeted drug delivery. Int J Nanomedicine 10:6199–6214
Dowdy SF (2017) Overcoming cellular barriers for RNA therapeutics. Nat Biotechnol 35(3):222–229
Duchardt F, Ruttekolk IR, Verdurmen WPR, Lortat-Jacob H, Bürck J, Hufnagel H, Fischer R, van den Heuvel M, Löwik DWPM, Vuister GW, Ulrich A, de Waard M, Brock R (2009) A cell-penetrating peptide derived from human lactoferrin with conformation-dependent uptake efficiency. J Biol Chem 284(52):36099–36108
El-Andaloussi S, Johansson HJ, Holm T, Langel U (2007) A novel cell-penetrating peptide, M918, for efficient delivery of proteins and peptide nucleic acids. Mol Ther 15(10):1820–1826
Elliott G, O'Hare P (1997) Intercellular trafficking and protein delivery by a herpesvirus structural protein. Cell 88(2):223–233
Elmquist A, Lindgren M, Bartfai T, Langel U (2001) VE-cadherin-derived cell-penetrating peptide, pVEC, with carrier functions. Exp Cell Res 269(2):237–244
El-Sayed A, Harashima H (2013) Endocytosis of gene delivery vectors: from clathrin-dependent to lipid raft-mediated endocytosis. Mol Ther 21(6):1118–1130
El-Sayed A, Futaki S, Harashima H (2009) Delivery of macromolecules using arginine-rich cell-penetrating peptides : ways to overcome endosomal entrapment. AAPS J 11(1):13–22
Ezzat K et al (2011) PepFect 14, a novel cell-penetrating peptide for oligonucleotide delivery in solution and as solid formulation. Nucleic Acids Res 39(12):5284–5298
Forsman H, Bylund J, Oprea TI, Karlsson A, Boulay F, Rabiet MJ, Dahlgren C (2013) The leukocyte chemotactic receptor FPR2, but not the closely related FPR1, is sensitive to cell-penetrating pepducins with amino acid sequences descending from the third intracellular receptor loop. Biochim Biophys Acta 1833(8):1914–1923
Frankel AD, Pabo CO (1988) Cellular uptake of the tat protein from human immunodeficiency virus. Cell 55(6):1189–1193
Futaki S, Suzuki T, Ohashi W, Yagami T, Tanaka S, Ueda K, Sugiura Y (2001) Arginine-rich peptides. J Biol Chem 276(8):5836–5840
Gao C, Mao S, Ditzel HJ, Farnaes L, Wirsching P, Lerner RA, Janda KD (2002) A cell penetrating peptide from a novel pVII-pIX phage-displayed random peptide library. Bioorg Med Chem 10:4057–4065
Gomez JA, Gama V, Yoshida T, Sun W, Hayes P, Leskov K, Boothman D, Matsuyama S (2007) Bax-inhibiting peptides derived from Ku70 and cell-penetrating pentapeptides. Biochem Soc Trans 35(4):797–801

Gomez JA, Chen J, Ngo J, Hajkova D, Yeh IJ, Gama V, Miyagi M, Matsuyama S (2010) Cell-penetrating penta-peptides (CPP5s): measurement of cell entry and protein-transduction activity. Pharmaceuticals 3(12):3594–3613

Guzman-Villanueva D, Weissig V (2016) Delivery of biologically active molecules to mitochondria. In: Buhlman L (ed) Mitochondrial mechanisms of degeneration and repair in Parkinson's disease. Springer, Cham, pp 255–267

Han K, Jeon M-J, Kim K-E, Park J, Choi SY (2000) Efficient intracellular delivery of GFP by homeodomains of Drosophila Fushi-tarazu and engrailed proteins. Mol Cells 10(6):0728–0732

Horton KL et al (2008) Mitochondria-penetrating peptides. Chem Biol 15(4):375–382

Huang J et al (2013) Mitochondria-targeted antioxidant peptide SS31 protects the retinas of diabetic rats. Curr Mol Med 13(6):935–945

Jean SR et al (2016) Peptide-mediated delivery of chemical probes and therapeutics to mitochondria. Acc Chem Res 49:1893–1902

Jia Y et al (2016) SS31, a small molecule antioxidant peptide, attenuates β-amyloid elevation, mitochondrial/synaptic deterioration and cognitive deficit in SAMP8 mice. Curr Alzheimer Res 13(3):297–306

Johansson HJ, El-Andaloussi S, Holm T, Mäe M, Jänes J, Maimets T, Langel Ü (2008) Characterization of a novel cytotoxic cell-penetrating peptide derived from p14ARF protein. Mol Ther 16:115–123

Joliot A et al (1991) Antennapedia homeobox peptide regulates neural morphogenesis. Proc Natl Acad Sci U S A 88:1864–1868

Kamei N et al (2013) Mechanistic study of the uptake/permeation of cell-penetrating peptides across a caco-2 monolayer and their stimulatory effect on epithelial insulin transport. J Pharm Sci 102(11):3998–4008

Kerkis A, Kerkis I, Rádis-Baptista G, Oliveira EB, Vianna-Morgante AM, Pereira LV, Yamane T (2004) Crotamine is a novel cell-penetrating protein from the venom of rattlesnake us. FASEB J 18(12):1407–1409

Kilk K, Magzoub M, Pooga M, Göran Eriksson LE, Langel U, Gräslund A (2001) Cellular internalization of a cargo complex with a novel peptide derived from the third helix of the islet-1 homeodomain. Comparison with the penetratin peptide. Bioconjug Chem 12(6):911–916

Kobayashi S, Takeshima K, Park CB, Kim SC, Matsuzaki K (2000) Interactions of the novel antimicrobial peptide buforin 2 with lipid bilayers: a proline as a translocation promoting factor. Biochemistry 39(29):8648–8654

Kumar S et al (2015) Peptides as skin penetration enhancers: mechanisms of action. J Control Release 199:168–178

Lehto T et al (2016) Peptides for nucleic acid delivery. Adv Drug Deliv Rev 106(2016):172–182

Levy G (1965) Pharmacokinetics of salicylate elimination in man. J Pharm Sci 54(7):959–967

Li J et al (2011) Mitochondria-targeted antioxidant peptide SS31 attenuates high glucose-induced injury on human retinal endothelial cells. Biochem Biophys Res Commun 404(1):349–356

Li K et al (2014) Targeting acute myeloid leukemia with a proapoptotic peptide conjugated to a toll-like receptor 2-mediated cell-penetrating peptide. Int J Cancer 134(3):692–702

Liberman EA et al (1969) Mechanism of coupling of oxidative phosphorylation and the membrane potential of mitochondria. Nature 222(5198):1076–1078

Lim S et al (2015) dNP2 is a blood–brain barrier-permeable peptide enabling ctCTLA-4 protein delivery to ameliorate experimental autoimmune encephalomyelitis. Nat Commun 6:8244

Lin YZ, Yao SY, Veach RA, Torgerson TR, Hawiger J (1995) Inhibition of nuclear translocation of transcription factor NF-κB by a synthetic peptide containing a cell membrane-permeable motif and nuclear localization sequence. J Biol Chem 270(24):14255–14258

Lindgren M et al (2006) Overcoming methotrexate resistance in breast cancer tumour cells by the use of a new cell-penetrating peptide. Biochem Pharmacol 71(4):416–425

Lundin P et al (2008) Distinct uptake routes of cell-penetrating peptide conjugates. Bioconjug Chem 19(12):2535–2542

Magzoub M, Sandgren S, Lundberg P, Oglęcka K, Lilja J, Wittrup A, Göran Eriksson LE, Langel U, Belting M, Gräslund A (2006) N-Terminal peptides from unprocessed prion proteins enter cells by macropinocytosis. Biochem Biophys Res Commun 348(2):379–385

Manczak M et al (2010) Mitochondria-targeted antioxidants protect against Abeta toxicity in Alzheimer's disease neurons. J Alzheimers Dis 20(Suppl 2):S609–S631

Mano M, Henriques A, Paiva A, Prieto M, Gavilanes F, Simões S, Pedroso de Lima MC (2006) Cellular uptake of S413-PV peptide occurs upon conformational changes induced by peptide-membrane interactions. Biochim Biophys Acta 1758:336–346

Margus H, Padari K, Pooga M (2013) Insights into cell entry and intracellular trafficking of peptide and protein drugs provided by electron microscopy. Adv Drug Deliv Rev 65(8):1031–1038

Marks JR, Placone J, Hristova K, Wimley WC (2011) Spontaneous membrane-translocating peptides by orthogonal high-throughput screening. J Am Chem Soc 133(23):8995–9004

Martín I, Teixidó M, Giralt E (2011) Design, synthesis and characterization of a new anionic cell-penetrating peptide: SAP(E). ChemBioChem 12(6):896–903

Milletti F (2012) Cell-penetrating peptides: classes, origin, and current landscape. Drug Discov Today 17(15–16):694–704

Mitchell DJ, Steinman L, Kim DT, Fathman CG, Rothbard JB (2000) Polyarginine enters cells more efficiently than other polycationic homopolymers. J Pept Res 56(5):318–325

Morris MC, Vidal P, Chaloin L, Heitz F, Divita G (1997) A new peptide vector for efficient delivery of oligonucleotides into mammalian cells. Nucleic Acids Res 25(14):2730–2736

Morris MC et al (2001) A peptide carrier for the delivery of biologically active proteins into mammalian cells. Nat Biotechnol 19:1173–1176

Murphy MP, Smith RAJ (2007) Targeting antioxidants to mitochondria by conjugation to lipophilic cations. Annu Rev Pharmacol Toxicol 47:629–656

Nakamura T et al (2013) The nanoparticulation by octaarginine-modified liposome improves α-galactosylceramide-mediated antitumor therapy via systemic administration. J Control Release 171(2):216–224

Nakase I, Hirose H, Tanaka G, Tadokoro A, Kobayashi S, Takeuchi T, Futaki S (2009) Cell-surface accumulation of flock house virus-derived peptide leads to efficient internalization via macropinocytosis. Mol Ther 17(11):1868–1876

Nasrollahi SA et al (2012) Cell-penetrating peptides as a novel transdermal drug delivery system. Chem Biol Drug Des 80(5):639–646

Oehlke J, Krause E, Wiesner B, Beyermann M, Bienert M (1997) Extensive cellular uptake into endothelial cells of an amphipathic β-sheet forming peptide. FEBS Lett 415(2):196–199

Oehlke J, Scheller A, Wiesner B, Krause E, Beyermann M, Klauschenz E, Melzig M, Bienert M (1998) Cellular uptake of an α-helical amphipathic model peptide with the potential to deliver polar compounds into the cell interior nonendocytically. Biochim Biophys Acta 1414(1-2):127–139

Oehlke J, Birth P, Klauschenz E, Wiesner B, Beyermann M, Oksche A, Bienert M (2002) Cellular uptake of antisense oligonucleotides after complexing or conjugation with cell-penetrating model peptides. Eur J Biochem 269(16):4025–4032

Oskolkov N, Arukuusk P, Copolovici D-M, Lindberg S, Margus H, Padari K, Pooga M, Langel U (2011) NickFects, phosphorylated derivatives of transportan 10 for cellular delivery of oligonucleotides. Int J Peptide Res Ther 17(2):147–157

Pooga M, Hällbrink M, Zorko M, Langel U (1998) Cell penetration by transportan. FASEB J 12(1):67–77

Qian Z, Dougherty PG, Pei D (2015) Monitoring the cytoslic entry of cell-penetratin peptides using a pH-sensitive fluorophore. Chem Commun (Camb) 51(11):2162–2165

Reddy PH, Manczak M, Kandimalla R (2017) Mitochondria-targeted small molecule SS31: a potential candidate for the treatment of Alzheimer's disease. Hum Mol Genet 26(8):1483–1496

Reynolds F, Weissleder R, Josephson L (2005) Protamine as an efficient membrane-translocating peptide. Bioconjug Chem 16(5):1240–1245

Rhee M, Davis P (2006) Mechanism of uptake of C105Y, a novel cell-penetrating peptide. J Biol Chem 281(2):1233–1240

Scheller A, Oehlke J, Wiesner B, Dathe M, Krause E, Beyermann M, Melzig M, Bienert M (1999) Structural requirements for cellular uptake of α-helical amphipathic peptides. J Pept Sci 5(4):185–194

Shin MC et al (2014) Cell-penetrating peptides: achievements and challenges in application for cancer treatment. J Biomed Mater Res A 102(2):575–587

Song W et al (2005) A potent opiate agonist protects against myocardial stunning during myocardial ischemia and reperfusion in rats. *Therapy and.* Prevention 16:407–410

Soomets U, Lindgren M, Gallet X, Hällbrink M, Elmquist A, Balaspiri L, Zorko M, Pooga M, Brasseur R, Langel U (2000) Deletion analogues of transportan. Biochim Biophys Acta 1467(1):165–176

Sparr C et al (2013) Improved efficacy of fosmidomycin against Plasmodium and Mycobacterium species by combination with the cell-penetrating peptide octaarginine. Antimicrob Agents Chemother 57(10):4689–4698

Stewart KM, Horton KL, Kelley SO (2008) Cell-penetrating peptides as delivery vehicles for biology and medicine. Org Biomol Chem 6(13):2242

Szeto HH (2006) Cell-permeable, mitochondrial-targeted, peptide antioxidants. AAPS J 8:E277–E283

Szeto HH (2008) Mitochondria-targeted cytoprotective peptides for ischemia-reperfusion injury. Antioxid Redox Signal 10(3):601–619

Taylor BN, Mehta RR, Yamada T, Lekmine F, Christov K, Chakrabarty AM, Green A, Bratescu L, Shilkaitis A, Beattie CW, Das Gupta TK (2009) Noncationic peptides obtained from Azurin preferentially enter cancer cells. Cancer Res 69(2):537–546

Tréhin R, Krauss U, Beck-Sickinger AG, Merkle HP, Nielsen HM (2004) Cellular uptake but low permeation of human calcitonin-derived cell penetrating peptides and Tat(47-57) through well-differentiated epithelial models. Pharm Res 21(7):1248–1256

Vivès E, Brodin P, Lebleu B (1997) A truncated HIV-1 tat protein basic domain rapidly translocates through the plasma membrane and accumulates in the cell nucleus. J Biol Chem 272(25):16010–16017

Weissig V (2015) DQAsomes as the prototype of mitochondria-targeted pharmaceutical nanocarriers: preparation, characterization, and use. In: Weissig V, Edeas M (eds) Mitochondrial medicine: vol II. Manipulating mitochondrial function. New York, NY, Springer New York, pp 1–11

Wisnovsky S et al (2016) Mitochondrial chemical biology: new probes elucidate the secrets of the powerhouse of the cell. Cell Chem Biol 23(8):917–927

Woldetsadik AD et al (2017) Hexokinase II-derived cell-penetrating peptide targets mitochondria and triggers apoptosis in cancer cells. FASEB J 31(5):2168–2184

Wu D et al (2002) A highly potent peptide analgesic that protects against ischemia-reperfusion-induced myocardial stunning. Am J Physiol Heart Circ Physiol 283(2):H783–H791

Yin X, Manczak M, Reddy PH (2016) Mitochondria-targeted molecules MitoQ and SS31 reduce mutant huntingtin-induced mitochondrial toxicity and synaptic damage in Huntington's disease. Hum Mol Genet 25(9):1739–1753

Zhao K et al (2004) Cell-permeable peptide antioxidants targeted to inner mitochondrial membrane inhibit mitochondrial swelling, oxidative cell death, and reperfusion injury. J Biol Chem 279(33):34682–34690

Zhao K et al (2005) Mitochondria-targeted peptide prevents mitochondrial depolarization and apoptosis induced by tert-butyl hydroperoxide in neuronal cell lines. Biochem Pharmacol 70(12):1796–1806

Zhao WY et al (2013) Mitochondria-targeted antioxidant peptide SS31 prevents hypoxia/reoxygenation-induced apoptosis by down-regulating p66Shc in renal tubular epithelial cells. Cell Physiol Biochem 32(3):591–600

Mitocans: Mitochondrially Targeted Anti-cancer Drugs

Stepana Boukalova, Katerina Rohlenova, Jakub Rohlena, and Jiri Neuzil

Abstract Mitochondria are intriguing organelles that are inherently present in most eukaryotic cells, with notable exceptions. They undertake multiple vital functions in a cell, including energy conversion, metabolite synthesis, regulation of the cellular redox state, production of reactive oxygen species, initiation of apoptosis, and buffering cellular Ca^{2+}. Although aberrant, mitochondria are indispensable in malignant cells for critical involvement in synthesis of vital precursors for a variety of metabolic pathways. Therefore, mitochondria have recently emerged as plausible, as yet underexploited targets for cancer therapy. Here we discuss why mitochondria may be clinically relevant anti-cancer therapeutic modalities and give examples of agents that act via mitochondria that we, collectively, refer to as mitocans. Some of these agents hold a great promise for making it to the 'bedside', entering clinical trials.

Keywords Mitochondrial targeting · Delocalised cations · Anti-cancer effect · Mitochondrial respiratory complexes

1 Introduction

Mitochondria are multi-functional cellular organelles with prominent semi-autonomous status. This is exemplified by the presence of a discrete mitochondrial genome and mitochondrial gene transcription/translation machinery enclosed

---

S. Boukalova · K. Rohlenova · J. Rohlena
Molecular Therapy Group, Institute of Biotechnology, Academy of Sciences of the Czech Republic, Prague-West, Czech Republic
e-mail: Stepana.Boukalova@ibt.cas.cz

J. Neuzil (✉)
Molecular Therapy Group, Institute of Biotechnology, Academy of Sciences of the Czech Republic, Prague-West, Czech Republic

Mitochondria, Apoptosis and Cancer Research Group, School of Medical Science and Menzies Health Institute of Queensland, Griffith University, Southport, QLD, Australia
e-mail: j.neuzil@griffith.edu.au

P. J. Oliveira (ed.), *Mitochondrial Biology and Experimental Therapeutics*,
https://doi.org/10.1007/978-3-319-73344-9_27

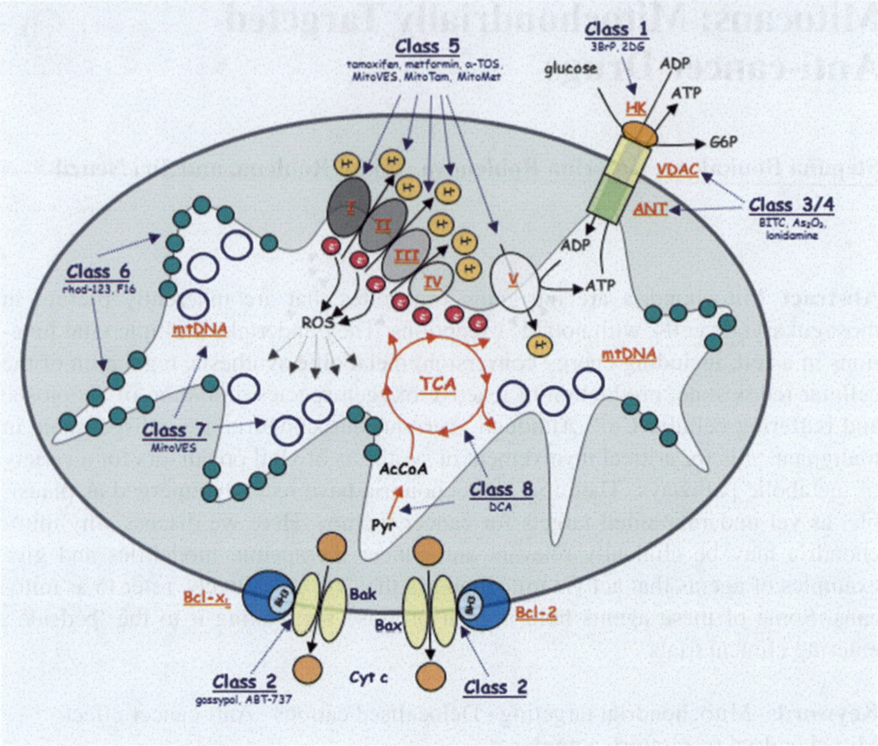

Fig. 1 Mitochondrial respiratory complexes and complex V. The cartoon shows the four respiratory complexes of OXPHOS and the adjacent complex V (ATP synthase), and also indicates the electron flow within the respiratory complexes. The inhibitors of individual complexes that were shown to have anti-cancer activity are depicted

in a complex two-membrane structure, remnants of an ancient endosymbiotic event preserved by evolution.

Mitochondria produce cellular energy in the form of ATP, using the process of oxidative phosphorylation (OXPHOS). Embedded in the inner mitochondrial membrane (IMM), the OXPHOS system consists of two parts: the electron transport chain (ETC) and the ATP synthase (Fig. 1). ETC accepts electrons from NADH and $FADH_2$ that are generated in the tricarboxylic acid (TCA) cycle, the central metabolic pathway, and transfers them in a coordinated manner to a final acceptor, molecular oxygen, in a reaction producing water. The individual steps of electron transfer are facilitated by four macromolecular complexes, known as complex I, II, III and IV (CI–CIV) (Scheffler 2008). The individual complexes further assemble into higher structures called supercomplexes (SCs), which optimize the process of electron transfer and minimize formation of potentially harmful reactive oxygen species (ROS), due to premature escape of electrons (Maranzana et al. 2013; Letts et al. 2016; Acin-Perez et al. 2008;

Schagger and Pfeiffer 2000). Coupled to the transit of electrons through the ETC complexes is an active transport of protons across the IMM, which generates electro-chemical gradient, also called IMM potential ($\Delta\Psi_{m,i}$). ATP synthase (i.e. complex V), acting as a molecular turbine, uses the proton gradient for the generation of ATP (Scheffler 2008).

Mitochondria in cancer cells were traditionally, but erroneously, viewed as OXPHOS-defective, and the role of mitochondria in cancer was therefore largely neglected. This notion stemmed from the work of a German biochemist Otto Warburg. Almost a century ago, Warburg observed that proliferating cancer cells consume large amounts of glucose and convert it into lactate even in the presence of oxygen, in a process called 'aerobic glycolysis' (reviewed in Koppenol et al. 2011). This was later misinterpreted by Warburg and others as evidence that mitochondria are 'injured' in cancer and therefore dysfunctional, explaining the apparent increase in glycolysis (Warburg 1956). Indeed, this seems a highly wasteful behavior, as employing aerobic glycolysis instead of reduction of oxygen by OXPHOS gives only minor fraction of ATP. While for 1 mole of glucose consumed the cell generates 2 moles of ATP in glycolysis, it would gain another 34 moles of ATP, should the pyruvate coming out of glycolysis be converted to acetyl-coenzyme A, and catabolized fully in mitochondria via the TCA cycle (Vander Heiden et al. 2009). From this point of view, it indeed made little sense to Warburg and other that cancer cell would use glycolysis in the presence of oxygen without the need to compensate for an unspecified mitochondrial injury that would preclude ATP generation in mitochondria.

Today it is assumed that the Warburg effect is related, in the majority of cases, to dysregulation of growth signaling due to the activation of oncogenes and inactivation of tumor suppressors in cancer, and is needed to gain building blocks for biosynthetic reactions for which the ATP is not limiting (Vander Heiden et al. 2009). Furthermore, glycolysis is kinetically much superior to OXPHOS and, consequently, the overall ATP production is comparable irrespective of the preferred pathway. Still, glycolysis cannot meet all the biosynthetic needs, and some of the necessary components can only originate in mitochondria. In fact, functional mitochondria are required for tumorigenesis, cells lacking mitochondrial DNA cannot form tumors (Dong et al. 2017; Tan et al. 2015), and ETC-derived ROS support transformation and tumor proliferation (Weinberg et al. 2010). Accordingly, tumors retain functional OXPHOS along with the usage of aerobic glycolysis (Hensley et al. 2016; Fan et al. 2009; Marin-Valencia et al. 2012), and both pathways generate indispensable metabolic intermediates to sustain rapid growth (Sullivan et al. 2015; Birsoy et al. 2015; Ying et al. 2012).

In summary, mitochondria in cancer are undoubtedly functional, and as central hubs of cellular homeostasis, cell death induction and ROS generation, they represent intriguing targets for anti-cancer therapy.

2 Classification of Mitochondrially Targeted Anti-cancer Agents

The structure of mitochondria, composing of several separate and functionally different compartments, allows for multiple targets for anti-cancer agents within these organelles (Canto et al. 2015; Bhola and Letai 2016; Vyas et al. 2016; Mills et al. 2017; Martinez-Outschoorn et al. 2017). The notion that mitochondria may be a plausible target for anti-cancer therapy stems from their difference between normal and malignant cells (Gogvadze et al. 2008).

There are hundreds of small molecules with anti-cancer activity that exert their effect via mitochondria. We have coined these compounds as 'mitocans', which is the acronym for mitochondria and cancer (Neuzil et al. 2006), encompassing compounds targeting various aspects of mitochondria. Consequently, we have proposed the classification of mitocans, which is based on the molecular targets of the agents within mitochondria, starting from the surface of the outer mitochondrial membrane (OMM), to the components of the mitochondrial matrix, such as the TCA cycle or the mitochondrial DNA (mtDNA; Fig. 2) (Neuzil et al. 2013). Individual classes of mitocans are described below.

3 Individual Classes of Mitochondrially Targeted Agents

3.1 Class 1: Hexokinase Inhibitors

Hexokinases (HKs) catalyze the first step of glucose metabolism phosphorylating glucose to glucose-6-phosphate (G6P), which is then processed to obtain energy and biosynthetic precursors. In mammals, there are four isoforms of HK, HKI–HKIV, with different tissue distribution and affinity to glucose (Wilson 2003). HKII is a predominant isoform in insulin-sensitive tissues (Heikkinen et al. 2000), but it is also upregulated in many types of tumors (Sato-Tadano et al. 2013; Chen et al. 2014; Wolf et al. 2011; Qiu et al. 2011). HKII is believed to promote the Warburg phenotype of cancer cells by increasing their glycolytic activity even in the presence of oxygen (Mathupala et al. 2009). A prominent feature of HKII is its association with OMM mediated by direct interaction with the voltage-dependent anion channel (VDAC) protein (Nakashima et al. 1986) (Fig. 2). The HK-mitochondria interaction is crucial for cancer cells as it facilitates the evasion of apoptosis by antagonizing apoptotic Bcl-2 family proteins and thus prevents the formation of mitochondrial outer membrane permeabilization (MOMP) (Pastorino et al. 2002; Gall et al. 2011; Majewski et al. 2004; Vyssokikh et al. 2002). Therefore, HKII has a dual role in tumor cells. First, its upregulation results in increased glycolysis rates, and, second, HKII association with OMM contributes to inhibition of apoptosis by counteracting the release of pro-apoptotic factors from mitochondria via MOMP. This makes it an attractive target for anti-cancer therapy.

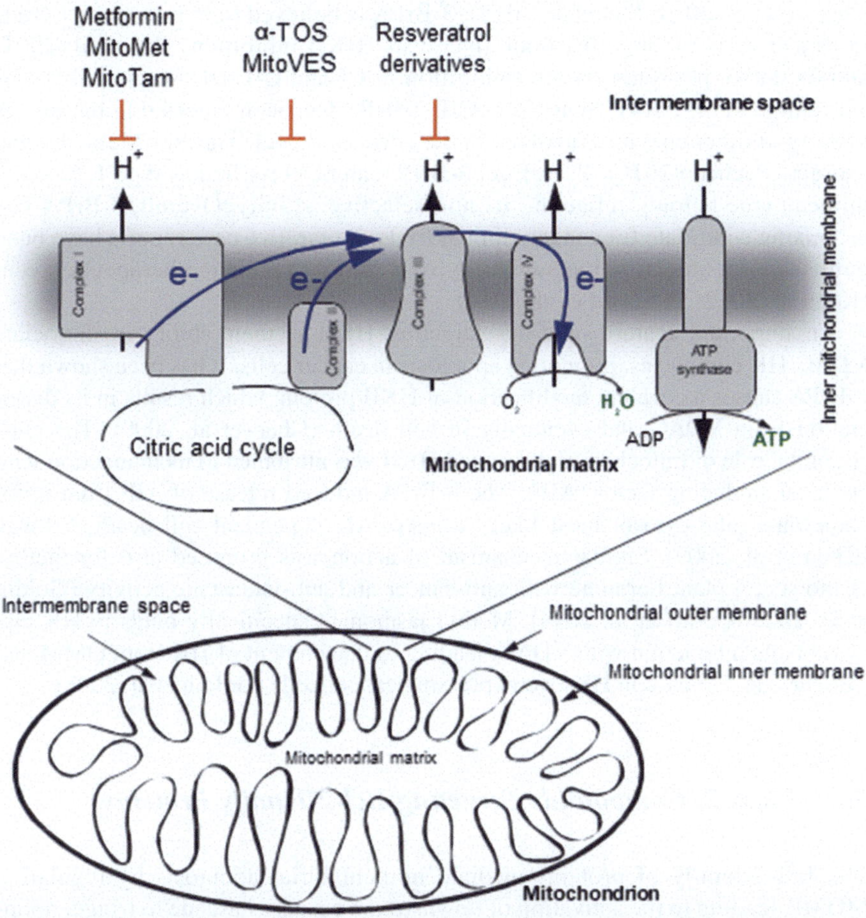

Fig. 2 Schematic indication of targets for mitocans within mitochondria. The cartoon indicates the eight individual classes of mitocans within mitochondria, also showing examples of some of the better studied agents acting via these sites. Adapted from Neuzil et al. (2013)

Several inhibitors of HK, such as 2-deoxyglucose (2-DG) or 3-bromopyruvate (3-BrPA), have been found to suppress cancer growth and are in pre-clinical or early phase clinical trials. 2-DG, a glucose analogue, is phosphorylated by HK to 2-DG-P, which cannot be further metabolized. 2-DG-P accumulates in the cell and inhibit the activity of HK and, consequently, glycolysis. When used as a single agent, 2-DG has only modest or no anti-tumorigenic activity (Maschek et al. 2004; Cheng et al. 2012; Kim et al. 2017). However, 2-DG was shown to increase the overall efficacy of chemotherapy or radiotherapy (Maschek et al. 2004; Coleman et al. 2008; Simons et al. 2007; Cheng et al. 2012). 3-BrPA, a halogenated derivative of pyruvate, is an alkylating agent with potent and selective anti-cancer activity (Geschwind et al. 2002; Ko et al. 2004;

Chapiro et al. 2014; Sun et al. 2015). 3-BrPA is believed to induce energy crisis in cancer cells by targeting both glycolysis (HKII inhibition) (Ko et al. 2001) and oxidative phosphorylation (inhibition of CI and CII) (Sanborn et al. 1971; Macchioni et al. 2014). Apart from CII, 3-BrPA has been reported to inhibit the activity of other enzymes involved in the citric acid cycle (Jardim-Messeder and Moreira-Pacheco 2016). Therefore, 3-BrPA can be classified as type 1, 5, and 7 mitocan (see below), reflecting its non-selective activity. Overall, 3-BrPA is a promising candidate for cancer clinical trials. So far, two case reports have been published documenting the suitability of 3-BrPA as a chemotherapeutic agent (Ko et al. 2012; El Sayed et al. 2014).

An important feature of drugs targeting HKII is their ability to dissociate VDAC-HK complexes, promoting apoptosis in cancer cells. It has been shown that 3-BrPA causes a covalent modification of HKII protein, which results in its dissociation from VDAC and eventually in cell death (Chen et al. 2009). The anti-apoptotic role of mitochondrially bound HKII was attributed to its interaction with apoptosis-inducing factor (AIF). The 3-BrPA-induced release of AIF from mitochondria to the cytosol most likely triggers AIF-dependent cell death pathway (Chen et al. 2009). Similar mechanism of action was proposed also for methyl jasmonate, a plant hormone with anti-cancer and anti-metastatic activity (Goldin et al. 2008; Cesari et al. 2014). Methyl jasmonate specifically binds to HK and disrupts its interaction with VDAC, leading to detachment of HK from OMM and cytochrome c release in HK-overexpressing cancer cells (Goldin et al. 2008).

3.2 Class 2: Compounds Targeting Bcl-2 Family Proteins

The Bcl-2 family of proteins mediate mitochondrial apoptosis by regulating MOMP, leading to the activation of downstream caspase cascade to trigger apoptosis. The family includes several pro-apoptotic and anti-apoptotic members, many of which interact with each other. The MOMP is mediated by oligomerization of effector proteins, BAX and BAK, in the OMM. This process is inhibited by pro-survival cell ‘guardians’, such as Bcl-2 itself and Bcl-X_L, while it is promoted by pro-apoptotic BH3-only proteins (Czabotar et al. 2014). The pro-survival Bcl-2 family proteins are often overexpressed in cancer cells supporting their resistance to apoptosis (Kelly and Strasser 2011). To counter this, a novel class of small-molecule therapeutics that mimic the pro-apoptotic BH3-only proteins, has been developed. These agents bind to and inhibit the pro-survival Bcl-2 family members, thereby sensitizing cancer cells to apoptosis (Opydo-Chanek et al. 2017). The prototypic BH3 mimetic, ABT-737 selectively targets the pro-survival proteins with sub-nanomolar range affinity, displaying anti-cancer activity in several pre-clinical models (Oltersdorf et al. 2005; Konopleva et al. 2006). However, the clinical application of ABT-737 has been limited due to low aqueous solubility and poor bioavailability. Nowadays, two orally available derivatives of ABT-737, navitoclax (ABT-263) and venetoclax (ABT-199), are being evaluated in clinical trials as a therapy for solid tumors as well as lymphoid malignancies either as single agents or in combination with other chemotherapeutics (Opydo-Chanek et al. 2017). Two other small molecule inhibitors of pro-survival Bcl-2 proteins, Obatoclax and Gossypol, have reached clinical testing. These compounds display lower affinity toward their targets compared to ABT-737 and its analogues (Konopleva et al. 2008; Kitada et al. 2003), and are associated with adverse effects in patients (Goard and Schimmer 2013; Kitada et al. 2008). However, several derivatives of gossypol with improved therapeutic efficacy and reduced toxicity have been developed, which raises hopes for the possible utilization of these agents in anti-cancer therapy (Kitada et al. 2003, 2008).

3.3 Class 3: Thiol Redox Inhibitors

Cancer cells display increased levels of reactive oxygen species (ROS), which facilitate their proliferation and promote chemo- and radioresistance. Even though dietary antioxidants are considered to be broad-spectrum anti-cancer agents, they have consistently failed to reduce tumor burden in prospective human clinical trials, possibly because they do not dampen mitochondrially localized ROS, which are believed to be the prime pro-tumorigenic mediators (Bjelakovic and Gluud 2007). On the other hand, tumor cells are highly vulnerable to the toxic effect of ROS overproduction (rather than scavenging), probably due to already saturated antioxidant capacity (Trachootham et al. 2009). Therefore, stimulation of ROS production (by the electron transport chain targeting drugs—class 5 mitocans; see below) or inhibition of the antioxidant system (by the thiol redox inhibitors) is a way to selectively kill cancer cells (Trachootham et al. 2009). Indeed, human cancers often display elevated levels of glutathione (GSH) and thioredoxin (Trx), which maintain the redox homeostasis by eliminating ROS-induced damage (Gamcsik et al. 2012; Iwasawa et al. 2011). Several agents known to inhibit GSH and Trx biosynthesis are undergoing trials for cancer treatment. The anti-arthritic drug auranofin, which suppresses Trx levels by inhibiting Trx reductase, is receiving increasing attention for its potential to be repurposed as a novel tumor-suppressing therapeutic. In pre-clinical testing, auranofin induced cytotoxicity in cancer cells by increasing intracellular ROS levels and was able to suppress tumorigenic potential in human chronic leukemia and gastric cancer models (Fiskus et al. 2014; Zou et al. 2015). The depletion of GSH by its conjugation with phenethyl isothiocyanate (PEITC) or inhibition of GSH synthesis using buthionine sulfoximine was shown to have cancer preventive effects and synergize with Trx suppression approach in cancer treatment (Gupta et al. 2014; Harris et al. 2015). Interestingly, PEITC and other isothiocyanates are naturally occurring dietary compounds, which are assumed to be responsible for the reduced risk of cancer in humans with high intake of cruciferous vegetables (Gupta et al. 2014).

3.4 Class 4: ANT Targeting Drugs

Adenine nucleotide translocase (ANT) is a transporter localized in the IMM, and together with the VDAC protein serves as a mode of transport for ATP/ADP, linking mitochondrial matrix and the cytosol (Zhivotovsky et al. 2009). ANT plays an important role in cell death regulation as it modulates the activity of the mitochondrial permeability transition pore (mPTP). Opening of the mPTP results in organelle swelling and rupture, and is inherent to necrotic cell death (Kwong and Molkentin 2015). Several thiol cross-linking agents that have been shown to induce mPTP opening and cell death owing to their direct modification of ANT are considered as potential anti-cancer drugs (McStay et al. 2002; Costantini et al. 2000). A non-toxic derivative of phenyl arsenoxide, 4-(N-(S-glutathionylacetyl)amino)phenylarsenoxide (GSAO), was shown to have anti-proliferative effects on tumor endothelial cells and to inhibit tumor angiogenesis in mice, which was attributed to GSAO-mediated ANT inactivation (Don et al. 2003). This compound was already tested in phase I clinical trial in patients with advanced solid tumors (Horsley et al. 2013). Moreover, arsenic trioxide, a thiol-oxidizing agent currently used as a chemotherapeutic agent for the treatment of acute promyelocytic leukemia, was proposed to activate mPTP possibly via direct modification of ANT (Larochette et al. 1999).

3.5 Class 5: Electron Transport Chain Targeting Drugs

Respiration mediated by the ETC is an important energy-transforming catabolic process, while it also provides access to electron acceptors for crucial anabolic reactions (Sullivan et al. 2015; Ahn and Metallo 2015; Birsoy et al. 2015). Indeed, functional ETC is indispensable for tumorigenesis (Tan et al. 2015; Weinberg et al. 2010) and, as the major source of ROS, promotes cancer initiation and progression (Sabharwal and Schumacker 2014). Recent evidence highlights ETC as a promising target for anti-cancer therapy. As noted above, cancer cells often show increased levels of oxidative stress and higher saturation of their antioxidant defenses, which makes them more sensitive to apoptosis induction by inhibitors of ETC complexes that are strong ROS inducers (Trachootham et al. 2009). Moreover, tumor-initiating cells, a small tumor cell subpopulation responsible for the metastatic propensity and recurrence of cancer, have been shown to be dependent on OXPHOS in several types of tumors (Sancho et al. 2016). By targeting ETC, this highly aggressive population of cancer cells can be eliminated allowing efficient suppression of tumorigenesis and metastases (Lonardo et al. 2013; Sancho et al. 2015; LeBleu et al. 2014; Yan et al. 2015a). Class 5 mitocans may thus present a valuable option for cancer therapy in the future. The next section provides a brief overview of selected anti-cancer agents acting on individual ETC complexes (Fig. 1).

3.5.1 Complex I

Intensive research interest is currently focused on metformin, a widely used anti-diabetic drug and a known inhibitor of CI. The repurposing of metformin as an anti-cancer agent was initiated by several reports showing a reduced risk of developing cancer and better response to chemotherapy in metformin-treated patients (Evans et al. 2005; Landman et al. 2010; Jiralerspong et al. 2009). Since then, many studies have demonstrated that metformin efficiently inhibits cancer cell proliferation *in vitro* and suppresses tumorigenesis *in vivo*, although the doses of the agent are very high. The anti-cancer effects of metformin are thought to be mediated primarily via inhibition of CI. The resulting decreased cellular energy charge leads to activation of AMP-activated protein kinase (AMPK), an important energy sensor, promoting cell cycle arrest (Foretz et al. 2014). Additionally, the decreased mitochondrial function associated with metformin treatment affects the NAD^+/NADH homeostasis and inhibits essential mitochondria-linked anabolic processes, and thus suppresses proliferation of tumor cells (Gui et al. 2016). Several other modes of action have been proposed to mediate anti-cancer effects of metformin, including regulation of DNA methylation and insulin signaling (Yan et al. 2015b; Dowling et al. 2012; Cuyas et al. under revision). The anti-cancer effects of metformin are currently being evaluated in a number of clinical trials. Preliminary results warrant the use of metformin in combination therapy and as an adjuvant treatment in patients with different types of cancer (Chae et al. 2016; Coyle et al. 2016). However, metformin may not be a viable option for advanced carcinomas, as it did not show favorable outcomes (Chae et al. 2016).

In search for more efficient analogue of metformin, we and others have modified the drug with triphenylphosphonium (TPP^+) group, which allows preferential accumulation in mitochondria (Boukalova et al. 2016; Cheng et al. 2016). The mitochondrially targeted metformin, MitoMet, inhibits CI-mediated respiration when administrated at much lower doses compared to the parental compound. Indeed, mitochondrial targeting strongly enhanced the anti-cancer efficacy of metformin in models of pancreatic cancer, reflecting its increased association with mitochondria. Moreover, tagging with TPP^+ shifted the activity of metformin from anti-oxidant to pro-oxidant, which may further augment the tumor suppressive function of MitoMet (Boukalova et al. 2016; Algire et al. 2012).

We have taken a similar approach to modify the activity of tamoxifen, another anti-cancer drug used in the clinics. Tamoxifen, a mixed agonist/antagonist of the estrogen receptor, is used as the first-line therapy in hormone-sensitive breast cancer, and was reported to inhibit CI at supra-pharmacological concentrations (Moreira et al. 2006). Recently, tamoxifen was shown to induce metabolic reprogramming in breast cancer cells via inhibition of CI, independently of the estrogen receptor (Daurio et al. 2016). Interestingly, the mitochondrially targeted tamoxifen (MitoTam) has greatly increased anti-cancer potential compared to the parental agent, broadening its applicability to breast cancer subtypes so far recalcitrant to treatment (Rohlenova et al. 2017). This was clearly related to its direct effect on mitochondria, as interference with the mitochondrial targeting group or the absence of the molecular target, CI, greatly reduced efficacy. In addition, MitoTam shows the highest efficiency in breast cancer subtypes with increased mitochondrial respiration, relating to increased assembly of respiratory SCs, particularly in malignant cells and tumors featuring high expression of the Her2 oncogene. Apparently, MitoTam association with CI can disrupt mitochondrial SCs, unleashing considerable ROS production in a self-amplifying feedback loop. This is a completely new mechanism of action, unparalleled in any other known anti-cancer agent, demonstrating that mitochondrial targeting can result in unexpected beneficial consequences. One interesting outcome of this work is the finding that oncogenes, such as Her2, may increase mitochondrial respiration by enhancing ETC assembly into SCs by as yet uncharacterized mechanism(s). This may be dependent on the fraction of Her2 that localizes into mitochondria, as the amount of mitochondrial Her2 was directly correlated with sensitivity to MitoTam (Rohlenova et al. 2017). MitoTam has already passed the pre-clinical testing, where it showed a very favorable toxicity profile, and continues to phase I clinical trial.

3.5.2 Complex II

CII (succinate dehydrogenase) directly links ETC with the TCA cycle, where it catalyzes oxidation of succinate to fumarate. Similarly to CI, inhibition of CII may lead to increased ROS production and to induction of cell death (Kluckova et al. 2015). Such a mechanism of action was proposed for the anti-cancer agent from the family of vitamin E analogues, α-tocopheryl succinate (α-TOS), that selectively kills tumor cells (Neuzil et al. 2001a, b; Weber et al. 2002; Dong et al. 2008, 2009). Besides, α-TOS displays other indirect anti-cancer effects. By means of ROS-dependent disruption of FGF2 pathway, the agent eliminates proliferating endothelial cells and thus suppresses angiogenesis, a vital process in tumor formation (Dong et al. 2007; Stapelberg et al. 2005). In analogy to MitoMet and MitoTam, mitochondrial targeting of α-TOS mediated by its modification with the TPP^+ tag resulted in increased apoptogenic activity in cancer cells and superior anti-cancer efficacy in animal models (Dong et al. 2011a, b). Moreover, mitochondrially targeted α-TOS (MitoVES) efficiently eliminated breast tumor initiating cells in a CII-dependent manner (Yan et al. 2015a), further supporting the therapeutic significance of TPP^+-tagged mitocans. The TPP^+ group localizes the agents to the interface of IMM and matrix, which is an ideal position for interaction with their ETC targets. Using this approach, the anti-cancer activity and selectivity of class 5 mitocans can be considerably enhanced (Biasutto et al. 2010).

3.5.3 Complex III and IV

CIII and CIV mediate the last step of the ETC, transporting electrons from ubiquinol (reduced by CI and CII) to molecular oxygen. CIII is regarded as an important site of ROS production and a possible target for anti-cancer therapy (Sabharwal and

Schumacker 2014). Resveratrol, a plant-derived polyphenol with chemoprotective activity, has been shown to inhibit several ETC complexes, especially CIII. However, the mechanism of action of resveratrol is still a matter of controversy, as it was reported to act either as an anti-oxidant with cytoprotective activity, or as a cytotoxic pro-oxidant (de Oliveira et al. 2016). On the other hand, mitochondrially targeted derivatives of resveratrol unequivocally induce oxidative stress by inhibiting CI and CIII, and selectively kill fast-growing cells (Sassi et al. 2014a, b). Yet the anti-cancer activity of resveratrol derivatives remains to be confirmed in animal models.

So far, CIV has been rarely evaluated as a target in anti-cancer therapy despite its important role in regulating the ETC activity. CIV, also known as cytochrome c oxidase, has been proposed to mediate the cytotoxic effects of photodynamic therapy (Wu et al. 2014). This procedure uses low-power laser irradiation of tumor tissue, which affects the activity of ETC and induces ROS bursts, leading to eradication of cancer cells. The efficacy of the process may be further increased by administration of photosensitizers such as porphyrin derivatives (Hilf 2007). Among the main advantage of photodynamic therapy is the possibility of its precise targeting on tumor tissue without affecting healthy cells.

3.6 Class 6: Lipophilic Cations Targeting the Inner Membrane

Delocalized lipophilic cations (DLCs) target the IMM due to large hydrophobic surface areas and a delocalized positive charge. The hydrophobic nature of these compounds allows them to rapidly pass the membranes, while the positive charge supports their accumulation in mitochondria driven by $\Delta\Psi_{m,i}$ across the IMM. Cancer cells have been documented to have considerably higher $\Delta\Psi_{m,i}$ than non-malignant cells, which makes them a selective target for DLCs (Modica-Napolitano and Aprille 2001). As mentioned above, DLCs such as TPP^+, have been employed to promote targeting of mitocans to mitochondria. However, DLCs have intrinsic cytotoxic activity, and thus act as mitocans in their own right. Several different targets inside mitochondria were reported to mediate the toxic effect of these agents, including CV, mtDNA, and the ETC (Baracca et al. 2003; Anderson et al. 1989). Anti-cancer activity of the lipophilic cation MKT-077 was tested in animal models with promising outcomes (Koya et al. 1996; Chiba et al. 1998). Nevertheless, clinical trials with MKT-077 were terminated owing to renal toxicity (Propper et al. 1999). Yet the search for DLCs with less non-specific toxicity profile continues (Yang et al. 2017).

3.7 Class 7: Drugs Targeting the TCA Cycle

The TCA cycle plays a central role in cellular metabolism. In this process, the end-product of glycolysis, pyruvate, is metabolized, producing reduced electron carriers, NADH and $FADH_2$, which serve as a source of electrons for ETC. Due to oncogenic signaling, the entrance of pyruvate to the TCA cycle is reduced, and a larger fraction of pyruvate is converted to lactate and excreted from the cell. This is mediated by upregulation of pyruvate dehydrogenase kinase (PDK), which negatively regulates conversion of pyruvate to acetyl-CoA by pyruvate dehydrogenase (Dang 2007). A PDK inhibitor, dichloroacetate (DCA), was reported to promote a shift to oxidative metabolism accompanied by a decrease in the $\Delta\Psi_{m,i}$, ROS generation and activation of the K^+ channel. These effects were specific for cancer cells, suggesting that DCA may have anti-cancer efficacy in patients (Bonnet et al. 2007; Michelakis et al. 2008). Phase I clinical trials showed that DCA was generally well tolerated, which supports future testing (Dunbar et al. 2014; Chu et al. 2015).

3.8 Class 8: Drugs Targeting mtDNA Transcription/ Translation

Mitochondria contain multiple copies of their own genome, which encodes several subunits of the ETC complexes. mtDNA is replicated independent of the cell cycle status by DNA polymerase γ (POLG), an enzyme specific for mitochondria. The indispensable role of mtDNA in tumorigenesis is documented by the fact that cancer cells devoid of mtDNA cannot form tumors (Tan et al. 2015). As OXPHOS is the major mitochondrial system maintained by mtDNA, interference with mtDNA function will therefore, in principle, be similar to direct OXPHOS or ETC inhibition. Recently, it has been proposed that inhibition of POLG and the resulting depletion of mtDNA may be an efficient therapeutic approach for treating acute myeloid leukemia (AML). The viability of AML cells was shown to be suppressed by administration of a nucleoside analogue 2′,3′-dideoxycytidine (ddC), which is in cells phosphorylated by nucleoside kinases to 2′,3′-dedeoxycytidine triphosphate (ddCTP), a selective inhibitor of POLG. ddC treatment resulted in suppression of mtDNA biosynthesis followed by ETC defects and decreased respiration. As the nucleoside kinases are overexpressed in AML cells, ddCTP was preferentially produced in cancer cells, while non-cancerous cells were resistant to ddC treatment. Importantly, ddC efficiently suppressed AML also *in vivo* (Liyanage et al. 2017). As ddC is an approved drug for the treatment of HIV, it has already passed safety tests and may be further repurposed as an anti-cancer agent. Interestingly, FDA-approved antibiotic tigecycline was highly effective in suppressing experimental AML, and its effect could be recapitulated by genetic interference with mitochondrial translation machinery (Skrtic et al. 2011). Later on, it was shown that several classes of antibiotics that interfere with mitochondrial transcription/translation have broad anti-cancer activity (Lamb et al. 2015).

Other compounds with anti-tumor properties were also shown to target mtDNA, via an indirect mechanism. MitoVES, a CII inhibitor discussed above, suppressed mitochondrial transcription factor A (TFAM) in a ROS-dependent manner, resulting in decreased mitochondrial gene transcription and significant alternations in mitochondrial function (Truksa et al. 2015). MitoVES amplifies its direct effect on CII by stimulating a long-term reduction of ETC function by interfering with mtDNA transcription. This provides an intriguing possibility for interfering with tumor progression by means of suppressing the proliferation of cancer cells without necessarily inducing apoptosis.

4 Mitochondrial Targeting: Broad Applicability with Selectivity for Cancer

Based on the considerable heterogeneity of cancer cells within a tumor (Gerlinger et al. 2012), it is becoming increasingly clear that targeting of a single oncogene or signaling pathway is unlikely to bring benefit on a large scale. In this respect, the ascendance of mitochondria as direct therapy targets may provide additional broad-range therapeutic options to be combined with the narrow-range personalized treatment. In fact, many of the activated oncogenic pathways targeted by kinase inhibitors and other narrow-range therapeutics lead to alterations in major metabolic systems further downstream, including mitochondria. Targeting mitochondria may therefore avoid the frequent appearance of resistance associated with the emergence of alternative signaling pathways that bypass the original upstream site of intervention. Furthermore, cancer cell populations resistant to conventional therapy are particularly dependent on mitochondrial activity for reasons that are not yet fully understood, and are sensitive to mitochondria-directed interventions (Wolf 2014). As mitochondria are indispensable for tumor initiation and progression (Tan et al. 2015), damage to mitochondria may cause severe condition, which cannot be readily overcome by other compensatory mechanisms.

One of the major challenges for translation of mitochondrially targeted therapy to clinical use is the specificity for tumor tissue. While kinase inhibitors target pathways specifically upregulated in cancer, mitochondria are ubiquitous. To achieve specificity using mitochondrially targeted agents, both cell autonomous and external aspects must be considered, as well as the rate of uptake of the bioactive compound by the tumor relative to non-cancerous tissue. For autonomous effects, the readiness of the cancer cell to initiate cell death in response to mitochondrial stress is likely an important parameter. This is particularly apparent after mitochondrial ROS induction, which effectively triggers cell death in proliferating cancer cells, while somatic, non-dividing cells are resistant. This stems from increased antioxidant capacity in mitochondria in non-dividing cells (Trachootham et al. 2009).

The external effects are represented by the specific tumor milieu. As the majority of cancer cells feature the Warburg phenotype with relatively high glucose consumption, glucose will become scarce commodity within the tumor (Hirayama et al. 2009; Rodrigues et al. 2014; Goveia et al. 2016). Cancer cells with functional mitochondria will gain advantage, being able to switch to alternative fuel sources, which the cells reliant solely on glycolysis (i.e. without functional mitochondria) cannot do. Interference with the respiratory function or mitochondrial ATP generation will therefore compromise survival in such environment by decreasing the flexibility of cancer cell metabolism. In this regard, TPP^+-modified compounds can be designed to efficiently enter mitochondria, and the molecular linker attaching TPP^+ moiety to the active group may be adjusted in order that the biologically active moiety of the agent can associate with its molecular target, as we have shown for analogues of MitoVES differing in the length of a carbon linker (Truksa et al. 2015). TPP^+ conjugation will also increase accumulation in tumor tissue, based on increased mitochondrial/plasma membrane potential of tumor cells, further improving specificity. Taking into consideration the above reasoning, mitochondrially targeted agents have the potential to be both broadly-applicable and selective to cancer.

5 Conclusion and Perspectives

Mitochondria are as yet underexploited targets for cancer therapy. The recently emerging notion that the organelles differ between normal and malignant cells provides a window of opportunity for selective targeting. This is supported by a number of recent publications as well as commanding review papers, some of which are referred to above. The issue at hand that appears to be complicating a straightforward progress towards clinical trials of mitocans stems from the ubiquitous presence of mitochondria in malignant as well as normal cells, indicating potential deleterious effects. It is therefore essential to design drugs that would be selective for cancer cells. For this, altered properties of mitochondria of cancer cells, such as their higher $\Delta\Psi_{m,i}$ and increased propensity to ROS-induced apoptosis, may be utilized for selective targeting. One strategy that appears plausible, is tagging of mitocans with a delocalized cationic tag, such as triphenylphosphonium. To this end, we and others have synthesized mitochondrially targeted analogues of agents targeting mitochondrial ETC, of which mitochondrially targeted tamoxifen is entering a clinical trial.

In conclusion, mitocans, agents that target mitochondria, may be the long sought after drugs that could be developed into broad-spectrum, efficient and selective anti-cancer modalities that will overcome recurring complications encountered during tumor management, such as development of resistance towards therapy or lack of response due to high tumor heterogeneity.

Acknowledgements This work was supported in part by grants from the Czech Science Foundation GA16-12719S to J.N. and GA17-20904S to J.R. and from Czech Ministry of Health grant AZV16-31604A to J.N. Further support was provided by BIOCEV CZ.1.05/1.1.00/02.0109 from the ERDF and by RVO: 86652036.

References

Acin-Perez R, Fernandez-Silva P, Peleato ML, Perez-Martos A, Enriquez JA (2008) Respiratory active mitochondrial supercomplexes. Mol Cell 32(4):529–539. https://doi.org/10.1016/j.molcel.2008.10.021

Ahn CS, Metallo CM (2015) Mitochondria as biosynthetic factories for cancer proliferation. Cancer Metab 3(1):1. https://doi.org/10.1186/s40170-015-0128-2

Algire C, Moiseeva O, Deschenes-Simard X, Amrein L, Petruccelli L, Birman E, Viollet B, Ferbeyre G, Pollak MN (2012) Metformin reduces endogenous reactive oxygen species and associated DNA damage. Cancer Prev Res (Phila) 5(4):536–543. https://doi.org/10.1158/1940-6207.CAPR-11-0536

Anderson WM, Patheja HS, Delinck DL, Baldwin WW, Smiley ST, Chen LB (1989) Inhibition of bovine heart mitochondrial and Paracoccus denitrificans NADH—ubiquinone reductase by dequalinium chloride and three structurally related quinolinium compounds. Biochem Int 19(4):673–685

Baracca A, Sgarbi G, Solaini G, Lenaz G (2003) Rhodamine 123 as a probe of mitochondrial membrane potential: evaluation of proton flux through F(0) during ATP synthesis. Biochim Biophys Acta 1606(1–3):137–146

Bhola PD, Letai A (2016) Mitochondria-judges and executioners of cell death sentences. Mol Cell 61(5):695–704. https://doi.org/10.1016/j.molcel.2016.02.019

Biasutto L, Dong LF, Zoratti M, Neuzil J (2010) Mitochondrially targeted anti-cancer agents. Mitochondrion 10(6):670–681. https://doi.org/10.1016/j.mito.2010.06.004

Birsoy K, Wang T, Chen WW, Freinkman E, Abu-Remaileh M, Sabatini DM (2015) An essential role of the mitochondrial electron transport chain in cell proliferation is to enable aspartate synthesis. Cell 162(3):540–551. https://doi.org/10.1016/j.cell.2015.07.016

Bjelakovic G, Gluud C (2007) Surviving antioxidant supplements. J Natl Cancer Inst 99(10):742–743. https://doi.org/10.1093/jnci/djk211

Bonnet S, Archer SL, Allalunis-Turner J, Haromy A, Beaulieu C, Thompson R, Lee CT, Lopaschuk GD, Puttagunta L, Bonnet S, Harry G, Hashimoto K, Porter CJ, Andrade MA, Thebaud B, Michelakis ED (2007) A mitochondria-K+ channel axis is suppressed in cancer and its normalization promotes apoptosis and inhibits cancer growth. Cancer Cell 11(1):37–51. https://doi.org/10.1016/j.ccr.2006.10.020

Boukalova S, Stursa J, Werner L, Ezrova Z, Cerny J, Bezawork-Geleta A, Pecinova A, Dong L, Drahota Z, Neuzil J (2016) Mitochondrial targeting of metformin enhances its activity against pancreatic cancer. Mol Cancer Ther 15(12):2875–2886. https://doi.org/10.1158/1535-7163.MCT-15-1021

Canto C, Menzies KJ, Auwerx J (2015) NAD(+) metabolism and the control of energy homeostasis: a balancing act between mitochondria and the nucleus. Cell Metab 22(1):31–53. https://doi.org/10.1016/j.cmet.2015.05.023

Cesari IM, Carvalho E, Figueiredo Rodrigues M, Mendonca Bdos S, Amoedo ND, Rumjanek FD (2014) Methyl jasmonate: putative mechanisms of action on cancer cells cycle, metabolism, and apoptosis. Int J Cell Biol 2014:572097. https://doi.org/10.1155/2014/572097

Chae YK, Arya A, Malecek MK, Shin DS, Carneiro B, Chandra S, Kaplan J, Kalyan A, Altman JK, Platanias L, Giles F (2016) Repurposing metformin for cancer treatment: current clinical studies. Oncotarget 7(26):40767–40780. https://doi.org/10.18632/oncotarget.8194

Chapiro J, Sur S, Savic LJ, Ganapathy-Kanniappan S, Reyes J, Duran R, Thiruganasambandam SC, Moats CR, Lin M, Luo W, Tran PT, Herman JM, Semenza GL, Ewald AJ, Vogelstein B, Geschwind JF (2014) Systemic delivery of microencapsulated 3-bromopyruvate for the therapy of pancreatic cancer. Clin Cancer Res 20(24):6406–6417. https://doi.org/10.1158/1078-0432.CCR-14-1271

Chen Z, Zhang H, Lu W, Huang P (2009) Role of mitochondria-associated hexokinase II in cancer cell death induced by 3-bromopyruvate. Biochim Biophys Acta 1787(5):553–560. https://doi.org/10.1016/j.bbabio.2009.03.003

Chen J, Zhang S, Li Y, Tang Z, Kong W (2014) Hexokinase 2 overexpression promotes the proliferation and survival of laryngeal squamous cell carcinoma. Tumour Biol 35(4):3743–3753. https://doi.org/10.1007/s13277-013-1496-2

Cheng G, Zielonka J, Dranka BP, McAllister D, Mackinnon AC Jr, Joseph J, Kalyanaraman B (2012) Mitochondria-targeted drugs synergize with 2-deoxyglucose to trigger breast cancer cell death. Cancer Res 72(10):2634–2644. https://doi.org/10.1158/0008-5472.CAN-11-3928

Cheng G, Zielonka J, Ouari O, Lopez M, McAllister DM, Boyle KA, Barrios CS, Weber J, Johnson BD, Hardy M, Dwinell MB, Kalyanaraman B (2016) Mitochondria-targeted analogs of metformin exhibit enhanced antiproliferative and radiosensitizing effects in pancreatic cancer cells. Cancer Res. https://doi.org/10.1158/0008-5472.CAN-15-2534

Chiba Y, Kubota T, Watanabe M, Matsuzaki SW, Otani Y, Teramoto T, Matsumoto Y, Koya K, Kitajima M (1998) MKT-077, localized lipophilic cation: antitumor activity against human tumor xenografts serially transplanted into nude mice. Anticancer Res 18(2A):1047–1052

Chu QS, Sangha R, Spratlin J, Vos LJ, Mackey JR, McEwan AJ, Venner P, Michelakis ED (2015) A phase I open-labeled, single-arm, dose-escalation, study of dichloroacetate (DCA) in patients with advanced solid tumors. Investig New Drugs 33(3):603–610. https://doi.org/10.1007/s10637-015-0221-y

Coleman MC, Asbury CR, Daniels D, Du J, Aykin-Burns N, Smith BJ, Li L, Spitz DR, Cullen JJ (2008) 2-Deoxy-D-glucose causes cytotoxicity, oxidative stress, and radiosensitization in pancreatic cancer. Free Radic Biol Med 44(3):322–331. https://doi.org/10.1016/j.freeradbiomed.2007.08.032

Costantini P, Belzacq AS, Vieira HL, Larochette N, de Pablo MA, Zamzami N, Susin SA, Brenner C, Kroemer G (2000) Oxidation of a critical thiol residue of the adenine nucleotide translocator enforces Bcl-2-independent permeability transition pore opening and apoptosis. Oncogene 19(2):307–314. https://doi.org/10.1038/sj.onc.1203299

Coyle C, Cafferty FH, Vale C, Langley RE (2016) Metformin as an adjuvant treatment for cancer: a systematic review and meta-analysis. Ann Oncol 27(12):2184–2195. https://doi.org/10.1093/annonc/mdw410

Cuyàs E, Fernández-Arroyo S, Verdura S, Stursa J, Werner L, Joven J, Viollet B, Neuzil J, Menendez JA. Metformin regulates global DNA methylation via one-carbon metabolism. Oncogene (in press).

Czabotar PE, Lessene G, Strasser A, Adams JM (2014) Control of apoptosis by the BCL-2 protein family: implications for physiology and therapy. Nat Rev Mol Cell Biol 15(1):49–63. https://doi.org/10.1038/nrm3722

Dang CV (2007) The interplay between MYC and HIF in the Warburg effect. Ernst Schering Found Symp Proc 4:35–53

Daurio NA, Tuttle SW, Worth AJ, Song EY, Davis JM, Snyder NW, Blair IA, Koumenis C (2016) AMPK activation and metabolic reprogramming by tamoxifen through estrogen receptor-independent mechanisms suggests new uses for this therapeutic modality in cancer treatment. Cancer Res 76(11):3295–3306. https://doi.org/10.1158/0008-5472.CAN-15-2197

Don AS, Kisker O, Dilda P, Donoghue N, Zhao X, Decollogne S, Creighton B, Flynn E, Folkman J, Hogg PJ (2003) A peptide trivalent arsenical inhibits tumor angiogenesis by perturbing mitochondrial function in angiogenic endothelial cells. Cancer Cell 3(5):497–509

Dong LF, Swettenham E, Eliasson J, Wang XF, Gold M, Medunic Y, Stantic M, Low P, Prochazka L, Witting PK, Turanek J, Akporiaye ET, Ralph SJ, Neuzil J (2007) Vitamin E analogues inhibit angiogenesis by selective induction of apoptosis in proliferating endothelial cells: the role of oxidative stress. Cancer Res 67(24):11906–11913. https://doi.org/10.1158/0008-5472.CAN-07-3034

Dong LF, Low P, Dyason JC, Wang XF, Prochazka L, Witting PK, Freeman R, Swettenham E, Valis K, Liu J, Zobalova R, Turanek J, Spitz DR, Domann FE, Scheffler IE, Ralph SJ, Neuzil J (2008) Alpha-tocopheryl succinate induces apoptosis by targeting ubiquinone-binding sites in mitochondrial respiratory complex II. Oncogene 27(31):4324–4335. https://doi.org/10.1038/onc.2008.69

Dong LF, Freeman R, Liu J, Zobalova R, Marin-Hernandez A, Stantic M, Rohlena J, Valis K, Rodriguez-Enriquez S, Butcher B, Goodwin J, Brunk UT, Witting PK, Moreno-Sanchez R, Scheffler IE, Ralph SJ, Neuzil J (2009) Suppression of tumor growth in vivo by the mitocan alpha-tocopheryl succinate requires respiratory complex II. Clin Cancer Res 15(5):1593–1600. https://doi.org/10.1158/1078-0432.CCR-08-2439

Dong LF, Jameson VJ, Tilly D, Cerny J, Mahdavian E, Marin-Hernandez A, Hernandez-Esquivel L, Rodriguez-Enriquez S, Stursa J, Witting PK, Stantic B, Rohlena J, Truksa J, Kluckova K, Dyason JC, Ledvina M, Salvatore BA, Moreno-Sanchez R, Coster MJ, Ralph SJ, Smith RA, Neuzil J (2011a) Mitochondrial targeting of vitamin E succinate enhances its pro-apoptotic and anti-cancer activity via mitochondrial complex II. J Biol Chem 286(5):3717–3728. https://doi.org/10.1074/jbc.M110.186643

Dong LF, Jameson VJ, Tilly D, Prochazka L, Rohlena J, Valis K, Truksa J, Zobalova R, Mahdavian E, Kluckova K, Stantic M, Stursa J, Freeman R, Witting PK, Norberg E, Goodwin J, Salvatore BA, Novotna J, Turanek J, Ledvina M, Hozak P, Zhivotovsky B, Coster MJ, Ralph SJ, Smith RA, Neuzil J (2011b) Mitochondrial targeting of alpha-tocopheryl succinate enhances its pro-apoptotic efficacy: a new paradigm for effective cancer therapy. Free Radic Biol Med 50(11):1546–1555. https://doi.org/10.1016/j.freeradbiomed.2011.02.032

Dong LF, Kovarova J, Bajzikova M, Bezawork-Geleta A, Svec D, Endaya B, Sachaphibulkij K, Coelho AR, Sebkova N, Ruzickova A, Tan AS, Kluckova K, Judasova K, Zamecnikova K, Rychtarcikova Z, Gopalan V, Andera L, Sobol M, Yan B, Pattnaik B, Bhatraju N, Truksa J, Stopka P, Hozak P, Lam AK, Sedlacek R, Oliveira PJ, Kubista M, Agrawal A, Dvorakova-Hortova K, Rohlena J, Berridge MV, Neuzil J (2017) Horizontal transfer of whole mitochondria restores tumorigenic potential in mitochondrial DNA-deficient cancer cells. Elife 6. https://doi.org/10.7554/eLife.22187

Dowling RJ, Niraula S, Stambolic V, Goodwin PJ (2012) Metformin in cancer: translational challenges. J Mol Endocrinol 48(3):R31–R43. https://doi.org/10.1530/JME-12-0007

Dunbar EM, Coats BS, Shroads AL, Langaee T, Lew A, Forder JR, Shuster JJ, Wagner DA, Stacpoole PW (2014) Phase 1 trial of dichloroacetate (DCA) in adults with recurrent malignant brain tumors. Investig New Drugs 32(3):452–464. https://doi.org/10.1007/s10637-013-0047-4

El Sayed SM, Mohamed WG, Seddik MA, Ahmed AS, Mahmoud AG, Amer WH, Helmy Nabo MM, Hamed AR, Ahmed NS, Abd-Allah AA (2014) Safety and outcome of treatment of metastatic melanoma using 3-bromopyruvate: a concise literature review and case study. Chin J Cancer 33(7):356–364. https://doi.org/10.5732/cjc.013.10111

Evans JM, Donnelly LA, Emslie-Smith AM, Alessi DR, Morris AD (2005) Metformin and reduced risk of cancer in diabetic patients. BMJ 330(7503):1304–1305. https://doi.org/10.1136/bmj.38415.708634.F7

Fan TW, Lane AN, Higashi RM, Farag MA, Gao H, Bousamra M, Miller DM (2009) Altered regulation of metabolic pathways in human lung cancer discerned by 13C stable isotope-resolved metabolomics (SIRM). Mol Cancer 8(1):41. https://doi.org/10.1186/1476-4598-8-41

Fiskus W, Saba N, Shen M, Ghias M, Liu J, Gupta SD, Chauhan L, Rao R, Gunewardena S, Schorno K, Austin CP, Maddocks K, Byrd J, Melnick A, Huang P, Wiestner A, Bhalla KN (2014) Auranofin induces lethal oxidative and endoplasmic reticulum stress and exerts potent preclinical activity against chronic lymphocytic leukemia. Cancer Res 74(9):2520–2532. https://doi.org/10.1158/0008-5472.CAN-13-2033

Foretz M, Guigas B, Bertrand L, Pollak M, Viollet B (2014) Metformin: from mechanisms of action to therapies. Cell Metab 20(6):953–966. https://doi.org/10.1016/j.cmet.2014.09.018

Gall JM, Wong V, Pimental DR, Havasi A, Wang Z, Pastorino JG, Bonegio RG, Schwartz JH, Borkan SC (2011) Hexokinase regulates Bax-mediated mitochondrial membrane injury following ischemic stress. Kidney Int 79(11):1207–1216. https://doi.org/10.1038/ki.2010.532

Gamcsik MP, Kasibhatla MS, Teeter SD, Colvin OM (2012) Glutathione levels in human tumors. Biomarkers 17(8):671–691. https://doi.org/10.3109/1354750X.2012.715672

Gerlinger M, Rowan AJ, Horswell S, Larkin J, Endesfelder D, Gronroos E, Martinez P, Matthews N, Stewart A, Tarpey P, Varela I, Phillimore B, Begum S, McDonald NQ, Butler A, Jones D,

Raine K, Latimer C, Santos CR, Nohadani M, Eklund AC, Spencer-Dene B, Clark G, Pickering L, Stamp G, Gore M, Szallasi Z, Downward J, Futreal PA, Swanton C (2012) Intratumor heterogeneity and branched evolution revealed by multiregion sequencing. N Engl J Med 366(10):883–892. https://doi.org/10.1056/NEJMoa1113205

Geschwind JF, Ko YH, Torbenson MS, Magee C, Pedersen PL (2002) Novel therapy for liver cancer: direct intraarterial injection of a potent inhibitor of ATP production. Cancer Res 62(14):3909–3913

Goard CA, Schimmer AD (2013) An evidence-based review of obatoclax mesylate in the treatment of hematological malignancies. Core Evid 8:15–26. https://doi.org/10.2147/CE.S42568

Gogvadze V, Orrenius S, Zhivotovsky B (2008) Mitochondria in cancer cells: what is so special about them? Trends Cell Biol 18(4):165–173. https://doi.org/10.1016/j.tcb.2008.01.006

Goldin N, Arzoine L, Heyfets A, Israelson A, Zaslavsky Z, Bravman T, Bronner V, Notcovich A, Shoshan-Barmatz V, Flescher E (2008) Methyl jasmonate binds to and detaches mitochondria-bound hexokinase. Oncogene 27(34):4636–4643. https://doi.org/10.1038/onc.2008.108

Goveia J, Pircher A, Conradi LC, Kalucka J, Lagani V, Dewerchin M, Eelen G, DeBerardinis RJ, Wilson ID, Carmeliet P (2016) Meta-analysis of clinical metabolic profiling studies in cancer: challenges and opportunities. EMBO Mol Med 8(10):1134–1142. https://doi.org/10.15252/emmm.201606798

Gui DY, Sullivan LB, Luengo A, Hosios AM, Bush LN, Gitego N, Davidson SM, Freinkman E, Thomas CJ, Vander Heiden MG (2016) Environment dictates dependence on mitochondrial complex I for NAD+ and aspartate production and determines cancer cell sensitivity to metformin. Cell Metab. https://doi.org/10.1016/j.cmet.2016.09.006

Gupta P, Wright SE, Kim SH, Srivastava SK (2014) Phenethyl isothiocyanate: a comprehensive review of anti-cancer mechanisms. Biochim Biophys Acta 1846(2):405–424. https://doi.org/10.1016/j.bbcan.2014.08.003

Harris IS, Treloar AE, Inoue S, Sasaki M, Gorrini C, Lee KC, Yung KY, Brenner D, Knobbe-Thomsen CB, Cox MA, Elia A, Berger T, Cescon DW, Adeoye A, Brustle A, Molyneux SD, Mason JM, Li WY, Yamamoto K, Wakeham A, Berman HK, Khokha R, Done SJ, Kavanagh TJ, Lam CW, Mak TW (2015) Glutathione and thioredoxin antioxidant pathways synergize to drive cancer initiation and progression. Cancer Cell 27(2):211–222. https://doi.org/10.1016/j.ccell.2014.11.019

Heikkinen S, Suppola S, Malkki M, Deeb SS, Janne J, Laakso M (2000) Mouse hexokinase II gene: structure, cDNA, promoter analysis, and expression pattern. Mamm Genome 11(2):91–96

Hensley CT, Faubert B, Yuan Q, Lev-Cohain N, Jin E, Kim J, Jiang L, Ko B, Skelton R, Loudat L, Wodzak M, Klimko C, McMillan E, Butt Y, Ni M, Oliver D, Torrealba J, Malloy Craig R, Kernstine K, Lenkinski RE, DeBerardinis RJ (2016) Metabolic heterogeneity in human lung tumors. Cell 164(4):681–694. https://doi.org/10.1016/j.cell.2015.12.034

Hilf R (2007) Mitochondria are targets of photodynamic therapy. J Bioenerg Biomembr 39(1):85–89. https://doi.org/10.1007/s10863-006-9064-8

Hirayama A, Kami K, Sugimoto M, Sugawara M, Toki N, Onozuka H, Kinoshita T, Saito N, Ochiai A, Tomita M, Esumi H, Soga T (2009) Quantitative metabolome profiling of colon and stomach cancer microenvironment by capillary electrophoresis time-of-flight mass spectrometry. Cancer Res 69(11):4918–4925. https://doi.org/10.1158/0008-5472.CAN-08-4806

Horsley L, Cummings J, Middleton M, Ward T, Backen A, Clamp A, Dawson M, Farmer H, Fisher N, Halbert G, Halford S, Harris A, Hasan J, Hogg P, Kumaran G, Little R, Parker GJ, Potter P, Saunders M, Roberts C, Shaw D, Smith N, Smythe J, Taylor A, Turner H, Watson Y, Dive C, Jayson GC, Cancer Research UKDDOPIct (2013) A phase 1 trial of intravenous 4-(N-(S-glutathionylacetyl)amino) phenylarsenoxide (GSAO) in patients with advanced solid tumours. Cancer Chemother Pharmacol 72(6):1343–1352. https://doi.org/10.1007/s00280-013-2320-9

Iwasawa S, Yamano Y, Takiguchi Y, Tanzawa H, Tatsumi K, Uzawa K (2011) Upregulation of thioredoxin reductase 1 in human oral squamous cell carcinoma. Oncol Rep 25(3):637–644. https://doi.org/10.3892/or.2010.1131

Jardim-Messeder D, Moreira-Pacheco F (2016) 3-Bromopyruvic acid inhibits tricarboxylic acid cycle and glutaminolysis in HepG2 cells. Anticancer Res 36(5):2233–2241

Jiralerspong S, Palla SL, Giordano SH, Meric-Bernstam F, Liedtke C, Barnett CM, Hsu L, Hung MC, Hortobagyi GN, Gonzalez-Angulo AM (2009) Metformin and pathologic complete responses to neoadjuvant chemotherapy in diabetic patients with breast cancer. J Clin Oncol 27(20):3297–3302. https://doi.org/10.1200/JCO.2009.19.6410

Kelly PN, Strasser A (2011) The role of Bcl-2 and its pro-survival relatives in tumourigenesis and cancer therapy. Cell Death Differ 18(9):1414–1424. https://doi.org/10.1038/cdd.2011.17

Kim EH, Lee JH, Oh Y, Koh I, Shim JK, Park J, Choi J, Yun M, Jeon JY, Huh YM, Chang JH, Kim SH, Kim KS, Cheong JH, Kim P, Kang SG (2017) Inhibition of glioblastoma tumorspheres by combined treatment with 2-deoxyglucose and metformin. Neuro Oncol 19(2):197–207. https://doi.org/10.1093/neuonc/now174

Kitada S, Leone M, Sareth S, Zhai D, Reed JC, Pellecchia M (2003) Discovery, characterization, and structure-activity relationships studies of proapoptotic polyphenols targeting B-cell lymphocyte/leukemia-2 proteins. J Med Chem 46(20):4259–4264. https://doi.org/10.1021/jm030190z

Kitada S, Kress CL, Krajewska M, Jia L, Pellecchia M, Reed JC (2008) Bcl-2 antagonist apogossypol (NSC736630) displays single-agent activity in Bcl-2-transgenic mice and has superior efficacy with less toxicity compared with gossypol (NSC19048). Blood 111(6):3211–3219. https://doi.org/10.1182/blood-2007-09-113647

Kluckova K, Sticha M, Cerny J, Mracek T, Dong L, Drahota Z, Gottlieb E, Neuzil J, Rohlena J (2015) Ubiquinone-binding site mutagenesis reveals the role of mitochondrial complex II in cell death initiation. Cell Death Dis 6:e1749. https://doi.org/10.1038/cddis.2015.110

Ko YH, Pedersen PL, Geschwind JF (2001) Glucose catabolism in the rabbit VX2 tumor model for liver cancer: characterization and targeting hexokinase. Cancer Lett 173(1):83–91

Ko YH, Smith BL, Wang Y, Pomper MG, Rini DA, Torbenson MS, Hullihen J, Pedersen PL (2004) Advanced cancers: eradication in all cases using 3-bromopyruvate therapy to deplete ATP. Biochem Biophys Res Commun 324(1):269–275. https://doi.org/10.1016/j.bbrc.2004.09.047

Ko YH, Verhoeven HA, Lee MJ, Corbin DJ, Vogl TJ, Pedersen PL (2012) A translational study "case report" on the small molecule "energy blocker" 3-bromopyruvate (3BP) as a potent anticancer agent: from bench side to bedside. J Bioenerg Biomembr 44(1):163–170. https://doi.org/10.1007/s10863-012-9417-4

Konopleva M, Contractor R, Tsao T, Samudio I, Ruvolo PP, Kitada S, Deng X, Zhai D, Shi YX, Sneed T, Verhaegen M, Soengas M, Ruvolo VR, McQueen T, Schober WD, Watt JC, Jiffar T, Ling X, Marini FC, Harris D, Dietrich M, Estrov Z, McCubrey J, May WS, Reed JC, Andreeff M (2006) Mechanisms of apoptosis sensitivity and resistance to the BH3 mimetic ABT-737 in acute myeloid leukemia. Cancer Cell 10(5):375–388. https://doi.org/10.1016/j.ccr.2006.10.006

Konopleva M, Watt J, Contractor R, Tsao T, Harris D, Estrov Z, Bornmann W, Kantarjian H, Viallet J, Samudio I, Andreeff M (2008) Mechanisms of antileukemic activity of the novel Bcl-2 homology domain-3 mimetic GX15-070 (obatoclax). Cancer Res 68(9):3413–3420. https://doi.org/10.1158/0008-5472.CAN-07-1919

Koppenol WH, Bounds PL, Dang CV (2011) Otto Warburg's contributions to current concepts of cancer metabolism. Nat Rev Cancer 11(5):325–337. https://doi.org/10.1038/nrc3038

Koya K, Li Y, Wang H, Ukai T, Tatsuta N, Kawakami M, Shishido, Chen LB (1996) MKT-077, a novel rhodacyanine dye in clinical trials, exhibits anticarcinoma activity in preclinical studies based on selective mitochondrial accumulation. Cancer Res 56(3):538–543

Kwong JQ, Molkentin JD (2015) Physiological and pathological roles of the mitochondrial permeability transition pore in the heart. Cell Metab 21(2):206–214. https://doi.org/10.1016/j.cmet.2014.12.001

Lamb R, Ozsvari B, Lisanti CL, Tanowitz HB, Howell A, Martinez-Outschoorn UE, Sotgia F, Lisanti MP (2015) Antibiotics that target mitochondria effectively eradicate cancer stem cells, across multiple tumor types: treating cancer like an infectious disease. Oncotarget 6(7):4569–4584. https://doi.org/10.18632/oncotarget.3174

Landman GW, Kleefstra N, van Hateren KJ, Groenier KH, Gans RO, Bilo HJ (2010) Metformin associated with lower cancer mortality in type 2 diabetes: ZODIAC-16. Diabetes Care 33(2):322–326. https://doi.org/10.2337/dc09-1380

Larochette N, Decaudin D, Jacotot E, Brenner C, Marzo I, Susin SA, Zamzami N, Xie Z, Reed J, Kroemer G (1999) Arsenite induces apoptosis via a direct effect on the mitochondrial permeability transition pore. Exp Cell Res 249(2):413–421. https://doi.org/10.1006/excr.1999.4519

LeBleu VS, O'Connell JT, Gonzalez Herrera KN, Wikman H, Pantel K, Haigis MC, de Carvalho FM, Damascena A, Domingos Chinen LT, Rocha RM, Asara JM, Kalluri R (2014) PGC-1alpha mediates mitochondrial biogenesis and oxidative phosphorylation in cancer cells to promote metastasis. Nat Cell Biol 16(10):992–1003., 1001–1015. https://doi.org/10.1038/ncb3039

Letts JA, Fiedorczuk K, Sazanov LA (2016) The architecture of respiratory supercomplexes. Nature 537(7622):644–648. https://doi.org/10.1038/nature19774

Liyanage SU, Hurren R, Voisin V, Bridon G, Wang X, Xu C, MacLean N, Siriwardena TP, Gronda M, Yehudai D, Sriskanthadevan S, Avizonis D, Shamas-Din A, Minden MD, Bader GD, Laposa R, Schimmer AD (2017) Leveraging increased cytoplasmic nucleoside kinase activity to target mtDNA and oxidative phosphorylation in AML. Blood. https://doi.org/10.1182/blood-2016-10-741207

Lonardo E, Cioffi M, Sancho P, Sanchez-Ripoll Y, Trabulo SM, Dorado J, Balic A, Hidalgo M, Heeschen C (2013) Metformin targets the metabolic achilles heel of human pancreatic cancer stem cells. PLoS One 8(10):e76518. https://doi.org/10.1371/journal.pone.0076518

Macchioni L, Davidescu M, Roberti R, Corazzi L (2014) The energy blockers 3-bromopyruvate and lonidamine: effects on bioenergetics of brain mitochondria. J Bioenerg Biomembr 46(5):389–394. https://doi.org/10.1007/s10863-014-9577-5

Majewski N, Nogueira V, Robey RB, Hay N (2004) Akt inhibits apoptosis downstream of BID cleavage via a glucose-dependent mechanism involving mitochondrial hexokinases. Mol Cell Biol 24(2):730–740

Maranzana E, Barbero G, Falasca AI, Lenaz G, Genova ML (2013) Mitochondrial respiratory supercomplex association limits production of reactive oxygen species from complex I. Antioxid Redox Signal 19(13):1469–1480. https://doi.org/10.1089/ars.2012.4845

Marin-Valencia I, Yang C, Mashimo T, Cho S, Baek H, Yang XL, Rajagopalan KN, Maddie M, Vemireddy V, Zhao Z, Cai L, Good L, Tu BP, Hatanpaa KJ, Mickey BE, Mates JM, Pascual JM, Maher EA, Malloy CR, Deberardinis RJ, Bachoo RM (2012) Analysis of tumor metabolism reveals mitochondrial glucose oxidation in genetically diverse human glioblastomas in the mouse brain in vivo. Cell Metab 15(6):827–837. https://doi.org/10.1016/j.cmet.2012.05.001

Martinez-Outschoorn UE, Peiris-Pages M, Pestell RG, Sotgia F, Lisanti MP (2017) Cancer metabolism: a therapeutic perspective. Nat Rev Clin Oncol 14(1):11–31. https://doi.org/10.1038/nrclinonc.2016.60

Maschek G, Savaraj N, Priebe W, Braunschweiger P, Hamilton K, Tidmarsh GF, De Young LR, Lampidis TJ (2004) 2-Deoxy-D-glucose increases the efficacy of adriamycin and paclitaxel in human osteosarcoma and non-small cell lung cancers in vivo. Cancer Res 64(1):31–34

Mathupala SP, Ko YH, Pedersen PL (2009) Hexokinase-2 bound to mitochondria: cancer's stygian link to the "Warburg Effect" and a pivotal target for effective therapy. Semin Cancer Biol 19(1):17–24. https://doi.org/10.1016/j.semcancer.2008.11.006

McStay GP, Clarke SJ, Halestrap AP (2002) Role of critical thiol groups on the matrix surface of the adenine nucleotide translocase in the mechanism of the mitochondrial permeability transition pore. Biochem J 367(Pt 2):541–548. https://doi.org/10.1042/BJ20011672

Michelakis ED, Webster L, Mackey JR (2008) Dichloroacetate (DCA) as a potential metabolic-targeting therapy for cancer. Br J Cancer 99(7):989–994. https://doi.org/10.1038/sj.bjc.6604554

Mills EL, Kelly B, O'Neill LAJ (2017) Mitochondria are the powerhouses of immunity. Nat Immunol 18(5):488–498. https://doi.org/10.1038/ni.3704

Modica-Napolitano JS, Aprille JR (2001) Delocalized lipophilic cations selectively target the mitochondria of carcinoma cells. Adv Drug Deliv Rev 49(1–2):63–70

Moreira PI, Custodio J, Moreno A, Oliveira CR, Santos MS (2006) Tamoxifen and estradiol interact with the flavin mononucleotide site of complex I leading to mitochondrial failure. J Biol Chem 281(15):10143–10152. https://doi.org/10.1074/jbc.M510249200

Nakashima RA, Mangan PS, Colombini M, Pedersen PL (1986) Hexokinase receptor complex in hepatoma mitochondria: evidence from N,N′-dicyclohexylcarbodiimide-labeling studies for the involvement of the pore-forming protein VDAC. Biochemistry 25(5):1015–1021

Neuzil J, Weber T, Gellert N, Weber C (2001a) Selective cancer cell killing by alpha-tocopheryl succinate. Br J Cancer 84(1):87–89. https://doi.org/10.1054/bjoc.2000.1559

Neuzil J, Weber T, Schroder A, Lu M, Ostermann G, Gellert N, Mayne GC, Olejnicka B, Negre-Salvayre A, Sticha M, Coffey RJ, Weber C (2001b) Induction of cancer cell apoptosis by alpha-tocopheryl succinate: molecular pathways and structural requirements. FASEB J 15(2):403–415. https://doi.org/10.1096/fj.00-0251com

Neuzil J, Wang XF, Dong LF, Low P, Ralph SJ (2006) Molecular mechanism of 'mitocan'-induced apoptosis in cancer cells epitomizes the multiple roles of reactive oxygen species and Bcl-2 family proteins. FEBS Lett 580(22):5125–5129. https://doi.org/10.1016/j.febslet.2006.05.072

Neuzil J, Dong LF, Rohlena J, Truksa J, Ralph SJ (2013) Classification of mitocans, anti-cancer drugs acting on mitochondria. Mitochondrion 13(3):199–208. https://doi.org/10.1016/j.mito.2012.07.112

de Oliveira MR, Nabavi SF, Manayi A, Daglia M, Hajheydari Z, Nabavi SM (2016) Resveratrol and the mitochondria: from triggering the intrinsic apoptotic pathway to inducing mitochondrial biogenesis, a mechanistic view. Biochim Biophys Acta 1860(4):727–745. https://doi.org/10.1016/j.bbagen.2016.01.017

Oltersdorf T, Elmore SW, Shoemaker AR, Armstrong RC, Augeri DJ, Belli BA, Bruncko M, Deckwerth TL, Dinges J, Hajduk PJ, Joseph MK, Kitada S, Korsmeyer SJ, Kunzer AR, Letai A, Li C, Mitten MJ, Nettesheim DG, Ng S, Nimmer PM, O'Connor JM, Oleksijew A, Petros AM, Reed JC, Shen W, Tahir SK, Thompson CB, Tomaselli KJ, Wang B, Wendt MD, Zhang H, Fesik SW, Rosenberg SH (2005) An inhibitor of Bcl-2 family proteins induces regression of solid tumours. Nature 435(7042):677–681. https://doi.org/10.1038/nature03579

Opydo-Chanek M, Gonzalo O, Marzo I (2017) Multifaceted anticancer activity of BH3 mimetics: current evidence and future prospects. Biochem Pharmacol. https://doi.org/10.1016/j.bcp.2017.03.006

Pastorino JG, Shulga N, Hoek JB (2002) Mitochondrial binding of hexokinase II inhibits Bax-induced cytochrome c release and apoptosis. J Biol Chem 277(9):7610–7618. https://doi.org/10.1074/jbc.M109950200

Propper DJ, Braybrooke JP, Taylor DJ, Lodi R, Styles P, Cramer JA, Collins WC, Levitt NC, Talbot DC, Ganesan TS, Harris AL (1999) Phase I trial of the selective mitochondrial toxin MKT077 in chemo-resistant solid tumours. Ann Oncol 10(8):923–927

Qiu MZ, Han B, Luo HY, Zhou ZW, Wang ZQ, Wang FH, Li YH, Xu RH (2011) Expressions of hypoxia-inducible factor-1alpha and hexokinase-II in gastric adenocarcinoma: the impact on prognosis and correlation to clinicopathologic features. Tumour Biol 32(1):159–166. https://doi.org/10.1007/s13277-010-0109-6

Rodrigues TB, Serrao EM, Kennedy BW, Hu DE, Kettunen MI, Brindle KM (2014) Magnetic resonance imaging of tumor glycolysis using hyperpolarized 13C-labeled glucose. Nat Med 20(1):93–97. https://doi.org/10.1038/nm.3416

Rohlenova K, Sachaphibulkij K, Stursa J, Bezawork-Geleta A, Blecha J, Endaya B, Werner L, Cerny J, Zobalova R, Goodwin J, Spacek T, Alizadeh Pesdar E, Yan B, Nguyen MN, Vondrusova M, Sobol M, Jezek P, Hozak P, Truksa J, Rohlena J, Dong LF, Neuzil J (2017) Selective disruption of respiratory supercomplexes as a new strategy to suppress Her2high breast cancer. Antioxid Redox Signal 26(2):84–103. https://doi.org/10.1089/ars.2016.6677

Sabharwal SS, Schumacker PT (2014) Mitochondrial ROS in cancer: initiators, amplifiers or an Achilles' heel? Nat Rev Cancer 14(11):709–721. https://doi.org/10.1038/nrc3803

Sanborn BM, Felberg NT, Hollocher TC (1971) The inactivation of succinate dehydrogenase by bromopyruvate. Biochim Biophys Acta 227(2):219–231

Sancho P, Burgos-Ramos E, Tavera A, Bou Kheir T, Jagust P, Schoenhals M, Barneda D, Sellers K, Campos-Olivas R, Grana O, Viera CR, Yuneva M, Sainz B Jr, Heeschen C (2015) MYC/PGC-1alpha balance determines the metabolic phenotype and plasticity of pancreatic cancer stem cells. Cell Metab. https://doi.org/10.1016/j.cmet.2015.08.015

Sancho P, Barneda D, Heeschen C (2016) Hallmarks of cancer stem cell metabolism. Br J Cancer 114(12):1305–1312. https://doi.org/10.1038/bjc.2016.152

Sassi N, Mattarei A, Azzolini M, Bernardi P, Szabo I, Paradisi C, Zoratti M, Biasutto L (2014a) Mitochondria-targeted resveratrol derivatives act as cytotoxic pro-oxidants. Curr Pharm Des 20(2):172–179

Sassi N, Mattarei A, Azzolini M, Szabo I, Paradisi C, Zoratti M, Biasutto L (2014b) Cytotoxicity of mitochondria-targeted resveratrol derivatives: interactions with respiratory chain complexes and ATP synthase. Biochim Biophys Acta 1837(10):1781–1789. https://doi.org/10.1016/j.bbabio.2014.06.010

Sato-Tadano A, Suzuki T, Amari M, Takagi K, Miki Y, Tamaki K, Watanabe M, Ishida T, Sasano H, Ohuchi N (2013) Hexokinase II in breast carcinoma: a potent prognostic factor associated with hypoxia-inducible factor-1alpha and Ki-67. Cancer Sci 104(10):1380–1388. https://doi.org/10.1111/cas.12238

Schagger H, Pfeiffer K (2000) Supercomplexes in the respiratory chains of yeast and mammalian mitochondria. EMBO J 19(8):1777–1783. https://doi.org/10.1093/emboj/19.8.1777

Scheffler I (2008) Mitochondria, 2nd edn. Wiley, Hoboken, NJ

Simons AL, Fath MA, Mattson DM, Smith BJ, Walsh SA, Graham MM, Hichwa RD, Buatti JM, Dornfeld K, Spitz DR (2007) Enhanced response of human head and neck cancer xenograft tumors to cisplatin combined with 2-deoxy-D-glucose correlates with increased 18F-FDG uptake as determined by PET imaging. Int J Radiat Oncol Biol Phys 69(4):1222–1230. https://doi.org/10.1016/j.ijrobp.2007.07.2343

Skrtic M, Sriskanthadevan S, Jhas B, Gebbia M, Wang X, Wang Z, Hurren R, Jitkova Y, Gronda M, Maclean N, Lai CK, Eberhard Y, Bartoszko J, Spagnuolo P, Rutledge AC, Datti A, Ketela T, Moffat J, Robinson BH, Cameron JH, Wrana J, Eaves CJ, Minden MD, Wang JC, Dick JE, Humphries K, Nislow C, Giaever G, Schimmer AD (2011) Inhibition of mitochondrial translation as a therapeutic strategy for human acute myeloid leukemia. Cancer Cell 20(5):674–688. https://doi.org/10.1016/j.ccr.2011.10.015

Stapelberg M, Gellert N, Swettenham E, Tomasetti M, Witting PK, Procopio A, Neuzil J (2005) Alpha-tocopheryl succinate inhibits malignant mesothelioma by disrupting the fibroblast growth factor autocrine loop: mechanism and the role of oxidative stress. J Biol Chem 280(27):25369–25376. https://doi.org/10.1074/jbc.M414498200

Sullivan LB, Gui DY, Hosios AM, Bush LN, Freinkman E, Vander Heiden MG (2015) Supporting aspartate biosynthesis is an essential function of respiration in proliferating cells. Cell 162(3):552–563. https://doi.org/10.1016/j.cell.2015.07.017

Sun Y, Liu Z, Zou X, Lan Y, Sun X, Wang X, Zhao S, Jiang C, Liu H (2015) Mechanisms underlying 3-bromopyruvate-induced cell death in colon cancer. J Bioenerg Biomembr 47(4):319–329. https://doi.org/10.1007/s10863-015-9612-1

Tan AS, Baty JW, Dong LF, Bezawork-Geleta A, Endaya B, Goodwin J, Bajzikova M, Kovarova J, Peterka M, Yan B, Pesdar EA, Sobol M, Filimonenko A, Stuart S, Vondrusova M, Kluckova K, Sachaphibulkij K, Rohlena J, Hozak P, Truksa J, Eccles D, Haupt LM, Griffiths LR, Neuzil J, Berridge MV (2015) Mitochondrial genome acquisition restores respiratory function and tumorigenic potential of cancer cells without mitochondrial DNA. Cell Metab 21(1):81–94. https://doi.org/10.1016/j.cmet.2014.12.003

Trachootham D, Alexandre J, Huang P (2009) Targeting cancer cells by ROS-mediated mechanisms: a radical therapeutic approach? Nat Rev Drug Discov 8(7):579–591. https://doi.org/10.1038/nrd2803

Truksa J, Dong LF, Rohlena J, Stursa J, Vondrusova M, Goodwin J, Nguyen M, Kluckova K, Rychtarcikova Z, Lettlova S, Spacilova J, Stapelberg M, Zoratti M, Neuzil J (2015) Mitochondrially targeted vitamin E succinate modulates expression of mitochondrial DNA transcripts and mitochondrial biogenesis. Antioxid Redox Signal 22(11):883–900. https://doi.org/10.1089/ars.2013.5594

Vander Heiden MG, Cantley LC, Thompson CB (2009) Understanding the Warburg effect: the metabolic requirements of cell proliferation. Science 324(5930):1029–1033. https://doi.org/10.1126/science.1160809

Vyas S, Zaganjor E, Haigis MC (2016) Mitochondria and cancer. Cell 166(3):555–566. https://doi.org/10.1016/j.cell.2016.07.002

Vyssokikh MY, Zorova L, Zorov D, Heimlich G, Jurgensmeier JJ, Brdiczka D (2002) Bax releases cytochrome c preferentially from a complex between porin and adenine nucleotide translocator. Hexokinase activity suppresses this effect. Mol Biol Rep 29(1–2):93–96

Warburg O (1956) On the origin of cancer cells. Science 123(3191):309–314

Weber T, Lu M, Andera L, Lahm H, Gellert N, Fariss MW, Korinek V, Sattler W, Ucker DS, Terman A, Schroder A, Erl W, Brunk UT, Coffey RJ, Weber C, Neuzil J (2002) Vitamin E succinate is a potent novel antineoplastic agent with high selectivity and cooperativity with tumor necrosis factor-related apoptosis-inducing ligand (Apo2 ligand) in vivo. Clin Cancer Res 8(3):863–869

Weinberg F, Hamanaka R, Wheaton WW, Weinberg S, Joseph J, Lopez M, Kalyanaraman B, Mutlu GM, Budinger GR, Chandel NS (2010) Mitochondrial metabolism and ROS generation are essential for Kras-mediated tumorigenicity. Proc Natl Acad Sci U S A 107(19):8788–8793. https://doi.org/10.1073/pnas.1003428107

Wilson JE (2003) Isozymes of mammalian hexokinase: structure, subcellular localization and metabolic function. J Exp Biol 206(Pt 12):2049–2057

Wolf DA (2014) Is reliance on mitochondrial respiration a "chink in the armor" of therapy-resistant cancer? Cancer Cell 26(6):788–795. https://doi.org/10.1016/j.ccell.2014.10.001

Wolf A, Agnihotri S, Micallef J, Mukherjee J, Sabha N, Cairns R, Hawkins C, Guha A (2011) Hexokinase 2 is a key mediator of aerobic glycolysis and promotes tumor growth in human glioblastoma multiforme. J Exp Med 208(2):313–326. https://doi.org/10.1084/jem.20101470

Wu S, Zhou F, Wei Y, Chen WR, Chen Q, Xing D (2014) Cancer phototherapy via selective photoinactivation of respiratory chain oxidase to trigger a fatal superoxide anion burst. Antioxid Redox Signal 20(5):733–746. https://doi.org/10.1089/ars.2013.5229

Yan B, Stantic M, Zobalova R, Bezawork-Geleta A, Stapelberg M, Stursa J, Prokopova K, Dong L, Neuzil J (2015a) Mitochondrially targeted vitamin E succinate efficiently kills breast tumour-initiating cells in a complex II-dependent manner. BMC Cancer 15:401. https://doi.org/10.1186/s12885-015-1394-7

Yan L, Zhou J, Gao Y, Ghazal S, Lu L, Bellone S, Yang Y, Liu N, Zhao X, Santin AD, Taylor H, Huang Y (2015b) Regulation of tumor cell migration and invasion by the H19/let-7 axis is antagonized by metformin-induced DNA methylation. Oncogene 34(23):3076–3084. https://doi.org/10.1038/onc.2014.236

Yang N, Weinfeld M, Lemieux H, Montpetit B, Goping IS (2017) Photo-activation of the delocalized lipophilic cation D112 potentiates cancer selective ROS production and apoptosis. Cell Death Dis 8(2):e2587. https://doi.org/10.1038/cddis.2017.19

Ying H, Kimmelman AC, Lyssiotis CA, Hua S, Chu GC, Fletcher-Sananikone E, Locasale JW, Son J, Zhang H, Coloff JL, Yan H, Wang W, Chen S, Viale A, Zheng H, Paik JH, Lim C, Guimaraes AR, Martin ES, Chang J, Hezel AF, Perry SR, Hu J, Gan B, Xiao Y, Asara JM, Weissleder R, Wang YA, Chin L, Cantley LC, DePinho RA (2012) Oncogenic Kras maintains pancreatic tumors through regulation of anabolic glucose metabolism. Cell 149(3):656–670. https://doi.org/10.1016/j.cell.2012.01.058

Zhivotovsky B, Galluzzi L, Kepp O, Kroemer G (2009) Adenine nucleotide translocase: a component of the phylogenetically conserved cell death machinery. Cell Death Differ 16(11):1419–1425. https://doi.org/10.1038/cdd.2009.118

Zou P, Chen M, Ji J, Chen W, Chen X, Ying S, Zhang J, Zhang Z, Liu Z, Yang S, Liang G (2015) Auranofin induces apoptosis by ROS-mediated ER stress and mitochondrial dysfunction and displayed synergistic lethality with piperlongumine in gastric cancer. Oncotarget 6(34):36505–36521. https://doi.org/10.18632/oncotarget.5364

SkQ1: The Road from Laboratory Bench to the Market

Boris A. Feniouk and Maxim V. Skulachev

Abstract SkQ1 and other mitochondria-targeted plastoquinone derivatives are promising drug candidates for the treatment of pathologies associated with mitochondrial dysfunction. Experiments in animals revealed that SkQ alleviates damage induced by ischemia-reperfusion injury, prevents autoimmune arthritis, suppresses the development of Alzheimer's disease signs, inhibits the development of ophthalmological disorders, including dry eye syndrome (DES), and has anti-inflammatory activity. The first SkQ-based drug (Visomitin eye drops) is already developed and registered in Russia; clinical studies of orally administered drug is in progress. In this review, we summarize recent experimental data on SkQ effects in animals, results of clinical trials and future perspectives of mitochondria-targeted antioxidants as therapeutic compounds.

Keywords SkQ1 · Mitochondria-targeted plastoquinone derivatives · Drug development · Dry eye syndrome · Visomitin · Clinical trials

1 Introduction

The history of mitochondria-targeted compounds dates back to late 1960s, when Skulachev and Liberman performed first experiments with membranophilic penetrating cations (MPCs) on mitochondria and submitochondrial particles (Liberman et al. 1969; Bakeeva et al. 1970; Feniouk and Skulachev 2018). At the

B. A. Feniouk
Faculty of Bioengineering and Bioinformatics, Lomonosov Moscow State University, Moscow, Russia

A.N. Belozersky Institute of Physico–Chemical Biology, Lomonosov Moscow State University, Moscow, Russia

M. V. Skulachev (✉)
Institute of Mitoengineering, Moscow State University, Moscow, Russia

A.N. Belozersky Institute of Physico–Chemical Biology, Lomonosov Moscow State University, Moscow, Russia
e-mail: max@mitotech.ru

P. J. Oliveira (ed.), *Mitochondrial Biology and Experimental Therapeutics*,
https://doi.org/10.1007/978-3-319-73344-9_28

end of 1990s the first MPC-conjugated antioxidants that were able to specifically accumulate in over 1000-fold excess in mitochondria (especially in those with high a $\Delta\psi$) and to mildly de-energize the mitochondrial membrane were developed. The obvious advantage of such antioxidants consists in extremely low effective concentrations due to their ability to selectively accumulate in the inner mitochondrial membrane. For potential drug candidates, this property allows to greatly diminish the dose applied and the risks of unwanted side effects.

At the beginning of this millennium another important step on the way of mitochondria-targeted antioxidants design was made: MitoQ, an MPC conjugated with ubiquinone, was developed by Murphy et al. (Kelso et al. 2001).

The important advantage of MitoQ was its ability to be regenerated from the oxidized form back to the reduced one by accepting electrons from the respiratory chain (James et al. 2005), i.e. it was a rechargeable mitochondrial antioxidant. This means that each MitoQ molecule can neutralize numerous ROS species, and therefore such drugs can be applied in even lower doses. It also allows to use the more stable oxidized form of the drug, because it will eventually be reduced in mitochondria, yielding the active anti-oxidant quinole form.

Later studies performed in the group of Skulachev on model systems, mitochondria, and cell cultures indicated that ubiquinone, although being a natural mitochondrial electron carrier, is not the best option for the antioxidant moiety for mitochondria-targeted ROS quenchers. MPCs conjugated with plastoquinone, an electron carrier from the chloroplasts, proved to be more promising antioxidants than MitoQ (Feniouk and Skulachev 2018).

It must be noted that all quinone-based antioxidants exhibit pro-oxidant activity at high concentrations. However, the anti- and pro-oxidant concentrations of MitoQ are rather close to each other (~300 and ~500 nM, respectively), while for the plastoquinone conjugates this difference was almost 32-fold (~25 and ~800 nM) (Antonenko et al. 2008; Bakeeva et al. 2008; Skulachev et al. 2009).

These observations were the starting point for the SkQ-project—a multidisciplinary study of MPCs (or "Sk"—Skulachev ions, as they were named in 1974 by Green (1974)) conjugated to plastoquinone (abbreviated to "Q") as potential drugs for treatment of pathologies related to oxidative stress, mitochondrial dysfunction, and aging.

At the initial stage of the project, we analyzed a broad spectrum of SkQ compounds properties and activities in artificial lipid membranes, in isolated mitochondria, in cell cultures (Antonenko et al. 2008), and in animal models (summarized in Skulachev et al. 2011; Pasyukova et al. 2017; Feniouk and Skulachev 2018).

The effects of SkQ compounds on age-related retinal pathologies proved to be the most striking results obtained in the pre-clinical studies. Oxidative stress is an important damaging factor in the retina, because it contains a lot of polyunsaturated fatty acids and because the oxygen concentration in that site is near-arterial, i.e., much higher than in the great majority of other tissues (Beatty et al. 2000; Kanda et al. 2007).

Numerous indications of a crucial role of ROS in age-related ocular pathologies, as well as the results we obtained in OXYS rats (an animal model suffering from increased sensitivity to oxidative stress and developing cataract and retinopathies as early as 2.5–3 months of age (Sergeeva et al. 2006; Kolosova et al. 2006)) suggested that eye pathologies are probably a promising line for SkQ application.

Experiments showed (Neroev et al. 2008; Skulachev et al. 2009) that 250 nmol SkQ1/kg per day consumed with food completely prevented the development of retinopathy and cataract in OXYS rats up to the age of 2 years (Skulachev et al. 2011; Saprunova et al. 2012). Later it was found that such effect can be achieved with local administration of SkQ1 in the form of eye-drops (Neroev et al. 2008). Further development of SkQ1 as potential pharmaceutical split into two major directions: ophthalmic (development of eye-drops) and systemic (development of oral and injectable forms).

2 SkQ1 in Ophthalmics

A stable eye-drop form of SkQ1 called «Visomitin» was developed in 2009. Preclinical studies confirmed safety and efficacy of Visomitin eye drops in several animal models of age-related eye diseases, such as cataract, retinopathy (Neroev et al. 2008), glaucoma (Iomdina et al. 2015) and dry eye syndrome (ORA inc, unpublished). Experiments performed on rats in the group of Prof. Kolosova revealed several aspects of SkQ1 mechanism of action in eye pathologies: it slowed down the lacrimal gland degeneration (Bakeeva et al. 2016), normalized the lens metabolome, and the chaperone crystalline activity (Yanshole et al. 2015) and expression (Muraleva et al. 2014; Rumyantseva et al. 2015), ameliorated the pathological gene expression alterations in retina in animals suffering from AMD-like retinopathy (Perepechaeva et al. 2014). Electron microscopy studies demonstrated that SkQ1 given daily with food at the dose of 250 nmol/kg for 5 months, starting from the age of 19 months, prevented such age-dependent destructive processes in retina and vascular layer of eyes as accumulations of lipofuscin granules, and flattening, overgrowing, and degradation of endothelial cells of choriocapillaries (Saprunova et al. 2012). Oral SkQ1 administration to OXYS rats led to an increase in the concentration of mRNA and in the expression of vascular endothelial growth factor A (VEGF, a key regulator of angiogenesis) in retina up to the levels corresponding to the Wistar rats (Markovets et al. 2011).

Further clinical studies with dry eye syndrome (DES) carried out on 240 patients for 45 days confirmed the beneficial effect of SkQ1 in humans (Brzheskiy et al. 2015). This double-blinded placebo-controlled study also included a 45 days post-treatment period. The study revealed that SkQ1 reduced such DES symptoms as eye dryness, burning, grittiness, and blurred vision. Importantly, SkQ1 promoted faster corneal wound healing, especially in the subgroup of patients with heavy initial DES (Fig. 1).

Similar results were obtained in US Phase II clinical trial that utilized novel system of anti-DES drugs investigation, Controlled Adverse Environment (CAE) challenge (Ousler et al. 2005). In this study, two variants of SkQ1-based eye-drops were tested versus placebo («artificial tear», same as in a previous clinical study) on 90 patients (30 subjects per arm). The first variant was the same «Visomitin» preparation containing 250 nM SkQ1 as in a previous DES study, the second contained tenfold more (2500 nM) SkQ1. The experiment showed higher tolerability (followed with corneal damage and DES symptoms) of CAE challenge in group of Visomitin comparing to placebo (Petrov et al. 2016).

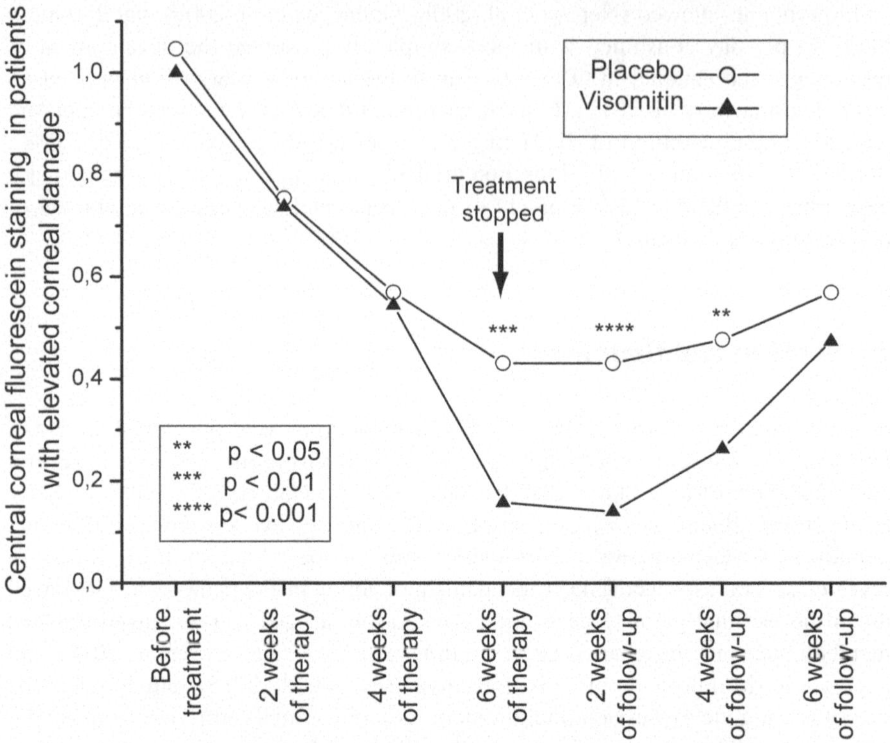

Fig. 1 Corneal damage dynamics evaluated by fluorescein staining in subgroup of patients with heavier initial symptoms of DES. Therapy was stopped after 6 weeks of treatment

Another double-blinded placebo-controlled phase II clinical study was performed on 80 patients with senile cataract. In this study, «Visomitin» administration period was 6 months. The trial confirmed long-term safety of ocular SkQ1 administration in humans. The study revealed statistically significant effect of SkQ1 on non-corrected visual acuity of patients in «Visomitin» group compared with placebo group. Importantly, measurements of tear antioxidant activity showed a strong, statistically significant increase in «Visomitin» group only (Erichev et al. 2016).

«Visomitin» is now approved and marketed as an Rx drug in Russia. Already over one million doses were sold here in the drug stores. FDA approval process is ongoing in the US.

3 Systemic Use of SkQ1

In parallel with the development of «Visomitin» in ophthalmics, several preclinical studies were performed with SkQ1 (and its close analog SkQR1) administered systemically. The results of these studies are summarized in Table 1.

Table 1 Results of pre-clinical studies of SkQ1 in animals

Pathology/Disease (animal models)	SkQ1 effect	Reference
Alzheimer disease	Protection of neurons from β-amyloid toxicity.	Kapay et al. (2011, 2013)
Alzheimer disease	Reduction of β-amyloid and tau protein hyperphosphorylation *in vivo*	Stefanova et al. (2016)
Alzheimer disease	In rats SkQ1 increased behavioral activity, decreased the percentage of animals with demyelination, reduced hippocampal Aβ1-40 and Aβ1-42 protein levels, and significantly reduced the destructive changes in mitochondria	Kolosova et al. (2017)
Inflammation-related diseases and pathology	Prevention of inflammation-triggered remodeling of blood vessel endothelium	Zinovkin et al. (2014) and Galkin et al. (2016)
Autoimmune arthritis	Reduction of arthritis symptoms in Wistar rats	Andreev-Andrievskiy et al. (2016)
Pyelonephritis	Suppression of inflammation, protection of kidney function	Plotnikov et al. (2013)
Kidney ischemia	Tissue protection, increase of animal survival	Plotnikov et al. (2011)
Brain focal ischemia	Brain tissue protection, reduction of neuro-motoric deficit	Isaev et al. (2012)
Cardiomyopathy, cardiac hypertrophy, and diffuse myocardial fibrosis	Lifelong SkQ1 treatment in mice retards senescence-associated myocardial disease (cardiomyopathy), cardiac hypertrophy, and diffuse myocardial fibrosis	Manskikh et al. (2015)
Spermatogenesis abnormalities	SkQ1 improved spermatogenesis, morphology of testicular seminiferous tubules, and significantly reduced the apoptosis of germ cells in old Immp2 l mutant mice	Jiang et al. (2017)
Systemic inflammatory response syndrome (SIRS)	SkQ1 down-regulated the expression of NFκB-regulated genes (VCAM1, ICAM1, MCP1, and IL-6), prevented the body temperature drop and lethality in mice injected with high doses of a SIRS inducer, tumor necrosis factor (TNF)	Zakharova et al. (2017)
Progeria induced by mitochondrial DNA-polymerase dysfunction	In mutator mice SkQ1 delayed the appearance of kyphosis, alopecia, lowering of body temperature, body weight loss, as well as ameliorated heart, kidney and liver pathologies	Shabalina et al. (2017)
Myocardial hypertrophy	SkQ1 prevented heart hypertrophy and mitochondrial oxidative stress in mice	Zhang et al. (2016)
Heart ischemia	Reduction of infarction area, reduction of arrhythmia symptoms in rats	Bakeeva et al. (2008)

Combined with toxicology and pharmacokinetics studies, these results allowed the start of Phase I (safety) study of systemic oral form of SkQ1 in healthy volunteers. This study was successfully completed in 2016 in Russia.

4 Conclusions

SkQ1 and other mitochondria-targeted plastoquinones are a promising and efficient means to alleviate several age-related pathologies. In case of dry eye syndrome and cataract it is already successfully implemented as a drug (eye drops); clinical trials of the form for oral administration are currently running.

Acknowledgements This work was supported by the Russian Science Foundation (Project No. 14-50-00029).

Conflict of Interest M.V.S. is the general director of Mitotech LLC, a biotech company which owns rights for compounds of SkQ type.

References

Andreev-Andrievskiy AA, Kolosova NG, Stefanova NA, Lovat MV, Egorov MV, Manskikh VN, Zinovkin RA, Galkin II, Prikhodko AS, Skulachev MV, Lukashev AN (2016) Efficacy of mitochondrial antioxidant plastoquinonyl-decyl-triphenylphosphonium bromide (SkQ1) in the rat model of autoimmune arthritis. Oxidative Med Cell Longev 2016:8703645

Antonenko YN, Avetisyan AV, Bakeeva LE, Chernyak BV, Chertkov VA, Domnina LV, Ivanova OY, Izyumov DS, Khailova LS, Klishin SS, Korshunova GA, Lyamzaev KG, Muntyan MS, Nepryakhina OK, Pashkovskaya AA, Pletjushkina OY, Pustovidko AV, Roginsky VA, Rokitskaya TI, Ruuge EK, Saprunova VB, Severina II, Simonyan RA, Skulachev IV, Skulachev MV, Sumbatyan NV, Sviryaeva IV, Tashlitsky VN, Vassiliev JM, Vyssokikh MY, Yaguzhinsky LS, Zamyatnin AA Jr, Skulachev VP (2008) Mitochondria-targeted plastoquinone derivatives as tools to interrupt execution of the aging program. 1. Cationic plastoquinone derivatives: synthesis and in vitro studies. Biochemistry (Mosc) 73:1273–1287

Bakeeva LE, Grinius LL, Jasaitis AA, Kuliene VV, Levitsky DO, Liberman EA, Severina II, Skulachev VP (1970) Conversion of biomembrane-produced energy into electric form. II. Intact mitochondria. Biochim Biophys Acta 216:13–21

Bakeeva LE, Barskov IV, Egorov MV, Isaev NK, Kapelko VI, Kazachenko AV, Kirpatovsky VI, Kozlovsky SV, Lakomkin VL, Levina SB, Pisarenko OI, Plotnikov EY, Saprunova VB, Serebryakova LI, Skulachev MV, Stelmashook EV, Studneva IM, Tskitishvili OV, Vasilyeva AK, Victorov IV, Zorov DB, Skulachev VP (2008) Mitochondria-targeted plastoquinone derivatives as tools to interrupt execution of the aging program. 2. Treatment of some ROS- and age-related diseases (heart arrhythmia, heart infarctions, kidney ischemia, and stroke). Biochemistry (Mosc) 73:1288–1299

Bakeeva LE, Eldarov CM, Vangely IM, Kolosova NG, Vays VB (2016) Mitochondria-targeted antioxidant SkQ1 reduces age-related alterations in the ultrastructure of the lacrimal gland. Oncotarget 7:80208–80222

Beatty S, Koh H, Phil M, Henson D, Boulton M (2000) The role of oxidative stress in the pathogenesis of age-related macular degeneration. Surv Ophthalmol 45:115–134

Brzheskiy VV, Efimova EL, Vorontsova TN, Alekseev VN, Gusarevich OG, Shaidurova KN, Ryabtseva AA, Andryukhina OM, Kamenskikh TG, Sumarokova ES, Miljudin ES, Egorov EA, Lebedev OI, Surov AV, Korol AR, Nasinnyk IO, Bezditko PA, Muzhychuk OP, Vygodin VA, Yani EV, Savchenko AY, Karger EM, Fedorkin ON, Mironov AN, Ostapenko V, Popeko NA, Skulachev VP, Skulachev MV (2015) Results of a multicenter, randomized, double-masked, placebo-controlled clinical study of the efficacy and safety of visomitin eye drops in patients with dry eye syndrome. Adv Ther 32:1263–1279

Erichev VP, Kozlova IV, Reshchikova VS, Alekseev VN, Levko MA, Zamyatnin AAJ, Gudkova EY, Kovaleva NA, Vygodin VA, Fedorkin ON, Ostapenko V, Senin II, Savchenko AY, Popeko NA, Skulachev VP, Skulachev MV (2016) Efficacy and safety of Visomitin® eye drops, in patients with age-related cataract: a randomized, double-blind, placebo-controlled clinical study. Nat.J.Glaucoma(Rus.) 1:61–69

Feniouk BA, Skulachev VP (2018) Studies on mitochondria directed plastoquinones. In: Oliveira PJ (ed) Mitochondrial biology and experimental therapeutics. Springer, New York

Galkin II, Pletjushkina OY, Zinovkin RA, Zakharova VV, Chernyak BV, Popova EN (2016) Mitochondria-targeted antioxidant SkQR1 reduces TNF-induced endothelial permeability in vitro. Biochemistry (Mosc) 81:1188–1197

Green DE (1974) The electromechanochemical model for energy coupling in mitochondria. Biochim Biophys Acta 346:27–78

Iomdina EN, Khoroshilova-Maslova IP, Robustova OV, Averina OA, Kovaleva NA, Aliev G, Reddy VP, Zamyatnin AA Jr, Skulachev MV, Senin II, Skulachev VP (2015) Mitochondria-targeted antioxidant SkQ1 reverses glaucomatous lesions in rabbits. Front Biosci (Landmark Ed) 20:892–901

Isaev NK, Novikova SV, Stelmashook EV, Barskov IV, Silachev DN, Khaspekov LG, Skulachev VP, Zorov DB (2012) Mitochondria-targeted plastoquinone antioxidant SkQR1 decreases trauma-induced neurological deficit in rat. Biochemistry (Mosc) 77:996–999

James AM, Cocheme HM, Smith RA, Murphy MP (2005) Interactions of mitochondria-targeted and untargeted ubiquinones with the mitochondrial respiratory chain and reactive oxygen species. Implications for the use of exogenous ubiquinones as therapies and experimental tools. J Biol Chem 280:21295–21312

Jiang Y, Liu C, Lei B, Xu X, Lu B (2017) Mitochondria-targeted antioxidant SkQ1 improves spermatogenesis in Immp2l mutant mice. Andrologia. https://doi.org/10.1111/and.12848

Kanda A, Chen W, Othman M, Branham KE, Brooks M, Khanna R, He S, Lyons R, Abecasis GR, Swaroop A (2007) A variant of mitochondrial protein LOC387715/ARMS2, not HTRA1, is strongly associated with age-related macular degeneration. Proc Natl Acad Sci U S A 104:16227–16232

Kapay NA, Isaev NK, Stelmashook EV, Popova OV, Zorov DB, Skrebitsky VG, Skulachev VP (2011) In vivo injected mitochondria-targeted plastoquinone antioxidant SkQR1 prevents beta-amyloid-induced decay of long-term potentiation in rat hippocampal slices. Biochemistry (Mosc) 76:1367–1370

Kapay NA, Popova OV, Isaev NK, Stelmashook EV, Kondratenko RV, Zorov DB, Skrebitsky VG, Skulachev VP (2013) Mitochondria-targeted plastoquinone antioxidant SkQ1 prevents amyloid-beta-induced impairment of long-term potentiation in rat hippocampal slices. J Alzheimers Dis 36:377–383

Kelso GF, Porteous CM, Coulter CV, Hughes G, Porteous WK, Ledgerwood EC, Smith RA, Murphy MP (2001) Selective targeting of a redox-active ubiquinone to mitochondria within cells: antioxidant and antiapoptotic properties. J Biol Chem 276:4588–4596

Kolosova NG, Shcheglova TV, Sergeeva SV, Loskutova LV (2006) Long-term antioxidant supplementation attenuates oxidative stress markers and cognitive deficits in senescent-accelerated OXYS rats. Neurobiol Aging 27:1289–1297

Kolosova NG, Tyumentsev MA, Muraleva NA, Kiseleva E, Vitovtov AO, Stefanova NA (2017) Antioxidant SkQ1 alleviates signs of Alzheimer's disease-like pathology in old OXYS rats by reversing mitochondrial deterioration. Curr Alzheimer Res 14(12):1283–1292

Liberman EA, Topaly VP, Tsofina LM, Jasaitis AA, Skulachev VP (1969) Mechanism of coupling of oxidative phosphorylation and the membrane potential of mitochondria. Nature 222:1076–1078

Manskikh VN, Gancharova OS, Nikiforova AI, Krasilshchikova MS, Shabalina IG, Egorov MV, Karger EM, Milanovsky GE, Galkin II, Skulachev VP, Zinovkin RA (2015) Age-associated murine cardiac lesions are attenuated by the mitochondria-targeted antioxidant SkQ1. Histol Histopathol 30:353–360

Markovets AM, Fursova AZ, Kolosova NG (2011) Therapeutic action of the mitochondria-targeted antioxidant SkQ1 on retinopathy in OXYS rats linked with improvement of VEGF and PEDF gene expression. PLoS One 6:e21682

Muraleva NA, Kozhevnikova OS, Zhdankina AA, Stefanova NA, Karamysheva TV, Fursova AZ, Kolosova NG (2014) The mitochondria-targeted antioxidant SkQ1 restores alphaB-crystallin expression and protects against AMD-like retinopathy in OXYS rats. Cell Cycle 13:3499–3505

Neroev VV, Archipova MM, Bakeeva LE, Fursova AZ, Grigorian EN, Grishanova AY, Iomdina EN, Ivashchenko ZN, Katargina LA, Khoroshilova-Maslova IP, Kilina OV, Kolosova NG, Kopenkin EP, Korshunov SS, Kovaleva NA, Novikova YP, Philippov PP, Pilipenko DI, Robustova OV, Saprunova VB, Senin II, Skulachev MV, Sotnikova LF, Stefanova NA, Tikhomirova NK, Tsapenko IV, Shchipanova AI, Zinovkin RA, Skulachev VP (2008) Mitochondria-targeted plastoquinone derivatives as tools to interrupt execution of the aging program. 4. Age-related eye disease. SkQ1 returns vision to blind animals. Biochemistry (Mosc) 73:1317–1328

Ousler GW, Gomes PJ, Welch D, Abelson MB (2005) Methodologies for the study of ocular surface disease. Ocul Surf 3:143–154

Pasyukova EG, Feniouk BA, Skulachev VP (2017) Mitochondria-targeted rechargeable antioxidants as potential anti-aging drugs. In: Vaiserman A (ed) Anti-aging drugs: from basic research to clinical practice. Royal Society of Chemistry, London

Perepechaeva ML, Grishanova AY, Rudnitskaya EA, Kolosova NG (2014) The mitochondria-targeted antioxidant SkQ1 downregulates aryl hydrocarbon receptor-dependent genes in the retina of OXYS rats with AMD-like retinopathy. J Ophthalmol 2014:530943

Petrov A, Perekhvatova N, Skulachev M, Stein L, Ousler G (2016) SkQ1 ophthalmic solution for dry eye treatment: results of a phase 2 safety and efficacy clinical study in the environment and during challenge in the controlled adverse environment model. Adv Ther 33:96–115

Plotnikov EY, Chupyrkina AA, Jankauskas SS, Pevzner IB, Silachev DN, Skulachev VP, Zorov DB (2011) Mechanisms of nephroprotective effect of mitochondria-targeted antioxidants under rhabdomyolysis and ischemia/reperfusion. Biochim Biophys Acta 1812:77–86

Plotnikov EY, Morosanova MA, Pevzner IB, Zorova LD, Manskikh VN, Pulkova NV, Galkina SI, Skulachev VP, Zorov DB (2013) Protective effect of mitochondria-targeted antioxidants in an acute bacterial infection. Proc Natl Acad Sci U S A 110:E3100–E3108

Rumyantseva YV, Ryabchikova EI, Fursova AZ, Kolosova NG (2015) Ameliorative effects of SkQ1 eye drops on cataractogenesis in senescence-accelerated OXYS rats. Graefes Arch Clin Exp Ophthalmol 253:237–248

Saprunova VB, Lelekova MA, Kolosova NG, Bakeeva LE (2012) SkQ1 slows development of age-dependent destructive processes in retina and vascular layer of eyes of wistar and OXYS rats. Biochemistry (Mosc) 77:648–658

Sergeeva S, Bagryanskaya E, Korbolina E, Kolosova N (2006) Development of behavioural dysfunctions in accelerated-senescence OXYS rats is associated with early postnatal alterations in brain phosphate metabolism. Exp Gerontol 41:141–150

Shabalina IG, Vyssokikh MY, Gibanova N, Csikasz RI, Edgar D, Hallden-Waldemarson A, Rozhdestvenskaya Z, Bakeeva LE, Vays VB, Pustovidko AV, Skulachev MV, Cannon B, Skulachev VP, Nedergaard J (2017) Improved health-span and lifespan in mtDNA mutator mice treated with the mitochondrially targeted antioxidant SkQ1. Aging (Albany NY) 9:315–339

Skulachev VP, Anisimov VN, Antonenko YN, Bakeeva LE, Chernyak BV, Erichev VP, Filenko OF, Kalinina NI, Kapelko VI, Kolosova NG, Kopnin BP, Korshunova GA, Lichinitser MR, Obukhova LA, Pasyukova EG, Pisarenko OI, Roginsky VA, Ruuge EK, Senin II, Severina II, Skulachev MV, Spivak IM, Tashlitsky VN, Tkachuk VA, Vyssokikh MY, Yaguzhinsky LS, Zorov DB (2009) An attempt to prevent senescence: a mitochondrial approach. Biochim Biophys Acta 1787:437–461

Skulachev MV, Antonenko YN, Anisimov VN, Chernyak BV, Cherepanov DA, Chistyakov VA, Egorov MV, Kolosova NG, Korshunova GA, Lyamzaev KG, Plotnikov EY, Roginsky VA, Savchenko AY, Severina II, Severin FF, Shkurat TP, Tashlitsky VN, Shidlovsky KM, Vyssokikh

MY, Zamyatnin AA Jr, Zorov DB, Skulachev VP (2011) Mitochondrial-targeted plastoquinone derivatives. Effect on senescence and acute age-related pathologies. Curr Drug Targets 12:800–826

Stefanova NA, Muraleva NA, Maksimova KY, Rudnitskaya EA, Kiseleva E, Telegina DV, Kolosova NG (2016) An antioxidant specifically targeting mitochondria delays progression of Alzheimer's disease-like pathology. Aging (Albany NY) 8:2713–2733

Yanshole LV, Yanshole VV, Snytnikova OA, Fursova A, Kolosova NG, Tsentalovich YP, Sagdeev RZ (2015) Effect of SkQ1 eye drops on the rat lens metabolomic composition and the chaperone activity of alpha-crystallin. Dokl Biochem Biophys 464:341–345

Zakharova VV, Pletjushkina OY, Galkin II, Zinovkin RA, Chernyak BV, Krysko DV, Bachert C, Krysko O, Skulachev VP, Popova EN (2017) Low concentration of uncouplers of oxidative phosphorylation decreases the TNF-induced endothelial permeability and lethality in mice. Biochim Biophys Acta 1863:968–977

Zhang YB, Meng YH, Chang S, Zhang RY, Shi C (2016) High fructose causes cardiac hypertrophy via mitochondrial signaling pathway. Am J Transl Res 8:4869–4880

Zinovkin RA, Romaschenko VP, Galkin II, Zakharova VV, Pletjushkina OY, Chernyak BV, Popova EN (2014) Role of mitochondrial reactive oxygen species in age-related inflammatory activation of endothelium. Aging (Albany NY) 6:661–674

Part III
COMIC: Mitochondrial Folies

Mitochondrial Follies: A Short Journey in Life and Energy

Anabela Marisa Azul, João Ramalho-Santos, Paulo Jorge Oliveira, and Rui Tavares

Abstract Communicating through comics can be an effective multi-layered strategy. The narrative of this comic transports the reader to a realistic, if abbreviated, view of mitochondrial biology blended with the changing daily routines of Lara. Lara is a fictionalized patient with an undisclosed metabolic disorder, who takes her health into her own hands, becoming interested in metabolism and mitochondria to better understand the processes by which living organisms convert food into energy at the cellular level. Moreover, her goal is also to communicate this fascinating world to friends and colleagues, in a way that may also be useful for scientists and the general public. The comic combines the discoveries of real individuals who have greatly contributed to the knowledge of mitochondria, namely Peter Mitchell, Hans Adolf Krebs and Lynn Margulis, with fictional characters, such as Lara and George, directly associated to the narrative. The interdisciplinary nature of graphic narration reflects the blending of text and scientific facts alongside visual information (both realistic and caricature-like) and critical-thinking-based dialogues and actions, that was enriched by the collaborative work between the researchers who developed the concept and the illustrator who brought it to life.

Keywords Mitochondria · Metabolism · Comic · Disease · Science communication

---

A. M. Azul · J. Ramalho-Santos · P. J. Oliveira (✉) · R. Tavares
CNC - Center for Neuroscience and Cell Biology, University of Coimbra, Coimbra, Portugal
e-mail: pauloliv@cnc.uc.pt

P. J. Oliveira (ed.), *Mitochondrial Biology and Experimental Therapeutics*,
https://doi.org/10.1007/978-3-319-73344-9_29

Mitochondrial Follies; A Short Journey in Life and Energy

[illegible]

Abstract [illegible]

[illegible]

[illegible]

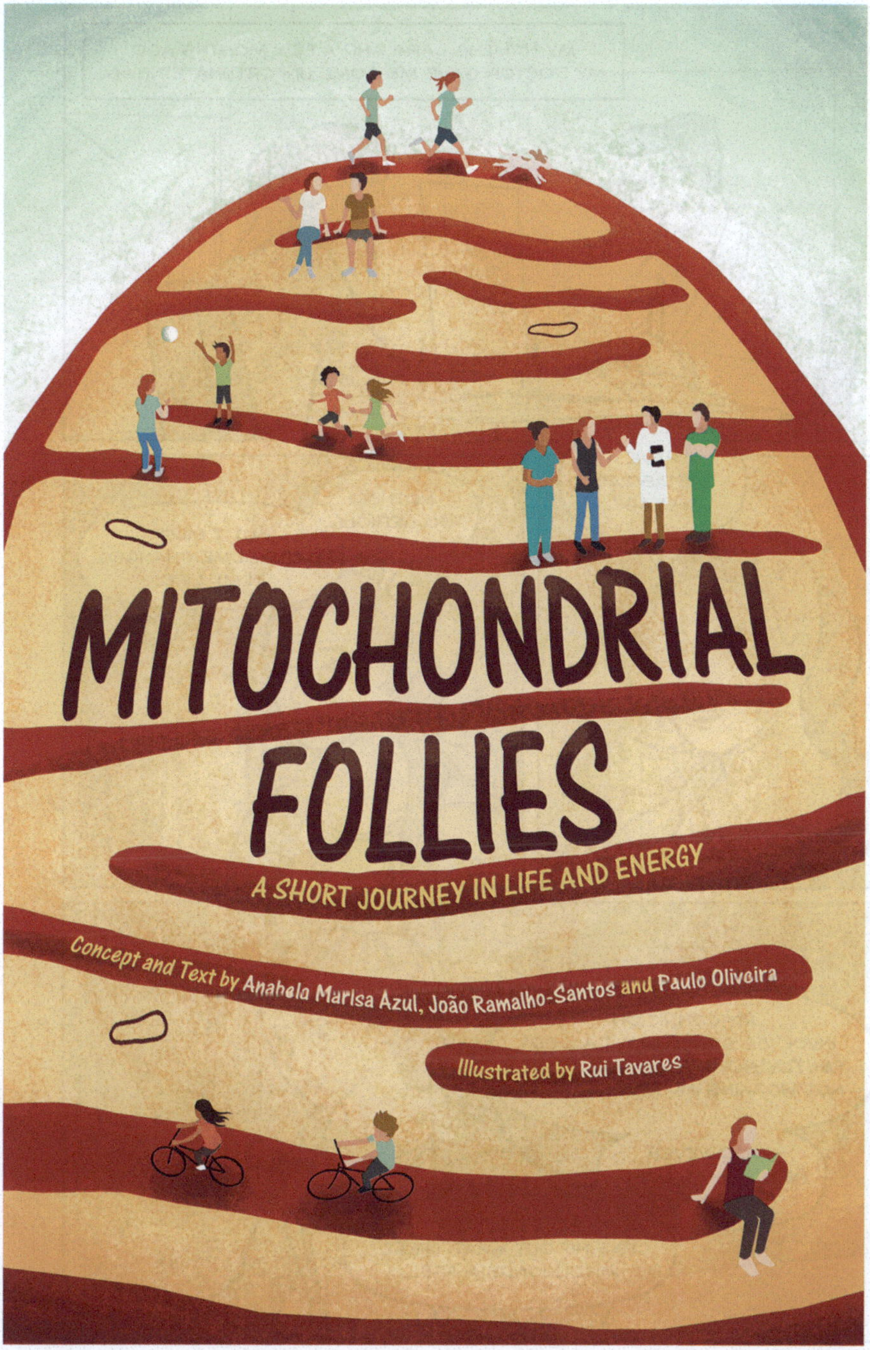
MITOCHONDRIAL FOLLIES
A SHORT JOURNEY IN LIFE AND ENERGY
Concept and Text by Anabela Marisa Azul, João Ramalho-Santos and Paulo Oliveira
Illustrated by Rui Tavares

MY NAME IS LARA AND A FEW MONTHS AGO MY DOCTOR GAVE ME SOME UNFORTUNATE NEWS.
I HAVE A METABOLIC DISEASE.
NOTHING TOO SERIOUS.
BUT I MUST MAKE SOME CHANGES IN MY LIFE.
MY CONDITION INVOLVES SOMETHING CALLED "*MITOCHONDRIA*", WHICH ARE NOT WORKING AS THEY SHOULD.
I HAD NEVER HEARD THAT WORD BEFORE.
SO I DECIDED TO FIND OUT WHAT THESE MITOCHONDRIA ARE...

IT'S FASCINATING, REALLY!
EXERCISING, PLAYING, READING, RESTING, GROWING UP...
...ALL THESE THINGS RUN ON ENERGY.

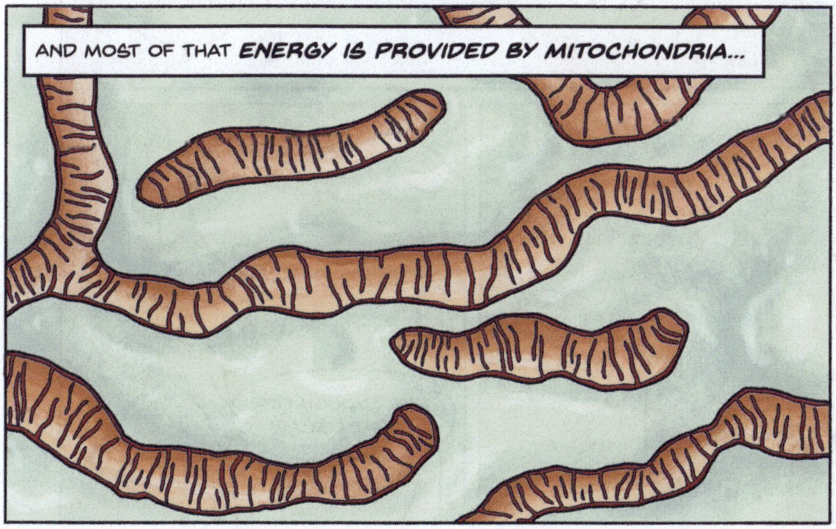
AND MOST OF THAT ENERGY IS PROVIDED BY MITOCHONDRIA...

...WHICH ARE STRUCTURES INSIDE OUR CELLS...
...THAT CAN HAVE A ROUND OR FILAMENTOUS SHAPE.
EXERCISING IMPROVES MITOCHONDRIA PERFORMANCE.
WOF! WOF!
IT'S ENOUGH FOR TODAY, OSCAR!

NOT ALL CELLS HAVE THE SAME NUMBER OF MITOCHONDRIA.
VRRR
YES, I'LL BE THERE NEXT WEDNESDAY.
GREAT! SEE YOU THEN.
"CELLS THAT NEED MORE ENERGY HAVE MORE MITOCHONDRIA.
A LIVER CELL MAY HAVE HUNDREDS OF MITOCHONDRIA.
MUSCLE CELLS MAY HAVE THOUSANDS.
ONLY RED BLOOD CELLS HAVE NO MITOCHONDRIA."

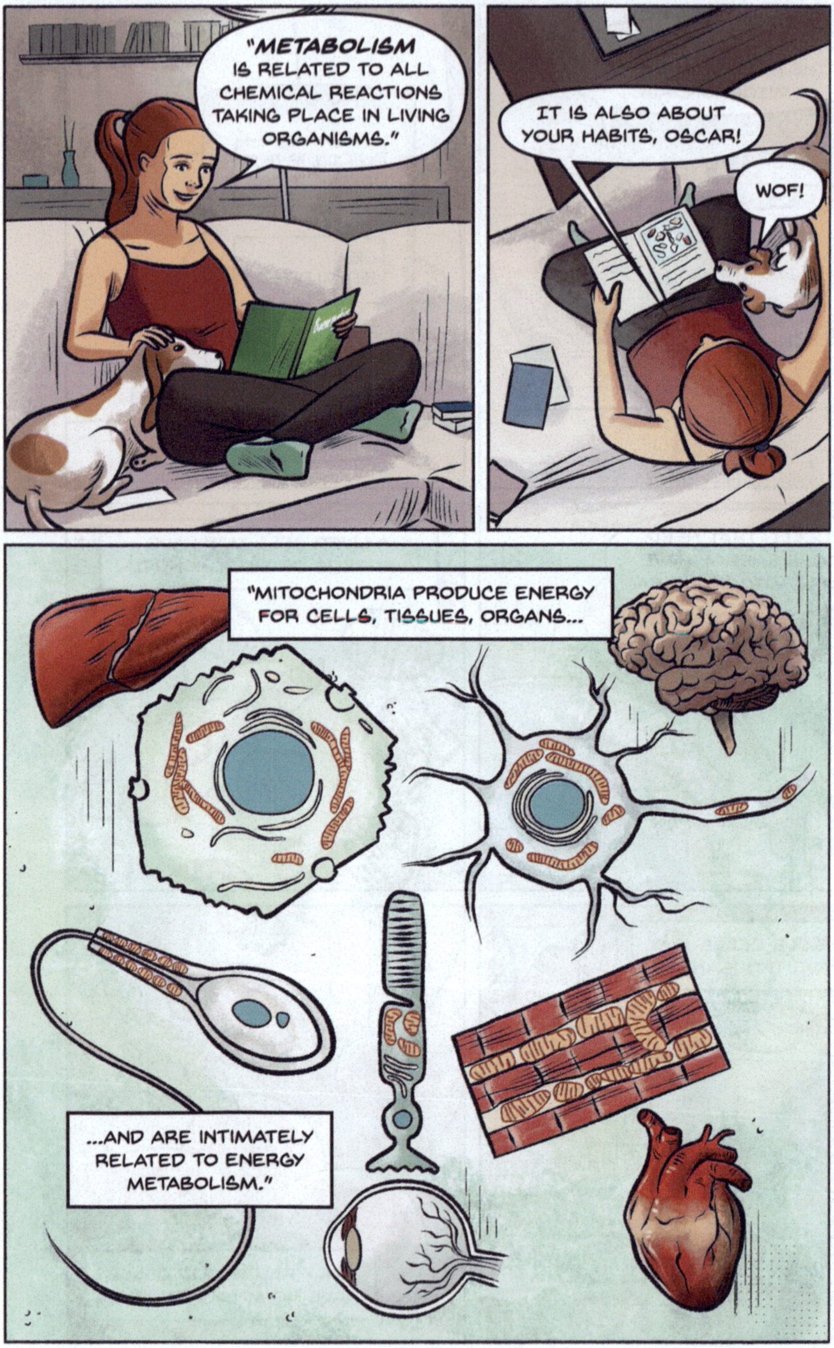
"METABOLISM IS RELATED TO ALL CHEMICAL REACTIONS TAKING PLACE IN LIVING ORGANISMS."
IT IS ALSO ABOUT YOUR HABITS, OSCAR!
WOF!
"MITOCHONDRIA PRODUCE ENERGY FOR CELLS, TISSUES, ORGANS...
...AND ARE INTIMATELY RELATED TO ENERGY METABOLISM."

ON WEDNESDAY...
IT'S NOT EASY CATCHING UP WITH YOU, LATELY!
SORRY GEORGE! I'VE BEING FASCINATED BY MITOCHONDRIA AND METABOLISM.

HMMM... LET'S PLAY?

SURE!
BUT WE CAN ONLY DO IT THANKS TO THE POWERHOUSE OF OUR CELLS: MITOCHONDRIA!

MITOCHONDRIA CONVERT FOOD INTO ENERGY.

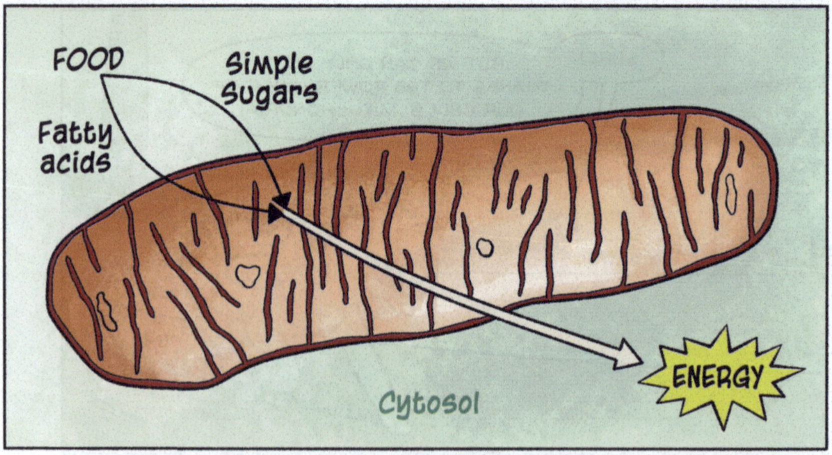
FOOD
Simple Sugars
Fatty acids
ENERGY
Cytosol

BUT ALSO PRODUCE COMPONENTS FOR MORE COMPLEX MOLECULES.
FOR EXAMPLE, FATTY ACIDS FROM FAT DIGESTION CONTINUE THEIR BREAKDOWN INSIDE MITOCHONDRIA.
THE CHEMICAL PATHWAY CALLED BETA-OXIDATION GENERATES MOLECULES, NAMELY COENZYMES, THAT HELP IN THE METABOLISM, SUCH AS ACETYL-COA, NADH, FADH2.
Fatty acids
CO.COA
BETA OXIDATION
acetylCOA
ACETYLCOENZYME A
NADH RIBOSYLNICOTINAMIDE 5'-DIPHOSPHATE COUPLED TO ADENOSINE 5'-PHOSPHATE BY PYROPHOSPHATE LINKAGE
$FADH_2$ REDUCED FLAVIN ADENINE DINUCLEOTIDE
H_2O WATER
WHAT?!
WHERE DID YOU LEARN THOSE WEIRD NAMES?
THE POINT IS TO BETTER UNDERSTAND OUR BODY AT THE CELL LEVEL.

ALSO SIMPLE SUGARS FROM SUGAR DIGESTION CONTINUE THEIR BREAKDOWN INSIDE MITOCHONDRIA
THROUGH THE CHEMICAL PATHWAY CALLED ***TRICARBOXYLIC ACID*** (TCA) CYCLE, OR KREBS CYCLE.
IT STARTS WITH PYRUVATE THAT IS CONVERTED TO ACETYL-COA, NEXT TO CITRATE, WHICH THEN ENTERS INTO A CYCLE THAT COMES BACK TO CITRATE. TCA CYCLE ALSO PROVIDES THE CO-ENZYME NADH.
Pyruvate
acetylCOA
Succinate
(via $FADH_2$)
TCA
cycle
Citrate
NADH
Sugar
Fatty Acids
Pyruvate
Beta Oxidation
AcetylCOA
Citrate
GERMAN-BORN BRITISH ***HANS ADOLF KREBS*** WON THE 1953 NOBEL PRIZE IN PHYSIOLOGY OR MEDICINE...
...FOR HIS PIONEERING EFFORTS IN CELLULAR RESPIRATION PATHWAYS TO PRODUCE ENERGY.

IF YOU SAY SO...
BUT HOW IS THIS RELATED WITH ENERGY PRODUCTION?
BEFORE WE GET INTO THAT, LET ME INTRODUCE A MOLECULE THAT STORES CHEMICAL ENERGY WITHIN CELLS: *ADENOSINE TRIPHOSPHATE (ATP).*
ATP
P
ENERGY
ADP
ATP IS PRODUCED THROUGH THREE MAIN CHEMICAL PATHWAYS, TWO OCCURRING IN MITOCHONDRIA.
FIRST, IN THE CYTOSOL, *GLYCOLYSIS* CONVERTS SIMPLE SUGARS (FRUCTOSE, GLUCOSE, LACTOSE) TO 2 MOLECULES OF PYRUVATE AND PRODUCES 2 ATP MOLECULES.
Simple Sugars
GLYCOLYSIS
2 Pyruvate
2 ATP

Pyruvate
acetylCOA
2 ATP
TCA cycle
2x
THE PYRUVATE MOLECULE ENTERS INTO A MITOCHONDRION TO INITIATE THE SECOND PATHWAY, THE TCA CYCLE. AND 1 ATP MOLECULE RESULTS FROM EACH CYCLE.
THAT CYCLE THAT GOES AROUND AND AROUND?
YES!

THEN COMES THE ELECTRON TRANSPORT CHAIN (ETC).
H+
ATP Synthase
Cytochrome C
Coenzime Q
e
I
II
III
IV
Succinate
TCA cycle
NADH
NADH
NADH AND SUCCINATE (VIA FADH2), BOTH PRODUCED BY THE TCA, ARE USED TO FUEL A METABOIC PATHWAY CALLED OXIDATIVE PHOSPHORYLATION (OXPHOS) WHICH PRODUCES ATP.
ELECTRONS (e−) FLOW FROM COMPLEX I TO COMPLEX IV...
...AND THE ENERGY RELEASED IS COUPLED TO THE PASSAGE OF PROTONS ACROSS THE INNER MEMBRANE.
BESIDES NADH, THE TCA ALSO PRODUCES SUCCINATE WHICH IS THEN OXIDIZED IN COMPLEX II.
THIS LEADS TO A DIFFERENCE IN H+ CONCENTRATION ACROSS THE INNER MEMBRANE, A PROTON GRADIENT.
THAT SOUNDS VERY COMPLICATED...

NOT REALLY.
BRITISH PETER MITCHELL WAS AWARDED THE NOBEL PRIZE FOR CHEMISTRY IN 1978...

...FOR PROPOSING A CHEMIOSMOTIC MECHANISM FOR THE GENERATION OF ENERGY THROUGH ELECTRON TRANSFER...
H+
...COUPLED TO THE TRANSLOCATION OF H+ BY SOME OF THE ETC PROTEINS.

ETC
THE PROTON GRADIENT IS USED TO DRIVE THE SYNTHESIS OF ATP, FORMED FROM ADP (ADENOSINE DIPHOSPHATE) AND PI (INORGANIC PHOSPHATE) BY THE ATP-SYNTHASE.
ATP Synthase
H^+
O_2
e^-
H_2O
THE FINAL ACCEPTOR OF THE e- IN THE ETC IS OXYGEN (O_2) THAT IS CONVERTED INTO WATER (H_2O).
ADP + P_i
34 ATP
THE H+ GRADIENT THAT IS BUILT BY THE ETC ACTS AS AN ENERGY RESERVOIR
JUST AS THE DIFFERENCE IN WATER LEVELS IS USED BY THE DAM TO SPIN THE TURBINE AND GENERATE ELECTRICITY.
I GIVE UP, FOR TODAY...

WHAT ARE YOUR PLANS NOW?
INVITE YOU TO A PICNIC!
SO, DID YOU HAVE A NICE TIME ON YOUR HOLIDAYS?
WELL, I DIDN'T STUDY AS MUCH AS YOU!

AHA!
WHAT DO WE HAVE HERE?
A PIECE OF PAPER TO DRAW A MITOCHONDRION!
Symbiotic Planet
IT'S BEEN A LONG TIME SINCE I'VE EATEN BLACKBERRIES.
THEY REMIND ME OF SUMMER HOLIDAYS WITH MY GRANDPARENTS AND COUSINS.
SO, OUR CELLS HAVE A DYNAMIC MEMBRANE SYSTEM
Organelles
Nucleus
Phospholipid Bilayer

MITOCHONDRIA HAVE 2 MEMBRANES, THE INNER CONNECTING TO TUBULAR STRUCTURES CALLED CRISTAE.
IF YOU SEE WHAT I MEAN.
INSIDE, THERE IS A GEL-LIKE MATRIX
WITH HUNDREDS OF PROTEINS AND THE MITOCHONDRIAL GENOMES.
ATP IS THEN EXPORTED THROUGH MITOCHONDRIAL MEMBRANES
(TRANSPORTERS LOCATED IN THE MITOCHONDRIAL MEMBRANES).
Outer Membrane
Inner Membrane
Matrix
Cristae
DNA
ATP
3 Metabolic Pathways
GLYCOLYSIS
2ATP
TCA
2ATP
ETC
34ATP
MITOCHONDRIA USES CHEMICAL ENERGY FOUND IN FOOD TO RELEASE ATP DURING CELLULAR RESPIRATION.
THAT'S WHY MITOCHONDRIA ARE CALLED THE POWERHOUSE OF OUR CELLS? I GET IT NOW!

LARA, THESE EGGS WITH MUSHROOMS ARE A CROWD-PLEASER!
BUT, THAT'S NOT THE END...
WITH YOU THERE IS ALWAYS A "BUT"...
Metabolism For Laypeople
DURING OXIDATIVE PHOSPHORYLATION, SOME ELECTRONS (e-) CAN BE DEVIATED FROM THE NORMAL FLUX
TO
PARTIALLY REDUCE OXYGEN, PRODUCING ***REACTIVE OXYGEN SPECIES (ROS).***
WAIT... CAN YOU EXPLAIN THAT SLOWLY?
O_2
e^-
$O_2^{\cdot -}$
ROS
H_2O_2
LOOK. ROS ARE OXYGEN DERIVATIVES, WHICH CAN BE VERY REACTIVE. SOME OF THEM ARE CALLED FREE RADICALS
INCLUDING SUPEROXIDE ANION OR THE HYDROXYL RADICAL CAN DAMAGE OR KILL THE CELL...
OXIDATION IS JUST WHAT HAPPENS WHEN YOU CUT AN APPLE AND BROWN COLOR APPEARS.

OUR BODY HAS A CELLULAR **ANTIOXIDANT DEFENSE SYSTEM** IN MITOCHONDRIA.
BUT NOT ALL IS BAD WITH ROS...
AT LOW TO MODERATE CONCENTRATIONS, ROS HAVE PHYSIOLOGICAL ROLES...
ROS
...NAMELY BY ACTIVATING THE EXPRESSION OF CERTAIN GENES.
MITOCHONDRIA HAVE ANTIOXIDANT PROTEINS TO CONTROL ROS LEVELS.
SUPEROXIDE DISMUTASE (SOD2) CONVERTS SUPEROXIDE ANION INTO HYDROGEN PEROXIDE (H2O2) AND GLUTATHIONE PEROXIDASE (GPX1), WHICH MAINTAINS LOW H2O2 LEVELS.
e^-
O_2
$O_2^{\cdot -}$
ROS
H_2O_2
SOD2
GPX1
WHEN ROS OVERWHELM THE CELLULAR ANTIOXIDANT DEFENCE SYSTEM, **OXIDATIVE STRESS** OCCURS.
DIET INFLUENCES THE PRODUCTION OF ROS.

OH, I FORGOT! I HAVE SOME CRISPY COOKIES AND JUICE.
THANKS. I SHOULD BE CAUTIOUS WITH PASTRIES, SUGARY DRINKS, AND SALTY FOODS...
ALL CAN TRIGGER INCREASED ROS FORMATION.
HIGH ROS AND FREE RADICALS ARE MAJOR CAUSES OF MITOCHONDRIA DYSFUNCTION, WHICH IN TURN IS ASSOCIATED WITH METABOLIC SYNDROME, TYPE 2 DIABETES, VASCULAR INFLAMMATION, ATHEROSCLEROSIS, AND RENAL, HEART AND LIVER DISEASE.
Anti Oxidants
ROS
FINE-TUNED CONTROL OF OXIDANTS AND ANTIOXIDANTS ARE KEY FOR CELLULAR HOMEOSTASIS.
AIR POLLUTANTS, HEAVY METALS, CIGARETTE SMOKE, PESTICIDES AND INDUSTRIAL CHEMICALS ALSO CAUSE OXIDATIVE STRESS.
ANTIOXIDANTS ARE PRESENT IN FRUITS AND VEGETABLES.

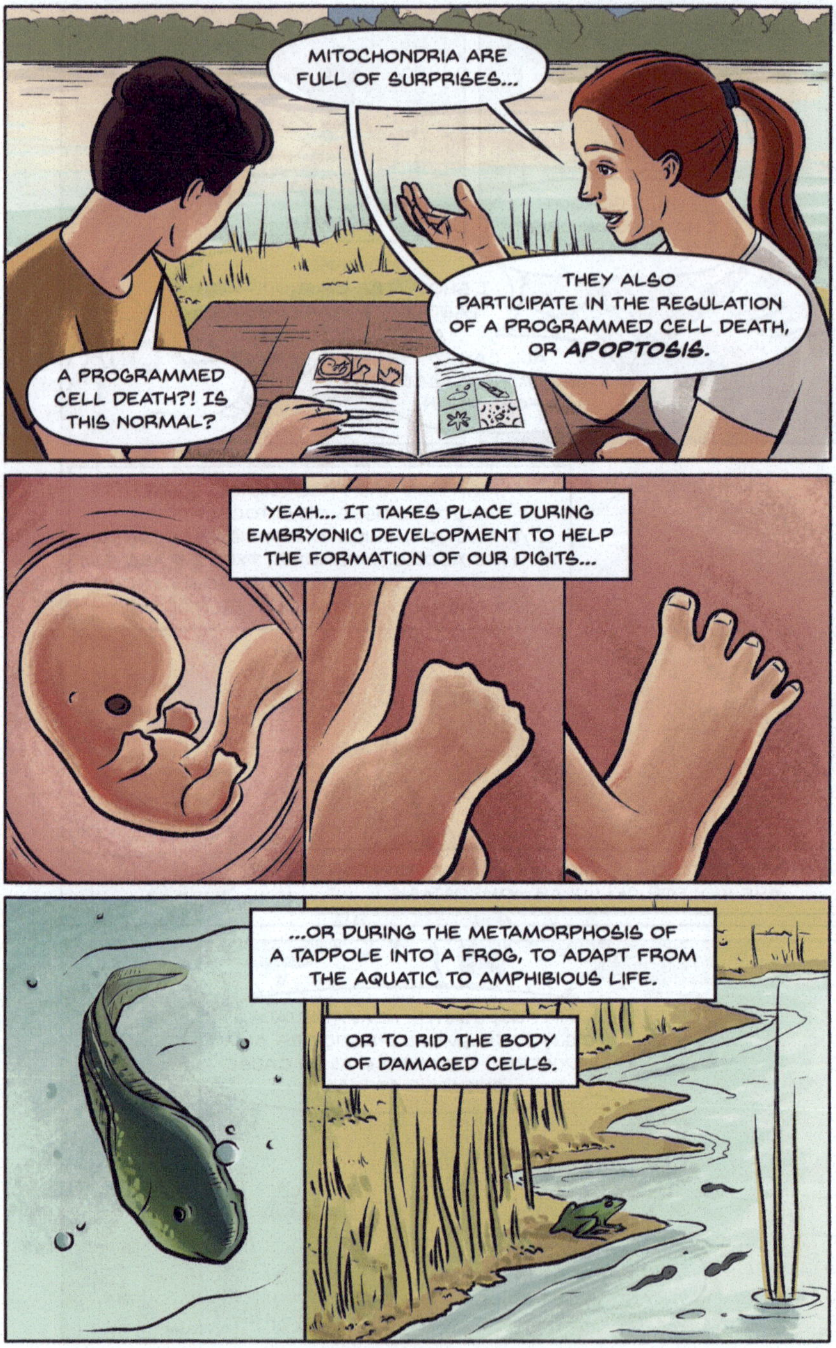
MITOCHONDRIA ARE FULL OF SURPRISES...
THEY ALSO PARTICIPATE IN THE REGULATION OF A PROGRAMMED CELL DEATH, OR APOPTOSIS.
A PROGRAMMED CELL DEATH?! IS THIS NORMAL?
YEAH... IT TAKES PLACE DURING EMBRYONIC DEVELOPMENT TO HELP THE FORMATION OF OUR DIGITS...
...OR DURING THE METAMORPHOSIS OF A TADPOLE INTO A FROG, TO ADAPT FROM THE AQUATIC TO AMPHIBIOUS LIFE.
OR TO RID THE BODY OF DAMAGED CELLS.

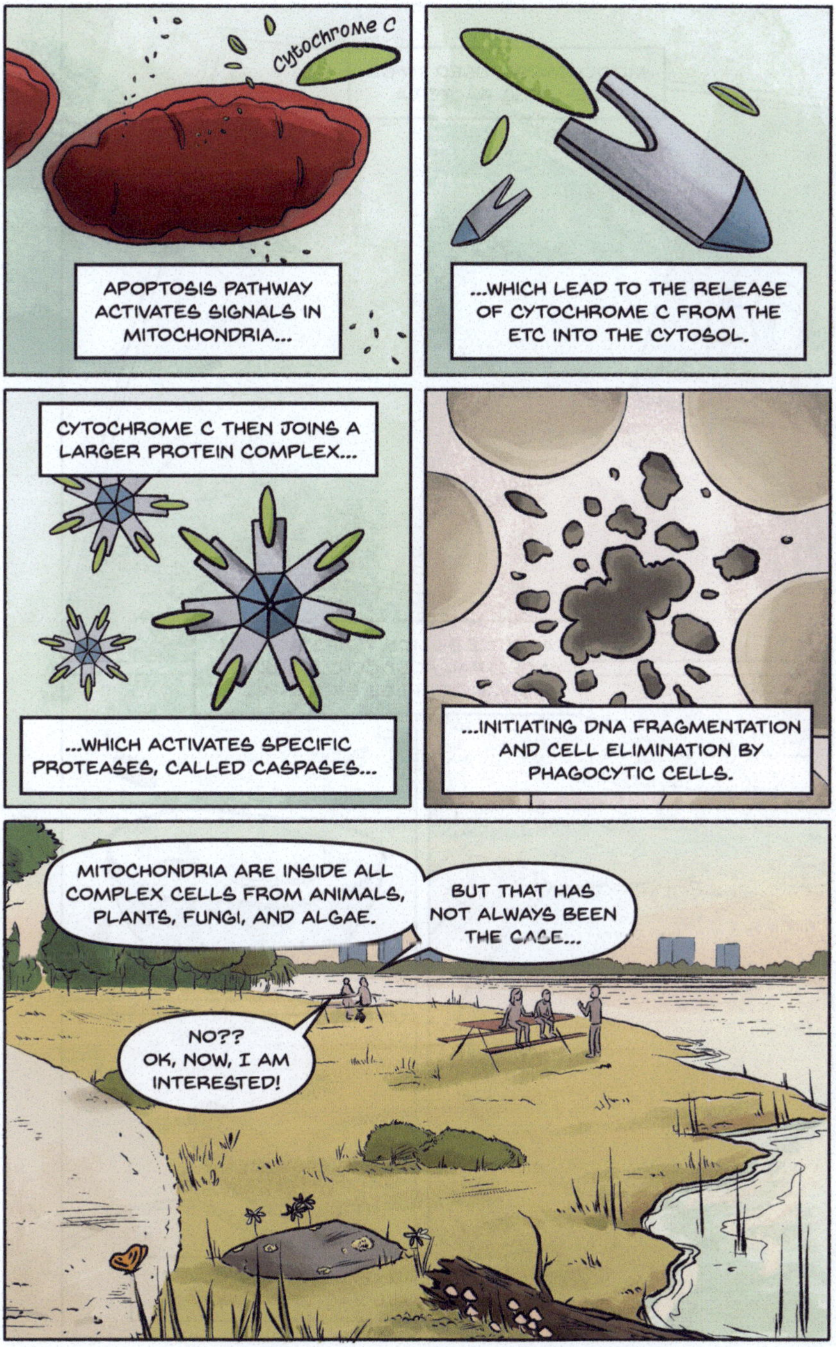
Cytochrome C
APOPTOSIS PATHWAY ACTIVATES SIGNALS IN MITOCHONDRIA...
...WHICH LEAD TO THE RELEASE OF CYTOCHROME C FROM THE ETC INTO THE CYTOSOL.
CYTOCHROME C THEN JOINS A LARGER PROTEIN COMPLEX...
...WHICH ACTIVATES SPECIFIC PROTEASES, CALLED CASPASES...
...INITIATING DNA FRAGMENTATION AND CELL ELIMINATION BY PHAGOCYTIC CELLS.
MITOCHONDRIA ARE INSIDE ALL COMPLEX CELLS FROM ANIMALS, PLANTS, FUNGI, AND ALGAE.
BUT THAT HAS NOT ALWAYS BEEN THE CASE...
NO?? OK, NOW, I AM INTERESTED!

MITOCHONDRIA USED TO BE FREE-LIVING BACTERIA.
Ancient Symbiosis
ABOUT 2 BILLION YEARS AGO, AN ANCESTRAL HOST CELL ENGULFED SOME OF THESE BACTERIA...
INSTEAD OF BEING DESTROYED, THE SMALL BACTERIA SURVIVED...

...AND THE SUBSEQUENT EVOLUTIONARY EVENTS CREATED WHAT WE NOW CONSIDER EUKARYOTIC CELLS, OUR CELLS.
ANIMAL
AMERICAN LYNN MARGULIS (1967) PROPOSED THAT MITOCHONDRIA AND CHLOROPLASTS DERIVED FROM FREE-LIVING BACTERIA...
...AND THAT EUKARYOTIC CELLS ARE THE RESULT OF THE EVOLUTION OF ANCIENT ENDOSYMBIOSES.
PLANT
SHE DEVELOPED AN IDEA INITIALLY PROPOSED BY RUSSIAN KONSTANTIN MERESCHKOWSKI.

MITOCHONDRIA MAINTAINED CERTAIN BACTERIAL CHARACTERISTICS
SUCH AS ONE OR MORE CIRCULAR ***MITOCHONDRIAL DNA*** (MtDNA) MOLECULES, SEVERAL COMMON PROTEINS AND THE CAPACITY FOR INDEPENDENT REPLICATION.
NOWADAYS, MITOCHONDRIAL PROTEINS ARE DERIVED EITHER FROM GENES ENCODED BY THE NUCLEAR GENOME, OR BY THE MtDNA, OR BY BOTH...
MITOCHONDRIA IMPORT MORE THAN 1,000 DIFFERENT PROTEINS FROM THE CYTOSOL, SOME OF WHICH PARTICIPATE IN THE ARCHITECTURE OF ELECTRON TRANSPORT CHAIN COMPLEXES THAT CARRY OUT THE CELULAR RESPIRATION.

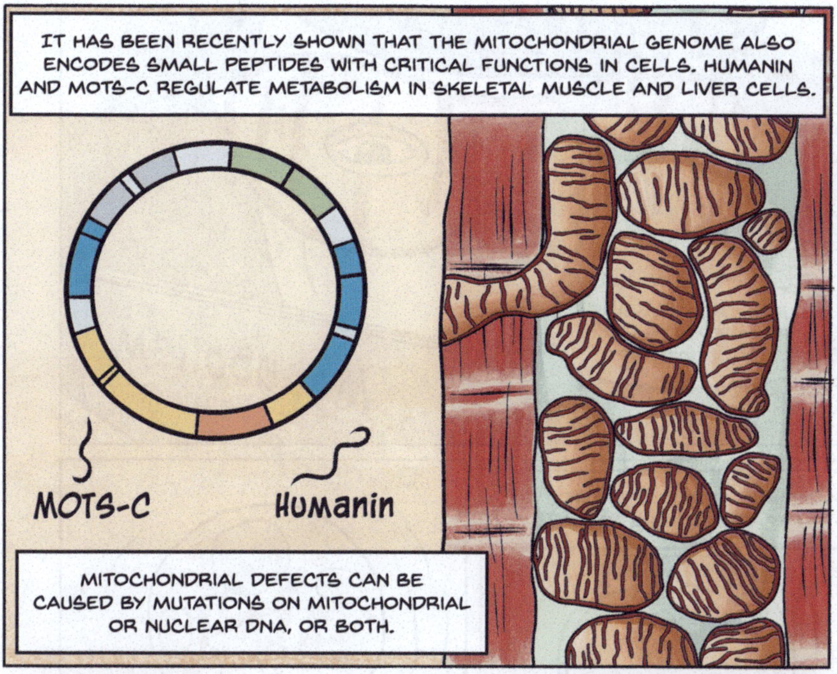
IT HAS BEEN RECENTLY SHOWN THAT THE MITOCHONDRIAL GENOME ALSO ENCODES SMALL PEPTIDES WITH CRITICAL FUNCTIONS IN CELLS. HUMANIN AND MOTS-C REGULATE METABOLISM IN SKELETAL MUSCLE AND LIVER CELLS.
MOTS-C
Humanin
MITOCHONDRIAL DEFECTS CAN BE CAUSED BY MUTATIONS ON MITOCHONDRIAL OR NUCLEAR DNA, OR BOTH.

AND THINGS FAIL SOMETIMES... CHILDREN CAN BE BORN WITH MITOCHONDRIAL DNA DEFECTS.
WICH IS A MAJOR PROBLEM, BECAUSE THEIR CELLS DON'T HAVE ENOUGH ENERGY TO FUNCTION PROPERLY.
THAT'S TERRIBLE...
YES, ESPECIALLY BECAUSE THERE IS NO CURE YET FOR THOSE MITOCHONDRIAL DNA DISEASES.

MITOCHONDRIA ARE RELATED WITH HEAT PRODUCTION?
Metabolism
for Laypeople
"MAMMALS HAVE DIFFERENT TWO TYPES OF ADIPOSE TISSUE...
THE WHITE ADIPOSE TISSUE STORES ENERGY AS TRIGLYCERIDES...
... AND THE BROWN TISSUE GENERATES HEAT, MAINTAINS BODY TEMPERATURE AND REGULATES ENERGY BALANCE."
THE BROWNISH COLOUR IS DUE TO THE HIGH NUMBER OF MITOCHONDRIA."
H+
UCP1
H+
H+
"UNCOUPLING PROTEIN 1 (UCP1), OR THERMOGENIN, IN THE INNER MEMBRANE OF MITOCHONDRIA, DISSIPATES THE PROTON ENERGY GRADIENT AS HEAT, RATHER THAN PRODUCING ATP, IN BROWN FAT TISSUE.

"THE THERMOGENIN PROTEIN CONTRIBUTES TO INCREASE BODY TEMPERATURE AT THE END OF THE HIBERNATION PERIOD IN SOME ANIMALS."
UCP1
BROWN ADIPOSE TISSUE IS ALSO IMPORTANT TO PROTECT NEWBORNS FROM COLD EXPOSURE.

I GUESS YOU'RE BECOMING FASCINATED BY MITOCHONDRIAL STORIES TOO.

CAROB CAKE... IS IT GOOD FOR MITOCHONDRIA?
IT'S A CAKE WITHOUT SUGAR. YES, WE MAY SAY IT'S MITOCHONDRIA-FRIENDLY...

AND THERE IS MORE...SEE HERE?

IT SAYS THAT OUR LITTLE FRIENDS ALSO REGULATE CALCIUM LEVELS IN OUR CELLS!
THEY HAVE DIFFERENT CHANNELS FOR CALCIUM (CA2+) TO ENTER AND EXIT. CALCIUM IS IMPORTANT TO REGULATE MANY SIGNALING EVENTS.
AND IT IS AMAZING HOW DIFFERENT CELL ORGANELLES TALK WITH EACH OTHER WITH CALCIUM WAVES!
Ca2+
Ca2+
AND I THOUGHT CALCIUM WAS GOOD ONLY FOR OUR BONES!

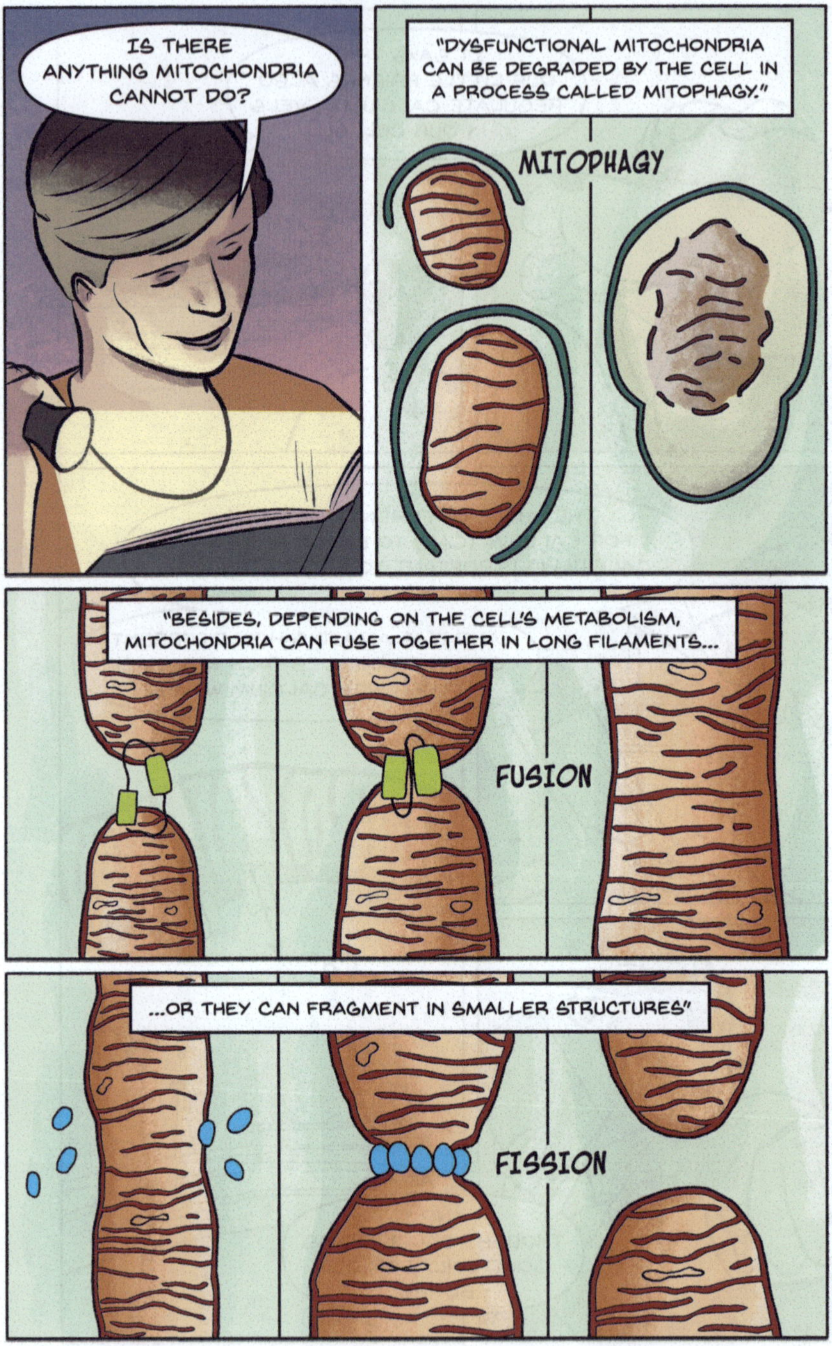
IS THERE ANYTHING MITOCHONDRIA CANNOT DO?
"DYSFUNCTIONAL MITOCHONDRIA CAN BE DEGRADED BY THE CELL IN A PROCESS CALLED MITOPHAGY."
MITOPHAGY
"BESIDES, DEPENDING ON THE CELL'S METABOLISM, MITOCHONDRIA CAN FUSE TOGETHER IN LONG FILAMENTS...
FUSION
...OR THEY CAN FRAGMENT IN SMALLER STRUCTURES"
FISSION

"MITOCHONDRIA CAN BE A VERY DYNAMIC ORGANELLE IN SOME CELLS, THEY OFTEN MOVE AROUND THE CELL THROUGH THE CYTOSKELETAL ELEMENTS."
WHAT'S THIS, LARA?
A MURAL REPRESENTING MITOCHONDRIA TRAVELING!
THEY CAN TRAVEL LONG DISTANCES, SUCH AS IN NEURONS.

TOMORROW, I HAVE MY ROUTINE MEDICAL CHECKUP.
IT´S SO PEACEFUL HERE AT SUNSET!
WHAT A JOURNEY TODAY!

THE NEXT DAY.
VRRR
MEDICAL CHECKUP
3:30 PM

DÉJÀ VU...

METABOLIC RUN
ROOM 2
WORK SESSIONS
VRRR
METABOLIC SYNDROME
BLOOD PRESSURE
FAT AROUND THE WAIST
TRIGLYCERIDE LEVELS
BLOOD SUGAR
LEVELS OF GOOD CHOLESTEROL (HDL)
HOW WAS YOUR VISIT?
IT WILL BE IN 10-15 MINUTES
HI, LARA!
HI, FRED! HI, TONY! NICE DRAWINGS!

YOU ARE DOING MUCH BETTER!
VRRR
CONGRATULATIONS ON THE PAPER, ANDREW!
THANK YOU!
MAYBE WE CAN THINK OF A CLINICAL TRIAL TO TEST THIS NEW TREATMENT FOR NON-ALCOHOLIC FATTY LIVER DISEASE?
Mitochondrial Network
YES. IT'S A GOOD OPPORTUNITY TO INVOLVE ACADEMICS, HEALTHCARE PROVIDERS, PATIENTS AND THE GENERAL PUBLIC.

BESIDES LIFESTYLE CHANGES MITOCHONDRIAL RESEARCH ALSO FOCUSES ON THE IDENTIFICATION OF COMPOUNDS AND SMALL NUTRITIONAL BIOACTIVE INGREDIENTS THAT DIRECTLY TARGET MITOCHONDRIA.
SEE YOU NEXT MONTH!
FATTY LIVER

ACTUALLY LARA, I THINK YOU'D BE A GOOD LEADER IN THE PATIENT ASSOCIATION. YOU WERE VERY KEEN TO LEARN ABOUT YOUR CONDITION!
THINK ABOUT IT.
AND LARA THOUGHT ABOUT IT...
THANKS FOR COMING ON BOARD, GUYS! LOOKING FORWARD TO WORKING WITH YOU.
VRRR
READY FOR A REMATCH? :)
SEE YOU ON WEDNESDAY :)

TITLE

MITOCHONDRIAL FOLLIES: A SHORT JOURNEY IN LIFE AND ENERGY

CONCEPT AND TEXT

ANABELA MARISA AZUL

JOÃO RAMALHO-SANTOS

PAULO JORGE OLIVEIRA

ILLUSTRATION

RUI TAVARES (ruidiastavares@gmail.com)

AKNOWLEDGEMENTS

AUTHORS AND ILLUSTRATOR ACKNOWLEDGE MIREIA ALEMANY I PAGÈS, PAULA TAVARES, SUSANA JORGE, SUSANA PEREIRA AND TERESA PAIS FOR THEIR ASSISTANCE AND HELPFUL DISCUSSIONS IN GRAPHICAL COMPOSITION AND TEXT. AUTHORS ACKNOWLEDGE FUNDING THROUGH THE EUROPEAN REGIONAL DEVELOPMENT FUND (ERDF), THROUGH THE COMPETE 2020 - OPERATIONAL PROGRAMME FOR COMPETITIVENESS AND INTERNATIONALISATION, AND THE PORTUGUESE FUNDING AGENCY FUNDAÇÃO PARA A CIÊNCIA E A TECNOLOGIA (FCT): I&D UID/NEU/04539/2013, POCI-01-0145-FEDER-007440, PTDC/DTP-FTO/2433/2014/POCI-01-0145-FEDER-016659 AND PTDC/DTP-DES/1082/2014/POCI-01-0145-FEDER-016657.

DISCLAIMER

THIS COMIC REFLECTS ONLY THE VIEWS OF THE AUTHORS. IN THIS COMIC REFERENCES ARE MADE TO REAL INDIVIDUALS WHO HAVE GREATLY CONTRIBUTED TOWARDS CURRENT MITOCHONDRIAL KNOWLEDGE, NAMELY P. MITCHELL, H. KREBS AND L. MARGULIS. ALL OTHER CHARACTERS ARE FICTITIOUS, AND ANY SIMILARITIES TO REAL PEOPLE (LIVING OR DEAD) IS PURELY COINCIDENTAL.

CNC
CENTER FOR NEUROSCIENCE AND CELL BIOLOGY
UNIVERSITY OF COIMBRA, PORTUGAL

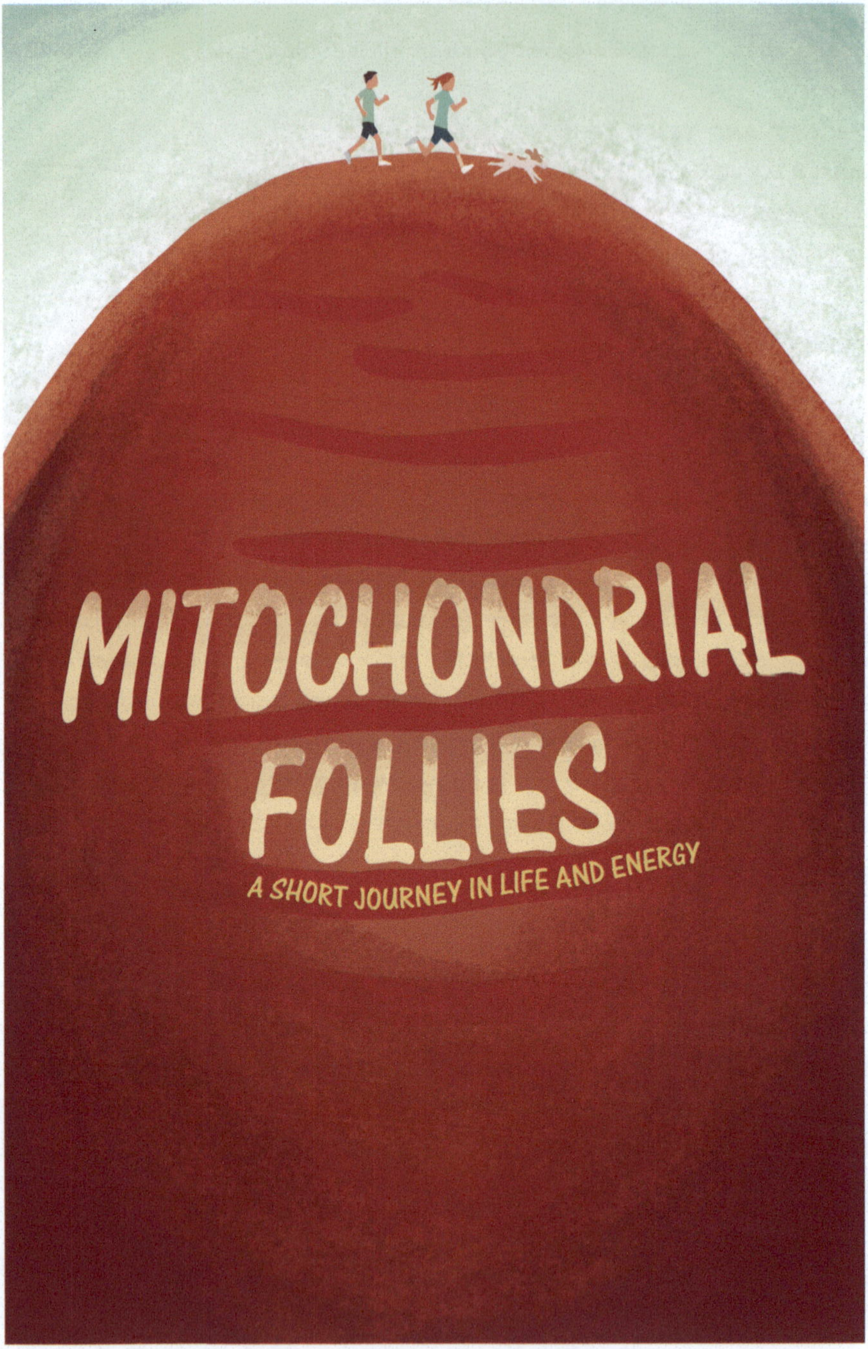
MITOCHONDRIAL
FOLLIES
A SHORT JOURNEY IN LIFE AND ENERGY

Index

P. J. Oliveira (ed.), *Mitochondrial Biology and Experimental Therapeutics*,
https://doi.org/10.1007/978-3-319-73344-9

The manufacturer's authorised representative in the EU is Springer Nature Customer Service Centre GmbH, Europaplatz 3, 69115 Heidelberg, Germany. If you have any concerns regarding our products, please contact ProductSafety@springernature.com

Printed and bound by CPI Group (UK) Ltd, Croydon, CR0 4YY
07/10/2026
02229059-0001